WALLACE'S LINE

WEBER'S LINE

EX

ECOLOGY AND FIELD BIOLOGY

# Robert Leo Smith

*West Virginia University*

# *ECOLOGY AND FIELD BIOLOGY*

## *Third edition*

**HARPER & ROW, PUBLISHERS, New York**

Cambridge, Hagerstown, Philadelphia, San Francisco,
London, Mexico City, São Paulo, Sydney

1817

*To Tricia and Gretchen*

**Credits for Part Opening and Chapter Opening Photographs**
**Part I:** NASA
*Chapter 1:* Charles Harbutt, Magnum Photos, Inc.
*Chapter 2:* Jen and Des Bartlett, Photo Researchers, Inc.
**Part II:** Ned Haines, Photo Researchers, Inc.
*Chapter 3:* © Hal H. Harrison from National Audubon Society, Photo Researchers, Inc.
*Chapter 4:* NCAR Photo
*Chapter 5:* NASA
*Chapter 6:* Karl H. Maslowski, Photo Researchers, Inc.
**Part III:** © Gordon Smith from National Audubon Society, Photo Researchers, Inc.
*Chapter 7:* Tom Wells, Photo Researchers, Inc.
*Chapter 8:* Ron Church, Photo Researchers, Inc.
*Chapter 9:* David Donoho, Photo Researchers, Inc.
*Chapter 10:* © Grant M. Haist from National Audubon Society, Photo Researchers, Inc.
**Part IV:** Charlie J. Ott from National Audubon Society, Photo Researchers, Inc.
*Chapter 11:* Des Bartlett, Photo Researchers, Inc.
*Chapter 12:* © Miguel Castro, 1972, Photo Researchers, Inc.
*Chapter 13:* Bruce Roberts, Photo Researchers, Inc.
*Chapter 14:* Des Bartlett, Photo Researchers, Inc.
*Chapter 15:* © Leonard Lee Rue III, Photo Researchers, Inc.
*Chapter 16:* Jen and Des Bartlett, Photo Researchers, Inc.
*Chapter 17:* © Frank Stevens II from National Audubon Society, Photo Researchers, Inc.
*Chapter 18:* Leonard Lee Rue III from National Audubon Society, Photo Researchers, Inc.
*Chapter 19:* Jen and Des Bartlett, Photo Researchers, Inc.
**Part V:** Andy Bernhaut, Photo Researchers, Inc.
*Chapter 20:* Jen and Des Bartlett, Photo Researchers, Inc.
*Chapter 21:* Ron Winch, Photo Researchers Inc.

Project editor: David Nickol
Designer: Gayle Jaeger
Senior production manager: Kewal K. Sharma
Compositor: The Clarinda Company
Printer and binder: The Murray Printing Company
Graphs and diagrams by Danmark & Michaels,
    Inc., and EWPIC, Inc.
Line drawings by Robert Leo Smith, Jr., and
    Ned Smith
Cover design by Robert Leo Smith, Jr.

**ECOLOGY AND FIELD BIOLOGY, Third Edition**
Copyright © 1980 by Robert Leo Smith

**Library of Congress Cataloging in Publication Data**

Smith, Robert Leo.
  Ecology and field biology.

  Bibliography: p.
  Includes index.
  1. Ecology.  2. Biology—Field work.  I. Title.
QH541.S6  1980          574.5          79-28390
ISBN 0-06-046329-5

# Contents

# *Preface*

This is my third time through *Ecology and Field Biology.* Many readers, I suspect, think that once a book has been written, revisions come easily. Actually a thorough revision is more difficult to write than the original text. An author must take out old material, incorporate new material, correct weaknesses, and try not to allow the knowledge explosion to fatten the text.

While this new edition is similar to the previous one in many ways, much of the material has been completely rewritten, an abundance of new material and many new illustrations have been added, and in some places the book has been reorganized. One major difficulty of writing an ecology text is the eclectic nature of the subject. As yet there is no firmly defined body of basic information that is regarded as making up a general ecology course. Introductory ecology is further distorted by the lack of time to present the material. Most general ecology courses are still only a quarter or semester in length, even though the subject has expanded enormously during the past 20 years.

Whether a general ecology course is taught in a quarter, a semester, or two semesters, with or without a laboratory, this text should fit admirably. If a general ecology course is lecture and discussion, this text provides the instructor and the student with the necessary additional material to develop topics for discussion, as well as a key to finding new sources of information. If ecology is a lecture and laboratory course, the section on comparative ecosystem ecology provides the students with an insight into the communities and ecosystems that may be involved in field studies, and the appendixes provide a guide to necessary field techniques. If the instructor requires individual projects, then *Ecology and Field Biology* provides the students with the basic information and techniques needed to undertake independent field studies. This versatility is not possible in a shorter text, nor is this versatility available in other ecology texts on the market. In addition the lists of journals, selected references, and general bibliographies provide a guide to the literature that should be invaluable to the instructor and student alike.

In *Ecology and Field Biology,* Third Edition, I have tried to present a balanced approach to ecology and to cover what appears to be that certain body of knowledge which should be common to all general ecology courses. This includes three broad areas: organismal or physiological ecology (the interaction of organisms and their adaptations to their physical environment); population and community ecology (hard to separate); and ecosystem ecology (primarily concerned with ecosystem structure and function). Some may argue that a fourth area should be ecological genetics. I consider that a part of population ecology. Evolution and genetics are not ecology, but rather integral parts of it. Interactions between organisms and their environments and among organisms and the function of ecosystems cannot be appreciated without considering genetics, natural selection, and evolution. It is a theme that has to run through all of ecology; but a general ecology course is not the place to teach genetics and evolution.

Ecosystem ecology is given extensive coverage in this edition, especially with the inclusion of Part III, "Comparative Ecosystem Ecology." However, I do not think it has been over-emphasized. Except for some selected items, much of this material may have to be omitted in shorter courses. But this is one place where the reader may acquire an appreciation of what ecosystems are like and how they differ. Without this material a student could go through an ecology course, be exposed to some basic principles, and yet have no concept of aquatic or terrestrial ecosystems.

I have attempted to carry through two themes in this text: systems and evolution. Obviously the ecosystems approach is the strongest in Parts II and III and the evolutionary theme in Part IV. The two merge in Part V. I have tried to keep a balance between theoretical and applied ecology. In some parts the applied may be more obvious than in other recent texts; but a common complaint of students is that ecology has become so conceptual and abstract that it has lost touch with the real world. Because most ecology students are going to live in the real world, where the task is to apply theory to actual problems, I have not skirted applied aspects. There should be no sharp division, however, between theoretical and applied, between abstract and real-world ecology; one complements the other.

I have retained the term *field biology* in the title, despite something of an old-fashioned ring, because it emphasizes that ecology is still a field science in spite of the deluge of paper ecology. All theories are for naught if they go untested in the field. Unfortunately, too many of them are not tested, but they become accepted as truth anyway.

I have endeavored to avoid two pitfalls in the text—dogmatism and North Americanism. It is sometimes difficult to avoid the impression of dogmatism when presenting the material in an introductory fashion. But in places I do point out where field data do not support theory and where theory is questioned. Not everything in this text is established truth. Some of today's ecological dogmas may become tomorrow's ecological myths.

The second edition was criticized by some for its North American outlook, which resulted mostly from the chapters on various ecosystems stressing North American examples. Throughout the third edition, examples are more global, especially in Part III. Overdependence on North American examples has been eliminated, and in Part V succession is treated as a worldwide phenomenon and is not restricted to North American coniferous and deciduous forest biomes.

While this edition in many ways appears similar to the second edition, there are major differences. Many of the changes reflect suggestions and criticisms obtained from a detailed questionnaire returned by many users and nonusers alike. I have deleted chapters on man and energy, man and biogeochemical cycles, ecosystem approach to resource management, and some behavioral material. (For those interested in ecology and the environment, see my *Ecology of Man: An Ecosystem Approach*.) I eliminated the organism and environment chapter and incorporated much of the material in the chapters on abiotic influences where the emphasis is adaptation. I expanded the population chapters and rewrote the ecosystem chapters to incorporate information made available by International Biological Program (IBP) studies. New and expanded material includes predation, niche theory, community structure, succession, foraging strategies, resource partitioning, vegetation-herbivore systems, plant population dynamics, phenology and seasonality, animal communication, mating systems, mutualism, evolution of seed dispersal systems, and Pleistocene ecology.

Two features which made *Ecology and Field Biology* unique among the many ecology texts have been retained: the illustration program and the Appendixes. Many of the illustrations from previous editions have found their way into numerous ecology, general biology, zoology, botany, and environmental texts. Some of these illustrations have been borrowed, modified, and reborrowed so many times they have in effect become generic. In this edition many new illustrations have been added and some old ones have been modified or eliminated. The number of photographs has been greatly reduced. Those with low information content relative to the space they occupy have been eliminated. Visual impact relies mostly on illustrations. To add interest to graphs, spot illustrations of the organisms concerned have been added, a technique initiated in the first edition and picked up by many biology texts since then.

The Appendixes have been updated to include such mathematical approaches as population dispersion, coefficient of community, community matrix, community ordination, niche measurement, life table construction, estimation of $r$, and others. The Appendixes, Aids to Identification, Selected References, and Journals of Interest make *Ecology and Field Biology* a source book for the user as well as a text.

The annotated lists of journals, bibliographies, and abstracts have been expanded greatly to call attention to the large number of new journals that have appeared during the past few years. Selected References has been thoroughly updated to include many of the specialized volumes on various aspects of ecology. These reference aids plus the Bibliography provide a major access to sources of literature in ecology. In addition a new feature, a glossary, has been added to this edition.

Like past editions, this text is divided into Parts. Part I consists of two introductory chapters. Chapter 1 discusses the nature of ecology with an emphasis on history. Chapter 2 introduces some basic concepts upon which the book rests: adaptation, natural selection, evolution, nature of systems, the ecosystem with emphasis on two basic functions, photosynthesis and decomposition. The purpose of introducing photosynthesis and decomposition so early in the text is to emphasize these two basic functional processes of ecosystems. In other texts, usually photosynthesis is considered in some detail with primary production, and decomposition is lost piecemeal elsewhere.

Part II considers the structure and function of ecosystems. Chapters 3 and 4 examine abiotic limitations on ecosystems. The approach is mostly from an organismal point of view. It is through the organism that these abiotic limitations are imposed on ecosystem structure and function. Chapter 5 examines energy flow through the ecosystem with emphasis on primary and secondary production, food chains, trophic levels, and energy budgets. Chapter 6 looks at nutrient cycles, nutrient budgets, and the intrusion of human activity on them. Nutrient cycles are discussed in more detail here than in other ecology texts, which for some reason or another are giving this topic scant attention. Yet nutrient cycling is an extremely important aspect of real-world ecology.

Part III in a way is a continuation of Part II. Drawing heavily on IBP biome studies, Chap-

ters 7, 8, 9, and 10 provide a comparative look at terrestrial and aquatic ecosystems, their structure and function. From these chapters the reader should gain some appreciation and perspective of lakes, streams, marine, grassland, shrubland, tundra, forest, and other major ecosystems.

Part IV is population ecology with an emphasis on an evolutionary and theoretical approach. This part involves some of the most extensive changes in the text. Plant population dynamics receives careful and considerable treatment. Chapter 11 is about ecological genetics, natural selection, and speciation. Chapter 12 examines social behavior with an emphasis on animal communication and on mating systems. Chapter 13 looks at population demography and 14 considers population growth. Chapter 15 looks at population regulation. Interspecific competition is treated in Chapter 16. I cannot agree with those who feel that interspecific competition and predation should be treated at equal length. Interspecific competition is largely abstract, is difficult to study in the field, and it therefore lacks the solid field data that predation possesses. To expand competition would result in verbosity; to compress predation would result in an inadequate treatment. Chapter 17 looks at theory of predation, and Chapter 18 considers predator–prey systems including vegetation–herbivore systems. Chapter 19 examines the coevolution of predator strategies and prey defense as well as the evolution of mutualistic systems.

Part V considers community ecology and succession. I moved these two topics to the end because it became obvious during the revision that such community concepts as niche, species diversity, and succession could not be discussed adequately without some knowledge of population theory and ecosystem functioning. Succession needs to be considered both from the trophic–dynamic and competitive–evolutionary points of view. The successional chapter has been expanded to include ecological changes through geological time. Present-day ecosystems have been strongly influenced by Pleistocene ecology, which has had a strong role in evolutionary change, extinctions, and speciation.

As I said of the previous edition, I have written this book to be read and used. I hope that I have been able to infuse into *Ecology and Field Biology* some enthusiasm for the subject and some feeling for the natural world. By necessity the reader will find in these pages some of the dull textbook stuff; but in it, too, I hope the user will find a feeling for the world outdoors. If that too can become as much a part of ecology as theory, mathematics, and computers, then perhaps posterity also will be able to study ecology.

*Robert Leo Smith*

# *Acknowledgments*

The author of a textbook depends on many people, mostly those researchers whose long hours in the field and laboratory have provided the raw material out of which textbooks are fashioned. Revisions depend heavily on the input of users of the text. I appreciate the comments, suggestions, and criticisms offered by a number of instructors and students. I am especially appreciative of the several hundred ecology instructors who responded to a rather lengthy questionnaire on the second edition. These responses served as a guide to the direction the third edition should take. In addition to these many anonymous individuals, Jerrold Zar provided many helpful suggestions in improving matters statistical in the second edition. Lewis Oring reviewed the material on behavior and suggested the animal communication approach. James Tanner and Eric Pianka provided incisive criticism of the population chapters that resulted in considerable improvement in that material. My colleague three office doors down the hall, Robert Whitmore, sharpened my approaches to predation, competition, and foraging strategies through numerous discussions. Richard Lee, author of a text on forest microclimatology and residing at the other end of the building, answered a lot of questions on thermal budgets and microclimates, while Stanislaw Tajchman provided access to otherwise inaccessible European literature on forest climatology. I am particularly indebted to James Karr, who reviewed the entire manuscript, suggested ways out of some organizational problems, provided new insights, and spent time at meetings and over the phone discussing various aspects of the material.

Again this edition is something of a family project. My son Thomas M. Smith, a graduate student in ecology at the University of Tennessee and Oak Ridge National Laboratory, revised much of the Appendix. While many the illustrations done by Ned Smith have been retained from the first edition, my son Robert Leo Smith, Jr., a natural history and scientific illustrator, did scores of new illustrations, prepared a majority of the new graphs and maps, and drew the cover design for this edition. My wife, Alice, took care of all the problems of everyday living while we were preoccupied with the revision.

The book could not have arrived at its present stage without the help of the publisher's staff. Dan Cooper, field representative, provided me with inputs from users. Judy Kahn, science writer and copy editor, provided great help in smoothing out the first draft and correcting rough spots in the second. David Nickol, Project Editor, had the difficult task of pulling typescript, illustrations, captions, bibliography and other pieces together from a manuscript that arrived in parts rather than in a whole.

I wish to acknowledge the following for permission to adapt, reprint, or redraw from their publications the figures listed below. Full information on source and citation will be found in the captions and Bibliography.

● Figures 3-8 © 1976, 3-11 © 1969, 4-29 © 1966, 9-9 © 1961, 9-14 © 1973, 11-1 © 1962, 13-7 © 1973, 13-13 © 1973, 15-1 © 1967, 15-10 © 1971, 16-6 © 1966, 16-7 © 1971, 16-11 © 1975, 17-12 © 1965, 18-6 © 1967, 20-19 © 1971, 21-6 © 1975, 21-11 © 1974, 21-14 © 1978. Copyright Ecological Society of America. Used with permission of Duke University Press.
● Figures 4-19 © 1970, 6-6b © 1966, 6-10 © 1972, 12-20 © 1977, 15-16 © 1973, 16-19 © 1977, 18-11 © 1973, 20-20 © 1965, 20-21 © 1965. From *Science*. Copyright by the American Association for the Advancement of Science.
● Figure 4-30, 4-31, 4-32, 4-33, 4-34, 4-36, 4-38, 9-19, 10-5, 10-20 a, b. Copyright Springer Verlag, New York and Berlin.
● Figures 7-6, 14-9, 14-10, 15-12, 15-15, 19-3, 19-5, 20,10, 20-16, 21-7, 21-8, 21-9, 21-10. From the *American Naturalist*. Copyright University of Chicago

## part I

# Introduction

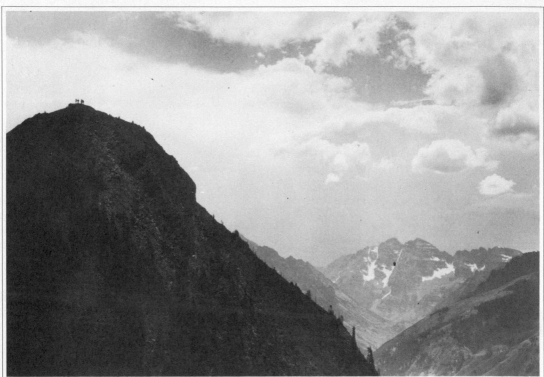

# *Ecology: its meaning and scope*
## CHAPTER 1

This is an age of ferment and change. A little more than 50 years ago Charles Lindbergh bravely crossed the Atlantic alone in a single-engine plane; today men have explored the surface of the moon and spacecraft have landed on Mars, photographed its surface, and sampled its environment. In the early 1920s the transmission of sound waves through radio had just been developed; today pictures along with voices are sent around the world by the way of communication satellites. In probing the secrets of the atom and its tremendous power, we have changed the direction of world history and the destiny of mankind. Our exploration of the secrets of the cell has revolutionized biology. The discovery of DNA and advances in biochemistry and genetics, including in vitro synthesis of genes, the ability to clone DNA, and the ability to transfer recombinant genes into bacterial and plant cells have profound implications for the future of human life.

### *Awareness of ecology*

While these technological and scientific advances were being made, another change has been taking place. The human environment has been deteriorating. Human populations over much of the world have been increasing explosively. The draining of nitrogenous wastes and excessive phosphorus from farmland and urban areas has been causing the pollution of natural waters and lowering of water quality. DDT and other chlorinated hydrocarbons, PCB, mercury and other heavy metals, and other chemicals, have been accumulating in some species of animals, impairing their reproduction or making them unfit as human food. The nonreturnable bottle, nonbiodegradable plastics, and forced obsolescence of appliances and automobiles have been creating massive solid waste problems and littering the countryside. The increasing demand for power to run industry and the modern conveniences of air conditioning and all-electric homes has increased sulfur dioxide and particulate pollution of the air. The search for fuel to supply this power has resulted in the upturning of the Appalachian mountains, midwest cornland, and western grazing land for strip-mined coal. A rapid multiplication of automobiles has been pouring increasing amounts of nitrous oxides and lead into the atmosphere, producing choking chemical smog. Roads slashed through the open country and suburban expansion have eaten into the hinterlands and farmland. Wilderness areas have been disappearing at an accelerating pace. The suburban and urban resident's increasing interest in outdoor recreation has been placing intolerable pressures on state and national parks. Even the oceans have not escaped, as human debris and chemicals have been deadening the seas. Warnings of deterioration have been sounded for years, but only now have people begun to awaken to the fact that planet Earth is in trouble. Suddenly the public has become aware of ecology.

In the late 1960s the general public was hardly aware of the term ecology. As a topic of public discussion, ecology stirred little interest. As a science it had none of the glamour of molecular biology. By 1970 ecology had become a household word, but it was misunderstood and misused. Too many failed to understand that ecology studies the interrelations of organisms and their environment and that human beings must be included among the organisms so studied. They only vaguely realized that the relationship is two-way, that just as the environment has an impact on the organism, so an organism has an impact on its environment. But at least a majority became aware of the environment. And the shattering view of earth from outer space forced on us the realization that the earth is finite and that what it is and what it contains are all we have.

Because it deals with life, ecology has been considered a part of biology. In the early part of the twentieth century the major introductory path to biology was natural history or, as it was more popularly known, nature study. This was a time when people were just awakening to the world about them. Nature had ceased to be an enemy. The fields were cleared, the forest subdued, and wildlife was no longer a threat. In fact many common animals—gray squirrels, beaver, deer, wild turkeys, and ducks—were on the edge of extinction. The conservation movement was building up full steam in the 1930s, and nature study was a part of nearly every school curriculum, even though more often than not it was poorly taught. Too often it consisted only of coloring bird pictures, but at least youngsters became aware that birds existed, that they were colorful and interesting, and that they were something more than living targets for BB guns. It was a time when John Burroughs was popular, the Reed *Bird Guides* were the last word in field guides, and the Comstock *Handbook of Nature Study* was the bible of natural history.

Out of this background of close contact with nature and an interest in life, biologists developed. But as the country became more urbanized and less rural, people lost this contact with nature. Interest in biology from a field approach declined, and research biologists became more concerned about the functioning of an organism than about its relationship to its environment. Modern biologists appeared at the doorways of chemistry, physics, and mathematics—disciplines not immediately related to the living environment. They looked upon biology as beginning and ending with a group of chemical compounds, and they thought that the answer to life lay within the realm of the physical sciences.

Part of the reason for the swing away from natural history lies in biology itself. For a long time traditional biology started and ended with the naming of organisms. Biology as taught in schools and colleges was an endless repetition of the study of types of organisms. It was largely descriptive, it was weak in quantitative data, and it lacked the strong conceptual foundation that marked physics, chemistry, and mathematics. At the popular level the mass of amateur naturalists who watched birds or collected insects rarely went beyond identification. They made little or no attempt to understand the organism, to find how it really lived or to question what its function was in nature. Even professionals fell into this trap and confined their work to descriptive biology. As a result natural history, once a rigorous subject, lost its position among the sciences and became equated with emotionalism and superficiality.

With the environmental awareness of the 1970s, interest in natural science began to revive. Books on natural history and ecology have become popular sellers; even the Reed *Bird Guides* and Comstock and Burroughs are back in vogue. Environmental study has returned to some classrooms; interest in wildlife and forestry has increased. Public outcries, wise or unwise, have been voiced against hunting and against environmental destruction by timber-cutting practices, highways, dams, power plants, and strip-mining. Many people are seeking a closer contact with the natural world. Some, especially the young, seek to return to the earth by establishing rural communes and attempting a subsistence agricultural way of life. Industry for the first time is finding itself on the defensive. Its uncontested right to pollute the air and water and to destroy the landscape for profit is being challenged.

Thus natural history evolved into ecology and ecology into a science that has entered the public consciousness. Where the old focal point was kinds of organisms, the new focal point is the nature of living systems. Just as molecular biology attempts to probe the secrets of living systems at the cellular level, so ecology probes the secrets of living systems at the levels of the organisms, the population, and the ecosystem.

## Development of ecology

The term ecology is derived from *okologie,* a word coined by the German zoologist Ernst Haeckel to mean "the relationship of the animal to its organic as well as its inorganic environment." The origin of the word is the Greek *oikos,* meaning "household," "home," or "place to live." Thus ecology deals with the organism and its place to live, that is, its environment, so ecology might well be called environment biology. An organism's environment includes its surroundings, other organisms of its own kind, and organisms of other kinds. Ecology is concerned with how the organism relates to all these factors. It studies the relationships and interactions among individuals within a population and with individuals of different populations. Because all organisms have become adapted to their surroundings and are always adjusting to a changing environment, natural selection and evolution are also a part of ecology.

Because of its involvements with so many fields, ecology is often regarded as a generality rather than a speciality. Indeed one ecologist, A. MacFadyen, in his book *Animal Ecology: Aims and Methods* (1963) wrote:

The ecologist is something of a chartered libertine. He roams at will over the legitimate preserves of the plant and animal biologist, the taxonomist, the physiologist, the behaviourist, the meteorologist, the geologist, the physicist, the chemist, and even the sociologist; he poaches from all these and from other established and respected disciplines. It is indeed a major problem for the ecologist, in his own interest, to set bounds to his divagations.

This statement nicely emphasizes that ecology is a multidisciplinary science. It has to be to reach the heart of the problems of environmental biology.

It is difficult to trace ecology back to any clear beginnings. The Greek scholar Theophrastus, a friend and associate of Aristotle, wrote of the interrelation between organisms and their environment. But modern impetus to the subject probably came from the plant geographers Humboldt, De Candolle, Engler, Gray, and Kerner. The development of ecology in America was greatly influenced by two early European ecologists, E. Warming through his book *Oecological Plant Geography* (1895) and A. F. W. Schimper through *Plant Geography on a Physiological Basis* (1898). Out of the roots of plant geography developed the study of the plant community which in time evolved into community ecology.

The study of plant communities developed separately in western Europe and the United States. Prominent among the early plant ecologists in the United States were J. M. Coulter of the University of Chicago and C. E. Bessey of the University of Nebraska. Coulter was the major professor of H. C. Cowles, who received his doctoral degree in 1897 for his study "Ecological Relations of the Sand Dune Flora of Northern Indiana." This work marked the beginning of pioneering studies of plant succession, one of the central concepts of modern ecology. At the same time F. C. Clements working under Bessey did a study on the plant geography of Nebraska. Clements quickly became the major theorist of early twentieth-century plant ecology, particularly in the area of plant succession. While American plant ecologists were concerned with the development and dynamics of plant communities, European plant ecologists, notably Braun-Blanquet (1932), concerned themselves with the composition, structure, and distribution of plant communities.

Early plant ecologists were concerned primarily with terrestrial plant communities. Studies on aquatic communities were stimulated by the work of two European aquatic ecologists, Thienemann and Forel. Thienemann introduced the concept of trophic or feeding levels in the terms of producers and consumers. Forel introduced the term *limnology* for the study of freshwater life in his monograph on Lake Leman. In America S. A. Forbes, an entomologist at the University of Illinois and the Illinois State Laboratory of Natural History, wrote what was later to become a classic of ecology, "The Lake as a Microcosm." Upon this foundation two American zoologists and aquatic biologists, E. A. Birge and C. Juday of the Wisconsin Natural History Survey, built the beginnings of limnology in America. Their detailed studies of photosynthetic processes, respiration, and decay and their measurements of energy budgets in lakes resulted in the concept of primary production, the fixation of the sun's energy by plants. Out of their studies came the trophic-dynamic concept of ecology. Introduced by R. Lindeman in 1942, this concept marked the beginning of modern ecology. Out of Lindeman's study came further pioneering work on energy flow and energy budgets by G. E. Hutchinson (1957, 1969) and H. T. and E. P. Odum in the 1950s. Work on the cycling of nutrients was done by Ovington (1962) in England and by Rodin and Bazilevic (1967) in Russia.

Meanwhile zoologists were becoming interested in the ecology of animals, but they pursued a course divergent to plant ecology. Early in the history of American ecology, a schism developed between botanists and zoologists over the term ecology. Botanists, responsible for the early development of ecology as a science, decided at the Madison (Wisconsin) Botanical Congress in 1893 to drop the *o* from oecology *(ökologie)* and adopt the anglicized spelling. Zoologists refused to recognize the term ecology. The entomologist William Morton Wheeler complained that botanists had usurped the word, had eliminated the letter *o* improperly, and had distorted the science. He urged that zoologists drop the term ecology and adopt the word ethology. The scism was widened by a more fundamental difference of approach in that plant ecologists ignored any interaction between plants and animals. In effect they viewed plants as growing in a world without parasitic insects and grazing herbivores. For years plant and animal ecologists went separate ways.

Early animal ecology was dominated in Europe by Charles Elton and R. Hesse. Elton's *Animal Ecology* (1927) and Hesse's *Tiergeographie auf oekologischer grundlage* (1924), later translated into English as *Ecological Animal Geography* (1937), were very strong influences on the development of animal ecology in America. Two early American animal ecologists were Charles Adams and Victor Shelford. Adams published the first text on animal ecology, *A Guide to the Study of Animal Ecology* (1913), and Shelford's *Animal Communities in Temperate America* (1913) was a landmark work because he stressed the relationship of plants and animals and emphasized the concept of ecology as a science of communities. The community concept had been developed much earlier by the geographer Alexander Humboldt and the marine biologist Karl August Mobius and was central to ecology until 1935 when A. Tansley, a British plant ecologist, advanced the concept of the ecosystem. The appearance in 1949 of the encyclopedic *Principles of Animal Ecology* by five second generation ecologists from the University of Chicago, W. C. Allee, A. E. Emerson, Thomas Park, Orlando Park, and K. P. Schmidt, pointed out the direction that modern ecology was to take with its emphasis on trophic structure and energy budgets, population dynamics, and natural selection and evolution.

The early twentieth century also witnessed a developing interest in the dynamics of population. The theoretical approaches of Lotka (1925) and Volterra (1926) stimulated experimental approaches by biologists (see Chapters 14, 16, and 17). In 1935 Gause investigated the interactions of predators and prey and the competitive relationships between species. At the same time A. J. Nicholson studied intraspecific competition. Later the work of Andrewartha and Birch (1954) and the field studies of Lack (1954) provided a broader foundation for the study of the regulation of populations. Animal ecologists explored such aspects as population density, reproduction and mortality, population growth, and community organization. Such population studies led to the concept of the niche (see Chapter 20) suggested by the plant ecologist H. Gleason (1917, 1926), advanced by J. Grinnell (1917) and C. Elton (1927), and expanded by G. E. Hutchinson (1957). The discovery of the role of territory in bird life by H. E. Howard in 1920 led to further studies by M. M. Nice in the 1930s and 1940s. Out of such studies came the field of behavioral ecology. In the 1940s and 1950s K. Lorenz and N. Tinbergen developed concepts of instinctive and aggressive behavior (see Chapter 13). The role of social behavior in populations has been explored by two controversial authors, V. C. Wynne-Edwards in his *Animal Dispersion in Relation to Social Behavior* (1962) and E. O. Wilson in *Sociobiology* (1975). Population ecology, because of its quantitative approach, led to theoretical mathematical ecology. First concerned with sampling techniques and statistical analysis of resulting data and studies of distribution of organisms, theoretical population ecology quickly moved into quantitative studies of competition, predation, community and population stability, cycles, community structure, community association, and species diversity.

The writings of Malthus (1798), who called attention to expanding populations and limited food supply, led indirectly to evolutionary studies and new systematics, for it was from Malthus' essay on populations that Darwin received the first inspiration for his theory of evolution (see Mayr, 1977). Out of the work of Darwin (1859) and the genetic theories of Mendel (1866), S. Wright (1931), and others grew the field of population genetics, the study of evolution and adaptation. Early ecologists were concerned with evolution as a central focus. For example, W. F. Ganong in 1904 recognized seven ecological principles all relating to adaptations of organisms to the environment. Allee et al. (1949) further emphasized the role of evolution in ecology, but the evolutionary viewpoint was more or less lost in the 1950s. It was strongly revived in the 1960s.

In the 1960s ecology took on new directions. While some ecologists continued in the earlier tradition, most new work, stimulated by large-scale funding to universities, by the development of new and expensive instrumentation in chemistry and physics, and by advances in statistical techniques and mathematical modeling, represented a significant departure from traditional ecology. With the use of electronic equipment and biotelemetric techniques, ecologists can sample and measure plant and animal populations without destroying them. Radioisotopes enable investigators to follow pathways of nutrients through ecosystems and determine the time and extent of transfer. Laboratory microcosms—samples of both aquatic and soil microecosystems taken from natural ecosystems—are useful in determining rates of nutrient cycling and other parameters of ecosystem functioning. The use of statistical procedures, such as correlation, multiple regression, and multivariate analysis, and the application of matrix algebra, calculus, and computer science to mathematical models simulating field conditions are providing new insights into population interactions and ecosystem functioning.

A source of dramatic influence on the development of ecology since the 1950s was the International Biological Program, known as IBP. The organization of the program in 1960 was stimulated by a growing concern over environmental problems facing the world. The American IBP, initiated in 1967, focused on a cooperative study and analysis of whole ecosystems, including the tundra, the coniferous forest, the eastern deciduous forest, the desert, and so on. The multiple goals of IBP, as summarized by McIntosh (1976), included: (1) understanding the interactions of the many components of complex ecological systems; (2) exploiting this understanding to increase biological productivity; (3) increasing the capacity to predict the effects of environmental impacts; (4) enhancing the capacity to manage natural resources; and (5) advancing the knowledge of human genetic, physiological, and behavioral adaptations.

IBP's greatest contribution was to increase our understanding of processes involved in the functioning of ecosystems, particularly photosynthesis and productivity, water and mineral cycling, decomposition, and the role of detritus. Although it may have lacked strong organization and coordinated direction in research it did more than any other organized program to advance modern ecology. Summaries of IBP research, being published in numerous volumes, will provide a base for future ecological research.

Whatever its successes, IBP did not provide any definitive concept of the scope and theoretical foundations of ecology. What constitutes modern ecology depends upon the ecologist or groups of ecologists to whom the question is addressed. Although it may be said that ecology has been identified as a separate field of scientific inquiry for about 50 years, it is a science that coalesced out of other fields and still lacks well-defined principles. In many ways its adherents are as divided as they were in the past. One group focuses on the structure and functioning of ecosystems, especially energy flow and nutrient cycling. Another group, closely related to population biologists, view ecology mostly in terms of mathematical models, sometimes resulting in untested hypotheses that often get accepted as dogma. Still another group views ecology largely as evolutionary studies and population genetics. Looked down upon as hopelessly old-fashioned is descriptive ecology, mostly because of its lack of theory and lack of a strong quantitative approach. Attempting to relate ecology to environmental and population problems and the management of natural resources is another group, the applied ecologists. Often they overlook and fail to utilize theories and models developed by mathematical and system ecologists.

## Divisions of ecology

Considered as objectively as possible, modern ecology might be viewed as centered on the study of the ecosystem and divided into three basic approaches—systems ecology, theoretical ecology, and evolutionary ecology. Systems ecology is concerned with the analysis and understanding of the structure and function of ecosystems by the use of applied mathematics, especially that developed by engineering and general systems. It involves the construction of models that represent the real system for the purpose of experimentation. To be valid the model has to mimic the real system at least over some restricted range, include the important variables, and be subject to mathematical expression. Models can be constructed to provide a simplified description of the system or to predict changes over time.

Theoretical ecology, involved in modeling of a somewhat different sort, is strongly oriented to populations (for example, see May, 1975) and focuses on such areas as predation, competition, niche theory, diversity, community structure, and community or ecosystem stability. It utilizes theories and equations developed in pure mathematics, physics, and even economics, and applies them to ecological problems. One of the strengths of theoretical ecology is its attempt to provide a substantial mathematical foundation to some ecological concepts upon which predictive ecology can be based. One of its weaknesses is that too many of its practitioners, trained in physics and mathematics but with a weak background in ecology, base their work on simplified assumptions, on poorly acquired or inadequate data, or on no data at all.

Evolutionary ecology is concerned with the interaction between organisms and their biological and physical environment, as expressed in adaptation. Adaptations of a species come about through natural selection, which favors those individual of the species able to reproduce successfully under given environmental conditions. Natural selection results in changes in relative frequencies of genotypes. And the changes brought about in the genetic constitution of a population is evolution.

The work of the three approaches obviously interrelates. The systems ecologist cannot explain how an organism or a set of organisms functions in the ecosystem without some appreciation of how they are adapted to changes over time. The evolutionary ecologist's study is incomplete if it views organisms outside of the system of which it is a part. The theoretical ecologist cannot begin to create adequate models of population change or community stability without giving some consideration to the organism's role in the ecosystem and to evolutionary change. Thus all facets of ecological investigation relate to the ecosystem. If each of the specialities would view its own interest within this context, much of the division in ecology would disappear.

Whatever the basic approach, ecologists are also grouped according to whether they work with plants or animals. Thus one might be a plant ecologist or an animal ecologist. Both plant and animal ecology can be further divided artificially into two parts, *synecology* and *autecology*. (For a good example of this division in plant ecology, see R. F. Daubenmire's *Plants and Environment: A Textbook of Plant Autecology, 1959,* and *Plant Communities: A Textbook of Plant Synecology,* 1968). Autecology is concerned with the study of the interrelations of individual organisms with the environment. Synecology is concerned with the study of groups of organisms—the community, Both have developed independently, although a knowledge of both is necessary for the understanding of the individual, population, or ecosystem. Autecology is experimental and inductive. Synecology is philosophical and deductive. Autecology, because it is concerned with the relationship of an organism to one or more variables such as light or salinity, is easily quantified and subject to experimental design both in the laboratory and in the field. Synecology is largely descriptive and not easily subject to experimental design. Autecology has borrowed techniques from physics and chemistry. With the development of such tools as computers and radioactive tracers synecology has entered a strong experimental phase.

Synecology is often subdivided into aquatic, terrestrial, and marine ecology. Terrestrial ecology, subdivided further into such areas as forest ecology, grassland ecology, and desert ecology, is concerned with terrestrial ecosystems—their microclimate, soil chemistry, nutrient and hydrological cycles, and productivity. Because they are biologically controlled and subject to much wider environmental fluctuation, terrestrial ecosystems are more difficult to study than aquatic ecosystems. Because aquatic ecosystems are affected more by the physical environment and have more stable environmental conditions (such as temperature), considerable attention is paid to chemical and physical characteristics. But as will be seen in the pages to come it is very difficult to separate aquatic from terrestrial ecology. Because one relates to the other in the form of inputs and outputs of nutrients and water, one cannot be studied without some references to the other.

## Some basic principles

One of the problems of ecology is the failure to arrive at some basic principles. Because the ecological world is by its very nature a rather messy place and because nature has so many exceptions to its "rules," ecology may never have any precise principles. Nevertheless, some ecological principles may be stated with the caution that they not be accepted as dogmas or laws as they often are.

1. The ecosystem is the major ecological unit. It contains both abiotic and biotic components through which nutrients are cycled and energy flows.
2. To permit these cycles and flows the ecosystem must possess a number of structured interrelationships among its components (soil, water, nutrients, producers, consumers, and decomposers).
3. The function of ecosystems is related to the flow of energy and the cycling of materials through the structural components of the ecosystem.
4. The total amount of energy that flows through a natural system depends upon the amount fixed by plants or producers. As energy is transferred from one feeding level to another, a considerable portion is lost for further transfer. This limits the number and mass of organisms that can be maintained at each feeding level.
5. Ecosystems tend toward maturity. In so doing they pass from a less complex to a more complex state. This directional change is called succession. Early stages are characterized by a relatively high energy flow per unit of biomass and high

net production. Later stages are characterized by low energy flow per unit of biomass and low net production, with most of the energy going into system maintenance.

6. When an ecosystem is exploited and that exploitation is maintained, the maturity of the ecosystem declines.

7. The major functional unit of the ecosystem is the population. It occupies a niche in the system, that is, it plays a particular role in energy flow and cycling of nutrients.

8. A niche within a given ecosystem cannot be simultaneously and indefinitely occupied by a self-maintaining population of more than one species.

9. Both the environment and the amount of energy fixation in any given ecosystem are limited. When a population reaches the limits imposed by the ecosystem, its numbers must stabilize or, failing this, decline (often sharply) from disease, strife, starvation, low reproduction, and so on.

10. Changes and fluctuations in the environment (exploitation and competition, among others) represent selective pressures upon the population.

11. Species diversity is related to the physical environment. An environment with a more complex vertical structure in general holds more species than one with a simpler structure. Thus a forest ecosystem will contain more species than a grassland system. Similarly, a benign, predictable environment holds more species than a harsher or more unpredictable environment. Tropical systems exhibit greater diversity than temperate or arctic systems.

12. In island situations immigration rates tend to balance extinction rates. The nearer an island is to a population source, the greater its immigration rate per unit time. The larger the island is, the higher its carrying capacity for each species. In any island situation immigration of species declines as more species become established, and fewer immigrants are new species.

13. The ecosystem has historical aspects; the present is related to the past, and the future is related to the present.

## Application of ecology

In the 1970s ecology was described as being a new science, and it became involved in social, political, and economic issues. This involvement came about primarily because the public was aware of environmental problems and was seeking answers to the problems of pollution, overpopulation, and degraded environments. Many have treated the issues as if they were new, as if ecology and what the science and its spokesmen had to say was very contemporary. Actually ecologists have been concerned with natural resource management and have been involved with environmental issues for years. The problem was people were not listening. For example, in 1932 H. L. Stoddard introduced the idea of fire in the control of plant succession in his book *The Bobwhite Quail*. Aldo Leopold expounded the application of ecological principles in the management of wildlife in his classic *Game Management* (1933). In *Forest Soils* (1954) H. L. Lutz and R. F. Chandler discussed nutrient cycles and their role in the forest ecosystem. And J. Kittredge pointed out the impact of forests on the environment in *Forest Influences* (1948). Even earlier, in 1864 George Perkins Marsh's dramatic book *Man and Nature* called attention to the effects of poor land use on the human environment. P. B. Sears' *Deserts on the March* (1935) was written in response to the dust bowl of the 1930s, and William Vogt's *Road to Survival* (1948) and Fairfield Osburn's *Our Plundered Planet* both called attention to the growing population-resource problem. Aldo Leopold's *A Sand County Almanac* (1949), which called for an ecological land ethic, was read largely by people interested in wildlife management, but it did not become widely read until 1970 when it became the bible of the ecology movement.

Rachel Carson probably did more than anyone to bring environmental problems to the attention of the public. Since the publication of her book *Silent Spring* (1962), we have become increasingly aware of the chemical poisons and other pollutants that are being cycled through the ecosystem. Once castigated as more fiction than fact, Carson's dire predictions came only too true. As urban developments eat away at the countryside, more and more people are becoming concerned about open spaces and are beginning to accept the work of such ecological land planners as Ian McHarg. The problem of population growth, once whispered about, is now widely debated in public. *The Population Bomb* (1968) by Paul Erlich and *Too Many* (1969) by George Bergstrom are widely read paperbacks; zero population growth is a goal toward which many in the United States have become committed. Population control, a subject once taboo, is now the topic of numerous books and newspaper and magazine articles. The public has become aware that humankind can exceed the limits of the environment and can deplete and contaminate it.

In spite of the widespread acceptance of the facts, we have had little real success in improving the environment or preventing its continued contamination. Population growth has been slowed only in some parts of the world. Industry has reluctantly accepted the fact that it must reduce pollution, but efforts to do so have been impeded by such unpopular yet inevitable results as higher costs, employment layoffs, and decreased production. Efforts to reduce pollution by pesticides and heavy met-

als are resisted by both industry and agriculture. The increasing concentration by agriculture upon single-crop farming has developed simplified ecosystems that create many difficulties and dangers. Pressures still mount against parklands and other semiwild places. Wildlife species decline in the face of "progress." Roads still cut across the landscape to make room for more cars to move faster from one place to another. Subdivisions still close in on prime agricultural lands while in more undeveloped regions forest and grasslands are being destroyed to make room for more agricultural lands.

This discourse could be continued, but enough has been written to press home the fact that the future of human life on earth demands more knowledge about ecosystems than we now possess. For the first time in the history of Earth *Homo sapiens* has become the completely dominant organism, changing the earth and its vegetation almost at will, with little regard for the consequences. It is little wonder then that some of today's most intellectually challenging problems are found in the area of environmental sciences to which the science of ecology has much to contribute.

# Ecological systems
## CHAPTER 2

Imagine yourself in a spacecraft far from Earth. Through the window the planet appears as a bluish ball suspended in the black void of space. Its surface is kaleidoscopic, patterns of white, blues, reds, and greens constantly changing across its surface as it turns and as swirls of clouds move across it. It hangs alone, self-contained, dependent on the outer reaches of space only for the energy of sunlight. It is close enough to the sun to be warmed by its radiant energy, yet it does not become overheated or overcooled. It is pro-

tected from damaging radiation by an atmosphere unlike that of any of its sister planets.

As the spacecraft approaches Earth, the changing patterns sharpen into broad outlines of mountains, deserts, plains, and seas. In the vast area from outer space to the core of Earth it is only at the narrow interface of land, air, and water that life exists. This thin blanket of life surrounding the earth is called the *biosphere* (Figure 2-1).

Supporting the biosphere is the *lithosphere* (Figure 2-1), the rocky material of Earth's

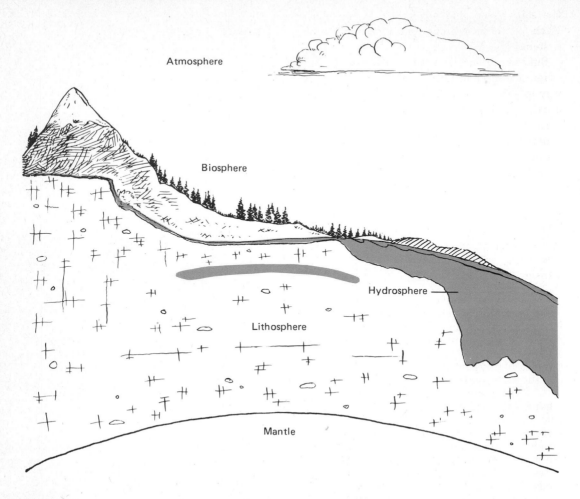

Atmosphere

Biosphere

Hydrosphere

Lithosphere

Mantle

**FIGURE 2-1**
*Schematic profile of the structural features of the outer part of the earth. The rocky material of the earth's outer crust makes up the* lithosphere. *Enveloping the earth is a layer of air, the* atmosphere. *Extending from below the crust in the form of groundwater to the surface in the form of rivers, streams, lakes, and seas is the earth's water, the* hydrosphere *(colored areas). The hydrosphere is closely associated with the moisture in the atmosphere. Extending from the top of the highest mountain to sea bottom is the* biosphere. *The biosphere for rooted plants ends at the snowline, but heterotrophic communities may extend beyond that point. The deep ocean floor also supports a specialized heterotrophic community. The atmosphere, the biosphere, the lithosphere, and the hydrosphere are often collectively called the* ecosphere.

outer shell from the surface to about 100 km deep. The thin, uppermost layer of the crust is the substrate of life—the foundation on which

it rests, the material of which it is made, and its primary source of nutrients.

The body of liquid on or near the surface of Earth is the *hydrosphere.* Most of the water is in the oceans, some is in the lithosphere as ground water, and a small fraction is in the atmosphere.

As the spaceship comes even closer to Earth, the mountains, deserts, plains, and seas focus into expanses of grasslands, forests, croplands, rivers, lakes, estuaries, and oceans. Each is physically and biologically different, and each is inhabited by different organisms that are well adapted to the environment in which they are found. Yet in spite of their differences, each functions in the same fashion. Energy is fixed by plants and transferred to animal components. Nutrients are withdrawn from the substrate, deposited in the tissues of plants and animals, cycled from one feeding group to another, released by decomposition to the soil, water, and air, and then recycled. Each of these regions constitutes an eco-system.

But the regions are not independent of each other. Energy and nutrients in one find their way to another, so that ultimately all parts of Earth are interrelated, each comprising a part of the total system that keeps the biosphere functioning. Each region is thus one element in the ecosystem that is the total planet.

## Definition of a system

The word *ecosystem* was coined by A. G. Tansley (1935) in an article that appeared in *Ecology.*

The more fundamental conception is . . . the whole *system* (in the sense of physics) including not only the organism-complex, but also the whole complex of physical factors forming what we call the environment.
. . . We cannot separate them (the organisms) from their special environment with which they form one physical system. . . . It is the system so formed which [provides] the basic units of nature on the face of the earth. . . . These *ecosystems,* as we may call them, are of the most various kinds and sizes.

Thus the *eco-* indicates environment and *-system* indicates a complex of coordinated units.

A system consists of a set or collection of interdependent parts or subsystems enclosed in a defined boundary. For example, in Figure 2-2, the system is represented by the area enclosed within the dashed lines. Within the system are a number of boxes representing subsystems. The lines between the objects represent interactions.

Each system exists in an environment which provides the input necessary for the functioning of a system. An input is any signal from outside the system to which it responds (Fig 2-3). The result of the response is the system's output, any attribute transmitted to the environment.

The system processes inputs in a set way with each subsystem performing set functions. Each part has a specific function, but its expression depends upon the proper functioning

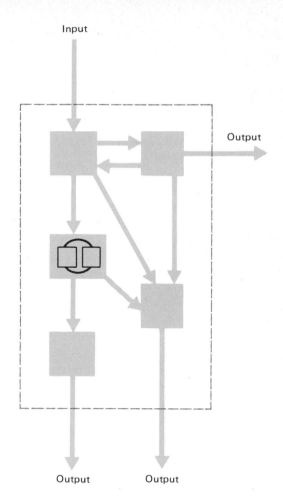

**FIGURE 2-2**
*Diagrammatic representation of a system. The dashed line represents the boundary of the system. The area outside of that line is the system's environment. The boxes inside the boundary represent the system's various components or subsystems. The smaller boxes within one of the boxes represent some of the elements of that subsystem. The lines connecting the boxes represent interactions between components. The lines crossing the boundary of the system represent inputs from and outputs to the environment.*

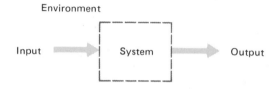

**FIGURE 2-3**
*Diagram of an open system showing inputs from and outputs to the system's environment.*

**13**

of all parts. The amount of output from the system is directly related to the input. If the input ceases, the system ceases to function.

A system may be open or closed. A closed system, of little interest to ecologists, has no exchange of energy or matter with its surroundings. For that reason closed systems tend to run down. An open system (Figure 2-3) depends upon an outside input. A radio can be used as an example of an open system. The boundary of the system is the radio itself. It consists of a number of subsystems—transistors, transductors, wires, a speaker, and various controls. The input upon which the system depends consists of electricity and electromagnetic waves. The output to the environment is sound waves and heat.

An open system with some sort of feedback mechanism that provides a degree of control over its functioning is called a cybernetic system. In a cybernetic system some of the output is fed back into the system to control some future output, that is, output becomes input determining the future functioning of the system. In Figure 2-4 the system of Figure 2-3 has been changed from open to cybernetic by the addition of a feedback loop. Feedback exists if any of the inputs are determined by the state of the system. To function, a feedback system has to involve some sort of an ideal state or set point toward which the system adjusts (Figure 2-5). If the set point is exceeded, then some internal mechanism is activated to reduce the input, slowing the tendency to exceed the set point. If input is too low, then the internal mechanism remains inactive, allowing increased input until the set point is reached.

A room dehumidifier is an example of a purely mechanical system in which a specific set point can be fixed. Suppose the desired humidity of a room is 50 percent. The dehumidifier set point is placed at 50 percent. When the humidity of the air exceeds that point, the moisture-sensitive device in the dehumidifier trips the switch to allow an inflow of electricity to turn the fan that pulls the air over the refrigerated coils on which the water condenses and drips off to be carried away. When the humidity falls below the set point, the moisture-sensitive device responds by shutting off the input of electricity to stop the fan.

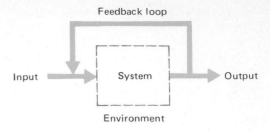

**FIGURE 2-4**
*The same system as Figure 2-3 with a feedback loop leading from output back to input. This makes the system cybernetic.*

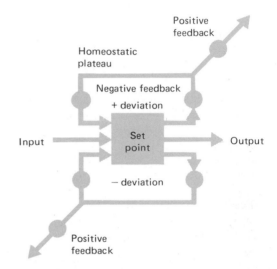

**FIGURE 2-5**
*Cybernetic systems have a set point about which the system maintains itself. The homeostatic plateau represents the maximum and minimum values or area within which the system can function by using feedback to regulate itself about the set point. When the system deviates from the set point a control mechanism brings it back toward the set point. This readjustment is termed negative feedback because it inhibits any strong movement away from the set point. If the system exhibits a continually increasing tendency away from the set point it is controlled by positive feedback which ultimately can destroy the system.*

The feedback of information that causes the dehumidifier to shut off or turn on when the humidity exceeds or falls below the set point is called a negative feedback loop. It halts or reverses a movement away from a set point. In other words the state of the system and its inputs are inversely related.

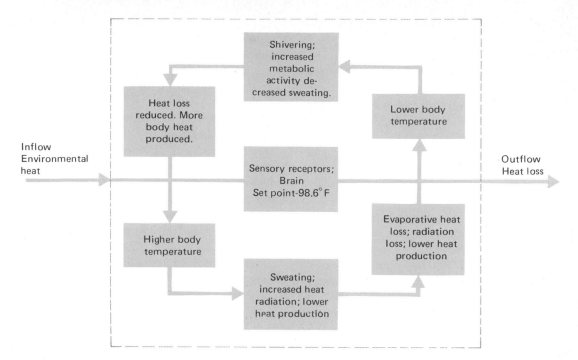

**FIGURE 2-6**
*Cybernetic control of body temperature in humans.*

Feedback which results in a continued movement away from a set point is called positive (Figure 2-5). The measure of this feedback is directly related to input. An example of positive feedback is the accumulation of compound interest or the geometric growth of population or even the growth of cancerous cells. Thus positive feedback can ultimately destroy the system.

A cybernetic system has a homeostatic plateau which represents the upper and lower limits within which the negative feedback system will work. (*Homeostasis* is the tendency of a system (an organism or group) to maintain internal stability.) When this plateau is exceeded, positive feedback takes over. It will ultimately destroy the system unless conditions can be corrected (Figure 2-5).

Cybernetic responses produce stable systems. Stable or homeostatic systems respond to stimuli by developing forces to maintain or restore the original condition of the system. In natural ecosystems, stability may mean that the system undergoes many changes and still preserves a similar structure or it may imply persistence, that is, the system remains the way it was.

An example of cybernetic control in a living system is temperature regulation in homeothermal animals such as humans (Figure 2-6). The normal temperature for humans is 98.6° F. If the temperature of the environment rises, sensory mechanisms, primarily in the skin, detect the change and transmit the information to the brain. The brain, acting on the information, sends a message to the effector mechanisms that increase the blood flow to the skin and induce sweating. Water excreted through the skin evaporates, cooling the body. If the environmental temperature falls below a certain point, a similar sequence takes place, this time reducing blood flow and causing shivering, an involuntary muscular exercise producing more heat. If environmental temperatures become extreme, the homeostatic plateau is exceeded and the cybernetic system breaks down. If the temperature becomes too hot, the body is unable to lose heat fast enough to hold the temperature at normal. Positive feedback sets in. Body metabolism speeds up, further increasing body temperature and eventually ending in heat stroke or death. If the environmental temperature drops too low, metabolic processes slow down, further decreasing body temperature and eventually resulting in death by freezing.

## Components of ecological systems

An ecosystem is basically an energy-processing system whose components have evolved together over a long period of time. The boundaries of the system are determined by the environment, that is, by what forms of life can be sustained by the environmental conditions of a particular region. Plant and animal populations within the system represent the objects through which the system functions.

Inputs into the system are both biotic and abiotic. The abiotic inputs are energy and inorganic matter. Radiant energy, both heat and light, imposes restraints on the system by influencing temperature and moisture regimes that determine what organisms can live in a system under a given set of environmental extremes and by affecting the productive capability of the system. Inorganic matter consists of all nutrients, water, carbon dioxide, oxygen, and so forth that affect the growth, reproduction, and replacement of biotic material and the maintenance of energy flow. Some of the materials in this chemical environment are necessary for the maintenance of the system, while others may be toxic or detrimental to its functioning.

The biotic inputs include other organisms that move into the ecosystem as well as influences imposed by other ecosystems in the landscape. No ecosystem stands alone. One is influenced by the other. A stream ecosystem, for example, is strongly influenced by the terrestrial ecosystem through which it flows. Island ecosystems are influenced by their distance from continental land masses.

Consider two ecosystems: the forest and the pond. The sun shining on the open pond supplies energy to warm the shallow water and for photosynthetic activity of the microscopic plants and rooted aquatics. These plants in turn support a variety and abundance of minute animal life. Both provide food for young sunfish, tadpoles, and aquatic insects. The insects are eaten by adult sunfish, frogs, and birds. Sunfish and frogs become food for bass and heron. Cattails, reeds, and waterlilies growing along the pond's shore furnish food and shelter for muskrats, nesting sites for ducks and red-winged blackbirds, and support for aquatic insects, snails, and flatworms. Remains of dead plants and animal sink to the bottom where they decompose and, mixed with the inflowing sediments add to the bottom muck, supply of organic and inorganic nutrients. If the water of the pond is drained, all life sustained by an aquatic environment is destroyed. If the cattails and reeds are covered by fill, the birds, muskrats, and many aquatic insects disappear. If insect life is destroyed, the food supply of frogs and sunfish is eliminated, and this in turn affects the bass and heron. If the bass are removed, the sunfish population may become so large that fish growth will be stunted. Thus all the organisms of the pond depend not only on water but also, directly or indirectly, upon one another for their well-being and existence.

The forest on the slope contains different elements but functions as a system in much the same way. Trees and plants capture energy from the sun and channel it to other members of the forest. Deer browse on leaves and twigs; earthworms and other soil organisms consume fallen leaves. Insects feed on leaves and plant juices. Woodland mice eat seeds and insects, and they in turn become food for weasels and hawks. The forest provides shelter for many forms of life and modifies the wind and temperature. The forest vegetation depends upon those organisms that break down organic matter and return the minerals to the soil. When trees are cut or burned, organisms dependent on the forest environment disappear; if deer become too numerous they overbrowse the forest and destroy young trees that serve as food and shelter for other animals. Just as organisms in a pond, all forest organisms depend, directly and indirectly, on each other for existence.

In simplest terms all ecosystems, aquatic and terrestrial, consist of three basic components which derive their inputs from the abiotic environment. Traditionally the abiotic inputs—$CO_2$, $O_2$, nutrients derived from the weathering of materials and from precipitation, and so on—are considered as abiotic components of the system. Here they are considered as inputs to the system rather than a part of the ecosystem (Figure 2-7).

The three components are the producers, the consumers, and the abiotic matter. The *pro-*

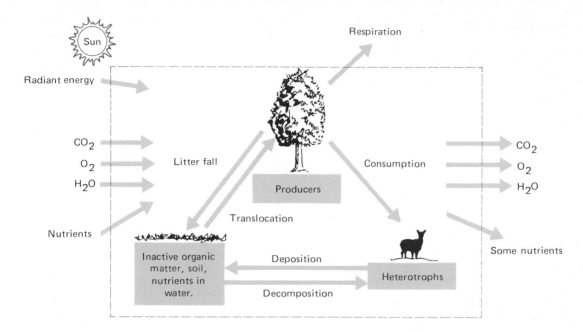

**FIGURE 2-7**
*Schematic diagram of an ecosystem. Again the dashed line represents the boundary of the system. The three major components are the producers, the consumers, and the abiotic elements—inactive organic matter, the soil matrix, nutrients in solution in aquatic systems, sediments, etc. The lines indicate interactions within the system. Inputs to the ecosystem from the environment include $CO_2$, $O_2$, $H_2O$, nutrients, and radiant energy. Outputs to the environment include $CO_2$, $O_2$, $H_2O$, some nutrients, and heat of respiration. (Adapted in part from O'Neill, 1976.)*

ducers or *autotrophs,* the energy-capturing base of the system, are largely green plants. They fix the energy of the sun and manufacture food from the simple organic and inorganic substances. Autotrophic metabolism is greatest in the upper layers of the ecosystem—the canopy of the forest and the surface water of lakes and oceans.

The *consumers* or *heterotrophs* utilize the food stored by autotrophs, rearrange it, and finally decompose the complex materials into simple, inorganic compounds again. In this role they function in regulating the rate of energy flow and nutrient flow and in stabilizing the system. The heterotrophic component is often subdivided into two subsystems, consumers and decomposers. The consumers feed largely on living tissue and the decomposers break down dead matter into inorganic substances. But no matter how they may be classified, all heterotrophic organisms are consumers and all in some way act as decomposers. Heterotrophic activity in the ecosystem is most intense where organic matter accumulates—in the upper layer of the soil and the litter of terrestrial ecosystems and in the sediments of aquatic ecosystems.

The third or *abiotic* component consists of the soil matrix, sediments, particulate matter, dissolved organic matter and nutrients in aquatic systems, and dead or inactive organic matter in terrestrial systems. Inactive organic matter is derived from plant and consumer remains and is acted upon by the decomposer subsystem of the heterotrophs. Such organic matter is important to the internal cycling of nutrients in the ecosystem.

The driving force of the system is the energy of the sun which causes all other inputs to circulate through the system. The various outputs, or more correctly outflows, from one subsystem become inflows to another. While energy is utilized and dissipated as heat of respiration, the chemical elements from the environment are recycled by organisms within the system. How fast nutrients turn over in the system is influenced by these consumers, which function as rate regulators in the ecosystem.

## Ecosystem stability

The ecosystem is an open system that receives energy from an outside source (the sun), fixes and utilizes it, and ultimately dissipates heat to space. If the sun's energy were to be shut off, the ecosystem (and life on earth) would cease to function. The ecosystem is also cybernetic. The regulatory feedback systems function through organisms to maintain homeostasis, that is, they constitute a mechanism for stability.

Ecosystems can respond to changes or perturbations only through the response of the populations that make up the system. Thus homeostasis in the ecosystem begins with the individual organism's ability to respond to a range of environmental conditions by adjusting its physiology or behavior. The success of the organism in maintaining homeostasis is reflected in its ability to leave reproducing offspring.

The size and success of a population of organisms depend upon the collective ability of the organisms to maintain their numbers in the environment. Population homeostasis involves the adjustment of reproductive, death, and immigration rates by interactions among members of the population (see Chapter 15) and with members of other populations.

Populations are the structural elements of the ecosystem through which energy flows and nutrients cycle. The pathways the inputs take through the system depend upon the interactions of the populations, their relationships to each other in terms of eating and being eaten. The pathway available in the simplest systems allows for only a linear flow of inputs and outputs (Figure 2-8). Each element derives its input from only one other element. In complex systems every element at some point receives inputs from or transmits outputs to every other element. Although systems with more complex pathways within the system are considered more stable, this may not necessarily be the case (May, 1973).

Some populations, such as annual plants, insects, and mice, respond rapidly to favorable environmental conditions and decline rapidly under unfavorable conditions. Populations of other organisms, such as trees and wolves, show less response to environmental changes,

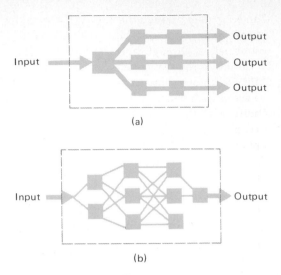

**FIGURE 2-8**
*Diagram of the pathway of energy flow in two systems. In a simple system interactions are minimal and linear. In a more complex system interactions are more numerous and may move in several directions.*

little variation in reproductive and death rates, and fluctuate little from one time frame to another. Such populations increase under favorable environmental conditions, but they are unable to exploit rapid, favorable fluctuations. At the same time they are less vulnerable to unfavorable environmental fluctuations. By containing elements of both types of populations, ecosystems achieve some level of homeostasis.

Ecosystem homeostasis or stability may be considered from two viewpoints. Stability may be judged in terms of the persistence of the ecosystem (Margalef, 1968; Preston, 1962). For example, a forest ecosystem is more persistent than a weedy field. Stability may also be judged in terms of the ability of a system to return to an equilibrium state after a temporary disturbance (Holling, 1973; May, 1973). The faster it returns to and the less it varies from its original state, the more stable is the system. By this standard the weedy field is more stable than the forest, because the site, once disturbed, quickly grows back to the same weed species. In either situation the stability of the system results from the ability of the population to react to perturbations to the ecosystem.

Stability in ecosystems involves persistence.

If an ecosystem through the ability of its populations is able to absorb changes, it tends to last or persist through time. Persistence may be viewed as the *resilience* of the system — its ability to absorb changes and still persist and the speed with which it returns to its original condition (Holling, 1973). Resilience does not necessarily imply high stability of individual populations. Populations within a system may fluctuate widely in response to environmental changes yet the system itself may be highly resilient.

For example, in the northern spruce-fir forest of North America the spruce budworm population under certain environmental conditions increases rapidly, escapes control of predators and parasites, and feeds heavily on balsam fir, killing many of the trees and leaving only the less susceptible spruce and birch. After the spruce budworm population collapses because of exhaustion of food supply, young balsam firs grow back in thick stands with spruce and birch. Between budworm outbreaks balsam fir outcompetes spruce and birch, but during outbreaks spruce and birch are favored over balsam fir. Thus interactions among populations of budworm, fir, spruce, and birch maintain ecosystem homeostasis (Holling, 1973). The spruce-fir forest ecosystem is resilient, even though the population elements involved exhibit low stability.

Related to resilience of an ecosystem is *resistance,* its ability to resist changes from disturbance. Often the resistance of an ecosystem results from its structure. Most resistant systems characteristically have large biotic structure such as trees and have nutrients and energy stored in standing biomass. A forest ecosystem is relatively resistant to disturbance. It can withstand such environmental perturbations as sharp temperature changes, drought, and insect outbreaks because the system is able to draw on stored reserves of nutrients and energy. For example, a late spring frost may kill the new leaves of trees, but the trees are able to draw on energy reserves to replace the leafy growth. But if the forest system is highly disturbed by fire or logging, the system exhibits low resilience; its return to its original condition is slow. Thus a sort of inverse relationship often exists between resistance and resilience.

Aquatic ecosystems which lack any long-term storage of energy and nutrients in biomass, exhibit little resistance. An influx of pollutants such as sewage effluents perturbs the systems by adding more nutrients to the ecosystem than it can handle. But because the system is limited in its capacity to retain and recycle nutrients, the system returns to its original condition relatively soon after the perturbation, such as the pollutants, is reduced or removed.

Lake Washington near Seattle is an example. Used as a basin for sewage disposal, the lake received a large input of excessive nutrients, especially phosphorus. The input eliminated certain diatom and algal populations and encouraged others, especially filamentous algae, to increase, destroying the clear water lake ecosystem. Once the sewage input was diverted from the lake, the phosphorus levels in the water declined, the algal populations shrunk, and the lake began to return to its original clear water condition. This lake system had high resilience, but low resistance.

A perturbation, if strong enough, may result in a different outcome. The system may be so greatly disturbed that it is unable to return to its original state and a different ecosystem may replace it. For example, when the forests of Scotch pine were cut in northern Britain, they were replaced by moorlands; and when the spruce forests of the central Appalachians in North America were cut and burned, they were replaced by stands of blueberry and stunted birch.

Ecosystem stability ultimately results from the adaptive responses of organisms to environmental conditions and changes. Adaptation of an organism is simply its conformity to its environment. Changes in the physiology, behavior, and morphology of species come about because some individuals in the population are better able to cope with environmental conditions. Those individuals best adjusted to environmental conditions make the greatest contribution to future generations; that is, they leave the greatest number of progeny in the population. This ability of an individual to leave offspring is termed its *fitness.*

Changes in the environment act as selective pressures on individuals, reducing the fitness of some and increasing the fitness of others. This differential reproduction is *natural selection*. It translates environmental patterns into adaptations. Natural selection results in *evolution,* the change in gene frequencies through time. Or to state it differently, evolution is the development of organisms through time.

Thus homeostasis of a system comes about through evolutionary changes in the elements that form it. (This is what is implied in the term ecosystem evolution or community evolution. Ecosystems and communities, of course, cannot evolve, because they are not evolutionary units on which natural selection can work.) Individuals in the population that make up the system respond through physiologic, behavioral, and genetic changes that restore or maintain the ability to respond to future unpredictable environmental changes. In variable environments natural selection acts to maintain flexibility by allowing variations in genetic types to be maintained in low numbers in the population. If environmental conditions or selective pressure are such that one genetic variant is more fit than others, that type is favored by natural selection and it increases in the population. If the environment changes so that a different type is more fit, that type is favored. Because the environment is always changing, genetic variety and variability allow populations to contribute to ecosystem stability.

## Essential processes in the ecosystem

The inputs of energy and nutrients into the ecosystem are handled by four functional processes: photosynthesis, herbivory, carnivory, and decomposition. Herbivory and carnivory serve to process nutrients and energy, move them along the routes, and retain them in the system as long as possible. But the two fundamental processes are photosynthesis—the fixing of energy and the incorporation of nutrients into active plant tissue—and decomposition—the final dissipation of energy and the reduction of organic matter into organic substances.

### PHOTOSYNTHESIS

The functioning of ecosystems is based on the fixation of energy and the production of organic compounds through the photosynthetic activity of autotrophs. Energy enters the ecosystem as visible light and is stored in plants during photosynthesis. From that point biochemical changes involve a series of rearrangements of matter into compounds of less chemical energy. These chemical rearrangements are accompanied by the production of heat that is lost to the environment. This loss of heat is accompanied by the loss of carbon dioxide, water, and nitrogenous compounds which are recycled through the ecosystem. Although some energy is irrevocably lost, some of it is stored in the system.

The forest and the pond ecosystem differ in energy fixation and flow. Energy fixation in the pond is accomplished by a number of plant populations consisting of many small individuals capable of rapid reproduction under certain optimal conditions. When environmental conditions change, one population disappears and is rapidly replaced by another. This rapid turnover in populations prevents any lasting accumulation of biomass, but seasonal replacements of plant populations maintain high productivity during the growing season. Energy fixation in the forest is accomplished by a plant population consisting of individuals of large size and slow reproductive rates. The individuals are long-lived (according to a human time scale) and can persist through environmental changes by utilizing stored energy reserves accumulated over time.*

*Time in ecology is relative, measured in spans comprehensible to humans but probably irrelevant to other organisms and certainly to ecosystems. A blue-green alga or a diatom is called short-lived from a human point of view. A few weeks to a human may be a century in alga time span. So it is with trees and animals. A tree with a life span of 100 years is called long-lived by humans, but such a tree is short-lived compared to a 2500-year-old redwood. The time frame of an alga, a tree, a squirrel, a warbler, or an elephant is quite different from that of a human. This should be kept in mind when such humanly relative terms as short-lived and long-lived are used in the text.

**The process of photosynthesis.** The familiar formulation of photosynthesis is

$$6CO_2 + 6H_2O + energy \rightarrow$$
$$C_6H_{12}O_6 + 6H_2O + 6O_2$$

Whereas the intermediate compounds of photosynthesis do include glucose and the by-products of water and oxygen, further related syntheses produce free amino acids, proteins, fatty acids and fats, vitamins, pigments, and coenzymes. All of these substances probably are synthesized in the chloroplasts by reaction involving photoelectron transport and photophosphorylation. The synthesis of various products may take place in different parts of the plant or under different environmental conditions. Mature leaves of certain species of plants may produce only simple sugars, whereas young shoots and rapidly developing leaves may produce fats, proteins, and other constituents.

Photosynthesis is carried out primarily by chlorophyll-bearing vascular plants and algae, both aquatic and terrestrial. Other minor contributions are made by photosynthetic bacteria that instead of water, use hydrogen, hydrogen sulfide, and various organic compounds as electron donors. Photosynthesis is a complicated process, the essential features of which can be outlined briefly.

When light energy strikes the chlorophyll molecule in a green plant, it is absorbed by pigments in the chloroplasts and its energy transferred to an electron of the chlorophyll molecule. The energy of the molecule is not raised to a higher state, and the electron is passed along through a series of chemical reactions. This results in the synthesis of adenosine triphosphate (ATP) from adenosine diphosphate (ADP), in the production of a strong reductant NADPH from nicotinamide adenine dinucleotide phosphate (NADP), and in the oxidation of water to H and OH ions. The H ions go to the reduction of NADP and the OH ions form water, release some molecular oxygen, and supply the electron to replace the one lost to the chlorophyll molecule. Further reactions involve the synthesis of carbohydrate from $CO_2$ and a series of reactions that incorporate the energy as well as the hydrogen and $CO_2$ into carbohydrates.

One of the recent discoveries of plant physiology is the difference in the way plants take up $CO_2$ and carry out the later reactions of photosynthesis. In most plants the photosynthetic process involves the reaction of $CO_2$ from the atmosphere with ribulose diphosphate (RuDP). This is a phosphorylated sugar with five carbon atoms. When a carbon dioxide molecule reacts with RuDP, two molecules of phosphoglyceric acid with three carbon atoms form. This product subsequently is converted into other products. Some of the phosphoglyceric acid is used to reform RuDP molecules that again act as $CO_2$ acceptors. This three-carbon photosynthetic pathway is known as the Calvin-Benson cycle and the plants using this method are known as $C_3$ plants.

In 1965 some plant physiologists discovered that sugarcane and other plants had a different pathway (Black, 1973). These plants have within the leaf acceptor molecules of phospho-enol-pyruvate (PEP), a three-carbon compound. $CO_2$ entering the leaf reacts with PEP to form malic acid and aspartic acid, each having four-carbon molecules. Both malic and aspartic acids are broken down by enzymes to release the fixed $CO_2$ to form phosphoglyceric acid with three-carbon molecules as in the Calvin cycle. At the same time the two acids also produce pyruvic acid used to form more PEP. Plants using the four-carbon method are called $C_4$ plants.

Thus the difference between the two types of plants is in the initial fixation of $CO_2$. $C_4$ plants have an extra step in $CO_2$ fixation which gives them a certain advantage. PEP is more reactive with $CO_2$ than RuDP, whose activity is somewhat inhibited by oxygen. In situations where the concentration of $CO_2$ is low and oxygen is high, the PEP system is more efficient than RuDP. PEP provides an initial and very efficient means of fixing $CO_2$ at very low concentrations, which exists when light saturation and temperatures are high and water is in short supply. The fixed $CO_2$ can then be fed to the $C_3$ or Calvin-Benson cycle.

Anatomically and physiologically $C_3$ and $C_4$ plants are different. $C_3$ plants have vascular bundles surrounded by a layer or sheath of large colorless cells and the cells containing the chlorophyll are irregularly distributed throughout the mesophyll of the leaf. The $C_3$ plants have an optimum temperature for photosynthesis of 16° to 25° C (Figure 2-9). The $CO_2$ compensation point (at which $CO_2$ uptake is equal to $CO_2$ evolution by the plant) is in the range of 30 to 70 ppm $CO_2$. $C_3$ plants also have a low light saturation threshold (Figure 2-10). Photosynthesis is inhibited when light falling on leaves reaches a range of 1000 to 4000 foot-candles.

*FIGURE 2-9*
*Effect of changes in light intensity on the photosynthetic rates of $C_3$ and $C_4$ plants grown under identical conditions. In this case the conditions are a 16-hour day, 25° C at day, 20° C at night, with ample water and nutrients. The $C_3$ species spear orache* (Atriplex patula) *exhibits a decline in the rate of photosynthesis, as measured by $CO_2$ uptake, as light intensity increases. The $C_4$ species red orache* (Atriplex rosea) *shows no such inhibition. (After Bjorkman, 1973.)*

Plants exhibiting the $C_4$ cycle have vascular bundles surrounded by a sheath of cells rich in chlorophyll, mitochondria, and starch. Mesophyll cells are arranged laterally around the bundle sheaths and have few chloroplasts. Compared to $C_3$ plants, $C_4$ plants have a higher optimum temperature for photosynthesis, 30° to 45° C. They have a lower $CO_2$ compensation point, 0 to 10 ppm $CO_2$, and exhibit less loss of reduced $CO_2$ through photorespiration, thus saving energy for other metabolic processes. $C_4$ plants, then, can carry on photosynthesis in very low concentrations of carbon dioxide. They have a high light saturation threshold; in fact it is difficult for the plants to reach light saturation even at full sunlight (10,000 to 12,000 footcandles). The net rate of photosynthesis for $C_4$ plants in full sunlight is 40 to 80 mg $CO_2/dm^2$ leaf area/hr, while the net rate of $C_3$ plants is 15 to 35 mg $CO_2/dm^2$ leaf area/hr (see Black, 1973).

These differences have ecological conse-

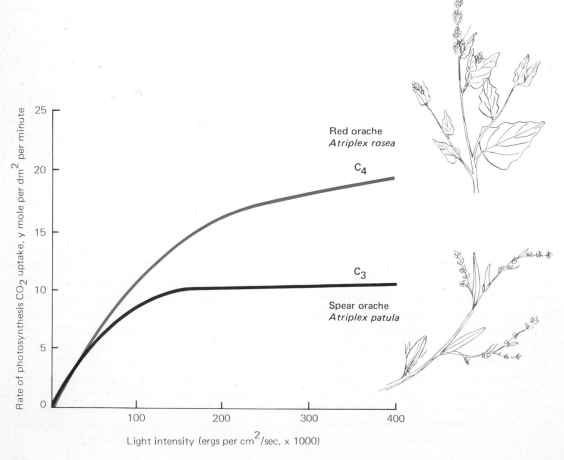

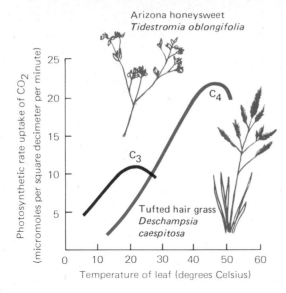

FIGURE 2-10
*Effect of changes in leaf temperature on the photosynthetic rates of $C_3$ and $C_4$ plants. $C_3$ plants, represented by the alpine tundra plant tufted hair grass (Deschampsia caespitosa), exhibit a decline in the rate of photosynthesis as the temperature of the leaf increases. They also reach maximum photosynthetic output at a lower temperature than $C_4$ species represented by Arizona honeysweet (Tidestromia oblongifolia). This desert species increases its rate of photosynthesis as the temperature of the leaf increases up to about 50° C. (After Bjorkman and Berry, 1973.)*

quences. Because they are more efficient photosynthetically, $C_4$ plants are more competitive than $C_3$ plants and more easily adjust to environmental stresses. Thus $C_4$ plants are characteristic of desert ecosystems, where light intensities and temperatures are high and water is scarce. Some desert plants have modified the $C_4$ and $C_3$ cycles into a crassulacean acid metabolism (CAM). They employ the $C_4$ cycle at night to minimize water loss through the stomata, taking up $CO_2$ and storing it as malic acid, and then switch to the $C_3$ cycle during the day, using the malic acid to complete photosynthesis.

The $C_4$ method is the more advanced photosynthetic process. It is not found in algae, bryophytes, ferns, gymnosperms, or more primitive angiosperms. Grasses hold about half of the known $C_4$ species. In North America most of these are subtropical in distribution. From Florida to Texas 65 to 82 percent of the grass species are $C_4$ plants, in the central plains 31 to 61 percent, and in the northern part of the continent 0 to 23 percent (Terri and Stowe, 1976). To date no $C_4$ grasses are known to grow on the tundra.

Some agricultural crops, such as sugarcane, are $C_4$ plants, but many agricultural crops are $C_3$ species with a low photosynthetic rate (see Bjorkman and Berry, 1973; Bassham, 1977), while their weed competitors are $C_4$ species. Agricultural plants can escape competition only by human imposition of cultural practices. The fact that $C_4$ plants appear to be inferior food sources for grazing herbivores and are often avoided by them has evolutionary significance (Casswell et al., 1973).

*Limits on rates of photosynthesis.* The rate of photosynthesis in an ecosystem is limited by both the environmental conditions and the nature of community structure. These limits include such environmental variables as light intensity, temperature, moisture, atmospheric gases, soil nutrients, and competition and such plant variables as leaf area, the geometry of the vegetational canopy, and the photosynthetic productive capacity of plants. To further complicate the picture, all of these variables interact with each other.

The intensity of light that reaches the plant is influenced by the local light climate, the modification of that light by the structure of the vegetation, and the position of the leaf in the vegetational profile (Figure 2-11). The relationship of the leaf to light interception can be described by the leaf area index (LAI). This is the ratio of leaf area per unit of ground area. For some plants there is an optimum LAI, the point at which there is minimal shading of one leaf by another (Pearce et al., 1965). As the LAI increases beyond this value, there is a decrease in photosynthesis. Other plants do not show this response (Loomis et al., 1967). The angle of the leaf influences light interception. Leaves that are perpendicular to incoming light (i.e., horizontal leaves) intercept the most light, but a number of layers of horizontal leaves reduce the amount of light reaching lower leaves (Loomis et al., 1967). Leaves growing at an angle require a much higher LAI to intercept the same quantity of light as

**23**

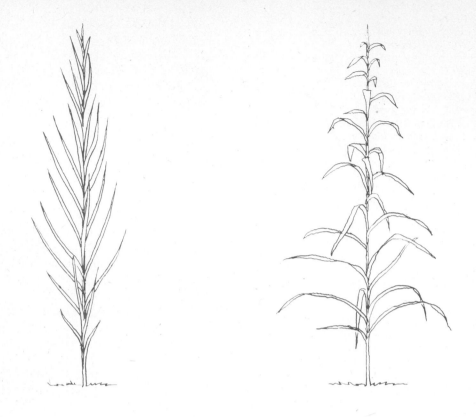

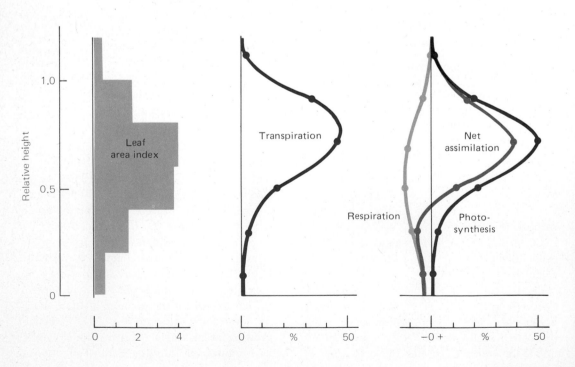

Leaf
area index

Transpiration

Net
assimilation

Respiration

Photo-
synthesis

Relative height

1.0

0.5

0

0    2    4

0         %         50

−0 +      %         50

FIGURE 2-11
*Relation of photosynthesis to leaf orientation, leaf area, and canopy structure. Although horizontal leaves capture the most sunlight, the upper layers shade the lower, reducing the overall interception of light. An upright orientation of leaves is more efficient in capturing the sun's energy. That type of leaf arrangement is typical of communities in which the individual plants are growing closely as among the grasses.*

*The amount of photosynthetic area is measured in terms of leaf area index (LAI). This is the ratio of the surface area of a leaf (the upper or lower surface only) to a given surface area of land as measured in hectares or acres. An LAI value of 1 means 1 hectare of leaf surface to 1 hectare of ground surface; value of 2 means 2 hectares of leaf surface to 1 hectare of ground surface. The LAI varies with the type of plant, leaf orientation, type of leaf, and canopy structure. Note that the LAI index varies with the vertical structure of the canopy. As indicated in the graphs for a forest ecosystem, the greatest amount of transpiration and photosynthesis takes place at the height where the LAI is the highest. (b after Baungartner, 1968).*

horizontal leaves. As their LAI increases above a certain value, the angled leaves carry on photosynthesis at a faster rate. Such an adaptation in leaf position to obtain maximum photosynthetic rate and still possess a high LAI is characteristic of corn, grasses, beets, turnips, and other row crops (Loomis et al., 1967).

For an individual leaf, at low light intensity, photosynthesis increases linearly as the light intensity increases. At higher intensities this rate increases more slowly until a maximum rate is achieved (Kok, 1965). This is particularly true for $C_3$ plants. In $C_4$ plants photosynthesis increases even at light saturation, although at a slower rate.

**Efficiency of photosynthesis.** The photosynthetic efficiency of converting the energy of the sun to organic matter can be assessed from two viewpoints: (1) the amount of energy required for the evolution of a molecule of oxygen and (2) the ratio of calories per unit area of harvested vegetation to the total solar radiation intercepted. In terms of energy input photosynthesis is a rather efficient process. To release 1 mole of oxygen and to fix 1 mole of

carbon dioxide a green plant needs an estimated 320 kcal of light energy. For each mole of oxygen evolved, approximately 120 kcal of energy are fixed. This corresponds to an efficiency of approximately 38 percent. Efficiencies calculated for isolated chloroplasts and for some algae amount to 21 to 33 percent (Kok, 1965; Wassink, 1968; Bassam, 1965).

When considered from the standpoint of calories stored in relation to light energy available, the efficiency is considerably less. The usable spectrum, 0.4 to 0.7 $\mu$ wavelengths, is only about half the total energy incident upon vegetation. Highest short-term efficiency measured over a period of weeks of active growth may amount to 12 to 19 percent (Wassink, 1968). In most instances, however, photosynthetic efficiency is computed either on an annual or a growing-season basis. Mean photosynthetic efficiency of a Puerto Rican tropical rain forest has been estimated at 7 percent (H. T. Odum, 1970). Sugarcane fields in Java have an efficiency of around 1.9 percent (Hellmers and Bonner, 1959). The photosynthetic efficiency of land areas is about 0.3 percent, of the ocean about 0.13 percent. Total yields of solar energy on earth amount to about 0.15 to 0.18 percent (Wassink, 1968).

A sampling of ecological efficiencies is given in Table 2-1. However, these figures cannot be accepted with a great deal of confidence nor are they really comparable. All of them involve a number of assumptions in their determination. There is no standard method for determining photosynthetic efficiency. In many cases the radiant energy incident upon the vegetation studied was estimated from tables or from measurements taken in the region but not in the study area. Temperature, moisture, nutrient status of the soil, and other variables are not held constant. Estimates are often based on the total energy available for the year rather than for the growing season alone. This lowers the estimates considerably. If based on the growing season, estimates might be as high as 10 percent instead of 1 to 3 percent. Error is also introduced if energy fixation is based on estimates of community peak standing crops rather than on the determination of peak standing crop of individual species within the community. Because different species reach peak standing crop at different

TABLE 2-1
*Comparative yields and photosynthetic efficiencies of some plant communities*

| Community | Location | Growing season (days) | Dry matter ($g/m^2/day$) | Efficiency (%) | Reference |
|---|---|---|---|---|---|
| Corn | Minnesota | 92 | 11.00 | 2.10 | Ovington and Lawrence, 1967 |
| 1-year weed field | New Jersey | 120 | 19.50 | 3.80 | Botkin and Malone, 1968 |
| Meadow steppe | West Siberia | 100 | 15.00 | 3.40 | Rodin and Bazilevic, 1965 |
| Short-grass steppe | Colorado | 115 | 22.00 | 3.40 | Moir, 1969 |
| Tall-grass prairie | Colorado | 140 | 8.00 | 1.20 | Moir, 1969 |
| Tall-grass prairie | Colorado | 218 | 3.85 | 0.77 | Sims and Singh, 1971 |
| Short-grass prairie | Colorado | 266 | 5.38 | 0.80 | Sims and Singh, 1971 |
| High mountain grassland | Colorado | 92 | 10.74 | 1.20 | Sims and Singh, 1971 |
| Desert grassland | Colorado | 267 | 1.58 | 0.16 | Sims and Singh, 1971 |
| Oak-pine forest | Long Island | 250 | | 0.90 | Whittaker and Woodwell, 1969 |
| Polar desert | Devon Island | 50–60 | 0.59 | 0.03 | Bliss, 1975 |
| Sedge-moss meadow | Devon Island | 50–60 | 12.36 | 0.79 | Bliss, 1975 |
| Lichen heath | Norway | 105 | 10.80 | 0.70 | Wielgolaski and Kjelvik, 1975 |
| Wet meadow | Norway | 120 | 36.30 | 2.30 | Wielgolaski and Kjelvik, 1975 |
| Birch forest | Norway | 150 | 28.20 | 1.70 | Wielgolaski and Kjelvik, 1975 |

periods of the growing season, estimates made only at one period would seriously underestimate photosynthetic efficiency.

### DECOMPOSITION

In the November woods the leaves of summer lie brown and withered. They had formed the green canopy of the forest, and in the course of a summer they pumped water and minerals from the soil, traded oxygen for carbon dioxide in the atmosphere, and trapped the energy of the sun by photosynthesis, converting it into carbohydrates and other energy-storing compounds. What energy the trees did not need for their own maintenance and production of fruit and seed, the leaves sent back to the trunk, branches, and roots to be stored as woody growth.

As the days shortened, the process slowed and finally stopped. A delicate layer of cells formed between the leaf and the twig, causing the stem of the leaf to fall off. As many as 10 million leaves per acre may flutter to the

ground in fall in our hardwood forests, an equivalent of 2000 to 3000 lb.

It is the activity of the organisms of the forest floor that ensures that next spring will be green. Unless the tons of leafy material and other debris are disposed of, the forest will smother in its own litter. And if the mineral matter it contains is not released to the soil to be recycled to the trees again, forest plants will face a shortage of nutrients necessary to life. Thus in the leaves underfoot is an almost unbelievable activity as billions of organisms reduce the debris to chemical elements and mix the organic matter with the soil.

*Role of the decomposers.* In effect what the organisms of decay accomplish is the reversal of the process of photosynthesis—the reduction of organic matter to the inorganic compounds (carbon dioxide, water, and oxygen) from which it was synthesized and the utilization of the energy this matter contains. The process of decomposition is a complex of many processes, including fragmentation, mixing, change in physical structure, ingestion, egestion, concentration, and action of enzymes, all accomplished by a wide diversity of organisms.

The organisms involved in this process are the decomposers. The use of a single term to include all the organisms involved may be misleading if it is taken to mean that they all function in the same manner, all somehow mysteriously converting organic matter back to inorganic compounds. Actually, the world of the decomposers is one of complex food webs, involving herbivory, carnivory, parasitism, and, to a limited extent, producers. In one way or another all contribute to the breakdown of organic matter.

In a popular sense decomposition is associated with the death of an organism—the withered vegetation lying on the ground or an animal carcass lying in a field. However, the process begins long before the detritus falls to the ground or to the bottom of a pond. Decomposition is occurring when any herbivore consumes vegetation. Not only does the animal extract minerals and nutrients from the plant for its own nutrition, but it also deposits a substantial portion as partially decomposed fecal material. A portion of what is not consumed is left as fragments exposed to microbial action. In this sense all animals are both consumers and decomposers.

*Microbial decomposition.* Microorganisms, the bacteria and fungi, are the organisms most commonly associated with decomposition. The bacteria may be autotrophic or heterotrophic. The autotrophic forms, commonly considered producers, may oxidize such compounds as ammonia, nitrates, and sulfides as a source of energy for the conversion of carbon dioxide into living matter. In utilizing this source of energy, they are aiding in decomposition. The heterotrophic bacteria are the true decay organisms. They may be *aerobic,* requiring oxygen as the electron acceptor, or they may be *anaerobic,* able to carry on their metabolic functions without oxygen by using some inorganic compound as the oxidant. Anaerobic bacteria are commonly found in aquatic muds and sediments and in the rumen of ungulate herbivores. Many are facultative anaerobes; they use oxygen when it is present, but in its absence they can use inorganic compounds as their energy source. The fungi are the

other group of major decomposers of plant materials.

The bacteria and fungi produce enzymes necessary to carry out specific chemical reactions. Enzymes are secreted into the plant and animal material, some of the products are absorbed as food, and the remainder is left for other organisms to utilize. Once one group has exploited the material to its capabilities, another group of bacteria and fungi move in to continue the processes. Thus a succession of microorganisms occurs in the detritus until the material is finally reduced to elemental nutrients.

Microbial decomposition of plant leaves begins while the leaves are still on the trees. Living plants produce varying quantities of exudates that support an abundance of surface microflora. These organisms feed on the exudates and on any cellular material that sloughs off. The same exudates account for the source of nutrients leached from the leaves during a rain. In tropical rain forests leaves are heavily colonized by bacteria, actinomycetes, and fungi (Ruinen, 1962).

While microbes are utilizing exudates of the leaves, other organisms are utilizing organic material from the roots of living plants. The soil region immediately surrounding the roots, known as the rhizosphere, and the root surface itself, known as the rhizoplane, support a whole host of microbial feeders on root litter and root exudates. Root exudates may consist of simple sugars, fatty acids, and amino acids. In fact some 10 sugars, 21 amino acids, 10 vitamins, 11 organic acids, 4 nucleotides, and 11 miscellaneous compounds have been identified in the rhizosphere. Obviously, not all such exudates occur in the rhizosphere of all plants. The absence of certain exudates in the rhizosphere can influence the quantitative and qualitative differences in the microflora of rhizospheres of different plant species (F. E. Clark, 1969a, b).

When the plant body becomes senescent, decomposition is accelerated. The plant is invaded by microbial decomposers. If moisture is sufficient, fungi colonize decaying culms of grass plants. A favorite point of invasion is the internode where the leaf is attached to the stem. The species of fungal flora involved is influenced by the distance of the internodes and leaves from the ground (the closer to the ground the more humid the habitat) and by the species of grass (Hudson and Webster, 1958; J. Webster, 1956–1957). Fungi infect pine needles 5 to 6 months before the needles fall (Burges, 1963). Destruction of the palisade layers of deciduous leaves by leaf miners opens up the affected leaves to microbial attack while they are still hanging on the tree. But the bulk of decomposition does not take place until the dead vegetation comes in contact with the soil.

Once on the ground, plant debris is subject to attack by other bacteria and fungi. The rate at which these organisms feed on the debris depends upon the moisture and temperature. Higher temperature favors more rapid decomposition, and continuous moisture is more favorable than alternate wetting and drying. The microbial mass is largely fungal mycelia. Among the first to invade the material are sugar-consuming fungi and bacteria that rapidly utilize the readily decomposable organic compounds (see Stewart et al., 1966). Once the glucose is utilized, the debris is invaded by other bacteria that feed on cellulose.

As the bacteria and fungi work on the plant debris, they assimilate the nutrients and incorporate them into their tissues. As long as these nutrients are a part of the living microbial biomass, they are unavailable for recycling. This is known as *nutrient immobilization*. The amount of mineral matter that can be tied up by microbes varies greatly, but many exhibit luxury consumption, that is, they consume more than they need for maintenance and growth. Such consumption of nutrients such as potassium, calcium, and nitrogen can affect primary production.

Such bacteria and fungi are short-lived. They die or are consumed by litter invertebrates. Death and consumption as well as the leaching of soluble nutrients from the decomposing substrate, release minerals contained in the microbial and detritus biomass. This process, known as *mineralization,* makes nutrients available for use by primary producers and microbes. Thus a cycle of immobilization and mineralization takes place within the soil. Nutrients are temporarily immobilized in microbial tissue. As microbes die, the nutrients are released or mineralized and become available for uptake again. Microbial uptake occurs simultaneously with mineralization. The amount of nutrients available for primary producers depends in part on the magnitude of uptake by the microbial decomposers.

*Decomposition by invertebrates.* The process of decomposition is aided by the fragmentation of detritus by litter-feeding invertebrates, collectively known as the *reducer-decomposers*. They feed on plant and animal remains and on fecal material. In terrestrial ecosystems this group includes such organisms as earthworms, millipedes, isopods, and the larvae of beetles and flies. In aquatic and semiaquatic situations mollusks, crabs, and scuds are among the detritus feeders.

In terrestrial situations litter-feeders consume parts of the leaves serving to open up the material to microbial invaders. The action of such litter-feeders as millipedes and earthworms may increase exposed leaf area to 15 times its original size (Ghilarov, 1970). Because the net assimilation of plant debris by litter-feeders is on the average less than 10 percent, a great deal of the material passes through the gut of these organisms. They utilize only the easily digested proteins and carbohydrates. Mineral matter in the material is often concentrated. This fecal matter is readily attacked by microbes. Some litter-feeders such as earthworms enrich the soil with vitamin $B_{12}$. In addition they mix organic matter with soil, thus bringing the material in contact with other microbes. Although the mineral pool in the contained biomass and the contribution to energy flow (van der Drift, 1971) by the reducers is relatively small (about 4 to 8 percent), they still make a major but indirect contribution to decomposition.

*Humus formation.* As the amount of energy decreases with time, the least decomposable material, largely derived from lignin, is left behind as humus. *Humus* is a structureless, dark-colored, chemically complex material whose characteristic constituents are humin, a complex of unchanged plant chemicals, and other organic compounds such as fulvic acid and humic acid. The latter is derived from lignin and plant flavonoids, which undergo degradation and conjugation with amino compounds, carbohydrates, and silicate materials. Because of the chemical complexity of the material, this process is accomplished in part by such fungi as *Penicillium* and *Aspergillus* and such bacteria as *Streptomyces* and *Pseudomonas*. Decomposition proceeds so slowly that the amount of organic matter in soil changes little each year. Annual loss by decomposition is balanced by the formation of new humic material. Carbon-dating indicates that humus in podzol soils (See Chapter 3) has a mean residence time of 250 ±60 years, that in chernozems 870 ±50 years (Campbell et al., 1967).

The importance of the reducer's role has been demonstrated in several experiments. These employed nylon litter bags to exclude reducers from the litter sample or used naphthalene, which drives away arthropods and macrofauna but does not inhibit the activity of bacteria and fungi. Witkamp and Olsen (1963) placed white oak leaves, some confined in litter bags and some unconfined, in pine, oak, and maple stands in November. By the following June both the unconfined and confined leaves showed a similar loss in weight. But after June the unconfined leaves showed a sudden increase in breakdown. Before June both types of leaves lost weight by the breakdown of easily decomposable substances through the action of microflora and microfauna. Because the unconfined leaves were broken into fragments by the reducer-decomposers as well as by birds, mice, wind, and rain, they were more available to microorganisms for further decay. Experiments by Kurcheva (1964), Edwards and Heath (1963), and Witkamp and Crossley (1966) show that suppression of activities of the reducer-decomposers results in a marked slowdown in the rate of microbial decomposition. In the absence of these invertebrates, the decomposition of wood is slowed by half. Not only do these organisms physically fragment the substrate but they also inoculate it with fungi and bacteria (Ghilarov, 1970). Without the activity of the reducers, nutrient elements could stagnate in the litter, bound energy in the ecosystem would increase, and both primary and secondary productivity would decrease.

Thus the decomposition of plant and animal material involves a complicated succession and interaction of different organisms, each of which contributes a part.

*Microbial consumers.* Feeding on the bacteria and fungi is another group, the *microbial consumers.* These consist of such diverse groups as the nematodes, collembolans or springtails, larval forms of beetles (Coleoptera), flies (Diptera), and mites (Acarina). Nematodes feed heavily on bacteria, fungi, and algae in the soil. Their impact on the microbiotic is not known. They may so reduce the microbial population that they delay ordinary decomposition. Or perhaps they promote microbial activity by preventing the microbial population from overproducing and by maintaining it at its level of maximum productivity or rate of division. Thus nematodes and other grazers may prevent aging and senescence of the bacterial populations. Collembolans and mites may have the same effect on fungi. If they do reduce the fungal population, they also stimulate its growth by dispersing fungal spores. By consuming fungi these microbial grazers may hasten recirculation of nutrients by releasing the nutrients locked up in the biomass of microbial populations.

*Influences on decomposition.* The rate of decomposition is influenced by a number of environmental and biotic variables. Among these are moisture, temperature, exposure, altitude, type of microbial substrate, and vegetation. Both temperature and moisture greatly influence microbial activity by affecting metabolic rates. Alternate wetting and drying and continuous dry spells tend to reduce both activity and populations of microbes. Slope exposure, especially as it relates to temperature and moisture, and type of vegetation can increase or decrease decomposition. Witkamp (1963) found that bacterial counts from north-facing slopes were nine times higher in hardwoods than in coniferous stands, but counts from hardwood and coniferous stands on south slopes did not differ. This was undoubtedly due to drier conditions on the south slopes.

The species composition of leaves in the litter had the greatest influence. Easily decomposable and highly palatable leaves from such species as redbud, mulberry, and aspen support initially higher populations of decomposers than litter from oak and from pine, which is high in lignin. Earthworms have a pronounced preference for such species as aspen, white ash, and basswood, take with less relish and do not entirely consume sugar maple and red maple, and do not eat red oak at all (Johnston, 1936). In a European study (Lindquist, 1942) earthworms preferred the dead leaves of elm, ash, and birch, ate sparingly of oak and beech, and did not touch pine or spruce needles. Millipedes likewise show a species preference (van der Drift, 1951). Thus decomposition of litter from certain species proceeds more slowly than litter from others. On easily decomposable material initially high populations of microbes decline with time as energy is depleted. But on more resistant oak and pine litter, initially low population densities increase as decomposition proceeds (Witkamp, 1963).

*Aquatic decomposers.* In terrestrial ecosystems bacteria and fungi take the major role in decomposition. The same cannot be said for aquatic ecosystems. Although the role of bacteria in aquatic environments is poorly understood, evidence suggests that bacteria and, to a limited extent, fungi act more as converters than as regenerators of nutrients, whereas phytoplankton and zooplankton take a major part in the cycling of nutrients.

Phytoplankton, macroalgae, and zooplankton furnish dissolved organic matter, with algae being main contributors. Phytoplankton and other algae excrete quantities of organic matter at certain stages of their life cycles, particularly during rapid growth and reproduction. During photosynthesis the marine algae *Fucus vesticulosus* produces as exudate on the average of 42 mg C/100 g dry weight of algae/hr. Total exudate accounts for nearly 40 percent of the net carbon fixed (Sieburth and Jensen, 1970). Johannes (1968) points out that 25 to 75 percent of the regeneration of nitrogen and phosphorus takes place in the presence of microorganisms by autolysis and solution rather than by bacterial decomposition. In fact 30 percent of the nitrogen contained in the bodies of zooplankton is lost by autolysis within 15 to 30 minutes after death, too rapidly for any significant bacterial action to occur.

Bacteria, phytoplankton, and zooplankton utilize inorganic nutrients as well as such organic nutrients as vitamin $B_1$ (necessary for the growth of both phytoplankton and zoo-

plankton) and organic sources of nitrogen and phosphorus. In effect they tend to concentrate these nutrients by incorporating them into their own biomass. Important in this concentration of nutrients are the bacteria. Dissolved organic matter is a substrate for the growth of bacteria. Both dissolved and colloidal matter condense on the surface of air bubbles in the water, forming organic particles on which bacteria flourish (R. T. Wright, 1970; Riley, 1963). Fragments of cellulose supply another substrate for bacteria. Bits of plant detritus, bacteria, and phytoplankton are consumed by both bacteria and planktonic animals (Strickland, 1965). As in terrestrial ecosystems the utilization of these organic nutrients by bacteria results in both the increase and the immobilization of nutrients. Bacteria can use nutrients such as phosphorus in excess of their need. Such luxury consumption can reduce the supply of available nutrients to phytoplankton, thus reducing algal blooms.

Bacteria are consumed by ciliates and zooplankton that in turn excrete nutrients in the form of exudates and fecal pellets in the water. Zooplankton, too, in the presence of an abundance of food, consumes more than it needs and can reduce microbial population. In the presence of abundance zooplankton will excrete half or more of the ingested material as fecal pellets, which make up a significant fraction of suspended material. These pellets are attacked by bacteria that utilize the nutrients and growth substances they contain. Thus the cycle starts over again.

Aquatic muds are largely anaerobic habitats. Fungi are absent, and the decomposer bacteria are largely facultative anaerobes. Incomplete decomposition often results in the accumulation of peat and organic muck. But the particulate matter from dead vegetation and animals nevertheless supports a rich bacterial flora. Plant fragments are colonized by bacteria, and both become food for snails and mollusks. Newell (1965) points out that the snail *Hydrobia* feeds on the detritus found in the mud flats. Its fecal pellets are devoid of nitrogen but rich in carbohydrates, suggesting that the snail cannot digest cellulose and other complex carbohydrates. If the pellets are held in filtered seawater, the nitrogen content quickly rises. The rise in nitrogen is accomplished by the growth of marine bacteria that colonize on the fecal material and utilize nitrogen dissolved in seawater. The fecal pellets, enriched with nitrogen in the form of bacterial protein, are reingested by the snail. The snail then digests the bacterial bodies, and the resultant fecal pellets again are devoid of nitrogen. Recolonization of fecal pellets is repeated.

Thus bacteria function primarily to concentrate nutrients rather than to release them to the environment by decomposition. That task is accomplished largely by the algae, zooplankton, and detritus-feeding animals that release certain nutrients and metabolites into the water by physical and chemical breakdown of plant and animal tissues, releasing cellular contents and excreting organic matter into the water. (For further details on decomposition in aquatic systems, see Chapter 7.)

## SUMMARY

The various biotic units which make up the biosphere, the thin layer of life on earth, are known as ecosystems. Ecosystems may be as large as vast unbroken tracts of forest or as small or smaller than a pond. They are, as the name implies, systems. A system is a collection of interdependent parts enclosed in a defined boundary. It receives inputs from the environment and processes them, resulting in outputs to the environment. If the system has a feedback mechanism in which some of the output is returned to the system as input to provide a degree of control over its functioning, it is cybernetic.

The ecosystem is an energy-processing cybernetic system, receiving abiotic and biotic inputs. The driving force is the energy of the sun. Abiotic inputs include oxygen, carbon dioxide, and nutrients from the weathering of the earth's crust and from precipitation. Biotic inputs include organic materials from surrounding ecosystems.

The ecosystem itself consists of three components: (1) the autotrophs, producers that fix the energy of the sun; (2) the heterotrophs, consumers and decomposers that utilize the energy and nutrients fixed by the producers and return nutrients back to the system; and (3) dead organic material and inorganic substrate that act as a short-term nutrient pool and maintain internal cycling of nutrients in the system.

Feedback mechanisms involve both organismic and population homeostasis that function to maintain ecosystem homeostasis or stability. Ecosystem stability can be considered as the persistence of a system or as its ability to return to an equilibrium state. Ecosystems that are able to absorb changes or perturbations, return to their original conditions, and still persist are considered resilient. This does not imply that such ecosystems are highly stable. Ecosystems that are able to remain relatively unchanged through changing conditions or perturbations are said to be resistant. A strong perturbation may result in the displacement of an ecosystem and its replacement by another.

Ecosystem stability ultimately results from adaptative responses of organisms in the ecosystem. Adaptation is the conformity of a organism to its environment. As the environment changes, adaptations of organisms change through natural selection or differential reproduction. Some organisms and populations leave behind more offspring than others. This ability to leave future progeny is known as fitness. Natural selection results in evolution, the development of organisms through time. Thus ecosystem homeostasis comes about through evolutionary changes in populations that comprise it.

The most basic processes in the maintenance of the ecosystem are photosynthesis and decomposition. Photosynthesis is the process by which green plants utilize the energy of the sun to convert carbon dioxide and water into carbohydrates. Plants possess one of at least two different photosynthetic processes. Most plants utilize the Calvin cycle which involves the formation of a three-carbon phosphoglyceric acid used in subsequent reactions. Other plants utilize a four-carbon process in which carbon dioxide taken into the leaf reacts to form malic and aspartic acids. The $CO_2$ fixed in these compounds is then released to the Calvin cycle. These two groups of plants, $C_3$ and $C_4$ respectively, possess structural and physiological differences that are important ecologically. $C_4$ plants can carry on photosynthesis at higher leaf temperatures, at higher light intensities, and at lower $CO_2$ concentrations than $C_3$ plants.

Involved in the return of nutrients to the ecosystem and the final dissipation of energy are the decomposer organisms. True decay organisms are the heterotrophic bacteria and the fungi, but the reducer-decomposers also play an important role by making organic matter more available to the decay organisms. They feed on plant and animal remains and on fecal material. Feeding on the bacteria and fungi are the microbial grazers, which reduce microbial populations and thus influence microbial activity. In terrestrial ecosystems bacteria and fungi take the major role in decomposition; in aquatic ecosystems bacteria and fungi act more as converters, whereas phytoplankton and zooplankton take a major part in the cycling of nutrients.

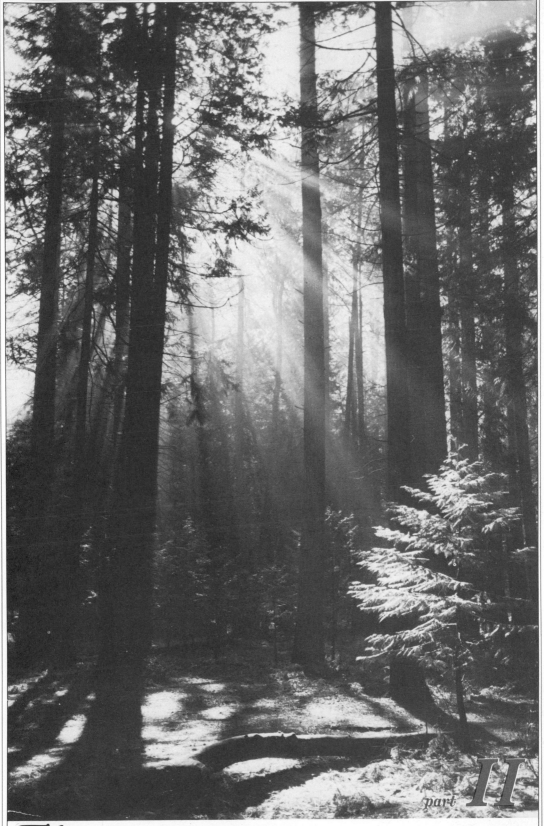

part *II*

# The ecosystem

# Abiotic limits to the system I
## CHAPTER 3

### Adaptability of the organism

The nature of the ecosystem, ecosystem functioning as characterized by energy flow and nutrient cycling, and the responses of the plants and animals that comprise the system are influenced by such environmental variables and inputs as nutrients, temperature, moisture, wind, and fire. The role of environmental limitations in the responses of organisms to their environment was recognized as early as 1840 by the German organic chemist Justus von Liebig. In his book *Organic Chemistry and Its Application to Agriculture and Physiology* Leibig described his analysis of surface soil and plants and set forth his conclusions in a simple statement, rev-olutionary for his day: "The crops of a field diminish or increase in exact proportion to the diminution or increase of the mineral substances conveyed to it in nature." Essentially what he said was that each plant requires certain kinds and quantities of nutrients and food materials. If one of these food substances is absent, the plant dies. If it is present in minimal quantities only, the growth of the plant will be minimal. This became known as the *law of the minimum*.

Continued investigations through the years disclosed that not only nutrients, but other environmental conditions, such as moisture and temperature, also affect the growth of plants. Animals, too, are limited by food, essential nutrients, water, temperature, and humidity among many factors. Eventually the law of the minimum was extended to cover all environmental requirements of both plants and animals.

Later studies of environmental influences on plants and animals showed that the success of an organism is limited not only by too little of a requirement, but also by too much. Organisms, then, live within a range between too much and too little, the limits of tolerance. This concept of maximum substances or conditions limiting the presence and success of organisms was incorporated by V. E. Shelford in 1913 into the *law of tolerance.*

The law of tolerance is summarized in a bell-shaped tolerance curve (Figure 3-1). This curve represents the range of a particular environmental factor. The peak is the optimal state for a particular physiological process and the tails are the states in which that process

functions poorly or not at all. For many organisms the tolerance curves are broad. The organisms are able to maintain homeostasis over a wide range of values for a particular environmental factor, such as salinity or humidity. Other organisms have a narrow range of tolerance. Organisms can be described by their tolerance curves for specific conditions using the prefix *eury,* meaning wide, and *steno,* meaning narrow. For example, a species tolerant to a wide range of salinities would be called a euryhaline species, while one with a narrow range would be called stenohaline.

The range of tolerance is not a fixed point. As seasons and conditions change, organisms may alter the tolerance curves to the right or left as they become acclimated. Tolerance curves reflect the ability of the organisms to adjust to environmental changes (Figure 3-2).

Organisms are limited by and must adjust to a number of conditions and interactions among the conditions. An organism, for example, may have a wide range of tolerance for one

*FIGURE 3-1*
*The law of tolerance.*

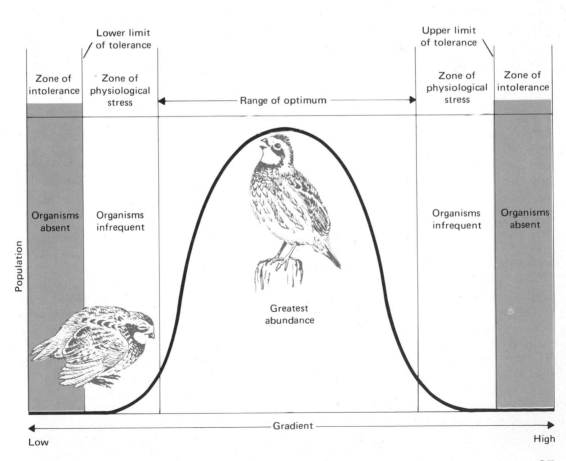

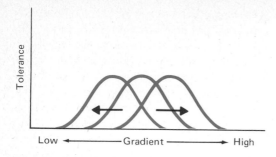

**FIGURE 3-2**
*Change in range of tolerances in response to environmental change. The tolerance range may shift within the physiological limits of the organism in response to changed conditions. For example, as water warms in the spring, the tolerance of fish for higher temperatures gradually shifts upward; at the same time the tolerance level for lower temperatures also shifts upward. Similarly, as the water cools in fall and winter, the tolerance for low temperatures increases, while the tolerance for high temperatures decreases. Thus, a temperature that would be lethal for a fish in winter, if the fish were suddenly exposed to it, can be tolerated in summer.*

substance or condition, but a narrow range for another and will thus be limited by the condition for which it possesses a narrow range. In some cases a less than optimum state in one condition may lower the limits of tolerance for another condition. Or the organism may achieve best fitness at some intermediate point between interacting factors. Some organisms may utilize an item in surplus as a substitute for another that is deficient; for example, some plants respond to sodium when the potassium supply is inadequate (Reitemeier, 1957). The range of tolerance determines distribution; organisms which exhibit a wide range of tolerances for all environmental influences will be widely distributed.

The individual reflects the adaptation of the population or the species to the environment in which it is found. The ability of plants and animals to live in the desert, for example, results from adaptations to heat and a limited water supply. The ability of plants to exist in the Arctic results in part from adaptations to make maximum use of a short growing season. The timing of daily and seasonal activities of populations of plants and animals is mediated through the photoperiodic responses of individuals.

Responses of animals to the environment are more easily observed than those of plants because animal reactions to sudden changes involve behavior. Behavior is intrinsic to the ecology of the animal. It is involved in the success of the animal to survive and to reproduce.

The behavior of an animal, like its structure, is the result of natural selection. Perhaps more often than not structure and behavior evolve together, structure influencing behavior and behavior in turn influencing the development of structure. To respond to environmental changes, the animal must first receive a stimulus from the environment through its sensory system—sight, taste, smell, touch, hearing. The stimulus is transmitted through the central nervous system, and finally the motor organs and the muscular system respond. The kind of sensory apparatus the animal possesses, the complexity and organization of its central nervous system, and the type and development of motor organs determine the manner in which the animal is able to respond to its environment. Because of this the world appears differently to other animals than it does to humans. What is sensed and is important to some animals may be unperceivable to us. Each animal lives, as Jacob von Uexkull put it, in its own phenomenal or self world, its *Umwelt*. Behavior becomes a mechanism for survival, and through a period of time natural selection favors the best adapted behavior patterns for a species to meet the contingencies of the environment.

## Nutrients

### IMPORTANT NUTRIENTS IN THE ECOSYSTEM

Living organisms require at least 30 to 40 elements for their growth and development. Most important of these are carbon, hydrogen, oxygen, phosphorus, potassium, nitrogen, sulfur, calcium, iron, magnesium, boron, zinc, chlorine, molybdenum, cobalt, iodine, and fluorine.

Molecular *oxygen*, $O_2$, which makes up 21 percent of the earth's atmosphere, is a byproduct of photosynthesis. Three major pools of oxygen are carbon dioxide, water, and molecular oxygen, all of which interchange atoms with one another. Other sources include nitrate and sulfate ions, which release oxygen upon their decomposition and supply oxygen for a number of living organisms. Molecular oxygen is a building block of protoplasm and is necessary for biological oxidation, in which it serves as a hydrogen acceptor, producing water. Plants and animals use large amounts of oxygen in respiration. In terrestrial habitats the supply of oxygen is rarely inadequate for life, except at high altitudes, deep in the soil, and in soils saturated with water. In aquatic communities, however, oxygen may be limited. Here the supply comes from photosynthesis and from the diffusion of atmospheric oxygen into the water, the rate of which is proportional to the surface area of water exposed to the air. The total quantity of oxygen that water can hold at saturation varies with temperature, salinity, and depth. It is depleted by the respiration of aquatic plants and animals. Oxygen content in aquatic environments fluctuates daily from a low at night, when respiration is greatest, to a high at midday, when photosynthesis is the highest.

*Carbon* is the basic constituent of all organic compounds. In the ecosystem it occurs as carbon dioxide, carbonates, fossil fuel, and as part of living tissue and it makes up only 0.03 percent of the atmosphere. The amount of carbon dioxide in natural waters is highly variable, for it occurs in free and combined states. The pH of aquatic media and of the soil has a pronounced influence on the proportions of the two types. Carbon dioxide combines with water to form a weak carbonic acid, $H_2CO_3$, which dissociates:

$$CO_2 + H_2O \rightleftharpoons H_2CO_3 \rightleftharpoons H^+$$
$$+ HCO_3^- \rightleftharpoons H^+ + CO_3^{2+}$$

Carbon dioxide in solution and carbonic acid make up free carbon dioxide; the bicarbonate ($HCO_3^-$) and the carbonate ($CO_3^{2+}$) ions are the combined forms. The presence of bicarbonate and carbonate ions in soil or water helps to buffer or maintain a certain pH of that medium. This is significant in ecological systems because an increase in soil pH lowers the availability of most nutrients to plants. The amount of carbon dioxide fluctuates during the day, but in a manner opposite to daily oxygen fluctuations. The carbon dioxide concentration is lowest at midday and highest at night, when only respiration is taking place.

*Nitrogen* makes up 78 percent of the atmosphere as molecular nitrogen, $N_2$, but most plants can utilize it only in a fixed form, such as in nitrites and nitrates (the exceptions are nitrogen-fixing bacteria and blue-green algae). Most of the nitrogen in the soil is found in organic matter. Nitrates leached from the soil and transported by drainage water are an important source of nitrogen for aquatic communities. During the summer the nitrogen supply in these may be utilized completely by phytoplankton and nitrates may disappear from the surface water. As a result phytoplankton growth, or "bloom," is reduced greatly in late summer. Nitrates build up again in winter. On the other hand, nitrogenous wastes draining from agricultural fields, sewage disposal plants, and other sources often overload aquatic ecosystems with nitrogen. This results in massive plankton growth and other undesirable changes in community structure.

All life processes depend upon nitrogen. Even chlorophyll is a nitrogenous compound. Nitrogen is a building block of protein and a part of enzymes. It is needed in an abundant supply for reproduction, growth, and respiration.

Oxygen, carbon, and nitrogen are considered the energy elements and are needed in relatively large quantities. Other elements and compounds, called macro- and micronutrients, are needed in smaller amounts. The macronutrients include calcium, magnesium, phosphorus, potassium, sulfur, sodium, and chlorine. The micronutrients are the trace elements. These include copper, zinc, boron, manganese, molybdenum, cobalt, vanadium, and iron. Some are essential to all organisms; others appear to be essential only to animals. If micronutrients are lacking, plants and animals fail as completely as if they lacked nitrogen, calcium, or any other major element.

Two elements needed in appreciable quantities are *calcium* and *phosphorus,* which are closely associated in the metabolism of animals. Together the two make up 70 percent of the total ash present in the animal body and nearly 1 percent of the wet body mass. In animals calcium is necessary for proper acid-base relationships, for the clotting of blood, for contraction and relaxation of the heart muscles, and for the control of fluid passage through the cells. It gives rigidity to the skeleton of vertebrates and is the principal component in the exoskeleton of insects and the shells of mollusks and arthropods. A number of mollusks and bivalves are restricted to hard water because there is insufficient calcium in soft water to harden the shells. In plants calcium is especially important in combining with pectin to form calcium pectate, a cementing material between cells. Plant roots need a supply of this element at the growing tips to develop normally.

Phosphorus not only is involved in photosynthesis, but also plays a major role in energy transfer within the plant and animal. It is a major component of the nuclear material of the cell, where it is involved in cellular organization (DNA) and in the transfer of genetic material. Animals require an adequate supply of calcium and phosphorus in the proper ratio, preferably 2:1 in the presence of vitamin D. An inadequate supply of either may limit the nutritive value of other elements. The lack of either in animals is associated with rickets, a condition involving the improper calcification of bones. When the supply of phosphorus in plants is low, growth is arrested, maturity delayed, and roots stunted. An excessive intake of calcium can inhibit the assimilation of other mineral elements.

Accompanying calcium and phosphorus is *magnesium*. In animals most of it is in the bones. It is an integral part of chlorophyll, without which no ecosystem could function, and the element is active in the enzyme systems of plants and animals, especially those enzymes transferring phosphate from ATP to ADP. Low intake of magnesium by grazing ruminants causes a serious disease, grass tetany, which may result in death.

*Potassium* is utilized in large quantities by plants, and if it is readily available and growing conditions are favorable, the uptake (in crop plants at least) may be above that of their average total requirements (Reitemeier, 1957). The formation of sugar and starches in plants, the synthesis of proteins, normal cell division and growth, and carbohydrate metabolism in animals all depend upon an adequate supply of potassium. Unlike calcium, phosphorus, and magnesium, potassium occurs in body fluids and soft tissues. In animals it is readily absorbed metabolically, and excess over needs is immediately excreted. Because plants usually contain an adequate supply, deficiencies rarely occur among grazing animals.

*Iron* and *manganese* are involved in the production of chlorophyll. Iron is part of complex protein compounds that serve as activators and carriers of oxygen and as transporters of electrons. Over half the iron present in the animal body is in the hemoglobin of the blood. Lack of iron results in anemia. A low level of manganese in animals can result in malformation of bones, in delayed sexual maturity, and in impaired reproduction.

*Boron* and *cobalt* are two micronutrients whose deficiency effects, the one in plants and the other in animals, are notable. Some 15 functions have been ascribed to boron, including pollen germination, cell division, carbohydrate metabolism, water metabolism, maintenance of conductive tissue, and translocation of sugar in plants. Plants with boron deficiency are stunted both in leaves and roots. This condition is most common in the croplands of eastern and central North America, where vegetation is continuously being removed. Without cobalt, an element not required by plants, animals become anemic and waste away. Cobalt deficiency is most pronounced in ruminants such as deer, cattle, and sheep, which require the element for the synthesis of vitamin $B_{12}$ by bacteria in the rumen. The quantity of cobalt required is very small. An acre of grassland supporting seven sheep needs to supply only 0.01 oz/yr.

In addition to producing deficiency symptoms when undersupplied, some micronutrients can be toxic when they are in excess. Among these are *copper* and *molybdenum*. Molybdenum acts as a catalyst in the conversion of gaseous nitrogen into a usable form by free-living, nitrogen-fixing bacteria and blue-green algae. But high concentrations of molybdenum cause teart disease in ruminants such as cattle and deer. This disease is characterized by diarrhea, debilitation, and permanent fading of the hair color. In plants copper is concentrated in the chloroplasts, where it affects the photosynthetic rate, is involved in oxidation-reduction reactions, and acts as an enzyme activator. When present in excess, copper interferes with phosphorus uptake in plants, depresses iron concentration in the leaves, and reduces growth. Copper can interact with molybdenum effectively to "tie up" the copper and produce a copper deficiency. In animals this may cause poor utilization of iron. As a result, iron concentrates in the liver, anemia develops, and calcification of bones decreases.

*Sulfur,* like nitrogen, is a basic constituent of protein, and many plants may utilize as much of this element as they do phosphorus. The sulfur supplied by rainwater and by organic matter in soil is sufficient to meet plant needs. Considerable quantities are released to the atmosphere in industrial areas and carried to the soil by rainwater. Excessive sulfur can be toxic to plants. Exposure for only an hour to air containing 1 ppm (part per million) of sulfur dioxide is sufficient to kill vegetation. Plants, especially in arid and semiarid country, are affected by high concentrations of soluble sulfates in soils, which limit the uptake of calcium.

*Sodium* and *chlorine* are on the borderline between micro- and macronutrients. They are required in minute quantities by plants, but in much greater quantities by animals. (Sodium can substitute for potassium to satisfy the plant's need for that element.) Both sodium and chlorine are indispensable to vertebrate animals. Obtained from common salt, these elements are important for the maintenance of acid-base balance of the body, the total osmotic pressure of extracellular fluids, and for the formation and flow of gastric and intestinal secretions.

*Zinc* usually is abundant in the soil, but it may be unavailable to plants. Zinc may exist as insoluble compounds in the soil when the pH is around 7. Zinc is needed in the formation of auxins (elements in plant growth substances or hormones), is a component of several plant enzyme systems, and is associated with water relations in plants. In animals zinc functions in several enzyme systems, especially the respiratory enzyme carbonic anhydrase in red blood cells, where it plays an essential role in the elimination of carbon dioxide. Insufficient zinc in the diet of mammals can cause dermatitis.

*Iodine* and *selenium* are two other micronutrients of note. Iodine, the natural deficiency of which occurs in the soils of the northeast region of North America, the Andes Mountains of South America, and the mountainous parts of Europe, Asia, and Africa, is involved in thyroid metabolism. In animals lack of iodine results in goiter, hairlessness, and poor reproduction. Selenium, closely related to vitamin E in its function, is required to prevent white muscle disease in the newborn ruminant. The amount required is on the order of 0.1 mg/kg of ration. The borderline between requirement level and toxicity level is very narrow. Too much selenium results in the loss of hair, sloughing of hooves, liver injury, and death by starvation.

### NUTRIENTS AND PLANT LIFE

Sixteen elements are essential for plant growth. Not all plants require these elements in the same quantities or in the same ratios, but all do require a certain minimal amount of essential ions for growth, and the requirements are specific.

Each species of plant has a specific ability to exploit the nutrient supply in some manner that may not be duplicated by other species. This enables different plants growing in the same environment to exploit slightly different nutrient sources. Shallow-rooted plants, for example, may utilize the nutrient supply on the upper surface soil, while others with deep tap roots may draw on deeper supplies of nutrients. Some species growing on soils poor in nutrients have become adapted to low nutrient levels, whereas species growing on more fertile soil have become adapted to higher levels of nutrition.

Nutrient levels in the soil have a pronounced influence on the local distribution of plants. Nowhere has this been better demonstrated than on the long-term grassland fertilizer trial plots at the Rothamsted Experimental Farm in England. Among the plots, established in 1856, are some containing natural vegetation that probably resembles the original. The unfertilized, unmanaged plots contain some 60 species of higher plants

(Thurston, 1969), representing not only every plant found in all other plots, but some restricted to the natural plots as well. Species diversity on the plots is high, and no one species is clearly dominant. The vegetation is short and the yield of hay low. On plots that received applications of phosphorus, potassium, sodium, and magnesium, but no nitrogen, legumes became dominant at the expense of other species. The addition of nitrogen discouraged the legumes, reduced their growth, and encouraged the grasses.

Soil acidity, by affecting nutrient uptake by plants, also has a strong influence on plant distribution. Some plants require a relatively high pH (over 6.5); others can get along on more acidic soils. Plants that are tolerant of acidic conditions often are tolerant of aluminum, a toxic ion, and are able to grow in spite of impaired calcium uptake. In fact, such plants are sensitive to calcium and may experience a yellowing of the leaves (chlorosis), a condition that can occur among rhododendrons and other ornamental evergreens. Plants adapted to a high pH and a high level of calcium are sensitive to ions of aluminum. Other ions such as sodium, manganese, nickel, and sulfur also have specific toxic effects. (For examples, see Marchland, 1973; Whittaker et al., 1954; Whittaker, 1960; Franklin and Dyrness, 1973.)

### NUTRIENTS AND ANIMAL LIFE

Because all animals depend directly or indirectly on plants for food, plant quality and quantity affect the well-being of animals. In the face of limited quantity, animals may suffer from malnutrition, starve, or be forced to leave the area. In other situations the quantity of food available may be sufficient to allay hunger, but the quality is such that it affects reproduction, health, and longevity. For example, the size of deer, their antler development, and their reproductive success all relate to nutrition (Figure 3-3). Deer on diets low in calcium, phosphorus, and protein have stunted growth, and bucks on such a low quality diet develop only thin spike antlers (French et al., 1955). Reproductive success of does is highest where food is abundant and nutritious (Cheatum, 1950; Severinhaus and Taber, 1953).

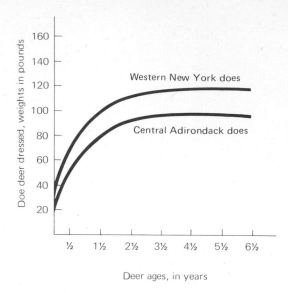

**FIGURE 3-3**
*Growth patterns and quality of range. Differences in the dressed weights of doe deer reflect differences in quality of forage and nutrient status of the soil. Deer in western New York occupy a range characterized by fertile soils and agricultural development. Deer from the Adirondacks occupy a range characterized by nutrient-poor soils, mixed coniferous-hardwood forests, and low quality browse. (After Severinghaus and Gottlieb, 1959.)*

The nutritional value of a food depends upon the balance and nature of its proteins, vitamins, and mineral elements, as well as the dietary requirements of the animals. Different animals have different requirements; what may be an acceptable food for some may be poor quality food for others. Except in overbrowsed or overgrazed areas, herbivores usually do not lack a sufficient quantity of food, but they often lack quality food, which can result in malnutrition.

The need for quality food differs among herbivores. For example, ruminant animals can synthesize requirements such as vitamin $B_1$ and certain amino acids from simple nitrogenous compounds; therefore they can subsist on rougher or lower quality forage than nonruminants who must ingest more complex proteins.

Among the carnivores quantity is more important than quality. Carnivores rarely have a dietary problem because they consume other animals that have stored up and resynthesized proteins and other nutrients from plants. What they fail to obtain from animal sources carnivores can obtain by consuming some plant concentrates such as berries and fruit.

All other organisms require a generous, continuous supply of protein throughout life for growth, proper development, weight gain, milk secretion, and other functions. Different periods of the life cycle have different protein requirements, with the greatest need during the period of rapid growth. The white-tailed deer, for example, can maintain life on a diet of 6 to 7 percent protein, but requires 13 to 16 percent protein for optimum growth (Deitz, 1965).

## Soil

Soil is the weathered superficial layer of the earth's surface that is composed of mineral and organic matter and is capable of supporting plant growth. It harbors bacteria, fungi, and other organisms that fix atmospheric nitrogen, break down organic matter, and incorporate it with mineral matter. Roots occupy a considerable portion of the soil. They serve to tie the vegetation to the soil and to pump water and its dissolved minerals to other parts of the plant for use in the photosynthetic process.

### SOIL FORMATION

The formation and development of soil begins with the weathering of rocks and their minerals. Exposed to the combined action of water, wind, and temperature, rock surfaces peel and flake away. Water seeps into crevices, freezes, expands, and cracks the rock into smaller pieces. Accompanying and continuing long after this disintegration is the decomposition of the minerals themselves. Water and carbon dioxide combine to form carbonic acid, which reacts with calcium and magnesium in the rock to form carbonates. These slightly soluble carbonates either accumulate deeper in the rock material or are carried away, depending upon the amount of water passing through. Primary minerals that contain aluminum and silicon, such as feldspar, are converted to secondary minerals, such as clay. Because iron is especially reactive with water and oxygen,

iron-bearing minerals are prone to rapid decomposition. Iron remains oxidized in the red ferric state or may be reduced to the gray ferrous state. Fine particles, especially clays, are shifted or rearranged within the mass by percolating water and on the surface by runoff, wind, or ice. Eventually the rock is broken down into loose material. This may remain in place, but is more often lifted, sorted, and carried away. Materials remaining in place are called residual. Material transported from one area to another by wind is known as loess; that transported by water as alluvial, lacustrine (lake), and marine deposits; and that by glacial ice as till. In a few places soil materials come from accumulated organic matter such as peat.

This mantle of unconsolidated material is called the regolith. It may consist of slightly weathered material with fresh primary minerals or it may be intensely weathered and consist of highly resistant minerals such as quartz. Because of variations in slope, climate, and native vegetation, many different soils can develop in the same regolith. The thickness of the regolith, the kind of rock from which it was formed, and the degree of weathering affect the fertility and water relations of the soil.

Eventually plants root in this weathered material. Frequently intense weathering goes on under some plant cover, particularly in glacial till and water-deposited materials, which already are favorable sites for some plant growth. Thus soil development often begins under some influence of plants. They root, draw nutrients from mineral matter, reproduce, and die. Their roots penetrate and further break down the regolith. The plants pump up nutrients from its depths and add them to the surface, and in doing so recapture minerals carried deep into the material by weathering processes. By photosynthesis plants capture the sun's energy and add a portion of it, as organic carbon—approximately 18 billion metric tons, $1.7 \times 10^{17}$ kcal—to the soil each year. This energy source, the plant debris, enables bacteria, fungi, earthworms, and other soil organisms to colonize the area.

The breakdown of organic debris into humus is accomplished by decomposition and finally mineralization. Higher organisms in the soil—millipedes, centipedes, earthworms, mites, springtails, grasshoppers, and others—consume fresh material and leave partially decomposed products in their excreta. This is further decomposed by microorganisms, the bacteria and fungi, into various compounds of carbohydrates, proteins, lignins, fats, waxes, resins, and ash. Through the process of mineralization these compounds are then broken down into simpler products such as carbon dioxide, water, minerals, and salts.

The fraction of organic matter that remains is called humus. It is a stage in the decomposition of soil organic matter and therefore it is not stable. New humus is being formed as old humus is being destroyed by mineralization. The equilibrium set up between the formation of new humus and the destruction of the old determines the amount of humus in the soil. (See Chapter 2.)

## SOIL PROFILES AND PROCESSES

*Horizons.* Activities of soil organisms, the acids produced by them, and the continual addition of organic matter to mineral matter produce profound changes in the weathered material. Rain falling upon and filtering through the accumulating organic matter picks up acids and minerals in solution, reaches mineral soil, and sets up a chain of complex chemical reactions. This continues further in the regolith. Calcium, potassium, sodium, and other mineral elements, soluble salts, and carbonates are carried in solution by percolating water deeper into the soil or are washed away into streams, rivers, and eventually the sea. The greater the rainfall the more water moves down through the soil and the less moves upward. Thus high precipitation results in heavy leaching and chemical weathering, particularly in regions of high temperatures. These chemical reactions tend to be localized within the regolith. Organic carbon, for instance, is oxidized near the surface, whereas free carbonates precipitate deeper in the rock material. Fine particles, especially clays, also move downward.

These localized chemical and physical processes in the parent material result in the development of layers in the soil, called *horizons,* which impart to the soil a distinctive profile. Within a horizon, a particular property of the soil reaches its maximum intensity and away from this level decreases gradually in both directions. Thus each horizon varies in thickness, color, texture, structure, consistency, porosity, acidity, and composition.

In general soils have four major horizons: *O,* an organic layer, and *A, B,* and *C,* the mineral layers. Below the four may lie the *R* horizon, the consolidated bedrock. Because the soil profile is essentially a continuum, often there is no clear-cut distinction between one horizon and another, and the content of each horizon varies considerably (Figure 3-4).

The *O* horizon is the surface layer, formed or forming above the mineral layer and composed of fresh or partially decomposed organic material. It is usually absent in cultivated soils. This and the upper part of the *A* horizon constitute the zone of maximum biologic activity. They are subject to the greatest changes in soil temperatures and moisture conditions and contain the most organic carbon and are the sites where most or all decomposition takes place. The *A* horizon is characterized by an accumulation of organic matter, by the loss of clay, iron, and aluminum, and by the development of granular, platy, or crumb structure. *B* is characterized by a concentration of all or any of the silicates, clay, iron, aluminum, and humus, alone or in combination, and by the development of blocky, prismatic, or columnar structure. *C* contains weathered material, either like or unlike the material from which the soil is presumed to have developed.

*Differentiation into types.* The differentiation of the soil profile into horizons and the nature of the soil material, its contents and distribution of organic matter, its color, and its chemical and physical characteristics are influenced over large areas by the combined action of vegetation and its prime determinant, climate (Figure 3-5).

The subhumid-to-arid and temperate-to-tropical regions of the world—the plains and prairies of North America, the steppes of Russia, and the veldts and savannas of Africa—support grassland vegetation. Dense root systems may extend many feet below the surface. Each year nearly all of the vegetative material above ground and a part of the root system are turned back to the soil as organic matter. Al-

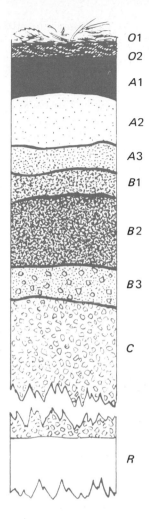

O1
O2
A1
A2
A3
B1
B2
B3
C
R

**FIGURE 3-4**
*Generalized profile of the soil. Rarely does any one soil possess all of the horizons shown. O₁: loose leaves and organic debris. O₂: partly decomposed or matted organic debris. A₁: dark-colored horizon with a high content of organic matter mixed with mineral matter. A₂: light-colored horizon of maximum leaching; prominent in podzolic soils and faintly developed or absent in chernozemic soils. A₃: transitional to B, but more like A than B; sometimes absent. B₁: transitional to B, but more like B than A; sometimes absent. B₂: a deeper colored horizon of maximum accumulation of clay minerals or of iron and organic matter; maximum development of blocky or prismatic structure or both. B₃: transitional to C. C: the weathered material, either like or unlike the material from which the soil presumably formed; a glei layer may occur, as well as layers of calcium carbonate, especially in grasslands. R: consolidated rock.*

though this material decomposes rapidly the following spring, it is not completely gone before the next cycle of death and decay begins. Soil inhabitants mix the humus with mineral soil, developing a soil high in organic matter. Because the amount of rainfall in grassland regions generally is insufficient to remove calcium and magnesium carbonates, they are carried down only to the average depth that the percolating waters reach. Grass maintains a high calcium content in the surface soil by absorbing large quantities from the lower horizons and redepositing them on the surface. Likewise little clay is lost from the surface. This process of soil development is called *calcification,* and the soil itself is called *mollisol* (see Table 3-1).

Soils developed by calcification have a distinct *A* horizon of great thickness and an indistinct *B* horizon, characterized by an accumulation of calcium carbonate. The *A* horizon is high in organic matter, giving a deep, dark, color, and in nitrogen, even in tropical and subtropical regions.

Humid regions support forests. Here the cycle of organic matter accumulation differs from that of the grassland. Only part of the organic matter—leaves, twigs, and some trunks—is turned over annually. Leaves, the largest source of organic matter, and vegetation of the ground layer remain on the surface. Dead roots add relatively little to soil organic matter because they die over an irregular period of time and are not uniformly concentrated near the surface. Because only the leaves are returned annually to the soil and much of the mineral matter and energy is tied up in trunk and branches, most of the currently available nutrients turned back to the soil come from annual leaf fall.

Rainfall in forested regions is sufficient to carry away many elements, especially calcium, magnesium, potassium, iron, and aluminum. Because trees generally return an insufficient amount of bases back to the surface soil, it becomes acid, although the degree of acidity will vary according to the forest's composition and its site (see Chapter 5). Increased acidity may cause dispersion and downward movement of organic and clay colloids. This process of soil development is called *podzolization* and the soil itself is called

**Chernozem**
**Mollisol**

**Tundra**
**Inceptisols**
**Histosols**

**Prairie (Brunizem)**
**Molisol**

**Podzol**
**Spodosol**

**Mountain**
**soils**
**Entisol**

**Chestnut**
**and Brown**
**Mollisol**

**Gray-brown**
**Podzolic**
**Alfisol**

**Sierozem**
**and**
**Desert**
**Aridisols**

**Red-yellow**
**Podzolic**
**Ultisol**

**Laterite**
**Oxisols**

**FIGURE 3-5**
*The great soil groups. This map of North
America shows the general distribution of
the important zonal or great soil groups of
the continent and points out the general
relationships of soils to vegetation and
climate. The majority of the soils illustrated
here (the exception is the tundra) are those
that develop on well-drained sites. In the
humid regions bases do not accumulate in
the soils because of the leaching processes
associated with high rainfall. Podzol soils, or
spodosols, characterized by a very thin
organic layer on top of a gray, leached soil
overlying a dark brown horizon, generally
develop in a cool, moist climate under
coniferous forests. The gray brown podzolic
soils, or alfisols, develop under deciduous
forests in a cool-temperate, moist climate.
These differ from podzols in that the
leaching is not as great and beneath the
organic layer is a horizon of grayish brown,
leached soil. The red and yellow soils, or
ultisols, occur in the warm-temperate, moist
climate of southeastern North America.
Developed through podzolization with some
laterization, yellow soils are characterized by
a grayish yellow, leached horizon over a
yellow one; the red by a yellowish brown,
leached soil over a deep red horizon. In the*

*TABLE 3-1*
*New soil orders and approximate equivalents in old classification*

| New soil order[a] | Formative syllable | Derivation and meaning | Approximate equivalent in old classification[b] |
|---|---|---|---|
| Entisol | ent | Coined from recent | Azonol and some low humic gley soils |
| Vertisol | ert | L. *verto*, "inverted" | Grumusols |
| Inceptisol | ept | L. *inceptum*, "young" | Ando, sol brun acids, some brown forest, low humic gley, and humic gley soils |
| Aridisol | id | L. *aridus*, "arid" | Desert, reddish desert, sierozem, solonchak; soils; some brown and reddish-brown and associated solonetz soils |
| Mollisol | oll | L. *mollis*, "soft" | Chestnut, chernozem, brunizem (prairie) rendzinas, some brown, brown forest, and associated solonetz and humic gley soils |
| Spodosol | od | Gk, *spodos*, "ashy" | Podzols, brown podzolic soils, and ground-water podzols |
| Alfisol | alf | Coined from Al-Fe | Gray-brown podzolic, gray wooded, noncalcic brown, degraded chernozem, and associated planosols and half-bog soils |
| Ultisol | ult | L. *ultimus*, "last" | Red-yellow podzolic, reddish-brown lateritic (of the U.S.), and associated planosols and half-bog soils |
| Oxisol | ox | Fr. *oxide*, "oxidized" | Lateritic soils, latosols |
| Histosol | ist | Grk. *histos*, "organic" | Bog soils |

*Source:* [a]Soil Survey Staff, 1960; [b]Baldwin, Kellogg, and Thorp, 1938.

*tall-grass prairie region with a temperate, moist climate the prairie or brunizem soils, or mollisols result from calcification. They are very dark brown in color, grading through lighter brown with depth. West of the brunizem lie the chernozem soils, another mollisol, black soil high in organic matter, some 3 to 4 feet deep, which grade into lime accumulations. The chernozem developed under tall- and mixed-grass prairie. Closely related are the chestnut and brown soils, also mollisols, dark brown and grading into lime accumulations at 1 to 4 feet. These soils developed under mixed- and short-grass prairie. In desert regions are sierozem and desert soils, or aridisols. They are pale grayish in color, low in organic matter, and closely underlain with calcareous material. Lateritic soils, or oxisols, typical of tropical rain forest where decomposition is rapid, have a thin organic layer over a reddish, leached soil. In high mountains are a variety of soils, here vaguely classified as mountain soils, entisols. Many of them are stony and lack any well-developed horizons. Tundra soils are variable, but the common one is a glei, subject to considerable disturbance from frost action and underlain with a permanently frozen substrate.*

*spodosol*. Both names refer to the ashy gray, leached horizon of strongly podzolized soils.

In the humid subtropical and tropical forested regions of the world, where rainfall is heavy and temperatures high, the soil-forming process is much more intense. Because temperatures are uniformly high, the weathering process in these regions is almost entirely chemical, brought about by water and its dissolved substances. The residues from this weathering—bases, silica, alumina, hydrated aluminosilicates, and iron oxides—are freed. Because precipitation usually exceeds evaporation, the water movement is almost continuously downward. With only a small quantity of electrolytes present in the soil water because of continual leaching, silica and aluminosilicates are carried downward, while sesquioxides of aluminum and iron remain behind. The reason for this is that these sesquioxides are relatively insoluble in pure rainwater, but the silicates tend to be precipitated as a gel in solutions containing humic substances and electrolytes. If humic substances are present they act as protective colloids about iron and aluminum oxides and prevent their precipitation by electrolytes. The end product of such a process is a soil composed of silicate and hydrous oxides, clays, and residual quartz, deficient in bases, low in plant nutrients, and intensely weathered to great depths. Because of the large amount of iron oxides left, these soils possess a variety of reddish colors and they generally lack distinct horizons. Below, the profile is unchanged for many feet. The clay has a stable structure and unless precipitated, iron is hardened into a cemented laterite. It is very pervious to water and is easily penetrated by plant roots. This soil-forming process is termed *laterization* or *latosolization.* Soils so formed are called *oxisols.*

Arid and semiarid regions have relatively sparse vegetation. Because plant growth is limited, little organic matter and nitrogen accumulate in the soil. Scant precipitation results in slightly weathered and slightly leached soils high in plant nutrients. The horizons are usually faint and thin. In these regions occur areas where soils contain excessive amounts of soluble salts, either from parent material or from the evaporation of water draining in from adjoining land. The infrequent rain penetrates the soil, but soon afterward evaporation at the surface draws the salt-laden water upward. The water evaporates, leaving saline and alkaline salts at or near the surface to form a crust, or *caliche.* Soils of arid and semiarid regions are called *aridisols.*

Calcification, podzolization, and laterization are processes that take place on well-drained soil. Under poorer drainage conditions a different soil-development process is at work. The slope of the land determines to a considerable extent the amount of rainfall that will enter and pass through the soil, the concentration of erosion materials, the amount of soil moisture, and the height at which the water will stand in the soil. The amount of water that passes through or remains in the soil determines the degree of oxidation and breakdown of soil minerals. The iron in soils where water stays near or at the surface most of the time is reduced to ferrous compounds. These give a dull gray or bluish color to the horizons. This process, called *gleization,* may result in compact structureless horizons. *Gley* soils are high in organic matter because more organic matter is produced by vegetation than can be broken down by humification, which is greatly reduced because of the absence of soil microorganisms. On gentle to moderate slopes, where drainage conditions are improved, gleization is reduced and occurs deeper in the profile. As a result the subsoil will show varying degrees of mottling of grays and browns. On hilltops, ridges, and steep slopes, where the water table is deep and the soil well drained, the subsoil is reddish to yellowish-brown from the oxidized iron compounds (Figure 3-6).

In all, five drainage classes are recognized (Figure 3-7): (1) well-drained soils are those in which plant roots can grow to a depth of 36 in. without restriction due to excess water; (2) on moderately well-drained soils plant roots can grow to a depth of 18 to 20 in. without restriction; (3) in somewhat poorly drained soils, plant roots cannot grow beyond a depth of 12 to 14 in.; (4) poorly drained soils are wet most of the time and usually are characterized by the growth of alders, willows, and sedges; (5) on very poorly drained soils water stands on or near the surface most of the year.

### TIME SPAN OF DEVELOPMENT

The weathering of rock material, the accumulation, decomposition, and mineralization of organic material, the loss of minerals from the upper surface and gains in minerals and clay in lower horizons, and horizon differentiation all require considerable time. Well-developed soils in equilibrium with weathering and erosion may require 2,000 to 20,000 years for their formation. But soil differentiation from parent material may take place in as short a time as 30 years. Certain acid soils in humid regions develop in 100 years because the leaching process is speeded by acid materials. Parent materials heavy in texture require much longer to develop into "climax" soils, because of impeded downward flow of water. Soils develop more slowly in dry regions than in humid ones. Soils on steep slopes often remain young regardless of geological age because rapid erosion removes the soil nearly as fast as it is formed. Flood-plain soils age little through time because of the continuous accumulation of new materials. Young soils are not as deeply weathered as and are more fertile than old soils because they have not been exposed to the leaching process as long. The latter tend to be infertile because of long-time leaching of nutrients without replacement from fresh material.

**FIGURE 3-6**

*Effect of topography and native vegetation on the soil. This diagram shows the normal sequence of eight representative soil types from the Mississippi to the uplands in Illinois. The drawing also illustrates how bodies of soil types fit together in the landscape. Boundaries between adjacent bodies are gradations or continuums, rather than sharp lines.*

*The lower part of the diagram pictures the profiles of seven of the soils, showing the color and thickness of the surface horizon and the structure of the subsoil. Note how the natural vegetation that once covered the land (trees for forest areas, grass clumps for grass) influenced surface color. The diagram also shows how topographic position and the distance from the bluff influence subsoil development.*

*Profile A (Sawmill) is a bottomland soil formed from recent sediments and has not been subject to much weathering. Profile B (Worthen) on the foot slope also developed from recent alluvial material and shows little structure. Profile C (Hooper) on the slope break developed from a thick loess on top of leached till, while the soil on the bottom of the slope developed directly from the till. Profile D (Seaton) is an upland soil formerly covered with timber. It possesses a light surface color and lacks structure, the result of rapid deposition of loess during early soil formation, holding soil weathering to a minimum. Profile E (Joy) represents an upland soil developed under grass. Note the dark surface and lack of structure, again the result of rapid deposition of loess. Profile F (Edgington) is a depressional wet spot. Extra water flowing from adjacent fields increased the rate of weathering, resulting in a light, grayish surface and subsurface and a blocky structure to the subsoil. This indicates a strongly developed soil. The depth of subsoil suggests that considerable sediment has been washed in from the surrounding area. Profile G (Sable) represents a depressional upland prairie soil. Note the deep, dark, surface and coarse, blocky structure. Abundant grass growth produced the dark color. (After Veale and Wascher, 1956.)*

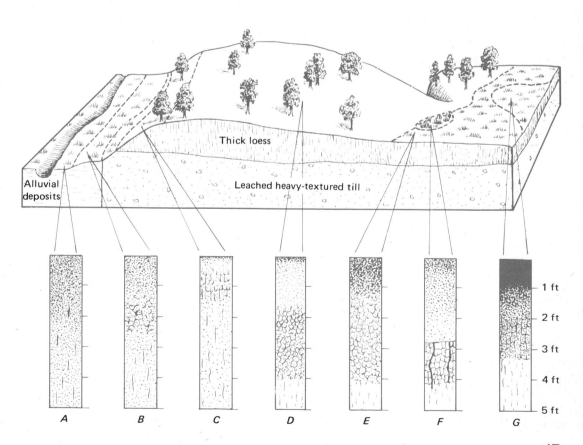

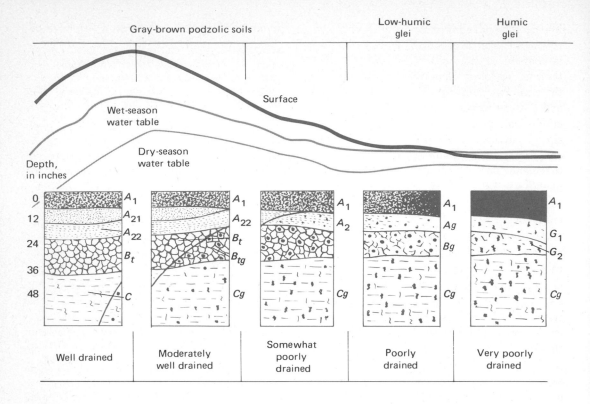

**FIGURE 3-7**
*Effect of drainage on the development of gray-brown podzolic soils. Wetness increases from left to right. This diagram represents the topographic position the profiles might occupy. Note that the strongest soil development takes place on well-drained sites where weathering is maximum. The least amount of weathering takes place on very poorly drained soils where the wet season water table lies above the surface of the soil. G or g indicates mottling; t indicates translocated silicate clays. (Adapted in part from Knox, 1952.)*

## Temperature

Environmental temperatures in any area fluctuate from sunlight to shade, from daylight to dark, and from season to season. Surface temperatures of soil may be 30° C higher in the sunlight than in the shade. Daytime temperatures may be 17° C higher than nighttime temperatures; on deserts this spread may be as high as 40° C. Temperatures on tidal flats may rise to 38° C when exposed to the sun and fall to 10° C within a few hours when the flats are covered by water. Seasonal fluctuations may also be extreme. In North Dakota where the annual mean temperature is between 3° and 9° C, temperatures fluctuate from a low of −43° C in winter to 49° C in summer. In West Virginia, where the mean annual temperature is 12° C, temperatures range from −37° C to 44° C.

The ability to withstand extremes in temperature varies widely among plants and animals, but there are temperatures above and below which no life can exist. A temperature of 52° C is about as high as any animal can withstand and still grow and multiply. Among plants some hot-spring algae can live in water as warm as 73° C under favorable conditions (Brock, 1966), and some arctic algae can complete their life cycles in places where temperatures barely rise above O° C. Nonphotosynthetic bacteria inhabiting hot springs can actively grow at temperatures greater than 90° C (Bott and Brock, 1969).

### THERMAL ENERGY EXCHANGE

Living organisms are intimately associated with an energy environment. The same influx of energy upon which photosynthesis depends is also the source of heat energy that characterizes the physical environment of life. What impact this heat influx or the thermal environment has on an organism depends on how it is adapted to cope with thermal budget and thermal stresses. The relationship involves the exchange of energy between the organism and the environment and the balance of heat energy gained and heat energy lost. In effect energy absorbed from the environment plus energy produced must equal energy lost from the body and the energy stored. This statement can be clarified by considering the sources and processes of energy gains and energy losses.

The source of heat absorbed from the environment is solar radiation. It may be direct radiation transmitted through the atmosphere, diffused radiation scattered by moisture and dust particles in the atmosphere, or reflected radiation bounced off objects in the environment. In addition there is infrared thermal radiation or reradiation of heat from the soil, rocks, vegetative surfaces, and the sky. The source of produced energy is the heat of metabolism of the organism. This includes basal heat production necessary to sustain life plus heat added by the physical act of digestion and by activity. In addition one has to consider gains and losses in energy in relation to changes in body temperature or heat storage (see Figures 3-8, 3-9, 3-10).

Just as the organism gains heat from the environment, so it loses heat to the environment. One source of loss is infrared radiation. Organisms are continually emitting radiant energy in the form of electromagnetic waves. The rate at which a surface (in this case the organism's surface) emits infrared radiation depends upon its surface temperature and its ability to give off radiant energy. The difference between the radiant energy lost from an animal and that absorbed from the environment represents the net energy exchange. Although it is simple to calculate net energy exchange for a one-dimensional flow between plane surfaces, as in a leaf, it is very complex for such an irregular object as an animal's body.

Another source of heat transfer is conduction, resulting from collision between oscillating molecules. The amount of heat lost by conduction varies with the area of surface exposed and the distance and temperature difference between the two surfaces which collide (molecularly). Thus, for an animal the heat loss varies with its body position (sitting, standing, curled) and with its movement from place to place.

Heat may also be transferred by convection, the movement of molecules in a fluid due to density differences resulting from temperature differences. Convection may occur naturally in the fluid (air or water) surrounding an object; or it may be forced, which involves external pressures from fluids passing by or over the object. The amount of heat transferred by convection depends on the shape and area of the organism, the velocity of the fluid, and the physical properties of the fluid and the organism.

Forced convection can speed up another process, evaporation. Evaporation depends upon a vapor pressure gradient existing between air and the object, and boundary (surface) resistance. Every organism is coupled to air temperature by a boundary layer of air adhering to the surface. The resistance this layer offers to the absorption and loss of heat depends upon the size, shape, texture, and orientation of the organism, wind speed, and the temperature difference between the surface of the organism and the air. In plants one has to add resistance by the leaf as influenced by the size and density of stomata and the resistance of the cuticle to loss of water (see Gates, 1962).

The relationship of all of these components in heat exchange can be summarized as: the net heat gain by solar radiation, infrared radiation, and metabolism plus or minus gains or losses in energy storage must equal the total heat lost by radiation, conduction, evaporation, and convection. In animals mechanisms of thermoregulatory adaptation have evolved to exploit or decrease the effects by one or more of the components in the heat equilibrium equation on the organism.

One aspect remains important. Heat produced continuously by plants and animals is lost passively to the environment. Such loss can take place only when the ambient or surrounding temperature is lower than the core body temperature (Figure 3-11). When ambient temperature equals core temperature the route for passing heat off to the environment is lost. When ambient temperature exceeds body temperature, the flow is reversed,

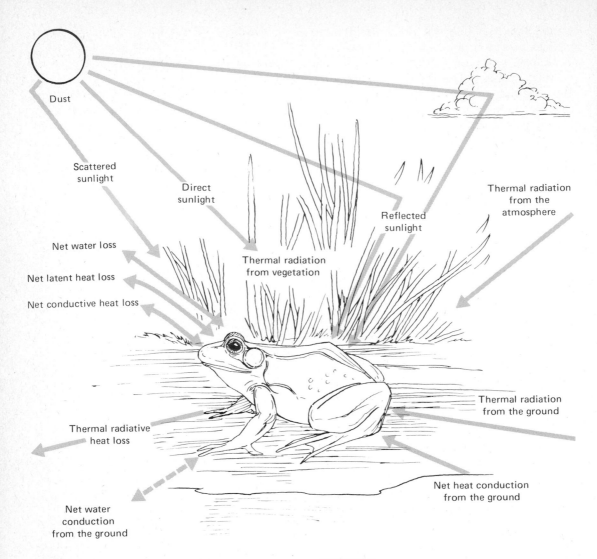

Dust

Scattered
sunlight

Direct
sunlight

Reflected
sunlight

Thermal radiation
from the
atmosphere

Net water loss

Net latent heat loss

Net conductive heat loss

Thermal radiation
from vegetation

Thermal radiation
from the ground

Thermal radiative
heat loss

Net heat conduction
from the ground

Net water
conduction
from the ground

**FIGURE 3-8**
*Exchange of energy and water between a frog and its environment. The flow of energy to the frog's upper surface includes (1) short-wave radiation directly from the sun and solar radiation scattered from clouds and other compartments of the environment; and (2) long-wave thermal radiation emitted from substrate, vegetation, and atmosphere. The frog loses energy (1) by emitting long-wave infrared radiation; (2) by exchanging energy to and from ambient air; and (3) by convection, evaporation, and condensation. It will also lose or gain energy by conduction of heat to or from its body core and will gain or lose heat by conduction and thermal radiation to the ground. (After Tracy, 1976.)*

**FIGURE 3-9**
*Energy exchange between a mammal and its environment. The energy balance is similar to that of the frog. (After Gates, 1961.)*

**FIGURE 3-10**
*Energy exchange—absorption and emission—in a meadow on a sunny day. (a) Q = net radiation; V = evaporation; L = sensible heat convection; B = soil heat flux; figures are cal/cm². The active layer in day lies between 30 and 55 cm; it absorbs 45 percent of net radiation. The second most active layer, the lowermost, which absorbs 28 percent. During input 80 percent of the radiant energy is used for evaporation of water, 15 percent for sensible heat convection, 5 percent to raise soil temperature. At night net radiation as well as heat exchange is reversed. (b) (After Cernusca, 1976.)*

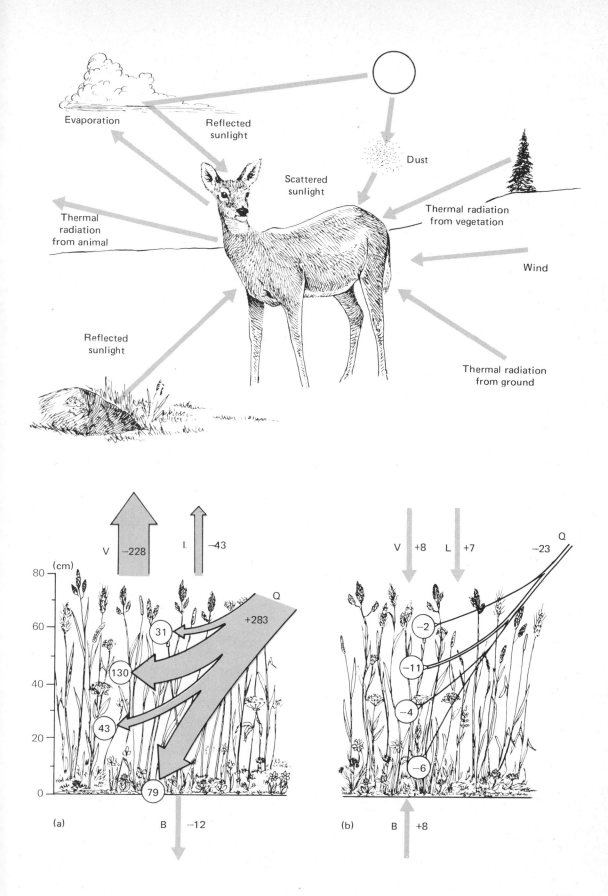

Evaporation

Reflected
sunlight

Dust

Scattered
sunlight

Thermal radiation
from vegetation

Thermal
radiation
from animal

Wind

Reflected
sunlight

Thermal radiation
from ground

(cm)

V   −228      L   −43          V   +8   L   +7         −23   Q

80

Q
+283

31

130                                −2

43                                 −11

−4

79                                 −6

0

(a)                              (b)

B   −12                           B   +8

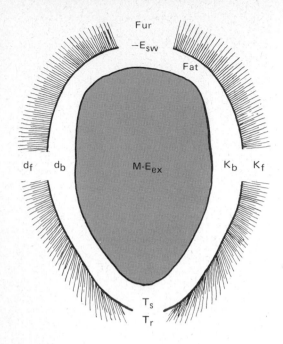

*FIGURE 3-11*
*Energy relations in the body temperature process. M = metabolism; $E_{ex}$ = respiratory moisture loss; $E_{sw}$ = moisture loss by sweating (if applicable); $T_b$ = body temperature; $T_s$ = skin temperature; $T_r$ = radiant surface temperature; $K_b$ = conductivity of fat; $K_f$ = conductivity of fur or feathers; $d_b$ = thickness of fat; $d_f$ = thickness of fur or feathers. The temperature of an animal is determined by metabolic energy (M) generated in the body core. Respiratory moisture loss expells a quantity of energy ($E_{ex}$) from the body cavity. The difference between M and $E_{ex}$ is body temperature ($T_b$). The central body core is surrounded by a layer of fat of a certain, but variable thickness ($d_b$) and of a certain conductivity ($K_b$). The amount of body heat lost or retained is influenced in part by these two variables. The fatty layer ends at the skin, which has a certain temperature ($T_s$). A layer of fur or feathers covering the skin represents another cylinder with a certain thickness ($d_f$) and a certain conductivity ($K_f$). The amount of energy M − $E_{ex}$ is conducted through the fatty layer across a temperature difference $T_b$ − $T_s$. The amount of energy flow depends upon the conductivity of the fat and its thickness: $M - E_{ex} = \dfrac{K_b}{d_b}$ ($T_b$ − $T_s$). At the skin's surface a certain amount of energy may be lost through sweating or evaporation of moisture ($E_{sw}$). The amount of energy conducted through the fur or feathers across which the temperature difference is $T_s$ − $T_r$ must be the same as*

*the amount of energy conducted across the fat less energy lost by sweating. The whole body temperature process is described by the equation*

$$(T_b - T_r) + (T_b - T_s) + (T_s - T_r) =$$
$$\frac{d_b}{K_b} (M - E_{ex}) + \frac{d_f}{K_f} (M - E_{ex} - E_{sw})$$

*(After Porter and Gates, 1969.)*

and heat moves from the environment to the organism.

Depending upon the values of the several components discussed, organisms are continually either emitting or absorbing radiation. Within a few minutes or even seconds an organism may undergo great temperature changes on the surface. Energy gained at one moment is lost at another. For example, Moen (1968b) measured the fluctuations on the surface of a ring-necked pheasant. Under a bright sun and free convection condition the surface temperature of the bird was 45° C, 4° to 5° higher than the core temperature. Thus heat flow was from the outside surface to the interior of the animal. With a cloud cover and a slight breeze the bird's surface temperature decreased nearly to air temperature (21° C), and heat flow was from the inside of the animal to the outside. These variations took place in a matter of seconds, while the air temperature remained the same. Extreme temperatures in the bird were buffered by the insulation of the feathers.

*Plant responses.* Although both plants and animals experience and exist within the same external energy environment, there are fundamental differences in their response to thermal pressures. Plants cannot move to escape adverse effects of heat or cold, and their metabolic heat is derived from absorbed solar radiation.

Except for the time when they sprout from seeds or vegetative cuttings, plants receive all their heat from radiation in the environment and from convection. To balance the input plants lose heat by radiation, convection, and evapotranspiration (Figure 3-10). The role of evapotranspiration is modified by the opening and closing of the stomata and by changes in the shape and position of the leaf. A notable example is the response of the evergreen leaves of rhododendron and laurel. In winter the leaves droop and curl. The colder the temperature, the tighter the curl. In fact it is possible to estimate winter temperature by the curl of the leaf.

The temperature of the leaf results from a combination of climatic variables such as wind speed, internal diffusion resistance, radiation absorbed, relative humidity, and air temperature. Wind speed is important. Winds as low as 1 mi/hr affect the temperature. Loss of water from the leaf increases if the air is relatively dry.

Heat affects the physiological processes of plants, particularly rate processes that depend upon temperature, such as photosynthesis and respiration. On a hot summer day a sunlit leaf can become too warm for photosynthesis. Following an early morning burst of activity, net photosynthesis ceases and respiration becomes dominant (Figure 3-12). Photosynthesis resumes later in the day after the leaf has cooled, but at this point light intensity has fallen to a suboptimal level. Because a plant's leaves face in different directions, photosynthesis of north-, south-, east- and west-facing leaves will differ throughout a summer day (see Gates, 1968). In arctic and alpine regions plants have a low threshold for photosynthesis. Starting growth while snow is still on the ground, the plants possess basal leaves and prostrate forms that make use of the heat layer of air next to the soil surface. These organs start photosynthesis at air temperatures unfavorable for taller shoots (Scott and Billings, 1964).

In addition to possessing optimum temperatures for photosynthesis, plants also have an optimum temperature range for other aspects of their life cycle. To germinate, seeds of many plants in cold regions require chilling under moist conditions after a period of maturation. Fluctuation in temperatures is necessary for best germination of most seeds. Some plants require low temperatures during or shortly after germination. And the temperature necessary to stimulate flowering may be lower than the temperature that favors flower development.

When temperatures drop below the minimum for growth, a plant becomes dormant even though respiration and photosynthesis may continue slowly. Low temperatures further affect the plant by precipitating the protein in leaves and tender twigs and by dehydrating the tissues. Intercellular ice draws water out of the cells, and rapid freezing causes ice to form within the protoplasts.

The ability to endure low temperatures varies among plants and among stages in the life cycle of the plant. Flowers are more sensitive to low temperatures than fruits or leaves, and young leaves are more resistant than old ones. Trees may be more severely injured by frost than herbaceous plants.

Adaptations to endure low temperatures are primarily protoplasmic. Major exceptions are plants with waxy blooms or dense hairs on leaves and stems. Cold hardiness is acquired

*FIGURE 3-12*

*The effect of temperature on net photosynthesis is illustrated in these graphs: (a) solar radiation incident on an exposed and shaded leaf; (b) air temperature in relation to exposed and shaded leaf temperature; (c) transpiration rate of an exposed and shaded leaf; (d) relative metabolic rates for an exposed and shaded leaf. The four graphs show that the exposed leaf is inhibited photosynthetically when its leaf temperature is high. (From Gates, 1968b.)*

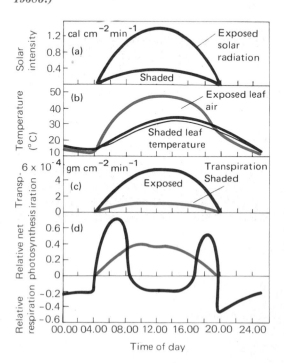

by a hardening process in which the protoplasm develops low structural viscosity. Free-water content decreases and water-soluble sugars and proteins increase. Both of these changes lower the freezing point of tissues. Hardening of perennials in cool and cold climates takes place in the fall. In spring cold hardiness is lost with the renewal of activity. A sudden drop in temperature after hardiness is lost may kill tissues that were able to survive far lower temperatures during the winter.

*Animal responses.* Animals respond to the thermal environment both physiologically and behaviorally. Animals produce an often considerable amount of heat by their own metabolism and maintain body temperature through various regulatory processes, but also they move about to seek a favorable temperature regime.

Terrestrial animals are more subject to radical and potentially dangerous changes in their thermal environment than aquatic animals because air has a lower specific heat than water and absorbs less solar radiation. Incoming solar radiation can produce lethally high temperatures and radiational loss to space can result in lethally low temperatures. Aquatic organisms live in a more stable energy environment, but at the same time they have a lower tolerance of temperature change.

*Thermoregulation and acclimation.* Body temperatures of animals are the result of heat gains and heat losses. Although all living things produce heat continuously, the magnitude and importance of this heat production depend upon the particular evolutionary strategy of each species. Some animals exhibit high rates of thermal conductance and low rates of heat production (usually less than $2°$ C). These animals exploit sources of heat energy other than metabolism (solar radiation, reradiation, etc.) and are called *ectotherms* because their principle source of heat is external. If the environmental temperature varies appreciably during the day the body temperature of these animals also varies. In terms of this variation in body temperature, such animals are categorized as *poikilotherms*. Within the range of temperatures that poikilothermic animals can tolerate, the rate of metabolism and thus oxygen consumption increase according to van't Hoff's rule. For

every temperature rise of $10°$ C, the rate of oxygen consumption doubles. Oxygen consumption follows temperature, but its rate of change is exponential. Lacking any homeostatic devices, poikilotherms of terrestrial environments in particular must depend upon some behavioral control over body temperatures. Most aquatic poikilotherms generally encounter temperature fluctuations of lesser magnitude and usually have more poorly developed behavioral and physiological thermoregulatory capabilities than terrestrial forms. They are, in effect, "trivially homoiothermic" (McNab, 1978).

A relatively small group of animals, largely birds and mammals but also a few reptiles to some extent, possess a sufficiently high rate of oxidative metabolism and a sufficiently low rate of thermal conductance to accomplish thermogenesis (that is, to generate heat). The body temperature is a product of the animal's own oxidative metabolism. Such animals are *endotherms*. Because the body temperature of endothermic animals remains constant regardless of environmental temperatures, they are called *homoiothermic*. Homoiotherms maintain an appreciable difference between body temperature and external environmental temperature through internal heat production. Their oxygen consumption decreases linearly with increasing temperature to some critical point where it is independent of environmental temperature. In other words, the lower the temperature, the more metabolic work the animal does to maintain a uniform temperature, but within a certain range of temperatures the amount of energy needed to maintain body temperatures becomes minimal. This level of minimum work is the *zone of thermal neutrality*. The lower limit of this zone represents the critical air temperature, the lowest ambient temperature at which a bird or mammal can maintain its body temperature at basal metabolism at rest (BMR). At all ambient temperatures, homoiotherms maintain their metabolic rate and body temperature by means of a closely regulated feedback system mediated by the hypothalamus. The metabolic costs of such a system, however, are high. Homoiotherms use some 80 to 90 percent of the oxidative energy to maintain thermal homeostasis.

How an animal responds to the energy environment depends upon whether it is a poikilotherm or a homoiotherm, whether it is living in a warm or cold environment, and whether it is in a terrestrial or aquatic situation.

Aquatic poikilotherms cannot maintain any significant difference between their body temperature and the environment. Their tissues are completely permeated by the circulatory system. Any heat produced in the muscles is transferred to the blood flowing through them and is carried away to the gills or skin, where it is lost to the water by convection and conduction. Fish are therefore ideal poikilotherms. Because of this close relationship between body temperature and environmental temperature, fish are readily victimized by any rapid changes in environmental temperatures. When temperature changes are not so rapid, the fish may escape to more favorable waters, for they can perceive minute differences in temperature—less than 1° C.

One can view the total range of temperature environment of a fish or any aquatic poikilotherm as a group of zones (Fig. 3-13). The central zone of *thermal tolerance* is the range of temperatures within which the fish is most at home. Within this range fish may have certain *preferred temperatures* that they will seek. The zone of thermal tolerance is bounded by an upper and lower zone of *thermal resistance,* temperature ranges within which the organism can survive for an indefinite period of time. The upper and lower bounds of thermal resistance are marked by the upper and lower *incipient lethal temperature.* This is the temperature at which a stated fraction of a fish population (generally 50 percent), when brought rapidly to it from a different temperature, will die within an indefinitely prolonged exposure (Fry et al., 1946).

The incipient lethal temperature is not fixed. Its value is affected by the previous thermal history of the organism. Poikilothermic animals are, within limits, able to adjust or *acclimate* to higher or lower temperatures. At this point it is best to define two terms, acclimation and acclimatization, similar but with different meanings. *Acclimation* refers to differences in the physiology of an animal after exposure to environments differing in

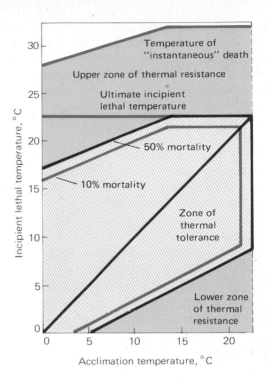

**FIGURE 3-13**
*Thermal tolerance of a hypothetical fish in relation to thermal acclimation. (From C. C. Coutant, 1970.)*

one or two well-defined variables as temperature and oxygen level. It is largely a laboratory phenomenon. *Acclimatization* refers to the differences that appear in the physiology of an animal after exposure to different natural environments, which vary in a number of parameters simultaneously. If an organism lives at the higher end of the tolerable range, it acclimates itself sufficiently so that the lethal temperatures will be somewhat higher and the lower limits will be somewhat higher than if it were living within the cooler range of the tolerable environment. Once acclimated to a given temperature, fish acclimate more readily to an increase than to a decrease in temperature. Acclimation to a higher or lower temperature both increases the length of time an organism can survive an elevated or lower temperature and raises the maximal or minimal temperature it can survive for a given period of time. However, acclimation has upper and lower limits and a temperature ultimately will be reached that will be lethal to an organism. This is the *ultimate incipient le-*

*thal temperature.* (For an excellent discussion of acclimation, see Coutant, 1970.) Because adjustment or acclimatization to varying environments takes place from one season to the next, a population is able to adjust to prevailing temperatures each season.

Amphibians present a somewhat different situation. Permanently aquatic forms, such as many salamanders, maintain body temperatures in the same manner as fish. Their temperature control is primarily one of seeking preferred temperatures within their habitat. But for semiterrestrial frogs and salamanders the problem of temperature control is more complex. The avoidance of temperature extremes is probably the most important mechanism, but within this framework amphibians are able to exert considerable control over their body temperature, which does not, as is so often assumed, simply follow the air temperature (see also Brattstrom, 1963). By basking in the sun *(heliothermism)* frogs can raise their body temperature as much as 10° C above the ambient temperature. Because of associated evaporative water losses, such amphibians must either be near water or sit partially submerged in it (Figure 3-14). Forms that live near water also use evaporative cooling through the skin to reduce body heat loads. By changing position or location or by seeking a warmer or cooler substrate, amphibians can maintain body temperatures within a narrow range of variation (see Lillywhite, 1970).

Relatively few reptiles are aquatic. Most are terrestrial and lack the buffering effects of water. Exposed to the widely fluctuating temperatures of the terrestrial environment, reptiles must possess more refined means of temperature regulation.

Although poikilothermic, reptiles exhibit little relationship between their core body temperature and the ambient temperature. By behavioral means they are able to maintain a relatively stable body temperature (Figure 3-15). Reptiles utilize evaporative cooling by panting and by water loss through the skin to prevent body temperature from reaching the *critical thermal maximum* (CTM). This is the temperature at which the animal's capacity to move becomes so reduced that it cannot escape from thermal conditions that will lead to its death (Cowles and Bogert, 1944). Thus reptiles

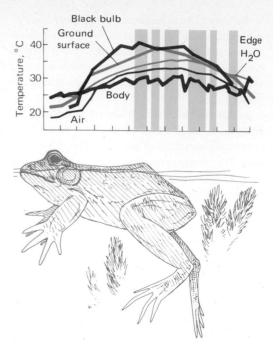

**FIGURE 3-14**
*The body temperature of a bullfrog measured telemetrically. Dips in the black bulb temperature indicate effects of cloud cover, convection, or both. Water temperature around the pond's edge varies from one location to another by as much as 2 to 3° C. Thus while in shallow water a frog may show a higher body temperature than that recorded for the edge water. Note the relative uniformity of temperature the bullfrog maintains by moving in and out of the water. (From Lillywhite, 1970.)*

possess some of the basic physiological mechanisms so characteristic of and so highly developed in the endotherms.

The reptile is able to regulate body temperature in a number of ways, the simplest of which is heliothermism. Remaining in the sun raises the reptile's core body temperature. When the temperature reaches the preferred level, the animal moves into the shade. It remains there until the body temperature drops below the preferred range and then moves back out into the sun again.

More elaborate behavior common to many lizards is *proportional control.* If the air temperature is lower than the preferred, the lizard can spread its ribs, flatten its body, and orient itself so that its body is at right angles to the sun. In this manner it gains the maximum amount of heat. If the temperature is too high, the lizard can appress its ribs and orient its body parallel to the sun, thereby decreasing the surface area exposed to radiant energy.

Other behavioral methods involve burrowing into the substrate, panting, and possibly changing color. Thus reptiles have at their command a variety of behavioral mechanisms useful for temperature regulation.

In addition to behavior reptiles possess some physiological mechanisms that permit them some degree of control over the maintenance of preferred body temperature. At least four families of lizards (Iguanidae, Gekkonidae, Varanidae, and Agamidae) can control the

rate of change in body temperature by changing the rate of heartbeat and by varying the rate of metabolism.

Birds and mammals escape the constraints of the environment by becoming homoiothermic. Instead of depending completely upon the environment for heat, they produce it by metabolic oxidation and regulate the gradient between body and air temperatures by changes in insulation (the type and thickness of fur, the structure of feathers, and layer of fat), by evaporative cooling, and by changing body temperature. Vascularization is another device among birds and mammals for conserving or getting rid of heat. Some animals, especially those in the arctic, have extensive areas in the extremities where the veins are closely juxtaposed to the arteries. Much of the heat lost by the outgoing arterial blood is picked up by the returning venous blood. As a result the venous blood is warmed on its return and reenters the body core only

**FIGURE 3-15**
*Behavioral mechanisms involved in the regulation of body temperature by the horned lizard* (Phrynosoma coronatum). *(From Heath, 1965.)*

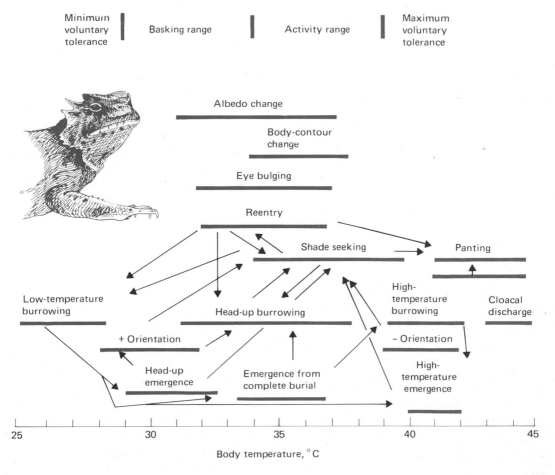

slightly cooler than the outgoing arterial blood. This arrangement prevents excessive heat loss through the extremities and reduces their temperature to prevent freezing. For example, tissue temperature in arctic gulls declines rapidly along the tibia under the covering of feathers. Heat from the warm arterial blood is transferred to cool venous blood returning from the foot which may be close to 0° C (Irving, 1960). In addition fat of low melting point is selectively deposited in those parts of the extremities subject to excessive cooling. Fat in the foot pads of the arctic fox and the caribou has a melting point of 0° C, about 30° C lower than the fat of the body core, yet the fat remains soft and flexible (for details, see Irving, 1972; Hill, 1976).

Within limits warm-blooded animals can maintain their basal heat production by changing insulation. But with declining temperatures there is a point beyond which insulation is no longer effective and body heat must be maintained by increased metabolism. The temperature at which this takes place is called the *critical temperature,* and it varies greatly between tropical and arctic animals (Scholander, Hock, Walters, Johnson, and Irving, 1950). Tropical birds and mammals exposed to temperatures below 23.5° to 29.5° C increase their heat production. If the air temperature is lowered to 10° C, the tropical animal must triple its heat production; and if lowered to freezing, the animal is no longer able to produce heat as rapidly as it is being lost. Arctic small mammals, on the other hand, do not increase their heat production until the air temperature has fallen to −29° C. Large arctic mammals can sustain the coldest weather without heat beyond that produced by normal basal metabolism. Eskimo dogs and arctic foxes can sleep outdoors at temperatures of −40° C without stress. This is not due to any difference in metabolism itself but to effective insulation and cold acclimation. A number of mammalian species significantly increase their basal metabolism during cold acclimation without the intervention of shivering.

When body temperature falls below the critical level metabolism may be further increased by shivering. Among birds exposed to cold, shivering and muscular activity are primary sources of extra heat.

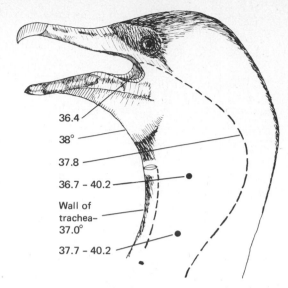

Body temperature–41.5° C

*FIGURE 3-16*
*Representative evaporative surface temperatures from two 6.5-week-old double crested cormorants* (Phalacrocorax auritus). *Ambient temperature was held at body temperature (41.5°C) during measurements. The lower temperatures in the buccal and pharyngeal regions, tracheal walls, and the portion of the esophagus moved during gular fluttering indicate that these areas are sites of evaporation. (After Lasiewski and Snyder, 1969.)*

Mammals acclimated to cold temperatures can increase their heat production by nonshivering thermogenesis. Nonshivering thermogenesis is the generation of heat from the metabolism of brown fat, a highly vascular brown adipose tissue localized around the head, neck, thorax, and major blood vessels. It is found in the young of most species and in animals that hibernate, and it increases in mass when animals are chronically exposed to low temperatures (for detailed reviews, see Smith and Horwitz, 1969; Chaffee and Roberts, 1971). Heat generated from the metabolism of this fat is transported to the heart and brain. Nonshivering thermogenesis allows animals to be active at lower temperatures than if they had to depend on shivering alone. This form of heat production and exercise are additive, so the two together can be important in maintaining body temperatures.

Physiologically it is more difficult for a homoiotherm to adapt to high temperatures than to cold. Endogenously produced heat must be transferred to the atmosphere, and for the most part this involves an evaporative loss of water, largely through sweating and panting. Because birds possess a body temperature some 4° to 5° C higher than mammals, they are more tolerant of intense heat. Birds do not sweat, and water loss through the skin is inhibited by the insulating covering of feathers. Body heat is lost largely by radiation, conduction, and convection. But when conditions demand it, birds can decrease their heat load by evaporative cooling through panting. However, panting requires work and work only adds more metabolic heat. Some groups of birds, particularly the goatsuckers, owls, pelicans, boobies, doves, and gallinaceous birds get around this by *gular fluttering* (see Figure 3-16), movements of parts of the gullet. Evaporative cooling by gular fluttering uses less energy than the heavy breathing of panting.

Another approach to the heat problem, especially in birds, is *hyperthermia*. A rise in body temperature adjusts the difference between the body and the environment, and thermal homeostasis is reestablished at a higher temperature. (Hyperthermia among mammals is considered below in the discussion of moisture.)

Parts of the body may serve as thermal windows in the conservation or radiation of heat. Horns and antlers of Artyodactyla, such as deer and goats, are vascular and only incompletely shielded by hair. Dilation of the blood vessels in response to heat stress allows increased radiation of heat, and constriction in the cold reduces radiation. The large ears of such desert mammals as the kit fox and jack rabbit function as efficient radiators to the cooler desert sky, which on clear days may have a radiation temperature of 25° C below that of the animal (Schmidt-Nielsen, 1964). By seeking shade, where the ground temperatures are low and solar radiation is screened out, or by sitting in depressions, where radiation from hot ground surface is obstructed, the jack rabbit through two large ears (400 cm²) could radiate 5 kcal/day. This is equal to one-third of the metabolic heat produced in a 3-kg

rabbit. Such a radiation loss alone may be sufficient to take care of the necessary heat loss without much loss of water.

Some animals of arid regions simply avoid heat by adopting nocturnal habits and remaining underground or in the shade during the day. Some desert rodents that are active by day periodically seek burrows and passively lose heat through conduction by pressing their bodies against burrow walls. Some birds, such as the whippoorwill and certain hummingbirds, and some bats go into a daily torpor.

*Dormancy.* One way to avoid both heat and cold is to go into a state of dormancy during a period of environmental stress. Many species of insects as well as certain crustaceans, mites, and snails, enter *diapause,* a period of dormancy and arrested growth. Eggs, embryonic larvae, or pupal stages may be involved. In addition to tiding the animal over unfavorable periods, diapause also synchronizes the life cycle of a species with the weather and ensures that active stages will coincide with the climatic conditions and food supply that favor rapid development and high survival.

Some poikilothermic animals become dormant during periods of temperature extremes. In winter amphibians and turtles bury themselves in the mud of pond bottoms; snakes seek burrows and dens in rocky hillsides. There they remain in a state of suspended animation until the temperature warms or cools again. Winter dormancy is called *hibernation;* summer dormancy is *estivation.*

Hibernation and estivation are not confined to poikilothermic animals. The phenomena also occur among a few homoiothermic animals, particularly bats, ground squirrels, woodchucks, and jumping mice. Like dormancy in poikilotherms, hibernation in homoiotherms is characterized by a reduction in the general metabolism to a degree never found in the deepest everyday sleep. But there is a difference. When reptiles and amphibians are exposed to cold, the animal cools because it has no way to stay warm. As the poikilotherm's system becomes cooler, the heart rate and metabolic rate decline. But when a mammal enters hibernation, the heart rate and metabolic rate decline, and *then* the body temperature drops.

The onset of dormancy in warm-blooded

animals may take place with one decline in temperature, as in the hamster and pocket mouse, or it may come in a series of steps, as in the groundhog and ground squirrel. With the onset of winter the California ground squirrel enters periods of incomplete dormancy. Each day its temperature drops a number of degrees, remains at this point for a while, then rises to a normal body temperature again (Strumwasser, 1960), but with each day the body temperature drops a few degrees lower. This continues until the body temperature is just a few degrees above that of the environment. This apparently conditions the body and the brain to lower temperatures. After a few such fitful starts at complete dormancy, the animal finally curls up to sleep in its retreat.

In this state of suspended animation, only the vital life of the animal continues, maintained by shallow breathing, slow circulation, and continual digestive absorption of stored fat. The reduced metabolism is characterized by great differences in body temperature, pulse rate, and breathing. The breathing rate, rapid in small mammals, is reduced to less than one breath a minute, and these breaths usually occur in a series of two or three gasps with long intervals between. The body temperature in the ground squirrel drops from 18.5° to 4.2° C. The heart rate is reduced to two to three beats a minute, although the blood pressure remains relatively high. The optimum environmental temperature for hibernators is 4.5° C (40°F); the body temperature passively follows the fluctuation in environmental temperatures between 3.3° to 12.8° C. If the environmental temperature reaches the freezing point of water, the metabolic rate of the hibernating animal increases. This increase may be able to keep the body temperature above freezing without the animal awakening. Mammalian hibernators, however, do not remain in a state of complete inertia all during the period of dormancy. They awake from time to time, eat a little, void, and return to their deep sleep.

Arousal from hibernation is explosive and certainly the most dramatic aspect of the hibernation cycle (Figure 3-17). As the animal starts to come out of the sleep, its body temperature rises rapidly, perhaps from 4° to 17.5° C in an hour and a half (Mayer, 1960). In the

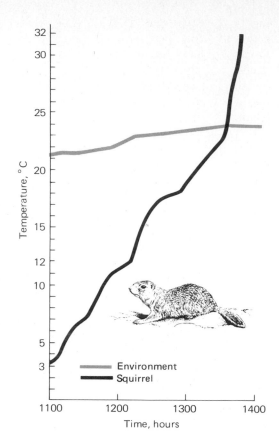

**FIGURE 3-17**
*Rise in body temperature of the arctic ground squirrel upon arousal from hibernation. With the second rapid rise, at 24° C, the animal opens its eyes and sits up. (After Mayer, 1960.)*

case of the ground squirrel, when the anterior part of the body has reached about 36.5° C, the temperature of the hind parts rises rapidly due to dilatation of the blood vessels. The animal's hair is erect, and its body shakes. At around 16° C, the animal tries to right itself, and at 17.5° C the shivering stops and the squirrel moves its tail. At 24° C the animal opens its eyes and suddenly sits up. At this time the temperature rises rapidly again. When the body temperature reaches 24° to 25.4° C, about 3 hours from the time of initial arousal, the ground squirrel is sitting up and flicking its tail. The animal is warm and active again.

The black bear and skunk are not true hibernators; in both, neither body temperature nor pulse is lowered significantly, and breathing is more frequent. During warm spells both will leave their dens and move about. Rarely are adult skunks inactive for over a month nor the young for more than four months. During these deep sleeps the animals may lose from 15 to 40 percent of body weight (Hock, 1960).

Estivation is common among some desert rodents. Not only does this prolong the period during which the animal can live on energy reserves through times of severe heat, but it also saves considerable amounts of water because the reduced ventilation of the lungs lowers the pulmonary evaporation of water. In addition, the lower body temperature of the animal reduces the amount of water which is needed to saturate the expired air (Schmidt-Nielsen, 1964).

## TEMPERATURE AND DISTRIBUTION

Because the optimum temperature for the completion of the several stages of the life cycle of many organisms varies, temperature imposes a restriction on the distribution of species. Optimal temperatures for some species are so different from those of others that the animals cannot inhabit the same area. Some organisms, particularly plants that are growing under suboptimal temperatures, cannot compete with the surrounding growth, a situation that would not exist under optimal conditions.

Generally the range of many species is limited by the lowest critical temperature in the most vulnerable stage of its life cycle, usually the reproductive stages. Although the Atlantic lobster will live in water with a temperature range of 0° to 17° C, it will breed only in water warmer than 11° C. The lobster may live and grow in colder water, but a breeding population never becomes established there.

A classic example of temperature limitation on animal distribution is found among three species of ranid frogs (J. A. Moore, 1949a). The wood frog breeds in late March, when water temperature is about 10° C. Its eggs can develop at temperatures as low as 2.5° C. Larval stages transform in about 60 days. This frog ranges into Alaska and Labrador, further north than any other North American amphibian or reptile. The leopard frog breeds in late April, when the water temperature is about 15° C, and the larvae require around 90 days to develop. As a result the northern limit of its range is southern Canada. The southernmost species of the three, the green frog, does not breed until the water is about 25° C, and the eggs will develop at 33° C, a lethal temperature for the others. Its eggs, however, will not develop at all until the temperature exceeds 11° C. The range of the green frog extends only slightly above the northern boundary of the United States.

A number of examples can be found among plants. The northern limit of the sugar maple closely parallels the 2° C mean annual isotherm. The paper birch, a cold-climate species, is found as far north as the 12° C July isotherm and seldom grows naturally where the average July temperature exceeds 21° C. The distribution of the black spruce follows a similar pattern. Its southern distribution is approximately the same as that of the paper birch, and its northernmost outliers are seldom far from the mean July isotherm of 10° C.

Some plants, such as blueberries, will not flower or fruit successfully unless chilled. Other plants may grow in a region during the summer, but are unable to reproduce or grow to normal size because the twigs freeze back in winter or are killed by late spring frosts. These plants are restricted within their natural range to areas where temperatures are favorable for growth and reproduction.

## Moisture

Water is essential to all life. Means of obtaining and conserving it have shaped the nature of terrestrial communities. Means of living within it have been the overwhelming influence on aquatic life. Because of its enormous importance, water and its properties need some further discussion.

### WATER

*Structure.* Because of the physical arrangement of its hydrogen atoms and hydrogen bonds, liquid water consists of branching chains of oxygen tetrahedra. The physical state of water, whether liquid, gas, or solid, is determined by the speed at which hydrogen bonds are being formed and broken. Heat increases that speed; hence weak hydrogen bonds cannot hold molecules together as they move faster. The thermal status of water in a liquid state is such that hydrogen bonds are being broken as fast as they form. At low temperatures the tetrahedral arrangement is almost perfect. When water freezes, the arrangement is a perfect lattice with considerable open space between ice crystals, and thus a decrease in density (this is the reason ice floats). As the temperature of the frozen water is increased, the molecular arrangement becomes looser and more diffuse, resulting in random packing (because of the continuous breaking and reforming of hydrogen bonds) and contraction of molecules. The higher the temperature, the more diffuse the pattern becomes, until the whole structure (and the hydrogen bonds) breaks down and the water melts. Upon melting, water contracts, and its density increases up to a temperature of 3.98° C. Beyond this point, the loose arrangement of molecules means a reduction in density again. The existence of this point of maximum density at approximately 4° C is of fundamental importance to aquatic life.

Seawater behaves somewhat differently. Seawater is defined as water with a minimum salinity of 24.7 ‰ (‰ = parts per thousand). Its density, or rather its specific gravity relative to that of an equal volume of pure water (sp. gra. = 1) at atmospheric pressure, is correlated with salinity. At 0° C the density of seawater with a salinity of 35 ‰ is 1.028. The lower its temperature, the greater becomes the density of seawater; the higher its temperature, the lower the density. No definite freezing point exists for seawater. Ice crystals begin to form at a temperature that varies with salinity. As pure water freezes out the remaining unfrozen water is even saltier and has an even lower freezing point. If the temperature can continue to decrease to reach a low enough point, ultimately a solid block of ice crystals and salt is formed.

*Physical properties.* Water is capable of storing tremendous quantities of heat with a relatively small rise in temperature. It is exceeded in this capacity only by liquid ammonia, liquid hydrogen, and liquid lithium. Thus water is described as having a high *specific heat,* the number of calories necessary to raise 1 g of a substance 1° C. The specific heat of water is given the value of 1.

Not only does water have a high specific heat, but it also possesses the highest heat of fusion and heat of evaporation—collectively called *latent heat*—of all known substances that are liquid at ordinary temperatures. Large quantities of heat must be removed before water can change from a liquid to its solid form, ice; and conversely ice must absorb considerable heat before it can be converted to a liquid. It takes approximately 80 cal of heat to convert 1 g of ice at 0° C to a liquid state at 0° C. This is equivalent to the amount of heat needed to raise the same quantity of water from 0° to 80° C.

Evaporation occurs at the interface between air and water at all ranges of temperature. Here again considerable amounts of heat are involved; 536 cal are needed to overcome the attraction between molecules and convert 1 g of water at 100° C into vapor. This is as much heat as is needed to raise 536 g of water 1° C. When evaporation occurs, the source of thermal energy may come from the sun, from the water itself, or from objects in or around it. Rendered latent at the point of evaporation, the heat involved is returned to actual heat at the point of condensation (the conversion from vapor to liquid). Such phenomena play a major role in worldwide meteorological cycles.

The property of *viscosity* also has biological importance. The viscosity of water is high because water molecules interact with neighboring molecules by forming hydrogen bonds. Vicosity can be visualized best if one imagines or observes liquid flowing through a clear glass tube. The liquid moving through the tube behaves as if it consisted of a series of parallel concentric layers flowing over one another. The rate of flow is greatest at the center, but because of the amount of internal friction between layers, the flow decreases toward the sides of the tube. This type of resistance between layers is called *lateral* or *laminar viscosity.*

This phenomenon can be observed along the side of any stream or river with uniform banks. The water at streamside is nearly still, whereas the current in the center may be swift.

In the flow of water viscosity is complicated

by another type of resistance, *eddy viscosity,* in which water masses pass from one layer to another, creating turbulence both horizontally and vertically. Biologically important, eddy viscosity is many times greater than laminar viscosity.

Viscosity is the source of frictional resistance to objects moving through the water. Since this resistance is 100 times that of air, animals must expend considerable muscular energy to move through the water.

Another property of water with significance for biological processes is surface tension. Within all substances particles of the same matter are attracted to one another. Molecules of water below the surface are symmetrically surrounded by other molecules and thus the forces of attraction are the same on all sides of the molecules. But at the water's surface, the molecules exist under a different set of conditions. Below is a hemisphere of strongly attractive similar water molecules; above is the much smaller attractive force of the air. Since the molecules on the surface, then, are drawn into the liquid, the liquid surface tends to contract and become taut. This is *surface tension.*

In the aquatic ecosystem surface tension is a barrier to some organisms and a support for others. It is the force that draws liquids through the pores of the soil and the conducting networks of plants. Aquatic insects and plants have evolved structural adaptations that prevent the penetration of water into the tracheal systems of the former and the stomata and internal air spaces of the latter.

## ADAPTIVE RESPONSES TO MOISTURE

On a hot summer afternoon a catbird sits perched in the shade of a shrub, its wings half dropped, its bill open. At the same time the leaves of the shrub hang drooped and slightly curled on the stem. Both are responding to the stress of heat, which is affecting the water economy. In a practical sense it is difficult to separate the water budget from the heat budget, for organisms utilize water to maintain their temperature. However, we can examine the physiological mechanisms both plants and animals have evolved to maintain water and solute balance.

*Plant responses.* Plants adapt to the moisture environment in a number of ways. Some plants adopt a life cycle in which the population survives dry periods as dormant seeds ready to sprout quickly when the rains come. They complete the active stages of their life cycle in a short period before moisture is completely gone, and then return to seed dormancy. The cycle may be based on annual wet and dry seasons, or it may extend over a period of several years, as in the case of some desert plants which may remain dormant for 10 years.

Another means of conserving water is succulence, an adaptation common to plants of salt marshes and deserts. Notable are the cacti. Because their root systems are shallow, cacti are able to draw up considerable amounts of water during periods of rain. The plants swell rapidly as they store water in the enlarged vacuoles of the cells. As they draw on this store of water during periods of moisture stress, the plants gradually shrivel. To further conserve water, cacti open the stomata at night and take in and fix carbon dioxide for use during the day when their stomata are closed (see $C_4$ photosynthesis, Chapter 2).

Plants may meet moisture stress by means of various leaf adaptations. In some the leaf size is small, the cell walls are thickened, the vascular system is denser, and the palisade tissue is more highly developed than the spongy mesophyll. This increases the ratio between the internal exposed cellular surface area of the leaf and the external surface. If the plants still do not obtain enough water, they simply shed their leaves to reduce transpiration. This is also the wintertime response of deciduous broad-leafed trees of temperate regions. Plants that have small leaves or shed them to conserve water maintain their productive capacity by increasing photosynthetic activity in the stem. Plants may also react to drought conditions by rolling, curling, or folding their leaves to reduce the unit area exposed to drying sun and wind. Water loss may also be reduced by the secretion of a waxy or resinous substance over the surface of the plant.

Root adaptations provide still other means to maintain water balance. An extensive root system that either penetrates deeply or covers a large shallow surface area is able to secure the maximum amount of water when moisture is present. Mesquite, for example, has a root system that penetrates to a depth of 54 m (W. S. Phillips, 1963).

Some desert plants have the ability to allow only certain ions to pass across cellular membranes and to keep others out. This allows the plant to absorb essential nutrients from the soil and helps maintain internal osmotic pressures higher than those of the surroundings.

Although plants of the salt marsh are surrounded by an abundance of water, they grow in a physiologically dry substrate because the salinity of the water limits the amount of moisture they can absorb by osmosis and the water they do absorb contains a high level of solutes. These plants compensate for an increase in sodium and chlorine uptake with a corresponding dilution of internal solutions with water stored in the tissues. In addition the plants exhibit high internal osmotic pressure many times that of freshwater and terrestrial plants, possess salt-secreting glands, secrete heavy cutin on the leaves, and are succulent (Chapman, 1960).

*Animal responses.* Adaptations to moisture stresses are more complex in animals than in plants. A more or less universal mechanism is the excretory system, designed to rid the body of water and solute wastes or to conserve them.

Aquatic organisms maintain their water balance by regulating osmotic pressure. Freshwater organisms live in an environment where the surrounding water has a lower salt concentration than their bodies. Their problem is to rid the body of excess water taken in through the permeable membranes. Each organism has an organelle, specialized cells, or organ that functions as an excretory system. For example, protozoans dispose of excess water through the contractile vacuole. Other organisms have flame cells, nephridia, or kidneys. Freshwater fish maintain their salt concentration by salt absorption through special cells in the gills. Amphibians have skin that is permeable to water, and through it they take up water by osmosis. Water produced by the kidney flows to the bladder where it is stored. If circumstances require it, aquatic amphibians cease urine production and conserve water and solutes for metabolic purposes by increasing reabsorption through the bladder.

Terrestrial animals have three means to gain water and solutes: through drinking, through food, and through production of metabolic water from food. They can lose water and solutes through urine, feces, evaporation over the skin, and respiration.

Birds and reptiles have three pathways through which they can physiologically control water losses: the gut, the kidney, and in some a salt gland. The kidneys of birds and reptiles are similar and both possess a cloaca, the chamber into which intestinal, urinary, and generative canals all discharge. After urine leaves the kidney the cloaca modifies its volume and composition, converting it into a semisolid paste, consisting almost entirely of uric acid. In birds, however, most concentration seems to occur in the renal tubules of the kidney; uric acid then appears in the collecting ducts in a precipitated form (Sturkie, 1976). The salt gland, found mainly in marine species, produces copious secretions containing high concentrations of sodium chloride.

Mammals have kidneys capable of lowering water losses by producing higher urinary osmotic and inorganic ion concentration. Although they lack salt glands and cloaca, many species of mammals do possess sweat glands. Among terrestrial animals a major source of water loss is through the skin and respiratory system. Birds and mammals have relatively high losses through this avenue and mammals through sweating and urine excretion.

If an organism is not to dehydrate, the input of water must equal the losses. For the most part moisture becomes a problem in only two environments, the saltwater environment and the desert. The saltwater environment is a physiological desert, in which the concentration of salts outside of the body of the organism can dehydrate it osmotically. In the true desert an absolute lack of moisture exists. Interestingly many organisms of different environments use the same strategy to overcome the problem.

In the marine and brackish environments the problem is one of inhibiting the loss of water through the body wall by osmosis and preventing an accumulation of salts in the system. Algae and invertebrates get around the problem by possessing body fluids that have the same osmotic pressure as seawater. In a way this adds to the problem of marine invertebrates because they cannot obtain fresh water through their food. Marine teleost fish absorb the water with the salt into the gut. They secrete magnesium and calcium through the kidneys and pass these ions off as a partially crystalline paste. The fish excrete sodium and chlorine by pumping the ions across membranes of special cells in the gills. This pumping process is called *active transport*. It involves the movement of salts against a concentration gradient at the cost of metabolic energy. Sharks and rays retain urea to maintain a slightly higher concentration of salt in the body than in surrounding seawater. Birds of the open sea are able to utilize seawater, for they possess special salt-secreting glands located on the surface of the cranium (Schmidt-Nielsen, 1960). Gulls, petrels, and other sea birds excrete from these glands fluids in excess of 5 percent salt. Petrels and other tube-nosed swimmers forcibly eject the fluid through the nostrils; in other species the fluid drips out of the internal or external nares.

Among marine mammals the kidney is the main route for the elimination of salt. Porpoises have highly developed renal capacities to eliminate salt loads rapidly (Malvin and Rayner, 1968). In marine mammals the urine has a greater osmotic pressure than blood and seawater (hyperosmotic); the physiology is poorly understood.

Vertebrates in the Arctic and Antarctic have special problems. As seawater freezes it becomes colder and more salty. The only alternative for most organisms is to increase the solute concentration in the body fluids to lower the body temperature. Some species of fish in the Antarctic possess in their blood a glycoprotein antifreeze substance that enables the animals to exist in temperatures below the freezing point of blood.

Among the reptiles the diamondbacked terrapin, an inhabitant of the salt marsh, lives in an environment in which the salinity is variable. It retains its osmotic pressure when the water is dilute, yet it possesses the ability to accumulate substantial amounts of urea in the blood through the functioning of salt glands when it finds itself in water more concentrated than 50 percent seawater.

Animals in arid environments possess adaptations suggestive of those of arid land plants. One adaptation is to become dormant during periods of environmental stress. The animal may adopt an annual cycle of active and dormant stages. For example, for eight or nine months the spadefoot toad estivates in an underground cell lined with a gelatinous substance that reduces evaporative losses through the skin. It appears when rainfall saturates the earth, moves to the nearest puddle, mates, and lays eggs. Young tadpoles hatch in a day or two. They rapidly mature and metamorphose into functioning adults that are capable of digging their own retreats in which to estivate until the next rainy period.

The flatworm *Phagocytes vernalis,* which occupies ponds that dry up in summer, encysts. The flatworm reacts to rising water temperature by detaching small pieces from the posterior end of the body, until the entire animal is reduced to a number of fragments. Each fragment rounds up and secretes a layer of slime, which hardens into a cyst highly resistant to drying. The cysts remain in the debris of the pond bottom until the pond fills up again. When water is again present, the cysts hatch. Other aquatic or semiaquatic organisms retreat deep into the soil until they reach groundwater level. Still others spend the dry season in estivation. Many insects enter diapause.

The desert also contains many animals that are active the year round, all of which have evolved ways of circumventing the lack of water by physiological adaptations or by activity habits. The kangaroo rat of the southwestern North American desert and its ecological counterparts, the jerboas and gerbils of Africa and the Middle East and the marsupial kangaroo mice and pitchi-pitchi of Australia, feed on dry seeds and dry plant material even when succulent green plants are available. These mammals obtain water from their own metabolic processes and from the water content of their food. To conserve water these rodents remain by day in sealed burrows, they possess no sweat glands, their urine is highly concentrated (24 percent urea and salt, twice the concentration of seawater), and their feces are dry.

Large desert mammals, such as the camel, can use water effectively for evaporative cool-

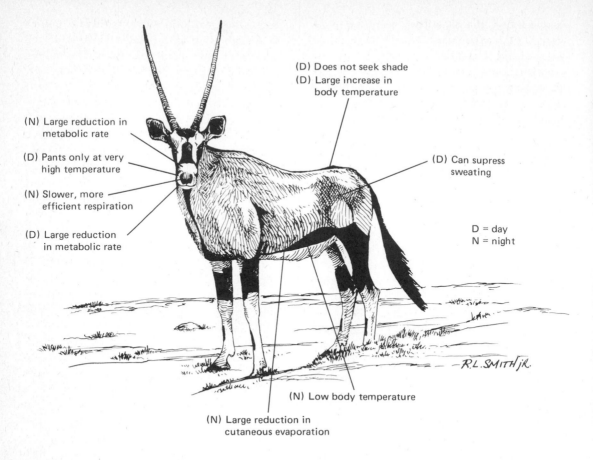

(D) Does not seek shade
(D) Large increase in
    body temperature

(N) Large reduction in
    metabolic rate

(D) Pants only at very
    high temperature

(N) Slower, more
    efficient respiration

(D) Large reduction
    in metabolic rate

(D) Can supress
    sweating

D = day
N = night

(N) Low body temperature

(N) Large reduction in
    cutaneous evaporation

*R.L.SMITH jR.*

**FIGURE 3-18**
*The physiological adaptations to heat and
aridity of the African ungulate, the oryx.
D-day; N-night. (Adapted from M. S. Gordon,
1972; based on data from C. R. Taylor, 1969.)*

ing through the skin and respiratory system
because their low ratio of surface area to body
size and low internal heat production result in
slower accumulation of heat. The camel not
only excretes highly concentrated urine, but
also can withstand dehydration up to 25 percent of body weight and it loses water from
body tissues rather than from the blood
(Schmidt-Nielsen, 1959). The camel's body
temperature is labile, dropping to 33.8° C
overnight and rising to 40.6° C by day, at
which point the animal begins to sweat. The
camel accumulates its fat in the hump rather
than over the body which speeds heat flow
away from the body. Its thick coat prevents the
flow of heat inward to the body.

Perhaps a more extreme adaptation to aridity exists among some of the African antelope.

Outstanding is the oryx (Figure 3-18). Many
African ungulates migrate to escape the heat
and dryness, but the oryx remains. During the
day it stores heat in its body, causing *hyperthermia,* that is, a substantial rise in body
temperature, thus reducing evaporative losses. Because evaporative loss is due to the difference between body and air temperature, a
high body temperature means less loss of
moisture. The oryx further reduces daytime
evaporative losses by suppressing sweating,
and panting is triggered only at very high
temperatures. It reduces its metabolic rate;
the amount of evaporative cooling necessary is
reduced by this lowered rate of internal production of calories. By night the oryx reduces
its nonsweating evaporation across the skin
by reducing its metabolic rate below that of
the daytime. The oryx's respiratory rates are
proportional to its respiratory efficiency and
inversely proportional to body temperature. A
cool animal breathes more slowly than a warm
one, thus using a greater proportion of in-
spired oxygen. With a lowered nighttime body

temperature, the saturation level for water vapor in the exhaled air is lower. The animal normally does not drink water; it exists on metabolic water and by feeding on grasses and shrubs, many of them succulent. In fact the oryx can obtain all the water it needs by eating food containing an average of 30 percent water.

Some desert birds, like marine birds, utilize a salt gland to help maintain a water balance. Otherwise, both marine and desert birds and those of the salt marsh have the same basic adaptations for the conservation of water and the elimination of sodium and chlorine ions. The black-throated sparrow *(Amphispiza bilineata)* of the deserts of western North America is not only restricted to arid regions and commonly occurs far from water in some of the most extreme desert habitat, but it also feeds largely on seeds. If the bird has some access to green vegetation it does not need to drink water (Smyth and Bartholomew, 1966). When it feeds largely on seeds and water is available, it will drink up to 30 percent of its body weight per day. But if water is not available, the sparrow can survive indefinitely without drinking by reducing the water content of its excreta from about 81 percent to 57 percent. At the same time it is able to concentrate $Cl^-$ in the urine. Among birds with similar adaptations are the budgerigar and zebra finch of Australia and the gray-backed finch lark of Africa (see Dawson and Bartholomew, 1968; Serventy, 1971). Because birds normally excrete semisolid uric acid as an adaptation related to weight reduction for flying, they are well fitted for water conservation in arid situations.

RAINFALL AND HUMIDITY

Moisture relationships within an ecosystem closely relate to the distribution of rainfall. Seasonal distribution of rainfall is more important than average annual precipitation. A world of difference exists between a region receiving 50 in. of rain rather evenly distributed throughout the year and a region in which nearly all of the 50 in. falls during a several-month period. In this situation, typical of tropical and subtropical climates, organisms must face a long period of drought. An alternation of wet and dry seasons influences the reproductive and activity cycles of organisms as much as light and temperature in the temperate regions. For example, the coming of the dry season to the African plains initiates the migrational movements of the large herbivores. In the southwestern United States the lack of winter rains fails to stimulate the growth of green vegetation, needed by Gambel's quail as a source of vitamin A. This in turn inhibits the reproductive activity of the bird, which is reflected in reduced numbers of offspring.

The moisture content of the air is usually expressed as *relative humidity,* the percent of moisture relative to the amount of water that the air holds at saturation at the existing temperature. If the air is warmed while its moisture content remains constant, the relative humidity drops, because warm air can hold more moisture than cool air. As the relative humidity drops, the vapor pressure deficit—the difference between the partial pressure of water at saturation and the prevailing vapor pressure of the air—increases, and increased evaporation takes place.

The relative humidity varies during a 24-hour period. Generally it is lower by day and higher by night. Relative humidity under a closed forest canopy is higher than it is on the outside during the day, and it is lower than the outside during the night.

Over normal surfaces relative humidity during the day usually increases with height because of the decrease in temperature with height. This contrasts with absolute humidity, which decreases with height. But if a temperature inversion occurs, especially at night, the relative humidity decreases upward to the top of the inversion, then changes little or increases only slightly.

In any one area relative humidity varies widely from one spot to another, depending upon the terrain. Variations in humidity are most pronounced in mountain country. Low elevations warm up and dry out earlier in the spring than high elevations, and soil moisture becomes more depleted later in summer. The daily range of humidity is greatest in the valley and decreases at higher elevations. Because daytime temperatures decrease with altitude, as does the dew point, relative humidities are greater on the tops of mountains than in the valleys. As nighttime cooling begins, the temperature change with altitude is reversed. Cold air rushes downslope and accumulates at the bottom. Through the night, if additional cooling occurs, the air becomes saturated with moisture and fog or dew forms by morning. These differences in humidity can produce vegetative differences on mountain slopes. They are most pronounced on the slopes of the mountains along the Pacific coast.

Temperature and wind both exert a considerable influence on evaporation and relative humidity. An increase in air temperature causes convection currents. This sets up an air turbulence, which mixes surface layers with drier air above. Wind movements associated with cyclonic disturbances also mix moisture-laden air with drier air above. As a result, the vapor pressure of the air is lowered and evaporation from the surface increases.

### MOISTURE AND PLANT DISTRIBUTION

Moisture or the lack of it influences the distribution of plants on both a geographic and local basis. For example, the western red cedar and the western hemlock grow in western North America where the average annual rainfall is around 23 cm (Little, 1971). The influence of moisture on the local distribution of plant communities is well illustrated by the grassland vegetation of the central plains. In a study of the prairie, meadow, and marsh vegetation of Nelson County, North Dakota, Dix and Smeins (1967) divided the soils into ten drainage classes from excessively drained to permanent standing water. They determined the indicator species for each drainage class and then divided the vegetational display into six units corresponding to the drainage pattern (Figure 3-19). The uplands fell into high prairie, mid prairie, and low prairie, and the lowlands into meadow, marsh, and cultivated depressions. The high prairie, with an excessive drainage pattern, was characterized by stands of needle-and-thread grass, western wheatgrass, and prairie sandweed. The mid prairie, considered to be the climax or true prairie, was dominated by big and little bluestem, porcupine grass, and prairie dropseed. Low prairie, on soils of moderate moisture, was characterized by big bluestem, little bluestem, yellow Indian grass, and muhly.

Lowlands, occupying soils in which the drainage was sluggish and the water table within the rooting depth of most plants, were characterized by canarygrass, sedge, and rivergrass. Meadows, on even wetter soils, were dominated by northern reed grass, wooly sedge, and spikerush. Marshes, with standing water, contained stands of reed, cattails, and tule bulrush. Cultivated depressions were usually colonized by spikerush and water plantain.

Although each drainage class supported a characteristic stand of vegetation, no one species was associated solely with another species. Each species had its own set of optimal drainage requirements and behaved independently of other species. At any particular site the drainage conditions influenced the combination of dominant plants and each community blended into the other. The only sharp breaks came where the drainage differences were sharp and severe.

In a similar manner soil moisture influences the distribution of woody plants in the northern Rocky Mountains. In northern Idaho, warm, dry lowland soils, droughty in summer, support Idaho fesque and snowberry. As the ground slopes up, soil moisture increases and the fesque and snowberry are replaced by ponderosa pine and Douglas fir; on midslope grand fir appears; and still higher western red cedar, western hemlock, and subalpine fir appear. On the upper parts of south-facing slopes, where the soil may become as droughty as that of the lowlands, wheat grass and Idaho fesque replace wood plants (Daubenmire, 1968a).

## Wind

Because the atmosphere is constantly in motion, winds are a continual and highly variable influence on the environment. Winds, especially those near the earth's surface, are strongly affected by the topography and by local heating and cooling.

Winds that move with the leading edge of the air mass or frontal system or are carried by the general circulation winds aloft are *general winds*.

In addition to broad general winds, there are local convective winds caused by temperature

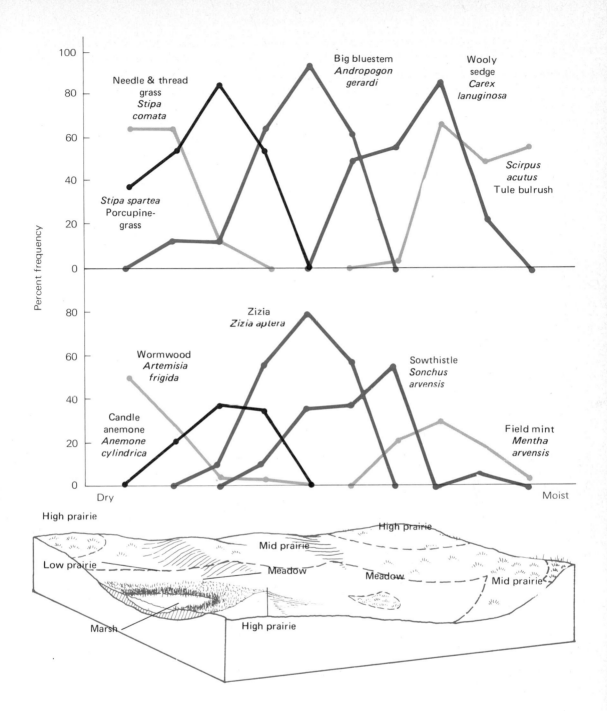

**FIGURE 3-19**
*The influence of topography and drainage regimes on prairie vegetation. The hypothetical block diagram of North Dakota landscape shows the relative positions of vegetational units. The distributional curves of selected species are based on a drainage gradient. The high prairie is the vegetational unit of excessive drainage. (From Dix and Smeins, 1967.)*

differences within a locality. The most familiar of these are the land and sea breezes experienced along ocean shores and large inland lakes and bays. Sea breezes flow inland from the water and bring in cool, moist marine air, often accompanied by morning fog. Land breezes at night are the reverse of daytime sea breezes. Air in contact with land becomes cooler at night, gains in density, and flows from land to water.

In mountainous topography local winds can be exceedingly complex. Differences in heating of air over mountain slopes and canyon and valley bottoms result in several wind systems. Because of the larger heating surface, air in mountain valleys and canyons becomes warmer during the day. Similarly, a larger cooling area of the valleys causes a reversal of this situation at night. The resulting pressures cause a flow of air upslope by day and downslope by night. Combined valley and upslope winds exit at the ridgetops by day. As the slopes go into the shadows of late afternoon, the slope and valley winds shift direction and become downslope winds.

Local winds affect soil moisture and humidity and have an important bearing on forest fire conditions and local drought situations. Drying action of high, warm winds in winter when soil moisture is low or unavailable causes drought. The wind removes humid air about the leaves and increases transpiration. Losing more water than they are able to absorb, evergreens in particular dry out and their foliage turns brown.

In areas of frequent and regular high winds plants that in another location grow tall become low and spreading. On high, wind-swept ridges cushion plants with small, uniform, crowded branches are most common. Because of constant desiccation, cells of plants growing in these places never expand to normal size, and all organs are dwarfed. Terminal branch shoots are killed by desiccation, by blasting of ice particles, and, along the ocean, by the effects of salt spray. As a result terminals are replaced by strong laterals, which form a mat close to the ground.

Strong and persistent winds blowing from a constant direction bend the branches of trees around to the windward side until, like a weathervane, they point out the direction of the prevailing wind. Often the wind may kill all the twig-forming buds, so that no limbs develop on the windward surface.

Shallow-rooted trees and trees with brittle woods, such as the willows, cottonwoods, and maples, are thrown or broken by strong winds. Windthrow is most prevalent among trees growing in dense stands that, through logging or natural damage, are suddenly exposed to the full force of the wind. Hurricanes and other violent windstorms sweeping over forested areas may uproot and break trees across a considerable expanse of land.

Hurricanes and strong storms cause deaths among animals also, and often carry individuals far from their normal environment and set them down elsewhere. Winds are an important means of dispersal for seeds and small animals such as spiders, mites, and even snails (Andrewartha and Birch, 1954; Darlington, 1957).

Wind may also play a secondary role in the distribution of small mammals. The deeper accumulation of litter and snow in areas sheltered from the wind support more small mammals than exposed areas (Vose and Dunlap, 1968).

## SUMMARY

**The physical and chemical conditions of the environment influence the well-being and distribution of plants and animals and the functions of the ecosystem. The kind and quantity of elements and nutrients available for circulation in the ecosystem affect the growth and reproduction of plants and animals that vary in their requirements and tolerances for different elements. Some elements, the macronutrients, are required in relatively large quantities by all living organisms. Others, the trace elements or micronutrients, are needed in lesser and often only minute quantities; yet without them plants and animals will fail, as if they lacked one of the major nutrients.**

**A major pool of nutrients in the ecosystem is the soil. Soil is the site of decomposition of organic matter and of the return of mineral elements to the system. It is the home of animal life, the anchoring medium for plants, and the source of water and nutrients for plants. Soil begins with the weathering of rocks and minerals, which involves the leaching and carrying away of mineral**

matter. Its development is guided by slope, climate, original material, and native vegetation. Plants rooted in the weathering material further break down the substrate, pump up nutrients from deeper layers, and add important organic material. Through decomposition and mineralization this material converts into humus, an unstable product that is continuously being formed and destroyed. As a result of the weathering process, accumulation and breakdown of organic matter and the leaching of mineral matter, horizons or layers are formed in the soil. There are four basic horizons: *O*, the organic layer; *A*, characterized by an accumulation of organic matter and a loss of clay and mineral matter; *B*, in which mineral matter accumulates; and *C*, the underlying material. Of all the horizons none is more important than the humus layer which plays a dominant role in the life and distribution of plants and animals, in the maintenance of soil fertility, and in much of the soil-forming process.

Soil profile development is influenced over large areas by vegetation and climate. In grassland regions calcification is the chief soil-forming process, in which calcium accumulates at the average depth reached by percolating water. In forest regions podzolization, involving the leaching of calcium, magnesium, iron, and aluminum from the upper horizon and retention of silica, take place. In tropical regions laterization, in which silica is leached and iron and aluminum oxides are retained in the upper horizon, is the major soil-forming process. Gleization takes place in poorly drained soils. Organic matter decomposes slowly and iron is reduced to the ferrous state.

Thus soils develop under different regimes of temperature and moisture which influence weather and biological activity. Life exists in an energy or thermal environment. Plants and animals are continually receiving and losing heat to the environment. Heat gains come from solar radiation, diffuse and reflected radiation, and infrared thermal radiation. Heat losses result from infrared radiation from the organism, from conduction, convection, and evaporation.

Response to the energy environment differs between plants and animals. Plants receive their energy from the environment. Heat affects the physiological processes of plants, especially photosynthesis and respiration. High temperatures may inhibit photosynthesis and various phases of a plant's life cycle have their optimum range of temperature. Plants possess certain adaptations to variations in temperature, such as reduced transpiration and reduced size of leaf.

Animal responses are more complex. Body temperatures of animals are maintained either from an external heat source or from metabolic regulation. Animals that depend upon environmental heat are poikilothermic. They gain most of their heat from the environment and regulate body temperature behaviorally by moving in and out of areas of different temperatures. Animals that are heated metabolically are homoiothermic. They maintain their body temperature by means of a closely regulated feedback system, mediated by the hypothalamus.

Both types respond to temperature changes by acclimatization. Within limits poikilotherms adjust or acclimate to higher or lower temperatures either by avoiding temperature extremes, by adjusting tolerances to given temperature ranges, by heliothermism, and in some instances by evaporative cooling. Some reptiles change their body temperatures by changing the rate of heart beat and metabolism. Acclimation in homoiotherms is accomplished by changes in body insulation, vascularization, evaporative heat loss, and nonshivering thermogenesis. Homoiotherms find it more difficult to adapt to warm temperatures than to cold.

One method of evading periods of stress utilized by both poikilotherms and homoiotherms is undergoing a period of dormancy or torpor. Animals going into dormancy in periods of heat and dryness are said to estivate. Those who become dormant in winter hibernate. Both involve a sharp reduction in physiological functions and at the same time conserve energy.

Plants and animals also face moisture problems, most acute in saline and arid environments. Similar strategies are employed by each group. A number of animals secrete excess salts through the salt glands, and thus conserve body moisture. Others concentrate their urine, use metabolic water, and reduce daytime evaporative losses by reducing the amount of sweating. Some plants and animals of arid regions have annual life cycles which involve surviving dry periods as seeds, eggs, or cysts. Plants in both saline and arid environments may possess certain adaptations as small leaves, waxy blooms, and succulence. In one way or another both plants and animals exhibit adaptations to handle variations in the thermal and moisture conditions of the environment.

# Abiotic limits to the system II
## CHAPTER 4

## Climate

The abiotic environment ultimately is controlled by climate, the prevailing weather conditions of a region. Climate determines the availability of water and the degree of heat over the earth, and thus it influences the development of soil, the nature of vegetation, and the type of biological community. Because climate determines moisture and temperature, it influences nutrient cycling, photosynthesis, and decomposition. The major structural and functional aspects of an ecosystem relate directly and indirectly back to climate.

### GLOBAL PATTERNS

*Solar radiation: determiner of climate.* The major determiner of climate is solar radiation Variations in the heat budgets of the earth along with the earth's daily rotation and path around the sun produce the prevailing winds and move the ocean currents. Winds and ocean currents in turn influence rainfall patterns over the earth.

The amount of solar radiation that reaches the outer limit of the atmosphere as measured on a surface held perpendicular to the sun's rays is 2.0 cal/cm²/min. This value, called the *solar constant,* fluctuates slightly, about 15

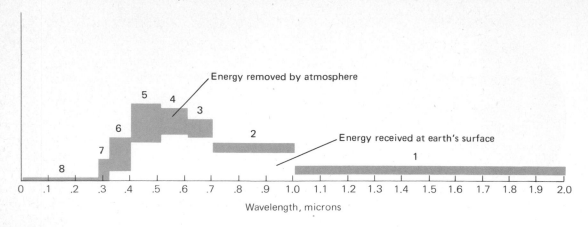

**FIGURE 4-1**

*Energy in the solar spectrum before and after depletion by the atmosphere from a solar altitude of 30°. The figures above the bars indicate: 1, near infrared with wavelength over 1 micron; 2, near infrared, 0.7–1.0 microns; 3–5, visible light; 3, red; 4, green, yellow and orange; 5, violet and blue; 6–7, ultraviolet. Note the strong reduction in ultraviolet. Nearly all wavelengths are absorbed by ozone at high levels. The region of peak energy is shifted toward the red end of the spectrum. Visible light in blue wavelengths is scattered rather than absorbed producing the blue light of the sky. (From Reifsnyder and Lull, 1965.)*

percent, through the year. The amount of solar radiation that reaches the earth's surface is one-half of the solar constant or 1.0 cal/cm²/min.

In passing through the atmosphere, the energy of certain wavelengths is absorbed so that the spectrum of energy that reaches the earth's surface is different from that of incident solar radiation (Figure 4-1). Ultraviolet wavelengths are nearly all removed by the atmosphere. Molecules of atmospheric gas scatter the shorter wavelengths, giving a bluish color to the sky and causing the earth to shine out in space. Water vapor scatters radiation of all wavelengths, so that an atmosphere with much water vapor is whitish; thus the grayish appearance of a cloudy day. Dust scatters long wavelengths to produce the reds and yellows in the atmosphere. Because of the scattering of solar radiation by dust and water vapor, part of it reaches the earth as diffuse light from the sky, called *skylight*. Infrared radiation that reaches the earth and is felt as

heat (sensible heat) is absorbed and a portion is reradiated back as far infrared (4 to 100 $\mu$). What we see as light is visible radiation which can be broken down into the spectrum that ranges from violet to red.

Taking the amount of solar radiation that penetrates the earth's atmosphere as 100 percent, 21 percent is reflected from the clouds back to space, 5 percent is reflected by dust and aerosols, and 6 percent is reflected by the earth. The total reflective loss to space by earth and atmosphere is 32 percent. Another portion of solar radiation, 3 percent, is absorbed by clouds, and 15 percent is absorbed by dust, water vapor, and carbon dioxide. The absorptive loss is 18 percent. Thus reflection and absorption account for the 50 percent loss of solar radiation as it passes through the earth's atmosphere to its surface.

The amount of solar radiation that reaches a specific point on the surface depends upon the atmosphere, moisture concentration of ozone and dust particles, altitude, and location on the planet. It varies seasonally and daily.

The solar radiation absorbed as short-wave energy by the earth and reradiated as long-wave radiation is responsible for heating the earth's atmosphere. Long-wave radiation prevented from escaping into space by the ozone layer of the stratosphere warms the upper atmosphere. Thus the atmosphere receives most of its heat directly from the earth and only indirectly from the sun.

***Creation of air and ocean currents.*** Sunlight does not strike the earth uniformly (Figure 4-2). Because of the earth's shape the sun's rays strike more directly on the equatorial

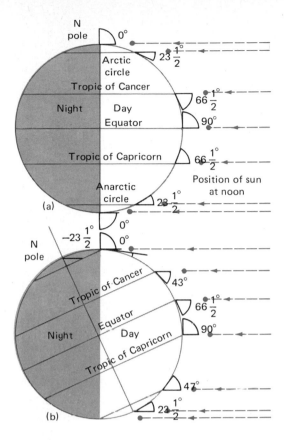

Pole. At the summer solstice the position of the sun is 90° at the Tropic of Cancer. To visualize this, turn the diagram upside down and change North to read South. At this time of year the sun remains above the horizon at the North Pole for the entire 24 hours. (From A. Strahler, 1971.)

**FIGURE 4-2**

*Altitude of the sun at the equinoxes and solstices. Latitude and inclination of the earth's axis at an angle of 66¹/₂° and the rotation of earth about the sun determine the amount of solar radiation reaching any point on the earth at any time. Earth's surface at all times lies half in the sun's rays and half in shadow marked by a dividing line, the circle of illumination. The circle of illumination is bisected by the equator and always lies at right angles to the sun's rays. Two times a year at vernal and autumnal equinoxes (March 20 or 21, September 22 or 23), the circle of illumination passes through the poles. At the time of summer and winter solstices (June 21 or 22, December 22 or 23) the circle of illumination is tangent to the Arctic and Antarctic Circles (66° N and S latitudes.). Thus at the time of fall and spring equinoxes the sun's rays fall directly on the equator, so that at noon the altitude of the sun at the equator is 90°. At the North Pole the sun is at the horizon and keeps that position as the earth rotates. At the time of the winter solstice the noon altitude of the sun is 66¹/₂° at the equator and 90° at the Tropic of Capricorn. The sun remains below the horizon at the North Pole and above the horizon at an altitude of 23¹/₂° at the South*

than on the polar region. The lower latitudes are heated more than the polar ones. Air heated at the equatorial regions rises until it eventually reaches the stratosphere, where the temperature no longer decreases with altitude. There the air whose temperature is the same or lower than that of the stratosphere is blocked from any further upward movement. With more air rising the air mass is forced to spread out north and south toward the poles. As the air masses approach the poles, they cool, become heavier, and sink over the arctic regions. This heavier cold air then flows toward the equator, displacing the warm air rising over the tropics (Figure 4-3).

If the earth were stationary and without any irregular land masses and oceans, the atmosphere would flow in this unmodified circulatory pattern. The earth, however, spins on its axis from west to east. This spinning produces a force, called the Coriolis force, that deflects the air flow in the Northern Hemisphere to the right and in the Southern Hemisphere to the left and prevents a simple direct return from the poles to the equator. The result is a series of belts of prevailing east winds, known in the polar regions as the polar easterlies and near the equator as the easterly trade winds. In the middle latitudes is a region of west winds known as the westerlies (see Figure 4-3).

These belts break the simple flow of air toward the equator and the flow aloft toward the poles into a series of cells. The flow is divided into three cells in each hemisphere. The air that flows up from the equator forms the equatorial zone of low pressure. The equatorial air cools, loses its moisture, and descends. By the time the air has reached 30° latitude north or south, it has lost enough heat to sink, forming a cell of semipermanent high pressure encircling the earth and a region of light winds known as the horse latitudes. The air, warmed again, picks up moisture and rises once more.

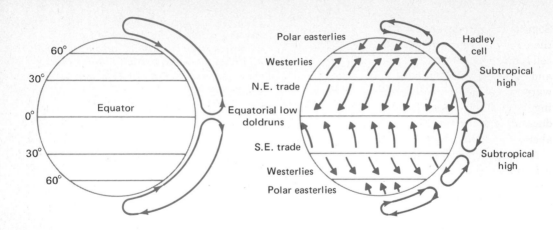

**FIGURE 4-3**
*Circulation of air cells and prevailing winds. On an imaginary, nonrotating earth air heated at the equator would rise, spread out north and south, and, upon cooling at the two poles, would descend and move back to the equator. On a rotating earth the air current in the Northern Hemisphere starts as a south wind that is moving north, but is deflected to the right by the earth's spin and becomes a southwest or west wind. A north wind is deflected to the right and becomes a northeast or east wind. Because the northward air flow aloft just north of the equatorial regions becomes a nearly true westerly flow, northward movement is slowed, air piles up at about 30° north latitude and loses considerable heat by radiation. Because of the piling up and heat loss, some of the air descends, producing a surface high-pressure belt. Air that has descended flows both northward to the poles and southward to the equator at the surface. The northward flowing air current turns right to become the prevailing westerlies; the southward flowing current, also deflected to the right, becomes the northeast trades of*

*the low latitudes. The air aloft gradually moves northward, continues to lose heat, descends at the polar region, gives up additional heat at the surface, and flows southward. This flow of air deflected to the right becomes polar easterlies. Similar flows take place in the Southern Hemisphere, but the air flow is deflected to the left by the rotating earth. The pattern of rising and descending air forms tubes around the earth called Hadley cells, after the English meteorologist who described them.*

**FIGURE 4-4**
*A schematic diagram of the oscillations in precipitation over tropical regions that result in intertropical convergences producing rainy seasons and dry seasons. Note that as the distance from the equator increases, the longer the dry season and the less rainfall involved. These oscillations result from changes in the altitude of the sun between the equinoxes and the summer and winter solstices as diagramed in Figure 4-1. (From H. Walter, 1971.)*

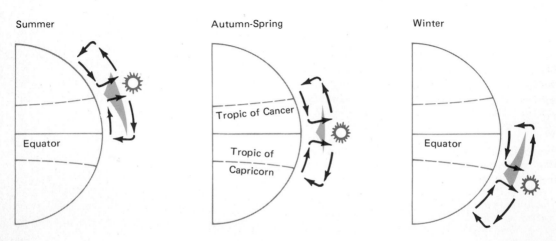

Some of it flows toward the poles, some toward the equator. Meanwhile at each pole, another cell builds up and flows outward in a southerly and northerly direction to meet the rising warm air at approximately 60° latitude flowing toward the poles. This convergence produces a semipermanent low-pressure area at about 60° north and south latitude.

Although tropical regions about the equator are always exposed to warm temperatures, the sun is directly over the equator only two times a year, March 21 and September 21. On June 21, summer in the Northern Hemisphere, the sun is directly over the Tropic of Cancer; on December 21, summer in the Southern Hemisphere, it is directly over the Tropic of Capricorn. Aside from causing seasonal changes, the sun also affects the heating of land, water, and air masses as its vertical orientation changes north and south. This results in the movement of the junction of rising and falling air masses over the equatorial regions, the *intertropical convergence* (Figure 4-4). Because air and land heat slowly a time lag exists between the change in the vertical orientation of the sun and the intertropical convergence by about one month and does not move as far north and south as does the sun (about 15° north and south instead of 30°). The rainy season in the tropics moves with the intertropical convergence.

Because the masses of land and water are not uniform and because land heats and cools more rapidly than the oceans, the surface of the earth experiences uneven heating and cooling. At any given time the temperature changes are much greater over continental areas than over oceans. Oceans act as heat reservoirs; continents affect the circulation. In winter, for example, the west coast of a continent is warmer than the east coast because the air reaching the west coast has traveled over warmer ocean areas. Also produced are monsoon winds, dry winds that blow from continental interiors to the oceans in summer and winds heavy with moisture that blow from the oceans to the interior in winter. Last are moving air masses with cyclonic and anticyclonic frontal systems.

These circulatory patterns and the moisture regimes they influence are mostly but not entirely responsible for forests, grasslands, and deserts, which in turn reflect the climatic patterns of the planet (Figure 4-5).

The turning of the earth, solar energy, and winds produce ocean currents, the horizontal movements of water around the planet. In absence of any land masses oceanic waters could circulate unimpeded around the globe, as does the flow of water around the Antarctic continent. But land masses divide the ocean into two main bodies, the Atlantic and Pacific. Both oceans are unbroken from high latitudes, north and south, to the equator; and both are bounded by land masses on either side (Figure 4-6).

Each ocean is dominated by two great circular water motions or *gyres,* each centered on a subtropical high pressure area. Within each gyre the current moves clockwise in the Northern Hemisphere and counterclockwise in the Southern Hemisphere. The movements of the currents are caused partly by the prevailing winds, the trades or tropical easterlies on the equator side and prevailing westerlies on the pole side. The two gyres, north and south, in both oceans are separated by an equatorial countercurrent that flows eastward. This current results from the return of lighter (less dense) surface water piled up on the western side of the ocean basin by the equatorial currents.

As the currents flow westward they become narrower and increase their speed. Deflected by the continental basin they turn poleward, carrying warm water with them. Two major currents in the Northern Hemisphere are the Gulf Stream in the Atlantic and the Kuroshio Current in the Pacific. Their counterparts in the Southern Hemisphere are the Australian Current in the Pacific and the Brazil Current in the Atlantic. As an example, the Gulf Stream flows north from the Caribbean, presses close to Florida, swings along the southeast Atlantic coast of North America, flows northeastward across the North Atlantic, and divides. One part becomes the Norway Current (or North Atlantic Drift), carrying warm water past Scotland and along the Norwegian coast. The other part swings south to become the Canary Current, completing the gyre (Figure 4-6). The counterpart of the Canary Current in the Southern Atlantic is the Benguela Current which flows north along the African coast. In

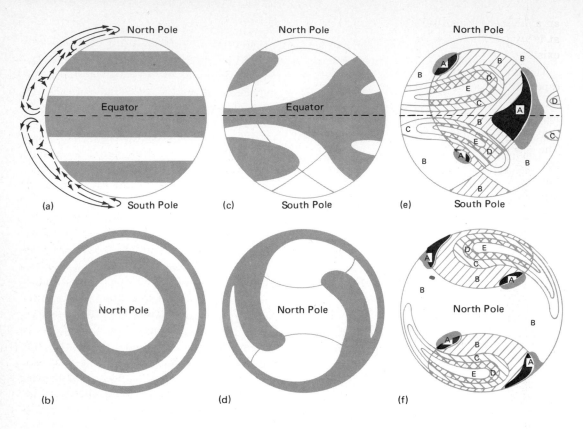

**FIGURE 4-5**

*Patterns of air currents and rainfall and associated pattern of vegetation. If the earth were uniform, without irregular masses of lands and oceans, the general circulation of the atmosphere and the rainfall belts would appear as in (a) looking down on the equator and (b) looking down on the pole. Because of the differential heating of the earth and its rotation air flow is broken into a series of three cells in each hemisphere, described in detail in Figure 4-3. This air flow influences rainfall. Shaded portions are areas of maximum rainfall; unshaded portions are dry areas.*

*The earth is not uniform, and unequal masses of land and water prevent the development of rain belts that correspond to the belts of ascending air. Land heats and cools more rapidly than water, the oceans act as heat reservoirs, and the continents affect the pattern of circulation of air. The generalized rainfall pattern as modified by oceans and continents is shown in (c) and (d). Again the shaded portions are areas of maximum rainfall; unshaded areas are dry. The loop formed by the black line in (c) is a generalized continental area. It roughly represents North and South America in the in the Western Hemisphere and Europe,*

*Asia, and Africa in the Eastern Hemisphere. In (d) the continents are represented by an egg-shaped loop in each hemisphere. The dry areas form an S in the Northern Hemisphere, the center being at the pole.*

*The circulation pattern can be carried one step further. Rainfall varies greatly over the earth from less than 1 inch to over 900 inches. This variation is reflected in the climatic regions and vegetation types. The letters and shadings in (e) and (f) represent the climatic conditions of the earth according to Thornwaite's classification:*

| Climatic region | Vegetation type | Code |
|---|---|---|
| *superhumid* | *rainforest* | *A* |
| *humid* | *forest* | *B* |
| *subhumid* | *grassland* | *C* |
| *semiarid* | *steppe* | *D* |
| *arid* | *desert* | *E* |

the Pacific the two currents are the California and the Peruvian or Humboldt. These cold currents coming from the Arctic and Antarctic regions result in upwellings that recharge surface waters with nutrients from the deep and in fogs that move over coastal areas providing an abundance of moisture and cooling the air at latitudes where the world's hottest deserts exist inland.

These ocean currents influence ecosystems, especially in coastal regions. The Gulf Stream, for example, brings milder temperatures and more moisture to Great Britain and Norway than would normally occur at their latitudes. The cold California Current is responsible for the mild, wet climate of the northwestern Pacific coast of North America that holds the tall dense coniferous rain forests dominated by Douglas fir and Sitka spruce. The fogs and low clouds of summer reduce moisture stress during that dry season.

*FIGURE 4-6*
*Ocean currents of the world. Note how the circulation is influenced by continental land masses and how oceans are inter-connected by the currents. 1, Antarctic West Wind Drift; 2, Peru Current (Humboldt); 3, South Equatorial Current; 4, Counter Equatorial Current; 5, North Equatorial Current; 6, Kuroshio Current; 7, California Current; 8, Brazil Current; 9, Benguela Current; 10, South Equatorial Current; 11, Guinea Current; 12, North Equatorial Current; 13, Gulf Stream; 14, Norwegian Current; 15, North Atlantic Current; 16. Canaries Drift; 17, Sargasso Sea; 18, Monsoon Drift (summer east, winter west); 19, Mozambique Current; 20, West Australian Current; 21, East Australian Current. (Dashed arrows represent cool water.) (From Coker, 1947.)*

## REGIONAL CLIMATES

*Interaction of temperature and moisture patterns.* The climate of any given region is the combination of patterns of temperature and moisture which are determined not only by latitude, but also by the location of the region in the continental land mass. Nearby bodies of water and geographical features such as mountains affect these patterns. In turn, the interaction of temperature and moisture patterns are reflected in the continental distribution of vegetation.

Zonation of vegetation in North America more usually follows the pattern of moisture than that of temperature. In Europe the zonation of natural vegetation more nearly follows that of temperature, resulting in broad belts of vegetation running east and west. Only in the far north in North America do the vegetation zones (tundra and boreal coniferous forest) stretch in these directions. Below this, the vegetation is mostly controlled by precipitation and evaporation, the latter influenced considerably by temperature. Because the available moisture becomes less from east to west, vegetation follows a similar pattern with belts running north and south. Humid regions along the coast support natural forest vegetation. This zone is broadest in the east. West of the eastern forest region is a subhumid zone where precipitation is less and evaporation is higher. Here the ratio of precipitation to evaporation is about 60:80 percent, and the land

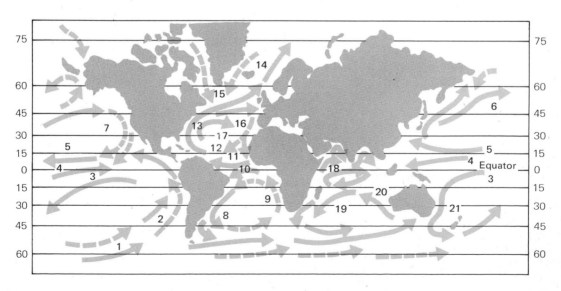

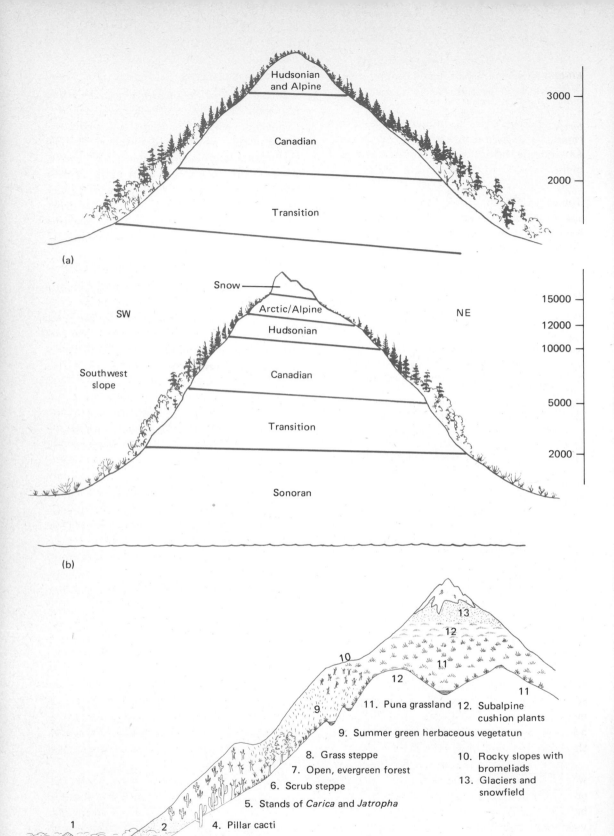

(a)

3000

2000

Hudsonian and Alpine

Canadian

Transition

(b)

SW

NE

Snow

Arctic/Alpine

Hudsonian

Southwest slope

Canadian

Transition

15000

12000

10000

5000

2000

Sonoran

(c)

13

12

11

10

12

11

11. Puna grassland   12. Subalpine cushion plants

9

9. Summer green herbaceous vegetatun

8. Grass steppe        10. Rocky slopes with bromeliads

7. Open, evergreen forest

6. Scrub steppe        13. Glaciers and snowfield

5. Stands of *Carica* and *Jatropha*

1

2

4. Pillar cacti

3. Rootless Tillandsias in desert

2

3

2. Desert

1. Lomas: fog green slope

**FIGURE 4-7**
*Altitudinal zonation of vegetation in mountains. On Mount Marcy in the Adirondack Mountains of New York the forest on the lower slope is Transition, consisting chiefly of northern hardwoods. The forest on the middle slope is Canadian, with paper birch, red spruce, and balsam. The upper slope is Hudsonian and Alpine, characterized by dwarf spruce and willows and heaths. In the southern Appalachians oaks and hickory replace the northern hardwoods, the northern hardwoods replace the spruce, and the spruce replaces willows and stunted spruce.*

*In the Rocky Mountains the Sonoran zone is characterized by grassland and shrubby vegetation–such as chaparral and juniper; the Transition by oaks and, higher up, lodgepole pine. The Canadian zone contains lodgepole pine, Engelmann spruce, red fir, and silver fir; the Hudsonian, mountain hemlock, western white pine; the Arctic-Alpine, willows, etc. Note that the life zones extend higher on the southwest slope than on the northeast slope. Elevations are approximate only.*

*Zonation of vegetation in the Andes Mountains is more complex. This cross-section of the Andes is at the latitude of Lima, Peru, from the coast to the high Andean region. Moving up the mountains one passes from deserts in the lowlands, the fog green slopes or lomas of the coastal zone, stands of rootless* Tillandsias *in the desert, zone of pillar cacti, stands of* Carica *and* Jatropha, *valley slopes with summer green herbaceous cover, scrub steppe, open evergreen forest, rocky slopes with bromeliads, grass steppe with scattered shrubs, the puna grasslands, subalpine cushion plants, and finally glaciers and snowfields. (After Walter, 1971.)*

supports a tall-grass prairie. Beyond this is semiarid country, where the precipitation-evaporation ratio is 20:40 percent. It supports a short-grass prairie. To the west of this and on the lee of the mountains is the desert.

In the mountainous country, both east and west, vegetation zones reflect climatic changes on an altitudinal gradient (Figure 4-7). These belts often duplicate the pattern of latitudinal vegetation distribution. In general the belts include the land about the mountain's base that has a climate characteristic of the region. Above the base region is the montane level, which has greater humidity and temperatures

that decrease as the altitude increases. Here the forest vegetation changes from deciduous to coniferous. Beyond this is the subalpine zone that includes coniferous trees adapted to a more rigorous climate than the montane species. Above the subalpine lies the krumholtz, a land of stunted trees. Above that is the alpine or tundra zone where the climate is cold. Here trees are replaced by grasses, sedges, and small tufted plants. On the very top of the highest mountains is a land of perpetual ice and snow.

Mountains influence regional climates by modifying the patterns of precipitation. Mountain ranges intercept air flows. As an air mass reaches the mountains it ascends, cools, becomes saturated (because cool air holds less moisture than warm air), and releases much of its moisture on the windward side. As the cool and dry air descends on the leeward side, it warms and picks up moisture (Figure 4-8). As a result the windward side of a mountain range supports more lush vegetation than the leeward side where dry, often desertlike conditions exist. Thus in North America the westerly winds that flow over the Sierra Nevada and the Rockies drop their moisture on the west-facing slopes that support excellent forest growth, while the leeward sides are desert or semidesert. Similar conditions exist in the Himalayas and the mountain ranges in Europe.

A picture of regional climates and the progression of temperature and precipitation can be obtained from a climograph. A *climograph,* a composite picture of the climate of an area, is a plot of the mean monthly temperatures against the mean monthly relative humidities or precipitation (Figure 4-9). The points for each month connected together form an irregular polygon which can be compared with the polygon for another area. The climograph is simpler to use than tables in comparing one region with another.

***Patterns of atmospheric movements.*** An important aspect of regional and local climates is the daily heating and cooling of air masses. Earth's atmosphere is a gas. A gas allowed to expand becomes cooler; the same volume of gas compressed into a smaller space becomes warmer. The change in temperature results from the change in degree of crowding of gas

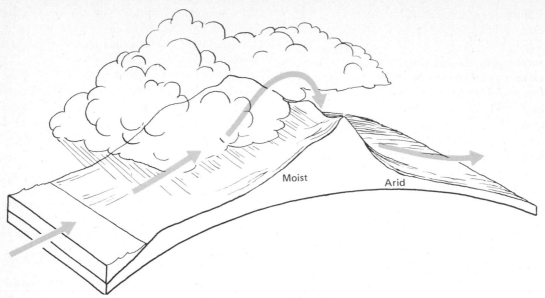

**FIGURE 4-8**
*Formation of a rain shadow. When an air
mass encounters a topographical barrier,
such as a mountain range, it is forced to go
over it. As it rises the air mass cools and
loses its moisture on the windward side. The
abundant moisture stimulates heavy
vegetational growth. The relatively dry air
descending the leeward slope picks up
moisture, creating arid to semiarid
conditions.*

**FIGURE 4-9**
*Temperature-moisture climographs. Note
how the hot, dry desert climate differs
graphically from the cool, temperate, moist
climate of the east. The data for the graph
are mean temperature and precipitation from
1941 to 1950 for Yuma, Arizona, and Albany,
New York. The second climograph compares
conditions on the rain shadow side and the
high rainfall side in the Appalachian
Mountains of West Virginia. (Numbers in
parentheses indicate month.)*

molecules. Expanded into a larger space the
molecules are more widely separated and collide less frequently, resulting in a drop in sensible heat. When the same volume of gas is
compressed, the molecules are closer together
and collide more frequently, raising sensible
heat. Such a process, in which heat is neither
lost to or brought in from the outside is termed
an *adiabatic process*. The temperature change
is called the *dry adiabatic temperature change*.

The principle cause of important and sustained decrease and increase in temperature of
vast masses of air is the adiabatic process in
the sinking and rising air mass. The atmospheric pressure at high altitudes is less than
near the earth's surface. As a mass of unsaturated air rises, it moves into a region of lower
pressure; its temperature decreases at the rate
of 1° C per 100 meters. As the air mass moves
down to lower elevations, the air is compressed
and warmed at the same rate. This rate of
change in temperature with change in height
is called the *dry adiabatic lapse rate*.

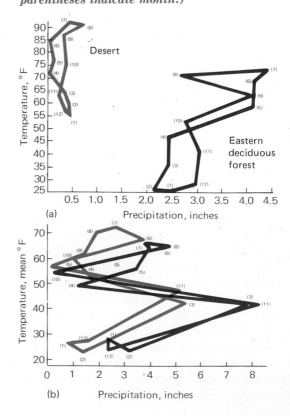

If a mass of air is cooled at the dry adiabatic rate, it will rise; if it becomes immersed in warmer air, it will fall to a level where the surrounding air has the same temperature. If an air mass descends, but the region it encounters is cooler and denser, it will rise to its original level. When this condition prevails the atmosphere is said to be stable. If a parcel of air moves up and tends to remain at the new level, the air mass is termed neutral. A layer of such air in which the temperature increases with height is extremely stable. Such a layer is called an *inversion*. But if the air mass moves up and down on its own accord and continues to rise or fall, the atmosphere is unstable.

As air rises and adiabatic warming and cooling continue, the air temperature approaches dew point (the temperature at which water content reaches saturation and condenses). The dew point temperature decreases at the rate of 0.2° C per 100 meters. Dew point and air temperature approach each other at the rate of about 0.9° C per 100 meters. As the air mass reaches dew point, the water it contains condenses to form clouds or fog.

As a parcel of air moves up and down and condenses its moisture as it cools, it gains some heat of condensation, generating sensible heat, and warms the air in which the condensation is taking place. This production of sensible heat reduces the adiabatic lapse rate which takes on a reduced value known as the *wet adiabatic lapse rate*. The wet adiabatic lapse rate has no single value because the rate varies with the temperature of and the moisture in the air. In general this rate falls somewhere between 1° C and 3° C per 100 meters.

The surrounding air is judged to be stable, neutral, or unstable by comparing its lapse rate with the wet adiabatic rate. In mountain country it is not uncommon for the air to change its lapse rate from dry to moist as the air cools sufficiently to condense its moisture.

Instability of the atmosphere is created by differential heating of the earth and lower atmosphere. The earth by day is heated by short-wave solar radiation (Figure 4-10), which is absorbed in different amounts over the land depending on vegetation, slope, soil, season and so on. Lower layers of the atmosphere, heated by the earth's surface by radiation, conduction, and convection, rise in small volumes; colder air falls. This turnover of the

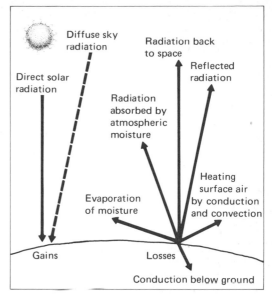

(a) Daytime surface heat exchange

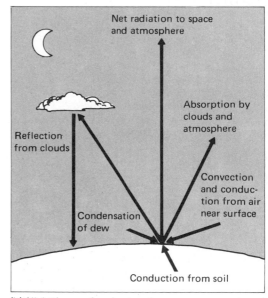

(b) Nighttime surface heat exchange

**FIGURE 4-10**
*Radiant heating of the earth. Solar radiation that reaches the earth's surface in the daytime is dissipated in several ways, but heat gains exceed heat losses. At night there is a net cooling of the surface, although some heat is returned by various processes. (Adapted U.S. Department of Agriculture, 1970.)*

atmosphere produces a turbulence that is increased by winds.

After the sun goes down, the earth begins to lose heat (Figure 4-10b). The layer of surface air is cooled, while the air aloft remains near warmer daytime temperatures. In mountainous or hilly country, cold dense air flows down slopes and gathers in the valleys. The cold air then is trapped beneath a layer of warm air

*FIGURE 4-11*
*Formation of inversions. Topography plays an important role in the formation and intensity of nighttime inversions. At night air cools next to the ground, forming a weak surface inversion in which the temperature increases with height. As cooling continues during the night, the layer of cool air gradually deepens. At the same time cool air descends down slope. Both cause the inversion to become deeper and stronger. In mountain areas the top of the night inversion is usually below the main ridge. If air is sufficiently cool and moist, fog may form in the valley. Smoke released in such an inversion will rise only until its temperature equals that of the surrounding air. Then smoke flattens out and spreads horizontally just below the thermal belt. (Adapted from U.S. Department of Agriculture, 1970.)*

(Figure 4-11). Such radiational inversions trap impurities and other air pollutants. Smoke from industry and other heated pollutants rise until their temperature matches the surrounding air. Then they flatten out and spread horizontally. As pollutants continue to accumulate, they may fill the entire area with smog. Such inversions are most intense if the atmosphere is stable. Inversions break up when surface air is heated during the day to create vertical convections up through the inversion layer or when a new air mass moves in.

Similar but more widespread inversions occur when a high-pressure area stagnates over a region. In a high-pressure area the air flow is clockwise and spreads outward. The air flowing away from the high must be replaced, and the only source for replacement air is from above. Thus surface high-pressure areas are regions of sinking air movement from aloft, called *subsidence*. When high-level winds slow down, cold air at high levels in the atmosphere tends to sink. The sinking air becomes compressed as it moves downward, and as it warms it becomes drier. As a result, a layer of warm air develops at a higher level in the atmosphere (Figure 4-12). Rarely reaching the ground, it hangs several hundred to several thousand feet above the earth, forming a sub-

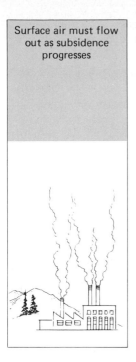

Surface air must flow out as subsidence progresses

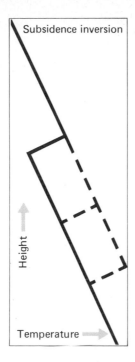

Subsidence inversion

Height

Temperature

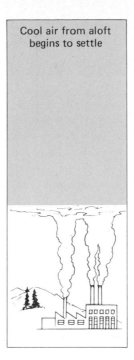

Cool air from aloft begins to settle

Warm, very dry air approaches the surface

**FIGURE 4-12**
*Descent of a subsidence inversion. The movement of the inversion traced by successive temperature measurements as shown by the dashed line. The nearly horizontal dashed lines indicate the position of the descending base of the inversion. The solid line indicates temperature. The temperature lapse rate in the descending layer is nearly dry adiabatic. The bottom surface is marked by a temperature inversion. Two features, temperature inversion and a marked decrease in moisture, identify the base of the subsiding layer. Below the inversion is an abrupt rise in the moisture content of the air.*

sidence inversion. Such inversions tend to prolong the period of stagnation and increase the intensity of pollution. The subsidence inversion that brings about our highest concentrations of pollution is often accompanied by lower-level radiation inversions.

Along the west coast of the United States, and occasionally along the east coast, the warm seasons often produce a coastal or marine inversion. In this case cool, moist air from the ocean spreads over low land. This layer of cool air, which may vary in depth from a few hundred to several hundred thousand feet, is topped by warmer, drier air, which also traps pollutants in the lower layers.

## MICROCLIMATES

When the weather report states that the temperature is 24° C and the sky is clear, the information may reflect the general weather conditions for the day. But on the surface of the ground in and beneath the vegetation, on slopes and cliff tops, in crannies and pockets, the climate is quite different. Heat, moisture, air movement, and light all vary radically from one part of the community to another to create a range of "little" or "micro" climates.

On a summer afternoon the temperature under calm, clear skies may be 28° C at 1.83 m, the standard level of temperature recording. But on or near the ground—at the 5-cm level—the temperature may be 5° C higher; and at sunrise, when the temperature for the 24-hour period is the lowest, the temperature may be 3° C lower at ground level (Biel, 1961). Thus in the middle eastern part of the United States, the afternoon temperature near the ground may correspond to the temperature at the 1.83-m level in Florida, 700 mi to the south; and at sunrise the temperature may correspond to the 1.83-m level temperature in southern Canada. Even greater extremes occur above and below the ground surface. In New Jersey March temperatures about the stolons of clover plants 1.5 cm above the surface of the ground may be 21° C, while 7.5 cm below the surface the temperature about the roots is −1° C (Biel, 1961). The temperature range for a vertical distance of 9 cm is 2° C. Under such climatic extremes most organisms exist.

The chief reason for the great differences between temperature at ground level and at 1.83 m is solar radiation. During the day the soil, the active surface, absorbs solar radiation, which comes in short waves as light, and radiates it back as long waves to heat a thin layer of air above it (Figure 4-13). Because air flow at ground level is almost nonexistent, the heat radiated from the surface remains close to the ground. Temperatures decrease sharply in the air above this layer and in the soil below. The heat absorbed by the ground during the day is reradiated by the ground at night. This heat is partly absorbed by the water vapor in the air above. The drier the air, the greater is the outgoing heat and the stronger is the cooling of the surface of the ground and the vegetation. Eventually the ground and the vegetation are cooled to the dew point, and water vapor in the air may condense as dew. After a heavy dew a thin layer of chilled air lies over the surface, the result of rapid absorption of heat in the evaporation of dew.

By alternating wind movement, evaporation, moisture, and soil temperatures, vegetation influences or moderates the microclimate

**FIGURE 4-13**
*Idealized temperature profiles in the ground and air for various times of day.* C = *heat transport by convection;* G = *heat transport by conduction;* L = *heat transport by latent heat of evaporation or condensation. (From Gates, 1962.)*

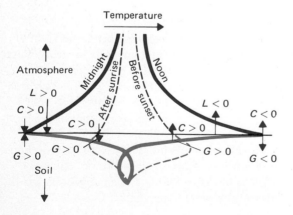

of an area, especially near the ground. Temperatures at the ground level under the shade are lower than those in places exposed to the sun and wind. Average maximum soil temperatures at the 2.5-cm level below the surface in a northern hardwoods forest and an aspen-birch forest in New York State were 15.5° C from mid-May to late June and 20° C (absolute maximum 25° C) from early July to mid-August (Spaeth and Diebold, 1938). Mean differences between maximum temperatures for the forest and adjacent fields ranged from 5° to 12° C. Maximum soil temperatures at 2.5 cm in open chestnut-oak forests were 19° C from mid-May to late June and 24° C (absolute maximum, 34° C) from July to mid-August. On fair summer days a dense forest cover reduces the daily range of temperatures at 2.5 cm by 10° to 18° C, compared with the temperature in the soils of bare fields.

Vegetation also reduces the steepness of the temperature gradient and influences the height of the active surface, the area that intercepts the maximum quantity of solar insolation. In the total absence of or in the presence of very thin vegetation, temperature increases sharply near the soil; but as the plant cover increases in height and density, the leaves of the plants intercept more solar radiation (Figure 4-14). The plant crowns then become the active surface and raise it above the ground. As a result temperatures are highest just above the dense crown surface and lowest at the surface of the ground. Maximum absorption of solar radiation in tall grass occurs just below the upper surface of the vegetation, whereas in the short grass maximum temperatures are at the ground level (Waterhouse, 1955). As the grasses grow taller the level of maximum temperature falls into and rises with the upper level of the grass stalks until the temperature eventually reaches an approximate equilibrium with the air above. (Among broad-leafed plants daily maximums occur on the upper leaf surfaces.) At night minimum temperatures are some distance above the ground because the air is cooled above the tops of plants and the dense stalks prevent the chilled air from settling to the ground.

Within dense vegetation air movements are reduced to convection and diffusion (Figure 4-15). In dense grass and low plant cover, com-

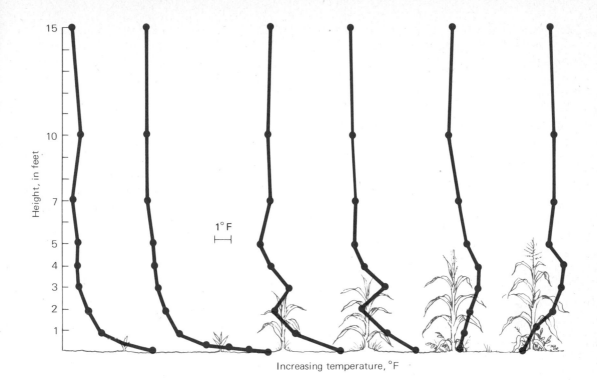

**FIGURE 4-14**
*Vertical temperature gradients at midday in a cornfield from the time of seeding stage to the time of harvest. Note the increasing height of the active surface. (Adapted from Wolfe et al., 1949.)*

plete calm exists at ground level. This calm is an outstanding feature of the microclimate near the ground because it influences both temperature and humidity and creates a favorable environment for insects and other animals.

Vegetation also deflects wind flow up and over its top (Figure 4-16). If the vegetation is narrow, such as a windbreak or a hedgerow, the microclimate on the leeward side may be greatly affected. Deflection of wind produces an area of eddies immediately behind the vegetation, in which the wind speed is low and small particles such as seeds are deposited. Beyond this is an area of turbulence, in which the climate tends to be colder and drier than normal. If some wind passes through the barrier and some goes over it, no turbulence develops, but the mean temperature behind the barrier is high in the morning and lower in the afternoon.

Humidity differs greatly from the ground up. Because evaporation takes place at the surface of the soil or at the active surface of plant cover, the vapor content (absolute humidity) decreases rapidly from a maximum at the bottom to atmospheric equilibrium above. Relative humidity increases above the surface, since actual vapor content increases only slowly during the day whereas the capacity of the heated air over the surface to hold moisture increases rather rapidly. During the night little difference exists above and on the ground. Within the growing vegetation, however, relative humidity is much higher than above the plant cover. In fact near-saturation conditions may exist.

Soil properties, too, enter the microclimatic picture. In a soil that conducts heat well, considerable heat energy will be transferred to the substratum, from which it radiates to the surface at night. On such soils surface temperatures are lower by day and higher by night than the surface temperatures of poorly conducting soils. This influences the occurrence of frost. Moist soils are better conductors of heat than dry soils. Light-colored sandy soils increase reflection and reduce the rate at which heat energy is absorbed. Dark soils, on the other hand, absorb more heat.

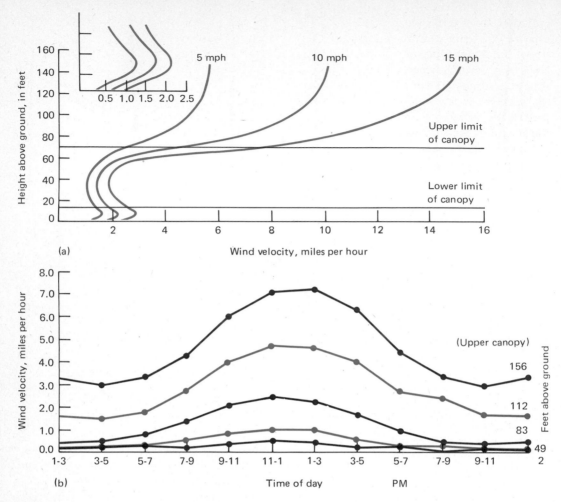

(a)

(b)

**FIGURE 4-15**
*Comparative wind velocities. The graphs show the distribution of wind velocities with height as affected by the timber canopy of coniferous forests for wind velocities of 8, 16, 24 km/hr at 43 m above ground and the average wind velocity during a June day (based on 1938, 1939, 1940 dates) inside a coniferous forest with a cedar understory in northern Idaho. Note decrease in velocity near the ground. Numbers 2, 49, 83, 112 and 156 are distances in feet from ground into canopy.*

**Valleys and frost pockets.** The widest climatic extremes occur in valleys and pockets, areas of convex slopes, and low concave surfaces. These places have much lower temperatures at night, especially in winter, much higher temperatures during the day, especially in summer, and a higher relative humidity. Protected from the circulating influences of the wind, the air becomes stagnant. It is heated by insolation and cooled by terrestrial

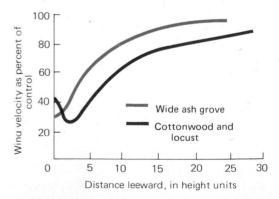

**FIGURE 4-16**
*Effect of vegetational form on wind velocity. The graph compares wind velocity at 16 in. above the ground to the leeward side of an open ash grove and a cottonwood and locust shelterbelt. The curve for the ash grove illustrates the effect of a open stand, which allows the wind to sweep under the crowns. Zero is the lee face of the barrier. (After Stoeckeler, 1962.)*

radiation, in sharp contrast to the wind-exposed, well-mixed air layers of the upper slopes. In the evening cool air from the uplands flows down the slope into the pockets and valleys to form a lake of cool air. Often when the warm air in the valley comes into contact with the inflowing cold air, the moisture in the warm air may condense as valley fog.

A similar phenomenon takes place on small concave surfaces. Like the larger valley, these concave surfaces radiate heat rapidly on still, cold nights and cold air flows in from surrounding higher levels. On such sites the air temperature near the ground may be 8° C lower than the surrounding terrain. This results in a temperature inversion. Because low ground temperatures in these areas tend to result in late spring frosts, early fall frosts, and a subsequent short growing season, these depressions are called frost pockets. The pockets need not be deep. Minimum temperatures in small depressions only 1 to 1.2 m deep were equivalent to those of a nearby valley 60 m below the general level of the land (Spurr, 1957). Such variations in temperature due to local microrelief can strongly influence the distribution and growth of plants. Tree growth is inhibited; and because the low surfaces more often than not accumulate water as well as cold air, such sites may contain plants of a more northern distribution. Frost pockets may also develop in small forest clearings. The surface of the tree crowns channels cold air into the clearings as terrestrial radiation cools the layer of air just above.

***North-facing and South-facing slopes.*** The greatest microclimatic differences exist between north-facing and south-facing slopes (Figure 4-17). South-facing slopes in the Northern Hemisphere receive the most solar energy. The amount of energy is maximal when the slope grade equals the sun's angle from the zenith point. North-facing slopes receive the least energy, especially when the slope grade equals or exceeds the angles of sun-ray inflection. At latitude 41° north (about central New Jersey and southern Pennsylvania) midday insolation on a 20° slope is, on the average, 40 percent greater on the south-facing slopes than on the north-facing slopes during all seasons. This has a marked effect on the moisture and heat budget of the two sites.

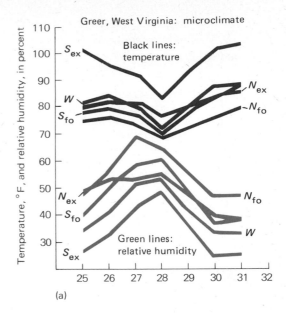

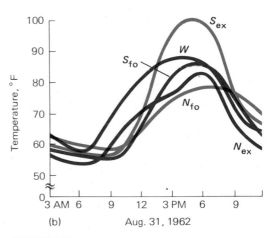

**FIGURE 4-17**
*Microclimates of north-facing and south-facing slopes. Daily maximum temperature and minimum relative humidity were recorded by four weather stations measuring microclimate and a single standard weather station during a week in August, 1962, at Greer, West Virginia. W = standard weather station on ridge; $N_{fo}$ = microclimate station on forested north-facing slope; $N_{ex}$ = station on exposed north-facing slope; $S_{fo}$ = station on forested south-facing slope; $S_{ex}$ = station on exposed south-facing slope. Note extremes recorded at exposed sites $S_{ex}$ and $N_{ex}$ and differences in readings of microclimate stations and the standard station. Among temperatures recorded at the five stations in August on a sunny day those recorded at the exposed sites showed the greatest variation. Contrast these with the forested north-facing slope $N_{fo}$. (Data courtesy Dr. W. A. van Eck.)*

**89**

High temperatures and associated low vapor pressures induce evapotranspiration of moisture from soil and plants. The evaporation rate often is 50 percent higher, the average temperature higher, the soil moisture lower, and the extremes more variable on south-facing slopes. Thus the microclimate ranges from warm and dry (xeric) conditions with wide extremes on south-facing slopes to cool and moist (mesic), less variable conditions on north-facing slopes (Figure 4-18). Xeric conditions are most highly developed on the top of south-facing slopes where air movement is greatest, while the most mesic conditions are at the bottom of the north-facing slopes.

The whole north-facing and south-facing

**FIGURE 4-18**
*Influence of microclimate on the distribution of vegetation on north-facing and south-facing slopes. Diagrams show the distribution of trees and shrubs in the hill country of southwestern West Virginia and the vegetation in the Point Sur area in California. In each case note the similarity of the vegetation on the lower parts of north-facing and south-facing slopes.*

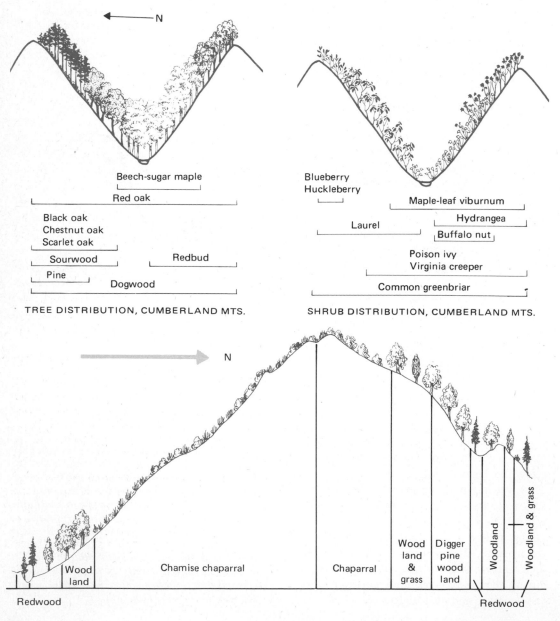

slope complex is the result of a long chain of interactions: solar radiation influences moisture regimes; the moisture regime influences the species of trees and other plants occupying the slopes; the species of trees influence mineral recycling, which is reflected in the nature and chemistry of the surface soil and the nature of the herbaceous ground cover.

Being mobile, few if any animals are typical of only north-facing or south-facing slopes, as far as we now know. However, their movements may be limited to some extent by the differences in conditions and food supplies on the slopes. Deer tend to use south-facing slopes more heavily in winter and early spring and north-facing slopes in summer (Taber and Dasmann, 1958). In the central Appalachians the red-backed vole, throughout its range normally an inhabitant of cool fir, spruce, aspen, and northern hardwood forests, is restricted in its local distribution to forested, mesic, north-facing slopes. Those species of soil invertebrates intolerant of humidity, such as some mites, can exist only in a dry habitat and therefore are confined to the xeric conditions of the south-facing slope.

*Microclimate of the city.* The building of cities has altered not only the slope, soil, and vegetation of the land they occupy, but also has altered the atmosphere, creating a distinctive urban microclimate.

The urban microclimate is a product of the morphology of the city and the density and activity of its occupants. In the urban complex, stone, asphalt, and concrete pavement and buildings with a high capacity for absorbing and reradiating heat replace natural vegetation with low conductivity of heat. Rainfall on impervious surfaces is drained away as fast as possible, reducing evaporation. Metabolic heat from masses of people and waste heat from buildings, industrial combustion, and vehicles raise the temperature of the surrounding air. Industrial activities, power production, and vehicles pour water vapor, gases, and particulate matter into the atmosphere in great quantities. The effect of this storage and reradiation of heat is the formation of a heat island about cities (Figure 4-19) in which the temperature may be 6° to 8° C higher than the surrounding countryside (see Landsberg, 1970; SMIC, 1971).

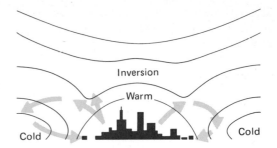

**FIGURE 4-19**
*Idealized scheme of nighttime circulation above a city in clear, calm weather. A heat island develops over the city. At the same time a surface inversion develops in the country. This results in a flow of cool air toward the city, producing a country breeze in the city at night. Lines are temperature isotherms; arrows represent wind. (From H. Landsberg, 1970.)*

Heat islands are characterized by high temperature gradients about the city. Highest temperatures are associated with areas of highest density and activity; temperatures decline markedly toward the periphery of the city (Figure 4-20). Although detectable throughout the year, heat islands are most pronounced during summer and early winter and are most noticeable at night when heat stored by pavements and buildings is reradiated to the air. The magnitude of the heat island is influenced strongly by local climatic conditions such as wind and cloud cover. If the wind speed, for example, is above some varying critical value, a heat island cannot be detected.

During the summer the buildings and pavement of the inner city absorb and store considerably more heat than does the vegetation of the countryside. In cities with narrow streets and tall buildings the walls radiate heat toward each other instead of toward the sky. At night these structures slowly give off heat stored during the day. Although daytime differences in temperature between the city and the country may not differ noticeably, nighttime differences become pronounced shortly after sunset and persist through the night. The nighttime heating of the air from below counteracts radiative cooling and produces a positive temperature lapse rate while an inversion is forming over the countryside. This, along

**91**

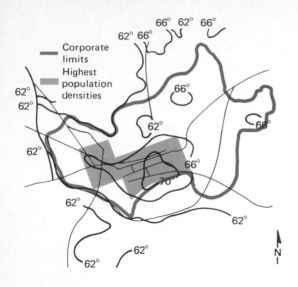

*FIGURE 4-20*
*Thermal pattern of night air in a small city,*
*Chapel Hill, North Carolina. Note that the*
*highest temperatures are inside the*
*corporate limits where the population and*
*activity are the greatest. (From Kopec,*
*1970.)*

with the surface temperature gradient, sets
the air in motion, producing "country breezes"
to flow into the city.

In winter solar radiation is considerably less
because of the low angle of the sun, but
heat accumulates from human and animal
metabolism and from home heating, power
generation, industry, and transportation. In
fact heat contributed from these sources is 2.5
times that contributed by solar radiation. This
energy reaches and warms the atmosphere directly or indirectly, producing more moderate
winters in the city than in the country.

Urban centers influence the flow of wind.
Buildings act as obstacles, reducing the velocity of the wind up to 20 percent of that of the
surrounding countryside, increasing its turbulence, robbing the urban area of the ventilation it needs, and inhibiting the movement of
cool air in from the outside. Strong regional
winds, however, can produce thermal and pollution plumes, transporting both heat and particulate matter out of the city and modifying
the rural radiation balance a few miles down
wind (Clarke, 1969; Oke and East, 1971).

Throughout the year urban areas are blan-
keted with particulate matter, carbon dioxide,
and water vapor. The haze reduces solar radiation reaching the city by as much as 10 to 20
percent. At the same time, the blanket of haze
absorbs part of the heat radiating upward and
reflects it back, warming both the air and the
ground. The higher the concentration of pollutants, the more intense is the heat island.

The particulate matter has other microclimatic effects. Because of the low evaporation
rate and the lack of vegetation, relative
humidity is lower in the city than in surrounding rural areas. But the particulate matter
acts as condensation nuclei for water vapor in
the air, producing fog and haze. Fogs are much
more frequent in urban areas than in the
country, especially in winter (Table 4-1).

Another consequence of the heat island is
increased convection over the city. Updrafts,
together with particulate matter and large
amounts of water vapor from combustion processes and steam power, lead to increased
cloudiness over cities and increased local rainfall both over cities and over regions
downwind. An evidence of weather modification by pollution is the increase in precipitation and stormy weather about La Porte, Indiana, downwind from the heavily polluted

*TABLE 4-1*
*Climate of the city compared to the country*

| Elements | Comparison with rural environment |
|---|---|
| Condensation nuclei and particles | 10 times more |
| Gaseous admixtures | 5–25 times more |
| Cloud cover | 5–10 percent more |
| Winter fog | 100 percent more |
| Summer fog | 30 percent more |
| Total precipitation | 5–10 percent more |
| Relative humidity, winter | 2 percent less |
| Relative humidity, summer | 8 percent less |
| Radiation, global | 15–20 percent less |
| Duration of sunshine | 5–15 percent less |
| Annual mean temperature | 0.5°–1.0° C more |
| Annual mean wind speed | 20–30 percent less |
| Calms | 5–20 percent more |

*Source:* Adapted from H. E. Landsberg, 1970.

areas of Chicago, Illinois, and Gary, Indiana, and close to moisture-laden air over Lake Michigan. Since 1925 there has been a 31 percent increase in precipitation, a 34 percent increase in thunderstorms, and a 240 percent increase in the occurrence of hail (Changnon, 1968).

## Fire

In the early 1930s and 1940s many ecologists ignored the importance of fire as an environmental influence, just as many ignored the role of animals as an influence on plant communities. Today ecologists recognize fire as an important ecological force that is a part of the natural environment, along with moisture, temperature, wind, and soil. (The literature on fire as an ecological force is enormous. For reviews, see Komarek, 1967–1976; Kozlowski and Ahlgren, 1974).

### CONDITIONS FOR FIRE

Three conditions are necessary for fire to assume ecological importance: (1) an accumulation of organic matter sufficient to burn; (2) dry weather to render the material combustible; and (3) a source of ignition, the main sources being lightning and human beings.

A fire climate is one that combines a wet period during which quantities of organic matter can grow with a dry period during which the matter can become dry enough to burn readily. The wetter the growing season and the longer and hotter the dry season, the more chances there are for lightning-set fires.

Globally certain regions possess conditions conducive for burning and the spread of fires set off by lightning. Africa is ideally located for the development and occurrence of thunderstorms and lightning (J. Phillips, 1965; Batchelder, 1967). Southern and western Australia experience hot, dry summers with low humidity and drying westerly winds (Cochrane, 1968). Long before the evolution of man fire periodically swept these regions (T. M. Harris, 1958). Thus each is characterized by vegetation that evolved under the influence of fire: the grasslands of North America, the chaparral of the southwestern United States, the maquis of the Mediterranean, the South African fynbos (J. Phillips, 1965), the African grasslands and savannas (Batchelder, 1967), the southern pinelands of the United States, and the even-aged stands of coniferous forests of western North America.

Lightning storms are not universally accompanied by precipitation. In the temperate regions weather patterns at the end of a drought are often characterized by cloud thunderstorms with no rain. The same is true at the beginning and the end of dry periods in the tropical regions. When lightning strokes hit the ground, the dry material is kindled, and fires are set off. In the western United States 70 percent of the forest fires are caused by dry lightning during the summer. Because of the seasonal nature of lightning, fires so caused are most numerous during the summer in the midst of the growing season, when they would have the greatest impact as a selection force.

When human beings appeared on the scene, fire became an even more powerful influence on vegetation, for they added a new dimension to it. Whereas lightning fires are random and often periodic, fires set by human beings are often a deliberate attempt to modify or change the environment. Fires became more numerous through the years, and their pattern was and is adjusted to the season, to agricultural calendars, and to religious beliefs. Fires were set to clear ground for agricultural use, to improve conditions for hunting, to develop grass and shrubby vegetation attractive to game, to improve forage for grazing, to open up the countryside, to reduce enemy cover, to develop areas for wild fruits and berries and other desirable plants, and to make travel easier. Other fires were set simply for excitement or revenge or to burn trash or by mistake. Whatever the reason, most fires burned in the nongrowing seasons of fall and spring, when damage can be much more severe.

As human beings spread from the fire-evolved grasslands and savannas to the more humid forested areas, they introduced fire into less fire-resistant vegetational areas, such as hardwood forests. Further, they caused fires when the forests were the most inflammable—fall and early spring. Indian fires undoubtedly produced the open heaths of the northeast United States and the glades of Kentucky, developed oak stands in the central hardwoods (R. L. Smith, 1978), and maintained large areas of blueberries (Thompson and Smith, 1970). Lumbering operations left massive piles of debris that fed extremely hot fires that swept across much of the logged-over country. In many places fire burned into the deep layers of organic matter and peat to rock and mineral soil, destroying any opportunity for former forest types to return.

## EFFECTS OF FIRE

As a selective pressure, fire has a pronounced impact on the ecosystem. It reduces dead and dry organic matter to soluble ash and releases phosphorus, calcium, potassium, and other elements for rapid recycling, stimulating new growth. Although the flush of new vigorous growth is attributed to an increased availability of nutrients, this may not necessarily be true for all ecosystems. Daubenmire (1968b) suggests that the response of grasslands to fire may reflect the release of new shoots from the competition of old tillers and increased root activity. The response of animals to vegetation on burns may be due to the increased availability of both tender shoots and biomass previously unavailable because litter and old stems were a hindrance to grazing. Considerable nitrogen may be lost to the atmosphere, but unless the litter is converted to white ash, some nitrogen will remain in the charred litter, increasing the total nitrogen content of the surface soil. Further increases in nitrogen come about by a marked increase in nitrogen-fixing legumes following a fire.

Fire exposes the mineral soil, stimulates the germination of certain seeds, and may encourage erosion, changing the character of the site. In the western United States erosive forces favor such species as the knobcone pine (Vogl, 1967). Hot fires may heat the soil sufficiently to kill the roots of some plants and change the nature of the soil faunal community. The dark color of burned-over lands absorbs more heat from the sun, warming up the soil in temperate regions and reducing the depth of permafrost in arctic regions.

Fire modifies the vegetational community. Crown and severe surface fires in a forest can destroy all existing vegetation. Light and moderate surface fires may destroy only the undergrowth, kill some thin-barked fire-sensitive trees by heating the cambium, and damage others. Heat-damaged trees are susceptible to attack from insects and disease.

Fire can free grass from woody competition and maintain grass as dominant vegetation. However, in some communities fire may have no effect on the composition of vegetation. Through long evolution in the presence of fire, many species have become fire-adapted. Communities containing such species may experience no loss of species, but with the stand opened up, opportunistic species may invade the area, increasing the richness of the community. In such a manner fire can diversify the mosaic of vegetation over the landscape.

The effect of fire on animal life has received considerably less attention than the effect of fire on plant life. In general the heat from moderate fires has much less effect on animal populations than do changes brought about by fire. Soil fauna, such as earthworms, snails, spiders, centipedes, and millipedes, are reduced markedly, not so much from the heat of the fire as from the xeric conditions brought about by increased evaporation and post-fire loss of soil moisture (Ahlgren, 1974). Mites and Collembola are reduced by the heat of the fire and return to normal in 3 to 5 years. Losses of birds and mammals are confined largely to nest destruction. Apparently most vertebrates are able to escape destruction by fleeing or remaining underground in burrows. Death among mammals and birds is attributed more to smoke inhalation and suffocation than to burning (Bendell, 1974).

Once humans recognized the destructiveness of their fires, they moved to the other extreme, the attempt to exclude all fire, which can have an effect on the ecosystem as deleterious as too frequent or too hot fires. The lack of periodic fires allows the accumulation of trash and litter. Then if fires do start, they are much hotter and more destructive than more frequent cool fires. Exclusion of fire alters the composition of forest stands by permitting the dominance of fire-sensitive species, and by encouraging the spread of woody vegetation into grasslands. It inhibits the regeneration of fire-resistant species and causes deterioration of forage and range, stagnation of forest stands, and decline in the habitats of certain wildlife.

Carefully used, fire can be an important tool in the regulation and manipulation of vegetation. It can be used to improve forage stands and increase net productivity of grasslands. It can reduce the hazard of destructive forest fires by removing litter before it accumulates to a great degree, and can improve seed bed for regeneration of certain forest types and tree species. It can be used to improve wildlife habitats, to maintain certain fire-controlled ecosystems, to maintain the naturalness of wilderness areas, and even to improve the esthetics of the natural landscape.

### ADAPTATIONS TO FIRE

Because of its pervasive effects, fire acts as a strong selection agent, resulting in a number of fire-adapted species. Some plants are fire resistant. They are characterized by underground stems and buds, aerial parts that die off annually, dormant buds at the soil surface, or thick fire-resistant bark. Fire-resistant species also are usually intolerant of shade and require mineral soil and full sunlight for the germination of seeds and growth of seedlings. Some plant species, although not fire-resistant as individual plants, require fire in their life cycle to release seeds from cones or provide a seedbed of mineral soil. Among such species are jack pine, knobcone pine, white pine, paper birch, aspen, and eucalyptus.

An example of a highly fire-adapted ecosystem is chaparral, a Mediterranean vegetation type that is highly flammable and depends upon recurring fires for restoration and optimum growth. Natural selection has favored characteristics that make chaparral plants more flammable and at the same time more fire tolerant. One such plant is chamise *(Denostema fasciculatum),* a shrub that grows rapidly after fire. It develops dead branches on which are produced lateral twigs 15 to 30 cm back from the tip producing a fine, dry fuel. With nearly half of its stems less than 1 cm in diameter, with highly resinous leaves, and with a high surface area to volume ratio in leaves and twigs, the shrub is highly flammable. After a fire new plants resprout or grow from high heat-resistant seeds that may have been buried in the litter for three to five years (Biswell, 1974).

Some animals may be considered fire adapted to the extent that they depend upon fire environment to maintain habitat and to increase their numbers. For example, Kirtland's warbler and blue grouse may respond to fire by expanding their population into burned areas (Redfield et al., 1970; Redfield, 1972). Moose respond to fires because burns produce succulent forages such as aspen, willow, and birch that become most abundant about 15 years after a fire (Bendell, 1974).

## Light

Light influences ecosystems in two ways. It affects photosynthetic activity and it influences the daily and seasonal activity patterns of plants and animals. The process of photosynthesis is discussed in Chapter 2, but a few points should be emphasized here, especially the relation of light to the structure of ecosystems.

### ADAPTATIONS TO LIGHT

The light intensity below which plants can no longer carry on sufficient photosynthesis to maintain themselves is called the *compensation intensity.* At this point photosynthesis balances respiration and plants are just able to replace material loss in respiration day and night. Only a few plants have a compensation intensity of less than about 100 footcandles or 1 percent of sunlight.

The optimum condition of light intensity varies among plants ranging from full sunlight to dense shade. To fully exploit light gradients, plants exhibit a range of light requirements (Table 4-2). Plants meet their specific requirements by certain modifications in structure and function, such as reducing respiratory losses to a minimum, increasing the photosynthetic rate, and changing the leaf area ratio, which in the extreme results in the development of sun and shade leaves. Sun leaves tend to be thicker with a greater development of the palisade layer; shade leaves tend to be thinner with a single palisade layer and less respiratory tissue.

In general plants can be divided into two groups, shade-intolerant plants (also called sun plants) and shade-tolerant plants (also called shade plants). Shade-intolerant plants are adapted to high light intensities, rarely reach

TABLE 4-2
*Adaptations of seedlings in five different light and habitat situations*

| Type of habitat | Illumination 0–40 cm above ground surface | Adaptative features |
|---|---|---|
| Dry, severe | Continuously high through growing season | Annual life cycle; small seeds efficiently dispersed; creeping shoots; compact rosettes; vertical leaves; high resistance to wilting |
| Recently cleared, nutrient-rich, moist | Initially high | Prolific seed production; efficient dispersal; tall stature; high growth rate; competitive; rapid extension of stem on shading; high leaf area index; minimal mutual shading. |
| Grassland | Vertical variation, small increase in intensity with increase in height above ground | Tall stature; rapid extension of growth on shading |
| Open woodland | High intensity early in growing season before canopy closes (spring); moderately high late in season after canopy thins (fall) | Growing and flowering before leaf canopy closes; flowering after canopy thins; tall stature; low resistance to wilting; large leaf area with minimal mutual shading; horizontal leaves; rapid extension of growth on shading; potential for high rates of photosynthesis |
| Dense woodland | Low intensity over most of growing season; small increase in intensity with increase in height above ground | Low respiration; horizontal leaves; limited extension of growth on shading; high resistance to fungal attack |

*Source:* After J. P. Grime, 1966.

light saturation ($C_4$ plants), and have a high rate of respiration. Under low light conditions sun plants tend to grow rapidly in height with a reduction in the thickness of the cell wall. Lacking effective supporting tissue, the plant collapses. Under such conditions the plant is highly susceptible to fungal infection (Grime, 1966).

Shade-tolerant plants have lower photosynthetic rates than sun plants, lower respiration and metabolic rates, and lower growth rates, all of which lead to conservation of nutrients and energy. The photosynthetic system of shade plants becomes light saturated at lower intensities than sun plants, but reaches maximum photosynthesis for longer periods of time during the daily cycle. Such plants cannot compete with sun plants under high light intensities. However, shade plants, because of their low metabolism and slow growth under low light conditions, can sustain themselves

under such a condition for a long period of time. This ability is further strengthened by the shade plant's inherent resistance to fungal infection (Grime, 1966). Typical among the shade species are sugar maple, white cedar, and hemlock that exist successfully under a dense forest canopy at low light intensities and reach maximum growth rates only when the canopy is opened sufficiently to allow considerably more light to enter. Thus shade plants and sun plants are adapted to live under certain environmental conditions and have evolved certain characteristics that enable them to colonize certain habitats successfully and prevent them from growing in others.

In aquatic situations the growth of rooted aquatic plants is limited by the amount of light penetration. Turbid water caused by silt or by phytoplankton growth prevents the light from reaching deeper water. Visible light that penetrates water becomes limited more and more to a narrow band of blue light as the water depth increases from 0.1 to 100.0 m. This is in part the reason why water of deep, clear lakes looks blue. Eventually blue light is filtered out and the remaining green light is poorly absorbed by chlorophyll. Depths at which green light occurs are occupied by red algae which possess supplementary pigments that enable them to utilize the energy of green light.

## CIRCADIAN RHYTHMS

One aspect of communities with which everyone is familiar is rhythmicity, the recurrence of daily and seasonal changes. Bird song signals the arrival of dawn. Butterflies, dragonflies, and bees become conspicuous, hawks seek out prey, and chipmunks and tree squirrels become active. At dusk light fades and daytime animals retire, the blooms of waterlilies and other flowers fold, and animals of the night appear. Fox, raccoon, flying squirrel, owls, and moths take over niches occupied by others during the day. As the seasons progress, day length changes and with it other conspicuous activities. Spring brings the migrant birds and initiates the reproductive cycles of many animals and plants. In fall trees of temperate regions become dormant, insects and herbaceous plants disappear, the summer-resident birds return south, and winter visitors arrive. On the ocean shore the tide rises and falls about 50 minutes later each day and affects all life on the edge of the sea.

Underlying these rhythmicities are the movements of the earth relative to the sun and moon, its rotation, the tilt of its axis, and its annual revolution around the sun. Life evolved under the influences of daily and seasonal environmental changes, so it is natural that plants and animals would have some rhythm or pattern to their lives that would synchronize them with fluctuations in the environment.

For years biologists have been intrigued by the means by which organisms keep their activities in rhythm with the 24-hour day, including such phenomena as the daily pattern of leaf and petal movements in plants, the emergence of insects from pupal cases, the sleep and wakefulness of animals. At one time biologists thought that these rhythmicities were entirely exogenous, that is, that the organisms responded only to external stimuli such as light intensity, humidity, temperature, and tides. Laboratory investigations, however, indicate that this is not the complete answer (Figure 4-21).

At dusk in the forests of North America a small squirrel with silky fur and large, black eyes emerges from a tree hole. With a leap the squirrel sails downward in a long sloping glide, maintaining itself in flight with broad membranes stretched between its outspread legs. Using its tail as a rudder and brake, it makes a short, graceful, upward swoop that lands it on the trunk of another tree. This is *Glaucomys volans* the flying squirrel, perhaps the commonest of all our tree squirrels. But because of its nocturnal habits, this mammal is seldom seen by most people. Unless it is disturbed, the flying squirrel does not come out by day. It emerges into the forest world with the coming of darkness; it returns to its nest with the first light of dawn.

The squirrel's day-to-day activity forms a 24-hour cycle. This correlation of the onset of activity with the time of sunset suggests that light has a regulatory effect on the activity of the squirrel. If the flying squirrel is brought indoors and kept under artificial conditions of night and day, the animal will confine its periods of activity to darkness, its periods of inactivity to light. Whether the conditions under which the animal lives are 12 hours of darkness and 12 hours of light or 8 hours of darkness and 16 hours of light, the onset of activity always begins shortly after dark.

But the photoperiodism (response to changing light and darkness) exhibited by the squirrel is not quite so simple. There is more to it than the animal becoming active because darkness has come. If the squirrel is kept in constant darkness, it still maintains a relatively constant rhythm of activity from day to day (DeCoursey, 1961). But in the absence of any external time cues, the squirrel's activity rhythm deviates from the 24-hour periodicity exhibited under light and dark conditions. The daily cycle under constant darkness varies from 22 hours, 58 minutes, to 24 hours, 21 minutes, the average being less than 24 hours (most frequent 23:50 and 23:59) (DeCoursey,

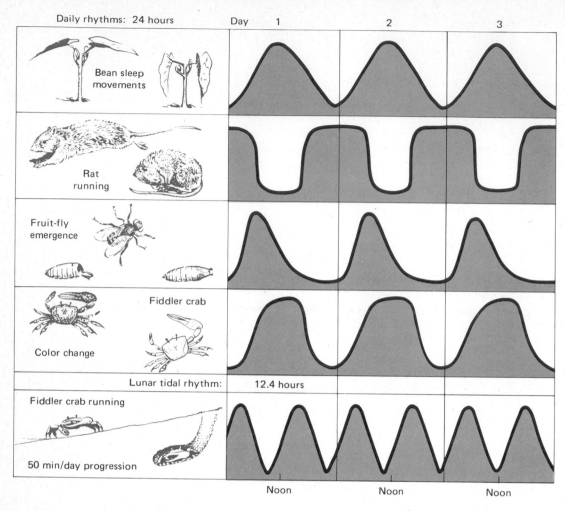

FIGURE 4-21
*Examples of rhythmic phenomena. These phenomena have been experimentally demonstrated to persist under constant conditions in the laboratory, illustrating diagrammatically the natural phase relationships to external physical cycles. (From Brown, 1959.)*

1961). If the same animal is held under continuous light, a very abnormal condition for a nocturnal animal, the activity cycle is lengthened, probably because the animal, attempting to avoid running in the light, delays the beginning of its activity as much as it can. The length of the activity cycle including periods of activity and inactivity, maintained under a given set of conditions is an individual characteristic. Because of the deviation of the average cycle length from 24 hours, each indi-

vidual squirrel gradually drifts out of phase with the day-night changes of the external world (Figure 4-22).

Like the flying squirrel, many forms of life, including humans, possess a rhythm of activity that under field conditions exhibits a periodicity of 24 hours (Aschoff, 1969; Menaker, 1971). When these organisms are brought into the laboratory and held under constant conditions of light, darkness, and temperature away from any external time cues, they still exhibit a rhythm of activity of approximately 24 hours. Because these rhythms approximate, but seldom match, the periods of the earth's rotation, they are called *circadian* (from the Latin *circa,* "about," and *dies,* "day"). The period of the circadian rhythm, the number of hours from the beginning of a period of activity one day to the beginning of activity on the next, is referred to as a free-running cycle. In other words, the rhythm of activity exhibits a self-sustained oscillation under constant conditions. The length of this cycle is usually a function of the intensity of light provided under constant conditions. Increasing the light intensity causes a lengthening of the free-running cycle in organisms active by night and a shortening of the cycle in organisms active by day (Hoffman, 1965). Circadian rhythms apparently are endogenous, that is, they are determined by internal mechanisms. They are affected very little by temperature changes, are insensitive to a great variety of chemical inhibitors, and are not learned from or imprinted upon the organisms by the environment.

Thus many plants and animals are influenced by two periodicities, the external rhythm of 24 hours and the internal circadian rhythm of approximately 24 hours. If the two activity rhythms are to be brought into phase, some external environmental "time-setter" must adjust the endogenous rhythm to the

exogenous. The most obvious timekeepers, cues, synchronizers, or *Zeitgebers* (Aschoff, 1958) are temperature and light. Of the two light is the master *Zeitgeber*. It brings the circadian rhythm of many organisms into phase with the 24-hour photoperiod of the external environment.

Although in the field one might have difficulty proving the role of light in synchronizing the circadian rhythm with the environment, it can be demonstrated in the laboratory (see Bruce, 1960). If an animal or plant is kept under constant conditions, such as continuous darkness or continuous light, the circadian rhythm drifts out of phase with natural light and dark and eventually may fade away. The length of time required for this to happen depends upon the organism and the conditions of light and darkness (Kramm, 1975). The activity rhythm of rodents may continue for several months in constant darkness. Other rhythms, such as the leaf movements of plants, may fade much more quickly. Once a rhythm has faded, a new one can be started by some exposure to light or dark. This may be the interruption of continuous darkness by a short flash of light, the interruption of continuous light by darkness, the change from continuous darkness to continuous light, or vice versa. With some organisms, a change in temperature may start a new rhythm.

The activity period of organisms shows an entrainment to light-dark cycles, that is, the phase or period of their circadian rhythm is determined or can be modified by light. The flying squirrel, both in its natural environment and in artificial day-night schedules, synchronizes its daily cycle of activity to a specific phase of the light-dark cycle. This was demonstrated in DeCoursey's (1960a, 1961) experiments. Flying squirrels were held in constant darkness until their circadian rhythms were no longer in phase with the natural environment. Then they were subjected to a light-dark cycle that was out of phase with their free-running cycle. If the light period fell in the animal's subjective night (natural period of activity), it caused a delay in the subsequent onset of activity. Synchronization took place in a series of stepwise delays, until the animal's rhythms were stabilized with the light-dark change (Figure 4-23). If the light

FIGURE 4-22
*Drift in the phase of activity of a flying squirrel held in continuous darkness at 20° C for 25 days. (After DeCoursey, 1960.)*

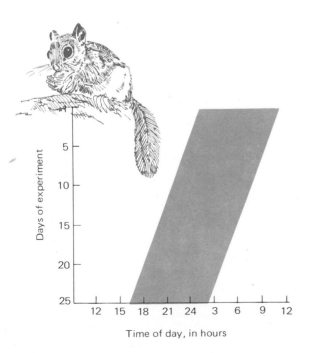

Days of experiment

5
10
15
20
25

12  15  18  21  24  3  6  9  12

Time of day, in hours

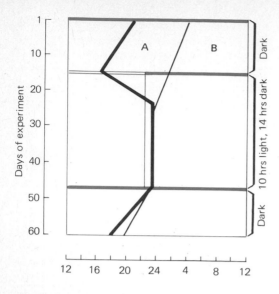

**FIGURE 4-23**

*Representation of synchronization for flying squirrels with circadian rhythm in constant darkness of less than 24 hours to a cycle of 10 hours of light, 14 hours of darkness. For squirrel A rephasing light fell during the squirrel's subjective night, synchronization was accomplished by a stepwise delay, and the onset of activity was stabilized shortly after light-dark change. For squirrel B light fell in the subjective day, and the free-running period continued unchanged until the onset drifted up against "dusk" light change. This prevented it from drifting forward by a delaying action of light. When returned to constant darkness, the onset of activity continued a forward drift. (After DeCoursey, 1960.)*

period fell at the subjective dawn or at the end of the dark period (when the animal's activity was about to end), it caused an advance of activity toward the dusk period. And if the light fell in the animal's inactive day period, it had no effect. The flying squirrels did not need to be exposed to a whole light-dark cycle to bring about a shift in the phase of the activity rhythm. A single 10-minute light period was sufficient to cause a phase shift in the locomotory activity, provided it was given during the squirrel's light-sensitive period (DeCoursey, 1960*b*).

A somewhat similar response is exhibited by two diurnal squirrels, the antelope ground squirrel *(Ammospermophilus leucurus)* and the red squirrel *(Tamiasciurus hudsonicus).*

When the experimental animals held under conditions of constant darkness were exposed to light in their subjective night (period of inactivity), the animals shortened the activity period of their cycle. When exposed to light in their subjective day (activity period), the animals lengthened their activity period. In effect, if the animal's activity were too late in the day, the activity period would be phase shifted toward dawn; and if the activity period were too early in the day, the phase shift would be delayed (Kramm, 1975, 1976).

Light and dark may be the *Zeitgebers* that control the phase of an organism's circadian activity rhythm, but the rhythms may relate more directly to other aspects of the environment, which, ecologically, are more significant to the organism than light and dark per se. The transition from day to night, for example, is accompanied by such environmental changes as a rise in humidity and a drop in temperature. Woodlice, centipedes, and millipedes, which lose water rapidly when exposed to dry air, spend the day in a fairly constant environment of darkness and dampness under stones, logs, and leaves. At dusk they emerge, when the humidity of the air is more favorable. In general, these animals show an increased tendency to escape from light as the length of time they spend in darkness increases. On the other hand, their intensity of response to low humidity decreases with darkness. Thus they come out at night into places too dry for them during the day; and as light comes they quickly retreat to their dark hiding places (Cloudsley-Thompson, 1956, 1960). Among some animals the biotic rather than the physical aspects of the environment may relate to the activity rhythm. Deer undisturbed by man may be active by day, but when they are hunted and disturbed they become strongly nocturnal. Predators must relate their feeding activity to the activity rhythm of the prey. Moths and bees must visit flowers when they are open and provide a source of food. And the flowers must have a rhythm of opening and closing that coincides with the time when the insects that pollinate them are flying. The entrainment of the phase of its activity rhythm to a natural light–dark cycle means more to an organism than simply an adjustment to a precise 24-hour period. More important, the entrainment serves to time the activities of plants and animals to a day–night cycle in a manner that is appropriate to the ecology of the species.

The possession of a circadian rhythm that can be entrained to environmental rhythms provides plants and animals with a biological clock, which probably is an integral part of cellular structure. With this, organisms can not only determine the time of day, they can also use the clock for time measurement. The clock, as already suggested, is not simply an hourglass or stopwatch. It does not start on some given signal, such as dawn, run until stopped by another signal, such as darkness, and then start up again on another. The clock runs, or oscillates, continuously, but it must be regulated or reset by environmental signals. It is this latter characteristic of the clock that makes the 24-hour photoperiod possible. The environmental rhythm of daylight and dark is the signal by which the biological clock is set to the correct local time each day.

## PHOTOPERIODISM IN PLANTS AND ANIMALS

As the seasons turn, the daily periods of daylight and darkness change. The activities of plants and animals are geared to this seasonal rhythm of night and day. The flying squirrel starts its daily activity with nightfall and maintains this relation through the year. As the short days of winter turn to the longer days of spring, the squirrel begins its activity a little later each day (Figure 4-24).

Most animals and plants of the temperate regions have reproductive periods that closely follow the changing day lengths of the seasons. For most birds the height of the breeding season is spring; for deer the mating season is the fall. Brook trout spawn in the fall; bass and bluegills in late spring and summer. The trilliums and violets bloom in the short days of early spring before the forest leaves are out and while an abundance of sunlight reaches the forest floor. Asters and goldenrods flower in the shortening days of fall.

Based on photoperiodic responses, plants can be classed as short-day, long-day, and day-neutral. Day-neutral plants are those whose flowering is not affected by day length, but rather is controlled by age, number of nodes, previous cold treatment, and the like.

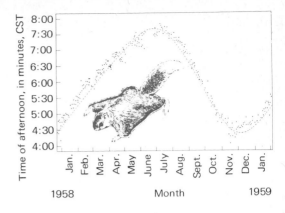

1958　　　　　Month　　　　1959

**FIGURE 4-24**
*Onset of running wheel activity for one flying squirrel in natural light conditions throughout the year. The graph is the time of local sunset through the year. (From DeCoursey, 1960.)*

Short-day and long-day plants both are influenced by the length of day. When the period of light reaches a certain portion of the 24-hour day, it inhibits or promotes a photoperiodic response. The length of this period so decisive to the response is called the *critical day length*. It varies among organisms but usually falls somewhere between 10 and 14 hours. Throughout the year plants (and animals) "compare" this time scale with the actual length of day or night. As soon as the actual length of day or night is greater or smaller than the critical day length, the plant may flower or cease to flower, expand its leaves, or lengthen its shoots. Short-day plants are those whose flowering is stimulated by day lengths shorter than the critical day length and thus bloom early in the growing season. Long-day plants are those whose flowering is stimulated by day lengths longer than a particular value and thus usually bloom in late spring and summer.

In the cotton fields of the southern United States lives the pink cotton bollworm, the larva of a tiny moth. Except for a few hours directly after hatching, the larva spends its life in the flower buds or bolls of cotton. At the fourth larval instar stage, the insect goes into diapause. The onset of diapause comes in late August, but not until near the autumnal equinox, September 21, when the night be-

**101**

comes equal to or longer than the day, does the number of diapausing larvae sharply increase. By the end of October virtually all the larvae are in diapause. The larva remains in this state of arrested development through the winter; then in late winter, as the days begin to lengthen, the insect comes out of diapause and continues its growth. The emergence from diapause reaches its maximum just after the spring equinox, when the days are just slightly longer than those that induced diapause.

When the larva of the pink bollworm was exposed to regimes of light and dark in the laboratory, the insect would go into diapause only when the light phase of the 24-hour day was 13 hours or less (Adkisson, 1966). If the larva was exposed to a light period of 13.25 hours, the insect was prevented from going into diapause. So precise is the time measurement in the insect that a quarter-hour difference in the light period determines whether the insect goes into diapause or not. Once growth is arrested, the exogenous rhythms do not cease but continue through the diapause until the day length becomes longer than 13 hours. Diapause terminates most rapidly under photoperiods of 14 hours, less rapidly at 16 and 12. Thus to the pink bollworm, the shortening days of late summer and fall forecast the coming of winter and call for diapause; and the lengthening days of late winter and early spring are the signals for the insect to resume development, pupate, emerge as an adult, and reproduce.

That increasing day length increases gonadal development and spring migratory behavior in birds was experimentally demonstrated some 40 years ago when Rowan (1925, 1929) forced juncos into the reproductive stage out of season by artificial increases in day length. Results of subsequent experimental work with a number of species have shown that the reproductive cycle is under the control of an exogenous seasonal rhythm of changing day lengths and an endogenous physiological response timed by a circadian rhythm.

After the breeding season the gonads of birds studied to date have been found to regress spontaneously. This is the *refractory* period, a time when light cannot induce gonadal activity, the duration of which is regulated by day length (see Farner, 1959, 1964*a*,

*b*; Wolfson, 1959, 1960). Short days hasten the termination of the refractory period; long days prolong it. However, Hamner (1968) makes the suggestion, based on experimental work with house finches, that the refractory period consists of two distinct periods of physiological states. One is an absolute refractory period, the length of which is independent of photoperiod. This is followed by a relative refractory period when photosensitive birds will not respond to day lengths equal to or shorter than the one to which they were previously exposed. As winter approaches and the natural day shortens, there is a continual temporal readjustment of the birds' timing mechanism so that the progressively shorter days become photoperiodically stimulatory. After the refractory phase is completed, the *progressive* phase begins in the late fall and winter. During this period the birds fatten, they migrate, and their reproductive organs increase in size. This process can be speeded up by exposing the bird to a long-day photoperiod. Completion of the progressive period brings the bird into the *reproductive* stage.

A similar photoperiodic response exists in the cyprinid fish, the minnows. The annual sexual cycle among the minnows consists of a sexually inactive period, followed by the reproductive period (Harrington, 1959). Underlying this is an intrinsic sexual rhythm that consists of a long responsive period, alternating with a shorter refractory period. Long days occurring within the responsive period start the prespawning period, characterized by mating and territorial behavior. The prespawning period ends with the laying of the first eggs. The subsequent spawning period is consummatory and ends sometime before the days shorten to the critical length that initiated the prespawning activity. Following this is the refractory, or postspawning period, in which light fails to stimulate gonadal development. The prespawning period, from about mid-November to mid-July, is the phase in which the annual sexual period is timed.

Seasonal cycles of photoperiodism influence the breeding cycles of many mammals. In the northeastern United States the flying squirrel has two peaks of litter production, the first in early spring, usually April, and the second in late summer, usually August. To produce litters in April, the flying squirrel must be in breeding condition in January and February. Muul (1969) investigated the responses of gonadal development to changing photoperiod under laboratory conditions. He found that in the flying squirrel the descent of the testes into the scrotum (in nonbreeding condition, the testes are held in the body cavity) occurred in January under short-day and long-night conditions. An accelerated decrease in day length hastened the descent. The experimental animals held under natural photoperiod came into reproductive condition and produced litters at the same time as squirrels in the wild in absence of temperature cues. After the maximum day length in summer, the testes regressed and remained in that condition. If the photoperiod was altered so that the animal's minimal day length came 2 months early and then increased, the testes descended 2 months early.

Muul subjected one group of squirrels to a photoperiod 6 months out of phase with the natural world. Squirrels were exposed in July to a photoperiod characteristic of January. Their testes descended in July and regressed in January, a time when the testes of males exposed to natural photoperiod were descending. Likewise females produced litters out of phase in January, 6 months later than births observed in nature. Thus in the flying squirrel the testes of the male descend and ovulation in females takes place when day length increases from 11 to 15 hours, and ovulation ceases and testes regress when the photoperiod decreases.

The food-storing behavior of the flying squirrel in fall also appears to be photoperiodically controlled (Muul, 1965). Squirrels held in the laboratory under seasonal photoperiods and controlled constant temperature exhibited an intensity of food storing similar to that of animals held under natural conditions. Squirrels exposed to seasonal temperatures and a controlled photoperiod of 15 hours of daylight showed no intense food-storage activity through the winter (Figure 4-25). When the light was reduced to 12 hours in March, there was a marked rise in food storage. Another group was held under constant temperature and a controlled photoperiod of 15 hours of daylight, which was reduced to 13 hours in mid-December. Within a week this group in-

creased food-storage activity sharply and continued it from January through March. In still another experiment the squirrels were subjected in mid-October to a photoperiod typical of mid-November. The intensity of food storage increased more rapidly than normal and reached an equivalent of that of mid-November under natural conditions. Further decreases in the length of day increased the performance of squirrels. By the beginning of November, when the squirrels were subject to a photoperiod equivalent to that of late December, the storage peak was maximum. Immediately a long day of 15 hours of light and 9 hours of dark, equivalent to midsummer con-

*FIGURE 4-25*
*Food storage activity of flying squirrels held under controlled photoperiods compared to natural photoperiods. (After Muul, 1965.)*

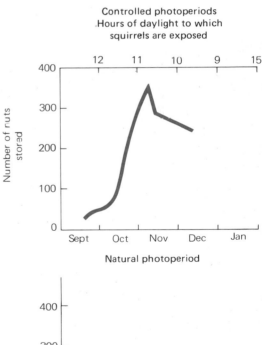

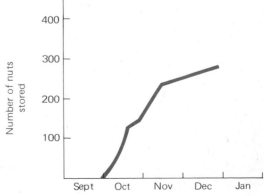

ditions, was imposed on the squirrels. Some squirrels showed a sudden decrease in storage intensity; others showed a gradual decrease. But among the squirrels held under natural conditions, storage of food was still increasing. These experiments demonstrate that the food-storage activity of the flying squirrel is photoperiodically controlled. Such a control synchronizes exploratory and storing behavior with a ripening of the mast crop (nuts and acorns) and prevents a premature harvest.

*Physiological Mechanisms: The Bünning Model.* The photoperiodic responses of plants and animals are not dependent upon the length of day or night as such. Instead, what seems to be involved is a circadian rhythm of sensitivity to light as the inducing or inhibiting agent. Current experimental evidence suggests that a time-measuring process starts at the beginning of the light period or the beginning of the dark. This induces, after a certain length of time from the beginning of light or dark, a sensitive stage that responds specifically to light.

When plants are held under short-day or long-night conditions, the short-day plants are stimulated to flower, and the long-day plants are inhibited. When day length is increased, flowering is inhibited in the short-day and stimulated in the long-day plants (Figure 4-26). If the dark period of the short-day and long-day plant is interrupted, each reacts as if it had been exposed to a long day. The long-day plant flowers; the short-day plant does not.

A similar response occurs in animals. The diapause of insects, a short-day phenomenon, is inhibited by light breaks in the dark period (Bünning and Jaerrens, 1960). The breeding period of the ferret can be initiated by exposing the animal to 12 hours of light each day for a month if 1 hour of light is given from midnight to one o'clock (Hart, 1951). This contrasts with 18 hours of light required daily if the animal is exposed to light in a continuous period. In fact, 6 hours of light are sufficient to bring the ferret into breeding condition if the cycle includes 4 hours of continuous light and 20 hours of darkness interrupted between the seventeenth and nineteenth hours (J. Hammond, 1953).

That such a circadian rhythm of light sensitivity exists in birds has been demonstrated

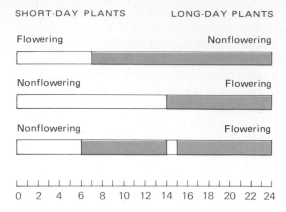

SHORT-DAY PLANTS     LONG-DAY PLANTS

Flowering     Nonflowering

Nonflowering     Flowering

Nonflowering     Flowering

0   2   4   6   8   10   12   14   16   18   20   22   24

*FIGURE 4-26*
*The influence of photoperiod on time of flowering in long-day and short-day plants. If exposure to light is experimentally controlled; short-day plants are stimulated to flower under short-day conditions, inhibited from flowering under long-day conditions, and respond to an interruption of a long dark period as though they had been exposed to long-day conditions. Long-day plants do not flower under short-day conditions, flower under long-day conditions, and flower under interrupted short-day conditions.*

in a series of experiments by Hamner (1963). House finches were held under six experimental light regimes, all involving 6 hours of daylight: short-day cycles of 24 hours (6L/18D), 48 hours (6L/42D), and 72 hours (6L/66D); and long-day cycles of 12 hours (6L/6D), 36 hours (6L/30D), and 60 hours (6L/54D). The finches held under the first three responded as though they had received a short-day treatment and exhibited no enlargement and maturation of the testes and no production of sperm. The birds held on the other three treatments responded as if under long-day conditions. The testes enlarged and spermatogenesis began. Since all the birds were subject to the same length of light but to varying lengths of darkness, the experiments showed that neither the length of the light period nor the length of the dark was critical. The results indicate that the house finch has an endogenous rhythm with a periodicity of about 24 hours. During the long dark period the birds endogenously reached light-sensitive states about 24 hours apart. When light is given at the proper phase of the rhythm, gonadal enlargement and maturation take place, but when light is given at the light-insensitive phase of the rhythm, no response occurred. Thus photoperiod-controlled reproductive cycles of birds are timed by a circadian rhythm of light sensitivity.

Further experimentation (Hamner, 1968) suggests that this response has a rhythm of two approximately 12-hour phases of different sensitivity to light. The basic rhythm can be phase shifted by and entrained to artificial lighting cycles. Thus as days shorten, the timing mechanism readjusts to increase the absolute duration of the second sensitive phase of the rhythm. As the days of spring begin to lengthen, additional light interacts with the readjusted sensitive phase of the photoperiodic clock and stimulates the growth of the testes.

How this rhythm of light sensitivity in plants and animals might function is suggested in a hypothesis advanced by Dr. Edwin Bünning of Germany to explain the short-day and long-day reactions of plants. According to the Bünning model (Figure 4-27), the circadian rhythm of light sensitivity goes through at least two half-cycles of more or less opposite sensitivity to light: a tension or photophil phase, in which development is favored by light, and a relaxation or scotophil phase, in which development is inhibited. Each phase of the rhythm is about 12 hours long, but to function, the light phase must precede the dark. Short-day plants are considered scotophil in the second half of the cycle; if exposed to light at that time, their flowering is inhibited. In the same half-cycle long-day plants are photophil; their flowering is stimulated. In addition, light sensitivity itself is rhythmic and seems to fluctuate in half-cycles within the light and dark periods. There is a time of maximum sensitivity during the dark (Figure 4-28), which in some organisms is about 14 to 16 hours from the beginning of the main light period. In the flying squirrel the sensitive periods are at the beginning of the subjective night and at subjective dawn (De-Coursey, 1960a, b). Light exerts its maximum effect in the pink bollworm either 10 hours after dark or 10 hours before dawn (Adkisson, 1964). Additional light offered a few hours after the beginning of the light period has a stimulatory effect on the flowering of short-day plants (Bünning, 1964). Among many or-

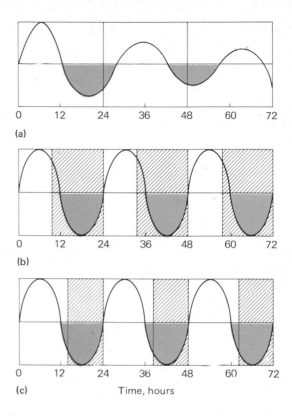

(a)

(b)

(c)

Time, hours

FIGURE 4-27
*The Bünning model. Oscillations of the clock cause an alternation of half-cycles with quantitatively different sensitivities to light (white vs. black). The free-running clock in continuous light or continuous darkness tends to drift out of phase with the 24-hour photoperiod. Short-day conditions allow the dark to fall into the white half-cycle; in the long day the light falls into the black half-cycle. (From Bünning, 1960.)*

ganisms the beginning of the previous light period may have a stronger influence on the timing of the point of maximum sensitivity than the beginning of the dark. At any rate both light and dark periods of a certain duration are required for a proper response. These rhythms in light sensitivity are endogenous. Light controls or sets the biological clock, and the clock in turn controls the light sensitivity of the organism.

Although the circadian rhythm is often considered as a single oscillation of approximately 24 hours, the fact is that the activity periods of most animals exhibit two peaks, occasionally more. The first, representing major activity, is followed by a second peak, which is much

**105**

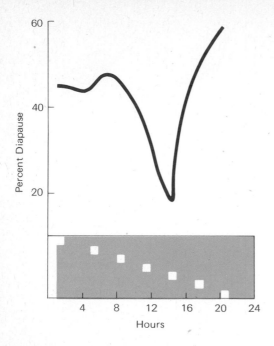

**FIGURE 4-28**
*Effect of light on diapause in the cabbage butterfly* (Pieris brassicae). *An additional 1 hour of light offered in the first half of the 24-hour cycle promotes diapause, but if offered in the second half of the cycle it inhibits diapause. Diapause is a short-day response. (Adapted from Bünning and Joerrens, 1960.)*

smaller, more variable in its position in time, and limited to about a half hour. The two peaks are separated by a trough of relative inactivity. In many animals this period is usually associated with some environmental condition adverse to the organism, such as high temperature or low humidity. Because the period of inactivity in nocturnal animals comes after midnight and in diurnal animals after noon, and the first peak of activity is associated with the dim light of dawn and dusk, some consider these peaks to be directly caused by environmental stimulus.

If this were so, the peaks should disappear under constant conditions of temperature and humidity and under artificial cycles of light and dark, with a sharp change between light and dark. Aschoff (1966) subjected three different species of finch to light cycles interposed with artificial twilight and then light cycles with a sharp change from light to dark. Under

both regimes the activity pattern remained the same (Figure 4-29). The birds were further subjected to light and dark cycles without twilight and then to constant illumination. Even under constant light the bimodal pattern of activity remained, indicating that the rhythm has two peaks that are endogenous, that are a persistent property of the circadian systems, and that do not depend on any concurrent change in environmental stimulus. Regardless of the length of the activity time, the period between the two peaks remains proportionately the same.

In all cases the responses of the organism are mediated by hormones and enzymes. Exactly how the hormonal and enzymatic systems function in periodism is the subject of much current research. (A discussion of this is beyond the scope of the book, but for an elementary and enlightening discussion see Hastings, 1970; Palmer 1974, 1976.)

**TIDAL AND LUNAR CYCLES**

For some organisms environmental time setters associated with tidal and lunar rhythms are of greater ecological importance than light-dark cycles. Animals that inhabit the intertidal zones of the sea show rhythms in their behavior that coincide with cycles of high and low tides. These endogenous timing processes also show persistent internal rhythms that are comparable to circadian rhythms of animals from nonintertidal environments. Among animals showing activity rhythms entrained to tidal cycles are the European shore crab *(Carcinus maenas)* and the fiddler crabs *(Uca minax* and *U. crenulata)* (for reviews see Enright, 1975; Palmer, 1974).

Reproduction in some marine organisms is restricted to a period that bears some relationship to tides. These rhythmic phenomena occur every lunar cycle of 28 days or in some instances every semilunar cycle of 14 to 15 days. Among some species as the grunion *(Leuresthes tenuis),* a small California fish that swarms in from the sea to lay eggs on sandy beaches, and the intertidal midge *(Clunio marinus)* the periodicities are so exact the activities of these animals can be predicted ahead of time. Laboratory studies confirm the entrainment of activity cycles to moonlight (see Enright, 1975).

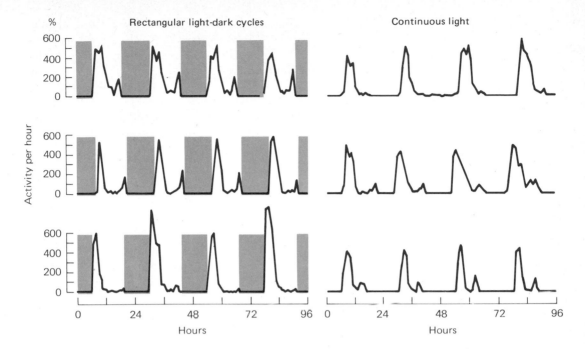

**FIGURE 4-29**
*Activity of three greenfinches* (Chloris chloris) *kept first in artificial light-dark cycles, thereafter in continuous illumination. Ordinate: Deviation of hourly activity from mean activity per 24 hours, the average being expressed as 100 percent. (From J. Aschoff, 1966.)*

## Seasonality

Seasonal periodicities, "the occurrence of certain obvious biotic and abiotic events or groups of events within a definite limited period or periods of the astronomic year" (Leith, 1974), are influenced by more than just light and photoperiod. The unfolding of leaves in spring and the dropping of leaves in fall, the blooming of flowers and the ripening of seeds, the migration of birds, and other biological events reoccurring with the passage of seasons are influenced by an interaction of light, temperature, and moisture. The study of the timing of recurring biological events, the causes of their timing with regard to biotic and abiotic forces, and the interrelations among phases of the same or different species is called *phenology* (Leith, 1974).

Seasonality in temperate regions is due largely to changes in light and temperature. Seasonality in tropical regions is keyed to rainfall. In a very broad way seasonal changes in temperature and light regimes result in alternate warm and cold periods. However, the progression is gradual and in temperate zones seasons can be identified as early or late spring, early or late fall, and so on. In the tropics the seasons are alternately wet and dry and their onset is abrupt. The beginning of the rainy season is a dependable environmental cue by which plants and animals of the tropics can become synchronized to seasonal changes. The dry season in the wetter tropics is a period with less than 100 cm rainfall per month. In extreme conditions, of course, no rain falls. The wet season is the period with more than 100 cm rainfall per month. The onset of the rainy season, which may last up to six months, varies with the movement of the intertropical convergence (see Figure 4-4). As a result the wet season and the dry season are highly predictable.

### SEASONALITY IN PLANTS

Phenological responses reflect altitudinal and latitudinal changes in light, temperature, and moisture. The advance of spring in the temperate regions is marked by progressively later timing of flowering of the same species of trees and herbs across a region. Although these progressive changes are the most pronounced

**107**

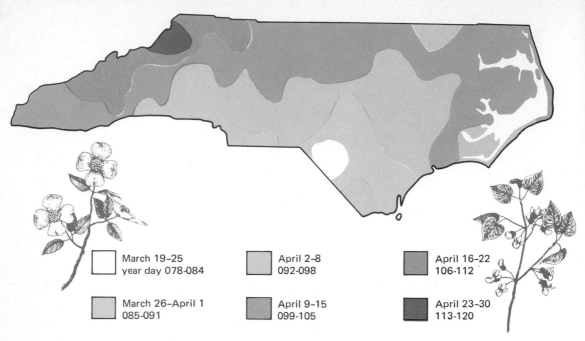

**FIGURE 4-30**
*Arrival of spring, 1970, across North Carolina as indicated by the opening of the flowers of dogwood* (Cornus florida) *and redbud* (Cercis canadenis). *(From Leith, 1974.)*

across broad geographical areas (Figure 4-30), distinct variations exist within a given region. These variations reflect local microclimates, which act as selective pressures on local populations resulting in ecotypic variation in environmental response. In tropical regions wet and dry seasons show a similar advance. For example, seedlings of northern origin of a tree species stop growth sooner than seedlings of southern origin of the same species, and seedlings from the highest altitudes stop growing before those from lower altitudes (Flint, 1974).

Although the temperate seasonal cycle is determined by an interaction of factors, some seasonal responses can be related to specific abiotic changes. Some responses result from photoperiod, others from temperature, and still others from both. The response to temperature by plants is related to temperature sums—the summation of mean daily temperatures. In general the breaking of seed dormancy depends upon chilling by an accumulation of low temperatures. Induction of dormancy and seedling growth are photo-

periodic responses as is the cessation of shoot growth in the fall. Depending upon the species seed germination requires either chilling or long days accompanied by high temperatures.

The full range of responses to seasonal abiotic changes by plants of the temperate region results in the seasonal aspects of the ecosystem. In the forests of eastern North America spring brings a flush of flowering and vegetative growth (Figure 4-31). The first major peak is composed largely of understory species which bloom before the canopy closes. A second peak comes in mid to late summer when plants of thickets, roadsides, woodlands openings, and forests where sufficient light penetrates come into bloom. Selection favors those species that bloom early before light reaching the forest floor diminishes or those that are able to compete for space in the openings. Some species, such as the late goldenrods and woodland asters, bloom when the canopy thins in the fall.

The timing of flowering and fruiting for the spring or fall has certain ecological and evolutionary advantages. Plants can flower without experiencing heavy vegetative demand for energy resources. Flowering and fruiting in spring is over or nearly so before much vegetative growth takes place and the plants can draw on energy and nutrient re-

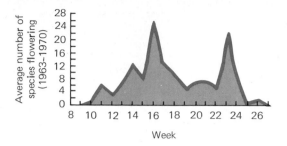

**FIGURE 4-31**
*Phenogram of the oak-hickory forest association in eastern Tennessee depicting flowering seasons as distinguished by spring and early summer flora. Data include flowering of 36 trees and shrubs and 97 herbaceous species and represent the mean date of the first flower between 1963 and 1970. (After Taylor, 1974.)*

**FIGURE 4-32**
*Phenological spectrum and seasonality in productivity of the mayapple (Podophyllum peltatum), a common spring flower in the deciduous forest of eastern North America. Productivity is maximum during the reproductive phase. Reduced productivity through fruit consumption by small mammals and insects coincides with the period of ripe fruits and seeds and seed dispersal. (After Taylor, 1974.)*

serves in the roots. Seeds get to the ground early allowing seedlings to develop a strong root system before winter. Flowering and fruiting in the fall takes place after vegetative growth is over and energy can be diverted to reproduction. Seeds produced usually require chilling before they can germinate.

Seasonal flux in plant activity influences biomass accumulation and energy fixation. Production is maximum during the reproductive phases from leaf budding to flowering. During these phases daytime temperatures appear to be the most important influence. Production decreases rapidly during the phases after flowering to fruit setting. Night temperatures are most important when plants are transferring energy from storage organs such as roots in the spring or to storage organs in the fall. In dry areas precipitation is more important than temperature in influencing these phases of activity.

Biomass production as influenced by seasonality varies among species, although studies on such seasonality are few. An example of phenological development for the deciduous forest is the mayapple *(Podophyllum peltatum)*. The mayapple grows in colonies in eastern North American woodland and develops rapidly in spring before the forest canopy closes (Figure 4-32). Shoot develop-

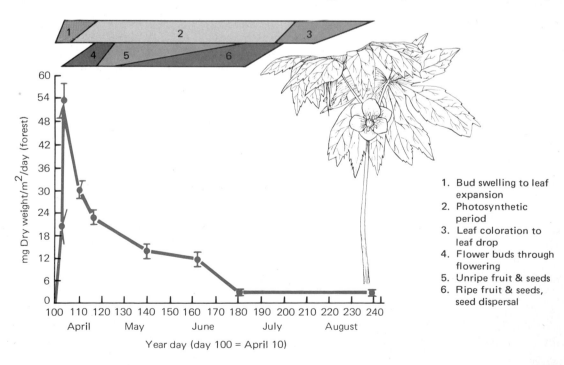

1. Bud swelling to leaf expansion
2. Photosynthetic period
3. Leaf coloration to leaf drop
4. Flower buds through flowering
5. Unripe fruit & seeds
6. Ripe fruit & seeds, seed dispersal

**109**

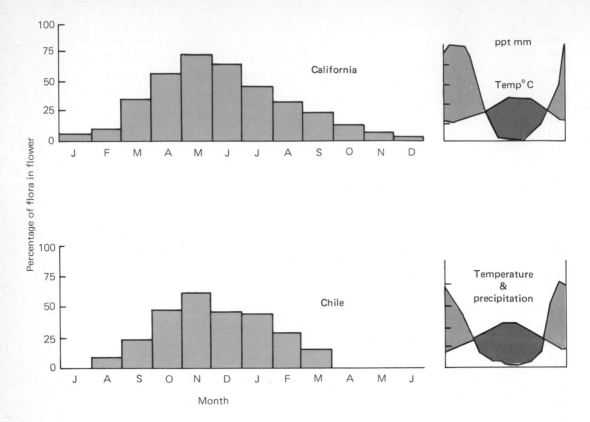

**FIGURE 4-33**
*Seasonal flowering activity of plants growing in the Mediterranean climatic regions of California and Chile. Climatic data are for the nearest station to the phenological observations. The California data are for the flora of San Dimas Experimental Forest, southern California. Chile phenological data are for shrubs only in the Santiago region. (After Mooney and Parsons, 1974.)*

ment and flowering are spring events; rhizome growth is a summer activity. Biomass of leaves, stems, rhizomes, roots, flowers, and fruit increases during the period when light is most available. It reaches maximum in early summer during the phase of unripe seeds and fruit (Taylor, 1974), just prior to the onset of senescence. In a one-year study the rhizome-root compartment increased 21 percent in dry weight above the standing crop estimate at leaf emergence. Leaf and above-ground stems accounted for 57 percent of the biomass; unripe fruits and seeds, 17 percent; and underground shoots, 4 percent. At the onset of senescence leaves change color and then the leaves and

fruit drop. At the end of the growth period fruits and seeds are consumed and distributed by small mammals and insects.

In mediterranean-type climates, characteristic of southern California, the Mediterranean region, parts of Chile, and Australia (see Chapter 9), with dry summers and wet winters, plants are most active metabolically in late spring when temperatures become warm but the stresses of summer drought have not yet developed (Figure 4-33). Summer sees a sharp drop in flowering and plant activity and in the presence of pollinating insects which are active when the food base is available (Mooney and Parsons, 1973).

Energy apportionment and biomass accumulation in plants in mediterranean climate ecosystems are exemplified by the evergreen shrub *Heteromeles arbutifolia* (Figure 4-34). Although the plant can fix carbon throughout the year, peak activity takes place during the winter and spring when the amount of moisture available is not limiting (Mooney, Parsons, and Kummerow, 1974). This energy fixed early is used for the growth of new leaves during late spring and early summer when temperatures are more favorable. Stem growth ceases when energy demands for reproductive functions are high.

In tropical regions flowering and fruiting and leafy growth of plants reflect the alternation of wet and dry seasons. The coming of the rainy season is marked by a flush of vegetative growth just as the warming spring temperatures trigger leafy growth in the temperate regions. Over much of the seasonal tropics, flowering and fruiting coincide with the dry season (Jansen, 1967). Some species flower at the end of the rainy season when soil moisture is still abundant. Other species flower at the end of the dry season. Although there may be definite peaks, flowering and fruiting continue through the dry season (Figure 4-35). In a Costa Rican tropical forest, for example, some 60 trees that flower during the dry period spread their flowering over 3.5 months. This results in a minimal overlap in flowering and reduces competition for pollinators. Because the trees are self-incompatible, they require the assistance of animals for the transfer of pollen. Thus competition for pollinators acts as a selective force leading to a sequence of blooming periods and unsynchronized flowering. The sequence of flowering provides a flow of floral food to a large number of pollinating animals (Frankie et al., 1974).

There are other ecological and evolutionary advantages to plant reproduction in the dry season. It reduces or eliminates conflict in the plant between energy demands for leaf growth and energy demands for reproduction. Plants are able to draw on nutrient and energy reserves stored in the roots during the rainy season. Flowers blooming in leafless trees are visible to pollinating insects, birds, and mammals, particularly bats, and fruits are conspicuous and accessible to fruit-eating animals. that aid in the dispersal of seeds.

Not all plants, of course, flower and fruit during the dry season. In the Costa Rican forest shrubs bloom shortly before and just after the maximum flowering of trees. Vines and herbaceous plants flower during the wet season. Plants with large-seeded fruits, whose seeds may be damaged by fruit-eating animals exhibit synchronized fruiting, the selective advantage being the escape of some seeds from

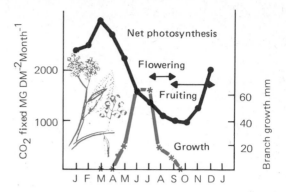

**FIGURE 4-34**
*Seasonality in photosynthesis, flowering, fruiting, and growth in the California sclerophyll shrub Christmas berry* (Heteromeles arbutifolia). *(After Mooney and Parsons, 1974.)*

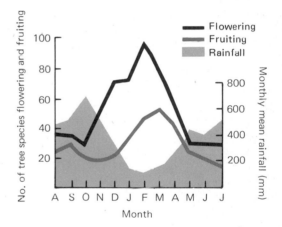

**FIGURE 4-35**
*Synchronization of flowering and fruiting in rain forest tree species in Golfito, Costa Rica, with mean monthly rainfall. Note that flowering and fruiting reach their highest during the dry season, the months of January, February, and March. (Adapted from D. H. Jansen, 1967.)*

excessive damage. Plants with smaller seeded fruits ripen throughout the year. By doing so they are able to utilize animal agents of dispersal with minimum competition for their services. In the Arima Valley of Trinidad grow around 20 common species of melastomes in the genus *Miconia*. Among them they disperse their fruiting season throughout the year. The animal agents have a year-long supply of food, and the plants are assured of seed dispersal (Snow, 1966, 1976).

## SEASONALITY IN CONSUMERS

Seasonal activity of consumers tends to correspond to seasonal variations in food supply. Appearance and disappearance of many insects is usually associated with the termination or initiation of diapause. In temperate regions this is usually controlled by photoperiod. Insects appear during periods of abundant food supply and go into diapause during unfavorable periods. Growth after diapause is influenced by temperature and moisture.

In a similar manner photoperiod influences the seasonality of birds in temperate regions (Figure 4-36). As with plants energy is apportioned according to the pattern of seasonal demands. Seasonal phases such as migration, reproduction, and molt usually do not overlap. In winter energy demands are channeled into survival and thermoregulation. Within this pattern some seasonal environmental mechanisms may modify bird activity. Migration, triggered by physiological processes, is highly correlated with weather patterns. Spring migration occurs with the onset of a warm front with an air flow to the north or northeast. The birds move on the wind and have a good chance of arriving in favorable weather at the end of the flight. When a cold front arrives, migratory movements stop, not

to resume again until high-pressure areas have passed. In fact radar studies show that a cold front in spring may immediately start a reverse southward movement (Drury et al., 1961; Eastwood, 1971). Fall migrations to the south appear to be timed to start after the passage of a cold front with its flow of continental polar air from a northerly direction.

Although the timing of nesting period in birds is controlled by photoperiod, the initiation of nesting across a geographic range is influenced by climatic gradients as well as the seasonal development, availability of nesting material, and food for young. In a study of the phenology of nesting in the robin, James and Shugart (1974) used a step-down multiple regression analysis involving seven climatic variables. They found that the combination of April wet-bulb and dry-bulb temperatures was the best predictor of the beginning of the nesting period: "If the mean noon relative humidity is near 50 percent in April, the beginning of the nesting period will be in late April or early May, regardless of the dry-bulb temperature; if the mean noon relative humidity is either higher or lower than 50 percent in April, the beginning of the nesting period will be later."

Among mammals periodicity influences reproductive behavior. The reproductive cycle of the white-tailed deer, for example, is initiated in the fall and the young are born in spring when the highest quality of food for lactating mother and young is available (Figure 4-37). The phenology of reproduction varies with altitude and latitude, especially among small mammals. Northern mammals have shorter breeding seasons and the young are born in the summer months (Figure 4-38). Further southward, the breeding season lengthens until the animals breed throughout the year.

A close interrelationship exists between flowering phenology and reproductive activities in the tropics. The breeding season and seasonal changes in bird populations are related to the availability of food (see Karr,

**FIGURE 4-36**
*Generalized diagram of the events in the annual life cycle of a migratory song bird. Note the strong seasonal periodicity of the various events in the cycle. (From C. M. Weise, 1974.)*

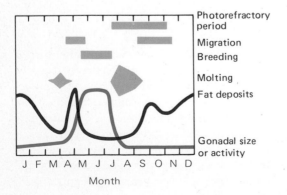

**FIGURE 4-37**
*The seasonal cycle of the white-tailed deer. The annual cycle is attuned to the decreasing day length of fall during which the breeding season begins.*

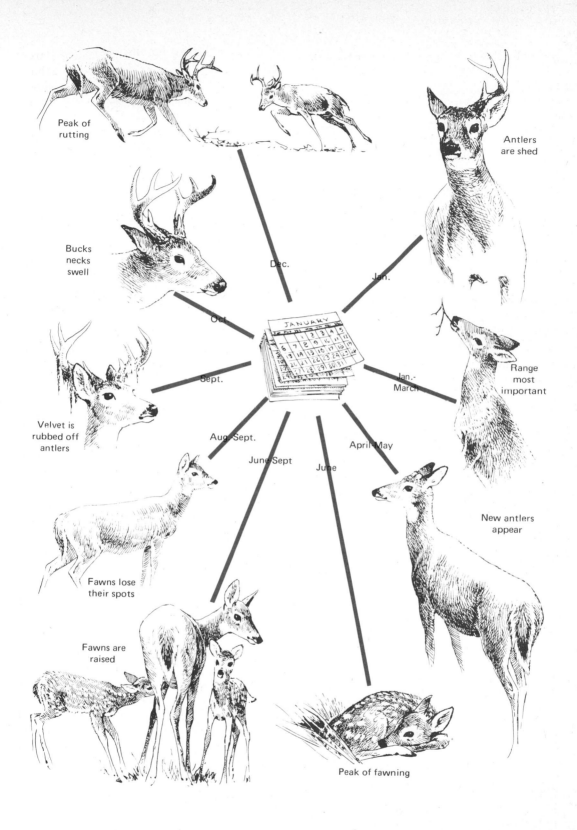

Peak of rutting

Antlers are shed

Bucks necks swell

Dec.

Jan.

Oct.

Range most important

Sept.

Jan.-March

Velvet is rubbed off antlers

Aug.-Sept.

June-Sept

June

April-May

New antlers appear

Fawns lose their spots

Fawns are raised

Peak of fawning

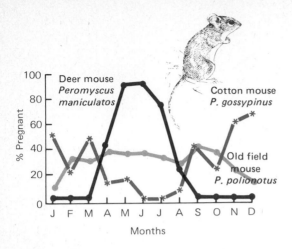

## Seasonality in decomposers

Decomposer activity is influenced by temperature, moisture, carbon:nitrogen ratio of leaves, and litter fall. In tropical rain forests and temperate forests litter fall continues through the year, but in seasonal tropical forests and temperate forests litter fall is seasonal. In temperate forests this seasonal litter fall is mostly in the fall with some additions from seasonal herbaceous growth. In evergreen forests, both broadleaf and needle-leaf, litter fall is less seasonal. With a heavy deposition of litter in the fall in temperate forests, microbial growth and the amount of carbon immobilized in bacterial and fungal tissue increases rapidly then declines when cold weather comes (Figure 4-39). At that time $CO_2$

**FIGURE 4-38**
*Seasonality in reproduction in woodland mice. The graph shows seasonal changes in the percentage of pregnant females in the deer mouse* (Peromyscus maniculatus) *from a northern locality characterized by extreme temperature fluctuations, the cotton mouse* (P. gossypinus) *from a southern locality with a mild climate, and the old field mouse* (P. polionotus) *from an intermediate location and climate. Note the strong seasonality in reproduction in the deer mouse of the northern climate and in the cotton mouse of the hot climate. (From M. H. Smith, 1974.)*

**FIGURE 4-39**
*Seasonal microbial activity in the eastern deciduous forest. The graph show microbial immobilization of carbon in the soil, in $O_1$ litter, and $O_2$ litter calculated monthly through the year. Carbon pools of $O_1$ and $O_2$ litter are indicated monthly. Note the sharp rise in carbon pool of $O_1$ litter in the fall. Periods of maximum litterfall and root sloughing are indicated by vertical arrows and maximum and minimum $CO_2$ efflux by horizontal arrows. (After Burgess and O'Neill, 1976.)*

1976). In some tropical regions nesting is greatest during the rainy season, while in areas with pronounced wet and dry seasons breeding is more evenly distributed through the year. In some tropical regions of South America influx of northern migrants during the temperate winter depresses the breeding of local bird populations and shifts most nesting to the dry season. Reproduction reaches a maximum at the beginning and at the end of the dry season.

Insects and other arthropods reach their greatest biomass early in the rainy season in the Costa Rican forests (Buskirk and Buskirk, 1976) and spiders are most abundant in the early rainy season, decline shortly before the end of the rains, and continue to decline steadily until the end of the dry season (Robinson and Robinson, 1970). The tropical dry season initiates other seasonal activities as estivation, diapause, and local and regional migrations to moister areas, exemplified by the large scale migrations of African ungulates.

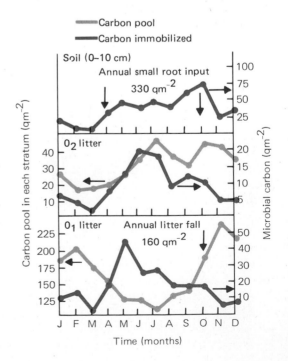

evolution from the litter declines and fungal propagules and accumulation of mycelium reach a maximum. As the temperature rises in spring, growth of bacteria and fungi resumes, and soil animals increase their activity with a corresponding rise in $CO_2$ evolution and microbial immobilization of carbon in soil and litter. While total availability of carbon is not limiting to decomposers during any period of the year, the biochemical form of carbon limits the rate of microbial and faunal catabolism. Similarly microbial activity results in immobilization of nitrogen. In fact nearly all of the nitrogen in the $O_1$ litter layer of the eastern deciduous forest in late spring may be in microbial cells (Ausmus et al., 1975).

## SUMMARY

Solar radiation is not only a source of light energy; it is also the major source of thermal energy. Thus solar radiation is the major determinant of climate. Variations in the heat budgets and the earth's daily rotation produce the prevailing winds, move ocean currents, and influence rainfall patterns over the earth. The climate of any given region is a combination of patterns of temperature and moisture which are influenced by latitude and location of the region in a continental land mass. In mountains climatic changes on an altitudinal gradient are reflected in zonation of vegetation. At the same time mountain ranges influence regional climate by intercepting air flow, causing its moisture to drop out on the windward side and creating a rain shadow on the leeward side.

Daily heating and cooling of air masses cause their rising and sinking above the earth. Under certain conditions the temperature of air masses increases with height rather than decreases. Such an air mass is very stable, creating an inversion which can trap atmospheric pollutants and hold them close to the ground. Inversions break up when air close to the ground heats up, causing it to circulate through the inversion, or when a new air mass moves into the area.

Although regional climates determine conditions over an area, the actual climatic conditions under which organisms live vary considerably from one area within the region to another. These variations, or microclimates, are influenced by topographic differences, height above the ground, exposure, and other factors. Most pronounced are environmental differences between ground level and upper strata and between north-facing and south-facing slopes. Other microclimates exist over urban areas. A city is characterized by the presence of a heat island. Compared to surrounding rural areas, a city has a higher average temperature, particularly at night, more cloudy days, more fog, more precipitation, a lower rate of evaporation, and lower humidity.

Another important abiotic influence closely related to climatic conditions is fire. Lightning-set fires have been part of the natural environment since the emergence of terrestrial plant life. Many plant communities, such as grasslands and certain forest types, evolved under a regime of fire. With the coming of humans fire became an even more important influence because it became more frequent, occurred intermittently throughout the year rather than only during the natural fire season of summer, and was often deliberately set to modify or change the environment. Because of the destructiveness of fires caused by humans, we moved to another extreme, the exclusion of fires, which also has adverse ecological effects. Properly handled fire is an important tool in the regulation and manipulation of vegetation.

Changes in light, temperature, and precipitation over the year bring about seasonal periodicities in recurring biological events, the study of which is phenology. In temperate regions seasonality is influenced by changes in light and temperature and in tropical regions by rainfall. Such periodicities are reflected in the occurrence of leaf growth, flowering, and fruiting of plants and in the breeding cycles, feeding activities, and migration of animals.

Daily periodicities of plants and animals are under the influence of day-night cycles. The timing of daily activities is controlled by an internal physiological biological clock, whose basic structure is probably chemical and is involved in the makeup of the cell. It is free-running under constant conditions with an oscillation or fluctuation that has its own inherent frequency. For most organisms the inherent clock deviates more or less from 24 hours. Under natural conditions this clock is set or entrained to 24 hours by external time cues which synchronize the activity of plants and animals with the environment. Because the most dependable external time setter is light (day and night), most of the selected species studied so far are entrained to a 24-hour period. The onset and cessation of activity are usually synchronized with dusk and dawn, the response depending upon whether the organisms are diurnal (light-active) or nocturnal (dark-active). The biological clock is useful not only to synchronize the daily activities of plants and animals with night and day consistent with the ecology of the species, but also to time the activities with the seasons of the year. The possession of a self-sustained rhythm with approximately the same frequency as that of the environmental rhythms enables organisms to "predict" such advance situations as the coming of spring. It brings plants and animals into a reproductive state at a time of year when the probability for survival of offspring is highest; it synchronizes within a population such activites as mating and migration, dormancy and flowering. The acquisition and refinement of a physiological timekeeper, geared to cues that provide organisms with distinct and species-specific or population-specific synchronization with the environment, are results of natural selection. The secrets of the clock, an understanding of how it works and where it resides in the organism, have yet to be discovered.

Light also influences the structure and function of ecosystems. In general plants can be classified as sun (shade-intolerant) or shade (shade-tolerant) plants. Each group is characterized by adaptations to certain light regimes. Shade-adapted plants have low photosynthetic, respiratory, metabolic, and growth rates and are resistant to fungal infections; sun tolerant plants have a high rate of respiration, are adapted to high light intensity, rarely reach light saturation, and are highly susceptible to fungal infections under shady conditions.

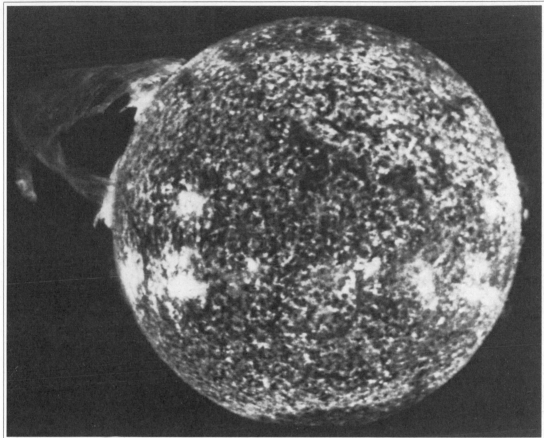

# Energy flow in ecosystems
## CHAPTER 5

The sunlight that floods the earth is a source of two forms of energy that keep the planet functioning: *heat energy,* which warms the earth, heats the atmosphere, drives the water cycle, and provides currents of air and water; and *photochemical energy,* which is used by plants in photosynthesis, fixed in carbohydrates and other compounds, and becomes fuel for cool-burning cellular furnaces of living organisms.

## The nature of energy and the laws of thermodynamics

The solar energy incident upon the outer atmosphere of the earth that eventually penetrates the atmosphere and reaches the earth's surface has wavelengths of of 0.1 to 10.0 $\mu$ (Figure 5-1). Of this approximately 4 percent is ultraviolet (wavelengths of 0.1 to 0.4 $\mu$), about 44 percent is visible light (wavelengths of 0.4 to 0.7 $\mu$), and the remaining 52 percent is infrared or long-wave radiation (wavelengths of 0.7 to 10.0 + $\mu$). What we see as light is visible radiation, broken down into a spectrum that ranges from violet to red.

Energy is measured in several units. For ecologists the gram calorie (g cal), the amount of heat necessary to raise 1 gram of water 1° C at 15° C, is the most convenient unit of energy. When large quantities of energy are involved the kilogram calorie (kcal) is more appropriate. A kcal is the amount of heat required to raise 1 kilogram of water 1° C at 15° C. Because ecologists are concerned with energy flow for a given area they measure energy per

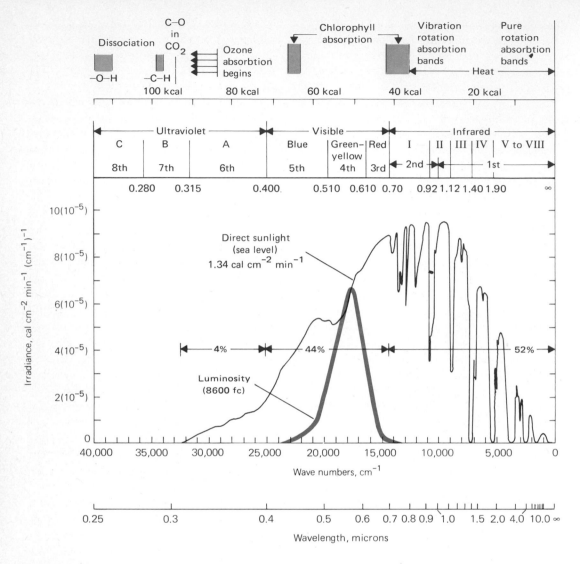

**FIGURE 5-1**
*Spectral distribution of solar radiation showing segmentation of the spectrum involved in plant processes. Shown at the top are the basic photochemical and radiation processes that may occur with the absorption of radiation by plants and other objects. The quantum content, expressed on a scale as kcal, is indicated for each frequency of incident radiation. (From Gates, 1965.)*

square centimeter or square meter. Often used units of measure are the *langley*, 1 g cal/cm²; the *joule*, 0.24 g cal; the *watt*, 1 joule/sec or 14.3 g cal/m²; the *British thermal unit* (Btu), amount of heat necessary to raise 1 pound of water 1° F or 252 g cal.

The transfer of radiation involves the movement of units of energy from one point to another. This flow of energy is known as a *flux*. The condition of flux requires that there be an energy source, the point from which energy flows, and an energy sink, the point to which it flows or a receiver. Without a sink for the flow of thermal energy, the sun could not be an energy source. The earth receives energy from the sun, absorbs a part of it, and gives up energy as heat to a sink, outer space.

The flow of energy is mediated at the molecular level. Characteristically thermal energy is distributed rapidly among all molecules in a system without necessarily causing any chemical reaction. The effect of thermal energy is to set the molecules into a state of random motion and vibration. The hotter the object, the more its molecules are moving, vibrating, and rotating. These motions tend to spread from a hot body to a cooler one, transferring energy from one to the other.

The energy of light waves, on the other hand, especially the red and blue wavelengths, causes electronic transitions within atoms and molecules, called *excitations*. These excitations can lead to photochemical reactions. The energy of light sends one electron of a pair of bonding electrons to a higher state, or orbit. Uncoupled from its partner, it is free to be involved in photochemical reactions.

There are two kinds of energy, *potential* and *kinetic*. Potential energy is energy at rest. It is capable of and available for work (defined as a force that causes a particle or other body to be moved or displaced). Kinetic energy is due to motion and results in work. Work that results from the expenditure of energy can both store energy (as potential energy) and arrange or order matter without storing energy.

The expenditure and storage of energy is described by two laws of thermodynamics. The first law states that energy is neither created nor destroyed. It may change forms, pass from one place to another, or act upon matter in various ways, but regardless of what transfers and transformations take place, no gain or loss in total energy occurs. Energy is simply transferred from one form or place to another. When wood is burned the potential present in the molecules of wood equals the kinetic energy released, and heat is evolved to the surroundings. This is an *exothermic* reaction. On the other hand, energy from the surroundings may be paid into a reaction. Here, too, the first law holds true. In photosynthesis, for example, the molecules of the products store more energy than the reactants. The extra energy is acquired from the sunlight, but again, there is no gain or loss in total energy. When energy from outside surroundings is put into a system to raise it to a higher energy state, the reaction is *endothermic*.

Although the total amount of energy involved in any chemical reaction, such as burning wood, does not increase or decrease, much of the potential energy stored in the substance undergoing reaction is degraded during the reaction into a form incapable of doing any further work. This energy ends up as heat, serving to disorganize or randomly disperse the molecules involved, thus making them useless for further work. The measure of this relative disorder is named *entropy*.

The second law of thermodynamics makes an important generalization about energy transfer. It states that when energy is transferred or transformed, part of the energy assumes a form that cannot be passed on any further. When coal is burned in a boiler to produce steam, some of the energy creates steam that performs work, but part of the energy is dispersed as heat to the surrounding air. The same thing happens to energy in the ecosystem. As energy is transferred from one organism to another in the form of food, a large part of that energy is degraded as heat and as a net increase in the disorder of energy. The remainder is stored as living tissue. But biological systems seemingly do not conform to the second law, for the tendency of life is to produce order out of disorder, to decrease rather than increase entropy.

The second law, theoretically, applies to the isolated, closed system in which there is no exchange of energy or matter between the system and its surroundings. An isolated system approaches thermodynamic equilibrium, that is, a point at which all the energy has assumed a form that cannot do work. A closed system tends toward a state of minimum free energy (energy available to do work) and maximum entropy, whereas an open system maintains a state of higher free energy and lower entropy. In other words, the closed system tends to run down; the open one does not. As long as there is a constant input of matter and free energy to the system and a constant outflow of entropy (in the form of heat and waste), the system maintains a steady state. Thus life is an open system maintained in a steady state.

Energy enters the biosphere as visible light that is stored in energetic covalent bonds during photosynthesis (see Chapter 2). From that point biochemical changes involve a series of rearrangements of matter into compounds of less chemical potential energy. These chemical rearrangements are accompanied by the production of heat that eventually goes into the energy sink. This loss of heat is accompanied by a loss of carbon dioxide, water, and nitrogenous compounds, which are recycled through the biosphere. Although some energy is irrevocably lost from the biosphere, some of it is stored in the system. The more organized the system, the longer energy is stored.

## Energy storage

### PRIMARY PRODUCTION

The flow of energy through the community starts with the fixation of sunlight by plants, a process that in itself demands the expenditure of energy. Plants rely on the food stored in the seed for energy until their own production machinery is working. Once mobilized, the green plants begin to accumulate energy. Energy accumulated by plants is called *production,* or more specifically, *primary production,* because it is the first and basic form of energy storage in an ecosystem. The rate at which energy is stored by photosynthetic activity is known as *primary productivity.* All of the sun's energy that is assimilated, i.e., total photosynthesis, is *gross primary production.* Like other organisms, plants require energy for reproduction and maintenance. The energy required for these needs is provided by a reverse of the photosynthetic process, *respiration.* The energy remaining after respiration and stored as organic matter is *net primary production,* or plant growth.

Production is usually expressed in kilocalories per square meter per year (kcal/m²/yr). However, production may also be expressed as dry organic matter in grams per square meter per year (g/m²/yr). If either of these two measures is employed to estimate efficiencies and other ratios, the same unit must be used for both the numerator and denominator of the ratio. Only calories can be compared with calories, dry weight with dry weight.

**Biomass.** Net primary production accumulates over time as plant biomass. Part of this accumulation is turned over seasonally through decomposition. Part is retained over a longer period as living material. The accumulated living organic matter found on a given area at a given time is the *standing crop biomass.* Biomass is usually expressed as grams dry weight of organic matter per unit area (for example, grams per square meter, kilograms per hectare, or calories per square meter). Thus biomass differs from production, which is the rate at which organic matter is created by photosynthesis. Biomass present at any given time is not the same as production. Nor does a high biomass necessarily imply high production.

Plants allocate net production to leaves, twigs, stem, bark, roots, flowers, and seed. How much is allocated to each compartment is difficult to determine because of the tedious chore of cutting sample trees, weighing the various components, and determining both the energy content and nutrient content. Such estimates have been made for several forest stands (for a summary of some eastern United States forests, see Whittaker et al., 1974). For example, in a young oak-pine forest on Long Island, New York, 25 percent of the net primary production was allocated to stem wood and bark, 40 percent to roots, 33 percent to twigs and leaves, and 2 percent to flowers and seeds. Among the shrubs, 54 percent of the net primary production was allocated to roots, 21 percent to stems, and 23 percent to leaves (Whittaker and Woodwell, 1968).

The proportionate allocation of net production to below-ground and above-ground biomass, to root and shoot, tells much about different ecosystems and about different components within the ecosystem. A high root-to-shoot ratio (R/S) indicates that most of the net production goes into the supportive functions of plants. Plants with a large root biomass are more effective competitors for water and nutrients and can survive more successfully in harsh environments, because most of their active biomass is below ground. Plants with a low R/S ratio have most of their biomass above ground and assimilate more light energy, resulting in higher productivity. Tundra sedge and grass meadows, characteristic of an environment with a long, cold winter and a short growing season, have R/S ratios ranging from 5 to 11. Tundra shrubs may range from 4 to 10. Further south, midwest prairie grasses have a R/S ratio of about 3, indicative of perhaps cold winters and limited moisture supply. In forest ecosystems, with their high above-ground biomass, the R/S ratio is very low. For the Hubbard Brook forest in New Hampshire the

R/S ratio (based on data of Gosz et al., 1976) for trees is 0.213, for shrubs 0.5, and for herbs 1.0. As one could predict, the R/S ratio increases through the vertical strata from canopy to forest floor.

Increasing competition can increase the R/S ratio. For example, consider the Borassus palm *(Borassus aethiopum)* which grows in the Ivory Coast savanna under three different situations: an open shrub savanna, a dense shrub savanna, and a tree savanna. The R/S ratio for trees and shrubs 2 to 8 m high, other than palms, in each of the three situations is 0.46, 0.42, and 0.53, respectively. For the palm the R/S ratio is 0.78, 4.80, and 0.78. In the presence of intense competition from other woody growth, the palm increases root biomass (data from Lamotte, 1975).

The above-ground biomass is distributed vertically in the ecosystem. The vertical distribution of leaf biomass or in aquatic systems the concentration of plankton influences the penetration of light, which in turn influences the distribution of production in the ecosystem. The region of maximum productivity in the aquatic ecosystem is not the upper sunlit surface (strong sunlight inhibits photosynthesis), but some depth below, depending upon the clarity of the water and the density of the plankton growth. As depth increases, light intensity decreases until it reaches a point at which the light received by the plankton is just sufficient to meet respiratory needs and production equals respiration (Figure 5-2). This is known as the *compensation level*. In the forest ecosystem a similar situation exists. The greatest amount of photosynthetic biomass as well as the highest net photosynthesis is not at the top, but at some point below maximum light intensity. In spite of wide differences in plant species and in types of ecosystems, the vertical profiles of biomass of the various ecosystems appear to be quite similar.

Within the vertical profile, biomass varies seasonally and even daily. In grasslands and old-field ecosystems much of the net production is turned over every year. The standing crop of living material in an old field in Michigan amounted to about $4 \times 10^3$ kg/ha (kilograms per hectare) in late summer, compared to 80 kg/ha in late spring. But at this time the standing crop in dead matter was nearly $3 \times$ $10^3$ kg/ha (Golley, 1960). The above-ground biomass of a tall grass prairie that included both living and dead material was approximately twice that of the standing crop, the living material added during the growing season (Kucera et al., 1967). The above-ground biomass has a turnover rate of approximately 2 years, the below-ground biomass of roots a turnover rate of 4 years. In a salt marsh the standing crop in autumn was $9 \times 10^3$ kg/ha; in winter it was just one-third of this. In a forest ecosystem a considerably greater proportion of the net production is tied up as wood. In an oak-pine forest, leaves, fruits, flowers, dead wood, and bark contributed 342 g/m²/yr to the organic horizon, and the roots 311 g/m²/yr, for a total of 653 g/m² or about 58 percent of the net primary production (Woodwell and Marpels, 1968).

*Net productivity.* Little of the energy assimilated by plants goes into organic production, as discussed in Chapter 2. Most of the light absorbed by plants is converted to heat and lost by convection and radiation. What fraction of light energy is used in photosynthesis goes into gross production and what is left over after respiration goes into net production. The production efficiency of plants is rather high. They fix energy continually during the daylight hours of the growing season and respiration requires a minimum amount of the assimilated energy. The ratio of net primary production to gross production (NPP/GPP) ranges between 40 and 80 percent. The most efficient are those plants which do not maintain a high supporting biomass, such as grass, large algae, and phytoplankton.

The productivity of ecosystems is strongly influenced by temperature and rainfall (Figure 5-3) and the productivity of various types of ecosystems varies widely over the globe (Figure 5-4). These variations in net production for a variety of ecosystem types are summarized in Table 5-1. The most productive terrestrial ecosystems are tropical forests with high rainfall and warm temperatures; their net productivity ranges between 1000 and 3500 g/m²/yr. Temperate forests, where rainfall and temperature are lower, range between 600 and 2500 g/m²/yr (Whittaker and Likens, 1975). Shrublands such as heath balds and tall-grass prairie have net productions of 700 to 1500 g/m²/yr

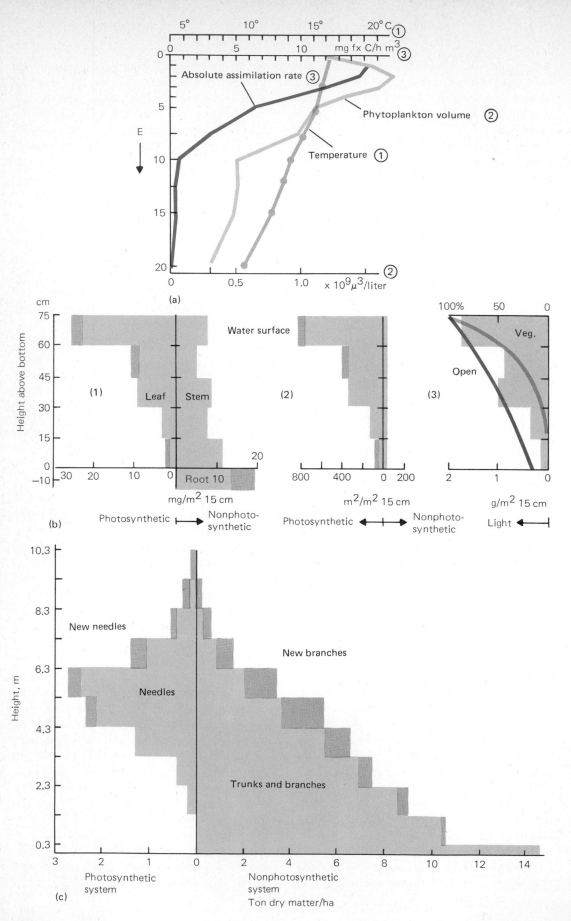

(a)

(b)

(c)

**FIGURE 5-2**

*Vertical distribution of production and biomass in aquatic and terrestrial communities. The three graphs for the pondweed community are (1) division of biomass into leaf, stem, and root; solid area represents winter buds; (2) concentration of chlorophyll in the plant community; (3) leaf area and light profile. The solid line is light in the community; the broken line represents light in open water. (After Ikusima, 1965.) (b) Structure and productive systems of a pine-spruce-fir forest in Japan. (After Monsi, 1968.)*

(Whittaker, 1963; Kucera et al., 1967). Desert grasslands produce about 200 to 300 g/m²/yr, whereas deserts and tundra range between 100 and 250 g/m²/yr (Rodin and Bazilevic, 1967). Net production of the open sea is generally quite low. The productivity of the North Sea is about 170 g/m²/yr, the Sargasso Sea 180

**FIGURE 5-3**

*A climatic-geographical distribution of primary production, biomass, and radiation input. P = primary production (tn/ha); B = biomass (tn/ha); R = solar radiant input 51 (kcal/m/yr 0.3-3.0 microns). (From Etherington, 1975.)*

g/m²/yr. However, in some areas of upwelling, such as the Peru Currents, net production can reach 1000 g/m²/yr. These differences in net primary productivity from tropic to arctic regions, are reflected in litter production, which in tropical forests ranges between 900 and 1500 g/m²/yr, in temperate forests 200 and 600 g/m²/yr, and in arctic and alpine regions 0 and 200 g/m²/yr (Bray and Gorham, 1964).

Some ecosystems have consistently high production. Such high productivity usually results from an additional energy subsidy to the system. This subsidy may be a warmer temperature, greater rainfall, circulating or moving water that carries food or additional nutrients into the community or in the case of agricultural crops the use of fossil fuel for cultivation and irrigation, the application of fertilizer, and the control of pests. Swamps and marshes, ecosystems at the interface of land and water, may have a net productivity of 3300 g/m²/yr. Estuaries, because of input of nutrients from river and tides, and coral reefs, because of input from changing tides, may have a net productivity between 1000 and 2500 g/m²/yr. Among agricultural ecosystems sugarcane has a net productivity of 1700 to 1800 g/m²/yr,

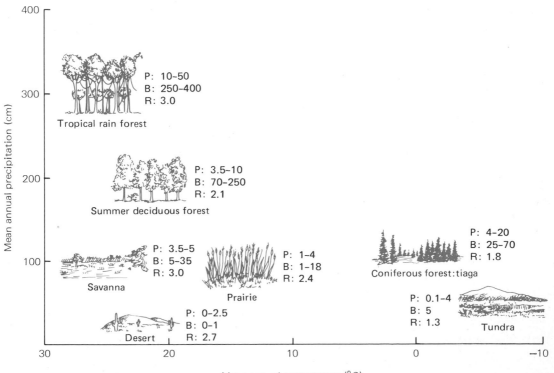

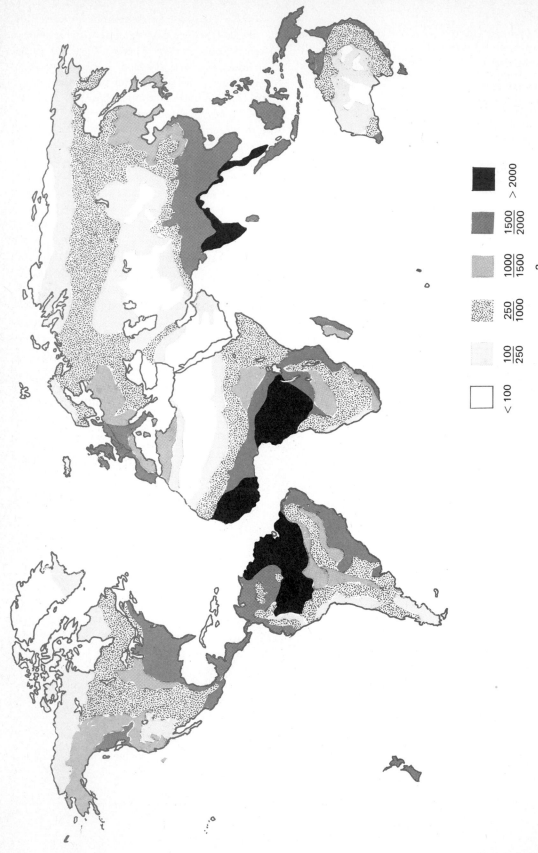

Productivity ranges — $9/m^2$/year, dry matter

| | |
|---|---|
| ⬛ | > 2000 |
| ▨ | $\dfrac{1500}{2000}$ |
| ▦ | $\dfrac{1000}{1500}$ |
| ⚬ | $\dfrac{250}{1000}$ |
| ▢ | $\dfrac{100}{250}$ |
| ☐ | < 100 |

TABLE 5-1
*Net primary production and plant biomass of world ecosystems*

| Ecosystems (in order of productivity) | Area ($10^6 km^2$) | Mean net primary production per unit area ($g/m^2/yr$) | World net primary production ($10^9 mtr/yr$) | Mean biomass per unit area ($kg/m^2$) |
|---|---|---|---|---|
| **CONTINENTAL** | | | | |
| Tropical rain forest | 17.0 | 2000.0 | 34.00 | 44.00 |
| Tropical seasonal forest | 7.5 | 1500.0 | 11.30 | 36.00 |
| Temperate evergreen forest | 5.0 | 1300.0 | 6.40 | 36.00 |
| Temperate deciduous forest | 7.0 | 1200.0 | 8.40 | 30.00 |
| Boreal forest | 12.0 | 800.0 | 9.50 | 20.00 |
| Savanna | 15.0 | 700.0 | 10.40 | 4.00 |
| Cultivated land | 14.0 | 644.0 | 9.10 | 1.10 |
| Woodland and shrubland | 8.0 | 600.0 | 4.90 | 6.80 |
| Temperate grassland | 9.0 | 500.0 | 4.40 | 1.60 |
| Tundra and alpine meadow | 8.0 | 144.0 | 1.10 | 0.67 |
| Desert shrub | 18.0 | 71.0 | 1.30 | 0.67 |
| Rock, ice, sand | 24.0 | 3.3 | 0.09 | 0.02 |
| Swamp and marsh | 2.0 | 2500.0 | 4.90 | 15.00 |
| Lake and stream | 2.5 | 500.0 | 1.30 | 0.02 |
| Total continental | 149.0 | 720.0 | 107.09 | 12.30 |
| **MARINE** | | | | |
| Algal beds and reefs | 0.6 | 2000.0 | 1.10 | 2.00 |
| Estuaries | 1.4 | 1800.0 | 2.40 | 1.00 |
| Upwelling zones | 0.4 | 500.0 | 0.22 | 0.02 |
| Continental shelf | 26.6 | 360.0 | 9.60 | 0.01 |
| Open ocean | 332.0 | 127.0 | 42.00 | 1.00 |
| Total marine | 361.0 | 153.0 | 55.32 | 0.01 |
| World total | 510.0 | 320.0 | 162.41 | 3.62 |

*Source:* Adapted from Whittaker and Likens, 1973.

hybrid corn 1000 g/m²/yr, and some tropical crops 3000 g/m²/yr.

In any ecosystem annual net production changes with time and age. For example, mean annual net primary production of a Scots pine plantation achieved maximum production of $22 \times 10^3$ kg/ha at the age of 20; it then declined to $12 \times 10^3$ kg/ha at 30 years of age (Ovington, 1961). Woodlands apparently achieve their maximum annual production in the pole stage, when the dominance of the trees is the greatest and the understory is at a minimum (Figure 5-5). The understory makes

**FIGURE 5-4**
*A map of world primary production. (Based on Golley and Leith, 1972.)*

its greatest contribution during the juvenile and mature stages of the forest. As age increases, more and more of the production is needed for maintenance, and very little gross production is left for growth.

Just as net production of a given community declines with age, so net production relative to gross production declines from a young ecosystem such as a weedy field or an agricultural crop to a mature plant community such as a forest. As plant communities approach a stable or steady-state condition (see Chapter 9), more of the gross production is used for the maintenance of biomass and less goes into newly added organic matter. Thus the ratio of gross production to biomass declines through time (Figure 5-6).

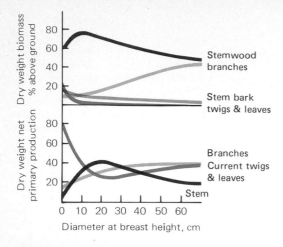

FIGURE 5-5
*Relation of above-ground biomass and production to size of tree, in the Hubbard Brook forest. The trends in above-ground biomass and production percentages in relation to diameter breast height (dbh) involve 63 sample trees of three major species—sugar maple, yellow birch, and beech. Note that as the trees increase in size the ratio of branches to stems increases. In larger trees branches account for a greater proportion of net primary production than stems. In smaller trees current leaves and twigs account for the greater percentage of primary production. This percentage declines rather rapidly as trees approach pole stage, then increases as the trees mature. (From Whittaker et al., 1974.)*

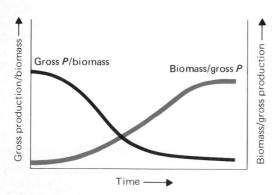

FIGURE 5-6
*Model showing the change through time in ratios between gross community photosynthesis and biomass, or production efficiency, and between biomass and gross community photosynthesis, or maintenance efficiency. Note the early high production efficiency and the later accumulation of biomass. (After G. D. Cooke, 1967.)*

Productivity may vary considerably not only among different types of ecosystems, but also among similar systems and within one system from year to year. Productivity is influenced by such factors as nutrient availability, moisture, especially precipitation, temperature, length of the growing season, animal utilization, and fire. For example, the herbage yields of a grassland may vary by a factor of eight between wet and dry years (Weaver and Albertson, 1956). Overgrazing of grasslands by cattle and sheep or defoliation of forests by such insects as the saddled prominent and gypsy moth can seriously reduce net production. Fire in grasslands may result in increased productivity if moisture is normal, but in reduced productivity if precipitation is low (Kucera et al., 1967). An insufficient supply of nutrients, especially nitrogen and phosphorus, can limit net productivity as can the mechanical injury of plants, atmospheric pollution, and the like.

Although the size of the standing crop is not synonymous with high productivity, the size of the standing crop does influence the capacity to produce. A pond with too few fish or a forest with too few trees does not have the capacity to utilize the energy available. On the other hand, too many fish or too many trees means less energy available to each individual. This lowers the efficiency of use and influences the storage and transfer of energy through the ecosystem. For example, Hayne and Ball (1956) found that the production of bottom-dwelling invertebrates in a pond amounted to nearly 17 times standing crop when predatory fish were present. In the absence of fish the production rate of bottom fauna declined and finally stopped with a large standing crop. Thus, in the presence of fish the size of the standing crop of bottom fauna was depressed, but production increased. In ponds with fish the annual production amounted to 811 lb/acre of bottom fauna and 181 lb of fish. This represents the efficiency of 18 percent in energy conversion from bottom fauna to fish.

### SECONDARY OR CONSUMER PRODUCTION

Net production is the energy available to the heterotrophic components of the ecosystem. Theoretically at least, all of it is available to the grazers or even to the decomposers, but rarely is it all utilized in this manner. The net production of any given ecosystem may be dispersed to another food chain outside of the ecosystem by humans, wind, or water. For example, about 45 percent of the net production of a salt marsh is lost to estuarine water (Teal, 1962). Much of the living material is physically unavailable to the grazers—they cannot reach many plants. The organic matter of live organisms is unavailable to decomposers and detritus-feeders, and that of dead plants may not be relished by grazers. The amount of net production available to herbivores may vary from year to year and from place to place. The quantity consumed varies with the type of herbivore and the density of the population. Once consumed, a considerable portion of the plant material, again depending upon the kind of plant involved and the digestive efficiency of the herbivore, may pass through the animal's body undigested. A grasshopper assimilates only about 30 percent of the grass it consumes, leaving 70 percent available to the detritus-decomposer food chain (Smalley, 1960). Mice, on the other hand, assimilate about 85 to 90 percent of what they consume (Golley, 1960; R. L. Smith, 1962).

Energy, once consumed, either is diverted to maintenance, growth, and reproduction or is passed from the body as feces and urine (Figure 5-7). The energy content of feces is transferred to the detritus food chain. Part of the energy is lost through urine, and, depending upon the nature of the organism, the loss of energy can be variable and often quite high (Coo et al., 1952). Another portion is lost as fermentation gases. Of the energy left after losses through feces, urine, and gases, part is utilized as *heat increment,* which is heat required for metabolism above that required for basal or resting metabolism. The remainder of the energy is *net energy,* available for maintenance and production. This includes energy involved in capturing or harvesting food, muscular work expended in the animal's daily routine, and energy needed to keep up with the wear and tear on the animal's body. The energy used for maintenance is lost as heat. Maintenance costs, highest in small, active, warm-blooded animals, are fixed or irreducible. In small invertebrates energy costs vary

with temperature, and a positive energy balance exists only within a fairly narrow range of temperatures. Below 5° C spiders become sluggish, cease feeding, and have to utilize stored energy to meet metabolic needs. At approximately 5° C assimilated energy approaches energy lost through respiration. From 5° to 20.5° C spiders assimilate more energy than they respire. Above 25° C, their ability to maintain a positive energy balance declines rapidly (Van Hooke, 1971). Energy left over from maintenance and respiration goes into the production of new tissue, fat tissue, growth, and new individuals. This net energy of production is *secondary production.* Within secondary production there is no portion known as gross production. What is analogous to gross production is actually assimilation. Secondary production is greatest when the birth rate of the population and the growth rates of the individuals are the highest. This usually coincides, for evolutionary reasons, with the time when net primary production is also the highest.

*FIGURE 5-7*
*Relative values of the end products of energy metabolism in the white-tailed deer. Note the small amount of net energy gained (body weight) in relation to that lost as heat, gas, urine, and feces. The deer is a herbivore, a first-level consumer. (After Cowan, 1962.)*

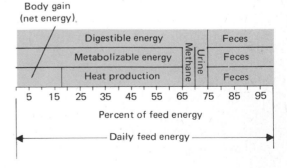

Body gain
(net energy)

| | Digestible energy | | Feces |
| Metabolizable energy | Urine / Methane | Feces |
| Heat production | | Feces |

5  15  25  35  45  55  65  75  85  95
Percent of feed energy

Daily feed energy

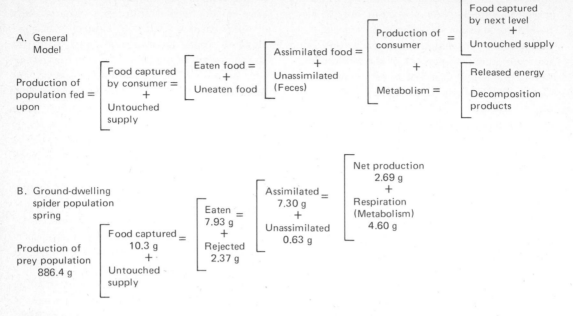

**FIGURE 5-8**
*General model and field example of components of energy metabolism in secondary production. (Data from Moulder and Reichle, 1972.)*

This scheme is summarized in Figure 5-8. It is applicable to any consumer organism, herbivore or carnivore. The herbivore represents the energy source of the carnivore, and like the plant food of the herbivore, not all of the energy contained in the body of the herbivore is utilized by the carnivore. Part of it, such as hide, bones, and internal organs, is unconsumed, and the same metabolic losses can be accounted for. At each transfer considerably less energy is available for the next consumer level.

The energy budget of a consumer population can be summarized by the following formula:

$$C = A + FU$$

where $C$ is the energy ingested, $A$ is the energy assimilated, and $FU$ is the energy lost through feces, urine, gas, and other rejecta.

The term $A$ can be further refined as

$$A = P + R$$

where $P$ is secondary production and $R$ is energy lost through respiration. Thus

$$C = P + R + FU$$

or secondary production is

$$P = C - FU - R$$

Just as net primary production is limited by a number of variables, so is secondary production. The quantity, quality (including nutrient status and digestibility), and availability of net production are three limitations. So is the degree to which primary and available secondary production are utilized.

The latter can be examined from the viewpoint of two different ratios. One is the ratio of assimilation to ingestion, $A/C$. This is the measure of the efficiency of the consumer population in extracting energy from the food it consumes. The other is the ratio of productivity to assimilation, $P/A$. This is a measure of the efficiency with which the consumer population incorporates assimilated energy into new tissue, or secondary production.

The ability of the consumer population to convert the energy it consumes varies with the species and the type of consumer (Table 5-2). Vertebrates utilize about 98 percent of their assimilated energy in metabolism and only about 2 percent in net production; invertebrates utilize 79 percent in metabolism. Thus invertebrates convert a greater proportion of their assimilated energy into biomass than vertebrates. A clearer distinction can be made if one considers, as Engelmann (1968) does,

TABLE 5-2
*Secondary production of selected consumers (kcal/m²/yr)*

| Species | Inges- tion (I) | Assimila- tion (A) | Respira- tion (R) | Produc- tion (P) | A/I | P/I | R/A | Reference |
|---|---|---|---|---|---|---|---|---|
| Harvester ant (h) | 34.50 | 31.00 | 30.90 | 0.10 | 0.1 | .0002 | .99 | Odum, Connell, and Davenport, 1962 |
| Plant hopper (h) | 41.30 | 27.50 | 20.50 | 7.00 | .67 | .169 | .75 | Wiegert, 1964 |
| Salt marsh grass- hopper (h) | 3.71 | 1.37 | 0.86 | 0.51 | .37 | .137 | .63 | Teal, 1962; Smalley, 1960 |
| Spider, small < 1 mg (c) | 12.60 | 11.90 | 10.00 | 1.90 | .94 | .151 | .84 | Moulder and Reichle, 1972 |
| Spider, large > 10 mg (c) | 7.40 | 7.00 | 7.30 | −3.00 | .95 | — | 1.04 | Moulder and Reichle, 1972 |
| Savannah sparrow (o) | 4.00 | 3.60 | 3.60 | 0 | .90 | 0 | 1.0 | Odum, Connell, and Davenport, 1962 |
| Old field mouse (h) | 7.40 | 6.70 | 6.60 | 0.1 | .91 | .014 | .98 | Odum, Connell, and Davenport, 1962 |
| Ground squirrel (h) | 5.60 | 3.80 | 3.69 | 0.11 | .68 | .019 | .97 | Wiegart and Evans, 1967 |
| Meadow mouse (h) | 21.29 | 17.50 | 17.00 | — | .82 | — | .97 | Golley, 1960 |
| African elephant (h) | 71.60 | 32.00 | 32.00 | 8 | .44 | | 1.0 | Petrides and Swank, 1966 |
| Weasel (c) | 5.80 | 5.50 | — | — | .95 | — | — | Golley, 1960 |

Note: h = herbivore; o = omnivore; c = carnivore.

the animal world divided into two broad energetic groups, the thermoregulators or homoiotherms and the nonregulators or poikilotherms. In this context the non-regulators are more efficient producers than the thermoregulators. However, a major difference exists in their assimilation efficiency. Poikilotherms have an efficiency of about 30 percent in digesting food, whereas homoiotherms have an efficiency of around 70 percent (see Table 5-2). Thus the poikilotherm has to consume more calories than a homoiotherm to obtain sufficient energy for maintenance, growth, and reproduction.

## Food chains

Net production theoretically represents the energy available either directly or indirectly to the consumer organisms and is the base upon which the rest of life on earth depends. This energy stored by plants is passed along through the ecosystem in a series of steps of eating and being eaten known as a *food chain*.

Food chains are descriptive. When worked out diagrammatically, they consist of a series of arrows, each pointing from one species to another, for which it is a source of food. In Figure 5-9, for example, the marsh vegetation is eaten by the grasshopper, the grasshopper is consumed by the shrew, the shrew by the marsh hawk or the owl.

But no one organism lives wholly on another; the resources are shared, especially at the beginning of the chain. The marsh plants are eaten by a variety of invertebrates, birds, mammals, and fish; and some of the animals are consumed by several predators. Thus food chains become interlinked to form a *food web*, the complexity of which varies within and between ecosystems.

**129**

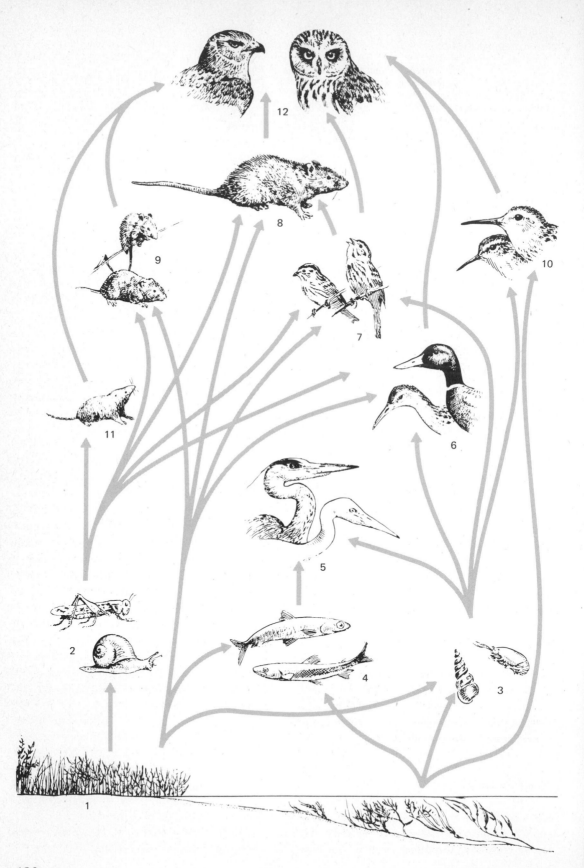

FIGURE 5-9

*A mid-winter food web in a* Salicornia *salt marsh (San Francisco Bay area). Producer organisms are the terrestrial and aquatic plants (1). The plants are consumed by terrestrial herbivorous invertebrates, represented by the grasshopper and snail (2), and by herbivorous marine and intertidal invertebrates (3). Fish, represented by the smelt and anchovy (4), feed on vegetative matter from both ecosystems. The fish in turn are eaten by first-level carnivores, represented by the the great blue heron and the common egret (5). Continuing through the web we have the following omnivores: clapper rail and mallard ducks (6), savanna and song sparrows (7), Norway rats (8), California vole and salt marsh harvest mouse (9), the least and western sandpipers (10). The vagrant shrew (11) is a first-level carnivore; the top carnivores (second level) are the marsh hawk and short-eared owl (12). (Adapted from R. F. Johnson, 1956a.)*

At each step in the food chain a considerable portion of the potential energy is used for maintenance and lost as heat. This limits the number of steps in any food chain to four or five. The longer the food chain, the less energy is available for the final members.

## COMPONENTS OF THE FOOD CHAIN

*Herbivores.* Feeding on plant tissues is a whole host of plant consumers, the herbivores, which are capable of converting energy stored in plant tissue into animal tissue. Their role in the community is essential, for without them the higher tropic levels could not exist. The English ecologist Charles Elton, in his classic book *Animal Ecology,* suggested that the term *key industry* be used to denote animals that feed on plants and are so abundant that many other animals depend upon them for food.

Only the herbivores are adapted to live on a diet high in cellulose. The structure of the teeth, complicated stomachs, long intestines, a well-developed cecum, and symbiotic flora and fauna enable these animals to use plant tissues. For example, ruminants, such as deer, have a four-compartment stomach. As they graze, these animals chew their food hurriedly. The material consumed descends to the first and second stomachs (the rumen and reticulum), where it is softened to a pulp by the addition of water, kneaded by muscular action, and fermented by bacteria. At leisure the animals regurgitate the food, chew it more thoroughly, and swallow it again. The mass again enters the rumen, where the coarse particles remain behind for further bacterial digestion. The finer material is pulled into the reticulum and from there forced by contraction into the third compartment, or omasum. There the material is further digested and finally forced into the abomasum, or true glandular stomach.

The digestive process in ruminants relies heavily on bacterial fermentation in the rumen, reticulum, and omasum. Millions of microorganisms attack various digestive materials such as cellulose, starch, pectin, and hemicellulose sugars and convert part of them to short-chain volatile fatty acids. These are rapidly absorbed through the wall of the rumen into the bloodstream and are oxidized to form the animal's chief source of energy. Part of the material is converted to methane and lost to the animal and part remains as fermentation products. Many of the microbial cells involved are digested in the abomasum to recapture still more of the energy and nutrients. In addition to fermentation, the bacteria also synthesize B-complex vitamins and essential amino acids.

Another outstanding group of herbivores is the lagomorphs—the rabbits, hares, and pikas. In contrast to ruminants, these herbivores have a simple stomach and a large cecum. During digestion part of the ingested plant material enters the cecum and part enters the intestine to form dry pellets. In the cecum the ingested material is attacked by microorganisms and is expelled into the large intestine as moist soft pellets surrounded by a proteinaceous membrane. The soft pellets, much higher in protein and lower in crude fiber than the hard fecal pellets, are reingested (coprophagy) by the lagomorphs. The amount of feces recycled by coprophagy ranges from 50 to 80 percent. The reingestion is important because it provides bacterially synthesized B vitamins and ensures a more complete digestion of dry material and a better utilization of protein (see McBee, 1971).

*Carnivores.* Herbivores are the energy source for carnivores, the flesh-eaters. Organisms that feed directly upon the herbivores are termed first-level carnivores or second-level consumers. First-level carnivores represent an energy source for the second-level carnivores. Still higher categories of carnivorous animals feeding on secondary carnivores may exist in some communities. As the feeding level of carnivores increases, their numbers decrease and their fierceness, agility, and size increase. Finally, the energy stored at the top carnivore level is utilized by the decomposers.

*Omnivores.* Not all consumers confine their feeding to one level. The omnivores consume both plants and animals. The red fox feeds on berries, small rodents, and even dead animals. Thus it occupies herbivorous and carnivorous levels, as well as acting as a scavenger. Some fish feed on both plant and animal matter. The basically herbivorous white-footed mouse also feeds on insects, small birds, and bird eggs. The food habits of many animals vary with the seasons, with stages in the life cycle, and with the size and growth of the organism (Figure 5-10).

*Decomposers.* Conventionally, the decomposers make up the final consumer group. Actually the decomposers are much more than that. Their basic function is to release the nutrients contained in the plant and animal biomass back into the mineral cycles. The work of the decomposers then is the opposite of that of the producers, who fix nutrients and energy into plant biomass. The process of decomposition involves a complex series of food chains in which the organisms of decay utilize the energy and materials of dead plant and animal matter.

There are two groups of decomposer organisms, macroorganisms and microorganisms. The macroorganisms or detritivores include small detritus-feeding animals such as mites, earthworms, millipedes, and slugs in the terrestrial ecosystems and crabs, mollusks, and worms in the aquatic ecosystems. The microorganisms or true decomposers are the bacteria and fungi. The detritus-feeders act as reducer-decomposers. They ingest organic matter, break it down into smaller pieces, mix it with soil, excrete it, spread spores, break down microbial antagonisms, and even add substances to the material that stimulate microbial growth. The same reducer-decom-

*Micropogon undulatus,* Atlantic Croaker

*Food categories*

| | Young | Juvenile | Adult |
|---|---|---|---|
| *Fishes* | | ▨ | ▨ |
| *Macrobottom animals* | | ▨ | ▨ |
| *Microbottom animals* | ▨ | ▨ | ▨ |
| *Zooplankton* | ▨ | ▨ | ▨ |
| *Phytoplankton* | | | |
| *Vascular plant material* | | | |
| *Organic detritus and undertermined organic material* | ▨ | ▨ | ▨ |

**FIGURE 5-10**
*Trophic spectra for young, juvenile, and adult stages of the Atlantic croaker from Lake Pontchartrain, Louisiana. The young fish subsist chiefly on zooplankton, the juveniles on organic debris, the adults on bottom animals and fish. (After Darnell, 1961.)*

posers also consume bacteria and fungi associated with the detritus, as well as protozoans and small invertebrates that cling to the material. Some of these organisms are microbial grazers, feeding on bacteria and fungi. These grazers may reduce bacterial and fungal populations, inhibiting the effects of increased population density and accelerating the division of soil microbes, thus speeding up microbial activity. The reducer-decomposers then reduce detritus to smaller pieces, condition the material for microbial action, and at the same time utilize the energy and nutrients stored in the biomass of bacteria and fungi.

The final breakdown of detritus is accomplished by bacteria and fungi. These bacteria may inhabit the digestive tracts of the reducer-decomposers, producing the enzymes necessary for the digestion of the material. Bacteria and hyphae of fungi secrete enzymes necessary to carry out the specific chemical action on the detritus. These organisms then absorb a portion of the product as food, and the unconsumed material is available as food for other organisms. In addition the bacteria and fungi become food for the microbial grazers. In this manner food chains involving decomposers reach up into the traditional herbivore-carnivore food chains. In fact, the decomposers, so frequently considered as some distant feeding group unclassifiable in the general scheme of food chains, actually function as herbivores or carnivores, depending upon whether their food is plant or animal material. At any rate, the end result of decomposition is the transformation of organic matter into nutrients available for plants.

*Other feeding groups.* Several other feeding groups are also involved in energy transfer. *Parasites* spend a considerable part of their life cycle living on or in and drawing their nourishment from their hosts. Most do not kill their hosts, but some, *the parasitoids,* do draw nourishment from the host until it dies. At this point the parasitoid transforms into another stage of its life cycle, becoming independent of the host. Functionally, parasites are specialized carnivores and herbivores.

*Scavengers* are animals that eat dead plant and animal material. Among these are termites and various beetles that feed in dead and decaying wood, and crabs and other marine invertebrates that feed on plant particles in water. Botflies, dermestid beetles, vultures, and gulls are only several of many animals that feed on animal remains. Scavengers may be either herbivores or carnivores.

*Saprophytes* are plant counterparts of scavengers. They draw their nourishment from dead plant and animal material, chiefly the former. Because they do not require sunlight as an energy source, they can live in deep shade or dark caves. Examples of saprophytes are fungi, Indian pipe, and beech drops. The majority are herbivores, but others feed on animal matter.

### MAJOR FOOD CHAINS

Within any ecosystem there are two major food chains, the grazing food chain and the detrital food chain (Figure 5-11). In most terrestrial and shallow-water ecosystems, with their high standing crop and relatively low harvest of primary production, the detrital food chain is dominant. In deep-water aquatic ecosystems, with their low biomass, rapid turnover of organisms, and high rate of harvest, the grazing food chain may be dominant.

The amount of energy shunted down the two routes varies among communities. In an intertidal salt marsh less than 10 percent of living plant material is consumed by herbivores and 90 percent goes the way of the detritus-feeders and decomposers (Teal, 1962). In fact most of the organisms of the intertidal salt marsh obtain the bulk of their energy from dead plant material. In a Scots pine plantation 50 percent of the energy fixed annually is utilized by decomposers (Figure 5-12). The remainder is removed as yield or is stored in tree trunks (Ovington, 1961). In some communities, particularly undergrazed grasslands, unconsumed organic matter may accumulate and remain out of circulation for some time, especially when conditions are not favorable for microbial action. The decomposer or detritus food chain receives additional materials from the waste products and dead bodies of both the herbivores and carnivores.

*Detrital food chain.* The detrital food chain is common to all ecosystems, but in terrestrial and littoral ecosystems it is the major pathway of energy flow, because so little of the net production is utilized by grazing herbivores. Of

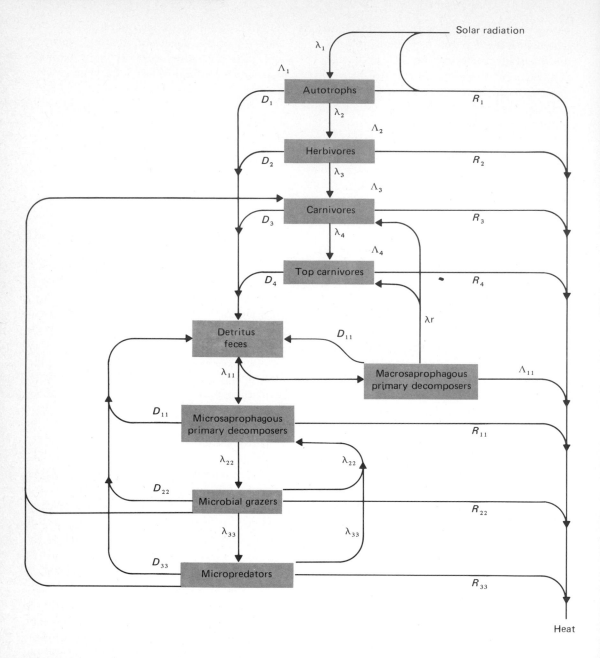

**FIGURE 5-11**
*Model of a detrital and grazing food chain.*
*Two pathways lead from the autotrophs, one*
*to the grazing herbivores, the other to the*
*detritus feeders. Note the interrelations*
*between the two food chains. (After Paris,*
*1968.)*

the total amount of energy fixed in a tulip poplar forest *(Liriodendron)* 50 percent of the
gross production goes into maintenance and
respiration, 13 percent is accumulated as new

tissue, 2 percent is consumed by herbivores,
and 35 percent goes to the detrital food chain
(Edwards, unpublished, as cited by O'Neill,
1975). Two-thirds to three-fourths of the
energy stored in a grassland ecosystem ungrazed by domestic animals is returned to the
soil as dead plant material, and less than one-
fourth is consumed by herbivores (Hyder,
1969). Of the quantity consumed by herbivores, about one-half is returned to the soil as
feces. In the salt marsh ecosystem the domi-

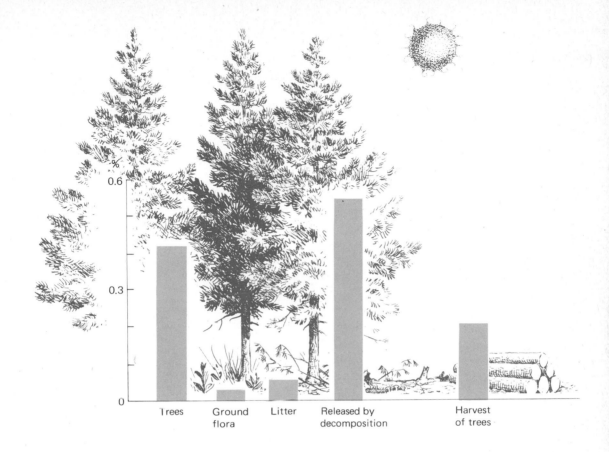

**FIGURE 5-12**
*The fate of the 1.3 percent of solar energy assimilated as net production by a 23-year-old Scots pine plantation. (After J. D. Ovington, 1961.)*

nant grazing herbivore, the grasshopper, consumes just 2 percent of the net production available to it (Smalley, 1960).

Enough has been said about the functional role of decomposers (see Chapter 2) in the detrital food chain. A quantified budget of energy flow through a detrital food chain is difficult to measure, although the use of radioactive tracers gives some idea of energy flow through selected food webs. Andrews et al. (1975), basing their estimates on a number of studies, determined that microbial activity accounted for 99 percent of total saprophytic assimilation in a short-grass prairie ecosystem. Saprophagic grazers accounted for the remaining 1 percent. A summary of energy flow in ungrazed and heavily grazed ecosystems is given in Table 5-3.

Gist and Crossley's (1975) study of a selected invertebrate population living in forest litter provides an example of a detrital food chain (Figure 5-13). Although no data for energy flow per se were collected, the flux of radioactive calcium and phosphorus permitted the construction of a food web. The quantity of elements involved in the flux provided some idea of the energy flow involved. The litter is utilized by five groups: millipedes (Diploda), orbatid mites (Cryptostigmata), springtails (Collembola), cave crickets (Orthoptera), and pulmonate snails (Pulmonata). Of these the mites and springtails are the most important litter-feeders. These herbivores are preyed upon by small spiders (Araneidae) and predatory mites (Mesostigmata). The predatory mites feed on annelids, mollusks, insects, and other arthropods. The spiders feed on the predatory mites. Springtails, pulmonate snails, small spiders, and cave crickets are preyed upon by carabid beetles, while medium-sized spiders feed upon cave crickets and other insects. The medium-sized spiders, in turn, be-

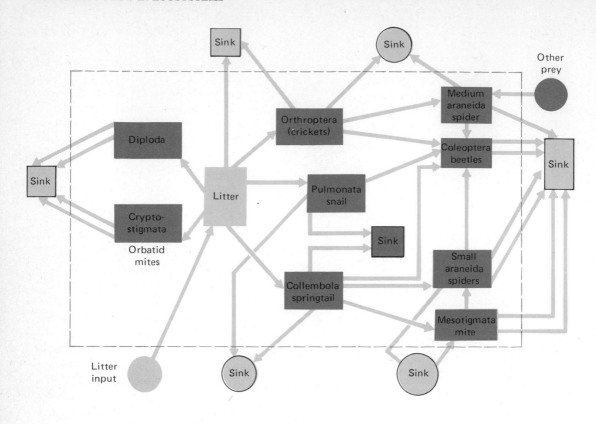

TABLE 5-3
*Primary production and energy flow through saprovores on a short-grass prairie, 1972 (kcal/m²)*

| Energy pathways | UNGRAZED | LIGHTLY GRAZED | HEAVILY GRAZED |
|---|---|---|---|
| **PRIMARY PRODUCTION (154-day measurement period)**[*] | | | |
| Gross production | 5838 | 6508 | 5596 |
| Net production | 3852 | 4296 | 3694 |
|   net above-ground production | 562 | 958 | 728 |
|   net below-ground production | 3290 | 3338 | 2966 |
|     net root production | 2256 | 2594 | 2539 |
|     net crown production | 1034 | 744 | 427 |
| Net respiration | 1986 | 2212 | 1902 |
|   net above-ground respiration | 936 | 1134 | 944 |
|   net below-ground respiration (model) | 1050 | 1078 | 958 |
|   net below-ground respiration (lab) | 598 | —[†] | 461 |
| **SAPROPHAGIC ACTIVITY (200-day measurement period)** | | | |
| Microbial respiration | 2990 | — | 2304 |
| Microbial production | 1594 | — | — |
| Microbial assimilation | 4584 | — | — |
| Saprophagic-grazing consumption | 45.0 | — | 35.0 |
| Saprophagic-grazing respiration | 16.2 | — | 12.5 |
| Saprophagic-grazing production | 3.7 | — | 3.0 |
| Saprophagic-grazing assimilation | 19.9 | — | 15.5 |

[*]Solar, radiation for 154-day plant growing season: received = $8.28 \times 10^5$ kcal/m², photosynthetically active = $3.89 \times 10^5$ kcal/m².
[†]Dashes indicate item not measured on this treatment.
*Source:* Andrews et al., 1975.

FIGURE 5-13
*Model of a detrital food web involving litter-dwelling invertebrates. The dashed line represents the boundaries of the system. Note that the detrital food chain involves a herbivorous component—millipedes and mites to the left and crickets and springtails to the right. The herbivores support a carnivorous component. The detrital food web, like the grazing food web, can become quite complex. (After C. S. Gist and D. A. Crossley, Jr., 1975.)*

come additional items in the diet of the beetles. The beetles, spiders, and snails are eaten by birds and small mammals, members of the grazing food chain not shown on the original web. Detrital food webs through predation are often linked to the grazing food chain at higher consumer levels.

Food chains involving saprophages may take two directions (as shown in Figure 5-11), toward the carnivores or toward microorganisms. The role of these feeding groups in the final dissipation of energy has already been mentioned. But they also are food to numerous other animals as well. Slugs eat the larvae of certain flies and beetles, which live in the heads of fungi and feed on the soft material. Mammals, particularly the red squirrel and chipmunks, eat woodland fungi. Dead plant remains are food sources for springtails and mites, which in turn are eaten by carnivorous insects and spiders. These in turn are energy sources for insectivorous birds and small mammals. Blowflies lay their eggs in dead animals and within 24 hours the maggot larvae hatch. Unable to eat solid tissue, they reduce the flesh to a fetid mass by enzymatic action in which they feed on the proteinaceous material. These insects are food for other organisms.

*Grazing food chain.* The grazing food chain is the one most obvious to us. Cattle grazing on pasture land, deer browsing in the forest, rabbits feeding in old fields, and insect pests feeding on garden crops represent the basic consumer groups of the grazing food chain. In spite of its conspicuousness, the grazing food chain, as pointed out earlier, is not the major food chain in terrestrial and many aquatic ecosystems. Only in some aquatic ecosystems do the grazing herbivores play a dominant role

in energy flow. Although voluminous data exist on phytoplankton productivity, filtration rates by grazing zooplankton, and production efficiencies of zooplankton (for summary, see Wetzel, 1975), few data are available on the flow of energy, rate of grazing, biomass turnover rates for phytoplankton, and turnover of zooplankton biomass within the same aquatic system. Carter and Lund (1968) found that certain grazing protozoans feeding on certain planktonic algae consumed 99 percent of the populations in 7 to 14 days. Hillbricht-Ilkowska (1975) studied the relationship between primary production of phytoplankton and consumer production in freshwater lakes. He found that the direct utilization of primary production by filter-feeding zooplankton was intense and its efficiency of energy transfer was high when the size and structure of phytoplankton were favorable for filter feeding. In lakes where the net form of phytoplankton was dominant, direct utilization was low and energy transfer inefficient. As an example, in lakes in which the production of phytoplankton ranged from 200 to 1200 kcal/m²/yr, production of filter-feeding zooplankton amounted to 10 to 250 kcal/m²/yr with a transfer efficiency of 2 to 30 percent. In lakes in which filter-feeding zooplankton production ranged from 50 to 150 kcal/m²/yr, the production of predacious zooplankton ranged from 2 to 50 kcal/m²/yr with an energy transfer efficiency of 5 to 40 percent.

In terrestrial systems a relatively small proportion of primary production goes by way of the grazing food chain. Over a three-year period only 2.6 percent of net primary production of a tulip poplar forest was utilized by grazing herbivores, although the holes made in the growing leaves resulted in a loss of 7.2 percent of photosynthetic surface. In a study of energy flow through a short-grass prairie ecosystem, involving ungrazed, lightly grazed, and heavily grazed plots, Andrews et al. (1975) found that on the heavily grazed prairie cattle accounted for only 15 percent of the net above-ground primary production, or 3 percent of total net primary production. On such grasslands, however, cattle grazing may account for the consumption of 30 to 50 percent of above-ground net primary production. The short-grass prairie, however, did respond to grazing stress by concentrating more of the net production to roots. In the heavily grazed plots grasses alloted 69 percent of net primary production to roots, 12 percent to crowns, and 19 percent to shoots. In contrast, the lightly grazed prairie allotted 69 percent of net primary production to roots, 18 percent to crowns, and 22 percent to shoots. Ungrazed plots allocated only 14 percent of production to shoots. This suggests that light grazing stimulates primary production above ground. In both lightly and heavily grazed plots about 40 to 50 percent of energy consumed by the cattle is returned to the ecosystem and the detrital food chain as feces (Dean et al., 1975).

Although the above-ground herbivores are the conspicuous grazers, below-ground herbivores can have a pronounced impact on primary production and the grazing food chain. Andrews et al. (1975) found that below-ground herbivores consisting mainly of nematodes (Nematoda), scarab beetles (Scarabaeidae), and adult ground beetles (Carabidae) accounted for 81.7 percent of total herbivore assimilation on the ungrazed short-grass prairie, 49.5 percent on the lightly grazed prairie, and 29.1 percent on the heavily grazed prairie. Ninety percent of the invertebrate herbivore consumption took place below ground, and 50 percent of the total energy was processed by nematodes. On the lightly grazed prairie cattle consumed 46 kcal/m² during the grazing season and the below-ground invertebrates consumed 43 kcal/m². Thus below-ground herbivorous consumption can impose a greater stress on the ecosystem than the above-ground herbivores. When a nematicide was added to a midgrass prairie, above-ground net production increased 30 to 60 percent.

Palatability and body size impose some limits on energy flow through the grazing food chain. Caswell et al. (1973) have hypothesized that plants possessing a $C_4$ dicarboxylic acid pathway of energy flow (see Chapter 2) are a poorer food source for herbivores than plants possessing the $C_3$ Calvin cycle. Because of anatomical structure and the pattern of translocation and accumulation of photosynthesis, $C_4$ plants may provide a less palatable, less nutritious food. Evidence of a tendency of herbivores to avoid feeding on a $C_4$ species has been found in some food habit studies.

No linear correlation exists between energy flow and body weight or surface area per se, but there is a correlation with the metabolically effective body weight. An increase of 100 percent in body weight means a 70 percent increase in metabolic weight. As the body size increases, the neuroendocrine system, which controls metabolism, increases proportionally with body surface rather than with weight. Thus metabolic rate per gram of weight rises exponentially as the weight of the individual declines.

Size affects energy flow in another way. It has considerable influence on the direction a food chain takes because there are upper and lower limits to the size of food an animal can capture. Some animals are too fleet to be caught; others are large enough to defend themselves successfully. Some foods are too small to be collected economically because it takes too long to secure enough to meet the animal's metabolic needs. Thus the upper limit to the size of an organism's food is determined by its ability to handle and process the item, and the lower limit by the animal's ability to secure large enough food to meet its needs.

There are exceptions, of course. By injecting poisons, spiders and snakes kill prey much larger than themselves; wolves hunting in packs can kill an elk or a caribou. The idea that food chains involve animals of progressively larger sizes is true only in a general way. In the parasitic chain the opposite situation exists. The larger animals are at the base and, as the number of links increases, the size of the parasites becomes smaller. The only animal that can deal with food of any size is a human being.

An example of a grazing food chain that has been rather carefully worked out (Golley, 1960) involves old field vegetation, meadow mice, and weasels (Figure 5-14). The mice are almost exclusively herbivorous, and the weasels live mainly on mice. The vegetation converts about 1 percent of the solar energy into net production, or plant tissue. The mice consume about 2 percent of the plant food available to them, and the weasels about 31 percent of the mice. Of the energy assimilated, the plants lose about 15 percent through respiration, the mice 68 percent, and the weasels 93 percent. The weasels use so much of their assimilated energy in maintenance that a carnivore preying on weasels could not exist.

In a very general way, in the transformation of energy through the ecosystem by way of the grazing chain, the energy is reduced in magnitude by 10 from one level to another. Thus if an average of 1000 kcal of plant energy is consumed by herbivores, about 100 kcal are converted to herbivore tissue, 10 kcal to first-level carnivore production, and 1 kcal to second-level carnivores. The amount of energy available to second- and third-level carnivores is so small that few organisms could be supported if they depended on that source alone. For all practical purposes, each food chain has from three to four links, rarely five. The fifth link is distinctly a luxury item in the ecosystem.

***Supplementary food chains.*** Other feeding groups, such as the parasites and scavengers, form supplementary food chains in the community. Parasitic food chains are highly complicated because of the life cycle of the parasites. Some parasites are passed from one host to another by predators in the food chain. External parasites (ectoparasites) may transfer from one host to another. Other parasites are transmitted by insects from one host to another through the bloodstream or plant fluids.

Food chains also exist among parasites themselves. Fleas that parasitize mammals and birds are in turn parasitized by a protozoan, *Leptomonas*. Chalcid wasps lay eggs in

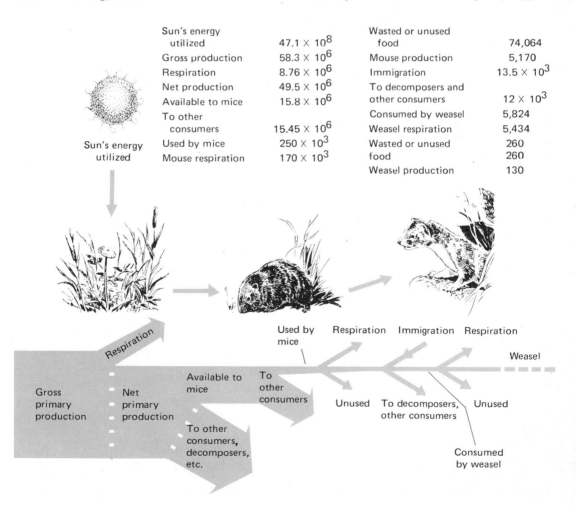

| | | | |
|---|---|---|---|
| Sun's energy utilized | $47.1 \times 10^8$ | Wasted or unused food | 74,064 |
| Gross production | $58.3 \times 10^6$ | Mouse production | 5,170 |
| Respiration | $8.76 \times 10^6$ | Immigration | $13.5 \times 10^3$ |
| Net production | $49.5 \times 10^6$ | To decomposers and other consumers | $12 \times 10^3$ |
| Available to mice | $15.8 \times 10^6$ | Consumed by weasel | 5,824 |
| To other consumers | $15.45 \times 10^6$ | Weasel respiration | 5,434 |
| Used by mice | $250 \times 10^3$ | Wasted or unused food | 260 |
| Mouse respiration | $170 \times 10^3$ | Weasel production | 130 |

**139**

the ichneumon or tachinid fly grub, which in turn is parasitic on other insect larvae.

In these parasitic food chains, the members, starting with the host, become progressively smaller and more numerous with each level in the chain.

## Trophic levels and ecological pyramids

Feeding relationships in an ecosystem as shown in food webs may also be classified as trophic or feeding groups. Such a classification involves the functions of organisms rather than the organisms themselves. Obviously a given species may belong to more than one feeding group. If all organisms that obtain their food in the same number of steps (that is, all those that feed wholly or in part on plants, wholly or part on herbivores, and so on) are superimposed, the structure can be collapsed into a series of single points representing the feeding or trophic levels of the ecosystem. Each step in the food chain represents a trophic level. Thus the first trophic level belongs to the producers, the second to the herbivores, the third to the first-order carnivores, and so on. Animals at the lower level, such as the grasshopper and the snail, may occupy a single trophic level, but most of the animals at the higher levels participate simultaneously in several trophic levels (see Figure 5-9).

Although trophic levels as considered by most ecological references do not include the decomposer and supplementary chains, they logically should. As already mentioned, decomposers and parasites should be considered as herbivores or carnivores, depending upon the nature of their food source. Decomposers feeding on the dead plant material, as well as bacteria occupying the rumen of ungulate animals, should be considered functional herbivores. Decomposers feeding on the dead bodies of animals should be considered carnivores. In this manner all the various steps in energy transfer in an ecosystem can be placed in some trophic level.

By adding all of the biomass or living tissue contained in each trophic level and all of the energy transferred between levels, one can construct pyramids of biomass and energy for the ecosystem (Figure 5-15). The pyramid of

biomass indicates by weight or other measurement of living material the total bulk of organisms or fixed energy present at any one time—the standing crop (Figure 5-15b). Because some energy or material is lost in each successive link, the total mass supported at each level is limited by the rate at which energy is being stored below. In general the biomass of the producers must be greater than the biomass of the herbivores they support, and the biomass of the herbivores must be greater than that of the carnivores. This usually results in a gradually sloping pyramid, particularly for terrestrial and shallow-water communities where the producers are large,

**FIGURE 5-15**
*Ecological pyramids. (Data from W. E. Pequegnat, 1961; H. T. Odum, 1957; Harvey, 1950; Teal, 1962.)*

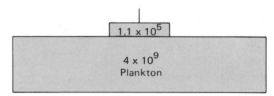

(a) Pyramid of numbers

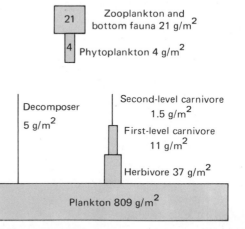

(b) Pyramids of biomass

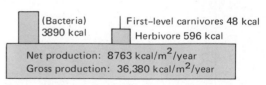

(c) Pyramid of energy

accumulation of organic matter is large, life cycles are long, and the rate of harvesting is low.

However, for some ecosystems the pyramid of biomass is inverted. Primary production in aquatic ecosystems such as lakes and open seas is concentrated in the microscopic algae. The algae have a short life cycle, multiply rapidly, accumulate little organic matter, and are heavily exploited by herbivorous zooplankton. At any point the standing crop is low. As a result, the base of the pyramid is much smaller than the structure it supports.

When production is considered in terms of energy, the pyramid indicates not only the amount of energy flow at each level, but, more important, the actual role the various organisms play in the transfer of energy. The base upon which the pyramid of energy is constructed is the quantity of organisms produced per unit or, stated differently, the rate at which food material passes through the food chain (Figure 5-15c). Some organisms may have a small biomass, but the total energy they assimilate and pass on may be considerably greater than that of organisms with a much larger biomass. On a pyramid of biomass these smaller organisms would appear much less important in the community than they really are.

These graphic representations of production and transfer always form pyramids because, in accord with the second law of thermodynamics, less energy is transferred from each level than was paid into it. In instances where the producers have less bulk than the consumers, particularly in open-water communities, the energy they store and pass on must be greater than that of the next level. Otherwise the biomass that producers support could not be greater than that of the producers themselves. This high energy flow is maintained by a rapid turnover of individual plankton, rather than by an increase of total mass.

Another pyramid commonly found in ecological literature is the pyramid of numbers (Figure 5-15a). This pyramid was advanced by Charles Elton (1927), who pointed out the great difference in the numbers of organisms involved in each step of the food chain. The animals at the lower end of the chain are the

most abundant. Successive links of carnivores decrease rapidly in number until there are very few carnivores at the top. The pyramid of numbers often is confused with a similar one in which organisms are grouped into size categories and then arranged in order of abundance. Here the smaller organisms again are the most abundant, but such a pyramid does not indicate the relationship of one group to another.

The pyramid of numbers ignores the biomass of organisms. Although the numbers of a certain organism may be greater, their total weight may not be equal to that of the larger organisms. Neither does the pyramid of numbers indicate the energy transferred or the use of energy by the groups involved. And because the abundance of members varies so widely, it is difficult to show the whole community on the same numerical scale.

## Model of energy flow

The concept of energy flow in ecological systems is one of the cornerstones of ecology. The model was first developed by Raymond Lindeman in 1942. It is based on the assumptions that the laws of thermodynamics hold for plants and animals, that plants and animals can be arranged into feeding groups or trophic levels, that at least three trophic levels—producer, herbivore, and carnivore—exist, and that the system is in equilibrium.

In the Lindeman trophic-dynamic model, the energy content or standing crop of any trophic level is designated by the Greek capital lambda, $\Lambda$. This letter is followed by a numerical subscript to denote trophic level. Thus $\Lambda_1$ represents the energy content of the producers, $\Lambda_2$ that of the herbivores, and so on. And $\Lambda_n$ indicates any designated trophic level. Energy is continuously entering and leaving (the dynamic aspect), and the contribution of energy through time from one trophic level to another is indicated by a lower-case lambda, $\lambda$. $\lambda_1$ represents the proportion of energy the organisms on any one trophic level, $\Lambda_n$, receive from the trophic level below. The loss of heat or respiration, $R$, plus the energy passed on to the next trophic level, $\Lambda_{n+1}$, is symbolized as $\Lambda_{n'}$. Thus for any trophic level designated as

**141**

$\Lambda_n$, $\lambda_{n'}$ represents $\lambda_{n + 1} + R$, and $\lambda_{n + 1}$ is the amount of energy passed on to the next trophic level, $\Lambda_{n + 1}$.

From this model one can write a generalized formula for energy flow from one trophic level to another:

$$\frac{\Delta \Lambda_n}{\Delta t} = \lambda_n - \lambda_{n'}$$

**FIGURE 5-16**

*Models of energy flow through the individual organism, through the population, and through the ecosystem. Note the losses and the portion of energy accumulated as growth in the organism. The ecosystem model considers the decomposers as occupying one of the several trophic levels rather than as a separate trophic pathway. For energy flow models based on field data, see examples in Part III.*

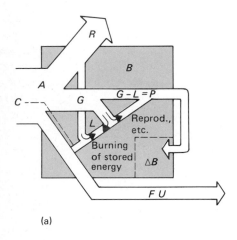

(a)

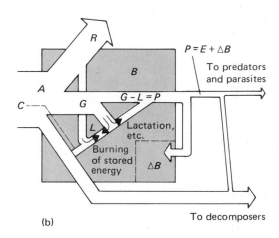

(b)

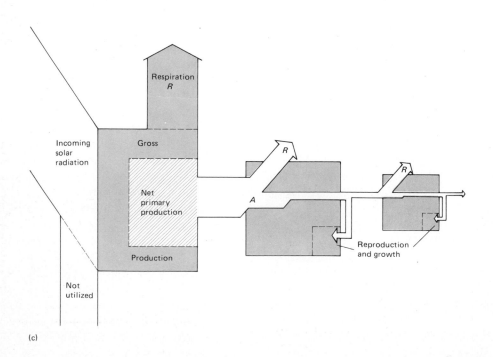

(c)

What this equation states is that the rate of change of the energy content of a trophic level is equal to the rate at which energy is assimilated minus the rate at which energy is lost from it. In the formula $\lambda_n$ is positive and represents the contribution of energy from the previous trophic level, $\Lambda_{n-1}$, and $\lambda_{n'}$ represents the sum of energy lost from $\Lambda_n$.

In the Lindeman model the amount of energy $\lambda_n$ transferred to the next trophic level $\Lambda_{n+1}$ is considered true productivity. The assumption, of course, is that all of this production must represent the gross primary production at the first trophic level and assimilation at the several consumer levels. As used in many studies, this has not been the case for the model because assimilation by decomposers at each level has been ignored, and obviously all of the production has not been assimilated by those organisms usually considered to be involved in the various trophic levels.

Whether at the producer or consumer level, energy flow through the ecosystem is mediated at the level of the individual (Figure 5-16a). A quantity of energy of food is consumed. Part of it is assimilated, and part is lost as feces, urine, and gas. Part of the assimilated energy is used for respiration, and part is stored as new tissue which can be utilized to some extent as an energy reserve. Part is used for growth or the production of new individuals.

A model of energy flow through a population (Figure 5-16b) obviously exhibits many of the characteristics of the model for the individual, but with some additions. Growth in the individual becomes changed in standing crop. Part of the biomass goes to predators and parasites, and there are gains and losses of energy and biomass from and to other ecosystems.

The population boxes can be fitted into several trophic levels and linked to form a model of energy flow through the ecosystem (Figure 5-16c). In the model there is no attempt to separate out the detritus-feeders and the decomposers. Contrary to what occurs in the Lindeman model, they must fall into one of several trophic levels when the food web is collapsed.

To provide for the grazing and detritus paths of energy transfer, Wiegert and Owens (1971) propose a somewhat different model of energy flow in which transfers through the two channels beyond the autotrophs are considered (Figure 5-17). They define all organisms utilizing living material as biophages, and all organisms utilizing nonliving matter as saprophages. First-order biophages utilize living plants and are the traditional herbivores, whereas first-order saprophages feed on dead plant material as well as organic material egested by first-order biophages. First-order biophages in turn are utilized at death by second-order saprophages, which are really functional carnivores. In turn, first-order saprophages may be utilized by second-order biophages. Although this model, like others, cannot separate out organisms that occupy several trophic levels, it does modify the Lindeman model so that the decomposers are broken down into functional components.

**FIGURE 5-17**
*Model of two-channel energy flow. The model separates energy flow into two pathways, one utilized by organisms that feed on living material, the biophages, and the other utilized by organisms that feed on nonliving material, the saprophages. (After Weigert and Owen, 1971.)*

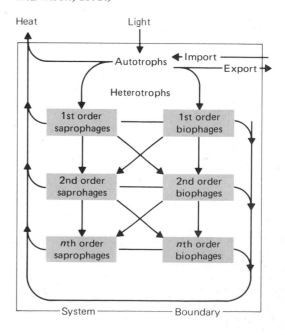

## Ecological efficiencies

The Lindeman model stimulated the study of energy flow through populations and communities and has given rise to various estimations of ecological efficiencies to the point that a great deal of confusion exists over the real meaning of the concept. Ecological efficiency is the ratio of any of the various parameters of energy flow in or between trophic levels, populations, or individual organisms (Table 5-4) (Kozlovsky, 1968). Efficiencies of individual organisms are more physiological than ecological.

Although the value of many ecological efficiencies might be dubious, a few are of real interest (see Tables 5-1, 3-3). Before they can be compared, the parameters involved must be defined. One parameter, ingestion $(I)$ is the quantity of food or energy taken in by an organism, be it a consumer, producer, or saprophage. In the case of producers $I$ is the amount of light available to or absorbed by the photosynthetic pigments. Assimilation $(A)$ is the amount of food absorbed in the alimentary canal by consumers, the absorption of extracellular products by decomposers, and the energy fixed by plants in photosynthesis. Respiration $(R)$ is all of the energy lost in metabolism and in activity, including loss in urine. Net productivity $(NP)$ is the energy accumulated and represents that left over after respiration: $NP = A - R$. Production $(P)$ is that portion of productivity a trophic level passes on to the next trophic level. It is that portion actually available and does not include losses to decomposers (if decomposers are not considered a part of the trophic level), losses to other systems, or increases and decreases to the standing crop. In this context production must be considered yield.

These efficiencies vary among species, populations, and trophic levels (see Table 5-3). Growth efficiencies among larger animals appear to be less than among small animals, and greater among young animals than older ones. Assimilation efficiencies appear to be higher among carnivores than among herbivores, but at the same time respiration in proportion to both ingestion and assimilation also increases at higher trophic levels. As a result, net productivity and production decreases in proportion to ingestion at carnivore levels. Although it may vary widely, the ratio of assimilation between one trophic level and another, $A/A_{n-1}$, is about 10 percent, and the net productivity

TABLE 5-4
*Definitions of selected energetic efficiencies*

| | | | |
|---|---|---|---|
| Assimilation efficiency | $\dfrac{A_n}{I_n}$ | $\dfrac{\text{energy fixed by plants}}{\text{light absorbed}}$ | (for producers) |
| | | $\dfrac{\text{food absorbed (assimilation)}}{\text{food ingested}}$ | (for consumers) |
| Ecological efficiency | $\dfrac{P_n}{I_n}$ | $\dfrac{\text{energy passed to } n+1}{\text{energy ingestion at } n}$ | |
| Ecological growth efficiency | $\dfrac{NP_n}{I_n}$ | $\dfrac{\text{net production at } n}{\text{ingestion at } n}$ | |
| Growth efficiency | $\dfrac{NP_n}{A_n}$ | $\dfrac{\text{net production at } n}{\text{assimilation at } n}$ | |
| Trophic-level production efficiency | $\dfrac{A_n}{NP_{n-1}}$ | $\dfrac{\text{assimilation at } n}{\text{net production at } n-1}$ | |
| Utilization efficiency | $\dfrac{I_n}{NP_{n-1}}$ | $\dfrac{\text{ingestion at } n}{\text{net production at } n-1}$ | |

*Source:* After Kozlovsky, 1963.

of one level compared to the net productivity of the previous level is virtually constant.

## Community energy budgets

Because energy flow involves both inputs and outputs, the efficiency and production of an ecosystem can be estimated by measuring the quantity of energy entering the community through the various trophic levels and the amount leaving it. Such information can give *net ecosystem production* (or net community production, as it is often called), which can be expressed as follows:

Net ecosystem production (biomass accumulation) = gross primary production − plant respiration − animal respiration − decomposer respiration‚

A balance sheet for energy flow and production with debit and credit sides can be drawn up for a community (Table 5-5). Few communities have been studied intensively enough to present such a broad picture, but some studies are available. One is a salt marsh, an autotrophic community (Teal, 1962); and another a cold spring, a hetero-trophic community (Teal, 1957), whose major energy source was plant material fallen into the water.

Of the energy transformed by organisms in the spring, 76 percent entered as leaves, fruit, and branches of terrestrial vegetation. Photosynthesis accounted for 23 percent, and 1 percent came from immigrating caddisfly larvae. Of this total input, 71 percent was dissipated as heat, 1 percent was lost through the emigration of adult insects, and 28 percent was deposited in the community. In the salt marsh the producers themselves were the most important consumers, for plant respiration accounts for 70 percent of gross production, an unusually high figure. (Scots pine, for example, utilizes only about 10 percent in respiration.) Plants are followed by the bacteria, which utilize only one-seventh as much as the producers. Primary and secondary consumers come in a poor third, using only one-seventh as much energy as the bacteria.

Researchers at present may be able to construct such energy budget sheets, but these examples point out how incomplete and fragmentary is our knowledge of energy flow through ecosystems. To understand something of energy flow through ecosystems, one has to

TABLE 5-5
*Community energy balance sheets for an autotrophic community and a heterotrophic community*

| AUTOTROPHIC COMMUNITY: THE SALT MARSH[a] | |
|---|---|
| Input as light | 600,000 kcal/m²/yr |
| Loss in photosynthesis | 563,620, or 93.9% |
| Gross production | 36,380, or 6.1% of light |
| Producer respiration | 28,175, or 77% of gross production |
| Net production | 8,205 kcal/m²/year |
| Bacterial respiration | 3,890, or 47% of net production |
| First-level consumer respiration | 596, or 7% of net production |
| Second-level consumer respiration | 48, or 0.6% of net production |
| Total energy dissipation by consumers | 4,534, or 55% of net production |
| Export | 3,671, or 45% of net production |
| HETEROTROPHIC COMMUNITY: TEMPERATE COLD SPRING[b] | |
| Organic debris | 2,350 kcal/m²/year, or 76.1% of available energy |
| Gross photosynthetic production | 710 kcal/m²/year, or 23.0% of available energy |
| Immigration of caddis larvae | 18 kcal/m²/year, or 0.6% of available energy |
| Decrease in standing crop | 8 kcal/m²/year, or 0.3% of available energy |
| Total energy dissipation to heat | 2,185 kcal/m²/year, or 71% of available energy |
| Deposition | 868 kcal/m²/year, or 28% of available energy |
| Emigration of adult insects | 33 kcal/m²/year, or 1% of available energy |

*Source:* [a]Teal, 1962; [b]Teal, 1957.

know something of energy flow through populations within the ecosystem and then relate this information to the flow of energy from one trophic level to another. Herein lies the weakness of the model. The energy budgets drawn to date are based in part on assumptions rather than on known values of energy flow. Too little is known of energy flow through any population to draw a clear picture of energy flow through an ecosystem. And what knowledge we do possess often is unreliable. If this knowledge is to be used, one has to assume that energy flow through a population is constant or, if fluctuating, at least predictable. But probably it is not, because energy flow through populations varies with ecological conditions. For example, growth, and thus energy storage by salmonid fish, is related to the size of food (Paloheimo and Dickie, 1966). When particle size is small, fish expend more energy obtaining food, and a smaller proportion is used in growth. Variations in temperature affect the rate of food assimilation, and variations in salinity and nitrogenous wastes in the water affect both energy turnover and efficiency of utilization.

Within any ecosystem animal populations may have a pronounced influence on the rate of energy fixation of plants. Overgrazing and overbrowsing reduce the amount of primary production in grassland and forest ecosystems. Overexploitation of a prey species either by humans or by predators can affect the amount of secondary production. The nutrient composition of soils or plants can limit energy fixation and storage in both primary and secondary production. Lack of boron in the soil, for example, can severely depress the growth of alfalfa; and the low nutrient status of a soil, and thus of the plants, can affect the production of animals. We lack enough detailed studies of trophic efficiency to give us any clear picture of the structure of ecosystems or to give strong support to the Lindeman model. Because they involve populations rather than trophic levels, studies to date neither support nor refute the model of energy flow through the ecosystem. But the concept of energy flow is valuable as a guideline for future studies and as a basis for understanding some of the relationships and interactions of humans and their environment.

## Our place in the food chain

Humans, like all other organisms, are intimately tied to the food chain and energy flow. However, we have moved out of the natural food chain into an artificial one that is manipulated toward our own ends and is dependent upon a large input of energy in addition to that supplied directly to producers by the sun.

Humans, the dominant consumer organism on earth through their history, have altered their role in energy flow and relationships in the food chain. For over 99 percent of their 2-million-year life on earth, humans existed as hunter-gatherers, a part of the natural ecosystem in which they existed. Energy input consisted largely of muscular energy needed to hunt game and gather seeds and fruits. About 10,000 years ago humans modified natural food chains and energy source to meet their own needs. They discovered ways of raising the plants whose seeds and roots they gathered and of domesticating the animals they hunted. In time they used two animals, oxen and horses, as a source of energy input into their agricultural efforts and developed irrigation techniques to supplement natural rainfall. After the industrial revolution humans largely replaced raising a number of crops together (polyculture) with raising single crops (monoculture). This shift came about largely by the development of planting and harvesting machinery and by the replacement of draft animals as an energy input with fossil-fueled tractors. The development of hybrid varieties of plants, advances in animal breeding and nutrition, and use of fertilizers have made agriculture ecologically artificial and placed it largely outside of natural food chains. (For a review of our changing place in the food chain, see R. L. Smith, 1976.)

Today 88 percent of our global food supply comes from plants. It is significant that in only four countries, Canada, the United States, Australia, and New Zealand, do humans obtain more calories from meat, milk, and eggs than from starchy food (Brown and Finsterbusch, 1972). The remainder of the world, to varying degrees, lives on the first consumer level.

Although humans have been growing food

for thousands of years and although they have applied knowledge of plant breeding and genetics to the improvement of food sources, they still rely on the basic crops that have been with them since the early days of agriculture.

The dominant staple foods are wheat, rice, and corn. Wheat is the principal staple in almost all high-income countries. It is the world's most widely cultivated and traded cereal, it supplies 20 percent of human calorie intake, and it provides a third of the world's population with food.

Rice is the principal food of the world's poor. Although a staple food in only 16 countries, those lands contain half the world's population. Providing 20 percent of calorie intake, rice yields twice as much per acre as wheat. This higher yield supports a much higher population.

The third ranking staple food is corn. The leading grain of 14 countries, including most of Latin America and many African countries, corn supplies about 5 percent of humans' energy needs.

An assortment of plants provide the rest of the energy. Millet, sorghum, and rye provide about 7 percent of energy intake. Starchy roots, particularly potatoes, sweet potatoes, and yams, provide another 5 percent; cassava (known in developed nations as tapioca) supplies about 2 percent, mainly in several South American and African countries. Fruits, sugar, and vegetables supply another 9 percent; fats and oils provide a little less than 9 percent; and livestock products and fish supply the remaining 12 percent (Brown and Finsterbusch, 1972).

The poorer nations of the world, particularly those in southeastern Asia, live close to primary production. Cereal crops are consumed directly; primary production is channeled directly to humans. In richer countries more and more of the grains are fed to livestock, and man moves up one step in the trophic level. Corn that provides the staple diet in Latin America is converted to pork and beef in the United States and Canada. Thus the direction of the food chain differs in various parts of the world. In monsoon Asia the major link is rice to human, in Latin America it is corn to human, and in Europe and North America it is wheat to human.

Peoples of poorer regions subsist on monotonous diets of cereal grains. In Africa the poorest groups live on even cheaper starchy food, such as potatoes, sweet potatoes, and yams. In the richer nations the diet becomes more diversified and there is greater dependence on animal protein. In North America 31 percent of daily food energy is in the form of meat, milk, eggs, and fish, in western Europe the proportion is about 22 percent, and in Oceania it reaches 36 percent. This is in sharp contrast to eastern Europe and Russia, where animal products contribute only 15 percent, and Asia, where the proportion drops to 5 percent.

These statements do not imply that diets high in animal proteins and fats are necessarily superior. There is increasing evidence that high protein diets may be somewhat maladaptive.

## HUMANS, FOOD, AND THE FUTURE

The number of organisms supported at any position in a food chain theoretically depends upon the limits of the energy supply available. For humans the supportable levels of populations depend upon the limits of photosynthesis.

World primary production amounts to approximately $6.9 \times 10^{17}$ kcal/yr (Leith, 1973, 1975). In 1970 approximately $14 \times 10^6$ km$^2$ of arable land in the world yielded $11,000 \times 10^6$ tn/yr dry weight of net production. Of this 12 percent was available as harvested food ($1000 \times 10^6$ tn/yr of cereal grains, $200 \times 10^6$ tn/yr of other food crops). In addition about $30 \times 10^6$ km$^2$ of pasture and range land produced in dry weight $20 \times 10^6$ tn/yr of meat, $4.7 \times 10^6$ tn of eggs, and $49 \times 10^6$ tn of milk. World harvest of aquatic organisms amounted to $17 \times 10^6$ tn dry weight. Of this 88 percent came from the oceans, which represents only 0.027 percent of total marine productivity. Total food harvest of dry weight of $1.22 \times 10^9$ plant matter and $80 \times 10^6$ of animal matter is only 0.75 percent of the net primary production of the world (Whittaker and Liken, 1975).

deWit (1967) has estimated the maximum sustainable human population that might be supported by usable agricultural net primary production (Table 5-6). The calculations are based on a number of selected variables: (1) a leaf scattering coefficient of 0.3; (2) a photo-

TABLE 5-6
*Potential productivity of Earth and the population it could support*

| North latitude (degrees) | Land surface ($10^8$ ha) | Number months above $10°$ C | Carbohydrate/ha/yr ($10^3$ kg) | M²/HUMAN TO SUPPORT LIFE | | | | Agricultural land (%) |
|---|---|---|---|---|---|---|---|---|
| | | | | NO ALLOWANCE FOR URBAN AND RECREATIONAL NEEDS | | 750 M²/HUMAN ALLOWANCE FOR URBAN AND RECREATIONAL NEEDS | | |
| | | | | $m^2/$ human | Number human ($\times 10^9$) | $m^2/$ human | Number human ($\times 10^9$) | |
| 70 | 8 | 1 | 12 | 806 | 10 | 1556 | 5 | 52 |
| 60 | 14 | 2 | 21 | 469 | 30 | 1219 | 11 | 38 |
| 50 | 16 | 6 | 59 | 169 | 95 | 919 | 17 | 18 |
| 40 | 15 | 9 | 91 | 110 | 136 | 860 | 18 | 13 |
| 30 | 17 | 11 | 113 | 89 | 151 | 839 | 20 | 11 |
| 20 | 13 | 12 | 124 | 81 | 105 | 831 | 16 | 10 |
| 10 | 10 | 12 | 124 | 81 | 77 | 831 | 11 | 10 |
| 0 | 14 | 12 | 116 | 86 | 121 | 836 | 17 | 10 |
| −10 | 7 | 12 | 117 | 85 | 87 | 835 | 9 | 10 |
| −20 | 9 | 12 | 123 | 81 | 112 | 831 | 11 | 10 |
| −30 | 7 | 12 | 121 | 83 | 88 | 833 | 9 | 10 |
| −40 | 1 | 8 | 89 | 113 | 9 | 863 | 1 | 14 |
| −50 | 1 | 1 | 12 | 833 | 1 | 1583 | 1 | 53 |
| Total | 131 | | | | 1022 | | 146 | |

*Source:* From deWitt, 1968.

synthetic rate of 20 kcal/ha/hr; (3) a leaf area index of 5.0; (4) a leaf display similar to that of young grass or grain; (5) photosynthesis taking place on a clear day; (6) latitudinal variations. He further assumes that each human requires $10^6$ kcal/yr. Thus the number of people who can exist on a hectare equals the total photosynthesis of the hectare expressed in tons. Based on these assumptions, the productivity of the earth is calculated for each 10° of latitude.

But human populations also need space. To meet this need, deWit adds 750 m² per person to support urban needs, recreation, and the like. The calculation cannot take into account the areas of land not suitable for agriculture, urban use, and so on. Of the earth's total land area only about 15 percent is available for agricultural purposes.

The calculations assume that humans will occupy the first consumer level in the food chain. If, however, they desire animal protein at the rate of 200 g/day, then about 5000 kcal

of forage will be needed per 500 kcal of meat, requiring twice the amount of land for agricultural purposes. If the need for space is doubled, the population would have to be halved from 146 billion to 73 billion. Of course, all of this is highly theoretical, pointing out how many people under ideal photosynthetic conditions the earth might possibly support.

On a more realistic level the supportable population, in terms of energy, depends upon the limitations of land and nutrient resources and upon the care that humans give to the planet. To date the care of the planet has not been aimed at increasing productivity. The basic support system of agriculture is soil. Wherever humans have cultivated the earth, they have caused a deterioration of the soil. Agricultural activities are increasing the amount of desert and wasteland, especially in North Africa and western India. In spite of technological ability to prevent soil erosion, it continues. Add to losses from erosion the loss of land to uses other than agriculture. In the

United States thousands of acres of farmland are lost each year to highways, urban expansion, and surface mining for coal. In fact three-fourths of the land invaded by urban expansion had been high quality cropland.

Today virtually all of the land that can be cultivated economically is under cultivation. Land only marginally suited to cultivation requires expensive and superior technology.

Much of the marginal land available requires irrigation, but irrigation is limited by the availability of fresh water and is energy demanding. Use of desalinized seawater is too expensive. The use of fossil energy to remove the salt, the transport of seawater inland, the disposal of waste brine create great problems in utilizing the ocean as a source of irrigation water. Today the cost of using fresh water for irrigation is greater than its value to agriculture. And in parts of the world irrigation results in poorly drained salty soils that eventually have to be flushed or abandoned.

Air pollution threatens to reduce the productivity of agricultural lands. As air pollution increases, the injury to crops increases, cutting yields (Howell and Kremer, 1970). Because of increased sulfur dioxide and nitrogen oxides in the air, rainwater is becoming more acid; this could have adverse effects on crop production as well as on forest growth in various parts of the world (see Chapter 6).

Much of the recent increase in world food production has come about not by increasing land already cultivated. The yield increases are the result of input of fossil fuel, labor, and fertilizer into croplands and the use of improved crop varieties. The replacement of open pollinated varieties of corn with hybrids and the use of herbicides have increased the yield of corn per acre in the United States nearly threefold since 1940, and the yield of rice per acre in Japan has more than doubled in that time (Figure 5-18). In these instances the maximum production per acre has been achieved. As nonrecurring sources of productivity are exhausted and as inputs of fertilizer and cultural techniques no longer yield significant returns, increases in the rate of yield per acre begin to slow. This results in a typical S-shaped growth curve (Figure 5-19). Although yield increases are continuing at a

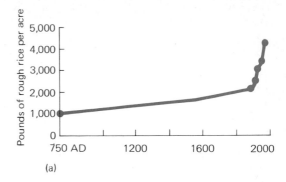

(a)

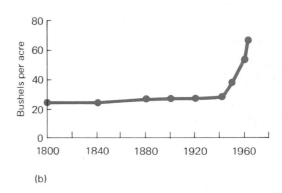

(b)

**FIGURE 5-18**
*Increases in yield of rice in Japan and corn in the United States. Expansion in rice production resulted largely from the use of fertilizers and the hand-planting of rice in rows. Expansion of corn production resulted from the replacement of open-pollinated varieties with hybrid corn and the use of herbicides, reducing competition from weeds. (From U.S. Department of Agriculture, 1970).*

rapid rate in some countries, they are slowing for some crops, such as wheat in the United States and rice in Japan. Eventually all crop production will reach this upper-yield plateau.

The problem then becomes one of attempting to achieve new plateaus. This might come about through the development of new food plants. Of the 700,000 plants in the world, nearly 80,000 are edible, but only 3,000 are considered crop plants, and of these only 300 are in abundant use. Only 12 species or genera supply man with 90 percent of his food. New sources of food may be found among unexploited plants.

As the world's population increases, man tends to look to the sea for food. But the sea

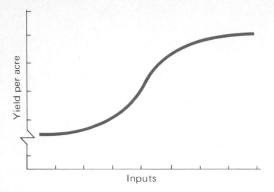

*FIGURE 5-19*
*Growth curves for yield increases. An*
*S-shaped curve results when nonrecurring or*
*"one-shot" sources of increased production*
*(such as the replacement of open-pollinated*
*varieties with hybrid varieties) are reduced*
*until eventually the rate of increased*
*production slows and levels off. Further*
*inputs of resources do not result in increased*
*yields. (From U.S. Department of*
*Agriculture, 1970.)*

has only a limited potential. This may seem rather surprising when one considers that the sea contains 90 percent of the world's vegetation and fixes from 18 to 20 billion tons of carbon a year. The productivity of the sea varies greatly. The open sea fixes approximately 50 g $C/m^2$, shallow waters and estuaries twice as much, and areas of upwelling perhaps six times as much. Occupying only 1 percent of the oceans, the regions of upwelling supply half the world's ocean fish production; the coastal waters and estuaries furnish the other half. In terms of fishing, the open sea is a watery desert.

Fish production has increased steadily over the years, reflecting intensified fishing effort and improvement in fishing gear and methods. It is the one resource in which the increase in production is higher than the rate of population growth. Today ocean fisheries supply 18 percent of the world's protein. But most fish protein is used to supplement livestock and poultry feeds in the richer countries of the world.

Based on the photosynthetic capacity of the sea, 240 million tons of fish is the maximum production, of which 100 million tons could be harvested on a sustained-yield basis (Ryther,

1969). To make the most of fish production humans will have to move to a lower trophic level. As it is now, the harvest comes from the upper trophic levels. Flounder and herring are on level three; tuna and halibut on level four. Every 1.5 oz of tuna fish requires 500 lb of phytoplankton. To obtain a larger harvest, humans will have to utilize the less palatable, less attractive species that now are thrown away. Yet a product of those unpalatable fish, fish flour or fish protein concentrate, has not been widely accepted, for no way has been found to incorporate it successfully into other foods (Holden, 1971).

The productivity of the sea is further limited by man himself. A major problem is over-exploitation. The rapid increase in fish yields has been at the expense of the stock of the fish. Of the 30 major fish stocks, 14 have been over-fished, and some have been eliminated as commercial species. Their populations are too low to be harvested economically. Among these are the Pacific sardine, the Atlanto-Scandinavian herring, the Berents Sea cod, the menhaden, and the haddock. Because the ocean is extraterritorial, no enforceable international regulations exist to control the take. Each nation takes as much as possible, with no regard for the future. The estimated 100-million-ton production was based on a sustained-yield management.

Pollution endangers much of the productive coastal waters. Although the region of upwelling may escape for the time any serious pollution, the coastal waters are becoming heavily polluted from sewage and industrial wastes, and are being destroyed by dredging and filling. These coastal waters are the nursery ground for many of our most valuable species—flounder, striped bass, and bluefish, as well as the oyster, clam, shrimp, and other shellfish. Highly vulnerable to pollution, the coastal waters are rapidly declining as the nursery of commercial species. Unless abated, pollution can eventually destroy half the world's ocean fisheries.

## SUMMARY

**A basic functional characteristic of the ecosystem is the flow of energy. The**

energy of sunlight is fixed by the autotrophic component of the ecosystem, green plants, as primary production. This energy is then available to the heterotrophic component of the ecosystem of which the herbivores are the primary consumers. The herbivores in turn are a source of food for the carnivores. At each step or transfer of energy in the food chain a considerable amount of potential energy is lost as heat, until ultimately the amount of available energy is so small that few organisms can be supported at that source alone. The animals high on the chain often utilize several sources of energy, including plants, and thus become omnivores. All food chains eventually end with the decomposers, mainly bacteria and fungi, that reduce the remains of plants and animals into simple substances. Energy flow in the ecosystem may take two routes: One goes through the grazing food chain, the other through the detritus food chain, in which the bulk of the production is utilized as dead organic matter by the decomposers.

The loss of energy at each transfer limits the number of trophic levels or steps in the food chain to four or five. At each level the biomass declines; if the total weight of individuals at each successive tropic level is plotted, a sloping pyramid is formed. In certain aquatic situations, however, where there is a rapid turnover of small aquatic consumers, the pyramid of biomass may be inverted. Energy, however, always decreases from one trophic level to another and is pyramidal.

The ratio of energy flow in or between trophic levels of natural communities or in or between individual organisms is ecological efficiency. Because efficiencies are dimensionless, any number of ratios can be determined. Among some of the most useful are assimilation efficiencies, growth efficiencies, and utilization efficiencies.

Although knowledge of energy flow in ecosystems is fragmentary and difficult to come by, the concept of energy flow is a valuable guide for the study and understanding of both ecosystem functioning and the relationship of humans to their environment.

The most significant change in the place of humans in the food chain came when they switched from a hunting-gathering society, depending mostly on animals for food, to an agricultural society, depending largely on plants. Maintenance of agricultural ecosystems depends upon heavy input of fossil fuels for production and distribution of food stuffs and for protection from pests and for fertilization. As in the past human populations obtain 80 percent of global food energy from plants, largely rice, wheat, and corn. Humans live mostly at the herbivore level.

# Biogeochemical cycling in the ecosystem
## CHAPTER 6

The existence of the living world depends upon the flow of energy and the circulation of materials through the ecosystem. Both influence the abundance of organisms, the metabolic rate at which they live, and the complexity and structure of the ecosystem. Energy and materials flow through the community as organic matter; one cannot be separated from the other. Certain elements are essential for energy fixation and flow; mineral cycling cannot function without the input of energy. One feeds upon the other in an apparent positive feedback loop; the availability of nutrients limits primary productivity; the more nutrients available the more energy flows; increased energy flow produces greater cycling of nutrients. What prevents the system from running away is a negative feedback loop of limits on available nutrients, the energy requirements of resource utilization, and limits on the ability of the system to regenerate nutrients. In essence two forces are at work. One is the tendency within ecosystems for living organisms to store resources in the form of growth. This results in the biotic structure of the ecosystem. The other force is dissipative,

the degradation of organic matter into inorganic compounds.

## Model of nutrient flow

A general model of nutrient flow, Figure 6-1, describes the relationship among the various components of the ecosystem. The food base depends upon: (1) the available nutrients, which in turn depend upon inputs into the detrital pool; (2) the rate at which detritus is decomposed; (3) the amounts of detritus and available nutrients that go into the storage in soil, humus detrital sediment, and the like; and (4) the release of nutrients from the reserve.

How nutrient utilization takes place in an ecosystem depends upon the the size of the abiotic nutrient reserve, the proportions of nutrients stored in the biotic component and the abiotic environment, and the rate of turnover of the recycling pool of nutrients, that is, how fast nutrients are passed from the detrital pool to available form through leaching and decomposition and their subsequent uptake. Because of nutrient limitation and the necessity to control the rate of nutrient flow to prevent overutilization, two mechanisms of nutrient utilization relative to energy flow exist (O'Neill et al., 1975).

One mechanism is to put energy conversion in a number of plant populations of low biomass with different responses to environmental conditions. These plant populations are capable of rapid reproduction when conditions are optimal. One or more populations can expand rapidly and take up nutrients quickly. A succession of plant populations through the season ensures continuous energy fixation and utilization of nutrients.

A second mechanism places energy conversion in individuals of great bulk and with slow reproductive rates. Large individuals are able to survive unfavorable conditions and large quantities of nutrients are stored in biomass.

Although the first mechanism is characteristic of aquatic ecosystems and the second of terrestrial ecosystems, especially forests, neither mechanism operates exclusively in only one type of ecosystem. Lakes have rooted aquatics with biomass accumulation, and forests have a seasonal parade of herbaceous understory plants.

Each mechanism has provisions for the conservation of nutrients. One provision is a large pool of organic matter. Litter and soil organic matter and the structural mass of autotrophs form such a pool in terrestrial ecosystems. In aquatic ecosystems the pool is particulate and dissolved organic matter. Turnover of the organic pool is relatively slow, roughly one magnitude lower than the turnover of the vegetative component (O'Neill et al., 1975). Organic matter has a key role in recycling nutrients because it prevents rapid losses from the system. Large quantities of nutrients are bound tightly in organic matter structure and are not readily available, but their release can be activated by decomposer organisms. This takes place mostly in those ecosystems in which the persistent standing crop of organic matter in the presence of adequate moisture and energy is determined by the supply of nutrients and the recycling mechanisms. If energy and especially water are limited, as they are in desert ecosystems, then recycling is minimal because the ecosystem cannot attain a standing crop large enough to deplete the nutrient supply.

**FIGURE 6-1**
*Model of nutrient flow of an ecosystem. Arrows indicate direction of flow (F) of nutrients between compartments represented by numbered blocks. The subscripts 0,1,2, etc., represent compartment numbers. Thus, $F_{3,1}$ indicates flow of nutrients from compartment 1 to compartment 3; $F_0$ indicates outflow to the environment; and $F_{0,3}$ is outflow to the environment from compartment 3. (From J. R. Webster, J. B. Waide, and B. Patten, 1975)*

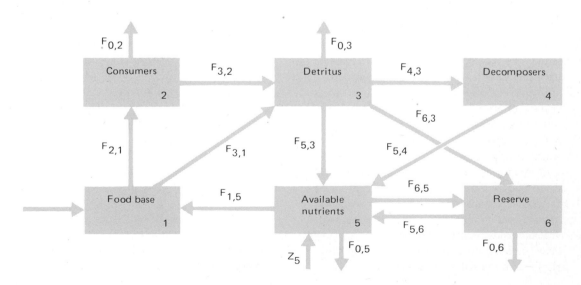

Another mechanism is to partition the nutrient reserve between short-term and long-term nutrient pools. For example, of the structural components of individual plants—wood, bark, twigs, and leaves—leaves are recycled the fastest and wood the slowest. The leaves represent a short-term nutrient pool and wood a long-term reservoir. Day and McGinty (1975) studied nutrient cycling by several tree species on the Coweeta Watershed at Franklin, North Carolina. Among the forest tree species studied were chestnut oak *(Quercus prinus)*, rhododendron *(Rhododendron maxima)*, and flowering dogwood *(Cornus florida)*. Although chestnut oak had the largest total standing crop biomass and the largest total standing crop of nutrients, nutrients were apportioned among the various structural components of the different species in a manner that influenced nutrient cycling. The largest standing crop of potassium was in the wood of chestnut oak and the largest standing crop of calcium and magnesium was in the bark of chestnut oak. The evergreen rhododendron had the highest concentration and the highest nutrient standing crop of nitrogen, calcium, magnesium, and potassium in the twig compartment, which was slowly recycled. Rhododendron also had the highest standing crop of leaf biomass. Chestnut oak, however, had the highest standing crop of nitrogen and potassium in that compartment. Rhododendron leaves had the highest standing crop of calcium and magnesium, but because the leaves are evergreen, the nutrients are recycled over a period of 7 years instead of 1 year. The leaves of dogwood concentrate over three times as much calcium per unit leaf biomass as oak and one and a half times as much per unit leaf biomass as rhododendron. Chestnut oak, because of its greater total biomass of leaves, recycled the most calcium through that compartment. However, the amount of calcium recycled by dogwood through its low leaf biomass—175 kg/ha compared to 845 kg/ha for chestnut oak—was 66 percent of that cycled by chestnut oak and 150 times more that that cycled by rhododendron. Because of its high wood to leaf ratio and its high concentration of calcium, the understory dogwood tree is very important in short-term cycling. Thus in the mature forest nutrients stored in vegetation are recycled at various time intervals from 1 to 100 years. This partitioning prevents excessive losses of nutrients and releases nutrients slowly to biogeochemical cycles.

## The water cycle

Leonardo da Vinci wrote, "Water is the driver of nature." Perceptive as he was, even he could not have appreciated the full meaning of his statement based on the scientific knowledge of his time. Without the cycling of water, biogeochemical cycles could not exist, ecosystems could not function, and life could not be maintained. Water is the medium by which materials make their never-ending odyssey through the ecosystem.

### DISTRIBUTION OF WATER

Although one views water as something of a local phenomenon, such as a stream or autumn rains, it forms a single worldwide resource distributed in land, sea, and atmosphere and unified by the hydrological cycle. It is influenced by solar energy, by the currents of the air and oceans, by heat budgets, and by water balances of land and sea. Through historical time the balance of free water has remained relatively stable although the balances between land and sea have fluctuated. According to the Russian hydrologist Shinitnikov (Kalinin and Bykov, 1969), at present we are passing from a humid period in earth's history to a dry one, in which the land areas are losing water at the rate of 105 mi³/yr, and the oceans are gaining that amount, rising on the average of 0.05 in./yr.

Oceans cover 71 percent of the earth's surface (Table 6-1). With a mean depth of 3.8 km (2.36 mi), they hold 93 to 97 percent of all the earth's waters (depending on the estimate used). Fresh water represents only 3 percent of the planet's water supply. Of the total fresh water on earth, 75 percent is locked up in glaciers and ice sheets, enough to maintain all the rivers of the world at their present rate of flow for the next 900 years. Thus about 2 percent of the world's water is tied up in ice. This leaves less than 1 percent of the world's water available as fresh. Freshwater lakes contain 0.3 percent of the freshwater supply, and at

*TABLE 6-1*
*World's water resources*

| Resource | Volume (W) ($10^3 \ km^3$) | ANNUAL RATE OF REMOVAL (Q) Volume ($10^3 \ kg^3$) | Process | Renewal period (T = W/Q) |
|---|---|---|---|---|
| Total water on earth | 1,460,000.0 | 520.0 | evaporation | 2,800 years |
| Total water in the oceans | 1,370,000.0 | 449.0 | evaporation | 3,100 years |
| | | 37.0 | difference between precipitation and evaporation | 37,000 years |
| Free gravitational waters in the earth's crust to a depth of 5 km in the zone of active water | 60,000.0 | 13.0 | underground runoff | 4,600 years |
| exchange | 4,000.0 | 13.0 | underground runoff | 300 years |
| Lakes | 750.0 | — | — | — |
| Glaciers and permanent snow | 29,000.0 | 1.8 | runoff | 16,000 years |
| Soil and subsoil moisture | 65.0 | 85.0 | evaporation and | 280 days |
| Atmospheric moisture | 14.0 | 520.0 | underground runoff precipitation | 9 days |
| River waters | 1.2 | 36.3* | runoff | 12 (20) days |

*Note:* Average error is probably 10–15 percent.
*Not counting the melting of Antarctic and Arctic glaciers.
*Source:* Kalinin and Bykov, 1969.

any one time rivers and streams contain only 0.005 percent of it. Soil moisture accounts for approximately 0.3 percent. Another small portion of the earth's water is tied up in living material.

Groundwater accounts for 25 percent of our fresh water. Groundwater fills the pores and hollows within the earth just as water fills pockets and depressions on the surface. Estimates, necessarily rough and inaccurate, place renewable and cyclic groundwater at $7 \times 10^6$ km³ (Nace, 1969), or approximately 11 percent of the freshwater supply. Some of the groundwater is "inherited," as in aquifers in desert regions, where the water is thousands of years old. Because inherited water is not rechargeable, heavy use of these aquifers for irrigation and other purposes is depleting the supply. In the foreseeable future the supply could be exhausted. A portion of the groundwater, approximately 14 percent, lies below 1000 m. Known as fossil water, it is often saline, and does not participate in the hydrological cycle.

The atmosphere, for all its clouds and obvious close association with the water cycle, contains only 0.035 percent fresh water. Yet it is the atmosphere and its relation to land and oceans that keeps the water circulating over the earth.

THE LOCAL WATER CYCLE

Outside it is raining. The rain strikes the house and runs down the windows and walls into the ground. It disappears into the grass, drips from the leaves of trees and shrubs, and trickles down the trunks. When the rain stops the windows and walls dry. The entire episode—the spring shower, the infiltration into the ground, the throughfall in the trees and bushes, the runoff from the walks, the evaporation—epitomizes the water cycle on a local scale (Figure 6-2).

*Precipitation* is the driving force of the water cycle. Whatever its form, precipitation begins as water vapor in the atmosphere. When air rises it is cooled adiabatically and when it rises beyond the temperature level at which condensation takes place, clouds form. The condensing moisture coalesces into droplets 1 to 100 $\mu$ in diameter and then into rain droplets with a diameter of approximately 1000 $\mu$ (1 mm). Where temperatures are cold enough, ice crystals may form instead. Particulate matter smaller than 10 m$\mu$ in the atmosphere may act as nuclei on which water

**155**

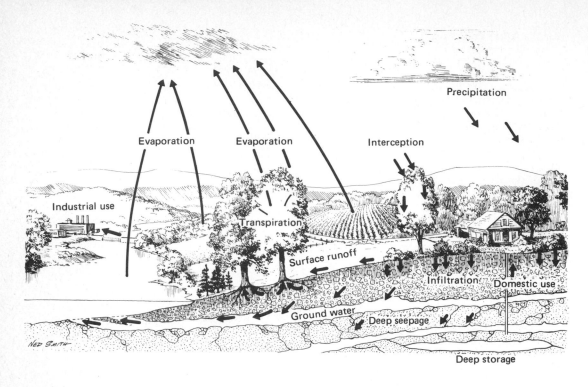

**FIGURE 6-2**
*The water cycle, showing major pathways through the ecosystem.*

vapor condenses. At some point the droplets or ice crystals fall as some form of precipitation. As the precipitation reaches the earth, some of the water reaches the ground directly, and some is intercepted by vegetation, litter on the ground, and urban structures and streets. It may be stored, run off, or in time infiltrate the soil.

Because of *interception,* various amounts of water evaporate into the atmosphere without ever reaching the ground. Grass in the Great Plains may intercept 5 to 13 percent of the annual precipitation (average loss, 7.9 percent), and litter beneath a stand of grass may intercept 2.8 to 8.0 percent of annual precipitation (average loss, 4.3 percent) (Corbett and Crouse, 1968). Precipitation striking forest trees or forest canopy must penetrate the crowns before it reaches the ground. A forest in full leaf in the summer can intercept a significant portion of a light summer rain (see Figure 6-3). The amount of rainfall intercepted depends upon the type and the age of the forest.

In general conifers intercept more rainfall on an annual basis than hardwoods. Mature pine stands, for example, will intercept 20 percent of a summer rainfall and 14 to 18 percent of the annual precipitation, whereas an oak forest will intercept 24 percent of a summer rain, but only 11 percent of the total annual precipitation (Lull, 1967).

In a deciduous forest in summer a relatively greater proportion of rainfall is intercepted during a light shower than during a heavy rain. During a light shower the rain does not exceed the storage capacity of the canopy and the water held by the leaves subsequently evaporates. Water exceeding the storage capacity of the canopy either drips off the leaves as *throughfall* or runs down the stem, twigs, and trunk as *stemflow.* The water then

**FIGURE 6-3**
*Yearly moisture regime of a 200-year-old upland oak forest in Russia. 1, interception of moisture by the crowns; 2, evaporation and transpiration by ground cover; 3, surface runoff; 4, transpiration by trees; 5, soil percolation; 6, loss to groundwater. (After A. A. Molchanov, 1960.)*

100%

14%

1

43%

14%

2

4

3

5

6%

6

19%

enters the soil in a relatively narrow band around the base of a tree. In winter deciduous trees intercept very little precipitation.

In urban areas a great portion of the rain falls on roofs and sidewalks, which are impervious to water. The water runs down gutters and drains to be hurried off to rivers. The percentage of rainfall striking impervious surfaces varies with the nature of the urban area. In downtown sections 100 percent of the surface may be impervious to rain. In residential areas imperviousness varies with the size of the lot. For lots of about 6,000 ft$^2$ about 80 percent of the area is impervious; for lots 6,000 to 10,000 ft$^2$, 40 percent; and for those 15,000 ft$^2$ and over, 25 percent (Antoine, 1964). In suburbia, with its expanse of lawns, impervious areas are smaller. On a 1.8-acre lot about 92 percent is lawn, 8 percent pavement and roof. A typical 0.4-acre lot is 24 percent roof and pavement (Felton and Lull, 1963). The largest paved areas in the suburbs are shopping centers, which require three to four times as much area for parking as for stores.

City streets may intercept, store, and eventually lose by evaporation 0.04 to 0.10 in. of water, and buildings 0.04 in. (Viessman, 1966). Estimates place interception by urban areas at about 16 percent, approximately the same amount intercepted by forest trees in summer leaf. Residential areas, surprisingly, intercept less; lawn grass 2 in. high intercepts 0.01 in. of water, and storage in leaf depression is about 0.04 in. (Felton and Lull, 1963). Total annual interception by residential areas comes to about 13 percent of a 45-in. rainfall (Lull and Sopper, 1969).

The precipitation that reaches the soil moves into the ground by *infiltration*. The rate of infiltration is governed by soil, slope, type of vegetation, and the characteristics of the precipitation itself. In general the more intense the rains, the greater rate of infiltration, until the infiltration capacity of the soil, determined by soil porosity, is reached. Because water moves through the soil by the action of two forces, capillary attraction and gravity, soils with considerable capillary pore space have initial rapid infiltration rates, but the pores rapidly fill with water. Water will infiltrate longer into soils with a high proportion of noncapillary pore space. Vegetation

that tends to roughen the surface retards surface flow and allows the water to move into the soil. Slope, impeding layers of stone or frozen soil, and other conditions influence the rate at which water moves into the soil.

Long wet spells and heavy storms may saturate the soil and intense rainfall or rapid melting of snow can exceed the infiltration capacity of the soil. When the soil can no longer absorb the precipitation, the water becomes overland flow. Sheet flow over the surface is changed to channelized flow as the water becomes concentrated in depressions and rills. This process can be observed even on city streets as water moves in sheets over the pavement and becomes concentrated in streetside gutters. Again the amount of runoff and associated erosion of soil depends upon slope, texture of the soil, soil moisture conditions, and the type and condition of vegetation.

In the undisturbed forest infiltration rates usually are greater than intensity of rainfall, and surface runoff does not occur. In urban areas infiltration rates may range from zero to a value exceeding the intensity of rainfall on certain areas where soil is open and uncompacted. Because of low infiltration, runoff from urban areas might be as much as 85 percent of the precipitation (Lull and Sopper, 1969). Because they are so compacted by frequent tramping and mowing, lawns have a low infiltration rate. Felton and Lull (1963) have demonstrated that water infiltrates in lawns at an average rate of 0.01 in./min, compared to 0.58 in./min in forest soil.

Water entering the soil will *percolate* or seep down to an impervious layer of clay or rock to collect as groundwater. From here the water finds its way into springs, streams, and eventually to rivers and seas. A great portion of this water is utilized by humans for domestic and industrial purposes, after which it reenters the water cycle by discharge into streams or into the atmosphere.

A part of the water is retained in the soil. The portion held by capillary forces between the soil particles is called *capillary water*. Another portion adheres as a thin film to soil particles. This is *hygroscopic water* and is unavailable to plants. The maximum amount of water that a soil can hold at one-third atmosphere of pressure after gravitational water is drained away is *field capacity*. Highly porous sandy soils have a low field capacity, while fine textured and humic soils have a high field capacity. Humus may retain 100 to 200 percent of its own weight in water. For each inch of humus about 0.8 in. of water is stored for as long as two days after a rain and is slowly discharged into streams (Lull, 1967). Storage in urban areas obviously is considerably less, almost nothing for paved areas.

Water remaining on the surface of the ground, in the upper layers of the soil, and collected on the surface of vegetation and water in the surface layers of streams, lakes, and oceans return to the atmosphere by *evaporation*. Evaporation is the movement of water molecules from the surface into the atmosphere at a rate governed by how much moisture the air contains (vapor-pressure deficit).

As the surface layers of the soil dry out, a dry barrier through which little soil water moves is created and evaporation ceases. Further water losses from the soil take place through the leaves of plants. Plants take in water through the roots and lose it through the leaves in a process called *evapotranspiration*. In the presence of sufficient moisture leaves remain turgid with the stomata fully open, which permits an easy inflow of carbon dioxide, but at the same time permits a large leakage of water. This leakage continues as long as moisture is available for roots in the soil, as long as the roots are capable of removing water from the soil, and as long as the amount of energy striking the leaf is enough to supply the necessary latent heat of evaporation. Thus plants can continue to remove water from the soil until the capillary water is exhausted. Some plants, known as phreatophytes, have roots that can reach and tap groundwater. For example, annual evapotranspiration in forests may range from 58 to 74 cm of water, but in urban areas it may be only 10 to 12 cm.

The temperate deciduous forest and the urbanized areas of the northeastern United States represent two environmental extremes in water cycling. In comparison to the forest, urbanized areas are characterized by reduced interception, less infiltration, much less soil moisture storage, less evapotranspiration, and reduced water quality. Urban areas also exhibit increased overland or surface flow, increased runoff, and increased peak flows of streams and rivers.

## THE GLOBAL WATER CYCLE

Once evaporated into the atmosphere or carried away by surface runoff, the water involved in the local hydrological cycle enters the global water cycle. The molecules of water that fell in the spring shower might well have been a part of the Gulf Stream a few weeks before and perhaps spent some time in the Amazon tropical rain forest before that. The local storm is simply a part of the mass movement and circulation of water about the earth, a movement suggested by the changing cloud patterns over the face of the earth. The atmosphere, oceans, and land masses form a single gigantic water system that is driven by solar energy. The presence and movement of water in any one part of the system affects the presence and movement in all other parts.

The atmosphere is one key element in the world's water system. At any one time the atmosphere holds no more than a 10- to 11-day supply of rainfall in the form of vapor, clouds, and ice crystals. Thus the turnover rate of water molecules is rapid. The source of water in the atmosphere is evaporation from land and sea, and therefore the amount of moisture in the atmosphere at any given point around the globe depends upon global variations in the rate of evaporation. Evaporation is considerably greater at lower latitudes than at higher latitudes, reflecting the greater heat budgets produced by the direct rays of the sun. The amount of moisture in the atmosphere over oceans is greater than over land, not only because there is more free water to evaporate, but also because oceans contain well over 90 percent of the world's water. Oceans provide 84 percent of total annual evaporation, considerably more than they receive in return from precipitation. Land areas contribute 16 percent, intercepting more water than evaporates from them.

Moisture in the atmosphere moves with the general circulation of the air. Air currents, hundreds of kilometers wide, are in fact giant unseen rivers moving above the earth. Only a part of this moisture falls as precipitation in any one place. For example, in a year's time the United States receives an unequally distributed 6000 km³ or 75 cm in depth of precipitation, but the liquid equivalent of water vapor passing over the country is 10 times that much. Atmospheric precipitation for the earth as a whole is approximately 100 cm in depth, and the average resident time for a water molecule in the atmosphere is approximately 10 days.

Variations in evaporation and precipitation follow the pattern of the global air currents. The trade winds move moisture-laden air toward the equator, where it is warmed, rises, cools, and drops its moisture as rain. Thus the equatorial regions are areas of maximum precipitation. The air that rises over the equator descends earthward in two subtropical zones around 30° N and 30° S latitude. As the air descends, it warms, and picks up moisture from land and sea. The highest annual losses to evaporation occur in the subtropics of the western North Atlantic (the Gulf Stream) and the Northern Pacific (the Kuroshio Current). North of this are two more zones of ascending air and low pressure that produce the west-coast areas of maximum rainfall. In high latitudes the air descends again in the polar regions, where it remains dry.

Global detention of precipitation varies with region and season. Maximum detention occurs in the boreal regions of the Northern Hemisphere. In tropical regions the period of maximum detention time is the early part of the rainy season. But whatever the location or season, the residence time of water on land is 10 to 120 days.

Precipitation on land in excess of evaporation is eventually transported to the sea by rivers. Rivers are the prime movers of water over the globe and carry many more times the amount of water their channels hold. By returning water to the sea they tend to balance the evaporation deficit of the oceans. Sixteen major rivers discharge 13,600 cm annually, 45 percent of all water carried by rivers. Adding the next 50 largest rivers brings the total to 17,600 cm, 60 percent of all water discharged to the sea.

Evaporation, precipitation, detention, and transportation maintain a stable water balance on the earth. Consider the amount of water that falls on the earth in terms of 100 units (Figure 6-4). On the average 84 units are lost from the ocean by evaporation, while 77 units are gained from precipitation. Land areas lose 16 units by evaporation and gain 23 units by precipitation. Runoff from land to oceans makes up 7 units, which balances the evaporative deficit of the ocean. The remaining 7 units circulate as atmospheric moisture.

In its global circulation the water also influences the heat budgets of the earth. The highest heat budgets are in the low latitudes, the lowest are in the polar regions, and a balance between incoming and outgoing cold and hot is achieved at 38° to 39° latitude. Excessive cooling of higher latitudes is prevented by the north and south transfer of heat by the atmo-

**FIGURE 6-4**
*Global water budget. The mean annual global precipitation of 83.6 cm in depth has been converted into 100 units.*

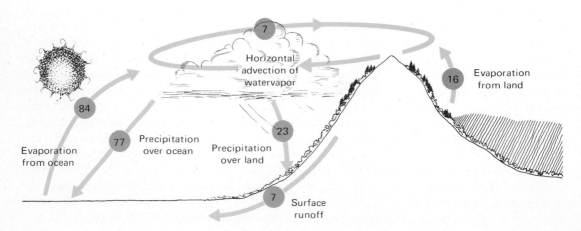

sphere in the form of sensible and latent heat in water vapor and by warm ocean currents.

Examined from a global point of view, the water cycle emphasizes the close interaction between the physical environment of and geographical locations on the earth. Thus the water problem often considered in local terms is actually a global problem, and local water management schemes can affect the planet as a whole. Problems result not because an inadequate amount of water reaches the earth, but because it is unevenly distributed, especially relative to human population centers. Because humanity has strongly interjected itself into the water cycle, the natural usable water resources have decreased, and water quality has declined. The natural water cycle has not been able to compensate for the deteriorating effects on water resources of human actions.

## Gaseous cycles

There are two types of biogeochemical cycles, the *gaseous* and the *sedimentary*. In gaseous cycles the main reservoir of nutrients is the atmosphere and the ocean. Because they are closely linked to the atmosphere and the ocean, they are pronouncedly global. In sedimentary cycles the main reservoir is the soil and the sedimentary and other rocks of the earth's crust. Both types involve biological and nonbiological agents, are driven by the flow of energy, and are tied to the water cycle. Three gaseous cycles basic to biological life are the cycles of oxygen, carbon dioxide, and nitrogen.

### THE OXYGEN CYCLE

Oxygen, the by-product of photosynthesis, is involved in the oxidation of carbohydrates with the release of energy, carbon dioxide, and water. Its primary role in biological oxidation is that of a hydrogen acceptor. The breakdown and decomposition of organic molecules proceeds primarily by dehydrogenation. Hydrogen is removed by enzymatic action from organic molecules in a series of reactions and is finally accepted by the oxygen, forming water.

Oxygen is very active chemically. It can combine with a wide range of chemicals in the earth's crust and is able to react spontaneously with organic compounds and reduced substances. Thus oxygen, necessary for life, can also be toxic, as it is to anaerobic bacteria. Higher organisms have evolved a system to protect themselves from the toxic effects of molecular oxygen. This system involves organelles called peroxisomes, which are contained within the cells and which produce peroxide. The oxidative reactions they mediate result in the production of hydrogen peroxide, which in turn is used through the mediation of other enzymes as an acceptor in oxidizing other compounds. The energy evolved is not utilized by the cells.

Higher organisms have also evolved elaborate mechanisms to ensure a supply of oxygen. Some animals obtain oxygen by diffusion through the skin from air or water; other animals have lungs and gills. Plants have stomata. Organisms have evolved an array of catalysts such as iron-containing molecules, the cytochromes and copper-containing enzymes, and cytochrome oxidases to mediate the transfer of hydrogen to oxygen molecules. In higher organisms this oxidative system is contained in the mitochondria of the cell, which acts as a low-temperature furnace where organic molecules are slowly burned with oxygen and the energy evolved is used to form the high-energy bonds of ATP.

The major supply of free oxygen that supports life is in the atmosphere. There are two significant sources of atmospheric oxygen. One is the photodissociation of water vapor in which most of the hydrogen released escapes into outer space. If the hydrogen did not escape, it would oxidize and recombine with the oxygen. The other source is photosynthesis, active only since life began on earth. Because photosynthesis and respiration are cyclic, involving both the release and utilization of oxygen, one would seem to balance the other, and no significant quantity of oxygen would accumulate in the atmosphere. At some time in the earth's history the amount of oxygen introduced into the atmosphere had to exceed the amount used in the decay of organic matter and the amount tied up in the oxidation of sedimentary rocks. Part of the atmospheric oxygen represents that portion remaining from the unoxidized reserves of photosynthesis—coal, oil, gas, and organic carbon in sedimentary rocks. The amount of stored carbon in the earth suggests that $150 \times 10^{20}$ g of oxygen has been available to the atmosphere, over 10 times as much as now present, $10 \times 10^{20}$ g (F. S. Johnson, 1970). The main nonliving oxygen pool consists of molecular oxygen, water, and carbon dioxide, all intimately linked to each other in photosynthesis and other oxidation-reduction reactions, and all exchanging oxygen with each other. Oxygen is also biologically exchangeable in such compounds as nitrates and sulfates, utilized by organisms that reduce them to ammonia and hydrogen sulfide.

Because oxygen is so reactive, its cycling is quite complex. As a constituent of carbon dioxide, it circulates freely throughout the biosphere. Some carbon dioxide combines with calcium to form carbonates. Oxygen combines with nitrogen compounds to form nitrates, with iron to form ferric oxides, and with many other minerals to form various other oxides. In these states oxygen is temporarily withdrawn from circulation. In photosynthesis the oxygen freed is split from the water molecule. This oxygen is then reconstituted into water during plant and animal respiration. Part of the atmospheric oxygen that reaches the higher levels of the troposphere is reduced to ozone ($O_3$) by high-energy ultraviolet radiation.

There is some concern that world oxygen will be depleted by the increased burning of fossil fuels and by a decreased biomass of photosynthetic plants caused by deliberate destruction and pollution. The earth is enveloped in a cloak of oxygen approximately 60,000 moles/m². To this, photosynthesis adds 8 moles/m² annually. Most of the oxygen released to the atmosphere by photosynthesis is consumed by animals and bacteria. The bulk of the remainder is utilized by the oxidation of geologic materials. Only a small fraction, about 1 part per 10,000, escapes oxidation and remains in the atmosphere. Thus the amount of oxygen in the atmosphere remains rather stable and resistant to short-time changes of 100 to 1,000 years. If we were to burn our fossil fuel reserves completely, we would still use up less than 3 percent of our oxygen reserve. And if photosynthesis should cease and all organic matter were decomposed, we would still have a large reserve of molecular oxygen on which to

draw (Broecker, 1970). However, before oxygen is depleted, some other catastrophe might wipe out man. The major problem is not oxygen depletion, but the accumulation of harmful gases and dusts in the world's atmosphere.

## THE CARBON CYCLE

Because it is a basic constitutent of all organic compounds and a major element involved in the fixation of energy by photosynthesis, carbon is so closely tied to energy flow that the two are inseparable. In fact the measurement of productivity (see Appendix C) is commonly expressed in terms of grams of carbon fixed per square meter per year. The source of all fixed carbon both in living organisms and fossil deposits is carbon dioxide, $CO_2$, found in the atmosphere and dissolved in the waters of the earth. To trace its cycling through the ecosystem is to redescribe photosynthesis and energy flow (Figure 6-5).

The cycling of carbon dioxide involves its assimilation by plants and its conversion to glucose. From glucose plants synthesize polysaccharides and fat and store them in the form of plant tissue. In digesting plant tissue herbivores synthesize these compounds into other carbon compounds. Meat-eating animals feed on herbivores and the carbon compounds are again redigested and resynthesized into other forms. Some of the carbon is returned directly by both plants and animals in the form of $CO_2$ as a by-product of respiration. The remainder for a time becomes incorporated in the living biomass.

Carbon contained in animal wastes and in the protoplasm of plants and animals is released eventually by assorted decomposer organisms. The rate of release depends upon environmental conditions such as soil moisture, temperature, and precipitation. In tropical forests most of the carbon in plant detritus is quickly recycled, for there is little accumulation in the soil. The turnover rate of atmospheric carbon over a tropical forest is about 0.8 year (Leith, 1963). In drier regions, such as grasslands, considerable quantities of carbon are stored as humus. In swamps and marshes, where dead material falls into the water, organic carbon is not completely mineralized; it is stored as raw humus or peat and is circulated only slowly. The turnover rate of atmo-

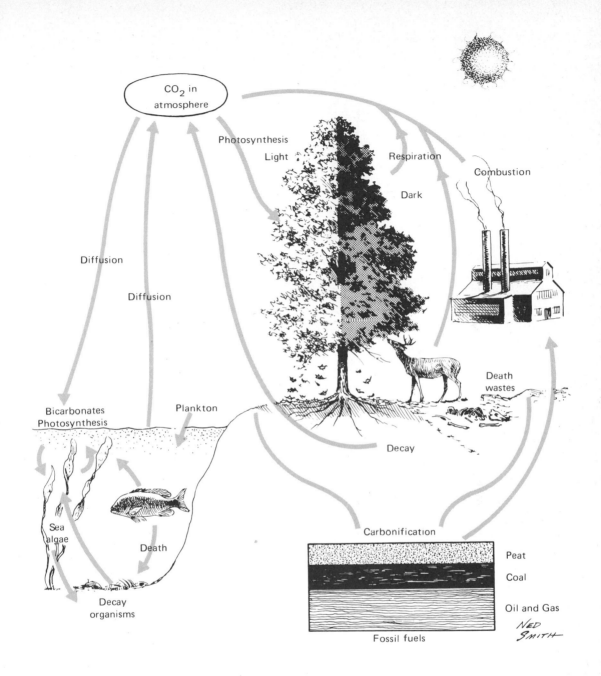

**FIGURE 6-5**
*The carbon cycle in the ecosystem.*

spheric carbon over peat bogs is somewhere on the order of 3 to 5 years (Leith, 1963).

Similar cycling takes place in the freshwater and marine environments. Phytoplankton utilizes the carbon dioxide that has diffused into the upper layers of water or is present as carbonates and converts it into carbohydrates. The carbohydrates so produced pass through the aquatic food chains. The car-

bon dioxide produced by respiration is reutilized by the phytoplankton in the production of more carbohydrates. Under proper conditions a portion is reintroduced into the atmosphere. Significant portions of carbon bound as carbonates in the bodies of shells, snails, and foraminifers become buried in the bottom mud at varying depths when the organisms die. Isolated from biotic activity, that carbon is removed from cycling and becomes incorporated into bottom sediments, which

**163**

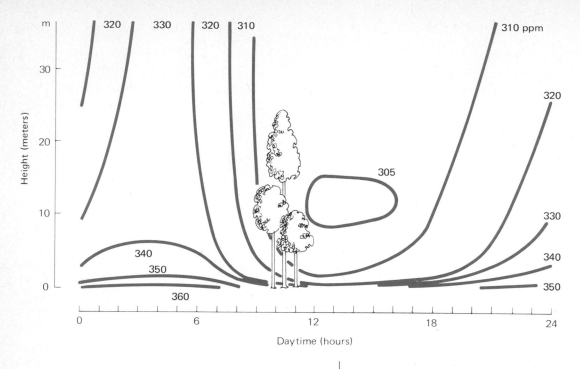

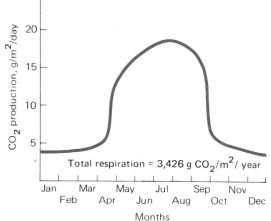

**FIGURE 6-6**
*Diurnal and seasonal pattern of carbon cycling. The respiration rates throughout the year show a high increase in production in the summer, when photosynthesis and decomposition are highest. Data are based on respiration rates during inversions, and rates are corrected to mean monthly temperatures.*

through geological time may appear on the surface as limestone rocks or as coral reefs. Other organic carbon is slowly deposited as gas, petroleum, and coal at an estimated global rate of 10 to 13 g/m²/yr.

***Diurnal and seasonal patterns in carbon cycling.*** The concentration of carbon dioxide in the atmosphere around plant life fluctuates throughout the day. At daylight, when photosynthesis begins, plants start to withdraw carbon dioxide from the air and the concentration declines sharply. By afternoon, when the temperature is increasing and the humidity is decreasing, the respiration rate of plants is increased, the assimilation rate of carbon dioxide declines, and the concentration of carbon dioxide in the atmosphere increases. By sunset, when the light phase of photosynthesis ceases, carbon dioxide is no longer being withdrawn from the atmosphere, and its concentration in the atmosphere increases sharply. A similar diurnal fluctuation takes place in aquatic ecosystems.

Not only is there a diurnal pattern of carbon dioxide concentration in the atmosphere, but there is also an annual course in the production and utilization of carbon dioxide (Figure 6-6). This seasonal change relates both to temperature and to growing seasons. In the spring, when land is greening and phytoplankton is actively growing, the daily production of carbon dioxide is high. As measured by nocturnal accumulation in spring and summer, the rate of carbon dioxide production may be two to three times higher than winter rates at the same temperature. The transition from lower to higher rates increases dramatically about the time of the opening of buds and falls off just as rapidly about the time the leaves of deciduous trees start to drop in the fall.

*The global carbon cycle.* Like water, the carbon budget of the earth is closely linked to the atmosphere, land, and oceans and to the mass movements of air around the planet. The carbon pool involved in the global carbon cycle amounts to an estimated 55,000 Gt (Gt is a gigaton, equal to 1 billion or $10^9$ metric tons). Fossil fuels and shale account for an estimated 12,000 Gt. The oceans contain 93 percent of the active carbon pool, that is, over 39,000 Gt, mostly as bicarbonate ions ($HCO_3^-$) and carbonate ions ($CO_3^-$) (Figure 6-7). Dead organic matter in the oceans accounts for 1,650 Gt of carbon, and living matter, mostly phytoplankton, 1 Gt. The terrestrial biosphere contains an estimated 1,456 Gt as dead organic matter and 826 Gt as living matter. The atmosphere, the major coupling mechanism in

the cycling of $CO_2$, holds about 702 Gt of carbon (1974 data).

Carbon cycling in the sea is nearly a closed system. The surface water acts as the site of main exchange of carbon between atmosphere and ocean. The ability of the surface waters to take up $CO_2$ is governed largely by the reaction of $CO_2$ with the carbonate ion to form bicarbonates:

$$CO_2 + CO_3^- + H_2O \rightleftharpoons 2HCO_3^-$$

In the surface water carbon circulates physi-

**FIGURE 6-7**
*Global compartments of $CO_2$. The diagram shows the interconnected subsystems of the atmosphere, oceans, and main continental organic matter pools. Shown are the major fluxes (in Gt/yr) and pool sizes (in Gt) of carbon. Fluxes include gross primary production (GPP), green plant respiration ($R_g$), net primary production (NPP = GPP/$R_g$), respiration by heterotrophs ($R_h$) and fires (F). (From Baes et al., 1977.)*

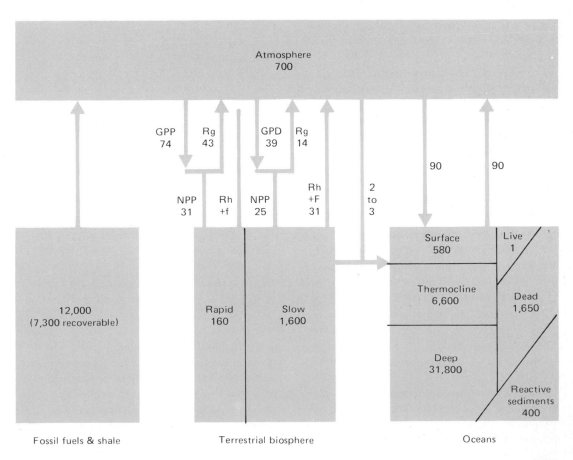

cally by means of currents and biologically through assimilation by phytoplankton and movement through the food chain. Only about 10 percent of the carbon that comes up from the deep in upwellings is used in organic production; the other 90 percent goes back to the deep (Broecker, 1973). The thermocline (a layer between the warmer upper zones and colder zones; see Chapter 7) separates the surface pool of carbon from deep waters. However, some mix between the surface and the deep takes place by eddy diffusion and 10 to 20 percent of the particulate matter sinks to the bottom. Because 80 percent of the ocean floor is bathed in water high in $CO_2$ and unsaturated in carbonates, about 85 percent of the calcium carbonate produced in the sea is destroyed by dissolution.

$$CaCO_3 + CO_2 + H_2O \rightleftharpoons Ca^{2+} + 2HCO_3^-$$

About 15 percent of the calcium carbonate becomes incorporated in deep sediments. Carbon in the sediments is trapped for about $10^8$ years, but carbon atoms in the water have a residence time of about $10^5$ years, which means that in 100,000 years carbon atoms are completely replaced (see Baes et al., 1977).

Most of the carbon in the land mass is in slowly exchanging matter, about 1456 Gt in humus and recent peat and about 600 Gt in large stems and roots. A small amount, 160 Gt, is in rapidly exchanging material. Exchanges between the land mass and the atmosphere are nearly in equilibrium. Photosynthesis by terrestrial vegetation removes about 113 Gt. Plant decomposition, plant respiration, and fires return about the same amount. Forests are the main consumers, fixing about 36 Gt of an estimated total world net production of 65 Gt/yr (Olsen, 1970). Forests are also a major reservoir of the terrestrial organic carbon pool, containing 1485 Gt of the estimated total 2216 Gt.

Of considerable importance in the terrestrial system of the carbon cycle is the proportion of detrital carbon, that is, carbon contained in the dead organic matter on the soil's surface and in the underlying mineral soil, to the living organic pool. Recent estimates place the amount of detrital carbon at 1456 Gt compared to 826 Gt for total world terrestrial biomass (Schlesinger, 1977). This estimate of

the total carbon pool is larger than previous ones because other estimates failed to include the carbon in the lower soil profile. This pool exceeds carbon in the surface soil by a factor of 25. The average amount of carbon per unit soil profile increases from the tropical regions poleward to the boreal forest and tundra (Table 6-2). Low values for the tropical forest reflect high rates of decomposition which compensate for high productivity and litterfall. Frozen tundra soil and waterlogged soils of swamps and marshes have the greatest accumulation of detritus because decay is inhibited by moisture, low temperature, or a combination of both. In boreal forests and tundra detritus greatly exceeds biomass. Detritus likewise exceeds biomass in temperate grasslands, but in tropical savannas the reverse situation exists, probably because of recurring fires. In forest ecosystems carbon loss in soil respiration averages two times the total annual detrital input of carbon from aboveground and below-ground sources. The world output of carbon from soil respiration in terrestial ecosystems amounts to an estimated 75 Gt a year.

The global cycling of carbon dioxide shows the same seasonal variations as the cycle exhibits on a more local basis. In January the concentration of carbon dioxide at the North Pole is near 313 ppm, at the South Pole 318 ppm. In July, as spring in the north progresses, the concentration of carbon dioxide from 20° north latitude to the North Pole declines sharply until it reaches its lowest concentration in August. By October the level increases again. The decline during the arctic summer suggests a heavy and rapid removal of carbon dioxide from the atmosphere by photosynthetic activity of arctic plant life. The scrubbing effect is at a maximum in August, when arctic ecosystems are most active and polar ice is at its minimum. A similar but much less pronounced seasonal change takes place at the South Pole.

***Intrusion into the carbon cycle.*** The $CO_2$ flux among land, sea, and atmosphere has been disturbed by the rapid injection of carbon dioxide into the atmosphere from the burning of fossil fuels and from the clearing of forests and other land use changes. This input, which has been increasing at an exponential rate since

TABLE 6-2
*Distribution of detritus and biomass by ecosystem types*

| Ecosystem type | Mean total profile detritus (kg C/m$^{-2}$) | CV* (%) | World area (10$^9$ ha) | Total world detritus (10$^9$ mtn C) | Total world biomass (10$^9$ mtn C) |
|---|---|---|---|---|---|
| Woodland and shrubland | 6.9 | 59 | 8.5 | 59 | 22.0 |
| Tropical savanna | 3.7 | 87 | 15.0 | 56 | 27.0 |
| Tropical forest | 10.4 | 44 | 24.5 | 255 | 460.0 |
| Temperate forest | 11.8 | 35 | 12.0 | 142 | 175.0 |
| Boreal forest | 14.9 | 53 | 12.0 | 179 | 108.0 |
| Temperate grassland | 19.2 | 25 | 9.0 | 173 | 6.3 |
| Tundra and alpine | 21.6 | 68 | 8.0 | 173 | 2.4 |
| Desert scrub | 5.6 | 38 | 18.0 | 101 | 5.4 |
| Extreme desert | 0.1 | — | 24.0 | 3 | 0.2 |
| Cultivated | 12.7 | — | 14.0 | 178 | 7.0 |
| Swamp and marsh | 68.6 | 63 | 2.0 | 137 | 13.6 |
| Total | | | 147.0 | 1456 | 826.9 |

*CV = coefficient of variation = standard deviation/mean × 100.
*Source:* Adapted from W. H. Schlesinger, 1977.

the beginning of the industrial revolution some 100 years ago, has increased the $CO_2$ concentration in the atmosphere from an estimated 260 to 300 ppm to 330 ppm. Up to 1950 two-thirds of the carbon dioxide injected came from biospheric sources and one-third from the burning of fossil fuels. Since 1950 most has come from the burning of fossil fuels. The amount retained in the atmosphere represents only about one-half of the annual input of 20 Gt (see Stuiver, 1978).

Because $CO_2$ in the atmosphere acts as a shield over the earth, allowing incoming short-wave radiation to penetrate the atmosphere, but holding outgoing long-wave radiation in, an increase in $CO_2$ can result in a warming of the earth (for a recent discussion, see Siegenthaler and Oeschger, 1978). Each doubling of $CO_2$ concentration results in a doubling of the average temperature of the trophosphere, with a greater increase at higher latitudes and a smaller increase at lower latitudes. Such increases could ultimately cause a rise in the sea level, resulting in the drowning of coastal regions; an increase in evaporation, resulting in a lowering of freshwater lakes and streams; an increase in respiration and decay; a decrease in ocean circulation, resulting in weakened upwelling and nutrient replacement; and the spread of microbial and insect pests from the tropics. If climatic changes were rapid enough, they could reduce the biological fitness and productivity of natural and agricultural plants and crops.

Because only about 50 percent of the total $CO_2$ injected into the atmosphere is retained, the other 50 percent must be stored in two possible sinks, the terrestrial biomass and the ocean. In view of increasing dependence on fossil carbon, the ability of the sinks to take care of excess carbon becomes increasingly important.

The terrestrial sink is accumulated plant biomass. Greenhouse experiments have shown that plant growth is stimulated by increased $CO_2$ concentrations provided that other nutrients are not limited. A 12 percent increase in atmospheric $CO_2$ theoretically could result in a 2.7 Gt/yr increase in carbon biomass. Such an increase could counterbalance $CO_2$ injected from the burning of fossil fuel. But the ability of terrestrial vegetation to act as a sink has been questioned (see Botkin, 1977). There is evidence that a 10 to 20 percent increase in $CO_2$ in the atmosphere would not result in any significant increase in plant biomass because of the different responses of plants to carbon enrichment. More importantly, although terrestrial vegetation does remove $CO_2$ from the atmosphere, it is also an additional source of $CO_2$ to the atmosphere (Stuiver, 1978; Woodwell et al., 1978). The destruction of natural vegetation probably accounts for a greater input of $CO_2$ to the atmosphere than is removed by biomass accumulation (Botkin, 1977; Baes, 1977; Woodwell et al., 1978).

This leaves the ocean as the major carbon sink (Stuiver, 1978). The storage of excess $CO_2$ by oceans depends upon eddy current circulation between the surface and deep water. Evidence based on oceanic studies suggests that the major $CO_2$ sink is the thermocline region of large ocean gyres. Thirty-four percent of excess $CO_2$ generated is stored in the surface and thermocline gyres, 13 percent is carried to deep seas, leaving 53 percent in the atmosphere. The future ability of the ocean to absorb $CO_2$ may be influenced by the amount of $CO_2$ in the atmosphere. As atmospheric $CO_2$ increases, the buffering factor of seawater decreases, decreasing the absorption of $CO_2$ and the effectiveness of the ocean as a sink. Because of the many variables involved, it is difficult to predict the future accumulation of $CO_2$ in the atmosphere. Computer models suggest that 100 years from now the cumulative fraction of airborne $CO_2$ will be between 46 and 80 percent (Siegenthaler and Oeschger, 1978). These figures are based on a realistic assumption of carbon dioxide production from burning of fossil fuels.

The carbon cycle is further altered by the heavy and widely distributed input of the air pollutant, carbon monoxide (Jaffe, 1971). The major atmospheric source is the gasoline engine. Lesser amounts come from the burning of coal. The annual worldwide input amounts to $27.17 \times 10^6$ ton, 95 percent of which is produced in the Northern Hemisphere. The concentration over the Northern Hemisphere is 0.1 to 0.2 ppm, but in urban areas local concentrations build up to levels as high as 50 to 100 ppm. In heavy automobile traffic drivers may be exposed to 650 ppm. The concentration in slow-moving traffic on an expressway may reach 140 ppm. At the level of 30 ppm, normal concentration for some urban atmospheres, carbon monoxide binds up about 5 percent of the hemoglobin of the blood.

Unlike carbon dioxide, carbon monoxide is an extremely stable gas and is not easily removed from the atmosphere. It may be scrubbed by atmospheric circulation, by photochemical oxidation to carbon dioxide, by plants and soil bacteria, and by oceanic absorption.

### THE NITROGEN CYCLE

*Processes involved.* Nitrogen is an essential constitutent of protein, a building block of all living material. It is also the major constituent, about 79 percent, of the atmosphere. The paradox is that in its gaseous state, abundant though it is, nitrogen is unavailable to most life. It must first be converted to some chemically usable form, and getting it into that form comprises a major part of the nitrogen cycle. The processes involved are fixation, ammonification, nitrification, and denitrification.

*Fixation* is the conversion of nitrogen in its gaseous state to ammonia or nitrate. Ammonia is the product of biological fixation; nitrate is the product of high-energy fixation by lightning or occasionally cosmic radiation or meteorite trails. In high-energy fixation nitrogen and oxygen in the atmosphere combine into nitrates which are carried to the earth in rainwater as nitric acid, $H_2NO_3$. Estimates suggest that less than 8.9 kg N/ha is brought to the earth by high-energy fixation.

Biological fixation, the more important method, makes available 100 to 200 kg N/ha, roughly 90 percent of the fixed nitrogen contributed to the earth each year. In biological fixation molecular (or gaseous) nitrogen, $N_2$, is split into two atoms:

$$N_2 \rightarrow 2N$$

This step requires an input of 160 kcal for each mole (28 g) of nitrogen. The free N atoms can combine with hydrogen to form ammonia, with the release of about 13 kcal of energy:

$$2N + 3H_2 \rightarrow 2NH_3$$

This fixation is accomplished by symbiotic bacteria living in association with leguminous and root-noduled nonleguminous plants, by free-living aerobic bacteria, and by blue-green algae. In agricultural ecosystems approximately 200 species of nodulated legumes are the preeminent nitrogen fixers. In nonagricultural systems some 12,000 species of plants, from free-living bacteria and blue-green algae to nodule-bearing plants, are responsible for nitrogen fixation.

Legumes, the most conspicuous of the nitrogen-fixing plants, have a symbiotic relationship with members of the bacterial genus *Rhizobium*. Rhizobia are aerobic, non–spore-forming rod-shaped bacteria. They exist in the immediate surroundings of the plant roots, called the rhizosphere, where, stimulated by secretions from the legumes, they multiply. The secretions, together with enzymes secreted by the legumes in response to the exudates of the bacteria, loosen the fibrils of the root hair walls. Swarming rhizobia enter the root hair tips and penetrate the inner corticular cells where they multiply and increase in size, resulting in swollen infected cells in which hemoglobin develops. These cells make up the central tissues of the nodules. Inside the nodules the bacteria change from rod-shaped to a nonmobile form that carry on nitrogen fixation.

A large number of nonleguminous nodule-bearing plants, most of them associated with early pioneering species on sites where the soil is usually low in nitrogen, make significant contributions of nitrogen to wildlands. Among such plants are *Alnus, Ceanothus,* and *Elaeagnus*.

Also contributing to the fixation of nitrogen are free-living soil bacteria. The most prominent of the 15 known genera are the aerobic *Azotobacter* and the anaerobic *Clostridium* (see Nishustin and Shilnikova, 1969). *Azotobacter* prefers soils with a pH of 6 to 7 that are rich in mineral salts and low in nitrogen. Less efficient at fixing nitrogen, *Clostridium* is ubiquitous, found in nearly all soils. Although it prefers a neutral pH, it is more tolerant of acidic conditions than *Azotobacter*. Both genera produce ammonia as the first stable end product. The free-living and symbiotic bacteria both require molybdenum as an activator and are inhibited by an accumulation of nitrates and ammonia in the soil.

Blue-green algae are another important group of largely nonsymbiotic nitrogen fixers. Of the some 40 known species the most common are in the genera *Nostoc* and *Calothrix,* found both in soil and aquatic habitats. Blue-green algae are well adapted to exist on the barest requirements for living. They are often pioneers on bare mineral soil. Especially successful in waterlogged soil, they appear to be nitrogen fixers in rice paddies of Asia (Singh, 1961), where studies indicate that they annually fix 30 to 50 kg N/ha. Blue-greens are perhaps the only fixers of nitrogen over a wide range of temperatures in aquatic habitats from arctic and antarctic seas to freshwater ponds and hot springs. In the hot springs of Yellowstone the blue-greens, which are responsible for the bluish-green color, fix nitrogen at a temperature of 55° C (W. D. P. Stewart, 1967). As with bacteria, blue-greens also require molybdenum for nitrogen fixation.

Other plants may be involved in nitrogen fixation. In humid tropical forests epiphytes growing on tree branches and bacteria and algae growing on leaves may fix appreciable amounts of nitrogen. Certain lichens *(Collema tunaeforme* and *Peltigera rufescens)* have been implicated in nitrogen fixation (Henriksson, 1971). Lichens with nitrogen-fixing ability possess nitrogen-fixing blue-green species as their algal component (see Chapter 19).

In addition to fixation nitrogen is also made available through the breakdown of organic matter by ammonification, nitrification, and denitrification, three other processes in the nitrogen cycle. In *ammonification* amino acids are broken down by decomposer organisms to produce ammonia, with a yield of energy. It is a one-way reaction. Ammonium, or the ammonia ion, is absorbed directly by plant roots and incorporated into amino acids, which are subsequently passed along through the food chain. Wastes and dead animal and plant tissues are broken down to amino acids by heterotrophic bacteria and fungi in soil and water. Amino acids are oxidized to carbon dioxide, water, and ammonia, with a yield of energy:

$$CH_2NH_2COOH + 1\tfrac{1}{2}O_2 \rightarrow 2CO_2 + H_2O \\ + NH_3 + 176 \text{ kcal}$$

Part of the ammonia is dissolved in water, part is trapped in the soil, and some is trapped and fixed in both acid clay and certain base-saturated clay minerals near the point where first broken down or introduced into the soil (Nommik, 1965).

*Nitrification* is a biological process in which ammonia is oxidized to nitrate and nitrite, yielding energy. Two groups of microorganisms are involved. *Nitrosomonas* utilize the ammonia in the soil as their sole source of energy because they can promote its oxidation to nitrite ions and water:

$$NH_3 + 1\frac{1}{2}O_2 \rightarrow HNO_2 + H_2O + 165 \text{ kcal}$$
$$HNO_2 \rightarrow H^+ + NO_2^-$$

Nitrite ions can be oxidized further to nitrate ions in an energy-releasing reaction. This energy left in the nitrite ion is exploited by another group of bacteria, the *Nitrobacter,* which oxidizes the nitrite ion to nitrate with a release of 18 kcal of energy (M. Alexander, 1965):

$$NO_2 + \frac{1}{2}O_2 \rightarrow NO_3^-$$

Thus nitrification is a process in which the oxidation state (or valence) of nitrogen is increased. *Nitrosomonas* oxidizes 35 mols of nitrogen for each mol of $CO_2$ assimilated; *Nitrobacter* oxidizes 100 mols.

Although nitrification is generally considered a beneficial process, it may in some situations have deleterious effects. Nitrification involves the conversion of slowly leached forms of nitrogen into readily leached nitrates. If quantities of nitrates are large enough and sufficient water percolates through the soil, the nitrates can be removed faster than they can be taken up by the roots, a situation that results in pollution of water. An abundance of nitrates also leads to increased losses of gaseous nitrogen.

Nitrates are a necessary substrate for *denitrification,* in which the nitrates are reduced to gaseous nitrogen by certain organisms to obtain oxygen. The denitrifiers, represented by fungi and the bacteria *Pseudomonas,* are facultative anaerobes. They prefer an oxygenated environment, but if oxygen is limited, they can use $NO_3^-$ instead of $O_2$ as the hydrogen acceptor and can release $N_2$ in the gaseous state as a by-product:

$$C_6H_{12}O_6 + 4NO_3^- \rightarrow 6CO_2 + H_2O + 2N_2$$

Nitrification and denitrification both require certain conditions: a sufficient supply of organic matter, a limited supply of molecular oxygen, a pH range of 6 to 7, and an optimum temperature of 60° C.

***Cycling of nitrogen.*** With the basic and necessary processes described, the nitrogen cycle can be followed briefly (Figure 6-8, Table 6-3). The sources of inputs of nitrogen under natural conditions are the fixation of atmospheric nitrogen, additions of inorganic nitrogen in rain from such sources as lightning fixation and fixed "juvenile" nitrogen from volcanic activity, ammonia absorption from the atmosphere by plants and soil, and nitrogen accretion from windblown aerosols, which contain both organic and inorganic forms of nitrogen. In terrestrial ecosystems nitrogen, largely in the form of ammonia or nitrates, depending upon a number of variable conditions, is taken up by plants, which convert it into amino acids. The amino acids are transferred to consumers, which convert them to different types of amino acids. Eventually their wastes (urea and excreta) and the decay of dead plant and animal tissue are broken down by bacteria and fungi into ammonia. Ammonia may be volatilized (lost as gas to the atmosphere), may be acted upon by nitrifying bacteria, or may be taken up directly by plants. Nitrates may be utilized by plants, immobilized by microbes, stored in decomposing humus, or leached away. This material is carried to streams, lakes, and eventually the sea, where it is available for use in aquatic ecosystems. There nitrogen is cycled in a similar manner, except that the large reserves contained in the soil humus are largely lacking. Life in the water contributes organic matter and dead organisms that undergo decomposition and subsequent release of ammonia and ultimately nitrates.

Tracer studies with $^{15}N$, a short-lived nonradioactive isotope, show that in marine ecosystems, ammonia is recycled rapidly and preferentially by phytoplankton (Dugdale and Goering, 1967). As a result there is little ammonia in most natural waters, and nitrate is utilized only in the virtual absence of ammonia. In addition to biological cycling there are small but steady losses from the biosphere to the deep sediments of the ocean and to sedimentary rocks. In return there is a small addition of "new" nitrogen from the weather-

*FIGURE 6-8*
*The nitrogen cycle in the ecosystem.*

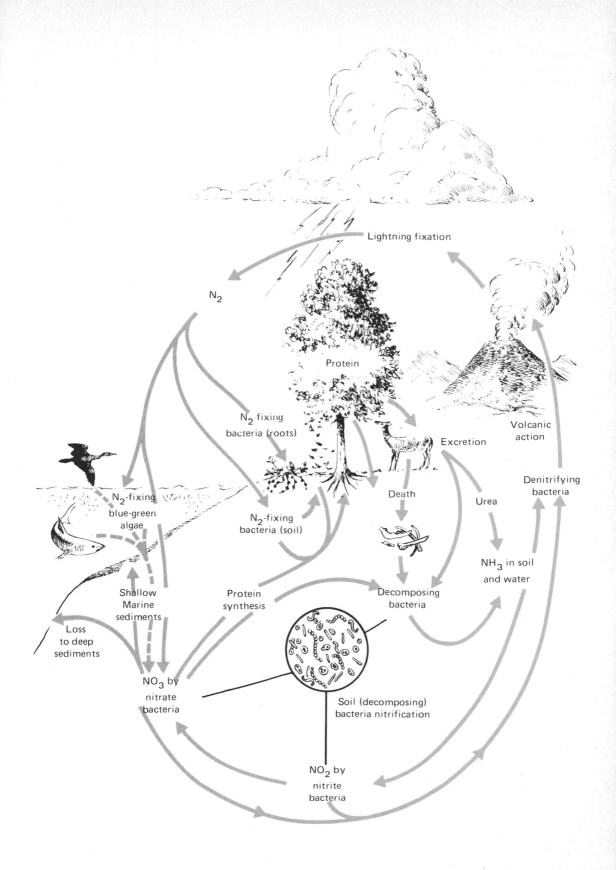

Lightning fixation

$N_2$

Protein

$N_2$-fixing
bacteria (roots)

Volcanic
action

Excretion

$N_2$-fixing
blue-green
algae

$N_2$-fixing
bacteria (soil)

Death

Urea

Denitrifying
bacteria

$NH_3$ in soil
and water

Shallow
Marine
sediments

Protein
synthesis

Decomposing
bacteria

Loss
to deep
sediments

$NO_3$ by
nitrate
bacteria

Soil (decomposing)
bacteria nitrification

$NO_2$ by
nitrite
bacteria

171

TABLE 6-3 *Budget for the nitrogen cycle*

| | LAND Rate ($10^6$ mtn/yr) | error (%) | SEA Rate ($10^6$ mtn/yr) | error (%) | ATMOSPHERE Rate ($10^6$ mtn/yr) | error (%) |
|---|---|---|---|---|---|---|
| **INPUT** | | | | | | |
| Biological nitrogen fixation | — | — | | | — | — |
| Symbiotic (31) | 14 | 25 | 10 | 50 | — | — |
| Nonsymbiotic (31) | 30 | 50 | — | — | — | — |
| Atmospheric nitrogen fixation (31) | 4 | 100 | 4 | 100 | — | — |
| Industrially fixed nitrogen fertilizer (31) | 30 | 5 | — | — | — | — |
| N-oxides from combustion | 14 | 25 | 6 | 25 | 20 | 25 |
| Return of volatile nitrogen compounds in rain | ? | — | ? | — | — | — |
| Riven influx (31) | — | — | 30 | 50 | — | — |
| $N_2$ from biological denitrification (31) | — | — | — | — | 83 | 100 |
| Natural $NO_2$ | — | — | — | — | ? | — |
| Volatilization ($HN_3$) | — | — | — | — | ? | — |
| Total input | 92+ | | 50 | | 103+ | |
| **STORAGE** | | | | | | |
| Plants (31) | 12,000 | 30 | 800 | 50 | — | — |
| Animals (31) | 200 | 30 | 170 | 50 | — | — |
| Dead organic matter (31) | 760,000 | 50 | 900,000 | 100 | — | — |
| Inorganic nitrogen (31) | 140,000 | 50 | 100,000 | 50 | — | — |
| Dissolved nitrogen (31) | — | — | 20,000,000 | 10 | — | — |
| Nitrogen gas(31) | — | — | — | — | 3,800,000,000 | 3 |
| $NO + NH_4$ (25) | — | — | — | — | Less than 1 | 50 |
| $NH_3 + NH_4$ (17) | — | — | — | — | 12 | 50 |
| $N_2O$ (33) | — | — | — | — | 1,000 | 50 |
| Total storage | 912,200 | | 21,000,970 | | 3,800,001,013 | |
| **LOSS** | | | | | | |
| Denitrification (31) | 43 | — | 40 | 100 | — | — |
| Volatilization | ? | — | ? | — | — | — |
| River runoff (31, 32) (includes enrichment from fertilizers) | 30 | 50 | — | — | — | — |
| Sedimentation (31) | — | — | 0.2 | 50 | — | — |
| $N_2$ in all fixation processes | — | — | — | — | 92 | — |
| $NH_3$ in rain (17) | — | — | — | — | Less than 40 | 50 |
| $NO_2$ in rain | — | — | — | — | ? | — |
| $N_2O$ in rain | — | — | — | — | ? | 50 |
| Total loss | 73 | | 40.2 | | 132+ | |

*The error columns list plus-or-minus probable errors as a percentage of the estimate.
*Source:* Inger et al., 1972. © 1972 by the Board of Regents of the University of Wisconsin System.

ing of igneous rocks and juvenile nitrogen from volcanic activity.

Under natural conditions nitrogen lost from ecosystems by denitrification, volatilization, leaching, erosion, windblown aerosols, and transportation out of the system is balanced by biological fixation and other sources. Both chemically and biologically, terrestrial and aquatic ecosystems constitute a dynamic equilibrium system in which a change in one phase affects the other.

*Intrusion into the nitrogen cycle.* Human intrusion into the nitrogen cycle often results either in a reduction in the nitrogen cycled or in overloading the system.

Cultivation of grasslands, for example, has resulted in a steady decline in the nitrogen content of the soil (Jenny, 1933). Mixing and breaking up of the soil exposes more organic matter to decomposition and decreases the amount of root material, thus decreasing new additions of organic matter. And the removal of nitrogen through harvested crops or grazing causes additional losses. Harvest of timber results in a heavy outflow of nitrogen from the forest ecosystem not only in timber removed, but also in short-term nitrate losses from the soil. In such situations outputs exceed inputs, impoverishing the system.

On the other hand, excessive amounts of nitrogen may be added to the system and the excess creates various problems. In agricultural systems heavy addition of commercial fertilizer, especially in the form of anhydrous ammonia, increases the amount of nitrogen in cropland ecosystems. However, if fertilizer is not properly applied a considerable portion of the added nitrogen may be lost as nitrate nitrogen to the groundwater. The amount of nitrates lost from the soil by the way of percolating water varies with the type of crop, rate and time of fertilizer application, soil permeability, the ratio of precipitation or irrigation to evaporation, hydrology of the area, the portion of watershed in crops, and general climate.

Another source of agricultural nitrate pollution is animal waste. Only recently has this become an environmental problem. Prior to the use of concentrated livestock feeding yards, animal wastes were recycled to croplands, and nutrients were eventually returned to the animals as grain and hay. Inputs of feed and fertilizer were converted into outputs of meat and milk. Increasing costs of chemical fertilizers may result in the return of feedlot wastes to the land.

A third source of nitrate pollution is human waste, particularly in the form of sewage. In spite of the magnitude of water pollution from agricultural sources, human effluents contribute even heavier loads. Municipal sewage treatment facilities contribute 25 percent of the nitrogen and 50 percent of the phosphorus found in the surface waters of Wisconsin (Corey et al., 1967, cited by Frink, 1971). Groundwater from rural areas contributes 50 percent of the nitrogen and only 2 percent of the phosphorus. In the Potomac River estuary waste water accounts for 51 percent of the nitrogen compared to 31 percent from agricultural runoff (Table 6-4) (Jaworski and Helling, 1970). As with agricultural wastes, the potential solution is to return human wastes to the land. This is an ecologically sounder solution than diluting sewage effluents in streams. The major drawback in such use of sewage effluent and sludge is the frequent high concentrations of such heavy metals as cadmium.

Automobiles and power plants are the major sources of other nitrogenous pollutants, nitrogen oxides. The major type of nitrogenous air pollutants is nitrogen dioxide, $NO_2$. In the atmosphere nitrogen dioxide is reduced by ul-

TABLE 6-4
*Estimates of sources of plant nutrients entering Wisconsin surface waters*

| Source | N (%) | P (%) |
|---|---|---|
| Municipal treatment facilities | 24.5 | 55.7 |
| Private sewage systems | 5.9 | 2.2 |
| Industrial wastes | 1.8 | 0.8 |
| Rural sources | | |
|     manured lands | 9.9 | 21.5 |
|     other cropland | 0.7 | 3.1 |
|     forest land | 0.5 | 0.3 |
|     pasture, woodlot, and other | | |
|       lands | 0.7 | 2.9 |
|     groundwater | 42.0 | 2.3 |
| Urban runoff | 5.5 | 10.0 |
| Rainfall on water areas | 8.5 | 1.2 |

*Source:* C. R. Frank, 1971.

traviolet light to nitrogen monoxide and atomic oxygen:

$$NO_2 \rightarrow NO + O$$

Atomic oxygen reacts with oxygen to form ozone:

$$O_2 + O \rightarrow O_3$$

Ozone, in a never-ending cycle, then reacts with nitrogen monoxide to form nitrogen dioxide and oxygen:

$$NO + O_3 \rightarrow NO_2 + O_2$$

This cycle illustrates only a few of the reactions that nitrogen oxides undergo or trigger. In the presence of sunlight atomic oxygen from nitrogen dioxide also reacts with a number of reactive hydrocarbons to form radicals. These radicals then take part in a series of reactions to form still more radicals that combine with oxygen, hydrocarbons, and nitrogen dioxide. As a result, nitrogen dioxide is regenerated, nitrogen monoxide disappears, ozone accumulates, and there form a number of secondary pollutants, such as formaldehydes, aldehydes, and peroxyacytnitrates, known as PAN (see Am. Chem. Soc., 1969). All of these collectively form photochemical smog.

Nitrogen dioxide and the secondary pollutants are harmful to both man and plants. Nitrogen dioxide, a pungent gas that produces a brownish haze, causes nose and eye irritations at 13 ppm and pulmonary discomfort at 25 ppm. Ozone irritates the nose and throat at 0.05 ppm and causes dryness of the throat, headaches, and difficulty in breathing at 0.1 ppm. All of these, along with other contaminants, such as sulfur dioxide and carbon monoxide, usually act synergistically. The total effect is much greater than individual effects alone. Thus air pollution has been positively correlated with increase in asthma, bronchitis, emphysema, and lung cancer (Lave and Seskin, 1970).

Ozone, PAN, and nitrogen dioxide severely injure many forms of plant life. PAN is extremely toxic. By destroying some of the lower epidermal cells of the leaves and by damaging the chloroplasts, it interferes with the plant's metabolic processes. Although the mechanism has not been defined, PAN also interferes with enzymes important in providing the energy necessary to split the water molecules in photosynthesis (see Treshow, 1970).

A number of important leafy vegetable plants are extremely sensitive to PAN. Among them are spinach, endive, and tobacco. Other sensitive crops are oats, alfalfa, beets, beans, corn, celery, and peppers. Affected plants usually show some form of glazing, silvering, or bronzing in irregular patches on the undersides of the leaves (see Hill and Heggestad, 1970).

Ozone is especially prevalent in the air near populated regions. Reactions to it vary among plants. Highly sensitive tobacco is flecked with white lesions, bean leaves show stippling and bleached areas, and leaves of woody plants may have reddish-brown lesions. Plants especially sensitive to ozone include white and ponderosa pines, alfalfa, oats, spinach, tobacco, and tomato. Nitrogen dioxide too causes direct injury to plants. Symptoms of damage from nitrogen dioxide are white or brown collapsed lesions or tissues between the veins and near the margins of the leaves. These symptoms closely resemble reactions to both ozone and sulfur dioxide.

## Sedimentary cycles

Mineral elements required by living organisms are obtained initially from inorganic sources. Available forms occur as salts dissolved in soil water or in lakes, streams, and seas. The mineral cycle varies from one element to another, but essentially it consists of two phases: the salt-solution phase and the rock phase. Mineral salts come directly from the earth's crust by weathering. Soluble salts then enter the water cycle. With water they move through the soil to streams and lakes and eventually reach the seas, where they remain indefinitely. Other salts are returned to the earth's crust through sedimentation. They become incorporated into salt beds, silts, and limestones; after weathering they again enter the cycle.

Plants and many animals fulfill their mineral requirements from mineral solutions in their environments. Other animals acquire the bulk of their minerals from plants and animals they consume. After the death of living organisms the minerals are returned to the soil and water through the action of the organisms and processes of decay.

There are as many different kinds of sedimentary cycles as there are elements. Two may serve as examples. The sulfur cycle is a sort of hybrid of the gaseous and sedimentary processes because sulfur reservoirs are found not only in the earth's crust, but also in the atmosphere. Phosphorus, on the other hand, is wholly sedimentary; it is released from rock and deposited in both the shallow and deep sediments of the sea.

### THE SULFUR CYCLE

The sulfur cycle is both sedimentary and gaseous. It involves a long-term sedimentary phase in which sulfur is tied up in organic and inorganic deposits, is released by weathering and decomposition, and is carried to terrestrial and aquatic ecosystems in salt solution. The gaseous phase of the cycle permits the circulation of sulfur on a global scale.

Sulfur enters the atmosphere from several sources: the combustion of fossil fuels, volcanic eruptions, the surface of the oceans, and gases released by decomposition. It enters the atmosphere initially as hydrogen sulfide, $H_2S$, which quickly oxidizes into another volatile form, sulfur dioxide, $SO_2$. Atmospheric sulfur dioxide, soluble in water, is carried back to

earth in rainwater as weak sulfuric acid, $H_2SO_4$. Whatever the source, sulfur in a soluble form is taken up by plants and is incorporated through a series of metabolic processes, starting with photosynthesis, into such sulfur-bearing amino acids as cystine. From the producers the sulfur in amino acids is transferred to the consumer groups (Figure 6-9).

Excretions and death carry sulfur in living material back to the soil and to the bottoms of ponds, lakes, and seas where the organic material is acted upon by bacteria, releasing the sulfur as hydrogen sulfide or sulfate. One group, the colorless sulfur bacteria, both reduces hydrogen sulfide to elemental sulfur and oxidizes it to sulfuric acid. The green and purple bacteria, in the presence of light, utilize hydrogen sulfide as an oxygen acceptor in the photosynthetic reduction of carbon dioxide. Best known are the purple bacteria found in salt marshes and in the mud flats of estuaries. These organisms are able to carry the oxidation of hydrogen sulfide as far as sulfate, which

**FIGURE 6-9**
*The sulfur cycle. Major sources are burning of fossil fuels and acid mine water from coal mines. o = oxidation; r = reduction; m = mobilization; im = immobilization.*

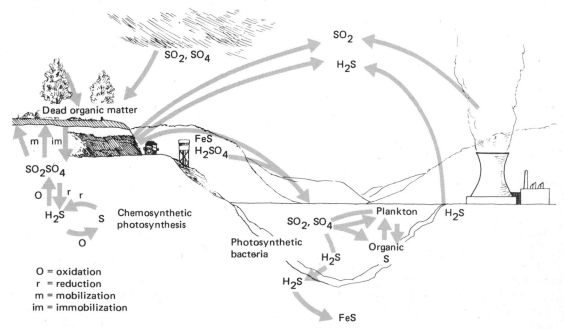

may be recirculated and taken up by the producers or may be used by sulfate-reducing bacteria. The green sulfur bacteria can carry the reduction of hydrogen sulfide to elemental sulfur.

Sulfur, in the presence of iron and under anaerobic conditions, will precipitate as ferrous sulfide, $FeS_2$. This compound is highly insoluble under neutral and alkaline conditions and is firmly held in mud and wet soil. Sedimentary rocks containing ferrous sulfide (called pyritic rocks) may overlie coal deposits. Exposed to the air in deep and surface mining, the ferrous sulfide oxidizes and in the presence of water produces ferrous sulfate ($FeSO_4$) and sulfuric acid:

$$2FeS_2 + 7O_2 + 2H_2O \rightarrow 2FeSO_4 + H_2SO_4$$

In other reactions ferric sulfate ($Fe_2SO_4$) and ferrous hydroxide ($FeOH_3$) are produced:

$$12FeSO_4 + 3O_2 + 6H_2O \rightarrow$$
$$4Fe_2(SO_4)_3 + 4Fe(OH)_3 \downarrow$$

In this manner sulfur in pyritic rocks, suddenly exposed to weathering by human activities, discharges heavy slugs of sulfuric acid, ferric sulfate, and ferrous hydroxide into aquatic ecosystems. These compounds destroy aquatic life and have converted hundreds of miles of streams and rivers in the eastern United States to highly acidic water.

*The global sulfur cycle.* As with nitrogen, oxygen, and other gaseous cycles, the biosphere plays an important role in the sulfur cycle. However, the sedimentary phase makes the total cycle more complex (Figure 6-10). Sources of sulfur include the weathering of rocks, especially pyrites, erosional runoff, industrial production and decomposition of organic matter. The bulk of sulfur appears first as a volatile gas, hydrogen sulfide. In the hydrosphere, the soil, and the atmosphere hydrogen sulfide is oxidized to sulfides and sulfates, the forms in which sulfur is most readily circulated. The atmosphere contains sulfate particles, sulfur dioxide, and hydrogen sulfide. The latter is most abundant over continents. The concentration of sulfur as hydrogen sulfide in the unpolluted atmosphere is estimated at 6 $g/m^3$; as sulfur dioxide at 1 $g/m^3$. Part of the sulfur in the atmosphere is recirculated to land and sea by rainwater. The concentration of sulfur dioxide in rain falling over land has been estiamted as 0.6 mg/liter and over sea as 0.2 mg/liter, excluding sea spray.

It is almost impossible to estimate the biological turnover of sulfur dioxide because of

**FIGURE 6-10**
*Sources and sinks of atmospheric sulfur compounds. Units are $10^6$ tons calculated as sulfate per year. (After Kellogg et al., 1972.)*

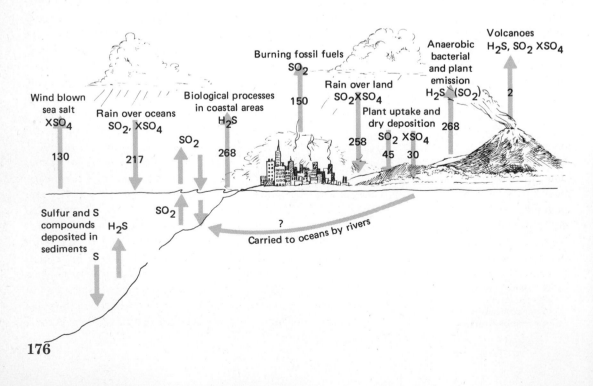

TABLE 6-5
**Budget of sulfur in nature** ($10^6$ tn/yr)

| Item | ATMOSPHERE | | LITHOSPHERE (SEDIMENT ROCKS) | | PEDOSPHERE | | (OCEANS) | |
|---|---|---|---|---|---|---|---|---|
| | To | From | To | From | To | From | To | From |
| River discharge | | | | | | 80 | 80 | |
| Weathering | | | | 15 | 15 | | | |
| Fertilizers | | | | 10 | 10 | | | |
| Precipitation | | 165 | | | 65 | | 100 | |
| Sea spray | 45 | | | | | | | 45 |
| Dry deposition | | 200 | | | 100 | | 100 | |
| Sedimentation | | | 15 | | | | | 15 |
| Industrial | 40 | | | 40 | | | | |
| Increase in sea | | | 50* | | | | | 50 |
| Balance | | | | | | | | |
|   soils—atmosphere | 110 | | | | | 110 | | |
|   oceans—atmosphere | 170 | | | | | | | 170 |
|     Total | 365 | 365 | 65 | 65 | 190 | 190 | 280 | 280 |
| Specification | | | | | | | | |
|   as $SO_2$ sulfur | 45 | 165 | | | 90 | 80 | 180 | 95 |
|   as $SO_2$ sulfur | 40 | 200 | | | 100 | | 100 | |
|   as $H_2S$ sulfur | 280 | | | | | 110 | | 170 |
|   as other forms of sulfur | | | 65 | 65 | | | | 15 |

*For the balance this has to be treated as an item borrowed by the ocean from the lithosphere.
*Source:* E. Eriksisson, 1963. Copyright by American Geophysical Union.

the complicated cycling within the biosphere. Eriksson estimates that net annual assimilation of sulfur by marine plants is on the order of 130 million tons. Added to the anaerobic oxidation of organic matter, this brings the total to an estimated 200 million tons. Both industrially emitted sulfur and fertilizer sulfur are eventually carried to the sea; therefore these two sources probably account for the 50-million-ton annual increase of sulfur in the ocean. The balance sheet of sulfur in the global cycle is summarized in Table 6-5.

***Intrusion into the sulfur cycle.*** To produce energy to keep our technological civilization running, we have to burn fossil fuels, and in doing so we inject into the environment a heavy dose of sulfur. Globally each year we pour into the atmosphere 147 million tons of sulfur dioxide. Seventy percent of it comes from the combustion of coal. Once in the atmosphere, sulfur dioxide does not remain in the gaseous state, but reacts with the moisture to form sulfuric acid.

Sulfuric acid in the atmosphere has a number of effects. It is irritating to respiratory tracts in concentrations of a few parts per million. In a fine mist or absorbed in small particles, it can be carried deep into the lungs to attack sensitive tissue. High concentrations of sulfur dioxide (over 1000 micromilligrams/m³) have been implicated as a prime cause in many air pollution disasters characterized by higher than expected death rates and increased incidence of bronchial asthma. Among such disasters are the Meuse Valley in Belgium in 1930, Donora, Pennsylvania, in 1938, London in 1952, and New York and Tokyo in the 1960s.

Plants exposed to atmospheric sulfur are injured or killed outright. Injury to plants is caused largely by acidic aerosols during periods of foggy weather, during light rains, or during periods of high relative humidity and moderate temperatures. Pines are more susceptible than broadleaf trees and react by partial defoliation and reduced growth. Exposure of plants to sulfur dioxide with as low a concentration as 0.3 ppm for 8 hours can produce both acute and chronic injury. Acute injury is characterized by dead tissue between the veins and along the margins of plant leaves; chronic injury is marked brownish-red or blackish areas in the blade of the leaf.

Not only does the emission of sulfur into the atmosphere cause injury to plants and problems in public health, it also produces acid rainfall over parts of the earth. In Scandinavia and downwind from the industrial centers of Britian and the Ruhr Valley (Figure 6-11), acidity of the rainfall has increased 200-fold since 1966; pH values as low as 2.8 have been recorded (Oden and Ahl, 1970). This acid rainwater is increasing the acidity of Scandinavian streams, interfering with salmon reproduction, and destroying salmon runs. It reduces forest growth and increases the amount of calcium and other nutrients leached from agricultural soils.

Current studies of the chemistry of surface water and rainfall in the northeastern United States, also downwind from major industrial centers, show that in recent years the acidity of rainfall has increased 10 to 100 times and that acidity of streams has also increased (Woodwell et al., 1972). Rain and snowfall at the Hubbard Brook Experimental Forest in New Hampshire has had a pH of 4.1 with weekly samples as low as pH 3.0. Contemporary rainfall over much of New England and eastern Canada has a pH of 4.1 to 4.4 compared to a more natural pH of 5.7 (Likens and Bormann, 1974; Bemmish, 1975).

As a result the chemistry of stream waters of the northeastern United States is changing from bicarbonates to sulfates. Such waters are detrimental to fish and other aquatic life. Thus the burning of high-sulfur coal and other fuels can easily convert nonacid streams into acid water without the input of mine acid drainage and the like. Owing to global air currents, a stream even relatively remote from industrial centers is not protected from that input of sulfuric acid.

### THE PHOSPHORUS CYCLE

The phosphorus cycle differs from the sulfur cycle in that the element is unknown in the atmosphere and none of its known compounds have any appreciable vapor pressure. Thus phosphorus can follow the hydrological cycle only partway from land to sea (Figure 6-12).

Under undisturbed natural conditions phosphorus is in short supply. It is freely soluble only in acid solutions and under reducing conditions. In the soil it becomes immobilized as phosphates of either calcium or iron. Even superphosphate (a soluble mixture of phosphates) applied to croplands may be rapidly converted to unavailable inorganic compounds. Its natural limitation in aquatic ecosystems is emphasized by the almost explosive growth of algae in water receiving heavy discharges of phosphorus-rich wastes.

The main reservoirs of phosphorus are rock and natural phosphate deposits, from which the element is released by weathering, by leaching, by erosion, and by mining for agricultural use. Some of it passes through terrestrial and aquatic ecosystems by way of plants, grazers, predators, and parasites; and it is returned to the ecosystem by excretion, death, and decay. In terrestrial ecosystems organic phosphates are reduced by bacteria to inorganic phosphates. Some are recycled to plants, some become immobilized as unavailable chemical compounds, and some are immobilized by incorporation into bodies of microorganisms. Some of the phosphorus of terrestrial ecosystems escapes to lakes and seas, both as organic phosphates and as particulate organic matter.

In marine and freshwater systems the phosphorus cycle involves four fractions: particulate phosphorus, the largest reservoir involving both living and dead particulate matter (including phytoplankton); inorganic phosphates, mostly soluble orthophosphate ($PO_4^{3-}$) with an extremely short turnover time; an organic phosphorus compound, XP, with low molecular weight (*ca* 250), excreted by organisms; and a soluble macromolecular colloidal phosphorus. The major exchange is between the inorganic phosphate and the particulate fractions. The organic compound fraction converts to the colloidal compound fraction, and both, especially the colloid, release phosphate to the soluble inorganic fraction. This is rapidly recycled through the plankton (Lean, 1973a, 1973b).

The phosphorus in the phytoplankton may be ingested by zooplankton or detritus-feeding organisms. Zooplankton in turn may excrete as much phosphorus daily as is stored in its

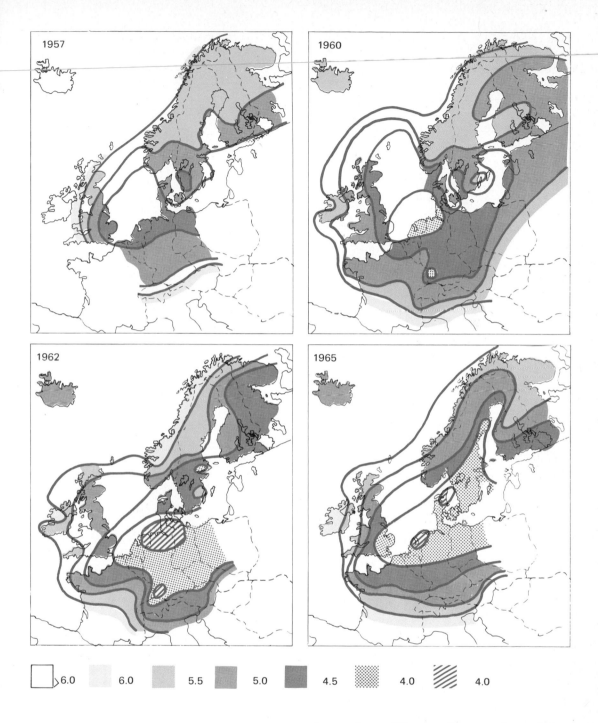

**FIGURE 6-11**
*Trends in the pH of rainfall over northern Europe from 1957 to 1965. Note the* *increasing acidity over the area downwind from the industrial heart of Europe. (From S. Oden, 1970.)*

179

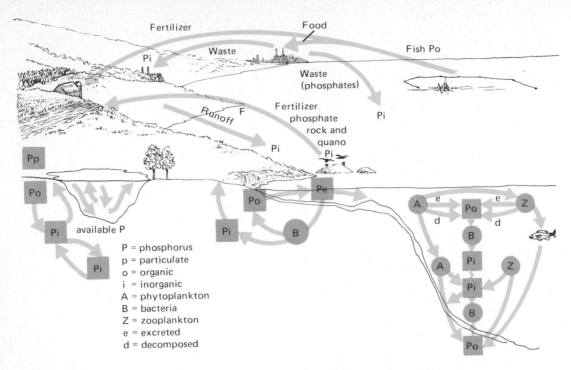

**FIGURE 6-12**
*The phosphorus cycle in terrestrial and aquatic ecosystems. P = phosphorus; p = particulate; o = organic; i = inorganic; A = phytoplankton; B = bacteria; Z = zooplankton; e = excreted; d = decomposers.*

biomass (Pomeroy et al., 1963). By excreting phosphorus it is instrumental in keeping the cycle going. More than half of the phosphorus zooplankton excretes is inorganic phosphate, which is taken up by phytoplankton. In some instances 80 percent of this excreted phosphorus is sufficient to meet the needs of the phytoplankton population. The remainder of the phosphorus in aquatic ecosystems is in organic compounds that may be utilized by bacteria, which fail to regenerate much dissolved inorganic phosphate. Bacteria are consumed by the microbial grazers, which then excrete the phosphate they ingest (Johannes, 1968). Part of the phosphorus is deposited in shallow sediments and part in the deep. In the ocean some of the latter may be recirculated by upwelling, which brings the phosphates from the unlighted depths to the photosynthetic zones, where it is taken up by phytoplankton. Part of the phosphorus contained in the bodies of plants and animals is deposited in shallow sediments and part in the deeper ones. As a result

the surface waters may become depleted of phosphorus and the deep waters saturated. Because phosphorus is precipitated largely as calcium compounds, much of it becomes immobilized for long periods in the bottom sediments. Upwelling returns some of it to the photosynthetic zones, where it is available to phytoplankton. The amount available is limited by the insolubility of calcium phosphate.

In the intertidal salt marshes of the southeastern United States marsh grass *(Spartina alterniflora)* withdraws phosphorus from the subsurface sediments. Half of the phosphorus withdrawn is fixed in plant tissue. The other half is leached from the leaves by rain and tides and carried out by tidal exchange (Reimold, 1972). When the grass dies the abundant animal population in marsh and adjacent waters and tidal creeks uses the detritus as food. The excretions of these animals add phosphorus to the water, which is taken up by the phytoplankton and eventually by the zooplankton (Pomeroy et al., 1969).

One of the detritus-feeders is the ribbed mussel *(Modiolus dimissus)*. It plays a major role in the turnover of phosphorus through the ecosystem (Kuenzler, 1961). To obtain its food, which consists of small organisms as well as particles rich in phosphorus suspended in the tidal waters, the mussel must filter great quantities of seawater. Some of the particles are ingested, but most are rejected and deposited as sediment on the intertidal mud. These particles, rich in phosphorus, are then retained in the salt marsh instead of being carried out to sea. Each day the mussel removes from the water one-third of the phosphorus found in suspended matter. Or to state it a little more precisely, every 2.6 days the mussel removes as much phosphorus as is found on the average in the particles in the water. The particulate matter deposited on the mud is utilized by the deposit-feeders, which release the phosphate back to the ecosystem. Thus the ribbed mussel, although of little economic importance to man and relatively unimportant as an energy consumer in the salt marsh, plays a major role in the cycling and retention of phosphates in the salt marsh. The worth of an animal cannot always be measured in terms of economic values or ignored because it contributes little to energy flow. It may serve some other important ecological function in the community that remains to be discovered.

Phosphorus, like other elements, may take diverse pathways through an ecosystem. Its movements through the several levels can be followed by the use of $^{32}P$ (phosphorus 32) as a tracer. By adding carefully regulated amounts of $^{32}P$, Ball and Hooper (1963) followed the movements of this element through a trout-stream ecosystem. The uptake was rather rapid, the material traveling from 414 to 10,298 m downstream, depending upon conditions, before being finally removed. Microscopic plants (periphyton) growing on the rocks and other substrate and three species of larger plants, *Potamogeton,* the alga *Chara,* and water moss *Fontanalis,* were responsible for most of the uptake. Maximum amounts of radioactive phosphorus appeared in plant tissues shortly after the material passed through the area, and the rate of loss was the greatest shortly thereafter. Losses decreased with time for almost 15 to 20 days, when equilibrium was achieved. This suggested that the plants were recycling phosphorus.

Concentration of phosphorus in consumer organisms reflected differences in both metabolic turnover rates and in food relationships. Small filter feeders, especially the black fly

larvae, reached the highest level of concentration in the shortest time, followed by the animals that scraped periphyton from the rocks. At the same time these organisms lost $^{32}P$ quite rapidly. The material persisted longer in such omnivorous feeders as the scud and the caddisfly appeared latest and was retained longest in the invertebrate and vertebrate predators. The investigators found considerable variation in the uptake and retention of $^{32}P$ by different plants and animals from year to year. This suggested that the major differences in the cycling of radioactivity were related to the way $^{32}P$ distributed itself between soluble and particulate phases in the stream water.

Seasonal changes in phosphorus sources have been described for a marine ecosystem (Ketchum and Corwin, 1965; Ketchum, 1967). In the prebloom period 28 percent of the phosphorus supply necessary for phytoplankton production was supplied by the inorganic fraction in solution in the euphotic zone (upper water layer) and 72 percent was supplied by vertical mixing and transport of nutrients from deeper water (Table 6-6). During the period of bloom 86 percent of the phosphorus came from the inorganic phosphorus dissolved in solution, 12 percent from vertical mixing, and 2 percent from regeneration by biological cycling. When the system was nearly in equilibrium, 43 percent of the phosphorus was supplied by regeneration and 57 percent by vertical mixing. During the prebloom period

TABLE 6-6
*Sources of supply of phosphorus under three environmental conditions*

| | Prebloom | Bloom | Steady state |
|---|---|---|---|
| Total phosphorus cycle (mg at/m²/day) | 0.89 | 1.65 | 0.54 |
| Source (%) | | | |
| Removal | 28 | 86 | 0* |
| Vertical mixing | 72 | 12 | 57 |
| Regeneration | 0 | 2 | 43 |

*Inorganic phosphate increased in the euphotic zone by 0.22 mg at m²/day. This amount was also supplied by vertical mixing.
*Source:* B. H. Ketchum, 1967.

13 percent of the phosphorus assimilated was located on the particulate matter about equally divided between the upper and lower layers; during the bloom 92 percent was located in particulate matter, of which about 47 percent remained in the upper water layers; and during the steady-state condition only 3 percent was in particulate form. Thus during the bloom stage much of the phosphorus used is tied up in organic matter and only by a rapid turnover of phytoplankton populations could the phosphorus requirements be met. The importance of turnover is further emphasized by the fact that under steady-state conditions only a small part of the phosphorus assimilated in production is in living matter; most of it is supplied by regeneration and vertical mixing from the deep.

The phosphorus requirements of marine phytoplankton can be met only by biological regeneration of nutrients and vertical mixing from the deep. Regeneration can come about only by a rapid recycling in the aquatic ecosystem, which in coastal waters may occur six to ten times a year.

Turnover time in the upper layers of lakes is 20 to 45 days (Rigler, 1973). However, phosphorus is rapidly recycled between plankton and the inorganic fraction. In summer turnover time between the two averages 0.9 to 7.5 minutes (Rigler, 1956). The average turnover time in summer seas ranges from 1 to 56 hours (Pomeroy, 1959).

*Intrusions into the phosphorus cycle.* As with other biogeochemical cycles, human activities have altered the phosphorus cycle. Because the cropping of vegetation depletes the natural supply of phosphorus in the soil, phosphate fertilizers must be added. The source of phosphate fertilizer is phosphate rock. Because of the abundance of calcium, iron, and ammonium in the soil, most of the phosphate applied as fertilizer becomes immobilized as insoluble salts. In 1968, for example, 50 percent more fertilizer was added to cropland than was lost to the oceans from global runoff from all sources.

Part of the phosphorus used as fertilizer is removed in crops when harvested. Transported far from the point of fixation, this phosphorus in vegetables and grain eventually is released as waste when foodstuffs are processed or consumed. Concentration of phosphorus in wastes of food-processing plants and of feedlots adds a quantity of phosphates to natural waters. Greater quantities are supplied by urban areas, where phosphates are concentrated in sewage systems.

Most of the phosphorus enrichment of aquatic ecosystems comes from sewage disposal plants. Primary sewage treatment removes only 10 percent of the total phosphorus. Secondary treatment removes only 30 percent at best, and even some of this ultimately finds its way into streams. Feedlots may contribute some from runoff. However, phosphorus has such a strong affinity to the soil particles that the amount coming from agricultural lands is insignificant. Thus sewage, with its heavy load of phosphate detergents, contributes nearly all of the phosphorus reaching lakes and rivers.

Phosphorus is perhaps more intimately involved in the eutrophication of water than nitrogen. Of the three nutrients required for aquatic plant growth, potassium is usually present in excess, nitrogen is supplemented by fixation, and phosphorus tends to be precipitated in the sediments and cannot be supplemented naturally (Am. Chem. Soc., 1969). Thus phosphorus is usually limiting, and in the presence of a luxury supply, algae respond with luxurious growth.

### CYCLING OF HEAVY METALS AND HYDROCARBONS

Bringing biogeochemical cycles forcibly to the attention of the public has been the accumulation of heavy metals in the biosphere. Among these are lead and mercury. Heavy metals have always been present in the environment and have entered into natural cycles of the ecosystem. However, human activities have markedly increased their concentration in the environment, and thus the passage of these metals through the food chain has markedly increased their concentration in the upper trophic levels.

*Lead.* Ninety-eight percent of the lead in the biosphere comes from automobile emissions, the burning of lead alkyl additives to gasoline. In 1968 more than $2.3 \times 10^8$ kg of lead was emitted to the atmosphere, which roughly amounts to 1 kg per person (Chow and Earl, 1970; Commoner, 1971). Minor sources are metal smelting plants and agricultural areas where lead arsenate is used as an orchard spray. Introduced as a fine aerosol lead eventually falls out either in precipitation or in dust to contaminate the soil. Once on or in plants, lead enters the food chain (Chow, 1970). Plant roots take up lead from the soil and leaves take it from contaminated air or from particulate matter on the leaf surface. There is evidence that microbial systems are also capable of taking up and immobilizing substantial quantities of lead (Tornabene and Edwards, 1972).

Atmospheric lead pollution poses a serious health problem in industrialized areas of the earth (Goldsmith and Hexter, 1967; Wessel and Dominski, 1977). The long-term increase in concentrations of atmospheric lead has resulted in a significant increase in the concentrations of lead in humans. The average body burden of lead among adults and children in the United States is 100 times greater than the natural burden and existing rates of lead absorption are 30 times the levels in preindustrial society.

*Mercury.* The danger of mercury-contaminated ecosystems has been tragically emphasized by the deaths and impaired lives of the Japanese who live in the villages around Minamata Bay and along the Agano River. The fish in these waters, a major part of these people's diet, became contaminated with methylmercury when neighboring industrial plants discharged mercuric wastes into the waters. The methylmercury ingested was assimilated and transported by the blood to the brain. Toxic accumulations resulted in severe neurological symptoms and death. Called Minamata Disease, mercury poisoning is characterized by blindness, deafness, incoordination, and intellectual deterioration. In some instances mercury can be transferred from a pregnant woman through the placenta to the fetus, even though the mother herself exhibits no symptoms of mercury poisoning.

The toxic properties of mercury have been recognized for centuries, but only in recent times has the element accumulated in the ecosystem. The discovery of mercury concentrations in wild birds and fish signaled its widespread distribution throughout the global ecosystem. High mercury contents in the feathers and beaks of grain-eating birds and their predators in Sweden gave the first scientific evidence that the terrestrial environment was contaminated (Berg et al., 1966; Borg, 1969). The source of contamination was alkyl mercury used as a fungicide on seeds. It was consumed directly by seed-eating birds and passed on to their predators. In Sweden 41 percent of seed-eating birds and 67 percent of predatory birds contained 2 mg/kg and over of mercury. However, since mercurial fungicides on seeds have been banned, the levels of mercury in seed-eating birds have declined.

More highly contaminated than terrestrial environments are aquatic ecosystems, which receive heavy local influxes of mercuric wastes from the manufacturers of chlorine, caustic soda, and electrical equipment, from pulp and paper mills, and from the burning of fossil fuels (Joensuu, 1972). The problem in aquatic environments is compounded by the chemical nature of mercury. It is unique in its ability to form stable compounds with organic radicals such as the methyl group. These compounds can be classified into three groups: (1) alkyl, as represented by methylmercuric hydroxide; (2) alkoxyalkyl, such as methoxyethylmercuric hydroxide; and (3) aryl, such as phenylmercuric acetate. Mercury is usually discharged into the aquatic environment as inorganic mercury compounds or as phenylmercury.

In the aquatic environment mercury may precipitate out as a highly insoluble sulfide. Under anaerobic conditions this compound may remain in the sediments for an indefinite period. Under aerobic or partially anaerobic conditions it can oxidize to the sulfate form and be subject to methylation (Figure 6-13). In the bottom sediments microbes capable of synthesizing vitamin $B_{12}$ can transform inorganic, aryl, and alkoxyalkyl compounds into both monomethyl and dimethyl forms. The less volatile monomethyl is usually formed under acidic conditions; the highly volatile dimethyl form is favored under neutral and alkaline pH. Methylmercury dissolves in the water. The more volatile dimethylmercury evaporates into the atmosphere and enters the global cycle. Fish in the methylmercury-contaminated water pick up this pollutant either by the food chain or by diffusion across the gills.

The concentration of methylmercury in fish inhabiting locally contaminated water may be

**183**

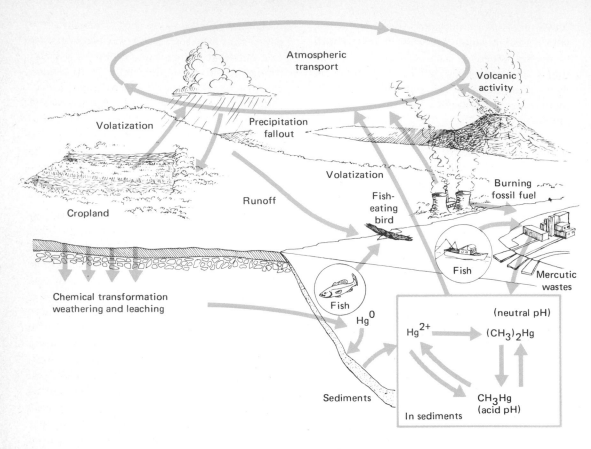

**FIGURE 6-13**
*The mercury cycle in nature. The portion of the cycle in methylation is shown in the insert.*

much too high. In Sweden average concentrations range from 0.2 to 5.0 mg/kg; a level of 0.05 mg is considered acceptable. Not only do these levels have a direct and often lethal effect on fish, but mercury is passed along to consumer organisms, with tragic results.

Because mercury is a natural substance, it has probably always been present at some level in oceanic and freshwater fish. In spite of bans on tuna and swordfish because of higher than permissible levels in their flesh, levels have changed little in time. Mercury levels of museum specimens of tuna and swordfish caught 93, 64, and 25 years ago are in the same range as specimens taken in recent years, all above permissible levels. This suggests that mercury levels now found in wide-ranging ocean fish come from natural sources rather than from man-made pollutants (Weiss

et al., 1971). In fact one authority (A. L. Hammond, 1972) estimates that if the total amount of mercury processed by man since 1900 were introduced and well mixed in the world's oceans, the concentration would be increased by only 1 percent. In freshwater ecosystems the story is different. D'Itri (1971) analyzed museum specimens of fish caught in the St. Clair River System before it received mercurial pollution and compared concentrations of mercury in those fish with ones recently caught. Concentrations of mercury in muskellunge and sea lamprey increased threefold from 1965 to 1970; concentrations in sturgeon, pike, sauger, small-mouthed bass, walleye, and white crappie increased five times during this period.

On a global basis the cycling of mercury appears to present no particular hazard, although as yet relatively little is known about the mercury cycle. Inorganic mercury and methylmercury are highly volatile at ordinary temperatures and are interjected into the atmosphere as an aerosol or vapor or both. Estimates place the atmospheric burden of mercury as 80,000 metric tons, but little is known of actual concentrations. The few measurements taken in the United States are well below the threshold limit of 0.1 mg/m³ for metallic vapor and 0.01 mg/m³ for organic mercury.

The major dangers of mercury are concentrations in local aquatic environments, where it accumulates in fish, and in terrestrial ecosystems, where it is introduced as a fungicide on seeds. Its greatest hazard to health is limited to those people for whom fish is the major part of the diet. For wildlife mercury is a hazard to seed-eating birds feeding directly on treated seeds and to predatory birds at the top of the aquatic food chain. For all animals, intake of mercury can induce chromosome breakage and other genetic risks not fully understood.

***DDT and other chlorinated hydrocarbons.*** Of all man's intrusions into biogeochemical cycles, none has done more to call attention to nutrient cycling than the widespread application of DDT. During World War II this pesticide began to be used in huge quantities to control disease-carrying and crop-destroying insects. As early as 1946 Clarence Cottam called attention to its damaging effects on ecosystems and nontarget species. But the impact of pesticides on ecosystems remained obscure until Rachel Carson's *Silent Spring* exposed the dangers of hydrocarbons. The detection of DDT in the tissues of animals in the Antarctic, far removed from any applied source of the insecticide, emphasized the fact that DDT does indeed enter the global biogeochemical cycle and becomes dispersed around the earth.

DDT (along with other chlorinated hydrocarbons) has certain characteristics that enable it to enter global circulation. Because it is highly soluble in lipids or fats and not very soluble in water, it tends to accumulate in the lipids of plants and animals. It is persistent and stable, undergoes little degradation (largely from DDT to DDE), and has a half-life of approximately 20 years. It has a vapor pressure high enough to ensure direct losses from plants. It can become adsorbed by particles or remain as a vapor; in either state it can be transported by atmospheric circulation and then return to land and sea with rainwater.

The major input of DDT to the biosphere is from its manufacture and subsequent application to croplands, forest, and marshes for insect control (Figure 6-14). In 1963 the maximum production of DDT was $8.13 \times 10^{10}$ g. Production declined sharply in 1970. The major areas of application have been and continue to be humid temperate and tropical regions.

Insecticides are applied on a large scale by aerial spraying. Half or more of a toxicant applied in this manner is dispersed to the atmosphere and never reaches the ground. If the vegetation cover is dense, only about 20 percent reaches the ground. In a massive spraying of DDT over 66,000 acres of forest in eastern Oregon, only about 26 percent reached the forest floor (Tarrant, 1971).

On the ground or on the water's surface the pesticide is subject to further dispersion. There is apparently little movement of DDT from the surface soil to the subsoil. In the Oregon study the concentration of DDT in the prespray samples was 0.006 ppm at a depth of 0 to 3 in. Twelve months after spraying, the concentration was 0.029 ppm; 36 months later it was back to 0.006 ppm again. There was no significant gain in the subsoil levels. Input from litterfall likewise declined from 11.32 ppm at 0 to 6 months after spraying to 3.08 ppm 3 years later. Throughfall precipitation contained insignificant amounts. Woodwell et al. (1971) estimate that agricultural soils of the United States contain 0.168 g/m² (1.50 lb/acre) of DDT, and nonagricultural soils 0.0045 g/m². Mean lifetime of DDT in the soil is about 4.5 years. Pesticides reaching the soil are lost through volatilization, chemical degradation, bacterial decomposition, runoff, and the harvest of organic matter, which can amount to about 1 percent of the total DDT used on the crop.

In flowing water pesticides are subject to further distribution and dilution as they move downstream. Insecticides released in oil solutions penetrate to the bottom and cause mortality of fish and aquatic invertebrates (see reviews in Pimentel, 1971a; and Cope, 1971). Trapped in bottom rubble and mud, the insecticide may continue to circulate locally and kill for some days.

In lakes and ponds emulsifiable forms of DDT tend to disperse through the water, but

**185**

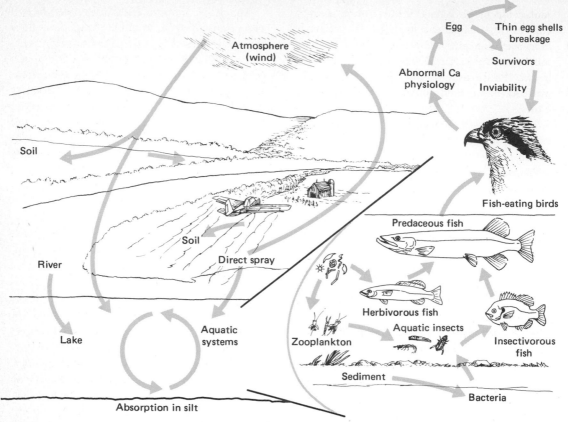

*FIGURE 6-14*
*The chlorinated hydrocarbon cycle. The
initial input comes from spraying. A
relatively large portion fails to reach the
ground and is carried on water droplets and
particulate matter through the atmosphere.*

not necessarily in a uniform way. DDT in oil
solutions tends to float on the surface and
move about in response to the wind.

Eventually the pesticides reach the ocean,
where they may concentrate on the surface
slicks where the pesticide concentration is
10,000 times greater than in lower waters.
These slicks, which attract plankton, are carried across seas by ocean currents. In the
oceans part of the DDT residues may circulate
in the mixed layer. Some may be transferred to
below the thermocline to the abyss. More may
be lost through the sedimentation of organic
matter.

Although considerable amounts of DDT and
other pesticides are transported by water, they
are relatively insignificant from a viewpoint of
global circulation. Estimates based on con-

centration of pesticides in river water and annual runoff amount to about 0.1 percent of the
amount of DDT produced per year in the United States.

The major transport of pesticide residue
takes place in the atmosphere. Not only does
the atmosphere receive the bulk of pesticidal
sprays (well over 50 percent of that applied),
but it also picks up that fraction volatilized
from soils, vegetation, and water. The vapor
pressure of DDT is such ($1.5 \times 10^{-7}$ mm Hg at
20° C) that the equilibrium concentration is
about 2 ppb by weight. If DDT remained as a
vapor alone, the saturation capacity of the atmosphere to the troposphere would hold as
much DDT as produced to date. But the capacity of the atmosphere to hold DDT is increased
greatly by the adsorption of residues to particulate matter. Thus the atmosphere becomes
a large circulating reservoir of DDT and other
chlorinated hydrocarbons.

Residues are removed from the atmosphere
by chemical degradation, by diffusion across
the air-sea interface, and mostly by rainfall
and dry fallout (SCEP, 1970). The mean con-

centration of DDT in rainfall in England in 1966-1967 was 80 ppt. Rainfall in southern Florida between June 1968 and May 1969 contained an average of 1000 ppt of DDT residues. If the total annual precipitation over the world's oceans contained an average of 80 ppt, a total of $2.4 \times 10^4$ metric tons of DDT would be carried to the oceans annually, about one-fourth of the world's annual production. Pesticides carried to land and sea are subject to volatilization and subsequent return to the atmosphere.

Although the quantity of residues of DDT and other chlorinated hydrocarbons may be relatively small, amounting to 1/30 or less of the amount produced each year, the concentrations still are sufficient to have a deleterious impact on marine, terrestrial, and freshwater ecosystems. DDT and related compounds tend to concentrate in the lipids of living organisms where they undergo little degradation (see Menzie, 1969; Bitman, 1970; and Peakall, 1970).

The high solubility of DDT in lipids allows the magnification of its concentration through the food chain. Most of the DDT contained in the food is retained in the fatty tissues of the consumer organism. Because it breaks down slowly, DDT accumulates to high and even toxic levels. The concentrated DDT is passed on to the consumer at the next trophic level where again it is retained and accumulated. The carnivores on the top level of the food chain receive massive amounts of pesticides.

There are a number of examples of species concentrating this pollutant. Eastern oysters held for 40 days in flowing seawater containing only 0.1 ppb of DDT concentrated the pesticide some 70,000 times that contained in the water (Butler, 1964). Four species of algae in water containing 1.0 ppm for 7 days concentrated the pesticide 227-fold (Vance and Drummond, 1969). Daphnia, a genus of zooplankton, concentrated DDT 100,000-fold during a 14-day exposure in water containing 0.5 ppb of DDT (Preuster, 1965). Slugs and worms in a cotton field concentrated DDT 18 and 11 times that of the level in the soil. Pesticides concentrated by first-level consumers are then passed on to second-level consumers, carnivores, who in turn magnify the pesticide even more.

The concentration of DDT in a Long Island salt marsh sprayed for mosquito control over a period of years was 13 lb/acre. The actual concentration of DDT in the water was 0.00005 ppm. The residue in consumers showed the increase in concentration along the food chain: 0.04 ppm in plankton, 0.16 ppm in shrimp, 0.28 in eel, 2.07 ppm in predacious fish, and 75.0 ppm in ringbilled gulls. Thus the ringbilled gull, a predacious bird near the top of the food chain, contained a level of DDT a million times greater than in the water (Woodwell et al., 1967).

In Clear Lake, California, the concentration of DDT in the tissue of organisms over that of water was 265-fold in plankton, 500-fold in small fishes, 85,000-fold in predacious fishes, and 80,000-fold in grebes, fish-eating birds (Rudd and Genelly, 1956).

High concentrations of DDT in the tissues often result in death or impaired reproduction and genetic constitution of organisms. Laboratory populations of zooplankton, shrimps, and crabs are killed outright by exposure to DDT in parts per billion. The continuous exposure of shrimp to DDT in 0.2 ppb, a concentration that has been detected in waters flowing into shrimp nursery areas, killed the entire population in less than 20 days (SCEP, 1972). A residue level of 5.0 ppm in the lipid tissues of the ovaries of freshwater trout causes 100 percent die-off of the fry, which pick up lethal doses as they utilize the yolk sac. High levels of DDT are correlated with the decline of such fish as sea trout and California mackerel.

DDT and its degradation product DDE interfere with calcium metabolism in birds. Chlorinated hydrocarbons block ion transport by inhibiting the enzyme ATPase, which makes available the needed energy. This reduces transport of ionic calcium across membranes and can cause death of organisms. DDE also inhibits the enzyme carbonic anhydrase (Bitman, 1970; Peakall, 1970), essential for the deposition of calcium carbonate in the eggshell and for the maintenance of a pH gradient across the membrane of the shell gland.

There is evidence that DDT can depress the growth of certain commercial vegetables, can cause significant changes in macro- and microelements in the above-ground parts of plant tissues (see Pimentel, 1971a), and can reduce the productivity of phytoplankton (Wurster, 1968). In most instances, however, the concentrations necessary to affect these changes were much higher than normally applied or higher than might occur in natural ecosystems.

*Polychlorinated biphenyls.* Another recently discovered contaminant widespread in the environment is polychlorinated biphenyl (PCB). PCB is a generic name used for a number of synthetic organic compounds characterized by biphenyl molecules containing chlorine atoms. PCBs are a mixture of different chlorinated biphenyls, each having a different percentage of chlorine atoms. They are widely used as dielectric fluids in capacitors and transformers, in plastics, solvents, and printing inks. The production of PCBs increased steadily from their introduction in 1930 to a high of 34,000 tons in 1970. Production and use are now declining since the discovery that PCBs are an environmental contaminant. (For a good review of PCBs, see Ahmed, 1976.)

A Swedish scientist, S. Jensen (1966), first recognized the threat of PCBs to the environment when he discovered them in fish and birds he was examining for the presence of DDT. Since then biologists have found PCBs in human tissue, food stuffs, and many species of fish and birds. Like DDT, PCBs have an affinity for fatty tissue, degrade slowly, and accumulate in the food chain.

Except for the source, the pathway of PCBs is similar to that of DDT. PCBs are discharged into the environment through sewage outfalls and industrial discharges. Rivers are a means of transport, but the atmosphere appears to be the major mode (see A. L. Hammond, 1972). PCBs accumulate in bottom sediments, adsorbed in silt and fine particles which hold them in the aquatic environment. Fish take up PCBs and concentrate them in their tissues, where the chemical residue is often higher than DDT. Typical concentrations of PCBs in fish, both ocean and fresh water, range from 0.01 to 1.0 ppm away from heavily polluted industrial areas. In polluted waters, as parts of the Great Lakes and the Hudson River in New York State, PCBs range from 10 to 85 ppm, with individual fish carrying as

much as 400 ppm. Federal Drug Administration acceptable tolerance levels of PCBs in food range from 0.5 ppm to 5.0 ppm, depending upon the food item. Predatory fish-eating birds, particularly ospreys and cormorants, have concentrations that range from 300 to 1000 ppm. Traces have been found in human tissue. Like DDT, PCBs appear to cause thinning of egg shells, deformities in newly hatched birds (Gilbertson et al., 1976), and reduction in growth rates of certain marine diatoms (Mosser et al., 1972).

## RADIONUCLIDES

Ever since the atomic bomb ushered in the atomic age, the impact of nuclear radiation on life on earth has been a major concern. Involved are high-energy, short wavelength radiations, known as ionizing radiations. They are so-called because they are able to remove electrons from some atoms and attract them to other atoms, producing positive and negative ion pairs. Of greatest interest is ionizing electromagnetic or gamma radiation, which has a short wavelength, travels a great distance, and penetrates matter easily.

Sources of gamma radiation are atomic blasts from weapon testing, nuclear reactors, and radioactive wastes. By-products of both weapons testing and nuclear reactors are radioactive such as zinc-65 ($^{65}Zn$), strontium-90 ($^{90}Sr$), cesium-137 ($^{137}Cs$), iodine-131 ($^{131}I$), and phosphorus-32 ($^{32}P$). When the uranium atom is split or fissioned into smaller parts, it produces, in addition to tremendous quantities of energy, a number of new elements, or fission products, including strontium, cesium, barium, and iodine. Some of these fission products last only a few seconds; others can remain active for several thousand years. These radioactive elements enter the food chain and become incorporated in living organisms.

In the same atomic reaction some particles with no electrical charges, called neutrons, get in the way of high-energy particles. Nonfission products are the result, and they include the radioisotopes of such biologically important elements as carbon, zinc, iron, and phosphorus, which are useful in tracer studies.

Both fission and nonfission products are released to the atmosphere by nuclear testing and by wastes from nuclear reactors unless carefully handled. Later these return to earth along with rain, dust, and other material as atomic fallout. In the case of weapons testing in the atmosphere, the fallout can be worldwide. Once the isotopes reach the earth, they enter the food chain and become concentrated in organisms in amounts that exceed by many times the quantities in the surrounding environment. This, in effect, produces local radiation fields in the tissues of plants and animals.

Of particular concern is the radioactive output of nuclear power plants. Pressurized water reactors, commonly used in nuclear power plants, release low levels of radioactivity to the air and condenser water. Radioactive contamination of water from nuclear power plants results when water passes through the intense neutron flux of the reactor. Trace elements in the water are activated, producing radioisotopes. Added to this are radioactive corrosive products from the surface of metal cooling tubes. Except for tritium, most of the radioisotopes are removed in a radioactive waste

removal process. Those left have a short half-life and rapidly decay below detection levels.

Dispersion of radionuclides in terrestrial ecosystems comes from gaseous, particulate, and aerosol deposition and from liquid and solid wastes. Plants both intercept particulate contaminants and absorb radionuclides through the foliage. Additional input into vegetation may come by way of uptake of radionuclides from soil and litter. From the plants radionuclides move through the ecosystem by way of the food chain.

Examples are strontium-90 and cesium-137, two of the most important radioactive materials released into the biogeochemical cycle (Figure 6-15). Ecologically they behave like

*FIGURE 6-15*
*Radionuclide cycling through the food chain.*
*Strontium and cesium are transferred*
*through fallout from the source to cows to*
*humans. (Courtesy Oak Ridge National*
*Laboratory.)*

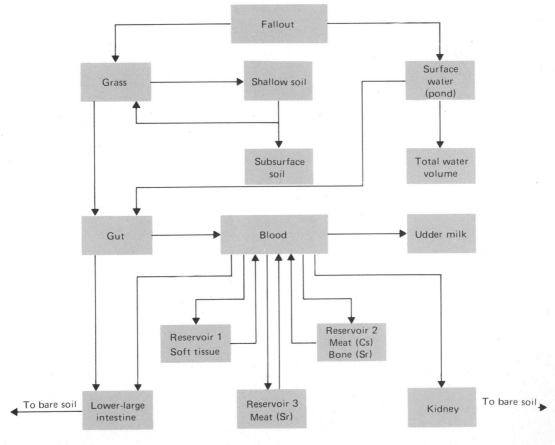

189

TABLE 6-7
*Aquatic and terrestrial food-chain concentration of elements*

ELEMENT CONCENTRATION FACTORS*

| Trophic level | Ca | Sr | K | Cs | Na | Co | Zn | Mn | Ru | Fe | P | Ra | I |
|---|---|---|---|---|---|---|---|---|---|---|---|---|---|
| **AQUATIC** | | | | | | | | | | | | | |
| Water | 1.0 | 1.0 | 1.0 | 1.0 | 1.0 | 1.0 | 1.0 | 1.0 | 1.0 | 1.0 | 1.0 | 1.0 | 1.0 |
| Algae and higher plants | 1–400 | 10–3,000 | | 50–25,000 | | 2,500–6,200 | 140–33,500 | 700–35,000 | 80–2,000 | 2,400–200,000 | 36,000–50,000 | 0.5 | 60–200 |
| Invertebrates | | | | | | | | | | | | | |
| saprovores | 16 | 10–4,000 | | 60–11,000 | | 325 | | 6,000–140,000 | 130 | 125 | 2–100,000 | 0.5 | 20–1,000 |
| herbivores | | 1 | | 600 | | | 150 | | | | 2,000 | | 2,000 |
| carnivores | | | | 800 | | | | | | | | | |
| Fish | | | | | | | | | | | | | |
| omnivores | | 1 | 300–2,500 | 125–6,000 | | | | | | 10,000 | 3,000–100,000 | 0.5 | 25–50 |
| carnivores | 0.5–300 | 1–150 | 400–2,700 | 640–9,500 | | | 4–40 | | | | | 1.5 | |
| **TERRESTRIAL** | | | | | | | | | | | | | |
| Plants | 1.0 | 1.0 | 1.0 | 1.0 | 1.0 | 1.0 | 1.0 | 1.0 | 1.0 | 1.0 | 1.0 | 1.0 | 1.0 |
| Invertebrates | | | | | | | | | | | | | |
| saprovores | 0.1–18 | 0.1 | 3.5 | 0.2 | 17 | 0.4 | | | 0.4 | | 11 | | 0.5 |
| herbivores | 0.1 | 0.1 | 3.0 | 0.3–0.5 | 21 | 0.5 | | | 1.2 | | 17 | | 0.2 |
| carnivores | 0.1 | | 2.0 | 0.1–0.5 | 27 | | | | | | 18 | | 0.1 |
| Mammals | | | | | | | | | | | | | |
| herbivores | 0.5–4.5 | 0.5–4.5 | | 0.3–2.0 | | 0.3 | | | 0.4 | 0.8 | | 0.01 | |
| omnivores | | | | 1.2–2.0 | | | | | | 0.2 | | | |
| carnivores | | | | 3.8–7.0 | | | | | | | | | |

*Ratio of element level in consumer to element level in food-chain base, with base value normalized at 1.0.
*Source:* D. E. Reichle et al., 1970.

calcium and follow it in the cycling of materials. Both enter the food chain most easily through the grazing food chain, especially in regions with relatively high rainfall or abundant soil moisture and with low levels of calcium and other mineral nutrients in the soil. One such region, the arctic tundra, has been subject to rather heavy atomic fallout from weapons testing, especially in the past. Lichens, the dominant plants, absorb virtually 100 percent of the radioactive particles falling upon them. From lichens the main contaminants, $^{90}$Sr and $^{137}$Cs, travel up the food chain from caribou and reindeer to wild carnivores and humans. Because caribou feed all winter on lichens, their flesh in spring contains 3 to 6 times as much $^{137}$Cs as it does in the fall. The caribou is a major food source for northern Alaskan natives who kill the animals during the northward migration in spring and stockpile the meat for late spring and early summer food. Thus in the spring the Eskimos show a corresponding rise in $^{137}$Cs level, which often increases by 50 percent. This level decreases when the people change to a diet of fish

in the summer (Figure 6-16). Nevertheless, humans who depend upon reindeer and caribou for their protein already have accumulated from one-third to one-half of their permissible amounts (Palmer et al., 1963; W. C. Hanson, 1971).

In vertebrate food chains strontium and cesium usually accumulate relative to whole-body levels at higher trophic levels, but cobalt, ruthenium, iodine, and some other radionuclides do not. Some, such as iodine, may concentrate in certain tissues. In arthropod food chains potassium, sodium, and phosphorus accumulate, whereas strontium and cesium do not (see Table 6-7).

*FIGURE 6-16*
*Cesium-137 concentrations in lichens, caribou flesh, and Eskimos at Anaktuvuk Pass, Alaska, during the period 1962-1968. Note the relationship between the concentration of $^{137}$Cs in caribou flesh and the amount in Eskimos. As the concentration in caribou declines seasonally, so does the concentration in humans. (From W. C. Hanson, 1971.)*

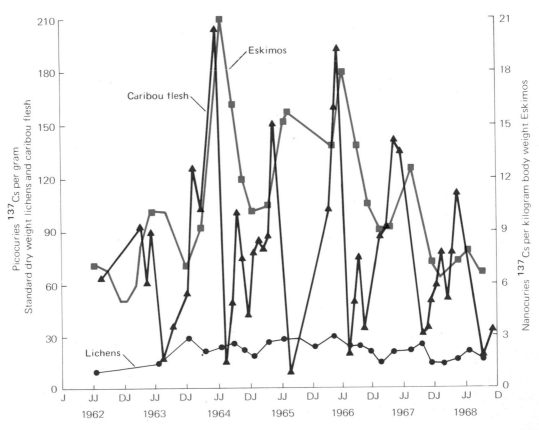

Radionuclides contaminate aquatic ecosystems largely through waste from nuclear power plants and from nuclear processing industry. Radioactive materials that enter the water become incorporated in bottom sediments and circulate between mud and water. Some become absorbed by bottom-dwelling insects and fish downstream from the source. In fact they may be exposed to chronic low-level radiation. Under such conditions a sort of equilibrium is established between retention in the organisms, the bottom sediments, daily input, and decay. As in terrestrial ecosystems concentrations of radionuclides in food chains vary with the system and the species involved. In aquatic environments $^{32}P$ concentrations increase at higher trophic levels; $^{90}Sr$ becomes concentrated in the bones of fish and in invertebrates possessing calcareous exoskeletons. Yet the concentration in fish flesh is less than that in plants and invertebrates. For example, cobalt-60 concentrations in algae (plants which absorb more nutrients physically through cell membranes than they take in and concentrate biologically) are 2500 to 6200 times that of the surrounding water; concentrations in fish are only 25 to 50 times that of the water. Thus the concentration of radionuclides does not necessarily increase consistently through the food chain as does the concentration of chlorinated hydrocarbons. In many situations the concentration of radionuclides decreases at higher trophic levels. Aquatic organisms tend to concentrate radionuclides the same as they do the stable element.

In general $^{137}Cs$, related biogeochemically to potassium, does not appear to increase at higher trophic levels in aquatic ecosystems. However, the fact that fish do accumulate some $^{137}Cs$ in their bones and clams $^{90}Sr$ in their shells serves as a monitor of low-level radiocontamination of the environment. Because they continuously feed on particulate matter in the aquatic environment and accumulate certain radionuclides, clams are excellent indicator species. Since there is no turnover of radionuclides deposited in the growth rings of the shell, their shells are a record of the radionuclide contamination in their environment (D. J. Nelson, 1962).

In spite of considerable study, we still know little about uptake, assimilation, distribution in tissues, turnover rates, and equilibrium levels of radionuclides for many taxonomic groups. We know even less about the role of the environment in the cycling of radionuclides through various ecosystems. As the number of nuclear power plants and the necessity of waste disposal increase, our knowledge of the behavior of radionuclides needs to be more sophisticated. The knowledge of hazards must extend not only to humans, but also to the biota upon which they depend.

## Nutrient budgets

Nutrients are constantly being added to or removed from ecosystems by artificial or natural processes (Figure 6-17). In woodland, shrub, and grassland ecosystems nutrients are imported by wind, rain, dust, and animal life and are returned annually to the soil by leaves, litter, roots, animal excreta, and the bodies of the dead. Released to the soil by decomposition, these nutrients again are taken up first by plants and then by animals. Some nutrients are retained in plant and animal biomass, some are stored in the soil, some are lost to ecosystems by leaching, erosion, and harvesting of plants and animals. In freshwater and marine ecosystems nutrients are imported by precipitation and runoff, taken up by phytoplankton, and consumed by animal life. The remains of plants and animals drift to the bottom where decomposition takes place. The nutrients again are recirculated to the upper layers by annual overturn and by upwellings from the deep. The inputs and outflows balanced one against the other make up the nutrient budget.

Knowledge of types of biogeochemical cycles, that is, how specific elements are cycled, enables us to construct nutrient budgets for particular ecosystems. A nutrient budget is the measure of the input and outflow of elements through the various components of an ecosystem. It involves tracing the pathways of each element through individuals, through parts of the ecosystem, and into and out of the system and measuring amounts of the elements so circulated.

Obtaining data for such a budget is difficult and time consuming. Inputs can be determined

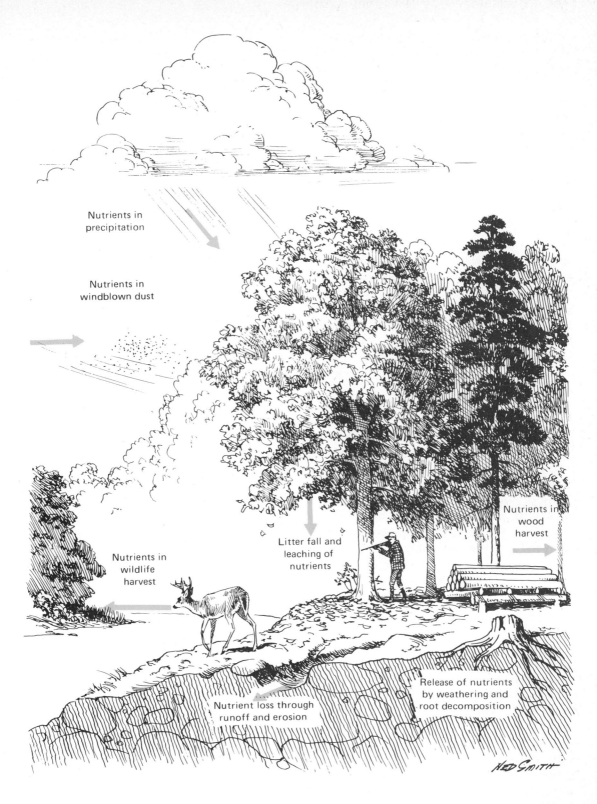

Nutrients in
precipitation

Nutrients in
windblown dust

Nutrients in
wood
harvest

Litter fall and
leaching of
nutrients

Nutrients in
wildlife
harvest

Release of nutrients
by weathering and
root decomposition

Nutrient loss through
runoff and erosion

NED SMITH

**FIGURE 6-17**
*Generalized nutrient cycle in a forest ecosystem. Input of nutrients is through precipitation, windblown dust, litterfall, and release of nutrients by weathering and root decomposition. Outgo is through wood harvest, wildlife harvest, runoff, erosion, and leaching.*

193

by measuring the quantities of nutrients carried in by precipitation and aerosols collected over the system. Estimates of the standing crop of nutrients can be obtained by sampling a number of trees of the various species, by determining the nutrient distribution within the biomass, and by measuring mineral content of leaf wash and stemflow. Transfers from the forest soil to the roots can be measured by collecting samples of soil solution. The uptake by vegetation can be estimated by sampling the mineral content of current years' growth, including foliage, branches, and bole. Pathways of nutrients within the system can be followed by means of radioactive tracers. Outflow from terrestrial systems can be determined by analyzing the nutrient content of streamflow from the system.

Because nutrient budgets of temperate forests have been studied more intensively than any other type of ecosystem, data from them can serve to illustrate nutrient budgets of ecosystems.

In the temperate forest ecosystem appreciable quantities of plant nutrients are carried in by rain and snow (Emanuelsson, Eriksson, and Egner, 1954; Eaton et al., 1973; Patterson, 1975) and by aerosols (White and Turner, 1970; Elwood and Henderson, 1975). One study of forested areas in western Europe showed the weight of nutrients carried in by precipitation to be roughly equivalent to the quantity removed by timber harvest (Neuwirth, 1957). For some elements the amount carried in by aerosols, known as dryfall, may exceed that carried in by precipitation. Estimates of annual income to an English mixed deciduous woodland by dryfall were 125.2 kg/ha of sodium, 6.3 kg/ha of potassium, 4.2 kg/ha of calcium, 16.2 kg/ha of magnesium, and 0.34 kg/ha of phosphorus. In an eastern United States deciduous forest ecosystem over 50 percent of the annual magnesium and potassium input and over 40 percent of the calcium input were dryfall (Elwood and Henderson, 1975). The income from aerosols is greater for elements known to occur as droplets, such as sodium, potassium, and magnesium. The income of calcium, associated with terrestrial sources, and phosphorus, associated with biological activity, is greater in rainfall (White and Turner, 1970).

Seventy to 90 percent of gross rainfall reaches the forest floor, mostly as throughfall and stemflow. On its way through the canopy and down the stems rainwater carries with it nutrients deposited as dust on the leaves and stems or leached from them. Such throughfall is richer in calcium, sodium, potassium, phosphorus, iron, magnesium, and silica than rainwater collected in the open at the same time (Tamm, 1951; Madgwick and Ovington, 1959; Eaton, Likens, and Bormann, 1975; Patterson, 1975). Throughfall of rain in an English oak woodland accounted for 17 percent of the nitrogen, 37 percent of the phosphorus, 72 percent of the potassium, and 97 percent of the sodium added by the canopy to the soil; the remainder was added by fallen leaves (Carlisle et al., 1967).

Although stemflow amounts to only 5 percent of the total rainfall reaching the forest floor, it is so concentrated about the trunk that with some species the moisture it supplies is 5 to 10 times as great as the rainfall. How much stemflow reaches the ground varies with the species. Smooth-barked beech has considerably more stemflow than oaks whose bark absorbs water (Patterson, 1975). Although throughfall provides more nutrients to the soil because of its large volume of flow, stemflow provides a more concentrated nutrient solution. Again the concentration of nutrients depends upon the species. Beech and hickories return considerably more calcium and potassium than oaks, and pines return smaller amounts of calcium, magnesium, potassium, and manganese than hardwoods. In northern North American forests of balsam fir, epiphytic lichens may remove nitrogen and add calcium and magnesium to stemflow (Lang et al., 1976).

The nutrients carried to the forest floor by throughfall, stemflow, and other sources are taken up in time by the surface roots of trees and translocated to the canopy. Such short-term internal cycling has been investigated by means of radioactive tracers. By inoculating white oak trees with 20 microcuries of $^{134}$Cs (cesium-134), Witherspoon and others (1962) were able to follow the gains, losses, and transfers of this radioisotope. About 40 percent of the $^{134}$Cs inoculated into the oaks in April moved into the leaves by early June (Figure

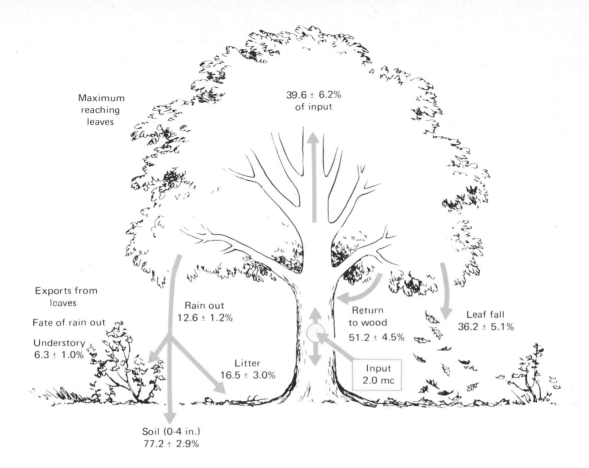

Maximum
reaching
leaves

39.6 ± 6.2%
of input

Exports from
leaves

Fate of rain out

Understory
6.3 ± 1.0%

Rain out
12.6 ± 1.2%

Return
to wood
51.2 ± 4.5%

Leaf fall
36.2 ± 5.1%

Litter
16.5 ± 3.0%

Input
2.0 mc

Soil (0-4 in.)
77.2 ± 2.9%

**FIGURE 6-18**
*Cycle of $^{134}$Cs in white oak as an example of nutrient cycling through plants. The figures are an average of 12 trees at the end of the 1960 growing season. (After Witherspoon et al., 1960; courtesy Oak Ridge National Laboratory.)*

6-18). When the first rains fell after inoculation, leaching of radiocesium from the leaves began. By September this loss amounted to 15 percent of the maximum concentration in the leaves. Seventy percent of this rainwater loss reached mineral soil; the remaining 30 percent found its way into the litter and understory. When the leaves fell in autumn, they carried with them twice as much radiocesium as was leached from the crown by rain. Over the winter, half of this was leached out to mineral soil. Of the radiocesium in the soil, 92 percent still remained in the upper 10 cm nearly 2 years after the inoculation. Eighty percent of the cesium was confined to an area within the crown perimeter and 19 percent was located in

a small area around the trunk. This suggests that cesium distribution in the soil was greatly influenced by leaching from rainfall and stemflow.

The use of $^{137}$Cs to trace the pathway of mineral cycling in the forest was carried one step further in a study of cesium cycling in a tulip poplar *(Liriodendron tulipifera)* stand (Olsen et al., 1970). Dominant trees were tagged with $^{137}$Cs. The fate of the tracer was continuously monitored for 5 years, and its pathway through the forest ecosystem was traced as illustrated in Figure 6-19. The leaves were richest in cesium in the spring and poorest in the fall. The reason for the decline in the leaves was the movement of $^{137}$Cs from the foliage to the woody tissue when the leaves were mature and senescent. There was a cumulative loss of the tracer from the foliage both from rainfall and litterfall, but the greatest transfer to the soil was through the roots. The $^{137}$Cs accumulated in the litter and organic layers of the forest floor and in the upper 10 cm of the min-

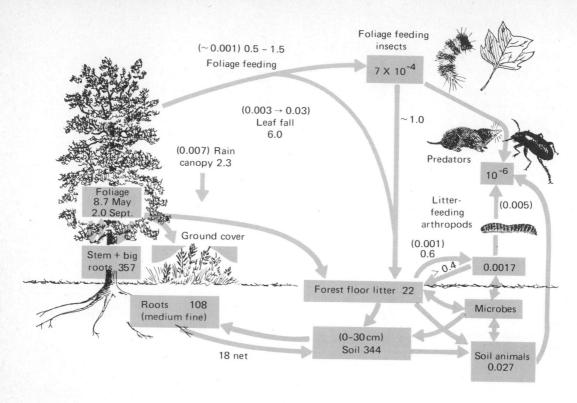

**FIGURE 6-19**

*The biogeochemical cycle of $^{137}Cs$ in a tulip poplar* (Liriodendron tulipifera) *forest ecosystem. The trees of this forest were originally tagged with a total of 467 microcuries of $^{137}Cs$ in May 1961. Continuous inventory followed the seasonal and annual distributions of and fluxes of $^{137}Cs$ among the abiotic and biotic components of the ecosystem. Data in this figure are for the 1965 season. Numbers in the boxes are microcuries of $^{137}Cs$ per square meter of ground surface. Arrows indicate the pathways of $^{137}Cs$ transfer between compartments. Numbers by the arrows are estimates of annual fluxes; numbers in parentheses are the averaged transfer coefficients (days) of $^{137}Cs$ for the seasons during which the process occurs. (Courtesy Oak Ridge National Laboratory.)*

eral soil. From here the cesium was picked up by decomposers, soil organisms, and litter-feeding arthropods and their predators.

Much of the nutrient pool in the forest ecosystem is involved in short-term cycling. Nutrients taken up by the trees are returned to the forest floor by litterfall, throughfall, and stemflow. But a portion of the nutrient uptake is stored in tree limbs, trunk, bark, and roots as accumulated biomass and additional quantities are stored in the living biomass in the forest floor and in consumer organisms. Much of this is effectively removed from short-term cycling. Additional losses come through leaching and the nutrients carried away by streamflow. If the forest is cut a considerable portion, up to 50 percent, of the accumulated nutrients may be withdrawn from the system in addition to increased losses through erosion and leaching. The nutrient cycle in the undisturbed system can continue and the nutrient budget can be balanced only by withdrawing elements from the soil.

Nutrient budgets for several ecosystems are discussed in a comparative way in Chapters 7, 8, 9, and 10. The budgets developed, especially for forest ecosystems, suggest several characteristics of mineral cycling. The accumulation of nutrients in the rooting zone of the soil and in the tree biomass makes them unavailable and slows down the biological cycle of minerals. Thus the rate of biological cycling can be maintained only by the pumping of nutrients from deep soil reserves and by the weathering of parent rock. Nutrient cycling is influenced by the climate, the nature of the the soil, and the nature of the plants occupying the area, including their ability to pump, transport, store, and return certain mineral elements. Seasonal variations in utilization, storage, and retention complicate the cycling. Except for a sharp peak in mineral uptake, which occurs when the productivity of the forest seems to be at its peak, age seems to have little influence on mineral cycling.

The gain to the ecosystem from precipitation, extraneous material, and mineral weathering is offset by losses. Water draining away removes more mineral matter than is supplied by precipitation (Viro, 1953; Whittaker and Woodwell, 1969; Likens et al., 1967).

Nutrient cycling is inseparable from energy flow and is in fact powered by it, as illustrated in Figure 6-20. Nutrients are pumped through

FIGURE 6-20
*Relationship of nutrient cycles and energy flow. The model stresses the fact that energy flow is unidirectional, whereas nutrient flow is cyclic (R = respiration). (From R. L. Smith, 1972.)*

the system and sped on their circular path by the action of photosynthesis. They are made available for recycling by the action of decomposers that carry on the complex chemical actions that set the nutrients free. Nutrients follow the same pathways as energy through the food chain, passing from one trophic level to another. The difference is that a given unit of energy is ultimately dissipated, whereas nutrients to varying degrees are recycled, that is, nutrients can be returned to their original chemical form. Some of the nutrients are involved in short-term cycles, some are tied up in organic storage temporarily, and some are locked up in deep sediments and rocks.

Because the medium in which all nutrients are eventually carried through the ecosystem is water, nutrient cycles are also inseparably tied to the water cycle. Foliar leaching, rainfall, percolation through the soil, decomposition, uptake, transport of nutrients in and out of the ecosystem—in fact the whole nutrient cycle—could not function without water. It is the movement of water, too, that ties the ter-

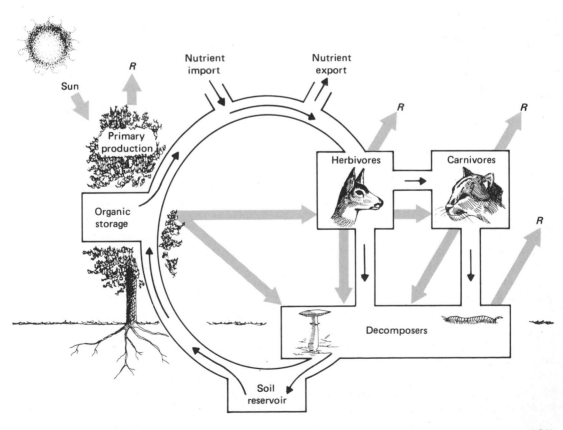

restrial ecosystem to the aquatic and in so doing relates the local ecosystem to the global ecosystem.

## SUMMARY

Materials flow from the living to the nonliving and back to the living parts of the ecosystem in a perpetual cycle. By means of these cycles plants and animals obtain nutrients necessary for their well-being.

There are two basic types of biogeochemical cycles, the gaseous represented by the carbon, oxygen, and nitrogen cycles and the sedimentary represented by the phosphorus cycle. The sulfur cycle is a combination of the two. The carbon cycle is so closely tied to energy flow that the two are inseparable. It involves the assimilation and respiration of carbon dioxide by plants, its consumption in the form of plant and animal tissue, its release through respiration, the mineralization of litter and wood, soil respiration, accumulation of carbon in standing crop, and withdrawal into long-term reserves. The carbon dioxide cycle exhibits both diurnal and annual curves. The equilibrium of carbon dioxide exchange between land, sea, and air has been disturbed by its rapid injection into the atmosphere by burning of fossil fuels, but over one-half is removed from the atmosphere by oceans and land vegetation.

The nitrogen cycle is characterized by fixation of atmospheric nitrogen by nitrogen-fixing plants, largely legumes and blue-green algae. Involved in the nitrogen cycle are the processes of ammonification, nitrification, and denitrification.

Human intrusion into the nitrogen cycle involves inputs of nitrogen dioxide into the atmosphere and nitrates into the aquatic ecosystems. The major sources of nitrogen dioxide are automobiles and burning of fossil fuels. Nitrogen dioxide is reduced by ultraviolet light to nitrogen monoxide and atomic oxygen. These react with hydrocarbons in the atmosphere to produce a number of pollutants including ozone and PAN. These make up photochemical smog, a pollutant harmful to plants and animals. Excessive quantities of nitrates are added to aquatic ecosystems by improper use of nitrogen fertilizer on agricultural crops, by animal wastes, and by sewage effluents. The latter accounts for the largest source. More closely involved with the pollution of aquatic systems is phosphorus which comes from sewage effluents.

The sedimentary cycle involves two phases, salt solution and rock. Minerals become available through the weathering of the earth's crust, enter the water cycle as salt solution, take diverse pathways through the ecosystem, and ultimately return to the sea or back to the earth's crust through sedimentation.

The phosphorus cycle is wholly sedimentary with reserves coming largely from phosphate rock. Much of the phosphate used as fertilizer becomes immobilized in the soil, but great quantities are lost in detergents and other wastes carried by sewage effluents.

The sulfur cycle is a combination of the gaseous and sedimentary cycles because it has reservoirs in both the earth's crust and the atmosphere. It involves a long-term sedimentary phase in which sulfur is tied up in organic and inorganic deposits, is released by weathering and decomposition, and is carried to terrestrial and aquatic ecosystems in salt solution. A considerable portion of sulfur is cycled in the gaseous state, which permits its circulation on a global scale. Sulfur enters the atmosphere from the combustion of fossil fuel, volcanic eruptions, the surface of the ocean, and gases released by decomposition. Entering the gaseous cycle initially as hydrogen sulfide, sulfur quickly oxidizes to sulfur dioxide. Sulfur dioxide, soluble in water, is carried to earth as weak sulfuric acid. Whatever the source, sulfur is taken up by plants and incorporated into sulfur-bearing amino acids, later to be released by decomposition. Injected into the atmosphere by industrial consumption of fossil fuels, sulfur dioxide has become a major pollutant, affecting and even killing plant growth, causing respiratory afflictions in humans and animals, and producing acid rainfall over parts of the world.

Industrial use of such heavy metals as lead and mercury, always present at low levels in the biosphere, has significantly increased their occurrence. Both pose potential and actual health problems, especially because they enter the food chain. Currently the major dangers of mercury are local rather than global.

Of more serious consequence globally are the chlorinated hydrocarbons. Used in insect control, these pesticides have contaminated the global ecosystem and entered food chains. Because they become concentrated at higher trophic levels, chlorinated hydrocarbons affect the predaceous animals most adversely. Fish-eating birds are endangered because chlorinated hydrocarbons interfere with reproductive capability.

Radioactive materials from nuclear weapons testing and from nuclear power plants have also been introduced. Some radioisotopes can enter and become concentrated in food chains. In many situations concentrations of radionuclides decrease at higher trophic levels.

Through our knowledge of biogeochemical cycles and the technique of using radioactive tracers, we can follow the pathways of specific nutrients to construct nutrient budgets for particular ecosystems. All nutrient cycles are inseparably tied to energy flow and the water cycle.

## part III

# Comparative ecosystem ecology

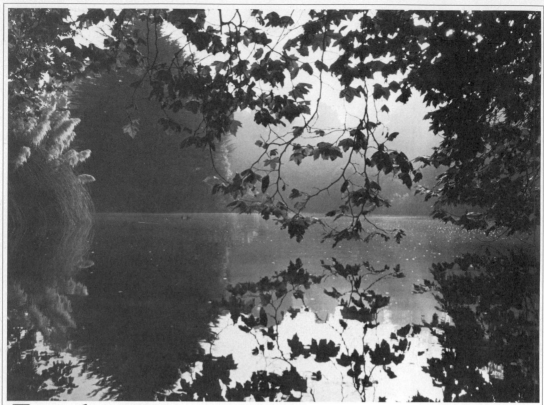

# *Freshwater ecosystems*
## CHAPTER 7

Global aquatic ecosystems fall into two broad classes defined by salinity—the freshwater ecosystem and the saltwater ecosystem. The latter may include inland brackish water, as well as marine and estuarine habitats. Freshwater ecosystems, the study of which is known as limnology, are conveniently divided into two groups, *lentic* or standing water habitats and *lotic* or running water habitats. Both can be considered on an environmental gradient. The lotic follows a gradient from springs to mountain brooks to streams and rivers. The lentic follows a gradient from lakes to ponds to bogs, swamps, and marshes.

## Lentic ecosystems

Lakes are inland depressions containing standing water. They may vary in size from small ponds of less than an acre to large seas covering thousands of square miles. They may range in depth from a few feet to over 5000 ft. Ponds are defined as small bodies of standing water so shallow that rooted plants can grow over most of the bottom. Most ponds and lakes have outlet streams; and both may be more or less temporary features on the landscape.

Lakes and ponds arise in many ways. Some North American lakes were formed by glacial erosion and deposition. Glacial abrasion of slopes in high mountain valleys carved basins, which filled with water from rain and melting snow to produce tarns. Retreating valley glaciers left behind crescent-shaped ridges of rock debris, which dammed up water behind them. Numerous shallow kettle lakes and potholes were formed on the glacial drift sheets that cover much of northeastern North America and northwestern Europe.

Lakes are also formed by the deposition of silt, driftwood, and other debris in the beds of slowflowing streams. Loops of streams that meander over flat valleys and flood plains often become cut off, forming crescent-shaped oxbow lakes.

Shifts in the earth's crust, either by the uplifting of mountains or by the breaking and displacement of rock strata, causing part of the valley to sink, develop depressions that fill with water. Craters of extinct volcanos may fill with water; and landslides can block off streams and valleys to form new lakes and ponds. In a given area all natural lakes and ponds have the same geological origin and the same general characteristics. But because of varying depths at the time of origin, they may represent several stages of succession.

Many lakes and ponds are formed by nongeological activity. Beavers dam up streams to make shallow but often extensive ponds. Man intentionally creates artificial lakes by damming rivers and streams for power, irrigation, and water storage or by constructing small ponds and marshes for water, fishing, and wildlife. Quarries and strip mines fill with water to form other ponds.

STRUCTURE

Unlike most terrestrial ecosystems, lentic ecosystems have well-defined boundaries— the shoreline, the sides of the basin, the surface of the water, and the bottom sediment. Within these boundaries environmental conditions vary. Light penetrates to a depth determined by turbidity produced by sediments and phytoplankton. Temperatures vary seasonally and with depth. Because only a relatively small proportion of the water is in direct contact with the air and because decomposition takes place on the bottom, the oxygen content of lake water is relatively low compared to that of running water. In some lakes oxygen may decrease with depth, but there are many exceptions. These gradations of oxygen, light, and temperature profoundly influence life in the lake, its distribution and adaptations.

*Temperature stratification.* Each year the waters of many lakes and ponds undergo seasonal changes in temperature (Figure 7-1). As the ice melts in early spring, the surface water, heated by the sun, warms up. When it

reaches 4° C and becomes more dense, a slight temporary stratification develops, which sets up convection currents. Aided by the strong winds of spring, these currents mix the water throughout the basin until the water in the lake is uniformly 4° C. At a uniform temperature even the slightest winds can cause a complete circulation of the water between the surface and the bottom. This is the spring overturn, when the nutrients on the bottom, the oxygen on the top, and the plankton within are mixed throughout the lake.

With the coming of summer the sun's intensity increases, and the temperature of the surface water rises. The higher the temperature of the surface water, the greater is the difference in the density between the surface and deeper layers. The thermal density gradient opposes the energy of the wind, and it becomes more difficult for the waters to mix. As a result, a mixing barrier is established. The freely circulating surface water, with a small but variable temperature gradient, is the *epilimnion* (Figure 7-1c). Below this is the barrier, the *metalimnion,* a zone characterized by a steep and rapid decline in temperature. Within the metalimnion is the *thermocline,* the plane at which the temperature drops most rapidly—1° C for each meter of depth. Below these two layers is the *hypolimnion,* a deep, cold layer, in which the temperature drop is gentle.

With the coming of autumn, the air temperature falls. The surface water loses heat to the atmosphere through evaporation, convection, and conduction. The temperature of the surface water drops and the thermocline sinks; the epilimnion increases until it includes the entire lake. The temperature once again is uniform from top to bottom, the lake waters circulate, and oxygen and nutrients are recharged throughout the lake. This is the fall overturn, which, through stirring actions caused by the slightest wind, may last until ice forms.

As the surface water cools below 4° C, it becomes lighter and remains on the surface. If the climate is cold enough, it freezes; otherwise, it remains close to 0° C. A slight inverse temperature stratification may develop, in which the water becomes warmer up to 4° C with depth. The water immediately beneath

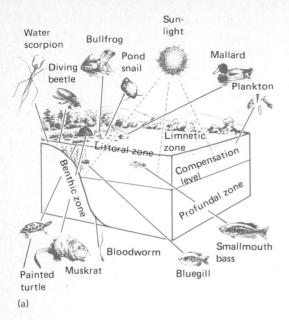

(a)

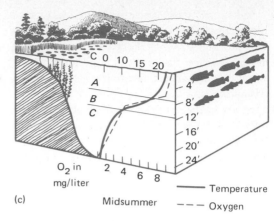

(c)        Midsummer

——— Temperature
– – – Oxygen

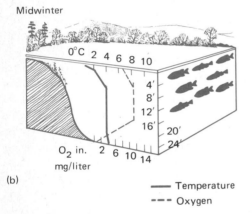

(b)

——— Temperature
– – – Oxygen

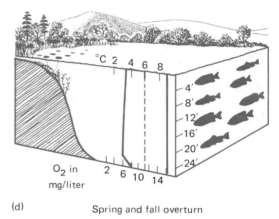

(d)        Spring and fall overturn

**FIGURE 7-1**
*Seasonal variations in oxygen and temperature stratification and distribution of aquatic life in lake ecosystems. The generalized picture of the lake in midsummer shows the major zones — the littoral, the limnetic, the profundal, and the benthic. The compensation level is the depth at which light is too low for photosynthesis. Surrounding the lake is a variety of organisms typical of the lake community. The distribution of oxygen and temperature in a lake during the different seasons affects the distribution of fish life. The narrow fish silhouettes represent trout, or the cold-water species; the wider silhouettes are bass, or the warm-water species. Note the pronounced horizontal stratification in midsummer and the almost vertical oxygen and temperature curves during the spring and fall overturns.*

the ice may be warmed by solar radiation through the ice. Because this increases its density, this water subsequently flows to the bottom, where it mixes with water warmed by heat conducted from bottom mud. The result of this is a higher temperature at the bottom, although the overall stability of the water is undisturbed. As the ice melts in spring, the surface water again becomes warm, currents pass unhindered, and the spring overturn takes place.

This is a general picture of seasonal changes in temperature stratification and must not be considered as a uniform condition in all bodies of water. In shallow lakes and ponds temporary stratification of short duration may occur; in others stratification may exist, but without a thermocline. In some very deep lakes the thermocline may simply descend during periods of overturn, but not disappear. In such lakes the bottom water never becomes mixed with the top layers.

In general, lakes of temperate regions and of high mountains in subtropical regions undergo two overturns during the year (they are *dimictic*). They are inversely stratified in winter and directly stratified in summer (Hutchinson, 1957). Lakes in warm, oceanic climates and in low mountains of subtropical latitudes undergo but one overturn a year *(monomictic)*. The water, never below 4° C at any level, cools sufficiently to allow complete circulation only during the coldest time of the year. Similarly, lakes in polar and arctic regions may rarely rise above 4°C and circulation takes place only in the summer. Freely circulating in summer, these lakes are inversely stratified in winter. In the high mountains of equatorial regions, where there is little seasonal change in temperature, lakes may circulate continually at a little above 4° C *(polymictic)*. These lakes lose just enough heat to prevent stable stratification. Some lakes in the humid tropics, whose waters are always well above 4° C, circulate only at irregular intervals.

*Oxygen stratification.* Oxygen stratification during summer nearly parallels that of temperature. In general the amount of oxygen is greatest near the surface where there is an interchange between the water and atmosphere and some stirring by the wind. The quantity decreases with depth, a decrease caused in part by the respiration of decomposer organisms feeding on the organic matter dropping down from the layers above. In some lakes oxygen may vary little from top to bottom; every layer is saturated relative to temperature and pressure. Water in some lakes is so clear that light penetrates below the depth of the thermocline and permits the development of phytoplankton. Because of photosynthesis, the oxygen content in these lakes may be greater in deep water than on the surface.

During the spring and fall overturns, when water recirculates through the lake, oxygen is replenished in the deep water and nutrients are returned to the top. In winter the reduction of oxygen in unfrozen water is slight because bacterial decomposition is reduced and water at low temperatures holds a maximum amount of oxygen. Under ice, however, oxygen depletion may be serious and result in a heavy winter kill of fish.

*Light stratification.* The energy source of the lake and pond ecosystem is sunlight. The depth to which light can penetrate is limited by the turbidity of the water and the absorption of light rays. On this basis lakes and ponds can be divided into two basic layers— the *trophogenic zone,* roughly corresponding to the epilimnion, in which photosynthesis dominates; and the *tropholytic zone,* corresponding to the hypolimnion, where decomposition is most active. The boundary between the two zones is the *compensation depth,* the depth at which photosynthesis balances respiration and beyond which light penetration is so low that it is no longer effective. Generally the compensation depth occurs where light intensity is about 100 footcandles, or approximately 1 percent of full noon sunlight incident to the surface (see Edmondson, 1956).

*Currents and seiches.* Oxygen and thermal stratification, depth and position of the thermocline, circulation of nutrients, and distribution of organisms all are influenced by currents. The most conspicuous water movements observed are traveling surface waves, the result of wind pressures on the surface of lakes and ponds. Except for the effects they have on the shoreline and shore organisms, surface waves are not too important biologically. More important are standing waves, or *seiches* (sāshes), a term that comes from the French and means dry, exposed shoreline. A seiche is produced by an oscillation of a structure of water about a point or node.

There are two kinds of seiches, surface and internal. Both are produced by the wind's movement across the water's surface, by heavy rain showers, or perhaps even by changes in atmospheric pressure (see Bryson and Ragotzkie, 1960; Vallentyne, 1957).

When the wind blows across a lake or pond, it piles up water on the leeward end and creates a depression on the windward end. When the wind subsides, the current flows back, but because the momentum of the returning currents is not broken on the shore, a depression is created on the former leeward side, and the water flows back again. Thus an oscillation or rocking motion is established about a stationary node. This continues until it is finally halted by friction on the lake basin or by such meteorological forces as an opposite wind or rain. Although surface seiches occur on all lakes, they are more easily observed on larger lakes.

Internal seiches, not observable on the surface, occur during the summer in thermally stratified lakes (Figure 7-2). They are much more pronounced and exert a greater influence on life in the lake than surface seiches. The internal seiche is caused not only by the action of wind on the surface waters, but also by density differences between warm and cold water. When the wind piles the water of the epilimnion up on the leeward side, the weight and circulation of the lighter water over the denser cold water tilts the thermocline and raises the hypolimnion on the windward side. When the wind stops, the raised hypolimnion, pushed down on and toward the leeward side, causes the epilimnion water to flow back toward the windward side. Thus an oscillation is established between the lighter water layer of the epilimnion and the denser water of the hypolimnion. In time the oscillations are

slowed by friction on the lake basin or by wind from an opposite direction.

These oscillations move about a point in the center of the lake where resistance is highest. Here vertical displacement movement is the least; the greatest displacement is on the windward and leeward ends. The position of an internal seiche can best be determined by charting the variations of the depth of a particular temperature over a period of hours. This movement makes difficult the determination of the true position of the thermocline because its position at any one point is determined by the amplitude of the oscillation at that point.

Internal seiches are important ecologically, because they distribute heat and nutrients vertically in the lake and transport them into shallow water. The seiche sets up two rhythmic but opposite current systems above and below the thermocline, whose speeds are greatest when the thermocline is level. A turbulence develops in the hypolimnion; without this turbulence the hypolimnion would become stagnant. In addition, plankton is moved up or down with the water mass. Even fish and other organisms are influenced by internal seiches.

The mixing of water and the transfer of nutrients are also carried on by *eddy currents,* small, turbulent currents whose energy is dissipated at right angles to the major currents. Because the major currents are largely horizontal and the turbulence is vertical, an interchange of adjacent water masses takes place. Oxygen may be transferred downward and heat upward; nutrients and even plankton are intermixed. The degree of intermixing depends upon the intensity of the turbulence, which changes gradually from one depth to another.

*Biological stratification.* Lentic ecosystems can also be subdivided into vertical and horizontal strata based on photosynthetic activity. The *littoral* or shallow-water zone is the zone in which light penetrates to the bottom. This area is occupied by rooted plants such as waterlilies, rushes, and sedges. Beyond this is the *limnetic* or open-water zone which extends to the depth of effective light penetration. It is inhabited by plant and animal plankton and the *nekton,* free-swimming organisms such as fish which are capable of moving about volun-

**FIGURE 7-2**
*Diagrammatic representation of the thermocline slope and the seiche.*

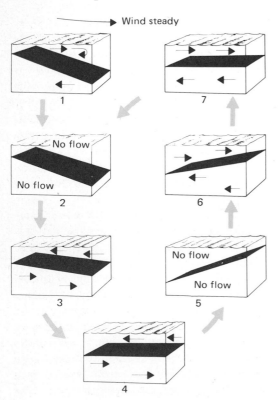

tarily. Beyond the depth of effective light is the *profundal* zone which depends upon the rain of organic material from the limnetic as the energy source. Common to both the profundal and littoral zones is the *benthic* zone or bottom region which is the zone of decomposition. Although these zones are named and often described separately, all are closely dependent upon one another in nutrient and energy flow.

*Limnetic Zone.* The open water is a world of minute, suspended organisms, the plankton. Dominant is the phytoplankton, a group of plant organisms containing the diatoms, desmids, and the filamentous green algae. Because photosynthesis in open water is carried on only by these tiny plants, they are the base upon which the rest of limnetic life depends. Suspended with phytoplankton is the zooplankton which grazes upon the minute plants. These animals form an important link in the energy flow in the limnetic zone.

Vertical distribution or stratification of plankton organisms is influenced by the physicochemical properties of water, especially temperature, oxygen, light, and current. Light, of course, sets the lower limit at which phytoplankton can exist. Because zooplankton feeds on these minute plants, it, too, is concentrated in the trophogenic zone. Phytoplankton, by its own growth, limits light penetration and thus reduces the depth at which it can live. As the zone becomes more shallow, the phytoplankton can absorb more light, and organic production is increased. But within these limits the depths at which the various species live is influenced by the optimum conditions for their development. Some phytoplankton live just beneath the water's surface; others are more abundant a few feet beneath, while those requiring colder temperatures live deeper still. Cold-water species, in fact, are restricted to those lakes in which phytoplankton growth is scarce in the upper region and in which the oxygen content of the deep water is not depleted by the decomposition of organic matter. Many of these cold-water species never move up through the metalimnion.

Because many species of zooplankton are capable of independent movement, animal plankton exhibits stratification that often changes seasonally. In winter some plankton forms are spread evenly to considerable depths; in summer they concentrate in the layers most favorable to them and to their stage of development. At this season animal plankton undertakes a vertical migration during some part of the 24-hour period. Depending upon the species and stage of development, zooplankton spends the night or day in the deep water or on the bottom and moves up to the surface during the alternate period to feed on phytoplankton (Figure 7-3).

Free, to a large extent, from the action of weak currents and capable of moving about at will are *nekton* organisms, the fish and some invertebrates. In the limnetic zone fish make up the bulk of the nekton. Their distribution is influenced mostly by food supply, oxygen, and temperature. During the summer, largemouth bass, pike, and muskellunge inhabit the warmer epilimnion waters, where food is abundant. In winter they retreat to deeper water. Lake trout, on the other hand, move to greater depths as summer advances. During the spring and fall overturn, when oxygen and temperature are fairly uniform throughout, both warm-water and cold-water forms occupy all levels.

*Profundal Zone.* Diversity and abundance of life in the profundal zone are influenced by oxygen, temperature, and the amount of organic matter and nutrients supplied from the limnetic zone above. In highly productive waters decomposer organisms so deplete the profundal waters of oxygen that little aerobic life can exist there. The profundal zone of a deep lake is relatively much larger, so the productivity of the epilimnion is low in comparison to the volume of water and decomposition does not deplete the oxygen. Here the profundal zone supports some life, particularly fish, some plankton, and such organisms as certain cladocerans, which live in the bottom ooze. Other zooplankton may occupy this zone during some part of the day, but migrate upward to the surface to feed. Only during the spring and autumn overturns, when organisms from the upper layers enter this zone, is life abundant in the profundal waters.

Easily decomposed substances floating down through the profundal zone are partly mineralized while sinking (Kleerekoper, 1953). The remaining organic debris, dead bodies of plants and animals of the open water

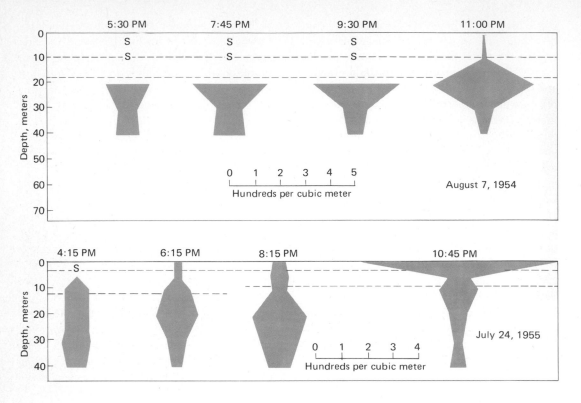

**FIGURE 7-3**
*Vertical distribution of the zooplankton*
copepod (Limnocalanus marcurus) *on two*
*midsummer days. The maximum number*
*reach the surface 1.5 to 4 hours before*
*sunset. Note that this organism inhabits the*
*deeper water. The broken lines represent the*
*metalimnion. (After Wells, 1961.)*

and decomposing plant matter from shallow-water areas, settles on the bottom. This, together with quantities of material washed in by inflowing water, makes up the bottom sediments, the habitat of benthic organisms.

*Littoral Zone.* Aquatic life is richest and most abundant in the shallow water about the edges of lakes (Figure 7-4). Plants and animals found here vary with water depth, and a distinct zonation of life exists from the deeper water to shore. A blanket of duckweed (*Lemna* spp.), supporting a world of its own, may cover the surface of the littoral waters. Submerged plants, such as pondweeds (*Potamogeton* spp.), grow at water depths beyond that tolerated by emergent vegetation. As sedimentation and accumulation of organic matter increase toward the shore and the depth of water decreases, floating rooted aquatics, such as pond

lilies and smartweed (*Polygonum* spp.), appear. Floating plants offer food and support for numerous herbivorous animals that feed both on phytoplankton and the floating plants. In the shallow water beyond the zone of floating plants grow the emergents, plants whose roots and stems are immersed in water and whose upper stems and leaves stand above the water. Among these emergents are plants with narrow, tubular or linear leaves, such as the bulrushes, reeds, and cattails. The distribution and variety of plants vary with the water depth and the fluctuation of the water level. Within the sheltering beds of emergent plants animal life is abundant. Although largely ignored in limnological studies, the littoral zone contributes heavily to the productivity of lentic ecosystems and provides a large input of organic matter to the system.

*Benthic Zone.* The bottom ooze is a region of great biological activity, so great that oxygen curves for lakes and ponds show a sharp drop in the profundal water just above the bottom. Because organic muck lacks oxygen completely, the dominant organisms there are anaerobic bacteria. Under anaerobic conditions, decomposition cannot proceed to in-

**FIGURE 7-4**
*Littoral vegetation. The heavy emergent growth of the littoral zone adds large amounts of detrital material that is important in the functioning of lake ecosystems.*

organic end-products. When the amounts of organic matter reaching the bottom are greater than can be utilized by the bottom fauna, odoriferous muck rich in hydrogen sulfide and methane results. Thus lakes and ponds with highly productive limnetic and littoral zones have an impoverished fauna on the profundal bottom. Life in the bottom ooze is most abundant in lakes with a deep hypolimnion in which oxygen is still available.

As the water becomes more shallow, the benthos changes. The bottom materials—stones, rubble, gravel, marl, clay—are modified by the action of water, by plant growth, by drift materials, and by recent organic deposits. Increased oxygen, light, and food result in a richness of species and an abundance not found on the profundal bottom.

Closely associated with the benthic community are the *periphyton* or *aufwuchs,* those organisms that are attached to or move upon a submerged substrate but do not penetrate it. Small aufwuchs communities are found on the leaves of submerged aquatics and on sticks, rocks, and other debris. The organisms found there depend upon the movement of the water, temperature, kind of substrate, and depth.

Periphyton found on living plants are fast growing, lightly attached, and consist primarily of algae and diatoms. Because the substrate is so short-lived, these rarely exist for more than one summer. Aufwuchs on stones, wood, and debris form a more crustlike growth of blue-green algae, diatoms, water mosses, and sponges.

**FUNCTION**

In many ways a lake might be considered a self-contained ecosystem, but, in fact, lentic ecosystems are strongly influenced by inputs of nutrients from sources outside of the basin (Figure 7-5). Nutrients and other substances move across the boundaries of the lentic system along biological, geological, and meteorological pathways (Likens and Bormann, 1975).

Wind-borne particulate matter, dissolved substances in rain and snow, and atmospheric

**209**

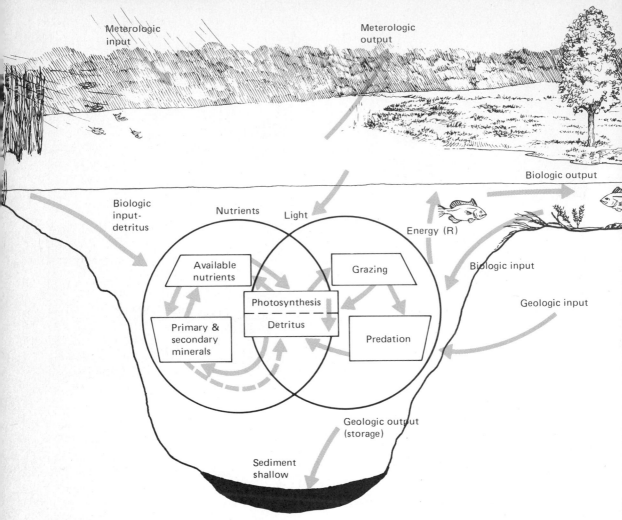

**FIGURE 7-5**
*Model for nutrient cycling and energy flow in a lake ecosystem. Meteprological, geological, and biological inputs enter the lentic system from the watershed that contains it. The nutrient and energy inputs as well as the nutrients and energy generated within the system move through a number of pathways. Part of the nutrients and energy fixed accumulate in bottom sediments. (After Likens and Bormann, 1974.)*

gases represent meteorological inputs to the system. Meteorological outputs are small, mainly spray aerosols and gases such as carbon dioxide and methane. Geological inputs include nutrients dissolved in groundwater and inflowing streams and particulate matter washed into the basin from the surrounding terrestrial watershed. Geological outputs include dissolved and particulate matter carried out of the basin by outflowing waters and nutrients incorporated in deep sediments which may be removed from circulation for a long period of time. Biological inputs and outputs, relatively small, include animals such as fish that move into and out of the basin. Energy input is largely sunlight, and energy output is heat. The lentic ecosystem receives its hydrological input from precipitation and the drainage of surface waters. Outputs involve seepage through walls of the lake basin, subsurface flows, evaporation, and evapotranspiration. Within the lentic ecosystem nutrients move among three compartments, dissolved organic matter, particulate organic matter, and primary and secondary minerals. Nutrients and energy move through the system by way of the grazing and detrital food chains.

Although studies of lake metabolism have

emphasized the phytoplankton-zooplankton grazing food chain, in reality lakes, like terrestrial communities, are dominated by the detrital food chain (Figure 7-6). The reason for the emphasis on the grazing food chain in lakes is the preoccupation with the open-water zone and the disregard of the littoral zone which adds significantly to lake productivity and supplies substantial quantities of detritus to the system. Detritus, by the way of recall, is all dead organic carbon. It includes particulate and dissolved organic matter derived from living organisms in the system, particulate and dissolved organic matter from external sources that enter and cycle within the system, and organic matter lost to a particular trophic level by such nonpredatory losses as egestion, excretion, and secretion (Rich and Wetzel, 1975; Wetzel, 1975).

Lake ecosystems function mostly within a framework of organic carbon transfer. The central pool is in dissolved form which comes from both internal (autochthonous) and external (allochthonous) sources. It represents the major flow through the system. Particulate organic carbon comes from three sources: (1) imports into the system from the outside; (2) the littoral zone; and (3) the lentic zone. The major areas of detrital metabolism are the benthic zone, where most of the particulate matter is decomposed and the open-water (or pelagic) zone during sedimentation.

An arctic tundra pond studied by Hobbie et al. (1972) provides an example of energy flow and carbon cycling in a lentic ecosystem. Although shallow, less than 40 cm deep, and frozen solid for nine months of the year, the pond's functional features during a midsummer day appear to be typical of lake ecosystems. The pond's surroundings are dominated by two sedges and a grass which are grazed by brown lemmings. Litter from clippings and feces supplies a considerable amount of particulate and organic matter to the pond. In the pond 50 percent of the primary production comes from emergent aquatic plants, 47 percent from benthic algae, and 3 percent from phytoplankton. As illustrated in Figure 7-7, most of the detrital organic carbon moves from aquatic plants to the sediments as particulate organic matter (POC). This pool provides a major source of dissolved organic matter (DOC), both in the sediments and in the water where DOC is augmented by inputs from allochthonous sources. Most of the carbon pool in the water portion of the system is DOC which is utilized largely by bacteria. The zooplankton are detritivores, depending largely on particulate matter rather than on phytoplankton. The grazing food chain is relatively unimportant. In the sediment POC and benthic algae are utilized by benthic animals. Part of the POC converts to DOC, utilized by bacteria. Carbon flux is much greater in the sediments than in the water column of the ecosystem.

The review of energy flow and nutrient cycling in lentic systems emphasizes the close relationship between land and water ecosystems. Primarily through the hydrological cycle one feeds upon the other. The water that falls on land runs from the surface or moves through the soil and deeper layers to enter springs, streams, and eventually lakes. The water carries with it silt and nutrients in solution, all of which enrich aquatic ecosystems. Human activities including road building, housing construction, mining, and agriculture add an additional heavy load of silt and nutrients, especially nitrogen, phosphorus, and organic matter. The outcome of these inputs is nutrient enrichment of aquatic ecosystems. This enrichment of aquatic systems is termed *eutrophication*.

The term *eutrophy* (from the Greek *eutrophus,* "well nourished") means a condition of being nutrient rich. The opposite of eutrophy is *oligotrophy,* the condition of being nutrient poor. The terms were introduced first by the German limnologist C. A. Weber in 1907 when he applied the terms to the development of peat bogs. E. Naumann later applied the terms to phytoplankton production in lakes: eutrophic lakes, found in fertile lowland regions, hold high populations of phytoplankton; oligotrophic lakes, common to regions of primary rocks, contain little plankton. This concept of oligotrophy and eutrophy ignores the inputs of highly productive littoral zones.

*Eutrophic systems.* A typical eutrophic lake (see Figure 7-8) has a high surface-to-volume ratio, that is, the surface area is large relative to depth. It has an abundance of nutrients, especially nitrogen and phosphorus, that stimulate a heavy growth of algae and other

**211**

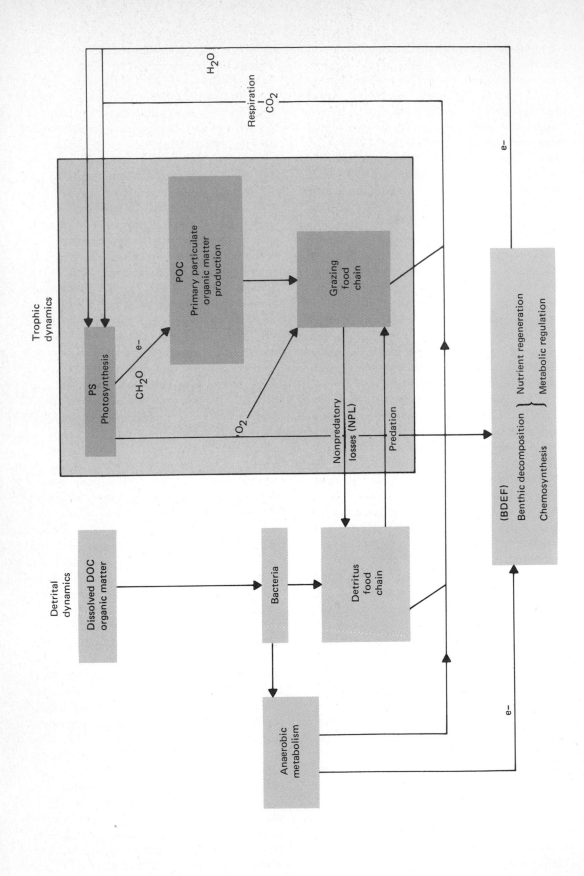

**FIGURE 7-6**

*Energy flow and nutrient cycling in a lake ecosystem. Heterotrophic, autotrophic, and detrital pathways are necessary for the operation of the whole lake ecosystem. This model shows the integration of the detrital and trophic structure involving herbivores, bacteria, and anaerobic metabolism. Herbivores and bacteria represent the grazing and detrital food chains which depend upon each other for prey and nonpredatory losses — the movement of organic matter to the decomposers. Anaerobic metabolism is a pathway for energy leaving organic substrates and the biota to interact more directly with abiotic factors. Export and uncoupled oxidation represent energetic losses, while import and net photosynthesis represent energy gains. $CO_2$ = oxidized carbon; $H_2O$ = reduced oxygen; PS = photosynthesis which reduces carbon and oxidizes water to create organic carbon, $CH_2O$, and molecular oxygen, $O_2$; DOC = dissolved organic carbon; POC = particulate organic carbon; NPL = nonpredatory loss; BDEF = benthic detrital electron flux; e = electron. (From Rich and Wetzel, 1978.)*

**FIGURE 7-7**

*Carbon flux among components of an tundra pond ecosystem representative of approximately average midsummer conditions at coastal Barrow, Alaska (lat 71° N). Compare this flow diagram with that of Figure 7-6. Note the important contribution of the detrital component to the overall functioning of the pond ecosystem. DOC = dissolved organic carbon; POC = particulate organic carbon; numbers in boxes = standing crops (mg C/m²); numbers on arrows = transfer rates ( mg c/m²/day). (Modified from Hobbie et al., 1972.)*

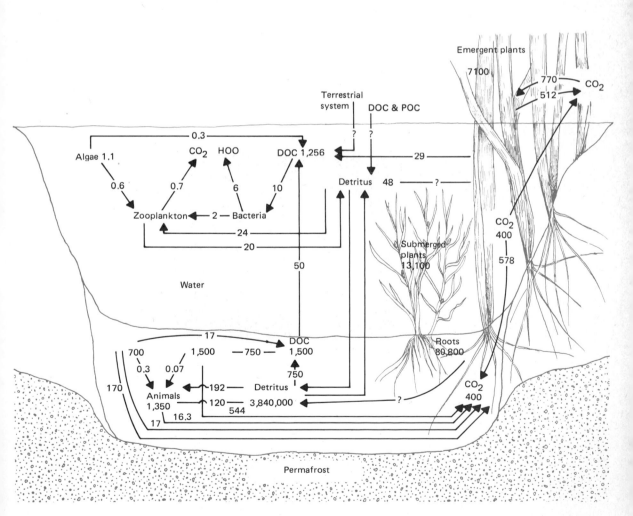

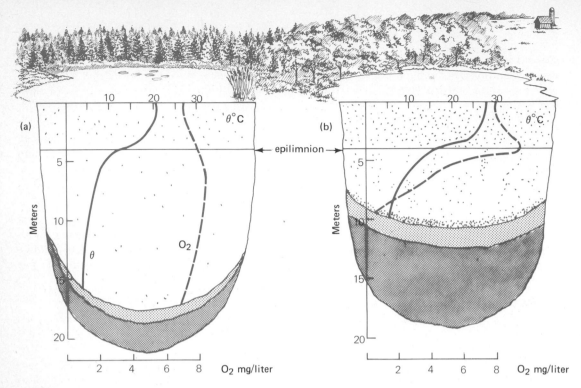

**FIGURE 7-8**
*Comparison of oligotrophic and eutrophic lakes. (a) The oligotrophic lake is deep and has relatively cool water in the epilimnion. The hypolimnion is well supplied with oxygen. Organic matter that drifts to the bottom falls through a relatively large volume of water. The watershed surrounding the lake is largely oligotrophic, dominated by coniferous forests on thin and acid soil. (b) The eutrophic lake is shallow and warm, and the oxygen in the deeper water is nearly depleted. The amount of organic detritus is large in relation to the volume of water. The watershed surrounding the lake is eutrophic, consisting of nutrient-rich deciduous forest and farmland.*

aquatic plants. Increased photosynthetic production leads to increased regeneration of nutrients and organic compounds, stimulating further growth. Phytoplankton becomes concentrated in the upper layer of the water giving it a murky green cast. The turbidity reduces light penetration and restricts biological productivity to a narrow zone of surface water. Algae, inflowing organic debris and sediment, and the remains of rooted plants drift to the bottom adding to the highly organic sediments. On the bottom bacteria partially con-

vert dead matter into inorganic substances. The activities of these decomposers depletes the oxygen supply of the bottom sediments and deep water to a point that the deeper parts of the lake are unable to support aerobic forms of life. The number of species declines, although the numbers and biomass of organisms may remain high. As the basin continues to fill, the volume decreases and the resulting shallowness speeds the cycling of available nutrients and further increases plant production. What develops is a sort of positive feedback mechanism that carries the system to eventual extinction—the filling in of the basin and the developement of a marsh, swamp, and ultimately a terrestrial community.

*Oligotrophic systems.* Oligotrophic lakes are characterized by a low surface-to-volume ratio, water that is clear and appears blue to blue-green in the sunlight, bottom sediments that are largely inorganic, and a high oxygen concentration through the hypolimnion. The nutrient content of the water is low; although nitrogen may be abundant, phosphorus is highly limiting. Low nutrient availability results from a low input of nutrients from external sources. This in turn causes a low production of organic matter, particularly phyto-

**214**

plankton. Low organic production results in low rate of decomposition and high oxygen concentration in the hypolimnion. These oxidizing conditions produce low nutrient release from the sediments. The lack of decomposable organic substances results in low bacterial populations and slow rates of microbial metabolism. Although the number of organisms in oligotrophic lakes may be low, the diversity of species is often high. Fish life is dominated by members of the salmon family.

When nutrients in moderate amounts are added to oligotrophic lakes they are rapidly taken up and circulated. According to Vallentyne (1974) tissues of aquatic algae contain phosphorus, nitrogen, and carbon in the ratio of 1 P : 7 N : 40 C per 500 g wet weight. If nitrogen and carbon are in excess and phosphorus is limiting, the addition of phosphorus would stimulate growth; if nitrogen is limiting, the addition of that element would do the same. In most oligotrophic lakes phosphorus rather than nitrogen is limiting. Because of its low ratio, 1 P per 500, an addition of even moderate amounts of phosphorus generates considerable growth of algae (for detailed discussions see Vallentyne, 1974; Wetzel, 1976). As increasing quantities of nutrients are added to the lake it begins to change from oligotrophic to *mesotrophic* (having a moderate amount of nutrients) to eutrophic. This has been happening at an increasing rate to clear oligotrophic lakes around the world.

In fact this "galloping eutrophication" has been changing naturally eutrophic lakes into *hypertrophic* ones. An excessive nutrient content results from a heavy influx of wastes, raw sewage, drainage from agricultural lands, river basin development, runoff from urban areas, and burning of fossil fuels. This accelerated enrichment, which results in chemical and environmental changes in the system and causes major shifts in plant and animal life, has been called *cultural eutrophication* (Hasler, 1969).

**Dystrophic systems.** Lakes that receive large amounts of organic matter from surrounding watersheds, particularly in the form of humic materials that stain the water brown, are called *dystrophic*. Although the productivity of dystrophic lakes is considered low, this refers only to planktonic production. Dys-

trophic lakes generally have highly productive littoral zones, particularly those that develop bog flora. This littoral vegetation dominates the metabolism of the lake ecosystem, providing both a source of dissolved and particulate organic matter (see Wetzel and Allen, 1970; Wetzel, 1975).

**Marl systems.** A fourth type of system, the *marl* lake, contains extremely hard water due to inputs of calcium over a long period of time. A hard-water lake is relatively unproductive and remains so because of the reduced availability of nutrients, even though carbonates remain high. Under certain conditions in these lakes, phosphorus, iron, magnesium, and other nutrients form insoluble compounds and are lost to the system. Sodium and potassium are low, but nitrogenous compounds are high. Because of low photosynthetic productivity, nitrogen remains unutilized. Calcium is often supersaturated, and carbonates, especially calcium carbonate, and humic dissolved organic matter precipitate to form marl deposits on the bottom. The metabolism of marl lakes is such that if carbonate inputs from drainage are reduced or depleted or if sediments build up high enough to support littoral vegetation and the growth of *Sphagnum,* the marl lake gradually develops into a bog (described later). (For details, see Wetzel, 1972, 1975.)

## Lotic ecosystems

Continuously moving water is the outstanding feature of streams and rivers. Current cuts the channel, molds the character of the stream, and influences the life and ways of organisms inhabiting flowing waters.

Streams may begin as outlets of ponds or lakes or they may arise from springs and seepage areas. Added to this in varying quantities is surface runoff, especially after heavy or prolonged rains and rapid snow melt. Because precipitation, the source of all runoff and subsurface water, varies seasonally, the rate and volume of streamflow may fluctuate widely from flood conditions to dry channels.

As water drains away from its source, it flows in a direction dictated by the lay of the land and the underlying rock formations. Its course may be determined by the original slope and its regularities; or the water, seeking the least resistant route to lower land, may follow the joints and fissures in bedrock near the surface and shallow depressions in the ground. Whatever its direction, water is concentrated into rills that erode small furrows, which soon grow into gullies. Water, moving downstream, especially where the gradient is steep, carries with it a load of debris that cuts the channel wider and deeper and that sooner or later is deposited within or along the stream. At the same time erosion continues at the head of the gully, cutting backward into the slope and increasing its drainage area. Just below its source, the stream may be small, straight, and often swift, with waterfalls and rapids. Further downstream, where the gradient is less and the velocity decreases, meanders become common. These are formed when the current, deflected by some obstacle on the floor of the channel, by projecting rocks and debris, or by the entrance of swifter currents, strikes the opposite bank. As the water moves downstream, it is thrown back to the other side again. These abrasive forces create a curve in the stream, on the inside of which the velocity is slowed and the water drops its load. Such cutting and deposition often cause valley streams to change course and to cut off the meanders to form oxbow lakes. When the water reaches level land, its velocity is greatly reduced, and the load it carries is deposited as silt, sand, or mud.

At flood time the material carried by the stream is dropped on the level lands, over which the water spreads to form flood-plain deposits. These flood plains, on which man has settled so extensively, are a part of the stream or river channel used at the time of high water, a fact that few people recognize. The current at flood time is swiftest in the normal channel of the stream and slowest on the flood plain. Along the margin of the channel and the flood plain, where the rapid water meets the slow, the current is checked and all but the fine sediments are dropped on the edges of the channel. Thus the deposits on the flood plain are higher on the immediate border and slope off gradually toward the valley side.

When a stream or river flows into a lake or sea, the velocity of the water is suddenly checked and the load of sediment is deposited in a fan-shaped area at the inlet point to form a delta. Here the course of the water is broken into a number of channels, which are blocked or opened with subsequent deposits. As a result the delta is characterized by small lakes and swampy or marshy islands. Material not deposited at the mouth is carried further out to open water, where it settles on the bottom. Eventually the sediments build up above the water to form a new land surface.

The character of a stream is molded by the velocity of the current. This velocity varies from stream to stream and within the stream itself, and it depends upon the size, shape, and steepness of the stream channel, the roughness of the bottom, the depth, and the rainfall.

The velocity of flow influences the degree of silt deposition and the nature of the bottom. The current in the riffles is too fast to allow siltation, but coarser silt particles drop out in the smooth or quiet sections of the stream. High water increases the velocity; it moves bottom stones, scours the stream bed, and cuts new banks and channels. In very steep stream beds, the current may remove all but very large rocks and leave a boulder-strewn stream.

Flowing water also transports nutrients to and carries waste products away from many aquatic organisms and may even sweep them away. Balancing this depletion of bottom fauna, the current continuously reintroduces bottom fauna from areas upstream. Similarly, as nutrients are washed downstream, more are carried in from above. Because of this, the productivity of primary producers in streams is 6 to 30 times that of those in standing water (Nelson and Scott, 1962). The transport and removal action of flowing water benefits such continuous processes as photosynthesis.

The temperature of a stream is not constant. In general, small, shallow streams tend to follow, but lag behind, air temperatures, warming and cooling with the seasons but never falling below freezing in winter. Streams with large areas exposed to direct sunlight are warmer than those shaded by trees, shrubs, and high steep banks (Figure 7-9). This is ecologically important because temperature affects the composition of the stream community.

The constant swirling and churning of stream water over riffles and falls result in

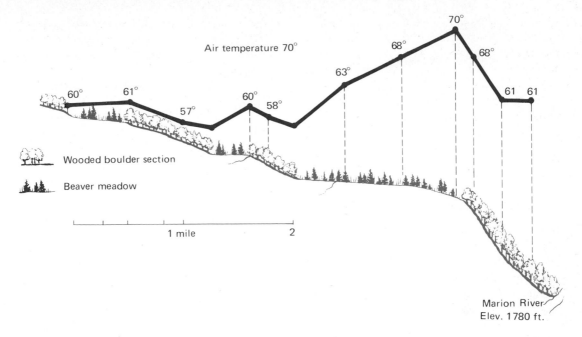

Air temperature 70°

60°     61°     57°     60°   58°     63°     68°     70°     68°     61   61

Wooded boulder section

Beaver meadow

1 mile     2

Marion River
Elev. 1780 ft.

**FIGURE 7-9**
*Profile of Bear Brook in the Adirondack Mountains of New York. The graph of its water temperatures shows the warming effect of open beaver meadows and the cooling effects of wooded boulder streams. (From the Biological Survey of Raquette Watershed, New York State Conservation Department, 1934.)*

greater contact with the atmosphere. Thus the oxygen content of the water is high at all levels and often is near the saturation point for the existing temperature. Only in deep holes or in polluted waters does dissolved oxygen show any significant decline.

Free carbon dioxide in rapid water is in equilibrium with that of the atmosphere, and the amount of bound carbon dioxide is influenced by the nature of the surrounding terrain and the decomposition taking place in pools of still water. Most of the carbon dioxide in flowing water occurs as carbonate and bicarbonate salts. Streams fed by groundwater from limestone springs receive the greatest amount of carbonates in solution. Because of a coating of algae and ooze on the bottom, little calcium carbonate is added by the action of carbonic acid on a limestone stream bed (Neel, 1951).

The degree of acidity and alkalinity of the water reflects the carbon dioxide content as well as the presence of organic acids and pollution. The higher the pH of stream water, the richer the natural waters generally are in carbonates, bicarbonates, and associated salts. Such streams support more abundant aquatic life and larger trout populations than streams with acid waters, which generally are low in nutrients.

STRUCTURE

*Fast water.* Fast or swiftly flowing streams are, roughly, all those whose velocity of flow is 50 cm/sec or higher (A. Nielsen, 1950). At this velocity the current will remove all particles less than 5 mm in diameter and will leave behind a stony bottom. The fast stream is often a series of two essentially different but interrelated habitats, the turbulent riffle and the quiet pool (Figure 7-10). The waters of the pool are influenced by processes occurring in the rapids above, and the waters of the rapids are influenced by events in the pool.

Riffles are the sites of primary production in the stream (see Nelson and Scott, 1962). Here the aufwuchs assume dominance and occupy a position of the same importance as the phytoplankton of lakes and ponds. The aufwuchs consist chiefly of diatoms, blue-green and green algae, and water moss. Extensive stands of algae grow over rocks and rubble on the stream bed and form a slippery covering.

**217**

**FIGURE 7-10**
*Two different but related habitats in a stream, the riffle (foreground) and the pool (background).*

**FIGURE 7-11**
*A fast mountain stream in a deep woods. The bottom is largely bedrock.*

Growth during favorable periods may be so rapid that the stream bottom is covered in 10 days or less (Blum, 1960). Many small algal species are epiphytes and grow on the tops of or in among other algae.

The outstanding feature of much of this algal growth is its ephemeral nature. Scouring action of water and the debris it carries tears away larger growth, epiphytes and all, and sends the algae downstream. As a result there is a constant contribution from upstream to the downstream sequence.

Above and below the riffles are the pools. Here the environment differs in chemistry, intensity of current, and depth. Just as the riffles are the sites of organic production, so the pools are sites of decomposition. They are the catch basins of organic materials, for here the velocity of the current is reduced enough to allow a part of the load to settle out. Pools are the major sites for free carbon dioxide production during the summer and fall.

Overall production in a stream is influenced in part by the nature of the bottom. Pools with sandy bottoms are the least productive because they offer little substrate for either aufwuchs or animals. Bedrock, although a solid substrate, is so exposed to currents that only the most tenacious organisms can maintain themselves (Figure 7-11). Gravel and rubble bottoms support the most abundant life because they have the greatest surface area for the aufwuchs, provide many crannies and protected places for insect larvae, and are the most stable (Figure 7-12). Food production decreases as the particles become larger or smaller than rubble. Insect larvae, on the other hand, differ in abundance on the several substrates. Mayfly nymphs are most abundant on rubble, caddisfly larvae on bedrock, and Diptera larvae on bedrock and gravel (Pennak and Van Gerpen, 1947).

The width of the stream also influences overall production. Bottom production in streams 6 m wide decreases by one-half from the sides to the center; in streams 30 meters wide it decreases by one-third (Pate, 1933). Streams 1.8 m or less in width are four times as rich in bottom organisms as those 5.8 to 7.3 m wide. This is one reason why headwater streams make such excellent trout nurseries.

*Slow water.* As the current slows, a noticeable change takes place in streams (Figure 7-13). Silt and decaying organic matter accumulate on the bottom, and fine detritus from upstream is the main source of energy. Faunal organisms are able to move about to obtain their food, and a plankton population develops. The composition and configuration of the stream community approaches that of standing water.

FIGURE 7-12
*Comparison of life in fast and slow streams. (1) blackfly larvae; (2) net-spinning caddisfly; (3) stone case of caddisfly; (4) water moss (Fontinalis); (5) alga (Ulothrix); (6) mayfly nymph (Isonychia); (7) stonefly nymph (Perla); (8) water penny; (9) hellgrammite; (10) diatoms (Diatoma); (11) diatoms (Gomphonema); (12) cranefly larva; (13) dragonfly nymph; (14) water strider; (15) damselfly nymph; (16) water boatman; (17) fingernail clam (Sphaerium); (18) burrowing mayfly nymph (Hexegenia); (19) bloodworm; (20) crayfish. The fish in the fast stream are brook trout (left) and redbelly dace (right). The fish in the slow stream are left to right, northern pike, bullhead, and smallmouth bass.*

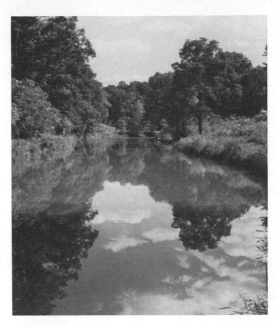

FIGURE 7-13
*A slow stream reflects the sky of a summer afternoon.*

(a)

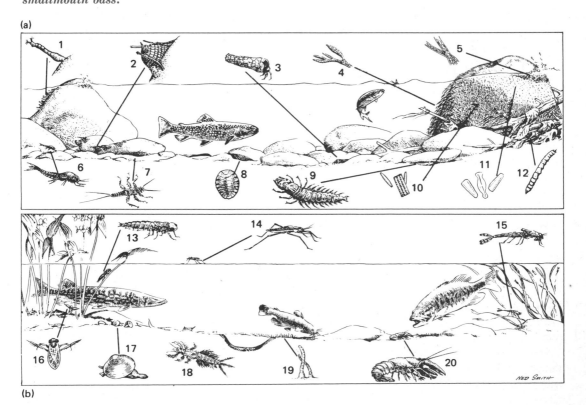

(b)

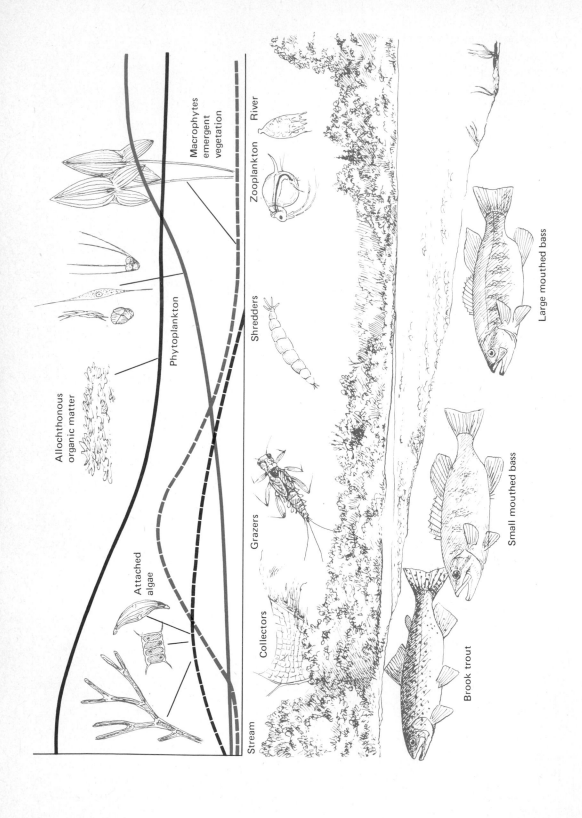

Allochthonous organic matter

Attached algae

Phytoplankton

Macrophytes emergent vegetation

Zooplankton

Shredders

Grazers

Collectors

Stream

River

Brook trout

Small mouthed bass

Large mouthed bass

With increasing temperatures, decreasing current, and accumulating bottom silt, organisms of the fast water are replaced gradually by organisms adapted to these conditions. Brook trout and sculpin give way to smallmouth bass and rock bass, the dace to shiners and darters. With current at a minimum many resident fish lack the strong lateral muscles typical of the trout and have compressed bodies that permit them to move with ease through masses of aquatic plants. Mollusks, particularly *Sphaerium* and *Pisidium,* and pulmonate snails, crustaceans, and burrowing mayflies replace the rubble-dwelling insect larvae. Only in occasional stretches of fast water in the center of the stream are remnants of headwater-stream organisms found.

As the volume of water increases, as the current becomes even slower, and as the silt deposits become heavier, detritus-feeders increase. Rooted aquatics appear. Emergent vegetation grows along the river banks, and duckweeds float on the surface. Indeed the whole aspect approaches that of lakes and ponds, even to zonation along the river margin.

The higher water temperature, weak current and abundant decaying matter promotes the growth of protozoan and other plankton populations. Scarce in fast water, plankton increases in numbers and species in slow water (Figure 7-14). Rivers have no typical plankton of their own. Those found there originate mainly from backwaters and lakes. In general, plankton populations in rivers are not nearly as dense as those in lakes. Time is too short for

### FIGURE 7-14
*Changes in a stream from headwater to river. The headwater stream is strongly heterotrophic, dependent upon allochthonous organic matter. The herbivores are largely collectors and grazers, and respiration exceeds production. The slow stream, halfway between the fast stream and the river, is dominated by shredders; phytoplankton, attached algae, and macrophytes increase. Diversity is the greatest at this point and production exceeds respiration. In the river phytoplankton increases, diversity decreases and again respiration exceeds production. (Adapted from Wetzel, 1975; Cummins, 1975).*

much multiplication of plankton because relatively little time is needed for a given quantity of water to flow from its source to the sea. Also occasional river rapids, often some distance in length, kill many plankton organisms by violent impact against suspended particles and the bottom. Aquatic vegetation filters out this minute life as the current sweeps it along.

### FUNCTION

Dependent upon energy input from the outside, stream ecosystems differ substantially from lakes and from terrestrial ecosystems whose main source of energy is primary production within the system. Energy is supplied mainly by allochthonous sources (Figure 7-15). Energy may come from leaf litter and branches dropped from vegetation overhead or blown in by the wind and from rainwater throughfall carrying with it nutrient-rich exudates from the leaves. Because the pathway for these nutrients is wind and rain, the inputs are considered meteorological.

Other inputs come by a geological pathway through subsurface seepage, which brings nutrients leached from the adjoining forest, agricultural and residential lands, and from upstream flow carrying both dissolved nutrients and particulate matter. Many streams receive inputs from mechanical pathways through the dumping of urban and industrial effluents into flowing water.

A minor biological autochthonous source of nutrients is primary production by diatomaceous algae growing on rocks and by rooted aquatics. Even in slow streams, where primary production may be substantial, rooted aquatics provide little immediate source of energy and become important only when production becomes available as detrital material. (But see Minshall, 1978, for importance of autotrophy in streams.)

Energy is lost along two pathways, geological through streamflow to downstream systems and biological from heat loss through respiration.

Energy flow in stream ecosystems is poorly documented compared to terrestrial and lake ecosystems. The only energy budget available is one for small, forested Bear Brook in the Hubbard Forest of northern New Hampshire

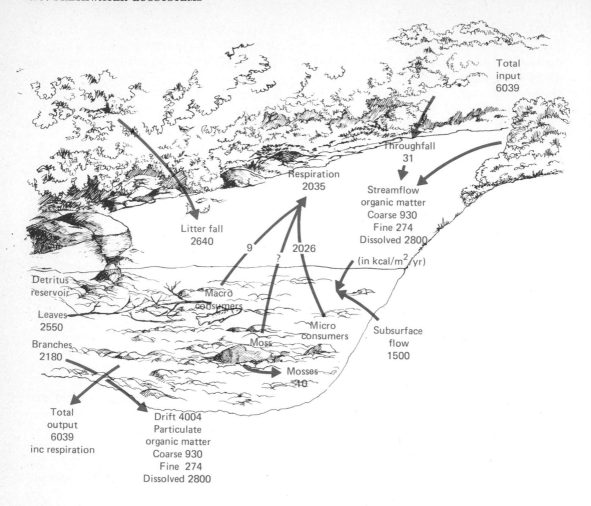

**FIGURE 7-15**
*Energy flow in stream ecosystems. Streams are open ecosystems with much of the energy input coming from outside sources. Note the great dependence on materials from terrestrial sources and inflow from upstream and the role of coarse particulate organic matter (CPOM), fine particulate organic matter (FPOM), and dissolved organic matter (DOM). Primary production contributes little to energy flow. Energy values are based on Bear Brook, New Hampshire. (Data from Fisher and Likens, 1973)*

(Table 7-1). Over 99 percent of the energy input comes from the surrounding forested watershed or from upstream areas. Primary production by mosses accounts for less than 1 percent of the total energy supply. Algae are absent from the stream. Inputs from litter and throughfall (meteorological) account for 44 percent of the energy supply and geological inputs account for 56 percent.

Energy is introduced in three forms: coarse particulate organic matter (CPOM), represented by leaves and other debris; fine particulate organic matter (FPOM), represented by drift and small particles; and dissolved organic matter (DOM). (For a detailed discussion of these sources, see Cummins, 1974.) In Bear Brook 83 percent of the geological input and 47 percent of the total energy is in the form of DOM. Sixty-six percent of the organic input is exported downstream, leaving 34 percent to be utilized locally.

Because most streams receive their basic energy supply from land, it is not surprising that the majority of first-level consumers are detritus-feeders (Figure 7-16) (Cummins, 1974, 1975). In the fall leaves drift down from overhanging trees, settle on the water, float downstream, and lodge against banks, debris, and stones. Soaked with water a leaf sinks to the bottom, where it quickly loses 5 to 30 per-

TABLE 7-1
## Annual energy budget for Bear Brook

| Item | kg (whole stream)* | kcal/m² | Percentage |
|---|---|---|---|
| **INPUTS** | | | |
| Litter fall | | | |
| Leaf | 1990 | 1370 | 22.7 |
| Branch | 740 | 520 | 8.6 |
| Miscellaneous | 530 | 370 | 6.1 |
| Wind transport | | | |
| Autumn | 422 | 290 | 4.8 |
| Spring | 125 | 90 | 1.5 |
| Throughfall | 43 | 31 | 0.5 |
| Fluvial transport | | | |
| CPOM | 640 | 430 | 7.1 |
| FPOM | 155 | 128 | 2.1 |
| DQM, surface | 1580 | 1300 | 21.5 |
| DOM, subsurface | 1800 | 1500 | 24.8 |
| Moss production | 13 | 10 | 0.2 |
| Input total | 8051 | 6039 | 99.9 |
| **OUTPUTS** | | | |
| Fluvial transport | | | |
| CPOM | 1370 | 930 | 15.0 |
| FPOM | 330 | 274 | 5.0 |
| DOM | 3380 | 2800 | 46.0 |
| Respiration | | | |
| Macroconsumers | 13 | 9 | 0.2 |
| Microconsumers | 2930 | 2026 | 34.0 |
| Output total | 8020 | 6039 | 100.2 |

*Note:* CPOM = coarse particulate organic matter; FPOM = fine particulate organic matter; DOM = dissolved organic matter.

*Budget in kg does not balance because of different caloric equivalents of budgetary components.

cent of dry matter as water leaches the soluble organic matter from its tissues. This leachate becomes part of the dissolved organic matter (DOM). Within a week or two, depending upon the temperature, the surface of the leaf is colonized by bacteria and fungi. As the microbes begin their work, the leaf is attacked by shredders, insect larvae that feed on coarse particulate organic matter such as leaves and needles. Among these shredders are craneflies (Tipulidae), caddisflies (Trichoptera), and stoneflies (Plecoptera). They break down the CPOM, feeding on the material not so much for the energy it contains, but for the bacteria and fungi growing on it (Cummins, 1974). Of the material they ingest, the shredders assimilate about 40 percent and pass off 60 percent as feces which becomes part of the fine particulate organic matter.

Broken up by the shredders and partially decomposed by the microbes, the leaf now becomes part of the FPOM, which also includes some precipitated DOM. Drifting downstream, the FPOM is picked up by another group of detritus-feeders, the collectors (fine particle detritivores), which include among others the larvae of black flies (Simulidae) and the net-spinning caddisflies (*Hydropsyche*). While the shredders and collectors feed on detrital material, another group of primary consumers, the scrapers, which include the water penny (*Psephenus* sp.) and a number of mobile caddisfly larvae, feed on the algal coating of stones and rubble. Much of the material they scrape loose enters the drift as FPOM.

Feeding on the detrital-feeders are predaceous insect larvae such as the powerful dobsonfly, in the larval form known as the hellgrammite *(Corydalis cornutus),* and such fish as the sculpin *(Cottus)* and trout. Even these predators do not depend solely on autochthonous material, the insects of the

**223**

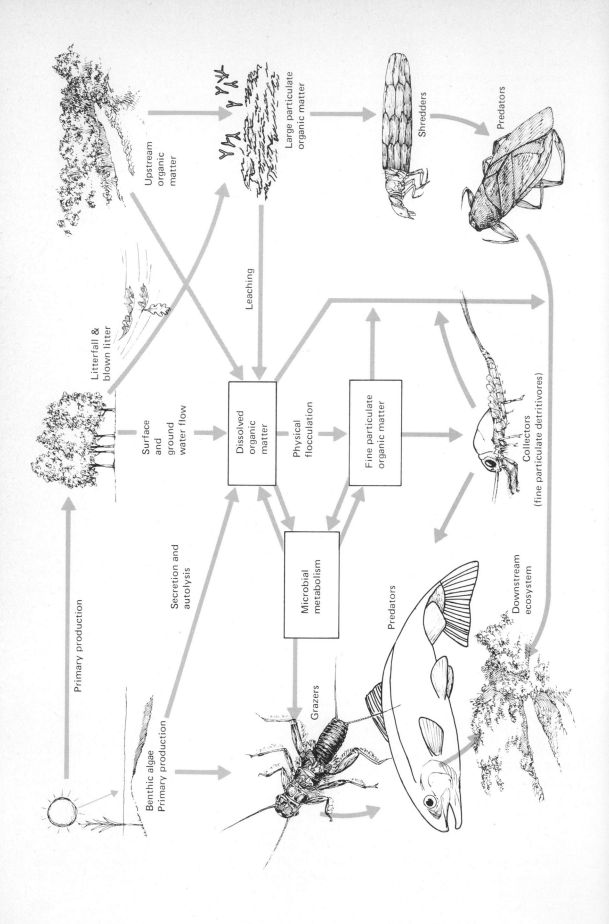

Primary production

Upstream organic matter

Large particulate organic matter

Shredders

Predators

Litterfall & blown litter

Leaching

Surface and ground water flow

Secretion and autolysis

Benthic algae Primary production

Dissolved organic matter

Physical flocculation

Fine particulate organic matter

Collectors (fine particulate detritivores)

Microbial metabolism

Grazers

Predators

Downstream ecosystem

FIGURE 7-16
*Model of a stream ecosystem emphasizing structure and function. On the right is the autotrophic component that supports the grazers or scrapers. On the left is the heterotrophic component which constitutes most of the stream ecosystem. Note the heavy input of materials and energy from terrestrial and upstream sources, the importance of coarse, fine, and dissolved particulate matter, and the role played by microbes, shredders or large particle detritivores, and the collectors or fine particle detritivores. (Adapted from Cummins, 1974.)*

stream; they also feed heavily on allochthonous material (Hunt, 1975). Brook trout in West Virginia feed largely on aquatic organisms from March through May, when stream invertebrates are most abundant. In June the fish shift to terrestrial invertebrates that fall into the stream; from July to September terrestrial insects account for 60 to 80 percent of the total food the fish consume (Redd and Benson, 1962).

## Wetlands

Associated with lentic and lotic ecosystems are wetlands, areas where water is near, at, or above the level of the land. Biologically they are among the richest and most interesting ecosystems. Yet they are also the least appreciated and the first to be destroyed by filling and draining. Wetlands are a half-way world between terrestrial and aquatic ecosystems and exhibit some of the characteristics of each. A wide variety of wetlands exists, and the problems of classifying them necessary for management and conservation has yet to be resolved (Sather, 1976). An older but useful classification is given in Table 7-2.

Wetlands dominated by emergent vegetation, plants with roots in soil covered part or all of the time by water and leaves held above water, are *marshes*. Growing to reeds, sedges, grasses, and cattails, marshes are essentially wet prairies. They develop along margins of lakes, in shallow basins with an inflow and outflow of water, and along slow-moving rivers and tidal flats. Wetlands in which consider-

TABLE 7-2
*Types of wetlands*

| Type | Site characteristics | Plant and animal populations |
|---|---|---|
| **INLAND FRESH AREAS** | | |
| Seasonally flooded basins or flats | Soil covered with water or waterlogged during variable periods, but well drained during much of the growing season; in upland depressions and bottomlands | Bottomland hardwoods to herbaceous growth |
| Fresh meadows | Without standing water during growing season; waterlogged to within a few inches of surface | Grasses, sedges, rushes, broadleaf plants |
| Shallow fresh marshes | Soil waterlogged during growing season; often covered with 6 in. or more of water | Grasses, bulrushes, spike rushes, cattails, arrowhead, smartweed, pickerelweed; a major waterfowl-production area |
| Deep fresh marshes | Soil covered with 6 in. to 3 ft of water. | Cattails, reeds, bulrushes, spike rushes, wild rice; principal duck-breeding area |
| Open fresh water | Water less than 10 ft deep | Bordered by emergent vegetation such as pondweed, naiads, wild celery, water lily; brooding, feeding, nesting area for ducks |
| Shrub swamps | Soil waterlogged; often covered with 6 in. or more of water | Alder, willow, buttonbush, dogwoods; nesting and feeding area for ducks to limited extent |

TABLE 7-2 (Con't)

| Type | Site characteristics | Plant and animal populations |
|------|---------------------|------------------------------|
| Wooded swamps | Soil waterlogged; often covered with 1 ft of water; along sluggish streams, flat uplands, shallow lake basins | North: tamarack, arborvitae, spruce, red maple, silver maple; south: water oak, overcup oak, tupelo, swamp black gum, cypress |
| Bogs | Soil waterlogged; spongy covering of mosses | Heath shrubs, sphagnum, sedges |

**COASTAL FRESH AREAS**

| Type | Site characteristics | Plant and animal populations |
|------|---------------------|------------------------------|
| Shallow fresh marsh | Soil waterlogged during growing season; at high tide as much as 6 in. of water; on landward side, deep marshes along tidal rivers, sounds, deltas | Grasses and sedges; important waterfowl areas |
| Deep fresh marshes | At high tide covered with 6 in. to 3 ft of water; along tidal rivers and bays | Cattails, wild rice, giant cutgrass |
| Open fresh water | Shallow portions of open water along fresh tidal rivers and sounds | Vegetation scarce or absent; important waterfowl areas |

**INLAND SALINE AREAS**

| Type | Site characteristics | Plant and animal populations |
|------|---------------------|------------------------------|
| Saline flats | Flooded after periods of heavy precipitation; waterlogged within few inches of surface during the growing season | Seablite, salt grass, saltbush; fall waterfowl-feeding areas |
| Saline marshes | Soil waterlogged during growing season; often covered with 2 to 3 ft of water; shallow lake basins | Alkali hard-stemmed bulrush, wigeon grass, sago pondweed; valuable waterfowl areas |
| Open saline water | Permanent areas of shallow saline water; depth variable | Sago pondweed, muskgrasses; important waterfowl-feeding areas |

**COASTAL SALINE AREAS**

| Type | Site characteristics | Plant and animal populations |
|------|---------------------|------------------------------|
| Salt flats | Soil waterlogged during growing season; sites occasionally to fairly regularly covered by high tide; landward sides or islands within salt meadows and marshes | Salt grass, seablite, saltwort |
| Salt meadows | Soil waterlogged during growing season; rarely covered with tide water; landward side of salt marshes | Cord grass, salt grass, black rush; waterfowl-feeding areas |
| Irregularly flooded salt marshes | Covered by wind tides at irregular intervals during the growing season; along shores of nearly enclosed bays, sounds, etc. | Needlerush, waterfowl cover area |
| Regularly flooded salt marshes | Covered at average high tide with 6 in. or more of water; along open ocean and along sounds | Atlantic: salt-marsh cord grass; Pacific: alkali bulrush, glassworts; feeding area for ducks and geese |
| Sounds and bays | Portions of saltwater sounds and bays shallow enough to be diked and filled; all water landward from average low-tide line | Wintering areas for waterfowl |
| Mangrove swamps | Soil covered at average high tide with 6 in. to 3 ft of water; along coast of southern Florida | Red and black mangroves. |

*Source:* Adapted from Shaw and Fredine, 1956.

able amounts of water are retained by an accumulation of partially decayed organic matter are *peatlands* or *mires*. Mires that are fed by water moving through the mineral soil and that are dominated by sedges are known as *fens*. Mires that depend upon precipitation for water supply and are dominated by *sphagnum* mosses are *bogs*. Wooded wetlands are *swamps*. They may be dominated by trees such as cypress, tupelo, and swamp oaks, or by shrubs such as alder and willow. Shrubby swamps are often know as *carrs*.

## Peatlands

Of all wetland types peatlands have been studied the most extensively. Because relatively little is known about the structure and function of freshwater marshes and swamps, bogs will serve as an illustration of the general features of wetlands. (For current review of freshwater marsh structure and function see Good et al., 1978.)

Peatlands are ecosystems in which the rate of production of organic matter by living organisms exceeds the rate at which the compounds are respired and degraded (Moore and Bellamy, 1974). As a result part of the production accumulates as organic deposits or peat. As the peat blanket thickens the surface vegetation becomes insulated from the mineral soil. Environmental conditions change bringing about a change in plant and animal life.

Although peatlands are most commonly associated with northern regions of the world, they occur worldwide wherever humidity and precipitation are relatively high or where hydrological situations encourage an accumulation of partly decayed organic matter.

Peat forms when the movement of water through a low wetland area becomes partially blocked. The inflowing water no longer carries sediments out of the basin and much of the water is retained in the system. When dead organic matter, its decomposition slowed by immersion in water, begins to act as an inert body displacing its own volume of water, peat formation begins (Moore and Bellamy, 1974). Peat formation continues until it reaches a level at which water drains from its surface. At this point peat no longer acts as an inert body but as its own reservoir holding quantities of water from drainage.

In tropical and subtropical regions peatlands are found mostly in mountains or in lowland estuarine and delta regions where enough inflowing water keeps the basin filled, retarding decomposition. Examples are the Everglades, the tropical coal swamps of the Carboniferous period, and the pocasins of the southern United States coastal plains. But peatlands are more characteristic of northern regions where inflow and precipitation balance outflow and retention, or where permafrost impedes drainage of water and the bulk of summer precipitation is retained. In northern regions peatlands are dominated by sphagnum moss and develop not only in basins, but also on mineral soil producing blanket mires, raised bogs, and patterned fens (see Moore and Bellamy, 1974; Heinselman, 1970).

### STRUCTURE

The formation and types of mires are complex. Peatland may develop from an aquatic ecosystem through the filling of the basin by sediments or from a terrestrial system through the swamping of wooded areas (paludification) (Malmer, 1975). In the process the original systems, characterized by rapid and fairly complete turnover of dead organic matter and a relatively high productivity, are slowly converted to a common ecosystem of low productivity, low plant biomass, low diversity, a slow and incomplete turnover of organic matter, and a storage of nutrients.

The classic example of bog development is the filling of a lake basin with sediment and organic matter carried into the area by inflowing water. As the open water, dominated by cations of calcium and carbonates, becomes more shallow, emergent vegetation begins to encroach upon and cover the basin. Excess organic matter produced by the marsh vegetation accumulates, building the bottom up to and eventually above the water table. The accumulated organic matter is continually saturated, inhibiting decomposition. If eutrophic conditions persist, as they often do, terrestrial vegetation may invade the area (see marsh succession, Chapter 21). But under oligotrophic and dystrophic conditions, calcium and sulphate ions dominate the water and humic

**227**

compounds dissolved from organic matter increase and pH decreases. The marsh vegetation is replaced by acid-tolerant sedges and grasses.

The buildup of sediment and peat diverts inflowing water first to the periphery of the basin and eventually away from the basin altogether. Now the only source of water is rainfall. Under conditions of high humidity and precipitation sphagnum mosses colonize the area. Sphagnum has the ability to absorb ions out of nutrient-poor water by exchanging hydrogen ions in its tissues for cations in solution. This exchange, which results in an increase in hydrogen and sulphate ions in the water, increases the acidity of the system, further decreasing the rate of decomposition and increasing the net accumulation of organic matter. Sphagnum, which has a spongelike ability to hold water, increases water retention on the site, and its habit of adding new growth onto the accumulating remains of past moss generations aids in thickening the organic mat. In time the water-saturated mat of peat and sphagnum expands and rises above the general water table level to form a raised bog (Figure 7-17). Between the mat of peat and surrounding higher land is a moatlike area of shallow water, often the remnant of flowing groundwater around the peat mat. These water areas, dominated by sedges, are called *laggs*.

Bogs may also develop by the filling of a basin from above rather than from below. Many of the sediments coming into the lake, especially the finer materials, remain suspended as colloids during their movement from the water's surface to the bottom. Often they remain in suspension until precipitated by bacterial, chemical, or photosynthetic action to form a fine, soft deposit called a false bottom. Meanwhile peat develops around the edges of the basin. Sphagnum invades the area and fills the open spaces between sedges and other emergent plants. When a consolidated mass of peat develops or when jutting rocks, logs, or even the leading edges of low growing plants allow a foothold, a mat extends outward over the water (Figure 7-18). As the mat thickens and advances toward the center of the lake, the older peat mat is colonized first by

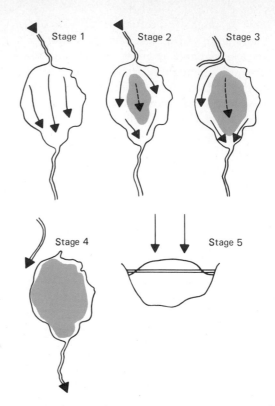

**FIGURE 7-17**
*Stages in the development of a mire. Stage 1. A flow of water brings allochthonous material into the system. If the water flow is large, the amount of outside material is heavy and the rate of peat growth is slow, producing a heavy peat above which the water flows. If the rate of water flow is low, growth of peat is fast, producing a light peat beneath which the water flows (see Figure 7-18). Stage 2. The accumulation of peat tends to channelize the main flow of water. Under certain conditions the whole peat mass may be inundated. Stage 3. Continued peat growth diverts inflow from the basin. The major source of water is now precipitation, surface runoff, or seepage from surrounding area. Stage 4. Additional accumulation of peat leaves large areas of the mire unaffected by moving water, but subject to inundation during periods of heavy rainfall. Stage 5. Continued peat growth raises the surface of the mire above the influence of groundwater. The mire now possesses its own water table fed by rain. (After Moore and Bellamy, 1975.)*

such shrubs as bog rosemary and leatherleaf, followed by forest trees as pine, spruce, or tamarack, producing the concentric concave rings of vegetation so characteristic of northern bog lakes. Because the pioneering mire vegetation often grows on a floating mat of peat over open water, such bogs are often termed quaking bogs or perhaps more descriptively *schwingmoor* in German.

Although basin filling is the textbook model of bog formation, most of the boreal peatland develops on higher ground on mineral soil. Covering large areas such mire development creates high moors and blanket mires (Figure 7-19). Such upland bogs start in one of several ways. Sphagnum and other mosses may in-

vade higher ground surrounding a lake basin by creeping into depressions. Beaver dams backing up water over streamside forests and along stream courses may initiate mire development. Logging of forests on wet sites where humus decomposition is incomplete may provide sites for colonization by mosses. Fresh mosses and sedges on the surface of developing peatland are porous and permeable to water. Partially decomposed remains of these plants beneath the new growth become compressed, increasing bulk density and reducing pore volume. Eventually these decomposed, compressed peats act much like impermeable clays. Water unable to move through the mass to mineral soil remains as a perched water table near peat surface. Far removed from flowing waters these blanket mires and raised bogs depend wholly upon precipitation for water and minerals. As a result the water is highly deficient in mineral salts and low in pH. Such bogs are termed *ombrotrophic*.

Upland bogs or paludified landscapes are not uniform, but rather are a mosaic of vegetation types ranging from blanket mires and raised bogs to fens fed by water that has moved

*FIGURE 7-18*
*Transect through a quaking bog showing zones of vegetation sphagnum mounds, peat deposits, and floating mats. A, pond lily in open water; B, buckbean and sedge zone; C, sweet-gale zone; D, leatherleaf; E, Laborador tea; F, black spruce; G, birch-balsam-black spruce forest. (Redrawn from Dansereau and Segadas-Vianna, 1952.)*

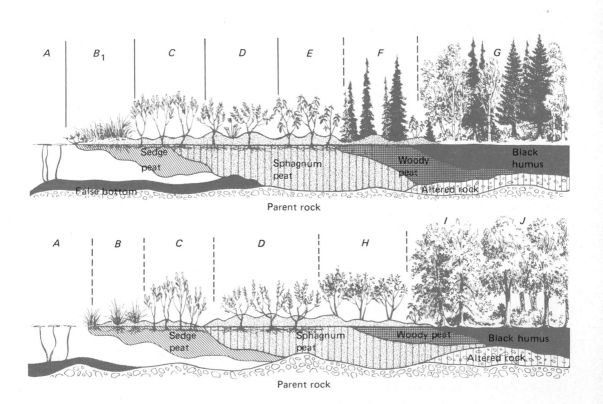

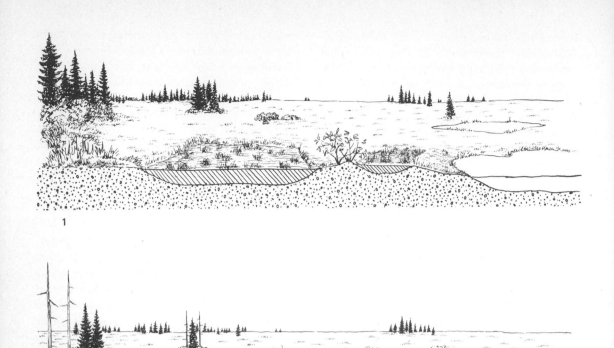

1

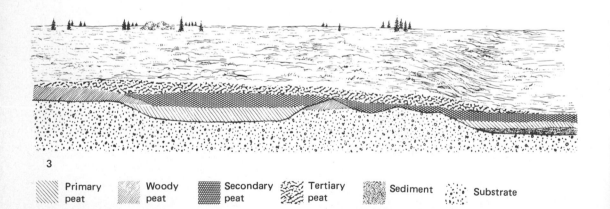

2

3

| | Primary peat | | Woody peat | | Secondary peat | | Tertiary peat | | Sediment | | Substrate |
|---|---|---|---|---|---|---|---|---|---|---|---|

**FIGURE 7-19**
*Development of a raised bog and blanket mire. The first stage involves the development of primary peat formed in basins and depressions. It reduces the surface retention of the reservoir. After peat fills the basin, it may continue to develop beyond the confines of the basin to form a raised bog. This is secondary peat which acts as a reservoir, increasing surface retention of water on the area. Peat that develops above the physical limits of ground water and blankets even terrestrial situations is tertiary peat. It acts as a reservoir holding a volume of water by capillary action above the level of main ground water drainage. The landscape acts as a perched water table, fed by precipitation falling on it. As the area becomes more wet and acid, tree growth dies out and the landscape is covered by sphagnum and sedges.*

through mineral soil and thus supports a richer vegetation of sedges and associated plants (see Moore and Bellamy, 1974; Heinselman, 1970, 1975). Although peatlands have been considered climax vegetation, it seems inaccurate to do so because their development or succession appears to be subject to random changes induced by such local events as fire, drought, flooding, and the like. For example, occasionally the peat mat becomes infiltrated with alkaline water which allows fermentation and decomposition of the peat. Pockets form beneath the surface; the peat mat sags and caves in. Patches of open water appear as bog pools. In the bog forest the evaporative power of the vegetation may become greater than the water-retaining capacity of the peaty material, upsetting the hydrological balance necessary for formation and maintenance of peatlands. Then succession tends toward a drier, more mesic condition. But lumbering, windthrow, or any action that destroys tree growth can reverse this process and convert the area to an open bog. Thus succession in peatlands may simply involve ceaseless change rather than directional development (Heinselman, 1975).

## FUNCTION

Because bog vegetation is not in contact with mineral soil and because inflow of groundwater is blocked, nutrient input is largely by the way of precipitation. Nutrient availability, especially nitrogen, phosphorus, and potassium, is low, and most of the nutrients fixed in plant tissue are removed from circulation in the accumulation of peat (Table 7-3). The amount of nutrients received by a bog depends upon the location of peatland. Bogs near the sea, such as English blanket mires, receive considerably more nutrients, especially magnesium and potassium, than inland bogs (Table 7-4). Few data exist on nutrient cycles in ombrotrophic bogs. But evidence suggests that some bog plants possess mechanisms to conserve nutrients.

In general nitrogen is available from three sources: (1) precipitation, the major source; (2)

TABLE 7-3
*Cumulative loss or gain of nutrients over the period February 1969 to January 1972 with an estimate of nutrients stored irrecoverably in deep peat each year ($mg/m^2$)*

| Element | 1969 | 1970 | 1971 | Total | Stored |
|---------|------|------|------|-------|--------|
| Ca | −525 | −329 | −1207 | −2061 | 6364 |
| Mg | +429 | +1950 | +964 | +2343 | 9091 |
| K | −52 | +361 | −84 | +225 | 4546 |
| $NH_4$ | −567 | −588 | −537 | −1692 | 3200 |
| $NO_3$ | −369 | −427 | −356 | −752 | |
| P | −5 | −2 | −13 | −20 | 2045 |
| Fe | −201 | −77 | −94 | −371 | nd* |
| Cu | +1 | +5 | −2 | +4 | nd |

*nd = not determined
*Source:* Moore et al., 1975.

TABLE 7-4
*Annual nutrient input by rainwater at various sites (kg/ha)*

| Site | Annual rainfall (cm) | Inorganic N | P | Na | K | Ca | Mg |
|------|----------------------|-------------|-----|------|------|-------|-------|
| Lerwick, Shetland Isles[34] | 57.3 | 2.10 | — | 133.00 | 5.52 | 6.70 | 19.20 |
| Lancashire UK[35] | 161.7 | 6.28 | 0.43 | 35.34 | 2.96 | 7.30 | 4.63 |
| Pennines, UK[33] | 186.1 | 6.89 | 0.27 | 32.14 | 2.27 | 9.53 | 4.48 |
| Kent, UK[46] | 84.0 | — | <0.4 | 19.30 | 2.80 | 10.70 | <4.20 |
| Hubbard Brook, USA[37] | 129.0 | — | — | 1.5 | 1.40 | 2.60 | 0.70 |

*Source:* Moore and Bellamy, 1974.

nitrogen fixation by blue-green algae living in close association with bog mosses and by bog myrtle, if the pH of the substrate is at least 3.5, below which nitrogen-fixing root nodule bacteria are inhibited; and (3) carnivorous habit by which certain plants as sundews and pitcher plants extract nitrogen from captured and digested insects. The blue-green algae may fix 0.2 to 0.9 g N/m²/yr. Rosswall (1975) has calculated a nitrogen budget and nitrogen cycle for an ombrotrophic bog dominated by the trailing ground plant *Rubus chamaemorus* (Figure 7-20). Yearly demand is about 1 to 2 g/m².

The phosphorus cycle holds losses to a minimum and is more closed than the nitrogen cycle (Figure 7-21). *Rubus chamaemorus* increases its uptake of phosphorus prior to bud break. After bud break the plant increases phosphorus in the stem, leaf, and root. This increase correlates with an increase in peat temperature. After the plant completes shoot growth in summer, its phosphorus levels decline in the shoots as it mobilizes phosphorus in the developing fruits or in roots and

rhizomes. In advanced senescence the plant rapidly loses phosphorus from the shoots and accumulates it in the roots and winter buds.

Energy flow in mire systems differs considerably from that of other systems (Figure 7-22) because the decomposer food chain is impaired. In most ecosystems material that enters the decomposer food chain eventually is recycled and the energy that enters the system is liberated or stored in living material. But in mire systems material resulting from primary

**FIGURE 7-20**
*Nitrogen budget for an ombrotrophic mire (Stordalen, Sweden). Quantities of nitrogen are expressed as g N/m², flows as g N/m²/yr. The vegetation has been divided into above-ground parts illustrated by* **Rubus chamaemorus** *and lichens (1.6 g N/m²), below-ground parts of vascular plants (3.0 g N/m²), above-ground litter (3.2 g N/m²), and bryophytes (4.4 g N/m²).*
$N_{org}$ = *organic nitrogen;* $N_{acc}$ = *nitrogen accumulated;* $N_{part}$ = *particulate nitrogen (dry deposition). (From Rosswall and Granhall, 1980)*

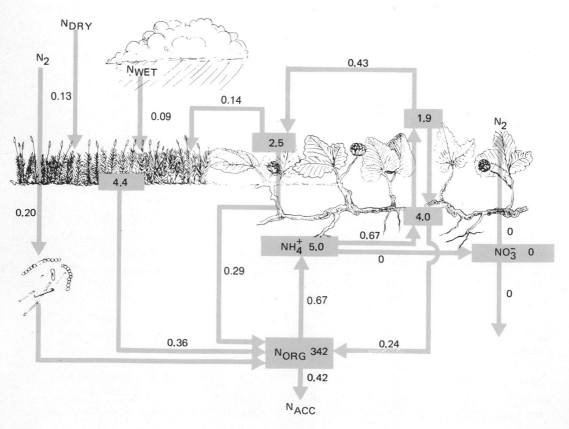

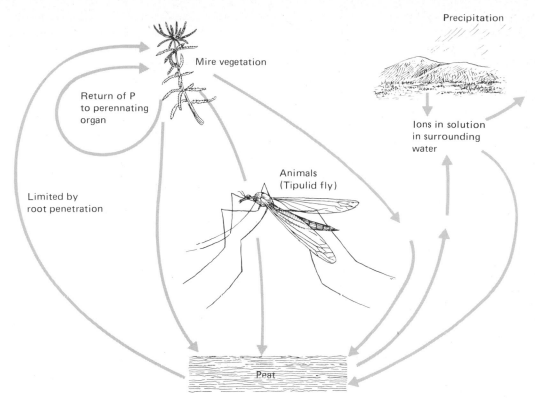

Precipitation

Mire vegetation

Return of P
to perennating
organ

Limited by
root penetration

Animals
(Tipulid fly)

Ions in solution
in surrounding
water

Peat

**FIGURE 7-21**
*Cycling of phosphorus in mire characterized
by a throughflow of water. (From Moore and
Bellamy, 1975.)*

production accumulates in an undecomposed state and energy is locked up in peat until environmental conditions change favoring decomposition or the material is burned.

Because of low temperatures, acidity, and nutrient immobilization primary production in peatlands is low. In the Stordalen mire in Sweden primary production amounted to 70 g/m²/yr by bryophytes and 83 gm/m²/yr by dwarf shrubs, forbs, and monocots collectively (Rosswall, 1975). In an English blanket mire average production was 635 g/m²/yr with sphagnum on wet sites contributing 300 g/m²/yr (Heal et al., 1975).

Herbivorous utilization of bog vegetation is low in part because of unpalatability of the vegetation. Herbivores may include red grouse, willow grouse, hares, bog lemmings, and sheep. With these herbivores consumption is selective and usually confined to current shoots, opening buds, and fruits. In most bogs the dominant herbivores are insects. In an English blanket mire psyllids and tipulids were

the main invertebrate herbivores, the former feeding on the phloem of heather and the latter feeding on liverworts (Heal et al., 1975; Moore et al., 1975). Predators include rodent-consuming weasels, harriers, short-eared owls, insect-consuming frogs, shrews, pipits, (Europe), Nashville warblers (North America), and invertebrates such as spiders and ground beetles.

Decomposer organisms are low in number and the dominant species vary from year to year. In a Swedish mire total fungal biomass was 58 g/m² and the bacterial biomass, largely anaerobic, was 22 g/m² (Rosswall et al., 1975). Invertebrate detritivores include rotifers, tartigrades, mites, and nematodes. In an English mire a single species of enchytraeid, *Cognettia sphagnetorum,* accounted for 70 to 75 percent of the total energy assimilated by the detritivores (Heal et al., 1975).

Rates of decomposition vary among bog vegetation. Litter of *Rubus chamaemorus* lost 2 percent of its weight in one year and 50 percent in three years. Shrub litter decomposes more slowly, while sphagnum decomposes hardly at all.

In an English mire (Moor House National

**233**

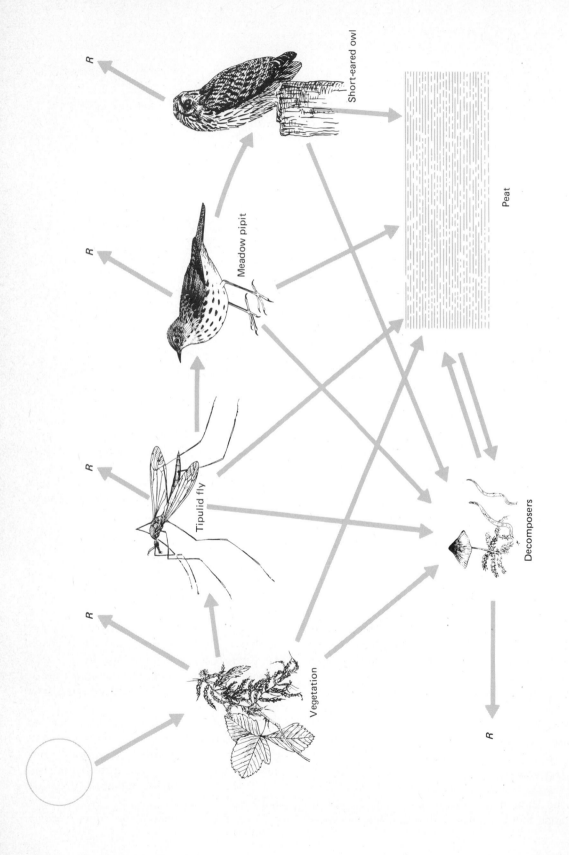

Short-eared owl

Meadow pipit

Tipulid fly

Vegetation

Decomposers

Peat

*R*

FIGURE 7-22
*FIGURE 7-22*
*Energy flow in a mire ecosystem. Note the distinct difference between this energy flow pattern and that of other ecosystems illustrated. Instead of being lost as respiration, the bulk of the energy is stored as peat which accumulates in the system. Such an accumulation in the geologic past resulted in the eventual formation of coal, oil, and gas deposits. (After Moore and Bellamy, 1975.)*

Nature Reserve) the vegetational standing crop has an annual turnover of about 0.3 percent and only about 1 percent of production is consumed by herbivores. Only 14 percent of the herbivore fauna is assimilated by predators. The remainder of the vegetation enters the decomposer food web where about 5 percent is assimilated by decomposers and 10 percent passes below the water table. About 85 percent of the production is decomposed by microflora. But the rate of decomposition is slow, with about 95 percent turnover time of 3,000 years. In the top 20 cm, 95 percent turnover time is about 70 years (Heal et al., 1975).

## SUMMARY

Lentic ecosystems are characterized by lakes and ponds, standing bodies of water that fill a depression in the earth. Geologically speaking, lakes and ponds are ephemeral features of the landscape. In time they fill in, grow smaller, and may finally be replaced by a terrestrial community.

A lake exhibits gradients in light, temperature, and dissolved gases. In summer its waters are stratified. It has a surface layer of warm, circulating water, the epilimnion, and a middle zone, the metalimnion, in which the temperature rapidly drops. Below this is the hypolimnion, a bottom layer of dense water approximately 4° C, often low in oxygen, and high in carbon dioxide. When the surface water cools in fall, the difference in density between the layers decreases and the waters circulate throughout the lake. A similar mixing of water takes place in the spring when the lake warms up. These seasonal overturns are important in recirculating nutrients and oxygen.

The area where light penetrates to the bottom of the lake, a zone known as the littoral, is occupied by rooted plants. Beyond this is the open water or limnetic zone, inhabited by plant and animal plankton and fish. Below the depth of effective light penetration is the profundal region where the diversity of life varies with temperature and oxygen supply. The bottom or benthic zone is a place of intense biologic activity, for here decomposition of organic matter takes place. Anaerobic bacteria are dominant on the bottom beneath the profundal water, while the benthic zone of the littoral is rich in decomposer organisms and detritus-feeders. Although lake ecosystems are often considered as autotrophic systems dominated by phytoplankton and the grazing food chain, lakes actually are strongly dependent on the detrital food chain. Much of the detrital input comes from the littoral zone.

Lakes may be classified as eutrophic, or nutrient rich; oligotrophic, or nutrient poor; or dystrophic, acidic and rich in humic material. Most lakes are subject to cultural eutrophication, which is the rapid addition of nutrients, especially nitrogen and phosphorus, from sewage and industrial wastes. Cultural eutrophication has produced significant biological changes, mostly detrimental, in many lakes.

From its source to its mouth the lotic ecosystem exhibits changes in its character. Upland headwater streams are small, shallow, usually swift, and cold. Production, although high, is dependent upon the watershed they drain for nutrients which do not remain in place. As nutrients are carried in, they are also removed and carried downstream by the current. To exist in the fast-flowing waters, organisms may be streamlined or flattened in shape so that they can escape the current in crevices and underneath rocks or they may attach themselves in one fashion or another to the substrate.

Downstream the volume of flow, augmented by tributaries, increases; the channel becomes wider and deeper; and the speed of flow decreases. In the lowlands the flow is slow and the bottom is soft with silt and mud. Aquatic plants and animals characteristic of ponds and lakes replace the life of the swift headwaters, but there is no clear-cut boundary between the fast- and slow-water communities. Like the vegetation continuum, one gradually blends into the other, reflecting a longitudinal gradient in temperature, velocity of current, and often pH. Certain conditions may reappear along the gradient and with them come organisms adapted to those conditions. For example, an area of increased current velocity downstream of a slow-water area will be inhabited by fast-water populations.

Streams and rivers are heterotrophic, functional parts of the landscapes in which they exist and obtain their characteristics from the landscapes. In headwater streams as much as 99 percent of the energy input comes from the outside. As streams develop into slow, deep rivers, there is somewhat less dependence on outside energy sources. Most of this external energy comes from leaves, throughfall of rain, surface drainage, subsurface seepage, and material carried in from upstream. Energy sources come in three factions: coarse particulate organic matter, fine particulate organic matter, and dissolved organic matter. Particulate matter is worked up by microbes and detrital shredders and collectors.

Closely associated with lakes and streams are wetlands, areas where water is near, at, or above the level of the ground. Wetlands dominated by grasses are marshes; those dominated by woody vegetation are swamps. Wetlands characterized by an accumulation of peat are mires. Those fed by water moving through mineral soil and dominated by sedges are fens; those dominated by sphagnum moss and dependent largely on precipitation for moisture are bogs. Bogs are characterized by blocked drainage conditions, an accumulation of peat, and low productivity. Most of the nutrients fixed in plants are removed from circulation and stored in accumulating peat; the stored energy remains locked up in the peat until environmental conditions change to favor decomposition or until the material is burned.

# Marine ecosystems
## CHAPTER 8

Freshwater rivers eventually empty into the oceans, and terrestrial ecosystems end abruptly at the edge of the sea. For some distance there is a region of transition. The river enters the saline waters of the oceans, creating a gradient of salinity. That gradient provides a habitat for organisms uniquely adapted to exist in the half-world between saltwater and fresh. The inhabitants of the region between land and sea are adapted to live in often severe environments dominated by tides. Beyond all this lies the open ocean—the shallow seas overlying the continental shelf and the deep ocean.

## Characteristics of the marine environment

The marine environment occupies 70 percent of the earth's surface. The volume of surface area lighted by the sun is small in comparison to the total volume of water involved. This and the dilute solution of nutrients limit production. It is deep, in places nearly 7 kilo. All of the seas are interconnected by currents, dominated by waves, influenced by tides, and characterized by saline waters.

### SALINITY

The salinity of the open sea is fairly constant, averaging about 35 ppt (0/00). This probably has not changed greatly since the earth was formed. Although quantities of salts are carried to the sea by rivers, they are removed at about the same rate as supplied by means of complex chemical reactions with sediments and particulate matter.

Two elements, sodium and chlorine, make up some 86 percent of sea salts. These along with other such major elements as sulfur, magnesium, potassium, and calcium comprise

99 percent of sea salts. Seawater, however, differs from a simple sodium chloride solution in that the equivalent amounts of cations and anions are not balanced against each other. The cations exceed the anions by 2.38 milliequivalents. As a result, seawater is weakly alkaline (pH 8.0 to 8.3) and strongly buffered, a condition that is biologically important.

The amount of dissolved salt in seawater is usually expressed as chlorinity or salinity. Because oceans are usually well mixed, sea salt has a constant composition, that is, the relative proportions of major elements change little. Thus the determination of the most abundant element, chlorine (Table 8-1), can be used as an index of the amount of salt present in a given volume of seawater. Chlorine expressed in 0/00 is the amount of chlorine in grams in a kilogram of seawater. Chlorinity can be converted to salinity, the total amount of solid matter in grams per kilogram of seawater. The relationship of salinity to chlorinity is expressed by

$$S (0/00) = 1.80655 \times chlorinity$$

TABLE 8-1
*Composition of sea water of 35 0/00 salinity, major elements (original)*

| Elements | g/kg | Millimole/kg | Milli-equivalent/kg |
|---|---|---|---|
| **CATIONS** | | | |
| Sodium | 10.752 | 467.56 | 467.56 |
| Potassium | 0.395 | 10.10 | 10.10 |
| Magnesium | 1.295 | 53.25 | 106.50 |
| Calcium | 0.416 | 10.38 | 20.76 |
| Strontium | 0.008 | 0.09 | 0.18 |
| Total | | | 605.10 |
| | | | |
| **ANIONS** | | | |
| Chlorine | 19.345 | 545.59 | 545.59 |
| Bromine | 0.066 | 0.83 | 0.83 |
| Fluorine | 0.0013 | 0.07 | 0.07 |
| Sulphate | 2.701 | 28.12 | 56.23 |
| Bicarbonate | 0.145 | 2.38 | — |
| Boric acid | 0.027 | 0.44 | — |
| Total | | | 602.72 |

*Note:* Surplus of cations over strong anions (alkalinity) : 2.38.
*Source:* Kalle, 1971.

The salinity of parts of the ocean is variable because physical processes change the amount of water in the seas. Salinity is affected by evaporation, precipitation, movement of water masses, mixing of water masses of different salinities, formation of insoluble precipitates that sink to the ocean floor, and diffusion of one water mass to another. Salinities are most variable near the interface of sea and air.

Salinity imposes certain restrictions on life that inhabits the oceans (see Kinne, 1971). Fish and marine invertebrates that inhabit marine, estuarine, and tidal environments have to maintain osmotic pressure under conditions of changing salinities. Most marine species are adapted to living in high salinities, and the number declines as salinity is reduced.

### TEMPERATURE AND PRESSURE

What has already been written about temperature and freshwater relations also applies to the sea. The range of temperature is far less than that on land, although as one would expect, the variation in temperature over the oceans is considerable. Temperatures range from arctic waters at $-3°$ C to tropical waters at $27°$ C. Currents are warmer or colder than the waters through which they flow. Seasonal and daily temperature changes are larger in coastal waters than on the open sea. The surface of coastal waters is the coolest at dawn and the warmest at dusk. In general, seawater is never more than $2°$ to $3°$ below the freezing point of fresh water or higher than $27°$ C. At any given place the temperature of deep water is almost constant and cold, below the freezing point of fresh water. Seawater has no definite freezing point, although there is a temperature for seawater of any given salinity at which ice crystals form. Thus pure water freezes out, leaving even more saline water behind. Eventually it becomes a frozen block of mixed ice and salt crystals. With rising temperatures the process is reversed.

Unlike fresh water, seawater (with a salinity of 24.7 0/00 or higher) becomes heavier as it cools and does not reach its greatest density at $4°$ C. Thus the limitation of $4°$ C as the temperature of bottom water does not apply to the sea. In spite of this, bottom temperatures of the sea rarely go below the freezing point of fresh water, generally averaging around 2° C, even in the tropics if the water is deep enough. The temperature of the ocean floor over 1 mi deep is 3° C.

Another aspect of the marine environment is pressure. Pressure in the ocean varies from 1 atm at the surface to 1000 atm at the greatest depth. Pressure changes are many times greater in the sea than in terrestrial environments and have a pronounced effect on the distribution of life. Certain organisms are restricted to surface waters, whereas others are adapted to the pressure at great depths. Some marine organisms, such as the sperm whale and certain seals, can dive to great depths and return to the surface without difficulty.

## ZONATION AND STRATIFICATION

Just as lakes exhibit stratification and zonation, so do the seas. The ocean itself is divided into two main divisions, the *pelagic,* or whole body of water, and the *benthic,* or bottom region (Figure 8-1). The pelagic is further divided

into the *neritic province,* water that overlies the continental shelf, and the *oceanic province.* Because conditions change with depth, the pelagic is divided into three vertical layers or zones. From the surface to about 200 m is the *photic zone,* in which there are sharp gradients of illumination, temperature, and salinity. From 200 to 1000 m is the *mesopelagic zone,* where little light penetrates and where the temperature gradient is more even and gradual and without much seasonal variation. It contains an oxygen-minimum layer and often the maximum concentrations of nitrate and phosphate. Below the mesopelagic is the *bathypelagic zone,* where darkness is virtually complete except for bioluminescence, temperature is low, and pressure is great.

The upper layers of ocean water exhibit a stratification of temperature and salinity. Depths below 300 m are usually thermally stable. At high and low latitudes temperatures

*FIGURE 8-1*
*Regions of the ocean.*

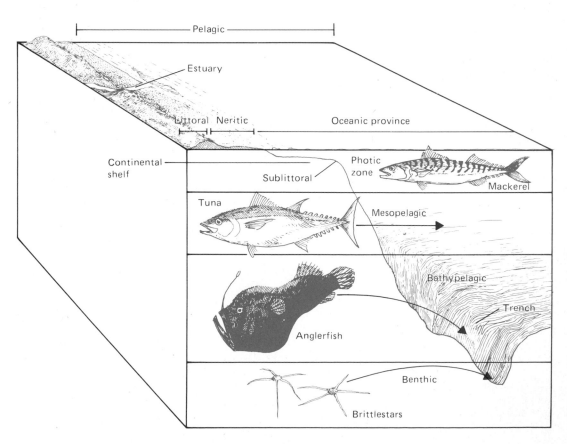

**239**

remain fairly constant throughout the year (Figure 8-2). At middle latitudes temperatures vary with the season. In summer the surface waters become warmer and lighter, forming a temporary seasonal thermocline. In subtropical regions the surface waters are constantly heated, developing a marked permanent thermocline. Between 500 and 1500 m a permanent but relatively slight thermocline exists.

Associated with a temperature gradient is a salinity gradient or *halocline,* especially at the higher latitudes (Figure 8-2). There the abundant precipitation reduces surface salinity and causes a marked change in salinity with depth. Thus, in the middle latitudes in particular, the two produce a marked gradient in density. Water masses form density layers with increasing depth. Because density of seawater does increase in depth and does not reach its greatest density at 4° C as with fresh water, there is no seasonal overturn. This results in a normally stable stratification of density known as the *pycnocline* (Figure 8-2). Because there are marked changes in temperature with depth in the open ocean and thus with density, the pycnocline often coincides with the thermocline.

Below the pycnocline is the *deep zone,* which comprises 80 percent of the ocean. Because of the contact of the deep ocean water with the cold surface waters of the high latitudes the deep ocean is cold and contains appreciable amounts of oxygen.

*FIGURE 8-2*
*Composite graph of temperature and salinity profiles of the sea. The side panel illustrates general temperature profiles of the ocean. At high latitudes colder, less saline surface waters overlie warmer, more saline waters. At the middle latitudes the ocean exhibits a seasonally varying surface temperature and a permanent thermocline. Waters of the lower latitudes possess stable surface temperatures and a permanent thermocline. (After R. V. Tait, 1968). The front panels show the relationship between temperature and salinity and the pycnocline. The back panel shows a schematic representation of the horizontal layering of ocean waters. Note the latitudinal differences. A, high latitudes; B, middle latitudes; C, low latitudes. (After M. Gross, 1972.)*

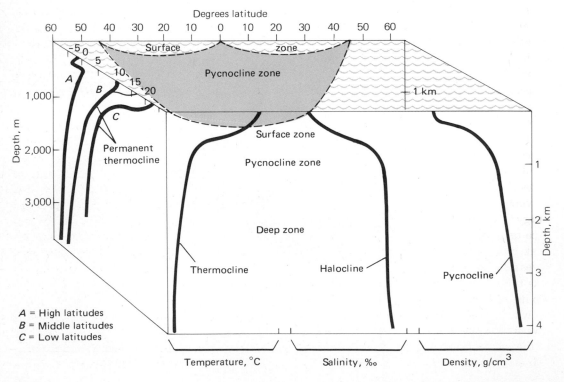

## WAVES AND CURRENTS

Waves on the surface of the sea are stirred by the wind. The frictional drag of the wind on the surface of smooth water ripples the water. As the wind continues to blow, it applies more pressure to the steep side of the ripple, and wave size begins to grow. As the wind becomes stronger, short, choppy waves of all sizes appear, and as they absorb more energy, they continue to grow in size. When the waves reach a point at which the energy supplied by the wind is equal to the energy lost by the breaking waves, they become the familiar whitecaps. Up to a certain point, the stronger the wind, the higher the waves.

Waves are generated on the open sea. The stronger the wind, the longer the fetch (the distance the waves can run without obstruction under the drive of the wind blowing in a constant direction) and the higher the waves. As the waves travel out of the fetch, or if the winds die down, sharp-crested waves change into smooth long-crested waves or swells that can travel great distances because they lose little energy as they travel. Swells are characterized by troughs and ridges, the height of which is measured from the bottom of the trough to the crest. The length of the wave is measured from the crest to the following wave and its period by the time required for successive crests to pass a fixed point. None of these is static, for all depend upon the wind and the depth of the wave.

The waves that break up on a beach are not composed of water driven in from distant seas. Each particle of water remains largely in the same place and follows an elliptical orbit with the passage of the wave form. As the wave moves forward with a velocity that corresponds to its length, the energy of a group of waves moves with a velocity only half that of individual waves. The wave at the front loses energy to the waves behind and disappears, its place taken by another. Thus the swells that break on a beach are distant descendents of waves generated far out at sea.

As the waves approach land, they advance into increasingly shallow water. The height of the waves rises higher and higher until the wave front grows too steep and topples over. As the waves break on shore they dissipate their energy against the coast, pounding rocky shores or tearing away at sandy beaches at one point and building up new beaches elsewhere.

There are internal as well as surface waves in the ocean. Similar to surface waves, internal waves appear at the interface of layers of water of different densities. In addition there are also stationary waves or seiches, already described in Chapter 7.

Just as there are internal waves, so there are internal currents in the sea. Surface currents are produced by wind, heat budgets, salinity, and the rotation of the earth. Water moving in surface currents must be replaced by a corresponding inflow from elsewhere. Because the surface waters are cooled and salinity changes, high-density water formed on the surface, largely at high latitudes, sinks and flows toward the low latitudes. These currents, subject to the Coriolis effect, are deflected or obstructed by submarine ridges and are modified by the presence of other water masses. The result is three main systems of subsurface water movements—the bottom, the deep, and the intermediate ocean currents—each of which runs counter to the other.

Also influencing subsurface currents is the Eckman spiral. As the wind sets the surface layer of water in motion, that layer in turn sets in motion a layer of water beneath, and that layer in turn another. Each layer moves more slowly than the layer above and is deflected to the right of it because of the Coriolis effect. At the base of the spiral the movement of water is counter to the flow on the surface, although the average flow of the spiral is at right angles to the wind.

In coastal regions the Eckman transport of surface water can bring deep waters up to the surface, a process called upwelling. Wind blowing parallel to a coast causes surface water to be blown offshore. This is replaced by water moving upward from the deep. Although cold and containing less dissolved oxygen, upwelling water is rich in nutrients that support a rich growth of phytoplankton. For this reason regions of upwelling are highly productive.

## TIDES

One of the fundamental laws of physics is Newton's law of universal gravitation. The law states that every particle of matter in the universe attracts every other particle with a force that varies directly as the product of their masses and inversely as the square of the distance between them. The gravitational pull of the sun and the moon each cause two bulges in the waters of the oceans. The moon, being much closer to the earth than the sun, exerts a tidal force twice as great as the sun's. The two bulges caused by the moon occur at the same time on opposite sides of the earth on an imaginary line extending from the moon through the center of the earth. The tidal bulge on the moon side is due to gravitational attraction; the bulge on the opposite side occurs because the gravitational force there is less than at the center of the earth. As the earth rotates eastward on its axis, the tides advance westward. Thus any given place on the earth will in the course of one daily rotation pass through two of the lunar tidal bulges, or high tides, and two of the lows or low tides, at right angles to the high. Because the moon revolves in a 29.5-day orbit around the earth, the average period between successive high tides is approximately 12 hours, 25 minutes.

The sun also causes two tides on opposite sides of the earth, but because they are less than half as high as the lunar tides, solar tides are usually partially masked by the lunar tides. Twice during the month, when the moon is full and when it is new, the earth, moon, and sun are nearly in line and the gravitational pulls of the sun and the moon are additive. The intensified force causes what are known as the spring tides, tides of the maximum rise and fall. When the moon is at either quarter, the gravitational pulls of the sun and moon interfere with each other, creating the neap tides of minimum difference between high and low tides.

Tides are not entirely regular nor are they the same all over the earth. They vary from day to day in the same place, following the waxing and the waning of the moon. They may act differently in several localities within the same general area. In the Atlantic semidaily tides are the rule. In the Gulf of Mexico the alternate highs and lows more or less efface each other, and flood and ebb follow one another at about 24 hour intervals to produce one daily tide. Mixed tides, combinations of semidaily and daily tides, are common in the Pacific and Indian oceans, with different combinations at different places.

Local inconsistencies of tides are due to many variables. The elliptical orbits of the earth about the sun and of the moon about the earth influence the gravitational pull, as does the declination of the moon—the angle of the moon in relation to the axis of the earth. Latitude, barometric pressure, offshore and onshore winds, depth of water, contour of shore, and internal waves modify tidal movements.

## The seashore

Where the land and sea meet there exists the fascinating and complex world of the seashore. Rocky, sandy, muddy, protected, or exposed to the pounding of incoming swells, all shores have one feature in common: they are alternatingly exposed and submerged by the tides. Roughly, the region of the seashore is bounded by the extreme high-water mark and the extreme low-water mark. Within these confines, conditions change from hour to hour with the ebb and flow of the tides. At flood tide the seashore is a water world; at ebb tide it belongs to the terrestrial environment, with its extremes in temperature, moisture, and solar radiation. In spite of all this, the seashore inhabitants are essentially marine, adapted to withstand some degree of exposure to the air for varying periods of time.

### STRUCTURE

*Rocky shores.* As the sea recedes at ebb tide, life hidden by tidal water emerges, layer by layer. The uppermost layers, those near the high-water mark, are exposed to air, wide temperature fluctuations, intense solar radiation, and desiccation for a considerable period of time, while the lowest fringes on the intertidal shore may be exposed only briefly before the flood tide submerges them again. These varying conditions result in one of the most striking features of the rocky shore, the zonation of life (Figure 8-3). Although this zonation may be strikingly different from place to place as a result of local variations in aspect, nature of the substrate, wave action, light intensity,

FIGURE 8-3
*The broad zones of life exposed at low tide on the rocky shore of the Bay of Fundy. Note the heavy growth of bladder wrack or rockweed on the lower portion, the white zone of barnacles above.*

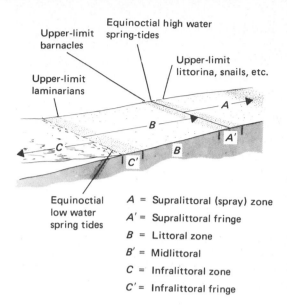

FIGURE 8-4
*Diagram of the basic zonation on a rocky shore. Use this as a guide when studying the subsequent drawings of zonation and the discussion in the text. A, supralittoral (spray) zone; A', supralittoral fringe; B, littoral zone; B', midlittoral; C, infralittoral zone; C', infralittoral fringe. (Adapted from Stephenson, 1949.)*

shore profile, exposure to prevailing winds, climatic differences, and the like, it possesses everywhere the same general features. All rocky shores have three basic zones, characterized by the dominant organisms occupying them (Figure 8-4).

Where the land ends and seashore begins is difficult to fix. The approach to a rocky shore from the landward side is marked by a gradual transition from lichens and other land plants to marine life, dependent in part at least on the tidal waters. The first major change from land shows up on the *supralittoral fringe* (Figure 8-5), where the salt water comes only every fortnight on the spring tides. It is marked by the black zone, a patchy or beltlike encrustation of Verrucaria-type lichens and Myxophyceae algae such as *Calothrix* and *Entrophsalis*. Capable of existing under conditions so difficult that few other plants could survive, these blue-green algae, enclosed in slimy, gelatinous sheaths, and their associated lichens represent an essentially nonmarine community, on which graze basically marine animals, the periwinkles (Doty, in

Hedgpeth, 1957). Common to this black zone is the rough periwinkle that grazes on the wet algae covering the rocks. On European shores lives a similarly adapted species, the rock periwinkle, the most highly resistant to desiccation of all the shore animals.

Below the black zone lies the *littoral zone,* a region covered and uncovered daily by the tides. It is universally characterized by barnacles, although often they are hidden under a dense growth of fucoid seaweeds or kelp (a brown alga, Phaeophyceae) and in the more northern reaches of the North American and European coasts by the red algae (Rhodophyceae).

The littoral tends to be divided into subzones. In the upper reaches the barnacles are most abundant. The oyster, blue mussel, and limpet appear in the middle and lower portions of the littoral, as does the common periwinkle.

Occupying the lower half of the littoral zone (midlittoral) of colder climates and in places

**243**

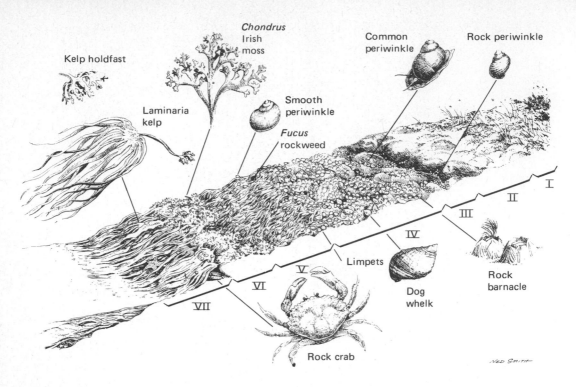

**FIGURE 8-5**
*Zonation on a rocky shore along the North Atlantic. Compare with the generalized diagram in Figure 8-4. I, land: lichens, herbs, grasses, etc.; II, bare rock; III, zone of black algae and rock periwinkles; IV, barnacle zone: barnacles, dog whelks, common periwinkles, mussels; V, fucoid zone: rockweed and smooth periwinkles; VI,* Chondrus *zone: Irish moss; VII, laminarian zone: kelp. (Zonation drawings based on data from Stephenson and Stephenson, 1954, and author's photographs and observations; all sketches drawn from life or specimens.)*

coast forms only a very narrow band, if present at all, for it is replaced by the more abundant bladder rockweed and, in sheltered waters only, the knotted rockweed.

The lower reaches of the littoral zone may be occupied by blue mussels (Figure 8-6) instead of rockweeds. This is particularly true on shores where the hard surfaces have been cov-

**FIGURE 8-6**
*A muddy tidal flat covered with blue mussels.*

overlying the barnacles is an ancient group of plants, the brown algae, more commonly known as rockweeds, or wrack *(Fucus)*. Rockweeds attain their finest growth on protected shores, where they may grow 2 m long; on wave-whipped shores they are considerably shorter. The rockweeds that live furthest up on the tidelands are channeled rockweeds, or wrack, a species found on the European but not the American shore. It is replaced in America by another, the spiral rockweed, a low-growing, orange-brown alga, whose short, heavy fronds end in turgid, roundish swellings. The spiral rockweed on the northeastern

ered in part by sand and mud. No other shore animal grows in such abundance; the blue-black shells packed closely together may blanket the area.

Near the lower reaches of the littoral zone, mussels may grow in association with a red alga, *Gigartina,* a low-growing, carpetlike plant. The algae and the mussels together often form a tight mat over the rocks. Here, too, grows another seaweed, Irish moss, which grows in carpets some 15 cm deep. Its color is variable, ranging from purple to yellow and green; its fronds are branched and tough; and when covered by water the plant has an iridescent sheen. Here, well protected in the dense growth from the waves, live infant starfish, sea urchins, brittle stars, and the bryozoan sea mats or sea lace *(Membranipora).*

The lowest part of the littoral zone, uncovered only at the spring tides and not even then if the wave action is strong, is the *infralittoral fringe.* This zone, exposed for relatively short periods of time, consists of forests of the large brown alga, *Laminaria,* one of the kelps, with a rich undergrowth of smaller plants and animals among the holdfasts.

Beyond the infralittoral fringe is the *sublittoral zone,* the open sea. This zone is principally neretic and benthic and contains a wide variety of fauna, depending upon the substrate, the presence of protruding rocks, gradients in turbulence, oxygen tensions, light, and temperature. This offshore zone is extremely interesting to explore by skin diving. Appropriately, this area is becoming known to marine biologists as the scuba zone.

*Tide pools.* The ebbing tide leaves behind pools of water in rock crevices, in rocky basins, and in depressions (Figure 8-7). These are tide pools, "microcosms of the sea," as Yonge (1949) describes them. They represent distinct habitats, which differ considerably from the exposed rock and the open sea, and even differ among themselves. At low tide all the pools are subject to wide and sudden fluctuations in temperature and salinity, but these changes are most marked in shallow pools. Under the summer sun the temperature may rise above the maximum many organisms can tolerate. As the water evaporates, especially in the smaller and more shallow pools, salinity increases and salt crystals may appear around

**FIGURE 8-7**
*Tidal pools, a rock pool nestled in a canyon on a rocky shore (top) and tiny rock pool on the ledge of a large rock (bottom). Note the rockweed and barnacles and the transparency of the water.*

the edges. When rain or land drainage brings fresh water to the pool, salinity may decrease. In deep pools such fresh water tends to form a layer on the top, developing a strong salinity stratification in which the bottom layer and its inhabitants are little affected. If algal growth is considerable, oxygen content of the water varies through the day. Oxygen will be high during the daylight hours but will be low at night, a situation that rarely occurs at sea. The rise of carbon dioxide at night means a lowering of pH.

Obviously pools near low tide are influenced least by the rise and fall of the tides; those that lie near and above the high-tide line are exposed the longest and undergo the widest fluctuations. Some may be recharged with seawater only by the splash from breaking waves or occasional high spring tides. Regardless of their position on the shore, most pools suddenly return to sea conditions on the rising tide and experience drastic and instantaneous changes in temperature, salinity, and pH. Life in the tidal pools must be able to withstand wide and rapid fluctuations in their environment.

*Sandy shores and mudflats.* Both the sandy shore and the mudflat at low tide appear barren of life, a sharp contrast to the life-studded rocky shores. But beneath the wet and glistening surface life exists, waiting for the next high tide.

Both in some ways are harsh environments, but the sandy shore is especially so. The very matrix of this seaside environment is a product of the harsh and relentless weathering of rock, both inland and along the shore. Through eons the ultimate products of rock weathering are carried away by rivers and waves to be deposited as sand along the edge of the sea. The size of the sand particles deposited influences the nature of the sandy beach, water retention during low tide, and the ability of animals to burrow through it. Beaches with relatively steep slopes usually are made up of larger sand grains and are subject to more wave action. Beaches exposed to high waves are generally flattened, for much of the material is transported away from the beach to deeper water, and fine sand is left behind (Figure 8-8). Sand grains of all sizes, especially the finer particles in which capillary action is

**FIGURE 8-8**
*A long stretch of sandy beach pounded by the surf. Although the beach appears barren, life is abundant beneath the sand.*

greatest, are more or less cushioned by a film of water about them, reducing further wearing away. The retention of water by the sand at low tide is one of the outstanding environmental features of the sandy shore.

Existence of life on the surface of the sand is almost impossible. It provides no surface for attachments of seaweeds and their associated fauna; and the crabs, worms, and snails so characteristic of rock crevices find no protection here. Life then is forced to exist beneath the sand.

Life on the sandy beach does not experience the same violent fluctuations in temperature as that on the rocky shores. Although the surface temperature of the sand at midday may be 10° C or more higher than the returning seawater, the temperature a few inches below remains almost constant throughout the year. Nor is there a violent fluctuation in salinity, even when fresh water runs over the surface of the sand. Below 25 cm, salinity is little affected.

### FUNCTION

The rocky shore is both autotrophic and heterotrophic. Many organisms, such as barnacles, depend upon the tides to bring them food; others, such as periwinkles, graze on the algal growth or primary production on the rocks. The sandy shore and mudflats, representative of the functioning of coastal intertidal ecosystems, are strongly heterotrophic.

For life to exist on the sandy shore, some organic matter has to accumulate. Most sandy beaches contain a certain amount of detritus from seaweeds, dead animals, feces, and material blown in from shore. This organic matter accumulates within the sand, especially in sheltered areas. In fact an inverse relationship exists between the turbulence of the water and the amount of organic matter on the beach, with accumulation reaching its maximum on the mudflats. Organic matter clogs the space between grains of sand and binds them together. As water moves down through the sand it loses oxygen both from the respiration of bacteria and oxidation of chemical substances, especially ferrous compounds. The point within the mud or sand at which water loses all its oxygen is a region of stagnation and oxygen deficiency, characterized by the formation of ferrous sulfides. The iron sulfides cause a zone of black whose depth varies with the exposure of the beach. On mudflats such conditions exist almost to the surface.

Life on sandy and muddy beaches consists of the *epifauna,* organisms living on the surface, and the *infauna,* organisms living within the substrate. Most infauna either occupy permanent or semipermanent tubes within the sand or mud or are able to burrow rapidly into the substrate. Multicellular infauna obtain oxygen either by gaseous exchange with the water through their outer covering or by breathing through gills and elaborate respiratory siphons.

The energy base for sandy beach and mudflat fauna is organic matter. Much of it becomes available through bacteria decomposition which goes on at the greatest rate at low tide. The bacteria are concentrated around the organic matter in the sand where they escape the diluting effects of water. The products of decomposition are dissolved and washed into the sea by each high tide which, in turn, brings in more organic matter for decomposition. Thus the sandy beach is an important site for biogeochemical cycling, supplying offshore waters with phosphates, nitrogen, and other nutrients.

At this point energy flow in sandy beaches and mudflats differs from that in terrestrial and aquatic systems because the basic consumers are bacteria. In other energy flow systems bacteria act largely as reducers responsible for conversion of dead organic matter into a form that can be utilized by producer organisms. In sandy beaches and mudflats bac-

teria not only feed on detrital material and break down organic matter, but they are also a major source of food for higher level consumers.

A number of deposit-feeding organisms ingest organic matter largely as a means of obtaining bacteria. Prominent among these on the mudflats are numerous nematodes and copepods (Harpacticoida), the polychaete clam worm *(Nereis),* and the gastropod mollusk *(Hydrobia).* Deposit-feeders on sandy beaches obtain their food by actively burrowing through the sand and ingesting the substrate to obtain the organic matter it contains. The most common among these is the lugworm *(Arenicola)* which is responsible for the conspicuous coiled and cone-shaped casts on the beach.

Other sandy beach animals are filter feeders obtaining their food by sorting particles of organic matter from tidal water. Two of these "surf fishers" who advance and retreat up and down the beach with the flow and ebb of tide are the mole crab and coquina clam.

Within the beach sand and mud live vast numbers of meiofauna, species 50 to 500 mm in size, including copepods, ostracods, nematodes, and gastrotrichs, all making up the interstitial life. Interstitial fauna are generally elongated forms with setae, spines, or tubercles greatly reduced. The great majority do not have pelagic larval stages. These animals feed mostly on algae, bacteria, and detritus. Interstitial life, best developed on the more sheltered beaches, shows seasonal variations, reaching maximum development in summer months.

Associated with these essentially herbivorous animals are the predators, always present whether the tide is in or out. Near and below the low-tide line live predatory gastropods, which prey on bivalves beneath the sand. In the same area lurk predatory portunid crabs such as the blue crab and green crab, which feed on mole crabs, clams, and other organisms. They move back and forth with the tides. The incoming tides also bring other predators such as killifish and silversides. As the tides recede, gulls and shorebirds scurry across the sand and mudflats to hunt for food.

Zonation of animal life exists on the beach but it is hidden and must be discovered by digging (Figure 8-9). Sand and mud shores can be divided roughly into supralittoral, littoral, and infralittoral zones based on animal organisms, but a universal pattern similar to that of the rocky shore is lacking. Pale, sand-colored ghost crabs and beach hoppers occupy the upper beach, the supralittoral. The intertidal beach, the littoral, is the zone where true marine life appears. Although it lacks the variety found in intertidal rocky shores, the populations of individual species of largely burrowing animals often are enormous. An array of animals, among them the starfish and related sand dollar, can be found above the low-tide line and in the infralittoral.

Because of their dependence upon organic matter imported to them and their essentially heterotrophic nature, the sandy beaches and

mudflats perhaps cannot be considered ecosystems, without including them as a part of the whole coastal ecosystem (Figure 8-10). Except in the cleanest of sands, some primary production does take place in the intertidal zone. The major primary producers are the diatoms confined mainly to fine grain deposits of sand containing a high proportion of organic matter. Productivity is low; one estimate places productivity of moderately exposed sandy beaches at 5 g $C/m^2/yr$. Production may be temporarily increased by phytoplankton carried in on high tide and left stranded on the surface. Again these are mostly diatoms. More important as producers are the sulfur bacteria in the black sulphide or reducing layer. These are chemosynthetic bacteria that use energy released as ferrous oxide is reduced. Some mudflats may be covered with algae *Enteromorpha* and *Ulva* whose productivity can be substantial.

In effect the shore and mudflats are part of a larger coastal ecosystem involving the salt marsh, the estuary, and coastal waters. They act as sinks for energy and nutrients because the energy they utilize comes not from primary production, but from organic matter that originates outside of the area. Many of the

**FIGURE 8-9**
*Life on a sandy ocean beach along the Atlantic coast. Although strong zonation is absent, organisms still change on a gradient from land to sea. I, supratidal zone: ghost crabs and sandfleas; II, flat beach zone: ghost shrimp, bristle worms, clams; III, intratidal zone: clams, lugworms, mole crabs; IV, subtidal zone. The dashed line indicates high tide.*

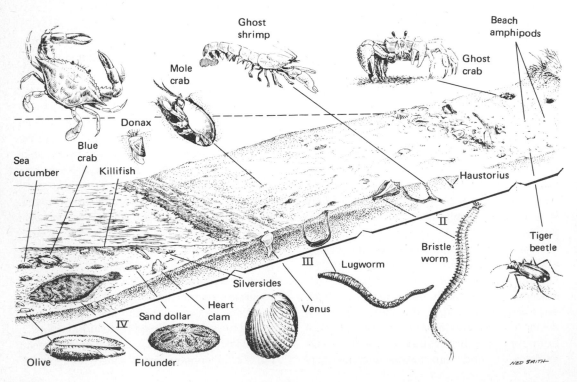

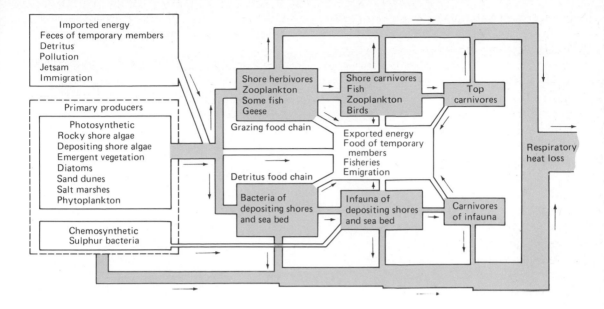

*FIGURE 8-10*
*Diagram of the coastal ecosystem. The coastal ecosystem is in effect a superecosystem consisting of the shore, the fringing terrestrial regions, and the sublittoral zones. It involves two food chains: the grazing food chain of rocky shore herbivores and zooplankton and the detrital food chain of bacteria of the depositing shore and sublittoral muds and the dependent detritivores and carnivores. Coastal ecosystems are extremely productive; because energy import exceeds export, the system is continuously gaining energy. (From S. K. Eltringham, 1971.)*

nutrient cycles are only partially contained within the borders of the shores.

## Tidal marshes

On the alluvial plains about the estuary and in the shelter of the spits and offshore bars and islands exists a unique community, the tidal marsh (Figure 8-11). Although to the eye tidal marshes appear as waving acres of grass, they are a complex of distinctive and clearly demarked plant associations. The nature of the complex is determined by tides and salinity. The tides perhaps play the most significant role in plant segregation, for twice a day the saltmarsh plants on the outermost tidal flats are submerged in salty water and then ex-

posed to the full insolation of the sun. Their roots extend into poorly drained, poorly aerated soil, in which the soil solution contains varying concentrations of salt. Only plant species with a wide range of salt tolerance can survive such conditions (Figure 8-12). Thus from the edge of the sea to the highlands, zones of vegetation, each recognizable by its own distinctive color, develop.

### STRUCTURE

*Plants.* Tidal salt marshes begin in most cases as sands or mudflats, first colonized by algae and, if the water is deep enough, by eelgrass. As organic debris and sediments accumulate, eelgrass is replaced by the first salt marsh colonists—the sea poa *(Puccinellia)* on the European coast and saltwater cordgrass *(Spartina alterniflora)* on the eastern coast of North America. Stiff, leafy, up to 3 m tall, and submerged in saltwater at every high tide, saltwater cordgrass forms a marginal strip between the open mudflat to the front and the higher grassland behind. As a wet grassland, the tall form of *Spartina alterniflora* is unique. No litter accumulates in the stand. Strong tidal currents sweep the floor of the *Spartina alterniflora* clean, leaving only black, thick mud.

As fine organic debris carried in and deposited by tides is buried by further deposition on top, an anaerobic environment is created.

**249**

*FIGURE 8-11*
*View of a tidal marsh on the Virginia coast showing the pattern of salt marsh vegetation. The mosiac is influenced by both water depth and salinity. The three dominant grasses are the coarse-leaved salt marsh cord grass* Spartina alterniflora, *the fine-leafed* Spartina patens, *salt marsh hay grass, and spike grass* Distichlis spicata.

Here bacteria and nematodes live on organic matter, utilizing it by parallel oxidizing and reducing reactions (Teal and Kanwisher, 1961). This results in the accumulation of such end products as methane, hydrogen sulfide, and ferrous compounds. Increasing degrees of reduction suppress biological activity. In fact, if the bacteria of the mud are supplied with oxygen, their rate of energy degradation increases 25 times. Thus the tidal marsh is a vertically stratified system in which free oxygen is abundant in the surface and absent in the mud. Between these two extremes is a zone of diffusion and mixing of oxygen.

*Spartina alterniflora* is well adapted to grow on the intertidal flats of which it has sole possession. It has a high tolerance for saltwater and is able to live in a semisubmerged state. It can live in a saline environment by selectively concentrating sodium chloride in its cells at a level higher than the surrounding salt water, thus maintaining its osmotic integrity. To rid itself of excessive salts, *Spartina alterniflora* has special salt-secreting cells in the leaves.

Water excreted with the salt evaporates, leaving behind sparkling crystals on the surface of leaves to be washed off by tidal water. To get air to its roots buried in anaerobic mud, *Spartina alterniflora* has hollow tubes leading from the leaf to the root through which oxygen diffuses.

Tall *Spartina alterniflora* occupies that portion of the tidal marsh between mean low water and mean high water (Figure 8-13). At the level of mean high water tall *Spartina alterniflora* rather sharply gives way to plants of the high marsh. One of the dominant plants of the lower reaches of the high marsh and the upper end of the intertidal zone is again *Spartina alterniflora,* but in the short form. Short *Spartina alterniflora* is yellowish, almost chloritic in appearance, contrasting sharply with the dark green tall form. Growing with the short *Spartina* are the fleshy, translucent glassworts (*Salicornia* spp.) that turn bright red in fall, sea lavender *(Limonium carolinianum),* spearscale *(Atriplex patula),* and sea blite *(Suaeda maritima).*

*FIGURE 8-12*
*Approximate ranges of species dominance among fresh-water and salt marsh plants in relation to water depth and salinity. (Adapted from U.S. Department of Agriculture, 1955.)*

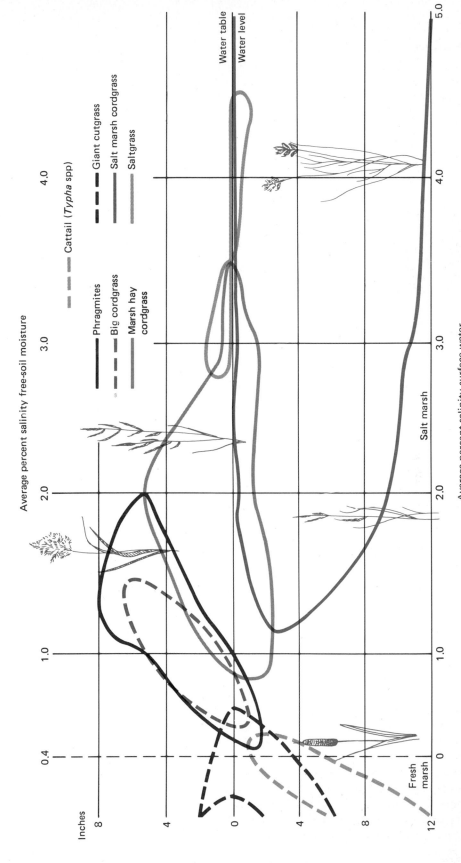

Average percent salinity free-soil moisture

4.0        3.0        2.0        1.0        0.4

Phragmites          Giant cutgrass
Big cordgrass       Salt marsh cordgrass
Marsh hay           Saltgrass
cordgrass
                    Cattail (*Typha* spp)

Water table
Water level

Salt marsh

Fresh
marsh

Inches    8    4    0    4    8    12

Average percent salinity surface water

5.0    4.0    3.0    2.0    1.0    0

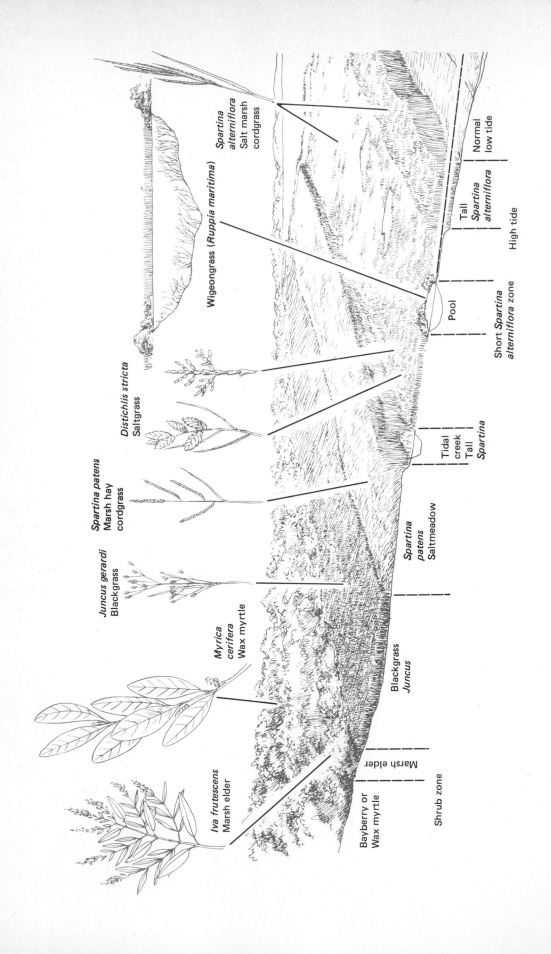

*Spartina alterniflora*
Salt marsh cordgrass

Wigeongrass (*Ruppia maritima*)

*Distichlis stricta*
Saltgrass

*Spartina patens*
Marsh hay cordgrass

*Juncus gerardi*
Blackgrass

*Myrica cerifera*
Wax myrtle

*Iva frutescens* Marsh elder

Normal
low tide

Tall
*Spartina
alterniflora*

High tide

Pool

Short *Spartina
alterniflora* zone

Tidal
creek
Tall
*Spartina*

*Spartina patens*
Saltmeadow

Blackgrass
*Juncus*

Marsh elder

Bayberry or
Wax myrtle

Shrub zone

FIGURE 8-13
*Stylized transect across part of the salt marsh shown in Figure 8-11 showing the relationship of plant distribution to microrelief and tidal submergence.*

Short *Spartina alterniflora* usually occurs on the lower, less well-drained high marsh. Where the microelevation is about 5 cm above the mean high water, it is replaced by a relative, marsh hay cordgrass *(Spartina patens),* and an associate, spike grass *(Distichlis spicata). Spartina patens* is a fine, small grass that grows so densely and forms such a tight mat that few other plants can grow with it. Dead growth of the previous year lies beneath current growth, shielding the ground from the sun and keeping it moist. Where the soil tends to have a higher salinity or to be waterlogged *Spartina patens* is replaced or shares the site with *Distichlis spicata.*

As the microelevation rises several more inches above mean high tide and if there is some intrusion of fresh water, *Spartina* and *Distichlis* may be replaced by two species of black grass *(Juncus roemerianus* and *Juncus gerardii),* so called because its very dark green color becomes almost black in the fall. Rarely are the rushes covered by ordinary high tides, but often they are submerged by the neap tides of spring and fall.

Beyond the black grass and often replacing it is shrubby growth of marsh elder *(Iva frutescens)* and groundsel *(Baccharis halimifolia).* These shrubs tend to invade the high marsh where a slight rise in microelevation exists, but such invasions are often short-lived as storm tides sweep in and kill the plants.

On the upland fringe grow bayberry *(Myrica pensylvanica)* and the pink flowered sea hollyhock *(Hibiscus palustris).* Where the water is fresh to brackish one finds the reed *(Phragmites communis),* spikerush *(Eleocharis* spp.), three-square bullrush *(Scirpus americanus),* and narrowleaf cattail *(Typha latifolia).*

Two conspicuous physiographic features of the salt marshes are the meandering creeks and the pond holes, called pannes or salt pans. The creeks are the drainage channels that carry the tidal waters back out to sea. The formation of these creeks is a complex process. In some cases the channels are formed by water deflected by minor irregularities on the surface. In estuarine marshes the river itself forms the main channel. Once formed, the channels are deepened by scouring and heightened by a steady accumulation of organic matter and silt. At the same time the heads of the creeks erode backward and small branch creeks develop. Where lateral erosion and undercutting take place, the banks may cave in, blocking or overgrowing the smaller channels. The distribution and pattern of the creek system plays an important role in the drainage of the surface water and the drainage and movement of the water in the subsoil.

Across the tidal marsh are many circular to elliptical depressions. At high tide they are flooded; at low tide the depressions remain filled with salt water. If shallow enough the water may evaporate completely, leaving an accumulating concentration of salt on the mud.

These pannes come about in several ways. Many of them are formed as the marsh develops. Early plant colonization is irregular and bare spots on the flat become surrounded by vegetation. As the level of the vegetated marsh rises, the bare spots lose their water outlet. If such a panne eventually becomes attached to a creek, normal drainage is restored and the panne eventually becomes vegetated. Other pannes are derived from creeks. Marsh vegetation may grow across the creek bottom and dam a portion of it, or lateral erosion may block the channel. With drainage no longer effective water remains behind after flood tide, inhibiting the growth of plants. Often a series of such pannes may form on the upper reaches of a single creek. Still another type is the rotten-spot panne, caused by the death of small patches of vegetation from one cause or another, such as inadequate drainage or a concentration of salt.

Pannes support a distinctive vegetation which varies with the depth of the water and salt concentration. Pools with a firm bottom and sufficient depth to retain tidal water support dense growths of wigeongrass *(Ruppia maritima),* with long threadlike leaves and small, black, triangular seeds relished by waterfowl. The pools are usually surrounded by those forbs such as sea lavender that add so much color to the marsh.

Shallow depressions in which water evaporates are covered with a heavy algal crust and crystallized salt. The edges of these "salt flats" may be invaded by *Salicornia, Distichlis,* or even *Spartina alterniflora.*

The exposed banks of tidal creeks that braid through the salt marshes support a dense population of mud algae, the diatoms and dinoflagellates, photosynthetically active all year. Photosynthesis is highest in summer during high tides and in winter during low tides when the sun warms the sediments (Pomeroy, 1959). Some of the algae are washed out at ebb tide and become part of the estuarine plankton available to such filter feeders as oysters. Thus the salt marsh functions both as a source of food and of fertilizer for the estuary.

The salt marsh described is typical of the North American Atlantic Coast, but many variations exist locally and latitudinally. Actually two groups of salt marshes occur in North America, those characteristic of the Atlantic and Gulf coasts and those characteristic of the Pacific Coast. Some salt marshes on the western coast of Europe and in Britain are similar to those of the northeastern coast of North America, but great differences exist among them (see Chapman, 1977). Tidal marshes of Europe and the east coast of North America form on a gently sloping continental coastal shelf. In western North America tidal marshes are confined to narrow river mouths, because the rivers flow directly onto a steep continental slope.

Although the species differ, *Spartina* is the dominant plant of salt marshes. From New England to New Jersey salt marshes exhibit a rather clear-cut zonation with much of the area in high marsh, growing to *Spartina patens* and associated plants. From New Jersey to North Carolina tidal marshes exhibit less distinctive zonation and much of the high marsh is dominated by the short *Spartina alterniflora.* From North Carolina south the salt marsh reaches its best development on the heavy silt deposits with large expanses of tall *Spartina alterniflora* (for a detailed review, see Cooper, 1974). On the west coast tidal land between mean sea level and mean high water is dominated by *Spartina foliosa* and the land between mean high water and the highest high tide is almost completely occupied by the perennial glasswort, *Salicornia virginica.*

**Consumers.** Animal life of the marsh, if not noted for its diversity, is certainly outstanding for its interest. Some of the inhabitants are permanent residents in sand or mud, others are seasonal visitors, and most are transients coming to feed at high or low tide.

Three dominant animals of the tall *Spartina alterniflora* are ribbed mussel *(Modiolus demissus),* fiddler crab *(Uca pugilator* and *Uca pugnax),* and marsh periwinkle *(Littorina* spp.). The marsh periwinkle, related to the periwinkles of the rocky shore, moves up and down the stems of *Spartina* and onto the mud as the tidal cycle changes. At low tide the periwinkle moves down the lower stems of *Spartina* and onto the mud to feed on algae and detritus.

Buried halfway in the mud is the ribbed mussel. At low tide the mussel is closed; at high tide the mussel opens to filter particles from the water, accepting some and rejecting others in a mucous ribbon known as pseudofeces.

Running across the marsh at low tide like a vast herd of tiny cattle are fiddler crabs. They earn their scientific name from their aggressive behavior and their common name from the highly developed single claw on the male. Among marsh animals fiddler crabs are the most adaptable. They have both gills and lungs. They can endure periods of high tides and cold winters without oxygen. They have a salt and water control system that enables them to move from diluted sea water to briny pools. The crabs are omnivorous feeders consuming animal remains, plant remains, algae, and small animals. Fiddler crabs live in burrows, marked by mounds of freshly dug, marble-sized pellets. The burrowing activity of the crabs is similar to that of the earthworms because in overturning the mud they bring nutrients to the surface.

Prominent about the base of *Spartina* stalks and under debris are sandhoppers. These detrital-feeding amphipods may be very abundant and are important in the diet of some of the marshland birds.

Two conspicuous vertebrate residents of the intertidal marsh of eastern North America are the diamond-backed terrapin *(Malaclemys terrapin)* and the clapper rail *(Rallus longiros-*

*tris).* The diamond-backed terrapin, which hibernates in the marsh mud, is carnivorous, feeding on fiddler crabs, small mollusks, marine worms, and dead fish. The rail finds its diet of fiddler crabs and sandhoppers along the marsh banks at ebb tide and the tall grass. It builds its nest in *Spartina alterniflora,* keeping it just above the level of high tide. Less conspicuous is the seaside sparrow *(Ammospiza maritima)* which, like the clapper rail, builds its nest just above the normal summer high-tide mark and feeds on the sandhoppers and other small invertebrates. It defends its territory by singing a short, buzzy song from isolated marsh elder bushes.

On the high marsh animal life changes almost as suddenly as the vegetation. The pulmonate marsh snail *Melampus* replaces the marsh periwinkle of the low marsh. At low tide the small, coffee-bean colored *Melampus* may be found by the thousands under the low grass where the humidity is high. Before high tide *Melampus* moves up the fine stalks of *Spartina patens.* Within the matted growth of *Spartina patens* can be found a maze of runways made by another high marsh inhabitant, the meadow mouse *(Microtus),* which feeds heavily on the grass. Replacing the clapper rail and the seaside sparrow on the high marsh are the willet *(Catoptrophorus semipalmatus)* and the seaside sharp-tailed sparrow *(Ammospiza caudacuta).*

Along the shrubby fringes of the marsh dense growths of marsh elder and groundsel give nesting cover for blackbirds and provide sites for heron rookeries. Remote stands of these shrubs support the nests of smaller herons and egrets, while the tall dead pines and man-made structures support the nests of the fish-eating osprey.

Low tide brings a host of predaceous animals onto the marsh to feed. Herons, egrets, gulls, terns, willets, ibis, raccoons, and others spread over the exposed marsh floor and the muddy banks of tidal creeks to feed. At high tide the food web changes as tide waters flood the marsh. Such fish as the killifish *(Fundulus heteroclitus),* silversides *(Menidia menidia),* and four-spined stickleback *(Apeltes quadracus),* restricted to channel waters at low tide, spread over the marsh at high tide, as does the blue crab *(Callinectes sapidus).* At high tide, too, ribbed mussels begin to strain the water for detrital material.

The two best known inhabitants of the salt marsh are the pesty salt marsh mosquito *(Aedes sollicitans)* and the biting greenhead fly *(Tabinus* spp.). Both are true marsh dwellers. Greenhead eggs are laid on the stems of *Spartina alterniflora* about a foot above the mud. Upon hatching, the larvae move down into the salty mud where they spend one to two years searching for any kind of prey including larvae of their own kind. Adult greenhead females seek their nourishment from the blood of any vertebrates they can attack. The males, lacking mandibles, feed on plant secretions. Both sexes live only several weeks, a sufficient time for mating and laying eggs. Salt marsh mosquitoes are most abundant in summer and fall. The female deposits eggs on damp mud where they begin to develop at once. At summer temperatures the larvae are ready to hatch in two days, requiring only the stimulus of water to emerge. When extra high tide or rains submerge the eggs, they hatch at once. The larvae exist in these small pools and develop into adult mosquitoes. The females, like the female greenhead fly, seek out prey across the marsh, sparing few vertebrates.

### FUNCTION

Two studies of salt marsh ecosystems, one done in Georgia by Teal (1962) and the other done in New England by Nixon and Oviatt (1973), provide contrasting insights on the functioning of salt marshes.

Teal confined his study to that area of the salt marsh of which 42 percent was in short *Spartina alterniflora* and 58 percent was in tall *Spartina alterniflora.* Average annual gross production of the marsh averaged about 34,600 kcal/m²/yr, resulting in an efficiency of about 6.1 percent (Figure 8-14). But respiration was high, averaging 28,000 kcal/m²/yr, over 75 percent of gross production. This reduced net production to 1.4 percent of incoming solar radiation. Average net production for *Spartina alterniflora* was 6,580 kcal/m²/yr (1,600 g/m²/yr), about 19 percent of gross production. Of this tall *Spartina* contributed 8,470 kcal/m²/yr and short *Spartina* 2,570 kcal/m²/yr. In addition to *Spartina,* mud algae contributed 1,620 kcal/m²/yr. Gross production

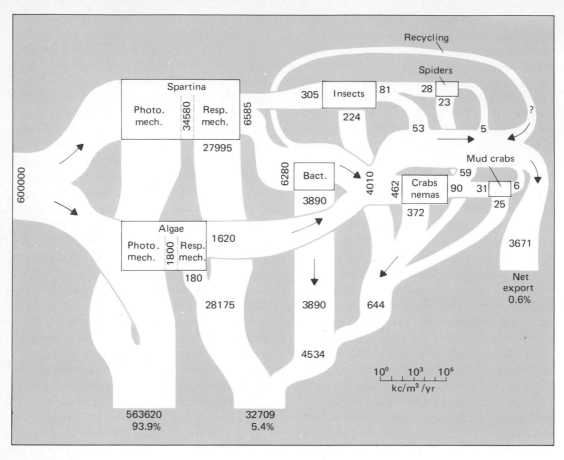

**FIGURE 8-14**
*Energy flow through a Georgia salt marsh.*
*Note the importance of the detrital food*
*chain. (From Teal, 1962.)*

amounted to 1,800 kcal/m²/yr and respiration 180 kcal/m²/yr. Total net production for the marsh averaged over 8,200 kcal/m²/yr (approximately 2,000 g/m²/yr).

Feeding on *Spartina alterniflora* were two groups of grazing herbivores: the salt marsh grasshopper *(Orchelimum),* which eats the plant tissues, and the plant hopper *(Prokelisis),* which sucks plant juices. Neither harvested significant portions of the standing crop. Grasshoppers injested about 3 percent of net production and assimilated 1 percent. Total annual net energy flow through the grasshoppers was 28 kcal of which production was 10.8 kcal. Plant hoppers accounted for an energy flow of 275 kcal/m²/yr of which 70 kcal was production. Thus most of the *Spartina* production went to the detrital food chain.

Before *Spartina* is available to most detrital feeders it must be broken down by bacterial action. Part of the *Spartina* particularly that in short growth, decomposes in place in the marsh. Most of the detritus from the tall *Spartina* is washed out to sea by tidal currents. Although bacterial decomposition consumes 82 percent of the organic matter, the remainder is enriched as animal food. Teal found that bacterial respiration accounted for 3890 kcal/m²/yr or 59 percent of the available *Spartina.*

Feeding on decomposing detrital material and the bacteria it supports are fiddler crabs. Where population densities are high, fiddler crabs can sweep over the surface of the marsh between successive tides. They accounted for an assimilation of 206 kcal/m²/yr. Mussels accounted for an assimilation of 56 kcal. Feeding on the herbivores were secondary consumers which largely went unstudied. The mud crab *(Eurytium)* assimilated 27 g/m²/yr of which 5.3 kcal was production.

**256**

Nixon and Oviatt (1973) studied a whole marsh system (16,800 m²) and its embayment, Bissel Cove (6,680 m²). Tall *Spartina alterniflora* made up about 7 percent of the marsh, short *Spartina alterniflora* 78 percent, and high marsh 15 percent. The New England marsh was considerably less productive than the Georgia marsh. Production for the tall *Spartina alterniflora* was 840 kcal/m²/yr and 432 g dry wt/m² for the short *Spartina*. Efficiency of production for the tall *Spartina* was 0.51 percent, one-half that of the tall *Spartina* of the Georgia marsh.

Growth in the New England marsh ceased in late fall (the Georgia marsh has some production in the winter) and dead grass remained until broken up by ice in late winter. Thus in New England marsh ice performed the task carried out in part by fiddler crabs in the Georgia marsh. Fiddler crabs are rare in New England salt marshes, as are the ribbed mussel and marsh snail, and there is no significant population of grasshoppers. Thus the entire production of the marsh goes into the detrital food chain.

The associated embayment supported dense, but patchy populations of wigeongrass *(Ruppia)* and sea lettuce *(Ulva)* that went through several periods of rapid growth, death, decay, and export. Maximum biomass of wigeongrass was twice that of tall *Spartina alterniflora*. Major consumers in the embayment were grass shrimp *(Palaemonetes pugio)* and fish, some 20 species including several mummichogs *(Fundulus)*, silversides, bluefish *(Pomatomus saltatrix)*, winter flounder, and stickleback. The fish were largely predatory; the grass shrimp were detritus-feeding herbivores. Feeding on fish were terns, gulls, black ducks, and mallards. The major fish predator, the herring gull, consumed less than 0.5 percent of the standing crop of fish. The ducks fed extensively on wigeongrass and sea lettuce.

The New England salt marsh is characterized by sharp seasonal changes in light, temperature, and salinity, all of which lower production. Most of the production of the salt marsh goes into embayment as detritus to enter a large sedimentary organic storage compartment as an energy source. Although the embayment is capable of considerable primary production over a short period of time, it is a semiheterotrophic system that depends upon the input of organic matter produced by the marsh grasses and algae. Over the year the annual energy budget for the embayment shows that consumption exceeds production.

The grass shrimp plays an import role in the embayment system at Bissel Cove by feeding on detrital material washed in from the salt marsh. It plucks away at the surface of dead leaves and stems, breaking them into smaller pieces that are colonized by bacteria and diatoms. Inefficient assimilation repackages the detrital material into fecal pellets and adds large quantities of ammonia and phosphates to the water. The grass shrimp also prevents the buildup of detritus from wigeongrass and sea lettuce by feeding on it and reducing it to fine sediment. By its action on the detrital material the grass shrimp makes nutrients and biomass available for other trophic levels. Although the grass shrimp is eaten by the mummichogs, the shrimp's ability to live in a low oxygen environment limits predation and competition and allows its population to reach the high levels necessary to function as the major detritivore.

Production of west coast salt marshes and British salt marshes is comparable to that of east coast salt marshes. Production of *Spartina foliosa* on California marshes ranged from a high of 1200 to 1700 g/m²/yr (Cameron, 1973) to 270 to 690 g/m²/yr (Mahall and Park, 1976) with about 56 percent of primary production exported as detritus (Cameron, 1973). Production of *Spartina townsendii* on British marshes ranged from 760 g/m²/yr (Ranwell, 1961) to 840 g/m²/yr (Jefferies, 1972). *Salicornia virginica* on California marshes produced 550 to 960 g/m²/yr (Mahall and Park, 1976); production of the annual *Salicornia strictissima* on British marshes was comparable, amounting to 876 g/m²/yr (Jefferies, 1972).

The salt marsh and its associated estuary together make up one of the most productive of all ecosystems (Figure 8-15). Marsh plants fix up to 6 percent of the available sunlight falling on them, more than most other ecosystems including intensive agriculture. There are several reasons for this high productivity. Tides continually carry out waste and bring in nutrients. The meeting of fresh and salt water traps

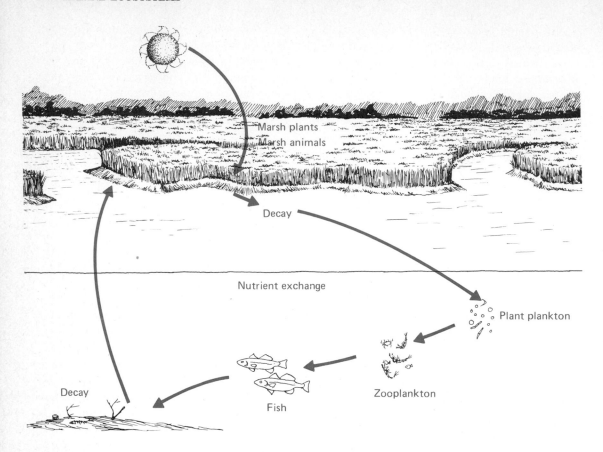

Marsh plants
Marsh animals

Decay

Nutrient exchange

Plant plankton

Fish

Zooplankton

Decay

**FIGURE 8-15**
*Relationship of the salt marsh to the estuary in nutrient cycling. Each contributes to the productivity of the other. (Adapted from C. N. Schuster, Jr., 1966.)*

and concentrates nutrients in the marsh, increasing natural fertility. Nitrogen-fixing blue-green algae in the marsh mud increase the supplies of needed nitrogen. And few nutrients are tied up in biomass for very long. Detrital material from the marsh plants is rapidly broken up and decomposed and algal and bacterial populations turn over rapidly. This production of the marsh is continually being fed into the coastal waters and estuaries making them the most productive waters in the world.

The relative contribution of salt marshes to coastal waters is influenced by latitude and the proportion of the marsh in tall *Spartina alterniflora* and other plants. Productivity studies along the east coast of North America indicate that annual net production of *Spar-*

*tina alterniflora* decreases with latitude (Table 8-2). Also influencing the amount of detritus supplied to the estuary is the proportion of high and low marsh. Salt marshes with extensive stands of *Spartina patens* or short *Spartina alterniflora* contribute much less than marshes with extensive stands of medium and tall *Spartina alterniflora*. Most of the production of such high marsh plants as *Spartina patens, Distichlis,* and *Juncus* ends up as peat accumulated in place.

## Estuaries

Waters of all streams and rivers eventually drain into the sea; the place where this fresh water joins the salt is called an *estuary*. Estuaries are semienclosed parts of the coastal ocean where the seawater is diluted and partially mixed with water coming from land. Estuaries differ in size, shape, and volume of water flow, all influenced by the geology of the region in which they occur. As the river

*TABLE 8-2*
*Biomass of* **Spartina alterniflora** *along the Atlantic Coast of North America* $(g/m^2)$

| Location | Tall and medium forms | Short form | Reference |
|----------|-----------------------|------------|-----------|
| Georgia | 1290–2200 | 643 | Teal, 1962 |
| North Carolina | 1100–1563 471 (medium) | — | Strond, 1969 |
| North Carolina | 640 (medium) | 471 | Williams, 1965 |
| Virginia | 1332 | — | Keefe, 1972 |
| Virginia | 990 | 490 | R. L. Smith, unpub. data |
| Maryland | 1270 | — | Johnson, 1970 |
| New Jersey | 325 (medium) | — | Good, 1965 |
| Connecticut | 840 (medium) | 432 | Nixon and Oviatt, 1973 |
| Nova Scotia | 580 | — | Mann, 1972, cited by Nixon and Oviatt, 1973 |

reaches the encroaching sea, the stream-carried sediments are dropped in the quiet water. These accumulate to form deltas in the upper reaches of the mouth and shorten the estuary. When silt and mud accumulations become high enough to be exposed at low tide, tidal flats develop, which divide and braid the original channel of the estuary. At the same time, ocean currents and tides erode the coast line and deposit material on the seaward side of the estuary, also shortening the mouth. If more material is deposited than is carried away, barrier beaches, islands, and brackish lagoons appear.

## STRUCTURE

Current and salinity, both complex and variable, shape life in the estuary. Estuarine currents result from the interaction of a one-direction stream flow, which varies with the season and rainfall, with oscillating ocean tides and with the wind (Ketchum, 1951; Burt and Queen, 1957). Because of the complex nature of the currents, generalizations about estuaries are difficult to make (see Lauff, 1967).

Salinity varies vertically and horizontally, often within one tidal cycle. Vertical salinity may be the same from top to bottom or it may be completely stratified, with a layer of fresh water on top and a layer of dense saline water on the bottom. Salinity is homogeneous when currents, particularly eddy currents, are strong enough to mix the water from top to bottom. The salinity in some estuaries may be homogeneous at low tide, but unstable at high tide. As the tide floods, a surface wedge of seawater moves upstream more rapidly than the bottom water creating a density inversion of salinity stratification. Seawater on the surface tends to sink, as lighter fresh water rises, and mixing takes place from the surface to the bottom. This phenomenon is known as tidal overmixing. Strong winds, too, tend to mix salt water with the fresh (Barlow, 1956) in some estuaries, but when the winds are still, the river water flows seaward on a shallow surface over an upstream movement of seawater that only gradually mixes with the salt.

Horizontally, the least saline waters are at the river entrance, and the most saline at the mouth of the estuary (Figure 8-16). The configuration of the horizontal zonation is determined mainly by the deflection caused by the incoming and outgoing currents. In all estuaries of the northern hemisphere, outward-flowing fresh water and inward-flowing seawater are deflected to the right because of the earth's rotation. As a result, salinity is higher on the left side.

Salinity also varies with changes in the quantity of fresh water pouring into the estuary through the year. Salinity is highest during the summer and during periods of drought, when less fresh water flows into the estuary. It is lowest during the winter and spring, when rivers and streams are discharging their peak loads. This change in salinity may happen rather rapidly. For example, early in 1957 a heavy rainfall broke the most severe drought in the history of Texas. The re-

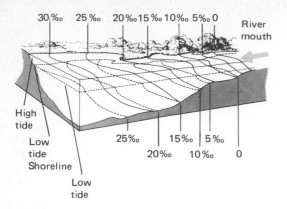

30‰  25‰  20‰ 15‰ 10‰ 5‰ 0

River mouth

High tide

Low tide Shoreline

Low tide

25‰    15‰    5‰
   20‰    10‰    0

**FIGURE 8-16**
*Generalized diagram of the estuary showing the vertical and horizontal stratification of salinity from the river mouth to the estuary at both high and low tide. At high tide the incoming seawater increases the salinity toward the river mouth; at low tide salinity is reduced. Note also how salinity increases with depth, because lighter fresh water flows over denser salt water.*

sultant heavy river discharge reduced the salinities in Mesquite Bay on the central Texas coast by over 30 ppt in a 2-month period. At the height of the drought salinities ranged from 35.5 to 50.0 ppt, but after the break in the drought they ranged from 2.3 to 2.9 ppt (Hoese, 1960). Such rapid changes have a profound impact on the life of the estuary.

The salinity of seawater is about 35 0/00; that of fresh water ranges from 0.065 to 0.30 0/00. Because the concentration of metallic ions carried by rivers varies from drainage to drainage, the salinity and chemistry of estuaries differ. The proportion of dissolved salts in the estuarine waters remains about the same as that of seawater, but the concentration varies in a gradient from fresh water to sea.

Exceptions to these conditions exist in regions where evaporation from the estuary may exceed the inflow of fresh water from river discharge and rainfall (a negative estuary). This causes the salinity to increase in the upper end of the estuary and horizontal stratification is reversed.

Temperatures in estuaries fluctuate considerably diurnally and seasonally. Waters are heated by solar radiation and inflowing and tidal currents. High tide on the mud flats may heat or cool the water, depending on the season. The upper layer of estuarine water may be cooler in winter and warmer in summer than the bottom, a condition that, as in a lake, will result in a spring and autumn overturn.

Mixing waters of different salinities and temperatures acts as a nutrient trap. Inflowing river waters more often than not impoverish rather than fertilize the estuary, except for phosphorus. Instead, nutrients and oxygen are carried into the estuary by the tides. If vertical mixing takes place, these nutrients are not soon swept back out to sea, but circulate up and down between organisms, water, and bottom sediments (Figure 8-17).

Organisms inhabiting the estuary are faced with two problems—maintenance of position and adjustment to changing salinity. The bulk of estuarine organisms are benthic and are securely attached to the bottom, are buried in the mud, or occupy crevices and crannies about sessile organisms. Motile inhabitants are chiefly crustaceans and fish, largely young of species that spawn offshore in high-salinity water. Planktonic organisms are wholly at the mercy of the currents. Because the seaward movement of stream flow and ebb tide transports plankton out to sea, the rate of circulation or flushing time determines the nature of the plankton population. If the circulation is too vigorous, the plankton population may be relatively small. Phytoplankton in summer is most dense near the surface and in low-salinity areas. In winter phytoplankton is more uniformly distributed. For any planktonic growth to become endemic in an estuary, reproduction and recruitment must balance the losses from physical processes that disperse the population (Barlow, 1955).

Salinity dictates the distribution of life in the estuary. Essentially the organisms of the estuary are marine, able to withstand full seawater. Except for anadromous fishes no freshwater organisms live there. Some estuarine inhabitants cannot withstand lowered salinities, and these species decline along a salinity gradient. Sessile and slightly motile organisms have an optimum salinity range within which they grow best. When salinities vary on either side of this range, populations decline. Two animals, the clam worm and the

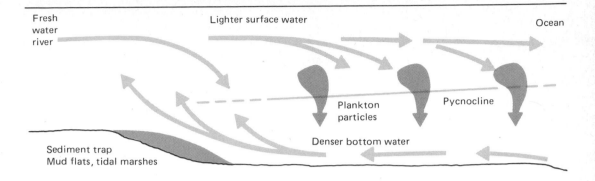

Fresh water river

Lighter surface water

Ocean

Plankton particles

Pycnocline

Denser bottom water

Sediment trap
Mud flats, tidal marshes

FIGURE 8-17
*A diagram showing how circulation of fresh
and salt water in an estuary creates a
nutrient trap. Note a salt wedge of intruding
seawater on the bottom, producing a surface
flow of lighter fresh water and a counterflow
of heavier brackish water. This
countercurrent serves to trap nutrients,
recirculating them toward the tidal marsh.
The same countercurrent also sends
phytoplankton up the estuary, repopulating
the water. When nutrients are high in the
upper estuary, they are taken up rapidly by
tidal marshes and mud flats. These areas
tend to trap particulate nitrogen and
phosphorus, convert them to soluble forms,
and export them back to open waters of the
estuary. Plants on the tidal marshes and
mud flats act as nutrient pumps between
bottom sediments and surface water. (From
Correll, 1978.)*

scud, illustrate this situation. Two species of
clam worm, *Nereis occidentalis* and *Neanthes
succinea,* inhabit the estuaries of the southern
coastal plains of North America. *Nereis* is
more numerous at high salinities and
*Neanthes* at low salinities. In European estuaries the scud *Gammarus* is an important
member of the bottom fauna. Two species,
*G. locusta* and *G. marina,* are typical marine
species and cannot penetrate far into the estuary. Instead they are replaced by a typical
estuarine species, *G. zaddachi*. This species,
however, is broken down into three subspecies,
separated by salinity tolerances. *G. zaddachi*
lives at the seaward end, *G. z. salinesi* occupies
the middle, and *G. z. zaddachi,* which can
penetrate up into fresh water for a short time,
lives on the landward end (Spooner, 1947;
Segerstrale, 1947).

Stage of development influences the range of
motile species within the estuarine waters.
This is particularly pronounced among estuarine fish. Some, such as the striped bass,
spawn near the interface of fresh and low-
salinity water (Figure 8-18). The larvae and
young fish move downstream to more saline
waters as they mature. Thus for the striped
bass the estuary serves both as a nursery and
feeding ground for the young. Anadromous
species, such as the shad, spawn in fresh water, the young fish spend the first summer in
the estuary, and then they move out to the
open sea. Other species, such as the croaker,
spawn at the mouth of the estuary, and the
larvae are transported upstream to feed in the
plankton-rich low-salinity areas. Others, such
as the bluefish, move into the estuary to feed.
In general marine species drop out toward
fresh water and are not replaced by freshwater forms. In fact the mean number of
species progressively decreases from the
mouth of the estuary to upstream stations
(H. W. Wells, 1961).

Salinity changes often affect larval forms
more severely than adults. Larval veligers of
the oyster drill *Thais* succumb to low salinity
more easily than the adults. A sudden influx of

FIGURE 8-18
*Relationship of a semi-anadromous fish, the
striped bass, to the estuary. (From Cronin
and Mansueti, 1971.)*

Fresh water    Estuarine                    Marine

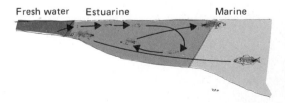

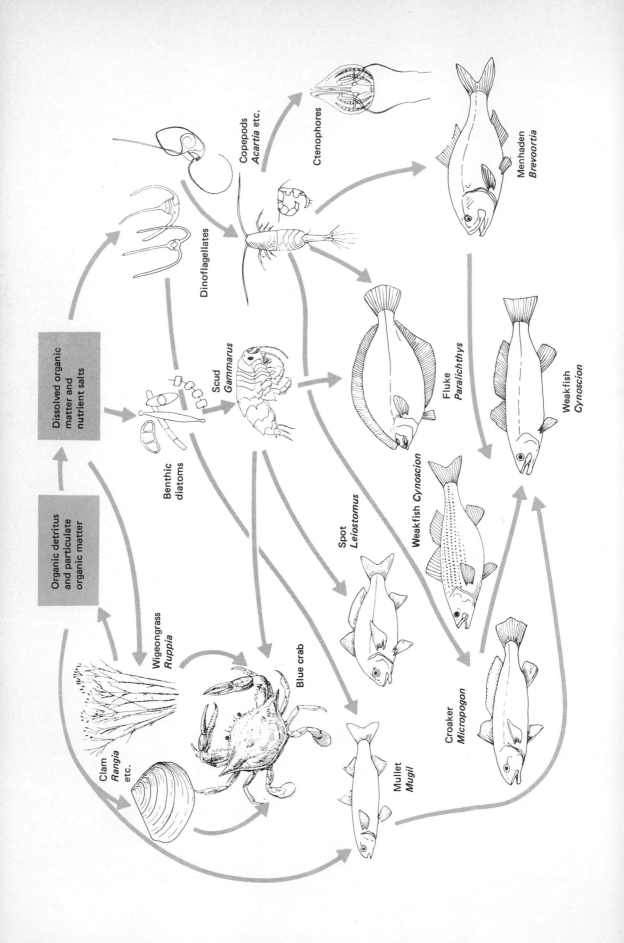

Dissolved organic matter and nutrient salts

Organic detritus and particulate organic matter

Copepods *Acartia* etc.

Ctenophores

Menhaden *Brevoortia*

Dinoflagellates

Scud *Gammarus*

Benthic diatoms

Fluke *Paralichthys*

Weakfish *Cynoscion*

Spot *Leiostomus*

Weakfish *Cynoscion*

Wigeongrass *Ruppia*

Blue crab

Clam *Rangia* etc.

Mullet *Mugil*

Croaker *Micropogon*

fresh water, especially after hurricanes or heavy rainfall, sharply lowers the salinity and causes a high mortality of oysters and their associates. When the drought-breaking heavy rainfall sharply reduced the salinities of Mesquite Bay in Texas, the high-salt-tolerant marine sessile and infaunal mollusks were completely wiped out. The high-salinity community of the oyster *Ostrea equestris* and the mussel *Brachidontes exustus* was replaced by the oyster *Crassostrea virginica* and the mussel *Brachidontes recurvus*. The rapid lowering of the salinity did not kill fish or other motile forms, which apparently moved out of the area (Hoese, 1960).

The oyster bed and the oyster reef are the outstanding communities of the estuary. The oyster is the dominant organism about which life revolves. Oysters may be attached to every hard object in the intertidal zone or they may form reefs, areas where clusters of living oysters grow cemented to the almost buried shells of past generations. Oyster reefs usually lie at right angles to tidal currents, which bring planktonic food, carry away wastes, and sweep the oysters clean of sediment and debris.

Closely associated with oysters are encrusting organisms such as sponges, barnacles, and bryozoans, which attach themselves to oyster shells and are dependent on the oyster or algae for food. The oyster crab strains food from the oyster's gills (Christensen and McDermott, 1958), and a pramidellid snail lives an ectoparasitic life by feeding on body fluids and tissue debris from the oyster's mouth (Hopkins, 1958). Beneath and between the oysters live polychaete worms, decapods, pelecypods, and a host of other organisms. In fact, 303 different species have been collected from the oyster bed (H. W. Wells, 1961).

**FUNCTION**

Estuarine systems function on both plankton-based and detrital-based food webs. The pro-ducer component, particularly in the middle and lower estuary, consists of dinoflagellates and diatoms. The latter convert some of the carbon intake to high-caloric fats and lipids rather than low-energy carbohydrates typical of most green plants. This provides a high-energy food base for higher trophic levels. In shallow estuarine waters rooted aquatics, particularly wigeongrass and eelgrass (*Zostera marina*), assume major importance (for review, see Phillips, 1974).

An example of energy flux in an estuary is provided by data for Pamlico River estuary in North Carolina (Copeland et al., 1974). The estuary is characterized by low salinity, high turbidity, and shallow water. The shallow water supports dense stands of wigeongrass, with associated attached algae and animal associates as the scuds and grass shrimp. The benthos is dominated by the clams *Rangia* and *Macoma*. The euphotic zone, the upper 2m, is dominated by dinoflagellates. Grazing on the phytoplankton are the zooplankters (*Acartia tonsa*) and the harpacticoid copepods that move up to the surface at night. Ctenophores crop each day 30 percent of the zooplankton population. At the same time considerable detrital material enters the estuarine system as dissolved organic matter and particulate organic matter, channeling much of the marsh production into estuarine production. Because the phytoplankton depends heavily on organic detritus, the complex food web can be considered as primarily detrital. Both the detritus and the phytoplankton base support a number of trophic levels that lead eventually to fish often harvested by man (Figure 8-19).

## Coral reefs

In subtropical and tropical waters one finds structures of biological rather than geological origin, coral reefs. Coral reefs are built by carbonate-secreting organisms of which coral (Coelenterata) may be the most conspicuous, but not always the most important. Also contributing heavily are the corralin red algae (*Porolithon*), foraminifera, and mollusks. Reefs are intergrown skeletons that withstand both the action of waves and the attack of coral-eating animals. Built only underwater at

*FIGURE 8-19*
*Simplified estuarine food web based on an estuary in southeastern United States. Note that the energy base is largely heterotrophic, supported by particulate matter and dissolved organic matter. (Adapted from Copeland et al., 1974.)*

**263**

shallow depths, coral reefs need a stable foundation to permit them to grow. Such foundations are provided by shallow continental shelves and submerged volcanos.

There are three kinds of coral reefs with many gradations among them. (1) *Fringing reefs* grow along the rocky shores of islands and continents. (2) *Barrier reefs* parallel shore lines along continents. (3) *Atolls* are coral islands which begin as horseshoe-shaped reefs surrounding a lagoon. Such lagoons are about 40 m deep and are usually connected to the open sea by breaks in the reef. Reefs build up to sea level. To become islands or atolls, the reefs have to be exposed by a lowering of the sea level or be built up by the accumulation of sediments and the piling up of reef material by the action of wind and waves.

Coral reefs are complex ecosystems involving close relationships between coral and algae. In the tissue of coral live zooxanthellae, endozoic dinoflagellate algae; on the calcareous skeletons live still other kinds, both filamentous and calcareous. At night the coral polyps feed, extending their tentacles to capture zooplankton from the water, thus securing phosphorus and other elements needed by the coral and its symbiotic algae. During the day the algae absorb sunlight, carry on photosynthesis, and directly transfer organic material to coral tissue. Thus nutrients tend to be recycled between the coral and the algae (Johannes, 1967; Pomeroy and Kuenzler, 1969). In addition the algae, by altering the carbon dioxide concentration in animal tissue, enable the coral to extract the calcium carbonate needed to build the skeletons (Goreau, 1963). The filamentous algae (Chlorophyta) living on the coral add to the primary production. Thus the symbiotic relationship between algae and coral, as well as the nutrient recharging by tidal waters, makes the coral reef one of the most productive ecosystems in the world.

Reflecting this high productivity and a wide variety of habitats provided by the coral structures is the great diversity of life about the coral reef. Thousands of kinds of exotic invertebrates, some of which feed on the coral animals and algae, hundreds of highly colored herbivorous fish, and a large number of predatory fish that lie in ambush for prey in the caverns that honeycomb the reef swarm about the coral. In addition there is a wide array of symbionts, such as the cleaning fish and crustaceans that pick parasites and detritus from larger fish and invertebrates.

## The open sea

Beyond the rocky and sandy shores and the estuaries lies the open sea. The open sea lacks any distinct communities because of the lack of any dominant large plant life and the freer movement of plankton and nekton that often follow ocean currents.

### STRUCTURE

*Phytoplankton.* As in lakes the dominant form of plant life or primary producers in the open sea is phytoplankton. The density of seawater is such that oceanic phytoplankton do not need well-developed supporting structures. In coastal waters and areas of upwelling phytoplankton 100 microns or more in diameter may be common, but in general plant life is much smaller and widely dispersed.

Because of its requirement for light, phytoplankton is restricted to the upper surface waters. The depth of its occurrence is determined by the depth of light penetration and so may range from tens to hundreds of feet. Because of seasonal, annual, and geographic variations in light, temperature, nutrients, and grazing by zooplankton, the distribution and composition of phytoplankton change with time and place.

Each ocean or region within an ocean appears to have its own dominant forms. Littoral and neritic waters and regions of upwelling are richer in plankton than mid-oceans. In regions of downwelling, dinoflagellates concentrate near the surface in areas of low turbulence. They attain their greatest abundance in warmer waters. In summer they may so concentrate in the surface waters that they color it red or brown. Often toxic to other marine life, such concentrations of dinoflagellates are responsible for red tides.

In regions of upwelling the dominant forms of phytoplankton are diatoms. Enclosed in a silica case, diatoms are particularly abundant in arctic waters.

Smaller than diatoms are the Coccolithophoridae, so small they pass through plankton nets (and so are classified as nannoplankton). Their minute bodies are protected by calcareous plates or spicules imbedded in a gelatinous sheath. Universally distributed in all waters except the polar seas, the Coccolithophoridae possess the ability to swim. They are characterized by droplets of oil which aid in buoyancy and storage of food.

In the equatorial currents and in shallow seas the concentration of phytoplankton is variable. Where both lateral and vertical circulation of water is rapid, the composition reflects in part the ability of the species to grow, reproduce, and survive under local conditions.

*Zooplankton.* Grazing on the phytoplankton is the herbivorous zooplankton, consisting mainly of copepods, planktonic arthropods that are the most numerous animals of the sea, and the shrimplike euphausiids, commonly known as krill. Other planktonic forms are the larval stages of such organisms as gastropods, oysters, and cephalopods. Feeding on the herbivorous zooplankton is the carnivorous zooplankton, which includes such organisms as the larval forms of comb jellies (Ctenophora) and arrowworms (Chaetognatha).

Because its food in the ocean is so small and widely dispersed, zooplankton has adapted ways more efficient and less energy demanding to harvest phytoplankton than filtering water through pores. It has evolved webs, bristles, rakes, combs, cilia, sticky structures, and even bioluminescence.

The composition of zooplankton, like that of phytoplankton, varies from place to place, season to season, year to year. In general zooplankton falls into two main groups, the larger forms characteristic of shallow coastal waters and the generally smaller forms characteristic of the deeper open ocean. Zooplankton forms of the continental shelf contain a large proportion of larvae of fish and benthic organisms. They have a greater diversity of species, reflecting a greater diversity of environmental and chemical conditions. The open ocean, being more homogeneous and nutrient-poor, supports less diverse zooplankton. In polar waters zooplankton species spend the winter in a dormant state in the deep water and rise to the surface during short periods of diatom blooms

to reproduce. In temperate regions distribution and abundance depend upon temperature conditions. In tropical regions, where temperature is nearly uniform, zooplankton is not so restricted, and reproduction occurs throughout the year.

Also like phytoplankton, zooplankton lives mainly at the mercy of the currents. But many forms possess sufficient swimming power to exercise some control Most species of zooplankton migrate vertically each day to arrive at a preferred level of light intensity. As darkness falls, zooplankton rapidly rise to the surface to feed on phytoplankton. At dawn the forms move back down to preferred depths.

By feeding in the darkness of night and hiding in the darkened waters by day, zooplankton avoids heavy predation, and by remaining in cooler water by day, conserves energy during the resting period. Surface currents move zooplankton away from its daytime location during feeding, but it can return home by countercurrents present in the deeper layers. Response to changing light conditions is useful in another way. As clouds of phytoplankton pass over the water above, zooplankton responds to the shadow of food by moving upward. This takes it out of the deep current drift and nearer the surface. At night it can move directly up to the food-rich surface water.

Zooplankton that lacks a vertical migration, and even some of those that do, drifts out of breeding areas with surface currents. Survival of a breeding population is assured by a complex cycle of seasonal migrations.

*Nekton.* Feeding on zooplankton and passing energy along to higher trophic levels are the small fish, squids, and ultimately the large carnivores of the sea. Some of the predatory fish, such as herring and tuna, are more or less restricted to the photic zone. Others are found in the deeper mesopelagic and bathypelagic zones.

Living in a world that lacks any sort of refuge as a defense against predation or as a site for ambush, inhabitants of the pelagic zone have evolved various means of defense and of securing prey. Among them are the stinging cells of the jellyfish, the remarkable streamlined shapes that allow speed both for escape and for pursuit, unusual coloration, advanced sonar, a highly developed sense of smell, and a social organization involving schools or packs. Some animals, such as the baleen whale, have specialized structures that permit them to strain krill and other plankton from the water. Others, such as the sperm whale and certain seals, have the ability to dive to great depths to secure food. Phytoplankton lights up darkened seas, and fish take advantage of that bioluminescence to detect their prey.

The dimly lighted regions of the mesopelagic and the dark regions of the bathypelagic zone depend upon a rain of detritus as an energy source. Such food is limited. The rate of descent of organic matter, except for larger items, is so slow that it is consumed, decayed, or dissolved before it reaches the deepest water or the bottom. Other sources include saprophytic plankton, which exist in the darker regions, particulate organic matter, and import of such material as wastes from the coastal zone, garbage from ships, and large dead animals. Because food is so limited, species are few and the populations low.

Residents of the deep also have special adaptations for securing food. Some, like the zooplankton, swim to the upper surface to feed by night. Others remain in the dimly lighted or dark waters. Darkly pigmented and weak bodied, many of the deep-sea fish depend upon luminescent lures, mimicry of prey, extensible jaws, and expandable abdomens (which enable them to consume large items of food) as means of obtaining sustenance. Although most fish are small (usually 15 cm or less in length), the region is inhabited by rarely seen large species such as the giant squid. In the bathypelagic region bioluminescence reaches its greatest development, where two-thirds of the species produce light. Bioluminescence is not restricted to fish. Squid and euphausiids possess searchlightlike structures complete with lens and iris, and squid and shrimp discharge luminous clouds to escape predators. Fish have rows of luminous organs along their sides and lighted lures that enable them to bait prey and recognize other individuals of the same species.

**Benthos.** There is a gradual transition of life from the benthos that exists on the rocky and sandy shores and that which exists in the ocean's depths. From the tide line to the abyss,

organisms that colonize the bottom are influenced by the nature of the substrate. If the bottom is rocky or hard, the populations consist largely of organisms that live on the surface of the substrate, the epifauna and the epiflora. Where the bottom is largely covered with sediment, most of the inhabitants, chiefly animals, are infauna and live within the deposits. Particle size of the substrate determines the type of burrowing organisms in an area because the mode of burrowing is often specialized and adapted to a certain type of substrate.

The substrate varies with the depth of the ocean and the relationship of the benthic region to land areas and continental shelves. Near the coast bottom sediments are derived from the weathering and erosion of land areas along with organic matter from marine life. Sediments of deep water are characterized by fine textured material, which varies with depth and the type of organisms in the overlying waters. Although these sediments are termed organic, they contain little decomposable carbon, consisting largely of skeletal fragments of planktonic organisms. In general, with regional variations, organic deposits down to 4,000 m are rich in calcareous matter. Below 4,000 m hydrostatic pressure causes some form of calcium carbonate to dissolve. At 6,000 m and lower sediments contain even less organic matter and consist largely of red clays, rich in aluminum oxides and silica.

Within the sediments are layers that relate to oxidation-reduction reactions. The surface or oxidized layer, yellowish in color, is relatively rich in oxygen, ferric oxides, nitrates, and nitrites. It supports the bulk of the benthic animals, such as polychaete worms and bivalves in shallow water, flatworms, copepods, and others in deeper water, and a rich growth of aerobic bacteria throughout. Below the oxidized layer is a grayish transition zone to the black reduced layer, characterized by a lack of oxygen, iron in the ferrous state, nitrogen in the form of ammonia, and hydrogen sulfide. It is inhabited by anaerobic bacteria, chiefly reducers of sulfates and methane.

In the deep benthic regions variations in temperature, salinity, and other conditions are negligible. In this world of darkness there is no photosynthesis, so the bottom community is strictly heterotrophic, depending entirely for its source of energy on what organic matter finally reaches the bottom. Estimates suggest that the quantity of such material amounts to only 0.5 g/m²/year (H. B. Moore, 1958). Bodies of dead whales, seals, and fish may contribute another 2 or 3 g.

Bottom organisms have four feeding strategies: (1) they may filter suspended material from the water, as do stalked coelenterates; (2) they may collect food particles that settle on the surface of the sediment, as do the sea cucumbers; (3) they may be selective or unselective deposit-feeders, such as the polychaetes; or (4) they may be predators, as are the brittlestars and the spiderlike pycnogonids.

Important in the benthic food chain are bacteria of the sediments (see Figure 8-20). Common where large quantities of organic matter are present, bacteria may reach several tens of grams per square meter in the topmost layer of silt. Bacteria synthesize protein from dissolved nutrients and in turn become a source of protein, fats, and oils for deposit-feeders.

## FUNCTION

The oceans occupy 70 percent of the earth's surface and their average depth is around 4000 m, yet their primary production is considerably less than the earth's land surface. They are less productive because only a superficial illuminated area up to 100 m deep can support plant life; and that plant life, largely phytoplankton, is patchy because most of the open sea is nutrient poor (Table 8-3). Much of this nutritional impoverishment results from a limited to almost nonexistent nutrient re-

serve that can be recirculated. Phytoplankton, zooplankton, and other organisms and their remains sink below the lighted zone into the dark benthic water. While this sinking supplies nutrients to the deep, it robs the upper layers.

The depletion of nutrients is most pronounced in tropical waters. The permanent thermal stratification, with a layer of warmer less dense water lying on top of a colder, denser, deep water, prevents an exchange of nutrients between the surface and the deep. Thus in spite of high light intensity and warm temperatures, tropical seas are the lowest in production, reaching an estimated $3.79 \times 10^9$ tn C/yr.

The temperate oceans are more productive, largely because a permanent thermocline does not exist. During the spring and to a limited extent during the fall temperate seas, like temperate lakes, experience a nutrient overturn. The recirculation of phosphorus and nitrogen from the deep stimulates a surge of spring phytoplankton growth. As spring wears on the temperature of the water becomes stratified and a thermocline develops, preventing a nutrient exchange. The phytoplankton growth depletes the nutrients and the phytoplankton population suddenly declines. In the fall a similar overturn takes place, but the rise in phytoplankton production is slight because of decreasing light intensity and low winter temperatures. Reduced production in winter holds down annual productivity of temperate seas to a level a little above that of tropical seas, $4.22 \times 10^9$ tn C/yr.

Most productive are coastal waters and regions of upwelling, whose annual production

TABLE 8-3
*Division of the ocean into provinces according to level of primary organic production*

| Province | Percentage of ocean | Area (km²) | Mean productivity (g c/m²/yr) | Total productivity ($10^9$ tn C/yr) |
|---|---|---|---|---|
| Open ocean | 90.0 | $326.0 \times 10^6$ | 50 | 16.3 |
| Coastal zone* | 9.9 | $36.0 \times 10^6$ | 100 | 3.6 |
| Upwelling areas | 0.1 | $3.6 \times 10^5$ | 300 | 0.1 |
| Total | | | | 20.0 |

*Includes offshore areas of high productivity.
*Source:* J. Ryther, 1969. © 1969 American Association for the Advancement of Science.

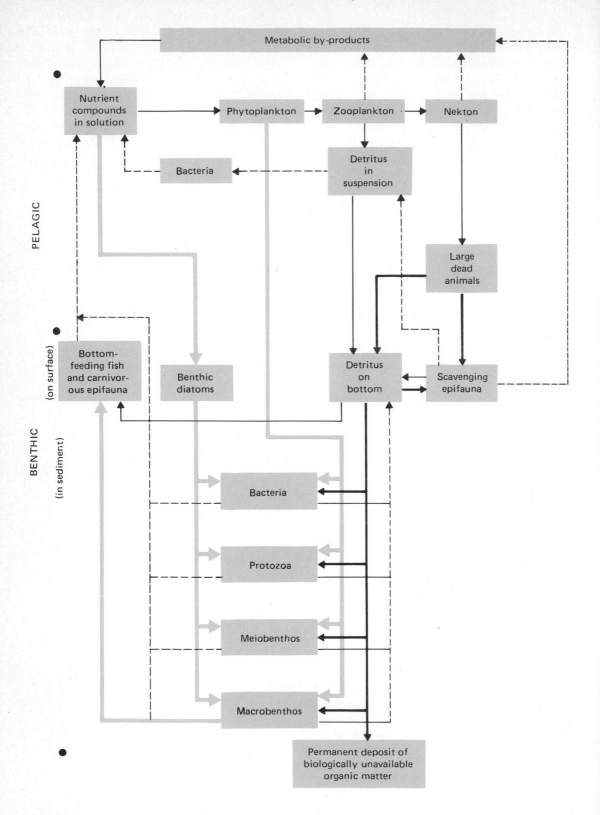

**FIGURE 8-20**
*Simplified marine food web. (After Raymont,*
*1963, from M. Gross, 1972.)*

may amount up to $6.90 \times 10^9$ tn C/yr. Major areas of upwelling are largely on the western sides of continents: off the coasts of southern California, Peru, northern and southwestern Africa, and the Antarctic. Upwellings result from the differential heating of polar and equatorial regions that produces the equatorial currents and the winds. As the water is pushed northward or southward toward the equator by winds, it is deflected away from the coasts by the Coriolis force. As the deflected surface water moves away, it is replaced by an upwelling of colder, deeper water that brings a supply of nutrients into the sunlit portions of the sea. As a result regions of upwellings are highly productive and contain an abundance of life. Because of their high productivity, upwellings support important commercial fisheries such as the tuna fishery off the California coast, the anchoveta fishery off Peru, and the sardine fishery off Portugal.

Other zones of high production are coastal waters and estuaries where productivity may run as high as 1000 g/m²/yr. Turbid, nutrient-rich waters are major areas of fish production. Thus between upwellings and coastal waters, the most productive areas of the seas are the fringes of water bordering the continental land masses. A great deal of the measured productivity of the coastal fringes comes from the benthic as well as the surface waters since the seas are shallow. Benthic production, largely unavailable, is not considered in the productivity of the open sea. Recent estimates (Koblentz-Mishke et al., 1970) of the total production for marine plankton is 50 Gt dry matter per year; if benthic production is considered, total production may be 55 Gt of dry matter per year.

Carbohydrate production by phytoplankton, largely diatoms, is the base upon which life of the seas exist. Conversion of primary production to animal tissue is accomplished by zooplankton, the most important of which are the copepods. To feed on the minute phytoplankton, most of the grazing herbivores must also be small, measuring between 0.5 to 5.0 mm. Most of the grazing herbivores in the oceans are the genera *Calanus, Acartia, Temora,* and *Metridia,* probably the most abundant animals in the world. The single most abundant copepod is *Calanus finmarchicus* and its close rela-

tive *Calanus helgolandicus.* In the Antarctic krill, fed upon by the blue whale, are the dominant herbivores.

The copepods then become the link in the food chain between the phytoplankton and the second-level consumers, as illustrated in the North Sea food web (Figure 8-21). The dominant primary producer is the diatom *Skeletoma* which is grazed by *Calanus.* The major predator on *Calanus* is the semiplanktonic sand eel *Ammodytes,* which in turn is food for the herring. The herring, however, can shorten the food chain by bypassing the sand eel and feeding directly on *Calanus.* In addition to

**FIGURE 8-21**

*Food web and energy flow for an area of the North Sea. The food web is based on the main groups of organisms. Note the low energy yield to humans which amounts to 6.6 kcal/m²/year from a primary production of 900 kcal/m²/year. (From Steele, 1971.)*

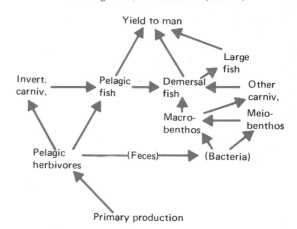

(a)

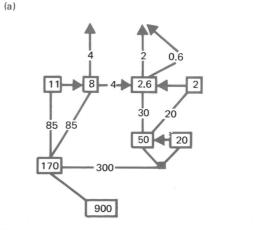

(b)

this major food chain, the herring is also involved in a number of side food chains that add stability to its food supply.

The anchoveta of the Peruvian upwelling shorten the food chain considerably by bypassing the zooplankton link completely and by feeding directly on the phytoplankton. The anchoveta possess modified gills that permit them to strain the phytoplankton from the water. Although these fish spawn throughout the year, their reproductive activity is concentrated in September in the Northern Hemisphere and in April and May in the Southern Hemisphere. Because the eggs have a negligible amount of stored food, the fry require a high density of food nearby (copepods) if they are to survive. Mortality is high. The pelagic eggs and larvae are a part of the plankton and subject to predation by a number of planktonic feeders, including the adult anchovetas themselves. Adults are preyed upon by squid, bonito, mackerel, and birds. Heavy predation comes from cormorants, pelicans, and gannets, all of which have throat pouches adapted for carrying a heavy load of fish. Ninety-five percent of the cormorant's diet and 80 percent of the pelican's and gannet's diet consists of anchoveta. If the fish population or availability drops, the bird population declines. In recent years the populations of these birds have dropped markedly, due in part perhaps to competition from humans. The birds harvest about 4 million metric tons of anchoveta a year; the Peruvian fishery has been harvesting 10.5 million metric tons. In 1971 the Peruvian fishery collapsed because of overfishing and because of shifts in the Humboldt Current. Recently the fishery has shown some recovery.

## SUMMARY

The marine environment is characterized by salinity, waves, tides, and vastness. Salinity is due largely to sodium and chlorine which make up 86 percent of sea salt. Although sea salt has a constant composition, salinity varies throughout the ocean. It is affected by evaporation, precipitation, movement of water masses, and mixing of water masses of different salinities. Because of its salinity, seawater does not reach its greatest density at 4° C, but becomes heavier as it cools.

Like lakes, the marine environment exhibits stratification and zonation. Because there is no seasonal overturn, a normally stable stratification occurs. This is known as the pycnocline. It often coincides with the thermocline.

The marine environment exhibits surface waves produced by winds, internal waves at the interface of layers of water, and subsurface currents that run counter to the surface.

Estuaries where fresh water meets the sea and their associated tidal marshes and swamps are a unit in which the nature and distribution of life are determined by salinity. In the estuary itself salinity declines from the mouth upriver. This decrease in salinity is accompanied by a decline in estuarine fauna because the fauna consists chiefly of marine species. The estuary serves as a nursery for marine organisms, for here the young can develop protected from predation and competing species unable to withstand the lower salinity. Tidal marshes add to the production of the estuary. Composed of salt-tolerant plants and flooded by daily tides, most of the primary production goes unharvested by herbivores. The organic matter is carried out to the estuary by the tides, where it is utilized by bacteria and detritus-feeders. If vertical mixing between fresh and saltwater occurs in the estuary, the nutrients are circulated up and down between the organisms and the bottom sediments. The importance of the tidal marshes as nutrient sources for the estuary and the role of the estuaries as a nursery for such marine species as the flounder and the oyster should be reason enough for their protection.

While the estuary is the place where fresh water meets the sea, the sandy shore and rocky coast are places where the sea meets the land. The drift line marks the farthest advance of tides on the sandy shore. On the rocky shore the tide line is marked by a zone of black algal growth on the stones. The most striking feature of the rocky shore, its zonation of life, results from the alternate exposure and submergence of the shore by the tides. The black zone marks the supralittoral, the upper part of which is flooded only every two weeks by spring tides. Submerged daily by tides is the littoral, characterized by barnacles, periwinkles, mussels, and fucoid seaweeds. Uncovered only at the spring tides is the infralittoral, which is dominated by large, brown laminarian seaweeds, Irish moss, and starfish. Left behind by outgoing tides are tidal pools. These are distinct habitats subject over a 24-hour period to wide fluctuations in temperature and salinity and inhabited by a varying number of marine organisms, depending upon the amount of submergence and exposure. Sandy and muddy shores, in contrast to rocky ones, appear barren of life at low tide, but beneath the sand and mud conditions are more amenable to life than on the rocky shore. Zonation of life is hidden beneath the surface. The energy base for sandy and muddy shores is organic matter made available by bacterial decomposition. They are important sites of biogeochemical cycling, supplying nutrients for off-shore waters. The basic consumers are bacteria, which in turn are a major source of food for both deposit-feeding and filter-feeding organisms. Sandy shores and mudflats are part of the larger coastal ecosystem including the salt marsh, estuary, and coastal waters.

Beyond the estuary and rocky and sandy shores lies the open sea. There the dominant plant life is phytoplankton and the chief consumers are zooplankton. Depending on these for an energy base are the nekton organisms dominated by fishes.

The open sea can be divided into three main regions. The bathypelagic is the deepest, void of sunlight and inhabited by darkly pigmented, weak-bodied animals characterized by luminescence. Above it lies the dimly lit mesopelagic region, inhabited by its own characteristic species such as certain sharks and squid. Both the mesopelagic and bathypelagic regions depend upon a rain of detritus from the upper region, the epipelagic, for their energy source.

Because of the currents and tides, the open sea can be considered one vast interconnected ecosystem that has its variants in the sandy beaches and rocky shores below the high tide mark.

The impoverished nutrient status of ocean water makes productivity low. This comes about because nutrient reserves in the upper layer of water are limited, phytoplankton and other life sink to the deep water, and a thermocline, permanent in deep water, prevents the recirculation of deep water to the upper layer. Most productive are shallow coastal waters and upwellings, where nutrient-rich deep water comes to the surface. The most productive fisheries are confined to these areas, but pollution and overexploitation are reducing the productivity of the seas.

# Terrestrial ecosystems I: distribution, soil, grasslands, shrublands, deserts
## CHAPTER 9

## Distribution of ecosystems

Human beings have always been interested in the plants and animals around them, but their knowledge for centuries was limited to life in their own immediate area. As adventurous explorers expanded man's horizons to new lands around the world, they brought back stories and specimens of new and often strange forms of life. As naturalists joined the explorers, they became more and more familiar with plant and animal life around the world, and began to note similarities and differences. But botanical explorers and zoological explorers looked at the world differently. Botanists in time noted that the world could be divided into great blocks of vegetation—deserts and grasslands, coniferous, temperate, and tropical forests. These divisions they called *formations* even though they had difficulty drawing sharp lines between them. In time plant geographers

attempted to correlate the formations with climatic differences, and found that blocks of climate reflected the blocks of vegetation and their life form spectra.

At the same time zoogeographers studying the distributions of animals found them much more difficult to map. By the beginning of the twentieth century, naturalists had accumulated the basic facts of worldwide animal distribution. All they needed to do—and this was a great enough task—was to arrange the facts and draw some general conclusions. This was done for birds in 1878 by Philip Sclater, who mapped them into six regions roughly corresponding to continents. Several different schemes of classification have been devised.

### BIOGEOGRAPHICAL REGIONS

The master work in zoogeography was done by Alfred Wallace, who is also known for reaching the same general theory of evolution as

Darwin. The *realms* of Wallace, with some modification, still stand today. There are six biogeographical regions, or realms, each more or less embracing a major continental land mass and each separated by oceans, mountain ranges, or desert (inside cover). They are the Palearctic, Nearctic, Neotropical, Ethiopian, Oriental, and Australian. Because some zoogeographers consider the Neotropical and the Australian regions to be so different from the rest of the world, they group the four other regions together to make a more general classification: Neogea (Neotropical), Notogea (Australia), and Metagea (the main part of the world). Each region possesses certain distinctions and uniformity in the taxonomic units it contains, and each to a greater or lesser degree shares some of the families of animals with other regions. Except for Australia, each at one time or another in the history of the earth has had some land connection with another across which animals and plants could pass.

Two regions, the Palearctic and the Nearctic, are quite closely related; in fact the two are often considered as one, the Holarctic. The Nearctic contains the North American continent south to the Tropic of Cancer. The Palearctic region contains the whole of Europe, all of Asia north of the Himalayas, northern Arabia, and a narrow strip of coastal North Africa. The regions are similar in climate and vegetation. They are quite alike in their faunal composition, and they share, particularly in the north, similar animals, such as the wolf, the hare, the moose (called elk in Europe), the stag (called elk in North America), the caribou, the wolverine, and the bison.

Below the coniferous forest belt the two regions become more distinct. The Palearctic is not rich in vertebrate fauna, of which few are endemic. Palearctic reptiles are few and usually are related to those of the African and Oriental tropics. The Nearctic, in contrast, is the home of many reptiles and has more endemic families of vertebrates. The Nearctic fauna is a complex of New World tropical and Old World temperate families; the Palearctic is a complex of Old World tropical and New World temperate families.

South of the Nearctic lies the Neotropical, which includes all of South America, part of Mexico, and the West Indies. It is joined to the Nearctic by the Central American isthmus and is surrounded by the sea. Isolated until 15 million years ago, the fauna of the Neotropical is most distinctive and varied. In fact about half of the South American mammals, such as the tapir and llama, are descendants of North American invaders, whereas the only South American mammals to survive in North America are the armadillo, opossum, and porcupine. Lacking in the Neotropical is a well-developed ungulate fauna of the plains, so characteristic of North America and Africa. The Neotropical, however, is rich in endemic families of vertebrates. Of 32 families of mammals, excluding bats, 16 are restricted to the Neotropical. In addition, 5 families of bats, including the famous vampire, are endemic.

The Old World counterpart of the Neotropical is the Ethiopian, which includes the continent of Africa south of the Atlas Mountains and Sahara Desert and the southern corner of Arabia. It embraces tropical forests in central Africa and savanna, grasslands, and desert in the mountains of East Africa. During the Miocene and the Pliocene, Africa, Arabia, and India shared a moist climate and a continuous land bridge, which allowed the animals to move freely between them. This accounts for some similarity in the fauna between the Ethiopian and the Oriental regions. Of all the regions the Ethiopian contains the most varied vertebrate fauna; it is second only to the Neotropical in endemic families.

Lush forests cover much of the Oriental region, which includes India, Indochina, south China, Malaya, and the western islands of the Malay Archipelago. It is bounded by the Himalayas, the Indian Ocean, and the Pacific Ocean. On the southeast corner, where the islands of the Malay Archipelago stretch out toward Australia, there is no definite boundary, although a line drawn by Wallace is often used to separate the Oriental from the Australian region. This line runs between the Philippines and the Moluccas in the north, then bends southwest between Borneo and the Celebes, then south between the islands of Bali and Lombok. A second line, Weber's, has been drawn to the east of Wallace's line; it separates the islands with a majority of Oriental animals from those with a majority of Australian ones. Because the islands between these two lines are a transition between the Oriental and the Australian regions, some zoogeographers call the area Wallacea (see inside cover).

Of the tropical regions the Oriental possesses the fewest endemic species and lacks a variety of widespread families. It is rich in primate species, including two families confined to the region, the tree shrews and the tarsiers.

Perhaps the most interesting and the strangest region, and certainly the most impoverished in vertebrate species, is the Australian. This includes Australia, Tasmania, New Guinea, and a few smaller islands of the Malay Archipelago. New Zealand and the Pacific Islands are excluded, for these are regarded as oceanic islands, separate from the major faunal regions. Partly tropical and partly south temperate, the Australian is noted for its lack of a land connection with other regions, the poverty of freshwater fish, amphibians, and reptiles, the absence of placental animals, and the dominance of marsupials. Included are the monotremes with two egg-laying species, the duck-billed platypus and the spiny anteater. The marsupials have become diverse and have evolved ways of life similar to those of the placental animals of others regions.

## LIFE ZONES

By the turn of the century some biologists were attempting to combine the regional distribution of both plants and animals into one scheme. C. Hart Merriam, then chief and founder of the United States Bureau of Biological Survey (later to become the Fish and Wildlife Service), proposed the idea of *life zones* (1894*a*, 1894*b*). Merriam divided the North American continent into three primary transcontinental regions, the Boreal, the Austral, and the Tropical. Differences among transcontinental belts running east to west, expressed by the animals and plants living there, supposedly are controlled by temperature. The Boreal region extends from the northern polar seas south to southern Canada, with extensions running down the three great mountain chains, the Appalachians, the Rockies, and the Cascade-Sierra Nevada Range. The Austral region embraces most of

the United States and a large part of Mexico. The Tropical region clings to the extreme southern border of the United States and includes some of the lowlands of Mexico and most of Central America. Each of these regions Merriam further subdivided into life zones.

The Boreal region he subdivided into three zones. The Arctic-Alpine zone, characterized by arctic plants and animals, lies north of the tree line and includes the arctic tundra as well as those parts of mountains further south that extend above the timber line. The Hudsonian zone, the land of spruce, fir, and caribou, embraces the northern coniferous forest and the boreal forests covering the high mountain ranges to the south. The Canadian zone includes the southern part of the boreal forest and the coniferous forests that cloak the mountain ranges extending south.

The Austral region is split into five zones. First is the Transition zone (called the Alleghanian in the east) which extends across northern United States and runs south on the major mountain ranges. It is a zone in which the coniferous forest and the deciduous forest intermingle. Extending in a highly interrupted fashion across the country from the Atlantic to the Pacific is the upper Austral zone. It is further subdivided into the Carolinian area in the humid east and the upper Sonoran of the semiarid western North America. The lower Austral embraces the southern United States from the Carolinas and the Gulf States to California. In the humid southeast, it is known as the Austroriparian area and in the arid west as the Lower Sonoran.

Once widely accepted, the life zones are rarely used today, although they creep now and then into the literature on the vertebrates. In the first place a life zone is not a unit that can be recognized continent-wide by a characteristic and uniform faunal or vegetational component. For example, the Transition zone includes the hardwoods of the East, the yellow pine of the Rocky Mountains, and the redwoods of California. Thus it covers too many types of vegetation and too many different animals to be useful. The life zones south of the Arctic and Canadian are not transcontinental. The Transition and Upper and Lower Austral of the east are totally different from those of the west. Then the temperatures at times of the year other than the season of growth and reproduction influence the distribution of plants and animals. In spite of all this there is something evocative about the life-zone terminology. The Arctic-Alpine zone recalls the cold, windswept mountains above the timber line; and the name Sonoran, slowly spoken, sings of the sun-baked desert, cactus, mesquite, horned lizards, and roadrunners.

## BIOTIC PROVINCES

A third approach to the subdivision of the North American continent into geographical units of biological significance was the biotic provinces concept defined and mapped by Dice in 1943. It differs from the others in that a province embraces a continuous geographic area that contains ecological associations distinguishable from those of adjacent provinces, especially at the species and subspecies level. Each biotic province is further subdivided into ecologically unique subunits, districts, or life belts, based largely on altitude, such as grassland belts and forest belts.

Basically the biotic province concept is an attempt to classify the distribution of plants and animals, especially the latter, on the basis of ranges and centers of distribution of the various species and subspecies. But the regions themselves and their subdivisions are largely subjective. The boundaries more often than not coincide with physiographic barriers rather than with vegetation types, and the regions never occur as discontinuous geographic fragments. Although a number of species may be confined to some biotic province, others occur over several provinces, because their distribution is determined more by the presence of suitable habitat, which is rarely restricted to a single region. Because the boundaries of biotic provinces and the ranges of subspecies of animals with a wide geographic distribution do coincide, this system is used at times by mammalogists, ornithologists, and herpetologists in the study of a particular group.

## BIOMES

As attempts at combining plant and animal distribution into one system, all of the above units of classification are unworkable, for plant and animal distributions do not coincide.

Another approach, pioneered by Victor Shelford, was simply to accept plant formations as the biotic units and to associate animals with plants. This approach works fairly well because animal life does depend upon a plant base. These broad natural biotic units are called *biomes* (Figure 9-1). Each biome consists of a distinctive combination of plants and animals in a climax community; and each is characterized by a uniform life form of vegetation, such as grass or coniferous trees. It also includes stages in the development of the community toward its final form, which may be dominated by other life forms. Because the species that dominate the developmental or seral stages are more widely distributed than those of the climax, they are of little value in defining the limits of the biome.

On a local and regional scale, communities are considered as gradients in which the combination of species varies as the individual species respond to environmental gradients. On a larger scale one can consider the terrestrial and even some aquatic ecosystems as gradients of communities and environments on a world scale. Such gradients of ecosystems are *ecoclines*.

If one were to sample vegetation and associated animal life along a transect cutting across a number of ecosystems, such as up a mountain slope or north to south across a continent, one would discover that communities change gradually just as species change along community gradients. If one were to run this transect across midcontinent North America beginning in the moist, species-rich forests of the Appalachians and ending in the desert, the transect would follow a gradient of ecosystems on a climatic moisture gradient—from the mesophytic forest of the Appalachians through oak-hickory forests, oak woodlands with grassy understory, tall-grass prairies (now cornland), mixed prairies (wheatland), short-grass plains, desert grasslands, and desert shrublands (Figure 9-2a). Likewise, a transect from southern Florida to the Arctic would move along a climatic temperature gradient from a subtropical forest through temperate deciduous forest, temperate mixed forest, to boreal coniferous forest and tundra (Figure 9-2b).

In addition to gradual changes in vegeta-

# LIFE AREAS OF NORTH AMERICA

| | |
|---|---|
| 1 | Arctic-Alpine |
| 2 | Open Boreal |
| 3 | Closed Boreal-Subalpine |
| 4 | Northern Hardwood-Conifer |
| 5 | Aspen Parkland |
| 6 | Montane Woodland-Brush |
| 7 | Pacific Rain Forest |
| 8 | Eastern Deciduous Forest |
| 9 | Grasslands |
| 10 | Oak-Savannah |
| 11 | Northern Desert Scrub |
| 12 | Southern Desert Scrub |
| 13 | Mesquite-Grassland |
| 14 | Piñon-Juniper-Oak |
| 15 | Chaparral-Oak Woodland |
| 16 | Southeast Evergreen |
| 17 | Mexican Pine and Pine-Oak |
| 18 | Tropical Areas Combined |

*FIGURE 9-1*
*Biomes of North America. (Map by John Aldrich, courtesy U.S. Department of the Interior, Fish and Wildlife Service.)*

tion, there are gradual changes in other ecosystem characteristics. As one goes from highly mesic and warm temperatures to xeric situations or cold temperatures, productivity, species diversity, and the amount of organic matter decreases. There is a corresponding decline in the complexity and organization of ecosystems, in the size of the plants, and in the number of strata to the vegetation. Growth forms change. The tropical rain forest is dominated by phanerophytes and epiphytes; the arctic tundra by hemicryptophytes, geophytes, and therophytes. Wherever similar environments exist on the earth, the same growth forms exist even though the species differences may be great. Thus different continents tend to have communities of similar physiognomy.

The various biomes of the world fall into a distinctive pattern when plotted on a gradient of mean annual temperature and mean annual precipitation (Figure 9-3). The plots are obviously rough. Many types intergrade with one another, and adaptations of various growth forms may differ on several continents. Climate alone is not responsible for biome types because soil and exposure to fire can influence which one of several biomes will occupy an area. Structure of biomes is further influenced by the nature of the climate, whether marine or continental. The same amount of rain, for example, can support either shrubland or grassland.

## HOLDRIDGE LIFE ZONE SYSTEM

Another useful approach to classification and study of ecosystems is the Holdridge life zone system (Holdridge, 1964, 1967; Holdridge et al., 1971). It is based on the assumption that (1) mature, stable plant formations represent physiognomically discrete vegetation types recognizable throughout the world and (2) geographical boundaries of vegetation correspond closely to boundaries between climatic zones (Holdridge et al., 1971). Vegetation is determined largely by an interaction of temperature and rainfall.

The Holdridge system divides the world into life zones arranged according to latitudinal regions, altitudinal belts, and humidity provinces (Figure 9-4). The boundaries of each zone are defined by mean annual precipitation and mean annual biotemperature. Altitudinal belts and latitudinal regions are defined in terms of mean biotemperatures and not in the usual terms of latitudinal degrees or meters of elevation.

The Holdridge system is based on three levels of classification: (1) climatically defined life zones, (2) associations which are subdivisions of life zones based on local environmental conditions, and finally (3) further local subdivisions based on actual cover or land use. The term association, as used by Holdridge, means a unique ecosystem, a distinctive habitat or physical environment and the naturally evolved community of plants and animals.

The subdivision into zones or associations is based on an interaction between environment and vegetation as defined by biotemperature, precipitation, moisture availability, potential evapotranspiration, and potential evapotranspiration ratio.

Biotemperature, as defined by Holdridge, is the mean of unit period temperatures with the substitution of zero for all unit period values below 0° C and above 30° C; temperatures below 0° and above 30° limit growth and physiological activities of plants. Mean biotemperature on a daily basis for an area is calculated by summing the hourly temperatures above 0° C and below 30° C and dividing by 24.

Precipitation is the total amount of precipitation for a region rather than seasonal distribution. Long-term average annual precipitation tends to be partially correlated with its seasonal distribution (in terms of soil moisture availability) and life zone boundaries as reflected in vegetation. This leads to a classification at the association level.

Moisture availability is the supply of moisture available for plant growth. It is a function of the relationship between precipitation and potential evapotranspiration. This determines the biologically significant condition of humidity. For all major components of vegetation, soil rather than direct precipitation supplies water necessary for physiological activity. In any given climatic situation moisture available to plants varies with those soil characteristics that influence filtration, storage, etc. These variations determine association classifications.

Potential evapotranspiration from land veg-

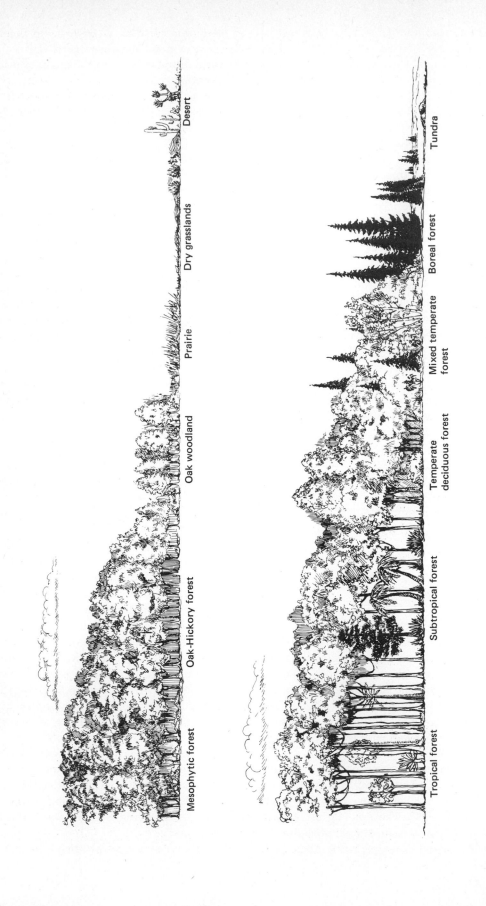

Mesophytic forest    Oak-Hickory forest    Oak woodland    Prairie    Dry grasslands    Desert

Tropical forest    Subtropical forest    Temperate deciduous forest    Mixed temperate forest    Boreal forest    Tundra

FIGURE 9-2

FIGURE 9-2
*Gradients of vegetation in North America from east to west and north to south. The east-to-west gradient runs from the mixed mesophytic forest of the Appalachians through the oak-hickory forests of the central states, the ecotone of burr oak and grasslands, to the prairie, short-grass plains, and the desert. The transect does not cut across the Rocky Mountains. This gradient is largely a result of variations in precipitation. The north-south gradient reflects temperatures. The transect cuts across tundra, boreal coniferous forest, the mixed mesophytic forests of the Appalachians, subtropical forests and Florida, and the tropical forests of Mexico.*

etation is essentially a function of biologically positive heat balance or biotemperature. Potential evapotranspiration from land vegetation is a hypothetical, not a directly measurable figure against which other moisture values may be compared. It is the amount of water that would be transpired under constantly optimal conditions of soil moisture and plant cover. It is calculated by multiplying mean annual temperature by an experimentally derived constant 59.93.

Potential evapotranspiration ratio is the ratio of the mean annual potential evapotranspiration to average total annual precipitation. It provides an index of biological humidity conditions. A ratio of 1.00 indicates precipitated moisture is exactly equal to potential evapotranspiration for a long-term average year. As the ratio increases, precipitation is progressively less than water lost and climates are increasingly arid. As the ratio decreases, the climates become progressively more humid.

FIGURE 9-3
*Pattern of world plant formation in relation to climatic variables of temperature and moisture. In certain areas where climate (maritime vs. continental) varies, soil can shift the balance between such types as woodland, shrubland, and grass. The dashed line enclosed a wide range of environments in which either grassland or one of the types dominated by woody plants may form the prevailing vegetation in different areas. (After Whittaker, 1970.)*

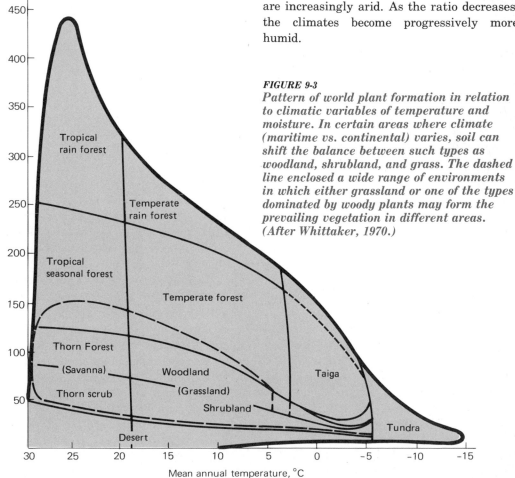

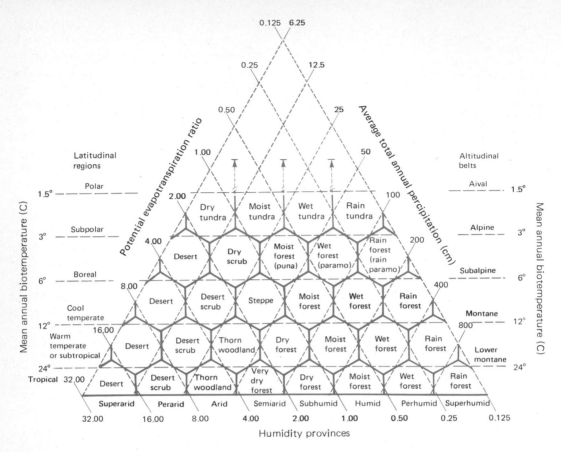

**FIGURE 9-4**
*The Holdridge life zone system for classifying plant formations. (After Holdridge, 1967.)*

## The soil system

Basic to all terrestrial ecosystems is soil. Populated by high numbers of species as well as individuals, the soil embraces another world with its whole chain of life, its predators and prey, its herbivores and carnivores, and its fluctuating populations. Because of their abundance, feeding habits, and ways of life, these small organisms have an important influence on the world a few inches above them. Because for all practical purposes it is a community separate from the one above, soil has been considered an ecosystem or biocenose, but it is not. Its energy source comes from dead bodies and feces from the community above (Figure 9-5). It is but a stratum of the whole ecosystem of which it is a part (see Castri, 1970; Kuhnelt, 1970; Ghilarov, 1970).

## SOIL AS AN ENVIRONMENT

Soil is a radically different environment for life than the one above the surface, yet the essential requirements do not differ. Like animals that live outside the soil, soil fauna require living space, oxygen, food, and water.

To soil fauna, the soil in general possesses several outstanding characteristics as a medium for life. It is relatively stable, both chemically and structurally. Any variability in the soil climate is greatly reduced compared to above-surface conditions. The atmosphere remains saturated or nearly so, until soil moisture drops below a critical point. Soil affords a refuge from high and low extremes in temperature, wind, evaporation, light, and dryness. This permits soil fauna to make relatively easy adjustments to unfavorable conditions.

On the other hand, soil has low penetrability. Movement is greatly hampered. Except to such channeling species as earthworms, soil pore space is very important, for this determines the nature of the living space, humidity, and gaseous condition of the environment.

**280**

Variability of these conditions creates a diversity of habitats, which is reflected by the diversity of species found in the soil (Birch and Clark, 1953). The number of different species, representing practically every invertebrate phylum, found in the soil is enormous.

Only a part of the soil litter is available to most soil animals as living space. Spaces between surface litter, cavities walled off by soil aggregates, pore spaces between individual soil particles, root channels and fissures—all are potential habitats. Most of the soil fauna are limited to pores and cavities larger than themselves. Distribution of these forms in different soils is often determined in part by the structure of the soil (Weis-Fogh, 1948), for there is a relationship between the average size of soil spaces and the fauna inhabiting them (Kuhnelt, 1950). Large species of mites inhabit loose soils with crumb structure, in contrast to smaller forms inhabiting compact soils. Larger soil species are confined to upper

layers where soil interstices are the largest (Haarløv, 1960).

Water in the spaces is essential because the majority of soil fauna are active only in this. Soil water is usually present as a thin film lining the surface of soil particles. This film contains, among other things, bacteria, unicellular algae, protozoa, rotifers, and nematodes. Most of these are restricted in their movements by the thickness and shape of the water film in which they live. Nematodes are less restricted, for they can distort the water film by muscular movements and thus bridge the intervening air spaces. If the water film dries up, these species encyst or enter a dormant state. Millipedes and centipedes, on the other hand, are highly susceptible to desiccation and avoid it by burrowing deeper into the soil (Kevan, 1962).

Excess water and lack of aeration are detrimental to many soil animals. Excessive moisture, which typically occurs after heavy rains, often is disastrous to some soil inhabitants. Air spaces become flooded with deoxygenated water, producing a zone of oxygen shortage for soil inhabitants. If earthworms cannot evade this by digging deeper, they are forced to the surface, where they die from excessive ultraviolet radiation. The snowflea (a collembo-

**FIGURE 9-5**
*Life in the soil. This drawing shows only a tiny fraction of the organisms that inhabit soil and litter. Note the fruiting bodies of the fungi which in turn are consumed by animals, invertebrate and vertebrate.*

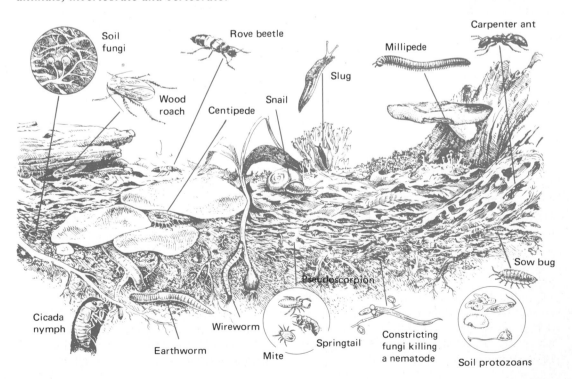

lan) comes to the surface in the spring to avoid excess soil water from melting snow (Kuhnelt, 1950). Many small species and immature stages of larger species of centipedes and millipedes may be completely immobilized by a film of water and unable to overcome the surface tension imprisoning them. Adults of many species of these organisms possess a waterproof cuticle that enables them to survive some temporary flooding.

Soil acidity long has been regarded as having an important effect on soil fauna. But because pH is readily measured, it has been overplayed in an attempt to correlate soil characteristics with the fauna. Bornebusch (1930) regarded a pH of 4.5 as inimical to earthworms, yet some earthworms, such as *Lumbricus rubellus,* are quite tolerant of relatively acid conditions. Although every species of earthworm has its optimum pH, and although some species, such as the *Dendrobaenus,* are characteristic of acid conditions, most of them seem to be able to settle in most soils, provided they contain sufficient moisture (Petrov, 1946). In northern hardwood forests earthworms are most abundant both in species an in numbers when the pH is between 4.1 and 5.5 (Stegeman, 1960). Mites and springtails (Collembola) can exist in very acid conditions (Murphy, 1953).

## STRUCTURE

The interrelations of organisms living in the soil are complex, but within the upper layers of the soil energy flows through a series of trophic levels similar to those of surface communities.

The primary source of energy in the soil community is the dead plant and animal matter and feces from the ground layer above. These are broken down by the microbial life—bacteria, fungi, protozoans. Upon this base rests phytophagous consumers, which obtain nourishment from assimilable substances of living plants, as do the parasitic nematodes and root-feeding insects; from fresh litter as do the earthworms; and from exploitation of the soil microflora. Some members of this consumer level, such as some protozoa and freeliving nematodes, feed selectively on the microflora. Others, including most earthworms, pot worms, millipedes, and small soil arthropods,

ingest large quantities of organic matter and utilize only a small fraction of it, chiefly the bacteria and fungi, as well as any protozoans and small invertebrates contained within the material.

On the next trophic level are the predators —the turbellaria, which feed on nematodes and pot worms, the predatory nematodes and mites, insects, and spiders. In such a manner does the community in the soil operate on an energy source supplied by the unharvested organic material of the world above.

Prominent among the larger soil fauna are the Oligochaetes, which include two common families, the Lumbricidae (earthworms) and the Enchytraeidae (white or pot worms). The small, whitish pot worms abound in the upper 8 cm of the soil if humidity is fairly constant. They are able to live under a greater variety of conditions than earthworms, but their numbers undergo violent fluctuations. Populations are at a maximum in winter and at a minimum in summer. They are not extensive burrowers and appear to divide the earth and humus more finely than earthworms. Little is known about the feeding biology beyond that they ingest organic debris, from which they may digest bacteria, protozoans, and other microorganisms (C. O. Nielson, 1961).

Earthworm activity in the soil consists mainly of burrowing, of ingestion and partial breakdown of organic matter, and of subsequent egestion in the form of surface or subsurface casts. Ingested soil is taken during burrow construction, mixed with intestinal secretions, and passed out either as aggregated castings on or near the surface or as a semiliquid in intersoil spaces along the burrow. Earthworms pull organic matter into their burrows and ingest some of it; it is then partially or completely digested in the gut. Casts of soil passed through the alimentary canal contain a larger proportion of soil particles less than 0.002 in. in diameter than uningested soil and a higher total nitrogen, organic carbon, exchangeable calcium and magnesium, available phosphorus, and pH.

Surface casting and burrowing slowly overturn the soil. Subsurface soil is brought to the top and organic matter is pulled down into and incorporated with the subsoil to form soil aggregates. These aggregates result in a more open structure in heavy soil and bind particles of light soil together.

Millipedes probably are the next most important group of litter-feeders. They and their somewhat similar associates, the centipedes, are essentially animals of the woodland floor. Millipedes occupy essentially three woodland habitats, the floor and aerial parts of vegetation, the litter and upper soil layer, and the areas beneath bark and stones and in rotten logs and stumps. The three most common forms are glomerids, or oval pill millipedes; flat-backed polydesmids with flattened lateral expansions; and large iuloids. The glomerids and polydesmids are not adapted to burrowing and must find refuge against both floods and drought in surface retreats. Iuloids, however, burrow extensively in the soil. Millipedes ingest leaves, particularly those in which some fungal decomposition has taken place, for lacking the enzymes necessary for the breakdown of cellulose, they live on the fungi contained within the litter. Different species of millipedes ingest varying quantities of litter, depending upon the tree species (van der Drift, 1951). *Iulus* consumes more red-oak litter, *Cylindroiulus* more pine.

The chief contribution of millipedes to soil development and to the soil ecosystem is the mechanical breakdown of litter, making it more vulnerable to microbial attack, especially by the saprophytic fungi.

Litter-feeders of importance are snails and slugs, which among the soil invertebrates possess the widest range of enzymes to hydrolyze cellulose and other plant polysaccharides, possibly even lignin (C. O. Nielsen, 1962). In Australian rain forests amphipods are a conspicuous part of the fauna and play a major part in the disintegration of the leaf litter (Birch and Clark, 1953).

Not to be ignored are termites (Isoptera), white, wingless, social insects. The termite, together with some dipteran and beetle larvae, is the only larger soil inhabitant that is able to break down the cellulose of wood. It accomplishes this with the aid of a symbiotic protozoan living in its gut. The termite has a mouth structure adapted to ingest wood; the protozoan produces the enzymes that effectively digest cellulose into the simple sugars that the termite can use. Together, the two organisms function perfectly. Without the protozoan, the termite could not exist; without the termite the protozoan could not gain access to wood.

Termites do not play a major role in the temperate soils, but in the tropics they dominate the soil fauna. In these regions they are responsible for the rapid removal of wood and other cellulose-containing materials, twigs, leaves, dry grass, structural timbers, etc., from the surface. In addition to removal of organic matter, termites are important soil churners. They move considerable quantities of soil, perhaps as much as 5,000 tons/acre, in constructing their huge and complex mounds. In semidesert country the openings and galleries of subterranean termites allow the infrequent rains to penetrate deep into the subsoil rather than to run off the surface (Kevan, 1962).

Of all soil animals, the most abundant and widely distributed are the mites (Acarina) and the springtails (Collembola). Both occur in nearly every situation where vegetation grows, from tropical rain forest to tundra. Flattened dorsoventrally, they are able to wiggle, squeeze, and even digest their way through tiny caverns in the soil. Here they browse on fungi or search for prey in the dark interstices and pores of the organic mass.

The more numerous of the two, both in species and in numbers, are the mites, tiny, eight-legged arthropods from 0.1 to 2.0 mm in size. The most common mites in the soil and litter are the Orbatei. In the pine-woods litter of Tennessee, for instance, they make up 73 percent of all the litter mites (Crossley and Bohnsack, 1960). These mites live largely on fungal hyphae that attack dead vegetation as well as the sugars digested by this microflora from evergreen needles.

The Collembola are the most generally distributed of all insects. Typically they may be brightly colored or completely white. Their common name, springtail, is descriptive of the remarkable springing organ at the posterior end, which enables them to leap comparatively great distances. The springtails are small, from 0.3 to 1 mm. They consist of two groups, the round springtails, or Symphypleona, and the long springtails, or Arthropleona. Neither have specialized feeding habits.

They consume decomposing plant materials, largely for the fungal hyphae they contain.

Small arthropods are the principal prey of spiders, beetles, especially the Staphylinidae, the pseudoscorpions, the mites, and the centipedes. The centipedes are one of the major invertebrate predators. The two most common groups are the nonburrowing type and those that burrow, earthwormlike, into the soil. Predacious Mesostigmata mites prey on herviborous mites, nematodes, enchytraeid worms, small insect larvae, and other small soil animals.

Most of the microorganisms of the soil, the protozoans and rotifers, myxobacteria and nematodes, feed on bacteria and algae. Nematodes are ubiquitous, found wherever their need for a film of water in which to move is met. Soil and freshwater nematodes form one ecological group, with many species in common. But in the soil they exist at much higher densities than in fresh water, up to 20 million/m$^2$. They are most abundant in the upper 5 cm in the vicinity of roots, where they feed on plant juices, soil algae, and bacteria. A few are predaceous.

These bacteria- and algae-feeders, in turn, are consumed by various predacious fungi. Among these, there are three groups: (1) the Zoopagales, an order of Phycomycetes that preys chiefly on protozoans, although a few species prey on nematodes; (2) the endozoic Hyphomycetes; and (3) the ensnaring Hyphomycetes. The latter two capture and digest nematodes, crustaceans, rotifers, and, to an extent, protozoans (Maio, 1958; Doddington, in Kevan, 1955). Zoopagales possess sticky mycelia, which capture the prey like flypaper. Endozoic Hyphomycetes release spores, which stick to the integument of nematodes. Germ tubes penetrate the tube of the animal and develop into internal mycelium.

The nematode-trapping Hyphomycetes possess unique morphological adaptations that enable them to capture their prey. One of the most common forms of traps is a network of highly adhesive loops which catch and hold nematodes on contact. Others possess sticky, knoblike processes to which the nematodes adhere. But the most unusual of all is the rabbit-snare trap, of which there are two types, nonconstricting and constricting. Both possess rings of filaments attached by short branches to the main filament. Each ring trap consists of three curved cells; and its inside diameter is just large enough to permit a nematode attempting to pass through to become wedged and unable to withdraw. In the constricting ring the friction of the nematode's body stimulates the ring cells to inflate to about three times their former volume and to grip the nematode in a stranglehold. The response is rapid; complete distention of the cells is accomplished within one-tenth of a second.

Other groups of animals, although feeding largely on the surface and contributing little to litter breakdown, are important as soil mixers. Ants are especially important as soil animals, for they are widely distributed, pioneer new sites, and bring up large quantities of soil from below ground. Prairie dogs raise earth from lower levels and deposit it at the surface, where it is broken down by weathering and incorporated with organic material. They carry surface soil down to plug passageways, and on clay soils they increase the proportion of fine soil particles on the surface. Moles, too, move considerable quantities of earth, although the amount has not been calculated. Their varied influences include improving the natural drainage and aeration of soil and increasing organic matter by burying surface vegetation and litter under their hills.

### FUNCTION

The soil community is largely heterotrophic, dependent upon energy fixed by the green plants it supports (Figure 9-6). Eighty to 90 percent of the energy bound in the litter and available to the community is captured by the microfloral decomposers, mostly the fungi, the most important functional group in the soil community. The remaining 10 to 20 percent is divided among the numerous and highly diverse groups of soil fauna.

How that energy is distributed is largely unknown and because of the complexity involved in studying the soil fauna, it will probably remain unknown for a long time. To discover how energy flow is partitioned one needs to know at least population density, biomass, fluctuations in both, and feeding, assimilation, and respiration rates for each population. En-

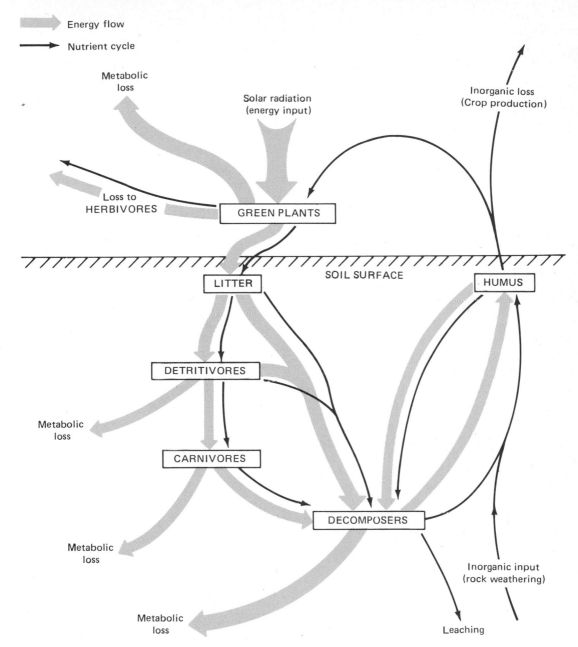

Energy flow

Nutrient cycle

Metabolic loss

Solar radiation (energy input)

Inorganic loss (Crop production)

Loss to HERBIVORES

GREEN PLANTS

SOIL SURFACE

LITTER

HUMUS

DETRITIVORES

Metabolic loss

CARNIVORES

Metabolic loss

DECOMPOSERS

Inorganic input (rock weathering)

Metabolic loss

Leaching

**FIGURE 9-6**

*Energy flow and nutrient cycling in the soil system. Note that the soil ecosystem, if it can properly be called one, is heterotrophic with its food chain dependent upon the autotrophic community above it. The diagram emphasizes the role of the soil as the site of decomposition and nutrient exchange. (Adapted from Wallwork, 1973.)*

gelmann (1961) investigated the functional role of soil arthropods, especially the herbivorous orbatid mites in an old field community. By gathering data on biomass, respiration, and caloric flow, he estimated that in 1 m² of soil 12.5 cm deep the mites consumed 10,248 calories of food a year and assimilated 2,085 calories or 20 percent of the food ingested (Figure 9-7). Respiration accounted for 96 percent of the energy assimilated, leaving little for production. Assuming the population in a steady state, Engelmann found that the mite biomass was replaced each year. The main role of the orbatid mites was to control the fungal and bacterial populations breaking down the dead litter.

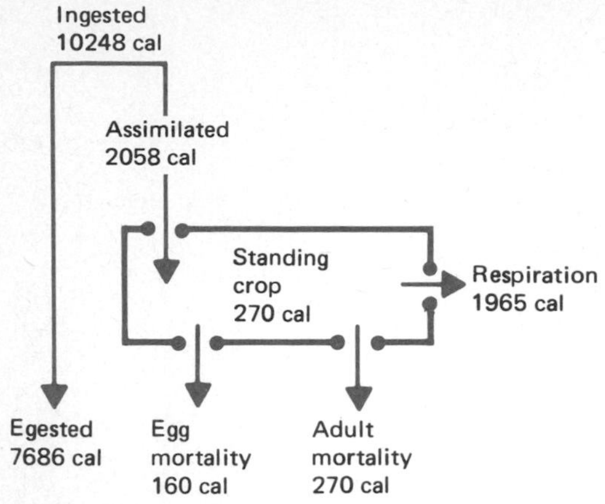


*FIGURE 9-7*
***Annual energy budget for a mite population of a square meter of old field grassland in Michigan. (Engelmann, 1961.)***

Feeding upon the soil herbivores are soil carnivores. Less is known of the flow of energy from herbivores to carnivores than from detritus to herbivores. Engelmann attempted to estimate this. He restricted his analysis to several groups of herbivores, two groups of carnivorous mites, and Japygidae (primitive, wingless insects with forcepslike structures at the caudal end of the body), which may be omnivorous or carnivorous. Using data on respiration rates for the various groups of herbivores and carnivores, Engelmann estimated ecological efficiency, the flow of energy from herbivores to carnivores, to range between 8 and 30 percent.

## *Grasslands*

When the explorers looked out across the prairies for the first time, they witnessed a scene they had never before experienced. Nowhere in all western Europe had they seen anything similar. Lacking any other name to call them, the explorers named these grasslands "prairie," from the French, meaning "grassland." This was the North American prairie and plains, the climax grassland that occupied the midcontinent (Figure 9-8). It was one of the several great grassland regions in the world, including the steppes of Russia, the pusztas of Hungary, the South African veld, and the South American pampas (see inside cover). In fact at one time grasslands covered about 42 percent of the land surface of the world, but today much of it is under cultivation. All have in common a climate characterized by high rates of evaporation, periodic severe droughts, a rolling-to-flat terrain, and animal life that is dominated by grazing and burrowing species. They occur largely where rainfall is between 10 and 30 in./yr, too light to support a heavy forest growth and too heavy to encourage a desert. Grasslands, however, are not exclusively a climatic formation—most of them require periodic fires for maintenance and renewal and for the elimination of incoming woody growth.

Grasses that make up the haylands, pastures, and prairies are either sod formers or bunch grasses. As the names imply, the former develop a solid mat of grass over the ground, and the latter grow in bunches (Figure 9-9), the space between which is occupied by other plants, usually herbs. Orchardgrass, broomsedge, crested wheatgrass, and little bluestem are typical bunch grasses, which form clumps by the erect growth of all shoots and spread at the base by tillers. Sod-forming grasses, which include such species as Kentucky bluegrass and western wheatgrass, reproduce and spread by underground stems. Some grasses may be either sod or bunch, depending upon the local environment. Big bluestem will develop a sod on a rich, moist soil and form bunches on a dry soil.

Associated with grasses are a variety of legumes and forbs. Cultivated haylands and pastures usually are planted to a mixture of grasses and such legumes as alfalfa and red clover. With these may grow unwanted plants, such as mustard, dandelion, and daisy. Seral grasslands often consist of a mixture of native grasses, such as timothy and bluegrass, and an assortment of herbaceous plants, including cinquefoil, wild strawberry, daisy, dewberry, and goldenrod. On the prairie legumes and forbs, particularly the composites, are important components of the climax grassland (J. E. Weaver, 1954). From spring to fall the color and aspects of the grassland change from pasqueflower and buttercups to goldenrod.

**FIGURE 9-8**
*Grasslands of central North America. The extensive, unbroken grasslands supported vast herds of bison and other ungulates, such as the pronghorn antelope and their associated predators. (Photo courtesy South Dakota Fish and Game Commission.)*

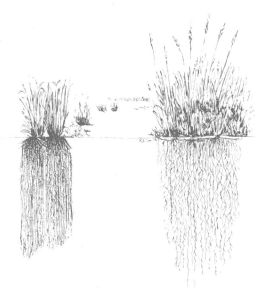

**FIGURE 9-9**
*Growth forms of a sod grass and two bunch grasses, crested wheatgrass and little bluestem (left). Also shown are root penetration and distribution in the soil (maximum depth about 2.5 meters.)*

## TYPES OF GRASSLANDS

***Tame and successional grasslands.*** Grasslands in normally forested regions are either tame or successional. In highly developed agricultural areas, such as eastern and central North America, Great Britain, and Europe, tame or cultivated grasslands are the major representatives of their class, although a few natural types do exist. By clearing the forests, humans developed tame grasslands principally as a source of food for livestock. In some agricultural regions, especially New England and the Lake States, grasslands were planted and then abandoned to revert to forest. In other regions, especially Britain, some grasslands have existed for centuries, becoming a sort of climax community supporting its own distinctive vegetation (see Duffey et al., 1974).

Tame grasslands can be classified as *permanent,* in grass over seven years and managed for hay or pasture; *temporary* or *rotational,* plowed every three to five years for crop production; and *rough,* marginal, unimproved, semiwild lands used principally for grazing. Many seral or successional grasslands fit the latter category.

Ecologically permanent hay fields and grazing lands differ from rotational hay fields. Permanent haylands, more common in Britain than in North America, and permanent grazing lands consist of species adapted to periodic defoliation by cutting and grazing. They consist chiefly of hemicryptophytes that produce their maximum leaf area in the early part of the season. Overgrazed pasturelands are poorer in species than are pasturelands subject to rotational grazing, in which part of the field is allowed to recover while another section is grazed. Fertilization reduces diversity by favoring more competitive grasses and reducing the proportion of forbs.

Rotational or temporary hay fields are dominated by two or three cultivated species, usually two grasses and a legume. Such hay fields are denser and ranker in growth than are permanent and seral grasslands. Management consists of fertilization, mowing, and plowing at regular intervals for growing other crops in the rotation. Such hay fields can provide an excellent habitat for grassland wildlife, but early season mowing destroys cover at the beginning or the height of the nesting season and exposes the ground surface to high solar radiation in late spring and early summer. Periodic destruction of grasslands every three to five years is less of a problem to wildlife if hay fields are established in adjacent fields while the original hay field is converted to cropland. The fauna simply moves from the plowed land to new hay fields.

*Tall-grass prairie.* Tall-grass prairie occupies, or rather occupied, a narrow belt running north and south next to the deciduous forest (see Figure 9-1). In fact, it was well developed within a region that could support forests. Oak-hickory forests did extend into the grasslands along streams and rivers, on well-drained soils, sandy areas, and hills. Prairie fires, often set by Indians in the fall, stimulated a vigorous growth of grass and eliminated the encroaching forest. When settlers eliminated fire, oaks invaded and overtook the grasslands (Curtis, 1959).

Big bluestem was the dominant grass of moist soils and occupied the valleys of rivers and streams and the lower slopes of the hills. Associated with bluestem were a number of forbs, goldenrods, compass plants, snakeroot, and bedstraw. Although grasses dominated the biomass, they were not numerically superior. Studies on remnant prairies in Wisconsin (Curtis, 1959) show the legumes comprised 7.4 percent of all species, grasses 10.2, composites 26.1. The high percentage of nitrogen-fixing legumes accounts in part for the annual production of 8,500 k/dry matter/ha.

Drier uplands in the tall-grass country were dominated by the bunch-forming needlegrass, side-oats grama, and prairie dropseed. Like the lowland, the drier prairie contained many species other than grass. In Wisconsin composites accounted for 27.5 percent of all species, butterflyweed and legumes 4.6 percent each, and grasses 13.7 (Curtis, 1959). The suggestion has been made that perhaps the xeric prairie might be more appropriately called "daisy-land."

The tall-grass prairie, as well as other types, is a continuously changing series of species ranging from those best adapted to wet, poorly aerated soils, such as slough grass, through those plants represented by big bluestem that flourish on mesic soils of high fertility, to those such as blue grama and a whole host of colorful forbs that dominate the xeric sites. Interestingly, no important genera of grasses or forbs have species that are at their optimum in all sections of the continuum nor have all their species with an optimum at any one particular point on the gradient (Curtis, 1959).

*Mixed-grass prairie.* West of the tall-grass prairie region is the mixed-grass prairie in which midgrasses occupy the lowland and short grasses the higher elevations. The mixed prairie, typical of the northern Great Plains, embraces largely the needlegrass–grama grass community, with needlegrass-wheatgrass dominating gently rolling soils of medium texture (Coupland, 1950). Because the mixed prairie is characterized by great annual extremes in precipitation, its aspect varies widely from year to year. In moist years midgrasses are prevalent, whereas in dry years short grasses and forbs are dominant. The grasses here are largely bunch and cool-season species, which begin their growth in early April, flower in June, and mature in late July and August.

*Short-grass plains.* South and west of the mixed prairie and grading into the desert are the short-grass plains, a country too dry for most midgrasses. The short-grass plains reflect a climate where the rainfall is light and infrequent (25 to 43 cm in the west, 51 cm in the east), the humidity low, the winds high, and the evaporation rapid. Shallow-rooted, the short grasses utilize moisture in the upper soil layers, beneath which is a permanent dry zone into which the roots do not penetrate. Sod-forming blue grama and buffalo grass dominate the short-grass plains. On wet bottomlands, switchgrass, Canada wild rye, and western wheatgrass replace grama and buffalo grass. Because of the dense sod, fewer forbs grow on the plains, but prominent among them is purple lupine.

Just as the tall-grass prairie was destroyed by the plow, so has much of the short-grass plains area been ruined by overgrazing and by plowing for wheat, which, because of low available moisture, the land could not support. Drought, lack of a tight sod cover, and winds turned much of the southern short-grass plains into the Dust Bowl, the recovery from which has taken years.

**Desert grasslands.** From southeastern Texas to southern Arizona and south into Mexico lies the desert grassland, similar in many respects to the short-grass plains except that triple-awn grass replaces buffalo grass (Humphrey, 1958). Composed largely of bunch grasses, the desert grasslands are widely interspersed with other vegetation types, such as oak savanna and mesquite. The climate is hot and dry. Rain falls only during two seasons, summer (July and August) and winter (December to February) in amounts that vary from 30 to 41 cm in the western parts to 51 cm in the east; but evaporation is rapid, up to 203 cm/year. Vegetation puts on most of its annual growth in August. Annual grasses germinate and grow only during the summer rainy season, whereas annual forbs grow mostly in the cool winter and spring months.

Like the tall-grass prairies on the eastern rim of the grasslands, the desert grasslands on the west exist because of fires, which periodically swept across them and eliminated the mesquite, the cacti, and low trees. Without fire the desert grasslands, long before their discovery by white man, would have been a land of low trees with an understory of grasses and small shrubs.

Separated from the midcontinent grasslands by the Rocky Mountains is the California prairie, composed largely of needlegrass and bluegrass. A region of winter rains, much of the California prairie is either under cultivation or is overgrazed.

**Tropical grasslands.** Around the world is a belt of tropical monsoon grasslands that extend from western Africa to eastern China and Australia. Within this belt monsoon grasslands fall into ecoclimatic gradients of arid to semiarid grasslands; medium rainfall grasslands found mainly in India, Burma, and northern Australia; the high-rainfall monsoonal or equatorial grasslands of southeast Asia; and grasslands whose species are adapted to a hot monsoonal summer and cool-to-cold winters (Whyte, 1968). The tropical grasslands of South America do not fall into this group because geographical conditions do not promote true or false monsoonal conditions. Instead the grasslands of Latin America are largely steppe, consisting almost entirely of bunch grass with no legumes and very few herbs, bushes, or trees (McIlroy, 1972).

Within this broad belt continental grasslands fall into their own ecoclines. In Africa, for example, climatic climax grasslands are confined to desert areas with prolonged drought. Grass cover, dominated by *Aristida*, is low and soil may be blown away by the wind. With a slight increase in wetness, desert scrub with scattered shrubs of the genera *Commiphora* and *Acacia* become the climax type. As rainfall increases, desert shrub gives way to desert-grass–Acacia savanna; tall-grass–Acacia savanna; tall-grass–low-tree savanna, and finally humid forests. Thus low rainfall areas are characterized by quick-growing tufted types of low ground cover and high rainfall areas by tall coarse grasses.

Much tropical grassland exists because of fires that prevent the intrusion of woody vegetation. Grass in turn reacts to burning by putting out new shoots and drawing on reserves of moisture and food in the rhizomes and roots, which then are depleted before the arrival of rain. If the grass is overgrazed, the plants are weakened and deteriorate and may be replaced by annual or unpalatable species. On the other hand, the elimination of fire is just as disastrous.

There are a number of differences between tropical and temperate grasses (Steward, 1970). Tropical grasses are lower in crude pro-

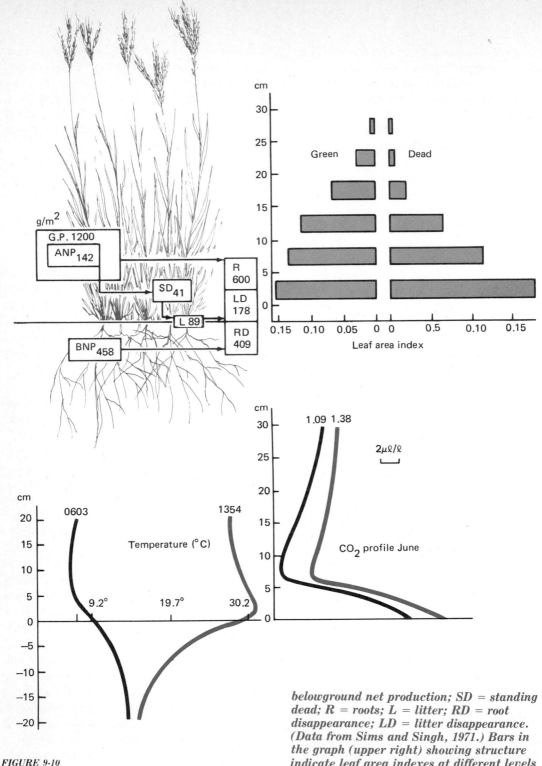

**FIGURE 9-10**

*Profile of a grassland showing energy flow, structure, and stratification of the physical environment. The energy flow diagram (upper left) indicates net primary production in an ungrazed short-grass prairie. Figures in the compartment represent net production during the growing season in g/m², ANP = above ground net production; BNP = belowground net production; SD = standing dead; R = roots; L = litter; RD = root disappearance; LD = litter disappearance. (Data from Sims and Singh, 1971.) Bars in the graph (upper right) showing structure indicate leaf area indexes at different levels for both green and dead plant structures. Note the change in scale for the two. In the temperature profile the actual temperature at soil surface is shown on each curve. In the $CO_2$ profile note the concentration of $CO_2$ near the soil surface, indicating flux from soil, litter, and lower canopy. (Adapted from Ripley and Redmann, 1977.)*

tein and higher in crude fibers. Tropical grasses have maximum photosynthesis at 30° to 35° C, temperate grasses at 15° to 20° C. Tropical grasses have maximum photosynthetic rates of 50 to 70 mg/hr compared with 20 to 30 mg/hr for temperate grasses. Individual leaves of temperate grasses reach light saturation at relatively low levels, whereas leaves of tropical grasses become saturated only at much higher levels. Temperate grasses have high rates of photorespiration; tropical plants do not.

## STRUCTURE

Grasslands possess essentially three strata—roots, ground layer, and herbaceous layer (Figure 9-10). The root layer is more pronounced in grasslands than in any other major community. Half or more of the plant is hidden beneath the soil; in winter this represents almost the total grass plant, a sharp contrast to the leafless trees of the forest. The bulk of the roots occupy rather uniformly the upper 1.6 dm of the soil profile and decrease in abundance with depth. The depth to which the roots of grasses extend is considerable. Little bluestem reaches 1.3 to 1.7 m and forms a dense mat in the soil to 0.8 m (J. E. Weaver, 1954). Roots of blue grama and buffalo grass penetrate vertically to 1 m. In addition, many grasses possess underground stems, or rhizomes, that serve both to propagate the plants and to store food. On the end of the rhizome, which has both nodes and scalelike leaves, is a terminal bud, which develops into aerial stems or new rhizomes. Rhizomes of most species grow at shallow depths, not over 10 to 14 cm deep. The exotic quack grass is notorious for its tough rhizomes. Forbs such as goldenrod, asters, and snakeroot possess large, woody rhizomes and fibrous roots that add to the root mat in the soil. Some, such as snakeroot, have extensive taproots 5 m long, and rushlike lygodesmia, common in many prairies, extends down over 6 m in mellow soil. Among hayland plants, alfalfa possesses a taproot that grows to a considerable depth.

All the roots of grassland plants are not confined to the same general area of the soil but develop in three or more zones. Some plants are shallow rooted and seldom extend much below 6 dm. Others go well below the shallow-rooted species, but seldom more than 1.5 m. Deep-rooted plants extend even further into the soil and absorb relatively little moisture from the surface soils. Thus plant roots absorb nutrients from different depths in the soil at different times, depending upon moisture (J. E. Weaver, 1954).

The ground layer is characterized by low light intensity during the growing season and by reduced windflow. Light intensity decreases as the grass grows taller and furnishes shade. Temperatures decrease as solar insolation is intercepted by a blanket of vegetation, and windflow is at a minimum. Even though the grass tops may move like waves of water, the air on the ground is calm. Conditions on grazed lands are different. Because the grass cover is closely cropped, the ground layer receives much higher solar radiation and is subject to higher temperatures and to greater wind velocity near the surface.

Grasslands, unmowed, unburned, and ungrazed, accumulate a layer of mulch on the ground surface. The oldest layer consists of decayed and fragmented remains of fresh mulch. Fresh mulch consists of residual herbage, leafy and largely undecayed. Three or four years must pass before natural grassland herbage will decompose completely (H. H. Hopkins, 1954). Not until mulch comes in contact with mineral soil does the decomposition process, influenced by compaction and depth, proceed with any rapidity. As the mat increases in depth, more water is retained, creating very favorable conditions for microbial activity (McCalla, 1943).

The amount of accumulated mulch often is enormous. On a relict of a climax prairie organic matter and other humic materials amounted to 885.4 g/m², 581.1 g of it fresh mulch and 50.1 g fresh herbage (Dhysterhaus and Schmutz, 1947). Another prairie supported 461.2 g/m² of fresh mulch and 830.1 g/m² of humus (Dix, 1960).

Grazing reduces mulch, as do fire and mowing. Light grazing tends to increase the weight of humic mulch at the expense of fresh (Dix, 1960); moderate grazing results in increased compaction, which favors an increase in microbial activity and a subsequent reduction in both fresh and humic mulch. Heavy grazing of a stand of bluestem reduced mulch from

1,014.7 g/m² to 100.8 g/m² (Zeller, 1961). An ungrazed North Dakota prairie averaged 441.2 g/m² of mulch compared to 241.5 g/m² for a grazed prairie. Burning reduces both, but the mulch structure returns two to three years after a fire on lightly grazed and ungrazed lands (Tester and Marshall, 1961; Hadley and Kieckhefer, 1963). Mowing greatly reduces fresh mulch and in a matter of time humic mulch also. An unmowed prairie accumulated 4 metric tn of humic mulch/acre; a similar prairie, mowed, had less than 1 metric tn (Dhysterhaus and Schmutz, 1947).

The influence of mulch on grasslands still is a point of controversy (see Tomanek, 1969). Mulch increases soil moisture through its effects on infiltration and evaporation; it decreases runoff, stabilizes soil temperature, and improves conditions for seed germination. In range management the question is how much natural mulch is needed for sustained yield of grass. Where mulch can accumulate, grassland maintains itself; but in areas of little or no accumulation, regression sets in, and the grassland deteriorates to weeds or mesquite. On the other hand, accumulation of mulch from one species can have a toxic effect on the germination and development of another. Heavy mulches can result in pure stands of some species or the invasion of forbs and woody vegetation. Some range ecologists maintain that a heavy mulch results in decreased forage production, smaller root biomass, lower caloric value of living shoots (Hadley and Kieckhefer, 1963), and affects the character and composition of grasslands by reducing understory plants (Weaver and Rowland, 1952).

The herbaceous layer may vary from season to season and from year to year, depending upon the moisture supply. Essentially the layer consists of three or more strata, more or less variable in height, according to the grassland type (Coupland, 1950). Low-growing and ground-hugging plants, such as wild strawberry, cinquefoil, violets, dandelions, and mosses, make up the first stratum. All of these become hidden, as the season progresses, beneath the middle and upper layers. The middle layer consists of shorter grasses and such herbs as wild mustard, coneflower, and daisy fleabane. The upper layer consists of tall grasses and forbs, conspicuous mostly in the fall.

All types of grasslands support similar forms of life. Much of the animal life exists within the several strata of vegetation: the roots, ground layer, and herb cover. Invertebrates, particularly insects, occupy all strata at some time during the year. During winter insect life is confined largely to the soil, litter, and grass crowns where these organisms exist as pupae or eggs. In spring soil occupants are chiefly earthworms and ants. Ants are the most prevalent in some eastern North American meadows, constituting 26 percent of the insect population (Walcott, 1937). Ground and litter layers harbor scavenger carabid beetles and predaceous spiders. Because this layer contains only limited supports for webs, there are more hunters than web builders among spiders.

Life in the herbaceous layer varies as the strata become more pronounced from summer to fall. Here invertebrate life is most diverse and abundant. Homoptera, Coleoptera, Orthoptera, Diptera, Hymenoptera, Hemiptera are all represented. Insect life reaches two highs during the year, a major peak in summer and a less well-defined one in the fall.

Mammals are the most conspicuous vertebrates of the grasslands and the majority of these are herbivores. A large and rich ungulate fauna evolved on the grasslands. The bison and antelope (*Antilocarpa americana*) of North America were equaled only by the richer and more diverse ungulate fauna of the East African plains. Today herds of cattle have replaced the buffalo and rodents and rabbits have the distinction of being the most abundant native vertebrate herbivores.

Grassland animals share some outstanding traits. Hopping or leaping is a common method of locomotion. Strong hind legs enable a variety of animals, such as grasshoppers, jumping mice, jackrabbits, and gazelles, to rise above the grass where visibility is unimpeded. Speed, too, is well developed. Some of the world's fastest mammals, such as antelopes and cheetahs, live in grasslands. Many of the rodents and rabbits are adapted to digging or burrowing. Because of dense grass and lack of trees for singing perches, some grassland birds have conspicuous flight songs that advertise territory and attract mates.

Animal life in tame and successional grasslands depends upon human management for the maintenance of habitat. But mowing for hay, a major management tool, also results in the destruction of habitat at a critical time of year. Nests of rabbits, mice, and birds are exposed at the height of the nesting season. Losses from both mechanical injury and predation are often heavy, although most species remain on the area to complete or reattempt nesting activity.

Pasturelands more often than not are so badly overgrazed that they support little in the way of vertebrate life. Rotation pastures may support more grassland life, but this still needs to be determined.

Successional grassland, because of infertility and plant cover of poverty grass (*Danthonia* spp.) and broomsedge, does not usually support as wide a variety of life as hayland. Poverty grass–dewberry fields are inhabited by grasshopper sparrows (*Ammodramus savannarum),* vesper sparrows (*Pooecetes gramineus),* and meadow mice, but deep grass species, such as meadowlarks (*Sturnella* spp.) and bobolinks (*Dolichonyx oryzivorus),* are few if not entirely absent. Broomsedge fields contain grasshopper sparrow, meadowlarks, and cotton rats. Both successional types offer poor quality food for such herbivores as cottontail rabbits and deer, although deer do feed on young broomsedge sprouts in early spring. Otherwise these dominant grasses are unpalatable to cattle and native herbivores alike.

## FUNCTION

North American grasslands vary in nature, type, species, species composition, and productivity as the pattern of rainfall changes (Table 9-1). Primary production is influenced by a number of environmental variables the most important of which appears to be available water (Figure 9-11). Sparse vegetation in arid regions

**FIGURE 9-11**
*Relationship between efficiency of water use and type of grassland. Efficiency is expressed in total primary production. (Adapted from NAS, 1974.)*

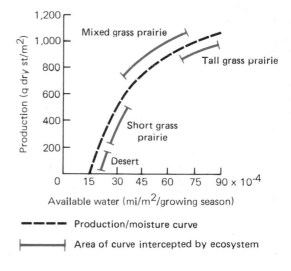

**TABLE 9-1**
*Mean standing crop of selected grasslands, growing season (g/m², oven dry)*

| Standing crop | Desert Ungrazed | Desert Grazed | Short-grass plains Ungrazed | Short-grass plains Grazed | Mixed prairie Ungrazed | Mixed prairie Grazed | True prairie Ungrazed | True prairie Grazed |
|---|---|---|---|---|---|---|---|---|
| Above ground | 81.2 | 49.5 | 69.0 | 42.9 | 154.2 | 94.6 | 256.1 | 220.8 |
|   Cool season grass | 0 | 0 | 6.6 | 2.6 | 101.6 | 9.5 | 4.6 | 24.2 |
|   Warm season grass | 51.3 | 9.4 | 42.6 | 27.1 | 45.8 | 82.1 | 227.1 | 152.6 |
|   Old dead | 0.9 | 1.1 | 25.9 | 18.3 | 80.2 | 58.5 | 204.0 | 44.0 |
|   Mulch | 68.7 | 52.3 | 97.5 | 89.3 | 448.3 | 239.2 | 111.2 | 206.5 |
| Below ground, alive and dead | 197.0 | 185.0 | 1600.0 | 1800.0 | 1213.0 | 2086.0 | 893.0 | 781.0 |
| Rainfall (cm) | 23 | | 30 | | 38 | | 94 | |

*Source:* Adapted from Lewis, 1971.

uses water inefficiently, but most of the water is evaporated from the bare surface of the soil. Adaptations among desert grasses are not for the efficient use of water, but for survival and persistence by other means. In semiarid country where water is limiting, plants apparently have evolved adaptations to make the maximum use of water. In humid, tall-grass country, the vegetative canopy is dense and intercepts most of the solar radiation. As a result adaptations in plants are for efficient capture of light at the expense of efficient use of water. This is reflected in mean standing crops of various grasslands (Table 9-1).

Productivity of grasslands is related to moisture availability. IBP studies of selected North American grassland sites reveal that mean standing crop of a tall-grass prairie was 256 g/m² with rainfall of 94 cm/yr. A mixed-grass prairie with a rainfall of 38 cm/yr had a standing crop of 154 g/m². A short-grass plain with a rainfall of 30 cm/yr had a standing crop of 69 g/m² (Lewis, 1971). For grassland ecosystems in general, data from IBP studies suggest a gross primary productivity of 983 g C/m²/yr; autotrophic respiration, 613 g C; net primary production, 370 g C; heterotrophic respiration, 199 g C; net ecosystem production, 171 g C; and ecosystem respiration 812 g C/m²/yr (NAS, 1974).

Grasslands support a conspicuous herbivore component that utilizes the vegetation in a complex manner. Each species to some degree utilizes the vegetational resource unavailable to or unused by the others. One of the better examples is the sharing of resources among a group of East African ungulates, the zebra, wildebeest, topi, and Thomson's gazelle (Bell, 1974). Resource division is based on the growth form and layering of grassland plants. At the start of the wet season the growing shoots of grass are above the ground, the growing point protected by unfolding leaves. The entire plant has thin cell walls and leaves are accessible to all grazers. In midseason the stems are below the growing point, raising the leaves from the ground. Protein level is reduced because of the structural nature of plants including high cellulose in cell walls. In late stages the growing point extends to the flowering culms of plants. The proportion of cell wall to cell contents is high and digestibility is reduced.

At the end of the wet season grasslands possess levels of different food values and different chemical composition available to the herbivores: low protein culms of grasses, stems and leaves of taller grasses, lower leaves and smaller grasses. Zebras, capable of digesting high cellulose foods, feed on the top herbaceous layer. Wildebeest and topi, requiring higher protein food, feed on the middle layers. Thomson's gazelle, requiring easily digested high protein food, feeds on small grasses and fruits on the ground. This description is highly simplified, for the relationship among the ungulates and between the ungulates and their food supply involves interactions among the metabolic needs of the mammals, the physiology of digestion, growth of grasses, drainage patterns, wet and dry seasonality, and migratory movements of the ungulates (see Bell, 1974).

In addition to large herbivores grasslands support a varied assortment of small mammals, birds, and insects. Small mammals of desert and mountain grasslands and tall-grass prairies are largely grazers or foliage eaters; those of mixed prairies are grazers and seed and fruit consumers; and those of short-grass plains are grazers. Rodents of semiarid desert grasslands are seed consumers (Lewis, 1971). Jackrabbits (*Lepus* spp.) are important consumers, ingesting about 6 percent of their body weight or about a pound of dry weight forage a day. They may waste 50 percent in addition to what they consume.

Among the array of herbivorous insects, grasshoppers and ants are the most important. Grasshoppers can occur in plague proportions and completely denude an area. Even a population of 3 per square yard in a relatively unproductive Wyoming short-grass plains can utilize up to 50 percent of the vegetation (Bullen, 1966). The degree of damage caused by grasshoppers is influenced by the amount of vegetation eaten daily, fluctuations of population size, and food preferences of each species. Some grasshoppers selectively feed on grass, others on forbs, and some on both. In addition to direct consumption, grasshoppers affect growth indirectly by their method of consumption. They eat close to the ground cutting the blades and stem near the crown. Such damage can reduce growth of the plant, prevent reseeding, and in extreme cases kill the plant.

Ants are ubiquitous to grasslands and can influence primary production by seed-gathering and leaf-cutting activities. Where their nests are abundant they can denude areas of grass. However, over a longer period of time the ant's activity of soil mixing and concentration of organic matter and nutrients in the rooting zone can increase net primary productivity.

Birds of grasslands are both granivorous (seed-eaters) and insectivorous. Insectivores feed largely on grasshoppers, Lepidoptera larvae, Hemiptera, and Hymenoptera. Horned larks are the most granivorous of grassland birds and meadowlarks one of the most carnivorous (see Wiens, 1973) (Figure 9-12).

The proportion of insectivorous to granivorous birds in grassland ecosystems is influenced by grazing intensity. The loss of grass cover reduces insect populations and nesting cover, thereby reducing the availability of food for insectivorous birds. Birds on grazed mixed prairie are largely seed-eaters, while those in ungrazed grasslands are insectivorous. Thus grazing reduces meadowlarks and grasshopper sparrows and increases horned larks.

Energy flow in breeding bird populations of grasslands is small, 1.01 to 2.33 kcal/m². Birds probably exert little influence on ecosystem structure, function, or dynamics "through their direct effects either on the flux rate or storage of energy and nutrients" (Wiens, 1973). Instead they may function as regulators of insect and plant populations through their consumption of insects and seeds. Or, Weins suggests, an intriguing possibility is that grassland birds are really "frills" in the ecosystem, "living and producing off its excesses without really influencing it in any way."

Feeding on herbivores are a host of carnivores: rattlesnakes, hawks, owls, badgers, coyotes, and once the wolf. The dominant mammalian carnivore on the North American grasslands is the coyote (Canis latrans), which feeds principally on jackrabbits. Other predators include the badger (Taxidea taxus), which feeds largely on burrowing mammals, and the endangered black-footed ferret (Mustela nigripes), which feeds on prairie dogs. Little information exists on energy flow through these feeding groups.

Less is known about a quantified functional role of decomposer organisms in grassland soils. Earthworms are abundant and their effect on grassland ecosystems as elsewhere is largely catalytic, reducing litter to smaller sizes to be worked over by microflora and fauna.

Grasslands evolved under the grazing pressure of ungulates. However, this grazing was accomplished by free-ranging and often migratory populations of animals whose numbers were checked by large carnivores. When wild ungulates were replaced by domestic ones, humans confined them with fences and often overstocked the range (Figure 9-13). This great change in the nature of grazing pressure had an impact on structure and function of grassland ecosystems.

Although the impact of grazing on grassland ecosystems has been observed over the years (Voight and Weaver, 1951; Weaver, 1954; Humphrey, 1958; Curtis, 1959) only recently have controlled experiments been undertaken to compare grazed with ungrazed grasslands. IBP grassland biome studies have examined grazed and ungrazed systems from a functional point of view. At IBP grassland sites total net primary production (TPN) is greater on grazed (647 g/m²) than on ungrazed sites (508 g/m²). However, grazed grasslands channel more organic matter below ground (406 g/m²) than do ungrazed grasslands (240 g/m²). For grazed grasslands this amounts to 63 percent of TPN going to below-ground net production (BNP) and 37 percent to above-ground net production (ANP). For ungrazed systems about 52 percent goes to BNP and 48 percent to ANP. Efficiency of BNP is greater in grazed than in ungrazed systems (Table 9-2). On cooler sites ungrazed grasslands have greater efficiency in ANP, while in warmer sites grazed grasslands are more efficient in ANP. In general, however, grazed grasslands balance less efficient ANP due to grazing to a greater efficiency of BNP (see Sims and Singh, 1971).

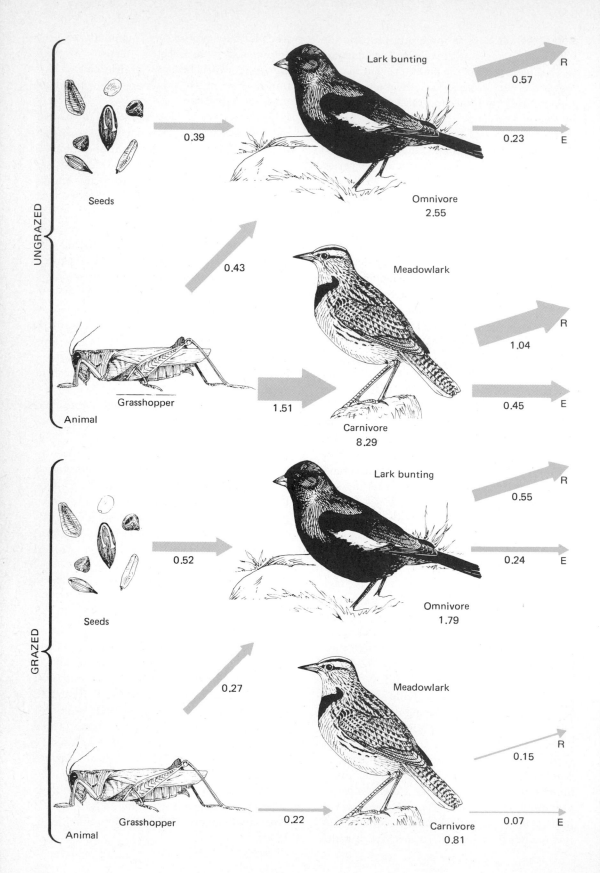

UNGRAZED

Seeds

0.39

Lark bunting

0.57    R

0.23    E

Omnivore
2.55

0.43

Animal

Grasshopper

Meadowlark

1.51

Carnivore
8.29

1.04    R

0.45    E

GRAZED

Seeds

0.52

Lark bunting

0.55    R

0.24    E

Omnivore
1.79

0.27

Animal

Grasshopper

Meadowlark

0.22

Carnivore
0.81

0.15    R

0.07    E

**FIGURE 9-12**
*Pattern and magnitude of energy flow
through a grassland breeding bird population
at the Cottonwood, South Dakota, IBS site,
April 1 to August 31, 1970. Omnivores are
represented by the lark bunting, carnivores
by the meadowlark, which feeds almost
exclusively on insects. Values of arrows are
for energy flow in kilocalories per square
meter per season. The values for omnivores
and carnivores are standing crop x days. R
= respiration; E = egestion. Note the
greater importance of the carnivore
component in ungrazed grassland compared
to grazed grassland. Grazing reduces the
insect population which then affects energy
flow through the carnivore component.
(Adapted from Wiens, 1973.)*

Grasslands respond to grazing by changing species composition. Some grasses and forbs are sensitive to intensive grazing pressure and tend to disappear, while other species increase (Table 9-3). On desert grasslands black grama is replaced by weedy species; on short-grass plains, which are the most stable under grazing pressure, blue grama and prickly pear increase; on mixed-grass prairies midgrasses decrease and short grasses and sedges increase. On tall-grass sites tall grasses disappear and

**FIGURE 9-13**
*Grazing by domestic herbivores. Domestic
ungulates, such as these sheep, have
replaced wild ungulates as the dominant
grazing animals on grasslands. Restricted in
their movements, these grazing herbivores,
whose density and intensity of grazing are
determined by human decisions, have a
tremendous impact on the structure and
function of grassland ecosystems. This is a
well-managed range on a high mountain
meadow in Idaho. (Photo courtesy Soil
Conservation Service.)*

*TABLE 9-2*
*Net primary productivity and efficiency of energy capture by ungrazed and grazed grasslands within the growing season*

| Type | ABOVE GROUND Net production (kcal/m²) | Efficiency (%) | BELOW GROUND Net production (kcal/m²) | Efficiency (%) | TOTAL Net production (kcal/m²) | Efficiency (%) |
|---|---|---|---|---|---|---|
| **GRAZED** | | | | | | |
| Desert | 456 | .06 | 625 | .09 | 1081 | .13 |
| Short-grass plains | 476 | .10 | 3285 | .70 | 3761 | .80 |
| Mixed prairie | 444 | .10 | 1810 | .41 | 2254 | .51 |
| True prairie | 2048 | .42 | 1701 | .35 | 3749 | .77 |
| **UNGRAZED** | | | | | | |
| Desert | 688 | .09 | 489 | .07 | 1177 | .16 |
| Short-grass plain | 568 | .12 | 2153 | .45 | 2721 | .57 |
| Mixed prairie | 788 | .18 | 1264 | .29 | 2052 | .47 |
| True prairie | 1348 | .27 | 872 | .17 | 2220 | .44 |

*Source:* Adapted from Sims and Singh, 1971.

*TABLE 9-3*
*Response of grassland species to grazing disturbance, grassland biome, 1970*

| | | % community standing crop | |
|---|---|---|---|
| *Grassland* | *Species* | *Ungrazed* | *Grazed* |
| Desert | Black grama *(Bouteloua eriopoda)* | 53 | 8 |
| | Broom snakeweed *(Xanthocephalum sarothrae)* | 15 | 26 |
| | Russian thistle *(Salsola kali)* | 3 | 25 |
| Short-grass plains | Blue grama *(Bouteloua gracilis)* | 63 | 72 |
| | Prickley pear *(Opuntia polyacantha)* | 5 | 12 |
| Mixed prairie | Wheat grass *(Agropyron smithii)* | 43 | 9 |
| | Blue grama *(Bouteloua gracilis)* | 9 | 37 |
| | Buffalo grass *(Buchloe dactyloides)* | 28 | 44 |
| True prairie | Little bluestem *(Andropogon scoparius)* | 69 | 23 |
| | Indian grass *(Sorghastrum nutans)* | 7 | t |
| | Japanese chess *(Bromus japonicus)* | 0 | 22 |
| | Tall dropseed *(Sporobolus asper)* | 2 | 37 |

*Source:* Lewis, 1971.

*FIGURE 9-14*
*Effects of overgrazing by cattle on short-grass range. Overgrazing has resulted in the replacement of grass by prickly pear and woody shrubs. The ground is bare, most of the moisture from rains is lost by runoff and evaporation, and erosion is serious in places. (Photo courtesy Soil Conservation Service.)*

little bluestem and tall dropseed increase; if grazing pressure is heavy the site may be invaded by the weedy Japanese chess (Lewis, 1971).

Overgrazing can cause serious deterioration of grassland systems. Overgrazing desert grasslands increases the spread of mesquite because of lessened competition from grass and the spread of seed by domestic stock (Figure 9-14). On other grasslands mulch deteriorates and disappears because little litter is added to the ground. Water runs off the surface taking topsoil with it. Lacking moisture and nutrients, the original plants cannot maintain themselves and the vegetative cover continues to decrease until only an erosion pavement remains. To maintain grasslands in good condition, at least one-third of the year's growth must be left to supply the annual addition to mulch (Dhysterhaus and Schmutz, 1947).

## Shrublands

Covering large portions of the arid and semiarid world is climax shrubby vegetation (see inside cover). In addition climax shrubland exists in parts of temperate regions, because historical disturbances of landscapes have seriously affected their potential to support forest vegetation (Eyre, 1963). Among such shrub-dominated plagioclimaxes are the moors of Scotland and the macchia of South America. But outside of these regions, shrublands are seral, a stage in land's progress back to forest. There they exist as second-class citizens of the plant world (McGinnes, 1972), unfortunately, given little attention by botanists, who tend to emphasize dominant plants. As a result, little work has been done on the seral shrub communities.

### CHARACTERISTICS OF SHRUBS

The success of shrubs depends largely on their abilities to compete for nutrients, energy, and space (West and Tueller, 1972). In certain environments shrubs have many advantages. They have less energetic and nutrient investment in above-ground parts than trees. Their structural modifications affect light interception, heat dissipation, and evaporative losses,

depending upon species and environments involved. The multistemmed forms influence interception and stemflow of moisture, increasing or decreasing infiltration into the soil (Mooney and Dunn, 1970b). Because most shrubs can get their roots down quickly and form extensive root systems, they can utilize soil moisture deep in the profile. This feature gives them a competitive advantage over trees and grasses in regions where the soil moisture recharge comes during the nongrowing season. Because they have a low shoot-to-root ratio, shrubs draw less nutrient input into aboveground biomass and more into roots. Their perennial nature allows immobilization of limiting nutrients and slows the nutrient recycling, favoring further shrub invasion of grasslands. Subject to strong competition from herbs, some climax shrubs, such as chamise (Adenostoma fasciculatum), inhibit the growth of herbs by allelopathy, the secretion of substances toxic to other plants (McPherson and Muller, 1969). Only when fire destroys mature shrubs and degrades the toxins do herbs appear in great numbers. As the shrubs recover, herbs decline. Seeds of herb species affected apparently have evolved the ability to lie dormant in the soil until released from suppression by fire.

### TYPES OF SHRUBLAND

*Mediterranean-type shrublands.* In five regions of the world lying for the most part between 32° and 40° North and South of the equator are areas characterized by a mediterranean climate: the semiarid regions of western North America, the regions bordering the Mediterranean, central Chile, the Cape region of South Africa, and southwestern and southern Australia. The mediterranean climate is characterized by hot, dry summers with at least one month of protracted drought and cool, moist winters. At least 65 percent of the annual precipitation falls during the winter months and for at least one month the temperature remains below 15° C (Aschmann, 1973).

All five areas (see inside cover) support physiognomically similar communities of xeric broadleaf evergreen shrubs and dwarf trees known as broad sclerophyll vegetation (for a detailed description, see McKell et al., 1972; de Castri and Mooney, 1973). Although vegetation in all the mediterranean-type ecosystems

**299**

shares certain characteristics, each has evolved its own distinctive flora and fauna. In the Northern Hemisphere the vegetation evolved from tropical floras and developed in dry summer climates that did not exist until the Pleistocene (Raven, 1973).

In addition to similar forms, vegetation in each of the mediterranean systems also shows similar adaptations to fire and to low nutrient levels in the soils. In these systems of the Northern Hemisphere annuals make up 50 percent of the species and 10 percent of the plant genera, and 40 percent of the species are endemics.

There are several variations of the basic mediterranean-type ecosystem. In the Mediterranean region shrub vegetation often results from forest degradation and falls into three major types. The *garrigue,* resulting from degradation of pine forests, is low, open shrubland on well-drained to dry calcareous soil. The *maqui,* replacing cork forests, is higher, thick shrubland in areas of more rainfall. The *mattoral,* which is also found in Chile, combines the other two types and appears to be equivalent to the North American chaparral (see Soriano, 1972).

In southwestern Australia the shrub country, known as *mallee,* is dominated by low-growing *Eucalyptus.*

In North America the sclerophyllic shrub community is known as *chaparral.* There are two types: the California chaparral, dominated by shrub oaks and chamise, and the inland chaparral of Arizona, New Mexico, Nevada, and elsewhere, dominated by Gambel oak and other species and lacking chamise.

For the most part these vegetation types lack an understory and ground litter, are highly inflammable, and are heavy seeders. Many species require the heat and scarring action of fire to induce germination.

For centuries periodic fires have roared through mediterranean-type vegetation, clearing away the old growth, making way for the new, and recycling nutrients through the ecosystem. When humans intruded into this type of vegetation, they changed the fire situation, either by attempting to exclude fire completely or by overburning. In the absence of fire chaparral grows tall and dense and yearly adds more leaves and twigs to those already on the ground. During the dry season the shrubs, even though alive, will nearly explode when ignited. Once set on fire by lightning or humans, an inferno follows.

After fire the land returns either to lush green sprouts coming up from buried root crowns or to grass, if a seed source is nearby. The grass and vigorous young sprouts are excellent food for deer, sheep, and cattle. But as the sprout growth matures chaparral becomes dense, the canopy closes, the litter accumulates, and the stage is set for another fire.

Because of the rough terrain characteristic of the mediterranean-type ecosystem some areas have remained relatively undisturbed, especially in California and South Africa. But in Australia and the Mediterranean basin human activity, especially animal raising and fruit and vegetable farming, have degraded the broad-leaf sclerophyllous vegetation.

**Sagebrush.** In the Great Basin of North America, the northern, cool, arid region lying east of the Rocky Mountains, is the northern desert scrub. Although this region is perhaps more appropriately considered a desert, it is one of the most important shrublands in North America (Figure 9-15). Its physiognomy differs greatly from the southern hot desert and the dominant vegetation is shrub. The vegetation falls into two main associations: sagebrush, dominated by *Artemisia tridentata* which often forms pure stands, and shad scale, dominated by *Atriplex confertifolia.* Inhabiting this shrubland are pocket and kangaroo mice, lizards, sage grouse, sage thrasher, sage sparrow, and Brewers sparrow, four birds that depend upon sagebrush for their continued existence.

**Successional shrublands.** On drier uplands shrubs rarely exert complete dominance over herbs and grass. Instead the plants are scattered or clumped in grassy fields, the open areas between filled with the seedlings of forest trees, which in the sapling stage of growth occupy the same ecological position as tall shrubs (Figure 9-16). Typical are the hazelnut, forming thickets in places, sumacs, chokecherry, and shrub dogwoods.

On wet ground the plant community often is dominated by tall shrubs and contains an understory intermediate between that of a meadow and a forest (Curtis, 1959). In north-

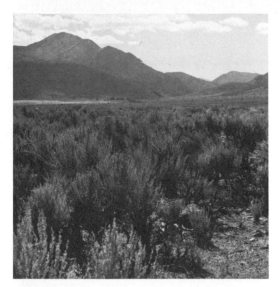

**FIGURE 9-15**
*Sagebrush in Wyoming. Although classified as cold desert, sagebrush country forms one of the most important shrub types in North America.*

**FIGURE 9-16**
*In eastern North America and in northern and western Europe shrublands are usually successional communities; but if the vegetation is dense shrub, these communities may persist for a very long time. This slope is claimed by St Johnswort (Hypcricum virginicum) and wild indigo (Baptisia tinctoria).*

ern regions the common tall shrub community found along streams and around lakes is the alder thicket, composed of alder or alder and a mixture of other species such as willow and red osier dogwood. Alder thickets are relatively stable and remain for some time before being replaced by the forest. Out of the alder country, the shrub carr community (carr is an English name for wet-ground shrub communities) occupies the low places. Dogwoods are some of the most important species in the carr. Growing with them are a number of willows, which as a group usually dominate the community.

Shrub thickets are valued as food and cover for game; and many shrubs, such as hawthorn, blackberry, sweetbrier, and dogwoods, rank high as game food. However, the overall value of different types of shrub cover, its composition, quality, and the minimum amounts needed, have never been assessed. There is some evidence (Egler, 1953; Niering and Egler, 1955) that even where the forest is the normal end of succession, shrubs can form a stable community that will persist for many years. If incoming tree growth is removed either by selective spraying or by cutting, shrubs eventually form a closed community resistant to further invasion by trees. This could have wide application to the management of power line rights-of-way.

### STRUCTURE

Shrubs are difficult to characterize. They have, as McGinnes (1972) points out, a "problem in establishing their identity." They constitute neither a taxonomic nor an evolutionary category (Stebbins, 1972). One definition is that a shrub is a plant with woody persistent stems, no central trunk, and a height of up to 15 to 20 ft. But size does not set shrubs apart because under severe environmental conditions many trees will not exceed that size. Some trees, particularly coppice stands, are multiple-stemmed, and some shrubs have large single stems. Shrubs may have evolved either from trees or herbs (for detailed discussion on evolution of shrubs, see Stebbins, 1972).

Shrub ecosystems, seral or climax, are characterized by woody structure, increased stratification over grasslands, dense branching on a fine scale, and low height. Dense growth of many shrub types, such as haw-thorne and alder, develop nearly impenetrable thickets that offer protection for such animals as rabbits and quail and nesting sites for birds. Deep shade beneath discourages any under-story growth. Many seral shrubs have flowers attractive to insects (and to man) and seeds that are enclosed in palatable fruits and are easily dispersed by birds and mammals.

Shrub communities have their own distinc-tive animal life, common to the shrubby edges of forests and shrubby borders of fields. In fact some species, such as the bobwhite quail and cottontail rabbit, depend heavily on seral shrub communities and disappear if they are destroyed. Alder flycatchers (*Empidonax traillii*) and swamp sparrows (*Melospiza geor-giana*) are typical of alder thickets; indigo buntings (*Passerina cyanea*), field sparrows (*Spizella pusilla*), and towhee (*Pipilo ery-throphthalmus*) are found in old field com-munities. In Great Britain some shrub com-munities, especially the hedgerows, have been stable for centuries and many forms of animal life, invertebrate and vertebrate, have become adapted to and dependent upon them. Among such animals are the whitethroat (*Sylvia communis*), the linnet (*Acanthis cannabina*), the blackbird (*Turdus merula*), and the reed bunting (*Emberiza citrinella*).

Climax shrub communities have a complex of animal life that varies with the region. North American chaparral and sagebrush communities support a variety of rodents, as well as jackrabbits, mourning doves, the California thrasher (*Toxostoma redivi-vum*), wren-tit (*Chamaea fasciata*), California quail (*Lophortyx californica*), mule deer, and coyote.

## FUNCTION

The functional aspects of shrubland ecosys-tems are poorly studied, but data from a few mediterranean-type systems provide some in-teresting insights into nutrient cycling.

In the California chaparral precipitation falls mostly in the cool winter months. About 8 percent of this precipitation is intercepted and is evaporated, 15 percent flows into streams, 40 percent penetrates substrate to flow as groundwater, and about 33 percent is lost as evapotranspiration (Mooney and Parsons,

1973). Most of the plant growth and flowering is concentrated in spring; 75 percent of flower-ing plants, half of them annuals, bloom in May at the end of the rainy period. However, the evergreen dominants fix carbon throughout the year. The amount fixed in summer depends upon the amount of rainfall received during the previous winter. Yearly accumulation of above-ground biomass for a sclerophyllous plant of intermediate age is about 1000 kg/ha (Mooney and Parsons, 1973).

The soils of mediterranean-type ecosystems are low in nutrients and are especially deficient in nitrogen and phosphorus. The cy-cling of these nutrients appears to be tight and conservative. Nitrates in the soil vary season-ally, reflecting the rainfall or the lack of it and the microbial activity which is relatively high during the winter. A flush of microbial activ-ity which involves decomposition of humus and mineralization of nitrogen and carbon follows wetting of dry soil. Nitrate accum-ulation depends upon a progressive drying period after rains. The topsoil gradually dries out to an increasing depth. The resulting im-proved soil aeration and increasing soil tem-perature favor rapid bacterial nitrification and the retention of nitrate ions in the soil. As the dry cycle continues, nitrates accumulate and remain fixed in the topsoil along with other soluble nutrients (Schaefer, 1973). When the rains arrive and wet the soil, the concentration of nutrients stimulates a flush of growth. If heavy rains suddenly enter dry topsoil, quan-tities of nutrients may be lost by leaching. In California chaparral much of the nitrogen re-turned to the soil is lost through erosion (Mooney and Parsons, 1973).

Some plants of the mediterranean systems exhibit some nutrient conservation mech-anisms. *Ceanothus*, an early successional species in the California chaparral, is a nitro-gen fixer (Mooney and Parsons, 1973). In the Australian ecosystem *Atriplex vesicana*, the dominant plant, lowers the nitrogen content of the surrounding soil and concentrates the ni-trogen through litterfall in the soil directly be-neath the plant. More nitrogen, however, is withdrawn from the soil than is returned by litterfall, which represents about 10 percent of total plant nitrogen. Litter has lower nitrogen and phosphorus content than fresh leaves, which suggests that the plant withdraws nitrogen and phosphorus from the leaf into the stem before the leaf falls.

Australian sclerophyllous shrubs also exhibit phosphorus conservation involving three mechanisms. First, phosphorus, like nitrogen, in aging leaves is recirculated through the plant with only minimal losses to litter. Second, a fine mat of proteoid or mycorrhizal roots penetrates the decomposing litter and takes up phosphorus. Third, polyphosphate forming and hydrolyzing enzymes are present in the roots of sclerophyllous plants. As orthophosphate is released from decomposing litter in spring, it is stored in roots as long-chain polyphosphate. When growth begins the polyphosphate seems to be hydrolyzed back to orthophosphate and transported to the growing shoots (Specht, 1973).

The best available data for mineral cycling in a mediterranean-type ecosystem are for a 17-year-old *Quercus coccifera* garrigue in southern France (Figure 9-17). The total above-ground mineral mass is 773 kg/ha; of this 629 kg is in wood, 144 kg in leaves. Calcium is the most important element, amounting to 485 kg/ha, followed by nitrogen, 159

kg/ha, and potassium, 85 kg/ha. The mean annual incorporation of biogenic elements into perennial organs amounts to 37 kg/ha, of which two-thirds is calcium. Annual returns of elements through the litter amounts to 75.2 kg/ha. Net primary productivity is on the order of 3.4 tn/ha/yr with a mean annual production of litter of 2.3 tn/ha (Lossaint, 1973).

## Deserts

Deserts are defined by geographers as land where evaporation exceeds rainfall. No specific rainfall can be used as a criterion, but deserts may range from extremely arid ones to those with sufficient moisture to support a variety of life. Deserts occur in two distinct belts about the earth, one confined roughly about the Tropic of Cancer, the other about the Tropic of Capricorn (see inside cover).

**FIGURE 9-17**
*Annual turnover of macronutrients in the* Quercus coccifera *garrigue at Le Puech de Juge near Montpellier in southern France. (After Lossaint, 1973.)*

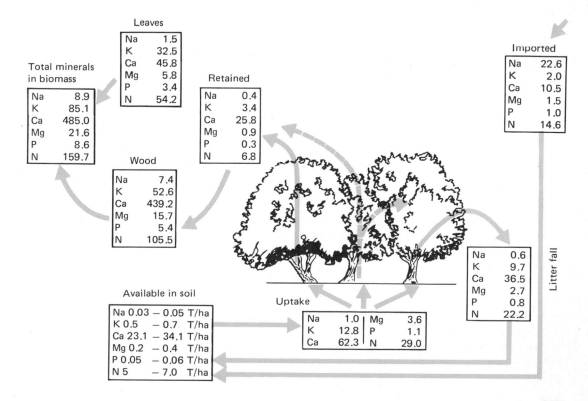

| Leaves | |
|---|---|
| Na | 1.5 |
| K | 32.5 |
| Ca | 45.8 |
| Mg | 5.8 |
| P | 3.4 |
| N | 54.2 |

| Total minerals in biomass | |
|---|---|
| Na | 8.9 |
| K | 85.1 |
| Ca | 485.0 |
| Mg | 21.6 |
| P | 8.6 |
| N | 159.7 |

| Retained | |
|---|---|
| Na | 0.4 |
| K | 3.4 |
| Ca | 25.8 |
| Mg | 0.9 |
| P | 0.3 |
| N | 6.8 |

| Imported | |
|---|---|
| Na | 22.6 |
| K | 2.0 |
| Ca | 10.5 |
| Mg | 1.5 |
| P | 1.0 |
| N | 14.6 |

| Wood | |
|---|---|
| Na | 7.4 |
| K | 52.6 |
| Ca | 439.2 |
| Mg | 15.7 |
| P | 5.4 |
| N | 105.5 |

| Litter fall | |
|---|---|
| Na | 0.6 |
| K | 9.7 |
| Ca | 36.5 |
| Mg | 2.7 |
| P | 0.8 |
| N | 22.2 |

| Available in soil | |
|---|---|
| Na | 0.03 – 0.05 T/ha |
| K | 0.5 – 0.7 T/ha |
| Ca | 23.1 – 34.1 T/ha |
| Mg | 0.2 – 0.4 T/ha |
| P | 0.05 – 0.06 T/ha |
| N | 5 – 7.0 T/ha |

| Uptake | | | |
|---|---|---|---|
| Na | 1.0 | Mg | 3.6 |
| K | 12.8 | P | 1.1 |
| Ca | 62.3 | N | 29.0 |

Deserts are the result of several forces. One that leads to the formation of deserts and the broad climatic regions of the earth is the movement of air masses over the earth (see Chapter 4). High-pressure areas alter the course of rain. The high-pressure cell off the coast of California and Mexico deflects rainstorms moving south from Alaska to the east and prevents moisture from reaching the southwest. In winter high-pressure areas move southward, allowing winter rains to reach southern California and parts of the North American desert. Winds blowing over cold waters become cold also; they carry very little moisture and produce little rain. Thus the west coast of California and Baja California, the Namib desert on coastal southwest Africa, and the coastal edge of the Atacama in Chile may be shrouded in mist, yet remain extremely dry.

Mountain ranges also play a role in desert formation by causing a rain shadow on their lee side. The High Sierras and the Cascade Mountains intercept rain from the Pacific and help maintain the arid conditions of the North American desert. The low eastern highlands of Australia effectively block the southeast trade winds from the interior of that continent. Other deserts, such as the Gobi and the interior of the Sahara, are so remote from the source of oceanic moisture that all the water has been wrung from the winds by the time they reach those regions.

## CHARACTERISTICS OF DESERTS

All deserts have in common low rainfall, high evaporation (7 to 50 times as much as precipitation), and a wide daily range in temperature, from hot by day to cool by night. Low humidity allows up to 90 percent of solar insolation to penetrate the atmosphere and heat the ground. At night the desert yields the accumulated heat of the day back to the atmosphere. Rain, when it falls, is often heavy and, unable to soak into the dry earth, rushes off in torrents to basins below.

The topography of the desert, unobscured by vegetation, is stark and, paradoxically, partially shaped by water. Unprotected, the soil erodes easily during violent storms and is further cut away by the wind. Alluvial fans stretch away from eroded, angular peaks of more resistant rocks. They join to form deep expanses of debris, the *bajadas*. Eventually the slopes level off to low basins, or *playas*, which receive waters that rush down from the hills and water-cut canyons, or *arroyos*. These basins hold temporary lakes after the rains, but water soon evaporates and leaves behind a dry bed of glistening salt.

The aridity of the desert may seem inimical to life, yet in the deserts life does exist, surprisingly abundant, varied, and well adapted to withstand or circumvent the scarcity of water.

The desert is not the same everywhere. Differences in moisture, temperature, soil drainage, alkalinity, and salinity result in variations in the amount of vegetation, the dominant plants, and the groups of associated plants. There are hot deserts and cool deserts, extremely dry deserts and ones with sufficient moisture to verge on being grasslands or shrublands. The Sahara desert, the largest, has minimal vegetation, mostly clustered about oases. The Arabian desert has its tamarisk, and the central Asian desert has saltbrush and saxual. The North American desert can be divided into two parts. One is the northern cool desert of the Great Basin. In terms of vegetation it can be considered a shrub steppe. The northern part is dominated by nearly pure stands of big sagebrush *(Artemisia tridentata)*. The southern part is dominated by shadscale *(Atriplex confertifolia)* and bud sage *(Artemisia spinescens)*. The deserts of the southwest—the Mohave, the Sonoran, and the Chihuahuan—are hot deserts. They are dominated largely by creosote bush *(Larrea divaricata)* and burro bush *(Franseria* spp.). Areas of favorable moisture and soil support tall growths of acacia *(Acacia* spp.), saguaro *(Cereus giganteus),* palo verde *(Cercidium* spp.), and ocotillo *(Fouquieria* spp.).

## STRUCTURE

Woody-stemmed and soft brittle-stemmed shrubs are characteristic desert plants (Figure 9-18). In the matrix of shrubs grows a wide assortment of other plants, the yuccas, cacti, small trees, and ephemerals. In the southwestern North American desert large succulents rise above the shrub level and change the aspect of the desert far out of proportion to

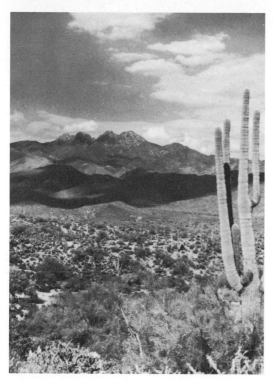

**FIGURE 9-18**
*The hot desert in southwestern United States. This photo shows a palo verde–saguaro community. Rainfall in the region is about 35 to 40 cm. In the distance the east-facing slopes are covered with interior chaparral, a mediterranean-type ecosystem. (Photo courtesy Arizona Fish and Game Commission.)*

their numbers. Like forest trees and prairie grasses, most desert species grow their best in certain topographical situations. The giant saguaro, the most massive of all cacti, grows on the bajadas of the Sonoran desert. Ironwood, smoketree, and palo verde grow best along the banks of intermittent streams, not so much because their moisture requirements demand it, but because their hard-coated seeds must be scraped and bruised by the grinding action of sand and gravel during flash floods before they can germinate.

The plants are adapted to the scarcity of water either by drought evasion or drought resistance. Drought evaders flower only in the presence of some moisture. They persist as seeds during drought seasons, ready to sprout, flower, and produce seeds when moisture and temperature are favorable. There are two periods of flowering in the North American desert, after winter rains come in from the Pacific northwest and after summer rains move up from the southwest out of the Gulf of Mexico. Some species flower only after winter rains, others only after summer rains, while a few bloom during both seasons.

Drought resisters have evolved means of storing water, locating underground water, or reducing the need for water, adaptations described in Chapter 3. These include reduction of leaf size, shedding of leaves, succulence, and fixation of $CO_2$ by night and the closing of stomata by day.

Desert animals are also drought resisters or drought evaders. Drought evaders, like ephemeral plants, adopt an annual life-style or go into estivation or some other stage of dormancy during the dry season. Included are many insects, some amphibians, such as the spadefoot toad, and small rodents. Birds nest during the rainy season when food is most abundant for the young. If extreme drought develops during the breeding season, some birds do not reproduce (Keast, 1959). A few birds, such as the poorwill, swift, and Allen and Anna's hummingbirds, become torpid when food is scarce (Cade, 1957).

Drought resisters are animals that are active the year round and have evolved ways to circumvent aridity by physiological adaptations, such as increasing radiation of heat from the body, by modifying feeding and activity patterns, and by adopting nocturnal habits and remaining underground or in the shade during the day.

### FUNCTION
The desert ecosystem differs from other ecosystems in at least one important way: the input of precipitation is highly discontinuous. It comes in pulses as clusters of rainy days 3 to 15 times a year and of these only 1 to 6 may be large enough to stimulate biologic activity. Thus the desert ecosystem experiences periods of inactive steady states broken by periods of production and reproduction. Both processes are stimulated by rain and continue through short periods of adequate moisture; when water is scarce again, both production and biomass return to some low steady state.

Primary production in the desert depends upon the proportion of available water used and the efficiency of its use. Data from various deserts in the world (for a summary, see Noy-Mier, 1973, 1974) suggest that annual net primary production of above-ground vegetation varies from 30 to 200 g/m².

The amount of biomass that accumulates and rate of turnover (ratio of production to biomass) depend upon the dominant type of vegetation. In deserts such as the Sonoran, where trees, shrubs, and cacti dominate, annual productivity is about 10 to 20 percent of the above-ground standing crop biomass of 300 to 1000 g/m². In deserts with perennial type vegetation annual production is 20 to 40 percent of biomass of 150 to 600 g/m². And, of course, annual or ephemeral communities have a 100 percent turnover of both roots and above-ground foliage; annual production is the same as peak biomass. These turnover rates are higher than those of forests and tundra. The ratio of below-ground biomass to above-ground (stems and foliage) for perennial grasses and forbs is between 1 and 20 and for shrubs between 1 and 3. In general deserts, unlike the tundra, do not have a high root-to-shoot ratio. The root biomass is relatively small, a characteristic of desert plants.

Adding to primary production in the desert are lichens and green and blue-green algae abundant as soil crusts. These blue-green algal crusts, whose biomass ranges up to 240 kg/ha, have the unusually high rate of nitrogen fixation of 10 to 20 g/m²/yr. In spite of high fixation rates only 5 to 10 g/m² of total nitrogen input becomes part of higher plants (Figure 9-19). Approximately 70 percent of the nitrogen is short-circuited back to the atmosphere as volatilized ammonia and as $N_2$ from denitrification speeded up by dry alkaline soils (Reichle, 1975).

Grazing herbivores of the desert are opportunists and generalists in their mode of feeding, that is, they consume a wide range of species, plant types, and parts. Desert sheep feed on succulents and emphemerals when available and then switch to woody browse during the dry period. As a last resort herbivores may consume dead litter and lichens. Small herbivores such as desert rodents and ants feed largely on seeds. One of the most notable small herbivores of the California desert is the harvester ant (*Pogonomyrmex occidentalis*) which lives on seeds gathered from the desert floor and stored in underground graneries. During periods of drought these ants gather mainly the seeds of two kinds of plants, wooly plantain and comb bur. During the period of winter rains, when annual plants flower and seed, ants gather these seeds and ignore seeds of plantain and comb bur. The utilization of different plants at different seasons provides a constant food source and enables the insect to remain active through the desert year.

Herbivores can have a pronounced impact on primary producers, especially if they are more abundant then the range's capacity to support them. Once grazers have utilized annual production, they consume plant reserves, especially in long dry periods. Overbrowsing can so weaken the plant that vegetation is destroyed or irreparably damaged. Areas protected from grazing, especially by goats and cattle, have higher biomass and a greater percentage of palatable species than grazed areas. In a study of the giant saguaro (*Cereus giganteus*) William Niering (1963) and his associates found that overgrazing was destroying the saguaro not only by physical damage from trampling, but also by the loss of nurse plants such as palo verde that provide the shade necessary for the survival of seedlings. Two desert herbivores the wood rat (*Nemotoma albigula*) and the black-tailed jackrabbit (*Lepus californicus*) are more abundant on grazed than on ungrazed desert. Wood rats tunnel into the cactus, increasing its susceptibility to a fatal bacterial disease. The jackrabbits feed on palo verde, destroying the nurse plant.

Herbivores in a shrubby desert, under most conditions, consume only a small part of the above-ground primary production, but seed-eating herbivores can consume most of the seed production. In one of the few studies of herbivory in the desert Chew and Chew (1970) found that small grazing herbivores (the jackrabbit and kangaroo rat) used only about 2 percent of the above-ground net primary production, but consumed 87 percent of the seed production, a rate of consumption that could have a pronounced effect on plant composition

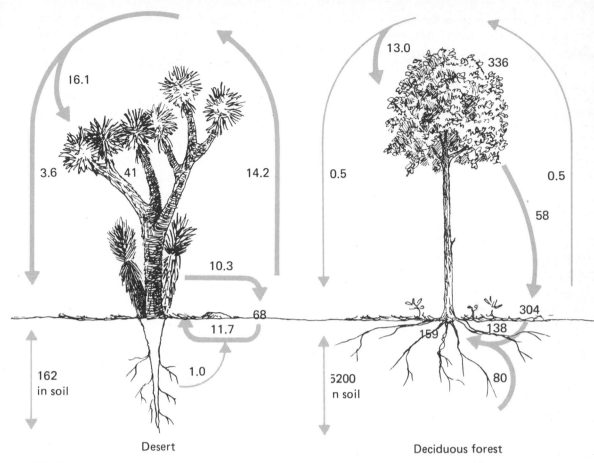

**FIGURE 9-19**

*Comparison of the nitrogen cycle of the desert and the temperate deciduous forest. In a forest a considerable portion of the nitrogen is cycled through the plant. In the desert although considerable quantities of nitrogen are fixed, most of it is lost through denitrification. (After Reichle, 1975.)*

and plant populations. Fifty-five percent of energy flow through small mammals in the shrub desert passed through the kangaroo rat, 22 percent through the browsing jackrabbit, and 6.5 percent through the insectivorous grasshopper mouse *(Onychomys torridus)*.

Carnivores, like the herbivores, are opportunistic feeders, with few specialists. Most carnivores, such as foxes and coyotes, have mixed diets that include leaves and fruits, and even insectivorous birds and rodents consume herbivorous foods. Omnivory rather than carnivory and complex food webs seems to be the rule in the desert ecosystem.

The detrital food chain seems to be less im-

portant in the desert than in other ecosystems. Although most functional and taxonomic groups of soil microorganisms exist in the desert, fungi and actinomycetes are prominent. Microbial decomposition, like the blooming of ephemerals, is limited to short periods when moisture is available. Because of this, dry litter tends to accumulate to a point where the detrital biomass is greater than above-ground living biomass. Most of the ephemeral biomass disappears through grazing, weathering, and erosion. Decomposition proceeds mostly through detritus-feeding arthropods such as termites that ingest and break down woody tissue in their guts. In some deserts considerable amounts of nutrients may be locked up in termite structures, to be released when the structure is destroyed. Other important detritivores are acarids and various isopods.

Because of limited production and decomposition in the desert ecosystem, nutrients, especially nitrogen, are limiting. The nutrient supply, confined largely to the upper surface of the soil is vulnerable to erosion and volatilization and can be rapidly depleted by the growth of annual plants. The sparse distribution of tree and shrub growth localizes nutrient return by litterfall and thereby concentrates nutrient recycling and microbial activity around the plants.

## SUMMARY

Deserts occupy about one-seventh of the land surface of the earth and are largely confined to a worldwide belt between the Tropic of Cancer and the Tropic of Capricorn. Deserts result largely from the climatic pattern of the earth, locations of mountain ranges, and remoteness of land areas from sources of oceanic moisture. Two types of desert exist in North America—the cool desert of the Great Basin dominated by sagebrush and the hot desert of the southwest dominated by creosote bush and cacti.

The desert is a harsh environment in which plants and animals have evolved ways to circumvent aridity and high temperature by becoming either drought resisters or drought evaders. Functionally deserts are characterized by low net production, by opportunistic feeding patterns for herbivores and carnivores, and by a detrital food chain that is relatively less important than in other ecosystems. Considerable quantities of nitrogen are fixed by crustlike blue-green algae on the desert floor, but most of the nitrogen is lost back to the atmosphere.

Semiarid regions support climax shrub ecosystems characterized by densely branched woody structure and low height. The success of shrubs depends upon their ability to compete for nutrients, energy, and space. In semiarid situations shrubs have numerous competitive advantages, including structural modifications that affect light interception, heat losses, and evaporative losses. Shrublands, which go by different names in different regions of the world,

are most typical of places where winters are mild and summers are long, hot, and dry. Successional shrublands occupy land in transition from grassland to forest. Such shrublands may remain stable for long periods.

Functional aspects of shrublands have not received the attention given to other ecosystems. Growth in mediterranean-type shrublands is concentrated at the end of the wet season when nutrients in solution and a relative abundance of moisture produce a flush of vegetation. Nutrient cycling, especially nitrogen and phosphorus, is tight, and some systems exhibit nutrient conservation such as the concentration of nitrogen in litterfall which the plants withdraw from the soil and the concentration of nitrogen and phosphorus in plant tissues in stems and roots to be recirculated through the plant.

Grasslands, once covering extensive areas of the globe, including midcontinent North America, have shrunk to a fraction of their original size because of human needs for crop and grazing lands. At the same time many of the native grazing herbivores have declined or disappeared, replaced in part by cattle and sheep. Clearing of forests and planting of hay fields and the development of successional grasslands on disturbed sites extended the range of grassland animals into once forested regions. Successional and climax grasslands consist of sod formers, bunch grasses, or both. When grasslands are undisturbed by grazing, mowing, or burning, they accumulate a layer of mulch that retains moisture, influences the character and composition of plant life, and provides shelter and nesting sites for some animals.

Grasslands vary in production according to environmental variables, especially the availability of water. Much of the above-ground net production goes to the grazing food chain which includes grazing ungulates, rodents, and insects. Although conspicuous, birds contribute little to grassland energy flow. Grazing has an important impact on grasslands, influencing both productivity and structure. Certain plant species become reduced or disappear under grazing pressure, while others increase. Uncontrolled grazing reduces productivity and may cause a collapse of the grassland ecosystem.

All ecosystems support and are supported by a heterotrophic soil community. Organisms in the soil, like all others, reflect their environment. Their abundance and composition in an area depend upon the nature of the soil, its nutrient status, the vegetation present, the kind of litter the vegetation produces, and the ability of plants to return calcium and other nutrients to the soil. Soil animals in turn play their role in influencing the future development of upper soil layers. The direct decomposition of plant litter is accomplished by microflora, bacteria, and fungi. Soil invertebrate fauna make organic matter more readily available to microflora by mechanical breakdown of litter, by exposing new areas for fungal invasion, by spreading fungal spores through their feces, and by increasing surface area exposed to attack by bacteria and fungi. But soil fauna also consume great quantities of fungi and depress bacterial and fungal populations. Predaceous species in turn influence the population levels of litter-feeding and decomposer organisms. Such is the chain of life in the world beneath the ground.

Through the years a number of attempts have been made to classify life into meaningful distributional units. Most of these attempts have been faunistic in approach. The first division of the world into distributional units was the biogeographical or faunal regions and realms. Three realms are recognized and are further subdivided into six regions. Each region is separated by a barrier of oceans, mountain ranges, or deserts which prevent the free dispersal of animals and each possesses its own distinctive forms of life. Each region is further subdivided by secondary barriers such as vegetation types and topography. Various classifications of vegetation and topography within the more general regions have been labeled life zones, biotic provinces, and biomes. The life zone concept, restricted to North America, divides the continent into broad transcontinental belts. The plant and animal differences between them are governed chiefly by temperature. The biotic province approach divides the North American continent into continuous geographic units that contain ecological associations different from those of adjacent units, especially at the species and subspecies level. The biome system groups plants and animals of the world into integral units characterized by distinctive life forms in a climax community, the stage of development at which the community is in approximate equilibrium with its environment. Boundaries of biomes, or major life zones as they are known in Europe, coincide with the boundaries of major plant formations of the world. By including both plants and animals as a total unit that evolved together the biome permits recognition of the close relationship that exists among all living things. More refined is the Holdridge life zone system which divides the world into zones defined by mean annual precipitation and mean annual biotemperature. Each system has its advantages and disadvantages. All reflect the fact that natural systems are complex and difficult to describe and classify.

# Terrestrial ecosystems II: forest and tundra
## CHAPTER 10

## Forest ecosystems

Of all the vegetation types of the world probably none is more widespread or more diverse than the forest (see inside cover). A map of world vegetation shows forest growth in distinct bands around the Northern Hemisphere. Below the tundra are consecutive belts of coniferous forest, temperate deciduous forest, and tropical forest. The tropical forest is most extensive in the northern part of the Southern Hemisphere. Within these bands of global forest are a diversity of forest types.

### STRATIFICATION

Regardless of type all forests possess large above-ground biomass. This biomass creates several layers or strata of vegetation which influence environmental conditions within the forest stand: light, moisture, temperature, wind, carbon dioxide (see Figure 10-2).

*Light.* Bathed in full sunlight, the uppermost layer of the canopy is the brightest part of the forest. Down through the forest strata light intensity is progressively reduced to only a fraction of full sunlight. In an oak forest only about 6 percent of the total midday sunlight reaches the forest floor; the brightness of light at the floor is about 0.4 percent of that of the upper canopy. Most pines form a dense upper canopy that excludes so much sunlight that the lower strata cannot develop. Pitch pine, Virginia pine, and jack pine have rather open crowns which allow more light to reach the forest floor than oak, hickory, beech, and maple. The upper crown of spruce and fir, a zone of widely spaced narrow spires, is open and well-lighted, while the lower crown is most dense and intercepts most of the light.

Light intensity within the forest varies seasonally. The greatest extremes exist within the deciduous forest. The forest floor receives its maximum illumination during March and April before the leaves appear; a second, lower peak occurs in the fall. The darkest period is midsummer. Light reaching the lower strata and the floor in the coniferous forest is approximately the same throughout the year, because the trees retain their foliage. Here illumination is the greatest in midsummer, when the sun's rays are most direct, and the lowest in winter, when the intensity of incident sunlight is the lowest.

*Humidity and moisture.* The forest interior has a high humidity because of plant transpiration and poor air circulation. During the day when the air warms and its water-holding capacity increases, relative humidity is lowest. At night when temperatures and moisture-holding capacities are low, relative humidity rises.

The lowest humidity in the forest is a few feet above the canopy where air circulation is the best. The highest humidity is near the forest floor, the result of evaporation of moisture from the ground and the settling of cold air from the strata above.

Variation of humidity inside the forest is influenced in part by the degree to which the lower strata are developed. Leaves add moisture to the immediate surrounding air; well-developed strata with more leaves have a higher humidity. Thus layers of increasing and decreasing humidity may exist from the forest floor to the canopy.

Humidity in the forest, of course, ultimately depends upon rainfall and its penetration of the forest canopy (Figure 6-3). The amount of penetration is determined by the degree of interception. Interception is determined by the type and age of forest, the amount of rainfall, associated stemflow and throughfall, and the season of the year (see Chapter 6).

*Temperature.* The highest temperatures in the deciduous forest are in the upper canopy because this stratum intercepts solar radiation. Temperatures tend to decrease through the lower strata. The most rapid decline occurs from the leaf litter down through the soil.

The temperature profile of a coniferous forest tends to be somewhat the reverse, particularly in those containing sprucelike trees. Here the coolest layer is in the upper canopy, perhaps because of greater air circulation, and the temperature increases down through the several strata to the forest floor.

The temperature profile changes through the 24-hour period. At night temperatures are more or less uniform from the canopy to the floor. This is due to the fact that radiation takes place most rapidly in the canopy; as the air cools it sinks and becomes slightly heated by the warmer air beneath the canopy. During the day the air heats up, and by midafternoon temperature stratification becomes most pronounced. On rainy days the temperatures are more or less equalized because water absorbs heat from warmer surfaces and transfers it to the cooler surfaces.

Temperature stratification varies seasonally (Christy, 1952). In fall, when the leaves drop and the canopy thins, temperatures fluctuate more widely at the various levels. Maximum temperatures decrease from the canopy downward, but rise again at the litter surface. The soil, no longer shaded by an overhead canopy, absorbs and radiates more heat than in summer. Below the insulating pavement of litter, temperatures decrease again through the soil. Thus there may be two temperature maximums in the profile, one in the canopy, the other on the surface of the litter. Winter temperatures decrease from the canopy down to the small tree layer, where in some forests they rise and then drop at the litter surface. From here the temperature increases rapidly down through the soil. During spring conditions are highly variable. Maximum temperatures are found on the leaf-litter surface, which at this season of the year intercepts solar radiation, and temperatures decrease upward toward the canopy.

*Wind.* Anyone who has entered a woods on a windy day is well aware of how effectively the forest reduces wind velocity. Overall wind velocities inside the forest may be reduced by 90 percent, and velocity near the ground usually ranges from 1 to 2 percent of wind velocity outside. The influence of forest cover on wind velocity varies with the height and density of the stand and the size and density of the crown (see Figure 4-16). Velocities in the open and cutover stands in wintertime deciduous forests are greater than in dense stands and in coniferous forests, and coniferous forests are the most effective in reducing the flow of wind.

The forest edge deflects the wind upward and over the trees, where the roughness of the canopy surface reduces the velocity. The velocity of wind in the forest is not a constant percentage of the speed of the wind above the canopy. During midday, for example, when the temperature decreases with height, this percentage decreases as the wind velocity above the crown increases.

## TYPES OF FOREST ECOSYSTEMS

*Tropical rain forests.* The tropical rain forest in several forms (the monsoon forest, the evergreen savannah forest, the montane rain forest) forms a worldwide belt about the equator (see inside cover). The largest continuous rain forest is found in the Amazon basin of South America. West and central Africa and the Indo-Malayan region are other major areas of tropical rain forest (see Richards, 1952; Odum, 1970; Whitmore, 1975).

The rain forest grows where the seasonal changes are minimal. The mean annual temperature is about 26° C, the mean minimum rarely goes below 25° C, and heavy rainfall occurs through much of the year. Under such perpetual midsummer conditions, plant activity continues uninterrupted, resulting in very luxurious growth. Tree species number in the thousands, with none dominant and each represented by a few individuals. The few communities with single dominants are limited to areas of particular combinations of soils and topography. The tree trunks are straight, smooth, and slender, often buttressed, and reach 25 to 30 m before expanding into large, leathery, simple leaves. Climbing plants, the lianas, long, thick, and woody, hang from trees like cables, and epiphytes grow on the trunks and limbs. Undergrowth of the dark interior is sparse and consists of shrubs, herbs, and ferns. Litter decays so rapidly that the clay soil, more often than not, is bare. The tangled vegetation, popularly known as the jungle, is a second-growth forest that develops where the primary forest has been despoiled.

*Tropical seasonal forests.* Monsoon forests and other deciduous and semideciduous forests grow where rainfall diminishes and where a pronounced dry season occurs. They differ from the rain forest in that they lose their leaves during the dry season. Such forests, most common in southeastern Asia, India, South America, and Africa, are commonly known as tropical seasonal forests. Similar stands occur along the Pacific side of Mexico and Central America.

*Tropical savannas.* Where forests merge with grasslands, a distinctive type of vegetation, the savanna, may exist. Natural savannas, the half-way world between grassland and forest, occur on an environmental gradient between a pure grassland and pure forest climate. Favorable conditions for the existence of savanna include a climate with a wet and dry season, fire, grazing by ungulates, domestic or wild, and particular soil conditions. Savannas are dominated by grasses and sedges and contain open stands of widely spaced, short trees. The amount of tree cover, however, may vary enormously.

Best known of all savannas is the high grass–low-tree savanna of Africa. It is dominated by elephant grass (*Pennisetum* spp.) and scattered deciduous trees 30 to 40 m high. A second type, referred to in Africa as the acacia–tall-grass savanna, but with analogous types elsewhere in the world, is dominated by tussock grasses and deciduous or evergreen trees. A third type, with discontinuous stands of desert grasses and scattered small, thorny trees, is found in Africa, Australia, India, South America, and southern North America. In some parts of the world savanna is increasing as forest cover is destroyed, while in other areas trees and shrubs are invading overgrazed grassland.

Savannas are usually characterized by a diversity of ungulate fauna. The African savanna contains the richest fauna in the world including the gazelles, impala, eland, buffalo, giraffe, zebra, and wildebeest. In contrast the successional and geologically more recent South American savannas lack these conspicuous mammalian herds.

*Temperate evergreen forests.* In several subtropical areas of the world are extensive mixed forests of both broadleaf evergreen and coniferous trees. Such forests include the eucalyptus in Australia, parana pine forests or araucaria forests of South America and New Caledonia, and false beech (*Northofagus* spp.) forests in Patagonia. Representatives of the temperate evergreen forest are also found in the Caribbean and on the North American continent along the Gulf Coast, in the hummocks of the Florida Everglades, and in the Florida Keys. Depending upon location, these forests are characterized by oaks, magnolias, gumbo-limbo, royal and cabbage palms.

*Temperate deciduous forests.* The temperate deciduous forest once covered large areas of Europe and China, parts of South America and the middle American highlands, and eastern North America (see inside cover). The deciduous forests of Europe and Asia have largely disappeared, cleared for agriculture. The dominant trees include European beech *(Fagus sylvatica),* pendunculate oak *(Quercus robor),* ashes *(Fraxinus* spp.), birches *(Betula* spp.), and elms *(Ulmus* spp.). Because of glacial history (see Chapter 21) the diversity of European deciduous forests does not compare with that of North America or China.

In eastern North America the temperate deciduous forest consists of a number of forest types, which intergrade into one another. The northern segment of the deciduous forest complex is the hemlock–white pine–northern hardwoods forest, which occupies southern Canada and extends southward through northern United States and along the high Appalachians into North Carolina and Tennessee. Beech, sugar maple, basswood, yellow birch, black cherry, red oak, and white pine are the chief components. White pine was once the outstanding tree of the forest, but because most of it was cut before the turn of the century, it now grows only as a successional tree on abandoned land and as scattered trees through the forest.

On relatively flat, glaciated country with its deep, rich soil grow two somewhat similar forests, the beech–sugar maple forest, restricted largely from southern Indiana north to central Minnesota and east to western New York; and the sugar maple–basswood forest, found from Wisconsin and Minnesota south to northern Missouri.

South of this is the extensive central hardwood forest. The central hardwood can be divided into three major types. (1) The cove, or mixed mesophytic, forest consists of an extremely large number of species, dominated by yellow poplar. This forest, which reaches its best development on the northern slopes and deep coves of the southern Appalachians, is one of the most magnificent in the world. Much of its original grandeur has been destroyed by high-grading and fire, but even in second- and third-growth stands, its richness is apparent. (2) On more xeric sites, the southern slopes and drier mountains, grows the oak-chestnut forest. The chestnut, killed by blight, has been replaced by additional oaks. (3) The western edge of the central hardwoods in the Ozarks and the forests along the prairie river systems are dominated by oak and hickory.

The southern pine forests of the coastal plains of the South Atlantic and Gulf States are considered part of the temperate deciduous forest because they represent a seral rather than a final stage. Unless maintained by fire and cutting, these forests are succeeeded by such hardwoods as oak, hickory, and magnolia. Magnolia and live oak dominate the climax forest of the southern Gulf States and much of Florida.

*Northern coniferous forests.* North of the deciduous forest in the northern hemisphere is a world-wide belt of coniferous forest. In Eurasia it begins in Scandanavia and extends across the continent to northern Japan. This forest is dominated in Europe by Norway spruce *(Picea abies),* in Siberia by Siberian spruce *(P. siberica),* and in Japan by the spruce *Picea janensis.* The northern coniferous forests extend southward along European mountain chains.

In North America the northern coniferous forest extends from New England, northern New York, and southern Canada north to the tundra, westward to the Pacific, and southward through the Rockies and Sierras into Mexico. The northern coniferous forest starts out as pine and hemlock mixed with hardwoods, a gradient or ecotone of the northern hardwood forest. Eastern hemlock, jack pine, red pine, white pine, and white cedar are characteristic. Originally the pine forests were most highly developed about the Great Lakes, but they were destroyed by exploitative logging in the 1880s and 1900s. The coniferous forest extends southward from New England through the high Appalachians. Here at high elevations the spruce and fir, the major forest type of the north woods, end. In the southern Appalachians, red spruce and Frasier fir dominate, but north, in the Adirondacks and White Mountains on into Canada and across the continent to Alaska, white spruce, black spruce, and balsam fir form the matrix of the forest. Occupying, for the most part, glaciated land, the northern coniferous forest is a region of cold lakes, bogs, rivers, and alder thickets.

*Temperate rain forests.* South of Alaska, the coniferous forest differs from the northern boreal forest, both floristically and ecologically. The reasons for the change are climatic and topographic. Moisture-laden winds move inland from the Pacific, meet the barrier of the Coast Range, and rise abruptly. Suddenly cooled by this upward thrust into the atmosphere, the moisture in the air is released as rain and snow in amounts up to 635 cm. During the summer, when the winds shift to the northwest, the air is cooled over chilly northern seas. Though the rainfall is low, the cool air brings in heavy fog, which collects on the forest foliage and drips to the ground to add perhaps 127 cm (50 in.) more of moisture. This land of superabundant moisture, high humidity, and warm temperatures supports the temperate rain forest, a community of lavish vegetation dominated by western hemlock, western red cedar, Sitka spruce, and Douglas fir. Further south, where the precipitation still is high, grows the redwood forest, occupying a strip of land about 724 km long.

*Montane coniferous forests.* The air masses that dropped their moisture on the western slopes of the Coast Range descend down the eastern slopes, heat up, and absorb moisture, creating the conditions that produce the Great Basin deserts already discussed. The same air rises up the western slopes of the Rockies, cools, and drops moisture again, although far less than it did on the Coast Range. Here in the Rockies develop several coniferous forest associations, influenced to a great extent by elevation. At high elevation, where the winters are long and snow is heavy, grows the subalpine forest, characterized by Engelmann spruce, alpine fir, and white-barked and bristlecone pines. Lower elevations support Douglas fir and ponderosa pine. The aridity-tolerant ponderosa pine more often than not has an understory of grass and shrubs.

Forests similar to those of the Rocky Mountains grow in the Sierras and Cascades. Alpine forests there consist largely of mountain hemlock, red fir, and lodgepole pine. At lower elevations grow the huge sugar pine, incense cedar, and the largest tree of all, the giant sequoia, which grows only in scattered groves on the west slopes of the California Sierras.

A deciduous seral stage, common to much of the western coniferous forest and the northern coniferous forest as well, is the aspen parkland, supporting trembling aspen, the most widespread tree of North America. The aspen is an important segment of the coniferous forest, for it is utilized by deer, grouse, bear, snowshoe, and beaver.

Montane coniferous forests dominated by pine also occur in Mexico and Guatemala. In Europe forests of Scotch pine extend across the mountain ranges to Spanish Sierra Nevada.

*Temperate woodlands.* In western parts of North America where the climate is too dry for montane coniferous forests, one finds the temperate woodlands. These ecosystems are characterized by open growth small trees with a well-developed understory of grass or shrubs. There are a number of types of temperate woodlands, which may consist of needle-leaved trees, deciduous broad-leaved trees, or sclerophylls, or any combination of these. An outstanding example is the pinyon-juniper woodland in which two dominant genera, *Pinus* and *Juniperus,* are always associated. This ecosystem is found from the front range of the Rocky Mountains to the eastern slopes of the Sierra Nevada foothills. One of the best examples stands on the Kaibab Plateau of northern Arizona. In southern Arizona, New Mexico, and northern Mexico occur oak-juniper and oak woodlands, and in the Rocky Mountains, particularly in Utah, one finds oak-sagebrush woodlands. In the Great Valley of California grows still another type—evergreen-oak woodlands with a grassy undergrowth.

## Temperate forest

### STRUCTURE

Highly developed, uneven-aged, deciduous forests usually consist of four strata (Figures 10-1, 10-2). The upper canopy consists of dominant or codominant trees. Beneath this is

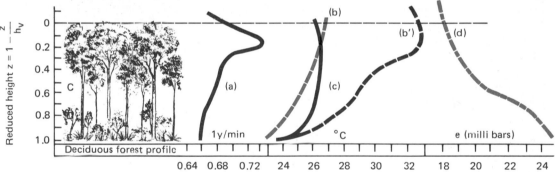

FIGURE 10-1
*A virgin stand of the mixed mesophytic forest in the Central Appalachians. Note the well-developed understory in this uneven-aged climax stand.*

FIGURE 10-2
*Microclimate profiles of a mixed oak, linden, and maple forest in central Russia. (a) Long wave radiation within the forest during summer. (b) Air temperature at midmorning during a sunny summer day. (b') Air temperature at midafternoon. (c) Idealized temperature profile in summer. (d) Water vapor pressure within the forest during warm weather. Height is scaled as reduced height determined by $Z = 1 - Z/h_v$ where $Z$ is some determined level and $h_v$ is maximum height of the stand. In this example height of the canopy is 12 meters. (After Raunder, 1972.)*

the lower tree canopy. Below is the shrub layer. The ground or field layer consists of herbs, ferns, and mosses.

Even-aged stands, the results of fire, clearcut logging, and other disturbances, often have poorly developed strata beneath the canopy because of the dense shade. The low tree and shrub strata are thin and the ground layer often is poorly developed, except in small, open areas.

In general the diversity of animal life is associated with the stratification of and growth forms of plants. Some animals are associated with or spend the major part of their life in a single stratum; others may range over two or more. Arthropods, in particular, confine their activities to one stratum. Dowdy, (1951) found

**315**

that 67 percent of all the arthropod species in Missouri oak-hickory forests inhabited one stratum only, and 78 percent of a red cedar forest were so restricted. None of the arthropod species were common to all strata and only 2 percent ranged through as many as four strata.

The greatest concentration and diversity of life are found on and just below the ground layer. Many animals, the soil invertebrates in particular, remain in the subterranean strata. Others, the mice, shrews, ground squirrels, and foxes burrow into the soil for shelter or food, but spend considerable time above ground. The larger mammals live on the ground layer and feed on herbs, shrubs, and low trees. Birds move rather freely among several strata, but even here there is some restriction. Ruffed grouse, hooded warblers, and ovenbirds occupy essentially the ground layer, but may move up into the tree to feed, to roost, or to advertise territory. Some invertebrates, such as the millipedes and spiders, move into the upper strata at night when humidity is favorable, a vertical migration that is somewhat similar to that of the plankton organisms in lakes and seas (Dowdy, 1944).

Other species occupy the upper strata—the shrub, low tree, and canopy. Red-eyed vireos inhabit the lower tree stratum of the eastern deciduous forest, the wood peewee the lower canopy. Blackburnian warblers and scarlet tanagers dwell in the upper canopy, where they are very difficult to observe. Flying squirrels and tree mice are mammalian inhabitants of the canopy, and the woodpeckers, nuthatches, and creepers live in the open space of tree trunks between the shrubs and the canopy.

Most of the intensive work on stratification of bird life in the temperate deciduous forest has been done in European woodlands, where one biologist (Turcek, 1951) found that 15 percent of the bird species in an oak-hornbeam forest nested on the ground, 25 percent in the herb and shrub strata, and 29 percent in the canopy. Thirty-three percent of the total population occupied the forest canopy, where more niches apparently were available, but its biomass was less than the ground and shrub populations. The ground layer, however, was the feeding area for 52 percent of the species; herb and shrubs, 9 percent; tree trunk, 10 percent; tree foliage, 23 percent; and the open spaces, 6 percent. In deciduous forests the diversity of bird species seems to be related to the height and density of foliage and resulting stratification, rather than to plant species composition (MacArthur and MacArthur, 1961).

## FUNCTION

Ecosystem function has been studied most intensively in forest systems. Energy flow and nutrient cycling have been assessed for several deciduous forest stands including the Hubbard Brook northern hardwoods forest in New Hampshire, Walker Branch mesic hardwoods forest at Oak Ridge, Tennessee, and an oak-pine forest at Brookhaven, Long Island.

Energy flow is most conveniently measured in terms of carbon pools and fluxes. The carbon budget for the mesic tulip poplar (*Liriodendron tulipifera*) and oak (*Quercus* spp.) at Oak Ridge is summarized in Table 10-1 and Figure 10-3. Estimates of above-ground and central root carbon pool are 8.03 kg with an annual above-ground accumulation rate of 0.166 kg $C/m^2/yr$ (Reichle et al., 1973). Mean standing below-ground lateral root carbon amounts to 0.76 kg $C/m^2$. Increment to lateral root biomass amounts to 8 percent per year. Root turnover from death occurs largely in roots less than 0.5 cm in diameter. Much of this root death takes place in late spring and late autumn. Similarly, most root production, which occurs mostly in late winter and midsummer, also takes place in roots less than 0.5 cm. Mean annual standing crop in the $O_1$ and $O_2$ litter layer is 237 g $C/m^2$. Total amount of soil carbon is 12.3 kg $C/m^2$ to a depth of 75 cm. Soil organic matter decreased from 4.6 percent dry weight of soil in the upper 10 cm to 1.3 percent at 21 cm depth. Litter invertebrates amounted to 520 mg $C/m^2$ and soil invertebrate fauna, largely earthworms, 6.4 mg $C/m^2$. Soil microflora, 65 percent fungi and 35 percent bacteria, totaled 58 g $C/m^2$.

Fluxes of carbon in photosynthesis and autotrophic respiration determined by means of gas exchange include an estimated gross carbon uptake (gross primary production) of 2.15 kg $C/m^2/yr$. Net primary production is 0.73 kg $C/m^2/yr$. The autotrophic respiration including contributions from the forest plants is 1.44 kg

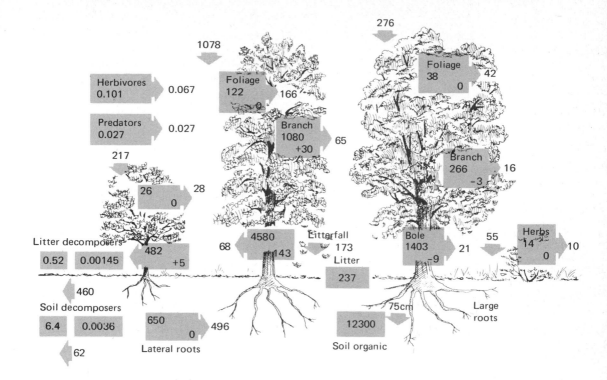

**FIGURE 10-3**
*Carbon budget in a mesic hardwood forest ecosystem at Oak Ridge, Tennessee. From left to right: the trees represent the understory, dominant tulip poplar* (Liriodendron tulipifera), *and other overstory trees. Structural components of the ecosystem have been abstracted as compartments with major fluxes. Vertical arrows represent photosynthetic fixation. Lateral arrows represent respiratory losses. Units of measure are gC/m² and gC/m²/yr for compartment increments and fluxes. (From Reichle et al., 1973.)*

C/m²/yr. Respiration from the forest floor was 1.04 kg C/m²/yr; decomposer respiration amounted to 0.21 kg C and summed respiration of canopy insects 0.094 kg C/m²/yr.

Annual litterfall amounts to 229 g C/m² of which 78 percent is leaves. Tree mortality accounts for 50 g C/m²/yr. Loss of photosynthetic surface through insect consumption varies from 1.9 percent to 3.4 percent, while actual reduction in photosynthetic surface amounts to 5.6 to 10.1 percent. Carbon flux due to actual consumption is 4.5 g C/m²/yr.

Another comprehensively analyzed temperate forest ecosystem is a young oak-pine forest at Brookhaven, New York, a relatively

xeric stand of pitch pine and scarlet, white, and bear oak with an understory of blueberry and huckleberry (Whittaker and Woodwell, 1969; Woodwell and Botkin, 1970). In some respects the oak-pine stand and the yellow poplar stands are similar even though they have different standing crops, 8.76 kg C/m² for yellow poplar–oak compared to 5.96 kg C/m² for the oak-pine (Table 10-1). Both have comparable net primary production and autotrophic respiration. Both lose approximately the same amount of photosynthetic surface to insects, about 3 percent of net annual production.

However, the two stands differ considerably at the heterotrophic level. Heterotrophic respiration for yellow poplar is more than twice that of the oak-pine forest. This large heterotrophic respiration plus the large autotrophic respiration result in total ecosystem respiration for yellow poplar over two times as great as that of the oak-pine forest. The main difference in heterotrophic respiration is in annual decomposition of 712 g C/m² for yellow poplar compared to 360 g C/m² for the oak-pine. Much of this difference is from a high root turnover rate in the yellow poplar stand, as well as the resistance of oak leaves and pine needles to rapid decomposition. Comparison of the ratios

**317**

*TABLE 10-1*
*Comparison of metabolism and structure of two terrestrial ecosystems*

|  | Parameter | Liriodendron Forest[a] | Quercus-Pinus Forest[b] |
|---|---|---|---|
| Total standing crop | TSC | 8.76 kg C/m² | 5.96 kg C/m² |
| Net primary production | NPP | 0.73 kg C/m²/yr | 0.60 kg C/m²/yr |
| Relative production | NPP/TSC | 8.3% | 10% |
| Autotroph respiration | $R_A$ | 1.44 | 0.68 |
|  | $R_A$/TSC | 0.16 | 0.11 |
| Heterotroph respiration | $R_H$ | 0.67 | 0.29 |
| Ecosystem respiration | $R_E = R_A + R_H$ | 2.11 | 1.01 |
| Net ecosystem production | $NEP = NPP - R_H$ | 0.06 | 0.28 |
| Annual decay |  | 0.70 | 0.36 |

Units of measure are kg C/m² and kg C/m²/yr[1] for compartments and fluxes unless otherwise noted.
*Source:* [a]Revised after Reichle et al., 1973a; [b]Woodwell and Botkin, 1970; in National Academy Science, 1974.

of net ecosystem production (NEP) to total standing crop (0.05 for oak-pine and 0.007 for the yellow poplar) indicates that the young oak-pine forest is accumulating carbon seven times as fast as the yellow poplar stand.

Mineral cycling has been studied extensively in recent years in several temperate forest ecosystems including oak-pine forest (Whittaker and Woodwell, 1969), a northern hardwoods forest (Likens et al., 1971; Likens, 1976), and European oak forests (Duvigneaud and Denaeyer-De Smet, 1970). One of the most intensively studied temperate forest ecosystems is an oak-beech-hornbeam forest in the Netherlands (Duvigneaud and Denaeyer-De Smet, 1970), 75 years old with trees up to 23 m. Total biomass of the stand is 156 metric tn/ha of which 80 percent is above ground. The dead organic matter in the soil amounts to 125 metric tn/ha, equal to the above-ground biomass. Net primary production amounts to 14.43 tn/ha. Forty percent of annual net productivity is returned to the soil as litter. The nutrient budget of the forest is presented pictorially in Figure 10-4 and a summary of the annual budget is given in Table 10-2.

Considerable amounts of nutrients are retained relative to total uptake. Five times more potassium is stored in tissues than is exchangeable in the soil pool, and as much magnesium is contained in the biomass as is in the soil pool. Oak is more important as a nutrient pool, but not if considered relative to biomass. Beech is the least important of the three dominant species.

Nutrient cycling may also be considered as a balance between inputs to the biological system from precipitation and outputs or losses from the system through streamflow from the watershed. The difference represents the amount of recharge to the system from the soil pool. In the oak-pine forest at Brookhaven,

*TABLE 10-2*
*Annual element balance of the Belgium mixed-oak forest (kg/ha)*

|  | K | Ca | Mg | N | P | S |
|---|---|---|---|---|---|---|
| Retained | 16 | 74 | 5.6 | 30 | 2.2 | 4.4 |
| Returned | 53 | 127 | 13.0 | 62 | 4.7 | 8.6 |
| Uptake | 69 | 201 | 18.6 | 92 | 6.9 | 13.0 |

*Source:* Data from Duvigneaud and Denaeyer-DeSmet, 1970.

*FIGURE 10-4*
*Annual mineral cycling (in kg/ha) of macronutrients in a mixed oak forest ecosystem at Virelles, Belgium. Retained: in annual wood and bark increment of roots and aerial parts of each species (total is hatched). Returned: by tree litter (tl), ground flora (gf), washing and leaching of canopy (w) and stem flow (sf). Imported: by incident rainfall (not included). Macronutrients contained in crown leaves when fully grown (July) are shown on the right side of the figure in italics; these amounts are higher (except for calcium) than those returned by leaf litter. Values for magnesium, nitrogen, and phosphorus in throughfall and stemflow. (From Duvigneaud and Denaeyer-De Smet, 1970.)*

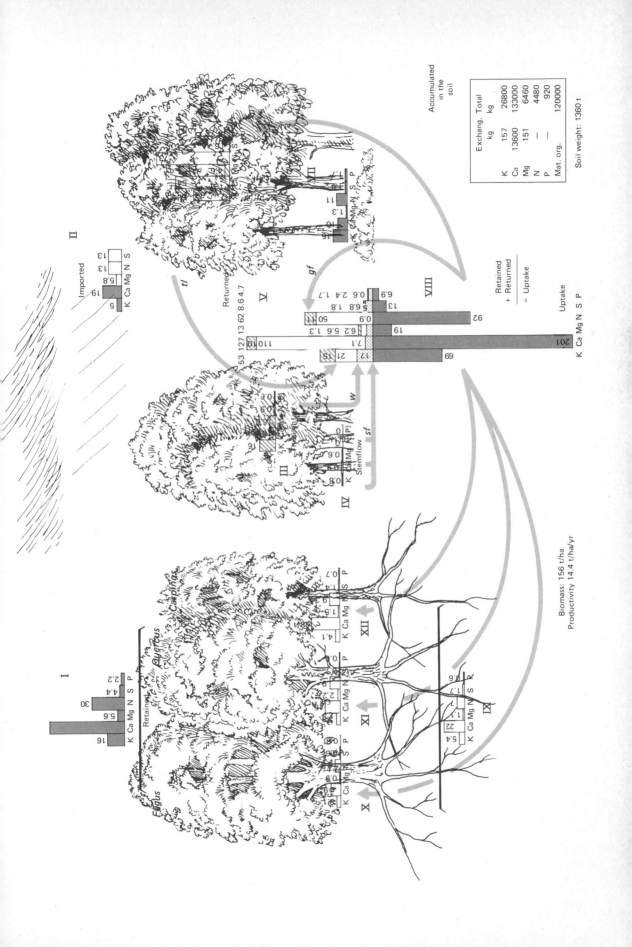

Biomass: 156 t/ha
Productivity 14.4 t/ha/yr

*TABLE 10-3*
**Input and losses of nutrients of the Hubbard Brook Forest (kg/ha)**

|  | Ca | Mg | Na | K |
| --- | --- | --- | --- | --- |
| Input | 3.0 ± 0 | 0.7 ± 0 | 1.0 ± 0 | 2.5 ± 0 |
| Output | 8.0 ± 0.5 | 2.6 ± 0.06 | 5.9 ± 0.3 | 1.8 ± 0.1 |
| Loss | 5.0 ± 0.5 | 1.9 ± 0.06 | 4.9 ± 0.3 | 0.7 ± 0.1 |

*Source:* Data from Likens et al., 1967.

*TABLE 10-4*
**Pool sizes and annual transfer rates of nitrogen, phosphorus, and potassium in natural oak-hickory stands on Walker Branch watershed**

|  | N | P | K |
| --- | --- | --- | --- |
| **NUTRIENT POOLS (KG/HA)** | | | |
| Vegetation | | | |
| Foliage | 53 | 5 | 42 |
| Branch | 120 | 10 | 51 |
| Bole | 164 | 11 | 110 |
| Stump | 37 | 2 | 24 |
| Root | 104 | 10 | 45 |
| Total | 478 | 38 | 272 |
| Forest floor | | | |
| Wood >2.5 cm | 28 | 2 | 6 |
| Twig <2.5 cm | 9 | <1 | 1 |
| $O_1$ horizon | 88 | 6 | 7 |
| $O_2$ horizon | 179 | 12 | 14 |
| Total | 304 | 20 | 28 |
| Mineral soil (to depth of 60 cm) | 4,000 | 950 | 32,500 (140)* |
| **NUTRIENT TRANSFERS (KG/HA/YR)** | | | |
| Litterfall (leaves, twigs <2.5 cm, and reproductive parts) | 36 | 3 | 20 |
| Bole and branch mortality | 1 | <1 | 1 |
| Root mortality | 72 | 7 | 32 |
| Throughfall | 13 | <1 | 24 |
| Incorporation in growth | 7 | <1 | 4 |

*140-exchangeable potassium on the basis of ammonium acetate extraction.
*Source:* National Academy of Science, 1975, and Oak Ridge National Laboratory.

Long Island, New York, total input from precipitation of the four cations potassium, calcium, magnesium, and sodium amounts to 2.48 g/m², while losses to the water table range from 3.68 to 5.18 g/m² (Woodwell and Whittaker, 1968). Output from the Hubbard Brook forest in New Hampshire, a northern hardwoods system, summarized in Table 10-3, exceeds the input of calcium, magnesium, and sodium, while potassium showed a gain through input (Likens et al., 1967).

These and other budgets suggest several characteristics of mineral cycling in temperate forests. Nutrients accumulate in the tree biomass to form a nutrient pool unavailable for short-term cycling (Table 10-4). In an oak-hickory forest in the central Appalachians, 65 percent of each pool of nitrogen, phosphorus, and potassium is incorporated in woody tissue, primarily bole, branches, and stump, 22 percent in roots, and 13 percent in foliage. Considering the whole forest system, 11 percent of the nitrogen, 4 percent of the phosphorus, and 1 percent of the potassium is stored in vegetation. In the forest floor the surface litter layer ($O_1$ and $O_2$ layers) is the most important nutrient pool, although the bulk of the nutrients in the system are in the mineral soil. Important in recycling nutrients from the vegetative pool is root mortality, followed by litterfall and foliage leaching. However, biological cycling can be maintained only if nutrients are pumped from soil reserves or are released through weathering of parent rock.

The role of the various components in nutrient cycling is illustrated by long term studies of the nitrogen cycle in deciduous forests at Hubbard Brook, New Hampshire (Bormann et al., 1977) and Walker Branch, Oak Ridge, Tennessee (Henderson et al., 1973). Both studies point out that natural forest ecosystems tend to accumulate and cycle large amounts of nitrogen. At both sites most of the nitrogen (87 to 90 percent) is incorporated in the mineral soil horizons. The remaining nitrogen is in the vegetation and forest floor Figure 10-5). The most important mechanism for

cycling nitrogen from vegetation to soil is lateral root turnover. In the Walker Branch forest this amounts to 56 kg/ha/yr. This amount includes only root mortality and not exudates. As measured at Hubbard Brook, root exudates release about 1 percent of the inorganic nitrogen made available by net mineralization. The second most important mechanism is litterfall, which at Walker Branch accounts for 37 kg/ha/yr and at Hubbard Brook 54.2 kg/ha/yr. Of this 91 percent consists of leaves and reproductive parts; the remaining 9 percent is bole and branch fall due to tree mortality. Foliar leaching, a third mechanism, contributes 4.4 kg/ha at Walker Branch, 9.3 kg/ha at Hubbard Brook. Vegetation retains 14.8 kg/ha of which bole and branch, stump, and lateral root components account for 47, 12, and 41 percent, respectively. Nitrogen released from forest litter decomposition amounts to an estimated 48 kg/ha. Atmospheric input accounts for 6.5 kg/ha/yr at Hubbard Brook, 23.5 kg/ha at Walker Branch. An additional estimated input of 14.2 kg/ha is added by nitrogen fixation, an input

*Representation of the nitrogen cycle in a mixed mesophytic forest watershed at Oak Ridge, Tennessee. Nitrogen pools in ecosystem components are shown on the right, while annual transfers are on the left. (After Auerbach et al., 1974; courtesy of Oak Ridge National Laboratory.)*

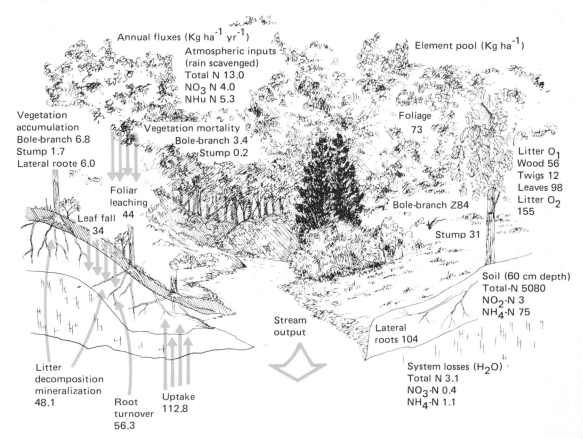

not estimated at Walker Branch. Uptake from mineral soil amounts to 112.8 kg/ha at Walker Branch, 79.6 kg/ha at Hubbard Brook.

The Hubbard Brook study emphasizes other important aspects of the nitrogen cycle in the forest. Of the nitrogen added to long-term storage about 54 percent is accumulated in living matter, and 46 percent is stored in organic matter in the forest soil. Of an estimated 20.7 kg/ha entering the system each year about 81 percent is held within the system. Of the estimated 119 kg/ha of nitrogen used in plant growth, 33 percent is withdrawn from storage location in the living plant and utilized in growth. A like amount is withdrawn from leaves before leaf fall and stored in the stems. In both forest ecosystems only a small fraction of the nitrogen added to the inorganic pool within the ecosystem is lost from the system in streamflow. At Hubbard Brook this leakage amounts to 5 percent; at Walker Branch, 1.9 percent. Thus an internal source of nitrogen not subject to loss through streamflow (providing the vegetation with a buffer against short-term fluctuations in available soil nitrogen), annual uptake by living vegetation, and annual additions of nitrogen to wood biomass are important in promoting a tight cycling of nitrogen.

How the substrate can influence mineral cycling is illustrated by the difference between the cation budgets of two forests of similar vegetation, one growing on a site underlain by granitic bedrock and one growing on a site underlain by dolomitic rocks rich in calcium and magnesium (Table 10-5). The forest growing at Walker Branch on dolomitic rocks shows an output of calcium and magnesium nearly 20 times greater than the output from the Coweeta, North Carolina, forest on granitic substrate. Net losses of the other two cations, sodium and potassium, are small and about the same magnitude.

The considerable quantities of nutrients tied up in forest tree biomass have certain implications. Assuming that all twigs, litter, and other parts are left behind to decay in the cutover forest, the harvest of oak timber from the 89-year-old Virelles stand would remove 51 percent of accumulated nutrients, potassium, calcium, magnesium, sulfur, and phosphorus (see Figure 10-4).

In addition to losses brought about by the actual removal of timber from the ecosystem there are additional losses from erosion and leaching. The magnitude of these losses is indicated by studies of timber removal at Hubbard Brook (Likens et al., 1969, 1978). An experimental watershed in the Hubbard Brook forest was clear-cut, the timber was left in place, but incoming vegetation was suppressed by herbicides in order to study losses of nutrients from the system without confounding uptake of nutrients by vegetation with increase of dissolved nutrients from accelerated decomposition. Evaporation decreased 68 percent and runoff increased 40 percent. An increased amount of water passing through and out of the ecosystem also meant an increased amount of nutrients lost from the clear-cut forest ecosystem. Because water was no longer being withdrawn from soil by trees during the growing season, more water was able to move through the soil, dissolving the nutrients and carrying them in solution. Loss of vegetation and subsequent increased surface temperatures increased action of decomposers. This enabled them to convert organic matter rapidly into mineral form, matter that was then dissolved in soil water. The amounts of calcium, magnesium, sodium, and potassium ions lost by subsurface runoff in the Hubbard Brook forest were 9, 8, 3, and 20 times greater, respectively, than the amounts lost from the undisturbed forest. Nitrogen losses were even greater. Once the vegetation was removed, ammonia converted rapidly to nitrate. This process caused the formation of both nitrate and hydrogen ions. The greater concentration of hydrogen ions increased the acidity of the soil, making calcium, magnesium, sodium, and potassium more soluble. The amount of nitrogen carried from the clear-cut forest in the form of nitrates was 100 times greater than nitrogen loss from the undisturbed forest.

Once the vegetation was permitted to grow, nutrient losses halted as the hydrological and biogeochemical parameters returned to previous levels. But the fact that losses declined before vegetation became fully effective suggests that readily available nutrients were becoming exhausted. During the ten years after cutting about 499 kg of nitrate nitrogen, 450 kg of calcium, and 166 kg of potassium per hectare

TABLE 10-5
*Average annual cation budgets for undisturbed watersheds at two sites in the Appalachian highlands, based on two water years 1969–1972 (kg/ha)*

| Vegetation type:<br>Geologic type: | COWEETA (NORTH CAROLINA)<br>*Mature Hardwoods*<br>*Granitic* | WALKER BRANCH (TENNESSEE)<br>*Mature Hardwoods*<br>*Dolomitic* |
|---|---|---|
| Calcium++ | | |
| Input | 6.16 | 28.6 |
| Output | 6.92 | 138.4 |
| Net loss or gain | −0.76 | −103.8 |
| Magnesium++ | | |
| Input | 1.26 | 3.2 |
| Output | 3.09 | 69.6 |
| Net loss or gain | −1.82 | −66.4 |
| Potassium+ | | |
| Input | 3.16 | 4.8 |
| Output | 5.17 | 5.6 |
| Net loss or gain | −2.02 | −0.8 |
| Sodium+ | | |
| Input | 5.40 | 9.2 |
| Output | 9.74 | 5.3 |
| Net loss or gain | −4.34 | +3.9 |

*Source:* National Academy of Science, 1974.

TABLE 10-6
*Comparative losses of calcium and nitrogen during the first two years after clear-cutting (kg/ha)*

| Origin of loss | HUBBARD BROOK EXPERIMENTAL DEFORESTATION | | GALE RIVER COMMERCIAL CLEAR-CUT | |
|---|---|---|---|---|
| | Ca | N | Ca | N |
| Dissolved substance in stream water | | | | |
| First year | 75 | 96 | 41[a] | 38[b] |
| Second year | 90 | 140 | 48[b] | 57[b] |
| Removed in wood products | 0 | 0 | 221* | 144* |
| Total removed | 165 | 236 | 310 | 239 |

*Estimated for an average clear-cut in second-growth northern hardwoods.
*Source:* [a]Likens et al., 1977; [b]Pierce et al., 1972.

were lost in stream water. In contrast only about 43 kg of nitrate nitrogen, 131 kg of calcium, and 22 kg of potassium per hectare were lost during the same period from an adjacent forested ecosystem. Loss of nitrogen from the clear-cut represents 28 percent of the total stored in the ecosystem prior to cutting, 6 times the amount taken up annually by vegetation. Loss of calcium represents 53 percent of the total amount stored in living biomass and dead organic matter, about 88 percent of the exchangeable calcium in the soil, 7 times an-

nual uptake. Potassium losses amount to 3 times the exchangeable potassium, 2.6 times annual uptake, and 58 percent of total stored.

These losses were compared to losses from a commercial clear-cut in which the timber was removed and the vegetation allowed to regrow without disturbance. Comparisons of the two areas are given in Table 10-6. Maximum nutrient losses through streamflow took place the first and second years after the commercial cutting. In spite of vegetative growth, cutting with regrowth resulted in accelerated loss of nutrients from the ecosystem through streamflow. This loss added to the wood removed was the same magnitude of nutrient losses as those from the experimental watershed. Nitrogen losses from the two watersheds were about the same, but calcium losses on the commercial clear-cut were 1.9 times greater, largely because of the removal of wood.

## Coniferous forest

### STRUCTURE

Coniferous forests (Figure 10-6) can be divided into three broad classes in respect to growth form and general growth behavior: (1) pines with straight, cylindrical trunks; whorled, spreading branches; and crown density that varies with the species, from the dense crown of red and white pine to the relatively open, thin crown of Virginia pine, jack pine, and Scotch pine; (2) spire-shaped evergreens, including spruces, firs, Douglas-fir, and (with some reservations) the cedars, with more or less tall pyramidal crowns, gradually tapering trunks, and whorled, horizontal branches; (3) deciduous conifers, with tall pyramidal open crowns and a deciduous habit. Growth form and behavior influence animal life and other aspects of the various coniferous ecosystems.

Stratification in coniferous forests is not well developed (Figure 10-7). Because of high crown density the lower strata are poorly developed and the ground layer consists largely of ferns and mosses with few herbs. The

**FIGURE 10-6**
*Spruce-fir forest in northern Maine. The tiaga or subalpine spruce and fir forests extend around the northern part of the world and down the high mountains of eastern north America.*

**FIGURE 10-7**
*Mean hourly profiles of temperature (a), humidity (b), carbon dioxide (c), and net radiation (d) in a coniferous forest, a Sitka spruce (Picea sitchensis) plantation in Scotland. Note the increase in CO₂ below the canopy and above the forest floor. The increase occurs because the soil is a source of CO₂. The hollow in the humidity profile in the lower canopy and beneath it probably represents a horizontal humidity flux caused by increased air flow below the canopy. Height of 12 meters is scaled as reduced height for comparison with the deciduous profiles in Figure 10-2. (After Jarvis et al., 1976.)*

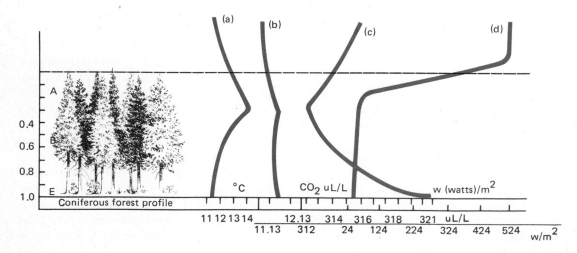

maximum canopy development in spire-shaped conifers is about one-third down from the open crown which gives such forests a foliage profile different from that of pines. Pine forests with a well-developed high canopy lack lower strata. However, old stand, open-crowned pines may have three strata, an upper canopy, a shrub layer, and a thin herbaceous layer. The litter layer in coniferous forests is usually deep, poorly decomposed, and rests on top of rather than mixed in with the mineral soil.

Animal life in the coniferous forest varies widely, depending upon the nature of the stand. Soil invertebrate litter fauna is dominated by mites. Earthworm species are few and their numbers low. Insect populations, while not diverse, are high in numbers and, encouraged by the homogeneity of the stands, are often destructive. Spruce budworm *(Choristonema fumiferana)*, found throughout the boreal forest, attacks balsam fir and spruces. Related species attack jack and red pine. Sawflies *(Neodriprion)* attack a wide variety of pines including pitch, Virginia, shortleaf, and loblolly.

A number of bird species are closely associated with coniferous forests. In North America these include such species as chickadees, kinglets, pine siskins, yellow-rumped warbler, purple finches, hermit thrush (Figure 10-8). Related species, the tits and grosbeaks, are common to European coniferous forests. Species diversity varies. In general coniferous forests of northeastern and southeastern North America and the Sierra Nevadas support the richest avifaunas (Wiens, 1975). Seventeen to 27 percent of all individuals present belong to a single dominant species. Bird densities are the highest in the Pacific northwest and lowest in immature northeastern forests. In northeastern coniferous forests more than 50 percent of the individuals are warblers, while in western coniferous forests less than 10 percent are warblers. Foliage-feeding insectivorous birds are dominant in all types of coniferous forests; in North America they are more dominant in the north, northeast, and southeast than in the west and northwest.

In extensive reforestation areas in New York State, where pine and spruce have been planted over thousands of acres, a definite species preference for certain conifer growth forms exists among birds (R. L. Smith, 1956). The pine forests attract the lowest number of species (8) and are preferred by the black-throated green warbler. Spruce plantings attract 22 species, of which the magnolia and Nashville warblers are the most common. The greater diversity of species in spruce undoubtedly reflects the greater abundance of niches offered by the spirelike crowns and the branches retained close to the ground. The low branches also provide excellent winter cover for grouse, snowshoe hares, and cottontail rabbits.

Except for strictly boreal species, such as the pine martin, much less affinity for coniferous forests exists among mammals. Most are associated with both coniferous and deciduous forests; the white-tailed deer, black bear, and mountain lion are examples. Their north-south distribution appears to be limited more by climate, especially temperature, than by vegetation. The red squirrel, commonly associated with coniferous forests, is quite common in deciduous woodlands in the southern part of its range.

## FUNCTION

The most intensively studied type of coniferous forest is Douglas-fir *(Pseudotsuga menziesii)*, although little information exists on primary production and energy flow through Douglas-fir stands. Wiens and Nussbaum (1975) have estimated energy flow through the breeding bird populations in six different stands of Douglas-fir in northwestern United States. They supported 7 to 15 breeding pairs of birds with standing crops of 223 to 526 g/ha. Total energy flow during the breeding season from April to October through the bird population varied with moisture conditions and altitude. The energy flow in a low elevation, moderately xeric stand was roughly 10 kcal/$m^2$/season; in a high elevation stand, 12 kcal/$m^2$; in a mesic flood plain stand, 17 kcal/$m^2$; and in a mid-elevational transitional stand, 21 kcal/$m^2$. Of the total intake, 13 to 19 percent, depending upon altitude, was used for thermoregulation; 15 to 16 percent was utilized for reproduction-related activities; and only 1 percent was channeled into production. This energy flow compares with estimates made for

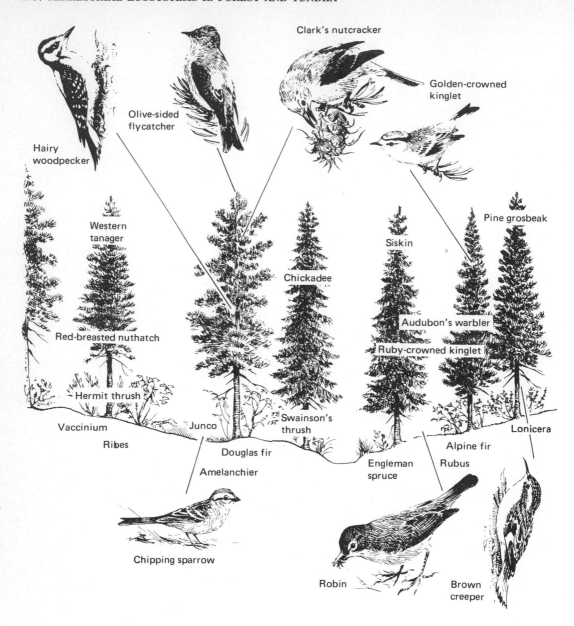

**FIGURE 10-8**
*Vertical distribution of some birds in a spruce-fir forest in Wyoming. (After Salt, 1967.)*

bird populations in forests of the midwestern United States and Panamanian tropics (Karr, 1971).

Better understood is mineral cycling in Douglas-fir stands. A nutrient budget for a Douglas-fir stand is illustrated in Figure 10-9 and summarized in Table 10-7. They show that 10 to 15 percent of the nutrients

held by the Douglas-fir are located in the root and stump below ground level. Mineral accumulation in the bark is relatively high. The understory plants contain a minimal proportion of the nutrients.

The budget also points out that the internal cycle is more important to nutrient budget than additions to or losses from precipitation. A portion of the nutrients is stored in the tree, then returned to the forest floor by the way of litterfall, stemflow, and leaf wash. Ten percent of the nitrogen, 50 percent of the phosphorus, 71 percent of the potassium, and 22 percent of

TABLE 10-7
**Budget of selected nutrients for second-growth Douglas fir forest**

| Ecosystem component | N (kg/ha) | % of total | P (kg/ha) | % of total | K (kg/ha) | % of total | Ca* (kg/ha) | % of total |
|---|---|---|---|---|---|---|---|---|
| **Trees** | | | | | | | | |
| Total | 320 | 9.7 | 66 | 1.7 | 220 | 44.6 | 333 | 27.3 |
| Foliage | 102 | 31.9 | 29 | 43.9 | 62 | 28.2 | 73 | 21.9 |
| Branches | 61 | 19.1 | 12 | 18.2 | 38 | 17.3 | 106 | 31.8 |
| Wood | 77 | 24.0 | 9 | 13.6 | 52 | 23.6 | 47 | 14.1 |
| Bark | 48 | 15.0 | 10 | 15.2 | 44 | 20.0 | 70 | 21.0 |
| Roots | 32 | 10.0 | 6 | 9.1 | 24 | 10.9 | 37 | 11.2 |
| **Subvegetation** | | | | | | | | |
| Total | 6 | 0.2 | 1 | 0.1 | 7 | 1.4 | 9 | 0.7 |
| **Forest floor** | | | | | | | | |
| Total | 175 | 5.3 | 26 | 0.6 | 32 | 6.5 | 137 | 11.2 |
| **Soil** | | | | | | | | |
| Total | 2,809 | 84.8 | 3,878 | 97.6 | 234 | 47.5 | 741 | 60.8 |
| Input by precipitation | 1.1 | | trace | | 0.8 | | 2.8 | — |
| **Uptake by forest from soil** | | | | | | | | |
| Total | 38.8 | | 7.2 | | 29.4 | | 24.4 | — |
| Foliage | 24.3 | | 4.7 | | 16.2 | | 17.8 | — |
| Branches | 4.2 | | 0.8 | | 2.7 | | 2.6 | — |
| Trunk | 10.3 | | 1.7 | | 10.5 | | 4.0 | — |
| **Return to forest floor** | | | | | | | | |
| Total | 15.3 | | 0.6 | | 15.0 | | 15.7 | — |
| Litter fall | 13.6 | | 0.2 | | 2.7 | | 11.1 | — |
| Stem flow | 0.2 | | 0.1 | | 1.6 | | 1.1 | — |
| Leaf wash | 1.5 | | 0.3 | | 10.7 | | 3.5 | — |
| Leached from forest floor | 4.8 | | 0.95 | | 10.5 | | 17.4 | — |
| Leached beyond rooting zone | 0.6 | | 0.02 | | 1.0 | | 4.5 | — |
| **Annual accumulation in ecosystem** | | | | | | | | |
| Forest | 23.5 | | 6.6 | | 14.4 | | 8.7 | — |
| Forest floor | 11.6 | | −0.4 | | 5.3 | | 1.1 | — |
| Soil | −34.6 | | −6.3 | | −19.9 | | −11.5 | — |

*Exchangeable components only.
Source: D. W. Cole et al., 1967.

the calcium are returned by this route. The annual accumulation of calcium, phosphorus, potassium, and nitrogen is considerable both in the trees and in the forest floor. The nutrient cycle depends upon the withdrawal of elements from the soil at the annual rate of 34.6 kg N/ha, 6.3 kg P/ha, 19.9 kg K/ha, and 11.5 kg Ca/ha. The time required for total soil depletion at this rate is summarized in Table 10-8. The depletion is speculative because it does not take into account change through time in transfer rate between components of the ecosystem or the addition of elements by the way of nitrogen fixation, mineral solubility, and accumulation of organic matter (Cole et al., 1969). This budget suggests that Douglas-fir is an accumulator plant (Fortesque and Martin, 1970). Accumulator plants are im-

TABLE 10-8
**Utilization of nitrogen, phosphorus, potassium, and calcium from soil by second-growth Douglas-fir**

| Element | Yearly uptake (% of total in system)* | Static supply (yr)† | Cyclic supply (yrs) |
|---|---|---|---|
| N | 1.4 | 73 | 125 |
| P | 0.2 | 537 | 582 |
| K | 12.5 | 8 | 12 |
| Ca | 3.3 | 30 | 64 |

*Includes input by precipitation and return by litter stem flow, and wash.
†Number of years supply will last.
Source: D. W. Cole et al., 1967.

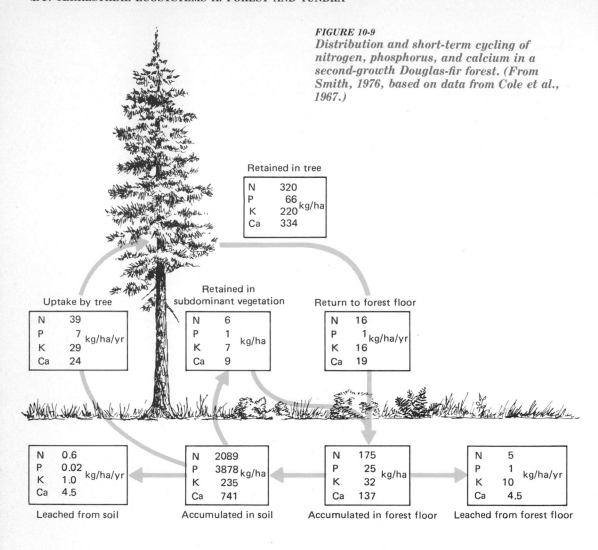

**FIGURE 10-9**
*Distribution and short-term cycling of nitrogen, phosphorus, and calcium in a second-growth Douglas-fir forest. (From Smith, 1976, based on data from Cole et al., 1967.)*

Retained in tree

| N | 320 | |
|---|---|---|
| P | 66 | kg/ha |
| K | 220 | |
| Ca | 334 | |

Uptake by tree

| N | 39 | |
|---|---|---|
| P | 7 | kg/ha/yr |
| K | 29 | |
| Ca | 24 | |

Retained in subdominant vegetation

| N | 6 | |
|---|---|---|
| P | 1 | kg/ha |
| K | 7 | |
| Ca | 9 | |

Return to forest floor

| N | 16 | |
|---|---|---|
| P | 1 | kg/ha/yr |
| K | 16 | |
| Ca | 19 | |

Leached from soil

| N | 0.6 | |
|---|---|---|
| P | 0.02 | kg/ha/yr |
| K | 1.0 | |
| Ca | 4.5 | |

Accumulated in soil

| N | 2089 | |
|---|---|---|
| P | 3878 | kg/ha |
| K | 235 | |
| Ca | 741 | |

Accumulated in forest floor

| N | 175 | |
|---|---|---|
| P | 25 | kg/ha |
| K | 32 | |
| Ca | 137 | |

Leached from forest floor

| N | 5 | |
|---|---|---|
| P | 1 | kg/ha/yr |
| K | 10 | |
| Ca | 4.5 | |

portant in the ecosystem because they can remove elements from the soil in quantities large enough to upset nutritional balance in the ecosystem.

Additional studies of nutrient budget of Douglas-fir stands on different sites in the Pacific Northwest support these earlier findings. Over a two-year study the average amount of water passing through the Douglas-fir ecosystems was 170 and 135 cm. Cation input through precipitation was less than 10 percent of that from mineral weathering, the chief source of cations. Average annual net losses of calcium, sodium, magnesium, and potassium were 47.0, 28.0, 11.0, and 1.5 kg/ha, respectively. Annual losses followed the same pattern as annual runoff, dominated by winter rainstorms from the Pacific Ocean. Both nitrogen and phosphorus tended to be conserved in

the system. From an average annual input of 1.0 kg/ha of dissolved nitrogen, largely organic (only 13 percent as nitrates and a trace ammonium) there was an average annual outflow of 0.5 kg/ha. Most of the soluble organic nitrogen was released by decomposition and exudates and incorporated into plant tissue. A similar small amount of phosphorus, 0.25 kg/ha, was lost as outflow. The higher loss of cations probably represents excess over need by the forest ecosystem (Fredrikson, 1972).

A study of nutrient cycling in a 450-year-old stand of Douglas fir indicated that nutrient input by throughfall was highest during summer and fall and lowest during winter. More nitrogen, phosphorus, and calcium were transferred to the soil by litterfall, mostly needles, than by throughfall. Throughfall added more potassium and magnesium.

The role of roots, fungi, and mycorrhiza in nutrient cycling in coniferous stands is poorly understood. Stark (1972, 1973) provides some interesting insights in the role of fungi in mineral cycling in a xeric stand of Jeffrey pine *(Pinus Jeffreyi)* in Nevada. Litterfall is largely pine needles which are infected by fungi a few months before they fall. During the first winter under snow on the ground basidiomycete fungi remove cell contents and most of the nitrogen and phosphorus, leaving mostly cellulose behind. In the second year the remains of the needles are incorporated in the fermentation zone of the litter to be acted upon by other fungi (Figure 10-10).

Stimulating fungal activity in the fermentation zone is the rain of pollen in spring. This pollen rain, which amounts from 0.9 to 3.0 kg/ha/yr, supplies among other nutrients, 38.0 g N, 15.0 g K, 8.5 g Mg, 5.2 g P, and 0.56 g Ca. While this level of nutrient input is inconse-

quential to tree growth, it is essential to fungi in the fermentation zone in summer. Nitrogen and phosphorus apparently stimulate the fungi to complete litter decay and release other elements to tree growth. Because of dry conditions during summer, only minimal amounts of nitrogen and phosphorus are leached from the fermentation layer; pollen is the main nutrient source for the fungi.

The fungal hyphae or rhizomorphs in turn

**FIGURE 10-10**
*Nutrient cycling in a Jeffrey pine forest. Mineral cycling involves decomposition of litter by free-living fungi assisted by bacteria and insects. Fungi store large quantities of elements from the litter and by a little understood mechanism pass them on to other parts of the system. Decomposition products of free-living organisms are leached into the soil and become available to roots. (After Stark, 1973.)*

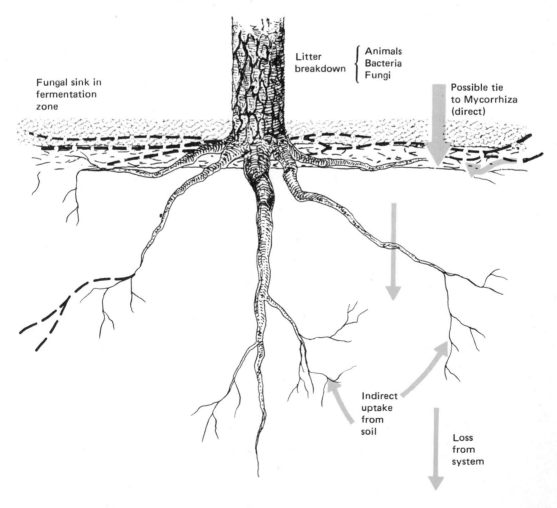

Litter breakdown { Animals Bacteria Fungi

Fungal sink in fermentation zone

Possible tie to Mycorrhiza (direct)

Indirect uptake from soil

Loss from system

concentrate nutrients in their tissues, especially the fruiting body. In comparison to the amount of nutrients found in pine needles, the basidiomycete fruiting bodies are low in calcium, magnesium, and manganese, but high in nitrogen, phosphorus, and sodium, and several trace elements. The rhizomorph tissue of fungi is high in calcium, nitrogen, phosphorus, sodium, and zinc. Thus fungi act as a living

sink of nutrients. Resistant to leaching, the fungal rhizomorphs hold biologically important elements in the litter. In fact in an experimental study, rhizomorph tissues were found to hold against a leaching force equivalent to one year's precipitation 99.9 percent of the 10 elements measured. Unprotected against leaching most elements would be lost to the ecosystem. Stark suggests that fungi

**FIGURE 10-11**
*Simplified representation of the annual biological cycle of nutrients in a mired oak (**Quercus**) forest at Virelles, Belgium (a), and a spruce forest (**Picea**) at Mirwart, Belgium (b). Productivity for both forests is 14.6 tn/ha/yr. (After Duvingneaud and Denaeyer-De Smet, 1975.)*

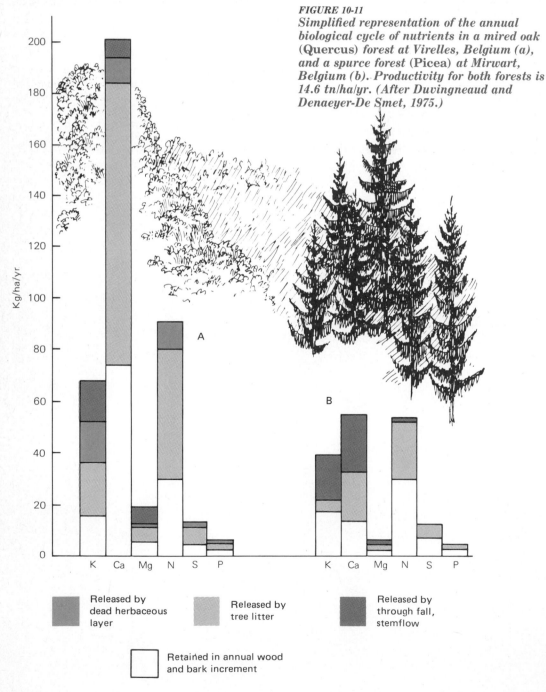

may release the elements slowly to the soil through exudates for recycling.

Coniferous forests appear to differ functionally from deciduous forests in the magnitude and nature of nutrient cycling. The differences are illustrated graphically in Figure 10-11, which compares nutrient cycling in a mixed oak forest in Belgium with a pine forest, both of which have an annual production of 14.6 tn/ha/yr. From these data it is evident that considerably more nutrients are cycled through the deciduous forest than through the coniferous forest and that the pine retains relatively more nutrients in its biomass than the oak. The pine retains more nitrogen, sulfur, and phosphorous than it returns through litterfall, throughfall, stemflow, and deal parts of the herbaceous layer. The oak forest on the other hand returns more of all elements than it retains. The deciduous forest is much more efficient at cycling calcium. The oak forest recycles calcium largely through litterfall, while the pine forest returns considerable quantities through stemflow and throughfall. Although the deciduous forest takes up more potassium than the pine forest, it retains less. The pine forest retains nearly half its uptake of potassium. This simply reemphasizes the suggestion that conifers tend to be accumulator plants.

## Tropical rain forest

### STRUCTURE

The tropical rain forest can be divided into five general layers, but the stratification is often poorly defined (Figure 10-12). The uppermost or emergent layer (A stratum) consists of trees over 60 m high whose deep crowns rise above the rest of the forest to form a discontinuous canopy. The second layer (B stratum), consisting of trees about 20 m high, forms another, lower discontinuous canopy. Strata A and B are not clearly separated from one another and together form an almost complete canopy. The third layer (C stratum) is the lowest tree layer; it is continuous, often the deepest layer, and, unlike A and B, is well defined. The fourth layer (D stratum), usually poorly developed in the deep shade, consists of shrubs, young trees, tall herbs, and ferns. The fifth (E stratum) is the ground layer made up of tree seedlings and low herbaceous plants.

A conspicuous part of the tropical rain forest is plant life dependent upon trees for support. Such plants include epiphytes, climbers, and stranglers. Epiphytes such as orchids and members of the Ericaceae are hemiparasites. They attach themselves to a tree and take up

**FIGURE 10-12**
*Microclimatic profiles in a tropical rain forest in November at Boseque de Florencia, Costa Rico. The height at the top of the canopy, 40 meters, is scaled as reduced height. $CO_2$ is expressed in relative values of 0 to 20; 0 represents a 0 gradient and values are deviations from an arbitrary standard. Note the increase in $CO_2$ near the ground and the increase in $CO_2$ in the morning and decrease in the afternoon, except near the ground. Temperature increases up through the canopy in the morning; in the late afternoon the temperature gradient is nearly 0. (After Allen et al., 1976.)*

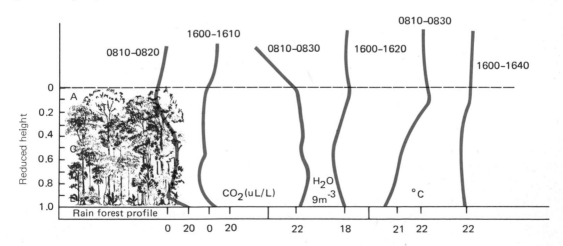

water, nutrients, and some photosynthate. Some of the epiphytes are important in recycling minerals leached from the canopy. Climbers are vinelike plants that reach to the tops of trees and expand into the form and size of a tree crown. Climbers grow prolifically in clearings, giving rise to the image of the impenetrable jungle. Stranglers start life as epiphytes. They then send roots to the ground and increase in number and girth, until they eventually encompass the host tree and claim the crown limbs as support for their own leafy growth.

The mature tropical forest, like the mature temperate forest, is a mosaic of continually changing vegetation. Death of tall trees creates openings in the forest which seedlings and saplings quickly fill. This is the gap phase in forest development. As the plants in the openings grow, the forest enters the building phase, dominated by pole-sized trees that occupy the B stratum. Most tropical rain forest trees complete height growth when they have achieved only about one-third to one-half of their final bole diameter. Thus stratification or layering results when a species or a group of species of similar mature height dominate a stand (Whitmore, 1975). Layering is also influenced by crown shape, which in turn is correlated with tree growth. Young trees still growing in height have a single stem and a tall narrow crown (monopodial). Mature trees have a number of large limbs diverging from the upper stem or trunk (sypodial). The limbs continue to grow, adding to crown width after the tree has reached mature height. Thus a pattern of gap, building, and mature phases in a stand results in poorly defined stratification in many tropical rain forests.

Stratification of animal life in the tropical rain forest, however, is pronounced. Harrison (1962) recognized six distinct feeding strata. (1) An insectivorous and carnivorous feeding group, consisting largely of bats and birds, works the upper air above the canopy. (2) Within the A-C canopy, a wide variety of birds, fruit bats, and mammals feed on leaves, fruit, and nectar. A few are insectivorous and mixed feeders. (3) Below the canopy, in the middle zone of tree trunks, is a world of flying animals, birds, and insectivorous bats. (4) Also in the middle zone of trunks are the scansorial

mammals, which range up and down the trunks, entering both the canopy and the ground zones to feed on the fruit of epiphytes growing on the tree trunks, on insects, and on other animals. (5) Large ground animals make up the fifth feeding group. This includes large mammals and a few birds, living on the ground and lacking climbing ability, that are able to reach up into the canopy or cover a large area of the forest. They include the herbivores, that eat leaves, fallen fruit, or root tubers, and their attendant carnivores. (6) The final feeding stratum includes the small ground animals, birds, and small mammals capable of some climbing, which search the ground litter and lower parts of tree trunks for food. This stratum includes insectivorous, herbivorous, carnivorous, and mixed feeders.

The tropical rain forest has been compared to a lake. The canopy, area of primary production, represents the phytoplankton, exploited directly by insects, comparable to the zooplankton and the large animals, comparable to the nekton. Food carried down into the deeper layer, the middle zone, and the ground, as fallen leaves, fruit, insect bodies, compares to the bodies of plankton organisms. This food source is utilized by the middle-zone birds and bats, corresponding to the nekton organisms of the deeper layer, and the middle-zone mammals which might be considered as periphytic organisms. The small ground mammals and birds are equivalent to the benthic organisms.

Animal life in the tropical rain forest is largely hidden, either by the dense foliage of the upper strata or by the cover of night. Birds are largely arboreal, and although brightly colored, remain hidden in the dense foliage. Ground birds are small and dark-colored, difficult to see. Mammals appear scarcer than they really are, for they are largely nocturnal or arboreal. Ground-dwelling mammals are small and secretive. Tree frogs and insects are most conspicuous at evening, when their tremendous choruses are at full volume. Insects are most diverse at forest openings, along streams, and at forest margins, where light is more intense, temperatures fluctuate, and air circulates freely. Highly colored butterflies, beetles, and bees are common. Among the unseen invertebrates, hidden in loose bark and in axils of leaves, are snails, worms, millipedes, centipedes, scorpions, spiders, and land planarians. Termites are abundant in the rain forest and play a vital role in the decomposition of woody plant material. Together with ants, they are the dominant insect life. Ants are found everywhere in the rain forest, from the upper canopy to the forest floor, although in common with other rain forest life, the majority tend to be arboreal.

Many specialized interactions among plants and animals exist in tropical rain forests. Plants, often widely dispersed in the forest, depend for pollination upon birds, bats, and insects, especially beetles, bees, moths, and butterflies (see Procter and Yeo, 1973) and for seed dispersal on fruit-eating birds, bats, rodents, and primates. Heavy predation on seeds by insects may result in wide dispersal of tree species. The further the seed is from the parent plant, the more likely it will escape predation and germinate (Jansen, 1971; see also Chapter 19). Other interactions involve repellent toxins to discourage predation by herbivores and even insect-plant mutualisms in which insects such as ants live in hollows of stem and prevent other insects from gaining entrance or feeding on the plant (Jansen, 1967).

## FUNCTION

Perhaps the most intensely studied tropical forest is the one at the Puerto Rico Nuclear Center. H. T. Odum (1970) and his associates worked up an energy budget for that forest. Incoming solar radiation amounted to 3,830 kcal/m²/day. Gross production amounted to 131 kcal/m²/day of which 116 kcal was used in respiration, leaving a net production of 15.2 kcal/m²/day as determined by gas analysis. Roots were responsible for 60 percent of the respiration, leaves for 33 percent, and trunks, branches, and fruit for the remainder. Net productivity as measured by biomass accumulation was 16.31 kcal/m²/day. Cumulative biomass addition through wood growth was 0.72 kcal/m²/day or 3.8 percent of net production. The remainder of net production passed through the grazing and detrital food chains.

Odum's energy budget is for a specific rain forest, and since site, soil, and other conditions of different tropical forests vary widely, productivity also varies widely (Figure 10-13). However, a few generalizations can be made

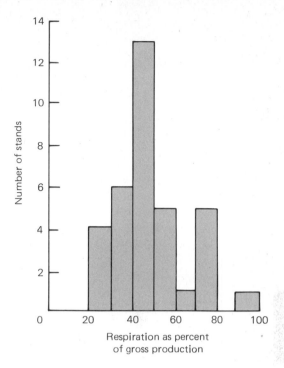

**FIGURE 10-13**
*Frequency of stands of tropical forest vegetation with autotrophic respiration (R) as a percent of gross primary production (GPP). All types of vegetation are represented in the figure. (From Golley, 1972.)*

about their energy budgets. Tropical forests use 70 to 80 percent of their energy intake in maintenance and 20 to 30 percent for net production. Average gross primary production is about 67 mtn/ha/yr or $28 \times 10^3$ kcal/m²/yr (Golley, 1972). Mean annual net production is about 21.6 mtn/ha/. This exceeds temperate forests, averaging 13 mtn/ha/yr, by a factor of 1.7 and boreal forests, averaging 8 mtn/ha/yr, by a factor of 2.7 (Golley and Farnsworth, 1973). However, tropical and temperate forests differ somewhat in their efficiency of production. Efficiency in this case is defined as the sum of energy stored in wood, leaves, fruit, and litter divided by total solar energy available to the community. Jordan (1971) found that rate of wood production in intermediate-aged stands was about the same for both tropical and temperate forests, but the rate of leaf and litter production was higher in the tropics. However, the efficiency of wood production was higher in temperate forests, probably because more selective pressure exists in temperate forests, where solar energy is not as abundant, to produce the maximum amount of wood.

High year-round temperatures and abundant rainfall in tropical rain forest areas produce rapid geologic cycling. Because geologic cycling is accelerated, biologic cycles apparently are modified to keep nutrients in the living portion of the system. Nutrients may be stored in living biomass where they are protected from leaching or the time the nutrient elements remain in the soil may be reduced to a minimum. This results in some basic differences between tropical forest and temperate forest ecosystems.

A large standing biomass is typical of tropical ecosystems. The tropical rain forest averages about 300 tn/ha compared to 150 tn/ha for a temperate forest. Concentrations of nutrients in the biomass differ (Figure 10-14). Tropical rain forests tend to concentrate proportionately more calcium, silica, sulfur, iron, magnesium, and sodium and less potassium

and phosphorus than temperate forests. In spite of these differences, however, rain forests hold larger quantities of nutrients simply because of their much larger biomass. Mineral concentrations do vary widely among tropical forests, influenced by site, soil, and climate (Golley, 1975). For example, Amazon rain forests are low in nutrient concentrations compared to Panamanian tropical forests.

As in the temperate forest a ratio of standing crop of available nutrients in the active part of the soil to the standing crop in vegetation provides some insight into nutrient cycling (Figure 10-15). In five Panamanian forests described by Golley and others (1975) potassium and phosphorus were held in large percentages in the vegetation. Much of the mineral recycling takes place through litterfall. The ratio of mineral elements held in bio-

**FIGURE 10-14**
*Comparison of concentration of selected elements in temperate and tropical forests. (From Golley, 1975.)*

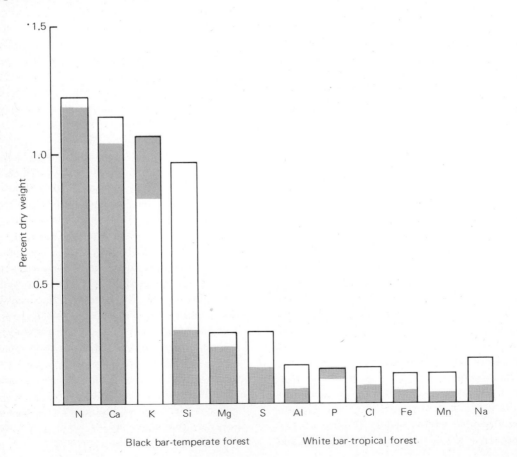

Black bar-temperate forest        White bar-tropical forest

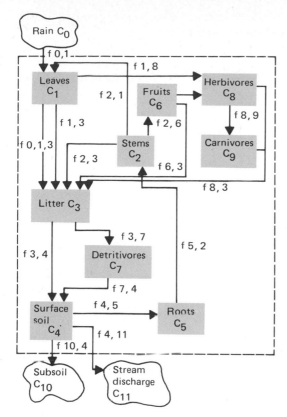

*FIGURE 10-15*
*Diagram of mineral cycling in a tropical moist forest. Dotted lines indicate system boundary. Boxes identify system components. Arrows indicate transfer functions between components and sources. (From Golley, 1975.)*

*TABLE 10-9*
*Turnover time of the mineral inventory in vegetation by litterfall in tropical forests (yr)*

| Forest | P | K | Ca | Mg |
|---|---|---|---|---|
| Tropical moist, Panama | 25 | 37 | 22 | 25 |
| Premontane wet, Panama | 9 | 28 | 17 | 10 |
| High forest, Ghana | 12 | 10 | 9 | 6 |
| Deciduous evergreen, Thailand | 5 | 6 | 9 | 8 |
| Montane, Peurto Rico | — | 84 | 9 | 10 |

*Source:* F. B. Golley, 1975.

calcium far exceeds the stream discharge, while the reverse is true for magnesium. The budget also suggests phosphorus and potassium might be limiting and that the elements are conserved by a rapid internal cycling.

Internal cycling may be aided by (1) the rapid return of nutrients leached by throughfall; (2) retention of nutrients by fungal rhizomorphs; and (3) uptake by mycorrhizal fungi. Data on throughfall in the rain forests of the Ivory Coast of Africa indicate that more than 60 percent of the potassium recycled and 15 to 56 percent of other nutrients such as calcium, magnesium, and nitrogen are supplied by throughfall. Nutrients leached from the leaves, especially at the end of the dry season when leaching is greatest, are apparently taken up efficiently by the soil-root system (Bernhard-Reversat, 1975).

In the temperate forest bacteria and fungi, the main agents of decay, release nutrients directly into the mineral soil, where they are subject to leaching. Feeder roots of these trees are woven into the matrix of mineral soil. Feeder roots of tropical trees, however, are concentrated in the well-aerated upper 2 to 15 cm of humus and only a few penetrate the upper layer of mineral soil (Cornforth, 1970). Symbiotically associated with the roots are mycorrhizal fungi. Some mycorrhizal fungi live around the roots. Others live partly within the cells of the root. Although their role as symbionts is not clearly defined, it appears the mycorrhizae are capable of digesting organic matter and passing phosphates and other minerals from the soil to the roots. During the process the mycorrhizae extract sugar and growth substances from the roots.

mass to the amount returned to the soil by litter provides some estimate of turnover time (Table 10-9). Phosphorus, magnesium, calcium, and potassium all appear to have turnover times of less than 100 years, and most are recycled in 20 years.

Nutrient budgets, like those of the temperate forest, can be assessed by comparing inputs from rainfall with outputs by stream discharge. The difference between the two represents inputs from weathering of the substrate. Data for a tropical forest of Panama indicate that inputs of phosphorus and potassium balance outputs (Table 10-10), but more calcium and magnesium are lost from the system than are gained by rainfall input. This suggests considerable input from the soil reservoir. Annual uptake of phosphorus, potassium, and

*TABLE 10-10*
*Comparison of biological and geological cycles in a tropical moist forest*

| | GEOLOGICAL CYCLE | | BIOLOGICAL CYCLE |
| | Input rain | Stream output | Annual uptake |
| Element | (kg/ha/yr) | /kg/ha/yr) | /kg/ha/yr) |
|---|---|---|---|
| P | 1.0 | 0.7 | 11 |
| K | 9.3 | 9.5 | 187 |
| Ca | 29.0 | 163.0 | 270 |
| Mg | 5.0 | 44.0 | 30 |

*Source:* F. B. Golley, 1975.

Retention of nutrients in the rhizomorphs of litter fungi was studied by Stark (1972). These fungi concentrate high levels of calcium, cooper, iron, potassium, nitrogen, sodium, phosphorus, and zinc, relative to the amount stored in rainforest leaves and especially dead wood, which forms the main substrate for the fungi. Rhizomorphs are highly resistant to leaching, holding better than 99.9 percent of the elements from such loss. Eventually these elements are released to the roots through exudates.

The roots of tropical forests support an abundance of mycorrhizal fungi that attach the feeder roots to dead organic matter by hyphae and rhizomorph tissue (Figure 10-16). As a result, considerable nutrient cycling in tropical rainforests appears to be through mycorrhizae (Went and Stark, 1968). They appear to cycle nutrients directly from dead organic matter to living roots with only a minimum of leakage into mineral soil. In such a direct mineral cycling, minerals remain tied up in living and dead organic matter and are transferred through hyphae from dead branches or leaves to living roots. Very little mineral matter becomes soluble and moves into the soil. This, according to Went and Stark (1968) may explain why feeder roots are concentrated mainly in the humus layer and why more mycorrhizal roots occur in poorer tropical soils.

## The tundra

North of the coniferous forest belt lies a frozen plain, clothed in sedges, heaths, and willows, which encircles the top of the world (see inside cover). This is the arctic tundra—the word comes from the Finnish *tunturi,* meaning a treeless plain (Figure 10-17). At lower latitudes similar landscapes, the alpine tundra, occur in the mountains of the world. But in the Antarctic a well-developed tundra is lacking. Arctic or alpine, the tundra is characterized by low temperatures, a short growing season, and low precipitation (cold air can carry very little water vapor).

The tundra is a land dotted with lakes and transected by streams. Where the ground is low and moist, extensive bogs exist. On high, drier areas and places exposed to the wind, vegetation is scant and scattered, and the ground is bare and rock-covered. These are the fell-fields, an anglicization of the Danish *fjoeld-mark,* or rock deserts. Lichen-covered, the fell-fields are most characteristic of the highly exposed alpine tundra.

### CHARACTERISTICS

Frost molds the tundra landscape. Alternate freezing and thawing and the presence of a permanent frozen layer in the ground, the *permafrost,* create conditions unique to the arctic tundra. The sublayer of soil is subject to annual thawing in spring and summer and freezing in fall and winter. The depth of thaw may vary from a few inches in some places to a foot or two in others. Below the thaw depth the ground is always frozen solid and is impenetrable to both water and roots. Because the water cannot drain away, flat lands of the Arctic are wet and covered with shallow lakes and bogs. This reservoir of water lying on top

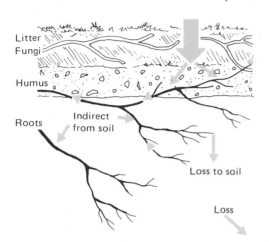

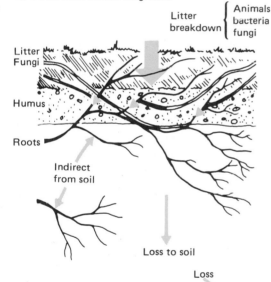

**FIGURE 10-16**
*Mechanisms of mineral cycling from soil in tropical forests. One type of direct nutrient cycling is characterized by a transfer of elements from litter to humus to roots without the aid of mycorrhizal fungi (except at root ends). This type of transfer occurs in tropical latisols where water penetration is slow. Another type of direct cycling involves a litter-humus-fungi-root pathway. It is characterized by the breakdown of litter by fungi, bacteria, and animals, uptake from humus by mycorrhizal fungi, and transfer of materials to living roots. (After Stark, 1973.)*

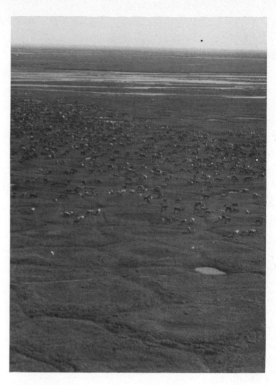

**FIGURE 10-17**
*Wide expanse of arctic tundra. This photo shows an area near the Sadlerochit River on the Arctic National Wildlife Refuge 5 miles from the Arctic Ocean. Note the frost polygons in the foreground. The caribou, a major arctic herbivore, are part of the Porcupine herd. (Photo courtesy U.S. Fish and Wildlife Service.)*

of the permafrost enables plants to exist in the driest parts of the Arctic.

The symmetrically patterned land forms so typical of the tundra result from frost. The fine soil materials and clays, which hold more moisture than the coarser materials, expand while freezing and then contract upon thawing. This action tends to push the larger material upward and outward from the mass to form the patterned surface.

Typical nonsorted patterns associated with seasonally high water tables are frost hummocks, frost boils, and earth stripes (Figure 10-18). Frost hummocks are small earthen mounds up to 1.5m in diameter and 1.3m high, which may or may not contain peat. Frost boils are formed when the surface freezes across the top, trapping the still unfrozen muck beneath. As this chills and expands, the mud is forced up through the crust. Raised earth stripes,

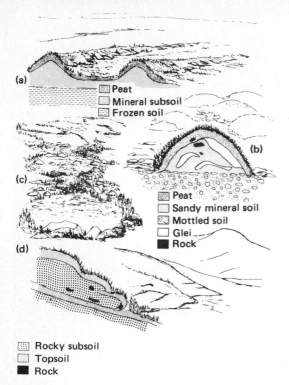

*FIGURE 10-18*
*Patterned ground forms of the tundra region:* (a) *unsorted earth stripes;* (b) *frost hummock;* (c) *sorted stone nets and polygons* (d) *solifluction terrace. (Diagrams adapted from Johnson and Billings, 1963.)*

found on moderate slopes, appear as lines or small ridges flowing downhill. They apparently are produced by a downward creep or flow of wet soil across the surface of the permafrost.

Sorted patterns are characteristic of better-drained sites. The best known of these are the stone polygons, the size of which is related to frost intensity and the size of the material (Johnson and Billings, 1962). The larger stones are forced out to a peripheral position, and the smaller and finer material, either small stones or soil, occupies the center. The polygon shape may result from an accumulation of rocks in desiccation cracks formed during drier periods. These cracks appear as the surface of the soil dries out, in much the same way as cracks appear in bare, dry, compacted clay surfaces in temperate regions. On the slopes, creep, frost-thrusting, and downward

flow of soil change polygons into sorted stripes running downhill. Mass movement of supersaturated soil over the permafrost forms solifluction terraces, or "flowing soil." This gradual downward creep of soils and rocks eventually rounds off ridges and other irregularities in topography. This molding of the landscape by frost action is called *cryoplanation* and is far more important than erosion in wearing down the arctic landscape.

In the alpine tundra permafrost exists only at very high elevations and in the far north, but the frost-induced processes—small solifluction terraces and stone polygons—are still present. The lack of a permafrost results in drier soils; only in alpine wet meadows and bogs do soil-moisture conditions compare with the Arctic. Precipitation, especially snowfall and humidity, is higher in the alpine than in the arctic tundra, but the steep topography results in a rapid runoff of water.

The arctic and the alpine regions share many features that characterize the tundra biome. The vegetation of the tundra is structurally simple (Figure 10-19). The number of species tends to be small, growth is low, and most of the biomass and functional activity are confined to a relatively few groups. Growing season and reproductive season are short. Most of the vegetation is perennial and reproduces vegetatively rather than by seed. Although it appears homogeneous, the pattern of vegetation is diverse. Small variations in microtopography result in a steep gradient of moisture. The combination of microrelief, snow melt, frost heaving, and aspect, among other conditions, produces an almost endless change in plant associations from spot to spot (Polunin, 1955).

Although the environment of the arctic and alpine regions is somewhat similar, the vegetation of the two differs in species composition and in adaption to light. Alpine sorrel *(Oxyria digyna),* found both in the arctic and alpine tundras, exhibits increased production of flowers and decreased production of rhizomes in the southern portion of its range (Mooney and Billings, 1960). Northern populations of the plant have a higher photosynthetic rate at lower temperatures and attain a maximum rate at a lower temperature. Alpine plants reach the saturation point for light at higher

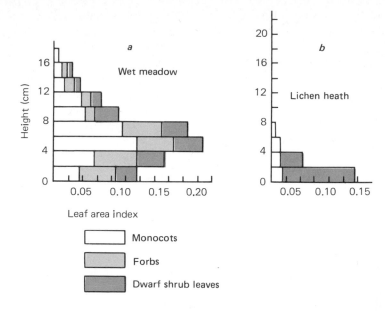

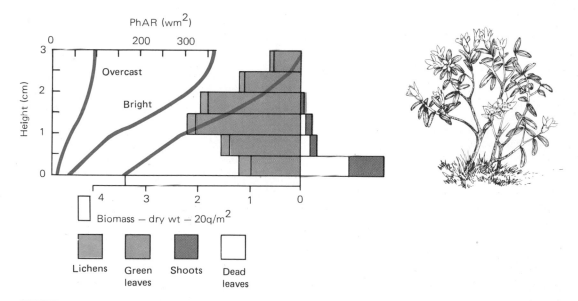

*FIGURE 10-19*
*Structure of vegetation in arctic tundra. (a)*
*Maximum leaf area indices (m²/m²) at*
*different plant heights (single surface)*
*measured by inclined point quadrats for wet*
*meadow. (b) Leaf area indices for lichen*
*heath. Note the structural simplicity of the*
*lichen heath (After Berg et al., 1975). (c)*
*Vertical structure of the* Loiseleuria *heath*
*and radiation profile. The instantaneous*
*values of photosynthetically active radiation*
*(W/m²) are plotted against stand height for a*
*bright day and an overcast day. The air and*
*soil temperatures are for a hot summer day.*
*(After Larcher et al., 1975.)*

intensities than arctic plants, which are
adapted at lower light intensities. Arctic
plants require longer periods of daylight than
alpine plants. The further north the geographic
origin of the plant, the more slowly the plant
grows under short photoperiod.

*Krumholtz.* At the tree line, where the forest
gives way to the tundra, lies an area of
stunted, wind-shaped trees, the krumholtz or
"crooked wood." The krumholtz in the North
American alpine region is best developed in
the Appalachians. In the west it is much less

marked, for there the timber line ends almost abruptly with little lessening of height; the trees for the most part are flagged, that is, branches remain only on the lee side (Figure 10-20). On the high ridges of the Appalachians, particularly in the White and Adirondack mountains, the trees begin to show signs of stunting far below the timber line. As the trees climb upward, stunting increases until spruces and birches, deformed and semiprostrate, form carpets 2 to 3 ft high, impossible to walk through but often dense enough to walk upon. Where strong winds come in from a constant direction, the trees are sheared until the tops resemble close-cropped heads, although the trees on the lee side of the clumps grow taller than those on the windward side.

Though the wind and cold generally are regarded as the cause of the dwarf and misshapen condition of the trees, Clausen (1965) has demonstrated that the ability of some species of trees to show a krumholtz effect is genetically determined. Eventually conditions become too severe even for the prostrate forms,

**FIGURE 10-20**
*Krumholtz in the Rocky Mountains. The tree line is sharply defined. Dwarf spruce and fir grow in narrow pockets that hold snow in winter. Although the low growth form is partly genetic, parts of the plants exposed above snow are broken and killed by wind and cold.*

and the trees drop out completely except for those that have taken root behind the protection of some high rocks. Tundra vegetation then takes over completely. On some slopes the trees might be able to grow on better sites at higher elevations, but they appear to be eliminated by competition from sedges (Griggs, 1946).

### STRUCTURE

*Arctic tundra.* In spite of its distinctive climate and many endemic species, the tundra does not possess a vegetation type unique to itself. Thus the word *tundra* does not imply a vegetation structure as do such terms as *prairie, deciduous forest,* or *tropical rainforest.* In effect, the tundra is structurally a grassland.

In the Arctic only those species able to withstand constant disturbance of the soil, buffeting by the wind, and abrasion from wind-carried particles of soil and ice can survive. On well-drained sites heath shrubs, dwarf willows and birches, dryland sedges and rushes, herbs, mosses, and lichens cover the land. On the driest and most exposed sites—the flat-topped domes, the rolling hills, and low-lying terraces, all usually covered by coarse, rocky material and subject to extreme action by the frost—vegetation is sparse and often confined to small depressions. Plant cover consists of scattered heaths and mats of mountain avens, as well as crustose and foliose lichens growing on the rocks. Willows, birch, and heath occupy well-drained soils of finer material, and between them grow grasses, sedges, and herbs.

But over much of the Arctic the typical vegetation is a cotton grass–sedge–dwarf heath complex (H. C. Hanson, 1953). Hummocks may support growths of lichens, willow, blueberry, and heaths. Depressions are covered with sedge-marsh vegetation, and over the rest grow tussocks of cotton grass. The spaces between the tussocks may be filled with sphagnum, on top of which dwarf shrubs grow; in other places sphagnum may overgrow the sedges and cotton grass. On mounds and hummocks in freshwater marshes, on well-drained knolls and slopes, in areas of late-melting snowbanks, along streams and on sandy and gravelly beaches, grassland types develop.

Topographic location and snow cover delimit a number of arctic plant communities. Steep, south-facing slopes and river bottoms support the most luxurious and tallest shrubs, grasses, and legumes, whereas cotton grass dominates the gentle north-facing and south-facing slopes, reflecting higher air and soil temperatures and greater snow depth. Pockets of heavy snow cover create two types of plant habitats, the snow patch and the snowbed. Snow-patch communities occur where wind-driven snow collects in shallow depressions and protects the plants beneath. Snowbeds, typical of both arctic and alpine situations, are found where large masses of snow accumulate because of certain topographic peculiarities. Not only does the deep snow protect the plants beneath, but the meltwater from the slowly retreating snowbank provides a continuous supply of water throughout the growing season. Snowbed plants, usually found only here, have an extremely short growing season but are able to break into leaf and flower quickly because of the advanced stage of growth beneath the snow.

The conditions unique to the Arctic result in part from three interacting forces: permafrost, vegetation, and the transfer of heat. Permafrost is sensitive to temperature changes. Any natural or man-made disturbances, however slight, can cause the permafrost to melt. Because the permafrost itself is impervious to water, it forces all the water to move above it. Thus the surface water becomes quite conspicuous even though precipitation is low (see Brown and Johnson, 1964; Brown, 1970). Vegetation protects the permafrost by shading, which reduces the heating of the soil. It retards the warming and thawing of the soil in summer and increases the average temperature in winter. If the vegetation is removed, the depth of thaw is 1.5 to 3 times that of the area still retaining the vegetation. Accumulated organic matter and dead vegetation further retard the warming of the soil in summer, even more than a vegetative cover. Thus vegetation and its organic debris impede the thawing of the permafrost and act to conserve it (see also Pruitt, 1970).

In turn permafrost chills the soil, retarding the general growth of both above-ground and below-ground parts of plants and the activity of soil microorganisms. It also impoverishes the aeration and nutrient content of the soil (Tyrtikov, 1959). The effect is more pronounced the closer the permafrost comes to the surface, where it contributes to the formation of shallow-root systems. The effect of permafrost on vegetation is so pronounced that vegetation can be used to map areas of permafrost.

The tundra world holds some fascinating animals, even though the diversity of species is low. The animals of the Arctic are mostly circumpolar, and although the species that inhabit the North American tundra are not the same as those that inhabit the Eurasian tundra, they are close relatives. The North American barren ground caribou *(Rangifer arcticus),* for example, is matched by the European reindeer *(Rangifer tarandus),* and some consider the two the same species. The muskox *(Ovibos moschatus),* arctic hare *(Lepus articus),* and arctic ground squirrel *(Alopex lagopus)* are or have been common to both. In addition some 75 percent of the birds of the North American tundra are common to the European tundra (Udvardy, 1958).

Muskox and caribou are the dominant large herbivores. Muskox are intensive grazers with low herd movements. In summer they feed on grasses, sedges, and dwarf willow in the valleys and plains; in winter they move up to the windswept ridges where snow cover is scant. Caribou are extensive grazers, spreading over the tundra in summer to feed on sedges and grasses, and in winter migrating southward to the taiga to feed on lichens.

Intermediate-sized herbivores include the arctic hares, which feed on willows. In winter the hares disperse over the range; in summer they tend to congregate in more restricted areas. The smallest and dominant herbivore over much of the arctic tundra is the lemming *(Lemmus* spp.*)* which feeds on fresh green sedges and grasses. Breeding throughout the year beneath the snow, the lemming has a three- to five-year population cycle. During the highs this rodent can reach densities as great as 125 to 250 per hectare and during the lows as few as 3 to 5 per hectare. Herbivorous birds are relatively few, dominated by ptarmigan and migratory geese.

The major arctic carnivore is the wolf *(Canis lupus).* (The polar bear is a marine predator.) The wolf preys on muskox, caribou, and when they are abundant, lemmings. Medium-sized predators include the arctic fox which feeds on the arctic hare. The smallest mammalian predators, the least weasel *(Mustela rixosa)* and the short-tailed weasel *(Mustela erminea)*, feed principally on lemmings and the eggs and nestlings of birds. Major avian predators, the snowy owl *(Nyctea scandiaca)* and the hawklike jaeger, feed heavily on lemmings. Except for the wolf, whose populations remain relatively stable when free from human persecution, the fortunes of most arctic predators rise and fall with the flood and ebb of lemming life.

The arctic tundra, with its wide expanse of ponds and boggy ground, is the haunt of myriads of waterfowl, sandpipers, and plovers, which arrive when the ice is out, nest, and return south before winter sets in.

Invertebrate life is scarce, as are amphibians and reptiles. Some snails are found in the arctic tundra about the Hudson Bay. Insects, reduced to a few genera, are nevertheless abundant, especially in mid-July. The insect horde is composed of black flies, deer flies, and mosquitos (Shelford and Twomey, 1941).

Animal activity in the arctic tundra is geared to short summers and long winters. The only hibernator is the ground squirrel, although the female polar bear does den up in the snow where she gives birth to her cub. The ground squirrel is active only from May to September. It mates almost as soon as it emerges from the burrow in spring, and the young are born in mid-June, after a 25-day gestation period. The young are self-sufficient by mid-July, attain adult weight and are ready to hibernate by late September and early October (W. Mayer, 1960). A similar speed-up in the life cycle exists among some of the arctic birds. Because of the long days, the northern robin feeds the young for 21 hours a day, and the young may leave the nest when they are slightly over 8 days old, in contrast to the 13 or more days when born in the temperate region (Karplus, 1949). The species that are unable to withstand severe cold migrate to warmer or more protected areas.

*Alpine tundra.* In general the alpine tundra is a more severe environment for plants than the arctic tundra, and the adaptation of plants to the physical environment is probably more important than the interrelations of one species with another. The alpine tundra is a land of strong winds, snow, and cold and widely fluctuating temperatures. During the summer the temperature on the surface of the soil ranges from 40° to 0° C (Bliss, 1956). The atmosphere is thin, and because of this, light intensity, especially ultraviolet, is high on clear days.

The alpine tundra of the Rocky Mountains is a land of rock-strewn slopes, bogs, alpine meadows, and shrubby thickets (Figure 10-21). In spite of the similarity of conditions, only about 20 percent of the plant species of the arctic and of the Rocky Mountain alpine tundra are the same, and these are different ecotypes. Heaths are lacking in the tundras of the Rockies, as well as the heavy growth of lichens and mosses between other plants. Lichens are more or less confined to the rocks, and the ground is bare between plants.

Cushion- and mat-forming plants, rare in the arctic, are important in the alpine tundra. Low and hugging the ground, they are able to withstand the buffeting of the wind. The cushionlike blanket traps heat and the interior of the cushion may be 20° warmer than the surrounding air, a microclimate that is

*FIGURE 10-21*
*Alpine tundra in the Rocky Mountains.*

utilized by insects. Thick cuticles, which increase plants' resistance to desiccation, and the abundance of epidermal hairs and scales are characteristic of alpine plants. The significance of this is still debated. The hairs appear to absorb and reflect the bright light of the alpine environment. At the same time they may act as a heat trap, perhaps preventing cold injury when air temperatures drop to freezing (Krog, 1955) and enabling the plants to develop and bloom while the air is still cold.

Alpine vegetation and its associated soils vary on a rather complex gradient of topographic site and snow cover, both of which interact with wind (Figure 10-22). The vegetational pattern, worked out for the Beartooth Plateau of Wyoming (Johnson and Billings, 1962), serves as an example of the high alpine areas of the Rocky Mountains. The high, windswept areas, rocky and free of snow, support only lichens, which may completely cover the sheltered side; but on the windward side they are short, no higher than the depth of snow, and they may be completely lacking on the most exposed sites. Below the lichen growth are the xeric cushion-plant communities, which extend further downslope on the windward side than they do on the lee. This land of rock, lichens, and cushion plants is the alpine rock desert. In somewhat more protected sites grows the geum turf, a sodlike covering of geum and associated plants, such as sedges, lupines, polygonums, and mountain avens. Alpine meadows develop on well-drained soils of sheltered uplands and lower mesic slopes and basins. Hairgrass (Deschampsia), often growing in pure stands, is the dominant species. These meadows are subject to considerable disturbance both from frost activity and from pocket gophers. Alpine bogs, communities quite similar to those of the arctic tundra, support a growth of sedge and cottongrass. Willow thickets, dense and uniform in height, grow along drainage channels and in alpine valley bottoms.

The alpine tundra of the high Appalachians is not nearly so cold and windswept as that of the Rockies. Tundra areas are small and lack the diversity of species found in the western mountains. Indeed little floristic similarity exists between the regions. There is a much closer affinity between the flora of the eastern

**FIGURE 10-22**
*Relationship of major vegetation types to slope cover and slope position in the western alpine tundra on Beartooth Plateau, Wyoming. (After Johnson and Billings, 1962.)*

alpine tundra and that of the arctic and the alpine communities of Scandinavia and central Europe.

Nine plant communities are recognized in the tundras of the Presidential Range in New Hampshire (Bliss, 1963). These occur on two gradients, one of increasing snow depth, the other of increasing moisture (Figure 10-23). On exposed windswept sites where winter snow cover is thin or nonexistent, *Diapensia,* a dwarf, tussock-forming shrub, grows. Over those widespread areas where snow cover is variable a dwarf heath–rush community occupies the sites. Dwarf heaths—bearberry, bilberry, Lapland rosebay—dominate where the deep snow cover melts early. Snowbank communities are most prevalent on the east- and southeast-facing slopes, in the lee of the prevailing winds.

The second gradient, one of increasing summer atmospheric and soil moisture and fog, is largely restricted to north- and west-facing slopes on the higher peaks. Sedge meadows at the highest elevations give way downslope to a sedge–dwarf heath community. At lower elevations this is replaced by a sedge–rush–dwarf heath type. Two other communities, the streamside and the bog, are common at low elevations.

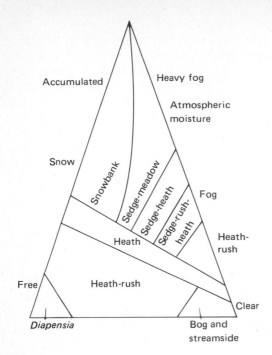

**FIGURE 10-23**
*Relationship of alpine communities to snow and atmospheric moisture in the Presidential Range of New Hampshire. (After Bliss, 1963.)*

The alpine tundra, which extends upward like islands in the mountain ranges of the world, is small in area and contains few characteristic species. The alpine regions of western North America are inhabited by pikas, marmots, mountain goat (not a goat at all, but related to the South American chamois), mountain sheep, and elk. Sheep and elk spend their summers in the high alpine meadows and winter on the lower slopes. The marmot, a mountain woodchuck, hibernates over winter, while the pika cuts grass and piles it in tiny haycocks to dry for winter. Some rodents, such as the vole and the pocket gopher, remain under the ground and snow during winter. The pocket gopher is important, for its activities influence the pattern of alpine vegetation. The tunneling gophers kill the sedge and cushion plants by eating the roots and by throwing the soil to the surface, smothering plant life. Other plants then take over on the wind-blown, gravelly soil. These pioneering plants are rejected by the gopher. The rodent moves on, the cushion plants move back in, organic matter accumulates again.

Slowly the sedges recover, and when they do, the gophers return once more.

The alpine regions contain a fair representation of insect life. Flies and mosquitos are scarce, but springtails, beetles, grasshoppers, and butterflies are common. Because of the ever-present winds, butterflies fly close to the ground; other insects have short wings or no wings at all. Insect development is slow; some butterflies may take two years to mature, and grasshoppers three.

**FUNCTION**

Low temperature, a short growing season ranging from 50 to 60 days in the high arctic tundra to 180 days in low latitude alpine tundras, and low availability of nutrients interact to keep primary production on the tundra low.

*Photosynthesis.* Plants are photosynthetically active on the arctic tundra about three months out of the year. The onset of photosynthesis occurs in spring as quickly as the snow cover disappears. Photosynthesis in early spring is limited because plant leaves are poorly developed with a limited ability to intercept light. Throughout the growing season photosynthesis is influenced by light and temperature (Johansson and Linder, 1975). Across the tundra plants exhibit ecotypic variation in their adjustment to these two influences. Maximum photosynthesis takes place when ambient temperatures are 5 to 15° C. Photosynthesis is inhibited when leaf temperatures rise above 40° C, which they can reach on bright, clear arctic summer days (Bliss, 1975). Tundra plants, which possess a $C_3$ photosynthetic cycle, have adapted to short growing seasons and low light intensities of the arctic summer in two ways: (1) they carry on photosynthesis throughout the 24-hour daylight period, even at midnight when the light is one-tenth that of noon; (2) the plants, especially the monocotes, possess a high leaf area index (0.5 to 1.0). The nearly erect leaves of some arctic plants permit the almost complete interception of the slanting rays of arctic solar radiation (Bunnell et al., 1975; Berg et al., 1975). In contrast some arctic plants, particularly somes mosses, are poorly adapted to 24-hour light and under this condition turn yellowish or brownish (Kallio and Valanne, 1975). Much of the photosynthesis goes into the production of new growth, but about one month before the growing season ends, plants cease to allocate photosynthate to above-ground biomass. They withdraw photosynthate from the leaves and move it to the roots and below-ground biomass.

*Primary production.* Net annual primary production varies markedly across the tundra, depending upon the plant community considered (Bliss, 1975; Wielgolaski, 1975). Because of microenvironmental conditions influenced by soil, slope, aspect, exposure to wind, snow depth, and drainage conditions plant communities change rapidly over relatively short distances. Primary production for selected tundra communities is summarized in Tables 10-11, 10-12, 10-13. In general alpine tundras are more productive than arctic tundras (Table 10-11). At the Devon Island IBP site in Canada total primary production above and below ground for a hummocky sedge-moss meadow was 183.3 g/m²/yr; for a wet sedge meadow, 279.9 g/m²/yr; and for a raised beach cushion plant–lichen community, 42.8 g/m²/yr for the total area and 131. g/m²/yr for the vegetated area only (Bliss, 1975). In contrast net primary production above ground only for two communities on an Austrian alpine heath, a *Vaccinium* heath and a *Loiseleuria* heath, was 422 g/m²/yr and 277 g/m²/yr, respectively (Larcher et al., 1975). Total above- and below-ground productivity for three Norwegian alpine tundra communities, a lichen heath, a wet sedge-moss meadow, and a dry meadow dominated by mountain avens, *Dryas,* forbs, and grass, was 276 g/m²/yr, 837 g/m²/yr, respectively.

Different components of the plant community make different contributions to net productivity, depending upon the site (compare Tables 10-11 and 10-12). In the hummocky sedge-moss meadow of the Devon Island IBP site, Canada, above-ground net primary production of vascular plants amounted to 44.7 g/m²/yr, while mosses contributed 33 g/m²/yr and lichens contributed nothing. In the raised beach community dominated by lichens and cushion plants vascular plants contributed 17.8 g/m²/yr, mosses 2 g/m²/yr, and lichens 25 g/m²/yr. In an alpine wet meadow vascular plants contributed 254 g/m²/yr and mosses 173 g/m²/yr.

Much of the primary production of vascular plants is below ground rather than above (Table 10-11). The data emphasize this important functional aspect of the tundra. The net annual above-ground primary production of vascular plants in a wet sedge meadow at Devon Island was 45.7 g/m² and below-ground 129.7 g.; a wet meadow at an alpine tundra at the Hardangervidda IBP site in Norway had a biomass of 254 g/m² above ground and 1316 g/m² below ground. On more xeric sites dominated by lichens there is less below-ground biomass than above.

Although total production of the tundra is low because of a short growing season, daily primary production rates of 0.9 to 1.9 g/m² in arctic tundra and 2.2 g/m² in some alpine tundras are comparable to some temperate grasslands. The efficiency of primary production of the Devon Island tundra (Table 10-14) ranges from 0.03 percent for the polar desert to 1.03 percent for the hummocky sedge-moss meadow (Bliss, 1975). Efficiency of a Norway tundra ranges from 0.6 percent for a dwarf heath community to 2.4 percent for a willow thicket (Wielgolaski and Kjelvik, 1975). This compares with temperate region ecosystems. Efficiency of grassland ecosystems ranges from 0.33 to 3.8 percent.

Net radiation in the tundra is somewhat higher compared to radiation in temperate regions, but photosynthetically active radiation appears to be lower. This effectively increases the percentage efficiency of primary production of tundra plants (Bliss, 1975).

*Biomass.* Biomass reflects the pattern of net production. At Devon Island, for example, biomass for the polar desert amounts to 270 g/m² and for the hummocky sedge-moss meadow 3,208 g/m². More live biomass exists below ground than above in most tundra communities (Table 10-11). Live roots make up about 60 percent of the total below-ground biomass. The ratio of above- and below-ground biomass depends upon the nature of the tundra community. On Devon Island the hummocky sedge-moss community has an A/B ratio of 1:12.6, the cushion plant lichen community 1:.06; the wet sedge communities of both the arctic and the alpine tundra 1:8.9.

Biomass on the tundra also accumulates below rather than above ground. The entire

TABLE 10-11
Comparison of primary production and biomass of vascular plants, selected tundra ecosystems (g/m²)

| Location and type | Vegetation | Primary production | Biomass | A/B ratio biomass | Reference |
|---|---|---|---|---|---|
| **Devon Island, Canada** | | | | | |
| arctic | hummocky sedge–moss meadow | 44.7 (103.6) | 86 (1085) | 1:12.6 | Bliss, 1975 |
| arctic | wet sedge–moss meadow | 45.7 (129.7) | 78 (691) | 1:8.9 | Bliss, 1975 |
| arctic | raised beach, cushion plant lichen | 17.8 (2.6) | 89 (57) | 1:.06 | Bliss, 1975 |
| **Taimyr, USSR** | | | | | |
| arctic | herb and grass meadow | 68.4 (100) | 71.5 (191) | | Matveyera et al., 1975 |
| arctic | mossy *Salix polaris* mesic frost boil | 31.8 | 20.2 | | Matveyera et al., 1975 |
| arctic | dry frost boil | 23.8 | 21.0 | | Matveyera et al., 1975 |
| **Hardangervidda, Norway** | | | | | |
| alpine | lichen heath | 88 (100) | 62 (191) | 1:3.08 | Ostbye et al., 1975 |
| alpine | wet meadow | 254 (1316) | 147 (410) | 1:8.95 | Ostbye et al., 1975 |
| alpine | dry meadow | 241 (545) | 161 (245) | 1:3.38 | Ostbye et al., 1975 |
| **Mt. Patscherkofel, Austria** | | | | | |
| alpine | *Loiseleuria* heath | 277 | 1084 (2213) | 1:2.04 | Larcher et al., 1975 |
| alpine | Vaccinnium heath | 422 | 1013 (2206) | 1:2.17 | Larcher et al., 1975 |
| **Macquarie Island, Tasmania** | | | | | |
| subantarctic | grassland | 1890 (3670) | 912 (1690) | 1:1.85 | Jenkin, 1975 |
| | herbfield | 314 (550) | 139 (670) | 1:4.82 | Jenkin, 1975 |

Note: Numbers in parentheses are below-ground production and biomass, without parentheses, above-ground production and biomass.

TABLE 10-12
*Comparison of primary production and biomass of mosses and lichens, selected tundra ecosystems*

| Location and type | Vegetation | Primary production (g/m²/yr) | | Biomass (g/m²) | | Reference |
|---|---|---|---|---|---|---|
| | | Mosses | Lichens | Mosses | Lichens | |
| Devon Island, Canada | | | | | | |
| arctic | hummocky sedge–moss meadow | 33.0 | 0 | 908 | 0 | Bliss, 1975 |
| arctic | wet sedge–moss meadow | 102.6 | 0 | 1097 | 0 | Bliss, 1975 |
| arctic | raised beach, cushion plant lichen | 2.0 | 25 | 15 | 49 | Bliss, 1975 |
| Hardangervidda, Norway | | | | | | |
| alpine | lichen heath | 10.0 | 78 | 7 | 370 | Ostbye et al., 1975 |
| alpine | wet meadow | 173.0 | 0 | 175 | 0 | Ostbye et al., 1975 |
| alpine | dry meadow | 48.0 | 4 | 31 | 19 | Ostbye et al., 1975 |
| Mt. Patscherkofel, Austria | | | | | | |
| alpine | *Loiseleuria*, heath | nd | | 0 | 136 | Larcher et al., 1975 |
| | Vaccinnium heath | nd | | 44 | 22 | Larcher et al., 1975 |

*TABLE 10-13*

*Live standing crop and net annual production of various communities on Truelove Lowland and plateau, (energy constant, kJ/m²)*

| | Vascular plants | | | | | |
| | Above-ground | Below-ground | Mosses | Lichen | Algae | Total |
|---|---|---|---|---|---|---|
| **STANDING CROP** | | | | | | |
| Cushion plant-lichen | 1,824 | 1,167 | 272 | 841 | 0 | 4,104 |
| Cushion plant-moss | 2,581 | 1,025 | 10,920 | 393 | nd | 14,919 |
| Frost-boil sedge–moss | 1,088 | 6,720 | 10,012 | 0 | 17 | 17,837 |
| Dwarf shrub–heath | 3,397 | 18,912 | 7,389 | 410 | nd | 30,108 |
| Hummocky sedge–moss | 1,674 | 20,656 | 16,527 | 0 | 67 | 38,924 |
| Wet sedge–moss | 1,519 | 13,154 | 19,966 | 0 | 33 | 34,672 |
| Polar desert (plateau) | 301 | 75 | 4,222 | 0 | 0 | 4,598 |
| Semidesert (raised beach) | 2,130 | 1,113 | 4,531 | 661 | 0 | 8,435 |
| Wet sedge tundra (all meadows) | 1,402 | 14,008 | 13,351 | 0 | 42 | 28,803 |
| Total Lowland | 1,602 | 8,422 | 9,506 | 205 | 21 | 19,756 |
| **NET PRODUCTION** | | | | | | |
| Cushion plant-lichen | 326 | 63 | 38 | 50 | 0 | 477 |
| Cushion plant-moss | 552 | 105 | 364 | 33 | nd | 1,054 |
| Frost-boil sedge-moss | 565 | 1,121 | 146 | 0 | 17 | 1,849 |
| Dwarf shrub heath | 423 | 1,904 | 364 | 59 | nd | 2,750 |
| Hummocky sedge–moss | 858 | 1,979 | 602 | 0 | 67 | 3,506 |
| Wet sedge–moss | 895 | 2,477 | 1,874 | 0 | 33 | 5,279 |
| Polar desert (plateau) | 42 | 8 | 92 | 0 | 0 | 142 |
| Semidesert (raised beach) | 418 | 79 | 167 | 46 | 0 | 710 |
| Wet sedge tundra (all meadows) | 724 | 1,619 | 460 | 0 | 42 | 2,845 |
| Total Lowland | 544 | 1,004 | 331 | 17 | 21 | 1,917 |

*Source:* Bliss, 1975.

*TABLE 10-14*

*Net annual production and efficiency of annual production for various tundra plant communities and system components based upon the length of the growing season (50 to 60 days)*

| Component | Net production (kJ/m²) | Total radiation (kJ/m²) | Efficiency (%) | |
| | | | Total radiation | PAR* |
|---|---|---|---|---|
| Polar desert (plateau) | 138 | $11.72 \times 10^5$ | 0.01 | 0.03 |
| Polar semidesert (raised beach) | 711 | $11.13 \times 10^5$ | 0.06 | 0.16 |
| Sedge–moss meadows (all meadows) | 2845 | $9.04 \times 10^5$ | 0.31 | 0.79 |
| Hummocky sedge–moss meadow | 3506 | $9.04 \times 10^5$ | 0.39 | 1.03 |
| Total lowland | 1916 | $9.62 \times 10^5$ | 0.20 | 0.50 |

*PAR (photosynthetically active radiation) = 40 percent of total radiation; efficiency was calculated on the basis of radiation received during the growing season for each component part and the contribution (%) which that component provides to the total lowland (raised beach types, meadow types, dwarf shrub heath, etc.).
*Source:* Bliss, 1975.

monocot portion of the vegetation grows and dies each season. Most of the herbaceous and woody plants accumulate little above ground and rapidly turn over the energy in the living portion of the system. The below-ground portion is more persistent with roots persisting for 2 to 10 years, reflecting low mortality and slow decomposition. Because decomposition is slowest in wet sites the difference between above- and below-ground biomass is greatest there. Decomposition is faster on well-drained and on nutrient rich sites. In fact greatest accumulation of below-ground biomass is associated with those sites where net productivity is the lowest.

*Decomposers.* Most of the production of the tundra goes by the way of the decomposers (Figure 10-24). At Devon Island, for example, grazing by lemmings at the most accounts for only 3 to 4 percent of aboveground standing crop. On the hummocky sedge-moss meadow 2 percent of the plant production is utilized by herbivores; 98 percent of the primary production is channeled to microbivores and saprovores. Comprising the latter is a wide variety of soil invertebrates, dominated by protozoans. Among important soil invertebrates are the Enchytraeidae or pot worms, nematodes, fly larvae, and various crustacea. Annual production and consumption among these animals is greater than the herbivore system even in a lemming high (Bunnell et al., 1975).

But the major detrital consumers are bacteria and fungi. Tundra soils contain a diversity of soil bacteria—iron oxidizers, nitrogen fixers, ammonia oxidizers, sulfate reducers, and fermenters. Bacteria occur in the same abundance as in temperate soils. The same is true for fungi. The amount of mycelia in tundra soils is as great or often greater than in temperate mull or mor soils. As in other terrestrial systems, fungi are more important than bacteria. But like the primary producers, the decomposers are restricted in their activities by the cold. Studies of decomposition rates for sedge indicate a dry weight loss of 19 percent for the first year, 12 percent the second year. For moss the rate was 1.3 percent per year (see Bliss et al., 1973; Bliss, 1975; Bunnell et al., 1975; Rosswall, 1975).

*Consumers.* Major herbivores of the tundra include waterfowl, ptarmigan, lemming, hares, muskox, caribou, and reindeer. In some parts of the arctic tundra lemmings are the dominant herbivores. Adult lemmings consume about 0.32 g dry material/g body weight and juveniles 0.53 g. Adults consume 170 percent and juveniles 200 percent of their body weight in summer with an assimilation efficiency of 47 percent. Lemmings during a high may consume over 25 percent of the above-ground primary production or 10 percent of total plant production. Consumption,

however is seasonal; heaviest consumption takes place in winter when during a high, lemmings may consume the entire aboveground vascular plant biomass. Of the food consumed lemmings return about 70 percent as feces which accumulate in winter and release quantities of nutrients during the spring melt (Bunnell et al., 1975). During periods of highs, lemmings can hold primary production to 3 to 48 g/m². By reducing the litter layer, these animals reduce insulation of the soil and increase the depth of thawing. In doing so. lemmings can influence the composition and nature of the tundra plant community.

In other parts of the arctic tundra lemmings are not significant grazing herbivores. Their place may be taken by the muskox and caribou. These two ungulates have an average standing crop of 0.17 kg/km² in the Canadian and Alaskan tundra. This is low compared to the 140 kg/km² of the African savanna. In summer muskox consume approximately 30 to 34 grams of vegetation per kilogram of body weight and have an assimilation efficiency of 56 percent. Muskox remove less than 1 percent of potential primary production. On the range at Devon Island, muskox grazed 14 percent of the total area available and removed 15 percent of the total herbage available. However, muskox are selective grazers and on the restricted areas they grazed the animals removed 80 to 85 percent of the herbage available. Dung decomposition is slow, requiring 5 to 12 years to be recycled (Bliss, 1975).

Preying on the herbivores are a number of carnivores. Arctic fox consume 23.3 to 44.7 kcal/kg body weight a day in summer and 54.5 to 72.2 kcal/kg body weight in winter. An efficient assimilator, the fox rarely passes as feces more than 5 percent of the total energy ingested. One investigator (Speller, cited by Bliss et al., 1973) estimated that between the first of May and the last of September, one fox would require the equivalent of 400 28-g lemmings, 400 30-g snow buntings, and 20 3.46-kg arctic hares. Weasels consume more than 1 28-g lemming a day. Extremely effective as predators, weasels may consume up to 20 percent of the lemming population. Snowy owl, another lemming predator, has a winter food requirement of 4 to 7 lemmings a day. Thus not only can the predators act as a force

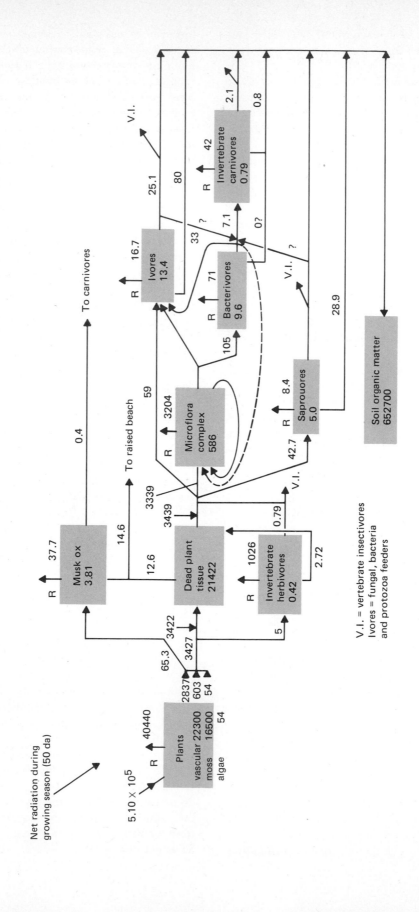

Net radiation during
growing season (50 da)

$5.10 \times 10^5$

R ← 40440

Plants
vascular 22300
moss 16500
algae 54

2837
603
54

65.3
3427
3422

5

R ← 37.7

Musk ox
3.81

0.4

To carnivores

12.6

14.6

To raised beach

Dead plant
tissue
21422

3439

3339

59

R ← 3204

Microflora
complex
586

105

R ← 16.7

Ivores
13.4

33

?

25.1

80

V.I.

R ← 9.6

Bacterivores
71

7.1

0?

R ← 2.1

Invertebrate
carnivores
0.79

42

0.8

1026

Invertebrate
herbivores
0.42

0.79

V.I.

2.72

42.7

R ← 5.0

Saprouores
8.4

V.I.

?

28.9

Soil organic matter
652700

V.I. = vertebrate insectivores
Ivores = fungal, bacteria
and protozoa feeders

*FIGURE 10-24*

*Energy flow diagram for a sedge-moss meadow system. Standing crop (boxes) and energy flow (arrows) are expressed in kJ/m². Respiration (R) is given for all components along with energy estimates for rejecta. Ivores are fungal, bacterial, and protozoan feeders. Insectivores are seed and insect-feeding birds. (From Bliss, 1975.)*

to drive the lemming population to a low, but predators themselves are strongly affected by a scarcity of lemmings. When lemming populations are low, arctic fox experience reproductive failures and snowy owls cannot exist on the tundra. When forced southward the owls face an uncertain fate.

*Nutrients.* Arctic and alpine tundras are low in nutrients because of a short growing season, cold temperatures, and restricted decomposition. Most of the nutrient capital in the soil is not directly available to plants. Pools of soluble soil nutrients, especially nitrogen and phosphorus, are small relative to exchangeable pools, which in turn are small compared to the nonexchangeable pools (Figure 10-25) (Bunnell et al., 1975). As a result

nutrient cycling in the tundra is conservative. Vascular plants have a strong internal cycling. They retain and reincorporate nutrients, especially nitrogen, phosphorus, potassium, and calcium, in their tissues rather than release them to decomposers. In forest ecosystems more nutrients are retained in new growth than released through litterfall, but in the tundra ecosystem death of tissues compensates for new growth. The system arrives at a steady state in which return equals uptake.

Leaching or removal of nutrients is minimal, occurring mostly in spring at the beginning of the growing season. As the snow melts, runoff increases carrying with it animal debris that accumulates over winter. Spring rains leach senescent plant material produced the previous growing season. Rising temperatures stimulate decomposition. Because 60 percent of the active roots are in the upper 5 cm of soil, the root mass once thawed takes up most of the

*FIGURE 10-25*

*Standing crops (g/m²) and flows (g/m²/yr) of nitrogen and phosphorus in major components of the Barrow tundra ecosystem. (From Bunnell et al., 1975.)*

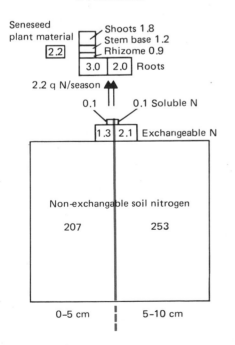

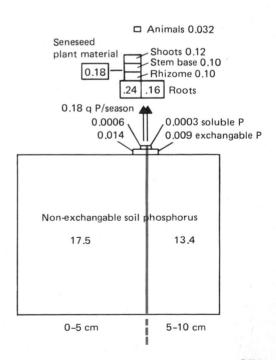

nutrients early in the season. In summer vascular plants leak nutrients, especially phosphorus and potassium, from the cuticle of the leaves to the surface where they are washed off by summer rains. These nutrients are often picked up by bryophytes which effectively absorb or adsorb nutrients before they reach the soil. The nutrients incorporated into the mosses are slowly released by them. Thus bryophytes function as a temporary nutrient sink.

In some tundra ecosystems lemmings become involved in nutrient cycling. Foraging in meadows and ridges, but building their nests and defecating in troughs, lemmings transport nutrients from one site to another. Higher levels of soil nutrients, especially phosphorus, in polygon troughs, and higher rates of decomposition in these microsites may reflect the activity of lemmings (Bunnell et al., 1975).

Two nutrients, nitrogen and phosphorus, are most limiting. Compared to temperate forest and grassland, the input of nitrogen is small. Two major sources are precipitation and biological fixation (Figure 10-26). Nitrogen fixation is accomplished by both anaerobic and aerobic free-living bacteria; by blue-green

algae in soil, in water, and especially in the moss where they live epiphytically; and by lichens, particularly in the Fennoscandian tundra (Granhall and Lid-Torsvik, 1975; Kallio and Kallio, 1975). A rough estimate places nitrogen fixation up to 1 to 2 g/m²/yr although values for blue-green algae range up to 5 to 6 g/m²/yr. Nitrogen fixation is controlled by light, temperature, and moisture.

Precipitation is the second important input of nitrogen. The input from precipitation is usually lower than the input from biological fixation, but at some sites the two are equal. At Hardangervidda, Norway, for example, input from precipitation equals that of fixation, about 200 mg/m²/yr (Granhall and Lid-Torsvik, 1975).

Nitrogen accumulates in tissues in greater

**FIGURE 10-26**
*Pools and fluxes in the nitrogen budget for the tundra ecosystem. Nitrogen fixation and ammonia deposition represent the largest inputs; nitrogen in runoff and denitrification represent major losses. Values are given as milligrams of nitrogen per square meter per year. (From Barsdate and Alexander, 1973.)*

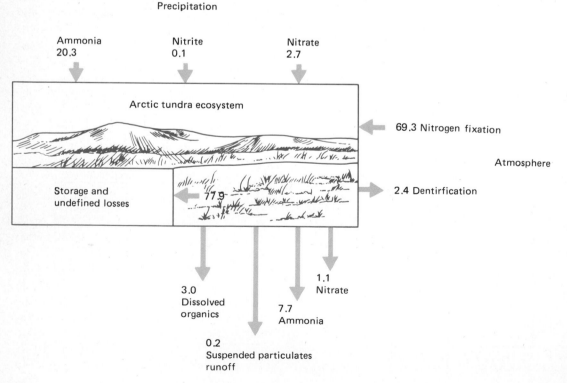

quantities than other nutrients, making much of it unavailable for recycling. At the Barrow IBP site in Alaska about 65 percent of the gross input, 59.2 mg/m², is stored in the system as living and dead organic matter (Bunnell, 1975). Circulation in the system is restricted by decomposers, while flux rates are more or less controlled by the nitrogen-fixing microorganisms. Relatively small quantities of nitrogen are lost through leaching, stream flow, and denitrification. The most significant leaching takes place during snowmelt in the spring.

Phosphorus can be very limiting in the tundra ecosystem. Phosphorus apparently controls the rate of production of new leaves, primarily by controlling the rate at which nutrients are removed from older leaves and translocated and incorporated into new leaves (Bunnell et al., 1975). The effect of phosphorus limitation is not nutrient deficiency, but a limitation of leaf area which lowers production. While the accumulation of nitrogen, potassium, and calcium is similar from site to site across the tundra, the accumulation of phosphorus is highest at the most productive sites.

Because pools of both nitrogen and phosphorus are small, a constant turnover between the exchangeable and soluble pools is necessary to replenish the quantity absorbed by plants. Bunnell et al. (1975) estimate that at the Barrow site soluble and exchangeable nitrogen must turn over 11 times during a growing season to meet plant needs and phosphorus must be replenished 200 times a season or 3 times a day during the growing season. Because the tundra soil does not store available nutrients in any great quantity, primary productivity depends upon release of nutrients from decomposition, the uptake of which is often aided by mycorrhizae. In this respect the tundra ecosystem is more similar in function to the tropical forest ecosystem than to the temperate forest ecosystem.

## Synthesis

A simple comparison of ecosystem functions provides a first step toward seeking some common properties of metabolic patterns in ecosystems and developing some general principles of ecosystem functioning. Table 10-15 presents comparative metabolic parameters for four types of terrestrial ecosystems: a deciduous forest (mesic hardwood), a coniferous forest, a short-grass prairie, and a tundra. The values are very general and, of course, selected ecosystems within the four types might present values different from those used. Data for the deciduous forest and the short-grass prairie are probably more accurate than data for the other two.

Although gross production of the deciduous forest ecosystem is 1.6 times that of the coniferous forest and 1.7 times that of the short-grass prairie, net primary production among the three is comparable, although the coniferous forest is somewhat less. Much of the GPP of the forest ecosystems goes to the maintenance of structure. This is reflected in a comparison of the deciduous forest with that of the short-grass prairie. Forest structure requires 4.5 times greater allocation of photosynthate to maintain structure than the grassland. Heterotrophic respiration in the deciduous forest and short-grass prairie is similar, but values for the coniferous forest are considerably lower, reflecting either an inaccurate data base or more likely the lower activity of soil microflora and microfauna, allowing accumulation of litter on the forest floor. All four systems are characterized by a large pool of organic detritus suggesting that a large, stable pool of organic matter is typical of ecological systems. Three of the systems are characterized by annual accumulation of large amounts of organic matter below ground.

This accumulation of organic matter is one aspect of capture, distribution, and conservation of elements essential to system persistence. Currency for translocation and retention of nutrient elements is carbon allocated for maintenance of primary producers and decomposers (Burgess and O'Neill, 1976). Retention and distribution of elements within an ecosystem may be dependent on the maintenance of the detrital pool in litter and soil and in the decomposer community. The cost to the ecosystem is the loss of carbon to heterotrophic respiration. Microbial immobilization of nutrients appears to be synchronized with likely periods of maximum elemental uptake by

TABLE 10-15

*Comparative metabolic parameters and metabolic ratios of four contrasting ecosystems* $(gC/m^2/yr)$

| | | Deciduous[a] forest | Coniferous[b] forest | Grassland[a] | Tundra[b] |
|---|---|---|---|---|---|
| **Metabolic parameters** | | | | | |
| Gross primary production | GPP | 2150 | 1320 | 983 | 208 |
| Autotrophic respiration | $R_A$ | 1440 | 680 | 430 | 120 |
| Net primary production | NPP | 720 | 600 | 840 | 88 |
| Heterotrophic respiration | $R_H$ | 660 | 370 | 670 | 85 |
| Net ecosystem production | NEP | 150 | 270 | 180 | 3 |
| Ecosystem respiration | $R_E$ | 2105 | 1050 | 1090 | 205 |
| **Metabolic ratios** | | | | | |
| Production efficiency | $R_A/GPP$ | 0.67 | 0.52 | 0.30 | 0.57 |
| Effective production | NPP/GPP | 0.33 | 0.45 | 0.70 | 0.42 |
| Maintenance efficiency | $R_A/NPP$ | 2.0 | 1.13 | 0.51 | 1.36 |
| Respiration allocation | $R_H/R_A$ | 0.46 | 0.54 | 1.55 | 0.71 |
| Ecosystem productivity | NEP/GPP | 0.02 | 0.20 | 0.14 | 0.01 |

Source: [a]Burgess and O'Neill, 1976; [b]National Academy of Science, 1974.

roots which results in a reduced amount of nutrients being lost from the ecosystem. Maximum growth of roots appears to take place at times of minimum nutrient immobilization by decomposers. This reciprocal interaction between plants and decomposers may be a major mechanism underlying nutrient conservation in ecosystems.

Among the four types of ecosystems the tundra stands out as possessing extremely low metabolic parameters. Gross production of the tundra, for example, is 10 times less than that of the deciduous forest. However, when all four systems are compared in terms of ratios of metabolic parameters, especially production efficiencies and effective production, ecosystems appear more uniform. Respiratory expenditures per unit production are about the same regardless of the environment. Thus certain parameters of ecosystem functioning analyzed in terms of ratios (such as the relative apportioning of resources among components of various ecosystems) yield some basic patterns common to most ecosystems.

## SUMMARY

The alpine tundra of the high mountain ranges in lower latitudes and the arctic tundra that extends beyond the tree line of the far north are at once similar and dissimilar. Both have low temperatures, low precipitation, and a short growing season. Both possess a frost-molded landscape and plant species whose growth rates are slow. The arctic tundra has a permafrost layer; rarely does the alpine tundra. Arctic plants require longer periods of daylight than alpine plants and reproduce vegetatively, while alpine plants propagate themselves by seed. Over much of the Arctic, the dominant vegetation is cotton grass, sedge, and dwarf heaths. In the alpine tundra cushion and mat-forming plants, able to withstand buffeting by the wind, dominate exposed sites while cotton grass and other tundra plants are confined to protected sites. Net primary production is low because of a short growing season, although daily primary production rates are comparable to those of temperate grasslands. Biomass accumulates below rather than above ground. In spite of an assemblage of grazing ungulates and rodents, most production goes to decomposers. Major detrital consumers are bacteria and fungi. Nutrient levels are low, and nutrients tend to accumulate and become stored in living and dead plant material unavailable for recycling. Circulation is restricted by limited activity of decomposers, while the flux rates are controlled by nitrogen-fixing organisms. Because pools of both nitrogen and phosphorus are small, constant turnover between exchangeable and soluble pools is necessary to replenish the quantity absorbed and retained by plants.

Coniferous, deciduous, and tropical rain forests are three of the dominant types of forest. The coniferous forest, which forms a vast belt encircling the northern portion of the Northern Hemisphere, is typical of regions where summers are short and winters are long and cold. The deciduous forest, richly developed in North America, western Europe, and eastern Asia, grows in a region of moderate precipitation and mild temperatures during the growing season. The tropical rain forest grows in equatorial regions where humidity is high, the rainfall heavy (especially at least during one season of the year), seasonal changes are minimal, and annual mean temperature is about 28° C. All three are more or less stratified into layers of vegetation. Accompanying this vegetative stratification is a stratification of light, temperature, and moisture. The canopy receives full impact of climate and intercepts light and rainfall; the forest floor is shaded through the year in most coniferous and tropical rain forests and in late spring and summer in the deciduous forest. The coniferous and deciduous forests hold different species of animal life, but animal adaptations are similar. The greatest concentration and diversity of life are on and just below the ground layer. Other animals live in various strata from low shrubs to the canopy. The tropical rain forest has pronounced feeding strata from above the canopy to the forest floor and many of its animals are strictly arboreal. Whatever the forest, the different trees that compose it create different environments that ultimately dictate the kinds of plants and animals that can live within it.

While gross primary productivity of forest ecosystems is high, so much of the GPP is allocated to the maintenance of forest structure that effective production is low: deciduous forest, 33 percent; coniferous forest, 45 percent; and tropical rain forest, 12 percent. Mineral cycling is tight. Nutrients accumulate in woody biomass to form a pool unavailable for short-term cycling. Although the bulk of nutrients is in mineral soil, the most important pools in mineral cycling are root mortality, litterfall, and foliar leaching. Internal cycling of some nutrients, particularly nitrogen, is important in nutrient conservation. In most forest systems only a small fraction of nutrients is lost from the system through streamflow.

Coniferous forests exhibit short-term cycling between litterfall and uptake by trees. At the same time conifers appear to be accumulators, removing elements from the soil in quantities large enough to upset nutritional balances in ecosystems. Nutrient cycling in conifers appears to depend on mycorrhizae, fungi, and root activity. Fungi act as nutrient sinks, while a symbiotic relationship exists between mycorrhizae and roots in the uptake of nutrients.

Tropical rain forests possess a large standing crop biomass that ties up great quantities of nutrients. Much of the mineral cycling takes place between a rapid decomposition of litterfall and rapid uptake of nutrients it contains. Roots are concentrated in the top of the ground in close contact with the litter where a symbiotic relation between roots and mycorrhizae facilitates nutrient uptake.

The tropical rain forest and coniferous forest contrast with the deciduous forest in which microbial decomposition releases nutrients to mineral soil which, in turn, provides the pathway through the roots for nutrient uptake.

## part IV

# *Population ecology*

# Ecological genetics, natural selection, and speciation
## CHAPTER 11

The many different groups that make up the ecosystem consist of populations of plants and animals. Considered ecologically, a *population* is a group of interbreeding organisms of the same kind occupying a particular space. Each population is a structural component of the ecosystem through which energy and nutrients flow. It is characterized by density, that is, the number of organisms occupying a definite unit of space. It has an age structure, the ratio of one age class to another. It acquires new members through birth and immigration and loses members through death and emigration. The difference between the gains and losses determines the rate of population growth.

Because it is composed of interbreeding organisms, the population is also a genetic unit. Each individual carries a certain combination of genes, a part of the population's total genetic information. The sum of all genetic information carried by all individuals of an interbreeding population is the *gene pool*. Gene

flow or the exchange of such information between populations comes about through immigration and emigration.

Evolution involves changes in the gene pool and physical expressions of genetic constitution. These changes are caused by selective pressures brought to bear by the environment upon individuals of the population. If the gene pool and its forms of expression contain enough variation and variability to change when necessary, individuals in the population over generation time become better adapted to the environment. Selection (anything that produces a systematic heritable change in a population) acts as a cybernetic system. Information from the environment is transmitted back to the gene pool, which responds with changes in its content and expression.

Evolutionary change of the population is the cumulative result of the adaptiveness of its individuals. The more fit individuals have more offspring than the less fit, and as a result, less adapted types in the population decrease and better adapted types increase. In other words, certain variations within the gene pool increase at the expense of other variations to produce a population better adapted to its environment. Because the environment is always changing, the population through the survival of certain individuals, is constantly changing adaptive characteristics. The rate at which individuals in the population respond in fitness to environmental changes determines the ability of that population to survive.

## Genetic variation

Wherever you go along the seashore, whether it be long stretches of beaches or harbors and docks, you see and hear the gulls, especially the ubiquitous herring gull. Even a moderately alert observer will detect obvious differences among the herring gulls. Most conspicuous are the adults with their bluish-gray back, their white head, neck, underparts and tail, their black-tipped primary wing feathers, and their yellow bill with a bright red spot near the tip of the lower mandible. Among the adult gulls are younger birds with a different pattern. Some are darkish brown-gray, mottled and barred on the back with white and grayish buff. Others are lighter in tone with some gray on the back. And still others may be similar to adults but with some dusky spotting on the tail and wings. Their bills may have only a suggestion of the red spot.

If you examine the gulls more closely, you will detect other differences. The size, the shades of gray on the back, the length of the bill, the shape of the red spot, the length of the wing, and other characteristics may vary among the birds. In fact so widespread are these smaller differences that a person who looks carefully at the birds and becomes acquainted with a colony can distinguish one bird from another.

### TYPES OF VARIATION

The most obvious type of variation among members of a population is discontinuous, that is, a variation in a specific character or set of characters which separates individuals into discrete categories. Thus the gulls can be divided by their plumage into first-year birds,

second-year birds, third-year birds, and mature adults. Another type of discontinuous variation is morphological, such as male and female or the red and gray phase of the screech owl. Other discontinuous differences may be biochemical, such as blood groups in humans, or even behavioral, such as song dialects in birds.

A second type of variation, the one commonly used in taxonomic and evolutionary studies, is the continuous variable, a variation in a character which can be placed along a range of values. Characters subject to continuous variation can be measured, for example, tail length of a species of mouse, number of scales on the belly of a snake, rows of kernels on ears of corn, and shape and size of sepals and petals. The measurements of a character or set of characters for several individuals in a population can be tabulated as a frequency distribution and arranged graphically as a histogram (Figure 11-1). The *mean* of the character for the population is the sum of all the values divided by the number of specimens. Many specimens will have approximately the same

FIGURE 11-1
*Histogram showing the frequency distribution of the hind tibia lengths of nymphal exuviae (shed skin) of the periodical cicada* Magicicada septendecim. *(After Dybas and Lloyd, 1962.)*

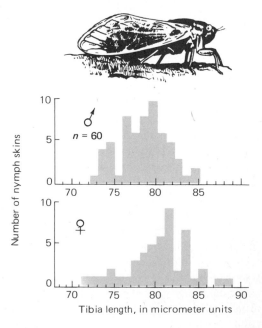

numerical value. The most frequent numerical value is the *mode* or modal class. Measurements of other specimens will vary above and below the mode. The *frequency* of occurrence of measured values will fall away from the mode with fewer and fewer individuals in the more distant classes. The *frequency distribution* of these variable characters tends to follow a bell-shaped curve, the normal frequency distribution.

Variations may also be typed according to whether they are genetic or nongenetic. The characteristics of a species and variations in individuals are transmitted from parent to offspring. The sum of the hereditary information carried by the individual is the *genotype*. The genotype directs the development of the individual and produces the characters that make up the morphological, physiological, and behavioral characteristics of the individual. The external or observable expression of the genotype is the *phenotype*. The expression of some genotypic characters may be influenced by external and internal conditions. For example, a seedling with the gene for the formation of chlorophyll will develop the normal green color if germinated in the light, but it will be white, that is, it will have a different appearance, if germinated in the dark. The gene directs the character of green color, but its expression is affected by environmental conditions. Thus, the phenotype of an individual (P) is determined by the genetic endowment (G) modified by the environment (E) and a factor of interaction between the genes and the environment or selective pressure.

The ability of a genotype to give rise to a range of phenotypic expressions under different environmental situations is known as *phenotypic plasticity*. Some genotypes have a narrow range of reaction to environmental conditions and therefore give rise to a fairly constant phenotypic expression. But many plants and animals that can survive under a wide range of environmental conditions may possess variable and diverse phenotypic responses. Some of the best examples of such phenotypic plasticity are found among plants. The size of plants, the ratio of reproductive tissue to vegetative tissue, and even the shape of the leaf may vary widely at different levels of nutrition, light, and moisture (Figure 11-2).

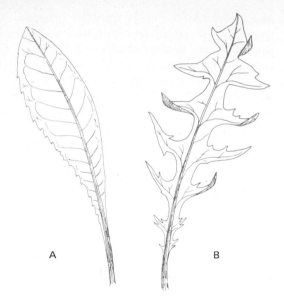

A                    B

*FIGURE 11-2*
*Plasticity of response to light by leaves of dandelion. A leaf growing in the shade exhibits minimal lobing of the leaf. Leaves growing in the sun are deeply lobed. Increased lobing of leaves growing in the open may relate to increased heating of the leaf in full sunlight. Lobed leaves present less surface area for absorption of heat and more edge per surface area which increases dissipation of heat.*

Lacking the mobility of animals, plants must possess more flexibility in their response to environmental conditions in order to survive. Phenotypic plasticity represents nongenetic variation. An environmentally induced modification of a character is not inherited. What is inherited is the ability of the organism to modify such a character.

### SOURCES OF VARIATION

The primary genetic control mechanism found within the nucleus of every cell in the organism is deoxyribonucleic acid, DNA. DNA, the information template from which all cells in the organism are copied, is a complex molecule in the shape of a double helix, resembling a twisted ladder in construction. The long strands, comparable to the uprights of the ladder, are formed by an alternating sequence of deoxyribose sugar and phosphate groups. The connections between the strands, or the rungs, consist of pairs of the nitrogen bases adenine, guanine, cytosine, and thymine. In the formation of the rungs adenine is always paired with thymine and cytosine is always paired with guanine. The DNA molecule is divided into smaller units, called nucleotides, consisting of three elements: phosphate, deoxyribose, and one of the nitrogen bases bonded to the strand at the deoxyribose. The information of heredity is coded in the sequential pattern in which the base pairs occur. According to current theory, each species is unique in that its base pairs are arranged in a different order and probably in different proportions from every other species.

DNA is present in larger units called chromosomes, which are found in most living organisms. Each species has a characteristic number of chromosomes in every cell and the chromosomes occur in pairs. When cells reproduce (a process of division called *mitosis*), each resulting cell nucleus receives the full complement of chromosomes, or the *diploid* number (for example, 46 in humans). In organisms that reproduce sexually the germ cells or gametes (sperm and egg) result from a process of cell division, *meiosis,* in which the pairs of chromosomes are split so that each resulting cell nucleus receives only one-half the full complement, or the *haploid* number (23 in humans). When egg and sperm unite to form a new individual the diploid number is restored. The chromosomes recombine in a great array of combinations. This segregation and recombination of chromosomes and the hereditary information they carry are the primary sources of variation.

Each chromosome carries units of heredity called *genes,* the informational units of the DNA molecule. Because chromosomes are paired, genes are also paired in the body cells. The position a gene occupies on a chromosome is known as a *locus*. Genes occupying the same locus on a pair of chromosomes are termed *alleles*. If each of the pair of alleles affects a given trait in the same manner, the two alleles are called *homozygous*. If each of the pair of alleles affects a given trait in a different manner, the pair is called *heterozygous*. During meiosis the alleles are separated as the chromosomes separate. At the time of fertilization the alleles, one from the sperm and one from the egg, recombine as the chromosomes recombine.

Although variation in individuals comes from differences in genetic materials, in macro- and microenvironments, and in gene-environment interaction, major interest lies in the source of new genetic variation. One source of differences in genetic material is the reassortment and recombination of existing genes both at the level of the gene and at the level of the chromosome. The other source is mutation or a change in the genetic material in the gene or chromosome.

*Recombination of genetic material.* When two gametes combine to form a zygote, the gene contents of the chromosomes of the parents are mixed in the offspring. Because the number of possible recombinations is infinitely large, recombination rather than mutuation is the immediate and major source of variation. Recombination does not result in any change in genetic information, as mutation does, but it does provide different combinations of genes upon which selection can act. Because some combinations of interacting genes are more adaptive than others, selection determines the variations or new types that will survive in the population. The poorer combinations are eliminated by selection and the better ones retained.

The amount or degree of recombination influencing the amount of variability in a population is limited by a number of characteristics of the species. One limitation is the number of chromosomes and thus the number of genes involved. Another is the frequency of crossing-over, the exchange of corresponding segments of homologous chromosomes during meiosis. Others include gene flow between populations, the length of generation time, and the type of breeding, for example, single versus multiple broods in a season in animals and self-pollination versus cross-pollination in plants.

*Mutation.* A *mutation* is an inheritable change of genetic material in the gene or chromosome. Organisms that possess such changes are called mutants.

*Micromutations,* or gene mutations, are alterations in the DNA sequence of one or a few nucleotides. During meiosis the gene at a given locus usually is copied exactly and eventually becomes part of the egg or sperm. On occasion the precision of this duplication process

**361**

breaks down and the offspring DNA is not an exact replication of the parent DNA. The alteration may be a change in the order of nucleotide pairs, a substitution of one nucleotide pair for another, a deletion of a pair, or various kinds of transpositions.

The rate of mutation in general is low. Most common mutations involve the change of one allele into another. Even in a population homozygous for *A*, for example, *A* eventually will mutate to *a* in some of the gametes; and in a population having both genes, mutations may be forward to *a* or backward to *A*. If *A* mutates to *a* faster than *a* to *A*, the frequency of allele *A* decreases. This rarely occurs to a point where one of the alleles is lost to the population, for reversibility prevents a long-term or permanent loss. Eventually such mutations arrive at an equilibrium. Even if one allele is lost from the population, it will usually reappear by mutation.

*Macromutations,* or chromosomal mutations, may result from a change in the number of chromosomes or a change in the structure of the chromosome.

A change in chromosomal number can arise in two ways: the complete or partial duplication of the diploid number rather than the transmission of the haploid number or the deletion of some of the chromosomes.

Polyploidy is the duplication of entire sets of chromosomes. It can arise from an irregularity in meiosis or from the failure of the whole cell to divide at the end of the meiotic division of the nucleus. The individual body cell is ordinarily diploid (2*n* or twice the haploid number). Forms of polyploidy are triploid (3*n* or three haploid sets), tetraploid (4*n*), etc.

Polyploidy exists mostly in plants. The condition is rare in animals because an increase in sex chromosomes would interfere with the mechanisms of sex determination and the animal would be sterile. Polyploid plants differ from normal diploid individuals of the same species in appearance and are usually larger, more vigorous, and occasionally more productive.

Another form of macromutation is the *duplication* or *deletion* of a part of a normal complement of chromosomes. Such deletions or duplications result in abnormal phenotypic conditions. (One such condition is Down's syndrome or mongolism in humans.)

A change in the physical structure of a chromosome may occur in the form of deletion, duplication, translocation, or inversion of segments of the chromosomes (Figure 11-3).

*Deletion* is the loss of a part of a chromosome; a definite segment and the genes thereon are missing in the offspring cell. Occasionally the functions of the genes at the missing loci will be assumed by genes in some other part of the chromosome, but if the segment lost is large, the individual dies. Also lethal is a deletion in individuals that are homozygous for the character involved. In heterozygous individuals deletion may permit the manifestation of characters determined by recessive genes. The loss of a short segment results in a marked effect on development.

*Duplication* involves an addition to a chromosome. In general duplication is less harmful than deletion and some cases may have little or no effect on the phenotype. Duplication may increase both the genetic material and the effect of certain genes on development or it may cause an imbalance of gene activity, reducing the viability of an organism.

*Translocation* is the exchange of segments between two nonpaired (nonhomologous) chromosomes. The genes in the translocated segment become linked to those of the recipient chromosome. If the translocation is reciprocal, all of the chromosome material is present in the individual even though it is rearranged. Individuals carrying such translocations are usually normal. If the translocation is not reciprocal, some genes will be transferred to completely different chromosomes and the linkage relationship becomes altered drastically. The effect of the translocation becomes evident during meiosis and the formation of gametes. During the segregation of chromosomes, the two within the translocated segments will produce balanced and unbalanced gametes. Some will have excessive deletions and others will have a duplication of material. If an unbalanced gamete fertilizes a balanced gamete the fertilized egg is not viable.

An *inversion* is an alteration of the sequence of genes in a chromosome. It may occur when a

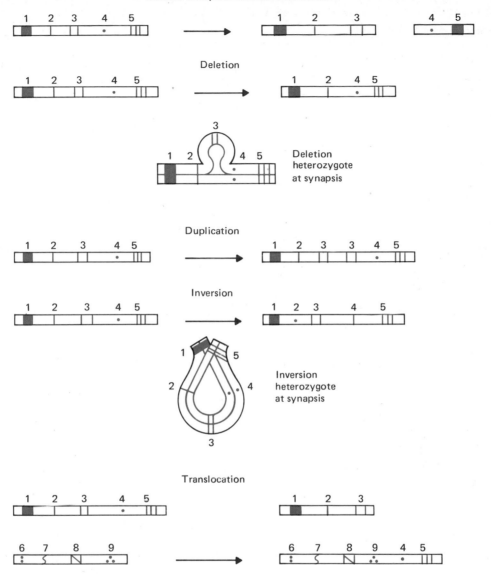

**FIGURE 11-3**
*Types of chromosomal aberrations. When these altered chromosomes join with normal homologs during the first meiotic division, they assume characteristic configurations that allow locus by locus matching. Synaptic configurations of deletion and inversion are shown. (Diagrams after Wilson and Bossert, 1971.)*

chromosome breaks in two places and the segment between the breaks becomes turned around, reversing the order of genes in respect to an unbroken chromosome. When the altered chromosome is paired with a normal chromosome in a heterozygous individual, the alteration interferes with pairings. The chromosomes must bend, loop, or in some way cross over so that each gene aligns itself with its homologue. This crossing-over between inverted and normal chromosomes usually produces abnormal, nonviable gametes.

## RECOMBINATION AND HARDY-WEINBERG EQUILIBRIUM

If genes occur in two forms, $A$ and $a$ then any individual carrying them can fall into three possible diploid classes: *AA, aa,* and *Aa.* Individuals in which the alleles are the same, *AA* or *aa,* are called homozygous; and those in which the alleles are different, *Aa,* are heterozygous. The haploid gametes produced by the homozygous individuals are either all *A* or all *a;* those by the heterozygous, half *A* and half *a.* These can recombine in three possible ways: *AA, aa, Aa.* Thus the proportion of gametes carrying *A* and *a* is determined by the individual genotypes, the genes received from the parents. Eggs and sperm unite at random, enabling the prediction of the proportion of offspring of different genotypes based on parental genotypes.

Assume that a population homozygous for the dominant *AA* is mixed with an equal number from a population homozygous for the recessive *aa.* Their offspring, the $F_1$ generation, then will consist of 0.25 *AA,* 0.50 *Aa,* and 0.25 *aa* (Figure 11-4). These proportions are called genotypic frequencies. The gene frequencies, of course, are 0.50 of *A* and 0.50 of *a.* This proportion will be maintained through successive generations of a bisexual population (Figure 11-5) if at least three conditions exist: (1) reproduction is random; (2) mutations either do not occur or they occur in equilibrium, that is, the rate of mutation from *A* to *a* is the same as *a* to *A;* and (3) the population is large enough so that changes by chance in the frequency of genes are insignificant.

The equilibrium of these three genotypes can be expressed as a general statistical law, known as the Hardy-Weinberg law, which can be stated simply as follows:

$p$ = frequency of allele *A*
$q$ = frequency of allele *a*
$p + q = 1.0$

Then

$p^2 + 2pq + q^2 = 1.0$

in which

$p^2$ = frequency of individuals homozygous
$q^2$ = frequency of individuals homozygous
$2pq$ = frequency of heterozygous individuals, *Aa*

In the hypothetical population above, the proportion of the genotypes in the $F_1$ generation will be $(0.5)^2 + 2(0.5 \times 0.5) + (0.5)^2$; and the same tendency can be demonstrated even

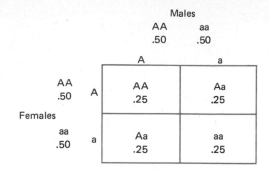

**FIGURE 11-4**
*Mixing two homozygous populations.*

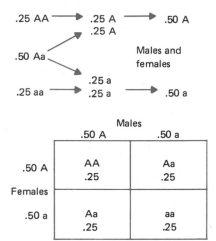

**FIGURE 11-5**
*Proportions in the $F_2$ generation.*

if the ratio is not the classical Mendelian 1:2:1. Imagine a population in which the ratio of *A* alleles *(p)* to the *a* alleles *(q)* is 0.6 to 0.4 (Figure 11-6). The frequency of the genotypes in the $F_1$ generation will be 0.36 *AA,* 0.48 *Aa* and 0.16 *aa,* and the gene frequency will be $(0.6)^2 + 2(0.6 \times 0.4) + (0.4)^2$. From this one can conclude that all succeeding generations will carry the same proportions of the three genotypes, provided that the assumptions mentioned earlier are fulfilled.

The stated assumptions are never perfectly fulfilled in any real population, so the Hardy-Weinberg law must be considered as wholly theoretical, a distribution against which actual observations can be compared with limitations (Wallace, 1968). Neverthe-

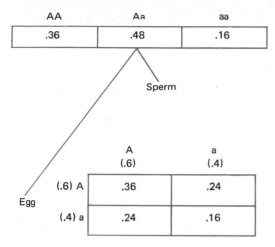

| AA | Aa | aa |
|:---:|:---:|:---:|
| .36 | .48 | .16 |

Sperm

|  | A<br>(.6) | a<br>(.4) |
|:---:|:---:|:---:|
| (.6) A | .36 | .24 |
| (.4) a | .24 | .16 |

Egg

FIGURE 11-6

*Illustration of the Hardy-Weinberg law.*

less, the Hardy-Weinberg law is of fundamental importance in theoretical population genetics.

## Natural selection

### NONRANDOM REPRODUCTION

Variation in a population seldom is constant from generation to generation. One reason is gene mutation, the ultimate source of genetic variation. But of more immediate consequence is nonrandomness of reproduction within a population. Not every individual is able to contribute its genetic characteristics to the next generation or to leave surviving offspring. It is this selectivity that is natural selection.

Before a given individual in a population can contribute to the succeeding generation, it must first survive to reproduce. Survival begins from the time of fertilization through the periods of development, growth, and sexual maturation. Fertilized eggs may fail to develop fully and die from physiological or environmental causes. Disease, predation, and accidents eliminate those young not quite as swift, as quick, or as strong as their siblings. In such survival genetic variation plays a key role, for natural selection influences the frequency of alleles in a population. If a mutation arises that places its carrier at a disadvantage, selective pressures eliminate the individual; on the other hand, an advantageous mutation is retained.

An example of such selection can be found among the flies. When DDT was first used as an insecticide against houseflies, the chemical was highly effective and destroyed the bulk of local populations. But among the flies were a few that did not die, that carried a mutation or a certain combination of genes that made them resistant to the spray. Resistance in one strain of flies was due to a recessive gene. Flies homozygous for this gene tolerated a high concentration of DDT, while homozygous dominants and heterozygotes were killed. These flies survived to multiply. Many of their offspring were as resistant to the sprays as the parents; some were even more resistant. The least resistant were selected against; the most highly resistant were retained in the reproductive population. Later applications of DDT continually selected for a combination of genes most resistant to the insecticide. As a result DDT became ineffective in fly control, and newer, stronger sprays were and are required. Eventually these sprays will select resistant strains of flies, which will become adapted to the new environmental conditions. But to acquire this resistance the flies pay a price. In the absence of DDT the resistant flies are inferior competitors to the nonresistant flies, which have a shorter development time (Pimentel et al., 1951). If the spraying is stopped, evolution will be reversed and the resistance will largely disappear from the fly population.

Once they reach reproductive age, more individuals are eliminated from the parental population. The maintenance of genetic equilibrium is based on random mating, but mating is not random. Many species of animals, particularly among birds, fish, and some insects, have elaborate courtship and mating rituals. Any courtship pattern that deviates from the commonly accepted pattern is selected against, and the deviating individual and its genes are eliminated from the reproductive population. On the other hand, animals possessing a color pattern or movement that accents the typical pattern are selected for. Any new mutations that improve on courtship, mating signals, and ritual would possess a favored position in subsequent generations. Among polygamous species, in particular, the majority of males go mateless, for the females mate with dominant males that tolerate no interference from younger or less aggressive males. States of psychological and physiological readiness also are involved in mate selection. Unless both male and female are of the same state of sexual readiness, mating will not occur.

Neither is fecundity random. Some families or lines increase in number through time; others fade away. Obviously those who produce more offspring increase the frequency of their genes in a population and affect natural selection. For example, if individuals with allele *A* produce 10 offspring to every 1 produced by those with allele *a,* the proportion of *A* in the population will increase. There is a limit, however, for natural selection does not always favor fecundity. If an increased number of young per female results in reduced maternal care, survival of offspring may be reduced. This is true particularly among those animals whose chances of individual survival are high. Those organisms whose chances of individual survival are low, for example, ground-nesting game birds and oceanic fish, have become very fecund. A sort of general rule applies to all organisms: high fecundity, low survival; low fecundity, high survival. Natural selection allows a wide range of interplay between these.

*Genetic drift.* Sexual reproduction is such that only a few of all the gametes produced are actually involved in the formation of a new generation. In general, all an individual's genes will be represented somewhere among its gametes, but not in any two of them. Yet, under conditions of stable population size, two gametes are all that an individual can leave behind. For a heterozygote, *Aa,* there is a 50:50 chance that these two gametes will either be *A, A* or *a, a,* assuming no natural selection (and with selection the chance is even greater). Thus there is a 50:50 chance that a heterozygote will fail to pass on one of its genes. In a whole population these losses tend to balance each other, so that the gene frequencies of the filial generation are a replica, but never an exact one, of the parents' gene frequencies. This is simply the familiar law of averages at work. The larger the population, the more closely the gene frequencies of each generation will resemble those of the previous one; the smaller the population, the greater the sam-

pling error, or *genetic drift* (S. Wright, 1931, 1935). If the deme is very small, there is a good chance that the whole population may become homozygous for a particular allele in only a few generations. This is called *genetic fixation.* Certain alleles are permanently lost until reintroduced by immigration or mutation. This loss of genetic variability is often maladaptive.

Theoretically, at least, the importance of genetic drift in natural selection and evolution may be considerable, especially because most species consist of partially isolated small populations. A special case of genetic drift, *founders principle* (Mayr, 1942) results when a small group of colonists from a species population establishes a new population in an unfilled habitat. The group carries with it a random but biased sample of the genetic variation of the parent population. The sample of genes carried by the founders becomes established in the new population, resulting in a gene frequency in the colony different from that in the parent population.

*Genetic assimilation.* From a mathematical point of view natural selection is regarded as a process that brings about changes in the frequency of genes within a population. But in actual operation natural selection does not act on genes per se, but on the individual organism, especially as it affects the individual's ability to leave viable offspring. The receiving end of selection, then, is the phenotype, which throughout its development is exposed to the rigors of the environment. Any change in the environment that requires some adaptive change in the species will be lethal unless at least some of the organisms, by some somatic change, are able to weather the period of environmental stress, either until the environment returns to its previous norm or until some appropriate genotypic change occurs (Bateson, 1963). Somatic or phenotypic flexibility will involve some of the genotypes in such a way that they will produce a phenotype suitable for the new conditions. If the period of stress is of long enough duration, then the somatic response in the form of acquired characteristics may, under appropriate conditions of selection, be replaced by similar characteristics that are genetically determined. Such replacement of acquired by genetic characters is called *genetic assimilation* (Waddington, 1957).

This genetic assimilation, which simulates Lamarckian inheritance, has considerable survival value when organisms must adapt to stress or change that remains constant over a generation. Through genetic assimilation a species acquires the ability to respond through somatic changes to changes in the environment. Upon return of the previous environmental norm the changes in the individual produced in response to specific environmental conditions will follow with a diminution or loss of characteristics (Waddington, 1957), but the ability to respond will be retained genetically.

## PATTERNS OF SELECTION

If an organism can tolerate a given set of conditions so that it can leave fertile progeny, thus contributing its genetic traits to the population gene pool, it can be said to be adapted to its environment. If an organism survives only as an individual and leaves few or no mature, reproducing progeny, thus contributing little or nothing to the gene pool of the population, it is poorly adapted. Those individuals that contribute the most to the gene pool are said to be the most fit, and those that contribute little or nothing are the least fit. The fitness of the individual is measured by its reproducing offspring. Natural selection is not a measure of individual survival, but of differential reproduction, the ability to leave the most offspring capable of further reproduction.

In simplest terms fitness is measured by comparing the number of offspring produced by one genotype to the number produced by another. Suppose genotype $AA$ produces 250 offspring and genotype $BB$ produces 200. The reproductive success of genotype $BB$ compared to $AA$ is reduced by 50, or expressed in fractional terms $50/250 = 0.20$. Obviously $AA$ is the superior genotype. In measuring selection or the adaptative value of a genotype, fitness is frequently designated as $W$, the value of which ranges from 1.00 for the most productive genotype to 0 for no reproduction (lethal genes). In our very simple example the value of $W$ (for $AA$) would be designated as 1.00; the value of $W$ for $BB$ would be $1.00-0.20$, or 0.80. The selective pressure acting on a genotype is designated as a *selection*

coefficient, s. It can be stated as the difference between 1.00 and the fitness value. In the example the selection coefficient for $AA$ is 0; for $BB$ it is 0.20. Thus, $W$ (fitness) $= 1 - s$; similarly, $s = 1 - W$. For $BB$, $W = 1 - 0.20 = 0.80$; $s = 1 - 0.80 = 0.20$. This extremely simple example and these figures are given here only to provide some appreciation of the term fitness. The calculations used to determine fitness values and selection coefficients are more complex, but not difficult. Good discussions are given in Haldane (1954), Wallace (1968), Ricklefs (1979), and Anderson (1977).

Within a population selection may act in three ways. Given an optimum intermediate genotype *stabilizing selection* favors the average expression of the phenotype at the expense of both extremes. *Directional selection* favors one extreme phenotype at the expense of all others. The mean phenotype is shifted toward the extreme, provided that heritable variations of an effective kind are present. *Disruptive selection* favors both extremes, although not necessarily to the same extent, at the expense of the average (Figure 11-7).

Disruptive selection is most apt to occur in a population living in a heterogeneous environment in which there is a strong selection for adaptability or phenotypic flexibility. Increased competition within the population may select for a closer adaptation to habitat, with the result that the population may subdivide. This would give rise either to a polymorphic situation or to separation into different populations with different characteristics. The latter is most likely to take place in areas where selection is intense and where optimum habitat adjoins or is penetrated by less than optimum habitat. Organisms settling in these habitats will adapt to the local environment. If disruptive selection is strong enough, it will lead to positive assortative mating and eventually genetic divergence of two or more groups.

An example of the development of a polymorphic species through selection is the case of the peppered moth *(Biston betularia)* in England. Before the middle of the nineteenth century the moth, as far as is known, was always white with black speckling in the wings and body (Figure 11-8). In 1850, near the manufacturing center of Manchester,

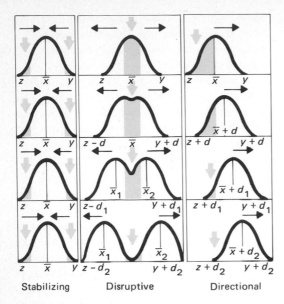

Stabilizing    Disruptive    Directional

**FIGURE 11-7**
*Three main kinds of selection. Stabilizing
selection favors organisms with values close
to the populational mean. Consequently little
or no change is produced in the population.
Directional selection favors one extreme and
tends to move the mean of the population
toward that extreme. Directional selection
accounts for most of the change observed
in evolution. Disruptive selection increases
frequencies of the extremes. In the figures
the curves represent the frequency of
organisms with a certain range of values
between z and y. The shaded areas are those
phenotypes that are being eliminated by
selection. The long arrows indicate the
direction and amount of evolutionary
change. (From O. T. Solbrig, 1970.)*

**FIGURE 11-8**
*Normal and melanistic forms of the
polymorphic peppered moth.* Biston
betularia *at rest on a lichen-covered tree.
The spread of the melanistic form,*
carbonaria, *in industrial areas is associated
with improved concealment of black
individuals on soot-darkened, lichen-free tree
trunks. Away from industrial areas the
normal color is most frequent because black
individuals resting on lichen-covered trunks
are subject to heavy predation by birds.*

a melanistic form of the species was caught for
the first time. The black form, *carbonaria,* increased steadily through the years until the
black form became extremely common, often
reaching a frequency of 95 or more percent in
Manchester and other industrial areas. From
these places *carbonaria* spread mostly westward into rural areas far from the industrial
cities. The black form has come about by the
spread of dominant and semidominant mutant
genes, none of which are recessive. This increased frequency and spread has been
brought about by natural selection. The normal form of the peppered moth has a color pattern that renders it inconspicuous when it
rests on lichen-covered tree trunks. But the

grime and soot of industrial areas carried
great distances over the English countryside
by prevailing westerly winds killed or reduced
the lichen on trees and turned the bark of the
trees a nearly uniform black. The dark form is
very conspicuous against the lichen-covered
trunk, but inconspicuous against the black. A
British biologist, H. R. D. Kettlewell (1961; see
also Kettlewell, 1965) experimentally demonstrated the role of natural selection in the
spread of the dark form. He reared, marked,
and released melanistic and light forms in polluted woods. The melanistic form had a better
survival rate than the light form and thus left
more offspring. To confirm the role of natural
selection, Kettlewell released light and dark
forms in unpolluted woods. There the light
form survived better. The reason was selective
predation. In woods with lichen-covered trees
the melanistic form was more easily seen by
several species of insect-feeding birds and was
therefore subject to heavier predation. In polluted woods the light form bore the brunt of
predation. This explained why the normal
form has virtually disappeared from the polluted country and why it is still common in the
unpolluted areas in western and northern
Great Britain.

Although selective predation is considered
to be the major influence in maintaining

melanic polymorphism in the peppered moth, the polymorphism may also be maintained by a heterozygous advantage of the dark-colored forms independent of selection by predation (Lees and Creed, 1975). The dark-colored individuals (those possessing the *carbonaria* allele) apparently have a physiological advantage over the nonmelanistic form in withstanding the effects of air pollution. Predation aside, the dark form will increase in those areas where it has the physiological advantage. In some cases the visual disadvantage is less than than the physiological advantage.

## GROUP AND KIN SELECTION

Selection pressures are generally considered as impinging upon individuals of a species. But the idea was suggested by Darwin in *The Origin of Species,* advanced by Sewell Wright (1931, 1935), and further expounded by Wynne-Edwards (1963) that a social group might be considered an evolutionary unit. Selection operating on a population as a unit is called *group selection*. It may be more precisely defined as differential genetic survival of unrelated individuals (Ball, 1976). Group selection results in inherited differences in fitness between groups just as individual selection results from inherited differences in fitness between individuals.

Group selection is one of the most controversial and divisive subjects in ecology and population biology today. It is considered by some population ecologists and geneticists (see Lewontin, 1965; J. Emlen, 1973; Gilpin, 1975; E. Wilson, 1975) and is strongly rejected by others (see M. Smith, 1976; Ghiselin, 1974; Williams, 1966), who argue that no characteristic of group selection exists that cannot be explained by natural selection at the individual level.

Group selection can be considered from the standpoint of group size; large diffuse units or small units or demes largely isolated from one another genetically. Wynne-Edwards (1963) advanced the idea that if a population is to persist, it must possess some homeostatic process that will dampen fluctuation while other populations die out. If a population grows too large, it will run out of food or other resources necessary to the survival of both individuals and the group. Through group selection individuals acquire certain behavioral calls and displays that serve to pass information, allowing other individuals to assess the size of population. This stimulates a tendency by members of the population to reduce reproductive efforts when the population size becomes too large. An example is territoriality in animals in which a large portion of the population may be prevented from reproducing because they are prevented from occupying breeding sites. The plausibility of selection operating at this level is strongly questioned.

More acceptable is selection in smaller units of unrelated individuals or interdemic selection. Originally advanced by Wright (1931, 1935), interdemic selection might result among those species with small local populations which are persistent and sufficiently isolated to permit some differentiation in sets of gene frequencies. In most cases there is also sufficient contact between demes to allow the gradual spreading of an advantageous genetic complex to other demes. However, gene flow is not enough to prevent local populations from acquiring some characteristics of their own.

Consider a deme possessing a high frequency of a gene which decreases mortality or increases reproduction that is surrounded by demes possessing the gene at lower frequencies. The deme with the adaptive gene in high frequency will develop some selective advantage over neighboring demes, produce a greater surplus population, emigrate to surrounding demes or into empty habitat patches (created perhaps by local extinctions of populations possessing the adaptive gene in low frequency), and in time improve its own selective advantage. This process can spread among less fit demes. In effect the successful take over from the less successful. Thus a gene which confers some benefit on the group enables the population that possesses it to succeed so that eventually the entire population consists of individuals possessing it.

Group selection, in effect, depends upon (1) more than one group within a large entity (a subpopulation in a regional population) and (2) different frequencies of adaptive alleles of genes in the groups. Further, in group selection the characteristic selected evolves because it increases the fitness of the group even though it may decrease the fitness of any individual bearing the trait. Traits that benefit the population at the expense of individual fitness are called *altruistic*. Such traits find their best expression in kin selection.

*Kin selection* acts on small groups such as parents and offspring or a group of closely related individuals. Theoretically kin selection increases the average genetic fitness of the group at the expense of the fitness of some individuals through aid given to reproducing individuals by others who do not mate (see Hamilton, 1972; E. Wilson, 1975). For example, 36 to 71 percent of the breeding pairs of Florida scrub jays *(Aphelocoma caerulescens)* may be assisted by nonbreeding helpers; nonbreeding birds assist parents, fathers and stepmothers, brothers and their mates, and very rarely an unrelated pair (Woolfenden, 1973, 1974). Helpers strongly prefer their closest kin. The helpers do not participate in nest construction or incubation, but do aid in the defense of nest and territory, attack predators, and help feed the young. Pairs with helpers increase the rate of reproduction two- to threefold over pairs without helpers. However, young helpers at the nest may improve their own fitness by eventually inheriting their parents' territories when suitable habitat is in short supply. Adults allow grown young to remain within their territory. The young in turn appease the parents by helping to feed the nestlings (Woolfenden and Fitzpatrick, 1978).

If the breeders and helpers are very close kin, the helpers serve to increase the fitness values of genetic traits they hold in common with the breeders. Replicate genes are found not only in an ancestor's direct descendents, but are also present to some degree in any relative. The closer the relative, the closer the replication. Thus while not leaving behind offspring of their own (at least for one breeding season), the helpers (immature birds or those whose nests have failed) do improve the fitness of their kin. Through such altruistic behavior traits they improve the fitness of their own genes (because they have a genetic relationship with the offspring).

Altruism is one of those terms that means what the user wants it to mean. As a result altruism has acquired many shades of meaning in ecology, sociobiology, and behavior that deviate from the dictionary definition. Altruism,

strictly defined, is the sacrifice of one's own well-being in the service of another or is any activity that decreases the fitness of the altruist or donor. But at least five different meanings of altruism exist in ecological literature that invoke both kin and individual selection (see Brown, 1975; Dawkins, 1976; Hamilton, 1971; Trivers, 1971). Brown suggests that from the standpoint of evolution and selection altruism might be defined as "the giving of aid in the form of arbitrarily defined goods and services to individuals of the same species who are not offspring or direct descendents of the donor and without direct benefit to the donor or its mate."

Altruism reaches its highest development in the community of social insects in which the bulk of the population are nonreproductive individuals who attend the reproductive unit, the queen, to insure successful reproduction and thus group survival. Altruism is also conspicuous in African lions, wild dogs, nonhuman primates, and birds, particularly those species with relatively little dispersal and with semi-isolated populations of kin groups. Successful groups are those that maintain their own genetic superiority in fitness while exporting genes to less successful groups.

The concepts of group selection, kin selection, and altruism are controversial and subject to considerable study and testing in the field. Some form of group selection probably does exist in some populations in a restricted way. D. Wilson (1975) suggests that no strong distinction can be made between individual and group selection. The two exist in a continuum of selection in which individual selection is one extreme and group selection is the other. This idea has been challenged by Maynard Smith (1976). (For a good review of current status of the concept see Alexander and Borgia, 1978.)

## Speciation

### TYPES OF SPECIES

*Morphological and sibling.* One has little difficulty distinguishing a robin from a wood thrush or a white oak from a red oak. Each has certain morphological characteristics that set them apart from other organisms. Each is a discrete unit to which a name has been given.

This was the way that Carl von Linné, who gave us our system of classification, saw the great number of plants and animals. He, as did others of his day, regarded the many organisms as fixed and unchanging units, the products of special creation. Differences and similarities were based on color pattern, structure, proportion, and other characteristics, and from these criteria the species were described, separated, and arranged into groups. Each species was monotypic, that is, it contained only those individuals that fairly well approximated the norm or type for the species, the specimens from which the species was described. Some variation was permissible, but these variants were considered accidental, and some slight changes within the species were admitted possible. This discrete unit is the *morphological species,* a classical concept still alive, useful, and necessary today for classifying the vast number of plants and animals.

The studies of Darwin on variation, of Wallace on geographical distribution, and of Mendel on genetics upset the idea of special creation and emphasized that variation was the rule. Naturalists explored new lands, collected new specimens, and observed plants and animals in their natural communities. They found that some species were quite distinct and easily identified by structural, color, or behavioral characteristics. But among some organisms the distinctions were much hazier. One species might differ from another in only minute, but constant morphological characteristics. Some, such as mosquitos, were morphologically indistinguishable as adults and were distinguishable only by the eggs. Others were structurally indistinguishable in all stages of life history but differed in their behavior (see R. D. Alexander, 1962), in ecology (see Dybas and Lloyd, 1962), in biochemistry (Sibley, 1960). These are *sibling species,* defined as "morphologically similar or identical natural populations that are reproductively isolated" (Mayr, 1963) (Table 11-1).

Examples of sibling species are widespread, especially among the insects. Some of the better known cases occur in the genus *Drosophila* and in the malaria mosquito complex of Europe. Sibling species can be found in almost every family of beetles and are quite common in the Orthoptera (see Table 11-1). They also occur in snakes, in amphibians, particularly the frogs, in birds to a limited extent, and in fish, especially the whitefish and the salmonids (see Neave, 1944; Ricker, 1940). A characteristic common to all is the lack of a sharp division between ordinary species and the sibling species. The latter are at the far end of a broad spectrum or continuum of increasingly diminishing morphological differences. To the human eye the species may be virtually indistinguishable, but the differences are apparent to the animals themselves.

*Biological species.* Biologists also realized, as Darwin observed, that many apparently closely related forms replaced each other geographically and in doing so intergraded with one another. So smooth and gradual is the transition that it is often difficult to separate precisely one from another. The question "When does a robin cease to be a robin and become something else?" becomes very real and important. The robin, of course, is quite distinct, but the same cannot be said about some other species.

Eventually some biologists realized that the classical morphological or typological concept of species, which emphasizes the individual, was incomplete. It was augmented by the population concept of the species—which is based on the premise that species are not fixed, static units, but are changing through long periods of time. The anatomical, physiological, and behavioral characteristics that clearly define a species on a local basis are now regarded as a part of the sum of characteristics found throughout the entire population. The pattern of characteristics can be graphed as a frequency distribution of variants of different characters present in the population at a given time. In other words, the species is multidimensional.

The species has been defined as "a group of actually or potentially interbreeding populations that are reproductively isolated from other such groups" (Mayr, 1942). This definition then, identifies a species as a group of interbreeding individuals living together in a similar environment in a given region and under similar ecological relationships. They interact with other species in the same environment. The individuals recognize each other as potential mates. They are a genetic unit in

**371**

TABLE 11-1
*Some differences in the sibling species of the 17-year and 13-year cicadas (Homoptera, Cicadidae, Magicicada)*

| | M. septendecim | M. tredecim | M. cassini* | M. tredecassini | M. septendecula* | M. tredecula |
|---|---|---|---|---|---|---|
| Size | NO STATISTICAL DIFFERENCE IN SIZE BETWEEN ANY OF THESE | | | | | |
| Body color | Black above, reddish below; appendages reddish; pronotum reddish yellow; prothoracic pleura reddish yellow | Same as *M. septendecim* except radial W in forewing heavily clouded | Black above, almost black below; appendages reddish; pronotum black; prothoracic pleura black | | Pronotum black; prothoracic pleura black; tibia reddish or with narrow black apical markings. | Same as *septendecula* except apical tarsal segments reddish all the way to tip; abdominal sternites with prominent reddish bands |
| Brood | 17 years | 13 years | 17 years | 13 years | 17 years | 13 years |
| Call | Low-pitched buzzing phrases; fairly even in intensity, ending with a drop in pitch; Phaaaaaaraoh; 1–3 sec | | Rapidly delivered tick series, alternated with high-pitched, sibilant buzzes; noticeable rise and fall in pitch and intensity; ticks 2–3 sec, buzz 1–3 sec | | High-pitched brief phrases in series of 20–40 at rate of 3–5 per sec; entire call 7–10 sec. | |
| Chorus | Even, monotonous roaring or buzzing; no regular fluctuations in intensity or pitch; individual males not synchronized; most intense in morning | | Shrill, sibilant buzzing; rise and fall in intensity due to synchronization among individual males; most intense in afternoon | | More or less continuous repeating of short, separated buzzes; no regular fluctuations in pitch or intensity; individual males not synchronized; most intense around midday | |

*Source:* Data from Alexander and Moore, 1962; M. Lloyd, personal communication.
*The only reliable way to distinguish *septendecula* from *cassini*, other than by song, is by morphometric characters. These color characteristics are very inconsistent. Where *cassini* occurs on the edge of its range in the absence of *septendecula*, some character displacement seems to take place. The abdominal sternites of *cassini* have reddish bands as prominent as those of *septendecula*.

which each individual holds for a short period of time a portion of the contents of an intercommunicating gene pool.

Such a definition of species is applicable only to bisexual organisms. It is also limited to *sympatric* species, those occupying the same area at the same time. *Allopatric* species, those occupying an area separated by time and space, cannot be involved because they never have the opportunity to meet other similar species and therefore there is no way of knowing whether they are reproductively isolated or not. The only direct evidence that an individual organism belongs to one species consists of observations in the wild that indicate that the organism is living with a specific population and functioning as a member of it. The individuals tend to possess the same morphological and physiological characteristics because they belong to the same evolutionary population. Thus, the biological species, in contrast to the static morphological concept, encompasses variation, is constantly changing, splitting up, and reuniting, and is almost impossible to define precisely.

The concept of the biological species arose mainly in vertebrate systematics, where it has the widest application. It is less generally accepted by botanists, for among plants, as well as among many small invertebrate animals, the biological species, as defined, is inadequate.

*Asexual species.* The biological species embrace only sexually reproducing organisms. But among plants and some invertebrates, asexual or vegetative reproduction is common. Many of the higher plants, even those in which cross-fertilization occurs, reproduce in this fashion and others rely on it, by means of stolons, root runners, bulbs, and corms. The blue flag along the streamside and water lilies in a pond occupy the habitat not primarily through sexually produced seeds but through asexual propagation. In a somewhat similar fashion, the same is true for many annual plants, especially those that occupy extreme environmental situations and pioneering communities. These annuals are self-fertilized and, because they are short-lived, they possess no effective means of vegetative reproduction. Selection has favored self-fertilizing forms most likely to produce a good crop of seeds.

Such organisms, which perpetuate their kind outside of sexual reproduction, are called *agamospecies* (an old, relatively unused but very useful term describing those forms that have no true sexual reproduction) (Cain, 1954). They possess, as one may well imagine, very little genetic variability, and they have for the most part lost their capacity to adapt to environmental changes. Variability has been sacrificed for the ability to take the maximum advantage of a given environmental situation. Here they are preeminently successful.

Agamospecies possess for many generations the genetic constitution derived from parent stock—in fact they are the parent stock. Disease, fire, climate, or age may destroy individual shoots, but separate vegetative parts of the same genetic individual remain. The buffalo grass of the western plains, some botanists believe, probably consists of the same genetic individuals that colonized the plains after the glacial retreat. They possess a sort of immortality.

Any changes in agamospecies are the result of either mutation or occasional reversal to sexual reproduction. Mutations involved are sudden and random changes with seemingly little relationship to the structure and needs of the individual in which they occur. Although small mutations may be tolerated, large ones have little chance of being beneficial. But if mutations do persist, then they are maintained by the same vegetative reproduction, and a different form of the same species has arisen.

Occasionally such plants may revert to sexual reproduction. Pollen from one may be transferred to another and some variability is introduced to the offspring, which propagate themselves either vegetatively or by self-fertilization. Eventually some forms may cross-pollinate and produce hybrids, which in turn are maintained and spread by self-pollination and vegetative reproduction. Hybridization may range from a slight blurring of distinctions between two or more species to the development of a huge complex of forms, in which the original species have become more or less lost. Such natural hybridization is the root of much of the "species problem" in plants and is involved in such "difficult" groups as hawthorns, blackberries, willows, and even some oaks. Certain groups of these are little more than aggregations of hybridizing semispecies (see Benson, 1962).

**373**

Agamospecies, however divergent from the biological species, still fit into the morphological species concept. They still can be associated typologically with their sexual relatives, yet they have no certain status.

## GEOGRAPHICAL VARIATION

Because of the widespread variation of many morphological, physiological, and behavioral characters in a widely distributed species, significant differences often exist among populations of different regions. One local group may differ, more or less, from other local populations, and the greater the distance between populations, the more pronounced the differences become. The geographic variants reflect the environmental selective forces acting on various genotypes, adapting each population to the locality it inhabits.

Two types of geographic variants are the cline and the geographical isolate.

*The cline.* The cline is the result of phenotypic response to environmental selection pressures that vary on a gradient or continuum. This continuous variation results from the intergradation of gene pools between local populations. Clines are most prevalent among organisms with continuous ranges over a continental area.

An example of a cline is the northern leopard frog *(Rana pipiens),* one of the most familiar of all North American amphibians. It has the largest range, occupies the widest array of habitats, and possesses the greatest amount of morphological variability of any North American ranid, and is the only ranid successfully established throughout the prairie country. But the variability and adaptability of the leopard frog are orderly, not haphazard. The species embraces a number of temperature-adapted races on a north-south gradient (J. A. Moore, 1949a,b). When the populations of the north and south extremes are compared, the differences are pronounced, yet between the two extremes, no break in variations occurs. Embryos of southern leopard frogs have an upper-limit temperature tolerance of 4° C above that of the northern embryos, although both survive equally well at low temperatures. Southern forms have smaller eggs and a slower rate of development at low temperatures. In fact so wide are these physiological differences that matings of individuals from the extremes on the north-south gradient result in defective or nonviable offspring (J. A. Moore, 1949a,b), although normal hybrids are produced when the leopard frog is crossed with either of the two gopher frogs *(R. areolata areolata* and *R. a. capito)* or the pickeral frog *(R. palustris).*

Clines have distinctive extremes, but because such gradual changes exist from one population to another, it is impossible to group the demes into separate entities. Clines are usually associated with an ecological gradient, such as temperature, moisture, altitude, and light. These changes may take place over a relatively short distance in response to some varying change in ecological conditions (ecocline) or they may take place over a much larger area, as is the case of the leopard frog. This phenomenon has resulted in a number of ecological rules, summarized in Table 11-2.

TABLE 11-2
*Rules correlating variations with environmental gradients (subject to frequent exceptions)*

| Rule | Statement |
| --- | --- |
| Bergmann's rule | Geographic races possessing a smaller body size are found in the warmer parts of the range; races of larger body size are found in the cooler climate. |
| Allen's rule | The extremities of animals, the ears, tail, bill, etc., are shorter in the cooler part of the range than in the warmer part. |
| Golger's rule | Among warm-blooded animals, black pigments are most prevalent in warm and humid areas, reds and yellows in the arid areas, and reduced pigmentation in the cool areas. |
| Jorden's rule | Fish living in warm waters tend to have fewer vertebrae than those living in cool waters. |
| — | Races of birds living in the warmer part of the range lay fewer eggs per clutch than those living in the cooler part of the range. |

Some species of plants exhibit clinal gradation in size and other structural characteristics, in time of flowering, in growth, or in other physiological responses to the environment. Clinal differences in plants can be demonstrated by transplant studies in which a series of populations from different climates are grown together under one uniform environment in the field and greenhouse. Further comparisons of differences can be obtained by growing such a series under several environmental conditions. Such studies have revealed that a number of prairie grasses, among them blue grama, side-oats grama, big bluestem, and switchgrass, flower earlier in northern and western communities and progressively later toward the south and east (McMillan, 1959). Goldenrod (Solidago sempervirens) flowers progressively later in the season from north to south along the Atlantic coast. Yarrow (Achillea) blankets the temperate and subarctic Northern Hemisphere with an exceptional number of ecological races. One species, Achillea lanulosa, exhibits considerable variation in response to different climatic environments at various altitudes. Individuals are progressively shorter at high altitudes, although considerable variation in height exists within each population (Clausen et al., 1948).

Until recently clinal variations were measured on such characteristics as wing length and body weight and correlated with environmental gradients such as temperature. Such measurements were often limited to statistical differences among subspecies. Since 1960 the computer has enabled biologists to employ multivariate biometry to measure simultaneously the relationships of a number of meristic characters to several environmental variables. This approach at last permitted biologists to demonstrate the obvious but at one time unmeasurable fact that geographic variation is not due to the adaptation of a few characters to a single environment. Rather it is a multidimensional process involving the adaptation of many characters to a variety of interdependent environmental variables whose gradients and ranges overlap in a complex way (Sokal and Rinkel, 1963).

One of the best examples is the study of the evolution of the house sparrow in North America by Johnston and Selander (1971). They studied 16 skeletal characters of 33 samples of North American male and female house sparrows. From correlations of skeletal characters they found that all 16 in both sexes were as geographically variable as such characteristics as wing length. Samples from central and eastern Canada, the Great Plains, and the Rocky Mountains had large characteristics; samples from southwestern United States and Texas were intermediates; and samples from the west coast, Gulf Coast, and Mexico had small characteristics. By feeding summaries of gross size variations into the computer, Johnston and Selander obtained a contour diagram of geographical variation in mean values of skeletal sizes (Figure 11-9). The contoured variations corresponded to the predictions of Bergmann's rule with the larger birds at the higher latitudes. Similarly, geographical variations of such characters as sternum and long-bone sizes agreed well with Allen's rule (see Table 11-2).

Not all species exhibit such clinal variations. Some show little or none at all. Ecological rules may be true in a general sort of way, but they are not universal. McNab (1971) argues that for many, if not all, animals the assumption underlying Bergmann's rule is not tenable. This rule assumes that large animals expend less energy to maintain body heat because of a small surface-area-to-volume ratio, and therefore it is more economical for them to live in a colder climate. McNab feels that the north-south gradient in size must be due to something else. He demonstrates that latitudinal variation in body size among carnivores and granivores reflects a change in the size of their prey. A latitudinal change in the size of the available prey is due either to the distribution of the prey or to the distribution of other predators utilizing the same prey species. Allen's rule postulates that animals living in warm climates have small bodies and relatively long appendages for effective thermoregulation at high temperatures. A regression of surface volume ratio of the bill against eight measures of the environment from latitude to four measures of temperature associated thinner bills with higher temperatures, in accordance with Allen's rule (Power, 1970). The regression of bill dimensions

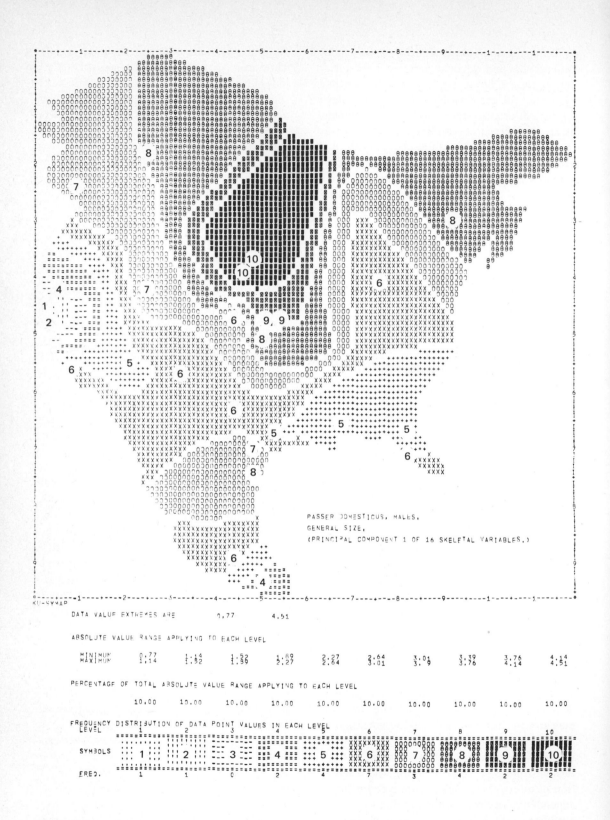

**FIGURE 11-9**
*Generalized contour diagram of the geographic variation in mean values of general size (based on a set of 16 skeletal size variables) of male house sparrows (Passer domesticus) in North America. (From S. J. Gould and R. F. Johnson, 1972.)*

376

against temperature in the red-eyed vireo follows Allen's rule, but in the sympatric Philadelphia vireo, tarsal length increases as temperature decreases, in opposition to Allen's rule (Barlow and Power, 1970). According to Gloger's rule, homeotherm animals are darker in warm and humid regions and lighter in cool areas. Among some insects, such as the collembolans, the higher the latitude, the darker the animal (Rapoport, 1969).

The nature of the variations in body size, in appendages, and in color may reflect the nature of a species' adaptive response to the environment. Species occupying broad ecological niches, but possessing high individual tolerance of the environment may show little clinal variation and have little tendency to speciate. Those occupying broad ecological niches, but showing low individual tolerance of the environment may show strong clinal variation and have a strong tendency to speciate (Levins, 1968). The latter is true of the house sparrow, for in a little over 100 years in North America it has evolved into a number of clinal variations comparable to those in its native Europe.

***The geographic isolate: The subspecies.*** In contrast to the cline, the geographic isolate is a population or a group of populations that is prevented by some extrinsic barrier from effecting a free flow of genes with others of the same species (Mayr, 1963). The degree of isolation depends upon the effectiveness of the extrinsic barrier, but rarely is the isolation complete (Figure 11-10). These geographical isolates, or races (or ecological races) and to some extent the clinal variants, taxonomically make up the *subspecies,* in itself a thorny problem (Figure 11-11).

The subspecies is defined by Mayr (1963) as "an aggregate of local populations of a species

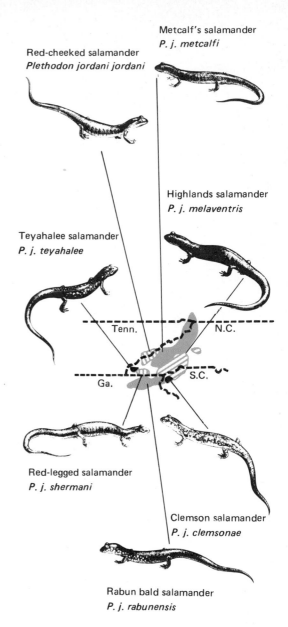

Metcalf's salamander
*P. j. metcalfi*

Red-cheeked salamander
*Plethodon jordani jordani*

Highlands salamander
*P. j. melaventris*

Teyahalee salamander
*P. j. teyahalee*

Tenn.    N.C.

Ga.    S.C.

Red-legged salamander
*P. j. shermani*

Clemson salamander
*P. j. clemsonae*

Rabun bald salamander
*P. j. rabunensis*

**FIGURE 11-10**
*Geographical races and subspeciation in the* **Plethodon** *salamanders of the Appalachian highlands. These salamanders of the* jordani *group resulted when the population of the salamander* **Plethodon** yonahlossee *became separated by the French Broad valley. In the eastern part the separated population developed into Metcalf's salamander, which spread northeastward, being the only direction in which any group*

*member could find suitable ecological conditions. South, southwest, and northwest the mountains end abruptly, limiting the remaining* jordani. *Metcalf's salamander is the most specialized and ecologically divergent and least competitive. Following separation of Metcalf's salamander, the isolation of red-cheeked salamander from the red-legged and the rest of the group resulted from the deepening of the Little Tennessee River. Remaining members are still somewhat connected, especially around the headwaters of the Little Tennessee. (Based on data from Hairston and Pope, 1948, and* **Checklist of Amphibians and Reptiles,** *1960.)*

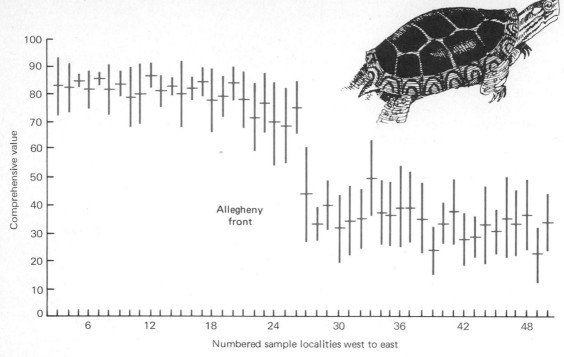

**FIGURE 11-11**

*Effect of mountain barrier on subspeciation of the painted turtle* (Chrysemys picta). *The painted turtle consists of two subspecies, the eastern painted turtle* C. p. picta *and the midland painted turtle* C. p. marginata. *North of the high Appalachians the two subspecies intergrade. In West Virginia the high Appalachians form an ecological barrier that effectively separates the two subspecies. This separation is illustrated in the plots of compositive values of such measurable characteristics as seam alignment, width of plastron, plastron figure, and others. The horizontal bar is the mean value; the vertical bar is standard deviation. (Data from L. Clack, 1975.)*

inhabiting a geographic subdivision of the range of a species, and differing taxonomically from other populations of the species." Here again the population is stressed, for although a whole population can be assigned to a subspecies, it is often difficult to assign an individual to a subspecies because of the inherent variability within a population and the overlap of the curves of variations of adjacent populations. In reality the subspecies is an artifact of man, set up for practical reasons; it is not a unit of evolution.

The geographical races of a continental species are, more often than not, connected by intermediate forms or intergrades, so that it is virtually impossible to draw a line that will separate them. The differences between the races are greatest at the geographical periphery of the species. Often these differences are so pronounced that the end races behave as perfectly distinct species. Such is the case in the cline of the leopard frog in which matings from neighboring locations produce normal offspring, but matings from the gradient extremes do not.

Sometimes a population of a species has diverged almost too far to be considered a race, yet not far enough to be considered a distinct species. This may result in a so-called ring of races. Sympatric populations that share the same habitat behave like distinct species, yet they are joined by a chain of allopatric races that smoothly intergrade. As a result, it is impossible to draw a line anywhere between the two, although for all practical purposes the end members are different species.

The classical example is the great tit *(Parus major)* of the Eurasian continent. The nominate race (the one first named) is *Parus major major,* and it has a dull green back and yellow belly. In southern Europe some variation becomes evident. The great tit of Spain, Portugal, and North Africa, *P. major excelsus,* has a more vivid yellow on the belly and a reduced amount of white on the outer tail feathers. The variant on the Balearics, Cyprus, Crete, and Greece, *P. m. aphrodite,* has a more grayish back and a paler yellow belly. The birds of Corsica and Sardinia are similar to *aphrodite,* except that they are more gray on the flanks and have been separated as *P. m. corsus.* The Palestinian subspecies, *P. m. terraesanctae,* is paler beneath and more yellowish on the back than any of the others mentioned, a condition even more pronounced in the birds of Persia and North Mesopotamia, which are known as *P. m. blanfordi.* However, all the forms tend to intergrade except where separated by the sea, and all have the same basic coloration of gray back and yellow belly.

In India, Japan, and Manchuria live two other forms of the great tit, whose color variation is so pronounced that by morphological concepts they could easily be called two other species. The Indian bird, *P. m. cinereus,* has a gray back and a white belly; and the Japanese bird, *P. m. minor,* has a green back and a white belly. Both, however, are regarded as subspecies of *Parus major,* for through south-central China and along the coast of Indochina lives another form, *P. m. commixus,* which resembles the gray-backed form except that the tail is gray and the back green tinted, and where this form meets the Indian and Japanese birds, they intergrade. Where they come in contact with the Persian race, the Indian birds intergrade with it and thus connect *P. m. major* of Europe with *P. m. cinereus* and ultimately with *P. m. minor.* The white-bellied, green-backed form of the extreme eastern Asian coast extends northward from Japan to the northern border of Manchuria, the Amur River, where *P. m. minor* meets *P. m. major.* Here they coexist and breed without intermixing as a separate species, although more recent information suggests that hybrid populations are formed (Delacour and Vaurie, 1950). Thus although two races *Parus m. major* and *Parus m. minor* may behave as separate species in the zone of overlap they are connected with one another by a whole chain of intergrading geographic races. Where does one draw a line between the two? Where would one species, if so classified, begin and the other end? This is the problem that modern systematists face when they attempt to define a species. This is what is meant by the species problem.

This problem is not so pronounced among insular populations, for here the distribution is discontinuous, and because of natural barriers, zones of intergradation do not exist.

### POLYMORPHISM

Variations, especially discontinuous ones, frequently arise within local populations. The occurrence of several distinct forms of a species in the same habitat at the same time is *polymorphism.* This may involve differences in morphological characters and in physiology. The important feature about polymorphism is that the forms are distinct and the characteristic involved is discontinuous. There are no intermediates.

Polymorphism may be caused by differences in major genes. An example is the snow goose, which is polymorphic with two clearly defined color phases, a blue one and a white one (Figure 11-12). The polymorphism appears to be determined by a single pair of alleles. Birds possessing genotype *BB* and *Bb* are blue, whereas birds possessing *bb* are white (Cooke and Cooch, 1968). But there are complications. Dominance appears to be incomplete, and a variable amount of white on the belly of the blue-colored birds indicates a heterozygous individual, whereas a dark-bellied blue-colored bird is homozygous. This polymorphism is further complicated by variations in plumage that are not due to the major alleles for polymorphism, but which have both genetic and nongenetic components. In addition there is assortive mating. Males select mates that have a plumage pattern similar to one of their parents, a behavior that could maintain the polymorphism. The goslings apparently become imprinted to the plumage of one of their parents. A similar polymorphism exists among goslings of the related Ross goose, in which the dominant allele confers a gray color to the down and the recessive a yellowish color (Cooke and Ryder, 1971). Since the gene is not manifested in the adult plumage, mating in terms of the dimorphism appears to be random.

B. Snow Goose
blue phase

A. Snow Goose
white phase

*FIGURE 11-12*
*Polymorphism in the lesser snow goose*
*(Anser caerulescens), showing white*
*plumage and dark plumage forms.*
*Imprinting seems to affect the choice of mate*
*in this complex, with white-plumaged birds*
*favoring the white phase and dark-colored*
*birds, the blue phase. At present the blue*
*morph is increasing at the expense of the*
*white.*

Polymorphism may be environmentally induced, but environmental modification of the action of genes is possible only when two environments, for example, background color, are present at the same time in the same place. Environmentally controlled polymorphism, favoring two or more forms, is the optimal expression of the characters concerned. All intermediates are at a disadvantage and usually are eliminated. The North American black swallowtail butterfly and the European swallowtail are good examples. Both swallowtails pupate either on green leaves and stems or on brown ones. Through natural selection both have acquired a genetic constitution that produces green pupal color in green environments, brown in brown. Green pupae would be quite conspicuous in winter, but butterflies emerge from the green in late summer; those in brown pupae do not emerge until the following spring (Sheppard, 1959).

There are times when environmental changes convert a disadvantageous allele or mutant gene to an advantageous one, permitting the latter to spread through the population. Polymorphism will exist until the new advantageous form has completely replaced the original or has so swamped it that the original can be maintained only by recurrent mutation. Such a situation is known as *transient polymorphism.* An excellent example is the industrial melanism of the peppered moth, discussed earlier.

The most common type of polymorphism, however, is *stable* or *balanced polymorphism,* a condition in which an apparent optimum proportion of two forms exists in the same habitat and any deviation in one direction or the other is a disadvantage.

### HOW SPECIES ARISE

The great diversity of living things in the world causes one to wonder how all these species arose. Each is adapted to an ecological niche in the community to which it belongs, and each is genetically independent. The process by which this has come about, by which one form becomes genetically isolated from the other, is *speciation,* the multiplication of species.

*Geographic or allopatric speciation.* Speciation in most organisms is accomplished by an interaction of heritable variation, chromosome rearrangement, natural selection, and spatial isolation. Species formation under spatial isolation is called allopatric or geographical speciation (see Mayr, 1963) and may arise through subdivision or founders effect.

*Subdivision.* The classic model of speciation is the subdivision of a widely distributed species into two large populations by some extrinsic barrier that interrupts gene flow between them (Figure 11-13). Under different arrays of selective pressures each population accumu-lates its own genetic differences. In time the accumulated differences acquired during isolation become great enough so that if contact does occur and hybrids are produced, the hybrids have low fitness and are eliminated from the population.

The first step in geographical speciation is the splitting of a single interbreeding population into two spatially isolated populations. Imagine an area of warm, dry land occupied by species *A*. Then geographical events create a mountainous, aquatic, or other ecological barrier which separates a segment of species *A* from the rest of the population. The newly isolated segment, *A'*, now occupies an area of cool, moist climate in our imaginary land.

Because the population *A'* is only a sample of species *A*, it will possess a slightly different ratio of genetic combinations. Because the climatic conditions are different, the selective forces are different. Natural selection will favor any mutation or any recombination of existing genes in population *A'* that results in better adaptation to a cool, moist climate, but selection for a warm, dry climate will continue in population *A*. With different selective forces acting upon them, the two populations will tend to diverge. Accompanying this genetic divergence will be changes in physiology, mor-

*FIGURE 11-13*

*Geographic or allopatric speciation. One species may become two as a result of isolation for a long period of time by a geographical barrier. If the barrier is removed before the separated groups have diverged from the parent stock, they simply rejoin (retrogression). If speciation is incomplete, the two races can interbreed (introgression). If genetic divergence is not great, potential hybrids are not at a disadvantage and the gene pools of the two races will mesh. If potential hybrids are at a disadvantage, isolating mechanisms are effective even though some interbreeding does occur (secondary introgression). If genetic divergence is complete, the two species can exist sympatrically after the barrier is removed without interbreeding. (Adapted from Mettler and Gregg, 1969.)*

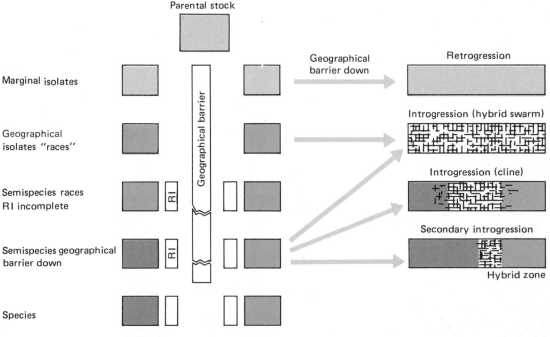

381

phology, color, and behavior, resulting in ever-increasing external differences, until *A'* becomes a geographical race but still a part of the species *A* population.

If geographical barriers break down at this point, interbreeding is possible and individuals produced by the cross are fully fertile and viable. If the offspring possess as high a reproductive potential as the parent stock and if neither *A* nor *A'* has a selective advantage, the two gene pools will merge. The final result will be a population with increased variability over the original population prior to the split.

If, however, the barrier remains, evolutionary diversification continues. *Isolating mechanisms,* factors that curtail or prevent gene interchange between populations (Dobzhansky, 1947) develop, and gradually become more established as diversification increases. Eventually the mechanisms become fully effective, and normal interbreeding is no longer possible if the two populations come together again. Population *A'* has now arrived at the species stage.

If the barrier fails at this stage or just prior to it, individuals of the two populations may interbreed and produce hybrid offspring. Such hybrids among animals are less fertile and less viable than the parent stock because they contain discordant gene patterns. Their reproductive potential, if they are fertile at all, is low; they produce fewer offspring. They are at a selective disadvantage, and any color pattern, voice, behavior, etc., in the parent stock or any mutation or genetic recombination that reinforces reproductive isolation will be selected. This selection against hybrids continues until gene flow between the two populations has stopped. Thus species *A* and new species *A'* can invade each other's territory, occupy suitable niches—in our example, a warm, dry environment and a cool, moist environment—and become wholly or partly sympatric.

Allopatric speciation by subdivision is characteristic of species having low reproductive rates, late sexual maturity, few offspring, long life span, high competitive ability, and high vagility. Such species exhibit outbreeding and little or no chromosome evolution, and they possess relatively large amounts of nongenetic DNA which protects them from mutagenic agents (Bush, 1975). Speciation re-

sults from an accumulation of adaptative changes in structural and regulatory genes. The changes are maintained and developed by reproductive isolation brought about by geographical isolation and by selection for the expression of those changes in mate selection should contact with other species occur.

*Founders effect.* A second and probably more common form of geographic speciation results from the establishment of a new colony by one (gravid female) or a small number of founders. Organisms most prone to establish founder colonies have high reproductive rates, early sexual maturity, large number of offspring, short life span, and low competitive ability. They are prone to inbreeding and often are characterized by small cohesive groups, such as herds or family units, with permanent pair bonds. They may have fairly restricted home ranges or be highly vagile and become established in distant unoccupied habitats. Founder individuals are often products of a population flush (Carson, 1968), "surplus" individuals resulting from a rapid population increase in which selection pressures are relaxed. Often founder populations exist on the periphery of the range and become isolated from the parent population or find unexploited areas suitable for invasion. Because they are away from the genetically stable conditions in the center of the population where variation in the gene pool is great, where much of the geome is tied up in coadaptive complexes, and where selection is for heterozygosity, founder populations have less genetic variation and experience greater selection for homozygosity.

When individuals or small groups become isolated from parent populations, they may become extinct, rejoin the parent population, or evolve into a new species, especially if selection is for homozygosity. In the latter situation some individual in the group, such as a dominant male, carries a major adaptive chromosomal macromutation. This adaptive rearrangement in a homozygous condition may become fixed rapidly in the population. Such major chromosome rearrangements could result in a rapid reorganization of regulatory mechanisms without a change at the gene level (see White, 1973), allowing the population to exploit new habitats (Bush, 1975). Once the founder population is well established in a new habitat, reproductive isolation arises by change. Postmating reproductive isolating mechanisms (discussed later in the chapter) will develop with the fixation of new chromosome arrangement. Premating reproductive isolating mechanisms evolve only if the population reestablishes contact with the parent population.

*Nongeographic speciation.* Although the idea is prevalent that speciation can take place only with geographical isolation, there is a rapidly growing body of evidence that speciation can occur without geographic isolation (White, 1973, 1974). Such parapatric and sympatric speciation involves chromosomal macromutations or rearrangements. Although adaptation, isolation, competition, and specialization are all part of the evolutionary process, evolution is basically a cytogenic process (White, 1973). Complex sequences of changes in chromosomes develop systems of cytogenic polymorphisms and genetic isolation that preadapt organisms to new habitats, different foods, or different environmental conditions and as a result bring about speciation.

Support for nongeographic speciation comes from studies of gene flow in plants, especially as it relates to the maintenance of purity of crop cultivars through isolation by distance to prevent inbreeding. Plants are nonvagile, so that even with seed dispersal by wind and animals the level of gene flow between plant populations is so low that effective population size can be measured in meters (Bradshaw, 1972). As a result plant populations consist of isolated breeding units of various sizes, each more or less adapted to narrow local environmental conditions. Such isolation and restricted gene flow provides excellent situations for nongeographic speciation in plants.

*Parapatric speciation.* Resembling speciation by founders effect is parapatric speciation (White, 1973; Bush, 1975). Parapatric speciation is the evolution of a species as a contiguous population in a continuous cline. The differentiated adjacent populations meet in a narrow zone of overlap; the area of overlap may be only a few hundred meters wide (Figure 11-14). It differs from founders effect in that (1) no spatial isolation is required during speciation, (2) the level of vagility is low, and (3) reproductive isolating mechanisms arise by

selection at the same time genetically unique individuals exploit or colonize a new environment.

Parapatric speciation is characteristic of short-lived, rapidly reproducing species and occurs in small- to medium-sized populations living on the periphery of the main population. As in founder populations, chromosome macromutations frequently initiate speciation with little or no genic differentiation even after speciation is complete. Such populations are characterized by high levels of homozygosity and lack of long range dispersal. Diverging populations are in close contact without the usual geographic isolation. Although gene flow does occur between populations, it rarely extends very far into the population, because individuals in the two populations experience strong directional selection in two different habitats or microhabitats (Mayr, 1974). The fixation of new karotypes results in major changes in the regulatory functions, releasing new chromosome races to occupy new habitats to which they were previously poorly adapted.

Examples are the flightless morabine grasshoppers endemic to Australia (White, 1973*a*, 1973*b*; Key, 1973) A subfamily of the Eumastaeidae widespread in the tropics and subtropics of the Old and New World, the morabine grasshoppers are found over a full range of temperature and rainfall experienced in Australia. Most are found on forbs, grasses, and shrubs up to 2 meters high; a few live on ferns. The distributional ranges of individual species are small, averaging on the long axis about 600 kilometers and ranging from 2,000 to 150 kilometers or less. Within their overall distribution the species are restricted by local habitat conditions, especially vegetation, although this is not limiting. The ranges appear to be limited by some form of interaction with species having adjacent ranges (Figure 11-14).

The morabine grasshoppers exhibit extensive intraspecific variability in the form of morph clines and geographical races. The most striking features of this variation are the widespread occurrences of chromosomal polymorphism and chromosomal races possessing unique karotypes. In the chromosomal races the chromosomal type has become fixed over the area occupied by the races and interracial hybrids, while usually fertile, are adaptively

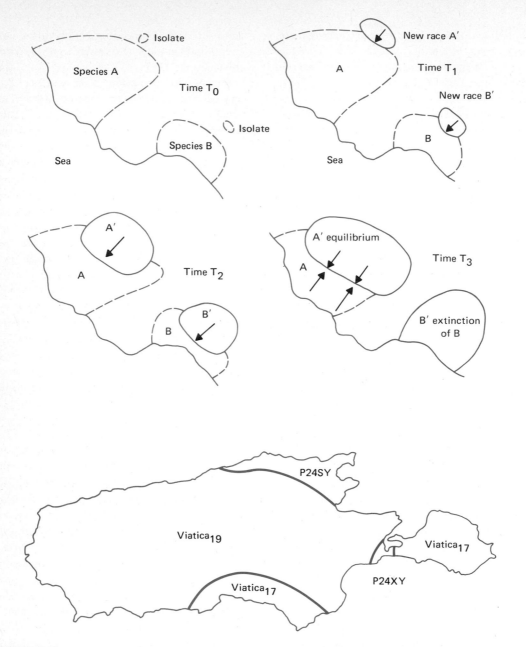

**FIGURE 11-14**

*Parapatric speciation and distribution of races in morabine grasshoppers in Australia. New races ($A^1$, $B^1$) arise from a peripheral isolate in each of two species (A, B). The new races from tension zones with the parent populations and advance against them until either equilibrium ecological conditions are reached under which the two homozygotes are equally adaptative, as in species A and $A^1$, or the parent form is eliminated, as in B and $B^1$.*

*The map of Kangaroo Island, South Australia, shows distribution of four parapatric races of morabine grasshoppers. The four races, existing with a radius of 15 km, are each confined to its own territory with different tension zones between the races. Parapatric morabine grasshopper species probably*
*began when a structural chromosome arrangement or a major gene mutation was fixed in an isolate of the species. The population isolate increased, expanded its range until it came in contact with the parent population and "hybridized." If the hybrids were not at a selective disadvantage, the low dispersal ability of the insects prevented individuals of the new race and their hybrids from penetrating the old population very far before they were eliminated. Thus a tension zone is established which prevents the new race from penetrating further into adjacent populations. After the tension zone has been established, further genetic changes may arise to further restrict or reduce compatability. (From Key, 1973.)*

inferior. Behavioral isolation is minimal or nonexistent.

Other organisms experiencing parapatric speciation include annual plants (Grant, 1971) and plants preadapted to colonize new habitats, such as soils heavily contaminated by heavy metals (mine spoils, for example) such as nickel and cadmium (Antonovics, 1971). Parapatric speciation apparently also takes place among such sessile animals as the marine snail *Partula* (Clarke and Murray, 1969), mole crickets (Nevo and Blondheim, 1972), mole rats *(Spalax)* (Nevo, 1969; Nevo and Shaw, 1972), pocket gophers *(Thomomys)* (Jevo et al., 1974; Thaeler, 1974), and endemic species of cichlid fish in East Africa (Greenwood, 1974).

*Sympatric speciation.* It may be difficult to draw a line between parapatric speciation and sympatric speciation. Sympatric speciation is defined by Mayr (1963) as the "origin of isolating mechanisms within the dispersal area of the offspring of a single cline." It differs from parapatric speciation in that premating reproductive isolation arises before the population shifts to a new niche (Bush, 1975). In parapatric speciation strong premating and postmating reproductive isolating mechanisms develop as the population enters a new environment. Sympatric speciation takes place in the center of a population in a patchy environment rather than on the periphery and chromosome macromutation is probably not involved. This results in the formation of a number of sibling species.

The first step in sympatric speciation is the formation of stable polymorphism. If stable polymorphism is accompanied by assortative mating, in which "individuals best adapted to a particular niche tend to mate with one another" (Bush, 1975), two reproductively isolated populations evolve. Groups of animals in which sympatric speciation is most likely to and probably does occur are the insect parasites of plants and animals. About 70 percent of the some 900,000 known species of insects are parasitic on plants and animals and many are host specific. These insects are small in size, have short life spans, possess high reproductive rates, have low competitive ability, adapt readily to new conditions, and experience sibmating and other forms of inbreeding.

Sympatric speciation, especially among insect parasites, depends upon several conditions (Bush, 1975). The genetic variation needed to establish a new host race has to be present in the population before the new host appears (preadaptation). Within the population some individuals must carry (1) new host recognition alleles permitting a parasite to move to a new host and (2) new survival genes to allow the parasite to counteract any chemical defenses of the new host. These preadaptive genes permit the insect parasite to recognize, move to, and survive on a new host. In a host-specific parasite mutations at one or two loci would be sufficient to ensure survival on a host closely related to the original one.

Let us assume that host recognition is controlled at a single locus and that a mutation of a host recognition allele $H_1$ to $H_2$ occurs. Through recombination the gene persists in the population. Homozygous individuals $H_1H_1$ recognize the original host plant. Heterozygous individuals $H_1H_2$ tend to stay and rarely, if ever, move to a new one. Homozygous individuals $H_2H_2$, maladapted for the original host, seek out a new host. If the $H_2H_2$ homozygotes locate a new host, they may become established; if not, they are eliminated from the population. However, the $H_2$ allele is still retained in the population in heterozygotes.

A similar situation holds for the survival gene $S$. Individuals homozygous for $S_1S_1$ and heterozygous individuals $S_1S_2$ have the selective advantage on the original host, while $S_2S_2$ individuals cannot survive. But on a new host plant $S_2S_2$ individuals would have a strong selective advantage, while $S_1S_1$ homozygosity is lethal.

Thus any individuals homozgygous for $H_2H_2S_2S_2$ are able to discover and survive on a new host plant. A host shift takes place, and if the new host is unexploited the new race spreads rapidly. In occupying a new host plant, the new race may be temporarily free from predation. Its parasites and predators have to change their host-seeking behavior before they can locate their former prey. As a result predatory pressure on the new race is relaxed, allowing time for the population to increase.

Several other conditions facilitate sympatric speciation. Ideally the new host plant should be closely related to the original one. Because chemical recognition cues and chemical defenses would be somewhat similar, great genetic changes in the new host race would not be necessary. The new host should grow abundantly with the original host within the dispersal range of the parasite. Then new genotypes could be tested on the hosts until the right combination for a successful host shift was "found." The host race should utilize the host as a site for courtship and mating. Mate selection would then depend upon host selection, and because only homozygous individuals could locate and survive on a new host, a strong ecological isolating mechanism would develop. Finally, the fruiting or leafing time of the new host plant must fall within the emergence, mating, and oviposition, (egg laying) period of the parasite.

An example of how sympatric speciation might come about is provided by a fruit-eating tephritid fly, the apple maggot *(Rhagoletis pomonella)*. The fly was once restricted to the hawthorn *(Crataegus),* but a new race that feeds on apples appeared in 1864 in the Hudson Valley and spread rapidly through the apple growing regions of North America. In 1970 in Door County, Wisconsin a new host race that feeds on cherries was described (Shervis et al., 1970). The new cherry race apparently developed from the apple race. Each race times its emergence, mating, and oviposition to coincide with the maximum amount of fruit available for oviposition. Thus the establishment of a population on a new host plant required some genetic alteration of emergence time to coincide with the availability of the appropriate fruit. Among *R. pomonella* the three host races have three emergence times. The cherry race emerges early to utilize the fruiting of the cherry in June. The original hawthorn race emerges late to coincide with the late summer and fall fruiting of the hawthorn. The apple race fills the gap between the two.

The new cherry race apparently developed when, because of changing market conditions, sour cherries were not picked. Fruit hanging on the trees in abandoned and unsprayed orchards provided new ovipostion sites for the insect. The apple maggot was able to exploit this new resource in part because its emergence time extended from late June (cherry time) to October (hawthorn time), although the life span of the individual fly is only 20 to 30 days. Dispersal of the insect depends upon the presence or absence of host fruit. If the fruit crop is heavy, the insect does not leave the host tree. If fruit is scarce, *R. pomonella* seeks out new host plants for distances up to a mile.

According to a model by Bush (1975) a *Rhagoletis* fly uses long-distance physical and chemical cues to locate a host plant (Figure 11-15). In the absence of chemical cues the insect orients to non–host-specific yellow and dark color masses. The principal chemical cue is olfactory—the fruit odor itself (not all fruit odors, even of apple varieties are equally attractive to the insects). Once the fly reaches the tree, tactile cues from the leaf hold the insect until it has found a fruit by visual cues. Then it moves slowly over the fruit, apparently assessing its color, size, and surface texture. For all *Rhagoletis* races red is the preferred color, but each has its own preferred range of fruit size. Hawthorn and cherry races apparently prefer slightly smaller fruits than the apple race.

Both males and females exibit the same behavior and follow the same cues in locating the preferred fruit. Once the site has been found, the female deposits eggs under the skin of the fruit and then drags the ovipositor around the fruit to lay down one or more marking pheromones. This inhibits other females from visiting the fruit and holds the male on the fruit. The male devotes his time on the fruit to mating, which takes place on the oviposition site. Because both male and female visit the same fruit, the host plant acts as an isolating mechanism.

The development of the apple and cherry races of *R. pomonella* probably would not have occurred without the unintentional help of humans. By the development of large acreages devoted to monocultural fruit production, providing an abundance of food and short dispersal distances, humans have provided ideal situations for sympatric speciation. Such speciation might not have occurred in the natural situation of a patchy environment. It is likely that this type of speciation has been repeated among many of the crop pests.

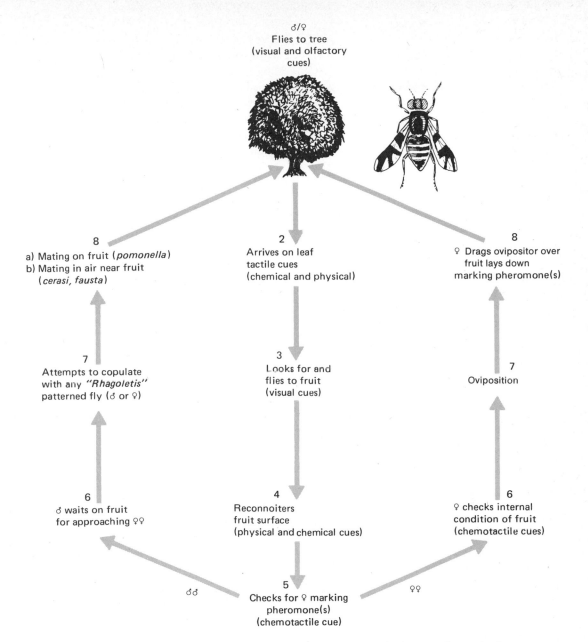

δ/♀
Flies to tree
(visual and olfactory
cues)

8
a) Mating on fruit (*pomonella*)
b) Mating in air near fruit
(*cerasi, fausta*)

2
Arrives on leaf
tactile cues
(chemical and physical)

8
♀ Drags ovipositor over
fruit lays down
marking pheromone(s)

7
Attempts to copulate
with any "*Rhagoletis*"
patterned fly (δ or ♀)

3
Looks for and
flies to fruit
(visual cues)

7
Oviposition

6
δ waits on fruit
for approaching ♀♀

4
Reconnoiters
fruit surface
(physical and chemical cues)

6
♀ checks internal
condition of fruit
(chemotactile cues)

δδ

5
Checks for ♀ marking
pheromone(s)
(chemotactile cue)

♀♀

**FIGURE 11-15**
*A possible example of sympatric speciation
among the true fruit flies, Tephritidae,
particularly the genus* Rhagoletis. *Involved
in sympatric speciation is the establishment
of a population on a new host plant,
development of specific olfactory cues
leading to the host tree, and reproductive
isolating mechanisms such as marking
pheromones. This figure illustrates the role
of host selection and mating behavior in the
apple maggot. (From Bush, 1975.)*

**Polyploidy.** Among plants new species can
arise spontaneously through polyploidy, the
alteration of the number of chromosomes.
There are two types of polyploids, both important from the standpoint of plant speciation—autopolyploids and allopolyploids.
The *autopolyploid* is formed by the doubling of chromosomes in any individual of the
species. Thus if the diploid species has *AA,
BB, CC,* and *DD* chromosomes, its autotetra-
ploid would have *AAAA, BBBB, CCCC,* and
*DDDD.* When gametes are formed by meiosis,
then groups of three and one, four, and two and
two may be formed by each type of chromosome. Such irregularity in the chromosomes of
the gametes results in few offspring. When a

diploid gamete unites with a normal haploid (1 *n*) gamete, a hybrid triploid results. Although incapable of sexual reproduction, the triploid can reproduce and spread vegetatively.

The *allopolyploid* arises something like this. Suppose that species *A* has four chromosome pairs: *AA, BB, CC, DD,* and a second species *B* nearby has chromosome pairs: *RR, SS, TT, UU.* If the two plants hybridize, the offspring will have a complement of A, B, C, D, R, S, T, U. Because the chromosome pairs are very dissimilar, the hybrid will be infertile. But if during the development of one of the individuals, the chromosome number should be doubled, if it should mutate to a polyploid, then an allopolyploid (allotetraploid) with chromosome pairs *AA, BB, CC, DD, RR, SS, TT, UU* will be formed. Each chromosome now has one definite partner; pairing at meiosis once again normal, and fertility is improved.

But there are other problems for the allopolyploid. Cross-fertilization with either parental species, both much more abundant, will produce sterile triploid offspring; and because of the small population of allopolyploids, few tetraploids will result, unless the plant is self-fertilized or relies on asexual reproduction. At such a competitive disadvantage, the polyploid is selected against and eventually disappears. If little cross-pollination takes place and asexual reproduction is the normal method of propagation, then the parental stock and the allopolyploid may coexist.

There is another alternative. If the allopolyploid colonizes an area unoccupied by either parent, then the allopolyploid may so completely dominate the area that the parents are excluded. The few that do gain a foothold are at a selective disadvantage, because, being less common, they will receive more pollen from the polyploid than from the diploid stock. Thus the offspring of the parental stock will be sterile triploids, whereas only a few allopolyploid offspring will be in this condition. This is more than a remote possibility because a number of polyploids thrive in areas different from the diploid ancestors. Because plants dispense with sexual reproduction under unfavorable conditions and multiply by rhizomes, bulbs, corms, etc., polyploidy is not at a disadvantage, particularly among perennial herbs. In fact this condition often enables plants to colo-

nize and to tolerate more severe environments. Thus the availability of new ecological niches favors the establishment of polyploidy.

From polyploidy have arisen many of our common cultivated plants—potatoes, wheat, alfalfa, coffee, grasses, to mention a few. It likewise is rather widespread among native plants, in which polyploidy produces a complex of species as in the blackberries. The common blue flag of northern North America is a polyploid and is believed to have originated from two other species, *Iris virginica* and *I. setosa,* when the two, once wide ranging, met during the retreat of the Wisconsin ice sheet. The sequoia is a relict polyploid, its diploid ancestors having become extinct, as are the willows and birches. The whole fascinating subject of polyploidy and species formation in plants is a complex one that cannot be pursued any further here. For this you should turn to such sources as Stebbins (1950) and Grant (1971).

### THE MAINTENANCE OF SPECIES

*Isolating mechanisms.* In spring there is a rush of courtship and mating activity in woods and fields, lakes and streams. Fish move into their spawning grounds, amphibians migrate to breeding pools, birds are singing. During this frenzy of activity each species remains distinct. Song sparrows mate with song sparrows, trout with trout, wood frogs with wood frogs, and few mistakes are made even between species similar in appearance. The means through which the many diverse species remain distinct are *isolating mechanisms.* These include any morphological characters, behavioral traits, habitat selection, and genetic incompatibility that enable different species to remain apart.

Isolating mechanisms fall into four broad classes: (1) ecological, (2) ethological or behavioral, (3) mechanical, and (4) reduction of mating success (Mayr, 1942, 1963).

*Ecological mechanisms* include habitat isolation and temporal or seasonal isolation. If two potential mates in breeding condition have little opportunity to meet, they are not likely to interbreed. *Habitat isolation* reduces the possibility of interbreeding. From North Africa and the Iberian Peninsula eastward through south-central Europe to Arabia and across central Asia to the Pacific are spread the *Alectoris* partridges, among them the chukar, well know as an exotic game bird in parts of North America (Figure 11-16). Most of the species are allopatric and separated by geographical distance. But in Thrace two very similar forms, the rock partridge, *A. graeca*, and the chukar, *A. chukar*, meet (G.E. Watson, 1962*a, b*). Here they are geographically sympatric. Both species inhabit rock-strewn hillsides and mountainsides with little cover and surface water. But the rock partridge is an alpine form found above 3,000 ft; the chukar lives below 3,000 ft. Otherwise their habitats, habits, and food are the same. Altitudinal allopatry separates the two forms. Thus among these partridges geographic sympatry exists because of habitat isolation.

Habitat isolation on a narrow local basis is highly important among frogs and toads (see

FIGURE 11-16
*Sympatry and allopatry in the* **Alectoris** *partridges. The overlap in ranges indicates regions of sympatry and ecological allopatry in the partridges. (Map redrawn from Watson, 1962.)*

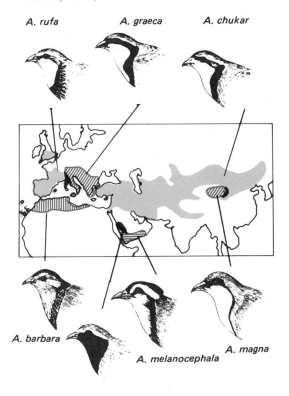

A. rufa          A. graeca          A. chukar

A. barbara

A. melanocephala          A. magna

A. philbyi

Bogert, 1960). Different calling and mating sites among concurrently breeding frogs and toads tend to keep the species separated. The barking tree frog typically calls while floating in open water, or while sitting on woody material projecting from the water. The closely related green tree frog calls from the ground or from low branches of trees or bushes near the breeding pool, but never in the water. Probably this is highly effective in preventing barking tree frog males from clasping green tree frog females. The upland chorus frog and the closely related southern chorus frog breed in the same pools, but ecological preferences tend to separate, partially at least, the calling aggregations of the species. The southern chorus frog calls from concealed positions at the base of grass clumps or among vegetational debris; the upland chorus frog calls from more open situations.

*Temporal isolation,* differences in the timing of the breeding and flowering seasons, effectively isolates some sympatric species. The American toad, for example, breeds early in the season, whereas the Fowlers toad breeds a few weeks later (W. F. Blair, 1942). Brown trout and rainbow trout may occupy the same streams, but the rainbows spawn in the spring, the brown trout in the fall. Fluctuations in environmental stimuli can time mating seasons. Among the narrow-mouthed toads, *Microhyla olivacea* breeds only after rain, whereas *M. carolinensis* is little influenced by rain (Bragg, 1950). Because temporal isolation is incomplete, call discrimination also is involved (W. F. Blair, 1955). Nevertheless some hybridization does occur.

*Ethological mechanisms,* differences in courtship and mating behavior, are the most important isolating mechanisms in animals. The males have specific courtship displays to which, in most instances, only females of the same species respond. These displays involve visual, auditory, and chemical stimuli (see Chapter 12). Some insects, such as certain species of butterflies and fruit flies, and mammals possess species-specific scents. Birds, frogs and toads, some fish, and such "singing" insects as the crickets, grasshoppers, and cicadas have specific calls that attract the "correct" mates. Visual signals are highly developed in birds and some fish. Species-specific color patterns, structures, and display, which give rise to a high degree of sexual dimorphism among such bird families as the hummingbirds and ducks, have apparently evolved under sexual selection (Sibley, 1957). Among the insects, the light flashes sent out by fireflies on a summer night are the most unusual visual stimuli. The light signals emitted by various species differ in timing, brightness, and color, which may range from white through blue, green, yellow, orange, and red (Barber, 1951).

*Mechanical isolating mechanisms* involve structural differences that make copulation or pollination between closely related species impossible. Evidence for such mechanical isolation among animals is very scarce. Even variations in body size and genitalic differences among insect species do not prevent cross-mating. Differences in floral structures and intricate mechanisms for cross-pollination within the species of many plants present mechanical barriers (see Grant, 1963). If hybrids do occur, especially among the orchids, they possess such unharmonious combinations of floral structures that they are unable to function together, either to attract insects to them or to permit the insects to enter the flower.

Ecological, ethological, and mechanical isolating mechanisms are significant in that they prevent the wastage of gametes, diminish the appearance of hybrids, and permit populations of incipient species to enter each other's ranges and become partly or wholly sympatric.

The fourth type of isolating mechanism, the *reduction of mating success,* does not prevent the wastage of gametes, but it is highly effective in preventing crossbreeding. Male gametes of animals that are liberated directly into the water, as is the case of most fish, amphibians, and marine and fresh-water invertebrates, either are unable to fertilize eggs other than of their own kind, or produce juveniles that fail to mature. If hybrids do mature, they may be sterile or at a selective disadvantage and thus eliminated from the breeding population. Isolation through sterility is important only in organisms in which ethological isolating mechanisms are poorly developed.

The different types of mechanisms act in various combinations to accomplish isolation among species. In the coastal sage communities of southern California grow two species of sage, *Salvia apiana* and *S. mellifera.* Although genetically compatible, the two species have established a number of partial barriers to maintain reproductive isolation. *S. mellifera* blooms earlier than *S. apiana* (temporal isolation), has smaller flowers (mechanical), and is pollinated by small bees, flies, and butterflies, whereas *S. apiana* is pollinated largely by the larger carpenter bees (ethological) (Epling, 1947). Two mosquitos, *Aedes aegypti* and *A. albopictus,* maintain reproductive isolation by means of at least five barriers: (1) mating behavior involving species differences in flight sounds and differential responses of males; (2) structural incompatibility, the result of differences in male genitalia; (3) sperm inactivation in interspecific crosses; (4) reduced oviposition due to differences in ovulation stimulus provided by substances from the male accessory glands; and (5) genetic incompatibility (Leary and Craig, 1967).

**Breakdown in isolating mechanisms: hybridization.** The gaps between sympatric species are absolute. There is no crossing them when the species occupy the same habitat and are adapted to different niches. Between allopatric species, on the other hand, the gaps are relative only. There are no assurances that they are even separate species; perhaps they are geographical races instead. The final test comes when the geographical barriers are broken and the two or more species in question meet. If the isolating mechanisms are not highly effective, if the populations are still diverging, then the organisms will hybridize freely. A more or less extensive hybrid zone is formed. Unlike the intergradation of subspecies, which involves a series of intermediate populations no more variable than neighboring populations, hybrid populations in the zone of secondary contact range from character combinations of species *A* to those of species *B.* Some hybrids may be indistinguishable from one or the other parent stocks; others may show a high degree of divergence.

If the hybrids are not at a selective disadvantage in competition with the parent populations, the genes of one species will be incorporated with the gene complex of the other.

This resulting introgression leads to a swamping of differences between parental forms. There will be a period of increased variability in the rejoined populations, and new adaptive forms will be established. Eventually the variability will decrease to a normal amount, and once again there is a single freely interbreeding population.

Introgressive hybrids are less common in animals than in plants. One of the most conspicuous examples in animals is the golden-wing–blue-wing warbler complex, in which the hybrid forms, the Lawrence and Brewster warblers, are distinctive in color and song. Prairie chicken–sharp-tailed grouse hybrids are rather common, but whether these hybrids are capable of producing offspring among themselves or with either parent species is not known. However, the variety of types (Figure 11-17) from one extreme to the other suggests that first generation hybrids crossed with either parent species have reproduced (Ammann, 1957).

There are numerous examples among plants. The white oak, for example, hybridizes with seven other members of the white oak group, including the chestnut, post, and overcup oaks. In plants, at least, two species may intergrade not only with each other but with still others to form chains or complex networks of partly segregated and partly interbreeding systems. These hybrids must compete with other plants selected by the environment and probably better adapted than they. Usually the hybrid swarms occupy permanent intermediate habitats somewhat removed from the parents or disturbed habitats where selective advantages to normal inhabitants are removed. If a hybrid population does survive, it may not retain all its members; through selection some members representing various genetic combinations may exploit special niches not suited for either parent.

If selected against, hybridization functions as a source of selection against individuals of both parental species that enter into mixed pairs. Any mechanism reducing the incidence of pairs then is selected for as long as the interaction continues. Any interaction between species that results in deleterious competition or in a wastage of gametes is selected against, reducing diversity in species-specific charac-

**FIGURE 11-17**
*Hybridization of sharp-tailed grouse and prairie chicken. Dorsal and ventral views of (1) sharp-tailed grouse, adult male, shot October 3, 1941, near Sidnaw, Houghton County, Michigan; (2) prairie chicken –sharp-tailed grouse, adult male, shot October 2, 1946, at Drummond Island, Chippewa County, Michigan; (3) prairie chicken – sharp-tailed grouse, immature male, shot October 17, 1947, near Ralph, Dickinson County, Michigan; (4) prairie chicken – sharp-tailed grouse, adult female, shot October 28, 1950, near Sharon, Kalkaska County, Michigan; (5) prairie chicken, adult male, from Michigan, no additional data. (Specimen 2, the University of Michigan Museum of Zoology collection; others, the Game Division collection, Michigan Department of Conservation; photos from Ammann, 1957; photos courtesy of Game Division, Michigan Department of Conservation.)*

ters and reinforcing isolating mechanisms (Sibley, 1957).

The North American plains serve or served as an effective barrier to woodland species of animals, a barrier that has existed at least since the Pleistocene, when the ice sheet separated the populations of many animals. (Mengel, 1970). Woodlands stretched like fingers along the rivers, but only a few, the Platte in particular, provided a continuous woodland connection across the grasslands. Even these river-bottom forests may have been broken by prairie fires, by floods, and by trampling from buffalo. When white settlers occupied the plains, they destroyed the woodlands for lumber. But importations of timber from the east, the control of fires, and the control of floods brought riparian woodlands back. Trees and shrubs were planted in shelter belts around towns, farms, fields, and even along the rivers. These changes, which resulted in the development of suitable breeding sites, as well as "islands" of suitable resting places for migrants, permitted the woodland species of birds to move both east and west, to colonize the new habitat, and eventually to come in contact with one another. The red-shafted flicker of the west and the eastern yellow-shafted flicker were two species that spread into an area of contact on the plains, as well as the lazuli bunting and the indigo bunting, the Bullock and the Baltimore orioles, and the black-headed and rose-breasted grosbeaks.

Because isolating mechanisms were not fully established, the two flickers and the two orioles have interbred extensively and each of the pairs are now considered single species.

The two grosbeaks illustrate well what happens when two closely related allopatric species that have not yet diverged sufficiently to possess strong isolating mechanisms come into contact (West, 1962; Kroodsma, 1974). Both the black-headed and rose-breasted grosbeaks have essentially the same habitat preferences, build the same type of frail nest from twigs and rootlets, lay eggs that are almost indistinguishable, possess nearly the same vocalizations, and have nearly identical female plumages. Only in the male plumage do the two species differ widely (Figure 11-18). On this basis the two have been considered separate species. Where the two come in contact, however, they interbreed, producing a variety of plumage combinations (Figure 11-19). The variation found in the hybrids suggests that back-crossing, producing second-generation

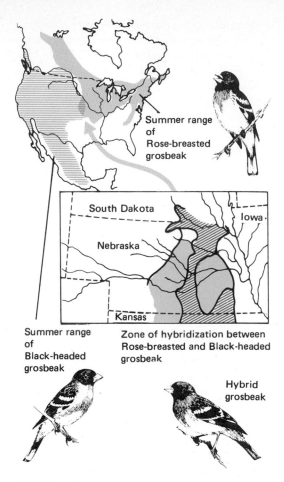

**FIGURE 11-18**
*Hybridization in the black-headed and rose-breasted grosbeaks. The map of North America shows the summer range of both species; inset map shows the zone of hybridization between the two. The drawing of the hybrid grosbeak shows characters somewhat intermediate between the two parent stocks. (Map based on West, 1962.)*

hybrids $(F_2)$, is taking place. Difference in plumage that has evolved in the two under spatial isolation is insufficient to prevent the two from interbreeding. At present there is no indication that any selection against hybrids is operating. Possibly either the differences will be swamped or hybridization will stabilize, with a hybrid zone between the two "pure" populations, since by hypothesis the two species are ecologically too similar to become sympatric.

A similar situation arises in plants when geographic barriers are no longer effective.

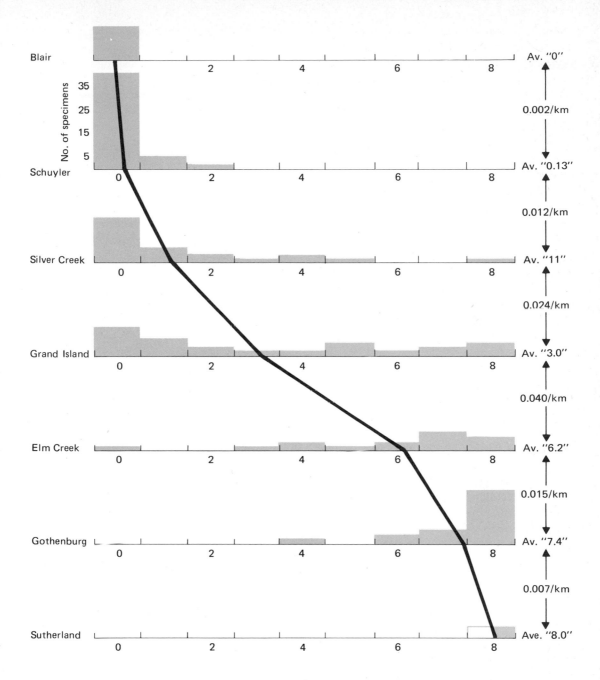

*FIGURE 11-19*
*Histogram of hybrid index scores for samples of grosbeaks from a Platte River transect. The shift per kilometer in average index is given. Localities are about 80 km (50 mi) apart. (From West, 1962.)*

The eastern red cedar and the Rocky Mountain juniper are related and quite similar in appearance. Where the two come in contact from North Dakota south and east into Kansas and Nebraska, they, like the grosbeaks, hybridize and exhibit to varying degrees the characteristics of both.

## ADAPTIVE RADIATION

The direction and degree to which a population diversifies is influenced by the preadaptability of the species population to a new situation, by selective pressures of climate and competition, and by the availability of ecological niches. All species of organisms are adapted to some particular environment, but because the environment is limited, overpopulation can result. This in itself is a selective force, for the time finally comes when those individuals that are able to utilize some unexploited environment and resource are at an advantage. Under lessened competition in the new environment, these individuals have some opportunity of leaving progeny behind. By eliminating disadvantageous genes, selection will strengthen the ability of the group to utilize the new niche or niches it has occupied.

Before a species can enter a new mode of life, it must first have physical access to the new environment. And having arrived there, the species must be capable of exploiting the niches, that is, be preadapted to new conditions. It must possess a level of physical and physiological tolerances high enough to enable it to gain a foothold in the new locality. Here animals, particularly the vertebrates, have an advantage over plants. Most plants, although easily dispersed, are rather exacting in their habitat requirements; animals, though they find it difficult to cross barriers, are better able to cope with a new environment.

Once the organism has established a beachhead, it must possess sufficient genetic variability to establish itself under the selective pressures of climate and competition from other organisms. The adaptations that permit the organism to gain a foothold are only temporary makeshifts, which must be altered, strengthened, and improved by selection before the organism can efficiently utilize the niche.

Finally, an ecological niche must be available for exploitation. Competition in a new habitat either must be absent or slight enough so that the new invader can survive in its initial colonization. Such niches have been available to colonists of some remote islands, such as the Galapagos, the Hawaiian Islands, and the archipelagos of the South Pacific. The abundant empty niches available in these diversified islands when the first colonists arrived encouraged the rapid evolution of species. Darwin's finches are classical examples of colonization and diversification in an unexploited environment. The original finch population probably consisted of a few chance migrants from South America. Because of the paucity of invaders, due to the great distance from the mainland, the successful immigrants were able to spread out in many evolutionary directions to exploit the islands' resources.

A similar development took place among the honeycreepers, Drepanididae, in the Hawaiian Islands, which evolved into finchlike, honey eaterlike, creeperlike, and woodpeckerlike forms. They so completely occupied the diverse niches on the island that they prevented similar adaptive expansion by later colonists of thrushes, flycatchers, and honey eaters.

The ancestor of the honeycreepers was probably a nectar-feeding coerebidlike bird with an insectivorous diet somewhat similar to *Himatione*. After the first arrivals had colonized one or two islands, stragglers undoubtedly invaded other surrounding islands. Because each group was under a somewhat different selective pressure, the geographically isolated populations gradually diverged. By colonizing one island after another and, after reaching species level, recolonizing the islands (double invasion) from which they came, the immigrants enriched the avifauna, especially on the larger islands with a more varied ecology. At the same time competition among sympatric forms placed a selective premium on divergence. The forms then became adapted to the somewhat different ecological niches available, survived by rigid specialization, or one or the other perished. Such divergence of one group into several different forms, each adapted to different ecological niches, able to exploit new environments or to tap a new source of food, is called *adaptive radiation* (Figure 11-20).

This principle is nicely illustrated by the genus *Hemignathus,* all members of which are primarily insectivorous. *Hemignathus obscurus,* whose lower mandible is about the same length as the upper mandible, uses its decurved bill like forceps to pick insects from crevices as it hops along the trunks and limbs of trees. The bill of *H. ludicus* is also decurved, but the lower mandible is much shorter and thickened. The bird uses the lower bill to chip and pry away loose bark as it seeks insects on the trunks of trees. In *H. wilsoni* the modification is carried even further. The lower mandible is straight and heavy. Holding its bill open

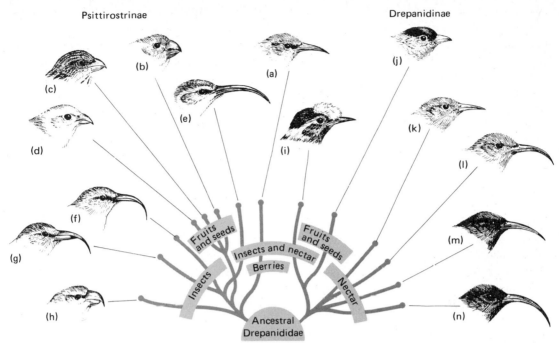

Psittirostrinae

Drepanidinae

Fruits and seeds

Insects and nectar

Berries

Fruits and seeds

Insects

Nectar

Ancestral Drepanididae

Adaptive radiation of Hawaiian honey creepers

**FIGURE 11-20**

*Adaptative radiation in the Hawaiian honeycreepers, Drepanididae. Selected representatives of two subfamilies Drepanidinae and Psittirostrinae illustrate how the family evolved through adaptive radiation from a common ancestral stock. Both subfamilies show a certain degree of parallel evolution based on diet. a, Loxops virens, probes for insects in the crevices of bark and folds of leaves and feeds on nectar and berries; b, Psittirostra kona, a seed-eater, extinct; c, Psittirostra cantans, feeds on a wide diet of seeds, insects, insect larvae, and fruit; d, Psittirostra psittacea, a seed-eater, feeding especially on the climbing screw pine (Freycinetia arborea); e, Hemignathus obscurus, an insect- and nectar-feeder; f, Hemignathus ludidus, an insect-feeder; g, Hemignathus wilsoni, an insect-feeder; h, Pseudonestor xanthophrys, feeds on larvae, pupae, and beetles of native Cerambycidae and grips branch by curved upper beak; i, Palmeria dolei, feeds on insects and the nectar of ohio (Metrosideros); j, Ciridops ana, a fruit- and seed-eater, extinct; k, Himatione sanguinea, feeds on caterpillars and the nectar of ohio; l, Vestiaria coccinae, feeds on loop caterpillars and the nectar from a variety of flowers; m, Drepanis funerea, a nectar-feeder, extinct; n, Drepanis pacifica, a nectar-feeder, extinct. (Drawings based on Amadon, 1947, 1950, and other sources; evolutionary data from Amadon.)*

to keep the slender upper mandible out of the way, the bird uses the lower mandible to pound, woodpeckerlike, into soft wood to expose insects. The bill of genus *Hemignathus* was specialized at the start to feed on insects and nectar. Although the result of modification through competition in *H. wilsoni* is grotesque, it is the only species among eight surviving in fair numbers.

The honeycreepers also exhibit parallel evolution, adaptive changes in different organisms with a common evolutionary heritage in response to similar environmental demands. The long, thin, decurved bill of *Hemignathus obscurus* is adapted to a diet of insects and nectar; so too are the bills of several members of the subfamily Drepanidinae (Figure 11-20).

Similar adaptive radiation took place among the whales. Arising from a carnivorous creodont stock, they eventually radiated into filter feeders, the whalebone whales, whose food is plankton, the great-toothed whales, which feed on deep-sea mollusks, and the fish-eating porpoises and dolphins.

**CHARACTER DISPLACEMENT**

Within an area of overlap between two closely related species the differences between them

**395**

may be accentuated, whereas outside of these areas the differences may be weakened or lost entirely (Brown and Wilson, 1956). This divergence in an area of overlap, known as *character displacement,* may be expressed in morphology, ecology, or behavior. Differences in feeding habits, in anatomical structures that assist in food gathering, in periods of activity, and in nesting sites reduce competition; differences in reproductive behavior prevent interspecific hybridization.

Although the concept of character displacement has been widely and uncritically accepted (for example, see Lack, 1971; Ricklefs, 1973; Emlen, 1973), little evidence for it exists. Evidence for character displacement requires that two closely related species can be shown to be similar in character in allopatric situations and more distinctive in sympatric situations. The characters under consideration should be displaced in relation to one another (Grant, 1973, 1975).

The classic example is the case of the rock nuthatches, *Sitta tephronota* and *Sitta neumayer.* In the zone of overlap in Iran the two species supposedly diverge in eye markings, bill size, and body size (Vaurie, 1950; Brown and Wilson, 1956). The ecological character displacement involving bill size (which relates to the utilization of a food resource) is invalid because of the variability of bill size in the allopatric populations. The differences probably existed before the two species met. Ecological character displacement may simply represent clinal variation. However, the larger size of the eye stripe in *S. neumayer* in the zone of sympatry might represent reproductive character displacement because differences in eye stripe serve to reinforce species recognition (Grant, 1975). Some evidence also exists for reproductive character displacement in frogs relative to mating calls (Blair, 1974). Overall evidence for ecological character displacement, including the more recent ones involving desert lizards (Huey et al., 1974) and marine snails (Fenchel, 1975), is weak and probably has been overemphasized as an evolutionary process. Evidence is much stronger for reproductive character displacement because in the zone of sympatry the need for reinforcing species-specific recognition is strong.

## CONVERGENCE

Instead of accentuating their differences in zones of overlap, some species will tend to reduce their differences and their characters converge. Instead of becoming less similar, they become more similar. This apparently happens among species in which reproductive isolation is complete. Selective pressures may tend to favor an increasing resemblance among species as long as reproductive isolation is not upset and the similarities are advantageous to the species (Moynihan, 1968; Cody, 1969). Such convergence can result because sympatric species evolve similar adaptations to the same environment and because it facilitates social reactions among species.

There are a number of possible situations. Sympatric species may become more or less cryptic against background color. Selection may favor dull or dark-colored forms, which might be a reason for the so-called sibling species. Or the animals may become more conspicuous against background color, a trait that can facilitate flocking among individuals of the same and very similar species such as the herons. Convergence in color pattern may be selected for if it is advantageous for a species to be a social mimic. Such mimicry to facilitate mixed flocking is common among birds of the mountains of neotropical regions (Moynihan, 1968). For example, in parts of the northern Andes most of the more common and conspicuous species in the mixed flocks of the humid temperate zone are predominantly brilliant blue or blue and yellow.

Similar convergence exists among plants. Within the chaparral vegetation of California, for example, the dominant plants belong to such diverse families as *Ericaceae, Rhamraceae,* and *Rosaceae,* yet all are deep-rooted, evergreen sclerophyllous shrubs (Mooney and Dunn, 1970). In fact throughout all areas with a Mediterranean type of climate the vegetation has a similar appearance and is dominated by woody, evergreen sclerophyllous species. Even though widely separated geographically and possessing different evolutionary histories, the vegetation has converged in both form and function. This convergence has been in response to similar selective forces, including fire, drought, high temperatures, and low rainfall (see Mooney, 1977; Orians, 1977).

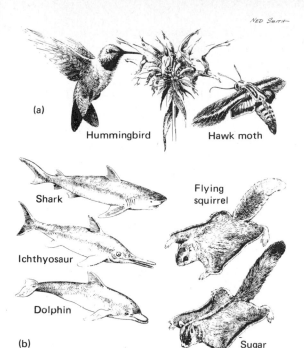

NED SMITH

(a)
Hummingbird      Hawk moth

Shark

Flying squirrel

Ichthyosaur

Dolphin

(b)

(c)

Sugar glider

**FIGURE 11-21**
*Convergent evolution in dissimilar organisms. The hummingbird and hawk moth both feed on the nectar of flowers, the bird by day, the moth by night. Both have adaptations of the mouth for probing flowers; both have the same rapid, hovering mode of flight. The prehistoric marine reptile, the ichthyosaur; the modern shark; and the marine mammal, the dolphin, all have the same streamlined shape for fast movement through the water. A North American rodent, the flying squirrel, and an Australian marsupial, the sugar glider, both have a flat bushy tail and an extension of skin between the foreleg and the hind leg that enables them to glide down from one tree limb to another.*

The influences of similar environmental selection has resulted in the convergence of wholly unrelated, geographically distant species that started out being quite dissimilar (Figure 11-21). Examples are numerous and include the marsupial phalanger of Australia and the placental flying squirrel of North America; the Tasmanian wolf and the true wolf. The sphinx moth and the hummingbird have converged in form, flying habit, and feeding procedure even though one is an insect and the other a bird.

## SUMMARY

Genetic variability and speciation result from natural selection, which is essentially nonrandom reproduction. This comes about in three ways: nonrandom mating, nonrandom fecundity, and nonrandom survival. This nonrandomness lies in the variations contained in the gene pool of the deme or local population of interbreeding individuals. Certain genetic combinations are more fit than others and these transmit more genetic information to future generations than less fit combinations. Two sources of genetic variation acted on by natural selection are mutations and recombination of genes provided by parents in a bisexual population. Theoretically, variations in biparental populations, as reflected in gene frequencies and genotypic ratios, remain in equilibrium if the conditions of random reproduction, equilibrium in mutation, and a relatively large population exist. In nature such conditions do not occur, and there is a departure from genetic equilibrium. This departure is evolution. The direction evolution takes depends upon the genetic characteristics of those individuals in the population that survive and leave behind viable progeny.

It is through a long evolutionary process that species arise by an interaction of heritable variation, natural selection, and, perhaps among most kinds, spatial isolation. As one segment of a species is separated from another, it carries a somewhat different sample of the gene pool and faces different selective pressures. Eventually the population diverges so far from the original parent stock that interbreeding cannot take place even if the geographical barriers are removed. At this point speciation is complete and the two populations could occupy the same geographical area (sympatry), yet remain distinct. This they accomplish by means of isolating mechanisms, which include any morphological character, behavioral trait, habitat selection, or genetic incompatability that is species specific. If isolating mechanisms break down, hybridization results. Species may also arise in the absence of geographical isolation through disruptive selection in which the phenotypes of both extremes are favored at the expense of the average. Speciation in plants may also come about through alteration in the number of chromosomes, or polyploidy.

Plants and animals are commonly classified in terms of the morphological species, a discrete entity encompassing little variation and containing only individuals that approximate the norm for the species, that is, the original specimen on which the description is based. But organisms are not static; they are continuously changing through long periods of time. This fact has given rise to the concept of the biological species, a group of interbreeding individuals living together in a similar environment in a given region under similar ecological relationships. This concept is limited to bisexual organisms and involves several problems, such as the inability to distinguish among some groups where one species begins and another ends. To get around this problem some evolutionary biologists suggest that species descriptions be retained to identify kinds of organisms, but as a concept the species should be regarded as evolutionary units.

# Behavior and social systems
## CHAPTER 12

Individual animals that make up a population interact among one another to varying degrees. These encounters act to space out individuals or to bring them together for security, sexual reproduction, and parental care. Any activity that directly relates to potential or actual encounters among individuals is *social behavior*.

Social behavior involves various forms of communication through sight, sound, and smell and utilizing color patterns, morphological structures, movement patterns, calls, and scents. Many behavioral patterns in individuals are genetically determined, and their appearance, retention, and further development are influenced by forces of natural selection. Other behavioral patterns develop through learning or are modified by experience. Whatever its original source behavior becomes a mechanism for survival and over time natural

selection favors the best adapted behavior pattern for the species.

The study of behavior of animals in their natural communities is the field of *ethology*, which is concerned primarily with determining how the behavioral processes work, how they develop during the life of an animal, and how they have evolved. Ethology has its own terminology and rapidly changing theory, much of which is of little direct concern to the ecologist and field biologist. But to understand aspects of behavior pertinent to the ecology of animals, some exposure to ethological concepts is necessary.

### Stereotyped behavior

In the bright sunlight of an April morning a male flicker taps out a drumming roll on the bone-gray limb of a dead oak tree. He is answered in the distance by another, and then suddenly he leaves his drumming perch and drops to the ground. Another male flicker has flown into the territory. The two face each other closely on the ground, point their bills in the air, wave them about for four or five seconds, utter a shrill *we-cup,* raise their bright red crests, and display the yellow-colored underwings. After several such dances, the birds fly into different trees.

In the quiet of a summer evening a field cricket in a dark crevice in the base of a dead oak tree calls. He occupies the crevice alone and sallies forth for short distances to find food and water, to challenge other males, and to court females. He is answered by other males, located in other dark corners, and the evening cricket chorus begins. Nearby female crickets respond to the chirps and travel in the direction of the source.

All the actions of these animals are stereotyped and automatic responses to stimuli. Each performed as other individuals of its kind would perform under similar circumstances. Each response is characteristic of the species and apparently is inherited or genetically fixed. It is this internally "programmed" behavior which ethologists refer to as innate or instinctive behavior or, more appropriately, stereotyped or species-typical behavior.

## STIMULUS AND RESPONSE

An animal performs a species-specific behavior only when it receives the appropriate external stimulus or cue. Not all of the information gathered by the sense organs of the animal is meaningful. Somewhere between the sense organs and the motor centers that control the behavioral action much of the sensory input is filtered out and the remainder is selectively admitted. What stimuli are admitted depends upon the internal condition of the animal. Only when a certain stimulus or stimuli are properly received by the animal is the behavior pattern discharged. This stimulus may be the appearance of a predator or prey, a sex partner, the presence or absence of a species-specific color pattern, morphological structure, or posture; it may be a chemical substance, such as the odor of a stream or a scent emitted

by an animal; or it may be a specific call or song. Often the stimulus is only a small part of the total object perceived, such as the mustache of a flicker (Noble, 1936) or the red breast feathers of a European robin (Lack, 1953). The essential feature of an object or situation that elicits a particular innate behavior pattern is called a *sign stimulus.* And a structure of an animal that serves exclusively to send out sign stimuli and results in a high degree of response is called a *releaser* (Figure 12-1).

The first phase of a behavioral pattern usually is spontaneous. Some internal stimulus in the animal, such as hunger or hormonal action, causes the animal to seek an external stimulus such as food or a sex partner. When the animal sights food or a potential mate, a series of behavioral acts is set in motion. The preliminary or searching phase is often called *appetitive behavior.* Internally controlled, this behavior in the right situation switches over to external stimuli. If the animal proceeds in a coordinated manner and in the right direction, the activity is self-stimulating and self-reinforcing until the releaser for the next stage is found.

**FIGURE 12-1**
*Mustache of the male flicker (below), the releaser that stimulates territorial defense in the male and courtship behavior in the female (above) that lacks a mustache.*

An example of such a reaction chain of behavior and its control is the courtship of the queen butterfly (L. P. Brower et al., 1965). The appearance of the female is the stimulus that sets off the behavioral pattern or program (Figure 12-2). When a female queen butterfly comes within the view of the male, he flies after her in pursuit. The female flies off, but eventually the male overtakes her. As he passes a few inches over her back, he extrudes

**FIGURE 12-2**
*Chain-reaction behavior in the courtship of the queen butterfly. The male behavior is shown on the right and the female behavior on the left. (From Brower et al.; 1965.)*

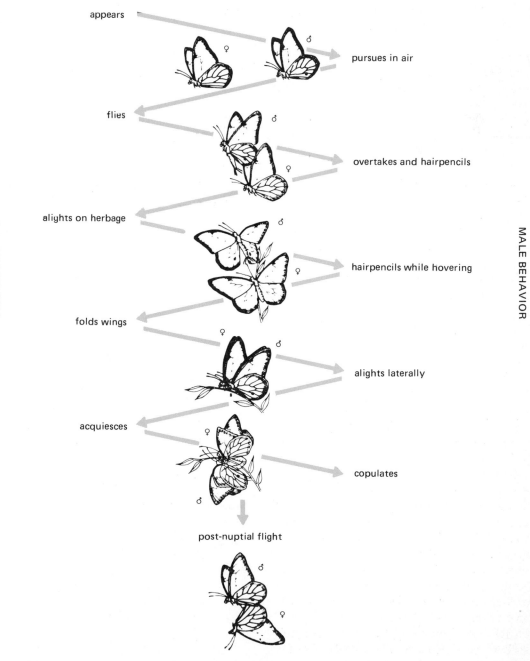

FEMALE BEHAVIOR

MALE BEHAVIOR

appears

pursues in air

flies

overtakes and hairpencils

alights on herbage

hairpencils while hovering

folds wings

alights laterally

acquiesces

copulates

post-nuptial flight

**401**

two bundles of hairs on either side of the abdomen. These are the abdominal hair pencils, which unfurled emit a strong, musky perfume. Changing to a bobbing flight, the male rapidly sweeps the hair pencils up and down the female's head and antennae. She responds to this by alighting in the herbage. The male then continues to "hair pencil" while still maintaining a bobbing flight. In response the female folds her wings tightly over her back. This stimulates the male to retract the hair pencils and alight alongside of the female. The female acquiesces and the male in turn attempts to copulate. If successful, the male and female engage in a postnuptial flight, in which the male flies off carrying the female suspended at the end of his abdomen. If unsuccessful, the male induces the female to fly up again by hovering over her and striking her on the back. As she flies up, the male pursues her and starts the courtship over again.

Internal mechanisms also exist to bring a series of behavioral acts to an end. A courtship behavior program ends with successful mating. Hunting behavior ends with the capture of a prey. And the entire program of feeding ends when the stomach of the animal is full and the animal's stretch receptors and blood glucose level registered through the hypothalamus turn off eating tendencies. The concluding or *consummatory act* of a behavioral program may act as a sort of a negative feedback mechanism that reduces the effect of a stimulus and brings, if necessary, the behavior to an end. It is followed by a period of quiescence during which there is no further appetitive behavior controlled by the same internal and external stimuli.

Although appetitive behavior and consummatory acts are useful as descriptive terms, it is almost impossible to break a particular behavior pattern into such discrete categories. Each subgoal of a particular behavior, with its own appetitive behavior and consummatory act, may be followed immediately by another sequence of appetitive behavior (see Manning, 1972).

The strong stimulation of one behavioral program apparently inhibits all others and allows only one particular behavior pattern to appear in an animal at one time. How this suppression of one pattern by another is ac-

complished is unknown, but in some way it may be controlled through connections in the central nervous system. However, when an animal is strongly stimulated in more than one way at the same time, for example, both to attack and to escape, on occasions neither behavioral program can suppress the other. Also, in some situations the dominant behavior program may be thwarted. The result is *conflict behavior*.

One form of conflict behavior is the emergence of a program that does not appear to be appropriate as a response to either of the conflicting stimuli. For example, the black-headed gull will preen itself or go through the actions of nest building when its brooding drive is thwarted (Moynihan, 1955a). During a fight with another male the pectoral sandpiper, like other waders, will suddenly turn its head around, put its bill under its scapulars, and act as if it were going to sleep (Hamilton, 1959). Such acts are called *displacement* or *irrelevant activities* (Figures 12-3 and 12-4).

The still controversial theory behind displacement is that the nervous energy of two opposing tendencies finds its outlet through another behavior pattern. Another theory is that two types of behavior are more or less evenly balanced and cancel each other out so that a third behavioral pattern emerges. A third theory is that if a hierarchy of behavior exists, a subordinate behavioral pattern appears when the dominant ones are absent. Under these circumstances, ethologists argue, such behavior may not be so inappropriate after all (see van Iersel and Bol, 1958; Sevenster, 1961; Rowell, 1961; Hinde, 1970). Another form of conflict behavior is *redirection activity*,

*FIGURE 12-3*
*Grass pulling, a displacement activity in the herring gull. (Based on photographs in N. Tinbergen, 1953.)*

(a)

(b)

*FIGURE 12-4*
*Posture of displacement feeding and normal feeding in the pectoral sandpiper. (After Hamilton, 1959.)*

behavior in which the object toward which an action is directed is suddenly changed (N. Tinbergen, 1952). Like displacement, redirection seems to occur when an activity or stereotyped behavior is thwarted or conflicts with another (see Bastock et al., 1953; van Iersel and Bol, 1958; Manning, 1972). When its nest is disturbed by a human intruder, a falcon may attack some passing bird. Both attack and escape tendencies are activated simultaneously by the intrusion. The escape drive may be strong enough to suppress attack behavior toward the intruder, but the thwarted attack drive still finds expression by redirection to another object.

Some of these displacement and redirection activities have become incorporated in display behavior patterns, both hostile and courtship; both involve considerable conflict and thwarting.

## Learning

Behavioral responses are adaptive. By the appropriate perception of, admission of, and reaction to stimuli, the flicker and cricket act successfully in a particular situation. The correct responses are built into their nervous system as part of the inherited structure. The behavior evolved gradually, modified by natural selection to fit the environment, but environment and the animals' experience of it play a minor role in their behavioral development. Their stereotyped responses represent one extreme in a continuum of behavioral responses.

At the other extreme are animals born with few inherited responses, but with the ability to modify behavior through learning in the light of experience, discovering what responses give best results and modifying behaviors accordingly. Environment plays a major role in their behavioral development. Between the two extremes is a gradient of behavior involving an interaction between stereotyped behavior and behavior influenced by environment.

Learning is important in perfecting skills, even stereotyped ones. Behaviors such as walking, flying, feeding, courtship, nest building, and vocal patterns appear in the animal in species-specific form at some stage of maturation; many of them appear suddenly in complete form. But with experience animals become more adept in performance. For example, nest building in birds is stereotyped. It appears at a certain hormonal level and improves as that level rises, but it also improves as the bird learns what materials are suitable for nest building such as size and type of twigs. Birds do not learn how to fly; they "know" how to fly once they have developed the proper motor coordination. But they do learn such things as recognizing a suitable landing place.

Learning is a large and complex subject which we cannot here pursue at length, but a few simple forms of learning important in behavioral responses should be recognized.

### HABITUATION

Probably the simplest form of learning is *habituation* which involves the loss of an old response rather than the acquisition of a new one. An animal repeatedly exposed to a given stimulus not associated with reward and punishment soon learns to ignore it. For example, a crow might initially respond to the appearance of a scarecrow by flight, but when over time the presence of the scarecrow results in neither reward nor punishment, the bird learns to ignore this intrusion into its environment. Learning to ignore repeated stimuli that have no accompanying consequences is important in the development of young animals. Because a young animal is threatened by a wide range of predators and other dangers, it is advantageous for it to respond with escape reactions to anything that moves. Eventually through habituation the animal is able to sort out relevant from irrelevant stimuli.

Habituation is common to all animals from protozoans to humans. It is an important learning process enabling an animal to adjust its behavior to its environment.

## OPERANT CONDITIONING OR TRIAL AND ERROR LEARNING

Operant conditioning or trial and error learning is a modification of the conditioned reflex or respondent conditioning in which a new stimulus becomes associated with positive or negative reinforcement. In respondent conditioning the existing stimulus-response reaction becomes attached to a novel stimulus. For example, an unconditioned stimulus, food, results in an unconditioned response, salivation. If a bell rings prior to each presentation of food to the animal, the animal associates the novel stimulus, the sound of the bell, with food and begins to salivate upon hearing the bell. In operant conditioning the animal may be motivated by hunger as in the conditioned reflex, but no unconditioned reflex stimulus exists to evoke an appropriate unconditioned reflex. The animal explores (appetitive behavior) its environment for food by looking, smelling, digging, etc. If any of its random motor patterns during the search is followed by the discovery of food, that pattern is reinforced, that is, it is more likely to occur again. Each positive reinforcement increases the likelihood of the appearance of the pattern, and, in this sense, the animal learns to follow the pattern repeatedly. In a similar way through negative reinforcement an animal learns to reject or ignore unpalatable food items. Such trial and error learning is fundamental in the development of new motor skills by young animals.

## IMPRINTING

Somewhat different from the conventional type of learning is *imprinting* (see Immelmann, 1975). Imprinting, a form of associative learning, is the rapid establishment of a perceptual preference for an object. It seems to occur only at a specific and brief highly sensitive or critical period in the life cycle of an animal and under a particular set of environmental conditions. The critical period for imprinting varies among species and is linked to the rate of physical development. This rapid form of learning is highly suited to those animals whose early and rapid development of motor abilities requires that contact with their parents and others of the same species be established early.

The term imprinting was first suggested by K. Lorenz (1933) to explain his experiences with greylag geese. Young geese without a mother attached themselves to Lorenz. Apparently because he, rather than the mother, was present at the critical period at which the goslings would have established perceptual preference for the mother, he was accepted as a substitute parent.

Experimental studies have stressed visual imprinting, but there is growing evidence that imprinting also involves other perceptual stimulation. Young wood ducks reared in a dark tree hole or nest box are exposed to the call of their mother for a relatively long period before they are exposed to the sight of her. The auditory stimulation apparently plays a major role in the ducklings' recognition of their parent (Gottlieb, 1963). Olfactory imprinting apparently exists among mammals that recognize both species and individuals by body odor.

The survival value of imprinting is obvious. Young animals, particularly those that move on their own power shortly after birth, have only a short time in which to learn important primary objects, particularly parents and food. Imprinting is important in species recognition, in behavioral isolating mechanisms, in special behavior such as the homing of salmon to their native streams to spawn (Hasler et al., 1978), and it may be partially or wholly responsible for certain evolutionary changes or equilibria in polymorphic species. (For discussions of imprinting, see Hess, 1971; Manning, 1972; Klopfer, 1973; Brown, 1975.)

## Habitat selection

Habitat selection may be one form of imprinting. There is some evidence, among birds at least, that imprinting is involved in the development of habitat preferences. (It is difficult to establish the nature of the imprinting because the characteristics of the habitat at the time of imprinting after hatching are different from its characteristics early the next spring when the bird settles on its territory.) Other evidence suggests that habitat selection among birds and mice is little influenced by early experience. The whole question of habitat selection cannot be dissociated from competition and natural selection, the forces that "fit" the organism to the environmental niche it occupies.

Habitat selection is partly a psychological process. Lack and Venerables (1939) and Miller (1942) suggest that birds recognize their ancestral habitat by conspicuous though not necessarily essential features. The Nashville warbler, a typical inhabitant of open heath edges of northern bogs, selects open stands of aspen and balsam fir and forest openings of blackberry and sweet-fern in the southern part of its range. These habitats are visually suggestive of bog openings (R. L. Smith, 1956). MacArthur and MacArthur (1961) and later MacArthur and many others (see MacArthur, 1972) demonstrated a strong correlation between structural features of the vegetation and the species of birds present. Basically birds of forest and grassland choose their habitat on the basis of the density of leaves at different elevations above the ground, irrespective of plant species composition. In effect it is the overall aspect that is important—the type of terrain, whether rolling or flat, open or grown with woody vegetation, homogenous or patchy.

There are specific features in addition to overall aspect that determine a habitat's suitability (Hilden, 1965). For example, the lack of singing perches may prevent some birds from colonizing an otherwise suitable area. The introduction of perches can mean the colonization of that area. For example, when telephone lines were strung across a treeless heath, tree pipits, birds that require an elevated singing perch, moved into the area. Woodcock will not utilize a singing ground unless it allows sufficient room for flight (Sheldon, 1967). A small opening surrounded by tall trees is not suitable, but an opening of the same size surrounded by low shrubs is.

An adequate nesting site is another requirement. Animals require sufficient shelter to protect parents and young against enemies and adverse weather. Selection of small island sites, such as muskrat houses, by geese provides protection against predators. Hole-nesting animals require suitable cavities or substrate in which they can construct such cavities. In areas where such sites are absent, populations of birds (Hartmann, 1956) and squirrels (Burger, 1969) can be increased dramatically by providing nesting boxes and den boxes.

There may be some relationship between food and habitat selection, although it is rarely a determinant. Some predatory birds nest only where food is abundant, which may account for local fluctuations in populations. Among such birds are owls, jaegers, and hawks. Perhaps the extension of the breeding range of the evening grosbeak has been encouraged by the widespread winter feeding of the birds south of their natural breeding range.

Once one or several animals have settled in an area, others may be stimulated to do likewise. Among colonial and semicolonial birds, the attractiveness of the colony is highly important. For herring gulls nesting for the first time, this is an important aspect in habitat selection (Dorst, 1958). Other birds are attracted to areas settled by different species. Thus the tufted duck (Aythya fuligata) has a strong social attraction to gull colonies. In fact this duck will not nest on small islets where gulls do not nest, and high-density populations have never been found where nesting gulls are few (Hildén, 1964). On the other hand, the presence of some animals may inhibit others from occupying otherwise suitable areas. Most pronounced in this respect is the presence of human activity.

Habitat selection in two subspecies of deer

mice, *Peromyscus maniculatus gracilis,* restricted to hardwood forests and brushy habitats, and *P. m. bairdi,* restricted to open field habitats, has been studied experimentally by Harris (1952) and Wecker (1963). Harris presented artificial forest and grassland habitats to the two subspecies, and found that both selected the appropriate artificial habitat. The physical conditions other than vegetation were essentially uniform throughout the experimental room, so he concluded the mice were reacting to the form of artificial vegetation. And because laboratory bred *Peromyscus,* that had no previous experience with either habitat, artificial or natural, exhibited a preference for the type of habitat they normally occupied in the wild state, he concluded that habitat selection was basically genetic.

The problem was probed more deeply by Wecker who investigated the role of early experience in determining the habitat preference of prairie deer mice *(P. m. bairdi)* under natural conditions. He used both wild stock and descendents of Harris's laboratory-bred mice 10 to 20 generations removed from any experience in the natural environment. Wecker demonstrated that the choice of fields by the wild strain prairie deer mice is reinforced by early experience in the field, but such experience is not essential. Early life in wooded and laboratory environments did not reduce affinity for the field habitat. The laboratory strain of prairie deer mice showed reduced affinity for the field habitat; early exposure to the field was necessary for the mice to develop a strong preference for the field. However, exposure to woods did not lead to a corresponding preference for woods. Wecker concluded that while habitat imprinting functioned in the mice, they possessed an inherent predisposition for the grassland habitat.

Habitat selection is not extremely rigid. Most species exhibit some plasticity; otherwise these animals would not spread into abnormal habitats. The ability or trait of some members of a particular species to select habitats that deviate from that of their companions must exist on both a phenotypic and genetic level. Hinde (1959) suggests that some animals may accept a normally inadequate habitat if some supernormal releaser, such as nest boxes, exists in the local environment. Plasticity in habitat selection is well illustrated by such birds as the chimney swift, which chooses chimneys over hollow trees as nest sites, and the magnolia warbler, which although partial to hemlock and spruce forests, is a typical inhabitant of the drier oak forests of West Virginia.

## Animal communication

The social organization of a population and in part the fitness of individuals depend upon the ability of individuals to communicate in some fashion with one another. Communication is the sending of a signal—visual, auditory, or chemical—by one individual and its reception by another. In the process the behavior of the sender influences the behavior of the receiver (Table 12-1). In some instances the behavior patterns involve signals changed from other adaptative functions to ones of signaling. In other instances certain movements which originally had a direct function evolved into ritual movements that function efficiently as social signals. For such signals to evolve, be they display, songs and calls, or scent, they must benefit the signaler. The signals may benefit or harm the receiver depending upon the social context.

### VISUAL COMMUNICATION

Because vision is the dominant means by which most animals relate to their environment, visual communication is probably the most widespread of all communication mechanisms (see Hailman, 1977). Visual communication involves displays: body postures and the use of certain morphological structures as patches of bright color. Often displays are ritualized, that is, they become evolutionarily specialized for communication (Brown, 1975). Such displays are involved in virtually every social context. Much visual communication is agonistic, involving tendencies of attack and escape (Moynihan, 1955b), and is associated with territoriality and dominance. Display advertises the presence of an individual and acts as an aggressive threat, as appeasement, as a defensive device and as an alarm mechanism. Visual display is involved

TABLE 12-1
*Basic types of information contained in animal communication signals*

| Type of information | Contents | Example expressed in words |
|---|---|---|
| Deictic | Signals or parts of signals which draw attention to an individual or object | *Look!* |
| Identification or indexical | Individual or class information, e.g. individual, sex, species, colony, age, relationship, etc. | I am *X*<br>I belong to *species X* |
| Spatial | Location in space, direction, distance | to the *right*<br>*90° to the right of the sun* |
| Response level | Information on the state of the emitter, on levels of probability of aggression, submission, sexual receptivity, etc., or on quantity or quality of exploitable resources such as food, nest sites, or shelter | The food source is *rich*<br>I am *very* angry |
| Temporal | Information on whether some event will occur, is occurring, or has already occurred | I will *soon* be ready |
| Event | Information on events taking place, such as the absence, presence, or need for food; presence of shelter, predators; or information on behavior which will take place | I *need food*<br>A *predator* is *nearby* |

*Note:* A signal may contain several types of information. Nonsignals (signs and symptoms) may contain the same kinds of information. The italicized words relate to the critical information involved.
*Source:* Adapted from D. Otte, 1974.

in pair formation, maintenance of pair bond, and parent-young relationships.

***Threat display.*** Under natural conditions hostile displays appear to be instigated by a direct response of one animal to the proximity of another individual. Agonistic behavior involves displays of threat that seem to have evolved because they conferred upon the animal the ability to achieve certain advantages without having to fight for them or risk physical injury. The most common of all ritualized forms of threat display is intimidation, of which most higher animals possess more than one kind. The primary function of intimidation display is to force the opponent to retreat or flee.

Threat display appears to involve attack and escape tendencies activated simultaneously. These tendencies appear to pass through stages in which the tendency to attack dominates and then gradually the tendency to escape dominates (if, for example, the threatening animal approaches another).

Threat displays have been derived from a number of sources, both hostile and nonhostile (Moynihan, 1955c). Of the hostile sources perhaps the most important are a whole series of unritualized intention movements indicating locomotion, retreat, or avoidance (Figure 12-5). Others undoubtedly were attack movements, such as pecking, flying or charging at

FIGURE 12-5
*Unritualized intention movement of flight in the hermit thrush. (After Dilger, 1956.)*

the opponent, or actual fighting. Out of this evolved those attack components—the color of bill or head, feather patterns, feather puffing, horns and antlers, bristling of body hair, and other morphological structures—that emphasize the visual conspicuousness of threat displays.

Agonistic displays vary from species to species, but among families and genera of animals a number of basic hostile displays appear, interspecific differences being a sort of "variations on a theme." Among passerine birds, in fact among most groups of birds, a common and perhaps universal aggressive posture is the head-forward threat (Figure 12-6). Another display commonly given is bill raising, the holding of the bill well above horizontal (Figure 12-7). It is probably derived from an upward flight intention movement and apparently serves as a distance-increasing display (N. Tinbergen, 1959).

Mammals, too, possess certain, basic aggressive displays. Among canids the baring of fangs, the curling of lips, erection of ears, and movement of tail all indicate degrees of aggressiveness (Figure 12-8). Aggressive behavior of black-tailed deer consists of several display components (Cowan and Geist, 1961). One is the crouch, a position in which hunching and partial flexing of the hind legs, shoulders, and elbow joints lower the body. The walk is slow, stiff, and stilted, with the head held in line with the body, the ears laid back (Figure 12-9). Other features are nose licking, circling, the snort, and finally the rush. The most extreme display consists of lowering the head to bring the antlers into contact with the object of aggression. Similar threat displays are found in the moose (Geist, 1963) and the barren-ground caribou (Pruitt, 1960).

*Appeasement display.* Appeasement displays, as one might expect, are almost as common as aggressive displays and are especially characteristic of encounters during the reproductive period. They serve to prevent attack without provoking escape, to reduce the strength of the opponent's attack tendencies, and to release other specific nonaggressive behavior (Moynihan, 1955b). Appeasement results when the escape tendency is much stronger than the attack. Like aggressive patterns, appeasement displays have derived

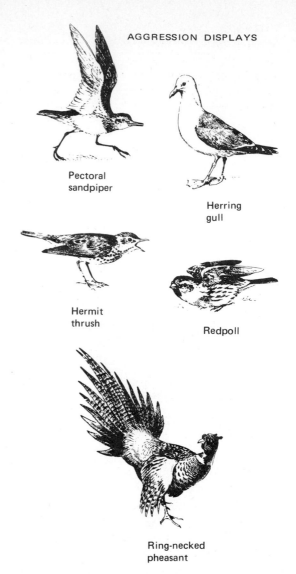

AGGRESSION DISPLAYS

Pectoral sandpiper

Herring gull

Hermit thrush

Redpoll

Ring-necked pheasant

**FIGURE 12-6**
*Agonistic displays among birds. Note the general similarity. (Pectoral sandpiper, after Hamilton, 1959; herring gull, based on photos in N. Tinbergen, 1953; hermit thrush, after Dilger, 1956; redpoll, after Dilger, 1960; ring-necked pheasant, after Collias and Taber, 1951.)*

from hostile sources, but the patterns involve most strongly the intention movements of escape. Usually appeasement behavior consists of withdrawal or avoidance movements, often specialized to hide offensive weapons, such as beaks, fangs, and any threat releasers. Appeasement displays, unlike intimidation, seldom have colors or structures evolved for this use alone.

**408**

A common appeasement display is the submissive pose, varying in appearance but still possessing some basic features in common (Figure 12-10): among birds, the lowered or in-drawn head, the crouched position, the fluffing or ruffling of feathers; among canids, the ears laid back, the eyes looking down, the tail curled between the hind legs (see Figure 12-8); among rabbits, a crouched position with ears laid back, the head pulled in toward the body, and the tail depressed (Marsden and Holler, 1964). Some gulls, such as the black-headed gull, turn their heads away from the opponent (head flagging) to hide the threat stimulus of the red bill and black face (Tinbergen and Moynihan, 1952). Occasionally ap-

(a)

(b)

**FIGURE 12-7**
*Bill raising, a courtship and aggressive display common to many birds.*

**FIGURE 12-8**
*Aggressive and submissive expressions in the American wolf, typical of canids. (Suggested by illustration in Schenkel, 1948.)*

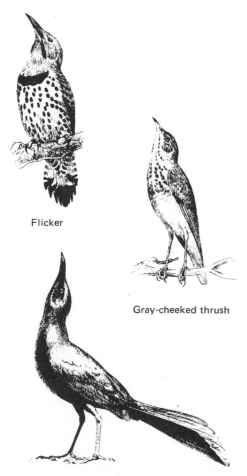

Flicker

Gray-cheeked thrush

Boat-tailed grackle

**FIGURE 12-9**
*Aggressive display in the black-tailed deer* (Odocoileus hemionus sitkensis). *The illustration shows details of the head during the snort that occurs when the buck is circling in the crouch position. Note the widely opened preorbital gland, curled upper lip, and bulged neck muscles. The snort is a sibilant expulsion of air through the closed nostrils, causing them to vibrate. (After Cowan and Geist, 1961.)*

**409**

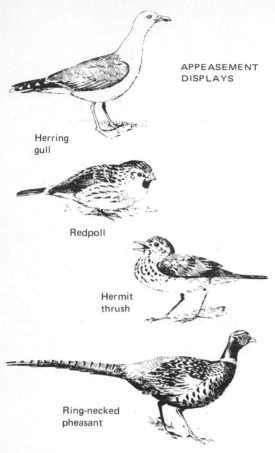

APPEASEMENT
DISPLAYS

Herring
gull

Redpoll

Hermit
thrush

Ring-necked
pheasant

*FIGURE 12-10*
*Appeasement displays among birds. (Herring*
*gull, based on photos and drawings in N.*
*Tinbergen, 1953; Moynihan, 1955b; redpoll,*
*after Dilger, 1956; ring-necked pheasant,*
*after Collias and Taber, 1951.)*

peasement displays may be superimposed on threat, and at times attack and escape may integrate, as they do in the "anxiety upright" of the gull (Figure 12-10). Although basically a threat display, it seldom elicits more than a low escape drive in the opponent.

*Courtship and reproductive behavior.* In the bright cold sunlight of a late winter day a flock of mallards is active on the open water of a pond. Suddenly some rear up out of the water, and some appear to be charging at others. Closer observation reveals a pattern to the activity. A number of green-headed males are swimming with their heads drawn in, the feathers ruffled, the body shaking repeatedly. As tension increases, a few drakes rear up out of the water and flick their head forward. On

occasions a drake rears up, arches his head forward, and rakes his bill across the water. Then with his bill pressed to his breast, he slowly sinks back to the water. At times this display may be accompanied by a low courtship call or followed by still another display, in which the male throws his head back in an arched position and jerks it abruptly upward. As the drake turns toward the female, he erects and spreads his tail feathers vertically and lifts his wing coverts to expose the irridescent metallic purple speculum. Then the drake lowers his head, stretches the head and neck forward just above the surface of the water and swims in rapid circles about the female. The female in turn is completely passive. The brown-feathered hens follow a mate or an intended one. With neck arched and head pointed toward the water, the hen moves her head back and forth from front to side away from the drake and toward the females. Often this display is accompanied by short dashes of attack. This courtship display is typical of the mallard (Figure 12-11). Other surface ducks have similar behavior patterns, but with a number of variations, omissions, or additions to the repertoire.

The ducks represent courtship behavior of a sexually dimorphic and polygamous group of animals. A somewhat different pattern is common to animals holding territories and to those in which sexual dimorphism is lacking. The female is attracted to the male by song or some other type of sound production, as in birds, frogs, and some insects, or by appearance, as in the three-spined stickleback. The male, in turn, sees an animal of the same species and reacts aggressively. If the intruder happens to be a male or an unreceptive female, the animal flees or threatens back. In the latter situation, a fight, even if a mock one, develops. If the intruder happens to be a receptive female, she remains. She may exhibit some hostility, but eventually she adopts a submissive pose, which tends to reduce or inhibit the male's aggressiveness. When the male's behavior becomes less aggressive, pair formation is accomplished. As times goes on, more generalized sexual elements enter courtship behavior. In contrast, females of some species become more aggressive after pair formation and may dominate the male. The

**410**

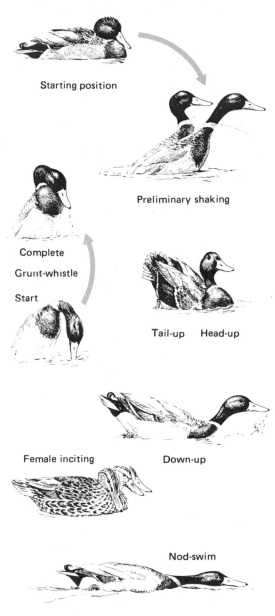

Starting position

Preliminary shaking

Complete

Grunt-whistle

Start

Tail-up    Head-up

Female inciting    Down-up

Nod-swim

*FIGURE 12-11*
*Courtship display of the mallard. Some of*
*these displays are common to many of the*
*river ducks.*

male's first reaction in courtship behavior then is to attack; the initial reaction of the female is to appease (escape) the male and elicit further courtship. The male in turn must suppress the escape tendencies of the female. Thus all courtship display contains elements of agonistic behavior (attack and escape) and sexual behavior. Progressive changes in the behavior of male and female toward each other depend in part on the stimulus situations presented by the partner: morphological, behavioral, vocal.

*Function of courtship display.* The function of courtship display is to attract the female to the male or vice versa. Once oriented to each other, neither may react to the partner's courtship until certain aggressive tendencies have been appeased or the reluctance of the male or female has been overcome. Later in courtship the behavior patterns are involved in the proper synchronization of the mating act, as exemplified by the courtship behavior of the queen butterfly (see Figure 12-2). Finally courtship displays serve to assure that males mate with the females of their own species and tend to reduce the error of mating with the wrong species. Thus the displays act as an isolating mechanism.

These functions—orientation, synchronization, persuasion, appeasement, and reproductive isolation—are achieved by species-specific signals or releasers. In fact the most highly developed specific signals and behavior have evolved around courtship, where errors would be most disadvantageous to the species.

Some basic patterns may be common to a group of animals (Table 12-2). Among nine species of surface ducks *(Anas)* sympatric to western North America eight possess many homologous displays, but there is a striking difference in male plumages. Some displays are almost universal among passerine birds: bill raising, feather sleeking, feather raising, lowering the bill, vibrating the wing or wings (Andrew, 1961).

Interspecific differences so necessary to reduce the chance of interspecific pairings set one species apart from the other. These include (1) differences in the color of plumage, skin, or other body structure, (2) variations in display movements, involving differences in relative strengths of tendencies to attack, flee, or behave sexually toward mates, (3) frequency in occurrence of several behavior patterns and their intensity, and (4) production of sound (Hinde, 1959).

In closely related sympatric species strong sexual dimorphism evolves in part because of

**411**

TABLE 12-2
*Comparison of presumably homologous behavior patterns in the*
*mergansers* (Mergus) *and goldeneyes* (Bucephala)

| | Buffle head | Common golden-eye | Barrow's golden-eye | Hooded mer-ganser | Smew | Red-breasted mer-ganser | Common mer-ganser |
|---|---|---|---|---|---|---|---|
| **MALE COURTSHIP** | | | | | | | |
| Upward stretch | × | × | × | × | × | × | × |
| Wing flapping | X | × | × | × | × | × | × |
| Crest raising | X | × | × | X | × | × | × |
| Head throw | — | × | × | × | × | — | — |
| Tail cocking | — | — | — | × | × | × | × |
| **FEMALE COURTSHIP** | | | | | | | |
| Inciting | × | × | × | × | × | × | × |
| **COPULATORY BEHAVIOR** | | | | | | | |
| Drinking (♂) | ? | × | × | × | × | × | × |
| Drinking (♀) | ? | × | × | × | × | × | × |
| Female prone | × | × | × | × | × | × | × |
| Upward stretch (♂) | — | — | — | × | × | × | × |
| Preen dorsally (♂) | × | — | — | — | × | × | × |
| Water twitch (♂) | × | × | × | × | — | — | ? |
| Preen behind wing (♂) | — | × | × | × | ? | — | — |
| Steaming to ♀ | — | × | × | × | — | — | — |
| Flick of wings (♂) | × | × | × | × | × | ? | — |
| Steaming from ♀ | — | × | × | × | — | — | — |

*Note:* An × indicates that the behavior pattern was observed; a large × indicates that the behavior pattern was exceptionally well developed.
*Source:* Johnsgard, 1961.

selective pressures brought to bear by the female: the choice of her own species based on innate and learned responses. Males among ducks, for example, will court females of any species. Females, on the other hand, in most situations will mate only with males of their own species, distinguished by color patterns and display movements. Competition for mates also increases the development of signal characters, particularly among polygamous species in which the males with the most pronounced or effective releasers are most successful in attracting a mate or mates. In such instances females choose males for the development of secondary sexual characteristics.

*Source of courtship displays.* Because many elements of agonistic behavior enter into courtship and because courtship is so closely related to other aspects of an animal's life, the fact that courtship displays have evolved from other behavior patterns is not surprising. Involved are intention movements, elements of aggressive behavior, ritualized dis-

placement or redirection activities, infantile behavior, and nest building. In fact among some species of birds, such as the wood thrush (Dilger, 1956) and redstart (Ficken, 1963), there are no special displays associated solely with pair formation. All displays involved are agonistic and appear in other situations. The same is true in many mammals.

The courtship display of the male caribou is simply a modified threat pose (Pruitt, 1960). The extended muzzle, grunts, and swift advance express to the doe the vigor of sexual drive in the buck. If the doe is not physiologically receptive, the buck's actions are "interpreted" as antagonistic, and the doe flees; if the doe is sexually receptive, the display does not release flight. On the other hand, in mountain sheep the ram is much more cautious, and the ewe employs the aggressive behavior (Geist, 1971). Her attacks and butting suggest the behavior of a smaller subordinate ram toward a larger one (Figure 12-12), except that the female dashes away whereas the young

**FIGURE 12-12**
*Courtship display of the estrous bighorn ewe. The ewe makes use of contact and aggression as she forcefully butts the ram on the shoulder. Through it all the ram maintains a stiff posture. (After V. Geist, 1971.)*

ram does not. Possibly by using aggressive behavior in her courtship, the ewe arouses sexual behavior in the ram by arousing aggressive impulses. This switch from female to male behavior at estrus is also common among mountain goats and old world deer (Geist, 1971).

One of the functions of courtship is to release the sexual response in the partner through specific signals. The male stimulates the female and in turn the female stimulates the male. The courtship is mutual. When an animal, especially the male, is under strong sexual impulses yet cannot mate because the partner has not given the final signal to release the mating act, a conflict situation arises, which finds its outlet in irrelevant activities.

Some irrelevant activities act as releasers in courtship just as they do in aggressive behavior. Thwarting of the sexual drive in some birds results in preening, which in ducks has become ritualized. The movements or components have become exaggerated to increase their efficiency as a signal. Brightly colored structures, such as the blue wing speculum of the mallard duck, are so located that they are conspicuous during the ritualized displacement activity. The mandarin duck, related to the North American wood duck, strokes a specialized secondary feather, which is extremely broad and bright orange in color, in contrast to the normal, narrow, dark green

secondary feathers. By such ritualized preening males call attention to the colored feather or feathers, thus making the movement more conspicuous and more species-specific. Ritualization of movements removes them further away from the original source; they become increasingly independent, which obscures the source or instinct from which they were borrowed.

*Maintenance of pair bond.* Once pair formation—the period from the initial meeting of the male until a bond is formed—has been accomplished, other behavioral elements more sexual in nature appear. Male birds may fluff the feathers of the scapulars, rump, and head, spread the tail, drop and wave the wings, and bow (Figures 12-13 and 12-14). These displays, so common among passerine birds in one form or another, probably are ritualized intention movements of flight, indicating either strong sexual tendencies to fly up and mount the female or tendencies to flee from the female. Common among paired gulls is head bobbing, regarded as solicitation or precopulatory display (N. Tinbergen, 1960).

Symbolic nesting and symbolic nest site selection are common in courtship behavior of birds, from grebes and cormorants to songbirds. Symbolic nesting involves either the manipulation of nesting material by that member of the pair, usually the male, who does not ordinarily help in nest construction or the unnecessary handling of nesting material by the other member (or both, if male and female together build the nest) prior to actual building.

Begging food by the female and reciprocal feeding of the female by the male are part of

**FIGURE 12-13**
*Courtship display of the male red-winged blackbird and invitational display of the female. (Drawn from photos in Nero, 1956.)*

Red-wing

**413**

King rail

**FIGURE 12-14**
*Invitational display of the king rail. This posture is assumed by a mated male upon close approach by the female. The bird displays the tail, points the bill downward, and slowly swings the bill from side to side. (After Meanley, 1957.)*

the sexual behavior patterns of many species of birds. Begging on the part of the female is regarded as a kind of infantilism, the reappearance of her behavior as a chick. Courtship feeding may serve to reduce or inhibit the aggressive behavior of the male and release sexual behavior, for courtship feeding often precedes or accompanies coition. It may also stimulate and maintain the pair bond. Ethologists still speculate on the origin of this ritualized feeding, but it appears most likely to have arisen in part from anticipatory feeding in which the male prematurely brings to the nest food intended for nestlings not yet hatched. Feeding on the part of the male belongs to parental behavior which normally appears long after sexual behavior wanes.

*Distraction display.* When danger threatens the nest or young of birds the adult may attempt to draw the attention of the intruder to itself. To do this some simulate injury, a distraction display in which the parent flutters along the ground, always keeping one jump ahead of the enemy. If the predator follows, the bird continues the act until the predator is no longer near the nest. If the enemy remains, the stimulus is increased, and the bird returns to repeat the display until the response tires.

Distraction displays may grade into aggres-

sive displays. The killdeer, for example, may run toward such intruders as a cow or horse, fluff its feathers, trail its wings, then strike the animal in the face. Other birds such as owls may remain on the nest, raise or spread the wings, and snap the bill. Some reptiles snap and hiss at intruders. Mammals become highly aggressive and lower the head, utter snorts or growls, snarl, bare the fangs, and so on. Other animals may change color as do some fish, or increase in size, as do some frogs, toads, and snakes, in an attempt to intimidate the enemy.

*Flight display.* Erratic flight display is common to many taxonomic groups such as mammals, birds, moths, grasshoppers, and cladoceran crustaceans. Flight, taking the forms of zigzagging, looping, spinning, wild bouncing, or leaping (Figure 12-15), disorients the predator's attack. Flight behavior among some species and groups may involve flashing of colored surfaces, such as the white tail of the white-tailed deer, the white tail feathers of birds, and the changing colors of fish. Such behavior may not only disorient the predator,

**FIGURE 12-15**
*Stotting, a form of flight behavior and predator avoidance in Thompson's gazelle. In stotting, the gazelle holds its front and hind legs stiffly stretched downward and bounces, springing from the pastern joints. The high level of flight excitation is caused by an approach of a predator. Stotting apparently conserves energy over other forms of flight behavior, and at the same time a number of gazelles stotting at different intervals would tend to confuse the predator. (Left, normal stotting gait; middle, paddling with hind legs in extremely high stotting; right, landing from high stotting.) (After Walther, 1969.)*

but may also stimulate escape tendencies in the predator.

Flight reaction is contagious. If one animal flees, others may be startled into doing the same. This results in stampedes among ungulates and massed flight among waterfowl. If danger is from above, such as from a hawk, bird flocks tend to draw close together to form a dense flock and perform sharp, swift turns with coordination and precision. Such behavior makes taking of an individual much more difficult.

### CHEMICAL COMMUNICATION

Communication by scent is widespread throughout the animal kingdom. It is accomplished through *pheromones,* scents or chemical releasers secreted from exocrine glands as liquids, transmitted as liquids or gases, and smelled or tasted by others. The pheromones, which function in intraspecific communication, may release an immediate behavioral response or they may alter the physiology of the receiving organism, usually through the endocrine system, stimulating a new set of behavioral patterns, particularly related to reproductive behavior.

Pheromones may be simple or complex chemical substances. Among insects pheromones are either a single component or a simple mixture, while among vertebrates they are complex chemicals which tend to be personal odors (Wilson, 1971).

Chemical communication has the advantage of being persistent in absence of its sender. Such signals, however, serve to transmit general messages only. They are important in communicating such information as the identity of an individual, its social rank and sex, territorial boundaries and trails, and location of food. They serve to stimulate sexual activity, to alert individuals to danger (alarm), and to defend against enemies (defense).

Among many insects, particularly moths, pheromones function largely as sex attractants. Virgin female moths ready to mate release a sex pheromone from glands at the tip of the abdomen. Released to the air, the attractant is carried by the wind, the distance dependent upon meteorological conditions. Males that come in contact with the pheromone locate "calling" females by flying upwind until they locate the source. Because of the weight of the female ovaries, it is energetically more efficient for the female to remain stationary and allow the male to find her. Because they are so effective, sex attractants have been used successfully as lures for males in insect pest control programs.

In some species pheromones serve as alarms, stimulating a flight or fight reactions. Some minnows release alarm substances in the water when tissue has been damaged. But alarm pheromones are most highly developed in such social insects as ants that release the substance as glandular secretions. (For a detailed review, see Whittaker and Feeney, 1971.)

Chemical signals are also useful in interspecific communication. Known as *allomones* they are valuable in situations involving defense. Many animals, including insects, other arthropods such as millipedes, amphibians, and mammals such as shrews and skunks, may eject or secrete defensive sprays and secretions highly irritating to the enemy.

Among some species the chemical communication repertoire involves different secretions and different glands. This is especially true among social insects and mammals, both of which rely extensively on chemical communication. For example, North American deer are equipped with a number of suboriferous glands and sebaceous glands that secrete pheromones (Figure 12-16). Each gland secretes its own distinctive scent (Muller-Schwarze, 1971). Scent from the tarsal gland is important in the mutual recognition of sex, age, and individuals. The metatarsal glands, located on the outside of the hind legs, secrete a scent associated with alarm and fear reaction. Scent from glands in the forehead is used for marking home range. And during aggressive posturing the black-tailed deer opens fully the oriface of the preorbital gland. Female urine attracts males. And rub-urinating in which deer of either sex and of all ages occasionally run their hocks together while urinating on them, serves as a distress signal in fawns and as a threat in adult males and females.

### AUDITORY COMMUNICATION

Auditory or sound communication is well developed in a diverse array of animals includ-

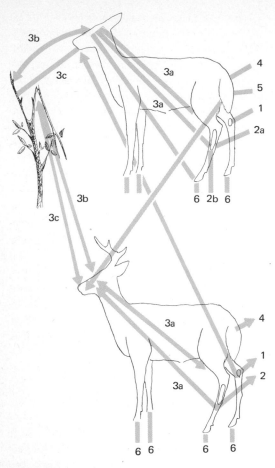

**FIGURE 12-16**
*Chemical communication through scent in North American deer. Various odors and secretions from subcutaneous glands released to the air and rubbed on ground and twigs have significant meanings in the world of the deer. Scents of the tarsal organ (1), metatarsal gland (2a), tail (4), and urine (5) are transmitted through the air. While the deer is reclining, the metatarsal gland touches the ground (2b). The deer rubs its hind legs over its forehead (3a), and rubs its forehead over dry twigs (3b). Marked twigs are sniffed and licked (3c). Interdigital glands leave scent on ground (6). (From Muller-Schwarze, 1971.)*

ing a few spiders and fiddler crabs; insects, especially grasshoppers, crickets, katydids, cicadas, flies, bees, mosquitos, and a few moths; many species of fish; frogs, toads, and a few salamanders; and crocodiles, geckos, and some turtles. Sound production is universal among birds and nearly so among mammals. Among mammals it is most highly developed

in the primates, although the carnivores, notably the canids and felids, have a rich repertoire of vocalizations.

Sound-producing mechanisms are often associated with non–sound-producing morphological structures. In some groups sound production evolved from respiratory structures. These include the larynx of mammals, the syrinx of birds, the resonating air sacs of frogs, toads, and some birds. Among insects sound production results from rubbing parts of the exoskeleton together or *stridulation*. Certain structures on the exoskeleton evolved that increase sound production. For example, crickets possess a sound-producing structure on the tegmina or modified wings which consists of two parts, a scraper or file and a vibrator or drum. Some birds possess secondary specialization of wing or tail feathers for sound production in aerial displays. Examples are woodcock *(Philohela)* and snipe *(Capella)*. Other animals produce sounds by beating the substrate. These include fiddler crabs, rabbits, beaver, and woodpeckers.

Type of sound production and of auditory reception influences the way sound is used. Insects are tone deaf and produce sound in pulses that vary little in frequency. A similar situation exists in frogs and toads, although they produce sound vocally. Sound production in both insects and anurans is influenced by temperature. Temperature changes affect the frequency of insect sound, and while frequency of sound in frogs and toads is more or less independent of temperature, the pulse rate is increased by an increase in temperature. In contrast, birds and mammals utilize a rapid change in frequency, resulting in complex sounds. Birds are unique in that they can produce two separate frequencies simultaneously. This ability increases their repertoire of both songs and calls. Although songs and calls are difficult to define and can be separated only by the behavioral context in which they are used, songs generally are longer and more complex than calls. Songs are associated with the breeding season and are often under hormonal control. Calls are short and simple and are given in a variety of social contexts particularly those involving contacts between members of a pair, family, etc. (see Thielcke, 1976).

The role of vocalizations depends upon the

social context in which they are used. The most conspicuous auditory efforts relate to courtship and aggressive behavior. The songs of birds and the calls of insects and anurans serve a dual role of attracting a potential mate and repelling rivals. Some species have vocalizations that are aggressive in nature and others that serve to attract a mate. Examples are the grasshopper sparrow (*Ammodramus savannarum*) (Smith, 1959) and its meadow associate the meadow grasshopper (*Tibicen* spp.) (R. D. Alexander, 1960).

Vocalizations serve as sexual recognition. Birds, frogs, and calling insects are among the animal groups where the ability to discriminate between the calls of their own species and those of others is widespread. That both wide and subtle differences exist between songs is readily apparent from spectrographs of songs (see Thielcke, 1976). These differences are important in maintaining reproductive isolation and in enabling close-range orientation of male to female, especially where a number of similar species assemble on a common breeding site as do frogs and toads. These subtle differences also enable birds to recognize neighboring individuals of their own species.

Sound production plays an important role in stimulation of the sexes during mating activity. Females of many species of birds give a solicitation call or note that releases copulatory behavior in the male. Males of some species, notably the blackbirds, likewise possess solicitation notes that indicate their readiness to mate (Nero, 1956a). When sexually responsive male and female crickets and cicadas are at close range, the male produces specialized courtship sounds, which stimulate the female to move forward and walk up the back of the male into position to receive the spermatophore (R. D. Alexander, 1960, 1961). During courtship display the male cod grunts to stimulate the female to display and to swim upward and spawn (Brawn, 1961).

More general and less species specific are warning and alarm calls. Although not given as conscious alarms to other animals, warning calls may have that effect. A warning call from a bird who has discovered some common enemy is sufficient to bring many birds to the scene, to fly about, and to scold and otherwise mob the animal. Such calls are cries of fear,

and the intensity and nature of the call may vary with the intensity of fear. Among birds in particular and rodents warning or alarm notes may vary either in intensity or in kind according to the situation. The tendency of birds to give a warning call increases as the reproductive cycle passes through stages of nest building, incubation, and brooding young. The intensity of response increases as the enemy comes nearer and nearer the young.

Like other behavioral traits sound production has been favored by natural selection. Its influence on the behavior of other animals has improved the fitness of the producer. Once a signal evolves the stage is set for it to give rise to several functionally different signals, the rate of change depending upon the signal function (Otte, 1975). The evolution of sound signals has been accompanied by the evolution of morphological structures which improve the ability of the animal to produce a diversity of communicative sounds.

## Social dominance and spacing

Social interactions among individuals in a population involve agonistic or aggressive behavior. Such behavior is the basis of social organization characterized in a very general way by dominance and territoriality. The difference between the two involves not only individual interactions but also division and utilization of space. Social organization based on individual distance (distance from another individual which provokes aggressive or avoidance behavior) and dominance relationships among members of a social unit result in space being shared among individuals. Social organization which results in space being divided among and occupied exclusively by a social unit with a defended boundary is territoriality.

### SOCIAL DOMINANCE

A tendency exists among many animals to band together in flocks, herds, schools, or loose colonies. Many of these animal groups have some form of social organization based on intraspecific aggressiveness and intolerance and on the dominance of one individual over another. Two opposing forces are at work at the same time: mutual attraction versus social intolerance, a negative reaction against crowding. Each individual occupies a position in the group based on dominance and submissiveness.

In its simplest form there is an alpha individual which is dominant over all others, a beta individual which is dominant over all but the alpha, and so on to the omega which is usually totally subordinate. This relationship was first described by Schjelderup-Ebbe (1922) for the domestic chicken. It is a straight-line or linear peck order, so named because pecking follows dominance, that is, birds peck others of lower dominance or rank (Figure 12-17). Even within peck order complexities may exist, such as triangular or nonlinear hierarchies. These are triplets of individuals whose pair relations are such that the first individual is dominant over the second, the second is dominant over the third, and the third is dominant over the first. In such a situation, an individual of a lower rank can peck an individual of higher rank.

In some groups peck order is replaced by peck dominance, in which social rank is not absolutely fixed. Threats and pecks are dealt by both members during encounters, and the individual that pecks the most is regarded as dominant. The position of the individual in the social hierarchy may be influenced by levels of male hormone, strength, size, weight, maturity, previous fighting experience, previous social rank, injury, fatigue, close associates, and environmental conditions.

In flocks made up of both sexes separate hierarchies may exist for males and females with dominance of males over females. Such peck orders are characteristic of flocks of red crossbills and ring-necked pheasants. Tordoff (1954) found that the top-ranking male in a crossbill flock was the most aggressive male (Tables 12-3 and 12-4), yet the low-ranking male was the most active in dominating females, followed closely by the top-ranking male. Collias and Taber (1951) observed a similar situation in a flock of ring-necked pheasants in which the males dominated the females, who had their own peck order. But at the onset of the breeding season, the dominance of cock over hen declined.

Dominance among white-tailed deer in a

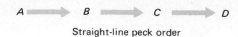

Straight-line peck order

Triangular peck order

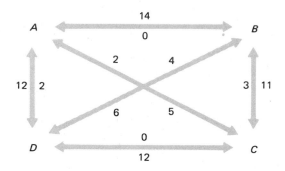

Triangular peck order

**FIGURE 12-17**
*Some examples of peck orders. (Upper) A straight-line peck order in which one animal is dominant over the animal below it. A is the alpha organism, D is the omega. (Middle) Triangular peck order in which A is dominant over B, B is dominant over C, yet C is dominant over A. (Lower) A more complex example of a triangular peck order. The double arrows indicate encounters between individuals. The numbers represent the numbers of wins of one individual over another. For example, A is clearly dominant over B with 14 wins. B never dominated A. A is also dominant over D with 12 wins out of 14. C is also strongly dominant over D with 12 wins to 0 for D. B is also dominant over D with 6 wins to D's 4 wins. D is clearly the omega individual. Other interactions can be read in a similar manner.*

wintering yard in Wisconsin (Kabat et al., 1953) was one in which the adult bucks dominated other deer, and the does dominated the fawns. Within this scheme the larger animals,

## TABLE 12-3
*Peck order in male red crossbills, based on 404 encounters*

|  | Dominates (December 31– January 12) | | | | | Dominates (January 14– March 25) | | | | |
|---|---|---|---|---|---|---|---|---|---|---|
| W | — | B | A | G | O | — | B | A | G | O |
| B |  | — | A | G | O |  | — | A | G | O |
| A |  |  | — | G | O |  |  | — | G | O |
| G | W |  |  | — | O |  |  |  | — | O |
| O |  |  |  |  | — |  |  |  |  | — |

*Source:* Tordoff, 1954.

## TABLE 12-4
*Social hierarchy of a mixed group of caged male and female house finches in winter*

|  | R | L | GB | WG | GR | WP | Y | O |
|---|---|---|---|---|---|---|---|---|
| R |  | 1 | 2 | 0 | 4 | 2 | 3 | 2 |
| L |  |  | 2 | 1 | 2 | 1 | 1 | 1 |
| GB |  |  |  | 7 | 7 | 4 | 6 | 10 |
| WG | 2 |  |  |  | 6 | 8 | 8 | 5 |
| GR |  |  | 1 |  |  | 3 | 2 | 5 |
| WP |  |  |  |  |  |  | 8 | 1 |
| Y |  |  |  |  |  |  |  | 3 |
| O |  |  |  |  |  | 9 |  |  |

*Note:* Read from left to right in horizontal rows. The number indicates the number of wins by the bird listed in the first column over other individuals in the group.
*Source:* W. L. Thompson, 1960.

whether buck, doe, or fawn, dominated the smaller animals. Wintering flocks of juncos exhibit no such dominance based on sex (Sabine, 1959). Females dominate the males in some species, such as the redstart (Ficken, 1963), while in other species, such as the snow goose (Jenkins, 1944) and jackdaws (Lorenz, 1931), the females appear to be equal to the male, especially where family ties are strong.

Once social hierarchies are well established within a group, newcomers and subdominant individuals rise in rank with great difficulty. Strangers attempting to join the group either are rejected or, as in the valley quail, are relegated to the bottom of the social order (Guhl and Allee, 1944). New birds in a wild covey remain a few yards behind the main group and do not mingle until after a period of acquaintanceship. Some individuals newly entering

the flock then may rise rather rapidly up the hierarchy, whereas others remain unassimilated. Several such individuals then may associate together and by coordinating their behavior maintain a mutual social rank against other members, as do some individuals in howler monkey society (Carpenter, 1934). Removal or injury of the dominant individual causes a scramble for the alpha position.

Rise in the hierarchy often is related to breeding and sexual activity and hormones. This is particularly true among those species that remain in flocks throughout the year. Breeding condition is important in determining social rank. Male house finches that occupied the bottom rung of the social ladder during the winter but came into breeding condition first rose near to the top of the hierarchy during the reproductive season (Thompson, 1960). The top male, however, was still dominated by his mate. Ring-necked pheasant cocks that occupied the dominant position in winter flocks usually were the first to crow and to establish individual territories in the spring. But subdominant males that came into breeding condition earlier rose higher in the hierarchy (Collias and Taber, 1951). Rise in hierarchy, then, appears to be related to a rise in male hormones.

Mating, too, improves the rank of lower individuals. Both members of a pair of house finches rise in hierarchy, but not always to the same level (Thompson, 1960). In the jackdaw, however, one member of the mated pair rises at the time of pair formation to the same level as the other member (Lorenz, 1931).

A rise in male hormone levels is not the only cause of increased aggressiveness. Shortages of food, space, and mates, among other things, will increase competition and hostility between individuals. The dominant individual has first choice of food, shelter, and space, and subdominant individuals may obtain less than the despots. When shortages are severe, the low-ranking individual may be forced to wait until all others have fed, to take the leavings, if any, to face starvation, or to leave the area.

### TERRITORIALITY

Territory is a defended, more or less fixed and exclusive area maintained by an individual or a social unit occupying it. Territory holders use aggressive behavior including advertisement by scent or song, threat and attack, to exclude intruders. Territory results in a uniform pattern of dispersion of individuals over an area (Figure 12-18). It is found among

**FIGURE 12-18**
*Territories of the grasshopper sparrow as determined by observations of banded birds. Dots indicate singing perches. Note changes between the two years among the same banded birds. (From R. L. Smith, 1963.)*

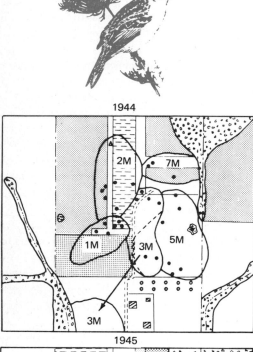

1944

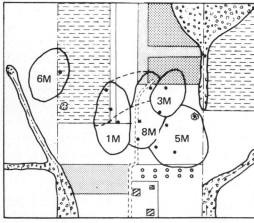

1945

(a)

birds, many fish and insects, and some mammals.

Territory should not be confused with home range. Home range is an area in which an animal normally lives and is not necessarily associated with any type of aggressive behavior (Figure 12-19). A home range may or may not be defended in part or in whole and may or may not overlap with those of other individuals of the same species. However, dominance may exist among individuals with overlapping home ranges and subordinate individuals tend to avoid contact with dominant individuals. Both use the same area by using different parts at different times.

Generally home ranges do not have a fixed boundary. Seldom is a home range rigid in its use, its size, and its establishment. It may be compact, continuous, or broken into two or more discontinuous parts reached by trails and runways. Irregularities in distribution of food and cover produce corresponding irregularities in home range and in frequency of animal visitation. The animal does not necessarily visit every part daily. Its movements may be restricted to runways or its activities may be concentrated in the most attractive parts or centers of activity.

Territories may be classified according to the activities that take place in them (Nice, 1941). One is the general purpose territory, common among songbirds and some mammals such as muskrats. All activities take place within territorial boundaries. Late in the breeding season and after it, territorial defense breaks down. A second type is mating and nesting territory with feeding done elsewhere. This is common among some hawks and black-capped chickadees. A third type is mating territory, exemplified by the leks of prairie grouse and the singing grounds of woodcock. A fourth type is a nesting territory in which only the nesting site is defended. Colonial birds such as puffins and swallows defend territories the size of which is determined by the distance the bird can strike from its nest with its beak. Some animals, particularly hummingbirds, defend only a food resource. This represents a feeding territory. Some birds may defend a roosting territory in situations where adequate roosting sites are scarce.

In many ways territoriality and social

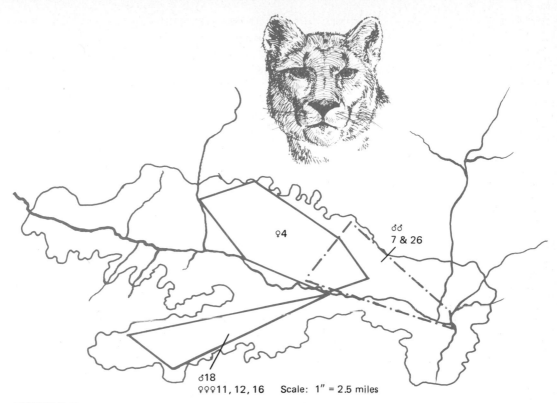

**FIGURE 12-19**
*Home ranges of female and male mountain lions in Idaho as determined by observations of marked animals. (After Hornocker, 1969.)*

Scale: 1″ = 2.5 miles

dominance represent degrees of manifestation of the same basic dominance behavioral pattern. It is difficult to draw a sharp line between the two. Depending upon the season and conditions of crowding territory can grade into social dominance. This behavior is perhaps best expressed in the behavior of the feral domestic fowl (McBridge et al., 1969). During the breeding season social behavior may range from extreme territorial to weak social hierarchy as illustrated by six classes of males. The first class, dominant territorial males, restrict the movements of all other males to the territory and have females as constant companions. The second class of males possess subordinate territories, small ones defended against even dominant neighbors. The dominant neighbor may be able to penetrate the periphery of the territory, but not over the whole extent. These males may have harem flocks. The third class, semiterritorial males, roost apart from the dominant males and make slight defense reactions against invasion by dominant males.

However, the territories of these males are within the territory of a dominant male. They may be attended by females, but not constantly. The fourth class, subordinante males, roost apart from dominant males. The areas they occupy are also usually occupied by females, but the birds do not possess any. The fifth class is a group of subordinate males who roost with a dominant male and do not possess females. The last class consists of the runts, who generally leave the territory. Where more than one male occupies an area, territorial succession is a matter of peck order position. When male domestic fowl are confined to smaller and smaller areas, dominant males adjust from territorial behavior to social dominance. When space is increased, the dominant males become territorial again (M. Schein, pers. comm.).

Another example is the dragonfly *Leucorrhini arubicunda*. Like other dragonflies, it maintains a territory along the edge of a pond. When the population density is relatively low and individuals are spaced 3 to 7 m apart, aggressive interaction is frequent. When density increases and individuals are spaced less than $1/2$ m apart, territoriality breaks down. As density increases, the level of territorial defense and attachment to site decreases (Pajunen, 1966).

## COLONIAL GROUPS

During the breeding season many animals typically are dispersed over an area as individuals and pairs. Other animals are highly clumped, nesting or breeding in colonies with individual territories restricted to small areas about the nest. Such colonial species congregate in relatively large groups in a limited area to breed, but feed and carry out other activities elsewhere. Examples are colonies of cliff and barn swallows, puffins and gannets on sea cliffs, and gulls and terns on small sea islands.

Selective pressures favoring colonial behavior could be limited nesting sites, predation, unpredictable food resources, or an interaction of any or all. In some instances scarcity of adequate breeding sites may have favored the exploitation of limited areas by large numbers of individuals and pairs. Because clumped distribution of animals would attract predators, selection would favor large numbers nesting together over small numbers. Large colonies would provide greater protection against intrusion by predators and decrease the odds that any individual nest of eggs or young would be taken. For example, Tenaza (1971) found that nests at the periphery of a nesting colony of Adelie penguins were more vulnerable to predation than those toward the center, because predators enter the colony from the edges. As the size of the colony increases, the percentage of nests on the periphery of the colony decreases and the percentage of successful predation decreases.

Most probably the greatest selective pressure favoring colonial nesting is the patchy distribution of food resources, which vary in location and abundance from year to year. By eliminating expenditure of energy for prolonged defense of territory (in which typically the food resource is uniformly abundant and predictable), colonial animals can direct their energies toward foraging some distance from the colony.

A comparative example can be found in two blackbirds, the red-winged and the tricolored (Orians, 1961). The red-winged blackbird is adapted for an adequate food supply, predictable from year to year and available for a sufficiently long time to feed young. It defends a territory in which the male's several mates can find sufficient food to feed the young. The tricolored blackbird, the male of which also acquires more than one mate, lives in large colonies. Males defend very small territories, and breeding begins simultaneously. Freed from energy demands of prolonged territorial defense, the male can accompany the female in foraging long distances from the colony for food. Unlike red-winged blackbirds, tricolor blackbirds depend upon an unpredictable food supply in the river valleys of California. The abundance of food depends upon the time and extent of spring floods and outbreaks of grassland and crop insects. Although an abundance of food may be available, the time and place of food availability cannot be predicted in advance. For this reason birds travel in nomadic flocks and delay their breeding until they discover an available rich supply of food. Thus selection has favored an opportunistic and synchronized nesting cycle in this species.

## COOPERATIVE SOCIAL GROUPS

Not all aggregations are based on social dominance. Some groups show no strong evidence of social rank among individuals and exhibit relatively strong group cooperation in intergroup aggressive situations.

The social structure in two animals, the harvester ant and the prairie dog, are excellent examples. Individual ants, possibly by imprinting, soon learn the odor of their own colony. Foraging ants returning to their own colony may stimulate some aggression on the part of their nest mates, but once their identity is clear, the ants are unmolested. But if an ant enters an alien colony, it is threatened, seized, and killed, for a strange taste or smell stimulates attack among ants in the colony. The colony acts as a unit, and conflicts between colonies may continue for days, leaving the ground littered with corpses.

Among vertebrates few social organizations are more fascinating than that of the prairie dog, aggregations of which are known as prairie dog towns. Some towns are subdivided by topography or vegetation into wards; further subdivisions reflect prairie dog behavior. Within each ward a group of prairie dogs is united into a cohesive, cooperative unit, known as a coterie, which defends a particular section of the ward against all trespassers (King, 1955). Territories of coteries, which cover less than an acre, may contain from 2 to some 30 members. Breeding coteries usually contain 1 male and 3 or 4 females; nonbreeding coteries may have all males, more males than females, or an equal number of both sexes. No social hierarchy exists, although one male, usually the most aggressive and the strongest defender of the territory, may dominate the rest. All coterie members use the same territory without conflict or threat, and social relations among most members are friendly and intimate, involving grooming, play activities, vocalizations, and the identification kiss—a recognition display in which each individual turns its head and opens its mouth to permit contact with the other. A coterie emphasizes its social unity by the defensive action of its members against invasion and by the inability of any member of the group to wander beyond the coterie territory without becoming involved in conflict (although members are ready to invade other coteries if the opportunity is present.) Members of the coterie do not drive away the young; instead the young are protected against the antagonism of neighboring coteries. Overpopulation in a coterie may force territorial expansion, in which the defending male of the adjacent coterie is vanquished and driven out; or it may result in social unrest and the eventual emigration of yearlings. Adults may leave perhaps to escape the demands of the pup or to seek more abundant food. Advantages of living in such a group are many, including a limited control over an area of the habitat, increased defense against predators, the prevention of overcrowding, results that can be achieved only by the combined activity of all individuals.

Prairie dogs exhibit little division of labor, but the most highly organized natural societies—those of ants, termites, and some bees and wasps—do. The insect society has one reproductive female, the queen. The queen, usually large in size, lays eggs from which other members of the group develop.

The workers among ants, bees, and wasps are sterile females; among termites workers are also sterile, but may be either male or female. In many colonies males, produced from unfertilized eggs, are rare. They serve only to fertilize the queen and having accomplished this either die or leave the group. The workers, as their name implies, perform the duties of their complicated society. Ants and termites have an additional caste, the soldiers; large and equipped with formidable mandibles, they defend the colony. Soldiers are males and females in the termites, males in the ants. Defense in bees and wasps, that lack a soldier caste, is a function of the worker.

All behavior in these societies is innate. There is no flexibility or variation in roles. In the societies with a soldier caste it is not the strongest individuals who assume the role of defense, but rather that the society produces a given number of soldier organisms. Although the population consists of physically separate individuals, the individual cannot survive if separated from the social group. Individuals are inextricably bound to the colony by behavior and physiology, more like the separate cells of an organism than like separate individuals.

## Parental behavior

Degree of parental care given to the young varies widely in the animal kingdom. Generally the greater the degree of parental care given, the lower is the fecundity.

Those invertebrates in which brood protection is highly developed lay relatively few eggs, and those that give eggs no protection whatsoever produce eggs in the millions (see Thorson, 1950, for example). Parental care is not highly developed among most invertebrates. Some retain eggs within the body until they hatch; others carry eggs externally. Invertebrate parental care is most highly developed in social ants, bees, and hunting wasps. The social insects provide all five functions of parental care: defense, food, sanitation, heat, and guidance.

Fish and amphibians either lay a few eggs and actively protect both them and the young or lay many eggs and give them no care at all.

Parental care is usually poorly developed among the amphibians, although a few salamanders remain with the eggs. Some frogs, notably the male midwife toad, carry eggs and subsequent young on their bodies and eventually place them in a suitable environment for further growth. Fish, likewise, may or may not care for eggs. The cod lays its eggs in the open sea; the trout constructs a gravel nest to ensure proper protection and ventilation of the eggs, but gives no care to the young. Other species, especially those that have highly developed courtship and mating patterns, build a nest, defend and aerate the eggs, and protect the young. This behavior is typical in the sticklebacks and catfish.

Internal fertilization and terrestrial reproduction is fully developed among reptiles. Relatively few eggs well supplied with yolk may be carried inside the mother's body until they hatch; or they may be placed in nests buried in the ground and given little subsequent care. Crocodiles, however, actively defend the nest and later the young for a considerable period of time.

It is among the homoiotherms that parental care reaches its highest development. Parental care is most complex among birds, since the young are hatched outside the body. Care must start with the nest, carry through the incubation of eggs and brooding and continue with the feeding of the young until they become independent. The female plays the major role in the care of eggs and young, but among some animals the male performs this function, and in others both sexes participate. All species defend the eggs, directly or indirectly, and many actively defend the young. Among the monotremes the mother plays the most significant role by carrying the young in the uterus until birth, by nourishing the young with milk, and by providing them with heat, sanitation, and guidance after birth. The male may play no part in the care of the young, as in the seals and deer; he may defend them, as does the musk-ox; or he may share in other parental duties, for example, by supplying food, as do the wolf and the fox.

### INCUBATION BEHAVIOR

Brooding or incubation behavior is that phase of parental activity that provides warmth and shelter for the eggs and young. As one would expect, it is best developed in birds. From the time of laying of the last eggs, and in some species from the laying of the first, through to the independence of the young, incubation and brooding dominate all other behavior. Sexual behavior is suspended, and even self-feeding, after the young have hatched, is reduced. The duration of the incubation period is influenced by a genetically fixed period of embryonic development and varies with the calories applied.

The start and continuation of incubation in birds and in those fish that tend the eggs are highly dependent upon nest and eggs, the sign stimuli that induce these animals to settle on the eggs. Effective incubation continues only when there is a proper feedback of the tactile and visual stimuli of the eggs and in some species (perhaps in the females of all) the thermoreceptors of the brood patch (Barends, 1959). Fanning of eggs by the three-spined stickleback is released by the appearance and fertilization of the eggs. If the eggs are removed, the cycle is broken; at the same time the tendency to fan inhibits sexual behavior (van Iersel, 1953).

Care and concern for the eggs increase as the incubation period progresses. Birds tend to desert nests less frequently after a disturbance as the time for hatching approaches. Ground-nesting birds, especially the geese, gulls, and terns studied, tend to retrieve any eggs accidently kicked from the nest, although it may take some time for the bird to respond to the egg-out-of-nest stimulus (N. Tinbergen, 1960). Then the bird rolls or attempts to roll the egg with its bill back into the nest. A broken or pecked egg does not release retrieval; rather the bird may eat the egg. No longer a normal shape the egg has become instead a bit of food. The three-spined stickleback, too, will retrieve eggs that happen to lie outside the nest (van Iersel, 1953). These the male sucks up and inserts back into the nest, but only if the clump of eggs is large enough, at least five or six. Single eggs are eaten. The male also removes or attempts to remove eggs that have become moldy.

The eggs are seldom left unattended among those birds in which both sexes incubate. As one bird arrives to relieve the other at the nest, one of several actions may happen. The signs or signals of broodiness in the mate may stimulate the sitting bird to rise from the nest and allow the other to take over. On the other hand, nest relief may arrive before the sitter is prepared to go. In this case the relief bird, unable to satisfy its brooding urge, may perform some irrelevant activity such as nest building. This activity seems to stimulate the sitting bird to rise. But if all else fails, the relief bird may force the sitter off the nest. Some birds, the herons for example, have a ritualized nest-relief ceremony; others may announce their approach with a call.

## CARE OF YOUNG

The hatching or birth of young ushers in another phase of parental behavior, the care of young. The extent and kind of care given to the young is influenced by the maturity of the young at the time of birth. Basically birds and mammals are either precocial or altricial at birth. Precocial animals are able to move about at or shortly after birth, although some time may elapse before they can fly or move about as adults. Altricial animals are born helpless, naked or nearly so, often blind and deaf. Between the two extremes there is a wide variation in the stage and nature of maturity at birth.

Nice (1962) has classified the maturity at hatching in birds (Table 12-5). Precocial and semiprecocial birds, hatched with eyes open

and completely covered with down, are mobile to some degree and leave the nest in a day or so. Most precocial birds are capable of feeding themselves on small invertebrates and seeds; others follow the parents and respond to their food calls; still others are fed by the parents. Semiprecocial birds are able to walk, but because of feeding habits of the parents, they are forced to remain in the nest. Semialtricial birds are hatched with a substantial covering of down and with eyes open or closed, but are unable to leave the nest. Altricial birds are completely helpless; their eyes are closed and they have little or no down.

A somewhat similar classification could be devised for mammals. Young mice, bats, and rabbits are born blind and naked and thus are altricial. Young of wolves, foxes, dogs, and cats are born with hair and are soon able to crawl about the nest or den, but are blind for several days. These might be called semialtricial. Deer, moose, and other ungulates, as well as horses and pigs, would fall into the semiprecocial category. They are very ungainly on their legs for several days after birth, and during this time they establish a nursing routine. They may be hidden alone by the mother or held in a "pool," characteristic of the elk (Altmann, 1960). The young wait for the dam to return to the hiding place to nurse and to be licked. Most precocial of all mammals are seals, which might well be a "precocial 4" according to the classification in Table 12-5. Not

TABLE 12-5
*Classification of maturity at hatching in birds*

| *Type* | *Characteristics* | *Example* |
|---|---|---|
| **FEED SELVES** | | |
| Precocials | Eyes open, down-covered, leave nest first day or two | |
|   Precocials 1 | Independent of parents | Megapods |
|   Precocials 2 | Follow parents but find own food | Ducks, shorebirds |
|   Precocials 3 | Follow parents and are shown food | Quail, chickens |
| **FED BY PARENTS** | | |
|   Precocials 4 | Follow parents and are fed by them | Grebes, rails |
| Semiprecocials | Eyes open, down-covered, stay at nest though able to walk | Gulls, terns |
| Semialtricials | Down-covered, unable to leave nest | |
|   Semialtricials 1 | Eyes open | Herons, hawks |
|   Semialtricials 2 | Eyes closed | Owls |
| Altricials | Eyes closed, little or no down, unable to leave nest | Passerines |

*Source:* Adapted from Nice, 1962.

only is delivery extremely rapid, approximately 45 seconds in the gray seal (Bartholomew, 1959), but movements and vocalization appear very shortly after birth. Newly born fur seals are able to raise up and call from 15 to 45 seconds after birth and are capable of shaky but effective locomotion a few minutes after birth (Bartholomew, 1959). Even while the umbilical cord is still attached, the pups are able to shake off water, bite and nip at each other, and scratch dog-fashion with the hind flippers. They attempt to nurse within five minutes after birth. Although pups continue to nurse the cows for some time, the cows are protective and attentive toward their young only between parturition and estrus. Thus this behavior is conspicuous only for a few hours to a day after the cow has given birth to the pup.

As a rule parental behavior is not well developed in the poikilotherms, but care for young by the sticklebacks and chiclid fish deserves some comment. At hatching, the young sticklebacks have rather large yolk sacs from which they draw nourishment, and so they tend to remain embedded in or lying on the nest material. After a while the young, moving in a series of jumps, attempt to swim out of the nest. The male tries to catch the jumping and swimming young, chasing them, sucking them into his mouth, and spitting them back into the nest pit. But in most instances the male is unsuccessful (van Iersel, 1953). In one small group of cichlid fish the young remain near their mother, and for about 6 days return to the mouth of their mother for protection (N. Tinbergen, 1952).

Nourishment of the young is chiefly a function of the female among mammals, for only she can provide milk. The males defend the young or ignore them. Among the canids, however, males hunt and bring prey to the den for the female and for the female and young once the young have started on solid food. No such behavior exists among the felids.

Among birds who feed their young, both male and female may participate whether the male assisted in the brooding or not. (There are exceptions, however, such as the hummingbirds.) Exactly how the male knows that the young have hatched when the female alone incubates the eggs is not really known, although changes in the female's behavior probably give the cue. In some species, such as the starling and prairie warbler, the male may start bringing food to the nest several days to a week prior to the hatching (Nolan, 1958). This is known as anticipatory feeding.

Providing sanitation, guidance, and heat are other parental functions. Brooding provides the necessary warmth for the young while altricial and semiprecocial mammals are sheltered in nests or dens; such precocials as the ungulates have temperature controls of their own from birth. Sanitation is no problem among precocial birds. Altricial birds usually deposit their wastes in fecal sacs, which are carried away by the parents. Mothers of some altricial mammals stimulate excretion in the young by licking their genital and anal regions and then swallowing the excreta.

Guidance by parents of young who have left the nest and are beginning to acquire some independence occurs among few species of invertebrates other than the social insects and among only a few species of amphibians, reptiles, and fish. The female caiman of Guiana keeps the young with her until the spring following their birth.

Guidance, again, is best developed among birds and mammals. Female ducks, especially river ducks, stay with the young until they can fly and set the rhythm for such activities as feeding, preening, and resting. Gallinaceous birds are highly dependent on parental guidance for food as well as warmth. Altricial birds also follow parents for several weeks after they leave the nest (see Nice, 1943).

Parental guidance is highly important among the ungulates. The young of moose and elk, for example, may follow the cow until they are yearlings (Altmann, 1960). A close bond exists between cow and calf, and there is considerable vocal communication. The moose cow makes the calf stay within her sight and will retrieve it if it strays. If the cow is killed during the first year, the calf rarely survives the winter. In fact during the rut season a moose cow will leave with her calf at once if the bull intimidates the young animal. Young raccoons and skunks follow the mother on nightly forays, and young wolves join their parents in the hunt. Young gray squirrels follow the female for a while after they have

**426**

left the nest, respond to her calls, and may even be groomed by her (Bakken, 1959).

A rather strong bond may exist between the young of the same brood or litter. Ducklings seemingly need the companionship of their fellows and may do poorly without it. The attachment between members of a brood often outlasts their bond with the parent. A strong bond also exists in broods of gallinaceous birds, but among some shore birds the bond between siblings is not especially strong (Nice, 1962).

As the young mature into adults, the bond between parent and offspring breaks, and agonistic behavior replaces it. This split more often than not is initiated by the parents themselves. The moose cow becomes hostile to the yearling, especially the female offspring, whom the old cow regards as a rival. The yearling bull is tolerated, but if near his mother he becomes a target for the courting bull (Altmann, 1960). Among elk the yearling cow is tolerated in the herd, but the young bull is chased out of the herd at breeding season by the dominant bull. Not only does he have to face the antagonism of mature cows and the dominant bull, he also has to face other free-roaming, unattached bulls, forcing the animal into a very insecure situation. Once rejected by parents and forced on their own, juvenile ungulates become as Altmann puts it "a rejected and most erratic non-conforming age group."

## Mating systems

The behavioral mechanisms and the nature of social organization involved in an organism's obtaining a mate is called a *mating system*. A mating system includes such aspects as the number of mates acquired, the manner in which they are acquired, the nature of the pair bond, and the pattern of parental care provided by each sex. The structure of mating systems ranges from monogamy through many variations of polygamy. Mating systems employed may vary even within a species, involving different degrees of pair bonds and relationships between the pair (Table 12-6).

The nature and evolution of male-female relationships and thus mating systems are influenced by ecological conditions, especially the availability and distribution of resources and the ability of individuals to control access to mates or resources. If the male has no role in the feeding and protection of young, no advantage accrues to the female to remain with the male. If the habitat is sufficiently uniform so that little difference in territorial quality exists, the number of young raised in the poorest habitat is only slightly less than the number reared in the best. Selection would favor monogamy because female fitness in both would be nearly the same. If the habitat is diverse with some parts more productive than others, competition among males may be intense and some males will settle on poorer territories. Under such conditions it may be more advantageous for a female to join another female in the territory of a male defending a rich resource than to expel the other female. Selection under those conditions will favor bigamous mating even though aid from the male in feeding the young is reduced or absent.

The various mating systems result in a differing fraction of the male population acquiring mates. Mating becomes competitive and selection works against those males deprived of mates. Competitive mating results in differential production of progeny by different genotypes, or *sexual selection*. Ultimately sexual selection results in the evolution of morphological and behavioral traits that influence both competitive mating and mating systems (for example, see Geist, 1978).

Not all structures and behaviors associated with mating necessarily relate to courtship and sexual selection. Such traits may also serve other functions, such as bringing individuals together and synchronizing breeding activities of colonial birds.

### MONOGAMY

Monogamy is the formation of a pair bond between one male and one female. It is most prevalent among birds. It is relatively rare among mammals except for humans, several carnivores, such as the fox and mustelids, and a few herbivores, such as the beaver.

Monogamy results when neither of the two sexes has the opportunity to monopolize the other either directly or through control of resources. Monogamy is most prevalent among

**427**

*TABLE 12-6*
*An ecological classification of mating systems*

| | |
|---|---|
| Monogamy | Neither sex has opportunity of monopolizing additional members of the opposite sex. Fitness often maximized through shared parental care. |
| Polygyny | Individual males frequently control or gain access to multiple females. |
|   Resource defense polygyny | Males control access to females *indirectly,* by monopolizing critical resources. |
|   Female (or harem) defense polygyny | Males control access to females *directly,* usually by virtue of female gregariousness. |
|   Male dominance polygyny | Mates or critical resources are *not economically monopolizable.* Males aggregate during the breeding season and *females select mates* from these aggregations. |
|     Explosive breeding assemblages | Both sexes converge for a short-lived, highly synchronized mating period. The operational sex ratio is close to unity and sexual selection is minimal. |
|     Leks | Females are less synchronized and males remain sexually active for the duration of the females' breeding period. Males compete directly for dominant status or position within stable assemblages. Variance in reproductive success and skew in operational sex ratio reach extremes. |
| Rapid multiple clutch polygamy | Both sexes have substantial but relatively *equal* opportunity for increasing fitness through multiple breedings in rapid succession. Males and females each incubate separate clutches of eggs. |
| Polyandry | Individual females frequently control or gain access to multiple males. |
|   Resource defense polyandry | Females control access to males *indirectly,* by monopolizing critical resources. |
|   Female access polyandry | Females do not defend resources essential to males but, through interactions among themselves, may limit access to males. Among phalaropes, both sexes converge repeatedly at ephemeral feeding areas where courtship and mating occur. The mating system most closely resembles an explosive breeding assemblage in which the operational sex ratio may become skewed with an excess of females. |

*Source:* Emlen and Oring, 1977.

those species in which cooperation by both parents is required to rear young successfully. For one member of the pair to spend time and energy courting and mating with other individuals would result in the loss of individual fitness. Monogamy also prevails in situations where resources are rather uniformly distributed and little opportunity exists for an individual to monopolize them. Maximum fitness for both individuals of the pair is achieved when both parents share in the care of the young (see Coulson, 1966).

## POLYGAMY

Polygamy is the acquisition by an individual of two or more mates, none of which is mated to other individuals. A form of pair bond exists between the individual and each mate. When one member of the pair is freed from parental duty, partly or wholly, the more time and energy the emancipated member of the pair can devote to intrasexual competition for mates and resources. The more such critical resources as food or quality habitat are unevenly distributed, the greater is the opportunity for such an individual to control the resource and thus available mates.

How many members of the other sex an individual can monopolize depends upon the degree of synchrony in sexual receptivity. For example, if females in the population are sexually active for only a brief period, the number a male can monopolize is limited. If females are receptive over a long period of time, the number a male can control depends upon the availability of females and the number of mates the male can energetically defend. Such variability in environmental and behavioral conditions results in various types of polygamy of which there are two basic forms: polygyny, in which an individual male gains control of or access to two or more females, and polyandry, in which an individual female gains control of or access to two or more males. A special form of polygamy is promiscuity, in which males and females copulate with one or many of the opposite sex but form no pair bonds. Emlen and Oring (1977) classify types of polygamous relationships according to the means by which individuals gain access to the limiting sex. This classification appears in Table 12-6.

*Resource defense polygyny.* Resource defense polygyny results when the male can defend resources essential to the female. How many females he can monopolize depends upon how the resources are distributed and how defendable they are. If resources are highly clumped, a male can control more of the resources and thus more of the females. In the choice of a male by the female the concomitant sexual selection pressure is influenced by the quality of the male-defended resource as well as by the quality of the male.

If both parents take care of the young, the female stands to lose if the male has more than one mate, unless the quality of the male or the quality of the resources he controls more than offsets the loss of his help at the nest. Under such a situation the female may have a greater chance of reproductive success and thus increase her fitness if mated polygamously to a male in high quality habitat than if mated monogamously to a male in poor quality habitat. Examples are the dickcissel (Zimmerman, 1971) and the red-winged blackbird (Howard, 1977), two species in which males defend optimal habitats, determined by volume of vegetation, attract multiple mates, while males in less suitable habitats are more apt to be monogamous or go unmated.

If the male takes no active interest in the nest and provides no care of the young and if the resources needed by the female are sufficiently clumped, he can spend his time defending the resources. Such a situation can lead to a strong development of polygyny, as it does in a number of species of hummingbirds (Stiles and Wolf, 1970; Gill and Wolf, 1975). The males defend high nectar-producing flowers, rather than territory per se (Wolf et al., 1976), at which they allow the females, but not other males to feed. By controlling a needed resource, food, the hummingbird also controls access to females, leading to polygyny.

*Female defense polygyny.* A second type of polygyny is defense of a group of females or a harem. This usually results when females are naturally gregarious and live in small herds for most of the year for reasons unrelated to reproduction. Such herds enable the dominant male to have access to and to control a number of females during the reproductive season. The male asserts his claim to the females by aggressive behavior toward other males and expends great amounts of energy defending the females from raiding activities of rivals. There is, of course, an upper limit to the number of females a male is able to defend; a male attempting to control too many will lose part of his harem to a rival.

Two types of harem defense exist. One, common to ungulates such as the wapiti and red deer and to wild horses, is defense of the females themselves. A second type, common to seals, is defense of access to areas where females congregate on the breeding ground.

Female defense polygyny leads to intense sexual selection and strong sexual dimorphism. The males of seals are many times larger than females; bull wapiti are considerably larger than cows and possess highly developed antlers. (For discussion of sexual dimorphism in this context, see Geist, 1977.)

*Male dominance polygyny.* On the mid-grasslands of North America where remnant populations of prairie grouse (prairie chickens and sharp-tailed grouse) still exist and on the sage lands of western North America where sage grouse live, males of these species congregate on communal display areas called *leks* in order to attract and mate with females. These leks, called booming grounds for prairie chickens, dancing or parade grounds for prairie chickens and sharp-tailed grouse, and strutting grounds for sage grouse, are located on areas of open ground somewhat elevated and so situated that they are visible to the surrounding area. The cocks gather from daybreak on through the afternoon on the lek to display. Early in the season behavior is largely aggressive and fighting is frequent. Territories are established as a result of this fighting, with dominant and subdominant birds obtaining central positions on the lek. Once fighting has subsided, females begin visiting the lek where they show a marked preference for centrally located males. When females arrive on the lek, courtship dominates with its bizarre displays and resonation of sounds from brightly colored air sacs (see Wiley, 1974).

Lek behavior is an example of male dominance polygyny. The male does not defend females nor does he defend a resource needed by the female. Rather, the males sort themselves out into dominance positions, excluding some or many from mating, and the females choose the male. Lek mating systems exist among a number of species of birds and some insects, frogs, and mammals.

Another type of male dominance polygyny, somewhat suggestive of lek behavior, is the explosive breeding assemblage. Males of such animals as singing insects and chorusing amphibians congregate on breeding sites to which females are attracted. During a short-lived, highly synchronous breeding period, mating is promiscuous. Breeding activity is frenzied, and because of female synchrony in breeding period, individual males have little opportunity to monopolize matings.

Male dominance polygyny results when individual males are completely freed from parental care, when the environment provides little potential for resource or mate control, and when it is economically (energetically) impossible for the male to control or monopolize resources necessary for acquiring females.

*Rapid multiple clutch polygyny.* Among some birds males assume the full burden of incubation and rearing of the brood. This frees the female from nesting duties and allows for more frequent matings. This reversal of parental duties could come about only if it increased male fitness. A lack of dependable breeding conditions could place a premium on the ability of the female to replace clutches lost to predation and other causes. From the male point of view, costs of assuming nesting duties are balanced by the ability of the female to produce new clutches rapidly. Among some shorebirds and gallinaceous birds the female lays the first clutch, which the male incubates, and then lays a second clutch, which she incubates. This doubles the reproductive effort in a short period of breeding time. Such an arrangement obviously increases the fitness of the female and it also increases male fitness if the female returns to her mate prior to laying of the second clutch and if the male remains sexually active as far into the breeding season as possible.

## POLYANDRY

The assumption of incubation and brood rearing roles by the male is a step toward polyandry. The female increases her fitness by producing multiple clutches, but this fitness can be increased only to the point that males are sexually receptive and available to assume responsibility for additional clutches.

*Resource defense polyandry.* In resource defense polyandry the female competes for and defends resources essential for the male. As in polygyny, the degree of polyandry depends upon the distribution and defensibility of resources, especially quality habitat. The production by females of multiple clutches that all require the brooding services of a male leads to competition among females for access to available males. After a clutch is laid and a male begins incubation, he becomes sexually inactive and is effectively removed from the male pool, resulting in a scarcity of available males.

Polyandry usually occurs when nesting failure is frequent, requiring the female to be free to produce multiple clutches, when the breeding season is sufficiently long to allow renesting, and when the males arrive asynchronously on the nesting grounds (in contrast to synchrony in female breeding condition in polyandry).

Polyandry is best developed in two bird orders, Gruiformes (cranes, rails, etc.) and Charadriiformes (the shorebirds). An example in eastern North America is the spotted sandpiper *(Actitis macularia),* a species in which losses to egg predation are high and in which the female has an exceptional ability to lay multiple clutches (Hayes, 1972; Oring and Knudsen, 1972; Oring and Marson, in press). An example of extreme polyandry is the American jacana of Central America. Because the breeding habitat of this bird is very limited, only a small fraction of the breeding population is active in any one year. Males defend small territories about ponds and lagoons, while females defend large territories that embrace the territories of several males. Females mate with more than one male, produce multiple clutches, and frequently have multiple mates incubating clutches simultaneously. Because nest predation is high, females provide replacement clutches for their males who assume parental duties. In effect female jacanas specialize in egg production, while the male specializes in brood rearing (Jenni and Collier, 1972; Jenni, 1974).

Among species exhibiting polyandry behavioral dimorphism and size dimorphism are evident. The females are larger, often considerably so, than the males (for example, the female jacana weighs 50 to 75 percent more than the male), are dominant over males in aggressive interactions, and provide minimal care of eggs or young.

*Female access polyandry.* An exceptional type of polyandry in which no resource defense is involved at all is exhibited by phalaropes (Emlen and Oring, 1977). Phalaropes, small shorebirds that nest on the arctic tundra and winter at sea in the Southern Hemisphere, are unique in that the female is more brightly colored than the male and the male performs all parental care. Because the food resource, aquatic larvae of mosquitos, midges, and beetles, is highly unpredictable and ephemeral, the birds shift courtship and nesting sites from year to year and even week to week. Males and females congregate at bodies of water where they feed, display, and mate. Pair bonds are brief. After the female has completed the first clutch of eggs, she may attempt to increase fitness by mating again. This is especially true for females that complete the first clutch early. If nesting failures are frequent and if sufficient time remains in the breeding season to allow renesting, females compete for males. If the density of birds is high, female interactions may become severe and some females may be prevented from breeding. Thus females may limit access of other females to males, with which the female must remain until the clutch is complete and incubation begins.

### PLASTIC MATING SYSTEMS

Mating systems often are not genetically fixed or species specific. Considerable variability may exist in mating systems employed by individuals within a species, depending upon the availability and distribution of needed resources, the energy cost of defending resources and mates, and population density. If environmental conditions or population densities change from year to year, individuals within local populations may be able to respond to those changes by shifting from monogamy to polygamy or from one form of polygamy to another.

How mating systems may change under differing environmental conditions has been out-

*FIGURE 12-20*
*General schema of the determinants of a plastic mating system. (From Emlen and Oring, 1977.)*

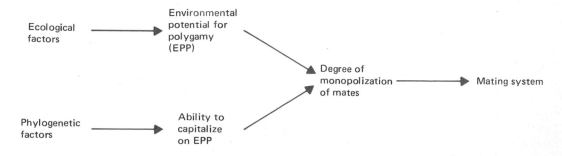

431

lined in a general schema by Emlen and Oring (1977) (Figure 12-20). Ecological factors influence the environmental potential (resource availability and distribution) for polygamy. Phylogenetic factors influence the ability of the animal to capitalize on environmental potential. The ability to capitalize involves the amount of parental care required for the successful rearing of young. If both parents are required, the parents must remain monogamous. If only one parent is required, the potential for polygamy exists.

## SUMMARY

Encounters and subsequent interactions among individuals involves social behavior. Such behavior includes both stereotyped responses and learning. In stereotyped behavior outside stimuli are selectively admitted, sensory data are processed and integrated within the nervous system, and the messages are sent out to the muscular system for the appropriate, species-specific responses. Environment contributes little input to the development of such behavior in an individual. Influenced considerably more by environment is learning. Learning, particularly habituation and operant conditioning, allows animals to modify their behavior, enabling them to adjust more finely to the changing environment in which they live. A special form of learning is imprinting, the rapid establishment of perceptual preference for an object. It appears to be important in species and sexual recognition and in habitat selection.

Social interaction requires a means of communication, a "language" or system of signals between individuals. These signals may be visual, involving simple to elaborate display; auditory, involving species-specific songs and calls; or chemical, involving pheromones. Communicative behavioral patterns are associated with such activities as territoriality and social dominance, courtship, pair formation, maintenance of pair bond, interaction between parents and young, and suppression of agonistic responses.

Basic social behavior is agonistic or aggressive, expressed as social dominance and territoriality which act as spacing mechanisms among individuals in the population. Territoriality, the defense of a fixed area, divides space among some individuals and excludes others. In contrast social dominance results in sharing of space, but with some individuals dominant over others. Subordinate individuals avoid contact with dominant ones by using different parts of a shared area and by using the area at different times.

Associated with territoriality and social dominance is competition for mates, which results in sexual selection, a form of natural selection. The behavioral mechanisms and social organization involved in the acquisition of mates is a mating system. Mating systems, which are influenced by sexual selection and the degree and nature of parental care given the young, include two basic types, monogamy and polygamy. Two general kinds of polygamy are polygyny, in which the male acquires more than one mate, and polyandry, in which the female acquires more than one mate. The potential for competitive mating is higher in polygamy than in monogamy. The nature of mating systems is influenced by both ecological and behavioral factors.

# Properties of populations
## CHAPTER 13

Individual populations can be described by such measurable properties as density, dispersion, age ratios, mortality, and natality. These characteristics of populations, their changes over time, and the prediction of future changes can be analyzed by certain quantitative techniques. The study of such vital statistics of populations is known as *demography*.

## Density and dispersion

### CRUDE VERSUS ECOLOGICAL DENSITY

The size of a population in relation to a definite unit of space is its density. Every ten years the census bureau counts the number of people living in the United States; wildlife biologists determine the number of game in a particular area; a forester determines the number and volume of trees in a timber stand. The measure of number of individuals per unit area is called *crude density*.

But populations do not occupy all the space within a unit because it is not all a suitable habitat. A biologist might estimate the number of deer per square mile, but the deer may not utilize the entire area because of such factors as human habitation and land use practices, lack of cover, or lack of food. A soil sample may contain 2 million arthropods per square meter, but these arthropods inhabit only the pore spaces in the soil, not the entire substrate. Goldenrods inhabiting old fields grow in scattered groups or clumps because of soil conditions and competition from other old field plants. No matter how uniform a habitat may appear, it is not uniformly habitable because of microdifferences in light, moisture, temperature, or exposure, to mention a few conditions. Each organism occupies only areas that can adequately meet its requirements, resulting in patchy distribution. Density measured in terms of the amount of area available as living space is *ecological density*.

**433**

Attempts have been made to make such ecologically realistic measurements. For example, one study in Wisconsin expressed the density of bobwhite quail as the number of birds per mile of hedgerow rather than per acre (Kabat and Thompson, 1963). However, ecological densities are rarely estimated because it is difficult to determine what portion of a habitat represents living space. Determining just what area is available to one particular kind of organism and just how suitable it is represents one of the most important problems in population biology.

The density of organisms in any one area varies with the seasons, weather conditions, food supply, and many other factors. However, an upper limit to the density is imposed by the size of the organism and its trophic level. Generally the smaller the organism, the greater its abundance per unit area. A 100-acre forest will support more woodland mice than deer and more trees 2 to 3 inches dbh (diameter breast height) than trees 12 to 14 inches dbh. The lower the trophic level, the greater the density of the organism. The same 100-acre forest will support more plants than herbivores, more herbivores than carnivores.

From a practical point of view, density is one of the more important parameters of populations. It determines in part energy flow, resource availability and utilization, physiological stress, dispersal, and productivity of a population. The density of human populations relates to economic growth and the expansion and management of towns, cities, regions, states, and nations. Increasing or decreasing populations can place strains on economic and social institutions. The distribution of humans in a given region affects certain land use and pollution problems. Wildlife biologists need to know about densities of game populations to regulate hunting and manage habitats. Foresters base timber management and evaluation of site quality in part on the density of trees.

### DISPERSION

Crude density provides only minimal information about a population. It gives no information on the evenness of concentration. A more important aspect of density is the dispersion of the population in an area, that is, how the individuals are distributed in terms of space and time.

*Spatial dispersion.* Individuals may be distributed randomly, uniformly, or in clumps (Figure 13-1). Distribution is considered random if the position of each individual is independent of the others. Random distribution is rare, for it can occur only where the environment is uniform, resources are equally available throughout the year, and interaction among members of the population produces no patterns of attraction or avoidance. Some invertebrates of the forest floor, particularly spiders (Cole, 1946; Kuenzler, 1958), the clam *Mulinia lateralis* of the intertidal mud flats of the northeastern coast of North America, and certain forest trees (Pielou, 1974) appear to be randomly distributed.

Uniform or regular distribution is the more even spacing of individuals than would occur by chance. Regular patterns of distribution result from intraspecific competition among members of a population. For example, territo-

*FIGURE 13-1*
*Uniform, random, and clumped patterns of distribution.*

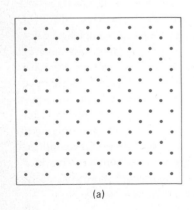

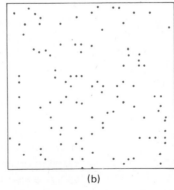

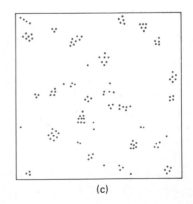

(a)                                    (b)                                    (c)

riality under relatively homogeneous environmental conditions can produce uniform distribution (Figure 13-2). Among plants it may result from severe competition for crown and root space among forest trees (see Gill, 1975) and for moisture among desert plants (Beals, 1968). Autotoxicity (the production of exudates toxic to seedlings of the same species), a characteristic common among plants of arid country, is another mechanism for achieving uniform distribution.

The most common type of distribution is clumped, also called clustered, contagious, and aggregated. This pattern of dispersion results from responses by plants and animals to habitat differences, daily and seasonal weather and environmental changes, reproductive patterns, and social behavior. The distribution of human beings is clumped or aggregated because of social behavior, economics, and geography.

There are various degrees and types of aggregated distribution. Groups of varying sizes and densities may be randomly or nonrandomly distributed over an area; individuals within the clumps may be distributed randomly or uniformly. Aggregations may be small or large. Population clusters may tend to concentrate around a geographical feature which provides nutrients or shelter (see Figure 13-3).

Aggregations among plants are often influenced by the nature of propagation and specific environmental requirements. Nonmobile seeds, such as those of oaks and cedar, are clumped near the parent plant or where they are placed by animals. Mobile seeds are more widely distributed, but even they tend to concentrate near the parent plant (Figure 13-4). Vegetative propagation produces clumping. Seed germination, survival of seedlings, and competitive relationships also affect the degree and type of aggregation.

Some animal aggregations represent separate responses of individuals to environmental conditions. They may be drawn together by a common source of food, water, or shelter. Moths attracted to light, earthworms congregated in a moist pasture field, barnacles clustered on a rock, all have low levels of or no social interaction. The individuals do not aid one another and only passively prevent other members of the same species from sharing the condition that brought each of them to the same location.

Aggregations on a higher social level reflect some degree of interaction among population members. Prairie chickens congregate for communal courtship; elk band together in herds with some social organization, usually with a cow as the head (Altmann, 1952); birds congregate on feeding grounds away from territorial sites, yet show intolerance for each other near the nest. Aggregations of the highest social structure are found among insect societies, such as ants and termites, in which individual members are organized into social castes according to the work they perform.

FIGURE 13-2
*Golden eagle territories in Scotland showing the even dispersion of breeding sites. (From L. Brown, 1976).*

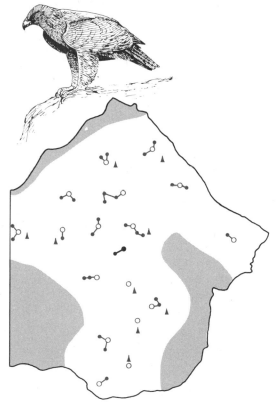

○—○ Group of sites belonging to one pair
○ Single site
●—● Marginal site not regularly occupied
▲  ▲ Breeding, year of survey 1967
░ Low ground unsuited to breeding eagles

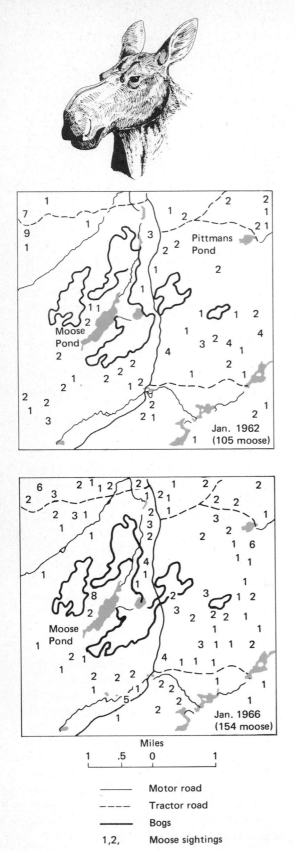

Miles
1  .5  0  1

—————  Motor road

- - - -  Tractor road

━━━━━  Bogs

1,2,  Moose sightings

**FIGURE 13-3**
*Random aggregate distribution of moose in the Lower Noel Paul River, Newfoundland. (From Bergerud and Manuel, 1969.)*

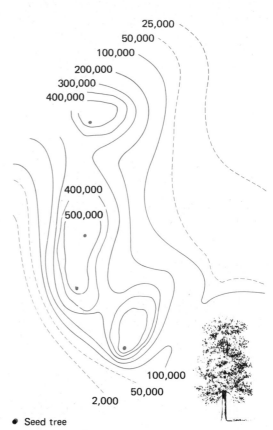

● Seed tree

**FIGURE 13-4**
*Pattern of annual seed fall of yellow poplar (Liriodendron tulipifers). Lines show equal seeding density. (After L. G. Engle, 1960.)*

The pattern of population distribution has two other features of note: intensity and grain (Pielou, 1974). The *intensity* of population dispersion is high if there is a wide range of densities (Figure 13-5), low if there is little variation in density. A pattern is *coarse grained* when clumps and the areas between them are large and *fine grained* when clumps and gaps between them are small. Although a coarse-grained pattern of dispersion relates more to relative than to absolute density, a change in density can affect both the intensity and grain of population dispersion. Fine-grained species

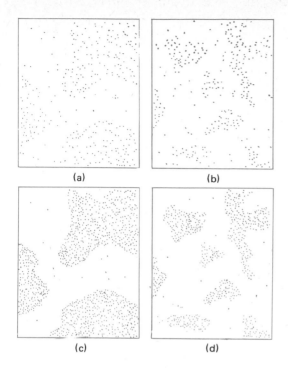

(a)          (b)

(c)          (d)

**FIGURE 13-5**
*Patterns of different grains and intensity.*
*(From E. Pielou, 1974.)*

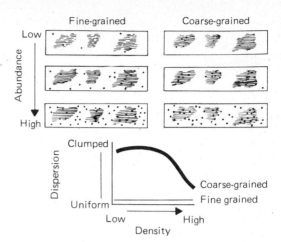

**FIGURE 13-6**
*Habitat or patch occupancy patterns for a fine-grained and a coarse-grained population as density increases from low to high levels. The fine-grained population may retain a uniform distribution pattern over a wide range of densities. The dispersion of a coarse-grained population is initially clumped, because individuals settle in preferred patches, but tends toward greater uniformity as density increases and individuals are forced to settle in less preferred patch types. (From Wiens, 1976.)*

that occupy more than one patch of habitat type may spread randomly over the area as density increases (Figure 13-6). Coarse-grained species that are restricted to certain patches of habitat become more aggregated as density increases. But as density becomes very high, individuals of coarse-grained species may be forced to occupy marginal patches of habitat in a fine-grained manner. As density increases, dispersion may become more uniform.

Populations of a species are not uniformly distributed over a region. While local distribution of populations may be defined by the spacing of individuals, on a larger scale the species are distributed as individual populations. In turn, individual populations may be concentrated in clusters within a given region. Regional distributions of populations make up the total range of a species (Figure 13-7).

The boundaries of a range are not fixed; they may fluctuate greatly. Habitat changes, competition, predation, and climatic changes can influence the extent of a species range, expanding it one year and contracting it another.

*Temporal dispersion.* Organisms in populations are distributed not only in space, but also in time. Temporal distribution may be circadian (relating to daily changes in light and dark) or it may relate to changes in humidity and temperature, seasons, lunar cycles, and tidal motions. Distribution may also be related to longer periods of time which encompass annual cycles, successional stages, and evolutionary changes. The environmental rhythm of daylight and dark is responsible for the movement of animal populations such as the attraction of nectar-feeding insects to open flower petals, the daily migration of plankton from deeper to upper layers of water, the withdrawal and emergence of nocturnal and diurnal animals. The presence or absence of organisms that inhabit intertidal zones coincides with tidal cycles. Seasonal changes result in changes in populations of wild flowers in forest and field and in the return and disappearance of migrant animals. The populations of forests and fields are quite different in spring, summer, fall, and winter.

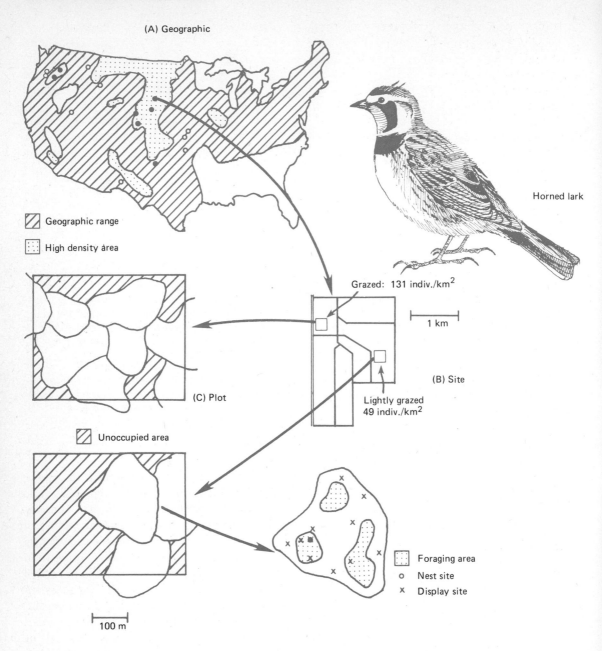

(A) Geographic

Geographic range

High density area

Horned lark

Grazed: 131 indiv./km$^2$

1 km

(B) Site

(C) Plot

Lightly grazed
49 indiv./km$^2$

Unoccupied area

Foraging area
o  Nest site
x  Display site

100 m

**FIGURE 13-7**
*Uneven population distribution within the range of the horned lark* (Eromophila alpestris). *Populations of an organism are not equally distributed over its range. Some areas hold higher densities than others, as indicated by the range map of the horned lark. Within a particular site the distribution of the bird is influenced by the availability of habitats. Within a plot of a given habitat the bird's distribution is influenced by territorial behavior. Within each territory space is allocated to different activities. (From Wiens, 1973.)*

## Mortality and survival

A major influence on population size is mortality. Mortality, which begins even in the egg, can be expressed either as the probability of dying (mortality rate) or as a death rate. The *death rate* is the number of deaths during a given time interval divided by the average population and is an instantaneous rate. The *probability of dying* is the number that died during a given time interval divided by the number alive at the beginning of the period.

The complement of the mortality rate is the *probability of living,* the number of survivors divided by the number alive at the beginning of the period. Because the number of survivors is more important to the population than the number dying, mortality is better expressed in terms of survival or in terms of *life expectancy,* the average number of years to be lived in the future by members of a population.

### THE LIFE TABLE

A clear and systematic picture of mortality and survival is best provided by a *life table,* a device first developed by students of human populations and widely used by actuaries of life insurance companies (Table 13-1). The life table consists of a series of columns, each of which describes an aspect of mortality statistics for members of a population according to age. Figures are presented in terms of a standard number of individuals, or cohort, usually, but not always, 1,000 at birth or hatching. The columns include: $x$, the units of age or age level; $l_x$, the number of individuals in a cohort that survive to the listed age level; $d_x$, the number of a cohort that die in an age interval, $x$ to $x + 1$ (from one age level listed to the next); $q_x$, the probability of dying or age-specific mortality rate, determined by dividing the number of individuals that died during the age interval by the number of individuals alive at the beginning of the age interval. Another column, $s$ or the survival rate, may be added. It can be calculated from the statement $1 - q$. In order to calculate life expectancy, given in column $e_x$, two additional statistics are needed, $L_x$ and $T_x$. $L_x$, the average years lived by all individuals in the population, is obtained by summing the number alive at the age interval $x$ and the number at age $x + 1$ and dividing the sum by 2. $T_x$ is calculated by summing all the values of $L_x$ from the bottom of the table to the top. Life expectancy $e_x$ is obtained by dividing $T_x$ for the particular age class $x$ by the survivors for that age as given in the $l_x$ column. (Construction of a life table is given in the appendix.)

Data for life tables are relatively easy to obtain for laboratory animals and for human beings. Data on mortality, survivorship, and age for organisms in the wild are much more difficult to obtain. Mortality $(d_x)$ can be estimated by determining the ages at death of a large number of animals born at the same

TABLE 13-1
*Life table for human beings in United States, 1966*

| MALE | | | | | FEMALE | | | |
|---|---|---|---|---|---|---|---|---|
| $x$ | $l_x$ | $d_x$ | $q_x$ | $e_x$ | $l_x$ | $d_x$ | $q_x$ | $e_x$ |
| 0 | 1000 | 26 | .02576 | 66.75 | 1000 | 20 | .01997 | 73.86 |
| 1 | 974 | 4 | .00405 | 67.51 | 980 | 3 | .00338 | 74.36 |
| 5 | 970 | 2 | .00253 | 63.78 | 977 | 2 | .00183 | 70.61 |
| 10 | 968 | 3 | .00260 | 58.93 | 975 | 1 | .00153 | 65.74 |
| 15 | 965 | 7 | .00730 | 54.08 | 973 | 3 | .00295 | 60.83 |
| 20 | 958 | 10 | .00992 | 49.46 | 971 | 3 | .00357 | 56.01 |
| 25 | 949 | 9 | .00938 | 44.93 | 967 | 4 | .00440 | 51.20 |
| 30 | 940 | 10 | .01088 | 40.33 | 963 | 6 | .00632 | 46.41 |
| 35 | 930 | 14 | .01520 | 35.74 | 957 | 9 | .00911 | 41.69 |
| 40 | 916 | 21 | .02345 | 31.26 | 948 | 13 | .01391 | 37.05 |
| 45 | 894 | 33 | .03716 | 26.94 | 935 | 20 | .02104 | 32.53 |
| 50 | 861 | 51 | .05956 | 22.88 | 915 | 28 | .03082 | 28.18 |
| 55 | 810 | 75 | .09216 | 19.16 | 887 | 40 | .04501 | 23.99 |
| 60 | 735 | 97 | .13260 | 15.84 | 847 | 56 | .06601 | 19.99 |
| 65 | 637 | 124 | .19505 | 12.86 | 791 | 84 | .10673 | 16.22 |
| 70 | 513 | 137 | .26772 | 10.35 | 707 | 114 | .16147 | 12.84 |
| 75 | 376 | 132 | .35064 | 8.22 | 593 | 146 | .24644 | 9.81 |
| 80 | 244 | 115 | .47188 | 6.33 | 447 | 170 | .38102 | 7.17 |
| 85+ | 129 | 129 | 1.00000 | 4.75 | 276 | 276 | 1.00000 | 5.05 |

*Source:* Data from Keyfitz and Flieger, 1971.

time. This can be done by marking or banding a considerable number of animals. One can record the ages at death of animals marked at birth, but not necessarily of those born during the same season of the year. Data from several years and several cohorts may be pooled to provide the information for the $d_x$ column. Another approach to gathering data is to determine the age of death of a representative sample of carcasses of the species concerned. Age can be determined by examining wear and replacement of teeth in deer, growth rings in the cementum of teeth of ungulates and carnivores, annual rings in the horns of mountain sheep, and weight of the lens of the eye. This information also goes into the $d_x$ column. Recording the ages at death of a sample of a population wiped out by some catastrophe could provide data for the $l_x$ series. Determining the age of animals taken during a hunting season provides information for the $l_x$ column because the sample is from a living population. But the data are often biased in favor of older age classes, especially if they are collected between breeding seasons.

There are three basic types of life tables. First, the *cohort, horizontal,* or *dynamic life table* records the fate of a group of animals all born at the same time (Table 13-2).

Second, the *time-specific* or *vertical life table* records the mortality of each age class in a given population over a year (Table 13-3). It is constructed from a sample of animals of each age class taken in proportion to their numbers in a population and age at death. It involves the assumption that birth and death rates are constant and that the population is stationary.

Third, the *dynamic-composite life table* records the same information as the dynamic life table, but takes as the cohort a composite of the numbers of animals marked over a period of years rather than at just one birth period (Table 13-4). For example, wildlife biologists may mark or tag a number of newly hatched young birds or young mammals each year over a period of several years. After following the fate of each year's group, they pool the data and treat all of the marked animals as one cohort (see Barkalow et al., 1970).

*Insect life tables.* The life tables above are typical of long-lived species in which generations overlap and in which different ages are

TABLE 13-2
*Dynamic life table for red deer on Isle of Rhum, 1957*

| $x$ | $l_x$ | $d_x$ | $100\ q_x$ | $e_x$ |
|---|---|---|---|---|
| **STAGS** | | | | |
| 1 | 1000 | 84 | 84.0 | 4.76 |
| 2 | 916 | 19 | 20.7 | 4.15 |
| 3 | 897 | 0 | 0 | 3.25 |
| 4 | 897 | 150 | 167.2 | 2.23 |
| 5 | 747 | 321 | 430.0 | 1.58 |
| 6 | 426 | 218 | 512.0 | 1.39 |
| 7 | 208 | 58 | 278.8 | 1.31 |
| 8 | 150 | 130 | 866.5 | 0.63 |
| 9 | 20 | 20 | 1000.0 | 0.50 |
| **HINDS** | | | | |
| 1 | 1000 | 0 | 0 | 4.35 |
| 2 | 1000 | 61 | 61.0 | 3.35 |
| 3 | 939 | 185 | 197.0 | 2.53 |
| 4 | 754 | 249 | 330.2 | 2.03 |
| 5 | 505 | 200 | 396.0 | 1.79 |
| 6 | 305 | 119 | 390.1 | 1.63 |
| 7 | 186 | 54 | 290.3 | 1.35 |
| 8 | 132 | 107 | 810.5 | 0.70 |
| 9 | 25 | 25 | 1000.0 | 0.50 |

*Source:* V. P. W. Lowe, 1969.

alive at the same time. However, a tremendous number of organisms have one annual breeding season and the generations do not overlap, so that all individuals belong to the same cohort or age class. Some insects and annual plants follow this pattern. In the conventional life table one can determine survivorship rates from the age distribution at one observation time. Among annual species one cannot because there is no age distribution, but there is a distribution of developmental stages. In species with no overlapping generations $l_x$ values can be obtained only by observing a natural population over a period of time, say a summer season, and estimating the size of population at each observation time. For many insects $l_x$ values can be obtained by estimating the size of the surviving population at each stage of development from egg to adult. One can estimate how many larvae and pupae transform to adults and, if records are also kept of weather conditions, abundance of predators, parasites, and disease, one can also estimate how many die from various causes.

An example is a life table for a sparce, stable gypsy moth population in Connecticut (Table 13-5). A brief description of the life cycle of this

TABLE 13-3
*Time-specific life table for red deer on Isle of Rhum, 1957*

| x | $l_x$ | $d_x$ | 1000 $q_x$ | $e_x$ |
|---|---|---|---|---|
| **STAGS** | | | | |
| 1 | 1000 | 282 | 282.0 | 5.81 |
| 2 | 718 | 7 | 9.8 | 6.89 |
| 3 | 711 | 7 | 9.8 | 5.95 |
| 4 | 704 | 7 | 9.9 | 5.01 |
| 5 | 697 | 7 | 10.0 | 4.05 |
| 6 | 690 | 7 | 10.1 | 3.09 |
| 7 | 684 | 182 | 266.0 | 2.11 |
| 8 | 502 | 253 | 504.0 | 1.70 |
| 9 | 249 | 157 | 630.6 | 1.91 |
| 10 | 92 | 14 | 152.1 | 3.31 |
| 11 | 78 | 14 | 179.4 | 2.81 |
| 12 | 64 | 14 | 218.7 | 2.31 |
| 13 | 50 | 14 | 279.9 | 1.82 |
| 14 | 36 | 14 | 388.9 | 1.33 |
| 15 | 22 | 14 | 636.3 | 0.86 |
| 16 | 8 | 8 | 1000.0 | 0.50 |

| x | $l_x$ | $d_x$ | 1000 $q_x$ | $L_x$ | $T_x$ | $e_x$ |
|---|---|---|---|---|---|---|
| **HINDS** | | | | | | |
| 1 | 1000 | 137 | 137.0 | 931.5 | 5188.0 | 5.19 |
| 2 | 863 | 85 | 97.3 | 820.5 | 4256.5 | 4.94 |
| 3 | 778 | 84 | 107.8 | 736.0 | 3436.0 | 4.42 |
| 4 | 694 | 84 | 120.8 | 652.0 | 2700.0 | 3.89 |
| 5 | 610 | 84 | 137.4 | 568.0 | 2048.0 | 3.36 |
| 6 | 526 | 84 | 159.3 | 484.0 | 1480.0 | 2.82 |
| 7 | 442 | 85 | 189.5 | 399.5 | 996.0 | 2.26 |
| 8 | 357 | 176 | 501.6 | 269.0 | 596.5 | 1.67 |
| 9 | 181 | 122 | 672.7 | 120.0 | 327.5 | 1.82 |
| 10 | 59 | 8 | 141.2 | 55.0 | 207.5 | 3.54 |
| 11 | 51 | 9 | 164.6 | 46.5 | 152.5 | 3.00 |
| 12 | 42 | 8 | 197.5 | 38.0 | 106.0 | 2.55 |
| 13 | 34 | 9 | 246.8 | 29.5 | 68.0 | 2.03 |
| 14 | 25 | 8 | 328.8 | 21.0 | 38.5 | 1.56 |
| 15 | 17 | 8 | 492.4 | 13.0 | 17.5 | 1.06 |
| 16 | 9 | 9 | 1000.0 | 4.5 | 4.5 | 0.50 |

characterized by different feeding behaviors. Larvae reach full size in late June or July, stop feeding, and begin to spin a few threads of silk to form the cradle that will hold the pupae on the tree. The insects spend about two weeks as pupae and then emerge as adult moths. Males emerge first, several days ahead of females. Females, who do not fly, attract strong flying males by means of a powerful sex attractant. After mating, the females lay an egg mass containing a variable number of eggs (the sparser the population, the greater the number of eggs).

The life table represents the fate of a cohort from a single egg mass. The age interval or *x* column indicates life history stages which are of unequal duration. The $l_x$ column indicates the number of survivors at each stage. The $d_x$ column gives a breakdown of deaths by causes in each stage. In this particular population dispersion and predation account for most of the losses. Note that no life expectancy is calculated because there is none. All the adult population will die in late summer. The uses of such a life table will be discussed later.

*Plant life tables.* Mortality of plants has not received the same kind of conceptual treatment as mortality of animals. Patterns of plant breeding, generations, and survivorship do not lend themselves easily to the kind of statistical summaries recorded in life tables. Survivorship in plants can be expressed as percent of germination of seeds and as mortality and survival of seedlings. Some studies consider the percent of mortality of trees and shrubs caused by drought, disease, and outbreaks of insects.

One approach to the use of the life table in the study of plant mortality was developed by R. R. Sharitz and J. F. McCormick (1973) for *Sedum smallii* (Table 13-6). The time of seed formation is considered the initial point in the life cycle. Life expectancy of these annual plants drops rapidly in seed stages and returns to a high level after seedling establishment. Although individuals that become established have a good chance of surviving, the high early mortality results in a low mean life expectancy. Because the stages are unequal in duration, life expectancy here does not mean the same as it does in the typical life table.

Another approach to the life table of plants

insect will give the tabulated statistics more meaning. In April and May gypsy moth larvae emerge from eggs deposited by females the previous summer. In response to light the larvae climb to the tops of trees and spin down from unfolding leaves by a silken thread. Because the thread is easily broken, larvae may be carried by the wind for long distances which often results in a mass redistribution of the population.

The larvae go through a number of developmental stages called instars. Male larvae pass through five instars, females through six. Each instar, particularly the last three, is

TABLE 13-4
*Dynamic-composite life table for gray squirrels marked as nestlings, 1956–1964 (sexes combined)*

| Year marked | Nestlings marked | YEAR OF RETURN 1957 | 1958 | 1959 | 1960 | 1961 | 1962 | 1963 | 1964 |
|---|---|---|---|---|---|---|---|---|---|
| 1956 | 40 | 8 | 4 | 3 | 2 | 0 | 0 | 0 | 0 |
| 1957 | 138 | | 60 | 30 | 28 | 13 | 9 | 4 | 3 |
| 1958 | 229 | | | 61 | 26 | 12 | 10 | 7 | 3 |
| 1959 | 193 | | | | 58 | 26 | 19 | 12 | 9 |
| 1960 | 162 | | | | | 19 | 13 | 8 | 6 |
| 1961 | 99 | | | | | | 4 | 1 | 1 |
| 1962 | 82 | | | | | | | 18 | 6 |
| 1963 | 80 | | | | | | | | 25 |

| Age (x) | Total known alive | Maximum available for recapture | Known live per 1000 available |
|---|---|---|---|
| 0–1 | 1,023 | 1,023 | 1,000.0 |
| 1–2 | 253 | 1,023 | 247.3 |
| 2–3 | 106 | 943 | 112.4 |
| 3–4 | 71 | 861 | 82.5 |
| 4–5 | 43 | 762 | 56.4 |
| 5–6 | 25 | 600 | 41.7 |
| 6–7 | 7 | 407 | 17.2 |
| 7–8 | 3 | 178 | 16.9 |

| $x$ (Age) | $l_x$ (known survival) | $d_x$ (Apparent mortality) | $q_x$ (Proportional mortality) | $L_x$ (Average years lived) | $T_x$ (Total years lived) | $e_x$ (Life expectancy) |
|---|---|---|---|---|---|---|
| 0–1 | 1,000.0 | 752.7 | .753 | 538.9* | 989.6 | .99 |
| 1–2 | 247.3 | 134.9 | .545 | 179.9† | 450.7 | 1.82 |
| 2–3 | 112.4 | 29.9 | .266 | 97.4 | 270.8 | 2.41 |
| 3–4 | 82.5 | 26.1 | .316 | 69.5 | 173.4 | 2.10 |
| 4–5 | 56.4 | 14.7 | .261 | 49.0 | 103.9 | 1.84 |
| 5–6 | 41.7 | 24.5 | .588 | 29.4 | 54.9 | 1.32 |
| 6–7 | 17.2 | .3 | .017 | 17.1 | 25.5 | 1.48 |
| 7–8 | 16.9 | 16.9 | 1.000 | 8.4 | 8.4 | .50 |

$$*n = \frac{1,000\,(l - e^{i})}{i}; \overline{qx} = 0.635.$$

$$†n = \frac{l_x + (l_{x+1})}{2}.$$

Source: Barkalow et al., 1970.

are yield tables developed in commercial forestry (Table 13-7). Like the life table for animals, the yield table considers age classes and numbers of trees in each class, with additional columns giving diameter, basal area, and volume. Yield tables indicate mortality of trees by the reduced number of trees in each class, but as numbers decline through competition or removal, basal areas and biomass increase. Mortality does not necessarily reflect a declining population, but rather a maturing one.

Like life tables, yield tables are not constant for a species; they are constructed for different site classes that take into consideration the different environmental conditions under which the species grows.

### MORTALITY AND SURVIVORSHIP CURVES

From life tables two kinds of curves can be plotted—mortality curves based on the $q_x$ column and survivorship curves based on the $l_x$ column.

TABLE 13-5
*Life table typical of sparse gypsy moth populations in northeastern Connecticut*

| x (age interval) | $l_x$ (number alive at beginning of x) | $d_{xf}$ (factor responsible for $d_x$) | $d_x$ (number dying during x) | 100 $q_x$ ($d_x$ as percent of $l_x$) |
|---|---|---|---|---|
| Eggs | 550.0 | Parasites | 82.5 | 15 |
| | | Other | 82.5 | 15 |
| | | Total | 165.0 | 30 |
| Instars I–III | 385.0 | Dispersion, etc. | 142.4 | 37 |
| Instars IV–VI | 242.5 | Deer mice | 48.5 | 20 |
| | | Parasites and Disease | 12.1 | 5 |
| | | Other | 167.3 | 69 |
| | | Total | 227.9 | 94 |
| Prepupae | 14.6 | Predators, etc. | 2.9 | 20 |
| Pupae | 11.7 | Vertebrate predators | 9.8 | 84 |
| | | Other | 0.5 | 4 |
| | | Total | 10.3 | 88 |
| Adults | 1.4 | Sex (SR = 30:70) | 1.0 | 70 |
| Adult, female | 0.4 | — | — | — |
| Generation | — | — | 549.6 | 99.93 |

TABLE 13-6
*Life table for a natural population of* **Sedum smallii**

| x | $l_x$ | $d_x$ | 100 $q_x$ | $L_x$ | $T_x$ | $e_x$ |
|---|---|---|---|---|---|---|
| Seed produced | 1000 | 160 | 160 | 920 | 4436 | 4.4 |
| Seed available | 840 | 630 | 750 | 525 | 756 | 0.9 |
| Seed germinated | 210 | 177 | 843 | 122 | 230 | 1.1 |
| Seedlings established | 33 | 9 | 273 | 28 | 109 | 3.3 |
| Rosettes | 24 | 10 | 417 | 19 | 52 | 2.2 |
| Mature plants | 14 | 14 | 1000 | 7 | 14 | 1.0 |

*Source:* Sharitz and McCormick, 1973.

TABLE 13-7
*Yield tables for Douglas-fir on fully stocked acre, total stand*

|  | SITE INDEX 200 | | |
|---|---|---|---|
| Age (years) | Trees (per acre number) | Av. DBH (in.) | Basal area (ft²) |
| 20 | 571 | 5.7 | 101 |
| 30 | 350 | 9.0 | 154 |
| 40 | 240 | 12.2 | 195 |
| 50 | 176 | 15.3 | 224 |
| 60 | 138 | 18.2 | 248 |
| 70 | 113 | 20.9 | 268 |
| 80 | 97 | 23.3 | 285 |
| 90 | 84 | 25.6 | 299 |
| 100 | 75 | 27.6 | 312 |
| 110 | 69 | 29.4 | 323 |
| 120 | 63 | 31.1 | 332 |
| 130 | 59 | 32.7 | 341 |
| 140 | 55 | 34.3 | 350 |
| 150 | 51 | 35.8 | 357 |
| 160 | 48 | 37.2 | 364 |

*Source:* McArdle et al., 1949.

*Mortality curve.* A mortality curve is derived by plotting data in the $q_x$ or mortality rate column of the life table against age. It consists of two parts: (1) the juvenile phase in which the rate of mortality is high and (2) the post-juvenile phase in which the rate first decreases as age increases and then increases with age after a low point in mortality is reached (Figures 13-8 and 13-9). For populations of mammals, a roughly J-shaped curve results.

Being a ratio of the number dying during an age interval to the number alive at the begin-

ning of the period (or surviving the previous age period), $q_x$ is independent of the frequency of the younger age classes. Thus it is freer of the biases inherent in the $l_x$ column and survivorship curves. Most life tables of wild populations are subject to bias because the 1-year age class is not adequately represented. This error in frequency of the 1-year age class distorts each succeeding $l_x$ and $d_x$ value. If the first values are inaccurate, all succeeding ones are inaccurate. But if the first values of $q_x$ are wrong, the error does not affect the other values. For this reason mortality curves, which indicate the rate of mortality indirectly by the slope of the line, are more informative than survivorship.

*Survivorship curves.* The survivorship curve depicts age-specific mortality through survivorship and is obtained by plotting the number of individuals of a particular age cohort against time. It may be constructed in two ways. One is to plot logarithms of the number of survivors, or the $l_x$ column, against

**FIGURE 13-8**
*Mortality curves for the human population in the United States, 1966, and for the red deer population on the Isle of Rhum, 1957. The curve for humans is the typical J-shaped characteristic of mammals. Note the break in the J-shape caused by the sharp rise in mortality for deer between the sixth and tenth years. Compare this with the survivorship curve for the same population in Figure 13-10.*

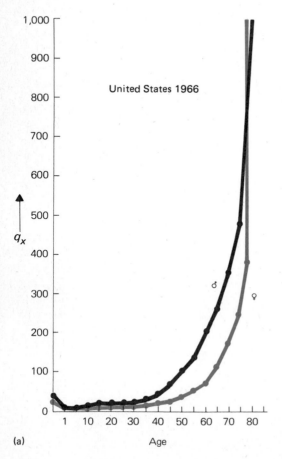

(a)

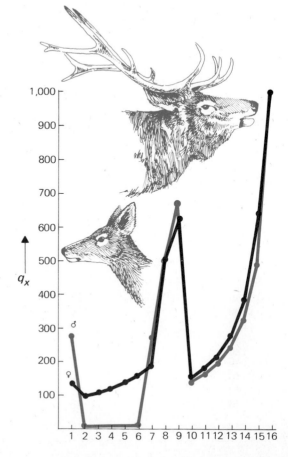

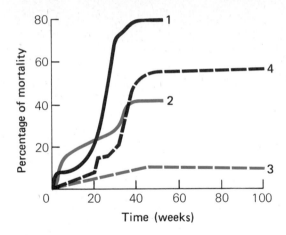

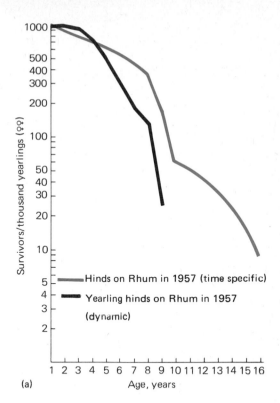

(a)

**FIGURE 13-9**
*Mortality curves for a population of dandelion* (Taraxacum officinale) *in a pure stand. The four curves show the variations resulting from conditions of high density and low light (1), low density and low light (2), low density and natural light (3), and unlimited water (4). (From Solbrig and Simpson, 1974.)*

time, with the interval on the horizontal coordinate and survivorship on the vertical coordinate (Figure 13-10). The other method is to plot the data for survivorship against the time interval scaled as percent deviation from the mean length of life (Figure 13-11). The validity of the survivorship curves depends on the validity of the life table and the $l_x$ column. Because life tables and thus survivorship curves are not typical of some standard population, but depict instead the nature of a population at different places at different times under different environmental conditions, survivorship curves are useful for comparing the population of one area, time, or sex with the population of another.

Survivorship curves are classified into at least three hypothetical types (Figure 13-12) (Deevey, 1947). The type I curve is concave and is typical of organisms with extremely

**FIGURE 13-10**
*Survivorship curves for red deer on the Isle of Rhum, 1957, as constructed from both time-specific and dynamic life tables. Both hinds and stags are harvested, which holds the survivorship curves of the two sexes to about the same level. (From Lowe, 1969.)*

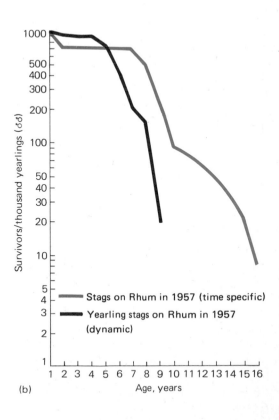

(b)

**445**

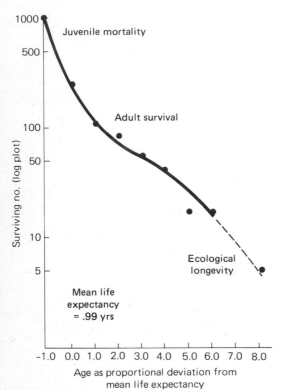

**FIGURE 13-11**
*Survivorship curve for 1,023 known-age gray squirrels. (From Barkalow, 1970.)*

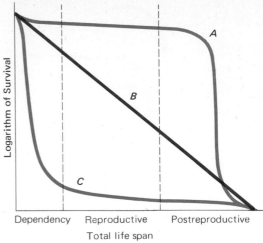

**FIGURE 13-12**
*Three basic types of survivorship curves. The vertical scale may be graduated arithmetically and logarithmically. If graduated logarithmically the slope of line will show the following rates of change: Type I, curve for animals with high mortality early in life; Type II, curve for animals in which the rate of change of mortality is fairly constant at all age levels, a more or less uniform percentage decrease in the number that survive; Type III, curve for animals living out the full physiological life span of the species. (After M. Alexander, 1958.)*

high mortality rates in early life, such as oysters, fish, many invertebrates, and some plants such as the *Sedum smallii* (Figure 13-13). The type II curve is linear and is typical of organisms with a constant mortality rate throughout all age classes. Such a curve is characteristic of hydra, adult stages of many birds (Figure 13-14) and rodents, and some perennial plants such as mustard, *Ranunculus* (Figure 13-15). The type III curve is convex and is typical of individuals that tend to live out their physiological life spans, that is, or-

ganisms with a high degree of survival throughout life and a heavy mortality at the end of the species' life span. Such a curve is typical of some plants (Figure 13-16) and of humans (Figure 13-17), mountain sheep, and other mammals.

These are conceptual models only and are not necessarily realistic, that is, not something to which survivorship curves must conform. But they do serve as a model to which survivorship of a species can be compared. Most survivorship curves are some sort of intermediate between two of the models. Consider, for example, the series of survivorship curves of the population of Sweden from the middle of the eighteenth century to the present (Figure 13-17). During the early eighteenth century the mortality of young individuals was high and life expectancy relatively low. With advances in medicine and modifications of the environment, the survivorship curves re-

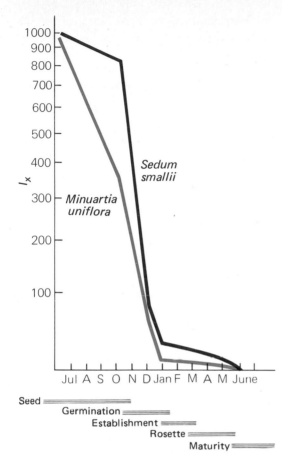

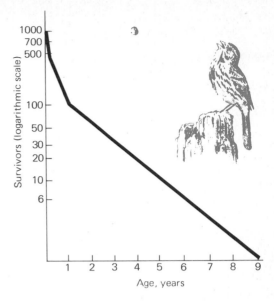

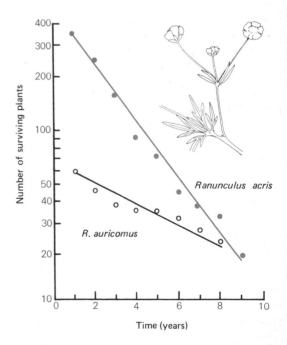

**FIGURE 13-13**
*Survivorship curves for a natural population of* Sedum smallii *and* Minuartia uniflora. *The curves suggest the type of survivorship curve typical of organisms experiencing early mortality. (From Scharitz and McCormick, 1973.)*

**FIGURE 13-14**
*Survivorship curve for the song sparrow* (Melospiza melodia). *The curve is typical of birds. After a period of high juvenile mortality, a concave or type I curve, the survivorship curve is linear or type II. (From Johnson, 1956.)*

**FIGURE 13-15**
*Survivorship curves of* Ranunculus acris *and* Ranunculus auricomus. *These curves are linear or type II. (From Sarukhan and Harper, 1974; based on data from Rabotnov, 1958.)*

flect a shift in mean life expectancy toward maximum.

## Natality

The greatest influence on population increase is usually natality, the production of new individuals in the population. Natality may be described as maximum or physiological natality and as realized natality. Physiological natality represents the maximum possible number of births under ideal environmental conditions, the biological limit. Because this is rarely achieved in wild populations, its measure is of little value to the field biologist, but it is a useful yardstick against which to compare

**447**

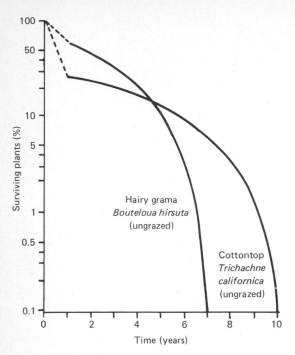

**FIGURE 13-16**
*Survivorship curves of two range grasses,*
*hairy grama* (Bouteloua hirsuta) *and*
*cottontop* (Trichachne californica). *These*
*curves are type III. (From Sarukhan and*
*Harper, 1974; data from Canfield, 1957.)*

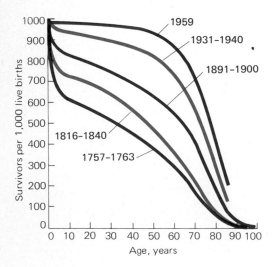

**FIGURE 13-17**
*Survivorship curves for the population of*
*Sweden over several centuries. Note that as*
*health conditions improved and the standard*
*of living increased, the survivorship curves*
*began to approach physiological life span,*
*changing from concave to convex, from type*
*I to type III. (From Clark, 1967.)*

realized natality. Realized natality is the amount of successful reproduction that actually occurs over a period of time. It reflects the type of breeding season (continuous, discontinuous, or strongly seasonal), the number of litters or broods per year, the length of gestation or incubation, and so on. It is influenced by environmental conditions, nutrition, and density of the population.

### NATALITY IN ANIMALS

Natality, measured as a rate, may be expressed either as crude birth rate or specific birth rate. *Crude birth rate* is expressed in terms of population size, for example, 50 births per 1,000 population. *Specific birth rate,* a more accurate measure, is expressed relative to a specific criterion such as age. The most usual form is an age-specific schedule of births, the number of offspring produced per unit time by females in different age classes.

Because population increase depends upon the number of females in the population, the age-specific birth schedule counts only females as giving rise to females. The age-specific schedule is obtained by determining the mean number of females born in each age group of females. This provides the statistic $m_x$. Given the $l_x$ or survivorship from the life table and the $m_x$ schedule one can obtain the *net reproductive rate, $R_o$*, the number of females left during a lifetime by a newborn female. In human demography the net reproductive rate is usually modified into a *fertility rate*, the number of births per 1,000 women 15 to 40 years of age.

An example of how net reproductive rates are determined is shown in the fecundity table for red deer (Table 13-8). This fecundity table uses the survivorship column $l_x$ from the life table (Table 13-3) and an $m_x$ column, the mean number of females born to females in each age group. For calculation of the net reproductive rate, the $l_x$ column is converted to a proportionality, in this case by dividing each age value by 1,000. Female red deer aged 1 and 2 years produced no young; therefore, their $m_x$ value is 0. The $m_x$ values for females 3 years old and above increase with age until 7 years, when a decline in mean number of births begins. If the $m_x$ values for each age group are summed, the sum 4.569 is the mean number of

**448**

TABLE 13-8
*Fecundity table of Red Deer*

| $x$ | $l_x{}^*$ | $m_x$ | $l_x m_x$ | $x l_x m_x$† |
|-----|-----------|-------|-----------|--------------|
| 1 | 1.000 | 0 | 0 | 0 |
| 2 | .863 | 0 | 0 | 0 |
| 3 | .778 | .311 | .242 | 0.726 |
| 4 | .694 | .278 | .193 | 0.772 |
| 5 | .610 | .308 | .134 | 0.920 |
| 6 | .526 | .400 | .210 | 1.260 |
| 7 | .442 | .476 | .210 | 1.470 |
| 8 | .357 | .358 | .128 | 1.024 |
| 9 | .181 | .447 | .081 | 0.729 |
| 10 | .059 | .289 | .017 | 0.170 |
| 11 | .051 | .283 | .014 | 0.154 |
| 12 | .042 | .285 | .012 | 0.144 |
| 13 | .034 | .283 | .010 | 0.130 |
| 14 | .025 | .282 | .007 | 0.098 |
| 15 | .017 | .285 | .005 | 0.075 |
| 16 | .009 | .284 | .003 | 0.048 |

$$R_o = \Sigma l_x m_x = 1.316 \qquad \Sigma x l_x m_x = 7.72$$

*Based on $l_x$ values from Table 13-3.
†This column represents generation time and is discussed in Chapter 14.

females that would be produced by each doe in a lifetime of 16 years if survival were complete. But obviously each doe does not live a full 16 years. To adjust for mortality in each age group the $m_x$ value for each age class is multiplied by the corresponding $l_x$ or survivorship value. The resulting value $l_x m_x$ gives the mean number of females born in each age group adjusted for survivorship; thus for age class 3, the $m_x$ value is 0.311, but adjusted for survivorship the value drops to 0.242. When these adjusted values are summed over all ages at which reproduction occurs the sum represents the number of females that will be left during a lifetime by a newborn female or $R_o$, the net reproductive rate. For red deer population with the survivorship indicated, the reproductive rate is 1.316.

Reproductive rate is measured on the basis of mature females and omits males and immature females. Useful for comparative purposes, the reproductive rate summarizes information on the frequency of pregnancy, the number of females born, and the length of breeding season. Because the simplest methods of obtaining reproductive rates involve counting embryos, placental scars, number of eggs, and unfledged young in birds, the reproductive rate often incorporates a measure of mortality for the original group of ova.

## NATALITY IN PLANTS

There are several distinct demographic differences between plant and animal populations. Plants, unlike animals, accumulate structural units including new shoots, leaves, and buds above ground, new roots below ground, and new clones or vegetative extensions of the parent plant. At the same time they produce new individuals through sexual reproduction. Thus a plant population exists at two levels, as individual plants and as colonies that develop asexually or grow as extensions of individual plants. Death and natality occur at both levels. A plant may acquire new structural units and such units may die while the plant remains alive. Or the whole plant may die or reproduce new individuals through sexual reproduction.

Sexual reproduction in plants involves two processes, the production of seeds and the germination of seeds. Except for annuals and biennials it is difficult to estimate seed production by individual plants and by populations. Perennial and woody plants, even within a population, vary in longevity and in seed production which is not necessarily an annual event. Individual plants vary widely in seed production from year to year and the amount of seed produced over a lifetime is largely unknown. Seeds usually undergo a period of dormancy until conditions are right for germination. Germination is the formal equivalent of birth in plants (for a discussion of these ideas see Harper and White, 1974).

## Life equation and key factor analysis

Two other approaches to an analysis of survivorship and fecundity are the life equation and key factor analysis.

### LIFE EQUATIONS

A picture of the limitations of the growth of a population, seasonal gains and losses, and other important events occurring throughout the year can be summarized in a life equation table (Table 13-9). Since slight changes in reproduction, survival, or sex ratios can influence the rate of increase considerably from year to year, wildlife biologists, especially, find the information summarized in the life equation highly useful.

The life equation is a modification of the life table. The life table is a mathematical expression of the vital statistics of a population. The life equation illustrates changes within the population. It involves a census or inventory of an identifiable population. Age in the life equation is referred to as stages in the life history of a population within one breeding cycle, instead of within a day, a month, or a year.

Life equations, like life tables, begin with a given population, usually 1,000, broken down into sex and age categories. If a game animal is involved, the table begins with a prehunting population. Hunting losses then are subtracted, according to sex and age, leaving a posthunting population. The number left after winter losses comprise the prebreeding season population. To this is added the breeding season gains. Finally, a new prehunting population estimate is obtained for one year later.

These tabular data of gains and losses can be presented as a sort of survival curve (Figure 13-18). Curves of several years can be joined to illustrate numerical changes over a period of years. In addition, changes in age structure can be indicated by showing the proportions of several age classes.

Life equations are not precise. Some categories in the equation cannot be measured accurately and must be estimated. However, the information which can be derived from these estimates may be quite accurate. Properly constructed, the life equation shows the magnitude of population losses due to several causes and where and when the heaviest losses occur. The life equation also indicates the extent and importance of production of young to the future of the population, gaps in knowledge of population behavior of the species involved, and the most important research problems for future study.

### KEY FACTOR ANALYSIS

A key factor (*k* factor) is any biological or environmental condition associated with mortality that is useful in predicting a future trend in a population, but that does not necessarily have a cause and effect relationship with mortality (Morris, 1959). It is utilized chiefly by some entomologists to assess the effects of any one environmental influence on the trend of a population. It requires a series of age-specific life tables or at least the type of data utilized in the life equation approach.

In order to understand how the key factor analysis works we can use the data from the life table of the gypsy moth (Table 13-5). The first figure in the *k* factor table (Table 13-10) is the maximum potential natality for each generation determined by multiplying the number of females of reproductive age by the maximum number of eggs per female. For our example we will simply consider the maximum fecundity for one female gypsy moth in a sparse population as 800 eggs. The $l_x$ value of each successive stage of the life cycle as it appears on the life table is entered and the values converted to logarithms. Each logarithm is subtracted from the previous one to give a *k* or mortality value for each age class. These values are added to give a total generation mortality *K*. This is done for a number of successive generations. To identify

*FIGURE 13-18*
*Diagrammatic life equation of a population of California quail* (**Lophortyx californica**). *The bars indicate the ratio of mature to immature birds in November. (After J. T. Emlen, 1940.)*

TABLE 13-9
*Life equation of a black-tailed deer population on a 36,000-acre area, 1949 to 1954*

| Year | Type of gain or loss | MALES | | | FEMALES | | | Total |
|------|----------------------|-------|-----------|-------|---------|-----------|--------|-------|
| | | *Adults* | *Yearlings* | *Fawns* | *Fawns* | *Yearlings* | *Adults* | *Total* |
| 1949 | Prehunting population | 312 | 140 | 456 | 475 | 274 | 1,003 | 2,690 |
| | Legal hunting kill | 204 | 3 | — | — | — | — | −207 |
| | Crippling loss and illegal kill | 10 | 11 | 8 | 8 | 8 | 38 | − 83 |
| | Winter losses | 13 | 19 | 304 | 229 | 40 | 135 | −740 |
| 1950 | Prefawning population | 85 | 107 | 144 | 238 | 226 | 860 | 1,660 |
| | Fawning season gain | 192 | 144 | +707 | +589 | 238 | 1,098 | 2,968 |
| | Summer mortality | 2 | 1 | 221 | 83 | 12 | 52 | −371 |
| | Prehunting population | 190 | 143 | 486 | 506 | 226 | 1,046 | 2,597 |
| | Legal hunting kill | 125 | 25 | 71 | 86 | 43 | 160 | −510 |
| | Crippling loss | 6 | 12 | 21 | 26 | 13 | 50 | −128 |
| | Winter losses | 2 | 6 | 80 | 61 | 10 | 37 | −196 |
| 1951 | Prefawning population | 57 | 100 | 314 | 333 | 160 | 799 | 1,763 |
| | Fawning season gain | 157 | 314 | +617 | +515 | 333 | 959 | 2,895 |
| | Summer mortality | 2 | 3 | 311 | 195 | 17 | 48 | −576 |
| | Prehunting population | 155 | 311 | 306 | 320 | 316 | 911 | 2,319 |
| | Legal hunting kill | 96 | 9 | — | — | — | — | −105 |
| | Crippling loss | 5 | 10 | 3 | 3 | 6 | 15 | − 42 |
| | Winter loss | 2 | 8 | 89 | 67 | 9 | 42 | −217 |
| 1952 | Prefawning population | 52 | 284 | 214 | 250 | 301 | 854 | 1,955 |
| | Fawning season gain | 336 | 214 | +762 | +601 | 250 | 1,155 | 3,318 |
| | Summer mortality | 3 | 2 | 436 | 260 | 12 | 58 | −771 |
| | Prehunting population | 333 | 212 | 326 | 341 | 238 | 1,097 | 2,547 |
| | Legal hunting kill | 178 | 80 | 62 | 64 | 55 | 205 | −644 |
| | Crippling loss | 9 | 20 | 19 | 20 | 17 | 70 | −161 |
| | Winter loss | 5 | 6 | 71 | 54 | 8 | 30 | −174 |
| 1953 | Prefawning population | 141 | 106 | 174 | 203 | 158 | 786 | 1,568 |
| | Fawning season gain | 247 | 174 | +608 | +506 | 203 | 944 | 2,672 |
| | Summer mortality | 2 | 2 | 338 | 225 | 10 | 47 | −624 |
| | Prehunting population | 245 | 172 | 270 | 281 | 193 | 897 | 2,058 |
| | Legal hunting kill | 85 | 26 | 38 | 36 | 18 | 70 | −273 |
| | Crippling loss | 4 | 9 | 8 | 8 | 7 | 27 | − 63 |
| | Winter loss | 5 | 7 | 71 | 53 | 7 | 29 | −172 |
| 1954 | Prefawning population | 151 | 130 | 153 | 184 | 161 | 771 | 1,568 |

*Source:* Adapted from E. R. Brown, 1961.

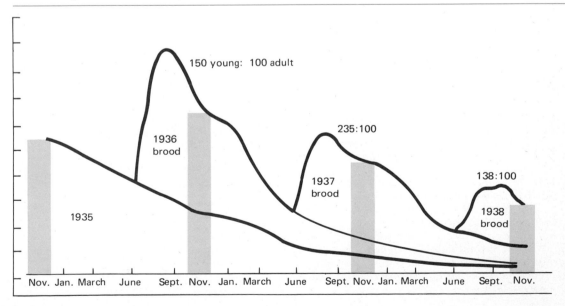

TABLE 13-10

k *Values for a Gypsy moth population based on life table for sparse population in Connecticut (Table 13-5)*

| x | $l_x$ | Logarithm of $l_x$ | k value |
|---|---|---|---|
| Maximum natality | 800.0 | 2.903 | — |
| Eggs laid | 550.0 | 2.740 | 0.163 |
| Larvae I–III | 385.0 | 2.585 | 0.155 |
| Larvae IV–VI | 242.0 | 2.385 | 0.200 |
| Prepupae | 14.6 | 1.164 | 0.221 |
| Pupae | 11.7 | 1.068 | 0.096 |
| Adult | 1.4 | 0.146 | 0.922 |
| | | | $K = 1.757$ |

the key factor which influences trends in adult populations, the *k* values for each successive generation are plotted along with *K*. The plot shows whether the mortality rate for one particular stage or age class consistently displays over the generations a strong correlation with total mortality (Figure 13-19). If there is some correlation between the *k* value of a particular stage and total *K*, then the analysis can be carried further to determine the *k* factor within that stage. (For details on use of *k* factor analysis, see Dempster, 1975.)

## Age structure

Populations may be divided into three ecological periods: prereproductive, reproductive, and postreproductive. The relative length of each period depends largely on the life history of the organism. Among annual species the length of

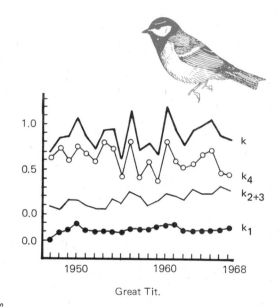

Great Tit.

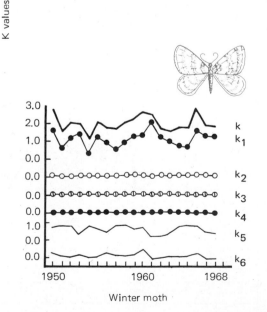

Winter moth

FIGURE 13-19

*Graphical key factor analysis applied to the great tit* (Parus major) *and the winter moth* (Operophtera brumata). *The key factor in the life cycle of the great tit in Marley Woods, Oxford, is* $k_4$ *mortality outside of the breeding season. In the rest of the annual life cycle variations in clutch size,* $k_1$, *and hatching success,* $k_2$, *are density dependent and are sufficient to regulate the population. Key factors in the life cycle of the winter moth are* $k_1$, *overwintering loss of winter moth eggs and larvae before the first larval census in spring, and* $k_5$, *a density dependent loss of pupae in the soil due to predation in the spring. (From Podoler and Rogers, 1975.)*

the prereproductive period has little significant effect on the rate of population growth, because all reproduction takes place in one season, except in multibrooded species. In longer lived plants and animals the length of the prereproductive period has a pronounced effect on the rate of growth. Organisms with a short prereproductive period often increase rapidly and have a short span between generations. Organisms with a long prereproductive period generally increase more slowly and have a long span of time between generations.

## AGE STRUCTURE IN ANIMALS

Theoretically, all continuously breeding populations tend toward a stable age distribution (see Chapter 14), that is, the ratio of each age group in a growing population remains the same if the age-specific birth rate and the age-specific death rate do not change. When mortality balances natality and the population is closed, that is, it experiences neither movements into nor out of the population, then the population has reached a stationary age distribution. If the stable condition is disrupted by any cause, such as a natural catastrophe, disease, famine, or emigration, the age composition will tend to restore itself upon the return to previous conditions, provided of course that the rate of birth and death are still the same.

Changes in age class distribution reflect changes in birth rates, survival rates, and age-specific death rates. Any influence that causes age ratios to shift because of changes in age-specific death rates affects the population birth rate. If life expectancy is low (high death rate for older age classes), a greater proportion of the population falls into the reproductive class, automatically increasing the birth rate. Conversely, if life expectancy is high, then a greater proportion falls into the postreproductive class, reducing the birth rate. Rapidly growing populations are usually characterized by declining death rates, especially in the very young age classes, which inflate the younger age groups. Declining or stabilized populations are characterized by lower birth rates with fewer young to rise into the reproductive age classes and by a larger proportion in older age classes.

Age structure is best visualized through age pyramids that show the percentage ratios among age groups (Figure 13-20). As a population grows and declines, the number of individuals and therefore the ratio in each age class fluctuates. Growing populations are characterized by a large number of young. The pyramid of a growing population has a broad base and a narrow pinched top. It indicates a youthful population, increasing numbers entering the reproductive age, and an increase in all age classes. If the population is neither growing nor declining (the number of births equal the number of deaths), then the number of individuals in each age class tends to remain the same (stationary age distribution). The pyramid of such a population is a narrow one and suggests a situation with no significant population growth. If a population is declining, few individuals are being added to the reproductive class and all other age classes in the population and a high proportion of the population may be in the older age classes. Pyramids of such populations usually have a narrow base of young. In some populations the age pyramid has a marked indentation in the reproductive age period and a relatively large percentage of its population in the pre- and postreproductive age periods.

From age structure one can determine a statistic useful in studies of human populations, the dependency ratio. This ratio measures the load placed on the reproductive portion of the population by the young and dependent old. The ratio is determined by dividing the sum of the number of people less than 20 years old and the number of people 65 years and over by the number of people 20 to 64 years old (Figure 13-20). The dependency ratio provides an assessment of the costs of education, health care, social security, and old-age assistance that must be provided by the productive age classes. One outcome of a stable age distribution and zero population growth is the large burden of social services the working group is forced to bear. This is especially true in a population that is moving from a decreasing number of young to an increasing number of old. Also bearing a heavy dependency burden are rapidly expanding populations with a large number of young.

*Age pyramids and population history.* The history of a population shows in its age dis-

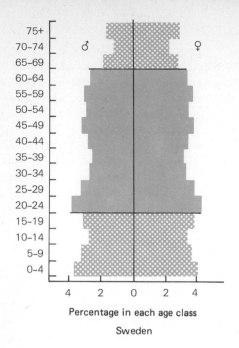

Percentage in each age class

Sweden

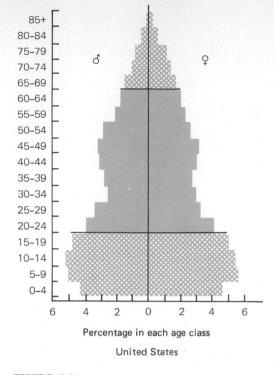

Percentage in each age class

United States

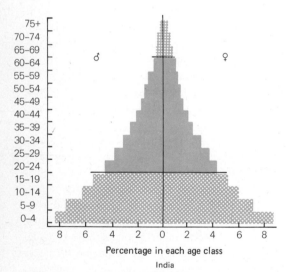

Percentage in each age class

India

**FIGURE 13-20**

*Types of age structures in populations as represented by age pyramids. The pyramids are divided into three dependency ratio components: young dependent, ages 0 to 14; economically active ages, 15 to 64; elderly dependents, 65 and over. The age pyramid for India shows a rapidly expanding population. It has a broad base of young age classes that will enter the reproductive age classes and a narrow peak of older age classes. This suggests a youthful population and foretells rapid population growth. The constricted shape of the age pyramid for the United States results from a declining birth rate. The youngest age classes are no longer numerically the largest. This type of age structure reflects declining fertility which in time will further distort the age distribution. The age pyramid of Sweden is characteristic of a country approaching zero population growth. These pyramids are applicable to human populations, but they should not be regarded as generalizations for all organisms. The pyramid for India, for example, may represent a stable population for a number of species with high reproductive rates and uniform mortality. Pyramids of growing, stable, and declining populations will vary among species. Compare these pyramids with those of Figure 13-23 for some game animals. (Data for India from UN Population Division, 1970. Data for United States from U.S. Census Bureau, 1970; diagram from R. L. Smith, 1976. Data for Sweden from UN Population Division, 1970; diagram from R. L. Smith, 1976.)*

tribution. An example is two contrasting areas in Appalachia, one in a county characterized by agriculture, small industry, and a stable population (Figure 13-21) and the other in a county characterized by an unstable, exploitative industry, coal mining, and a population which has been subject to an abrupt increase and decrease (Figure 13-22).

In the agricultural county with a relatively low population per square mile, the age pyramids for 1930 and 1940 have an expanded base of young, while the percentages shrink in successive age classes. Members of the young age classes obviously left the area as they en-

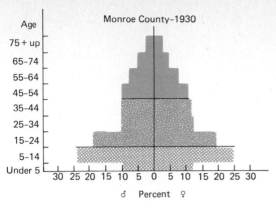

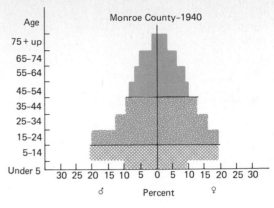

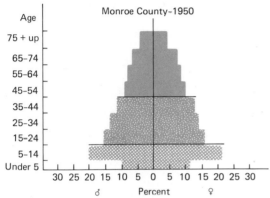

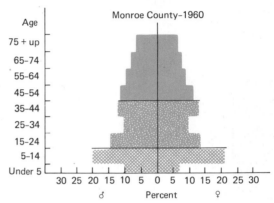

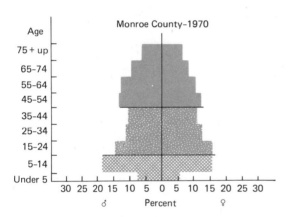

*FIGURE 13-21*
*Age pyramids for Monroe County, an agricultural region in southern West Virginia, from 1930 to 1970. Because of the variability in the way the Census Bureau broke down the age classes of county populations at different census periods, the age classes span 10 rather than 5 years; the 5-year-old age class is retained to show variation in births through the years. The pyramids reflect a population making a transition from growth to stability. Compare these pyramids with those in Figure 13-21 and study them in relation to the growth curve for the same region in Figure 14-7. (Data U.S. Census Bureau; pyramids from R. L. Smith, 1976.)*

tered the reproductive age, but not in numbers so great as to distort the 15- to-34-year-old age classes in the pyramid. The excess young were drained away, leaving a residual to maintain a fairly constant population level. Since 1950 the age pyramid has suggested a stable population (although in 1970 the crude birth rate was 13.4 and the crude death rate was 14.7).

Age pyramids for the coal mining county, by contrast, reflect a "boom and bust" type of population growth (see Figure 14-3). The age pyramid for 1930 reflects the heavy immigration of the previous decade. The pyramid has disproportionately large 25- to 44-year-old-age classes. The profile of the pyramid for 1940 is equilateral, characteristic of a population experiencing a high birth rate and a high death rate. Since the region was not experiencing a high death rate, the profile resulted from the movement of one age class into another over time.

The great migration of the 1950s shows in the pyramid for 1960. The sharply pinched middle emphasizes a large decrease in the 15- to 34-year-old age classes, a marked increase, in the older age classes, and only a moderate decline in the 0- to 15-year-old classes. The

**455**

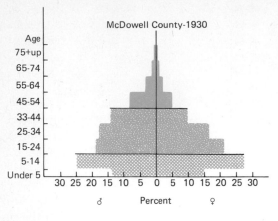

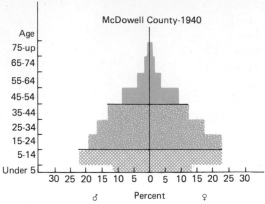

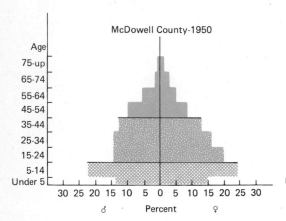

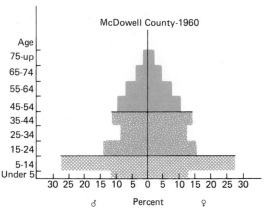

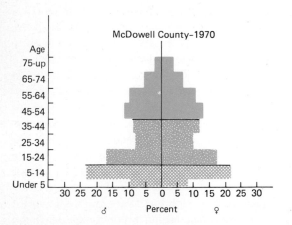

**FIGURE 13-22**
*Age pyramids for MacDowell County, a coal mining region in southern West Virginia, from 1930 to 1970. The pyramid for 1930 reflects the immigration of people into the mining region. A broad base of young suggests a future expansion of the population, and that expansion is reflected in the pyramids for 1940 and 1950. The 1960 pyramid shows a major change. The middle has become pinched and reflects the heavy emigration of younger elements of the reproductive age group. The emigration results in a high ratio of dependent young and old to the economically active. It also reflects a declining population. The extent of the decline is indicated in the age pyramid for 1970, which suggests an aging and declining population. Note, however, that young age classes, under 5 and 5 to 14, are still relatively high. Compare these pyramids with those in Figures 13-21 and 13-22 and study them in relation to the growth curve for this county in Figure 14-4. (Data from U.S. Census Bureau; pyramids from R. L. Smith, 1976.)*

pyramid for 1970 shows a further aging and decline of the population. The older age classes (45 to 65 and over) show an increase, the middle of the pyramid is still pinched, and the number of young is declining. In spite of a declining population the birth rate in this county is very high: 35.6 in 1970, 21.2 in 1972.

Basically a population can do three things: increase, decrease, or remain stable. The ratio of young to adults in a relatively stable population is approximately 2:1 in most populations of game animals (Figure 13-23). A normally increasing population should have an increasing number of young; a decreasing population, a decreasing number of young. Within this framework there are a number of variations. A population may be decreasing, yet show an increasing percentage of young. Or a population may be stable with a decreasing percentage of young. By combining information on population density, age ratios, and reproduction, a biologist can correlate changes in population structure with habitat changes and ecological and human influences.

### AGE STRUCTURE IN PLANTS

Age structure is rarely employed in the study of plants because in many instances plants do not lend themselves to that type of analysis (see White and Harper, 1970; Harper and White, 1974). In 1968 H. L. Kerster attempted to apply the procedures for measuring animal populations to plant populations. He worked up the age structure of the prairie forb blazing star, which can be aged by collecting (destructively) all the plants from sample quadrants and counting the annual rings on the freshly cut surfaces of the corms. The plants were sorted into two groups, rosettes only and rosettes bearing one or more flowering spikes. Thus the plants could be sorted into age categories and the reproductive members separated from the nonreproductive individuals.

The age pyramids developed from the data contrast two populations (Figure 13-24). The maximum age among the specimens is 35 years. The scarcity of plants over 20 years old can be accounted for by mortality from a hot prairie fire in the past, but desiccation and insect attacks were also involved. There is a greater variation in size of adjacent-year classes than one finds in animal populations because of variations in the germination of seeds and the survival of seedlings. Reproduction of flowering starts at about 9 years of age, and each "adult" reproduces rather regularly every other year. The average generation span is equal to the average age of the flowering plant. One population appears to be senescent and aging and the other shows recruitment of young or resurgence.

*FIGURE 13-23*
*Theoretical age pyramids, especially applicable to big game animals. Note how the pyramids flatten with increasing production of young. (After M. Alexander, 1958.)*

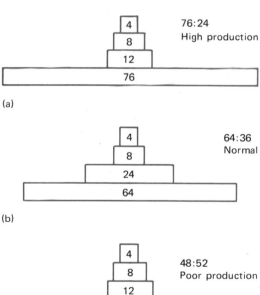

(a)    76:24 High production

(b)    64:36 Normal

(c)    48:52 Poor production

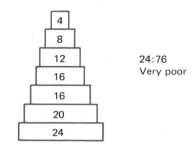

(d)    24:76 Very poor

## Sex ratio

An almost universal characteristic of plants and animals is sexual reproduction, although some organisms reproduce asexually at times. Even these species have some provision for sexual reproduction in their life cycle, for only by mixing and recombining genes can a population maintain genetic variability.

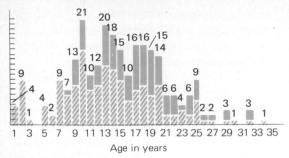

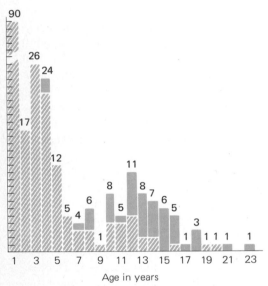

**FIGURE 13-24**

*Age pyramids for two populations of blazing
star* (Liatris aspera). *Numbers above bars
indicate the number in each year class.
Shaded portions of bars indicate plants not
flowering in 1965, the year of collection;
unshaded portions indicate flowering
specimens. The relatively low number of
young plants in population 1 suggests
recruitment is not adequate to maintain the
population. The low number of plants 7 to 11
years old in population 2 suggests population
senescence; however, the large number in
the seedling through 6-year-old classes
implies resurgence. (From Kerster, 1968.)*

Populations of most organisms tend toward
a 1:1 sex ratio (the proportion of males to
females in a population). The primary sex
ratio (the ratio at conception) tends to be 1:1.

(This may not be universally true and is, of
course, very difficult to determine.)

The secondary sex ratio (the ratio at birth)
among mammals is often weighted toward
males, but it shifts toward females in older age
groups. For example, the ratio among fetuses
of elk in western Canadian national parks was
113 males to 100 females (Flook, 1970). Be-
tween ages $1\frac{1}{2}$ to $2\frac{1}{2}$ the ratio of males to
females dropped abruptly and continued to do
so until it remained at about 85:100, although
in certain areas it dropped as low as 37:100.
The greatest decline in the number of males
occurred between the ages of 7 to 14. The decline of females was much less rapid. The difference in the rate of decline has an additional
effect on the higher ratio of females to males.
The loss of males allows more food and space
for the females and young, thus further increasing the female rate of survival.

Among humans, too, males exceed females
at birth, but as age increases the ratio swings
in favor of females. In 1965 the ratio of males
to females based on a stable age distribution
in the United States was: age 0 to 4, 104:100;
age 40 to 44, 100:100; age 60 to 64,88:100; age
80 to 84, 54:100. Lower mortality among females is characteristic of both advanced and
underdeveloped countries.

Among birds the sex ratios tend to remain
weighted toward the males (Bellrose, 1961).
Fall and winter sex ratios of prairie chickens
show a preponderance of males in adult
groups; similar ratios are characteristic of
other gallinaceous birds and some passerine
birds.

It is not easy to explain why sex ratios
should shift from an equal ratio at birth to an
unequal one later in life. Perhaps a partial
answer lies in factors related to the genetic determination of sex and the physiology and behavior of the two sexes.

Sex in organisms is determined by the X and
Y chromosomes. The XY combination of
chromosomes produces males in mammals and
females in birds and some insects, notably butterflies. Perhaps the chromosome combination
itself may be partially responsible for biased
losses of females in birds and males in mammals. Each gene on the X and Y chromosome
is expressed in the XY combination, while
in the XX combination the heterozygous com-

bination of alleles can mask the harmful effects of single recessive gene. Therefore, XY adults may be more susceptible to disease, physiological stress, and aging than XX organisms.

Physiological and behavioral patterns affect mortality. For example, during the breeding season the male elk battles other males for dominance of a harem, defends his harem from rivals, and mates with the females. These activities not only consume considerable energy, but leave little time for feeding, and the male often ends the breeding season in poor physical condition. Among birds the female may help the male defend territory, builds the nest, lays and incubates the eggs, broods the young, and, often with the help of the male, feeds the young. While incubating the eggs and brooding the young, the female is much more vulnerable to predation and other dangers than is the male. Thus adult males in mammals and adult females in birds apparently expend more energy, are subject to greater stress, and are more vulnerable to predation and other dangers. Their higher vulnerability and mortality is consistent with the imbalance in the sex ratio in the older age cohorts.

## SUMMARY

Living organisms exist in groups of the same species. These groups, or populations, occupy a particular space; the size of the population in relation to a definite unit of space is its density.

Populations are distributed in some kind of pattern over an area. Some are uniformly distributed, a very few are randomly distributed, and most exhibit a clumped or contagious distribution which results in aggregations. Some aggregations reflect a degree of sociality on the part of the population members, which may lead to cooperative or competitive situations. The pattern of clumped distribution is called coarse-grained when the clumps and areas between them are large and fine-grained when the clumps and areas between them are small.

Population density depends upon the number of individuals added to the group and the number leaving, the difference between the birth rate and death and the balance between emigration and immigration. Birth rate usually has the greatest influence on the addition of new individuals. Mortality is a reducer and is the greatest in young and old. Mortality and survivorship are best analyzed by means of a life table, which is an age-specific summary of mortality operating on a population. From the life table one can derive both mortality curves and survivorship curves that are useful in comparing demographic trends within a population and between populations living under different environmental conditions. Reproductive rates of populations can be best analyzed by means of a life table, relate reproduction with age and survivorship. Age and sex ratios influence natality and mortality rates of populations. Reproduction is limited to certain age classes; mortality is most prominent in others. Changes in age class distribution bring about changes in production of young and mortality. The sex ratio tends to be balanced between males and females at birth, but the ratio changes to favor females in mammals and males in birds as cohorts age.

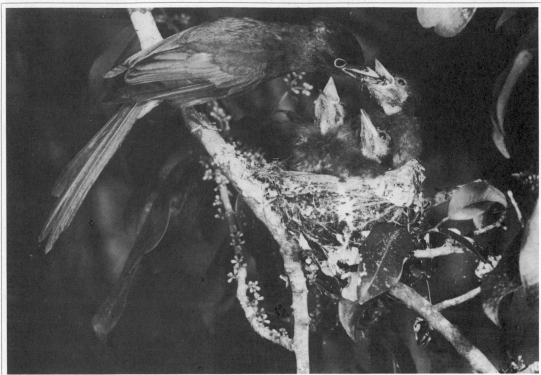

# Population growth
## CHAPTER 14

Populations are dynamic; they are constantly changing. Depending on the nature of the organism, populations may change from hour to hour, day to day, season to season, year to year. In some years certain organisms are abundant; in other years they are scarce. Local populations appear, expand, decline, and occasionally disappear into extinction. Eventually the area may be recolonized by individuals moving in from other populations. Changes are brought about by the interaction of organisms and their environment as it affects birth rates, death rates, and the movement of individuals in and out of populations.

### Rate of increase

Possessing sufficient data to construct a life table and knowing the age-specific fecundity (discussed in Chapter 13), one can determine some characteristics of population growth. By multiplying the age-specific birth rate, $m_x$, by survivorship, $l_x$, and summing the products, one obtains the net reproductive rate, $R_0$. If $R_0$ equals 1, the birth rate equals the death rate; individuals are replacing themselves and the population is remaining stable. If the value is greater than 1, the population is increasing; if it is less than 1, it is decreasing.

From the same information one can also chart the growth of the population through the construction of a life history table. To construct this table, one needs the data included in a life table and one more parameter, $p_x$, the proportion of animals surviving in each age class. The parameter $p_x$ is the complement of $q_x$ and is expressed as $1 - q_x$. It has an advantage over $l_x$, because like $q_x$ it is independent of survivorship of earlier age classes, is dependent only on the value of the previous age class, and expresses survivorship as a rate.

The construction of a life history table can be illustrated using a hypothetical squirrel population. The life table data and $p_x$ are given in Table 14-1. For ease in calculations, the $l_x$ column is expressed as a decimal fraction of 1 rather than in terms of 1,000. In the life history table (Table 14-2) the population begins with 10 females aged 1 year which will produce a litter the year of introduction, year 0. During year 0 the initial population of 10 1-year-old females gives birth to 20 females, which are added to age class 0 of that year. The survival of these two age groups is obtained by multiplying the number of each by the approximate $p_x$ value. Since the $p_x$ of the females in age 1 is 0.5, 5 individuals ($10 \times 0.5 = 5$) survive to year 1 (age 2). The $p_x$ value of age 0 is 0.3, so only 6 of the 20 in this age group in year 0 survive ($20 \times 0.3 = 6$) to year 1 (age 1). Survivorship is tabulated year by year diagonally down the table to the right through the various age groups. In year 1 the 6 1-year-olds and the 5 2-year-olds together contribute 27 young to age class 0. The $m_x$ value of the 6 1-year-olds is 2.0, so they produce 12 offspring. The 5

2-year-olds ($m_x$ value of 3.0) produce 15 offspring. The steps for determining the number of offspring in year $t$, $N_{to}$, is given by the equation

$$N_{to} = \sum_{x=1} N_{tx} m_x$$

where

$N_{to}$ = the number in the population at the given year $t$

$N_{tx}$ = the number of age $x$ in year $t$

For year 0 the calculation is

$$N_0 = (10)(2) = 20$$

and for succeeding years

$$N_1 = (6)(2) + (5)(3) = 27$$
$$N_2 = (8.1)(2) + (3)(3) + (4)(3) = 37.2$$

and so on. The number of offspring is obtained by multiplying the number in each age group by the $m_x$ value for that age and summing these values for all ages.

From such a life history table one can also calculate age distribution (Table 14-3). The age distribution for any one year can be obtained by dividing the number in each age group by the total population size for that year. The general equation is

$$c_{tx} = \frac{N_{tx}}{\sum_{y=0}^{\infty} N_{ty}}$$

where

$c_{tx}$ = the proportion of age group $x$ in year $t$

$N_{tx}$ = the sum of the number of each age group

$N_{ty}$ = the total population in year $t$

TABLE 14-1
*Life table, hypothetical squirrel population*

| Age | $q_x$ | $p_x$ | $l_x$ | $m_x$ | $l_x m_x$ |
|-----|-------|-------|-------|-------|-----------|
| 0 | 0.7 | 0.3 | 1.0 | 0 | 0 |
| 1 | 0.5 | 0.5 | 0.3 | 2.0 | .60 |
| 2 | 0.2 | 0.8 | 0.1 | 3.0 | .30 |
| 3 | 0.4 | 0.6 | 0.08 | 3.0 | .24 |
| 4 | 0.2 | 0.8 | 0.05 | 3.0 | .15 |
| 5 | 0 | 0 | 0.03 | 0 | 0 |

$$\sum m_x = GRR = 11.0$$
$$\sum l_x m_x = R_O = 1.29$$

*Note:* GRR = gross reproductive rate; $R_O$ = net reproductive rate.

TABLE 14-2
*Life history table, hypothetical squirrel population*

| Age | 0 | 1 | 2 | 3 | 4 | 5 | YEAR 6 | 7 | 8 | 9 | 10 |
|-----|----|----|-------|-------|-------|--------|--------|--------|--------|--------|--------|
| 0 | 20 | 27 | 37.2 | 48.87 | 60.10 | 77.25 | 98.99 | 126.28 | 161.02 | 205.70 | 262.71 |
| 1 | 10 | 6 | 8.1 | 11.16 | 14.66 | 18.03 | 23.17 | 29.69 | 37.88 | 48.31 | 61.71 |
| 2 | 0 | 5 | 3.0 | 4.05 | 5.58 | 7.33 | 9.02 | 11.58 | 14.84 | 18.94 | 24.16 |
| 3 | 0 | | 4.0 | 2.40 | 3.24 | 4.46 | 5.86 | 7.21 | 9.26 | 11.87 | 15.15 |
| 4 | 0 | | | 2.40 | 1.44 | 1.54 | 2.67 | 3.51 | 4.32 | 5.55 | 7.12 |
| 5 | 0 | | | | 1.92 | 1.15 | 1.55 | 2.13 | 2.81 | 3.45 | 4.44 |
| Total | 30 | 38 | 52.30 | 68.88 | 86.94 | 110.16 | 141.26 | 180.40 | 230.13 | 293.19 | 375.29 |
| $\lambda$ | 0 | 1.27 | 1.38 | 1.32 | 1.27 | 1.26 | 1.28 | 1.28 | 1.28 | 1.27 | 1.28 |

TABLE 14-3
*Approximation of stable age distribution*

| Age | | | | | | YEAR | | | | | |
| | 0 | 1 | 2 | 3 | 4 | 5 | 6 | 7 | 8 | 9 | 10 |
|---|---|---|---|---|---|---|---|---|---|---|---|
| 0 | .67 | .71 | .71 | .71 | .69 | .70 | .70 | .70 | .70 | .70 | .70 |
| 1 | .33 | .16 | .15 | .16 | .17 | .16 | .16 | .16 | .16 | .16 | .16 |
| 2 | | .13 | .06 | .06 | .06 | .07 | .06 | .06 | .06 | .06 | .06 |
| 3 | | | .08 | .03 | .04 | .04 | .04 | .04 | .04 | .04 | .04 |
| 4 | | | | .03 | .02 | .02 | .02 | .02 | .02 | .02 | .02 |
| 5 | | | | | .02 | 0.1 | .01 | .01 | .01 | .01 | .01 |

By comparing the age distribution of the hypothetical population in year 3 with the population in year 6, one observes that the population has attained a stable age distribution by the year 6. From that year on, the proportions of each age group in the population remain the same year after year.

The rate at which the population grows can be expressed as a graph of the numbers in the population against time. The graph of the growth of the hypothetical population indicates that it is growing exponentially. (Figure 14-1).

The rate of population growth also can be determined by dividing the size of the population in year $t$ by the total of the previous year, $t - 1$. For the hypothetical population this is 38/30 = 1.27 for year 1. (The total population and the rate of increase for each of the years is given in Table 14-2.) Once the population has reached a stable age distribution (Table 14-3), that is, the percentage of organism in each age group remains constant, the rate of growth becomes constant. The rate of increase attained with a stable age distribution is the finite rate of increase, $\lambda$. The finite rate of increase may also be expressed as a natural logarithm of $\lambda$, $\log_e \lambda$, which is $r$, the intrinsic rate of increase. (For a more accurate approximation of $r$, see Appendix C.) Thus $\lambda = e^r$ where $e$ is the base of natural logarithms.

Because $r$ is derived from $R_o$, which involves schedules of survivorship, $l_x$, and fecundity, $m_x$, $r$ includes both births and deaths. In a closed population, that is, one in which no individual enters or leaves, the intrinsic rate of increase is the instantaneous birth rate (per individual) minus the instantaneous death rate (per individual), or $r = b - d$. In an open population $r$ includes immigration $(i)$ and

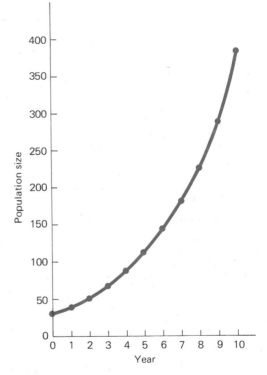

FIGURE 14-1
*Exponential growth curve of the hypothetical squirrel population.*

emigration $(e)$, or $r = (b + i) - (d + e)$. When births exceed deaths, the population increases; when deaths exceed births, the population decreases. Expressed somewhat differently, $r$ is positive when $R_o$ is greater than 1 and negative when $R_o$ is less than 1. When $R_o$ is unity, $r$ equals 0. When the environment is optimum, when $R_o$ is maximum, the population reaches its maximum rate of increase $r_{max}$ or $r_i$, the intrinsic rate of increase.

However useful $r$ may be, the parameter has certain limitations. It refers to a particular time interval; it is free from the effects of the population's own density and its own age structure. It is influenced by life history events such as frequency of reproduction; and it expresses the rate of increase under particular conditions before population density begins to affect birth rates and death rates. (For discussions, see Cole, 1954; Gadgil and Bossert, 1970; Mertz, 1971.)

## The growth curve

### EXPONENTIAL GROWTH

If a population were suddenly presented with an unlimited environment, as can happen when an animal is introduced into a suitable but unoccupied habitat, it would tend to expand geometrically. Assuming there were no movements into or out of a population and no mortality, then birth rate alone would account for population changes. Under this condition population growth would simulate compound interest. If you refer to a mathematics handbook, you will learn that if interest is compounded annually then

$$A = P(1 + r)^n$$

where

$A$ = the new amount at a given time
$P$ = the original amount or principal
$r$ = the rate of interest expressed as a decimal
$n$ = the number of years

If the interest is compounded several times a year, then

$$A = P\left(1 + \frac{r}{q}\right)^{nq}$$

where $q$ is the number of times interest is compounded during the year.

If interest is compounded continuously, then $q$ approaches infinity. Letting $r/q$ equal $x$, the expression can be written

$$A = P(1 + x)^{rn/x}$$

When $q$ approaches infinity, $x$ approaches 0, and from calculus it can be shown that

$$\lim_{x \to 0} (1 + x)^{1/x} = e$$

where $e$ is the base of natural logarithms,

which is approximately 2.7183. Thus if interest is compounded continuously, the expression reads

$$A = Pe^{rn}$$

In the symbols used in population ecology $A$ becomes $N_t$, the number in the population at the given year $t$; $P$ becomes $N_o$, the initial population; $r$ becomes the rate of increase, and $t$ is the unit of time. With these symbols the compound interest formula becomes the expression for logarithmic population growth, or the accumulation of compound interest on the population,

$$N_t = N_o e^{rt}$$

Expressed as a differential equation logarithmic growth is

$$\frac{dN}{dt} = bN - dN = (b - d)N = rN$$

where
$b$ = the instantaneous birth rate
$d$ = the instantaneous death rate
except that $d$ in the expression $dN/dt$ refers to an instantaneous rate of change in population growth.

The rate of growth at first is influenced by heredity or life history features, such as age at the beginning of reproduction, the number of young produced, survival of young, and length of reproductive period. Regardless of the initial age of the colonizers, the number of animals in the prereproductive category would increase because of births, whereas those in the older categories for a time would be stationary. As the young mature, more would enter the reproductive stage and more young would be produced, as has already been demonstrated in the hypothetical population example. If the number of animals is plotted against time, the points will fall into an exponential growth curve defined by the foregoing formula (see Figure 14-2). If the logarithm of the numbers of animals are plotted against time, the points fall into a straight line.

A population can increase exponentially until it overshoots the ability of the environment to support it. Then the population declines sharply or "crashes" through disease, starvation, or emigration. From the low point the population may recover to undergo

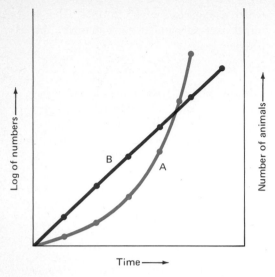

**FIGURE 14-2**
*Exponential growth curve plotted
arithmetically and logarithmically.*

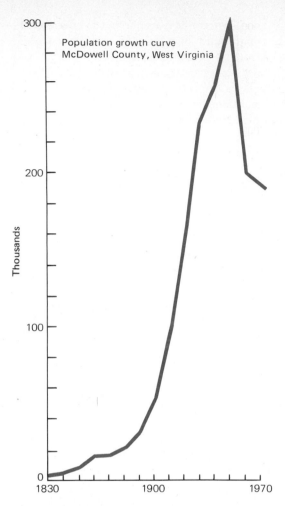

**FIGURE 14-3**
*Population growth curve for a southwestern
West Virginia coal region. Note exponential
growth followed by a sharp decline.
Resurgence of the coal industry will halt
temporarily the downward drop in
population. (Data from U. S. Census Bureau;
from R. L. Smith, 1976.)*

another phase of exponential growth, it may
decline to extinction, or it may recover and
fluctuate about some lower level. A J-shaped
curve is characteristic of many organisms introduced to a new and unfilled environment.

An example of exponential growth is the rise
and fall of the human population in a southwestern West Virginia county (Figure 14-3),
the age structure of which is presented in Figure 13-22. The growth curve from 1830 to
1950 is basically exponential. The population
reached its high in 1950; between 1950 and
1960 the population decreased dramatically.
The decline produces a curve typical of a population unresponsive to the carrying capacity of
the environment. Growth stops abruptly and
the population declines sharply in the face of
environmental deterioration. Another example is the reindeer herd on St. Paul one of the
Pribilof Islands, Alaska (Figure 14-4). Introduced on St. Paul in 1910, reindeer expanded
rapidly from 4 males and 22 females to a herd
of 2,000 in only 30 years. So severely did the
reindeer overgraze their range that the herd
plummeted to 8 animals in 1950.

### LOGISTIC GROWTH

As a population increases detrimental effects
of increased density begin to inhibit the rising
population until it reaches the carrying capac-

ity, the maximum number that can be supported in a given habitat.

Inhibition and slowing down of the growth
rate can be described mathematically by adding to the exponential equation $N_t = N_o e^{rt}$
some variable to describe the effects of density:

$$N = \frac{K}{1 + a e^{-rt}} \text{ where } a = \frac{K - N_o}{N_o}$$

$N_o$ is the number at time $t = 0$, and $K$ represents the asympototic level or balance point,
the density of organisms at which $R_o$ equals 1
and the rate of increase $r$ equals 0.

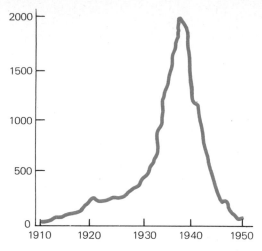

**FIGURE 14-4**
*Exponential growth of the St. Paul reindeer herd and its subsequent decline. (From V. C. Scheffer, 1951.)*

This equation is often written in the differential form

$$\frac{dN}{dt} = rN \frac{K - N}{K}$$

In words the equation says that the rate of increase of a population is equal to the potential increase of the population times the portion of the habitat that is still unexploited.

The equation was developed in 1838 by the French mathematician Verhulst as a model of population growth in a limited environment (see Hutchinson, 1978). In 1920 Pearl and Reed, in a classic paper, plotted the growth of the population of the United States by years and fitted it to the curve described by the Verhulst equation. Since then the model has been known as the Verhulst-Pearl equation.

The Verhulst-Pearl equation describes a logistic or sigmoid growth curve. The rate of increase is slow at first and then accelerates until it reaches a maximum (Figure 14-5). As density increases, the rate slows, marked by the inflection point in the curve. As the population reaches carrying capacity, the curve flattens out. Or to state it more precisely, as $K - N$ approaches 0, $dN/dt$ also approaches 0. The addition of the term $K - N/K$ makes the logistic curve depart from the exponential. It represents a feedback between the size of the population and the rate at which the population increases. If a population exceeds $K$, $r$ becomes negative and the population declines to $K$. If the population falls below $K$, $r$ is positive and the population increases to $K$. As a result populations tend to fluctuate around $K$.

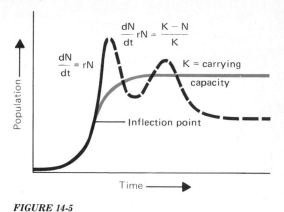

**FIGURE 14-5**
*Variations in growth curves. Curve A represents logistic growth which levels off at carrying capacity K. Curve B represents a hypothetical situation based on several small populations of white-tailed deer introduced into vacant habitats. The population expands exponentially (solid line of curve B). The broken line of curve B represents growth rate after the initial peak is reached. The population declines, experiences a resurgence in growth, and declines again. The population is then held below carrying capacity by controlled hunting.*

Logistic growth with all its fluctuations is common to many populations liberated in a new environment (Figure 14-6). Left undisturbed, they will usually exhibit a single eruptive oscillation (Caughley, 1970). Growth consists of four stages: (1) initial increase, the period between establishment of the population and attainment of the initial peak; (2) initial stabilization at peak population; (3) decline; (4) postdecline, in which the population adjusts to a lower level. As the population approaches carrying capacity, either upward or downward, growth departs from the logistic and fluctuates between some upper and lower limits.

An example of sigmoid growth is the population growth curve for an agricultural county in southeastern West Virginia (Figure 14-7), the age structure of which is given in Figure 13-21. The population reached an asymptote of 13,000 in 1900 and has fluctuated about that

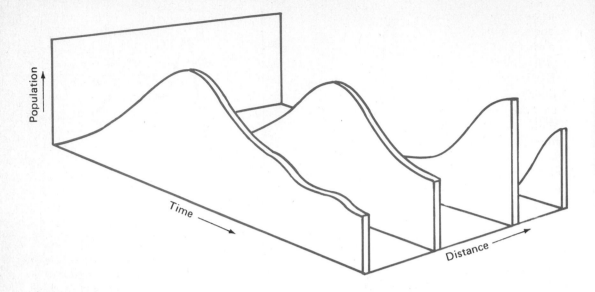

**FIGURE 14-6**
*Model of population increase of an ungulate population released in a vacant habitat. The curve on the extreme left shows population growth, eruptive fluctuation, decline, and postdecline stability at the point of initial colonization. The two curves in the middle illustrate the stage of growth at two successive fronts. The population at the advancing front (extreme right) is becoming established. (Adapted from Caughley, 1970; from Riney, 1968.)*

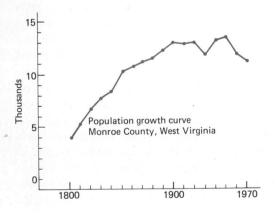

Population growth curve
Monroe County, West Virginia

**FIGURE 14-7**
*Population growth curve for Monroe County, West Virginia, dominated by agriculture and light industry. The growth curve is basically logistic. (From R. L. Smith, 1976.)*

level since. One dip in the population occurred in 1930; a period of recovery was followed by a second decline in the 1950s, the time of the great migration out of West Virginia.

**POPULATION DECLINE AND EXTINCTION**

Population decline is associated with a reduction in the rate of increase. Population size may become so low that $r$ becomes negative and the population dwindles to extinction. The heath hen is a classic example. Formerly abundant in New England, this eastern form of the prairie chicken was driven by excessive hunting to Martha's Vineyard off the Massachusetts coast and to the pine barrens of New Jersey. By 1880 it was restricted to Martha's Vineyard. Two hundred birds made up the total population in 1890. Conservation measures increased the population to 2,000 by 1917. But that winter a fire, gales, cold weather, and predation by goshawks reduced the population to 50. The number of birds rose slightly by 1920, then declined to extinction. The last bird died in 1932.

There are several causes for the decline of sparce populations. When only a few animals are present, females of reproductive age may have a small chance of meeting a male in the same reproductive condition. Many females may remain unfertilized, reducing average fecundity. A small population faces the prospect of an increased death rate. The fewer the animals the greater the individual's chances of succumbing to predation, because the odds of being taken by a predator increase as the number of exposed individuals declines.

Extinction is a natural process. Through millions of years of Earth's history, species have appeared and disappeared leaving a record of their existence in fossil bones and footprints in sedimentary rock. Some could not adapt to geological and climatic changes. Others diverged under selective pressures into new species while the parent stock disappeared. And still others could not withstand the competitive and predatory pressures of a relatively new species, *Homo sapiens,* who appeared to have the first great impact in the Pleistocene (Martin, 1973).

Whether or not recent extinctions have been precipitated by humans, the process continues today at an accelerated pace. The greatest number of extinctions has taken place since A.D. 1600. Well over 75 percent of the extinctions have been caused by humans through the alteration and destruction of habitat, introduced predators and parasites, predator and pest control, competition for the same resources, and hunting (see R. L. Smith, 1976, 1977).

Extinction is often thought of as taking place simultaneously over the full range of a species, but the process does not work in that fashion. It begins with isolated local extinctions when conditions so deteriorate in a given area that species of plants and animals disappear. Eventually, all the local extinctions reach a sum of total extinction.

The most important cause of extinction today is habitat alteration, which is a local phenomenon. Cutting and clearing away of forests, drainage and filling of wetlands, conversion of grasslands to croplands, construction of highways and industrial complexes, and urbanization and suburbanization greatly reduce available habitat for many species. When a habitat is destroyed, its plant life is eliminated, and animals either must adapt to changed conditions or leave the area and seek a new place to settle. Because of the rapidity of habitat destruction, no evolutionary time exists for a species to adapt to changed conditions. Forced to leave, the dispossessed usually find the remaining habitats filled and face competition from others of their kind or from other species. Restricted to marginal habitats, the animals may persist for a while as nonreproducing members of a population or succumb to predation or starvation. As the habitat becomes more and more fragmented, the species is broken down into small isolated or "island" populations, out of contact with other populations of its species. As a result, genetic variations in the isolated populations are reduced, making members of that population less able to withstand environmental changes. The maintenance of local populations often depends heavily on the immigration of new individuals. As distance between local populations or islands increases and as the size of the local populations declines, the continued existence of the local population becomes more precarious. As the number falls below some minimum level, $K,$ the local population may become extinct simply through random fluctuations.

## POPULATION GROWTH IN PLANTS

The rate of increase, $r,$ in plants is bipartite; increase occurs through seed production and through vegetative reproduction. (Some invertebrates fit the same pattern; for example, Daphnia reproduce by cloning in summer and resort to sexual eggs to overwinter.) Some populations, such as algae and duckweed (*Lemna,* spp.) exhibit growth rates that can be described by the formal equation of population growth. Other plant populations grow mostly in pulses, controlled in part by periodic seed production in the life cycle of the plant. A single event such as fire or harvest may eliminate a population and permit another to grow.

Harper and White (1974) point out the close relationship that exists between the number of plants in a population *(N)* and the weight of plants *(w),* including all subunits such as leaves and clones. When $N,$ density, is high, a relationship exists between mortality of individual plants and growth of surviving individuals; as mortality increases, growth of survivors increases. This observation gives rise to several generalizations about the growth of plant populations. (1) Rate of mortality is related to stress of population density caused by competition for light, moisture, nutrients, space, etc. (2) Growth of survivors is related to rate of mortality of individual plants in the population. (3) In populations characterized by stresses of density, more growth and weight becomes incorporated into a few individuals. As a result in a population of even-aged individuals a hierarchy of size in plants develops with relatively few dominants and a large class of suppressed individuals. (4) Risk of mortality tends to remain constant with time. (5) Density-dependent mortality of individual plants is preceded by death of components such as lower limbs and leaves.

## Reproductive strategies

Natural selection favors individuals that produce the maximum number of successful young in a lifetime. The size and number of offspring produced varies greatly among plants and animals. Some produce a large number of small offspring, while others produce a small number of large young. The nature of reproductive effort is determined by the manner in which parents invest energy in the production of offspring.

Parents have a finite amount of energy for reproduction at any given time. A certain minimum amount of energy is needed for offspring to be viable. As energy is apportioned among an increasing number of offspring, the fitness of individual offspring declines. Eventually a point is reached at which increases in reproductive effort per offspring results in declining payoffs in fitness (Smith and Fretwell, 1974).

Some of the energy available to and assimilated by parents is apportioned to growth and maintenence and some to reproduction. How this energy is allocated to growth and reproduction is central to the reproductive strategy an organism employs. If an organism directs more energy to reproduction, it has less energy to allocate to growth and maintenance. As a result the individual grows more slowly to the next age or fails to survive. For example, Lawlor (1976) found that reproductive females of the terrestrial isopod *Armadillidium vulgare* had a lower growth rate than nonreproductive females and that nonreproductive females devoted as much energy to growth as the reproductive females devoted to both growth and reproduction. And Tinkle (1969) has demonstrated that among lizards species with larger fecundity have poorer individual survival because more energy is channeled into egg production over a short period resulting in heavy physiological stress. In addition females that have large clutches or produce several clutches a year are less able to escape predation.

If parents are to contribute their maximum to future generations, they have to balance the profits of their immediate potential of reproductive success against the costs of future prospects. Basically two extreme options are available. One is *semilparity,* the sacrifice of future prospects by expending all energy in one reproductive effort in the life cycle, that is, to produce a large number of offspring and then die. This is typical of many insects and some Pacific salmon. The other option is *iteroparity,* to produce a few young at one time and repeat reproductive efforts through a lifetime (see Gadgil and Bossert, 1970).

Most organisms compromise between these two extreme options and make a trade-off between present progeny and future offspring. The nature of these trade-offs varies from species to species and even within species. If an individual grows rapidly while investing little if any energy into reproduction and is short-lived as an adult, its reproductive strategy will probably be an early single reproductive effort because any delays in producing offspring reduces the probability that the individual will survive to reproduce. Such a strategy results in a high individual fitness for the parent and a lower average fitness for the offspring (because a limited energy resource is apportioned among many young). But if an organism is longer lived and its chances for survival as an adult are relatively high, it can reproduce more than once during a lifetime. Under these conditions the strategy is usually late reproduction with more energy invested in fewer young. Parental fitness is reduced, but individual fitness of the young is increased. Among organisms that reproduce several times over a lifetime, reproductive effort should increase with age because older individuals require less energy for growth or maintenance (Williams, 1966; Gadgil and Bossert, 1970).

The proportion of total energy available to an organism that is invested in reproduction varies. Energy costs involve not only the weight of the progeny, but also the amount of energy expended in rearing the young. Harper et al. (1970) have estimated that herbaceous perennials invest between 15 and 25 percent of annual net production in production of new plants including vegetative propagation. Wild annuals (single stage reproducers) expend 15 to 30 percent, most grain crops 25 to 35 percent, and corn and barley 35 to 40 percent. The lizard *Lacerta vivipara* invests 7 to 9 percent of its annual energy assimilation in reproduction (Avery, 1975) and the female Allegheny mountain salamander *Desmognathus ochrophaeus* invests 48 percent of its annual energy flow into reproduction, including energy stored in the eggs and energy costs of brooding (Fitzpatrick, 1973).

## REPRODUCTIVE STRATEGIES IN ANIMALS

One method of apportioning energy to reproductive effort is the adjustment of clutch size. D. Lack (1954) proposed that among most birds clutch size evolved through natural selection to correspond to the average largest number of young that parents can feed. Thus clutch size is an adaptation to food supply. Temperate species have larger clutches because increasing day length allows parents a longer time to forage for food to support large broods. In the tropics, where day length does not change, food becomes a limiting resource.

M. Cody (1966) modified this concept by employing the principle of allocation of energy and proposed the hypothesis that clutch size results from different allocations of energy to egg production, avoidance of predators, and competitive ability. In temperate regions periodic local climatic catastrophes can hold the population below carrying capacity; natural selection favors an increase of *r* (rate of increase), and clutch size on the average is larger than in tropical regions. The stability and predictability of the climate in the tropics makes maintenance of carrying capacity more important than increasing production of offspring to replace losses due to environmental instability. More energy is expended in avoiding predation and in meeting intra- and interspecific competition which is keener than in temperate regions.

The predictions of this theory are met by a number of field examples. Birds in temperate regions have larger clutch sizes than in the tropics (Figure 14-8) and mammals at higher latitudes have larger litters than those at lower latitudes (Lord, 1960). Lizards exhibit a similar pattern. Those living at lower latitudes have smaller clutches, have higher reproductive success (or lower egg mortality), reproduce at an earlier age, and experience higher adult mortality than those living at higher latitudes (Tinkle and Ballinger, 1972; Andres and Rand, 1974).

FIGURE 14-8
*Relationship between clutch size and latitude in birds. Represented as examples are the family Icteridae (blackbirds, orioles, meadowlarks) distributed in North and South America and the genus* Oxyura *of the family Anatidae (ruddy and masked ducks) worldwide in distribution. (Adapted from Cody, 1966.)*

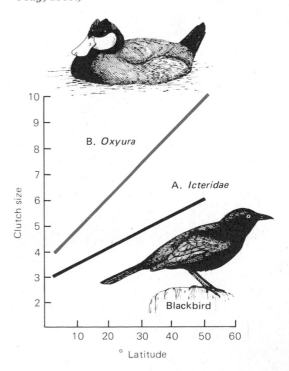

**469**

Insects, too, support this theory. An example is the milkweed beetle (Landahl and Root, 1969). When a population of the temperate species *Oncopeltus fasciatus* and one of the tropical species *O. unifasciatellus* were reared in the laboratory, both exhibited similar duration of egg stage, egg survivorship, developmental rate, and age of sexual maturity. Although clutch size for both species was the same, the temperate species produced eggs sooner and laid a larger number of eggs over time. The tropical species had fewer clutches, rather than smaller ones, and laid over a longer period. Its total egg production was only 60 percent of that of the temperate species.

The adjustment of reproductive effort to food supply also apparently has evolved among invertebrates (see Price, 1974, 1975). For example, Spight and Emlen (1976) provided two marine snails *Thias lamellosa* and *Thias emarginata* with an increased food supply. Adult *T. lamellosa* responded by increased growth which in turn resulted in an increase of average egg clutch size from 930 to 1,428 eggs per female. *T. emarginata* did not increase its growth, but it spawned many more times during the year.

Individuals of some species exhibit some flexibility to adjust to changing energy resources during the reproductive season. For example, the common grackle *(Quiscalus quiscula)* starts incubation before its clutch of five eggs is complete, a pattern of hatching ensuring survival of some young under adverse conditions (Howe, 1976). The eggs laid last are heavier and the young hatched from them grow fast, but if food is scarce, the late hatched are not fed and die of starvation. Asynchronous hatching favors the early hatched young at the expense of those hatched later. Although parents attempt to assure the survival of all young, their investment in young is protected by starvation of late-hatched birds if a food shortage arises.

Reproduction may also be so timed that offspring are produced when the environment is most favorable and food is most abundant. Among birds courtship is quickly followed by nesting and egg laying. Among mammals with long gestation periods the mating season is timed to ensure birth at a time of favorable conditions. Thus white-tailed deer mate in Oc-

tober and November and the young are born the following June. Some mammals mate shortly after the young are born. If the gestation period is short and if development of the egg were to continue, young would be born at an unfavorable time of year. A reproductive strategy of such mammals is delayed development of the fetus through delayed fertilization or delayed implantation.

Some species of bats live in temperate climates where their insect food is abundant only during summer. These bats time their reproductive processes with winter hibernation. They mate in late summer when the food supply is starting to decline, but at that time the female has not yet ovulated. The following spring one egg is released to be fertilized by sperm stored in the uterus of the female through the 6-month period of hibernation. Although the testes in male bats regress in fall and winter, sperm is stored in the epididymis permitting the male to inseminate any females immediately after hibernation that failed to mate the previous autumn.

Other mammals slow reproduction by delaying implantation of the blastocyst; it lies dormant in the uterine lumen until the following spring. After implantation the rest of pregnancy continues normally as if no delay had occurred. An example is the stoat *(Mustela erminea)*, which lives alone most of the year. The young are born in April or May. The young male stoat reaches sexual maturity a year after birth, but the young female stoat reaches sexual maturity before she is weaned at 6 to 8 weeks. This very early sexual maturity permits mating and impregnation before the litter is dispersed. However, the blastocyst does not become implanted until the following spring. Such early mating ensures that the female will be impregnated despite the stoats' solitary living habits, and delayed implantation ensures that she will put on sufficient growth to withstand the stress of pregnancy and produce viable young. Delayed implantation, which is common to many members of the weasel family as well as the armadillo, bear, roe deer, seals, and kangaroo-like marsupials such as the wallabys, does not usually involve such early sexual maturity. Other behavioral and climatic reasons are usually involved. (For a review, see Weir and Rowlands, 1973.)

## REPRODUCTIVE STRATEGIES IN PLANTS

Plants, like animals, must allocate a portion of their energy, their net annual assimilated income, to reproductive effort, and like animals, different types of plants employ different reproductive strategies. One group of plants allocates a larger portion of energy to vegetative structure than to flowers and seeds, thus possessing low reproductive effort. Among such plants are perennial and woody species occupying rather stable habitats and later stages of succession. By investing a greater proportion of energy in persistent vegetative structures, such as tree trunks and long-lived roots, these species are better able to compete in a crowded, resource-limited environment. Another group allocates a greater proportion of energy to reproductive effort, producing a large number of seeds. These are plants that occupy a relatively unstable or nonpersistent environment of early successional stages.

Plants can employ different strategies in the suballocation of energy invested in reproduction. They may increase or decrease the number and size of seeds. Or energy can be allocated to varying proportions of sexual and vegetative reproduction. The strategy employed is determined by such selection pressures as predation, dispersal, and environmental conditions.

Plants follow the general principle of allocation on a latitudinal basis as do many animals. S. McNaughton (1975) investigated the allocation of resources to reproduction in a series of greenhouse studies of the cattail *(Typha)*. He measured reproductive effort by the growth of rhizomes, the cattail's principal means of population growth within a given habitat. Cattails shed their very small seeds to the wind as a means of colonizing a new habitat. McNaughton considered three species of cattails on a climatic gradient: the common cattail *T. latifolia,* a climatic generalist that grows from the Arctic circle to the equator; the narrowleaf cattail *T. angustifolia,* restricted to northern latitudes of North America, and *T. domingenesis,* restricted to southern latitudes of North America. The cattails exhibited a climatic gradient in the allocation of energy to reproduction. *T. angustifolia* and northern populations of *T. latifolia* grew earlier and faster and produced a greater

number of rhizomes than *T. domingenesis* and southern populations of *T. latifolia,* but the southern plants produced larger rhizomes. Faced with stronger intra- and interspecific competition, southern cattails invested more energy into vegetative growth. However, cattail species exhibited a gradient in trade-offs between colonization potential and competitive potential (Figure 14-9). An inverse relation between production and rate of foliage production allows adaptation to site conditions. Where competition is intense, plants can produce more and taller foliage, but reduce rhizome production. Where competition is low, plants can invest more energy in rhizome production.

Plants exhibit a great diversity in size and number of seeds. Some plants, such as coconut palm, have seeds as large as 27,000 g; their main agent of dispersal is water. Other plants, such as some orchids and other saprophytes, have seeds as small as .000002 g that are easily carried by the wind and are able to become established in small, widely dispersed microhabitats (Harper, Lovell, and Moore, 1970). Between the two extremes exists a continuum of seed size. The optimum seed size for a plant relates to its dispersal ability, colonizing abil-

**FIGURE 14-9**
*Relationship between rate of rhizome production and foliage production rate in cattails* (Typha). *Line is best least squares fit. (From S. MacNaughton, 1975.)*

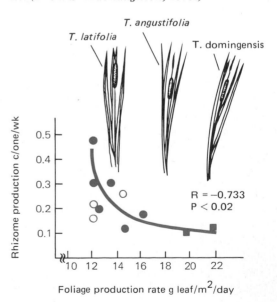

ity, and ability to escape predation. Plants that colonize new and often widely scattered habitats or harsh environments where competition is low invest their energy in large numbers of small seeds with minimal energy reserves for germination. Plants risk the loss of most of their energy investment by sacrificing most of the propagules to ensure the establishment of a few. Plants associated with more stable, less harsh environments where competition is keen produce a small number of large seeds with considerable energy reserves. Such plants sacrifice dispersal ability for strong, aggressive seedlings with high competitive ability.

These two strategies may be employed not only by different plant species, but also by populations of the same species living under different environmental conditions. Linhart (1974) studied two populations of *Veronica peregrina* growing in temporary pools in the Central Valley of California. In each population reproductive allocations differed from the center to the periphery of the pool. In the center of the pool, where the environment was predictable and moisture conditions were optimal, plants grew densely and intraspecific competition was intense. The *Veronica* growing there produced fewer, but heavier seeds which germinated rapidly. At the periphery of the pool, the environment was unpredictable, moisture was limiting, and interspecific competition was present. *Veronica* growing in the patchy environment of the periphery experienced higher mortality, produced more and lighter seeds, and grew taller, perhaps in response to competition from taller grasses. The smaller seeds of peripheral *Veronica* were better able to disperse and their greater abundance increased the probability of successful survival. The fact that the difference in the two populations of *Veronica* was maintained in the greenhouse suggests genetic polymorphism.

While alternate strategies of seed production among many plants appear to involve genetic polymorphism, some plants respond in a plastic manner. The annual species *Polygonium cascadense* grows on the dry upper slopes of many peaks in the western Cascade Mountains. Those plants that grow in more harsh, open habitats allocate proportionately

more of their resources (up to 71 percent) to reproduction than those in more moderate habitats. In the harsh environment plant density and seedling production per unit area were highest and plant size and total seedling mortality were lowest (Hickman, 1975). The difference between the two environmental responses was plastic rather than genetic, because no differences were detected in greenhouse populations. Similar responses have been observed in populations of dandelions (Solbrig and Simpson, 1974) and goldenrods (Abrahamson and Gadgil, 1973) (Figure 14-10).

Another selective pressure influencing seed size is predation. Seeds are a major source of food for many animals. If predation on seeds is heavy, energy invested in reproduction is lost. In response to predation many plants have evolved a strategy of producing small seeds in large numbers to increase the probability that some will escape the predators. An example is seed production in two groups of large leguminous trees in Central America (Jansen, 1969). Seeds of both are attacked by weevils. One group, which may lose 44 to 99 percent of its seed crop to predators, produces a large number of seeds with a mean weight of 0.269 g. The large number divides up the reproductive costs among many seeds and increases the probability that some will escape predation. The other group produces a smaller number of seeds with a mean weight of 3.0 g, but invests some energy into chemical defense which involves a toxic organic acid that repels weevils. Plants with large seeds increase the amount of protective tissue to reduce predation. Lodgepole pine in western North America responds to predation by squirrels by increasing protective tissue while maintaining a relatively constant number and size of seeds (Smith, 1970).

Closely associated with seed number and size is dispersal. Plants employ a great variety of mechanisms for seed dispersal (see Pijl, 1969). Plants with numerous light seeds, such as maple, tulip poplar, conifers, and tall weeds of old fields, employ the wind as a dispersal agent. Usually the seeds are borne on tall stems exposed to the wind. Animals may serve as the dispersal agents. Seeds of many herbaceous plants have hooks, burrs, and spines which catch on the fur of mammals and feath-

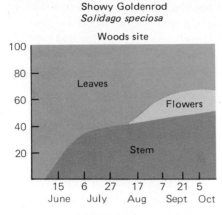

Showy Goldenrod
*Solidago speciosa*

Woods site

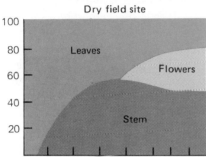

Dry field site

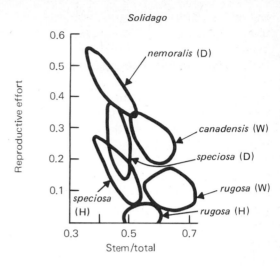

*Solidago*

**FIGURE 14-10**

*Allotment of energy to reproductive effort in goldenrods. The reproductive effort of six populations of goldenrod is measured by plotting the ratio of dry weight of reproductive tissue to total dry weight of above-ground tissue on the ordinate as a function of ratio of weight of stem tissue to total dry weight of above-ground tissue. Each closed curve embraces all points representing individuals included in a single population. D = dry field site; W = wet meadow; H = hardwood site. The reproductive effort of showy goldenrod (Solidago speciosa) growing in a hardwood site and on a dry field site is measured by plotting the percent of total biomass of stems, leaves, and flowers as a function of time during the growing season. Note that goldenrods growing on the dry site allocate a greater portion of energy to reproductive effort, probably because density-independent mortality is higher in that environmental situation. (Adapted from Abrahamson and Gadgil, 1973.)*

ers of birds. Other plants employ predation as a dispersal mechanism. Small seeds contained in highly palatable fleshy fruits such as blackberries and blueberries pass through the digestive tract and are dropped in suitable sites for germination. Heavy seeds of such plants as oak and hickory are consumed by food storing predators such as squirrels and jays. Many of the seeds cached by the animals are lost or forgotten and germinate. By sacrificing a large number of seeds to predators, plants achieve a measure of dispersal and planting of a portion of their reproductive effort.

## r-selection and K-selection

From the discussion of reproductive strategies among plants and animals two points are evident: (1) among populations living in a harsh or unpredictable environment in which mortality is largely independent of population density, individuals allocate more energy to reproduction and less to growth, maintenance, and the ability to compete; (2) among populations living in a stable or predictable environment in which mortality results mostly from density-related causes and competition is keen, individuals allocate more energy to nonreproductive activities. The former are known as *r*-strategists because they remain on the rising part of the logistic growth curve; the latter are known as *K*-strategists because they exist near the asymptote (MacArthur and Wilson, 1967).

Individuals known as *r*-strategists are typically short-lived, existing usually less than one year. They have high reproductive rates

**473**

and produce a large number of offspring with low survival, but rapid developmental rates. They are opportunistic, that is, have the ability to make use of temporary habitats, and often inhabit unstable or unpredictable environments where catastrophic mortality is environmentally caused and is relatively independent of the density of the population. For them environmental resources are not limiting and they are able to exploit relatively uncompetitive situations. Tough and adaptable, *r*-strategists have means of wide dispersal and are good colonizers; they respond rapidly to disturbances to the population. Such species are characteristic of early stages of succession.

*K*-strategists are competitive species with stable populations of long-lived individuals; they are relatively large and produce relatively few seeds, eggs, or young. Among animals parents care for the young; among plants seeds possess stored food that gives the seedling a strong start. *K*-strategists exist in environments in which mortality relates more to density-related causes than to unpredictability of conditions. They are specialists, efficient users of their particular environment, but their populations are at or near carrying capacity and are resource limited. *K*-strategists are typically long-lived, requiring a relatively long time to mature. These qualities combined with their lack of means of wide dispersal make *K*-strategists poor colonizers. They are characteristic of later stages of succession.

*K*-strategists and *r*-strategists are under different selection pressures. Among *r*-species selection favors those genotypes which confer the highest possible intrinsic rate of increase, rapid development, small body size, early and single-stage reproduction, large number of offspring, and minimal parental care. Among *K*-species selection favors genotypes that confer the ability to cope with physical and biotic pressures, the ability to tolerate relatively high population densities, delayed reproduction, large body size, slower development, and repeated reproduction. *r*-selection favors productivity; *K*-selection favors efficient use of the environment (see Pianka, 1970.)

Although *r*-selection and *K*-selection are described as two polar entities, selection forms a continuum from *r* to *K*. *K*-selection and *r*-selection are relative terms useful in comparing organisms, but there are degrees of strategies among species and within species, as discussed above. Under the condition of an unfilled environment an organism may behave as an *r*-strategist; as the environment fills it may behave as a *K*-strategist. When a population is subjected to intense mortality unrelated to density or crowding, individuals with a higher reproductive rate leave the most offspring and will dominate the area. Under intense mortality related to population density (see Chapter 15), individuals able to tolerate high densities at or near carrying capacity will be retained in the population.

The concept of *r*- and *K*-selection, however, has certain flaws that reduce its usefulness and create problems in its application. For example, many species exhibit relatively long life, repeated reproduction, high fecundity, and a high mortality of young, thus possessing features common to both *r*- and *K*-strategists. Much of the problem results because *K* cannot be expressed as a function of life history traits. It is a composite of population, its resources, and their interaction. On the other hand, *r* is a function of life history traits: age, survivorship, and fecundity. Only *r* is sensitive to strong selection pressure. Thus *r* and *K* are not equivalent terms; they "cannot be reduced to units of common currency" (Stearns, 1977). It may be more profitable to consider the response of individuals to selective pressure only in terms of *r*. For example, assume that the environment fluctuates or a population is near equilibrium *(K)*. If juvenile mortality or birth rate fluctuates and adult mortality does not, then selection should favor later maturity, smaller reproductive effort, and fewer young. If adult mortality fluctuates and juvenile mortality or birth rate does not, then selection should favor early maturity, larger reproductive effort, and more young (see Stearns, 1976, 1977).

### SUMMARY

**Population growth results from differences between additions to a population through birth and immigration and removals from a**

population by death and emigration. When births and immigration exceed death and emigration, a population increases. The difference between the two (when measured as an instantaneous rate) is the population's rate of increase, $r$. In an unlimited environment a population expands geometrically, a phenomenon that may occur when a small population is introduced to an unfilled habitat. Geometric increase is characterized by a constant schedule of birth and death rates, an increase in numbers equal to the intrinsic rate of increase, and the assumption of a fixed or stable age distribution which is maintained indefinitely. But because the environment is limited, such growth is not maintained indefinitely. Population growth eventually slows and arrives at some point of equilibrium with the environment or carrying capacity, $K$. However, natural populations rarely achieve such a stable level; instead a population fluctuates in numbers. When populations become quite small, chance events alone can lead to extinction.

Populations grow largely by additions through reproduction. The nature of reproductive effort—how many and how frequently offspring are produced—relates to allocation of energy to reproductive effort. One alternative is to invest a a maximal amount of energy into a single reproductive effort in a lifetime, as exemplified by annual plants and many insects. The other alternative is to allocate less energy to reproduction, but to have repeated reproduction. A single reproductive effort in which the amount of energy invested in each individual is minimal results in a large number of young with low individual fitness. Such a reproductive strategy is characteristic of situations where the mortality of young is very high. Repeated reproduction involves a large amount of energy invested in few young of high individual fitness. Because little difference in potential for population increase exists between single and repeated reproduction, repeated reproduction may be a response to an unpredictable survival of individuals from zygote to adult. By reproducing several times an organism is more likely to insure reproductive success.

Allocation of energy to reproduction is related in part to the nature of mortality, its relationship to population, and predictability of the environment. Organisms living in an unpredictable environment or subject to heavy environmentally induced mortality tend to allocate a greater proportion of energy to reproduction, to possess single reproduction, and to expand rapidly when conditions are favorable. Such organisms are said to be $r$-selected because selection favors high productivity. Organisms which occupy a more predictable environment, are more subject to density-related mortality, tend to allocate less energy to reproduction, and experience repeated reproduction are said to be $K$-selected because selection favors efficient use of the environment.

# Population regulation
## CHAPTER 15

No population increases indefinitely. Sooner or later growth slows and the population eventually achieves some kind of stability or equilibrium with its resources and environment.

A population may exhibit one of three basic types of response to growth. (1) As a population approaches some level of density of equilibrium relative to its environment, it may fluctuate above and below that point. Stability is likely to be governed by intrinsic influences. (2) A population may fluctuate widely without any reference to equilibrium size. Such fluctuations usually indicate the influence of extrinsic forces, such as weather, that may mask any hint of an equilibrium population level. (3) A population may oscillate between high and low points with some regularity between highs and lows.

### Density dependence versus density independence

Implicit in intraspecific population regulation is the influence of density dependence and density independence. Density-dependent influences on a population change as density of the population varies and the proportion of organisms affected changes with density. Density-independent influences in a population do not vary as a population changes. The same proportion of organisms is influenced at any density.

Density-dependent regulation of populations is a homeostatic process. As a population approaches a certain size some density-related effects act to reduce its rate of growth by decreasing the birth rate and increasing the

476

death rate. If a population falls below a certain level, the birth rate increases and the death rate decreases. Such a feedback mechanism results in oscillations or fluctuations. In general fluctuations will move about an equilibrium density, the level at which the production of offspring compensates for loss of individuals by death ($R_0 = 1$; $r = 0$). Any departure from this level brings regulation or compensating reactions into play which cease when equilibrium density is regained. Because a time lag is involved in response to density effects (organisms require time to reach the reproductive class of the population), equilibrium density is rarely attained.

Density-independent influences on the rate of increase do not regulate population growth per se because regulation implies a homeostatic feedback that functions with density. But density-independent influences can have considerable impact on changes in the population size and they can affect birth rates and death rates. Density-independent influences may so affect a population that they completely mask any effects of density-dependent regulation. A cold spring may kill the flowers of oaks, causing a failure in the acorn crop. Because of the failure, squirrels may experience widespread starvation the following winter. Although starvation relates to the density of squirrels and available food supply, weather is the major cause of decline. In general population fluctuations influenced by annual and seasonal changes in the environment tend to be irregular and correlated with variations in temperature and moisture.

Twenty or more years ago ecologists diverged widely in their views on the relative importance of density-dependent and density-independent influences on populations. (For density-dependent viewpoints, see Nicholson, 1954, 1956; Solomon, 1957; Lack, 1966; Hairston, Smith, and Slobodkin, 1960; for density-independent viewpoints, see Andrewartha and Birch, 1954; Milne, 1957; Thompson, 1956; for a historical overview of the debate, see Krebs, 1978.) The arguments were largely semantic, stemming from different approaches in different areas of study. (The density-independent school, for example, was dominated by insect population ecologists, while the density-dependent side was domi-

nated by vertebrate population ecologists.) Most ecologists now agree that numbers of organisms are determined by an interaction between density-dependent and density-independent influences which may vary among populations and within the same population.

Watt (1973), for example, suggests that the relative sensitivity of a species to various influences determines how its population is regulated. Fluctuations of a given population result from an interaction between the magnitude of fluctuations in weather and the sensitivity of a species to environmental changes. If weather fluctuates only a small amount, fluctuations of species insensitive to weather are regulated by density-dependent mechanisms. The more sensitive a species is to environmental fluctuations, the more important density-independent mechanisms become.

Because of interactions between density-dependent and density-independent influences, it is perhaps more instructive to discuss regulatory mechanisms in terms of influences that are intrinsic or extrinsic to the population.

## Extrinsic influences

### WEATHER

Among all the extrinsic forces impinging on a population, perhaps the most powerful is weather, particularly extremes of temperature and moisture. Conditions beyond the organisms' limits of tolerance can have a disastrous impact on the population by affecting growth, maturation, reproduction, survival, movements, and dispersal of individuals within a population and even by eliminating local populations.

In general the influence of weather tends to be irregular and unpredictable. Pronounced changes in population growth can often be correlated directly with variations in moisture and temperature. An example of a population reaction to extremes of temperature can be seen in the walnut aphid, an important pest in the walnut orchards of California. Sluss (1967) followed an aphid population over a period of time. He found that aphid population changes were influenced by (1) age of walnut leaflets, which determines the maximum number of aphids that can feed on a leaflet at any one time; (2) amount of prior aphid feeding, which damages leaves and reduces their carrying capacity; (3) predation; and (4) temperature.

Sluss found that sharp declines in the population were associated with high temperatures, especially when the temperature exceeded 100° F over a period of several days. This resulted in a J-shaped growth curve characteristic when a density-independent influence suddenly restricts the rapid growth of a population (Figure 15-1). The heavy feeding of the dense populations prior to the period of high temperature had significantly reduced the food supply and damaged the leaves, so the population was unable to resurge after temperatures became lower. Predation by coccinellid beetles, combined with high temperature, modified the J-shaped curve. The population grew more slowly and peaked at a lower level but still experienced a sharp decline. Because the population prior to the decline was high enough to reduce food and affect leaf condition, these aphids, too, were unable to resurge. In situations where predation was relatively heavy aphid populations reached carrying capacity more slowly, did not reach such high densities, and did not experience the same drastic decline at high temperatures. Because leaves were not left in an unfavorable condition by aphid attack, the insects were able to make a comeback. And because their food supply was not decimated, beetles that preyed on aphids remained in the area and responded to any new increase in the aphid population.

Populations of deer in the northern part of their range are sensitive to severe winters (Figure 15-2). Following severe winters (those that have had 60 days of 38 cm of accumulated snow, or 50 or more days of 61 cm of accumulated snow), populations of deer in the Adirondack Mountains decrease dramatically. Losses are due largely to the inability of deer to obtain sufficient food. Low-growing winter food is buried during periods of prolonged snow cover. Energy expenditures required by a fawn to move through 38 cm of snow and by an adult to move through 51 cm or more are greater than energy provided by available food and stored fat reserves of the deer. Often deer are unable to reach food supplies outside of the

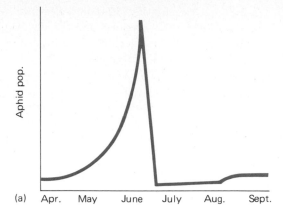

(a)

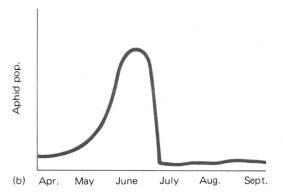

(b)

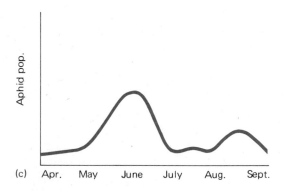

(c)

**FIGURE 15-1**

*Three generalized patterns of change in aphid populations found in northern Californian walnut orchards. The graphs indicate the effect on population change of high temperature and predation, operating as density-independent mechanisms. (From Sluss, 1967.)*

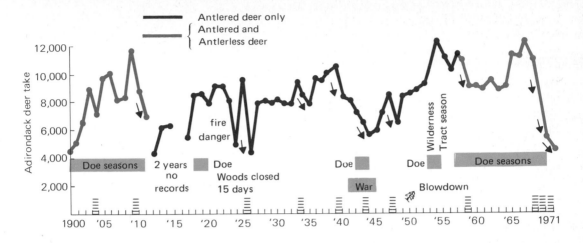

**FIGURE 15-2**
*Fluctuations in the Adirondack deer herd in New York. Changes in population appear to be closely related to severity of winters. Hard winters result in heavy mortality, a decline in reproduction, and a sharp decline in population. (After C. S. Severinghaus, 1972.)*

wintering area and an exceptionally high mortality of fawns results. For example, during the severe winters of 1968–1969, 1969–1970, and 1970–1971, characterized by prolonged deep snows and cold temperatures, approximately 5 out of every 6 fawns died (Severinghaus, 1972). These heavy winter kills were reflected in the deer population the following year.

In desert regions a direct relation exists between precipitation and rate of increase in certain rodents and birds (Figure 15-3). Merriam's kangaroo rat *(Dipodomys merriami)* occupies lower elevations of the Mojave Desert. Although the kangaroo rat has the physiological capacity to conserve water and survive long periods of aridity, it does require in its environment a level of moisture sufficient to stimulate the growth of herbaceous desert plants in fall and winter. The kangaroo rat becomes reproductively active in January and February when plant growth, stimulated by fall rains, is green and succulent. Herbaceous plants provide a source of water, vitamins, and food for pregnant and lactating females. If rainfall is scant, annual forbs fail to develop and production of kangaroo rats is low (Beatley, 1969; Bradley and Mauer, 1971). This close relationship to seasonal rainfall and relative success of winter annuals is also apparent in other rodents occupying similar desert habitat and in Gambel's quail and scaled quail (Francis, 1970).

### HUMAN ACTIVITY

Human activity is another extrinsic force with considerable impact on population growth. Human land use patterns radically alter habitats available for other organisms. Destruction of virgin forests in southern United States caused the extinction of the ivory-billed woodpecker; cutting of northern beech forests combined with market hunting brought about extinction of the passenger pigeon. Drainage of potholes in prairie regions has seriously reduced waterfowl populations. But clearing forests and subsequent cultivation of land in eastern North America increased habitat for animals of open fields and thickets. It permitted the eastward spread of such prairie life as the dickcissel and coyote.

Use of pesticides is a significant density-independent influence. Pesticides cause direct mortality of target organisms and of other species by contact or injection (Clawson, 1958; Rudd, 1964; Pimentel, 1971). They lower fecundity by interfering with reproductive processes or the survival of young. They also interfere with density-regulated processes by eliminating predators, parasites, and competitors of a target species and may thereby increase rather than decrease its numbers (Muir, 1965).

**479**

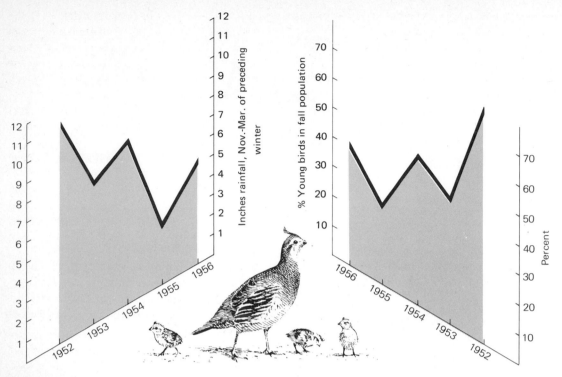

**FIGURE 15-3**
*Relationship of winter rainfall to age composition the following year of a Gambel's quail population in southern Arizona. (After Sowls, 1960.)*

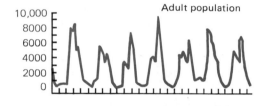

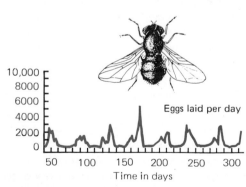

## RESOURCE AVAILABILITY

Abundance of resources such as food and reproductive sites directly or indirectly functions to regulate populations, largely through intraspecific competition. Competition exists when a resource is in short supply and results in the failure of many individuals in the population to survive or to reproduce.

In a long-term experiment involving sheep blowflies *(Lucilia cuprina)* Nicholson (1954) demonstrated the influence of intraspecific competition in a population. Although the experimental population lacked all the complex interactions one would expect to find in nature, the work does show what might happen.

In one experiment Nicholson fed to a culture of blowflies containing both adults and larvae a daily quantity of beef liver for the larvae and an ample supply of dry sugar and water for the adults. The number of adults in the cages varied with pronounced oscillations (Figure 15-4).

**FIGURE 15-4**
*Oscillations in the numbers of adult sheep blowflies (*Lucilia cuprina*) and in the daily rate of egg production in a laboratory population. Larvae received unlimited food; adults received a limited supply (0.5 g of liver daily). Peaks in the adult cycle alternate with those in the egg cycle. (After Nicholson, 1954.)*

When the population of adults was high, the flies laid such a great number of eggs that resulting larvae consumed all the food before they were large enough to pupate. As a result no adult offspring came from the eggs laid during that period. Through natural mortality the number of adults progressively declined, and few eggs were laid. Eventually a point was reached where the intensity of larval competition was so reduced that some of the larvae secured sufficient food to grow to a size large enough to pupate. These larvae in turn gave rise to egg-laying adults. Because of the developmental time lag between the survival of larvae and an increase in egg-laying adults, the population continued to decline, further reducing the intensity of larval competition and permitting an increasing number of larvae to survive. Eventually the adult population again rose to a very high level and the whole process started again.

Competition for limited food held this blowfly population in a stage of stability and prevented any continuing increase or decrease. The time lag involved in the addition of egg-laying adults to the declining population resulted in an alternate over- and undershooting of the equilibrium position, causing an oscillating population density.

In a second experiment Nicholson supplied the adults with a surplus of suitable food, which was unavailable to the larvae. As a result of the enormous quantity of eggs laid by the adults, larval competition intensified and eventually the density of adults decreased in a manner comparable to the other experiment.

In another variation the larvae were supplied with a surplus of food, and the adults were given a constant daily quota of protein food. Again the adult population oscillated. The adults produced a high number of eggs that, because of the lack of larval competition, nearly all developed into adults. The adults competed intensely for a limited amount of food. Lacking sufficient protein for the production of eggs, the adults laid fewer eggs; and for the lack of replacements the adult population declined. Competition was gradually relaxed to a point where some of the flies obtained enough protein to produce eggs. After a 2-week lag the adult population began to build up again.

From these results Nicholson felt that the magnitude of the oscillations would be reduced if the larvae and the adults competed for a limited quantity of food not available to the other. This assumption was confirmed experimentally. Under the conditions described not only were the fluctuations slight, but they lost their periodicity, and the mean population level was nearly quadrupled.

In these competitive situations the larvae and the adults were seeking food, the rate of supply of which was not influenced by the activity of the flies. In effect the resource, or food available, could be subdivided into many small parts to which the competitors, the larvae and the adult flies, had general access. The individuals "scrambled" for their food, which under gross crowding resulted in wastage because each competitor got such a small fraction of the food that it was unable to survive. "Scramble" competition tends to produce violent oscillations in the population not caused by environmental fluctuations. It limits the average density of the population far below that which the food supply could support.

In contrast to the scramble type of competition is the "contest" type, in which each successful individual claims a supply of requisites sufficient for self-maintenance and reproduction. Unsuccessful individuals are denied access to food or space by the successful competitors, so the deleterious effects of shortage are confined to a fraction of the population. This permits the maintenance of a relatively high population density and prevents violent oscillations in numbers.

In the blowfly experiment food was directly limiting, and its availability had a pronounced effect on the growth and size of the population (Figure 15-5). The response of the population was characterized by sharp fluctuations brought on by starvation or the lack of nutrients needed for reproduction.

It is not difficult to assess the role of food in limiting population growth and size in a laboratory situation, where variables can be controlled. It is much more difficult to do so in a natural situation, where food may interact with other limiting influences. There are field situations, however, that do illustrate that food, either directly or indirectly, influences both fecundity and mortality.

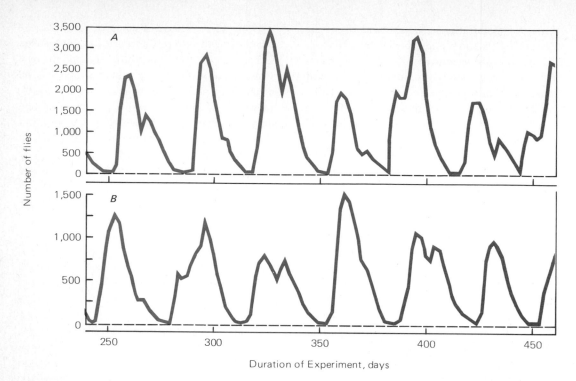

**FIGURE 15-5**
*Fluctuations in the number of adult blowflies in two cultures subjected to the same constant conditions, but restricted to different daily quotas of food. (After Nicholson, 1957.)*

A density-dependent relationship exists between many herbivores and carnivores and their food supply. For example, the mule deer and the white-tailed deer, inhabitants of the transitory early successional stages, do not appear to have developed a self-regulating mechanism characteristic of such ungulates as the Uganda kob and the mountain sheep that inhabit a relatively stable vegetation type.

If food quantity and quality are high, fertility is maximum, mortality is low, and the rate of increase is high unless the population is held down by hunting or predation. As the population increases to high levels, both the quantity and quality of available food decline. In periods of stress, such as a hard winter, starvation, especially among fawns, becomes common (Figure 15-6). But more commonly, too many deer survive and exist on inadequate nutrition. Poor nutrition results in the repression of growth, delayed sexual maturity, lower conception rates, increased intrauterine mor-

**FIGURE 15-6**
*Death of deer from winter starvation. Overshooting the carrying capacity of the environment, especially at a critical time of year when food resources are scarce and access is controlled by a strong dominance hierarchy, can result in widespread starvation, the fate of this deer. (Photo courtesy Michigan Conservation Department.)*

**482**

tality, increased fawn mortality, and an age structure distorted toward older age classes (see Klein, 1970; Caughley, 1970).

Carnivores are more subject to food shortages than herbivores and for many starvation and delayed reproduction result. Birds of prey may fail to nest in periods of low prey populations. Only 20 percent of a great horned owl (*Bubo virginiana*) population nested when the population of its prey, the snowshoe hare, was low (Rusch, 1972). When the snowshoe hare population was high, 100 percent of the owl population nested. Faced with the same shortage of prey, lynx in the same area continued to reproduce, but the young died from starvation.

Just as a shortage of food can limit the size of an animal population, so can a shortage of such resources as moisture and soil nutrients limit a plant population. Grime and Curtis (1976) planted naturally occurring bare patches of soil with seeds of two bunch grasses, sheep fescue (*Festuca ovina*) and tall oatgrass (*Arrhenatherum elatius*). One group of plots received no treatment. Other plots received various combinations of water, phosphorus, and nitrogen. On the untreated plots tall oatgrass experienced a slow, continuous seedling mortality through the winter and a sharp decline in the summer. Although some seedling loss was caused by herbivores, the catastrophic loss in summer was caused by a shortage of moisture. Sheep fescue, a drought-resistant grass, survived much better on the untreated plots. On the plots treated with both nitrogen and phosphorus, but no water, both fescue and tall oatgrass survived. With an adequate supply of soil nutrients, tall oatgrass was able to extend its roots deeper to reach moisture. On plots treated with water, nitrogen, and phosphorus tall oatgrass dominated and fescue declined. The shallow-rooted fescue could not compete with the tall oatgrass that possesses both deep and shallow roots.

## DISEASES AND PARASITES

Because the virulence and rate of spread of infectious diseases and certain parasites increase with the density of a population, they can act as density-dependent regulators. Disease, although prevalent, becomes important as a regulating mechanism only when it reaches an epidemic or epizootic level and this is more likely to occur when the density of the host population is high. Rampant disease can reduce populations in a density-dependent fashion, but it may not be the primary cause of population decline. An increasing population may result in a depression within the organism of antibody formation and other body defenses, an increase in inflammatory processes, and an increased susceptibility to disease. Thus disease may be the consequence of a high population rather than a cause of population decline.

Bacteria and viruses are causal agents of important animal diseases. They may be transmitted from one animal to another by direct contact or by insect vectors. An important disease associated with dense populations of wild mammals is rabies. A viral disease highly contagious to humans, rabies follows the nerves from the infection point to the spinal column and brain before symptoms appear. The symptoms are both behavioral and physical. Infected animals are restless and excitable, exhibit convulsions, wander aimlessly, and show no fear of humans. Physically they are emaciated, exhausted, and partially paralyzed. Among wild mammals rabies occurs most commonly in coyotes, foxes, skunks, and raccoons. Foxes and dogs are the primary causes of the spread of the disease.

Disease in plants is caused largely by fungi and spreads in a density-dependent fashion. One example is *Fomes annosus*, a disease of white pine which spreads rapidly through pure stands by way of root grafts (a situation in which roots of one tree become grafted onto roots of a neighbor). A second example is oak wilt which threatens oak forests of the upper Mississippi Valley and the southern Appalachians. The disease is caused by the fungus *Ceratocystes fagacearium*, which appears to be spread by root grafts and nitidulid beetles, especially *Glischorochilus* and *Coleopterus* (True et al., 1960).

Disease may be density-independent when it is introduced into a population lacking any resistance to the disease or when it results from an environmental change not brought about by the organism involved. Then disease can reduce populations, exterminate them locally, or restrict the distribution of the host.

**483**

Prior to the discovery and settlement of the Hawaiian Islands by Europeans, the avifauna of Hawaii inhabited all parts of the islands from the seashore to the upper limits of vegetation. In 1826 the tropical subspecies of the night mosquito was accidentally introduced. The mosquitos spread rapidly throughout the lowland areas of the high islands and spread bird pox and other unidentified diseases through the lowland bird populations. As a result several endemic species of birds disappeared from the lowlands to a 600 m elevation (Warner, 1968). A number became extinct and those of the species that still exist are restricted to regions above 600 m in spite of the availability of habitat at lower elevations. When they are carried to a lower elevation, they succumb to malaria or bird pox. Thus the high forests of the Hawaiian Islands are an ecological sanctuary from disease for the birds. A somewhat similar situation occurred with the American chestnut. An introduced fungal disease wiped out the chestnut as a timber tree in North America, but a residual population of root sprouts continues to persist, providing some hope that a resistant strain will evolve.

The relationship of diseases and parasites and their hosts is one of evolutionary response. As new defenses are built up by the host, new strains are evolved by the parasite. A dramatic example is the relation of hybrid corn, artificially developed or "evolved" by man, and corn blight. In breeding varieties of hybrid corn, hybridizers succeeded in developing a uniform cytoplasm. They did this to utilize factors within the cytoplasm that lead to male sterility, which eliminates the need for detasseling corn in seed fields. One particular corn cytoplasm, Texas male sterile, TMS, has been widely incorporated into many lines. But TMS conferred susceptibility to a virulent race T of southern corn blight (Tatum, 1971). When this blight did appear in the central corn belt in 1969 and in Florida, it became epidemic because 80 percent of field corn hybrids carried the highly susceptible TMS. The remaining hybrids containing normal male-fertile cytoplasm were resistant to race T. Such is the risk humans run with new varieties and races of cereal grains. The problem is to maintain a crop with a high degree of genetic diversity, so that when a disease develops a virulent strain,

a genetic bank exists from which resistant forms can be bred.

## PREDATION

Predation can act as a powerful extrinsic regulatory mechanism. Theoretically a predator can regulate or control a prey population if the predator increases or decreases its density or effectiveness as the abundance of prey increases or decreases (see Chapter 17). Or to state it somewhat differently, regulation results only if the average risk of each prey individual increases with an increasing density of prey (Nicholson, 1954). As long as the predator takes only recruitment to the prey population and does not reduce the parental stock, both predator and prey populations will tend to remain fairly stable. This is typical of compensatory predation. If the predator's rate of increase exceeds the prey's rate of increase predation cuts into parental stock and may limit or virtually exterminate local prey populations. The predator population tends to overshoot that of the prey and then the decrease in prey results in a sharp decline in the predator population through starvation. If a portion of the host or prey population is not available because of environmental discontinuity, then the oscillations will be damped, and under certain conditions the predator-prey system will be self-regulating. On occasions the prey overcomes the regulating influence of predation and reaches outbreak proportions. Under this condition it is unusual for a predator to overtake the prey and cause it to decline.

Data are accumulating on the influence of predation on selected populations. Waterfowl appear to be particularly vulnerable during the nesting season (Basler et al., 1968; Urban, 1970). The effects of predation have become more pronounced as available breeding habitat shrinks. In the past large expanses of nesting grounds and wide dispersion of breeding population compensated for predation.

Predation may depress populations of ungulates when the predator is one of the highly skilled canids, the wolf or the coyote, or a big cat, such as the mountain lion. In fact predation may be the major regulatory mechanism. Deer, moose, and caribou apparently have not evolved intrinsic population controls (Pimlott, 1967).

Just as animal predators have an impact on vertebrate and invertebrate prey, so do herbivores or plant predators have an impact on plant populations. Deer can adversely affect such plants as white cedar, birch, and American yew, all quite sensitive to overbrowsing. The American yew is nearly gone in the Adirondacks and central Applachians. Scattered vigorous stands exist only where populations of deer are low. Grazing of reef fish is so concentrated near the reefs that no sea grasses can sustain viable growth there (Randall, 1965).

In England the rabbit, introduced as a semidomesticated species in the twelfth century, has been part of the countryside since the mid-1800s. Although some early work by the English ecologist A. Tansley showed that rabbits grazing on the chalk flora produced a richer diversity of species, the real impact of rabbits was not noticed until after 1954. At that time myxomatosis nearly wiped out the rabbit population, reducing the predatory pressure on the grasslands. With the demise of the rabbits, there was a spectacular increase in the growth of grass and a profusion of flowering by perennial species whose existence had never been recorded. But after the initial flush the number of dicot species declined, the grasslands were dominated by a few species to the exclusion of others, and the number of vegetative shoots of both grasses and dicots were reduced as the plants grew into taller, self-shading vegetation (J. L. Harper, 1969).

## Intrinsic influences

### CROWDING

As population density increases through increased population growth or decrease of living space by environmental degradation, individuals become stressed. Evidence suggests that social stresses act on the individual through a physiological feedback involving the endocrine system (Figure 15-7). In vertebrates the feedback is most closely associated with the pituitary and adrenal glands (Christian, 1963; Christian and Davis, 1964). Increasing populations of mice held in the laboratory resulted in the suppression of somatic growth and curtailment of reproductive functions in both sexes. Sexual maturation was delayed or

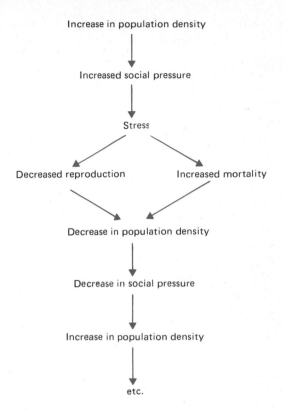

**FIGURE 15-7**
*Christian's stress hypothesis of population regulation. (From Krebs, 1964.)*

totally inhibited at high population densities, so that in some populations no females reached normal sexual maturity. Intrauterine mortality of fetuses increased, especially in the fetuses of socially subordinate females. Increased population density resulted in inadequate lactation and subsequent stunting of the nurslings. The same effect appeared again to a lesser degree in the animals of the next generation, even though the parents were not subject to crowding. Studies of some wild populations under stress seem to support the hypothesis that crowding can reduce populations. K. Myers and his associates (1971) studied confined populations of the Old World rabbit *(Oryctolagus cuniculus)*. They held the rabbits at several densities in different living spaces within confined areas of natural habitat. Those living in the smallest space, in spite of decline in numbers, suffered the most debilitating effects. Rates of sexual and aggressive behavior increased, particularly among the females. Reproduction declined, as

485

did the ovulation rate and the number of corpora lutea. Fat about the kidneys decreased, and the kidneys showed inflammation and pitting on the surface. The weight of the liver and spleen decreased, and adrenal size increased. Stressed individuals had abnormal adrenals. Young rabbits born to stressed mothers showed stunting in all body proportions and in organs. As adults they exhibited such behavioral and physiological aberrations as a high rate of aggressive and sexual activity and large adrenal glands in relation to body weight. Low body weight, the lack of lipids in the adrenals, abnormal adrenals, and poor survival indicated a lack of fitness of such rabbits. On the other hand, rabbits from low to medium density populations showed excellent health and survival.

Among fish there is a marked negative correlation between density and rate of body growth. This relationship, evident to anyone who has observed a pond overstocked with bluegills, is one of the best established examples of density-dependent relationships in populations (Figure 15-8). The decline in growth is not attributable only to the lack of food. There are more subtle influences at work, such as the conditioning of the water. The role of conditioners has been studied more intensively in frog populations, among which a high tadpole density apparently inhibits growth. Associated with inhibited growth is an algal cell found in the feces of tadpoles (Richards, 1958). The effects of this inhibitory cell are nonspecific. Water from the environment of crowded tadpoles of one species can inhibit growth of another (Licht, 1967).

Plant populations also become stressed at high densities, with differing responses in the vegetative and sexual modes of reproduction. In many ways the individual plant itself is a population that exhibits an increase in such basic components as leaves, flowers, and runners. As more and more units are added, interference between units increases and so does the demand on resources. As already discussed (see Chapter 13, life tables of plants), productivity of plants per unit area is independent of the number of plants over a wide range of densities. The biomass per unit area may be the same whether distributed among many small individuals or concentrated in a few large ones. In this respect plants and fish are similar in their growth responses to density.

One response, then, of plants to density is the reduction of the growth of individual plants. Botanists call it a plastic response. Individuals adjust their growth form, size, shape, number of leaves, flowers, and production of seeds in a scramble fashion to the limited resources available.

A second response is increased mortality. High mortality is characteristic of the seedling stage, just as high mortality is characteristic of young animals. The size of a population that develops from seed is the function of the availability of seed and sites. If sites are abundant and seeds are plentiful, density dependent mortality follows. As the number of seedlings increases, the chance that a seed will produce a mature plant decreases. No matter how great the densities involved, all finally converge to a certain number per unit area, a situation most obvious in forest trees. As the yield tables suggest, density varies with sizes of plants. A certain size has a similar level of survival at various densities. In some plants, reproduction by seed at high densities is almost nonexistent (Putwain et al., 1968); it is solely by vegetative means. As the density of plants increases, the number of vegetative offspring also declines.

A third response to the stress of density is a hierarchical exploitation of resources and space. A few individual plants become dominant and control the site. The remainder remain stunted and possess poor viability.

*FIGURE 15-8*
*Growth of fish as influenced by population density. The graph reflects a scramble type of competition. (From Backiel and LeCren, 1967.)*

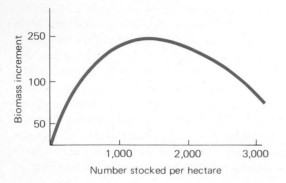

## BEHAVIOR

Social behavior may function in the regulation of animal populations by limiting the number of individuals to a habitat and reproductive activity. Two mechanisms are territoriality and social dominance discussed in Chapter 12.

Territorial behavior results in the exclusion of individuals from an area or habitat if all suitable space is taken. Territoriality can be considered from two extremes. The size of a territory may be a function of the number of animals who divide all available area among them. Each male receives his share of the area regardless of how high the density becomes. At the other extreme a territory has a rigidly fixed minimum size. The number of males an area can hold is determined by the total area available divided by the minimum size of territory. When the area is filled, the remaining animals are ejected or denied access. They become the surplus or floating population and live in groups in some neutral area or as individuals in areas too poor for breeding.

The real situation probably comes much closer to Julian Huxley's (1945) analogy which considers territory as an elastic disk compressible to a certain size. Territory size decreases as density increases, but when the territory reaches a certain size, the resident resists further compression and denies access to additional settlers. Because aggressive behavior is not the same among all individuals, the most aggressive have the advantage and the less aggressive individuals are forced to settle outside optimum habitat.

Fretwell and Lucas (1969) developed a model of territoriality involving habitat suitability (Figure 15-9). The model assumes that each individual will select the optimum habitat where its chances of survival and success at reproduction will be the highest. If all individuals choose a habitat of highest suitability, then from a point of view of unsettled individuals, all occupied habitats will be equally suitable. When all optimum habitat is filled, then the next most suitable or suboptimum habitat becomes for the remaining individuals the optimum habitat, and they begin to settle on it. Once that is filled, marginal habitat becomes most suitable. The model implies that as density increases, suitability of habitat decreases. Thus a habitat has its highest suitability

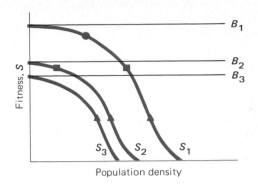

*FIGURE 15-9*
*Relationship between habitat suitability, population density, and settling of habitats of different quality. The suitability of habitat is ranked from $B_1$ to $B_3$. The curves $S_1$, $S_2$, and $S_3$ show the actual fitness of individuals in habitats at different population densities. At low population density all individuals will settle in habitat $B_1$. At intermediate density some individuals will settle in habitat $B_2$. At high density all three habitats will be settled. (After Fretwell and Lucas, 1969.)*

when density is 0. At low levels of population density only optimum habitat is occupied. At highest densities, all habitats from optimum to marginal are utilized. The behavior of settled individuals is a cue to unsettled ones that the area is completely populated.

Because territoriality spaces out individuals and restricts the number that occupy a suitable habitat, it is a potential population-regulating mechanism (Table 15-1). To demonstrate that territorial behavior does limit a population one has to show that a portion of a population, including both males and females, does not breed because they are excluded from suitable breeding sites by dominant or territorial individuals. This surplus population should be able to breed if a territory becomes available to them.

Ways in which territoriality influences the demography of a population and the degree to which it can bring about the existence of a surplus population have been demonstrated in several studies. In the arctic ground squirrel (*Spermophilus undulatus*) all females are allowed to nest but territorial polygamous males drive excess males from the colony into submarginal habitat where they exist as a nonbreeding floating population. The number of

TABLE 15-1
*Summary of effects of population surplus on reproduction in species defending large fractions of their home ranges*

| Species characteristic | Surplus Males | Females | Reproduction limited | Reproduction limited by territoriality | Reproduction limited by territoriality under natural conditions | Example |
|---|---|---|---|---|---|---|
| Polygynous | + | – | – | – | – | arctic ground squirrel (Carl, 1971) |
| Monogamous with excess males | + | – | – | – | – | many songbirds (Hensley and Cope, 1951) |
| Dependent on nest sites that are limiting | + | + | + | – | – | hole-nesting birds |
| Reaching abnormal densities in unnatural habitats | + | + | + | + | – | European blackbirds in Oxford Botanical Garden (Snow, 1958) |
| At normal densities in natural habitats | + | + | + | + | + | rock ptarmigan (Watson, 1965) |

*Source:* Brown, 1975.

breeding males remains constant because losses are continually replaced from the floating population, which in turn decreases drastically over the year largely from predation and weather. In this species territoriality stabilizes the number of breeding males but does not regulate the population because only males appear to be surplus.

The territorial response of the dickcissel *(Spiza americana)* conforms to the elastic disk analogy and model of optimum habitat (Zimmerman, 1971). Optimal habitat is characterized by a tall, dense cover of herbaceous vegetation and presence of song perches. Within this habitat, which is usually fully occupied, territorial size decreases to a minimum of approximately 0.9 acre, reached at a density of 60 to 70 per 100 acres. At greater densities surplus males are forced to less suitable vegetation where they attract fewer mates and experience poorer nesting success.

In Great Britain Krebs (1971) removed breeding pairs of great tits from their territories in an oak woodland. The pairs were replaced by new birds, largely first-year individuals, that moved in from territories in hedgerows, considered suboptimal habitat (Figure 15-10). The vacated hedgerow territories, however, were not filled, suggesting that a floating reserve of nonterritorial birds did not exist. In this case territorial behavior limits density of breeding birds in optimal habitat, but does not regulate the population because all birds are breeding in some habitat.

Territoriality results in a rather large floating population in Cassin's auklet *(Ptychoramphus aleuticus)*, which breeds on offshore islands from southern Alaska to central Baja California. Cassin's auklet lays one egg deposited in a crevice or burrow about which it defends a very small territory. Manuwal (1974) studied a breeding colony of approximately 105,000 birds on Southeast Farallon Island, California. Burrow density was high, from 1.09 to 0.02 burrows per square meter. Removal experiments showed a surplus population consisting equally of males and females, both capable of breeding. Fifty percent of the floating population was adult and 80 percent had no previous breeding experience. The

**FIGURE 15-10**
*Replacement of removed individuals and settlement of vacated territory. Six pairs of great tits* (Parus major) *were shot between March 19 and March 24, 1969 (left stippled area). Within three days four new pairs had taken up territories (right stippled area) and there was some expansion of residents' territories, so that after replacement territories again formed a complete mosaic over the woods. (From Krebs, 1971.)*

floaters roosted in areas where no territorial holders existed and actively searched for vacated burrows. When they found such a burrow floaters claimed it immediately. These successful floaters were usually paired before or immediately after occupying the burrow. Fifty percent of the floaters, who entered the population at a rate of about 19 percent per year, fledged young. Thus in Cassin's auklet territoriality in conjunction with high breeding density and shortage of nesting sites limits the net reproductive rate.

Nearly as effective is territorial behavior in the red grouse of the heather-dominated moors of Scotland. The red grouse has three social classes: (1) territorial cocks and their hens; (2) nonterritorial surplus birds that live as a floating reserve on the periphery of the breeding ground; and (3) nonterritorial transient birds. In fall and spring red grouse experience a sharp decline in numbers associated with territorial behavior and loss of nonterritorial birds from predation. If one of the territorial birds dies or disappears, its place is taken by a bird from the floating reserve. By late winter all surplus birds are removed from the moors

and the breeding population in spring is fixed by territorial behavior of the previous fall (Figure 15-11) (Watson and Jenkins, 1968; Watson and Moss, 1972). Competition for breeding territories sets an upper limit to the breeding numbers, but density of breeding population varies from year to year because of secondary variables that influence territorial behavior. Especially important is the abundance of heather *(Calluna vulgaris)* and its nutritional quality as measured by its nitrogen content (Moss et al., 1972; Lance, 1978). Size of grouse territories varies inversely with the proportion of ground occupied by patches of heather sward (Miller and Watson, 1978).

Although the existence of floating reserve populations is acknowledged little data exist on the social organization and behavior of surplus birds. Floaters may form flocks with a dominance hierarchy on areas not occupied by territory holders, as do the red grouse and Australian magpies. Or they may live singly and spend much time on the breeding territories of others. An example of this strategy is provided by the detailed studies by S. Smith (1978) of the rufous-collared sparrow *(Zonotrichia capensis)* in Costa Rica. By observing banded birds, both territorial and nonterritorial, and by selectively removing certain individuals, Smith was able to determine the role of the floater or "underworld" bird. Territorial sparrows on her study area occupied small territories ranging from 0.05 to 0.40 ha and made up 50 percent of the total population. The other 50 percent, underworld birds consisting of both males and females, lived in well-defined restricted home ranges within other birds' territories (Figure 15-12). Male home ranges, often disjoined, embraced three or four territories. Female home ranges were usually restricted to a single territory. Because home range boundaries of both sexes coincided with territorial boundaries, each territory held two single-sex dominance hierarchies of floaters, one male and one female. When a territorial owner, male or female, disappeared, it was quickly replaced by a local underworld bird of appropriate sex on the territory. These floaters usually entered the territories as young birds hatched some distance away and were tolerated by the owners.

The number of breeding birds in the rufous-

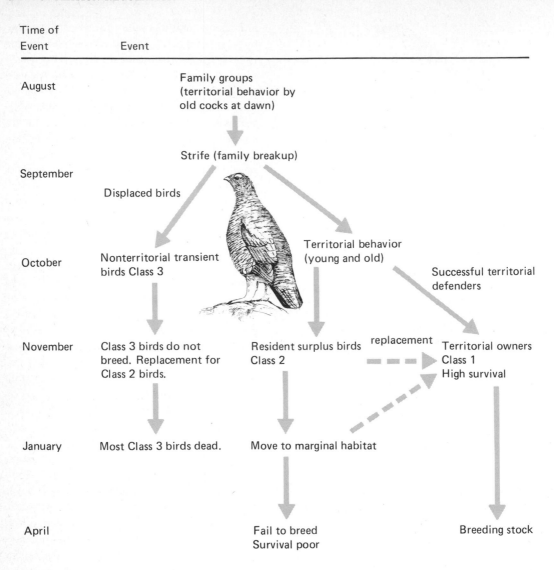

| Time of Event | Event |
|---|---|
| August | Family groups (territorial behavior by old cocks at dawn) |
| September | Strife (family breakup) |
| | Displaced birds |
| October | Nonterritorial transient birds Class 3 — Territorial behavior (young and old) — Successful territorial defenders |
| November | Class 3 birds do not breed. Replacement for Class 2 birds. — Resident surplus birds Class 2 — replacement → Territorial owners Class 1 High survival |
| January | Most Class 3 birds dead. — Move to marginal habitat |
| April | Fail to breed Survival poor — Breeding stock |

**FIGURE 15-11**
*Territoriality and social behavior as regulatory mechanisms in red grouse populations. In early winter the birds become divided into three classes, only one of which will breed the following spring. The other classes are eliminated. In general breeding success in any year is inversely proportional to breeding success the previous year: the greater the success was in the previous year, the poorer it is in the current one. Breeding success and thus population size is regulated through territorial behavior. (From Watson and Moss, 1971.)*

collared sparrow apparently was regulated by the territorial behavior of owners, because the population was not limited by nest sites, and since floaters could forage in territories, the birds apparently were not limited by food. Floaters seemingly were tolerated because the cost in energy to exclude them would be greater than the benefits accrued. Territorial owners could derive benefit by knowing their mate's potential replacement and in event of loss a new pair could be formed quickly. In addition because individual birds regulated the number of a hierarchy, floaters could reduce territorial defense costs by discouraging other intruders (S. Smith, 1978).

Territoriality may regulate populations of some species by restricting access to the food supply to a certain segment of the population (C. C. Smith, 1968). An example is found

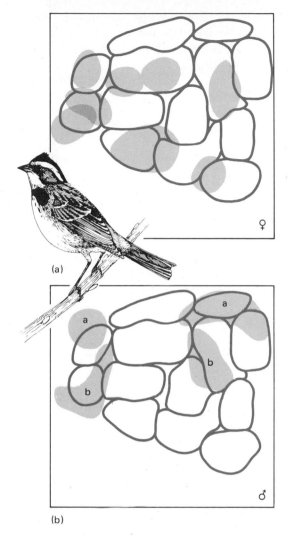

**FIGURE 15-12**
*Territorial boundaries of rufous-collared*
*sparrows. Home ranges of two "underworld"*
*females and males are superimposed on*
*occupied territories. Eventually some of the*
*females occupied the territory by replacing a*
*missing bird. The male home ranges a and*
*b are disjoined and each includes several*
*territories. (Adapted from S. Smith, 1978.)*

among the red squirrels *(Tamiascurius)* that inhabit the coniferous forests of the Pacific Northwest. Individual squirrels, male and female, each defend a territory through the year. Males enter the female's territory during a short period of one to two days when she is in heat. At that time the female ceases to defend her territory. The territory centers about the

food supply, and the size is adjusted according to its availability, which in general does not exceed three times the territorial owner's yearly requirements. The lower the overall food supply, which consists in part of cones and fungi that can be stored, the larger the territory. Thus territorial behavior allows each individual the optimal conditions for harvesting, defending, and storing a seasonal food supply, so that it will be available throughout the year. Kemp and Keith (1970) observed a similar type of territoriality in the populations of red squirrels in Alberta.

How territoriality might have evolved has been the subject of speculation (see Brown, 1969). Several hypotheses have been advanced including prevention of overpopulation (Wynne-Edwards, 1962) and need for individual food exploitation. Defense of a food resource, or rather space containing a food resource, may be important in some species such as the red grouse, the ovenbird (Stenger, 1958), and golden-winged sunbird (Gill and Wolf, 1975), because in those species territorial size is inversely related to food density. Brown has proposed a general theory that for territoriality to evolve some resource must be in short supply, it must be defensible, and it must be "worth fighting for" (Figure 15-13). Because space is more defensible than food, aggressive behavior is directed toward excluding other animals from space containing food resource.

Closely linked to the evolution of territoriality is aggressive behavior. Aggressive behavior is not likely to benefit an individual unless it provides something worth the risk. If an individual can obtain a mate, food, or nesting space without defending a territory, then territorial behavior would be disadvantageous. But if a resource is in short supply, an animal has much to gain in defense of a territory in spite of the risks of aggressive behavior, although risks are reduced if advertising and threat behavior are highly specialized (see Chapter 12).

### DISPERSAL

Overcrowding and associated increase in aggressive behavior are major forces in producing dispersal. Social hierarchy and territoriality can force low-ranking individuals to depart

**491**

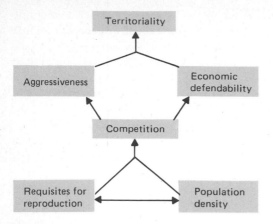

**FIGURE 15-13**
*Outline of a general theory of evolution of diversity in intraspecific territorial systems. This diagram represents a resource-centered theory. The value of territorial behavior for a species is determined by the supply of a resource that may be critical for that species and by the energy costs and gains associated with fighting for a share of the resource. (From Brown, 1964.)*

and to seek suitable habitat in areas that are unoccupied by higher-ranking individuals or competing species. Individuals forced to leave are largely the maturing young.

Dispersal is a constant phenomenon, regardless of whether a population is dense or sparse. It is most pronounced when densities are high. One-way movement out of a population with no return is *emigration*. It is closely associated with the one-way movement into the population, *immigration*.

If populations in an area are fairly stable and are at or near the carrying capacity, dispersal movements have little influence on population density and may only slightly influence age structure. But emigration movements induced by overpopulation and food shortages reduce populations and may greatly influence age structure and the reproductive rate of the remaining population. Immigration into rapidly growing populations can increase the growth rate not only by the addition of new members but also by an increase in the birth rate. Immigration into sparsely populated areas can increase the reproductive potential of a population by bringing in new genotypes.

Emigration or dispersal may be density

dependent and density independent to a greater or lesser degree. Environmental dispersal (Howard, 1960) or saturation dispersal (Lidicker, 1975) is density dependent. It is an outward movement of surplus individuals living at or near carrying capacity. These individuals, usually juveniles or subdominants, have two options; to stay and perish or at the very best not breed or to leave the area. If they move out, the odds are great that they will succumb to predators, disease, exposure, and accidents, although a few will arrive at some suitable area and settle down. An excellent example of such dispersal is that described for muskrats by Errington (1955). Because saturation dispersal is a symptom of approaching overpopulation, it has little influence on population regulation.

Of more importance is presaturation dispersal (Lidicker, 1975), which is basically density independent. Presaturation dispersal takes place before the population reaches carrying capacity and before resources are depleted. The emigrants are in good condition, consist of any sex and age group, possess a greater chance of survival than saturation dispersers, and have a high probability of settling in a new area.

Such dispersers may exhibit qualitative and genetic differences from residents. Charles Krebs and his associates (Myers and Krebs, 1971; Krebs et al., 1973) studied for five years the population dynamics, including aggressive behavior, dispersal and genetic differences, of two field voles, *Microtus pennsylvanicus* and *M. ochrogaster*. They established three large enclosures: a control area in which the population of mice was not disturbed and two removal areas in which the number of mice was continuously reduced by trapping. Mice dispersing from the control area into the removal areas were compared with resident control animals for several characteristics including age, sex, aggression, and genetic differences as determined by electrophoresis. The researchers found that dispersal into areas vacated by removal occurred largely during fall and winter. Dispersal was most common during the increase phase of population fluctuations in the control area and least frequent during the decline phase. During the period of population increase, 55 percent of the loss of females could be accounted for by emigration. Young females were more common in the dispersing population than young males, but adults of both sexes left the area. Dispersing males of *M. pennsylvanicus* were more aggressive than the males that remained. As density in the control areas increased, less tolerant individuals moved into less densely populated areas.

Presaturation dispersal may be one mechanism operating in the population regulation of some animals, particularly rodents. How such dispersal can function is suggested by several studies comparing responses of populations in which dispersal could occur with those in which dispersal was prevented. Krebs et al. (1969) enclosed three populations in such a way that immigration and emigration of voles was prevented, but predators had access. They compared these populations with a control population whose members were able to disperse. The enclosed populations increased in size; in one plot the population was three times as high as the control. Overpopulation in the enclosures resulted in overgrazing, habitat deterioration, and starvation. Stress was relatively unimportant because *Microtus* can exist at densities several times higher than those normally experienced by other voles.

While these experiments represented something of an artificial situation, Lidicker (1973) and Tamarin (1978) compared the population dynamics of naturally enclosed populations of *Microtus* on islands where a water barrier prevented emigration with populations living on the mainland. Lidicker followed the growth of the California vole *Microtus californicus* on Brooks Island in San Francisco Bay (Figure 15-14). He found that population growth rates were lower in the mainland populations from which dispersal could take place than in the island population in which emigration was greatly limited (for details, see Lidicker, 1973). Tamarin found that population of beach vole *(Microtus breweri)* on Muskeget Island near Nantucket Island, Massachusetts, did not undergo the degree of regular fluctuations characteristic of mainland populations of meadow vole *(M. pennsylvanicus)*.

In a way the results of the two studies seem somewhat contradictory, but the population of voles on Brooks Island represented only a

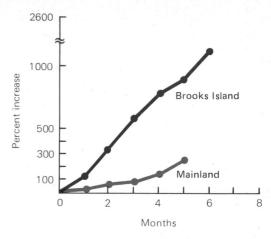

**FIGURE 15-14**
*Percentage increases (monthly intervals) from starting densities during periods of rapid increase to peaks in* Microtus californicus. *(Adapted from Lidicker, 1975.)*

period of rapid increase in the spring; the population was not followed over a period of several years. The Muskeget Island population was long established.

Based on his experimental results and those of other studies, Lidicker hypothesized that presaturation dispersal could play a key role in population growth of voles. To disperse individuals must be motivated, be physically able to leave the area, and have available a "dispersal sink," an empty or unfilled habitat or even marginal or unsuitable habitat in which the animals could survive for a short time. The dispersal sink must permanently remove animals from the population. If a barrier exists to prevent dispersal or if dispersal sinks are unavailable, the dispersers must return to their home area. Because voles are not strongly territorial, these "frustrated dispersers" are allowed back into the population, thus adding to population growth. Tamarin (1978) further hypothesizes that in situations in which emigration is inhibited the population would increase rapidly as it did in the experiments of Krebs and Lidicker, overshoot the carrying capacity, and decline. Without normal regulatory process, that is, dispersal, the population will oscillate until the voles adapt to a *K*-selected environment. (The beach voles on Muskeget Island exhibit stability probably because they have shifted more to-

**493**

ward *K*-selection.) In an unenclosed population a large portion leaves before the population reaches peak density, resulting in some degree of cycling (Figure 15-15). Because dispersers appear to possess behavioral and genetic differences from residents, cycles may be driven by a changing genetic composition of the population. The gene pool is changed by dispersal. Thus a mainland population that has a dispersal sink experiences high dispersal, density cycles, and *r*-selection, at least when the decline occurs. On islands or enclosed populations, dispersal is reduced, density cycles are virtually absent, and voles experience *K*-selection.

Some organisms experience an artificial dispersal by humans outside of the natural genotypic range and come into contact with different regional genotypes. An example is restocking game animals secured from one region and introduced into another with the objective of replenishing the species in depleted habitats.

*FIGURE 15-15*
*Population of female beach voles* (Microtus breweri) *on Muskeget Island compared to population of female meadow voles* (Microtus pennsylvanicus) *on the mainland (Barnstable, Mass.). Note the relative stability of the island population (solid circles) compared to the cyclic mainland population (open circles). (Adapted from Tamarin, 1978.)*

In northeastern United States there are two cottontail rabbits of the uplands. One is the eastern cottontail *Sylvilagus floridanus* of which there are 14 subspecies. The other species is the New England cottontail *Sylvilagus transionalis*. The eastern cottontail is a rabbit of open and brushy country. The New England cottontail is a woodland rabbit restricted to New England and the high Appalachians. Chapman and Morgan (1973) discovered that the introduction of different genotypes of *S. floridanus* from distant points of the species range had a distinct impact on the genetic structure and population dynamics of eastern cottontails of the region. The eastern cottontail benefited not only from the habitat changes, but also from increased genetic variability brought about by the introduction of new genotypes. The genetic variation added to the gene pool apparently enabled the eastern cottontail to colonize previously unsuitable habitats such as mature hardwood forests, mixed open woodland, coniferous forest, and rhododendron thickets. On the other hand, the New England cottontail, increasingly restricted to islands of habitats left after land clearing and lumbering and lacking an extensive pool of genetic variability, is declining to a point where it can be considered an endangered species.

Dispersal of individuals from a population and their colonization of a new area involves a number of variables. One is type of dispersal, active or passive. Plants for the most part are

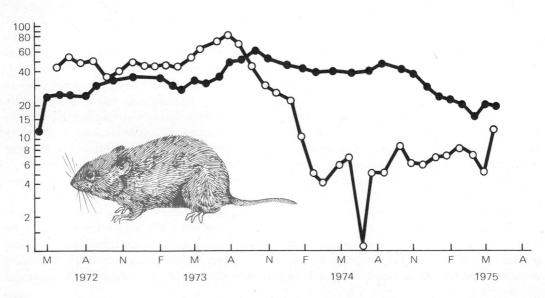

passively dispersed by currents of air or water or by being carried in the gut, fur, or feathers of animals. Protozoans, algae, and such invertebrates as spiders are also passively dispersed by air and water. Parasites may be carried in the gut of animals. Most animals, however, being mobile, are actively dispersed.

A second variable is frequency of dispersal. The more frequently a plant has a heavy seed crop or a population reaches a high density, the more often will new propagules be available for colonization, even though losses are enormous. Regardless of the means of dispersal, the number of disseminules per unit area decreases with distance, rapidly at first, then more slowly. If one plots the number of seeds against density, the curve approximates exponential decay. The reasons are rather obvious. Surviving exposure, escaping predation, moving through hostile habitat for great distances, deflection of seeds by barriers, loss of a means of transport—all of these and more influence the rate at which emigrants reach a given area.

Just as no species exists alone, so it does not disperse in the absence of other dispersing organisms. Successful colonization requires transport to the area, the arrival of at least one reproductive unit of an unrepresented species, the entry of that species into the community, and its survival for at least several generations. Once the colonists are settled on the area, they face such risks to survival as predation or stiff competition from members of another species that are also colonizing the area. Or they may be unable to survive because of physical or chemical conditions or be unable to give rise to a population because of biological conditions (see McGuire, 1972).

GENETIC CHANGE

In his studies of vole populations, Chitty (1960) found that a decline in the number of these rodents can take place even though the environment appears to be favorable. In fact a high density of population is insufficient to start a decline and a low density is insufficient to stop it. A majority of animals die from unknown causes, males more rapidly than females. However, the adult death rate is not abnormally high during years of maximum abundance. Because individuals in a declining vole population are intrinsically less viable than their predecessors, any changes in the cause of mortality are insufficient to account for the increased probability of death among voles.

Later, Chitty and Phipps (1967) suggested that the decline in vole populations resulted from the selective pressures of mutual interference. Although animals in declining populations may not necessarily be less viable than in increasing populations, some nevertheless are less fit than others and eventually disappear from the population. When population densities are low, Chitty reasoned (as they would be in newly colonized areas or in areas where the population was greatly reduced by mortality or some catastrophe), mutually tolerant individuals adapted to crowding would be most adaptable and have the selective advantage over aggressive individuals. As the population reaches peak density and begins to crowd in on available resources, selection would shift in favor of aggressive individuals. But when the population crashes, aggressive behavior in low populations is no longer an advantage. Aggressiveness may cause the population to continue to decline until aggressiveness is bred out of the population. As the number of tolerant individuals in the population increases, the population begins to build up again.

An example is provided by Wellington's (1960, 1964) studies of the population dynamics of the forest tent caterpillar (*Malacosoma pluviale*). He found that as the population increased, active adult females left the area to lay their eggs, leaving sluggish, adult females behind. As the local population dominated by relatively inactive individuals increased, quality of the population declined to the point where many larvae failed to survive even under the most favorable conditions. As environmental conditions changed, the inactive population sharply declined and the area was eventually recolonized by active moths.

The studies of Chitty and Wellington suggest that as animal populations increase in density, quality of the population deteriorates, preventing an indefinite increase. Genetic feedback apparently operates through density pressure, selection pressure, and genetic change within the population. Large increases

495

in population brought about by a changing environment increase the variability in the population and many "inferior" genotypes survive. When conditions become more rigorous, these ecologically inferior types are eliminated and the population is reduced, often abruptly (Figure 15-16).

How the genetic quality of a population can act on a population was demonstrated in a laboratory population of *Drosophila* by H. Carson (1968). He introduced a wild-type male into a population of an inbred line of females. This small change in the gene pool stimulated

*FIGURE 15-16*
*Modified version of Chitty's hypothesis to explain population fluctuations in small rodents. Density-related changes come about through natural selection. Selection through dispersal is highly important. Animals with the highest reproductive potential disperse; those that remain behind are less influenced by population densities. (From Krebs et al., 1973.)*

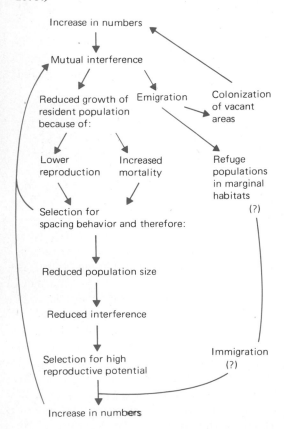

a population flush, that is, a rapid increase even though there were no changes in food and space. The population increased three times over the original number within nine generations, when the population abruptly crashed. Carson hypothesized that since other variables apparently were constant, the changes in the gene pool could be considered a stimulus sufficient to trigger a population flush and subsequent crash.

Carson recognized four stages to the flush. In phase 1 injection of new genetic material into a population promotes low mortality and high fecundity. In phase 2, a period of population increase beyond previous levels, selection is relaxed and all types of individuals survive. Because selection does not act to eliminate the less desirable new genetic combinations produced by high genetic variability, average fitness of the population declines. In phase 3 the population peaks, but the decline in fitness makes the population extremely vulnerable to environmental change. At the same time the high density creates harsher environmental conditions with increased pressure on available food and space. The population crashes, often to an extremely low level. In phase 4 the population is at a low level, lower perhaps than before the flush.

Genetic changes during a population flush are illustrated in the case of the marsh fritillary butterfly observed over a period of years by E. B. Ford (1964). For 19 years he followed a population of this butterfly and studied records of the insect left by collectors during the previous 36 years. In 1881 the marsh fritillary was unusually abundant and continued to be so until 1893. From 1893 to 1897 its numbers declined; between 1906 and 1912 the population was quite small; and between 1912 and 1917 the butterfly was actually rare. Between 1920 and 1924 the marsh fritillary increased dramatically, and by 1924 the species was very numerous again. In 1925 the population apparently stabilized at a high density and remained so until 1935 when observations ceased.

Variability in the phenotype was small in the population during the first period of abundance and continued to be throughout the phases of declining abundance and rarity. But when the numbers increased rapidly from 1920 to 1924, variability also increased greatly. Individuals differed in color pattern, size, and shape, and many individuals were deformed in various ways. After 1924 when population growth slowed and stabilized, deformed individuals virtually disappeared from the population. The population now consisted of individuals whose form was distinct from that characteristic of the butterflies in the first period of abundance.

## Population fluctuations and cycles

Populations are dynamic. Rarely, if ever, do they remain constant. They fluctuate because of changes in birth rates, death rates, and environmental conditions. The asymptote of the logistic curve represents a mean or a norm for a population about which density or numbers fluctuate. Fluctuations may be large or small, regular or irregular. Most population fluctuations are irregular (Figure 15-17), but some are more regular than one would expect by

chance. These are called oscillations or cycles. Some ecologists restrict their concept of a cycle to some endogenously caused fluctuations, particularly those affecting populations of two or more species simultaneously. In a strict sense any fluctuation with peaks regularly separated by the same time intervals is a cycle. The fluctuations of Nicholson's blowfly populations are oscillations (see Figures 15-4 and 15-5) brought about by delayed density responses.

The two most common intervals between oscillations are 3 to 4 years, typified by lemmings (Figure 15-18) and 9 to 10 years, typified by the lynx and snowshoe hare (Figure 15-19). These cyclic fluctuations are largely confined to the simpler ecosystems such as the northern coniferous forest and the tundra. Usually only local or regional populations are affected, although the 10-year cycle of the snowshoe hare and lynx are broadly synchronized from Quebec to the Northwest Territory (Keith, 1963). The cycle is most pronounced in midwest Canada and becomes

*FIGURE 15-17*
*Numbers of breeding pairs of great tits and blue tits in Marley Woods, 1947 to 1964. (From Perrins, 1965.)*

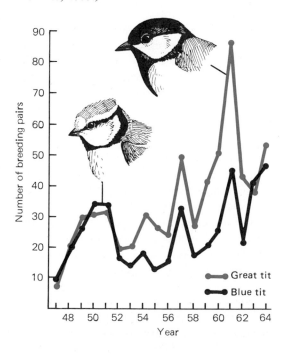

*FIGURE 15-18*
*3- to 4- year cyclic fluctuation of brown lemming population near Barrow, Alaska. (After Pitelka, 1957.)*

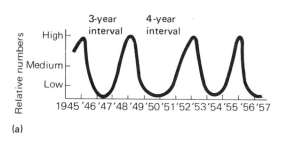

(a)

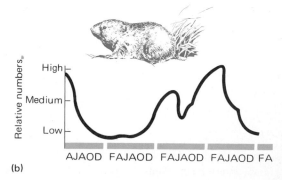

(b)

**497**

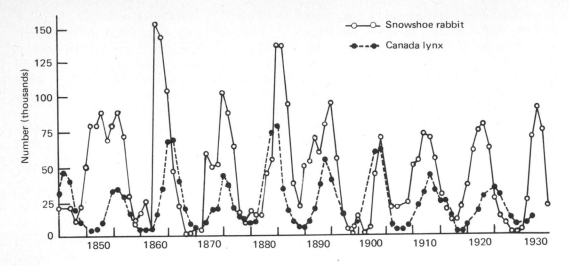

**FIGURE 15-19**
*9- to 10-year cyclic fluctuation of snowshoe hare and lynx population in northern North America. This cycle is illustrated in the fur returns from both snowshoe hare and lynx. (From MacLulich, 1937.)*

weaker as one moves away from the region. A similar cycle exists in the taiga of Russia.

A number of theories have been advanced concerning cycles. These can be divided into two main schools. One maintains that something in the physical environment, in the ecosystem, or in the population itself is able to cause rhythmic fluctuations. Predation has been singled out as a cause, but predators usually are not abundant enough when rodents are at a peak to bring about a decline. Food shortage, perhaps brought on by overpopulation, has been considered a cause. Lack (1954) suggests that cycles are caused by a combination of food shortage and predation. Other causes relate to mechanisms already discussed. A vegetation-herbivore hypothesis relates to a food shortage and loss of cover brought about by overpopulation of lemmings, followed by heavy predation and subsequent recovery of vegetation (Pitelka, 1957a, b). Endocrine malfunction has been suggested, as have changes in the quality and behavior of animals.

Another school, represented by Cole (1951, 1954b) and Palmgren (1949), holds that cycles cannot be distinguished statistically from random fluctuation. Populations are affected by a

variety of environmental forces, and random oscillations are the result. Cycles then could reflect random oscillations or fluctuations in environmental conditions. But the statistical reality of these cycles has been demonstrated by M. G. Bulmer (1974, 1975) for a number of northern animals (coyote, fisher, red fox, lynx, martin, mink, muskrat, skunk, wolf, wolverine, snowshoe hare, horned owl, ruffed grouse) during the period 1951 through 1969 and by R. May (1976) for the lemming. The main features of the cycles are the regularity of the period and the irregularity of the amplitude.

Fluctuations in populations are characterized by certain demographic parameters. L. Keith and his associates (Meslow and Keith, 1968; Keith and Windberg, 1978) followed a snowshoe hare population through two periods of decline and one of increase (15 years, 1.5 cycle periods) in the Rochester district of central Alberta. The decline, which actually set in prior to the peak winter population, was characterized by a high winter-to-spring weight loss, a decrease in juvenile growth rate, decreased juvenile overwinter survival, reduction of adult survival beginning one year after the population peak and continuing to the low, and decreased reproduction (characterized by changes in ovulation rate, third and fourth litter pregnancy rates, and length of breeding season). The upswing of the cycle, which set in about three years after the peak winter and slowed the rate of decline, was characterized by a lower winter-to-spring weight loss, increased juvenile growth rate, increased juvenile overwinter survival, and increased reproduction. Potential natality during the upswing was at least double that of the worst decrease years immediately preceding the low.

Based on his studies and the ideas of other investigators who studied animal cycles, Keith has developed a conceptual model of the 10-year cycle of the snowshoe hare and associated animals. The pattern of the cycle is shown in Figure 15-20. The cycle is generated intrinsically by a hare-vegatation interaction that triggers a population decline. As snow-

shoe hares reach peak densities, there is an increasing interaction between hares and their overwinter (September to May) food supply of browse. This essential browse consists largely of stems less than 3mm in diameter, largely new growth from the previous summer. The hare-vegetation interaction becomes critical when essential browse falls below that needed to support the population over winter. Excessive browsing and girdling reduces subsequent annual increases of woody growth, precipitating a food shortage, which causes a high winter mortality of juvenile hares and lowered reproduction the following summer. The decline in hares elevates the predator-hare ratio, intensifying a second interaction that extends the period of decline and drives the population to a level below that which could be sustained by the habitat. The decline in hares forces predators to turn to ruffed grouse, which experiences heavy predator-induced mortality. This predation drives a cyclic response in grouse. The low hare population allows vegetation to recover and accumulate biomass and causes a sharp decline in predator populations from re-

**FIGURE 15-20**
*Graphic model of the vegetation-herbivore-predalor cycle. Note the time lag between the cycle of vegetation recovery, growth and decline of snowshoe hare population, and the rise and fall of the predator population. Note the effect of the snowshoe hare decline on ruffed grouse population to which the lynx turns when hares decline. (Adapted from Keith, 1974.)*

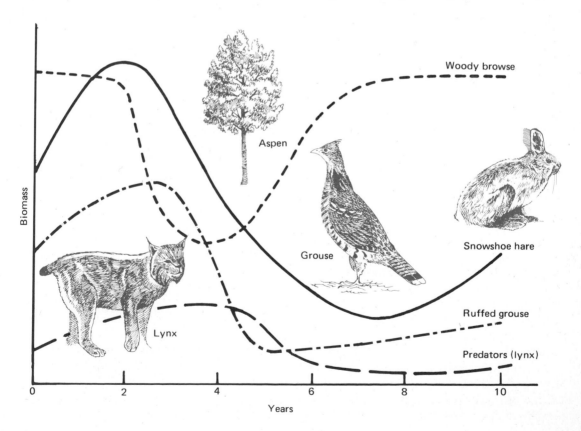

499

duced reproductive success (early loss of young is the chief cause of decline in mammalian predators), increased juvenile mortality, and mass emigration. With the decline of predators and a growing abundance of winter food, the hare population begins to rise sharply, initiating another cycle.

The periodicity of the 10-year cycle is determined mainly by hare-vegetation and predator-hare interaction. The length of the increase phase depends upon: (1) the average rate of population growth from low to high densities, and (2) average biomass of essential woody browse present. The length of the decrease phase is a function of (1) response of native species of woody browse to overutilization by hares, including a delayed density-dependent reduction in new browse production, and (2) intensity and duration of predation after the major initial decline of hares from a food shortage. The 3- to 4-year cycle of small mammals follows a similar pattern and characteristics (see Table 15-2).

The model does not explain observed synchrony between local hare populations or between regional populations. Local synchrony can be explained by widespread dispersal of hares at peak densities from better habitats into marginal and submarginal habitats (Keith, 1974, 1978). This dispersal insures rather uniform densities in acceptable habitats during cyclic peaks. When the hare population declines, movement from better habitats into marginal habitats ceases. Synchrony between regional hare populations is probably related to recurrent weather phenomenon of almost continental scope. The synchronizing weather phenomenon is probably a succession of two or more mild winters which would allow high populations to escape catastrophic mortality and persist at higher levels and allow lower populations to grow to peak or near peak densities. A return to average winter temperatures, combined with intense hare-vegetation interaction, could trigger a major decline. A severe winter following two mild ones would guarantee it. Once synchronized, regional hare populations would not need to be subjected to appropriately timed weather patterns for several decades.

TABLE 15-2
*Comparative demography of snowshoe hare and microtine cycles*

|  | *9- to 10-year cycle* | *3- to 4-year cycle* |
|---|---|---|
| Reproductive rate | highest during increase phase due to lower breeding age and longer breeding season | highest during increase phase due to lower breeding age and longer breeding season |
| Juvenile mortality | increase in rate and decrease in reproduction precipitates cyclic decline | high in peak summer; high in decline phase |
| Survival of all age classes | lower during decline than during increase phase | low during decline phase; mortality lowest during increase phase |
| Size of adults | greater proportion of large-sized adults in peak population | larger in peak populations than at other times in cycle |
| Juvenile growth rate | reduced rate accompanies onset of population decline | no information |
| Litter size and pregnancy rates | vary systematically through cycle | no systematic differences with decreasing and increasing populations |
| Prenatal mortality | no significant variation through cycle | slight variation over cycle, but not a serious loss in declining populations |
| Dispersal | throughout all available habitats during peak years | most frequent from increasing populations and infrequent from declining populations |

*Source:* Based on Keith, 1974; Krebs and Myers, 1974.

The cyclic model is applicable to hares in the boreal regions. However, among disjunct snowshoe hare populations south of the boreal forest some exhibit cyclic fluctuations and other do not (Brooks, 1955). Factors relating to the presence or absence of cyclic fluctuations have been studied by Dolbeer (1972) and Tanner (1975). In the coniferous forest and associated regions south of the boreal forest snowshoe hares exhibit cyclic fluctuations only in a uniform environment of spruce and fir (which, of course, is also characteristic of the boreal forest). But in regions where the environment is heterogeneous, where many kinds of vegetation patterns exist, cycles do not occur. Hares occupying high quality habitats are protected from predation, while in areas of poor cover hares are subjected to predation. Poor habitats are kept replenished by dispersal of animals from better habitats and these hares, in turn, are eliminated by predators. Such dispersal and predation tend to hold down the population of hares in better habitats to a level at which they do not overutilize their food supply, thus damping cyclic behavior.

## SUMMARY

The existence of regulatory mechanisms in plant and animal populations is a subject of considerable speculation and study. Some biologists have maintained that population fluctuations are most affected by density-independent influences. Others have argued that populations are regulated by density-dependent influences that relate to optimum population size. A more general idea is that local populations fluctuate between some upper and lower levels and the fluctuations are brought about by an interaction of population density, influences extrinsic to the population (which may act as density-dependent or density-independent forces), and influences intrinsic to the population. Extrinsic influences include weather, human activity, resource availability, disease and parasites, and predation. Intrinsic influences are crowding, behavior, dispersal, and genetic changes.

Density-independent influences affect, but do not regulate population. They can reduce local populations, even to the point of extinction, but their effects do not vary with population density. Regulation implies homeostatic feedback that functions with density.

Intraspecific competition for resources is a density-dependent mechanism in the regulation of population numbers. There are basically two types of competition. In scramble competition all individuals have access to the resource and as each attempts to get a part of it. Such scrambling may result in wastage if no individual or only a few can obtain an amount sufficient to ensure survival and reproduction. In contest competition successful individuals divide the resource and the unsuccessful are denied access to it. Contest competition is characteristic of territoriality and social dominance and hierarchy.

Social behavior, through the mechanisms of territoriality and dominance, reflects the influence of density. Territoriality may function in population regulation if it creates a surplus population consisting of sexually mature individuals prevented from breeding by territory holders.

Increased density may affect population growth through physiological responses of individuals. Crowding may produce stresses that can result in endocrine imbalances, especially in the pituitary-adrenal complex, and other responses which result in abnormal behavior, abnormal growth, and degeneration and infertility.

Social pressure and crowding may also induce emigration or dispersal. Dispersal may function in population regulation by acting as one factor accounting for the losses that occur when the growth rate becomes zero at carrying capacity. It may act as a major force in slowing or stopping population growth at carrying capacity, or, more speculatively, dispersal may prevent numbers from reaching carrying capacity. Dispersal may also function to increase population growth in areas where the immigrants settle. Dispersing animals often fill empty habitats or join populations low in numbers; and emigrating females are usually reproductively active. Thus emigrants can improve their fitness by leaving the home area, in spite of the risks.

Genetic feedback appears to operate through a combination of density pressure, selection pressure, and scenic variability with the population. These various mechanisms produce oscillations in populations which are damped or become irregular through the combined action of density-dependent and density-independent influences. When oscillations are more regular than one would expect by chance, they are called cycles. The most common intervals between oscillations are 3 and 4 years, typified by microtine rodents, and 9 to 10 years, typified by the lynx and snowshoe hare. Both cycles share many demographic characteristics.

# Relations among populations: interspecific competition
## CHAPTER 16

Animal and plant populations exhibit a wide range of relationships to one another. Some populations have little influence on one another except in indirect and often distant roles they play in energy exchange. Other populations, such as parasites and their hosts and predators and their prey, have a very direct and immediate relationship. From an individual standpoint these relations are often detrimental; from a population standpoint they may act as a depressant or stabilizer of population numbers. Such interactions influence the growth of populations.

## Population interactions

Interactions can be positive, negative, or neutral in their effect on population growth (Table 16-1). Neutral interactions (designated as 0 0) produce no effect on growth of the populations involved. Positive interaction (+ +) benefits both populations. It may take the form of *protocooperation,* in which the relationship is nonobligatory, that is, it is not essential to either population, or *mutualism,* in which the relationship is obligatory. Negative interactions (− −) produces an adverse effect on the populations. This includes *competition.*

In some situations one population may be favorably affected while the other is unaffected (+ 0); or one population may be negatively affected while the other remains unaffected (− 0). The former is *commensalism,* a one-sided relation between two species in which one benefits and the other is neither benefited nor harmed. Among such commensals are epiphytes, plants which grow in the

TABLE 16-1
*Population interactions, two species system*

| Type of interaction | Response | |
|---|---|---|
| | **A** | **B** |
| Neutral | 0 | 0 |
| Protocooperation (nonobligatory) | + | + |
| Mutualism (obligatory) | + | + |
| Commensalism | + | 0 |
| Amensalism | − | 0 |
| Parasitism | + | − |
| Predation | + | − |
| Competition | − | − |

Note: 0 = no direct effect; + = positive effect on growth of population; − = negative effect on growth of population.

branches of trees. They depend on trees for support only; their roots draw nourishment from humid air. The relationship in which one population is inhibited while the other remains unaffected is called *amensalism*. Amensalism probably involves some type of chemical interaction, such as the production of an antibiotic or allelochemical agent by one of the organisms involved. Amensalism is a nebulous relationship, most examples of which can probably be best described as a form of interspecific competition.

Other relationships may positively affect one population and be detrimental to the other (+ −). Such relationships involve predation and parasitism. *Predation* involves the killing and consumption of prey. *Parasitism*, which may be viewed as a weak form of predation, also involves the consumption of one organism by another, but either the host survives or its death occurs as a result of the exploitation by the parasite over a period of time. Parasites which cause the eventual death of the host, that is, which act as predators, are known as *parasitoids*. These relationships are summarized in Table 16-1. Of all these relationships three with important ecological consequences are mutalism, predation, and competition.

## TYPES OF COMPETITION

When two or more organisms in the same community seek the same resource which is in short supply in relation to the number seeking it, they compete with one another. Few concepts have had such an impact on ecological and evolutionary thinking, yet basically the nature of competition and its effects on the species involved are some of the least known and most controversial areas of ecology.

There are two basic categories of competition, scramble and contest (introduced in Chapter 15). *Exploitative* or *scramble competition* is a competitive situation in which organisms or groups of organisms have equal access to a limited resource. The outcome is determined by how effectively each of the competitors utilizes the resources. Exploitative competition often results in the reduction in growth of all competitors (see Haven, 1973; Werner and Hall, 1976). When distributed among all competitors, the limited resources may be sufficient only to maintain existence. The resources in effect are wasted because they are consumed, but none of the competitors secures enough for successful growth and reproduction. *Interference* or *contest competition* is a competitive situation in which a competitor is denied access to a resource, usually by some form of aggressive behavior. It permits an efficient use of a resource by some individuals and denies it to others, reducing wastage.

Competition theory developed out of studies of animal populations with little consideration of interactions of plant populations. Individual plants (and sessile animals) are fixed in space. Depending upon the degree of proximity an individual plant influences and is influenced by neighbors which may be individuals of the the same or different species. Each individual affects the environment of its neighbor by consumption of resources in limited supply, by modifying environmental conditions (as shading and protecting plants from the wind and predators), and by producing toxins. These changes can alter the rate of growth and the growth form of individual plants. Most competitive relations among plants can be considered interference.

Responses to competition are more complex among plants than among animals because of the population-like structure of plants: leaves, stems, twigs, rootlets, flowers. Each part has its own "birth rate" and "death rate." Plants may respond to interference competition by increasing mortality risk of the whole plant or parts of plants, reduced growth rate, reduced reproductive output, or delayed maturity and reproduction. Density-stressed populations tend to form a hierarchy of dominant and suppressed individuals with mortality risk concentrated in the suppressed class. Competitive advantage among individuals is often determined by early seedling emergence. The early seedlings gain a competitive advantage not necessarily because they have a longer time to grow, but because they claim a disproportionate share of environmental resources, depriving later emergents of those resources (for discussion of plant competition, see Harper, 1977; Grimes, 1977).

## Competition theory

### LOTKA-VOLTERRA MODELS

The idea of competition or the "struggle between species" was emphasized by Darwin. Later Lotka and Volterra separately developed mathematical equations describing relationships between two species utilizing the same food resource. They added to the differential logistic equation for the population of each species a constant to account for the interference of one species on the population growth of another:

$$\frac{dN_1}{dt} = r_1 N_1 \left( \frac{K_1 - N_1 - \alpha N_2}{K_1} \right)$$

and

$$\frac{dN_2}{dt} = r_2 N_2 \left( \frac{K_2 - N_2 - \beta N_1}{K_2} \right)$$

where

$r_1$ and $r_2$ = the rates of increase for species 1 and 2, respectively
$K_1$ and $K_2$ = equilibrium population size for each species in the absence of the other
$\alpha$ = a constant, characteristic of species 2, a measure of the inhibitory effect of one $N_2$ individual on the population growth of species 1
$\beta$ = a constant, characteristic of species 1, a measure of the inhibitory effect of one $N_1$ individual on the population growth of species 2.

In the absence of any interspecific competition $\alpha$ or $N_2 = 0$ in equation 1 and $\beta$ or $N_1 = 0$ in equation 2 and the population grow logistically to equilibrium at carrying capacity. Inherent in the logistic equations is the inhibitory effect of each individual in the population on its own population growth. This is represented by $1/K$ for species 1 and by $1/K_2$ for species 2. In two competing populations the inhibitory effect of each $N_2$ individual on $N_1$ is $\alpha (1/K)$ or $K_1$. Similarly, the inhibiting effect of each $N_1$ individual on the population growth of species 2 is $\beta/K_2$. The outcome of competition depends upon the relative values of $K_1$, $K_2$, $\alpha$, $\beta$. If $N_2 = K_1/\alpha$, $N_1$ can never increase, and if $N_1 = K_2/\beta$, $N_2$ can never increase.

Graphic models of interactions of two competing species and resulting outcomes are illustrated in Figure 16-1. In Figure 16-1$a$ the equilibrium conditions are represented by the line $K_1/\alpha$, $K_1$ obtained by plotting $N_1$ against $N_2$. The line represents a set of joint values of $N_1$ and $N_2$ along which $N_1$, the number of individuals in species 1, is neither increasing or decreasing. This set of joint values is represented by $dN_1/dt = 0$. All populations of species 1 inside the line, indicated by the shaded area, will increase in size until they reach the diagonal line which represents all points of equilibrium. All populations outside of the line (to the right) will decrease to the equilibrium points. A similar set of joint values exists in which $N_2$ the number of individuals in species 2, is stationary (Figure 16-2$b$). Outside the diagonal equilibrium line species 2 decreases; inside the line it increases. Values of $N_1$ and $N_2$ are considered jointly because of the effect of competition. An increase in $N_1$ diminishes the growth rate of species 2, and an increase in $N_2$ decreases the growth rate of species 1. In the presence of species 2 the higher the value of $N_2$, the lower is the value of $N_1$ at which species 1 stops growing.

Now consider situations in which species 1 and species 2 occupy the same space as competitors simultaneously. There are four possible outcomes in competitive situations according to the Lotka-Volterra model: (1) species 1 wins and species 2 becomes extinct locally, (2) species 2 wins and species 1 becomes extinct locally, (3) either species can win depending upon the ecological variables operative at any

**505**

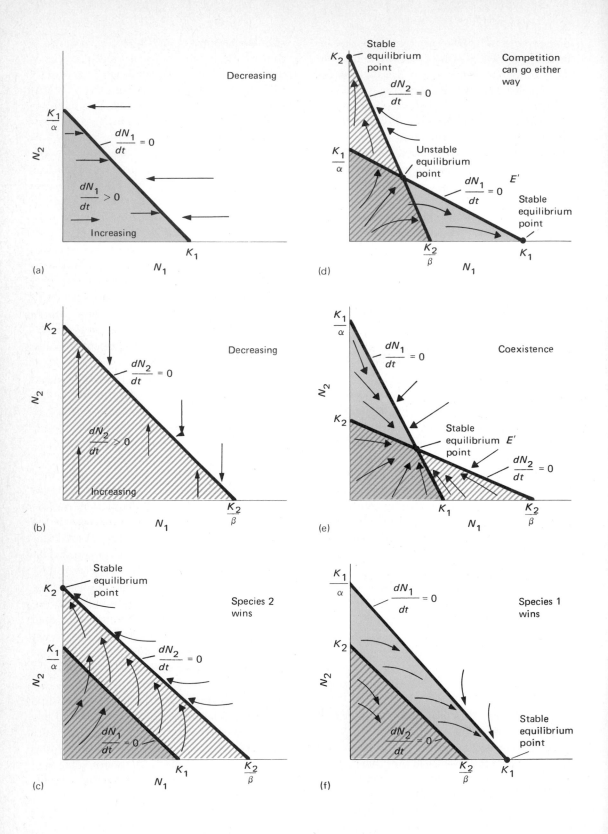

one particular time; (4) neither species wins and they eventually coexist dividing the resources between them in some manner.

To illustrate the first two outcomes, if species 1 and species 2 occupy the same space as competitors and species 2 is the stronger competitor, species 2 slows down the population growth of species 1 and eventually wins, leading to the extinction of species 1 ($\alpha > K_1/K_2$ and $\beta < K_2/K_1$). In Figure 16-1$c$ the plot of species 1 moves upward because the area on or above

FIGURE 16-1
*Models of competition between two species. In (a) and (b) populations of species 1 and 2 in absence of competition will increase in size and come to equilibrium at some point along the diagonal line or isocline. The line represents all equilibrium conditions (dN/dt = 0) at which the population just maintains itself (r = 0). In the shaded areas below the line r is positive and the population increases (as indicated by the arrows). Above the isocline r is negative and the population decreases (as indicated by the arrows). In (a) the intercepts of the $N_1$ isocline are $K_1$ and $K_1/\beta$ and the intercepts of the $N_2$ isocline are $K_2$ and $K_2/\alpha$. In (c) species 1 and 2 are competitive. Because the isocline, the zero growth curve of species 2, falls outside the isocline of species 1, species 2 wins, ultimately leading to the extinction of species 1, and the stable equilibrium point is at $K_2$ ($N_2 = K_2$, N = 0). In (f) the situation is reversed and species 1 wins, leading to the exclusion of species 2. Equilibrium is at $K_2$ ($N_1 = K_1$, $N_2 = 0$). In (d) and (e) the isoclines cross. Each species, depending upon the circumstances, is able to inhibit the growth of the other. In (e) neither species can exclude the other. Each by intraspecific competition inhibits the growth of its own population more than it inhibits the growth of the other population. There is one equilibrium point where the lines cross and both species coexist at densities below their respective carrying capacities. In (d) each species inhibits the growth of the other more than it inhibits the growth of its own population. Three possible equilibrium points exist. Two are stable: $N_1 = K_1$, $N_2 = 0$; $N_2 = K_2$, $N_1 = 0$. At these two points one species excludes the other. The other equilibrium point where the two lines intersect is unstable. What species wins often depends upon the initial proportion of the two species or changing environmental conditions.*

the line $K_1, K_1/\alpha$ (the carrying capacity of species 1) and the area below $K_2, K_2/\beta$ is below the carrying capacity of species 2. Species 2 will increase until it arrives at $K_2$; at that point only species 2 remains. Under a different set of conditions species 1 will win, as illustrated in Figure 16-1$f$.

In the third outcome (Figure 16-1$d$) the diagonal equilibrium lines cross each other. The equilibrium point is represented at their crossing, but it is unstable. The vectors are directed away from the equilibrium point indicating that the true equilibrium points are $K_1$ and $K_2$. In this situation equilibrium between competing species is unstable and either of the two species can win. Above the line $K_2, K_2/\beta$ species 2 is unable to increase, and above $K_1, K_1/\alpha$, species 1 is unable to increase. If the mix of the species is such that the point $N_1, N_2$ falls within the triangle $K_2, E'/K_1/\alpha$, species 1 is above its carrying capacity and species 2 is not. Species 2 will continue to increase and species 1 will decrease until it is gone. The reverse situation occurs in triangle $K_1, E', K_2/\beta$. What happens in parts of the diagram outside the triangles depends upon whether the starting value of $N_1$ is larger or smaller than that of $N_2$.

Finally, the two species might coexist with their populations in some sort of equilibrium (Figure 16-1$e$). As species 1 increases, species 2 may decrease and vice versa. Each species inhibits its own growth through density-dependent mechanisms more than it inhibits the growth of the other species. Neither species reaches a high enough density to bring about any serious competition between them and the population growth of each is not strongly controlled by the same limiting conditions. As long as each species is limited by a different resource and both are only weakly competitive, then the two species will continue to coexist. Thus, in Figure 16-1$e$, species 2 has the advantage in the area $K_1, E', K_2/\beta$ and the plot moves up and to the left to the equilibrium point $E'$. The two competing species will reach a stable equilibrium point and persist indefinitely when

$$\alpha < \frac{K_1}{K_2} \text{ and } \beta < \frac{K_2}{K_1}$$

**507**

The Lotka-Volterra equation and graphical models based on them apply well to animal populations, but not to plants. The models are based on an animal's potential for increase in numbers. In many plant populations the potential is for increase in biomass. Accordingly competitive relations among plants may be examined in terms of the influence of one species on the growth of another. Models developed from experimental studies of mixtures of two species sown at a variety of proportions detect changes in yield of dry matter, number of tillers, production of seed, etc. after a lapse of time. Proportions of the two species at the end of the period are plotted against proportions at the beginning. Such studies arrive at five basic types of interactions (Figure 16-2), suggestive of those predicted by the Lotka-Volterra equations (Harper, 1977).

1. Neutral interaction. Proportion of two species remains unaltered after a period of growth together. The balance of the two species is subject only to random variation. Such interactions are rarely, if ever, seen in nature.
2. Directional interaction in favor of species 1. Species 1 has the competitive advantage in the mixture at all proportions. The advantage is measured by the distance between the actual ratio line and the equilibrium line (45°) of no advantage. Parallel lines indicate that the advantage is independent of frequency. If the advantage is carried from planting to harvest and back to planting again, generation after generation of species 1 ultimately would bring about the extinction of species 2.
3. Directional interaction in favor of species 2. Species 1 is eliminated by species 2 at a speed dependent upon the distance between two parallel lines.
4. Stabilizing interaction. This is a frequency-dependent situation in which the minority component is always at an advantage. Which species will have the selective advantage depends upon the frequency at which the species are sown. The mix will always tend to change toward equilibrium frequencies and stability.
5. Disruptive interaction. This is a frequency-dependent situation in which the majority component is at an advantage. If a high proportion of species 1 is sown, species 2 goes extinct. Conversely, if a high proportion of species 2 is sown, species 1 goes extinct. This is a nonequilibrium mixture which is unstabilized and disruptive.

## EXPERIMENTAL EVIDENCE

Several experiments seem to provide at least partial support for the mathematical models of Lotka and Volterra. Gause (1934) set out to test the formulas experimentally. He used two species of *Paramecium, P. aurelia* and *P. caudatum. P. aurelia* has a higher rate of increase than *P. caudatum.* When both were introduced into one tube containing a fixed amount of bacterial food, *P. caudatum* died out. The population of *P. aurelia* interfered with the population growth of *P. caudatum* because of its higher rate of increase (Figure 16-3). In another experiment the two species used were *P. caudatum* and *P. bursaria.* In this experiment both species were able to reach stability because *P. bursaria* confined its feeding to the bacteria on the bottom of the tubes, whereas *P. caudatum* fed on bacteria suspended in solution. Although the two used the same food supply, they occupied different parts of the culture. In effect each utilized food essentially unavailable to the other. Park (1948) and Crombie (1947) carried out competition experiments involving several species of flour beetle and obtained results similar to those of Gause.

The British plant ecologist J. L. Harper (1961) performed similar experiments with duckweed *(Lemna).* When cultured separately in uncrowded conditions, the species *L. gibba* had a higher intrinsic growth rate than *L. polyrrhiza.* When cultured separately in crowded conditions, *L. polyrrhiza* had a higher growth rate than *L. gibba.* When the two species were grown together, *L. gibba* excluded *L. polyrrhiza* (see Figure 16-4).

## COMPETITIVE EXCLUSION

Experimental evidence and mathematical models indicate that in competitive situations if one species produces enough individuals to prevent the population increase of another, it

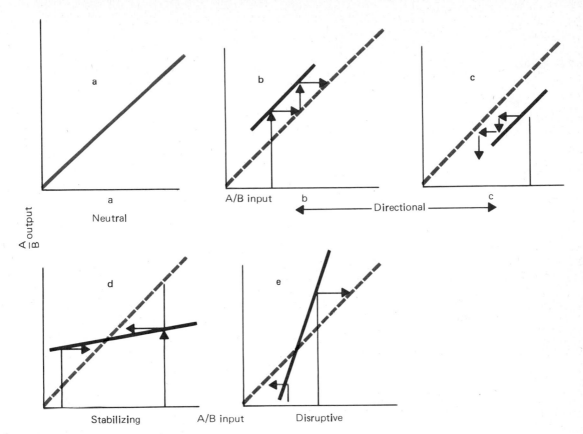

**FIGURE 16-2**
*Models of competitive interactions among plants. Compare these with Figure 16-1 which model competitive interactions in animal populations. These models are based on replacement series experiments in which the ratios of species present after a period of time are plotted against the ratio of the species sown or planted. (a) The proportion of two species remains unaltered after a period of growth together. This represents an ideal situation. (b) Species 1 gains an advantage over species 2 in all situations and Species 2 goes into extinction. The amount of advantage is measured by the distance of the actual ratio line from the theoretical diagonal line of no advantage. If the lines are parallel, the advantage is independent of frequency. In (c) the situation is reversed and Species 1 moves to extinction. (d) In this frequency-dependent situation the minority species is always at an advantage. Successional or selective advantage depends upon the frequency at which each species is sown or planted. For example, if a high population of species 1 is sown relative to species 2, species 2 gains in the mix. The mix tends toward a stable equilibrium frequency. (e) In this frequency-dependent situation the major species in the mix is at an advantage. If a high proportion of species 1 is sown, Species 2 will go extinct. Thus the outcome depends upon the starting frequency of the species. There is no equilibrium mix. (After Harper, 1977.)*

can reduce that population to extinction or exclude it from the area. These observations led to the development of the concept called Gause's principle: two species with identical ecological niches cannot occupy the same environment. The idea was far from original with Gause (and he laid no claim to it). For example, the ornithologist Joseph Grinnell in 1904 wrote: "Two species of approximately the same food habits are not likely to remain long evenly balanced in numbers in the same region. One will crowd the other out. The one longest exposed to local conditions, and hense best fitted, though ever so slightly, will survive to the exclusion of any less-favored would-be invader."

More recently the concept has been called the competitive exclusion principle (Hardin, 1960), which can be stated briefly as: "Com-

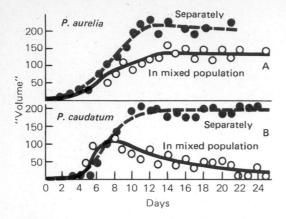

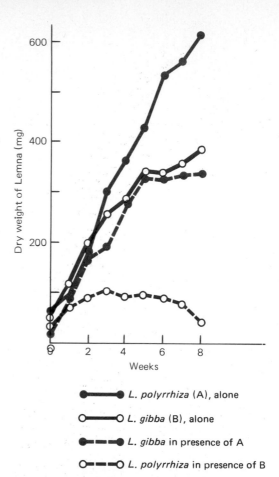

FIGURE 16-3

*Competition experiments with two species of*
**Paramecium.** *The graphs show the growth*
*of two related ciliated protozoans*
**Paramecium aurelia** *and* **P. caudatum**
*when grown separately and when grown in a*
*mixed culture. In the mixed culture* **P.**
**aurelia** *outcompetes* **P. caudatum** *and the*
*result is competitive exclusion. (From*
*Gause, 1934.)*

plete competitors cannot coexist. Two competing species with identical ecological requirements cannot occupy the same area." However, this so-called competitive exclusion principle hardly rates as a principle (Cole, 1960). It
is a little more than an ecological definition of
a species. A corollary of the statement is that if
two species coexist, they must possess ecological differences. Obviously, two separate
species cannot have identical requirements;
being different species they necessarily must
have somewhat different ecologies. However,
two or more species can compete for some essential resource without being complete competitors.

Pielou (1974) provides a set of conditions in
addition to utilization of resources in short
supply that should be met for competitive exclusion to take place. These conditions include:
(1) the competitors must remain genetically
unchanged for a long enough period of time for
one species to exclude the other; (2) immigrants from areas with different conditions
cannot move into the population of the losing
species; (3) environmental conditions must
remain constant; (4) competition must continue long enough for equilibrium to be
reached. While competitive exclusion might

FIGURE 16-4

*Competition experiments with species of*
*duckweed* **(Lemma).** *In this experiment*
*populations of two species were grown by*
*J. L. Harper. When grown alone, L.*
**polyrrhiza** *and* L. gibba *exhibit somewhat*
*similar rates of growth, although* L.
**polyrrhiza** *attains greater population levels.*
*When grown in a mixed culture,* L. gibba
*shows little difference in its growth rates,*
*but* L. polyrrhiza *in the presence of* L.
**gibba** *is eliminated. (After J. L. Harper,*
*1961).*

take place without fulfilling all these requirements, nonfulfillment usually results in
coexistence.

**COEXISTENCE AMONG MORE THAN TWO
SPECIES**

Coexistence and equilibrium have been considered from a two-species point of view. What
happens if a third species attempts to utilize
some portion of the resource? In natural com-

munities, of course, resources are shared by more than two species. Consider a resource gradient involving various sizes of food items as in Figure 16-5. Let species A occupy a lower end of the resource gradient and let species B occupy a position somewhat further up on the gradient, but with some overlap with species A. Now allow a third species to invade this resource gradient at a point between the utilization curves of A and B. Species C can successfully invade if A and B are relatively rare, if they are below carrying capacity, and if resources are abundant. Under these conditions competition will force each of the three to become more specialized in their resource utilization, to utilize optimal resources, and to space themselves more narrowly on the resource gradient (for theory, see MacArthur and Levins, 1967).

The overlap in resource utilization curves, used to quantify competitive interactions, does not necessarily imply competition. A large overlap may also indicate lack of competition. Resources may be sufficiently abundant so that two or more species can utilize the same

resources without competition. In fact where competition is strong, overlap may be minimal (Pianka, 1972, 1976; Connell, 1975).

Competition, for simplicity, is usually considered on a one-dimensional gradient. But in natural situations competition is spread over a number of resource gradients. A high competitive interaction on one gradient may be counterbalanced by low competitive interactions on other resource gradients. On the other hand, minimal competitive inhibitions on several gradients among a number of species can for some individual species be equivalent to strong competitive inhibitions from a few competing species. This relationship has been termed *diffuse competition* by MacArthur (1972). Diffuse competition can result in the inability of a species to fit into a community because it is excluded by competitive interactions with a specific combination of other species, rather than just one or two strong competitors. An example of diffuse competition may be that imposed by fugitive perennial and biennial species of plants settling on soil disturbed by badgers digging for ground squirrels in tall-grass prairie. Colonization of disturbed sites is determined by those adapted species whose seeds first arrive on the area. Subsequent immigrants do not become established although their seeds germinate (Platt and Weis, 1977).

### GENETIC CHANGES

If two or more closely related species are quite similar in their requirements and if the population densities are less than the carrying capacity of the habitat, each has to share a limited environmental resource. Natural selection will favor any new gene in the population that will eliminate or reduce sharing. It will promote the spread of genes that will enable a competing population to utilize a segment of resources unutilized by others.

The concept of competitive exclusion is based on the assumption that the competing species and their biotic environment remain constant. Because this is not the case a situation may arise in which two competitors coexist. The interspecific competitors may change genetically, so that both can live together and utilize the same food, space, and other necessary resources in the environment

**FIGURE 16-5**
*Theoretical resource gradient utilized by three competing species, A, B, and C. A and B share the resource gradient with minimal overlap (AB). A third species C, whose optimal resource utilization lies between A and B, competes with A and B (AC and BC). In response to selection pressures A and B narrow their range of resource utilization to the optimum and C utilizes that portion of the resource used at less than an optimal level by A and B.*

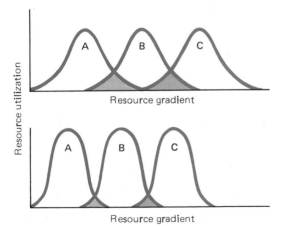

(Pimentel et al., 1965). Assume that species A and B are fairly evenly balanced in population, but species A is slightly superior. As the numbers of species A increase, species B declines and becomes sparse. At this point species A individuals contend largely with intraspecific competitive selection. Concurrently species B, still contending primarily with interspecific competitive pressure, evolves and improves its ability to compete with the more abundant and originally superior species A. As B improves as a competitor, its numbers increase until finally B becomes the abundant species. The original trend is reversed. After many such oscillations a state of relative stability should result.

## Examples of competition

Evidence of competition is difficult to obtain in the field. Data have been constructed to demonstrate competitive exclusion and other forms of competition when competition is probably only a fraction of the real situation.

### EXCLUSION

Some field evidence does seem to support competitive exclusion in part, although such competitive interactions are associated with other relations, both physical and biological.

An example is competition for space among two species of barnacles on the Scottish coast *Balanus balanoides* and *Chthamalus stellatus* (Connell, 1961*a*, 1961*b*). *Chthalmalus* is attached to rocks down to the mean tidal level. *Balanus,* with a higher growth rate than *Chthamalus,* occupies a lower zone up to about the level of high neap tide. It cannot survive above that line because it does not tolerate physical dessication as well as *Chthamalus*. *Chthamalus* cannot survive below that line because of competition from *Balanus* and predation from the whelk *Thias*. Larvae of both species are able to settle over a wider vertical range than occupied by either of the adults. But when young *Chthamalus* settle in the *Balanus* zone, they seldom survive unless the area is kept free of *Balanus*. *Balanus* colonizes an area in greater numbers and grows so much faster than *Chthamalus* that it undercuts, smothers, or crushes the latter. The *Chtha-*

*malus* that do survive are small and produce few offspring. In general the upper limit of distribution of barnacles is set up by physiological tolerances of dessication during low tide. The lower limit of *Chthamalus* is set up by competitive exclusion, and the lower limit of *Balanus* is set by predation by the whelk.

Another example of competitive exclusion may be the case of a feral house mouse *(Mus musculus)* in California (Lidicker, 1966). In a period of decline due to high density the house mouse population faced competition from a rapidly increasing population of voles. The house mice, experiencing reduced vitality, had a declining reproductive rate that was further reduced by persistent interference from aggressive voles, reduced food supply, and social disintegration caused by increased wandering and annoyance from voles. As a result the local house mouse population became extinct (Figure 16-6).

Exclusion frequently occurs along boundaries of contact between species and observed competition is a complex interaction of aggressive behavior, physiologial tolerances, food diversity, and other factors. In many instances one species simply denies another species access to a particular resource. The type of competition is interference or contest rather than exploitative.

On the eastern slopes of the Sierra Nevada live four species of chipmunks: alpine chipmunk *(Eutamias alpinus),* lodgepole chip-

*FIGURE 16-6*
*Graph depicting the decline and extinction of a house mouse population. The declining population was pushed to local extinction by competition and interference by aggressive meadow voles. (From Lidicker, 1966.)*

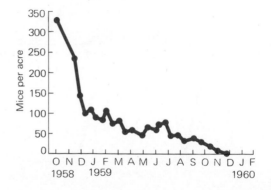

munk *(E. speciosus)*, yellow pine chipmunk *(E. amoenus)*, and least chipmunk *(E. minimus)*. Each occupies a different altitudinal zone (Figure 16-7), the line of contact partly determined by interspecific strife. The upper range of the least chipmunk is determined by aggressive interactions with the dominant yellow pine chipmunk (Heller, 1971). Although the least chipmunk is capable of occupying a full range of habitats from sagebrush deserts to alpine fell fields, in the Sierra Nevada it is restricted to the sagebrush habitat. The sagebrush desert is somewhat outside of the bounds of climatic adaptation or climate space of the least chipmunk, but it can exist there by the use of hyperthermia and daytime retreat to burrows (Heller and Gates, 1971). Because the sagebrush habitat is outside of the climate space of the yellow pine chipmunk, it cannot penetrate that part of the range of the least chipmunk. Aggressive behavior of the yellow pine chipmunk also determines the lower limit of the lodgepole chipmunk, restricted to open coniferous forests. There the abundance of food and cover as well as the secretive habits of the chipmunk do not bring it into aggressive contact with the yellow pine chipmunk. The upper limit of the lodgepole chipmunk is determined by the aggressive behavior of the alpine chipmunk which is limited to the alpine and Hudsonian zone of the mountains. Of the four species the distribution of two is, in part, influenced by the aggressive behavior of the other two. (See also Chappell, 1978.)

Aggressive behavior in competitive situations was probably selected for in the alpine and yellow pine chipmunks because in their habitat the seasonally limited food supply can be cached and economically defended. Aggres-

**FIGURE 16-7**
*Transect of the Sierra Nevada in California, latitude 38° N, showing vegetational zonation and altitudinal range of four species of Chipmunk* (Eutamias) *that inhabit the east slope. (Transect after Heller and Gates, 1971.)*

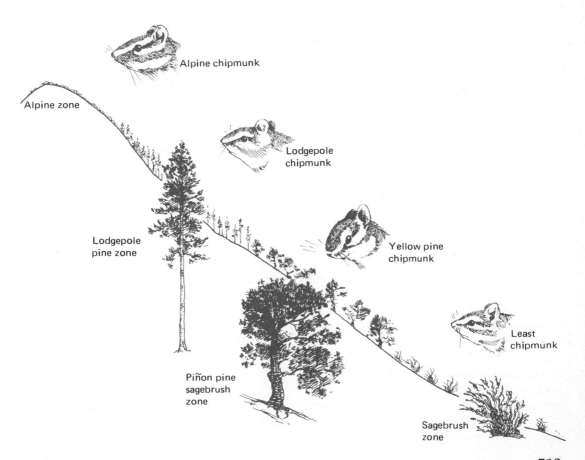

siveness was probably not selected for in the least chipmunk because such activity would not be metabolically feasible in the hot sagebrush desert. Likewise, aggressiveness was not selected for in the lodgepole chipmunk because in the presence of an abundance of food and vegetational diversity it does not have adaptive value (Heller, 1971). Possibly because of pressure from predators, the lodgepole chipmunk remains quiet and rather secretive in contrast to the noisy activity of the highly aggressive species.

## EQUILIBRIUM

Unstable equilibrium and possibly stable equilibrium may represent a stage on the road to competitive exclusion before sufficient time has elapsed to bring the interaction to conclusion.

Which species wins in unstable equilibrium more often than not is determined by the environmental conditions under which the competitors are living at a particular time. Subtle differences in the ecology of each species may favor one species over another. Such relationships are most evident in laboratory populations. Park (1948, 1954, 1955) found that the outcome of competitive interactions among flour beetles depended considerably on such environmental conditions as temperature and humidity, on the presence or absence of parasites, and on the fluctuation of the total number of eggs, larvae, pupae, and adults. Often the final outcome was not determined for generations.

An informative recent study was made by Neill (1975). He used replicated microcosm communities of four competing species of microcrustaceans along with associated bacteria and algae in a series of removal experiments involving different intensities of highly selective fish predation. Neill found that (1) the outcome of competition depended upon community composition; (2) the outcome among three species could not be predicted from separate interactions of the species in two-species systems; (3) each of the four species of microcrustaceans utilized slightly different microhabitats and food, and each excluded any additional species introduced; (4) the remaining species adjusted their food habits, but did not change their distribution in space; (5)

competition at an early stage in the life history can reduce or influence the outcome of competition at later stages.

In some ways unstable equilibrium is more easily observed in plants. From the time seedlings germinate and develop, the demand for growing space, light, moisture, and nutrients increases. Those plants that utilize resources in short supply most efficiently have the best chance for survival. Different species of plants with root systems in the same soil horizon may compete for limited moisture and nutrients. In western North America the shallow rooted annual grass *Bromus tectorum* grows early in spring and often reduces moisture so much that slower growing annuals, perennials, and even shrubs are unable to withstand the competition (Holmgren, 1956). In drier regions plants that develop roots rapidly after germination have the competitive advantage. Weaker plants are overtopped and gradually crowded out by more vigorous and aggressive individuals.

Stable equilibrium may come about for a number of reasons: immigration from outside populations, differences in regulating mechanisms, an abundance of resources, especially if the species populations are held below carrying capacity, and genetic changes in competitors.

Stabilizing situations have been observed in a number of competitive plant situations, largely experimental in nature. Leith (1960) found that mixtures of perennial ryegrass *(Lolium perenne)* and white clover *(Trifolium repens)* growing in pastures form moving mosaics. Patches dominated by ryegrass tend to be invaded by clover, and patches of clover by grass. Van den Bergh and De Wit (1960) seeded sweet vernal grass *(Anthoxanthum odoratum)* and timothy *(Phleum pratense)* together in field plots in different proportions. They then compared the ratio of tillers of the two species after the first winter with the ratios after the second winter (Figure 16-8). In those plots where sweet vernal grass was in excess, the proportion of timothy increased; and where timothy was in excess, vernal grass increased. This experiment points out that in some mixtures the species in the mix experience more intra- than interspecific interference. In a mix of two species the existence of a

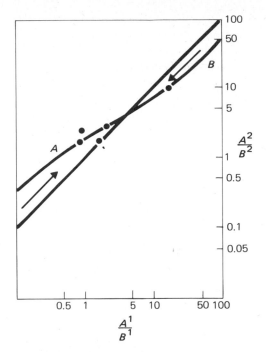

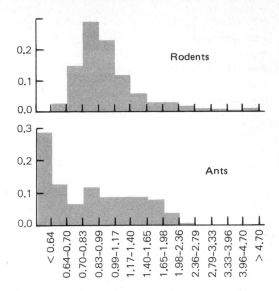

FIGURE 16-8
*Competitive relationship between two grasses, timothy (Phleum pratense) and sweet vernal grass (Anthoxanthum). The relationship is expressed as the ratio of the number of tillers of sweet vernal grass (A) and timothy (B) after the first winter ($A^1/B^1$) and after the second winter ($A^2/B^2$). In plots in which sweet vernal grass had been in excess, the proportion of timothy increased; where timothy had been in excess, sweet vernal grass increased. The point at which line AB intersects the 45 degree slope line represents the equilibrium point under a given set of environmental conditions. (From Harper, 1967, after Van den Bergh and De Wit, 1960.)*

FIGURE 16-9
*Sizes of native seeds harvested by coexisting ants and rodents in an Arizona desert. A total of 11,518 seeds was collected from the cheek pouches of 134 rodents representing five species, and 1,052 seeds were collected singly from ants representing seven species. Seeds were sized by passing them through a graded series of sieves. Although ants take some seeds smaller than those utilized by rodents, there is extensive overlap. Overlap of rodents on ants is 0.59 and overlap of ants on rodents is 0.50. (From Brown and Davidson, 1977.)*

stable equilibrium depends upon frequency-dependent competition.

Although stable equilibrium usually involves closely related species, it may also exist among distantly related organisms that utilize the same resources. One example is the competitive relationship between seed-eating rodents and ants in the scrub desert of Arizona (Brown and Davidson, 1977). Both rodents and ants are similar in their utilization of seed resources (Figure 16-9). The exclusion of either rodents or ants would increase the density of seeds available to the other taxon. When either rodents or ants were removed from experimental exclosures in the desert, the species of the remaining taxon that showed the greatest increase were those that tended to utilize densely distributed seed resources. Species showing the greatest increase were the kangaroo rat among the rodents and species of the genus *Pheidole* among the ants.

### ALLELOPATHY

A particular form of interference competition among plants is *allelopathy,* the production and release by one species of chemical substances that inhibit the growth of other species. These substances may range from acids and bases to relatively simple organic compounds that reduce competition for nutrients, light, and space. Produced in profusion in natural communities as secondary substances, most compounds remain innocuous

**515**

but a few influence community structure. For example, phenolic acids released by exudates and decay of dead roots may have an antibiotic effect on soil microorganisms. There is evidence that the grass prairie three-awn (*Aristida oligantha*) and other old field plants inhibit nitrogen-fixing bacteria and blue-green algae. In doing so they maintain a low level of nitrogen in the soil. Tolerant of low nitrogen concentrations, these plants may slow the invasion of grasslands by other species (Wilson and Rice, 1968).

In desert shrub communities a number of shrubs (*Larrea, Franseria,* etc.) release a variety of more or less toxic phenolic compounds to the soil through rainwater. Under laboratory conditions, at least, these substances inhibit germination and growth of seeds of annual herbs (McPherson and Muller, 1969). Other desert-type shrubs (*Artemisia* and *Salvia*) that commonly invade desert grasslands release aromatic terpenes such as camphor to the air. These terpenes are adsorbed from the atmosphere onto soil particles. In certain clay soils these terpenes accumulate during the dry season in quantities sufficient to inhibit the germination and growth of herb seedlings. As a result invading patches of shrubs are surrounded by belts devoid of herbs and by wider belts in which the growth of grassland plants is reduced (Muller et al., 1968).

But allelopathy may not be the only reason for the belts devoid of vegetation. Studies of plant-animal interactions suggest that while plants do produce toxins, the absence of plants may result from predation of plants by hares and consumption of seeds by rodents and birds. By means of exclosures and feeding experiments Bartholomew (1970) demonstrated that birds and rodents removed seeds at a far greater rate from the bare areas than from surrounding grassland. Grazing and trampling by hares reduced seedling development around the shrubs which the hares used as cover. Exclosures on bare areas supported over 20 times as much vegetation as areas open to rabbits. While toxins may be involved, the toxin hypothesis may not be necessary to account for the observed pattern in vegetation. In fact Harper (1977) suggests that toxic interactions in higher plants may not be common because (1) higher plants rapidly evolve tolerances to

such environmental toxins as zinc, nickel, and copper and have developed tolerances to herbicides and (2) complex organic molecules are broken down by soil microbial action and plant toxins probably experience the same fate.

Experimental work with toxins and other forms of interspecific competition suggests that competition plays an important role in community structure and function, but much of the data has been collected out of context of the whole system. It is difficult to design experiments in the field to test competition hypotheses. Much of the data already amassed is inconclusive because experiments did not consider other interactions such as nutrients other than those being studied, plant-animal interactions, sensitivity of each species to its own density, and other ways in which one species can affect the well-being of another. While competition may be a well-established laboratory phenonenon, there are too many interacting factors in the field environment to attribute an observed phenomenon to competition alone. As Harper (1977) admonishes: "There is room for much scepticism in the interpretation of competition experiments, particularly when they are applied to the field." This is not to downplay interspecific competition, for it undoubtedly plays an important role in species interactions and species evolution. But interspecific competition is a subtle interaction functioning on those occasions when resources are truly limiting and often hard to detect conclusively in the field.

## Resource partitioning

Among animals similar species or members of the same genus coexist in the same community because of ecological adaptations that permit them to occupy different habitat space, feed by different methods, utilize different kinds and sizes of food, or feed at different times and places. Among plants each species may occupy a different position on a soil moisture gradient, possess different nutrient requirements, or have different tolerances for light and shade. In some instances interspecific competition is avoided because normally the resource is abundant. Competition would be evident only when shared resources were in short supply.

For example, throughout Europe live nine species of tits *(Paridae)*. Three of these, the blue tit, the great tit, and the marsh tit, inhabit the broadleaf woods. The blue tit, the most agile of the three, works high up in the trees gleaning insects, mostly 2 mm in size or smaller, from leaves, buds, and galls. The great tit, which is large and heavy, feeds mostly on the ground and seeks prey in the canopy only when taking caterpillars to feed its young. Its food consists of large insects 6 mm and over, supplemented with seeds and acorns. The marsh tit feeds largely on insects around 3 to 4 mm, which it gleans in the shrub layer and in twigs and limbs below 6 m above the ground. It, too, feeds extensively on seeds and fruits. In the northern coniferous forests live the coal tit, the crested tit, and the willow tit (Figure 16-10). The more agile coal tit forages high up in the trees among the needles. There it seeks and feeds on aphids and spruce seeds. The willow tit consumes a high proportion of vegetable matter and feeds in the few available broadleaf trees. When in the conifers, it spends most of its time in the lower parts and on the branches rather than on the twigs. The crested tit is confined mostly in the upper and lower parts of the trees and on the ground, but the bird does not feed in the herb layer. Thus by feeding in different areas and on different size insects, as well as different types of vegetable matter, these species divide the resources among them.

A similar partitioning of resources reducing competition exists among plants. An example is three species of annuals growing together in a field on prairie soil abandoned one year after plowing. The annuals are bristly foxtail *(Setaria Faberii),* india mallow *(Abutilon Theophrasti),* and smartweed *(Polygonium pensylvanicum).* Each utilizes a different part of the soil resource (Figure 16-11). The foxtail has a fibrous, shallow root system that exploits a variable supply of moisture. It possesses the ability to recover rapidly from water stress, to obtain water rapidly after a rain, and to carry on a high rate of photosynthesis even when partially wilted. The mallow has a sparsely branched taproot extending to the intermediate depths where moisture is adequate during the early part of the growing season, but is less available later on. This annual is able to carry

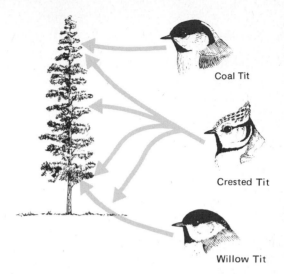

**FIGURE 16-10**
*Feeding areas and resource partitioning of coal tit, crested tit, and willow tit in a pine forest. The coal tit, more agile than the other two species, spends most of its time high up in the pines among the needles. Its smaller beak enables the coal tit to pick up and eat aphids in large quantities. The willow tit takes a higher proportion of vegetable food, mostly seeds, than the other two species, spends more time on lower than upper parts of the trees, and forages more on branches than on twigs and more on areas without needles. The crested tit divides its feeding time between upper and lower parts of trees, between areas with and without needles. It spends more time on the ground than the other two species. (Adapted from Lack, 1971.)*

on photosynthesis at a low water potential. The third species, the smartweed, possesses a taproot that is moderately branched in the upper soil layer and well developed at a depth below the rooting zone of the other species where the plant has a continuous supply of moisture (Wieland and Bazzaz, 1975).

## Competition and evolution

Interspecific competition as a mechanism of natural selection has been considered one of the strongest influences on the evolution of species and the structure of communities. Through the selective pressures of competition, species avoid competitive exclusion by par-

**517**

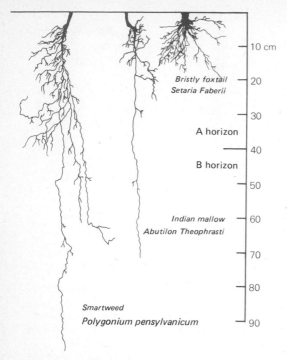

10 cm
20
30
40
50
60
70
80
90

Bristly foxtail
*Setaria Faberii*

A horizon

B horizon

Indian mallow
*Abutilon Theophrasti*

Smartweed
*Polygonium pensylvanicum*

**FIGURE 16-11**
*Partitioning of the soil resource by three species of annuals in a prairie soil in a one-year field. (From Wieland and Bazzaz, 1975.)*

titioning the resources among them through behavioral and morphological means. This differential utilization of resources and environment determines the structure and function of communities.

How well the theory of competition holds up in real world situations is basically an unanswered question. Many of the examples of mechanisms of coexistence are based in part on observed behavioral and morphological differences among closely related or associated species. Differences are then ascribed to selective pressures brought to bear by competition. But in some situations competition may not be as important as it is often considered (Wiens, 1977; Connell, 1975; Gilpin and Ayala, 1973).

Competition theory is based on several assumptions: (1) species populations are at carrying capacity; (2) the system is at equilibrium; (3) the environment remains constant and stable; (4) selection and thus competition is a continuous process; (5) all individuals are identical in resource needs and utilization; (6)

as resources increase in abundance, competitors become more specialized and utilize optimal resources.

Many of these assumptions have been accepted because they simplify the mathematical approach, but proofs that they are valid are mostly lacking. For example, no one has demonstrated that the resources are always in short supply, a condition necessary for continuous competition, that competition between two or more species is occurring, that a species population is at equilibrium or at carrying capacity, or that competitors become more specialized as resources increase in abundance. With an abundance of resources there may be no need to specialize.

The theoretical outcomes of competition also are based on the assumption that the environment is uniform, or relatively so, during the period when competition should be most pronounced, for example, the breeding season. The environment, however, is variable and unpredictable over both short and long intervals. The environment may range from very favorable to highly unfavorable, placing considerable stress on the population. This unpredictability of the environment may impose variable selection on the population (Wiens, 1974, 1977). During highly favorable periods resources are abundant, selective pressures are relaxed, with little loss of fitness across the range of phenotypes and little competitive interaction among species. Overlap in resource utilization may be considerable. But a period of stress or unfavorable environment could be a time of crisis for the population. Resources may be limited, competitive interactions intense, and competitive overlap minimal. Only phenotypes operating within the range of optimal resource utilization will survive. As a result overall genetic variation is reduced and the breeding population, consisting of a small number of individuals, is often slow to increase in favorable periods because of limitations on reproductive rates and energy allocations. If periods of environmental stress occur frequently, favorable periods may not be long enough to allow potentially competing populations to reach carrying capacity and experience competition. Some of the observed differences among species may not reflect the selective pressures of competition (because populations are rarely at equilibrium levels), but rather selective pressures imposed by predation and the physical environments. Selection through competition then becomes an intermittent process (Jaeger, 1971; Conley, 1976; Wiens, 1977) and other selective pressures may assume more importance than usually ascribed to them. The whole subject of competition is replete with unanswered questions, stimulating problems, and areas of controversy, which will be resolved only with careful and difficult experimental work in the field.

## SUMMARY

The concept of interspecific competition has had an important influence on the development of evolutionary and ecological theory. While competition between two species can be demonstrated in laboratory populations, it is more difficult to demonstrate competition in natural communities where evidence is more circumstantial than direct. However, some experimental work with plants in field situations indicates that competition can be very strong and influential.

One type of competition is exploitative in which organisms or groups of organisms have equal access to limited resources. The outcome is determined by how effectively each of the competitors utilizes the resource. A second type is interference in which one competitor is denied access to a resource by another, usually by some form of aggressive behavior.

Theoretically, competition can result in displacement or exclusion of one individual by another or in coexistence in stable or unstable equilibrium. A particular form of interference competition is allelopathy, the secretion of chemical substances that inhibit the growth of other organisms.

Cited as evidence of interspecific competition in the field is the amount of overlap in the utilization of resources. Large overlap of utilization curves on a resource gradient is often considered an indication of intense competition, but conversely such overlap might suggest minimal or no competition. Nonoverlapping curves or curves with minimal overlap instead may indicate strong competition. More important to a species may be diffuse competition, the cumulative effect of minimal competitive interactions with many species.

While competition theory suggests that interspecific competition is a continuous phenomenon, its role as a selective force may be intermittent, acting only during periods of resource scarcity or environmental crisis. At other times other selective forces may be more important.

# Relations among populations: predation
## CHAPTER 17

No phase of population interaction is more misunderstood (or hotly debated, especially by sportsmen) than predation. Predation in natural communities is a step in the transfer of energy. It is commonly associated with the idea of the strong attacking the weak, the lion pouncing upon the deer, the hawk upon the sparrow. But this idea must be modified, for predation grades into parasitism and vice versa. Between the two exists the broad gray area of the *parasitoid* and the host, which sometimes is called parasitism and sometimes predation. In this situation, one organism, the parasitoid, attacks the host (the prey) somewhat indirectly by laying its eggs in or on the body of the host. After the eggs hatch, the larvae feed on the host until it dies. Ultimately the effect is the same as that of predation, and the two can be considered the same. The concept of predation has been further extended to include the relationship between plants and herbivores. Grazing herbivores are considered predators upon plants and their impact on

plant populations as predation. A special form of predation is *cannibalism* in which predator and prey are the same species. Thus predation in its broadest sense can be defined as one organism feeding on another living organism.

From the viewpoint of the population ecologist, predation in its actions and reactions is more than just the transfer of energy. It represents a direct and often complex interaction of two or more species, of the eaters and the eaten. The numbers of some predators may depend upon the abundance of their prey, and predation may regulate the population of the prey. The ideas are debatable; certainly the same generalizations cannot apply to all groups of predators and prey.

## Models of predation

The influence of predation on population growth of a species received the attention of the mathematicians Lotka (1925) and Volterra

(1926); (for an excellent discussion of their work, see Hutchinson, 1978). Independently they proposed mathematical formulas to express the relationship between predator and prey populations. They attempted to show that as the predator population increases, the prey decreases to a point where the trend is reversed and oscillations are produced.

The Lotka-Volterra model involves one equation for the prey population and one for the predator population. The prey growth equation involves two components, the maximum rate of increase per individual and the removal of prey from the population by the predator:

$$\frac{dN_1}{dt} = r_1N_1 - PN_1N_2$$

where

$N_1$ = density of the prey population
$r_1$ = intrinsic rate of increase in absence of predation
$N_2$ = density of the predator population
$P$ = a coefficient of predation

The expression $N_1N_2$ assumes the removal of prey from the population is proportional to the chance encounter between predator and prey. In turn growth of the predator population is influenced by the density of the prey population:

$$\frac{dN_2}{dt} = P_2N_1N_2 - d_2N_2$$

where

$P_2$ = a coefficient expressing effectiveness of the predator
$d_2$ = density-independent mortality rate of the predator

The Lotka-Volterra model is graphically depicted in Figure 17-1. The ordinate $H$ is the number of predators; the abscissa P is the number of prey. In the area to the right of the vertical line, predators increase; to the left they decrease. In the area below the horizontal line, prey increase, above it prey decrease. The circle of arrows represents the joint population of predators and prey, and the size of the population of each changes with it. If a point or arrow falls in the region left of the vertical line, the prey population is not large enough to

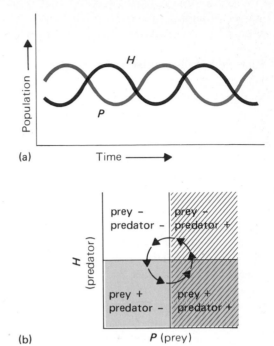

(a)

(b)

**FIGURE 17-1**
*Lotka-Volterra model of predator-prey interactions. The Lotka-Volterra equation can be depicted by the two graphs shown. The abundance of each population is plotted as a function of time. The model shows the joint abundances of species. The zero growth curves of both predator and prey are straight and intersect at right angles. The responses of the populations of each are indicated by a negative sign for population decline and a positive sign for population increase. Predators increase to the right of the vertical line; prey increases below the horizontal line.*

support the predators and the predator population declines. If arrows fall left of the vertical and above the horizontal, both populations are declining; the predator population decreases enough to permit the prey population to increase, moving the arrows left of the vertical below the horizontal. The increase in prey population now permits predators to increase and the arrows move right of the vertical below the horizontal. As the predator population increases, depressing the prey population, the arrows right of the vertical move above the horizontal. This interaction between predator and prey populations will oscillate through time as indicated in the graph.

The Lotka-Volterra model is based on a

**521**

number of simplifying assumptions that are difficult to justify in nature. It assumes that (1) predators move at random among a prey population that is distributed randomly; (2) every encounter of a predator with a prey results in capture and consumption; (3) in absence of predation the prey exhibits exponential growth; (4) all responses are instantaneous with no time lag for handling or ingestion of prey.

The model makes no allowance for age structure, for interaction of the prey with their own food supply, and for density-dependent mortality of the predator.

A decade later an ecologist, A. J. Nicholson, and a mathematician and engineer, V. A. Bailey, developed a mathematical model for a host-parasitoid relationship (Nicholson and Bailey, 1935). Nicholson and Bailey also based their model on the assumption that predators search randomly, but they assumed prey populations to be uniformly distributed in a uniform environment. They assumed that predators are insatiable, regardless of prey density, and that predators "sample" a certain proportion of the total prey population.

This feature of the Nicholson-Bailey model allows an estimate of prey in the next generation. If the number of hosts which the parasitoid removes within its sampling area is equal to a fraction of prey that represents recruitment, then the base parental stock remains. If the parental stock remaining is sufficient to replace the individual prey taken and if it is sufficient to maintain the density of parasitoids, then the two populations remain stable indefinitely. But if the parasitoid removes part of the parental stock along with recruitment, the prey and ultimately the predator populations decline. If much of the recruitment is left untouched, prey increase and predators may not be up to the task of removing increasing recruitment. In either of these two cases the two interacting populations will undergo oscillation with increasing amplitude, a feature of the Lotka-Volterra equations.

The Nicholson-Bailey model has a number of weaknesses. It assumes random search by the predator, which is seldom true in natural populations (but more convenient mathematically). The model ignores variations in searching efficiency of the predator. Also, predator

appetites are not insatiable, every encounter does not involve different age groups, and predator mortality is not density independent.

The Lotka-Volterra and Nicholson-Bailey models overemphasize the influence of predators on prey populations. Genetic changes, stress, emigration, aggression, availability of cover and hiding places, difficulty of finding prey as numbers become scarcer, and other attributes also influence fluctuations of populations. To make the Lotka-Volterra model more realistic, Rosenzweig and MacArthur (1963) developed a series of graphic models which consider a wider range of outcomes of predator-prey interactions. By modifying the zero growth curve of the prey to account for a low rate of growth at low and high population densities, they plot the growth curve as convex rather than horizontal. The basic components of increase and decrease as described in the Lotka-Volterra model remain the same. Thus, in the Lotka-Volterra model the growth curve of the prey is a horizontal straight line; the growth curve of the predator is the vertical line (see Figure 17-1). In the Rosenzweig-MacArthur model the growth curve of the prey is the convex line; the growth curve (isocline) of the predator is the vertical line (Figure 17-2).

The Rosenzweig-MacArthur model of a stable cycle of interaction is essentially the same as the Lotka-Volterra model. Predator and prey populations increase if the joint abundance of predator and prey falls inside the convex curve to the right of the vertical. If the joint abundance falls outside of the curve, populations of both decline. Because the prey curve intersects the predator curve at right angles, the curves produce neutrally stable cycles in populations of predator and prey.

If the predator curve is moved to the right, as in the damped cycle model, the curves no longer intersect at right angles. Interaction of predator with prey declines. The arrows do not form a closed circle, indicating that population fluctuations become damped, and the system eventually becomes stabilized.

If the predator curve is moved to the left, as in the unstable cycle model, more of the prey growth curve falls within the predator growth curve and interaction of predator and prey increases. The arrows now spiral outward, indicating unstable oscillations between high and low population sizes for both predator and prey populations (limit cycles). Such oscillations could lead to extinction of predator or prey or both.

These models represent situations in which the prey has no refuge from the predator. Growth rates of both predator and prey populations are a function of the frequency with which the predator comes in contact with the prey. A more accurate description of the situation in nature includes the removal of a portion of the prey population from contact with the predator, forcing the predator to turn to alternate prey (see switching, page 532). As the prey population increases in the refuge area, the surplus repopulates the surrounding area, as depicted in the stable cycle refuge model. Prey are cropped down to a certain level; predators then decline from emigration, starvation, or failure to reproduce.

However improved over the Lotka-Volterra model, even these models are unrealistic. The predator isocline is vertical, suggesting the probability that an encounter of a prey item by a predator will be constant at any given density regardless of the number of predators. This approach is based on several false assumptions: (1) an instantaneous change in reproduction of predators occurs without a change in prey density; (2) prey density is not affected by the predator death rate; (3) predators spend the same amount of time in search of prey regardless of prey density; (4) predators capture and eat the same proportion of prey they encounter; (5) prey are distributed at random and predators hunt at random; (6) there is no competitive interaction between predators while feeding. Nearly all these weaknesses are common to the Lotka-Volterra and Nicholson-Bailey models.

## Functional response

Models of interaction of predator and prey populations suggest two distinct responses of predator to changes in prey density. As prey density increases, each predator may take more prey or take them sooner. This is a functional response. Or predators may become more numerous through increased reproduction or immigration, a numerical response.

The idea of a functional response was introduced by Solomon (1949) and explored in detail by Holling (1959, 1961, 1966). He recognized three types of functional response (Figure 17-3): Type I in which the number of prey eaten per predator increases linearly to a maximum as prey density increases; Type II in which the number of prey eaten rises at a decreasing rate toward a maximum value; and Type III in which the number of prey taken is low at first and then increases in a sigmoid fashion approaching an upper asymptote.

### TYPE I RESPONSE

Type I response is a specialized type of functional response, the sort assumed in the simpler models just described (Lotka-Volterra, Rosenzweig-MacArthur, and others). In Type I predators of any given abundance take a fixed number of prey during the time they are in contact, usually enough to satiate themselves. Trout feeding on an evening hatch of mayflies would be an example of this type of functional response. Type I produces density-independent mortality up to satiation. Of more interest are Type II functional responses which produce inverse density-dependent mortality in prey, and Type III which produces changing density-dependent mortality.

### TYPE II RESPONSE

The Type II response, generally, but not exclusively associated with invertebrate predators, has attracted the most attention. It is described by the disk equation, named for an element in the experiment from which it was derived. In Holling's experiment the predator was represented by a blindfolded person and the prey by sandpaper disks 4 cm in diameter thumbtacked in different densities to a 3 ft square table. The predator tapped the table with a finger until a prey was found and then removed the disk. The predator continued the search and encounter (tapping, discovery, and removal) for one minute. Holling found that the number of disks the predator could pick up increased at a progressively decreasing rate as the density of disks increased. Predator efficiency rose rapidly as the density of disks increased until so much time was spent picking

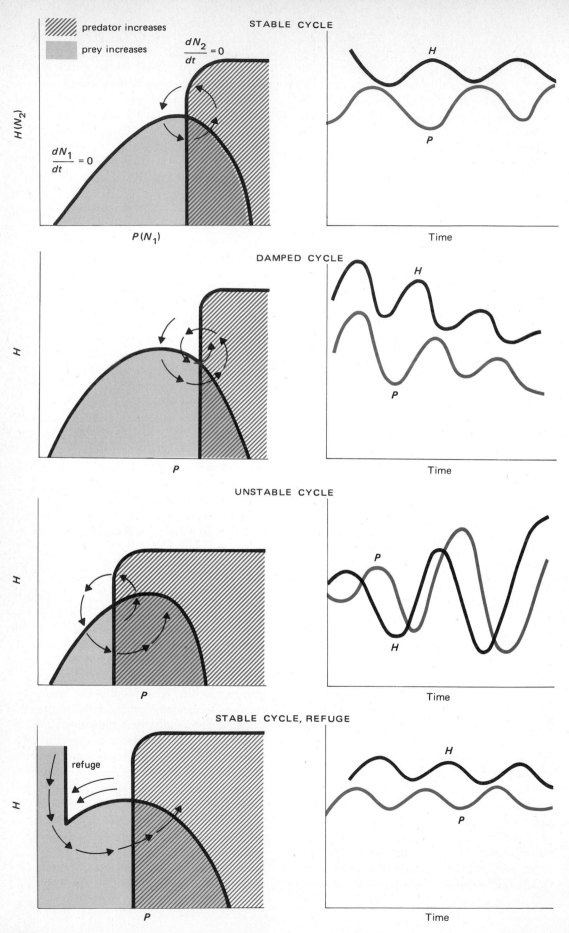

FIGURE 17-2

**FIGURE 17-2**

*Rosenzweig-MacArthur models of outcomes
of predator-prey interactions. In these
models growth curves of the predator and prey
are drawn more realistically. The prey curve
is convex rather than straight. In the stable
cycle model the prey population can increase
if the joint abundances of predator and prey
fall inside the area below the isocline of the
prey. It will decrease if it falls outside of this
area. In this model the position of the
growth curves are such that they intersect at
right angles as in Figure 17-1. This
generates a stable cycle.*

*In the damped cycle model the predator's
zone of increase intersects the prey's growth
curve at a point at which it is descending.
This dampens oscillations. In such situations
predators cannot successfully exploit prey
until the prey reaches carrying capacity.
Vectors spiral inward.*

*In the unstable cycle model the predator's
zone of increase intersects the ascending
part of the prey's growth curve. This
increases the oscillations producing an
unstable system. Vectors spiral outward and
the amplitude of the population increases
steadily until a limit cycle is reached. Such
situations can lead to extinction of either the
predator or prey or both.*

*In the stable cycle refuge model the
situation is somewhat similar to that in the
unstable cycle model except the prey has a
refuge where a portion of the population can
escape predation. This limits the oscillations.
(After MacArthur and Connell, 1966.)*

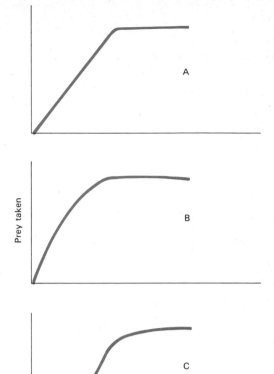

**FIGURE 17-3**

*Three types of functional response curves.
(A) Type I in which the number of prey
eaten per predator increases linearly to a
maximum as prey density increases. (B)
Type II in which the number of prey eaten
rises at a decreasing rate to a maximum
value. (C) Type III in which the number of
prey taken is low at first, then increases in a
sigmoid fashing approaching an upper
asymptote. (After Holling, 1959.)*

up and laying aside disks that the predator
could handle only a maximum number at a
time. The experiment demonstrated several
important components of predation: density of
prey, attack rate of predator, and handling
time, including time spent pursuing, sub-
duing, eating, and digesting prey.

Type II functional response is described by
the disk equation:

$$\frac{N_a}{P} = \frac{aNT}{1 + aT_h N}$$

where $T$ is determined by the equation

$$T = T_s + T_h N_a$$

and

$N_a/P$ = number of prey eaten per predator

$N_a$ = number of prey or hosts killed or
attacked

$P$ = number of predators or parasitoids

$a$ = a constant representing the attack
rate of the predator or the rate of
successful search

$N$ = number of prey

$T$ = total time predator and prey are
exposed

$T_n$ = handling time

$T_s$ = time spent by predator in search of
prey

Because handling time is the dominant
component, rise in the number of prey taken

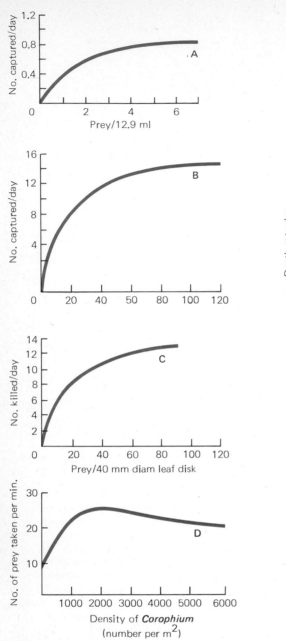

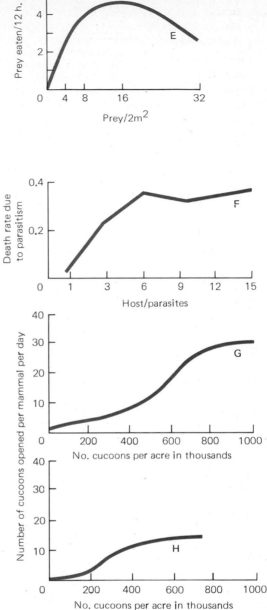

**FIGURE 17-4**

*Examples of Type II and Type III functional response curves. Graphs **A** through **E** are Type II curves for some predatory arthropods and a bird. (A) First instar* Linyphia triangulaus *(spider) feeding on* Drosophila. *(B) Second instar coccinellid* Harmonia axyridis *feeding on aphids. (C) Adult female* Phytoseiulus persimilis *(mites) feeding on nymphs of* Tetranychus urticae. *(D) Numbers of amphipod crustacean* Corophium *taken per minute by redshank in relation to density. (E) Dome-shaped Type II functional response of adult* Phytoseiulus persimilis *feeding on* Tetranychus urticae.

*The dome-shaped curve probably results from interference of predators. Female parasitoids discover many hosts are already parasitized or predators are discouraged from feeding in areas where a large number of individuals are already congregated and leave the area. Type III functional response curves are graphed in **F** through **H**. (F)* Encarsia formosa *parasitizing* Trialeurodes vaporariorum. *(G) Shrew* Sorex *feeding on sawfly larvae. (H)* Peromyscus *(deer mice) preying on sawfly larvae. (A, B, C, E, F from Hassel et al., 1976 and Beddington et al., 1976; D from Goss-Custard, 1977; G, H from Holling, 1964.)*

per unit time decelerates to a plateau (Figure 17-4) while the number of prey is still increasing. For this reason Type II functional response cannot act as a stabilizing force on prey population unless the prey occurs in patches (see page 531). Thus Type II response is destabilizing (for details see Murdock and Oaten, 1975).

The plateau may also be influenced by predator density of aggregation in response to patchy distribution of prey. Although the functional response equations assume predators searching at random in a uniform prey population, the usual situation in nature is an uneven distribution of prey. Aggregation of predators in areas of high prey density could well result in interference among predators. Encountering an individual of its own species, a predator may temporarily cease to hunt or may leave the area. Predator interference thus reduces the proportion of total prey or hosts the predator or parasite encounters because the predator's search time is reduced (Hassell et al., 1976).

Intermediate levels of aggregation may increase efficiency of predation through the social facilitation of locating prey. One or two members of the predator species discover and begin to feed on the prey item; other members of the species observe the feeding response and follow suit (see Curio, 1976). But at high levels of aggregation interference among predators may be so great that efficiency of predation declines or a number of predators leave the area.

Aggregative responses of predators to areas of high prey density can have a pronounced influence on the stability of predator-prey interactions. Hassell and May (1974) have presented an idealized general aggregative response curve for predators-to-prey distribution (Figure 17-5). The response curve exhibits a lower plateau of low prey density and an upper plateau of high prey density where predators do not distinguish between prey areas. But predators discriminate markedly in intermediate prey areas and tend to congregate in areas of higher density. An example of this type of distribution is the response curve for the redshank *(Tringa totanus)*, a shorebird which tends to concentrate in areas of its preferred food, the amphipod crustacean *(Corophium volutator)* (Figure 17-6).

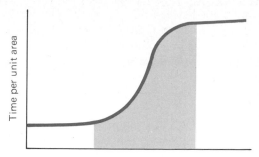

*Prey density per 1 unit area*

**FIGURE 17-5**
*Model of general aggregative response. At the lower plateau of prey density predators do not distinguish among relatively low (unprofitable) prey areas; at intermediate densities (shaded area) predators discriminate markedly; at the upper plateau predators do not discriminate among high (profitable) areas. (Hassel and May, 1974).*

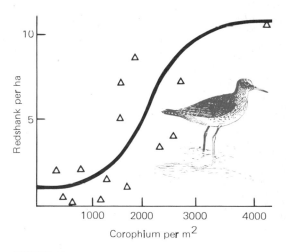

*Corophium per m²*

**FIGURE 17-6**
*Aggregative response in the redshank. The curve plots the density of the predator (the redshank) in relation to the average density of arthropod prey* (Corophium). *Compare in this curve with the profitability curve shown in Figure 17-9. (After Hassel and May, 1974.)*

**TYPE III RESPONSE**

Type III functional response is more complex than Type II. It has been associated with vertebrate predators that can learn to concentrate on a prey when it becomes more abundant, but recent studies by Hassell et al. (1977) show that it can be found among invertebrate predators as well. Because some vertebrate predators, especially feeding specialists, may also show Type II response, it is much wiser not to attempt to assign types to either invertebrate or vertebrate predators.

In Type III response the number of prey taken per predator increases with increasing density and then levels off to a plateau where the ratio of prey taken to prey available declines. Because the range of prey densities over which the death rate is imposed is an increasing function of density, that is, is density dependent, Type III functional response is potentially stabilizing.

Type II responses occur in situations of varying densities of one prey species. Type III responses invariably involve two or more prey species; the predator has a choice of prey. In the presence of several prey species, the predator may distribute its attacks among the prey in response to the relative frequency of the prey species. Predators take most or all of the individuals of a prey species that are in excess of a certain minimum number as determined, perhaps, by availability of prey cover and the prey's social behavior. The population level at which the predator no longer finds it profitable to hunt the prey species has been called the threshold of security (Figure 17-7) by Errington (1946). Type III responses have been called compensatory because as prey numbers increase above the threshold surplus animals become vulnerable to predation through interspecific competition (see Chapter 15). Below the threshold of security the prey

species compensates for its losses through increased litter size and greater survival of young. Functional response of the predator is very low below the threshold of security; above the threshold functional response is marked. An outstanding example of this type of predation is detailed by Errington (1963) in his notable study of the muskrat.

The reason for the sigmoidal shape of Type III response is the subject of much study and debate (see Royama, 1970; Croze, 1970; Murdock and Oaten, 1975; Curio, 1976). One explanation has been advanced by L. Tinbergen (1960), based on his studies of the relation between woodland birds and insect abundance. According to Tinbergen's hypothesis, when a new prey species appears in a given area, its risk of becoming prey is low at first. The birds have not as yet acquired a "searching image" for the species. Once the predator has secured a palatable item of prey, the predator finds it progressively easier to find others of the same kind. The more adept and successful the predator becomes at securing a particular prey item, the longer and more intensely it concentrates on the item. In time the numbers of the prey species become so reduced or its population so dispersed that encounters between it and the predator lessen. The searching image for that species begins to wane and the predator begins to react to another species. There are some problems with this hypothesis. For example, what "turns off" the searching image as the prey density decreases? Moreover, relations between the density of the prey species and its percentage in the food of a predator cannot be explained from the probability of encounters alone.

Also involved in the Type III response curve is the role of the facultative predator and alternate prey. Although the predator may have a strong preference for a certain prey, it can turn to an alternate, more abundant prey species which provides more profitable hunting. If rodents, for example, are more abundant than rabbits and quail, foxes and hawks will concentrate on the rodents. This idea was advanced early by Aldo Leopold in his book *Game Management* (1933) in which he described alternate prey species as buffer species because they stood between the predator on one hand and game species on the other. If the

*FIGURE 17-7*
*Compensatory predation as illustrated by a functional response curve. There is no response to the left of the vertical line which represents a threshold of security for the prey.*

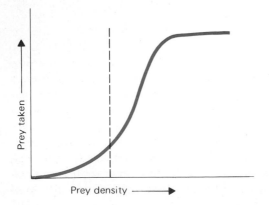

population of buffer prey is low, the predators turn to the game species; the foxes and hawks will concentrate on the rabbits and quail. This turning by a predator to an alternate, relatively more abundant prey has more recently been termed *switching* by Murdoch (1969). In switching the individual predator concentrates a disproportionate amount of attacks on the more abundant species and pays little attention to the rarer species. As the relative abundance of the two prey species changes, the predator changes its diet and turns to the alternate prey when it becomes more abundant.

When a predator switches depends a good deal on the threshold of security for the prey species involved. The threshold of security may be much lower for highly palatable prey and much higher for less palatable prey. Thus, a predator may hunt longer and harder for a palatable species before it turns to a less palatable alternate species. Conversely, the predator may turn from the less palatable species at a much higher level of abundance than it would if the palatable species were involved. In his laboratory experiments with sawfly larvae and sunflower seeds as prey and a white-footed mouse as predator, Holling found that the survival of the less palatable sawfly larvae was higher in the presence of the alternate prey, the more palatable sunflower seeds. Thus prey survival of one species is much greater in the presence of any more palatable species. For the palatable prey this situation is maladaptive. It is at a selective disadvantage when associated with less palatable prey. Such selective pressure favors temporal or spatial isolation of the more palatable from the less palatable species or it favors mimicry of the less palatable species by a palatable species (see Chapter 18).

Although the concepts of switching and alternate prey are valid, some predators deliberately seek out certain items, no matter how scarce. Peale's falcon, a subspecies of peregrine falcon, shows a marked preference for ducks or pheasants and will eat gulls and crows only when necessary (Beebe, 1960). Predators in a California grassland exhibited a distinct preference for the meadow vole over the harvest mouse and other rodents even though alternate prey was more abundant (Pearson, 1966). In fact the abundance of al-ternate prey apparently enables carnivores to maintain populations high enough to continue predatory pressure on the preferred species (Figure 17-8). Among herbivores deer exhibit a pronounced preference for certain species of browse (Klein, 1970). Meadow mice often concentrate on seeds of preferred grasses, even though seeds of other species are abundant and available (Batzle and Pitelka, 1970).

## Foraging strategy: concept of profitability of prey

An important point made by Hassell et al. (1976) is that although models of predator-prey interactions assume that predators search at random, the real world situation involves prey populations that are unevenly distributed, resulting in an aggregation of predators. In effect, predators tend to spend more time in areas where prey is most abundant.

Based on his study of hunting behavior and food selection by the great tit, Royama (1970) has proposed a model of hunting profitability. Basically Royama argues that it is energetically unprofitable for predators to spend time where prey density is low. As a result predators must discover the most productive way to allocate their hunting time among different prey species of different abundances in different niches or patches. Profitability is measured not by prey density, but by the amount of prey, preferably measured in terms of biomass, that a predator can harvest in a given time hunting the prey species. Improving food search efficiency would be adaptive (selected for) because it would reduce time allocated for foraging and would allow more time for other activities such as territorial defense. It would also conserve energy and reduce the time the predator itself is exposed to predation.

The profitability model assumes that (1) feeding rate within patches (niches) of uniform prey density rises at a decelerating rate to a plateau as density increases (Type II); (2) below densities of prey at which hunting or feeding levels off, the predator spends more time in areas not where prey density is greatest, but where profits are greatest; (3) predation rates may be depressed in areas of high prey density if the number of predators attracted to the area

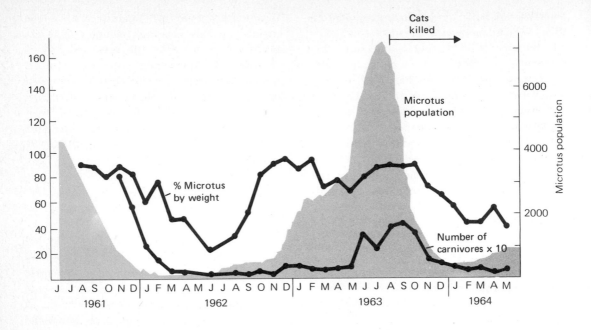

**FIGURE 17-8**
*Predatory pressure on a preferred species.
The amount by weight of Microtus, the
meadow vole, in the carnivores' diet is
compared with the number of carnivores and
the number of Microtus present. The left
scale divided by 10 gives the number of
carnivores. Note that the intensity of
predation does not decline measurably when
the vole population is low. This suggests that
Microtus is a preferred species and is
actively hunted regardless of prey density.
(From O. P. Pearson, 1966.)*

**FIGURE 17-9**
*Profitability curve plotting the profitability
of niche A in relation to density of two prey
species A and B and the predator's
movement between them. The density of
prey species B is fixed and the density of
prey species A changes relative to that
density. As the density of species A changes,
the profitability changes according to
Holling's disk equation. The model assumes
that species A and B have the same biomass,
palatability, and visibility. (From Royama,
1970.)*

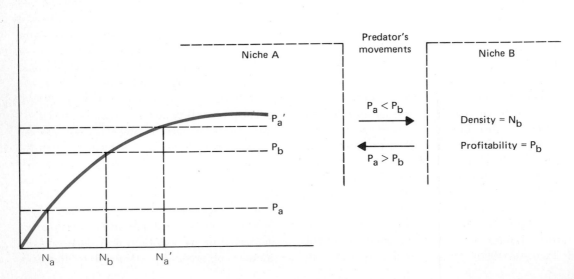

is large enough to produce predator interference; and (4) predators are capable of feeding on alternate prey.

Assume that a predator has two alternate prey of the same body size, nutritional value, and conspicuousness. Given these conditions, the profitability of a prey species concerned is described by $N/T$, the number caught per unit time. As the density of a prey species increases in its niche or patch, profitability of hunting increases, but gradually levels off (Figure 17-9). Suppose the predator can move freely between niches A and B inhabited by prey species A and B, respectively. In niche B the density of prey is fixed at $N_b$. In niche A the density of prey, given as $N_a$, is variable. If $N_a$ is smaller than $N_b$, then comparatively large differences exist between the profitability of A$(P_a)$ and B$(P_b)$. Predators should spend more time in B to increase their hunting efficiency. If density $N_a$ is increased to $N_b$, then profitability in niche A will increase to $P_b$. Now a predator no longer has any reason to spend more time in niche B than in niche A because it can take equal profits out of each. Under these circumstances the allotment of time between niches A and B would be a matter of chance, so on the average equal time is allotted to each. If density of prey is increased further from $N_a$ to $N_a'$, the profitability also increases from $P_a$ to $P_a'$. Although the profitability of $P_a'$ is now greater than $P_b$, the difference is much smaller than between $P_a$ and $P_b$. While the predator may spend more time in A than in B, profitability is much less than when density of prey $N_a$ was less than $N_b$. Further increase in prey density beyond $N_a'$ would not result in any worthwhile increase in profitability in niche A. From a predator's point of view, it is profitability rather than density of prey that is important.

The profitability hypothesis has been examined in relation to avian predators by a number of investigators (for example, see Davies, 1977; Smith and Dawkins, 1971; Smith and Sweatman, 1971; Goss-Custard, 1977). One of the most interesting and intensive studies is Zach and Falls' (1976a, b, c) report on the hunting and foraging behavior of the ovenbird (Seiurus aurocapillus). Captive ovenbirds were individually exposed to a patchy food supply presented in natural outdoor pens in typical habitat. In one experiment Zach and Falls presented four patch locations which were held constant although prey densities were interchanged.

The experimenters found that the ovenbird increased its search path exponentially with prey density; that the birds rapidly shifted their search efforts when prey densities were interchanged; and that because less search path was required per prey found in the dense patches, the birds concentrated their efforts in areas of high profitability and took a higher percentage of prey available in these sites.

Ovenbirds did not always visit all patch locations during the observation periods, but they always visited the densest patches. This suggests that the bird's discovery of one or more profitable feeding sites may restrict its sampling of other sites to assess profitability, as assumed by the hypothesis. The ovenbird's tendency to terminate its search after having encountered one or several profitable patches, together with its ability to learn the location of and to return to patches of high prey density may limit the number of patches it will exploit. The ovenbird may exploit other patches only if it discovers them by accident (random encounter).

Given a patchy food supply, the visiting pattern of the ovenbird appears to be determined by the rate of depletion of food owing to predation and the rate of renewal of food supply. Because food is depleted faster than it is replenished (because of the growth and movement patterns of the prey), the rate of decline in food supply determines whether the pattern of successive visits to patches is adaptative (see Stiles, 1976, in relation to hummingbirds). And at a given level of food consumption the rate of renewal determines the number of patches involved in an exploitation system. The number of patches in that system in relation to the number of patches the bird can remember may determine what proportion of visits are random or nonrandom. The fewer patches the bird has to remember, the less random are the visits.

In another set of experiments Zach and Falls (1976b) exposed ovenbirds to various sets of patches. When the birds were exposed to a single patch, they quickly concentrated their foraging in that patch. The birds took equal amounts of food at different prey densities and did not vary the amount of search path per visit. However, as the patch density increased, ovenbirds decreased the number of visits to the patch and the amount of search path per prey found. Also the birds reduced exploratory search outside the patch at higher prey densities, suggesting that with low depletion and high renewal rates, a single profitable feeding site might be sufficient. A day later the birds went directly to the old patch, now devoid of prey, and searched it thoroughly. In a second experiment Zach and Falls presented two prey patches successively. Prey in patch 1, presented on day 1, was not renewed on day 2. Instead a new patch with prey was presented. After an initial visit the birds quickly ignored the old patch location and after some exploratory behavior rapidly concentrated their search efforts in the new location. In a third experiment Zach and Falls presented ovenbirds with two patches simultaneously. The next day the birds exerted search efforts in both of these patches even though the prey was depleted.

These experiments suggest that regardless of prey density the birds search until they satisfy their needs. And, rather obviously, the birds have to exert more search effort as the prey density decreases. The ovenbirds respond quickly and in an adaptive fashion to a spatial pattern of food and learning is of major importance in this response. These experimental results are consistent with Royama's hypothesis, but, depending upon prey density, sampling of alternate feeding sites is less extensive than postulated.

Zach and Falls (1976c) also examined the question of whether ovenbirds hunt by expectation, that is, when a predator would give up search in one area and move on to the next. They discovered that the birds did not hunt by time or number expectation (in other words, the birds did not take as many prey as they expected to find by training and previous experience and then leave), but as in the previous experiments they did learn rapidly to find prey in patches. Their choice of feeding sites was nonrandom (all predator equations assume random search). They avoided areas of no food and patches previously visited. The ovenbirds improved their foraging efficiency by search-

ing nonrandomly within patches, avoiding areas already exploited. By doing so the ovenbirds were less likely to experience depletion of prey. If the birds followed a systematic search pattern, they gave up and left whenever a patch was completely covered. The time at which they gave up the search (see below, page 534) was unrelated to prey density. The conclusion was that birds do form expectations of where to find food, a subject explored further in the next section.

## Searching image, switching, and sigmoid response curves

As we have already stated, a long accepted explanation for the Type III functional response curve is the searching image (Tinbergen, 1960; Holling, 1959, 1965), the perception of certain prey species by predators in response to density levels of prey. As postulated by L. Tinbergen, when the frequency of random contacts with a prey species exceeds a threshold, the bird learns to see a prey. Characteristics of the prey perceived by the predator are associated with food and this association increases the likelihood that the predator will distinguish the prey from the background on subsequent contacts. The combination of increasing density of prey and establishment of a searching image results in a sudden increase of the perceived prey species in the predators diet, giving a sigmoid functional response curve. The predator adopts a searching image by random encounters and continues to concentrate on the species through random encounters until the overall composition of available prey shifts to give the predator an incentive to form another searching image.

The searching image is characterized by a restriction of the releasing stimuli to the visual properties of the prey and the visual properties of the background. The predator can acquire a searching image from remarkably few experiences (Croze, 1970; Dawkins, 1971; Curio, 1976). In losing an image the predator may simply not respond to the perceived stimulus or may in fact no longer perceive it, that is, no longer distinguish the properties of the prey from the background. The searching image is maintained by rewards in the form of the acquisition of food. When rewards are no longer there, the bird turns to another image. In effect the predator responds to changes in rewards. The extinction of an image tends to occur more slowly than its acquisition, and among some predators the searching image may be retained for some time, even in the absence of rewards. Croze (1970) found that carrion crows *(Corvis corone)* retained their searching image for eight days without reward. After that time the searching image declined rapidly, but was still retained.

The searching image as an explanation of sigmoid functional response has been criticized as inadequate or superfluous. Royama (1970) demonstrated that the sigmoid response curve could be explained by the hunting profitability hypothesis. Royama suggested three hypothetical curves based on profitability calculations (Figure 17-10), each assessing hunting time, $T$, as a function of density of the prey population, $N_o$, in a given patch. From these curves, values of $T$ as a function of $N_o$ can be obtained and the values of $N_a$, number of prey captured, then calculated from the disk equation. From these relationships, the number of captures can be assessed as a function of prey density. As Royama has shown, regardless of which of the three curves is chosen, the resulting $N_o - N_a$ curve is invariably sigmoid. If $N_a$ is plotted as a percentage of all other prey species taken, collectively considered as one species, as Tinbergen assumed, then the results follow the searching image trend that Tinbergen observed in his study of the great tit and that Royama obtained from his data. The use of the searching image is not needed to explain the sigmoidal functional response. The profitability of hunting is sufficient to produce a family of Type III response curves.

Murdock and Oaten (1975), taking a somewhat different approach, criticized the searching image as ambiguous and unconvincing from the data and showed that switching also yields an S-shaped curve. Switching, according to Murdoch and Oaten, is caused by the predator's (1) changing its preference toward the more abundant prey as it eats it more frequently by choice; (2) ignoring rare prey; (3) concentrating search in more rewarding areas.

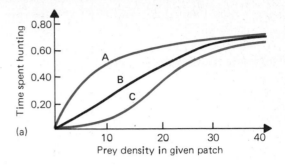

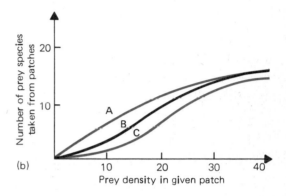

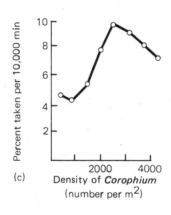

*FIGURE 17-10*
*Time-density response curves.* (a) *Hypothetical trends of time spent hunting* (T) *by a predator in relation to prey density* ($N_0$) *in a given patch of hunting range.* T *is scaled as a proportion of total hunting time in all patches involved.* (b) *Trends in number of prey taken* ($N_a$) *from patches concerned in relation to prey density (calculated from disk equation using values of* T *determined for curves in graph* a*).* (c) *Proportion of* Corophium *taken per unit time by redshanks in relation to prey density. (a, b from Royama, 1970; c from Goss-Custard, 1977a).*

Anyone of these three result in Type III response curves.

Notwithstanding the apparent close relationship between the switching and searching image, Murdoch and Oaten (1975) emphasize they are not the same. Searching image refers to an abrupt change in predator behavior in response to an increase in the absolute density of the prey of interest or to a sudden change in predator behavior over time as new prey appears. Searching image does not refer to a change in predator behavior in response to a change in the relative density of one prey in respect to another, a response which characterizes switching. However, Davies (1977), in a study of prey selection in wagtails *(Motacilla* spp.), found that a change in diet of these birds over a 10-day period was related to changes in the absolute rather than the relative abundance of the preferred food, midges (Chironomidae). As numbers of this prey decreased, the wagtails incorporated more alternate prey, *Drosophilia,* into their diet to maintain the feeding rate.

Although the searching image may not be sufficient or necessary to produce the sigmoidal response curve, it is still a valid observation of a behavioral phenomenon. It has been studied in some detail by a number of investigators (see Dawkins, 1971; Murton, 1971; Krebs, 1973; Croze, 1970). In his carrion crow studies Croze found that the bird did exhibit a searching image and that it needed only a few experiences with camouflaged prey to find it. In his experiments Croze placed meat bait under painted mussel and clam shells arranged on a sandy beach. The crows needed only 2.3 ± 0.5 experiences to become proficient in acquiring a searching image for shells of a particular color. After discovery of one prey, the crow tended to concentrate its efforts in that area.

Croze found two basic aspects to the crow's searching behavior. The first is area restricted search. Once it finds a prey, the crow will diligently search in a closely defined area and will stay in that vicinity as long as it is rewarded rather frequently with success. This behavior obviously relates to Royama's hypothesis of hunting profitability. Zach and Falls (1976) observed similar area restricted search in ovenbirds.

The second aspect concerns the predator's response to inter-catch distance and inter-prey distance (Figure 17-11). As prey density decreases, or as the distance between prey increases, the number of prey found decreases, or the distance between catches increases. The increased inter-prey distance forces the predator to spend more time seeking prey, thus lowering hunting success. As the inter-prey distance increases, the mean inter-catch distance also increases at a disproportionate rate until, in effect, the predator gives up. At this point the predator switches to another prey.

Such action, of course, implies that the predator would look for more than one prey item at a time. Because acquisition of a searching image involves visual cues both from the prey and the background, the predator is seeking camouflaged prey. Croze (1970) placed meat bait under empty mussel shells colored yellow, red, and black. In one experiment shells of only one color were baited, creating a monomorphic prey population. In another experiment shells of all three colors were baited, creating a trimorphic population. The trimorphic population suffered far less predation than the monomorphic. Out of 27 regularly arranged shells on the beach, a means of 10.5

*FIGURE 17-11*
*Relation of inter-catch and inter-prey distances. The inter-catch distance for carrion crows searching for hidden prey (A) increases more slowly than inter-prey distance (B) as the prey population becomes less dense. As prey distance increases the crow spends disproportionately more time hunting prey. (From Croze, 1970.)*

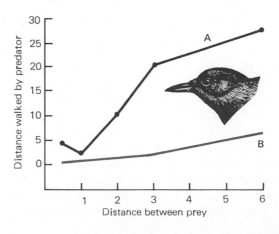

trimorphic and 3.5 monomorphic shells were left untouched by the crows. The crows needed 5 minutes to find the ninth trimorphic shell and only 3 minutes to find the ninth monomorphic shell. In the monomorphic population the crows got to the thirteenth shell in 5 minutes. After an inter-catch time of 5 minutes the crows gave up the search. Because morphs of a trimorphic population had a selective advantage over members of a monomorphic population, Croze was able to conclude that the ecological consequences of searching image behavior are: (1) camouflage is a prey's main defense against predation; (2) area restricted search increases selective pressure for prey behavior that scatters the population; and (3) searching image selects for polymorphism within a camouflaged prey species and for divergent coloration and behavior in sympatric species.

One can conclude that while the searching image itself may not result in a Type III response curve, it is an important and integral part of predation. Such a conclusion is supported by studies of food selection and foraging behavior of red-winged blackbirds (Alcock, 1973), white-crowned sparrows (Simon and Alcock, 1973), European thrushes (Smith, 1974), and titmice (Smith and Sweatman, 1974). In general these birds develop a searching image for the type of food fed upon previously and learn where they are most likely to find food (hunting profitability). Searching image and hunting profitability are complementary components of the Type III response curve.

## Numerical response

In addition to functional responses predators may exhibit numerical responses. One type of numerical response is a change of predator density through the movement of predators in and out of areas in response to prey density. Such movements of predators represent an aggregative response to prey patchiness (Hassell, 1966; Beddington et al., 1976), but can be considered a true numerical response only when predators move into an area from some distance. For example, Figure 17-12 shows a strong numerical response to increased prey

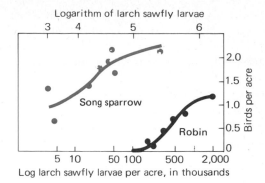

**FIGURE 17-12**
*Numerical response of song sparrows and robins to larval larch sawfly. (From Buckner and Turnock, 1965.)*

density by sparrows feeding on larch sawflies in certain Canadian bogs. Sparrow populations increased largely by immigration involving family flocks, adult and subadult birds, and premigratory flocks (Buckner and Turnovk, 1965). Aggregative response of local predators to local situations must be considered a part of functional response (Beddington et al., 1976).

Another and more important type of numerical response is the rate of change of a predator population (through birth and death rates) in response to prey death rate. The nature of this type of response is determined by the kind of predator involved, whether it is a parasitoid or a true predator.

For true predators, those that require several prey to complete their development, and for arthropod predators in particular, numerical response or overall rate of increase involves three components (Beddington et al., 1976): (1) duration of each predator instar; (2) survival rate within instars or survival rate of young in nonarthropod predators; and (3) fecundity of adults. All of these components, of course, depend upon the rate at which predators are able to locate and consume suitable prey (Lawton et al., 1975).

Duration of the instar is not influenced by prey density if the parasitoid feeds on only one host. But if the parasitoid feeds on more than one host during a developmental period, the amount of food intake and thus prey density influence development time. Growth takes

**535**

place only after metabolic energy needs are met; if only minimal energy is available after metabolic needs are met, growth is minimal. The less food available, the longer the development time and ultimately the slower the numerical increase of the predator.

Survival rates of arthropod instars and the young of nonarthropod predators are directly dependent on prey density, the size and availability of food. Too few prey means a lack of food, a lack of food results in poor survival of young or within instars, and a poor survival rate has a direct bearing on the numerical increase of a predator population.

For parasitoids in which the complete development of each larva requires only one host, adult fecundity is not limited by nutritional needs, but by the number of hosts females can find. The relationship between prey density and fecundity is linear (Hassell and May, 1973).

In all other situations nutrition controlled by the amount of prey eaten during the adult stage influences fecundity. Energy remaining after maintenance demands are met can be used in reproduction. If food and therefore energy is limited because of low prey density, fecundity is necessarily low (Figure 17-13) and positive numerical response is low. With increasing prey density fecundity increases and the numerical response is proportionately higher.

For example, the population of the great horned owl in a 62 mi² area in Alberta, Canada, increased over a 3-year period, 1966 to 1969, from 10 birds to 18 as the population of its prey, the snowshoe hare, increased sevenfold (Rusch et al., 1972). The proportion of owls nesting increased from 20 percent to 100 percent as biomass of snowshoe hare in the owl's diet increased from 23 to 50 percent (Figure 17-14). Coyotes inhabiting a 700 mi² area in Utah increased as the density of black-tailed jackrabbits increased, a species that made up three-fourths of the coyotes' diet (F. W. Clark, 1972).

As we have seen numerical response, positive or negative, is not immediate, especially in situations where fecundity depends upon the energy intake of adults. There is necessarily a time lag between adequate nutritional intake, developement and birth of young, and

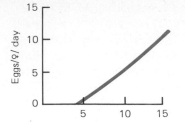

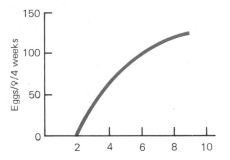

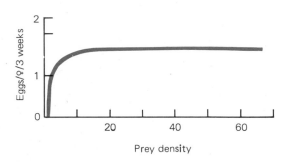

Prey density

**FIGURE 17-13**
*Fecundity in various predatory arthropods as a function of prey density. Note the shapes the curves can take. (From Beddington et al., 1976.)*

their maturation to reproducing individuals. Examples of the delayed numerical response can be found in a number of field studies.

Numerical response may involve both an aggregative response and an increase in fecundity. An example can be found among the "fugitive" warblers of northern forests, especially the Tennessee, Cape May, and bay-breasted warblers, whose abundance is dictated by outbreaks of spruce budworm. During such periods populations of the bay-breasted warbler have increased from 10 to 120 pair per 100 acres (Mook, 1963; Morris et al., 1958) and Cape May and bay-breasted warblers have larger clutches than associated warbler species (MacArthur, 1958). In fact Cape May

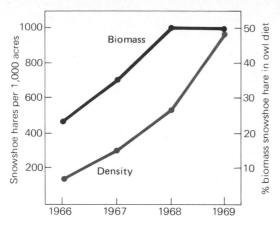

**FIGURE 17-14**
*Numerical response in the snowshoe hare–horned owl system. Density of the spring snowshoe hare in its habitat is plotted with the percentage of biomass of the hare in the diet of the great horned owl near Rochester, Alberta. (From Rusch et al., 1972).*

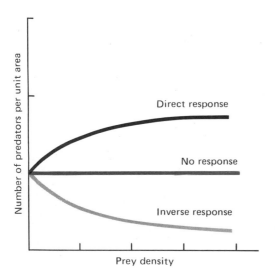

**FIGURE 17-15**
*Basic forms of numerical response. (From Hassell, 1966.)*

and possibly bay-breasted warblers apparently depend upon occasional outbreaks of spruce budworms for their continued existence. At those times these two species are able to increase more rapidly than other warblers because of extra large clutches. But during years between outbreaks they are reduced in numbers and are even extinct locally.

In general numerical response takes three basic forms (Figure 17-15): (1) direct or posi-

tive response, in which the number of predators per unit area increases as the prey density increases; (2) no response, in which the predator population remains proportionately the same; and (3) inverse or negative response, in which the predator population declines in relation to the prey population (Hassell, 1966).

## Total response

In analyzing the relationship between predator density and prey density, functional and numerical responses may be combined to give a total response, and predation may be plotted as a percentage. If this is done, predation falls into two types: (1) percentage of predation declines continuously as prey density increases (Figure 17-16) and (2) percentage of predation rises initially, then declines. The second type results in a dome-shaped curve (Figure 17-17) produced by the sigmoid (Type III) functional response to prey density and by direct numerical response.

*SUMMARY*

**Interactions between predator and prey have been described by the mathematical model of Lotka and Volterra and by subsequent modifications of their model by others. Essentially all these models predict oscillations of predator and prey populations. The oscillations may be stable, damped, or unstable. Relationships between predator and prey populations result in two distinct responses. As density of prey increases, predators may take more of the prey, a functional response, or predators may become more numerous, a numerical response.**

**There are three types of functional responses. In Type I the number of prey eaten per predator increases linearly to a maximum as prey density increases. In Type II the number of prey taken rises at a decreasing rate toward a maximum. In Type III the number of prey taken increases in a sigmoidal fashion.**

**Both Type II and Type III response**

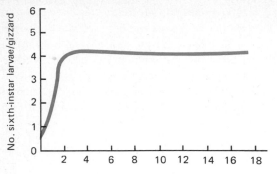

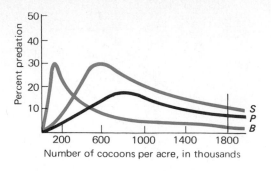

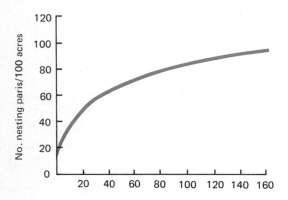

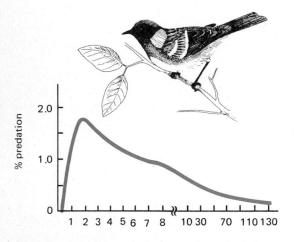

**FIGURE 17-17**
*Total response of predators to prey density expressed as percentage of predation to prey density. Total response includes both functional response and numerical response. S = Sorey shrew, B = Blarina shrew P = Peromyscus mouse. (From Holling, 1959.)*

**FIGURE 17-16**
*Functional response (a), numerical response (b), and total response (c) of bay-breasted warbler to changes in abundance of spruce budworm in New Brunswick. Total response is based on the assumption that larvae are available for 30 days, the average feeding day is 16 hours, and the digestive period is 2 hours. (After Mook, 1963.)*

curves may be found among invertebrates and vertebrate predators. Type II occurs in situations of varying densities of one species prey. Type III involves two or more species of prey. Inherent in Type III responses are searching image, in which the predator develops a facility for finding a particular prey item, and switching, in which the predator turns to an alternate, more abundant prey species for more profitable hunting. Because prey occurs in patches, the predator finds it more efficient to spend time in areas not necessarily where the prey is most abundant, but where hunting is most profitable in terms of time alloted.

Functional response views predation in terms of the relation of predator attack rates to prey density. Numerical response refers to the increase of predators resulting from an increased food supply. Numerical response may involve an aggregative response, the influx of predators to a food-rich area or, more importantly, a change in the rate of predator population growth through changes in developmental time, survival rates, and fecundity. Such changes produce a delayed numerical response, for a time lag necessarily exists between birth of young and maturation of reproducing individuals.

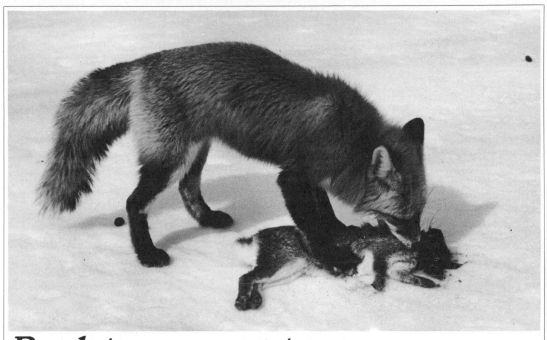

# Predator-prey systems
## CHAPTER 18

### Experimental systems

The interaction of predator and prey, particularly where an individual predator and an individual prey species are involved, is considered a predator-prey system. The predator directly influences growth and survival of the prey population, and the density of the prey population influences growth and survival of the predator population. It is such a system that Lotka and Volterra attempted to describe mathematically and the biologist G. F. Gause (1934) investigated experimentally. He reared together under constant environmental conditions a predator population *Didinium,* a ciliate, and its prey, *Paramecium caudatum* (Figure 18-1). The predator always exterminated the prey, regardless of the density of the two populations. After the prey was destroyed, the predator died of starvation. Only by periodic introductions of prey to the medium was Gause able to maintain the predator population and prevent it from dying out. In this manner he was able to maintain populations together and produce regular fluctuations in

both as predicted by the Lotka-Volterra model. The predator-prey relation was one of over-exploitation and annihilation, unless there was immigration from other prey populations.

In another experiment Gause introduced sediment in the floor of the tube. Here prey could escape from the predator. When the prey was eliminated from the clear medium, the predators died from the lack of food. The paramecia that took refuge in the sediment continued to multiply and eventually took over the medium.

The Gause experiments took place in a relatively simple environment. In a different type of experiment C. Huffaker (1958) attempted to learn if an adequately large and complex laboratory environment could be established in which a predator-prey system would not be self-exterminating. Involved were the six-spotted mite *(Eotetranychus sexmaculatus)* and a predatory mite *Typhlodromus occidentalis)*. Whole oranges, placed on a tray among a number of rubber balls the same size, provided food and cover for the spotted mite. Such an arrangement permitted the experimenter

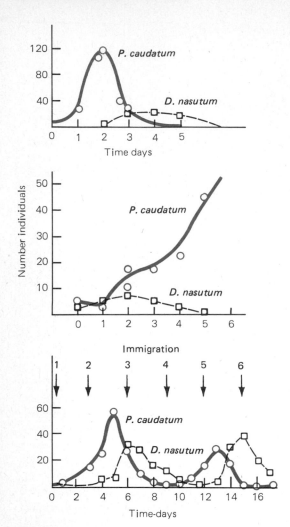

**FIGURE 18-1**
*Outcome of Gause's experiments of
predator-prey interaction between the
protozoans* **Paramecium caudatum** *and*
**Didinium nasutum** *in three microcosms:
oat medium without sediment, oat medium
with sediment, with immigration in oat
medium without sediment. (After Gause,
1934.)*

to control both the total food resource available and the pattern of dispersion by covering the oranges with paper and sealing wax to whatever degree desired and by changing the general distribution of oranges among rubber balls. The experimenter could manipulate conditions to simulate a simple environment where the food of the herbivore was concentrated or a complex universe where food was widely dispersed, partially blocked by barriers, and where refuge areas were lacking.

In both situations the two species found plenty of food available at first for population growth. Density of predators increased as the prey population increased. In the environment where food was concentrated and dispersion of the prey population was minimal, predators readily found the prey, quickly responded to changes in prey density, and were able to destroy the prey rapidly. In fact the situation was self-annihilative. In the environment where the primary food supply and the prey were dispersed, predator and prey went through two oscillations before the predators died out. The prey recovered slowly.

Several important conclusions resulted from the study. First, predators cannot survive when the prey population is low. Second, a self-sustaining predator-prey relationship cannot be maintained without immigration of prey. Third, the complexity of prey dispersal and predator-searching relationships, combined with a period of time for the prey population to recover from effects of predation and to repopulate the areas, had more influence on the period of oscillation than the intensity of predation.

The degree of dispersion and the area employed were too restricted in Huffaker's experiment to perpetuate the system. Pimental, Nagel, and Madden (1963) attempted to provide an environment with a space-time structure that would allow the existence of a parasite-host system. They chose as subjects a parasitic wasp *(Niasonia vitripennis)* and a host fly *(Musca domestica)* and provided for the environment a special population cage which consisted of a group of interconnected cells. A predator-prey system living in 16 cells died out, but a 30-cell system persisted for over a year. Increasing the system from 16 to 30 cells decreased the average density of parasites and hosts per cell and increased the chances for survival of the system. The lower density was due to the breakup and sparseness of both parasite and host populations. The greater number of individual colonies that remained following a severe decline of the host assured survival of the system, because these colonies provided a source of immigrants to repopulate the environment. Moreover, amplitude of the fluctuations of the host did not increase with time, as proposed by the model of Nicholson. Apparently the fluctuations were limited by intraspecific competition.

These laboratory experiments support studies made in the field. Sometime before 1839 prickly pear cactus *(Opuntia)* was introduced from America into Australia as an ornamental. As is often the case with introduced plants and animals, the cactus escaped from cultivation and rapidly spread to cover 60 million acres in Queensland and New South Wales. To combat the cacti, a South-American cactus-feeding moth *(Cactoblastis cactorum)* was liberated. The moth multiplied, spread, and destroyed the cacti until plants existed only in small, sparse, widely distributed colonies.

But the decline of the prickly pear also meant decline of the moth. Most of the caterpillars coming from moths that had bred on prickly pear the previous generation died of starvation. In areas where only a few moths survived not many plants were parasitized. As prickly pear increased, so did the moth until the cactus colony was again destroyed. In areas where no moths survived the colony spread once more, but sooner or later it was found by moths from other areas and was eventually destroyed. However, seed scattered into new areas established new colonies that maintained the existence of the species and thereby maintained the predator-prey system.

The rate of establishment of prickly pear colonies is determined by the time available for the colonies to grow before they are found by moths. As a result an unsteady equilibrium exists between cactus and moth. Any increase in the distribution and abundance of the cactus leads to an increase in the number of moths and subsequent decline in the cactus. The maintenance of this predator-prey, or more accurately herbivore-plant, system depends upon environmental discontinuity. The relative inaccessibility of host or prey in time and space limits the number of parasites and predators.

In further investigations of the moth and cactus relationship, J. Monro (1967) found that the moth may conserve food for succeeding generations of moths by limiting its own numbers. At high densities the moth clumps its egg sticks rather than laying them randomly on prickly pear. The clustering overloads certain plants of prickly pear with eggs, resulting in the destruction of the plants. In dense stands of prickly pear clustering initially has little influence on larval survival, for as an overloaded plant collapses, it falls on its neighbor and larvae can move to a new source of food. But, as dense stands become broken up into isolated plants, the relatively sedentary larvae are unable to cross the wide gaps and die of starvation. As mean density increases, the proportion of eggs wasted by clumping increases.

However, because the eggs are clustered rather than widely distributed, more plants escape infestation altogether or are subject to a lighter infestation than would be expected if eggs were laid at random. Monro found that this mechanism, which is employed most in the center of the range of the moths, acts to conserve food supply for succeeding generations and to maintain a constant level of both the food resource and the moth population.

These examples to some extent illustrate predator-prey interactions both at the plant-herbivore and at the herbivore-carnivore level. Although for simplicity they are considered separately, predator-prey interactions at one trophic level influence predator-prey interactions at the next trophic level. Interactions at two or more trophic levels are often involved in predator-prey stability.

## Plant-herbivore systems

Predation on plants by herbivores involves defoliation and consumption of fruits and seed. The results of the two forms of predation are different.

Defoliation is the destruction of plant tissue (leaf, bark, stem, and roots). Some plant predators, such as aphids, do not consume tissue directly, but, acting as parasites, tap plant juices without killing the plant. Other herbivores consume tissue directly, destroying parts or all of the plant. If grazers consume seedlings, they kill the plant outright. If they remove only part of a plant, its survival depends on the amount and continuation of grazing. Continued grazing may eventually kill the plant, but if grazing ceases, it may regenerate. Although grazed plants may persist and regenerate, defoliation still has an adverse affect. Plant biomass is decreased. Removal of leaves may damage the hierarchial position of the plant in the stand. Loss of foliage and subsequent death of some roots (root pruning) reduce the vigor of the plant, its competitive ability, and its reproductive effort (fitness). (Harper, 1977)

Although a plant may be able to compensate for the loss of leaves by increasing photosynethetic assimilation in the remaining leaves, it may be adversely affected by loss of nutrients, depending on the age of tissues removed. Young leaves are dependent structures, importers and consumers of nutrients drawn from reserves in roots and other plant tissues. As the leaf matures, it becomes a net exporter of nutrients, reaching its peak before senescence sets in. Grazing herbivores such as sawfly and gypsy moth larvae, deer, and rabbits concentrate on more palatable, more nutritious young leaves. They tend to reject older leaves because they are less palatable, are high in lignin, and often contain secondary compounds (tannin, for example). If grazers concentrate on young leaves they remove considerable quantities of nutrients. Plants respond to defoliation with a flush of new growth that drains additional nutrients from reserves that otherwise would have gone into growth or reproduction.

More important may be damage to the cambium and growing tip or apicial meristem. Deer, rabbits, mice, bark burrowing insects (as bark beetles) feed on bark destroying the cambium. Cambium destruction results in the death of a part of or the entire plant. Destruction of the growing tip by grazers or burrowing insects such as gall insects or shoot moths, often stimulates the development of dormant buds and increases branching, altering the growth form of the tree.

The impact of seed predation which results in elimination of individuals is difficult to assess. If density-dependent processes are such that few seedlings will survive, seeds removed by predators represent that portion of the population that has no future. In such instances seed predation has no real impact. But if predators remove seeds from an expanding population or from areas being colonized, predation reduces the rate of increase. On the other hand, if consumption of seeds is a mechanism for seed dispersal, as when seeds are contained within a palatable fruit and then carried in the gut of a fruit-consuming herbivore, predation can be to the plant's advantage.

The interrelations of plants and herbivores have been examined theoretically by May (1973), Caughley (1976a,b), and Noy-Meir (1975) all of whom present mathematical approaches and analyses.

The growth of vegetation as a function of plant biomass can be described by an expression comparable to the logistic growth equation:

$$aV\left(1 - \frac{V}{K}\right)$$

where
$V$ = biomass of vegetation
$K$ = maximum sustained biomass (carrying capacity)
$a$ = rate of increase

The rate of increase slows as competition for sunlight, moisture, nutrients, and self-interference increases (Figure 18-2a).

When ungrazed vegetation is subjected to grazing by a herbivore population, the vegetation's rate of growth is slowed by an amount proportional to the intensity of grazing or predation (number of herbivores consuming plants multiplied by the rate at which vegetation is consumed). When vegetation is at maximum sustained biomass *(K)*, herbivores can eat all they want, although the quantity is limited by the herbivores' intake capacity. If vegetation increases while the herbivore population remains the same, grazers increase consumption up to a point of saturation. If the vegetation declines, the amount herbivores consume also declines because intake is limited by the forage available. These conditions represent a Type II functional response curve (Figure 18-2a).

If herbivores increase, a numerical response, they may reach a level where they overgraze the vegetation, as frequently happens with deer, snowshoe hare, and lemmings. If the vegetation has no ungrazable reserve biomass

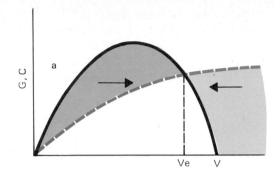

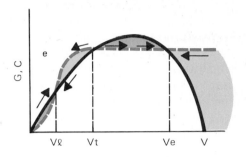

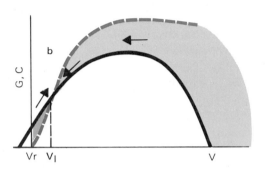

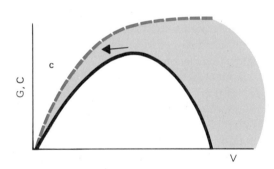

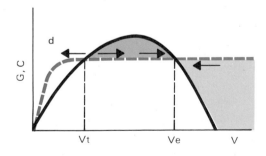

**FIGURE 18-2**

*Logistic plant growth (G) as a function of plant biomass (V) over which is imposed consumption per animal (C) as a function of plant biomass. Dark colored areas represent situations in which plant growth exceeds consumption; light-colored areas represent areas of curves where consumption exceeds growth. (a) Consumption curve is below the growth curve at all biomass levels. Intersection of the two curves indicates point of stable equilibrium between plant growth and herbivore consumption. Deviation from it in either direction will cause net changes in V tending to restore stable equilibrium. It is an undergrazed state as plant growth, animal consumption, and secondary production are below maximum levels possible. (b) Overgrazing to low biomass steady state. Vegetation has some ungrazable reserve biomass that prevents complete extinction. (c) Overgrazing to extinction. The consumption curve exceeds the plant growth curve at all levels of V. If no inaccessible plant reserves capable of producing new plant growth exist, vegetation becomes extinct. (d) Steady-state and unstable turning point to extinction. This situation occurs if the consumption curve is steeper than in (a). The two curves intersect at two points, one a steady state at high biomass and the other at low biomass. Any deviation can lead to extinction if V becomes lower than $V_t$. (e) Two steady states, one at high biomass and the other at low biomass. This occurs when an ungrazable plant reserve exists. The two curves intersect three times producing a stable steady state at high plant biomass. ($V_e$) and at a low plant biomass ($V_2$) and an unstable equilibrium or turning point between them ($V_t$). (After Noy-Meir, 1976.)*

or if it is grazed to a point where the green biomass is too sparse to maintain production, the plant population may go extinct (Figure 18-2*b, d*). If the vegetation has an ungrazed reserve, the reserve may be utilized to attain a low biomass steady state (Figure 18-1*c*). Depending upon the population density of the herbivore, the vegetation may stabilize at a high biomass *(V)* or reach an unstable equilibrium point at which the vegetation may be able to restore itself if grazing (predatory) pressure is reduced or at which the plant population may slip to extinction. In some situations, especially where a Type III functional response is involved, the vegetation may exhibit two stable steady points, one at a high plant biomass and another at a low plant biomass (Figure 18-2*e*). Between the two is an unstable equilibrium point.

Interactions between various vegetation growth curves and various herbivore densities can result in a number of plant-herbivore relations (for some detailed examples and discussion, see Noy-Meir, 1975). A typical interaction is one described earlier for vegetation and snowshoe hare. As herbivores increase, vegetation declines. In turn the herbivore population declines (Figure 18-3). The vegetation re-

covers, the herbivore population increases, and the two populations approach equilibrium, the vegetation with grazing pressure and the herbivore with its food supply. (Caughley, 1976*b*).

## Herbivore-carnivore systems

The herbivore-carnivore system can be described by the Lotka-Volterra equations modified by components of the disk equation including the attack rate and by the rate of predator decline (for details, see May, 1973). Like the plant-herbivore system, the herbivore-carnivore system can result in stability or instability. The general observation is that in an ecosystem stability does exist, more or less, between prey and predator. Rarely do predators exterminate their prey. On occasions the prey population may outstrip the ability of the predator to contain it and similarly predators can reduce or depress prey populations. There are three possible outcomes from such interactions: stable oscillation, stable limit cycle, and unstable oscillation (Figure 18-4). The unstable oscillation is more of a laboratory than a field phenomenon.

A stable oscillation results when a prey population is self-limiting and the predator therefore has little impact on the prey population. Some prey species possess strong limiting mechanisms such as territoriality. In such

**FIGURE 18-3**
*Trend of vegetation density and animal numbers after a herbivore eruption. Note that as the herbivore population increases, vegetation biomass decreases; then as herbivores decline, vegetation increases. Eventually vegetation growth and herbivore consumption reach a sort of steady state. (After Caughley, 1976.)*

**FIGURE 18-4**
*Diagram of the possible outcomes of a herbivore-vegetation system or a predator-prey system: (A) stable equilibrium, (B) stable limit cycle, (C) unstable equilibrium. (After Caughley, 1976.)*

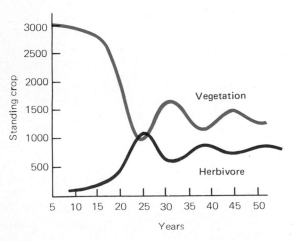

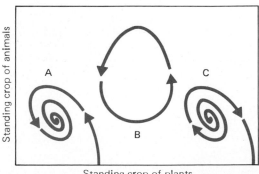

situations, as Tanner (1975) has demonstrated, the prey has a higher intrinsic rate of increase than the predator. An example is the mink-muskrat system (Errington, 1963). Muskrat populations are limited by territorial behavior. According to Errington, the predator takes only the surplus prey, the number in excess of $K$ as determined by availability of resources. However, Tanner suggests from his simulation models involving intrinsic growth rates of muskrats and mink that the predator can hold the prey at a density lower than the maximum attained in the absence of predation.

With prey populations that are not self-limiting, some stability in prey-predator interactions may be achieved because the growth rate of the prey species is less than that of the predator. Such prey species are many ungulates and hares, the populations of which appear to be predator controlled.

A stable limit cycle results when the growth rate of the predator and of the prey are approximately the same and the prey exhibits only a weak intraspecific population regulation, as exemplified by the snowshoe hare and lynx. In such cycles population numbers undergo well-defined cyclic changes in time (May, 1973). The amplitude of the cycle (maximum-minimum values the population reaches during the cycle) is determined by such variables as birth rate, death rate, predation rate, and so on. The period (time to complete one cycle) is also fixed. If the system is disturbed, a stable limit cycle will return to equilibrium.

Occupancy of two habitats by prey can dampen stable limit cycles if the second habitat provides poor cover from the predator and is inhabited by overflow or surplus animals from the first or optimal habitat. In some ways this is a refuge situation and is typical of snowshoe hare populations in the Rocky Mountains and in the central Appalachians (Tanner, 1975).

Ungulates such as the white-tailed deer have a lower intrinsic rate of increase than their major predators, the wolf and the mountain lion. Absence of self-limitation by the prey and a high $s/r$ ratio (intrinsic growth rate of the predator/intrinsic growth rate of the prey) result in some form of prey-predator equilibrium. In such cases the predator must be self-limiting through competitive and behavioral interactions. Removal of the predator from the system results in marked increases in prey population, particularly ungulates, with a resulting strong impact on vegetation. In many cases regulated hunting is an adequate substitute for natural predation. In other instances hunting can result in excessive kill and depressed populations of ungulates.

A number of properties within the system influence stability (see Murdoch and Oaten, 1975). Two of the most important appear to be prey self-limitation and predator searching time (Tanner, 1975). Predator searching time is influenced by a number of variables, especially spatial heterogeneity. The role of spatial heterogeneity in stability of predator-prey systems was suggested by the experiments of Huffaker and the field observations of Monro, discussed earlier. Spatial heterogeneity may involve refuges, as it does in some populations of snowshoe hares or in the ungrazed reserve of vegetation biomass. It may also involve barriers to predator dispersal and a mosaic of local prey populations out of growth phase with each other and fluctuating at different times. Stability also involves the presence of alternate prey which will lead to switching, removing predatory pressure from one population and allowing it to recover.

## Time lag and predator-prey stability

The models discussed fail to consider another aspect of predator-prey interaction, time lag. If the growth rate of a prey or predator species does not respond immediately to change in its own population or to changes in the interacting species, the result is a delayed logistic growth. Because the delayed response allows the predator to overshoot the prey or the prey to overshoot the predator, time lag results in an unstable system.

In comparing time lags in the different predator-prey systems, May (1973) argues that plant-herbivore systems are unstable, while plant-herbivore-carnivore systems are mostly stable. In the plant-herbivore system the time lag is relatively long and feedback is destabilizing. The herbivore birth rate time scale is less than the time scale for vegetation recovery from overgrazing. In the plant-herbivore-carnivore system the oscillating period of the herbivore-carnivore system in the absence of stabilizing density-dependent mechanism is greater than the time for vegetation recovery, resulting in a stable system (May, 1973).

Caughley (1976b), however, argues that in a vegetation-ungulate system, at least, there is no need to inject a time lag into the system because the time lag is little more than a mathematical nicety. Delayed logistic growth can be equated functionally with a numerical response term in the plant-herbivore equations.

## Population exploitation

A form of highly selective and intensive predation, often not related to the density of either predators or prey, is exploitation by humans. Overexploitation of wild populations, especially if coupled with the loss of habitat, has resulted in a serious decline and local if not global extermination of some species. The overharvesting of buffalo, great auk, African ungulates, whales, and many pelagic fish are examples of shortsightedness on the part of humans. On the other hand, such wildlife populations as the white-tailed deer are underharvested in many places, particularly since their natural predators have been eliminated. In contrast to these examples of destructive exploitation, the objective of the wise exploitation of any natural population is the maintenance of some sort of equilibrium between recruitment and harvest.

### BASIC CONCEPTS OF
### POPULATION EXPLOITATION

Although some of the terms used in defining exploitation of populations are similar to those used in productivity, the meanings are somewhat different. When fishery and wildlife biologists speak of *yield* they refer to the individuals or biomass removed when the population is harvested. *Biomass yield* is the product of the number harvested times the average weight. (Yield may indicate weight without numbers or vice versa.) The *standing crop* is the biomass present in a population at the time it is measured. *Productivity* is the differ-

ence between the biomass left in the population after harvesting at time $t$ and the biomass present in the population just before harvesting at some subsequent time $t + 1$.

The objective of regulated exploitation of a population is *sustained yield*. This means the yield per unit time is equal to productivity per unit time. In its simplest form sustained yield is described by an equation first proposed by E. S. Russell for fishery exploitation. Although the equation was specifically developed for fisheries, it is applicable to any exploitable population. With some minor modifications this equation is

$$B_{t+1} = B_t + (A_{br} + G_{bi}) - (C_{bf} + M_b)$$

where

$B_{t+1} =$ total biomass of exploitable stock just before harvesting, at time $t + 1$

$B_t =$ total biomass of exploitable stock just after the last harvest, at time $t$

$A_{br} =$ biomass gained by the younger recruits just grown to exploitable stock

$G_{bi} =$ biomass added by the growth of individuals in both $B_t$ and $A_{br}$

$C_{bf} =$ any biomass exploitatively removed during the harvest period

$M_b =$ biomass lost from exploitable stock by natural causes during the time $t$ to $t + 1$

Thus the equation stresses the fact that productivity also includes individuals that were born and individuals that died during the time interval from the end of one harvest period to the beginning of the next.

The equation is highly simplified. In an unexploited fish population $A_{br}$, $C_{bf}$, and $M_b$ are interdependent. For example, in a stable environment largely undisturbed by humans fish populations appear to be dominated by large species. In turn each species population appears to be dominated by large old fish (Johnson, 1972). When humans start to exploit such a population significant changes take place. To compensate for exploitation directed first toward the largest members of the population (organisms that under natural conditions are normally secure from predation), the population exhibits an increased growth rate, a reduced age of sexual maturity, increased number of eggs per unit of body weight, and reduced mortality of small members of the population (Regier and Loftus, 1972). Populations of other vertebrates react in a similar way.

Exploitation may also influence behavior of the species. Fishing or hunting techniques that are employed constantly because of their initial success gradually become less effective with time because of some conditioning or learning by members of the population. As harvest of the species begins to decline, the exploiters are forced to improve or change their methods of fishing and hunting. Also involved may be an interspecific competition. As both the numbers and larger members of a population decline, the niche may be occupied by unexploited, highly competitive, and closely related sympatric or introduced species. Thus, as the Pacific sardine populations declined, their place was taken by anchovies.

If exploitation is carried far enough, then the age classes in the population are too young to carry on reproduction and the population collapses. The principle behind sustained yield is to avoid that collapse of the population.

Exploitation and sustained yield of a population is clearly dependent on the rate of increase. Sustained yield does not imply holding a population at ecological carrying capacity *(K),* for obviously at that level the rate of increase equals zero. A population stable in the absence of harvesting can be harvested under sustained yield only after the plant-herbivore or the herbivore-carnivore system has been manipulated to raise *r.* This can be done in two ways: (1) improve food and cover to increase the carrying capacity by increasing available resources, fecundity, and survival; (2) lower the density by removing a certain number and then stabilize the population at some lower density (Figure 18-5). Within limitations the lower the density of a population is below the carrying capacity, the higher is the rate of increase. Thus a higher rate of harvest is needed to hold the population stable at some desired lower density.

The idea is to have the rate of harvest *H* equal to the rate of increase *r.* In effect the rate of harvest should hold the rate of growth at zero. *H* would have to equal the rate at which the population would increase if the harvest were stopped (Caughley, 1976). Con-

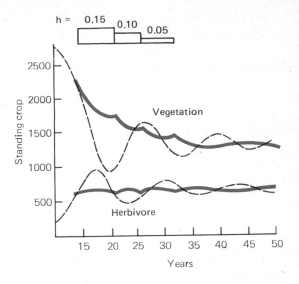

**FIGURE 18-5**
*Effect of harvesting on herbivore-vegetation system. Superimposed on the herbivore-vegetation interaction diagram in Figure 18-3 is the path of vegetation and herbivore standing crops after harvesting of herbivores over the intervals and at the rates per year as diagrammed in the rectangles above. As a result of management fluctuations are reduced and the system moves to equilibrium. Such equilibrium results if harvest is initiated at a rate of about one-half of the population's intrinsic rate of increase when the animal population is well below the peak. This rate of harvest must be maintained until plant density levels off and begins to increase. (After Caughley, 1976.)*

sider a deer population increasing at a rate of 20 percent a year. This gives a finite rate of increase, $e^r = 1.20$ and $r = 0.182$. To hold the population stable, the herd must be harvested at the instantaneous rate of $H = 0.182$. If the population is harvested only during a certain season of the year, as is usual with deer, then the sustained yield is calculated from an isolated rate of harvest $h$, defined as $h = 1 - e^{-H}$, in this case 0.167. This isolated rate of harvest would have take into account natural mortality to the population occurring between the period of birth and the time of harvest. Assuming a deer population after the fawning season of 1000 animals and allowing a natural finite rate of mortality of 0.25 per year, one could remove 151 animals in the fifth month. This would allow the number at the next fawning

**547**

season to climb back to 1000. The addition of 375 young would compensate for hunting and natural mortality (for details on carrying out calculations, see Caughley, 1976).

Although often considered as such, sustained yield is not of particular value for a given population. There may be a number of sustained yield values corresponding to different population levels and different management techniques. The level of sustained yield at which the population declines if exceeded is known as *maximum sustained yield.* Maximum sustained yield is not always the most efficient harvest level because of other considerations such as species interactions, esthetics, land use problems, and the like. Harvesting may be aimed at the *optimum sustained yield,* the level of sustained yield determined by consideration of these other factors as well as maximum sustained yield.

The higher the *r* of a species, the higher will be the rate of harvest that produces the maximum amount of biomass production. Species characterized by scramble competition (*r*-strategists) have a high wastage of production. To manage a population influenced by density-independent variables, such as climate or temperature, the objective is to reduce wastage by increasing the rate of exploitation. The role of harvesting is to take all individuals that otherwise would be lost to natural mortality. This type of exploitation is described by the expression (K. E. F. Watt, 1968):

$$\text{maximum yield} = B_t - \min(R_t)$$

where $B_t$ = biomass at time $t$ and $\min(R_t)$ = minimum number of reproducing individuals left at time $t$ in order to insure replacement of maximum yield at time $t + 1$.

Such a population is often difficult to manage because the stock can be severely depleted unless there is repeated reproduction. An example is the Pacific sardine (Murphey, 1966, 1967), a species in which there is little relationship between breeding stock and the subsequent number of progeny produced. Exploitation of the Pacific sardine population in the 1940s and 1950s shifted the age structure of the population to younger age classes. Prior to exploitation 77 percent of the reproduction was distributed among the first five years. In the fished population, 77 percent of the repro-

duction was associated with the first two years of life. The population approached that of single stage reproduction subject to pronounced oscillations (Figure 18-6). Two consecutive years of reproductive failures resulted in a collapse of the population from which it never recovered.

In populations characterized by density-dependent regulation (*K*-strategists) the maximum rate of harvest depends on age structure, frequency of harvest, number left behind after harvest, fluctuations in environment, and fluctuations in fecundity. It also depends on density of the population to be harvested and the rate of harvest needed to stabilize the density at that level.

This type of harvest is described by the expression:

$$(P_b) = \max B_{t+1} - B_t(X)$$

where

$P_b$ = biomass productivity from $t$ to $t + 1$
$B_t$ = biomass at time $t$
$B_{t+1}$ = biomass at time $t + 1$
$X$ = the several variables that influence biomass production over time $t$ to $t + 1$

This can be illustrated by means of some reproduction curves as shown in Figure 18-7. For

**FIGURE 18-6**
*Simulation of an exploited and an unexploited population of sardines, both subject to the random environmental variation in reproductive success. The dotted line indicates the asymptotic population size. Note how exploitation adds to instability and how dangerously low the population can get. (From Murphy, 1967.)*

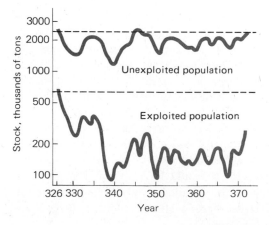

any position of the stock to the left of the 45°
line there is a rate of exploitation that will
maintain the stock at that position. Maximum
sustained yield does not necessarily require a
large standing crop. In the curves in Figure
18-7 let *a* be any position on the curve and *c* a
perpendicular line that cuts the 45° line at *b*.
At equilibrium the portion *bc* of the recruitment must be used for the maintenance of the
stock, for *bc = oc; ab* can be harvested. There
is, however, a limit to exploitation, a limit that
is influenced by the inflection point of the
curve. For curve *C,* the maximum rate of harvest is about 82 percent, for curve *A* 42 percent. In these curves the size of the reproductive stock that will give maximum sustained
yield will not be greater than half of the re-

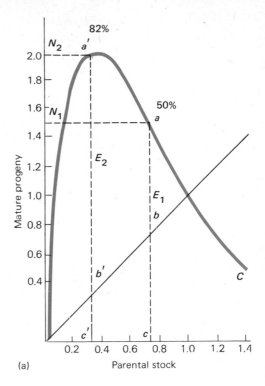

(a)

*FIGURE 18-7*
*Reproduction curves illustrating rates of*
*exploitation. Reproduction curves diagram*
*the relationship between recruitment or net*
*reproduction considered as mature progeny*
*and the density of parental stock. The 45°*
*line represents replacement level of the*
*stock, reproduction in which density*
*dependence is absent. The dome-shaped*
*curve is the plot of actual recruitment in*
*relation to the size of parental stock. The*
*apex of the curve, lying above and to the left*
*of the diagonal line, represents maximum*
*replacement reproduction. The curve must*
*cut the diagonal line at least once and*
*usually only once. Where the curve and*
*diagonal line intersect is the point at which*
*parents are producing just enough progeny*
*to replace current losses from reproductive*
*units.*
   *There are innumerable types of*
*reproduction curves. The reproduction curve*
*typical of r-strategists shows that low*
*parental stock can be very productive. On*
*curve C a perpendicular line ac cuts the 45°*
*line at b. Segment ab ($E_1$) is harvest; bc is*
*stock left for recruitment. Line ac*
*represents the point on the curve at which 50*
*percent of the mature population is*
*harvested each period. The line a'/c' is the*
*82 percent point with represents the*
*maximum surplus reproduction and*
*maximum rate of exploitation possible for*
*this population. Curve A is typical of*
*K-strategists among which a low density of*
*parental stock is not very productive. The*
*two points represent levels at which 20 and*
*42 percent, the maximum, can be harvested.*

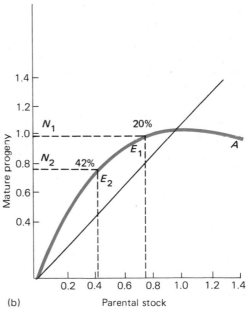

(b)

*Note the difference between curves C and A.*
*In curve c, $E_2 < E_1$ and $N_2 < N_1$. Under*
*these conditions the greater the standing*
*crop, the greater is sustained yield. In curve*
*A, where $E_2 < E_1$ yet $N_1 < N_2$, a high*
*standing crop does not result in greater*
*sustained yield. A knowledge of*
*parent-progeny relations is essential for the*
*exploitation of natural animal populations.*

**549**

placement of the reproductive population. The greater the area of the reproduction curve above the 45° line, the greater the optimum rate of reproduction.

In summary, Caughley (1976) gives six points applicable to harvesting of populations:

1. A population stable in numbers must be reduced below a steady density to obtain a croppable surplus.
2. There is an appropriate sustained yield for each density to which a population is reduced.
3. For each level of sustained yield there are two levels of density from which this sustained yield can be harvested (Figure 18-8).
4. Maximum sustained yield can be harvested at only one density (Figure 18-8).

**FIGURE 18-8**
*A reproduction diagram of an idealized population of K-strategists harvested for sustained yield under three regimes. In the first the population is harvested down from a steady state to a size $N_t = A$. The dashed line a represents the number that must be harvested (or replaced) each year to hold the population size stable at $N_t = A$. In the second regime the population is reduced to $N_t = B$ and a sustained yield represented by line b could be harvested each year to hold population stable at $N_t = B$. The yield in this case is a large proportion of a small population. Maximum sustained yield, M is obtained at the population size where the diagonal and the curve have maximum vertical separation, $N_t = M$. (From Caughley, 1976.)*

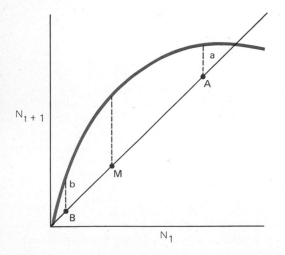

5. If a constant number is harvested from a population each year, the population will decline from steady density and stabilize at the upper population size for which that number is sustained yield. If this number exceeds the maximum sustained yield, the population declines to extinction.
6. If a constant percentage of the population is harvested each year (the percentage applying to the standing crop of that year), the population will decline and stabilize at a level at equilibrium with the rate of harvesting. This level may be above or below that generating maximum sustained yield.

One group of animals, notably fish and whales, is exploited commercially; another group is hunted for sport. Most game animals are characterized by contest type of competition and are largely, but not always, density regulated. It has been assumed by wildlife managers that regulated hunting and compensatory predation have similar effects on game animals. Just as a predator turns to another source of prey when the first source demands too great an expenditure of energy to hunt, so do sport hunters abandon the field when hunting success (animals taken per unit time) drops below a certain level. Hunting mortality supposedly replaces natural mortality. If the surplus were not harvested, the animals would succumb to disease, exposure, and the like. Thus for each individual removed by hunting natural mortality is reduced by one. It is assumed that population stability and sustained yield can be maintained if the population is harvested at the rate represented by the percentage of the young of the year or in some cases percentage of year-old animals in the population.

The weakness of this assumption is that if hunting mortality and natural mortality replace each other and if the rate of harvest applied to the population equals the rate of mortality in the absence of hunting, then the only cause of death in the post-young population is hunting. Similarly if the nonhunted population is below carrying capacity, no adult dies until $K$ is reached (Caughley, 1976). Obviously other mortality does take place among some hunted populations, and hunting and natural mortality are not compensatory. Part of the hunting mortality may be an addition to rather than a replacement of nonhunting mortality, as seems to be the case in some waterfowl populations (Geis et al., 1971).

## EXPLOITATION AND EXTINCTION

Few natural populations are really managed on a sustained yield basis, largely because enforceable regulatory mechanisms are neither available to control the harvest nor often desired by the exploiters. Exceptions are some game animal populations for which strong state and federal laws regulate the take. Many of these are underharvested rather than overharvested. Where their habitat is maintained, populations of these animals are largely controlled by nature and not man. Because of the lack of international regulations on the take, which could be handled by controlling gear, type of boats, fishing methods employed, and the like, most commercial fish populations are declining.

An example is the whaling industry (Figure 18-9). Early whalers with hand lances and harpoons sought only those whales that could be overtaken and killed from small boats. As whales became scarce along shore, whalers took to the high seas. In the sixteenth century whalers hunted off the Newfoundland coast and Iceland until stocks failed. The populations of Spitzenbergen and Davis Straight where next to go. The colorful New England whaling industry, which first exploited the stock of right whales, peaked in the first half of the nineteenth century, but as stocks of slow-moving, easily exploited species were depleted and as petroleum replaced whale oil as fuel, the New England whaling industry died. Then two developments put international whaling into business. One was the invention of the explosive harpoon in 1865 and the other was the development of more powerful, faster boats that could overtake the swifter whales. The revitalized industry began to concentrate on the plankton-feeding blue whale and its relatives the fin and the sei whales. Again stocks were overexploited. Blue whale fishery in Norway ended in 1904, followed by a decline of the species off Iceland, the Faeroes, the Shetlands, the Hebrides, and Ireland. When these areas failed, whalers sailing free-ranging factory ships turned to the Antarctic and the

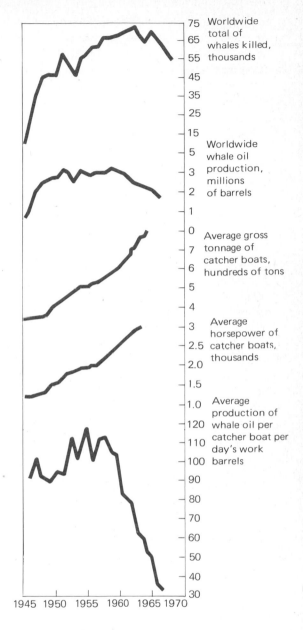

### FIGURE 18-9
*Relationship between a declining resource, whale stock, and the intensity of harvesting. Since 1945 more and more whales have been killed to produce less oil. At the same time boats have become larger and more powerful, but their efficiency has dropped greatly. (From D. C. Payne, 1968.)*

Falklands to concentrate on smaller fin whales. The catches rose and the stocks again collapsed. Antarctic whaling was over and most of the whaling nations were out of busi-

**551**

ness. Japan and the USSR developed a whaling industry and hunted the sei, sperm, and other whales wherever they could be found. Investments were made in large factory ships equipped with helicopters and accompanied by catcher boats that captured more whales but with less oil per effort and per ship.

Populations being overexploited exhibit certain easily discernible symptoms (K. E. F. Watt, 1968). Up to a certain rate of exploitation, the stock is able to replace itself. Beyond this critical point certain changes point to impending disaster. Exploiters experience decreased catch per unit effort as well as a decreasing catch of one species relative to the catch of related species. There is a decreasing proportion of females pregnant, due both to sparse populations and to a high proportion of young nonreproducing animals. The species fails to increase its numbers rapidly after harvest. A change in productivity relative to age and age-specific survival shows that the ability of the population to replace harvested individuals has been impaired. An outstanding example is the blue whale (Figure 18-10). After 1860 the blue whale became the most important commercial species. Catches peaked in 1931 at 150,000 animals and declined to 1,000 to 2,000 in 1963. For the past 40 years the average age of the blue whale caught in

the Antarctic has been 6 years, mostly immature females or females carrying their first calf. The species is near extinction.

## Cannibalism

A special kind of predation that exists within a species population is cannibalism. Cannibalism is common to a wide range of animals, both aquatic and terrestrial, from protozoans and rotifers through centipedes, mites, and insects to frogs and toads, fish, birds, and mammals, including humans. Interestingly about 50 percent, mostly insects, of terrestrial cannibals are normally herbivorous species. In fresh-water systems the bulk of cannibalistic species are predaceous, as they all are in marine ecosystems (Fox, 1975).

Cannibalism has been associated with stressed populations, particularly those facing starvation. While some animals do not become cannibalistic until other food runs out, others will do so when the relative availability of alternative foods declines and individuals in the population are disadvantaged nutritionally (Alm, 1952). This situation occurs among fish living in nutritionally poor water and among past human populations that lived in areas highly deficient in sources of animal protein, for example, New Guinea tribes (Dornstreich and Norren, 1974) and Aztec Indians (Harner, 1977).

Other conditions which may promote cannibalism are: (1) crowded conditions, even when food may be adequate, as in house mice and certain fish; (2) stress, especially when induced by low social rank, as in rats; (3) presence of vulnerable individuals such as nestlings and eggs, even though food resources are adequate, as in the carrion crow (Fox, 1975). Cannibalism of the crow can account for up to 75 percent of mortality of eggs and young (Yom-Tov, 1974). Whatever the cause, intensity of cannibalism is influenced by local conditions and the nature of local populations. In general cannibalism fluctuates greatly over both long and short periods of time. For example, among predaceous fishes such as walleye *(Stizostedium vitreum)* and yellow bass *(Roccis mississiensis)* cannibalism is most prevalent in summer (Fortney, 1974; Kraus, 1963).

*FIGURE 18-10*
*Rates of harvest of the blue and sperm whales. Since it has been intensively hunted, the great blue whale has slipped toward extinction. With the decline of the blue whale pressure has been exerted on the sperm whale. (Adapted from FAO Fisheries Tech. Paper No. 97.)*

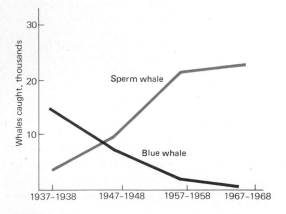

Demographic consequences of cannibalism depend upon the age structure of the population and the feeding rates of the various age classes. Even at very low rates cannibalism can produce demographic effects. For example, L. Cren (cited by Fox, 1975) has estimated that if small pike *(Esox lucius)* make up only 1 percent of the food of larger pike, cannibalism could account for all of the mortality of the young. Three percent cannibalism in the diet of adult walleye pike could account for 88 percent of the mortality among young (Chevalier, 1973). Fortney (1974) found that when walleyes sharply reduced their prey, yellow perch *(Perca flavescens),* the larger fish turned cannibalistic, reducing their own population. In a study of populations of the fresh-water backswimmer *(Notonecta hoffmanni)* isolated in small pools of water, Fox (1975) discovered that the insects ate each other over most of the growing season. Heaviest cannibalism came in early summer coinciding with a sudden decrease in food and a reduction of spatial refugees for the young. Mortality of young nymphs determined population sizes and age structures for the rest of the summer. The first nymphs to hatch were the most likely to survive and reproduce successfully.

Cannibalism in these instances became a mechanism of population control that rapidly decreases the number of intraspecific competitors as food became scarce. More young are produced by a few well-fed adults than by several malnourished individuals. Cannibalism is unlikely to bring about local population extinctions because it is short-term; it decreases as resources become relatively more available to survivors and as vulnerable individuals are scarcer or harder to find.

Cannibalism can be a selective advantage to survivors. Survivors gain a meal, eliminate a potential competitor for prey, and eliminate a conspecific (of the same species) predator. With the population reduced the survivor has more food, enhancing its chances for further survival, rapid growth, and greater reproductive success. Cannibalism can be a selective disadvantage if individual survivors become too aggressive and destroy their own progeny or genotype completely, reduce their genotype faster than those of conspecific competitors, or reduce their chances of successful reproduction by elimination of suitable mates. However, selection can balance advantages against disadvantages. In some situations the disadvantages of cannibalism are less severe than starvation and reproductive failure caused by inadequate nutrition. If a population is reduced by starvation, the survivors may be nutritionally stressed. If individuals are removed by cannibalism, density is reduced early and the per capita food supply remains high. Survivors have improved their fitness because, well fed as juveniles, they grow faster, survive better, and produce more young.

## Predation, competition, and community structure

When myxomatosis greatly reduced the rabbit population in England, it also released grasses and associated vegetation from grazing pressure (see page 558). The response of the vegetation, the rapid change from high diversity maintained by rabbit grazing to domination by a few grasses, suggested that predation coupled with interspecific competition may be a major influence on community structure. Further evidence on the role of herbivores in influencing the structure of grassland communities came from studies of controlled grazing of artificial and natural stands of grass. With sheep the primary grazing animal, rye grass and white clover varied in dominance as grazing pressures and fertility changed (Jones, 1933).

An example of the influence of long-term grazing on grassland is provided by a comparison of two grasslands near Oxford in the Thames Valley of England. One area has been grazed by cattle, sheep, and horses for 900 years. An adjacent area is cut regularly for hay late in summer. The grazed area is dominated by perennial grasses, laterally spreading dicots and some rosette-forming species without stolens or rhizomes. The hayland is also dominated by perennial grasses, but it has a high proportion of annual grasses and short-lived perennials. Its dicotyledonous flora are mostly plants with erect shoots and leaves on the stems. In the grazed land tall plants are selected against and the successful plants are ones that have the meristem at or below ground, extensive clonal growth, and leaf area close to the ground as protection against grazing. In the hayland tall plants have the selective advantage as well as those that flower early, while rosette and creeping plants are selectively disadvantaged.

Somewhat similar influences of herbivores on plant species have been discovered in fresh-water algal communities (Porter, 1977). Fresh-water phytoplankton includes a diverse association of blue-green and green algae, diatoms, dinoflagellates, and others. Feeding on phytoplankton is a group of grazing zooplankton dominated by cladocerans, mostly *Daphnia,* and cyclopoid and calanoid copepods. These grazers exhibit two types of feeding behavior, filter feeding and raptorial or grasping feeding. They discriminate among particles on the basis of size, shape, texture, and possibly on taste and abundance. *Daphnia* and a few copepods are filter feeders. They concentrate on small algae to an upper limit dictated by the size of the grazing zooplankton. Filter feeders reject large algae and those with rigid cell walls and spines. Grasping feeders include most copepods and some cladocerans. Ineffective at retaining small particles, these grazers feed on large particles which they can grasp.

Zooplankton grazing rates vary seasonally; they are lowest in winter and early spring and highest in summer. The impact grazing zooplankton have on the phytoplankton community varies with the season and with the type of phytoplankton. In general grazed algal species may decline, show no response, or increase. Algae that decline when the number of grazers is increased experimentally are small, naked green algae, nannoflagellates, and certain diatoms, all easily filtered, broken up, and digested. When zooplankton grazers are excluded, these phytoplankton species increase. Types that show no response to manipulations of grazing pressure are large unicellular desmids and dinoflagellates, filamentous diatoms, and colonial blue-green algae which apparently are rejected by zooplankton. Types of phytoplankton that increase are gelatinous green algae whose durable cell walls and gelatinous sheaths protect them from damage while passing through the gut of grazers. These algae pick up carbon and phosphorus compounds from digested material in the gut

which stimulate their increase after passing through the gut. Evidence, then, suggests that under certain conditions grazing zooplankton may control species composition of phytoplankton. Certain algae will be abundant in the presence of grazers, while others will be reduced in number (Porter, 1977).

Results from such studies can be reduced to some basic principles on the influence of grazers on plants (Harper, 1969). If the herbivore selectively grazes the major dominant, then plant diversity increases. Overgrazing in winter and spring followed by undergrazing in summer produces maximum floral richness. (This is usually the case in natural situations because deer, elk, and others can overutilize their food in winter, but undergraze it in summer.) The most species-rich communities are developed by continuous grazing with a maintained population of herbivores. If the dominant plant is highly palatable, overgrazing will reduce it and allow other, less palatable species to occupy the area. But if the dominant species are unpalatable, then grazing serves only to consolidate their dominance. If the herbivore is regulated by its food supply and not by predators, then it can reduce a plant species to a minor component and allow invasion by other species. As the intensity of grazing increases, more and more unpalatable species are grazed and eventually only wholly unpalatable species remain or move into the area. Thus the complex balance of species in a plant community is sensitive to rather precise control by the feeding activities of grazing animals.

By reducing certain prey species (plant or animal), predators may also alter community structure by ameliorating the intensity of competition for space and increase species diversity. The alteration of community structure was demonstrated by the experimental removal of a top carnivore, the starfish *Pisaster,* from an intertidal community (Paine, 1966, 1974). Under normal conditions, including predation by *Pisaster,* the mussel *Mytilus californianus,* the barnacle *Balanus cariosus,* and a goosenecked barnacle *Mitella polymerus* form a conspicuous band in the mid-intertidal zone. In areas where the predator *Pisaster* was removed *Balanus* occupied 60 to 80 percent of the available space. But within a year

*Balanus* was being crowded out by the rapidly growing *Mytilus* and *Mitella* and the benthic algae, chitons, and larger limpets disappeared from the lack of space. The removal of *Pisaster* caused a pronounced decrease in species diversity and an increase in the intensity of interspecific competition. Predation in this case was effective in preventing one or two species from monopolizing a given resource.

Predation can also influence community structure by selectively eliminating larger organisms, allowing smaller organisms to increase. This influence was found to be quite strong in experimental fresh-water systems in which fish were introduced (Brooks and Dawson, 1965; Hall, Cooper, and Werner, 1970). The change in relative abundance of zooplankton species may have come about for two possible reasons: (1) fish eliminated the larger zooplankton, allowing smaller species to utilize a limited resource previously unavailable to them; (2) fish removed not only the larger competing zooplankton, but also the predators of smaller species (Connell, 1975).

The impact of predation may be most pronounced when an exotic predator is introduced into an ecosystem. The peacock bass *(Cichla ocellatus),* a native of the Amazon, escaped from impoundments and entered Gatun Lake in the Panama Canal Zone. The peacock bass, a popular sport and food fish and a voracious predator, has profoundly affected community structure (Figure 18-11) and has had a devastating effect on the fish population (Table 18-1) (Zaret and Paine, 1973).

Predation by the peacock bass has seriously affected the key second-order consumer *Melaniris chagresi,* the lower trophic level on which it feeds, and the higher levels that feed on it. The peacock bass feeds mostly on older *Melaniris,* decreasing mature individuals and the breeding stock. Young *Melaniris* feed almost exclusively on zooplankton throughout the year. The predatory pressure they exert determines the composition and distribution of zooplankton. For example, one of the *Melaniris* prey species is *Ceriodaphnia cornata,* a species with morphologically distinct forms, horned and unhorned. Unhorned *Ceriodaphnia* have a reproductive advantage over the horned form and dominate lake areas where predation by *Melaniris* is absent. However, because *Mela-*

*TABLE 18-1*
**Change in fish species on Barro Colorado Island following appearance of Cichla**

| Family | Species | Percent change | |
|--------|---------|----------------|----------|
| | | Increase | Decrease |
| Atherinidae | *Melaniris chagresi* | | 50 |
| Characinidae | *Astyanax ruberrimus* | | 100 |
| | *Roeboides guatemalensis* | | 90 |
| Cichlidae | *Aequidens coeruleopunctatus* | | 100 |
| | *Cichla ocellaris* | 100 | |
| | *Cichlasoma maculicauda* | 50 | |
| Eleotridae | *Gobiomorus dormitor* | | 90 |
| Poeciliidae | *Gambusia nicaraguagensis* | | 100 |
| | *Poecilia mexicana* | | 100 |

Source: Zaret and Paine, 1973

*niris* prefers the unhorned forms, the horned forms predominate in areas where predation by *Melaniris* is present. With the decline of *Melaniris* due to peacock bass predation the two forms occur in equal numbers. Tertiary consumers that feed on *Melaniris,* such as Atlantic tarpon, black tern, and herons, have sharply declined. A complex community structure has been highly simplified. Six or eight common fish members of the community have been eliminated or seriously reduced, all by the introduction of a single top-level predator in the lake community.

Such studies suggest that predation and competition may be complementary selective pressures (Menge and Sunderland, 1976). When competing species are at carrying capacity or equilibrium, interspecific competition may be the major influence on community structure. But if predation is intensive enough it can reduce each of the competing species below its carrying capacity. In so doing predation would have the greatest influence on community structure by freeing certain prey

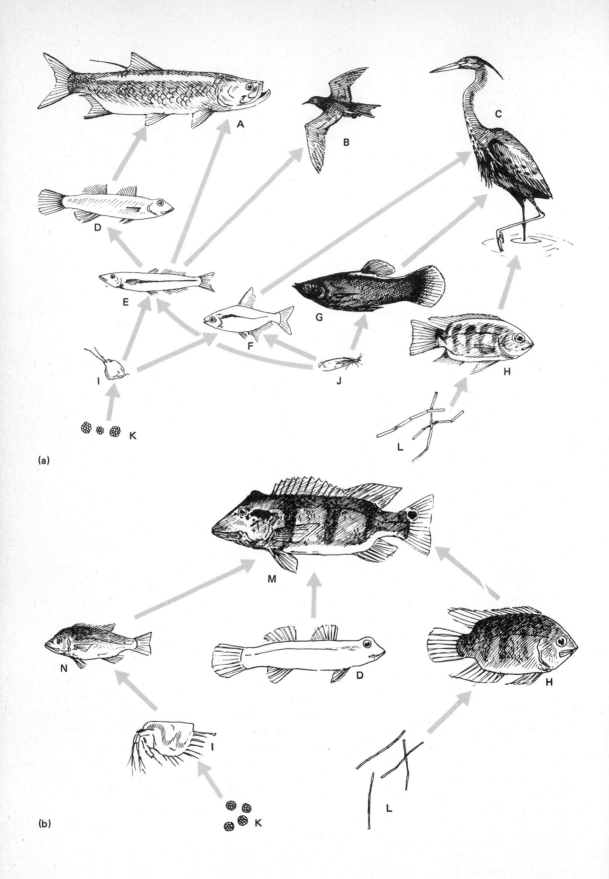

(a)

(b)

species from competition and increasing diversity. Predation as an influence on community structure is probably most important in ecosystems with complex food webs, while interspecific competition is most important in ecosystems with simple food webs. Competition is probably most intense among those species occupying higher trophic levels, while predation as a selective force is most important among species occupying lower trophic levels.

## SUMMARY

Interaction of predators and their prey is considered a predator-prey system. Predators influence growth and survival of prey populations and density of prey populations influence growth and survival of predator populations. Predator-prey interactions at one trophic level influence predator-prey interactions at the next trophic level. Involved are plant-herbivore systems and herbivore-carnivore systems.

Plant predation involves defoliation by grazers and consumption of seeds and fruits. Interactions between changes in plant biomass and herbivores result in oscillations of plant and herbivore populations or in equilibrium, the vegetation with the grazing pressure and herbivores with food supply. Herbivore-carnivore systems show similar interactions in nature: stable oscillations in which prey populations are self-limiting or stable limit cycles in which the rate of growth of predator and prey populations is approximately the same and the prey exhibits weak intraspecific population regulation. Stability may be the greatest in a three way plant-herbivore-carnivore system.

An artificial form of predator-prey interaction is commercial and sport hunting of animal populations by humans. Some attempt is made at sustained yield in which the yield per unit time is equal to production per unit time. Such an approach to management of exploited populations necessarily differs between $K$-selected and $r$-selected species. Too often commercial species are overexploited, resulting in a serious population decline that approaches extinction.

An unusual form of predation is cannibalism, a type that exists within a species. Often associated with stunted populations, cannibalism can result in pronounced demographic effects within a population including loss of younger age classes and lower reproduction.

Interaction of predation and interspecific competition can influence community structure and species diversity. Predation can so reduce populations of interspecific competitors that a number of different competitors can coexist. Interspecific competition unmodified by predation can result in dominance by a few species and a reduction in species diversity.

# Interaction between species: coevolution
## CHAPTER 19

Interdependent interaction between species within a given system—predator and prey, herbivore and plant, parasite and host, pollinator and plant, interspecific competitors—comes about through long periods of coevolution. *Coevolution* is the joint evolution of two interacting populations (taxa) in which selection pressures are reciprocal. Any evolutionary change in one component immediately changes selection forces acting on the other. Coevolution requires continuous genetic tracking by each component of genetic changes in the other. It is basically a game of adaptation and counteradaptation to changing selection pressures imposed on one taxon by the other.

The parasite-host interaction between the European rabbit and myxomatosis is an example of coevolution. To control the intro-

duced European rabbit, the Australian government introduced myxomatosis in the population. The first epizootic of the disease was fatal to between 97 and 99 percent of the rabbits; the second resulted in a mortality of 85 to 95 percent; the third, 40 to 60 percent (Fenner, 1953). The effect on the rabbit population was less severe with each succeeding epizootic, suggesting that the two populations were becoming integrated and adjusted to one another. In this adjustment attenuated genetic strains of virus, evolved by mutation, tended to replace virulent strains (Thompson, 1954). Also involved was passive immunity to myxomatosis, conferred to the young born of immune does. Finally, a genetic strain occurred in the rabbit population providing an intrinsic resistance to the disease.

The transmission of myxomatosis virus is dependent upon *Aedes* and *Anopheles* mosquitoes, which feed only on living animals. Rabbits infected with the virulent strain live for a shorter period than those infected with a less virulent strain. Because the latter live for a longer period, the mosquitoes have access to that virus for a longer time. This gives the nonvirulent strain a competitive advantage over the virulent. In those regions where the nonvirulent strains have a competitive advantage, the rabbits are more abundant because fewer die. This means that more total virus is present in those regions than in comparable areas where the virulent strains exist. Thus the virus with the greatest rate of increase and density within the rabbit population is not the one with the selective advantage. Instead the virus whose demands are balanced against supply has the greatest survival value in the ecosystem.

## Coevolution of predator-prey systems

The interaction between predator and prey involves a continual flux of adaptive genetic changes. As predators evolve more efficient ways of capturing prey, the prey evolves ways of avoiding or counteracting predation. If predators (parasites or pathogens) exert selective pressures for more effective defense in prey, then genetic changes will feed back as a selective force on the predator for a more effective means of exploiting prey.

### WARNING COLORATION, MIMICRY, AND CRYPTIC COLORATION

One approach to defense is the development of escape strategies to avoid predation, that is, strategies that make the predator's task of finding prey items more difficult. Certain color patterns and associated behaviors evolved by prey function to warn or confuse the predator or to hide the prey from the predator.

Animals that possess pronounced toxicity and other chemical defenses often possess warning coloration, bold colors with patterns that serve as a warning to would-be predators. The black and white stripes of the skunk, the bright orange of the monarch butterfly, and the yellow and black coloration of many bees

and wasps serve notice to their predators, all of whom must have some unpleasant experience with the animal before they learn to associate the color pattern with unpalatability or pain.

Together with these animals, other associated edible species have evolved a similar mimetic or false warning coloration. This phenomenon was described some 100 years ago by the English naturalist H. W. Bates in his observations on tropical American butterflies. The type of mimicry that he described, now called *Batesian,* is the resemblance of an edible species, the mimic, to an inedible one, the model. Once the predator has learned to avoid the model, it avoids the mimic also. An example among North American butterflies is the mimicry of the palatable viceroy butterfly to the monarch, definitely distasteful to birds (Brower, 1958). Both the model and the mimic are orange in ground color with white and black markings; they are remarkably alike. Yet the viceroy's nonmimetic relatives are largely blue-black in color.

Another group of models and mimics involve the inedible pipevine swallowtail and the palatable spicebush and black swallowtails (Figure 19-1). Tests involving the Florida scrub jay as the predator showed that this bird could not distinguish or else confused the color pattern of the mimic with that of the model (Brower, 1958).

Bumblebees and honeybees have their mimics among the flies. Of these perhaps the most interesting is the model bumblebee and the mimic robber fly *(Mallophora bomboides).* Not only does the robber fly benefit from reduced predation, but it also exploits the model for food. The robber fly preys on Hymenoptera by preference, and its resemblance to its prey allows it to escape notice of the bee until it is too late for the bee to defend itself or flee (Brower and Brower, 1962). This type of mimicry is called *aggressive.*

In another, less common type of mimicry, called *Mullerian* mimicry, both the model and the mimic are unpalatable. The pooling of numbers among the model and mimic reduces the losses of each, because the predator associates distastefulness with the pattern without having to handle both species. Mullerian mimicry differs from Batesian in that the feedback from handling either species is nega-

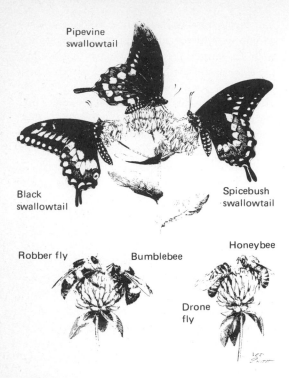

**FIGURE 19-1**
*Mimicry in insects. The model, the distasteful pipevine swallowtail, has as its mimics the black swallowtail and the spicebush swallowtail. The black female tiger swallowtail, not shown, is a third mimic. All these butterflies are found together in the same habitat. The robber fly illustrates aggressive mimicry. It is a mimic of the bumblebee on which it preys. The drone fly is a mimic of the honeybee.*

tive. Also, Batesian mimics and models are not of the same phylogenetic line, while Mullerian mimics include members of the same genus and family, probably because the ability to utilize or store poisonous substances from plants has become fixed in an evolutionary line.

Mimicry is usually considered an evolutionary response in animals, but animals in search of food may have stimulated mimicry in the plant kingdom also. Many examples undoubtedly go undetected. Gilbert (1975) in his study of the passionflower butterfly *(Heliconius)* and its food plant, the passionflower *(Passiflora),* found evidence of plant mimicry. *Passiflora,* a tropical vine of the New World comprising around 350 species, has a wide range of intra- and interspecific leaf and stipule shape. The number of *Passiflora* species found in any one

area is about 2 to 5 percent of the 350 species. *Passiflora* species are used as an egg laying site and a source of larval food by some 45 species of highly host specific species of *Heliconius,* each of which utilizes a limited group of the plants. The visually sophisticated butterflies learn the position of the vines and return to them on repeated visits. Within a habitat the leaf shapes of passionflowers vary among species. Under visual selection of butterflies passionflowers apparently evolved leaf forms that make them more difficult to locate. Because larval food niche is broader than that of the ovipositing females, there has been selective pressure for divergence among *Passiflora* species (Figure 19-2). Probably as a result of these selection pressures, *Passiflora* leaf shapes converge on those of associated tropical plants that *Heliconius* finds inedible. So close are the convergences that plant taxonomists have named some *Passiflora* species after the genus they resemble.

In addition two *Passiflora* species, *P. cyanea* and *P. auriculata,* have evolved glandular outgrowths on the stipules that mimic the size, shape, and golden color of *Heliconius* eggs at the point of hatching. Because *Heliconius* females detect and reject shoots that carry eggs and young, *Passiflora* achieves a measure of protection by egg mimicry.

Another evolutionary outcome of predation is cryptic coloration. Cryptic coloration involves patterns, shapes, postures, and movements that tend to hide the organism from its predators. Some animals are protectively colored so they blend into the background of their normal environment. Such protective coloration is common among fish, reptiles, particularly the lizard, and many ground nesting birds, such as the woodcock. Countershading, or obliterative coloration, in which the lower part of the body is light and the upper part is dark, reduces the contrast between the unshaded and shaded areas of the animal in bright sunlight. Object resemblance is common among insects. For example, the walking stick resembles a twig and some insects resemble leaves. Some animals possess eyespot markings which intimidate potential predators, attract their attention away from the animal, or delude them into attacking a nonvulnerable part of the body. Eyespot patterns in

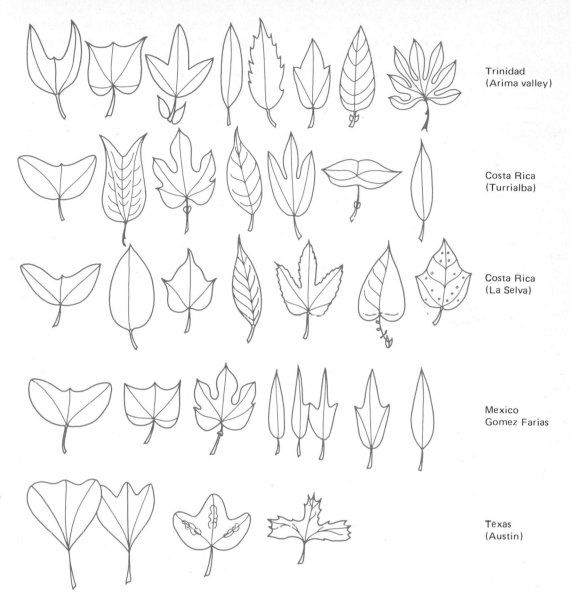

Trinidad
(Arima valley)

Costa Rica
(Turrialba)

Costa Rica
(La Selva)

Mexico
Gomez Farias

Texas
(Austin)

**FIGURE 19-2**
*Variation in leaf shape among groups of sympatric species of* Passiflora. *The leaf shapes tend to converge onto other common, but to Heliconicus butterflies inedible, species of a number of genera. (After Gilbert, 1975.)*

Lepidoptera seem to intimidate predators by imitating the eyes of a large avian predators that attack small insectivorous passerine birds.

Associated with cryptic coloration is flashing coloration. Some cryptic animals, particularly certain butterflies, grasshoppers, and birds, display extremely visible color patches when disturbed and put to flight. The flashing coloration distracts and disorients predators. When the animal comes to rest the bright colors vanish and the animal disappears into its surroundings.

**CHEMICAL DEFENSE**

Chemical defenses against predators are widespread among both animals and plants. They serve to warn potential prey and to repel or inhibit would-be attackers.

Some species of fish release pheromones from the skin into the water which act as alarm substances and induce a fright reaction in other members of the same or related species (Pfeiffer, 1962). The pheromone is produced in specialized epidermal cells which do not open to the surface so that it is released only when the skin is broken. Fish in the vicinity receive the stimulus through the olfactory organs. Only minute quantities are needed to drive a school of minnows from a feeding area (0.002 mg in a 14-liter aquarium) and the reaction may be transferred visually to individuals not exposed to the substance. The response to the pheromone is innate. Regardless of prior experience, at a certain age sometime after the alarm substance is produced in the skin, fish exhibit this reaction. The reaction affords a twofold protection. Not only does it shield fish from general predation, but also it protects the young from cannibalistic attacks of adults, for the would-be cannibals are frightened away by the alarm substance released from the skin of the young they attack. This alarm substance is most common among fish that are social, lack defensive structures such as spines, and are nonpredaceous.

Secretions are commonly employed by arthropods, amphibians, and snakes to repel predators. Strongly odorous, easily detected defensive substances are produced in often copious quantities by arthropods (Eisner and Meinwald, 1966; Eisner, 1970). The secretions are produced by glands containing large saclike reservoirs that are essentially infoldings of the body wall and are discharged through small openings. They may ooze onto the animal's body surface, as in millipedes, be aired by evagination of the gland, as in beetles, or be sprayed out for distances up to several feet, as in grasshoppers, earwigs, and stink bugs. The secretions effectively repel predators (birds, mammals, and insects alike) by their effect on sensitive areas of the predators' face and mouth. They are discharged during early phases of an attack to repel the assailant before bodily damage occurs.

Active components in the defensive secretions of many arthropods occur as toxic secondary substances, such as saponins, gossypol, and cyanogenic glycosides, used as chemical defense by plants. Such toxins inhibit herbivores from feeding on the plants. However, herbivorous arthropods are capable of incorporating toxic substances ingested from the plants into their own tissues. This in turn protects the herbivore from its enemies. The monarch butterfly, for example, feeds on milkweeds that contain a cardiac glycoside, a substance which causes illness in birds that eat the monarch.

Chemical defense involving secondary metabolic products is a first line of defense by plants against herbivores (see Levin 1976). The basis of chemical defense is the accumulation of secondary products ranging from alkaloids to terpenes, phenolics, and steroidal, cyanogenic, and mustard oil glycosides. Phenolics, a byproduct of amino acid metabolism, are ubiquitous to seed plants. Alkaloids, also amino acid derivatives, occur in several thousand species, and cyanogenic glycosides in a few hundred species. The secondary products may be stored within cells and released only when cells are broken or they may be stored and secreted by epidermal glands to function as a contact poison or a volatile inhibitor.

Production and storage of such metabolites are energy intensive processes that represent a trade-off between defense and reproductive effort. Over evolutionary time the processes involve responses between plant availability as prey or host and the presence of herbivores.

Chemical resistance to attack falls into two general types (Levin, 1976). One involves accumulations and changes in metabolites of the host plant that act as toxins at the wound site. This response is commonly used to resist attacks from bacteria, fungi, and nematodes. The other type of resistance is based on the presence of inhibitors prior to attack. This approach is commonly employed against animals feeding on plants as well as against fungi.

Inhibitors may function as warning odors, repellents, attractants, or in some cases direct poisons. Volatile components advertise substances that insects and other herbivores would find repellent if they touched the plant. Bitter tastes imparted by tannins and cardiac glycocides can deter further consumption of both seeds (Janzen, 1971) and foliage. Metabolites such as phenolic terpenes and saponins may be toxic and cause illness and occasionally death. Such repellents not only inhibit feeding on the plant possessing them, but also add a measure of protection to associated plants. For example, grazing by cattle on bentgrasses *(Agrostris)* and fesque *(Festuca)* is considerably reduced in the presence of buttercup *(Ranuculus bulbosus)* which contains a powerful irritant of skin and mucous membranes (Phillips and Pfeiffer, 1958). The presence of such plants can cause the predator to fail to locate its normal prey or to reject it along with the repellent plant (see Atsatt and O'Dowd, 1976).

Plants containing secondary metabolites may also function as attractant-decoys which cause the herbivore to feed on an alternate prey. Many attractant plants are not what they advertise and function as decoys and cause mortality or reduced fecundity because of the presence of toxins or deficiency in certain nutritive materials (Atsatt and O'Dowd, 1976). Coexisting toxic and nontoxic plants with similar attractant chemistry present a problem for host-specific herbivores. Some insects, for example, may be stimulated to lay eggs on a "wrong" but closely related plant which eventually results in larval death.

Although plants may possess powerful chemical defenses that work well against generalist herbivores, they can be breached by specialists. Many, if not most, toxic plants have specialist herbivores that can exploit their foliage or seeds. These specialists in some way absorb or metabotically detoxify foreign substances. As we have mentioned, some herbivores are able to store plant poisons and use them in their own defense or in the production of pheromones. Detoxification may be achieved by oxidation and conjugation reactions that yield inactive excretable products or by the action of gastrointestinal flora that degrade secondary metabolic products.

A general assumption is that plants respond to herbivore adaptations because their gene pools are diverse enough to counter any new herbivore breakthrough. Another view is that diversity of plant prey in the environment may be just as important (Atsatt and O'Dowd, 1976). Selection may favor individuals living in environments providing acceptable options for the herbivore. If the feeding environment provides little or no choice for the herbivore, selection favors specialization, forcing physiological adaptations of the herbivore to the chemical defense of the plant. But if the plant community is diverse enough, evolutionary tracking or specialization can be disrupted. By allowing the herbivore to switch between changing proportions of resistant and nonresistant hosts, such diversity, including intrapopulation variability or chemical defense, can impede selection for herbivores capable of detoxifying plant metabolites. If one variant becomes most common, the herbivore becomes adapted to it. As a result that variant becomes most susceptible to predation while other variants become less so. As the susceptible variant declines, the herbivore is forced to adapt to other variants. Situations that discourage herbivore specialization allow plants to maintain some stability in their populations and to conserve genetic variation for longer periods of time.

## PREDATOR SATIATION

A more subtle defense is the physiological mechanism of timing reproduction so that a maximum number of offspring are produced within one short period, thus satiating the predator and allowing a percentage of the reproduction to escape. For example, Schaller (1972) suggests that peak production of young wildebeest, highly susceptible to predation, is an antipredator mechanism. The young are so abundant at one short period of time that all predators become satiated. The remaining young quickly grow beyond a size handled by predators of the young. No young are available during the rest of the year.

Such predator satiation is a major strategy against predation in plants and is most prevalent in those species lacking strong chemical defenses. It involves four approaches (Janzen, 1971). The first is to distribute seeds so that all of a seed crop is not equally available to all members of a seed predator population. Seeds of most trees are concentrated near the parent and the number of seeds declines rapidly as the distance from the tree increases. The predators are attracted to the parent and many of the scattered seeds are missed by searching predators, in part because of searching image and unprofitability (see Chapter 18). These survivors produce most of the recruitment (Figure 19-3). A second approach is to shorten the time of seed availability. If all seed matures and is available at one time, seed predators will be unable to utilize the entire crop before germination. A third approach is the evolution

**563**

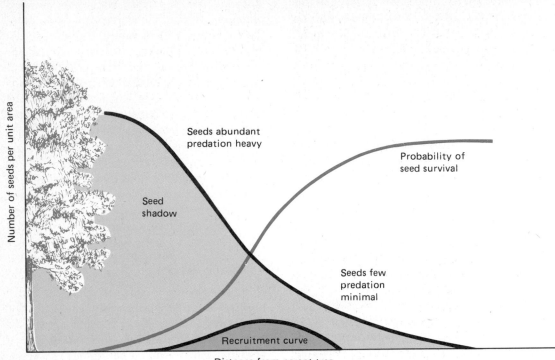

*Distance from parent tree*

**FIGURE 19-3**
*Model of seed dispersal in relation to distance from parent plant. All seeds close to the tree are eaten by seed predators. As distance from the parent plant increases the probablity of seed survival increases because density of both seeds and predators decreases. Although the number of seeds decreases rapidly with distance, recruitment or seed survival is higher well away from the parent plant. (After Janzen, 1970.)*

of a protective covering or mild toxicity so that seeds can be exploited only by predators that are partially to totally host specific. This reduces the number of different predators that can use the seed as food. A fourth approach is to produce a seed crop periodically rather than annually. The longer the time between seed crops, the less opportunity dependent seed predators have to maintain a large population between crops. Seed predators often experience local increases in density after good seed years, but decline rapidly when the food supply is depleted. This reduces the number of predators available to exploit the next seed crop.

The production of a periodic seed crop depends upon synchronization of seed production among individuals. This synchrony is usually achieved by weather events, such as late frosts or protracted dry spells. As individuals of a tree species in a community become synchronized, strong selection pressures build up against nonsynchronizing individuals, because they experience heavy seed predation between peak seed years. Such individuals either drop out of the community over evolutionary time or become synchronized.

Predator satiation may be further assured if during any fruiting season the timing of the seed crop of one species is influenced by the presence of seed crops of others. If two or more species synchronize seed production and share seed predators, those predators may be attracted away from one species to another, reducing predatory pressure on both species.

Of course, if the nature of predation or predators changes, synchronization or clumping of reproduction would be disadvantageous and natural selection would favor nonsynchrony. An example of an adaptive change in a prey species to meet a new type of predation is the caribou on the island of Newfoundland, effectively isolated from populations on the mainland of Canada. On Newfoundland in the past and on mainland Canada the principal predator of the caribou is the wolf. In response to very efficient predation by the wolf, the caribou evolved ways of keeping ahead of predation. These mechanisms include synchronized calving in which all are born in a couple of weeks, postparturient aggregation of does, and a strong perception of motion as a defense against an open approach by wolves (Bergerud, 1971).

The recent history of the caribou in Newfoundland differs from that of the caribou on the mainland. Predation by wolves, accompanied by heavy hunting pressure from both Indians and white settlers, reduced the caribou in the mid-1800s to 100 to 150 individuals. As the wolves were killed off and the Indians disappeared, the caribou reached peak numbers between 1895 and 1900. After the snowshoe hare was introduced to the island, it increased rapidly. With its increase came the rapid increase in the population of its predator, the lynx, an animal which had been quite rare in the 1800s. After reaching a peak population in 1896, snowshoe hares declined, leaving the abundant lynx without an adequate source of food. The lynx turned to caribou calves.

The caribou lacked any defense against a predator that travels by night and ambushes its prey. The caribou's habit of congregating during the calving season and keeping its young near cover is a good defense against the wolf, but a poor defense against the lynx. In recent years it appears that natural selection in the Newfoundland caribou is in the direction of defensive mechanisms against lynx predation. The caribou, especially on the Avalon Peninsula, exhibit three behavioral traits not common to caribou elsewhere: the caribou have no specific calving grounds, the females do not congregate after calving, and calving is spread over three weeks instead of two. Because the lynx, even though preying on caribou calves, still depends upon the hares, it cannot depart too far from those adaptations effective in catching them. Thus the constraints on the lynx are greater than those on the caribou. This change in behavior is reflected in a much improved rate of increase, 12 percent for the Avalon herd compared to 3 percent elsewhere (Bergerud, 1971).

## SEED DISPERSAL SYSTEMS

While some plants evolved defenses against plant predators, others evolved ways to exploit predators for seed dispersal. For such plants, dispersal agents are a resource that could be in short supply. A shortage would be reflected in the seeds produced being dispersed at something less than optimal rates. Selection pressure would favor some alternative to improve seed dispersal. One alternative is to produce an abundance of small light seeds that could be dispersed by some other mechanism such as wind, water, or passive transport by animals. By evolving small light seeds with "wings" or "parachutes" that allow seeds to be carried great distances by the wind, or bouyant seeds that can be carried by water currents, and by evolving devices on seeds such as bristles, barbs, and hooks that enable seeds to cling to fur of mammals and feathers of birds, plants escape competition for dispersal agents.

Another alternative is to increase predatory activity on fruit so that animals may serve as efficient agents of dispersal. To accomplish this, plants must evolve fruits or seeds that attract predators and at the same time discourage predispersal predation. Predispersal predation can be discouraged by cryptic coloration such as green unripened fruit among green leaves, by an unpalatable texture, repellent substances, and hard outer coats. When the seeds mature, plants can attract fruit-eating animals or *frugivores* by acquiring attractive odors, altering the texture of fruits and seed covering, improving the succulence, and acquiring a high content of sugar and oils. At the same time the plant must evolve some mechanism to protect seeds from mechanical and chemical action of the digestive tract.

Selection may favor one of two alternatives in response to competition for dispersal agents. Plants may evolve fruits that can be exploited by a large number of dispersal agents, opting for quantity dispersal, the scattering of large numbers of seeds at a high sustained rate to sites of variable quality. Or plants may evolve fruits attractive to a small number of specialized frugivores, favoring quality dispersal with a greater probability that fewer, larger seeds will be dispersed to a favorable site.

**565**

If a plant follows the first alternative of reducing competition for dispersers, it relieves the shortage of dispersers but at a cost of quality dispersal. Most animals feeding on such fruits are opportunistic, utilizing the fruits when available but switching to other food sources as insects when they appear in abundance.

Because opportunistic fruit-eaters do not depend on fruit for their basic sustenance, there is no need for the plants to provide a balanced diet. Their fruits are usually succulent and rich in sugars and organic acids, and contain small seeds that pass through the digestive tract unharmed. To accomplish such a passage, plants must evolve seeds with hard coats resistant to digestive juices. Seeds of such plants often will not germinate unless they have been conditioned or scarified by passage through the digestive tract. Large numbers of small seeds may be so dispersed, but relatively few are deposited on suitable sites.

The second alternative, to evolve fruit adapted to a relatively small group of specialized fruit eaters, sacrifices rate of dispersal for quality of dispersal. This involves a higher cost per seed because more energy must be invested in the production of nutritional fruit and a large seed containing high energy reserves for germinating seedlings.

If the plant depends on a small group of highly specialized frugivores, then the animals would have to depend upon the fruit crop for their sustenance to be effective dispersers of the plant. Rarely would they turn to an alternate source of food such as insects. The plant or combination of plants would have to provide a balanced diet, requiring expensively produced fruits rich in proteins and oils. This dependence on fruits by birds results in repeated visits to the plant insuring that seeds will be removed quickly after maturation and reducing chances that seed will be left to rot on the tree.

At the same time specialized frugivores must evolve efficient means of processing fruits. Because little mechanical breakdown is necessary for digestion of fruit, specialized frugivores are characterized by little-muscularized, thin-walled stomachs, and by a mechanism for regurgitation of large seeds before they pass into the intestinal tract.

Such nutritional-dispersal relationships between plants and fruiteaters require that the plant evolve a close relationship with frugivores. Such coevolution comes about if the plant becomes an important part in the diet of the frugivore. In the frugivore selection is for an efficient means of getting rid of seeds with minimum energetic costs before they reach the intestine. Evolution of seed regurgitation is essential for exploitation of large-seeded fruits, a feature independently evolved by frugivorous birds of several different families including toucans, cotingas, hornbills, and oilbirds. As the birds become more efficient at regurgitation, the plants no longer need to invest energy in hard-coated seeds to resist harsh treatment. Consequently, the plant can produce relatively large soft-coated seeds, a feature evolved independently by several unrelated families as Lauraceae, Myristiceae, and Palmae. Energy that would otherwise go to hard seed coats is invested in production of fruit nutritional enough to attract frugivorous birds capable of ingesting and regurgitating large seeds.

An exception to the coevolution of plants with large soft seeds and specialized frugivores is the mistletoe (Loranthaceae) which produces an abundance of small seeds that pass through the gut (Figure 19-4). Specialized frugivores are the most important dispersal agents for mistletoe throughout the Australian and Oriental regions, in the New World tropics, and in temperate regions. In North America mistletoes are eaten by waxwings (Bombycillidae) and phainopepla *(Phainopepla niteus)*. Mistletoe seeds lack a seed coat. They are chemically protected from digestive action of dispersal agents by a sticky, persistent viscin layer impregnated with chemicals that hasten the seed through the digestive tract. Treatment in the digestive tract by the mistletoe's associated frugivores is relatively gentle, resulting only in the removal of the outer layer of fruit pulp and leaving the viscid layer intact. When the birds defecate in trees, seeds stick to the limbs and germinate. Mistletoe seeds are often killed if fruits are consumed by birds with guts unspecialized for a frugivorous diet.

In addition to responding to competition for dispersal agents, plants must also respond to

**FIGURE 19-4**
*Seed dispersal of mistletoe and specialized frugivores. Specialized frugivores, ones that treat coevolved large soft seeds gently by expelling them before they reach the digestive tract, are found mostly in the tropics. A few specialized frugivores, however, notably certain tanagers of the genus* **Euphonia** *and certain species of flowerpeckers, Family Dicaeidae, are important agents of dispersal of mistletoe which has fruits with small, soft seeds lacking a seed coat. In temperate regions a very few specialized frugivorous birds, particularly waxwings (Bombycillidae) and the phainopepla (*Phainopepla nitens, Ptilogonatidae)*, feed on mistletoe berries. These birds have nonmuscular stomachs which allow seeds to pass through the digestive tract without being ground. Mistletoe seeds are given some chemical protection from destructive action in the frugivore's digestive tract by a viscid layer containing chemicals that speed the seed through the tract. This same layer also enables mistletoe seeds to stick to the limbs and twigs of host plants on which they are deposited by perching birds. Mistletoe seeds are likely to be destroyed by passage through the gut of unspecialized frugivores. Thus mistletoes and certain specialized frugivores are examples of coevolution involving a dispersal agent and a plant.*

time of fruiting in relation to associated plants. Again there are two possible alternatives.

One is to ripen fruit at a time when few other plants using the same dispersal agents are producing fruit. Because of a necessarily short fruiting season to avoid overlap and the need for mass production and mass dispersal of seeds, selection is for predation by a number of opportunistic fruit feeders.

The other alternative is to spread fruit production over a longer period of time so that each point in time places less stress on the dispersal agent. To do so the plant must present only a few mature plants at a time. Because such production may reduce the plant's ability to attract opportunistic dispersal agents, selection is for attracting a specialized group of frugivores dependent upon the plant for food. To support them, fruit production must be spread out over a long period to avoid overwhelming the frugivores with a large number of available fruit. Because fewer plants share the same dispersers, there is less selective pressure to avoid overlap. More important perhaps, because specialized frugivores must extract all their nutrition from fruit, the dispersal agents may require several species of fruit, each providing somewhat different levels of nutrition. Thus selection in those plants would be for broadly overlapping fruiting seasons.

### HUNTING ABILITY

As prey evolved ways of avoiding predators, predators necessarily had to evolve better ways of hunting and capturing prey.

Consider two examples of predatory activity. A Cooper's hawk takes up residence in a hemlock grove and so harasses a covey of quail that by spring not a single bird remains. Two predators, an alligator and a blue heron work independently side by side. The heron, during a relatively short period of observation, is quite successful, securing two fish. The alligator, so situated so that it can lunge at fish as they swim by, is unsuccessful.

The incidents appear to be quite simple, but each involves different approaches to hunting and capturing prey. The Cooper's hawk is a pursuer, the blue heron a stalker or searcher, and the alligator an ambusher. The alligator

**567**

can afford a low frequency of success because hunting by ambush requires a minimal amount of energy. The heron is a deliberate hunter, seeking suitable pools of water and procuring its prey with a sudden thrust of the beak. Its search time may be great, but pursuit time is minimal and therefore it can afford to take smaller prey. For the Cooper's hawk search time is minimal for it usually knows the location of prey, but pursuit time is relatively great and in general the hawk must secure relatively large prey. Searchers spend more time and energy to encounter prey. Pursuers, theoretically, spend more time to capture and handle prey once noticed.

Several different searchers cannot be common to an area unless they exhibit spatial separation of hunting sites, as herons do. This is not true for pursuers because they must specialize for size of prey on the foraging sites. Because pursuers are more restricted to a range of prey size, there is a closer match of predator to prey size.

The various species of predators have certain energy requirements that can be met only by profitable hunting. Predators cannot afford prey that is too small to meet their energy requirements unless the prey can be captured quickly and in abundance. Otherwise energy expended to procure food would exceed energy obtained. On the other hand, predators have an upper size limitation. The prey may be too big to consume or too difficult or dangerous to handle. Thus the predator has to select food on the basis of size and strength. Mountain lions, for example, avoid attacking large healthy elk, which they cannot successfully handle, and concentrate instead on deer (Hornocker, 1970). The alewife feeds on larger zooplankton and ignores smaller species (Brooks and Dodson, 1965).

Where a choice in prey is available, predators will choose the largest food items that their size and strength enables them to handle. Predators, especially arthropod instars, change the size of their prey as they grow larger. As the predator grows larger relative to the size of prey, handling time decreases because prey becomes relatively easier to handle, subdue, and consume. Decrease in handling time usually results in an increase in the attack rate (see Chapter 17).

Conversely, as prey size increases relative to predator size, handling time increases and attack rate decreases. Large predators can search fast and have a greater percentage of successful attacks, but as the prey grows larger they can more easily escape predation. In fact some become invulnerable to predation. Some predators have evolved methods to kill prey much larger and stronger than themselves. The predator may take the prey by surprise as a mountain lion ambushing a deer. Or predators may hunt in packs and by cooperative effort take down large prey, as wolves hunting down a moose. Certain snakes and arthropods may use venom to immobilize and kill prey very much larger than themselves. Predators very much larger than their prey have evolved ways of filtering organisms from their environment, particularly in aquatic communities. Examples are the net-spinning caddisflies and baleen whales that feed on krill. Intermediate-sized predators are usually hunters, while very small predators are usually parasitoids. The boundaries between the various predator-prey sizes are usually set by energy cost-benefit ratios. Predator-prey size relationships have been summarized graphically by Hespenhide (1974) and further modified by Enders (1976) (Figure 19-5).

Size and other characteristics of seeds may influence seed predation. Seed predators may discriminate between seed size taking larger or smaller seeds first. Or the seed predator may not discriminate in size, successfully taking all seeds on the correct species. In these three situations plants with smaller, more numerous seeds have a higher percentage of seeds escaping predation and will be under selection for smaller seeds. Some seed predators forage on the correct species, but some characteristic other than seed size, such as thickness of seed coat, determines what seeds the predator will succeed in taking (C. Smith, 1975).

An example of the interaction between seeds and seed predators is the relationship of pine squirrels and conifers in the Cascade Mountains of southwestern British Columbia (C. Smith, 1970, 1975). The squirrels feed principally on the seeds of lodgepole pine and Douglas-fir. Because the eastern slopes of the Cascades lie in a rain shadow, the forests there

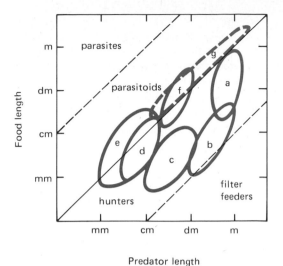

**FIGURE 19-5**
*Generalized pattern of relative sizes of predator and prey. Ellipses indicate relation of prey size for: a, hawks and owls; b. insectivorous birds; c, chewing insects; d, errant spiders and solitary wasps; e, web spiders; f, shrews; g, pack hunters as ants and wolves. Central dotted line indicates equality of size of predator and prey. Peripheral dotted lines represent prey or predator one-hundredth length of predator or prey. (From Enders, 1975, based on Hespenheide, 1973.)*

are dry and burn frequently. On these slopes the lodgepole pine is the common species, occurring in even-aged stands. Typical of fire-evolved conifers, this pine has serotinus cones which remain closed several years after they are mature and are usually opened only by fire. These cones offer the red squirrel *(Tamiasciurus hudsonicus),* which lives east of the Cascades, a rather constant food supply and stabilizes the squirrel population. The squirrels actually prefer the seeds of Douglas-fir. This tree, however, is not as common nor does it provide a dependable source of food. In years when the preferred Douglas-fir seed crop fails, the squirrels turn to seeds of lodgepole pine. On the western side of the Cascades moisture is abundant. There Douglas-fir is common and lodgepole pine is rare. The lodgepole pine in this area does not have serotinous cones and the resident Douglas squirrel *(T. douglassii)* experiences pronounced fluctuations. Due to the effect of selective pressure from the red

squirrel, the lodgepole pine on the eastern slopes has a harder texture, thicker cone scales, and only 1 percent of the weight of the cone is in seeds. The associated Douglas-fir produces fewer seeds per cone. These harder, thicker lodgepole pine cones in turn exert a selective pressure for squirrels with stronger jaw musculature. As a result red squirrels have stronger jaws and a different skull configuration than Douglas squirrels on the western slopes. In this situation coadaptive responses began with frequent fires that led to serotinous cones with fewer seeds and an all year round food supply for the squirrels. This led to high and constant squirrel predation and the evolution of hard cones in lodgepole pine and fewer seeds in Douglas-fir cones, increasing the work to be done per squirrel per seed extracted, which finally selects for squirrels with stronger jaws.

## Mutualism

*Mutualism* is a beneficial, obligatory relationship between two species; one population cannot live without the other. Facultative or nonobligatory relationships which are beneficial but not necessary for the existance of both species are often called mutualism, but are more accurately considered *protocooperation.* The term *symbiosis* is commonly used as another label for this type of mutually advantageous relationship, although the word more strictly defined means "living together" and thus includes other interactions such as commensalism and parasitism (see Henry, 1966). L. Gilbert (1975) defines mutualism more precisely as "those interspecific relationships for which benefits clearly outweigh costs for each species involved and for which net profits to individuals of each species can be translated into an increase in Darwinian fitness."

Some forms of mutualism are so permanent and obligatory that the distinction between the two interacting populations becomes quite blurred. A classic example is the lichen fungi, plants which consist of an algal and fungal population. The basic structure of the lichen is made up of a mass of fungal hyphae. Within this cellular formation is a thin zone of algae which usually forms colonies of 2 to 32 cells.

The hyphae support the plant while the algae supply the food. Without the algal component the fungal component probably would not survive (see Hale, 1971). Another well-known example of this kind of mutalism is the relationship of legumes with members of the bacterial genus *Rhizobium*. The bacteria infect the cells of root hairs whose exudates stimulate bacterial growth. Inside the nodule the bacteria assume the condition associated with nitrogen fixation.

Less well-known is mycorrhiza, a mutualistic relation of plant roots with the mycelium of fungi (a relationship already considered from other aspects in Chapters 5 and 10). Common to many trees of temperate and tropical forests, mycorrhizae produce shortened and thickened roots that suggest coral. The hyphae work between the root cells, extend into a network on the outside of the root, and act as a substitute for root hairs. They aid in decomposition of litter and translocation of nutrients, especially nitrogen and phosphorus, from the soil into the root tissue (Zak, 1964; Marx, 1971). Mycorrhizae increase the capability of roots to absorb nutrients (Voigt, 1971), provide selective ion absorption and accumulation, and mobilize nutrients in infertile soil, render unavailable substances available (particularly those bound up in silicate materials). In addition mycorrhizae reduce susceptibility of the host to invasion of pathogens by utilizing root carbohydrates and other chemicals attractive to pathogens. They provide a physical barrier to pathogens, secrete antibiotics, and stimulate the roots to elaborate chemical inhibitory substances (Marx, 1971). In turn the roots provide a constant supply of carbohydrates. A balanced association exists between the two. Any alteration in light or nutrient supplies creates a deficiency of carbohydrates and thiamin for the fungi; and interruption of photosynthesis causes a cessation of fruiting of the mycorrhizae. Thus any interruption or imbalance in the continuous supply of metabolites impairs or destroys the association.

Mutualistic relationships between plants and animals include interactions which involve shelter, food, transport, and reproduction. Some of the most interesting exist between ants and plants which range from the fungus-growing attine ants and a slow-growing fungus that cannot survive without them (Martin, 1970) to the ant-acacia relationship (Janzen, 1966; Hocking, 1975). The ants live in the swollen thorns of acacia (*Acacia* spp.) from which they derive a shelter and a balanced and almost complete diet for all stages of development. In turn the plants are protected from herbivores by the ants who at the least disturbance swarm out of their shelters emitting repulsive odors and attacking the intruder until it is driven away. Neither the ant nor the acacia can survive without the other.

One of the most important and complex mutualistic relationships is that between plants and their pollinators, insects, birds, and bats. The relationship between the common Central American plants *Heliconia* or wild plaintain and the hummingbird illustrates the many factors that may be involved in mutualism. Growing in openings of tropical forests or along the forest edge, *Heliconia* propagates vegetationally by rhizomes and usually forms large clumps. When two years of age or older, each individual *Heliconia* plant in the clump blooms. The bloom or inflorescence consists of several showy bracts, each of which encloses several flowers. The bracts open one after the other over a period of days or weeks with each flower within the bract lasting only a day. The flowers are tubular and vary in length and curvature depending upon the species of *Heliconia*. Some have long, curvaceous corollas, 33 mm or more in length; others have short, straight corollas, 32 mm or less (Stiles, 1975). Some species bloom either in the wet season or the dry season, while others bloom throughout the year but have a wet or dry seasonal peak in flowering. All are pollinated by insects or birds and offer a supply of sugar-rich nectar as an inducement to their pollinators.

Stiles (1975) found that in his Costa Rican study area nine species of hummingbirds visit the nine species of *Heliconia*. Just as the flowers of *Heliconia* have straight or curved corollas, Stiles found that the hummingbirds, too, could be divided into two groups: hermit hummingbirds with long, curved bills and nonhermits with shorter, straight bills. Stiles observed that the five *Heliconia* species with long, curved corollas are visited to a significantly greater extent by hermits than nonhermits, while three of the species with short corollas are visited mostly by straight-billed nonhermits.

The hermit and nonhermit groups differ in another way. The nonhermits frequently hold territories about clumps of *Heliconia* with short corollas. The hermits seldom held flower-centered territories and those that did defended them inconsistently. The difference in territoriality is directly related to a difference in feeding strategy that is in turn directly related to the energy-supplying attributes of different *Heliconia* species. The nonhermits defend large clumps of *Heliconia* that at the peak of flowering are able to supply the birds' total energy needs. The hermits feed on *Heliconia* that grow in scattered clumps, have a lower rate of flowering per inflorescence, and have a lower rate of nectar production. Even at the height of seasonal bloom clumps of these species cannot provide sufficient nectar to sustain a single bird. The hermits' feeding strategy involves "traplining." The birds travel between clumps of flowers, often following a definite route and visiting clumps in a particular sequence. Each clump produces sufficient nectar to warrent repeated visits, but not enough to be worth defending.

In return for nectar hummingbirds pollinate the respective flowers. *Heliconia* depend upon hummingbirds for pollen transfer. Hermit hummingbirds, probing into long, curved corollas, carry pollen at the base of the bill or on the head. The nonhermits or straight-billed hummingbirds carry pollen on the chin or mandibles. If the short corolla flowers are somewhat curved but the path to the nectar is short and straight allowing easy access, the hummingbird has pollen deposited on the bill.

Because of the number of types of plants and birds involved, some isolating mechanisms are essential. *Heliconia* select against hybridization by sequential and nonoverlapping peaks in flowering (wet or dry seasonal peaks in flowering), by spatial isolation (shady or sunny habitats), and by behavioral differences in hummingbirds. The behavioral differences include responses of hummingbirds to visual cues of flowers and to caloric content of nectar. The mutualism thus depends not only on the morphological fit between bird bill and flower corolla, but also on several other complementary relationships involving flowering phenology, energy content of nectar, and energy demands and behavioral responses of hummingbirds.

Even more complex mutualistic relationships exist between pollinating insects and plants. For example, Gilbert (1975) describes a coevolved mutualism between *Heliconius* butterflies, *Anguria* cucurbit vines, which the butterflies pollinate, and *Passiflora,* their larval host plant. *Passiflora* in turn has coevolved mutualistic defenses against its heliconine herbivores; predaceous ants and vespid wasps are attracted to the plant by nectar glands located on leaves and stems (for review of such extrafloral nectar production and mutualistic interactions, see Bentley, 1977). Such complex interactions suggest that mutualism, like predator-prey interactions and interspecific competition, is a powerful force in determining community structure. But because mutualism affects community structure by indirect pathways, its impact is less conspicuous and more difficult to study.

The examples of mutualism in the preceding discussion seem to support the suggestion that mutualistic relationships evolve from predator or parasite-host relationships (N. Smith, 1968). Initially one member of the interaction increases the stability of the resource level for the second. In time energy benefits accrue to the second member and perhaps its activities begin to improve the fitness of the first. Selection then favors the mutual interaction. In the *Heliconia*-hummingbird relationship the hummingbirds may have first exploited the nectar of the flowers. As a result of the exploitation some flowers were pollinated by the hummingbirds. The improved efficiency and success of pollination through the activity of the hummingbirds improved the fitness of the plants. Selection then favored the development of mechanisms to maintain the interaction, such as the production of sugar-rich nectar to keep the birds coming.

In spite of its importance and presumed relationship to other two-species interactions, mutualism has not received the same theoretical consideration as competition or predation. However, May (1976) and Christiansen and

Fenchel (1977) have suggested mathematical models involving Lotka-Volterra equations that might serve as a beginning. Because each species benefits the other in a mutualistic interaction, an increase in one population directly influences and is directly influenced by an increase in the other. Thus the population growth equation of each must include a term for the rate of increase of the other. However, since there is an upper limit to population growth based on carrying capacity, the influence of each has to have a saturation point. The environment may impose an upper limit on population growth. In addition in a plant-pollinator system a certain level of populations of both interacting species is necessary before any equilibrium is possible and both populations reach maximum stable densities. If plant density is too low and pollinators have difficulty finding plants, the pollinators will decline below replacement level.

If such a mutualistic system is in a predictable environment, it will persist. If it is in an unpredictable environment subject to perturbations, the population of one or the other is apt to fall below replacement levels with probable extinction. This suggests why mutualism is much more apparent in tropical than in temperate regions. As has been pointed out by Farnsworth and Golley (1974), "there are no obligate ant-plant mutualisms north of 24°, no nectarivorous or frugivorous bats north of 32° to 33° and no orchid bees north of 24°." The reason may be that in tropical regions insects and certain birds have a high and continual impact on plants throughout the year, which results in a stronger coevolutionary response between plant and herbivore.

## Parasitism

*Parasitism* is a condition in which two organisms live together but one derives its nourishment at the expense of the other. Parasites, strictly speaking, draw nourishment from the tissues of their larger hosts, a case of the weak attacking the strong. Typically parasites do not kill their hosts as predators do, although the host may die from secondary infection or suffer from stunted growth, emaciation, or sterility. Some parasitic larvae of insects

draw nourishment from tissues of their hosts and by the time of metamorphosis have completely consumed the soft tissues of the host. These parasites, known as *parasitoids*, essentially act as predators.

Parasites exhibit a tremendous diversity in ways and adaptions to exploit their hosts. Parasites may be plants or animals and they may parasitize plants or animals or both. They may live on the outside of the host or within its body. Some are full-time parasites, others only part-time. Part-time parasites may be parasitic as adults and free living as larvae or the reverse. Parasities have developed numerous ways to gain entrance to their hosts, even to the point of using several hosts as dispersal agents. They have evolved various means and degrees of mobility, ranging from free-swimming ciliated forms to forms totally dependent upon other organisms for transport. They have developed diverse ways of securing themselves to the host to maintain their position. Some such as the tapeworm have become so adapted to the host that they no longer require a digestive system. They absorb food directly through their body wall.

Parasites may be restricted to one host. Some parasites of birds, especially certain tapeworms, can live only on one particular order or genera (see Baer, 1951). Some parasites may live their entire life cycle on one host, while others require more than one host. For example, the brain worm (*Pneumostrongylus tenuis*) of the white-tailed deer has as a secondary host during its larval stage a snail or slug that lives in the grass. The deer picks up the infected snail while grazing. In the stomach the larvae leave the snail, puncture the deer's stomach wall, enter the abdominal membranes, and by the way of the spinal cord reach spaces surrounding the brain. Here the worms mate and produce eggs. Eggs and larvae pass through the blood stream to the lungs where larvae break into the air sacs, are coughed up, swallowed, and passed out through the feces. The larvae are ingested by the snail where they continue to develop to the infective stage. Through long years of coevolution the white-tailed deer has built up a resistance to the parasite.

Successful parasitism represents something of a compromise between two living populations. Parasites and hosts that have coevolved have developed a sort of mutual toleration with low-grade, widespread infection that remains so as long as conditions are favorable for the host. But if conditions become more favorable for the parasite or if the parasite finds a host which has developed no defense, relationships change.

Again, the brain worm of deer serves as an example. At one time the white-tailed deer and moose occupied nearly exclusive ranges. The deer inhabited deciduous forest, the moose the northern coniferous forest. When humans cut and burned the coniferous forest, aspen and birch replaced the conifers. The deer expanded into new range, came in close contact with the moose, and brought to it the brain worm. The moose, experiencing a new parasite against which it has evolved no defenses, succumbs to the infection.

## Social parasitism

Another form of parasitic relationship is social parasitism in which one organism is parasitically dependent on the social structure of another. Social parasitism may be temporary or permanent, facultative or obligatory, within a species or between species. Four forms of social parasitism can be defined in terms of these types of relationships (Wilson, 1975).

The first is temporary, facultative parasitism within a species. This type is rather well developed among the ants and wasps. For example, a newly mated queen of the wasp genus *Polistes* or *Vespa* will attack established colonies of her own species and displace the resident egg-carrying queen (Wilson, 1975).

A second type is temporary, facultative parasitism between species. An example of this occurs among the formicine ant genus *Lasius*. A newly mated queen of the species *L. reginae* will enter the nest of a host species *L. alienus* and kill its queen. The *alienus* workers will care for the *reginae* queen and her brood. In time the *alienus* workers, deprived of their own queen and thus replacements, die out, and the colony then consists of *reginae* workers. A somewhat parallel situation exists among birds. Twenty-one species of ducks are known

to lay eggs in nests other than their own (Weller, 1959). An estimated 5 to 10 percent of female redhead ducks are nonparasitic and nest early. All others lay eggs parasitically at one time or another. More than half of these are semiparasites and nest themselves. The remainder are completely parasitic.

A third type of social parasitism is temporary, obligatory parasitism between species. Although common in ants, the most outstanding examples are obligatory egg or brood parasitism in birds. Brood parasitism has been carried to the ultimate by the cowbirds and Old World cuckoos, both of which have lost the act of nest building, incubating the eggs, and caring for young. They are obligatory parasites who pass off these duties to the host species by laying eggs in their nests. The brown-headed cowbird of North America removes one egg from the nest of the intended host, usually the day she is to lay, and the next day lays one of her own as replacement. Some host birds counter by ejecting the egg from the nest. Others hatch the egg and rear the young cowbird, usually to the detriment of their own offspring. The host's young may be pushed from the nest or die from lack of food because of the more aggressive nature and larger size of the young cowbird.

A fourth type is permanent, obligatory parasitism between species. The parasitic form spends its entire life cycle in the nest of the host (Wilson, 1975). This type of social parasitism is common among ants and wasps. In most cases the species are workerless and queens have lost the ability to build nests and care for young. The queen gains entrance to the nest of the host and either dominates the host queen or kills her outright and takes over the colony.

Like parasitism, social parasitism can adversely affect a host experiencing its first contact with social parasite. Such a situation has developed with the Kirtland warbler *(Dendroica kirtlandii),* a relict species that inhabits extensive jack pine stands in a compact central homeland of about 100 mi² in northern lower Michigan (Mayfield, 1960). Before white settlers arrived, Kirtland's warbler apparently was isolated by 200 mi of unbroken forest from the parasitic brown-headed cowbird of the central plains, a bird closely associated with grazing animals (Figure 19-6). When settlers cleared the forest and brought grazing animals with them, cowbirds spread eastward and northward into jack pine country. Never associated with the cowbird, Kirtland's warbler has not evolved defenses against the social parasite, such as egg ejec-

**FIGURE 19-6**
*Maladaption of the Kirtland warbler to the social parasitism of the cowbird. The warbler apparently evolved outside the natural range of the cowbird and thus possesses no defense against egg parasitism. When the cowbird extended its range into Kirtland warbler country because of habitat changes brought about by human activity, it found the warbler an easy host to attack. As a result reproduction in the warbler declined sharply, barely reaching replacement levels. Control of cowbirds has reversed this trend.*

Kirtland's warbler
*Dendroica kirtlandii*

Brown-headed cowbird ♂
*Molothrus ater*

tion, building a new nest over the old, or rearing young successfully along with the cowbird. The warbler has a short nesting season and an incubation period a day or two longer than other songbirds, so the young cowbird is already out of the egg when the warblers hatch. This parasitism resulted in an alarming decline in the number of warblers fledged, a trend that has been reversed by a strong cowbird control program in the Kirtland warbler's breeding range.

Social brood parasitism can in some situations be mutually beneficial, as in the often cited case of the giant cowbird *(Scaphidura oryzivora)* and the hosts oropendolas *(Zarhynchus)* and cacique *(Cacicus),* grackle-like birds of the family Icteridae (N. Smith, 1968). Under certain conditions a greater number of host offspring survive from cowbird parasitized nests than from unparasitized ones. A botfly that burrows into the chick's body to feed and crawls out again to pupate on the bottom of the nest is the major cause of mortality among the nestlings. Host chicks in nests parasitized by the cowbird are almost never bothered by botflies, while chicks in unparasitized nests are and sustain heavy mortality. The reason is the aggressive parasitic cowbird nestling actually removed bots and eggs from its nestmates and at the same time protects itself by snapping at any moving object including adult botflies.

There is more to the story that can only be briefly told. Cowbird eggs vary in coloration and markings from colony to colony. At some colonies giant cowbirds produce eggs whose color and markings mimic the eggs of the host. In other colonies the cowbird lays nonmimetic eggs. Those cowbirds laying mimetic eggs secretly deposit only one egg in a host nest which already contains eggs. Cowbirds laying nonmimetic eggs openly leave two to three eggs in an empty nest. Hosts for mimetic eggs discriminate against mismatched eggs and throw them out. Hosts for nonmimetic eggs do not discriminate. A study showed that three-fourths of the nests of nondiscriminating hosts contained cowbird chicks; only one-fourth of the nests of discriminators had cowbirds.

Oropendula and cacique colonies are often clustered near nests of wasps and stingless but biting bees which in some way or another repel botflies. Nests near bees and wasps have a lower incidence of botfly parasitism. Interestingly, colonies protected by wasps and bees are composed of discriminator hosts and mimetic cowbirds, while unprotected colonies consist of nondiscriminating hosts and nonmimetic cowbirds.

The outcome of these complex interrelations which involve social parasitism, commensalism (birds and wasps), and protocooperation (nonmimetic cowbirds and nondiscriminatory hosts) is mixed. Reproductive success of discriminator oropendulas and caciques, who build nests near wasps and bees, is reduced by giant cowbird parasitism and gains no advantage from cowbird nestlings. However, nondiscriminatory hosts gain an advantage from an association with cowbirds. The average number of oropendolas and cacique chicks fledged per nest in colonies near wasps and bees was 0.39 and for nests distant from the insects was 0.43. The average number of cowbird chicks fledged in the two types of colonies was 0.76 and 0.73, respectively. Thus, reproductive success of both oropendolas and cacique and their cowbird parasites is about the same under both circumstances.

It is evident that oropendulas and caciques maintain polymorphisms in sets of populations which allow part of the population to accept cowbird eggs and others to reject them. At the same time cowbirds maintain polymorphisms of egg mimicry and of aggressive and secretive behavior to take full advantage of the host (for details, see N. Smith, 1968).

## SUMMARY

In exploitative or feeding situations interacting species often adjust their relationship to one another. Prey under the selective pressure of predation evolve some means of defense to counter the effects of predation. Predators must constantly improve their efficiency in exploiting prey in order to survive. Each in effect has a reciprocal influence on the evolution of the other. This reciprocal influence is called coevolution.

One response to selective pressures of predation is the development of color patterns and behavior to escape detection or inhibit predators. Warning coloration in prey serves to tell a predator that a prey item is distasteful or disagreeable in some manner. Usually the predator has to experience at least one encounter with such a prey to learn the significance of the color pattern. Often palatable species will mimic unpalatable species, thus acquiring some protection from predators. Batesian mimicry is the development of the appearance of a distasteful or dangerous model by an edible species. Müllerian mimicry is the development of a common warning coloration by different species, all of which are distasteful. Although mimicry is usually associated with animals, mostly insects, there is increasing evidence of mimicry among plants to hide from specialized herbivores or to attract seed dispersers. Cryptic coloration serves to blend the organism with its surroundings, making it hard to detect. Although warning and cryptic coloration and mimicry are usually associated with prey species, such mechanisms are also employed by predators to increase hunting efficiency.

Chemical defense, widespread among plants and animals, usually involves a distasteful or toxic secretion that repels, warns, or inhibits a would-be attacker. Plants employ secondary metabolic products such as alkaloids, phenolics, and cyanogenic glycosides. Chemical defense is most successful against generalist herbivores. Certain specialists breach the chemical defense and detoxify the secretions or sequester the toxins in their own tissues as a defense against their predators.

Another form of defense is predator satiation. Reproduction is so timed that offspring are so abundant that predators can take only a fraction of the very young, leaving a number to escape by growth to a size too large for the predator to handle.

Some plants have turned the situation around and utilize predators to disperse seeds. By packaging seeds in attractive fruits, plants encourage their consumption by animals that will transport the seeds in their digestive tracts. Because dispersal agents can be a resource in short supply, plants through selective pressures of competition for dispersers evolve different strategies of seed dispersal. One is to evolve small-seeded fruit adapted for exploitation by a large number of different dispersal agents. Selection is for small seeds that can tolerate the harsh treatment in the disperser's digestive tract, for a short fruiting season so that adequate dispersal can be obtained in a brief period of time, and for nonoverlapping fruiting seasons with competitors utilizing the same dispersal agents. Such a strategy results in a production of maximum number of seeds at a minimum per propagule cost of dispersal.

A second strategy is to produce large fruits, usually large-seeded, adapted for exploitation by specialized fruit-eaters. The frugivores in turn evolve mechanisms to regurgitate large, soft seeds with minimal damage and minimal expenditure of energy. Because such specialized frugivores necessarily depend on the plant for food, the plants evolve high-energy fruits rich in proteins and oils and a long fruiting period. This strategy, which reflects less competition among plants for dispersers, results in production of few seeds at a high per propagule cost of dispersal. Quality dispersal is favored over quantity dispersal.

Mutualism is a beneficial relation between two interacting species that cannot live separately. Like predator-prey interactions, mutualism is the result of coevolution over a long period of time. Some of the most complex mutualistic relationships evolved between plants and pollinators. These relationships probably developed out of some form of predator-prey relationship.

Another coevolved relationship is parasitism, a situation in which two animals live together, but one derives its nourishment at the expense of the other. Also rather common is social parasitism in which one species is dependent on the social structure of another. Social parasitism, like other forms of parasitism, may be temporary, permanent, facultative, or obligatory. A common example is brood parasitism among birds. Successful parasitism represents something of a compromise between two interacting populations. Hosts and parasites develop a sort of mutual tolerance with a low-grade, widespread infection.

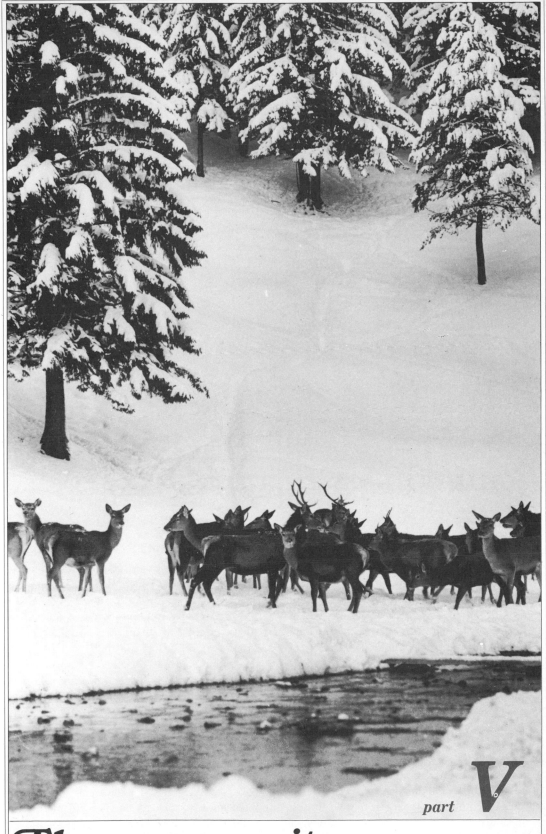

part *V*

# The community

# Community structure
## CHAPTER 20

A forest, grassland, hedgerow, lawn, stream, marsh—these and other habitats in the landscape are easily recognized as unique groupings of organisms or biotic communities. A *community* is a naturally occurring, mutually sustaining, and interacting assemblage of plants and animals living in the same environment and fixing, utilizing, and transferring energy in some manner. Like many ecological terms, the word community has several meanings. Some ecologists use it to describe certain groups of similar organisms, such as desert lizards or forest birds, in relation to habitat structure, latitudinal gradients, and the like. Others use the term in an even more restricted manner by employing it to refer to the relation of specialized groups to specific aspects of habitat, such as the relation of insect-feeding birds to the vertical structure of a forest or to differences between temperate and tropical habitats. In all cases, however, community does refer to an assemblage of interacting species.

A community can be considered not only as those interacting plant and animal populations that make up the biotic portion of an ecosystem, but also as those interacting organisms that inhabit a microenvironment, such as a fallen log, a spring, or a pool of water in a hollow tree.

A community may be autotrophic in the sense that it includes photosynthetic plants and gains its energy from the sun. Other communities, such as springs and caves, are heterotrophic; they depend upon the input of fixed energy, such as detritus, from the outside. Most autotrophic communities contain a number of microcommunities that are heterotrophic.

The nature of the community is determined

by the adaptations of its organisms to the physical environment—soil, temperature, moisture, light, nutrients, and the like—and the interactions among organisms—competition, predation, and symbiosis. These adaptations and interactions determine such attributes of the community as structure, dominance, species diversity, and niches. The community in effect is a product of the evolutionary processes that have adapted individual species to the environment and to each other.

## Physical structure

The most easily observed feature of a given community is its physical structure. There are pronounced differences between a grassland and a forest and between a stream and a lake. Differences in terrestrial communities are determined largely by vegetation. Differences in aquatic communities are determined largely by the depth and flow of water.

The form and structure of terrestrial communities are determined by the nature of the vegetation. Vegetation may be classified according to growth form. The plants may be tall or short, evergreen or deciduous, herbaceous or woody. One might speak of trees, shrubs, and herbs, and then further subdivide the categories into needle-leafed evergreens, broad-leafed evergreens, evergreen sclerophylls, broad-leafed deciduous, thorn trees and shrubs, dwarf shrubs, ferns, grasses, forbs, and lichens.

Perhaps a more useful system is the one designed in 1903 by the Danish botanist Christen Raunkiaer. Instead of considering plants' growth form, he classified plants by life form, the relation of their height above ground to their perennating tissue (tissue that survives from one growth season to the next, remaining inactive over winter or dry periods). Perennating tissue is the embryonic or meristemic tissue of buds, bulbs, tubers, roots, and seeds. Raunkiaer recognized five principal life forms (Table 20-1 and Figure 20-1). All the species in a region can be grouped into five life form classes: therophytes, cryptophytes, hemicryptophytes, chamaephytes, and phanerophytes. The ratio between these five life form classes

**FIGURE 20-1**
*Raunkiaer's life forms. 1, phanerophytes; 2, chamaephytes; 3, hemicryptophytes; 4, cryptophytes (geophytes); 5, therophytes. The parts of the plants that die back are unshaded; the persistent parts with buds are dark.*

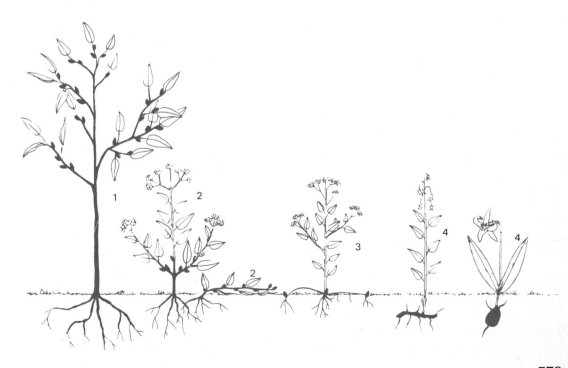

TABLE 20-1
*Raunkaier's life forms*

| Name | Description |
|---|---|
| Thero-phytes | Annuals survive unfavorable periods as seeds. Complete life cycle from seed to seed in one season. |
| Geophytes (Crypto-phytes) | Buds buried in the ground on a bulb or rhizome. |
| Hemicryp-tophytes | Perennial shoots or buds close to the surface of the ground; often covered with litter. |
| Chamae-phytes | Perennial shoots or buds on the surface of the ground to about 25 cm above the surface. |
| Phanero-phytes | Perennial buds carried well up in the air, over 25 cm. Trees, shurbs, and vines. |
| Epiphytes | Plants growing on other plants; roots up in the air. |

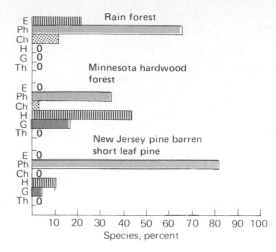

FIGURE 20-2
*Life form spectra of a tropical rain forest (adapted from Richards, 1952), a Minnesota hardwood forest (data from Buell and Wilbur, 1948), and a New Jersey pine barren (data from Stern and Buell, 1951).*

TABLE 20-2
*An example of an anlysis of life-forms spectra of two plant communities:*
*a New Jersey pine barren and a Minnesota jack pine forest*

| Basis of spectrum | Community | Number of species | Percentage | | | | |
|---|---|---|---|---|---|---|---|
| | | | *Ph* | *Ch* | *He* | *G* | *Th* |
| Species list | New Jersey | 19 | 84.2 | 0 | 10.5 | 5.2 | 0 |
| | Minnesota | 63 | 23.8 | 4.7 | 60.3 | 7.9 | 3.1 |
| Cover | New Jersey | 19 | 98.1 | 0 | 1.9 | 0 | 0. |
| | Minnesota | 63 | 11.8 | 2.5 | 55.6 | 28.7 | 1.4 |

*Source:* Stern and Buell, 1951.

expressed as a percentage provides a life form spectrum for the area that reflects the plants' adaptations to the environment, particularly climate (see Table 20-2 and Figure 20-2). A community with a high percentage of perennating tissue well above the ground (phanerophytes) would be characteristic of warm climates; a community consisting mostly of chamaephytes and hemicryptophytes would be characteristic of cold climates; and a community dominated by therophytes would be characteristic of deserts.

**VERTICAL STRATIFICATION**

A distinctive feature of the community is vertical stratification (Figure 20-3; see Chapters 7–10). Stratification of a community is determined largely by the life form of plants—their

size, branching, and leaves—which in turn influences and is influenced by the vertical gradient of light. The vertical structure of the plant community provides the physical structure in which many forms of animal life are adapted to live.

A well-developed forest ecosystem has a highly stratified structure with a large variety of components. It consists of several layers of vegetation, each of which provides a habitat for animal life in the forest. From top to bottom these layers are the *canopy,* the *understory tree,* the *shrub,* the *herb* or *ground* layer, and the *forest floor.* One can continue down into the root layer and the soil strata. The canopy is the major site of primary production and has a major influence on the structure of the rest of the forest. If it is fairly open, consid-

**FIGURE 20-3**
*Vertical section of communities from aquatic to terrestrial. All are structurally similar in that the zone of decomposition (D) and regeneration is in the bottom stratum, and the zone of energy fixation (P) is in the upper stratum. In succession from aquatic through terrestrial stages stratification and complexity of the community become greater. In the aquatic community there is little storage of materials in biomass. In terrestrial communities biomass storage increases as ecosystems develop from grass to forest.*

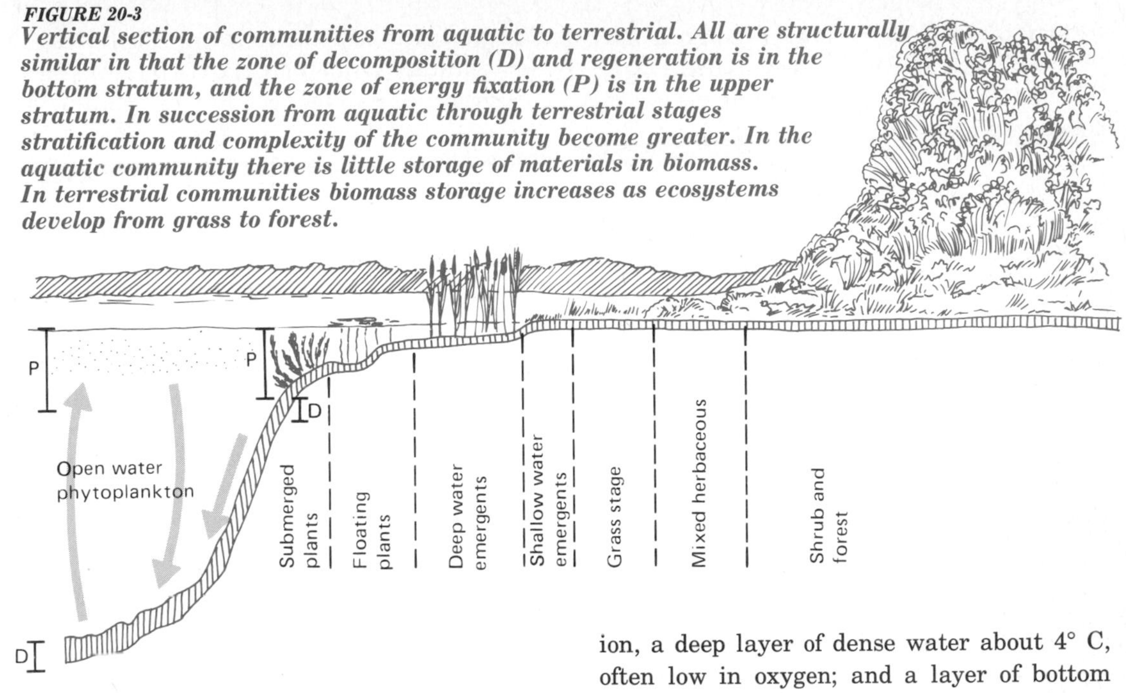

erable sunlight reaches the lower layers, and the shrub and understory tree strata are well developed. If the canopy is closed, the understory trees, the shrubs, and even the herbaceous layer are poorly developed. The nature of both the shrub and herb layers depends upon soil moisture conditions, slope position, density of the overstory, and aspect of slope, all of which can vary from place to place throughout the forest. The final layer, the forest floor (already discussed in Chapter 3) depends again on all these factors and in turn determines how and what nutrients are released for recycling for the growth of the other layers.

Other communities have a similar, if not as highly stratified, structure. Grasslands have a herbaceous layer that changes through the seasons, a ground or mulch layer, and a root layer. The root layer is more pronounced in grasslands than in any other ecosystem, and the mulch layer has a pronounced influence on plant development and animal life, especially insects and small mammals.

The strata of aquatic communities are determined by light penetration, temperature profile, and oxygen profile (for details, see Chapter 7). Well-stratified lakes in summer contain the epilimnion, a layer of freely circulating surface water; the metalimnion, characterized by a thermocline; the hypolimn-

ion, a deep layer of dense water about 4° C, often low in oxygen; and a layer of bottom mud. In addition two structural layers based on light penetration are recognized—an upper zone roughly corresponding to the epilimnion, dominated by plant plankton and the site of photosynthesis, and a lower layer in which decomposition is most active. The lower layer roughly corresponds to the hypolimnion and bottom mud.

Communities, whether terrestrial or aquatic, have similar biological structures. They possess an autotrophic layer which fixes the energy of the sun and manufactures food from inorganic substances. It consists of the area where light is most available: the canopy of the forest, the herbaceous layer of the grassland, and the upper layer of water of the lake and sea. Communities also possess a heterotrophic layer which utilizes the food stored by the autotrophs, transfers energy, and circulates nutrients by means of predation in the broadest sense and decomposition.

Each vertical layer in the community is inhabited by its own more or less characteristic organisms. Although considerable interchange takes place among several strata, many highly mobile animals confine themselves to only a few layers, particularly during the breeding season. Occupants of the vertical strata may change during the day or season. Such changes reflect daily and seasonal variations in humidity, temperature, light, oxygen

**581**

context of water, and other conditions (see Chapter 5) or the different requirements of organisms for the completion of their life cycles. For example, D. L. Pearson (1971) found that birds occupying the upper strata of a tropical dry forest in Peru moved to the lower strata during the middle of the day for several reasons: to secure food (insects move to lower levels), to escape heat stress, to escape a high degree of solar radiation, and to conserve moisture.

In general the finer the vertical stratification of a community, the more diverse is its animal life. The variety of life in a terrestrial community is in part a function of the number and development of layers of vegetation. If a certain layer is absent, then the animal life it normally shelters and supports is also missing. Thus grassland with few strata is poorer in species than a highly stratified forest ecosystem (see Karr and Roth, 1971). The distribution of life and biological activity in aquatic systems is to a large extent governed by vertical gradients of light, temperature, and oxygen; the greater the variation along these gradients, the greater the diversity of life the system supports.

## HORIZONTAL STRATIFICATION

Horizontal stratification relates to the distribution of organisms, principally plants, on the ground or across the canopy. Like vertical stratification, it can influence the presence and absence of certain forms of life. The pattern of plant distribution is mostly clumped (see Chapter 13). In terrestrial ecosystems the distribution of plant life is determined by a number of influences (Figure 20-4). Methods of plant reproduction produce characteristic patterns of distribution. Plants with airborne seeds may have a wider distribution than plants with heavy seeds (unless distributed by animals or water) or with pronounced vegetative reproduction. Allelopathic effects and shading may lead to suppression of some species and the development and growth of others. Soil structure, moisture conditions, and nutrient levels influence vegetational distribution. Runoff and small variations in topography and microclimate produce well-defined patterns of plant growth. Grazing animals have subtle but important effects on the spatial patterning of vegetation as do abiotic disturbances such as wind throw and fire. This mosaic of vegetation results in a patchy environment (see Wiens, 1976).

**FIGURE 20-4**
*General relationships of some of the major influences that govern vegetational patchiness in terrestrial environments. (From Wiens, 1976.)*

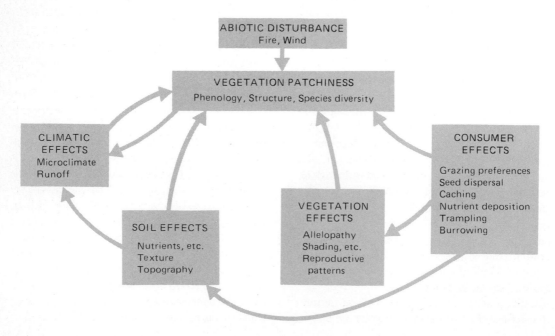

## ECOTONES: EDGE AND PATCHY ENVIRONMENT

The zone where two or more different communities meet and integrate is an *ecotone* or *edge* (Figure 20-5). This zone of intergradation may be narrow or wide, regional (for example, transition between prairie and forest), or local (for example, transition between a forest and grassy field). Of greatest ecological interest are local edges.

The concept of edge and patchy environment was first stated by Aldo Leopold (1933):

Game [wildlife] is a phenomenon of the edges. . . . [Wildlife] occurs where the types of food and cover which it needs comes close together, i.e., where their edges meet. . . . we do not understand the reason for all these edge effects, but in those cases where we can guess the reason, it usually harks back to the desirability of simultaneous access to more than one environmental type, or the greater richness of border vegetation, or both.

From this Leopold developed the law of dispersion and the law of interspersion. The *law of dispersion* states that potential density of wildlife species with small home ranges and requiring two or more habitat types is roughly proportional to the sum of the type of peripheries. The *law of interspersion* is that the abundance of resident species requiring two or more cover types appears to depend upon the degree of interspersion of numerous blocks of the same kind.

Patchy distribution of vegetation has a pronounced and important influence on animal life. As described in Chapter 13 animal populations may have a coarse-grained or a fine-grained pattern of dispersion. A fine-grained pattern results from an organism's ability to utilize the units (or grains) of the patchy environment in a random fashion, that is, to use a variety of patches in proportion to their frequency of occurrence. Organisms that utilize the patches nonrandomly, that is, show a distinct preference or need for certain types of patches, have a coarse-grained pattern of dispersion. Many species of birds and mammals found in edge situations are, in effect, coarse-grained species. They require a diversity of patches during the year which they use in a nonrandom fashion. For example, the ruffed

grouse requires relatively mature forests with an abundance of certain types of buds for winter food, pole timber stands for nesting cover, dense sapling stands for mating cover, and forest openings or forest borders with an abundance of herbaceous plants and low shrubs for brood cover and food in summer. Because the ruffed grouse spends its life in an area of 10 to 20 acres, this amount of land must provide all of its seasonal requirements (Gullion, 1970). Such species, therefore, are highly dependent on a very patchy environment.

Two basic types of edge or patchy environments are recognized: inherent and induced (Figure 20-5). Inherent edges result from permanent features in environmental conditions: soil type and soil drainage, extremes in topography, geomorphic features such as glacial deposits, and long-term microclimatic conditions that favor one type of plant community over another. Induced edges are temporary features in the landscape resulting from drastic short-term environmental disturbances such as floods, erosion, fire, timber harvesting, grazing, and land management practices. In general these disturbances result in a shift of vegetation patches to an earlier successional stage (see Chapter 21). Both inherent and induced edges may be either abrupt or mosaic. In the mosaic edge a continuous plant community becomes broken into islands and peninsulas that intermix with islands and peninsulas of other plant communities.

J. Ranney (1977) has reviewed the major features of the edge with emphasis on edges between forested and nonforested land. Environmentally such edges reflect steep gradients of wind flow, moisture, temperature, and solar radiation between the extremes of open land and forest interior. Greater wind velocity occurs at the forest edge, helping to create xeric conditions in and around the edge. Evaporation is increased resulting in increased demands on soil moisture by plants. But variations in solar radiation, both direct and reflected, probably have the most important physical influence on the forest edge, especially as it relates to north-facing and south-facing edges (Figure 20-6). A south-facing edge may receive 3 to 10 times more hours of sunshine a month during midsummer

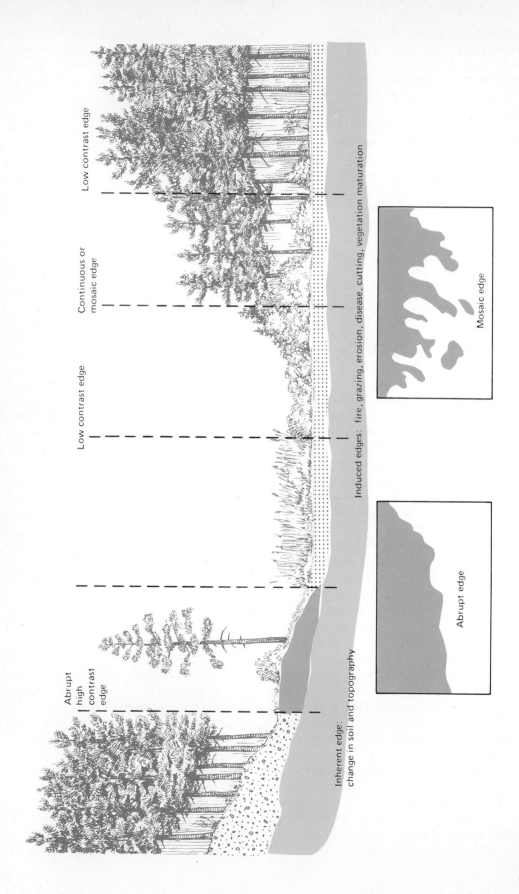

Abrupt
high
contrast
edge

Low contrast edge

Low contrast edge

Continuous or
mosaic edge

Low contrast edge

Inherent edge:
change in soil and topography

Induced edges:  fire, grazing, erosion, disease, cutting, vegetation maturation

Abrupt edge

Mosaic edge

*Inherent and induced edges. Edge is an area where two or more plant communities or successional stages within plant communities meet. There are two types. An inherent edge is an abrupt change in plant communities caused by changes in microclimate, soil types, geomorphic features, and topography. It is a long-term component of the landscape. An induced edge results from short-term environmental changes caused by natural or human disturbances such as flooding, fire, erosion, land treatment, grazing, disease, timber harvesting, and the like. Edges may differ in their degree of contrast. Edges of high contrast involve widely different adjacent communities, such as shrubs and mature forest. Edges of low contrast involve two closely related successional stages, such as shrubs and sapling growth. (After Thomas et al., 1977.)*

FIGURE 20-6

*Influence of solar radiation on the nature of the edge. Solar radiation does not affect all edge aspects equally. North-facing edges receive almost no direct sunlight and limited reflection from nearby fields, whereas south-facing edges receive considerable solar radiation. The depth to which solar radiation penetrates forest edge on level ground depends on the height of the canopy and the solar angle. Within the zone of radiation penetration light intensities and summer daytime temperatures are higher than forest interiors sheltered by tree canopies. (After Ranney, 1977.)*

as a north-facing edge, making it much warmer and drier. Also affecting the physical environment of the edge is the depth to which solar radiation penetrates the vertical edge of the forest. This penetration depends upon solar angle, edge aspect, density and height of vegetation, latitude, time of year, and time of day.

Because of increased exposure to wind and direct radiation, vegetation of the edge is subject to different environmental stresses than forest interior species. Trees on the forest edge are under greater heat stress and dissipate proportionately more heat by evaporation than by radiation; interior forest trees dissipate more heat by radiation. Higher light intensities result in sun scald of some tree species suddenly exposed to direct solar radiation and increased crown expansion and epicormic branching (development of new limbs on the bole of a tree) in others. High light intensity and xeric conditions favor those plant species that are tolerant of light and xeric soil moisture conditions, are capable of high root competition for moisture and nutrients, reproduce vegetatively, and depend on birds and mammals for distribution of seeds. Density and frequency of saplings and stump sprouts are greater in border areas.

As the edge canopy thickens, shade-tolerant plants appear. Eventually the edge is characterized by a mixture of both interior and edge species. Generally more interior species are found in the edge than edge species are found

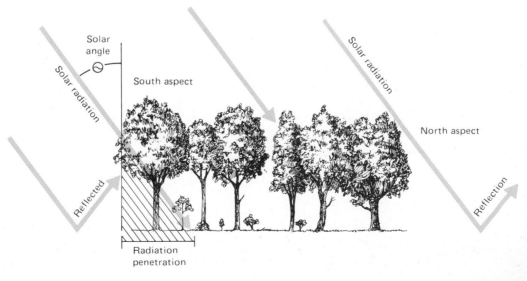

585

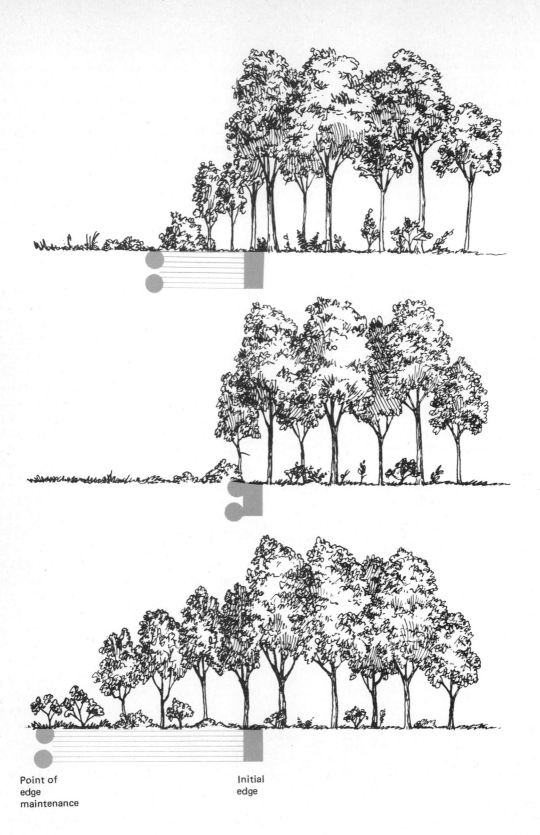

Point of
edge
maintenance

Initial
edge

586

in the interior, although the interior holds a greater proportion of rare species (see Gysel, 1951; Wales, 1972).

The highly variable nature of the edge results from certain attributes that influence the amount of edge and its richness. The amount of edge available, its length, width, and configuration (abrupt or mosaic), is often influenced by adjoining land use practices (Figure 20-7). Richness is a product of the combination of flora and fauna from the adjoining communities and the addition of species that favor edge situations. The degree of richness depends upon the contrast between adjoining plant communities and stand size. The greater the contrast between plant communities the greater may be the richness. Edge between forest and grassland, for example, should be richer than edge between shrub stand and sapling forest. As will be emphasized later in this chapter in the section on island biogeography, the size of the stand between certain upper and lower limits also influences richness. There is a limit to how small a vegetation patch can be and still support a rich edge. A large homogeneous vegetation can be so broken up by edge into smaller and smaller stands or patches that heterogeneity once provided by edge becomes itself a homogeneous community dominated by edge species. Small woodlands, for example, set in grasslands or other types of vegetation are essentially edge habitats rather than forest and edge. A minimum size of habitat block is needed to achieve a maximum number of species. The point at which maximum diversity tends to decrease is when the average size of the habitat block becomes smaller than the size required to support the maximum number of species present (see Galli et al., 1976).

The amount of edge relative to the size of forest interior or vegetation patch may have some impact on future trends in vegetation development. Exposed as they are to animals and wind, plant propagules of edge species have a higher probability of dispersal than forest interior species. As the ratio of cleared to forest land increases with a proportionate increase in edge, forest edge species may contribute more heavily to propagules available for transport between stands. Thus edge species may have a selective advantage in forest regeneration which could shift the composition of future forest stands toward edge rather than interior species.

## Species composition

### SPECIES DOMINANCE

The nature of a community may be controlled not only by physical or abiotic conditions such as the substrate, the lack of moisture, and wave action, but also to a large extent by some biological mechanism. Biologically controlled communities are often influenced by a single species or by a group of species that modify the environment. These organisms are called *dominants*.

It is not easy to delineate what constitutes a dominant species or, in fact, to determine the dominant species. The dominants in a community may be the most numerous, possess the highest biomass, preempt the most space, make the largest contribution to energy flow or mineral cycling, or by some other means control or influence the rest of the community.

Some ecologists assign the role of dominance to numerically superior organisms. But numerical abundance alone is not sufficient. A species of plant, for example, can be widely represented and yet exert little influence on the community as a whole. In a forest the small or understory trees may be numerically superior, yet the nature of the community is controlled by the few large trees that overshadow them. In such a situation dominance is measured not by number, but rather by biomass or basal area.

The dominant organism may be relatively scarce yet by its activity control the nature of the community. The predatory starfish *Piaster,* for example, preys on a number of similar species and thereby reduces competitive interactions between them so that these different prey species can coexist (Paine, 1966). If the predator is removed, a number of prey

**FIGURE 20-7**
*Effect of human management on edge structure. Edge may be abrupt or may extend outward for a considerable distance into adjacent plant community. (After Ranney, 1977.)*

species disappear and one of them becomes a dominant. In effect the predator controls the nature of the community and so must be regarded as the dominant.

Dominant species may not be the most essential species in the community from the standpoint of energy flow and nutrient cycling, although they often are. Dominant species achieve their status by occupying niche space that might potentially be occupied by other species in the community. For example, when the American chestnut was eliminated by blight from the oak-chestnut forests, the chestnut's position was taken over by oaks and hickory.

Although dominants frequently shape populations of other trophic levels, dominance necessarily relates to species occupying the same trophic level. If a species or a small group of species is to achieve dominance, it must relate to a total population of species, all of which possess similar ecological requirements. One or several become dominants because they are able to exploit the range of environmental requirements more efficiently than other species in the same trophic level. The subordinate species exist because they are able to occupy the niche or portions of it that the dominants cannot effectively occupy. Dominant organisms then are generalists capable of utilizing a wide range of physiological tolerances. The subdominants tend to be more specialized in their environmental requirements and more narrow in their physiological tolerances.

The degree of dominance expressed by any one species appears to depend in part on the position the community occupies on a physical or chemical gradient. At one particular point on a moisture gradient, species A and species B may be the dominants. As the gradient becomes drier, species B may assume a subdominant position in the community, and its place might be taken by species C. Nutrient enrichment can change the structure of the community. Lakes receiving excessive sewage discharges shift from a diverse assemblage of nutrient-thrifty diatoms to a few blue-green algae that are able to exploit a nutrient-rich system (see Edmundson, 1970).

To determine dominance ecologists have used several approaches. One can measure relative abundance of species, comparing the numerical abundance of one species to the total abundance of all species (see Appendix B). Or one can measure relative dominance, a ratio of the basal area occupied by one species to total basal area. Or one can use relative frequency as a measure. Often all three measurements are combined to arrive at an *importance value* for each species. This measurement is based on the fact that most species do not arrive at a high level of importance in the community, but those that do serve as an index, or guiding, species. Once species within a stand have been assigned importance values, the stands can be grouped by their leading dominants according to those values. Such techniques are useful in the study and ordination of communities on some environmental gradient such as moisture.

## SPECIES DIVERSITY

*Species richness and heterogeneity.* Communities vary in the number of species they contain and the species vary in the number of individuals they contain. Among the array of species that make up the community, relatively few are abundant and most contain only a small proportion of the total population of the community. Table 20-3 presents the structure of a mature woodland consisting of 24 species over 4 in. dbh (diameter breast height). Two trees, yellow poplar and white oak, make up nearly 44 percent of the stand. The next most abundant trees—sugar maple, red maple, and American beech—make up a little over 5 percent of the stand. Eight species range from 1.2 to 4.6 percent of the stand, while 10 remaining species as a group represent about 5 percent of the stand. Data in Table 20-4 show a woodland sample of somewhat different composition. That community consists of 10 species of which 2, yellow poplar and sassafras, make up 84 percent of the stand.

These two forest stands illustrate the pattern of a few common species associated with many rare ones. They also illustrate two other characteristics of distribution of species within a community—*species richness,* the number of species, often called species diversity, and *heterogeneity,* the relative abundance of species in the community. The greater the

TABLE 20-3

Structure of vegetation of a mature
deciduous forest in West Virginia

| Species | Number | Percent of stand |
|---|---|---|
| Yellow Poplar (*Liriodendron*) *tulipifera* | 76 | 29.7 |
| White Oak (*Quercus alba*) | 36 | 14.1 |
| Black Oak (*Quercus velutina*) | 17 | 6.6 |
| Sugar Maple (*Acer saccharum*) | 14 | 5.4 |
| Red Maple (*Acer rubrum*) | 14 | 5.4 |
| American Beech (*Fagus grandifolia*) | 13 | 5.1 |
| Sassafras (*Sassafras albidum*) | 12 | 4.7 |
| Red Oak (*Quercus rubra*) | 12 | 4.7 |
| Mockernut Hickory (*Carya tomentosa*) | 11 | 4.3 |
| Black Cherry (*Prunus serotina*) | 11 | 4.3 |
| Slippery Elm (*Ulmus rubra*) | 10 | 3.9 |
| Shagbark Hickory (*Carya ovata*) | 7 | 2.7 |
| Bitternut Hickory (*Carya cordiformis*) | 5 | 2.0 |
| Pignut Hickory (*Carya glabra*) | 3 | 1.2 |
| Flowering Dogwood (*Cornus florida*) | 3 | 1.2 |
| White Ash (*Fraxinus americana*) | 2 | .8 |
| Hornbeam (*Carpinus caroliniana*) | 2 | .8 |
| Cucumber Magnolia (*Magnolia grandiflora*) | 2 | .8 |
| American Elm (*Ulmus americana*) | 1 | .39 |
| Black Walnut (*Juglans nigra*) | 1 | .39 |
| Black Maple (*Acer nigrum*) | | |
| Black Locust (*Robinia pseudoacacia*) | 1 | .39 |
| Sourwood (*Oxydendrum arboreum*) | 1 | .39 |
| Tree of Heaven (*Ailanthus altissima*) | 1 | .39 |
| | 256 | 100 |

number of species, the higher is the species richness. Heterogeneity is based on a combination of species richness and equitability. *Equitability* is a measure of the evenness of allotment of individuals among the species. The more even the distribution of individuals among species, the greater is the equitability. In terms of these two characteristics of distribution of species, the stand described in

TABLE 20-4

Structure of vegetation of a deciduous forest
in West Virginia

| Species | Number | Percent of stand |
|---|---|---|
| Yellow Poplar (*Liriodendron*) *tulipifera* | 122 | 44.5 |
| Sassafras (*Sassafras albidum*) | 107 | 39.0 |
| Black Cherry (*Prunus serotina*) | 12 | 4.4 |
| Cucumber Magnolia (*Magnolia grandiflora*) | 11 | 4.0 |
| Red Maple (*Acer rubrum*) | 10 | 3.6 |
| Red Oak (*Quercus rubra*) | 8 | 2.9 |
| Butternut (*Juglans cinerea*) | 1 | .4 |
| Shagbark Hickory (*Carya ovata*) | 1 | .4 |
| American Beech (*Fagus grandifolia*) | 1 | .4 |
| Sugar Maple (*Acer saccharum*) | 1 | .4 |
| | 174 | 100 |

Table 20-3 is richer and has greater heterogeneity than the stand in Table 20-4.

In order to quantify species diversity for purposes of comparison, a number of indexes have been proposed (see Appendix C; Lloyd and Ghelardi, 1964; Pielou, 1971, 1975). One of the most widely used is the Shannon-Wiener Index, which includes both richness and heterogeneity.

The Shannon-Wiener Index measures diversity by the formula

$$H = -\sum_{i=1}^{s} (p_i)(\log p_i)$$

where

$H$ = diversity index

$s$ = number of species

$i$ = species number

$p_i$ = proportion of individuals of the total sample belonging to the $i$th species

This index is based on information theory. It was originally developed to enable communication engineers to predict the name of the next letter in a message. As such the index is a measure of uncertainty. The higher the value of $H$, the greater is the degree of uncertainty that the next letter or bit of information will be the same as the previous one. The lower the value, the less is the uncertainty. Thus, in the woodland with low diversity (84 percent yellow poplar and sassafras), the chances are great that in sampling the trees the next tree picked at random will be a yellow poplar or a sassafras. In the woodland with higher diversity, the chances that the next tree picked at random will be a yellow poplar or white oak are considerably less.

Diversity indexes assume that the more abundant a species is, the more important it is to the community. But the more abundant species are not necessarily the most important or the most influential. In communities embracing organisms possessing a wide range of sizes, the importance of fewer but larger individuals may be underestimated and the more common species are weighted more heavily than the many rare species. Thus one of the distinctive failures of the indexes is the inability to distinguish between the abundant and the important species. Nevertheless, diversity indexes do provide one measure for community comparisons. (For discussion of indexes, see Peet, 1974.)

***Diversity gradients.*** Species diversity can be used not only in comparing similar communities or habitats within a given region, but also in examining global ecosystems. Traveling north from the tropics to the Arctic, one finds the numbers of plants and animals decreasing on a latitudinal gradient. Species of nesting birds might approach 1395 in Colombia, drop to 630 in Panama, to 143 in Florida, to 118 in Newfoundland, and to 56 in Greenland (A. G. Fischer, 1960). The same pattern is found for mammals (Simpson, 1964), fish (Lowe-McConnell, 1969), lizards (Pianka, 1967), and trees (Monk, 1967). Diversity decreases from warm to cold climates.

But diversity is not restricted to a latitudinal gradient from the tropics to the Arctic. In oceans species diversity increases from the continental shelf where food is abundant but the environment changeable to deep water where food is less abundant but the environment more stable. Mountain areas generally support more species than flatlands, and peninsulas have fewer species than adjoining continental areas. From east to west in North America the number of species of breeding land birds (MacArthur and Wilson, 1967) and mammals (Simpson, 1964) increases. This increased diversity on an east-west gradient relates to an increased diversity of the environ-

ment both horizontally and altitudinally (Figure 20-8). Eastern North America has more uniform topography and climate and thus holds fewer species than western North America. However, because of more favorable moisture conditions, amphibians are more abundant and diverse in eastern North America than in the western part of the continent, while reptiles are more diverse in the hot arid regions of the west (Kiester, 1971) (Figure 20-9).

Small or remote islands have fewer species than large islands and those nearer continents (MacArthur and Wilson, 1967). The number of species on islands depends upon the rate at which replacement species reach an island and how long they survive (see page 603). Because of the tendency for immigration rates to balance extinction rates under equilibrium con-

**FIGURE 20-8**
*Latitudinal variation in distribution of mammals. Diversity of animals across a continent is influenced by temperature and moisture reflected in latitudinal and altitudinal variations. Species density of North American mammals from the Arctic to the Mexican border along the 100th meridian (a) and from the Arctic through Central America (b). (After Simpson, 1964.)*

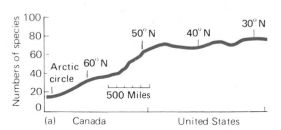

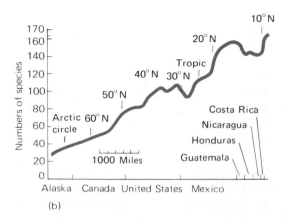

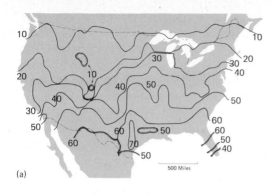

(a)

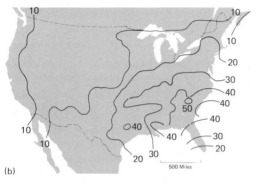

(b)

**FIGURE 20-9**
*Latitudinal variation in distribution of reptiles (a) and amphibians (b). The most pronounced latitudinal variations occur among reptiles and amphibians. Being poikilothermic and ectothermic, reptiles have their greatest diversity in hot desert regions and lower latitudes of North America. Being not only poikilothermic, but also highly sensitive to moisture conditions, amphibians reach their greatest diversity in the central Appalachians and decrease northward, southward, and westward. They are the lowest in species in the dry and the cold regions of the continents. (From Kiester, 1971.)*

ditions, one can, by knowing the size of an island and its distance from a continent, predict the diversity of its species (see Figure 20-23).

*Species diversity hypotheses.* Many hypotheses have been proposed to explain why the tropics should hold a greater abundance of species than the temperate region or why one island should hold more than another. Many of these hypotheses are similar, but not identical.

The *evolutionary time theory* (A. G. Fischer, 1960; Simpson, 1964) proposes that diversity relates to the age of the community. Old communities (in an evolutionary sense) hold a greater diversity than young communities. Tropical communities are older and evolve and diversify faster than temperate or arctic communities in part because the environment is more constant and climatic catastrophes less likely. There is some evidence to support this. For example, fossil planktonic Foraminifera from the Cretaceous period in the northern Hemisphere show a gradient in species abundance from the tropics to the Arctic similar to that found in living Foraminifera of today (Stehli, et al., 1969).

Considering a shorter time scale, the *ecological time theory* is based on the time required for a species to disperse into unoccupied areas of suitable habitat. Because there has not been enough time for many species to move into temperate zones, these areas are unsaturated as to the total number of species they now support. Many cannot move until barriers to dispersal are broken; others are already moving out of the tropics into temperate zones, as both the natural spread of the cattle egret in North America from Africa by way of South America and the northward spread of the armadillo indicate.

The *spatial heterogeneity theory* (Simpson, 1964) holds that the more complex and heterogeneous the physical environment, the more complex and diverse will its flora and fauna be. The greater the variation in topographic relief, the more complex the vertical structure of the vegetation, and the more types of microhabitats the community contains, the more kinds of species it will hold. This theory is supported by the fact that the more complex the vertical stratification of a community, the more species of birds it holds (MacArthur, 1972; Pearson, 1971).

The *climatic stability theory* (A. G. Fischer, 1960; Connell and Orias, 1964) holds that the more stable the environment, the more species will be present. Through evolutionary time the tropics, of all regions of the earth, has probably remained the most constant and has been relatively free from severe environmental conditions that could have catastrophic effects on a population. Under tropical conditions selection favors organisms with narrow niches and specialized feeding habits. Because each species uses a smaller fraction of the total resources, more species are able to exist in regions of more constant climate. At higher latitudes where the climate is severe and un-predictable selection favors organisms with broad tolerance limits for variations in the physical environment and with more general-ized food habits.

Another interesting hypothesis closely re-lated to climatic stability is the *productivity theory* advanced by Connell and Orias (1964). In brief, this hypothesis proposes that the level of diversity of a community is determined by the amount of energy flowing through the food web. The rate of energy flow is influenced by the limitation of the ecosystem and by the de-gree of stability of the environment.

If one assumes a hypothetical increase in the stability of the physical environment, then with increasing environmental stability, less energy is required for regulatory activities and more energy enters into net production. In-creased net productivity can support larger populations. Larger populations maintain greater genetic variety and increase the oppor-tunity for interspecific association. Greater productivity per unit area permits the less mobile animals to become even more seden-tary, and the species tend to be broken into many semi-isolated populations, which may bring about greater intraspecific genetic vari-ety. As a result speciation is favored, espe-cially if semi-isolated segments are exposed to new environments. Any new species that arise would tend to be more specialized and initially to have smaller populations.

In the early stages of the evolution of a community, positive feedback would increase the rate of speciation, resulting in a faster cycling of nutrients and an increase in net productivity. As the number of species in-creases, the food webs become more complex and the community becomes more stabilized. But the tendency toward overspecialization and the smaller population per species would tend to decrease community stability and act as a negative feedback on the whole system.

The productivity theory in effect says that the more food produced, the greater the diver-sity. While perhaps true in a general sense, there are too many exceptions. In some aquatic ecosystems increased productivity from en-richment of the system with sewage and other nutrients sources results in a decrease in di-versity. Marine benthic regions of low produc-tivity have a higher abundance of species than areas of high productivity (Sanders, 1968). In tropical bird communities it appears that the number of ways that energy is packaged rather than total energy is best correlated with diversity (Karr, 1975).

The *competition theory* (Dobzhansky, 1951; C. B. Williams, 1964) states that in environ-ments of high physical stress, such as the Arc-tic with its frigid cold and the temperate reg-ions with their wide fluctuations in annual temperatures, selection is controlled largely by the physical variables. In more benign climatic regions biological competition be-comes more important in the evolution of species and in the specialization of niches.

Out of his studies of animals of the rocky in-tertidal zones, Paine (1966) has proposed the *predation theory*. Because more predators and prey exist in the tropics than elsewhere, the predators tend to hold down the prey species to such a low level that competition among prey species is reduced. This theory supposes that more diverse communities support an in-creased proportion of predators, that predators are very efficient at regulating the abundance of prey, and that these assumptions are true at all trophic levels including the primary-producer level. Janzen (1971) suggests that this indeed does happen in the tropics. Be-cause seed-eating animals tend to cluster about the seed-producing tree, seed mortality is heaviest about the seed source, whereas the probability that seeds will be overlooked in-creases with the distance from the tree.

Sanders (1968) has combined the environ-mental stability hypothesis and the time hypothesis into still another one, the *stability-time hypothesis*. This assumes that two con-trasting types of communities exist, the physi-cally controlled and the biologically controlled.

In the physically controlled communities or-ganisms are subjected to physiological stress due to fluctuating physical conditions. The or-ganisms in time evolve adaptive mecha-nisms to meet these conditions. But at least some of the time the organisms are subject to severe physiological stress, the probabilities of reproductive success and survival are low. As a result diversity is low.

Low-diversity environments fall into three

categories: (1) new environments in which the number of organisms colonizing the area is increasing, but organisms are subject to environmental stress; (2) severe environments in which a slight environmental change such as an increase in temperature or salinity can eliminate life altogether; and (3) unpredictable environments in which the environmental properties vary widely and unpredictably about some mean value. A wide fluctuation from the mean can severely stress the population.

In the biologically controlled community the physical conditions are relatively uniform over long periods of time and are not critical in controlling the species. Evolution proceeds along the lines of interspecific competition, one species adapting to the presence of the other and sharing the resources with it. The environment is more predictable, the physiological tolerances of the organisms are low, and diversity is high.

But there is no such entity as a wholly physically controlled or a wholly biologically controlled community. Rather the community is influenced by the interaction of the two. In situations where physiological stress has been low, biologically accommodated communities evolved. As physiological stress increases due to increasing fluctuations in the physical environment the community changes from a biologically controlled to a physically controlled one. The number of species diminishes gradually along the gradient of stress. When

FIGURE 20-10
*Graphic representation of the stability-time hypothesis. The hypothesis states that species numbers diminish continuously along a gradient of physiological stress. The greatest diversity occurs among the predominantly biologically accommodated communities. (From Sanders, 1968.)*

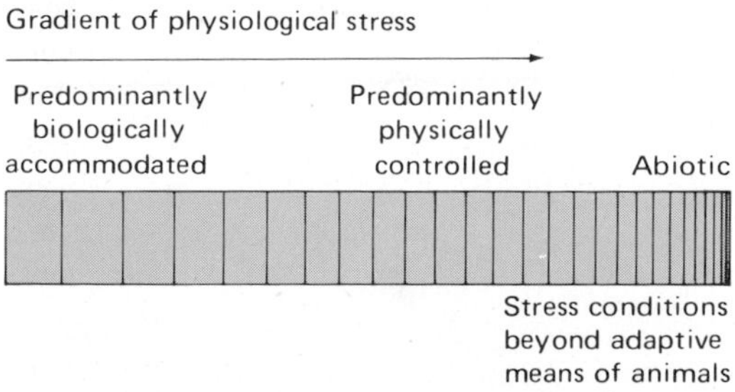

Gradient of physiological stress

| Predominantly biologically accommodated | Predominantly physically controlled | Abiotic |

Stress conditions beyond adaptive means of animals

the point is reached where stress is too severe, no species exist (Figure 20-10).

However intriguing these hypotheses may appear, it is difficult to test any of them in the field or even to put them into mathematical models. But in spite of this it is apparent that the diversity of species can be related to a number of such variables as the structure of the habitat, the diversity of microhabitats, the nature of the physical environment, climate and protection from its adverse effects, the availability of food, nutrient supply, and time. Also historical factors in other senses may also govern the detailed variation in community structure, including diversity, from place to place.

## The niche

The word niche has been used throughout the text. Like other ecological terms, niche is rather nebulous, its meaning colored by various interpretations that equate it with habitat, competition, functional roles, food habits, and morphological traits. The idea of the niche has been so important to ecology that it demands closer analysis.

One of the first to propose the idea of the niche was the ornithologist Joseph Grinnell (1917, 1924, 1928). In his study of the California thrasher *(Taxostoma redivivium)* and other birds, he suggested that the niche be regarded as a subdivision of an environment occupied by a species, "the ultimate distributional unit within which each species is held by its structural and functional limitations." Essentially, Grinnell was describing the habitat of the species.

Charles Elton (1927), in his classic book *Animal Ecology,* considered the niche as the fundamental role of the organism in the community—what it does, its relation to its food and its enemies. Basically this idea stresses the occupational status of the species in the community.

In 1959 G. E. Hutchinson proposed that the niche be considered the total range of environmental conditions under which the organism lives and reproduces. The Hutchinsonian concept was initially limited to environmental variables affecting a species, such as light, temperature, moisture, height of vegetation, and time. If biological variables are added to the environmental variables, this concept of the niche considers both the organism's environment and its function in the community.

E. P. Odum (1959) defined the niche as "the position or status of an organism within its community and ecosystem resulting from the organism's structural adaptations, physiological responses, and specific behavior (inherited and/or learned)." The habitat then becomes the organism's address and its niche is its profession.

More recently Whittaker et al. (1973) suggest that the term niche refer to the functional relationships of a species within a community; the habitat refers to the distributional response of an organism at different points in the landscape. If the two ideas must be considered together then the term ecotope, the habitat plus the niche, might be used. The ecotope would describe "the species response to the full range of environmental variables to which it is exposed" and "is the ultimate evolutionary context of the species."

Pianka (1978) has defined the niche as "all the various ways in which a given organismic unit conforms to its particular environment." (An organismic unit is an individual, a population, or a species.) Whatever definition one finally accepts it should be one that stresses the functional role of the organism in the community, the function it performs and the place in which it performs it.

## FUNDAMENTAL AND REALIZED NICHE

According to the Hutchinsonian concept, the community may be viewed as an aggregation of many environmental and functional variables, each of which can be considered to be a point in a multidimensional space. Hutchinson called this space the *hypervolume*. The niche for any one species would be the portion of the hypervolume it occupies as defined by the upper and lower limits of all the environmental and functional variables in which the species can live and replace itself. Within any one community a species free from interference from another species could occupy the full range of variables to which it is adapted. This is the idealized *fundamental niche* of the

species. Because it has an infinite number of dimensions and its environmental variables are not always linearly ordered as the model assumes, the fundamental niche cannot be determined completely.

The fundamental niche assumes the absence of competitors, but rarely is this the case in natural communities. Competitive relationships force the species to constrict the portion of the hypervolume it occupies. The set of conditions under which an organism actually exists in a community is its *realized niche* (Figure 20-11). In their studies ecologists usually confine themselves to one or two niche dimensions such as the feeding niche or a space niche.

Consider two examples. Root (1967) studied the exploitation of the feeding niche by the blue-gray gnatcatcher in California oak woodlands. He characterized the niche of the bird in part by the size of its food and by the height above the ground at which it captured food (Figure 20-12). Simplified for the sake of example, the bird's fundamental niche could be considered as characterized by a maximum range of size of prey between 1 and 14 mm in length and by a foraging area of ground level to 35 feet. The gnatcatcher's niche center, in-

**FIGURE 20-11**
*Fundamental and realized niches. The fundamental niche of a species represents the full range of environmental conditions, biological and physical, under which a species can successfully exist. However, under pressure of superior competitors the species may be completely displaced from part of the fundamental niche and forced to retreat to that portion of the fundamental niche hypervolume in which it is most highly adapted. The portion it occupies is its realized niche.*

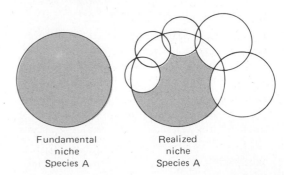

Fundamental
niche
Species A

Realized
niche
Species A

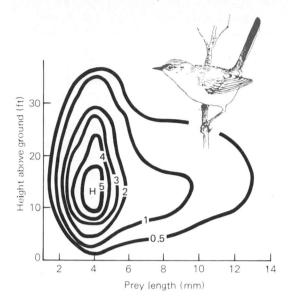

**FIGURE 20-12**

*Representation of niche space for the blue-gray gnatcatcher* (Polioptila caerulea) *based on a combination of two variables, size of prey and feeding height above ground. The contour lines map the feeding frequencies (in terms of total diet for the two niche axes) for adult gnatcatchers during the incubation period during July and August in California oak woodlands. The maximum response level is at H. This represents optimal conditions. Contour lines spreading out from this optimum represent decreasing response levels. For illustrative purposes the outer contour line may be considered the outer boundary of the fundamental niche, although under the field conditions in which the study was made it probably represents the outer boundary of the realized niche for these two variables. For a detailed discussion of such an analysis of the niche, see Maguire, 1973; Hutchinson, 1978. (Diagram from Whittaker et al., 1974, based on data from Root, 1967.)*

dicated by frequency of capture and stomach content analysis, consists of insects 3 to 5 mm taken 8 to 28 feet above the ground. The further the height and food dimensions diverge from this center, the more the gnatcatchers niche may overlap those of other species. The boundaries of the realized niche could be defined by the boundaries of any one of the contour lines according to the degree of competition.

Putwain and Harper (1970) studied the population dynamics of two species of dock,

*Rumex acetosa* and *R. acetosella,* each growing in hill grasslands in North Wales. *R. acetosa* grew in a grassland community dominated by velvet grass *(Holcus lanatus)* and red and sheep fesques *(Festuca rubra* and *F. ovina)*; *R. acetosella* grew in a community dominated by sheep fesque and bedstraw *(Galium saxatile)*. To determine interference and niches of the two dock species, Putwain and Harper treated the flora with specific herbicides to remove selectively in different plots (1) grasses, (2) forbs except *Rumex* species, (3) the *Rumex* species. All species except *R. acetosella* spread rapidly after the grasses were removed, but *R. acetosella* increased only after both grasses and nongrasses were removed.

The niches of these two plants are explained and diagrammed in Figure 20-13. The fundamental niche of *R. acetosella* (R) overlaps the fundamental niches of both grasses (G) and other forbs (D). Only when these competitors are eliminated does this dock realize its fundamental niche. *R. acetosa,* however, overlaps only with the grasses and only their removal is necessary to permit expansion of this dock throughout its fundamental niche.

Note that the niches of seedlings differ from those of the mature plant. The fundamental and realized niches of an organism can change with its growth and development. Insects with a complex life cycle may occupy one niche as a larva and an entirely different niche as an adult. In other organisms niche space can change as the organism matures because food and cover requirements change as the organism grows larger.

**NICHE OVERLAP**

The example of *Rumex* brings up the question of *niche overlap.* This occurs when two or more organisms use the same resource, such as food. Thus the concept of the niche is closely associated with the concept of competitive relationships among species (see Chapter 16).

The theoretical model of the niche assumes that competition is intense, that only one species can occupy a niche space, and that competitive exclusion takes place in areas of overlap. The amount of niche overlap is assumed to be proportional to the degree of competition for that resource. In a condition of lit-

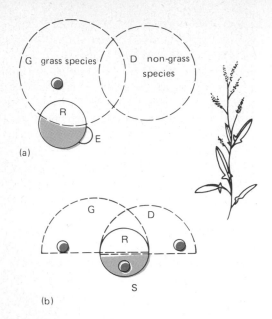

**FIGURE 20-13**

*Diagrammatic representation of niche relationships of* Rumex acetosa *and R. acetosella in mixed grassland swards. In each diagram the fundamental niches of grass species (G) and nongrass species (D) overlap. The fundamental niche of* Rumex *species (R) is shown as a continuous line and the realized niche is shaded. E is that part of the fundamental niche of R. acetosa which is expressed in the presence of nongrass species only and does not overlap the fundamental niches of G and D. The fundamental niches of seedlings, shown by the small dark colored circles, are contained within the fundamental niches of grasses, nongrasses, and mature* Rumex. *Compare this diagram with Figure 20-14. (From Putwain and Harper, 1970.)*

tle or no competition niches may be adjacent to one another with no overlap or they may be disjunct (Figure 20-14). At the other extreme, in a condition of intense competition the fundamental niche of one species may be completely within or correspond exactly to another, as in the case of the seedling *Rumex.* In such instances there can be two outcomes. If the niche of species 1 contains the niche of species 2 and species 1 is competitively superior, species 2 will be eliminated entirely. If species 2 is competitively superior, it will eliminate species 1 from the part of the niche space species 2 occupies. The two species then coexist within the same fundamental niche.

In most cases fundamental niches overlap. Some niche space is shared and some is exclusive, enabling the two species to coexist (Figure 20-14). Considerable niche overlap does not necessarily mean high competitive interaction. In fact the reverse may be true. Competition involves a resource in short supply. Extensive niche overlap may indicate that little competition exists and that resources are abundant. Pianka (1972, 1975) has suggested that the maximum tolerable overlap in niches should be lower in intensely competitive situations than in environments with low demand/supply ratios.

For simplicity niche overlap is usually considered one dimensional. In reality, of course, niche involves the utilization of many types of resources: food, a place to feed, cover, time, and so on. Rarely do two or more species possess exactly the same niche involving all requirements. While species may show overlap on one gradient, they may not on another. As with complete competition, species may exhibit little overlap (Figure 20-15).

Frequently more than two species, especially related genera, share a resource in a given community. A group or cluster of species that share a resource and thus occupy related or closely overlapping niches and strongly interact with one another but weakly with other members of the community is considered a *guild* (Root, 1967). For example, birds might be grouped to form a seed-eating guild, a flying-insect-feeding guild, or a nest hole guild (Figure 20-16).

### NICHE WIDTH

*Niche width* (or niche breadth or niche size) is the extent of the hypervolume representing the realized niche (see Van Valen, 1965; Van Valen and Grant, 1970; Vandermeer, 1972). Niche width is generally described as narrow or broad. This description is derived from plots of niches on a resource axis. The width is measured by the length of the axis intercepted by the curve (Figure 20-17). Measurements of niche width usually involve measures of some morphological trait such as bill size or some ecological variable as food size or habitat space. Obviously, determining what variables make up a niche is difficult, let alone measuring them and representing them graphically.

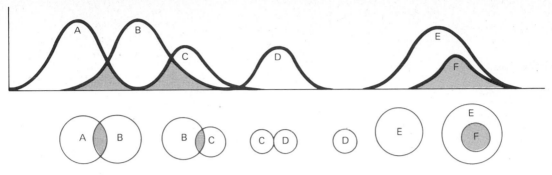

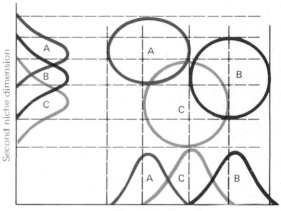

FIGURE 20-14

*Different types of niche relationships
visualized as graphs on a resource gradient
and as Venn diagrams. Species A and B have
overlapping niches of equal breadth but are
competitive at opposite ends for the resource
gradient. B and C have overlapping niches of
unequal breadth. Species C shares a greater
proportion of its niche with B than B does
with C. (In this example, however, B shares
its niche also with A at the other end of its
resource niche.) C and D occupy adjacent
niches with little possibility of competition.
D and E occupy disjunct niches and no
competition exists. Species F has a niche
contained within the niche of E. If F is
superior to E competitively, it persists and E
shares that part of its niche with F.
Compare this with Figure 20-13. (Adapted
from Pianka, 1978.)*

FIGURE 20-15

*Niche relationships based on two
gradients. Models of niche as a single
dimensional gradient do not indicate the
degree of true niche overlap where other
gradients are involved. Two species may
exhibit considerable overlap on one gradient
and little or none on another. When several
niche dimensions are considered, niche
overlap may be reduced considerably as
illustrated here. On resource gradient 1 A
and B exhibit no overlap and on resource
gradient 2, they overlap equally and
opposite. When both niches are considered
(circles) A and B do not overlap. C on
resource gradient 2 overlaps equally with B
and very little but equally with A. On
resource gradient 2 C overlaps with both A
and B. When both gradients are considered
C overlaps mostly with B and very little
with A.*

Multidimensional analysis allows many niche variables to be plotted as a single variable. The components are reduced to a single linear combination of the several resource axes (Harner and Whitmore, 1977).

More precisely niche width is "the sum total of variety of different resources exploited by an organism" (Pianka, 1975). The wider the niche, the more generalized the species is considered to be; the narrower the niche, the more specialized is the species. Generalist species have broad niches and sacrifice efficiency in the use of a narrow range of resources for the ability to utilize a wide range of resources; as competitors they are superior to specialists if resources are somewhat undependable. Specialists, equipped to exploit a specific set of resources, occupy narrow niches; as competitors they are superior to generalists if resources are dependable and renewable. A dependable resource supply is closely partitioned among specialists with low interspecific over-

lap (Roughgarden, 1974). If resource availability is variable, generalist species are subject to invasion and close packing during periods of resource abundance.

For simplicity, niche width can be considered in terms of space utilization, food re-

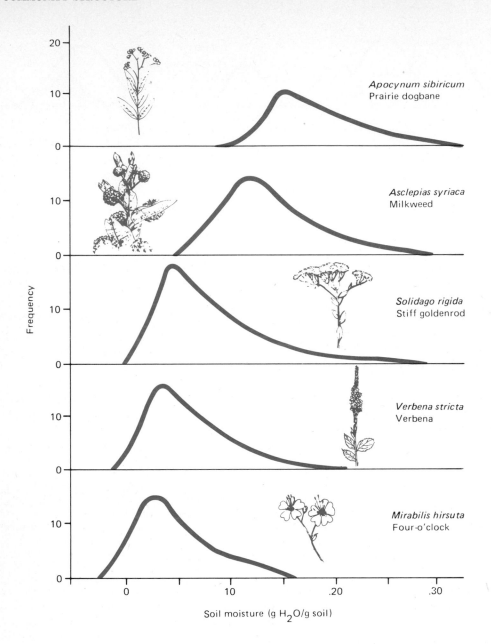

**FIGURE 20-16**
*Guild of wind-dispersed fugitive prairie plants that occur on localized disturbances, created largely by badgers, but rarely appear in the surrounding undisturbed prairie. This figure shows the distribution of these individuals along a soil moisture gradient. Abscissa is the percent moisture content of the soil at the layer of dead vegetation in badger disturbances. Ordinate is the number of disturbances at different moisture levels on which individuals were located. (Adapted from Platt and Weis, 1977.)*

source, or even morphological differences. Territorial animals, for example, parcel out bits of a given area of suitable habitat among relatively few individuals. Space demands become most critical among those species exhibiting interspecific territoriality.

Space can be divided into feeding niches. Organisms, especially birds and many insects, have restricted areas in the vertical vegetational profile in which they forage (Figure 20-18) (see MacArthur, 1958; Lack, 1971;

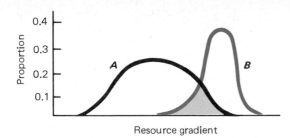

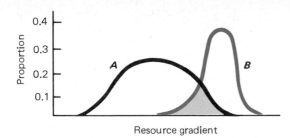

**FIGURE 20-17**
*Hypothetical distribution of a species (A) with a broad niche and of a species (B) with a narrow niche on a resource gradient. The niches overlap (shaded area). Species A overlaps species B more than species B overlaps A.*

Edington and Edington, 1972). The division of feeding space or the occupancy of somewhat different niches within species may be based on sex. The male red-eyed vireo, for example, gleans its insect food in the upper canopy, the female in the lower canopy and nearer the ground, with only about 35 percent overlap in feeding areas between the two (Williamson, 1971). Although similar foods may be utilized, each secures insects from different levels (Figure 20-19). Similar occupancy of separate niches exists between males and females of several of the woodpeckers (Ligon, 1968).

Morphological differences influencing food procurement may be a variable determining niche. For birds niches may be separated by food size dictated by the size of bill (see Lack, 1971). Differences in the length of bill are correlated with differences in foraging behavior. For example, a pronounced sexual difference in bill size exists between males and females of the Arizona woodpecker *(Dendrocopus arizonae);* the male forages on the trunk, while the female seeks food on the branches (Ligon, 1968).

## NICHE COMPRESSION, ECOLOGICAL RELEASE, AND NICHE SHIFT

Niche width provides some indication of resource utilization by a species. If a community consisting of a number of species with broad niches is invaded by competitors, intense competition may force the original occupants to restrict or compress their utilization of space.

The organisms may be forced to confine their feeding or other activities to those patches of habitat providing optimal resources. Competition which results in the contraction of habitat rather than a change in the type of food or resources utilized is called *niche compression* (MacArthur and Wilson, 1967).

Conversely, if interspecific competition is reduced, a species may expand its niche by utilizing space previously unavailable to it. Niche expansion in response to reduced interspecific competition is called *ecological release.* Ecological release may occur when a species invades an island which is free of competitors, moves into habitats it never occupied on the mainland, and increases its abundance (Cox and Ricklefs, 1977). Such expansion may also follow when a competing species is removed from a community, allowing a remaining species to move into microhabitats it previously could not occupy.

Associated with compression and release is another response, *niche shift.* Niche shift is the adoption of changed behavioral and feeding patterns by two or more competing populations to reduce interspecific competition. The shift may be a short-term ecological response or a long-term evolutionary response involving some change in a basic behavioral or morphological trait. For example, Werner and Hall (1976) demonstrated niche shift in three competing species of sunfish (Centrarchidae), the bluegill *(Lepomis maerochirus),* the pumpkinseed *(L. gibbosus),* and the green sunfish *(L. cynellus).* When the three species were stocked in separate replicated experimental ponds, the food habits of the three were similar and the average growth rate and food size increased. When the three species were stocked together, the bluegill concentrated on zooplankton prey in open water. The bluegill has long, fine, gill rakers which retain small prey. The pumpkinseed concentrated on prey associated with bottom sediments, mainly Chironominae. It has short, widely spaced gill rakers that do not become fouled when the fish sifts through bottom sediments. The green sunfish concentrated on insects found on the stems and leaves of vegetation and bottom sediments. The green sunfish is a more efficient forager in vegetation.

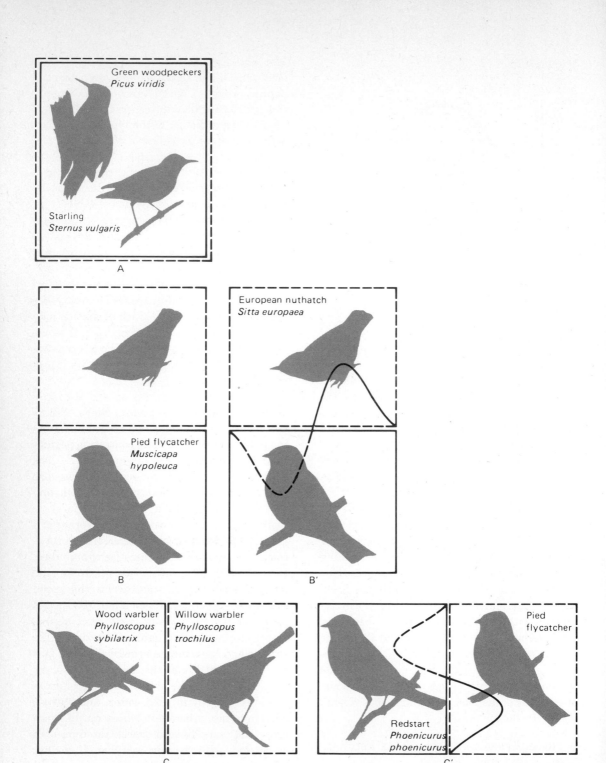

A

Green woodpeckers
*Picus viridis*

Starling
*Sternus vulgaris*

B

Pied flycatcher
*Muscicapa
hypoleuca*

B'

European nuthatch
*Sitta europaea*

C

Wood warbler
*Phylloscopus
sybilatrix*

Willow warbler
*Phylloscopus
trochilus*

C'

Redstart
*Phoenicurus
phoenicurus*

Pied
flycatcher

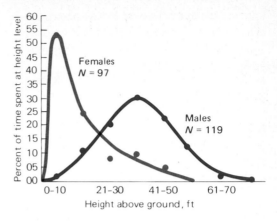

*FIGURE 20-18*
*Niche relationships based on spatial separation. Different species may share a resource, such as food, by dividing the single resource into units. The division may be based on food specialization without spatial separation. Food specialization (A) is exemplified by the green woodpecker and starling. Both are ground feeders, but the green woodpecker feeds on ants, the starling on insect larvae. Spatial separation results from behavioral and morphological specialization that enables each species to locate and utilize a particular part of the habitat. Vertical separation (B) or the use of different feeding zones depends upon the specialization of species to move and forage on the ground, tree trunks, branches, or twigs. The nuthatch forages on trunks and large branches of trees for insect larvae. The pied flycatcher hawks for flying insects from a vantange point. Horizontal separation (C) takes the form of mutually exclusive territories as in the case of the wood warbler and the willow warbler, or mutually exclusive feeding areas as exemplified by the redstart and pied flycatcher (C'). Involved in both vertical and horizontal separation are habitat selection and possibly competitive exclusion. The absence of one species may allow another to exploit a wider vertical range or horizontal range (B' and C'). In C' both the redstart and pied flycatcher catch flying insects and take insects from leaves and terminal twigs in the canopy. The pied flycatcher is more efficient at capturing flying insects. Where the two occur together the redstart occupies dry open parts of woods and gentle slopes. The pied flycatcher inhabits more shaded woods, woodland streams, and steeper slopes. But the absence of one allows the other to exploit a wider range of food or habitats. In B' the nuthatch and pied flycatcher are rather strongly separated by foraging sites. However the nuthatch does take many flying insects from the trunks and branches. The pied flycatcher utilizes the same flying insects when possible. (After Edington and Edington, 1972.)*

*FIGURE 20-19*
*Separation of male and female red-eyed vireos* (Vireo olivaeus) *by height of foraging. Mean height for males: 37.1 ft; standard deviation, 12; standard error (S.E.), 1.0; range, 9-75. Mean height for females: 14.2 ft; standard deviation, 10.8; S.E., 1.1; range 2.50. (From Williamson, 1971.)*

The variations in morphological traits among the three species of sunfish allow them to shift behavioral and feeding patterns as necessary in competitive and noncompetitive situations. Behavioral plasticity allows them to use to advantage the morphological differences. Thus niche shift is accomplished through a combination of the two and may occur over long-term environmental changes or with short-term seasonal patterns of availability of resources. In summer, when food resources dwindle, the three species become competitive and are forced into partitioning food resources to reduce competition.

### SPECIES ABUNDANCE HYPOTHESES

Some clue to the nature of niche relationships may be obtained from patterns of species abundance. Theoretically the way species divide up the hypervolume should be reflected in abundance of species. There are several hypotheses relating to species abundance, the random-niche-boundary hypothesis, the niche-preemption hypothesis, and the log-normal distribution of species.

The *random-niche hypothesis* (MacArthur, 1960) suggests that boundaries of the niche hypervolumes of several species can be treated as points cast at random on a line. The points then represent the niche boundaries. The length of each contiguous, nonoverlapping segment represents the niche size. If the segments, representing the importance value of the species (the percentage of total density, biomass, etc. of all species in a community that a single species represents), are plotted in sequence from the longest to the shortest, then a curve like that of Figure 20-20 will result. This hypothesis, also known as the broken-stick model (see Hairston, 1969), is realistic only in rare situations; it is still employed but generally considered obsolete. The curve is approached only by some small samples of taxonomically related animals with stable populations and relatively long life cycles occupying a small homogeneous community such as nesting birds in a forest.

The *niche-preemption hypothesis* supposes that the most successful or dominant species preempts the most space, the next most successful claims the next largest share of the space, and the least successful occupies what little space is left. If the relative importance of the species is plotted in species sequences on a log scale, the result is a straight line (Figure 20-20), and the distribution of the species forms a geometric series. Such a distribution is achieved only by a few plant communities containing relatively few species and occupying severe environments such as a desert. In most plant and animal communities, species overlap in the use of space and resources.

The *log-normal hypothesis* (Preston, 1962) supposes that the niche space occupied by a species is determined by a number of conditions such as resources, space, microclimate, and other variables that affect the success of one species in competition with another. The relative importance of each species is determined by the way the various variables affect each species. This results in a bell-shaped or normal distribution of importance values (Figure 20-21). Again, as in the random-niche theory, a line can be divided into segments representing the ranges in importance values, and the ranges are placed in frequency distribution classes. The central range of importance values will have the most segments or species in it, whereas fewer segments will fall into the ranges on either side. Again, the important species will be few in number, and many will be of intermediate importance. If

**FIGURE 20-20**
*Graphic representation of hypotheses suggested to delineate the niche. The hypotheses may be illustrated by graphs in which the importance values of the species (percentage of all species that particular species represents) expressed as total density, total net productivity, or some other measurement is plotted against the sequence of species. In the random niche-boundary hypothesis represented by graph line A boundaries are located at random positions in the hyperspace. In a geometric series (B) the size of the niche hypervolume is determined primarily by the ability of the dominant species to preempt part of the niche space, leaving the less successful species to occupy what is left. In a log-normal distribution (C) the niche space occupied by a species is determined by a large number of variables that affect the competitive abilities of the several species. (After Whittaker, 1965.)*

**FIGURE 20-21**
*Bell-shaped curve of importance values resulting from a log-normal distribution of plant species. The importance value in this example is determined by the percent of ground surface covered by the species. It is represented on the horizontal scale (logarithmic) by octaves in which the species are grouped by doubling the units of percentage of cover. The largest number of species occurs in the middle octaves. (After Whittaker, 1965.)*

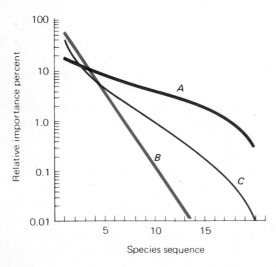

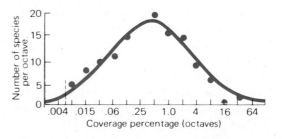

this distribution is plotted logarithmetically, the curve produced will fall somewhere between the random-niche-boundary and the niche-preemption curves. The log-normal distribution most closely approximates the distribution of importance values obtained from communities rich in species. All three approaches, however, are inadequate, and a way to define the boundaries of the niche still remains elusive.

## Island biogeography

The number of species found in a community is a function of immigration and extinction. Such a relationship is most easily observed in the study of the biogeography of islands where the number and diversity of species is often less than that of the mainland. F. W. Preston (1962) and MacArthur and Wilson (1963) suggested that the biota of an island reaches an equilibrium between the immigration of species new to the island and extinction of those already present. A new or uninhabited island would be colonized rapidly by those forms with the greatest dispersal ability. As the number of colonizing species increases, the number of new immigrants arriving on the island decreases. Immigration declines because as the number of species colonizing the island approaches the number of species in the source area, fewer immigrants belong to the new species. Also, later immigrants may be unable to establish populations because available habitats are filled or the resources are already utilized. At the same time populations of some island colonists go extinct. The extinction rate increases as more species arrive because of increased interspecific competition and reduced population size.

Eventually an equilibrium is established between immigration and extinction. The equilibrium level depends upon the number of species per island, the size of the island, and the distance of the island from the source of immigrants. The relationship of these factors is illustrated in Figure 20-22.

In general larger islands support more species than smaller islands. A ten-fold increase in area can double the number of species present. Larger islands are more likely

to have a more complex structure and thus provide habitats for a greater diversity of species. Also their larger area makes possible colonization by species with large range requirements, particularly the predators, usually missing on smaller islands. Extinction rates are lower on large islands than on small because of larger populations sizes, more specialized and diverse habitats, and greater availability of resources. Small islands by their very nature have less diverse habitats and hold smaller and thus more extinction-prone populations.

Islands close to a mainland or a source of immigrants support more species, especially at the higher trophic levels, than more distant islands, and more colonists arrive on them per unit time. Not only are near islands available to species with lower dispersal abilities, but also that availability permits recurrent colonization at a rate sufficient to offset the extinction of species. Because of the isolation of distant islands from the source, biota distance tends to suppress access of many potential plant and insect colonists. This results in a general impoverishment of insular habitats and resources, especially those needed by vertebrates (Johnson, 1976). As indicated in Figure 20-22 an island near a mainland may support as many species as a much larger island some distance from the mainland.

The replacement of extinct species by immigrant species or, more generally, the replacement of one species by another is called *turnover*. Because the species going extinct locally may be different from the immigrants replacing them, species composition constantly changes through turnover. Turnover can be determined by periodic censuses of island fauna, recording which species are breeding, the absence of species previously present, and the appearance of new species. But an accurate estimate requires a frequent monitoring of the population (Lynch and Johnson, 1974). A key to the determination of turnover and equilibrium is the interpretation of colonization and extinction. *Colonization* is defined as the persistence of an immigrant species on an island as a breeding species through at least one reproductive cycle (MacArthur and Wilson, 1967; Lynch and Johnson, 1974). Extinction, or more precisely local extinction, is

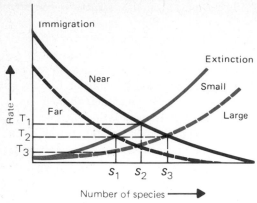

**FIGURE 20-22**
*Graphical representation of island biographical theory. Immigration rates decrease with increasing distance from source area. As a result distant islands reach species equilibrium with fewer species than near islands, all else being equal ($S_3 < S_2$ for large islands; $S_2 < S_1$ for small islands). Extinction rates increase as the size of island becomes smaller. As a result large islands have lower extinction rates than small islands. Considering both immigration and extinction rates one can hypothesize that near islands reach equilibrium with more species than distant islands ($S_2 < S_1$ for small islands; $S_3 < S_2$ for large islands). Small islands reach equilibrium with fewer species than large islands, all else being equal ($S_2$ for small islands, $S_3$ for larger islands) and large, far islands have an equilibrium density equal to that of near, small islands. Turnover rates are greater for near, small islands than for distant, small islands ($T_1 < T_2$) because near islands are closer to a source of immigrants. Similarly, turnover rates are greater for near, large islands than for distant, large islands ($T_2 < T_3$). Turnover rates are greater for near, small islands than for near, large islands because extinction rates on small islands are greater than on large islands ($T_1 < T_2$) and, all else being equal, turnover rates are greater on small islands than on large islands ($T_1 < T_3$).*

the disappearance or reduction from reproductive to nonreproductive status of a previous colonist (Simberloff, 1976).

Studies of island fauna tend to support much of island biogeography theory as it relates to the number of species present, distance from source of immigrants, and the size of the island. There is less empirical evidence, especially for vertebrates, to support equilibrium, because accurate data on species turnover are difficult to obtain.

A direct test of the equilibrium theory was performed by Simberloff and Wilson (1969, 1970; Simberloff, 1976). They eliminated the arboreal arthropod population (20 to 50 species) from a group of six small red mangrove islands in the Florida Keys by enclosing them in a large plastic tent and fumigating them with methyl bromide. They left two similar islands as controls. Simberloff and Wilson censused the areas carefully before fumigation and periodically afterwards. The number of species on the control islands remained the same. The "defaunated" islands were rapidly colonized and within 200 days the number of species on the islands stabilized. The turnover rate was high (0.5 to 1.0 species per day), but the number of species remained relatively constant over the two years of the study. On all islands but the most distant the number of species that recolonized the islands was greater than the predefaunation number, but it subsequently declined and oscillated about a lower number. This was probably caused by small population sizes in early colonization, allowing more species to coexist than would be possible in the untreated and more crowded islands. On the more distant islands the fewer species that were able to invade built up high populations early in the absence of predation and competition and retarded the attainment of equilibrium. Two of the defaunated islands reached equilibrium at a lower species density than their original equilibrium, probably because interspecific competition acted to keep the density lower and because the immigrants were not as well adapted to exploit the resources as the original inhabitants.

While the equilibrium theory originally applied to oceanic islands, not all islands are oceanic. Mountain tops, bogs, ponds, dunes, areas fragmented by human land use—all are essentially island habitats. "Ecologically any patch of habitat isolated from similar habitats by different, relatively inhospitable terrain traversed only with difficulty by organisms of the habitat patch may be considered an island" (Simberloff, 1974).

The equilibrium theory of island biogeography has direct application to the preservation and management of many continental and mainland species and communities (Terbrough, 1974, 1975; Wilson and Willis, 1975). The problem becomes one of maintaining a number of sufficiently large ecological systems adequately dispersed and interconnected by corridors to maintain faunal elements and biotic diversity.

The design of such islands is debated, some arguing for the largest possible areas, others for smaller areas connected by corridors (Wilson and Willis, 1975; Whitcomb et al., 1977). But because the amount of autecological and life history information available is small and because insufficient data exist to validate the island theory, such decisions are difficult to make (Simberloff and Abele, 1976). Some empirical evidence is being gathered in the field (Forman et al., 1976; Galli et al., 1976; MacClintoch et al., 1977). Such studies indicate that in eastern North America many of the neotropical migratory species once dominant in the forest interior tend to disappear from fragmented forests. Breeding bird censuses of eastern deciduous forests of North America indicate that neotropical species make up to 92 percent of the territorial birds in some forests (Whitcomb, 1977). How large a fragment has to be to hold such birds apparently depends upon the distance of the small woodland islands from large islands or continents of forests and the presence or absence of corridors between them. One Maryland study suggests that most forest interior species characteristic of the region are able to breed in forest fragments as small as 25 acres if the forest is subsidized by a major forest system (MacClintoch et al., 1977). It appears to be necessary to maintain large areas of habitat along with fragmented units connected by corridors. The problem of species preservation is a rich area for study in applied ecology and island biogeography.

## Nature of the community

The nature of the community has been the object of study and dispute for years. Is a community such as an oak-hickory forest a real entity that is definable, describable, and constant from one stand of oak-history to another? Or is it something of an abstraction, a collection of different populations that exist together because they have similar environmental requirements? To the last question one group of ecologists says yes, and another group says no.

### ORGANISMIC VERSUS INDIVIDUALISTIC CONCEPT

Involved in the concept of the community are two opposing philosophies. One is the *organismic concept* advanced by F. E. Clements (1916). This concept has had a powerful influence on the development of ecological thinking in America (McIntosh, 1976) and has persisted as a dominant concept until very recent years. The organismic concept regards the community as a sort of super organism, the highest stage in the organization of the living world—rising from cell to tissue, organs, organ systems, organism, population, community. Just as tissues have certain characteristics and functions above and beyond those of the cells that comprise them, so the community has characteristics and functions above and beyond the various populations it embodies. The distribution and abundance of one species in the community are determined by the species' interaction with others in the same community. Species making up the floristic community are organized into discrete groups. Groups of stands similar to one another form associations. Stands of one association are clearly distinct from stands of other associations. The community acts as a unit in seasonal activity, in competition with other communities, in trophic functions, and in succession. Clements (1916) considered the mature form of the community or the climax formation to be an

organic entity . . . As an organism the formation arises, grows, matures, and dies. Its response to the habitat is shown in the processes or functions and in structures which are the record as well as the results of these functions. Furthermore, each climax formation is able to reproduce itself, repeating with essential fidelity the stages of its development. The life history of a formation is a complex but definite process, comparable in its chief features with the life history of an individual plant. . . . Its most striking feature lies in the movement of populations, the waves of invasion which rise and fall through the habitat from initiation to climax. . . . all of these view points are summed up in that which regards succession as the growth or development of the climax.

The organismic approach was challenged by H. L. Gleason, who advanced an individualistic approach to the community. He argued:

the association represents merely the coincidence of certain plant individuals and is not an organic entity of itself. While the similarity of vegetation in two detached areas may be striking, it is only an expression of similar environmental conditions and similar surrounding plant populations. If they are for convenience described under the same name, the treatment is in no wise comparable to the inclusion of several plant individuals in one species (1917).

The sole conclusion we can draw from all the forgoing considerations is that the vegetation of an area is merely the resultant of two factors, the fluctuating and fortuitous immigration of plants and an equally fluctuating and variable environment (1926).

The individualistic approach was ignored by ecologists until the 1950s when the concept was advanced quantitatively by John Curtis and his students (1950, 1959) as the vegetation continuum and Robert Whittaker (1962, 1965) as gradient analysis (see McIntosh, 1975). The reintroduction of Gleason's individualistic concept coincided with the development of quantitative methods for analyzing vegetation. The individualistic concept emphasizes the species rather than the community as the essential unit in the analysis of interrelationships and distribution. Species respond independently to the physiological and biotic environment according to their own genetic characteristics. They are not bound together into groups of associates that must appear together. Instead, when species populations are plotted along an environmental gradient, long or short, the resulting graph suggests a normal or bell-shaped curve (Figure 20-23). The curves of many species overlap in a heterogeneous fashion. Thus the vegetation and its associated animal life exhibit a gradient or continuum from one extreme (e.g., dry conditions) to another (wet conditions). In this view, the community is regarded as a collection of populations of species requiring the same en-

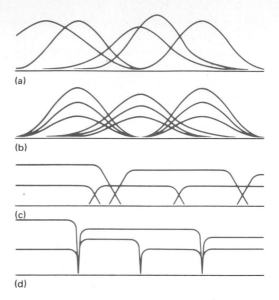

**FIGURE 20-23**
*Four models of species distribution along environmental gradients. (a) The abundance of one species on an environmental gradient is independent of the others. Thus the association of several species along the gradient changes with the response of individual species to that gradient. (b) The abundance of one species is associated with the abundance of another. Two or more species are always found in association with each other. (c) The distribution of one species is independent of another on an environmental gradient, but the abundance and distribution of each species is sharply restricted at some point on the gradient by interspecific competition. (d) The distribution of a species is sharply restricted by a change in some environmental variable. This is characteristic of an ecotone or edge.*

vironmental conditions. It is a continuous variable, not an integrated unit.

The individualist proponents grant that community discontinuities do exist, but that these reflect environmental discontinuities, such as breaks in soil type or sharp changes in moisture and salinity. If stands are as distinct as the organismic group considers them to be, then the boundaries between associations should be distinct. But they rarely are. Except where humans or some environmental catastrophe such as fire have interfered, it is rare for the vegetation to be a mosaic of discontinuous units. Instead, boundaries between units are more or less diffuse. If separate stands of an association are similar, all such associations should have similar distributions, and plants that comprise such associations should have distributions that coincide locally and over continental limits. If the association is a natural unit, then the species that comprise it should be bound together by obligate interrelationships. But most species relationships are not obligatory.

The organismic school argues that stands studied by gradient analysis are disturbed stands or ones that are not in equilibrium. If undisturbed, all stands develop to an endpoint in a few hundred years; and stands heading for an endpoint will naturally show a continuum of species. Besides, the technique of gradient analysis forces data into a continuum. These gradients are usually based on one variable such as moisture, but it is impossible to restrict a continuum to one variable alone because many interacting variables influencing plant distribution are involved. The continuum approach assumes that all species are equal when in fact some species are dominant.

Although the individualistic concept seems to be the more widely accepted of the two today, the organismic concept still has its strong supporters (Daubenmire, 1966, 1968a; Langford and Buell, 1969; Morrison and Yarranton, 1975) and it is still evident in much ecological thinking, especially in regard to succession. The ecosystem is still considered as a functional, biological unit that evolves and modifies its organization to store, conserve, and manage information. Obviously, communities cannot evolve and they are more of an abstraction than an entity.

## COMMUNITY PATTERN

The composition of any one community is determined in part by the species that happen to be distributed on the area and can grow and survive under prevailing conditions. Seeds of many plants may be carried by the wind and animals, but only those adapted to grow in the habitat where they are deposited will take root and thrive. The element of chance also is involved. One adapted species may colonize an area and prevent others equally as well adapted from entering. Wind direction and velocity, size of the seed crop, disease, insect and rodent damage all influence the establishment of vegetation. Thus the exact species that settle an area and the number of individual species that succeed are situations that seldom if ever are repeated in any two places at any two times. Nevertheless, there is a certain pattern, with more or less similar groups recurring from place to place. Only a relatively small group of species are potential dominants because a limited number are well adapted to the overall climate and soils of the region they occupy.

Communities are often regarded as distinct natural units or associations, especially for practical reasons of description and study, but more often than not community boundaries are hard to define. Some, such as ponds, tidal beaches, grassy balds, islands of spruce and fir within a hardwood forest, old fields and burns, have sharply defined boundaries. Here the vegetational pattern is discontinuous. Most often, however, one community type blends into another. The species comprising the community do not necessarily associate only with one another, but are found with other species where their distribution overlaps (Figures 20-24, 20-25). Some organisms will succeed only under certain environmental conditions and tend to be confined to certain habitats. Others tolerate a wider range of environmental conditions and are found over a wider area. Species shift in abundance and dominance, because of change in altitude, moisture, temperature, and other physical conditions. One

**FIGURE 20-24**
*Distribution of some forest trees on a continuum index. (Adapted from Curtis and McIntosh, 1951.)*

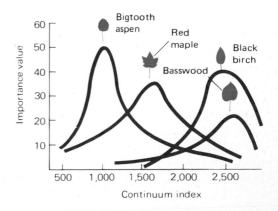

607

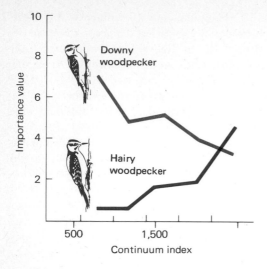

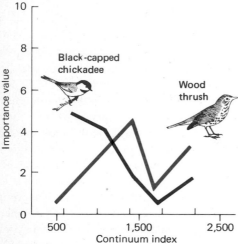

**FIGURE 20-25**
*Distribution of two bird species along a vegetational gradient. (Adapted from Bond, 1957.)*

boundaries grade continuously in either direction. Eventually the continuum must end, when environmental conditions favor a completely different group of organisms. A community in such a gradient can be described as a discrete area or point in the continuum, the point being defined by some given criteria.

The distribution of species along an environmental gradient is not confined to plants alone. The same phenomena also have been found in insects (Whittaker, 1952) and birds (Bond, 1957) (Figure 20-25).

### CLASSIFICATION

Although the major current concept of the community is that of a continuum, the plant and animal life of any large area is so complex that it must be separated into subdivisions. Thus the aggregation of organisms in any given locality or habitat must be regarded as a unit if the community is to be studied, described, or compared with similar community stands in other habitats. To give order to the study of communities, some system of classification is needed, even though the communities of a region often cannot be placed in discrete units.

There are a number of approaches to community classification, each arbitrary and each suited to a particular need or viewpoint (H. C. Hanson, 1958). The most widely used classification systems are based on physiognomy, species composition, dominance, and habitat.

Physiognomy, or general appearance, is a highly useful method of naming and delineating communities, particularly in surveying large areas and as a basis for further subdivision of major types into their component communities. Because animal distribution is most closely correlated to the structure of vegetation and not the species composition (cf., for birds, MacArthur and MacArthur, 1961), classification by physiognomy will relate both the animals and the vegetation of an area. Communities so classified are usually named after the dominant form of life, usually plant, such as the coniferous or deciduous forest, sagebrush, short-grass prairie, and tundra. A few are named after animals, such as the barnacle–blue mussel *(Balanus-Mytilus)* community of the tidal zone. One, of course, may

species may be dominant in one group, an associated species in another. This sequence of communities showing a gradual change in composition is called a *continuum* (Curtis, 1959). Each community is similar to, but slightly different from its neighbor, the difference increasing roughly with the distance between them. Even when the dominant plants change completely, the community may integrate in the understory vegetation. The continuum is much like a light spectrum. The end colors, red and blue, and other primary colors in the middle are distinguishable, but the grade into the other, so even here the classification may be based on arbitrary, although specific, criteria.

In areas where the habitat is well defined, physiography is used to classify and name communities. Examples of such are sand dunes, cliffs, tidal mud flats, ponds, and streams.

Finer subdivisions are often based on species composition, a system which works much better with plants alone than with animals or with both. Such a classification requires at first a detailed study of the individual community (see Appendix B). Such a system also involves a number of concepts: frequency (the regularity with which a species is distributed throughout a community), dominance, constancy, presence, and fidelity (the occurrence of a species in only a few community types).

A group of stands in which more or less the same combination of species occurs can be classified as the same community type, named after the dominant organisms or the ones with the highest frequency. Examples of such are the *Quercus–Carya* association, or oak-hickory forest, the *Stipa-Bouteloua* association, or mixed prairie, and the animal-dominated *Balanus-Mytilus,* or barnacle–blue mussel community of the tidal zone. European ecologists have developed the floristic classification with emphasis on dominance, constancy, and diagnostic species. They group communities into classes, orders, alliances, and associations (for a complete discussion on this, see Poore, 1962; Whittaker, 1962).

The floristic system is modified when the stands are treated as a continuum. The community complex of a major physiognomy is subdivided by species composition and correlated with an environmental gradient arbitrarily divided into five segments: wet, wet mesic, mesic, dry mesic, and dry (Curtis, 1959). Thus the deciduous forest in Wisconsin has been divided into southern and northern hardwoods and northern forest. These are further divided on a moisture gradient. The southern hardwoods, for example, include the dry southern hardwoods with bur, black, or white oak as the dominants; the dry mesic with red oak or basswood; the mesic with sugar maple and beech; the wet mesic with silver maple, elm, and ash; and the wet mesic with willow or cottonwood. Such a system recognizes the influences of habitat on community composition. Detailed studies on animal distribution may reveal similar influences of animal composition. A shift of species composition of birds occurs in the southern forests of Wisconsin (Bond, 1957). In the dry stands the most important species are the scarlet tanager, rose-breasted grosbeak, cardinal, blue jay, black-capped chickadee, downy woodpecker, and the red-headed woodpecker. In the dry mesic segment, however, the wood thrush, least flycatcher, blue-gray gnatcatcher, redstart, yellow-throated vireo, very, and ruby-throated hummingbird are the dominant species.

A large problem in community classification is to arrive at a system that will embrace animals as well as plants. Communities distinguished by plant composition indicate little about the animals of the community because animal distribution cannot be correlated with plant species distribution. As a result animal and plant communities usually are studied separately, which unfortunately obscures the wholeness of the community and limits our understanding of its functions.

To escape this dilemma in part, the distribution of animals can be related to the life form of plants and types of vegetation. This results in a more inclusive classification, which embraces several plant communities but includes all animal life associated with them; this classification is called the *biome*. The biome is a broad ecological unit characterized by the distinctive life forms of the climax species, plant or animal. The biome is further divided into smaller units, distinguished by uniformity and distinctness in species composition of the climax and its successional stages. Thus the life form of plants is given greater emphasis, rather than the taxonomic composition, which in the final analysis plays the most important role in dominance.

This brings up another concept: fidelity, or the "faithfulness" of a species to a community type. Species with low fidelity occur in a number of different communities and those with high fidelity in only a few. Seldom if ever are the latter found away from certain other plant and animal associates. The greater the ratio of the constant species to the total number of species, the more homogeneous is the community and the more sharply can it be delineated. Often, however, this simply reflects a group of species unable to grow successfully under a wide range of ecological conditions or with other species. Species, in general, can be grouped as *exclusive,* those completely or almost completely confined to one kind of community; *characteristic* (including the selective and preferential species of plant ecologists), those that are most closely identified with a certain community- and *ubiquitous* (or indifferent), those which have no particular affinity to any community. The characteristic species high in constancy and dominance are the ones that really characterize the community type.

## SUMMARY

However the community may be classified or whatever methods may be employed to distinguish one community from another, the basic concept remains unchanged. A biotic community is a naturally occurring assemblage of mutually sustaining and interdependent plants and animals living in the same environment, constantly utilizing and dissipating energy.

Communities are organized about dominant species, especially in the Temperate Zone. The dominant may be the most numerous, possess the highest biomass, preempt the most space, or make the largest contribution to energy flow. But the dominant species may not be the most important species in the community.

Each species in the community occupies a particular niche determined by a long process of natural selection and evolution. The niche is basically the functional role of the organism in the community. The more niches there are to occupy, the more specialized the occupants, the more complex the community, and the greater the diversity of species.

Species diversity implies both a richness in the number of species and equitability in the distribution of individuals among the species. A number of hypotheses have been proposed to explain species diversity. The evolutionary time theory proposes that diversity relates to the age of the community or the time available for speciation. The ecological time theory is based on the time available for species to disperse. The spatial heterogeneity theory relates diversity to the physical complexity of the environment. The climatic stability theory holds that the more stable the environment the more species will be present. The productivity theory suggests that the level of diversity is determined by the amount of energy flowing through the food chain. The competition theory says that where competition is keener, as in the tropics, the organisms are more specialized and thus more can coexist. The predation hypothesis states that predators reduce competition and permit more species to coexist. And the stability time hypothesis suggests that diversity is related to stress gradient with diversity lowest in environments subject to environmental stress and highest in those where the environment is benign.

All communities exhibit some form of layering or stratification, which largely reflects the life forms of the plants and influences the nature and distribution of animal life in the community. Communities most highly stratified offer the richest variety of animal life, for they contain a greater assortment of microhabitats and available niches.

Communities also exhibit horizontal stratification which is influenced by such environmental factors as soil type or disturbance. The area at the boundary between two plant communities is the edge or ecotone. Because of the additive effect of species from two adjoining plant communities and species characteristic of the edge itself, ecotones are noted for their species richness. Such edges create a patchy environment. Edge and patchy environment relate to island biogeography theory. If edge conditions result in fragmentation of homogeneous habitat stands, the community may become so small that it cannot support the maximum number of species associated with it. The habitat unit in effect becomes a small island separated from larger units of the same type. The smaller the area of an island, the fewer species it can hold. The more distant an island is from a mainland source of immigrants, the fewer species it will hold. Extinction rates are lower on larger islands than on small islands. Small islands have less diverse habitats and hold smaller and more extinction-prone species. Species diversity on islands represents a balance between immigration and extinction rates.

There are two opposing views concerning the nature of the community. According to the organismic school, the community is an integrated unit that has definable boundaries. The individualistic school argues that the community is a collection of populations that require similar environmental conditions. The makeup of a community is determined in part by the species that happen to be distributed on the area and that can grow and survive under prevailing conditions. The exact species that happen to settle on an area and the number that survive are rarely repeated in any two places at the same time, but there is a certain recurring pattern of more or less similar groups. Rarely can different groups of communities be sharply delimited because they blend together to form a continuum.

# Succession
## CHAPTER 21

Abandoned cropland is a common sight in agricultural regions, particularly in areas once covered with forests. No longer tended the lands grow up in grasses, goldenrod, and other herbaceous plants. Only the most unobservant would fail to notice that in a few years these same weedy fields are invaded by "brush"—blackberries, sumac, and hawthorne. The brush is followed by fire cherry, pine, and aspen. Many years later this abandoned cropland supports a forest of maple, hickory, oak, or pine. Thus over a period of years one community replaces another until a relatively stable forest or climax finally occupies the area (Figure 21-1).

## Succession defined

The changes involved in the return of the forest are not haphazard, but orderly, and bar-

ring disturbance by humans or natural events, the reappearance of the forest is predictable. This change in species composition and community structure over time is *ecological succession*. The whole series of communities from grass to shrub to forest that terminates in a relatively stable community is called a *sere,* and each of the changes that take place is a *seral stage*. Although each seral stage is a point in a continuum of vegetation through time, each is recognizable as a distinct community with its own characteristic structure and species composition, especially at the point of optimal development. Seral stages may last for a short time or for many years. Some stages may be missed completely or they may appear only in abbreviated or altered form. For example, when an old field grows up immediately in forest trees, the shrub stage appears to be bypassed, but its place is taken by the community of young trees.

(a)

(d)

(b)

(e)

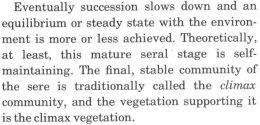

(c)

FIGURE 21-1
*Successional changes in an old field over 30 years. (a) The field as it appeared in 1942, when it was moderately grazed. (b) The same area in 1963. (c) A close view of the rail fence in the left background of (a). (d) The same area 20 years later. The rail fence has rotted away and white pine and aspen grow in the area (e) The same field in 1972. Note how aspen has claimed much of the ground.*

Eventually succession slows down and an equilibrium or steady state with the environment is more or less achieved. Theoretically, at least, this mature seral stage is self-maintaining. The final, stable community of the sere is traditionally called the *climax* community, and the vegetation supporting it is the climax vegetation.

Succession that takes place on areas devoid of or unchanged by organisms is called *pri-* mary (Figure 21-2). Succession that proceeds from a state in which other organisms were already present, for example, the abandoned cropland, is called *secondary*. Secondary succession arises on sites in which the natural vegetational cover has been disturbed by humans, animals, or natural forces as fires, wind storms, and floods. Its development may be controlled or influenced by the activities of humans or domestic and wild animals.

**FIGURE 21-2**
*Primary succession in the subalpine zone of the Wasatch Mountains, Utah. Here the early stages include the trees; the climax is a mixed herb community. Note the changes in soil depth from a rocky surface with fine soil only in the crevices to a well-defined solum essentially free from rocks. (Based on data from Ellison, 1954.)*

The fact that one aggregation of plants is replaced by another has been noted for many years (see Spurr, 1952). In 1863 Anton Kerner in a fascinating book, *Plant Life of the Danube Basin,* explained what he called the "genetical relationship of plant formations" as he described the formation of meadow from swamp, forest regeneration, and the forest edge. Within the pages of this old book lay the field of plant sociology in the embryonic state. In America Henry David Thoreau wrote about succession. Then in 1899 Henry Cowles published his classic description of plant succession on sand dunes of Lake Michigan. Sixteen years later the pioneer plant ecologist Frederick Clements published *Plant Succession,* a book which became the foundation of a system of studying and describing plant communities that still colors ecological thinking today (a point to be elaborated later).

## Kinds of succession

The most common approach to succession is one of a stereotyped progression of vegetational changes. Although such a structured approach is not very realistic because succes-

sion varies from site to site and region to region, a description of a general pattern of succession is useful as a point from which departures can be made.

### AQUATIC SUCCESSION

The transition of a pond to a terrestrial community can be observed in a limited area, often in one pond alone. The first step in succession is the *pioneer stage,* characterized by a bottom barren of plant life. Such a stage can be found in newly formed, human-created ponds and lakes. The earliest forms of life to colonize the area are plankton, which may become so dense as to cloud the water. This plankton consists of microscopic algae and animal life, which upon death settle to the bottom to begin to form a thin organic layer. If the plankton growth becomes rich enough, the pond may support other forms of life—small caddisflies that build cases of sand and feed on microorganisms living on the bottom, bluegills, green sunfish and large-mouthed bass.

At the same time the pond acts as a settling basin for sediments washed in from the surrounding watershed. These fine sediments and the accumulating organic matter form a layer of loose, oozy material that creates a substrate for rooted aquatics such as the branching green algae, *Chara,* pondweeds, and waterweeds. These plants bind the loose bottom into a firmer matrix and add materially to the deposition of bottom organic matter. Organisms common to the barren pond bottom cannot exist in the changed conditions of this second step in aquatic succession, the *submerged vegetation stage.* The caddisflies of the pioneering stage are replaced by other species able to creep over submerged vegetation and build cases from plant material. Dragonflies, mayflies, and small crustaceans appear.

Rapid addition of organic matter and sediments carried into the pond reduces water depth and provides nutrients for more demanding plants. Floating aquatics, with roots embedded in the bottom muck and leaves floating on the water's surface, invade the pond. Because these plants shut out the light from the pond depths, they tend to eliminate the submerged aquatic growth. This is the *floating aquatics stage,* one in which faunal living space is increased and diversified. Hydras, frogs, diving beetles, and a host of new insects capable of utilizing the undersurfaces of floating leaves appear.

Seasonal fluctuations in water levels alternately expose and cover the increasingly shallow bottom about the edge of the pond. Lacking the bouyancy and protection of water, the weak and soft-tissued floating plants cannot exist in the changing environment. Emergent plants— the cattails, sedges, bulrushes, and arrowheads, firmly anchored in the bottom muck by spreading fibrous roots and rhizomes—occupy the area to form the *emergent vegetation zone.* Because the plants extend above and lack the protection of water, they possess flexible leaves and wandlike stems that bend easily before the wind and water. Animals of the floating stage are replaced by those that inhabit the jungle of emergent plant stems. Gill-breathing snails give way to the lung breathers. Different species of mayflies and dragonflies spend their nymphal stages on submerged stems and climb to the surface when they are ready to emerge as adults. Redwinged blackbirds, ducks, and muskrats become common to the area. As the oxygen supply of the water decreases, because of the increasing quantities removed through respiration by organisms of decay breaking down the accumulating organic matter, only animals with low oxygen requirements can exist. Bullheads replace sunfish and annelid worms colonize the bottom muck.

Since the dense root system and the annual deposition of leaf growth add great quantities of organic matter to the bottom and bind inwashing sediments, the substrate builds up rapidly after the emergents have appeared. Much of the old open-water area now is covered by sedges, cattails, and associated plants to form a *marsh.* As the bottom rises above the groundwater level, the remnant of the open pond dries up in summer. It has now become a *temporary pond,* which contains only those organisms that can withstand drying in summer and freezing in winter.

Drainage improves as the land builds higher. The emergents disappear, the soil lies above the water table, and organic matter, exposed to the air, decomposes more rapidly. Meadow grasses, accompanied by land animals, invade to form a marsh meadow in forested regions and a prairie in the grass country. In forest regions these meadows may remain stabilized for years. In forested areas and in certain topographical situations between the prairie and the forest, alders, willows, and buttonbushes colonize the site. If the marsh is invaded directly by woody plants, the marsh-meadow stage never develops. Shrubs may give way to trees—aspen, elm, red and silver maples, and white pine. Root systems, limited by high water tables, spread horizontally instead of vertically in the soil. The substrate is rapidly but unevenly built up by the accumulation of fallen trunks and by upturned roots and soil. As the forest floor becomes drier and the crown closes, seedlings of intolerant forest trees are unable to develop, but seedlings of sugar maple, beech, hemlock, spruce, and cedar, able to grow in low light intensities, dominate the understory and subsequently replace the intolerant trees. These trees tolerate the environmental conditions they create, so the forest cover becomes stabilized (but see the description of peat bog development in Chapter 7).

### TERRESTRIAL SUCCESSION

A similar sequence takes place on dry areas. Barren areas, whether they are natural primary sites, such as rock and sand dunes, or disturbed areas, such as abandoned cultivated fields or roadbanks, are a sort of natural biological vacuum eventually to be filled by living organisms. Plants and animals that colonize such sites comprise the pioneer communities. On primary sites no soil exists initially and successive communities can become more complex as soils develop. Bare, disturbed areas, the secondary sites have some sort of soil present already. Both however, are characterized by full exposure to the sun, fluctuations in temperature, and rapid changes in moisture conditions.

*Primary succession.* Rocks and cliffs are common terrestrial primary sites. Burbanck and Platt (1964) and Shure and Ragsdale (1977) studied primary succession on granite outcrops in Georgia. Although crustose lichens colonize bare surfaces of rocks, primary succession is usually initiated in depressions resulting from exfoliation and weathering of rock surfaces. The dominant pioneering species is a small winter annual *(Sedum smalli)*. Lichens *(Cladonia)* washed or blown in colonize the depression and act to trap debris. The gradual increase in soil depth leads to greater soil moisture and more soil organic matter. These changes in physical condition favor the germination of several annual herb species *(Agrostis, Hypericum)*, forming a lichen-annual stage. As soil depth, soil organic matter, and soil moisture increase, perennial grasses, mosses, and herbs move in. On deeper soil islands low woody plants may follow, but extreme soil moisture stress in drought can cause high mortality of such plants, restricting the development of woody growth.

The sand dune is another severe primary site. A product of pulverized rock, sand is deposited by wind and water. Where deposits are extensive, as along the shores of lakes and oceans and on inland sand barrens, sand particles may be piled up in long windward slopes to form dunes (Figure 21-3) that move before the wind and often cover forests and buildings in their path, until stabilized by plants (Figure 21-3). With high surface temperatures by day and cold temperatures at night, dunes are rigorous environments for life to colonize. Grasses are the most successful pioneer and binding plants. When these, and such associated plants as beach pea, have at least partly stabilized the dunes, mat-forming shrubs invade the area.

From this point the vegetation may pass to pine and then oak, or to oak directly. The low fertility of the dunes favors plants with low nutrient requirements. Because these plants are inefficient in cycling nutrients, especially

(a)

(b)

FIGURE 21-3
*Sand dunes along the northeastern Atlantic coast.* (a) *Grass on a dune along the Virginia coast.* (b) *The foredune in the Cape Breton coast is claimed by marra grass, while the older dunes support a growth of white spruce.*

calcium, soil fertility remains low. Because of this infertility and the low moisture reserves in the sand, oak is rarely replaced by more moisture-demanding or more nutrient-demanding trees, except on leeward slopes and in depressions where microclimate and moisture conditions are more favorable. There a mesophytic forest (one that requires moderate levels of moisture) may become established (Olson, 1958).

Deposits left by retreating glaciers also are nutrient-poor primary sites. In Alaska newly exposed raw glacial till is invaded first by mountain avens *(Dryas)* whose feathery seeds are carried to the site by the wind (Lawrence, 1958). Mountain avens is followed closely by Sitka alder *(Alnus sinuata)*, a rapidly growing, nitrogen-fixing shrub. Because the rate of nitrogen accumulation from nitrogen compounds fixed in the roots and present in the leaves is high and because the cool moist climate prevents the utilization of nitrogen faster than it is fixed, nitrogen is stored in the soil. As a result alders and invading cottonwoods are stimulated to vigorous upright growth. As the alder thicket matures, cottonwoods and hemlocks rapidly emerge and eventually shade out the alder. With nitrogen no longer being fixed and added to the soil in quantity and with nitrogen reserves depleted, a carpet of moss and deep litter covers the forest floor.

Human-created primary sites include surface-mined lands on which the spoil has not been reclaimed by replacement of the upper layers of soil. Surface-mine spoils or regraded land may consist of an overburden of unweathered rocks and other materials. Elements such as nickel, zinc, and aluminum may be present in the overburden at levels toxic to plants. An overburden high in sandstone, shale, and pyritic materials is often highly acid and highly acid sites may go unvegetated for years. Sites that have pockets of less acidic material may support patches of vegetation. Generally the best sites on such spoils are invaded quickly and then the areas remain static for years. Increase in plant cover usually results from the spread of plants already established rather than any new invasion of bare areas.

Invading species vary with the region and the nature of the spoil. In deciduous forest regions of the eastern United States conditions on the spoils are usually too harsh for any of the heavy seeded forest species to colonize the area; invasion usually involves those species whose seeds are transported by animal or wind such as panic grasses and blackberries. On regraded sites where top layers of soil have been replaced, native vegetation tends to return to the mined area. In some cases this begins within three years because propagules in the soil are able to germinate and start growth before the grass cover becomes too competitive. Where sites are replanted, domestic agricultural grasses rather than native species are often sown, mainly because seed supplies are available. Thus tame grasslands rather than native prairie will dominate mined lands. On areas where the topsoil has not been replaced and grass seed has been added directly to the overburden, success is mixed. If well established, grass cover remains dominant for years with little invasion from native herbaceous or woody plants. Aggressive competition by grasses and lack of propagules on the site allows this continued dominance (R. L. Smith, in prep.). On other sites the presence of toxic and acid materials and low levels of nutrients cause slow growth or death of vegetation, exposing the mined surface to erosion. In mountainous regions grass cover is unable to stabilize the outslopes of the spoils and the area is subject to slides. In desert areas no reclamation has succeeded to date, for the aridity makes revegetation extremely difficult.

***Secondary succession.*** Secondary succession is most commonly encountered on abandoned farmland and noncultivated ruderal sites (waste places) such as fills, spoil banks, railroad grades, and roadsides, all artificially disturbed and frequently subject to erosion and settling movements.

Species most likely to colonize such places are the so-called weeds, species out of place from a human perspective. Although hard to define, weeds have two characteristics in common. They invade areas modified by human action; in fact a few are confined to such artificially modified habitats. And they are exotics, not natives to the region. Once native species invade the area, these "weed" plants eventually disappear.

Annual, biennial, or perennial, all plants that settle initially on disturbed areas possess a great tolerance for soil disturbance and partial defoliation. Their seeds remain viable for a long time and may remain in the soil for a number of years until the conditions are right for germination. Some weeds require an open seedbed and exposed mineral soil for germination. Their rapid and successful colonization is aided by an efficient means of dispersal. Some have light seeds that are carried by the wind; others spread by underground rhizomes. These vigorous pioneer plants grow rapidly under favorable conditions; in less favorable habitats they set seed even when small (Sorensen, 1954).

In spite of their vigor, these plants cannot maintain their dominance for long—3 to 4 years at most—if all disturbance ceases. Short life cycles, advantageous at first, are not adaptable to conditions imposed by incoming

**617**

plants with long life cycles, ones that begin growth earlier in the spring and persist throughout the summer.

The species composition of pioneering communities is highly variable. Infinite combinations exist, depending upon the seed source, the cultural practices prior to abandonment of the land, and moisture and soil conditions. Broad differences exist between the first-year communities that spring up in small grain fields and those that appear in cultivated row-crop fields. Because land in row crops has large areas of exposed soil, the number of annuals and biennials, such as ragweed, all of which do well in the hot weather of midsummer, is high. In the absence of cultivation, herbs are able to establish themselves in small grain fields unseeded to grass. In these the number of annuals and biennials is greatest the last year of use and decreases rapidly thereafter (Figure 21-4) (Beckwith, 1954). Moist soils rich in plant nutrients support such plants as burdock, catnip, and nettle; poorer, drier soils grow shepherd's-purse, chicory, and ragweed. On railroad grades and yards, fresh cinders, high in sulfur compounds and other toxic materials, compose a substrate suggestive of the saline soils of the western plains (Curtis, 1959). Here western plants carried eastward on rail cars may become established, among them western wheatgrass and sunflowers. In addition there is a wide assortment of exotic weeds.

Thus succession, whether primary or secondary, starts with the colonization of the area by pioneer species—plants able to grow on a substrate low in nutrients and organic matter, in an environment that is excessively dry, from strong solar radiation, and in wide variations of surface temperature. The number of plants colonizing and surviving on the site are few at first, but as conditions improve, more of them occupy the area. Through the deposition of organic matter and shading of the surface, they reduce the surface evaporation and modify the environment enough to permit more demanding plants to assume dominance. Better adapted to utilize the nutrients available, the new arrivals eventually take over and crowd out the pioneers by shading and by vigorous growth. Some new arrivals may take hold not because of changed environmental conditions but in spite of them. Many plants of an advanced seral stage could well have been the pioneers had they chanced to colonize the area earlier. Some pioneer species such as crabgrass, sunflower, and horseweed may produce chemicals that inhibit their own growth and the growth of other herbaceous species (Rice, 1965; Keever, 1950), paving the way for invasion by grasses that are not affected by the toxins (Parenti and Rice, 1969). Grasses in turn may inhibit nitrogen-fixing bacteria, slowing succession to the next stage (Rice, 1965). But eventually the stage is set for other plants and animals to invade and colonize the area and for a different community to develop.

### HETEROTROPHIC SUCCESSION

Within each major community and dependent upon it for an energy source are a number of microcommunities. Dead trees, animal carcasses and droppings, plant galls, tree holes, all furnish a substrate on which groups of plants and animals live, succeed each other, and eventually disappear, becoming in the final stages a part of the nutrient base of the major community itself. In these instances succession is characterized by early dominance of heterotrophic organisms, maximum energy available at the start, and a steady decline in energy as the succession progresses.

An acorn supports a tiny parade of life from the time it drops from the tree until it becomes a part of the humus (Winston, 1956). Succession often begins while the acorn still hangs on the tree. The acorn may be invaded by insects, which carry to the interior pathogenic fungi fatal to the embryo. Most often the insect that invades the acorn is the acorn weevil

*FIGURE 21-4*
*Succession of vegetation on cultivated fields. Note the early rise and rapid decline of annuals. (After Beckwith, 1954.)*

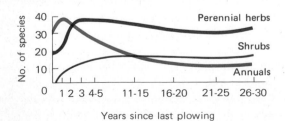

618

(*Curculio rectus*). The adult female burrows through the pericarp into the embryo and deposits its eggs. Upon hatching, the larvae tunnel through to the embryo and consume about half of it. If fungi (*Penicillium* and *Fusarium*) invade the acorn simultaneously with the weevil or alone, they utilize the material. The embryo then turns brown and leathery and the weevil larvae become stunted and fail to develop. These organisms represent the pioneer stage.

When the embryo is destroyed, partially or completely, by the pioneering organisms, other animals and fungi enter the acorn. Weevil larvae cut through the outer shell and leave the acorn. Through this exit hole fungi-feeders and scavengers enter. Most important is the moth *Valentinia glandenella,* which lays its eggs on or in the exit hole, mostly during the fall. Upon hatching, the larvae enter the acorn, spin a tough web over the opening, and proceed to feed on the remainder of the embryo and the feces of the previous occupant. At the same time several species of fungi enter and grow inside the acorn, only to be utilized by another occupant, the cheese mites (*Tryophagus* and *Rhyzozhphus*). By the time the remaining embryo tissues are reduced to feces, the acorn is invaded by cellulose-consuming fungi. The fruiting bodies of these fungi, as well as the surface of the acorn, are eaten by other mites and collembolans and, if moist, by cheese mites too. At this time predaceous mites enter the acorn, particularly *Gamasellus,* which is extremely flattened and capable of following smaller mites and collembola into crevices within the acorn. Outside on the acorn, cellulose and lignin-consuming fungi soften the outer shell and bind the acorn to twigs and leaves on the forest floor.

As the acorn shell becomes more fragile, holes other than the weevils' exits appear. One of the earliest appears at the base of the acorn where the hilum (the scar marking attachment point of seed) falls out. Through this, larger animals such as centipedes, millipedes, ants, and collembolans enter, although they contribute nothing to the decay of the acorn. The amount of soil in the cavity increases and the greatly softened shell eventually collapses into a mound and gradually becomes incorporated into the humus of the soil.

Thus microcommunities illustrate one concept of succession: that the change in the substrate is brought about by the organisms themselves. When organisms exploit an environment, their life activities make the habitat unfavorable for their own survival and create a favorable environment for different groups of organisms. Those responsible for the beginning of succession are all quite specialized for feeding in acorns, the later forms are less so, and the final group are generalized soil animals, such as earthworms and millipedes.

## Succession and animal life

As successional communities change, the animal life they support changes with them (Figure 21-5). Early terrestrial successional stages support animals of open fields such as meadowlarks and meadow pipits, meadow voles, and grasshoppers. As woody plants appear, stratification increases and habitat conditions change. Animals of annual plant and grass stages give way to animals of mixed herbaceous and shrubby habitats; woodland mice begin to replace grassland voles and rabbits and birds of thickets and shrubland appear. As the area passes into forest, vertical stratification increases. Animals of the forest edge and forest appear along with species that occupy the upper strata. As the forest matures shrub and edge species decline and are replaced by tree squirrels and birds and insects of the forest canopy. As the community matures the diversity of habitats associated with the earlier stages declines and the diversity of animal life declines with it. Each successional stage supports its own more or less distinctive forms of animal life and when these stages pass, their animal life also passes. The key to a diversity of wildlife is a variety of successional stages rather than a homogeneous climax stand and the maintenance of each stage in an area sufficiently large to hold its characteristic species.

## Nature of succession

While the change in species composition is the most obvious feature of succession, trophic-dynamic aspects and interspecific relations also change.

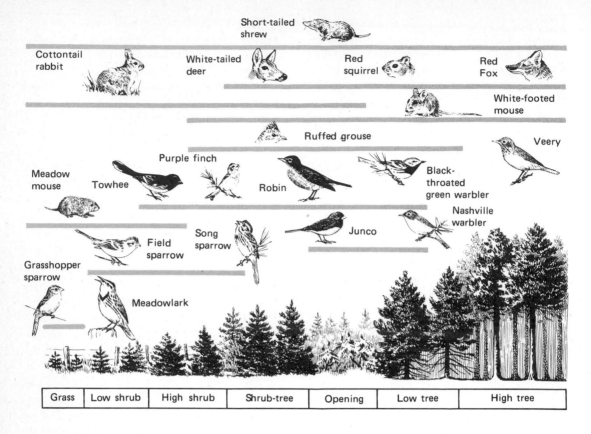

| Grass | Low shrub | High shrub | Shrub-tree | Opening | Low tree | High tree |

**FIGURE 21-5**

*Wildlife succession in a conifer plantation in central New York. Note how some species appear and others disappear as vegetation density and height change. Some species are common to all stages.*

### TROPHIC-DYNAMIC ASPECTS

In 1942 R. Lindeman postulated that productivity should increase as succession proceeds, that succession improves the efficiency of transfer and use of energy and nutrients. This postulate was modified and expanded by R. Margelef (1963, 1968) and E. P. Odum (1969) into the concept of maturity, that structural complexity and organization of an undisturbed ecosystem increase or mature with time. Although a number of their premises are debatable, the approach does serve to emphasize processes taking place during succession (Table 21-1).

One change already emphasized is increased structural complexity and organization of seral stages as succession proceeds. Succession may begin with a bare area colonized by small plants and terminate with large plants whose growth form results in increased vertical stratification and increased influence on environmental variables within the community. These changes and others, according to Margalef and Odum, represent differences between young or early successional ecosystems and mature or late successional ecosystems. In early successional stages community production exceeds community respiration *(P > R),* resulting in high net production which is stored as biomass. As long as gross production exceeds respiration, biomass continues to accumulate. As succession proceeds more of gross production is used in respiration, reflecting increased maintenance costs of the community. When production equals respiration *(P = R),* succession stops (Kira and Shidei, 1967) and production is utilized in maintenance of the system. This is reflected in the ratio of production to biomass *(P/B)* In early stages the ratio is high, but as the system matures, the ratio is low because most of the energy goes to support biomass.

Such changes in ratios of production to respiration and to biomass result because of

TABLE 21-1
*Proposed trends in ecological succession*

| Ecosystem attributes | Developmental stages | Mature stages |
|---|---|---|
| **COMMUNITY ENERGETICS** | | |
| 1. Gross production/community respiration ($P/R$ ratio) | Greater or less than 1 | Approaches 1 |
| 2. Gross production/standing crop biomass ($P/B$ ratio) | High | Low |
| 3. Biomass supported/unit energy flow ($B/E$ ratio) | Low | High |
| 4. Net community production (yield) | High | Low |
| 5. Food chains | Linear, predominantly grazing | Weblike, predominantly detritus |
| **COMMUNITY STRUCTURE** | | |
| 6. Total organic matter | Small | Large |
| 7. Inorganic nutrients | Extrabiotic | Intrabiotic |
| 8. Species diversity—variety component | Low | High |
| 9. Species diversity—equitability component | Low | High |
| 10. Biochemical diversity | Low | High |
| 11. Stratification and spatial heterogeneity (pattern diversity) | Poorly organized | Well-organized |
| **LIFE HISTORY** | | |
| 12. Niche specialization | Broad | Narrow |
| 13. Size of organism | Small | Large |
| 14. Life cycles | Short, simple | Long, complex |
| **NUTRIENT CYCLING** | | |
| 15. Mineral cycles | Open | Closed |
| 16. Nutrient exchange rate, between organisms and environment | Rapid | Slow |
| 17. Role of detritus in nutrient regeneration | Unimportant | Important |
| **SELECTION PRESSURE** | | |
| 18. Production | Quantity | Quality |
| **OVERALL HOMEOSTASIS** | | |
| 19. Internal symbiosis | Undeveloped | Developed |
| 20. Nutrient conservation | Poor | Good |
| 21. Stability (resistance to external perturbation) | Poor | Good |
| 22. Entropy | High | Low |
| 23. Information | Low | High |

*Source:* Odum, 1969.

changes in dominance and diversity. In young ecosystems energy is channeled through relatively few pathways to many individuals of a few species and production per unit is high. Food chains are short, linear, and largely grazing. In more advanced stages of succession energy is directed through many pathways and shared by many units. Food chains are complex and largely detrital. Average net productivity decreases as relative dominance declines and diversity increases. This is a result of apportionment of net production among many species, each reaching its maximum productivity at different periods during the growing season. However, total net community production remains relatively unchanged and appears independent of age, relative dominance, and diversity (Mellenger and McNaughton, 1975).

Nutrient cycles may also exhibit differences between young and mature ecosystems. In early successional stages nutrient cycles are characterized by a rapid exchange between organism and environment. Short-lived individuals return most nutrients accumulated to the soil and decomposers. In mature systems

**621**

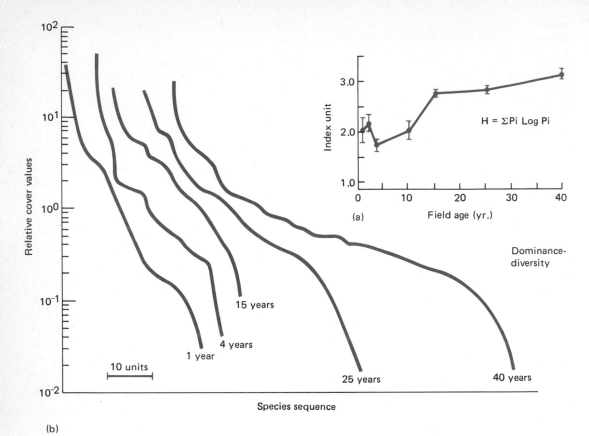

(b)

nutrients are locked up or hoarded in biomass of persistent and long-lived individuals. This reduces the amount of nutrients available for recirculation, another possible reason for decline in productivity. Also contributing to the decline may be an accumulation of excretion in the system that inhibits productivity (see Chapter 16).

Succession, as already hinted, also involves changes in diversity which increases as succession proceeds toward maturity and then decreases in the terminal stage. Increased diversity comes about largely through increased number of species while the total number of individuals in each successional stage remains about the same. Increase in diversity in part develops through decreased synchrony in the utilization of resources. Species which are nutrient demanding at different times are noncompetitive and can exist together. Thus spring flowering and fall flowering can be associated species, adding diversity to the community.

Diversity in succession can be considered from two points of view, diversity across an

**FIGURE 21-6**
*Relationship between plant species diversity and succession. (a) (Inset) Plant species diversity generally increases with succession and may reach a maximum in the forest stage (see Figure 21-7) when shade-tolerant and shade-intolerant guilds are present together in the community. Relatively low species diversity in a successional community (as in example graphed here) may result from development of strong dominance by a species with allopatric interference. High species diversity may result from a high degree of vertical and horizontal microenvironmental heterogeneity. (b) Dominance-diversity curves of successional community are geometric at first, suggesting the niche preemption hypothesis (see Figure 20-20). Dominance curves become less steep with time as more species are added and gradually a log normal distribution with an increase of species with intermediate relative importance values develops. (From Bazzaz, 1975.)*

environmental gradient and diversity across a temporal gradient (Auclair and Goff, 1971). Because the two are inversely interrelated no sweeping generalizations can be made concerning diversity and maturity. Diversity may or may not increase with advancing successional stages. Some old fields, for example, may have greater diversity in earlier stages than in later stages (Figure 21-6) (Bazzaz, 1975). Some later stages may be dominated by plants that produce strong allelochemicals that inhibit the growth of associated species reducing species diversity. A great deal depends upon the initial state of the site. Auclair and Goff (1971) concluded that pioneering forest communities on xeric sites increase in diversity through time (Figure 21-7), while successional forest communities on mesic sites decline in diversity from late successional to equilibrium forest. On intermediate sites diversity approaches on asymptote late in succession. Even within this framework diversity on a given site will be influenced by soil, microtopography, disturbance, and grazing by herbivorous mammals.

## COMPETITION: OPPORTUNISTIC AND EQUILIBRIUM SPECIES

Changes in species composition through succession are basically due to the inability of in-

**FIGURE 21-7**
*Species diversity in relation to time for xeric, intermediate, and mesic moisture conditions. Note that diversity increases with time on xeric and intermediate sites but decreases on mesic sites, which reach their maximum diversity in earlier stages of succession. Diversity in later stages is maximal on intermediate sites. (From Auclair and Goff, 1971.)*

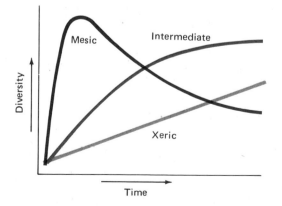

dividuals of early successional species to compete effectively against individuals of later successional species.

Early stages of succession, both primary and secondary, are characterized by low-growing plants that have short life cycles, are relatively small in size, reproduce annually by seeds or send out new photosynthetic growth from buds near the ground (geophytes), and generally produce an abundance of small seeds that possess means for wide dispersal. The seeds are able to persist in the soil for a long time until conditions are right for germination and growth. The plants are opportunistic species (see *r* strategists, Chapter 14), adaptable to a wide range of environmental conditions, especially the rigors of an open, disturbed site. Plants of early succession grow rapidly, they attain dominance quickly, and most of their production goes into photosynthetic and reproductive tissue. Plants of early stages maintain their dominance by suppressing the growth of any later successional plants that may exist as seedlings or sprouts on the site.

Because most pioneer plants of early successional stages have to renew their photosynthetic structure each year, they eventually lose their temporary dominance and in time are suppressed by taller, more vigorous plants that are able to initiate earlier seasonal growth. These gradually assume dominance because they are better able to exploit the site once stress conditions ameloriate and the environment becomes more stable and predictable. In later stages taller plants that carry biomass over from year to year have a competitive advantage, although this advantage may be gained by some form of external disturbance that reduces the vitality of early opportunistic species. Early stage plants, in effect, become fugitive species living in other disturbed and marginal areas, but able to disperse quickly to any newly disturbed area.

Plants of later stages of succession store and support more biomass and produce relatively few heavy seeds that are dispersed largely by animals or gravity. The large size of the seed provides an abundance of nutrients to get the seedlings started, but the seeds' vitality and longevity may be rather low. The plants are specialists, adapted to a narrow range of environmental conditions. They derive their persistence from their ability to hoard resources or to use them more efficiently. Plants of later stages grow more slowly and are relatively longer lived, thus able to dominate a site over a longer period of time. Much of their production goes into storage and maintenance; their root-to-shoot ratio is high, for well-developed root systems are needed to exploit restricted supplies of nutrients and water. Because resources are finely divided among many species, the dominance of any one species declines. Late successional species in effect are equilibrium species (see *K*-strategists, Chapter 14). They succeed in a more stable environment than opportunistic species, and are better able to weather temporary environmental stresses by drawing on energy reserves. Their populations through competition and other interactions reach some equilibrium level in the environment.

## Mechanisms of succession

Succession, demonstrated rather convincingly by H. C. Cowles (1899) in his study of vegetational development on sand dunes of Lake Michigan, was advanced as an ecological hypothesis by F. C. Clements (1916). His hypothesis had a powerful influence (or inhibition, as some critics say) on ecological thought. In fact the Clements theory of succession dominated ecology for years, almost becoming dogma, because it provided an orderly, logical explanation for the development of plant communities.

The Clementsian hypothesis centers about a developmental model of vegetational succession. The climax or terminal community is viewed as a superorganism and succession as the embryonic development of that organism. As an organism the climax arises, grows, matures, and dies. Furthermore, each climax formation is able to reproduce itself, repeating with essential fidelity the stages of its development. Thus Clementsian succession is linear or unidirectional, orderly, predictable, and developmental.

According to Clements's theory each climax owes its characteristic appearance to the species or dominants that control it. These dominants exhibit the same vegetation or life form (that is, trees, shrubs, or grass) and all belong to the highest type of life form possible under the prevailing climate. "Each climax is the direct expression of its climate; the climate is the cause, the climax is the effect, which in its turn reacts upon the climate" (Weaver and Clements, 1938).

Clement's concept of succession and the climax was challenged by Henry Gleason (1917). Gleason regarded a plant association or climax not as a superorganism, but as a community that developed from the random spread and establishment of individual plants. Overshadowed by Clements's theories, Gleason's ideas were largely ignored. The idea of individual development was later advanced by J. Curtis (1959) and R. A. Whittaker (1967) who viewed succession or community development as ending in a steady state community with its constituent plant populations in equilibrium with its position on an environmental gradient (see Chapter 20).

These two approaches to succession are important because they express different ways of considering the mechanisms of succession.

Succession can be considered as an expression of differences in colonizing ability, growth, and survival of organisms adapted to a particular set of conditions on an environmental gradient. The position of a species on the gradient is determined by the length of life cycle, time of reproduction, reproductive output, and other characteristics (Pickett, 1976). As certain environmental conditions and stresses change, populations of plants and sessile animals (such as barnacles and mussels) also change. The replacement of one of several species or groups of species by others results from interspecific competition and the interactions of herbivores, predators, and disease which permit one group of plants to suppress slower growing or less tolerant species.

Succession is initiated when a disturbance such as fire, land clearing, or floods, opens a relatively large area of ground or when a new area, such as glacial till or volcanic ash is available for colonization. Connell and Slatyer (1977) have developed models to represent suggested mechanisms of succession. The three types are presented here in somewhat modified form. All models assume no further

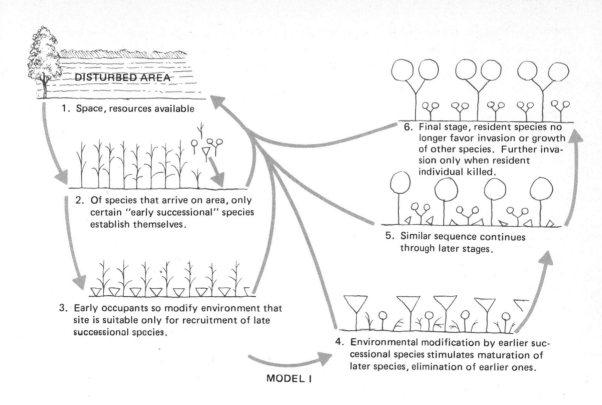

DISTURBED AREA

1. Space, resources available

2. Of species that arrive on area, only certain "early successional" species establish themselves.

3. Early occupants so modify environment that site is suitable only for recruitment of late successional species.

4. Environmental modification by earlier successional species stimulates maturation of later species, elimination of earlier ones.

5. Similar sequence continues through later stages.

6. Final stage, resident species no longer favor invasion or growth of other species. Further invasion only when resident individual killed.

MODEL I

**FIGURE 21-8**
*Model of Type I succession which assumes that later stage species depend upon early stage species to prepare a favorable environment for them. (Adapted from Connell and Slatyer, 1977.)*

significant changes in the abiotic environment after the initial perturbation. The models are similar in that all three pathways arrive at relatively stable communities. It is how this relatively stable community is achieved that makes the models different.

### TYPES OF SUCCESSION

Type I is essentially Clementsian. Only certain early successional species are able to colonize the site. These plants (or sessile animals) so modify environmental conditions that the site becomes less suitable for their own species and more suitable for a second species or group of species (Figure 21-8). The second group of species become dominant and in turn so modify the environment that they suppress the initial colonists and make way for the invasion of a third group which in turn also alter the environment. One group of species turns over the site to another until a relatively stable stage is achieved. In the stable stage resident species no longer modify the site and environment and they inhibit the invasion of a different species or group of species. Thus opportunistic species invade the area and in effect prepare the way for the invasion first of perennials, then shrubs

and trees. This sequence has been called relay floristics by Egler (1954) and is typical of primary succession.

Types II and III are based on the assumption that any species arriving on the area even before the period of abandonment or disturbance may be able to colonize the area. Because they grow and respond faster under initial environmental conditions, opportunistic or fugitive species usually (but not always) appear first. In type II succession these initial colonists modify the environment in a manner that makes it less suitable for their own recruitment but they neither increase or decrease the rate of recruitment or growth of later colonists (Figure 21-9). Propagules of some species probably were already on the site prior to disturbance and remain there after disturbance. Other propagules are carried to the site by wind or animals. Later stage species germinate and grow more slowly, but

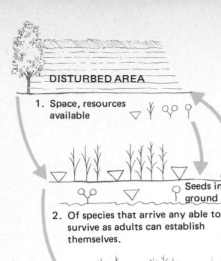

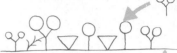

DISTURBED AREA

1. Space, resources available

Seeds in ground

2. Of species that arrive any able to survive as adults can establish themselves.

3. Early occupants modify environment so *less* suitable for recruitment of early successional species, little effect on recruitment of late species.

6. Further invasion or growth to maturity possible only when resident killed, releasing space.

5. Sequence continues until no species exists that can grow and reproduce in presence of resident.

4. Juveniles of later stage species that invade or are present grow to maturity despite early succession species. Earlier species eliminated.

MODEL II

FIGURE 21-9 (above)
*Model of Type II succession which assumes that succession leads to a community composed of those species most efficient in exploiting resources. (Adapted from Connell and Slatyer, 1977.)*

FIGURE 21-10 (below)
*Model of Type III succession which assumes that the species which colonizes a site holds it against all comers. When an empty space is filled, invaders can succeed only if former colonists are damaged or die, thus releasing resources. (After Connell and Slatyer, 1977.)*

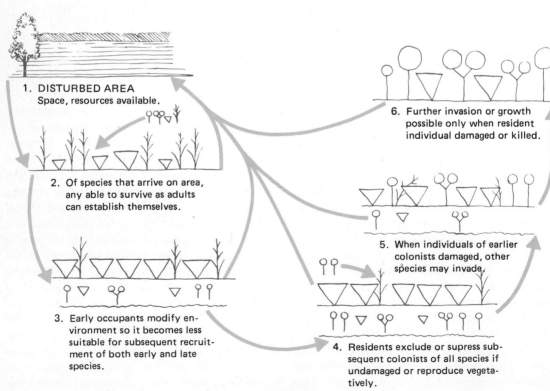

1. DISTURBED AREA
Space, resources available.

2. Of species that arrive on area, any able to survive as adults can establish themselves.

3. Early occupants modify environment so it becomes less suitable for subsequent recruitment of both early and late species.

6. Further invasion or growth possible only when resident individual damaged or killed.

5. When individuals of earlier colonists damaged, other species may invade.

4. Residents exclude or supress subsequent colonists of all species if undamaged or reproduce vegetatively.

MODEL III

are able to survive and grow in the presence of and under the environmental conditions imposed by the early colonists. Unable to compete, early stage plants are eventually replaced by perennials which assume dominance. As in Type I succession a point is reached where the species most tolerant of shading or other environmental conditions occupy and retain the site. (This is the initial floristic composition of Egler, 1954.) If no later stage species such as trees and shrubs have propagules in the soil, it may be years before woody vegetation invades the site. Not only may the seed source be some distance away, but competition from grasses and perennial forbs can inhibit the germination and survival of seeds and seedlings of trees and shrubs.

This leads to Type III succession in which early colonists, having secured space and other resources inhibit the invasion of subsequent colonists or suppress the growth of those species already present (Figure 21-10). New species can occupy the site only when some of the dominant residents are damaged or killed, usually by some local disturbance such as grazing by herbivores or by the action of pathogens. Because replacement individuals do not have to be adapted to or more tolerant of conditions imposed by previous residents, early successional species can replace themselves, essentially stopping succession. But if the replacement is a highly tolerant later stage species able to survive long periods of suppression, succession continues. Being short-lived, earlier stage species more often than not will be replaced, however slowly, by later stage species. In Type III the terminal species may be those most resistant to damage by herbivore, fires, and other hazards. This type is typical of secondary succession.

In each of the three types the process may be interrupted by some disturbance that reverts succession to some earlier stage.

## THE CLIMAX AND COMMUNITY STABILITY

According to tradition succession stops when the community or sere has arrived at an equilibrium state with the environment. At this point the community is stable and self-replicating and, barring disturbances, will persist indefinitely. This end point of succession is the climax.

There are three theoretical approaches to the climax. One is the monoclimax theory developed largely by Clements (1916). This theory recognizes only one climax whose characteristics are determined solely by climate. Successional processes and modifications of environment overcome the effects of differences in topography, parent material of the soil, and the like. All seral communities in a given region, if allowed sufficient time, will ultimately converge to a single climax and the whole landscape will be clothed with uniform plant community. All communities other than the climax are related to the climax by successional development and are recognized as subclimax, postclimax, disclimax, and so on. Several variations of the monoclimax theory have been advanced by a number of plant ecologists (for a review, see Whittaker, 1974).

Another approach is the polyclimax theory (Tansley, 1935). The climax vegetation of a region consists of not just one type but a mosaic of vegetational climaxes controlled by soil moisture, soil nutrients, topography, slope exposure, fire, and animal activity (see Daubenmire 1968a, b; Whittaker, 1974). The spatial pattern of habitats influences the spatial pattern of climax communities.

The third and perhaps most acceptable approach is the climax pattern hypothesis (Whittaker, 1953, 1974; McIntosh, 1958; Selleck, 1960). Composition, species structure, and balance of a climax community are determined by the total environment of the ecosystem and not by one aspect. Involved are the characteristics of each species population, their biotic interrelationships, availability of flora and fauna to colonize the area, the chance dispersion of seeds and animals, and soils and climate. The mosaic of climax vegetation will change as the environment changes; and the climax community represents a pattern of populations that corresponds to and changes with the pattern of environmental gradients to form ecoclines (see Chapter 20). The central and most widespread community in the pattern is the prevailing or climatic climax. It is the community that most clearly expresses the climate of the area. This mean of the prevailing climaxes relates the community to the climate in major ecoclines and provides a regional geographic pattern to vegetation.

The concept of the climax carries with it the idea of stability, the ability of a community to persist in spite of perturbations. The process by which communities recover from perturbations is succession. Ecosystem stability, discussed in Chapter 2, may involve both resistance and resilience to perturbations. If a community is resistant to perturbations, there is no succession. It achieves stability because individuals in the community resist disturbances by defenses against insects, fires, herbivores, invasion by competition, and other mechanisms. If a community is resilient, it returns to stable equilibrium through successional processes. Traditional theory of the climax assumes that following a disturbance the community through time returns to a similar species composition. There is one stable equilibrium point to which all successions converge. If communities return to different species compositions that vary from site to site, equilibrium may involve a number of stable points and communities do not necessarily converge to a composition similar to the original. A forest may return to the site, but the species composition may be somewhat different from the first.

To judge stability one needs to know how long and over what space a species composition must persist (Connell and Slayter, 1977). If one views stability as involving long-lived individuals whose populations are relatively constant over time, not subject to rapid replacement, and resistant to invasion from other species, then mature systems are stable ones. But early stage communities may also return quickly to original conditions, often to the same average species composition, and may even be relatively resistant to invasion. Plow a field of annual weeds and the next spring the field will be claimed by the same weeds again. Trample or tear up a grassy field and in a year or two the grass will claim the land again. But cut a mature forest and a 100 years may pass before any semblance of the original forest returns. Stability of a community can be a matter of a perspective of time.

Climax stability also carries with it the idea that every individual is replaced by another of the same kind; average species composition reaches equilibrium. If offspring of the same species are favored over others, then a dead mature individual may be replaced by a plant of its own kind. If offspring are concentrated about the mature parent, it may be replaced by its own progeny. This is most likely if the replacement is essentially the same unit—a root or stump sprout from the dead individual. But if conditions beneath the mature tree are less favorable for its own species than for other species, it will be replaced by one of another kind. And if conditions are neither more or less favorable for the offspring than for other species, replacement individuals depend upon the relative abundance of propagules arriving on the site, suppressed individuals already present, and competitive interactions among them.

Self-destructive biological changes are continually taking place in the climax community. Trees grow old and die and more often than not are replaced by individuals of a different species. Changes are constantly occurring in patches across the community and average species composition is slowly changing. Although succession may slow down it never ceases; and the stability of the climax reflects mostly the fact that dominant species are long lived in terms of a human time frame.

## Response to disturbance

Depending upon the size of the area and the severity of the disturbance, a perturbation to a community may result in either the growth of vegetation established following the disturbance or the reorganization of vegetation on the site established prior to the disturbance.

### LARGE-SCALE DISTURBANCE

A severe disturbance over a large area generally results in growth of vegetation established after the disturbance. Such areas are colonized by opportunistic species. What species will colonize the site depends upon local ecological conditions and the dissemination strategies employed. An opportunistic species may become established by a continuous input of seed from the outside or by storage of seed in the soil (Marks, 1974).

Successful colonization of a disturbed site through direct natural seeding from sources outside the site depends upon a number of conditions including distance of seed source, size of seed crop, timing of seed arrival, and favorable microclimate on the site. One opportunistic species that uses this dissemination strategy is yellow poplar *(Liriodendron tulipefera)*. To take advantage of the disturbed site seed trees must be nearby, seed must reach the site in sufficiently large numbers to insure some seed survival, and seed should land on exposed mineral soil to favor rapid germination and successful seedling survival.

Other opportunistic species colonize a disturbed area by means of propagules such as dormant seeds or roots on the area at the time of disturbance. One example is pin cherry *(Prunus pensylvanica)* whose seeds are carried to a forest by birds and small mammals or are deposited on the forest floor by an earlier stand of cherry. Pin cherry seeds can remain dormant for up to 50 years. When the forest canopy is removed and moisture, temperature, and light conditions become favorable, pin cherry seeds germinate and young trees quickly dominate the site crowding out the associated blackberry *(Rubus* spp.) that also colonize the area. If the seedling growth is dense, a pin cherry canopy can close in four years, eliminating other species except highly shade-tolerant seedlings of sugar maple and beech (Figure 21-11). If seedling growth is moderately dense, species with wind disseminated seeds, such as yellow birch and paper birch, will also occupy the site. Within 30 to 40 years pin cherry dies out allowing birch, sugar maple, and beech to dominate the gap. But during its period of tenure pin cherry contributes numerous seeds to the forest floor, ready to reclaim the site when another disturbance provides the opportunity.

A related species, the black cherry *(Prunus serotina),* functions in a similar manner in the oak forest of the Lake States and Appalachia. Like pin cherry, black cherry is an opportunistic species, but unlike pin cherry, it remains an integral component of the mature forest.

**FIGURE 21-11**
*Diagrammatic representation of the importance of different species along a gradient of time following a disturbance of a typical northern hardwoods forest. Immediately after disturbance, blackberries (a) dominate the site, but they quickly give way to yellow birch (b), quaking aspen (c), and pin cherry (d). Pin cherry assumes dominance early but with 30 years fades from the forest. Yellow birch assumes an early dominance which it retains into mature or climax stand. Trembling aspen begins to drop out after about 50 years. Meanwhile sugar maple (e) and beech (f), highly tolerant species, slowly gain dominance through time. In about 100 years the mature forest is dominated by beech, maple, and birch. (After Marks, 1974.)*

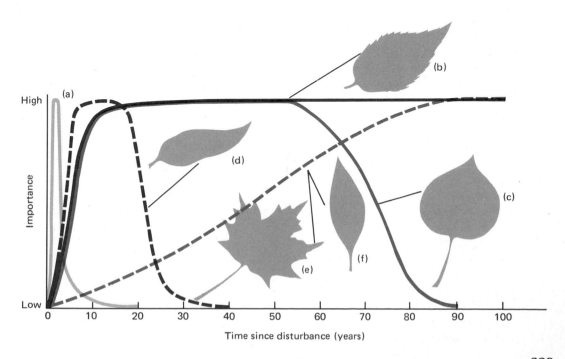

Black cherry has a high reproductive potential, a large number of seeds dispersed by birds and mammals, the ability to invade an area rapidly, a mechanism for delayed germination of seeds stored in the soil, and vigorous seedlings that have no strong seedbed preferences except freedom from competition. In addition black cherry tolerates regressive influences such as excessive shade, maintains a root system through repeated sprouting, and occupies a wide range of environments (Auclair and Cottam, 1971).

Most existing studies of responses to disturbance relate to tree species. Very few studies have investigated the response of the herbaceous (or field) layer of a forest to disturbance. Ash and Barkham (1974) studied the response of the herbaceous understory layer of an English coppice forest after cutting (A coppice forest is one in which stump sprouts or root suckers are maintained as the main source of regeneration with cutting rotations between 20 and 40 years.) Cutting coppice involves complete canopy removal, resulting in increased surface temperature on the forest floor and full exposure to light. Typically a number of open-habitat or opportunistic species germinate and become established (Figure 21-12), but are soon excluded by the developing canopy cover. In spite of the disturbance characteristic woodland species persist throughout the cycle. Adapted to a high light regime in spring before the leaves are out, these plants have the ability to tolerate high light intensities. At the same time they are able to coexist with annuals and open-habitat perennials which cast a shade on the ground like the tree cover. As the opportunistic species disappear, woodland herbs again assume dominance, often developing into monospecific stands.

Disturbances to forest ecosystems may result in the fragmentation of the ecosystems into isolated stands, islands of trees in an expanse of cropland, grassland, and suburban developments. Auclair and Cottam (1971) report that such fragmentation of oak woodland in southern Wisconsin radically changes environmental conditions in those woodlands compared to those found in larger units of vegetation. Small forest stands with unprotected borders experience increased penetration of light into the understory, increased exposure

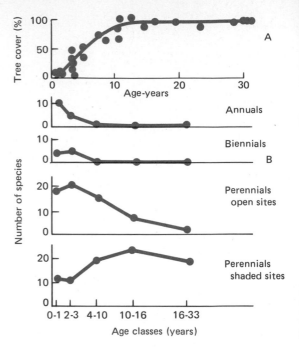

**FIGURE 21-12**
*Response of understory to disturbance in a coppice forest in England. The graphs show changes in the percentage of cover produced by the growth of the canopy and the number of herb and shrub species in the field or ground layer at different times after coppicing. Not all species were present throughout the 30-year period of growth and the total numbers involved for each type are also given. Note the rapid decline of annuals, biennials and open-site perennials as the canopy closes. Open-site perennials dominate the field layer shortly after coppicing. Perennials of shaded sites show a sigmoidal growth response as the canopy closes. (From Ash and Barkham, 1974.)*

to wind and erosion, and increasingly favorable environmental conditions for the dispersal and invasion of opportunistic species no more shade tolerant than the oaks of the canopy. Because of the absence of tolerant species in the understory succession in the stands cannot advance toward a more tolerant terminal stage. Succession in effect is arrested in something of a pioneering stage.

### SMALL-SCALE DISTURBANCE

If the area involved is relatively small or the intensity of disturbance is relatively slight, the response to disturbance entails reorganization of vegetation on the site prior to the disturbance. Reorganization may take the form of canopy expansion, growth of new branches on trunks of remaining trees, stump sprouting as in oaks and maples, root suckering as in aspen, and the growth of suppressed advanced regeneration (seedlings) to fill in the newly created gaps.

*Gap phase replacement.* Gaps, large or small, formed in the forest represent sites of a temporary reduction in the utilization of light, moisture and nutrients. With these in excess incoming growth is quickly stimulated. If the gaps are small, such as those created by the death or removal of individual trees or small groups of trees, the response is typically vegetational reorganization. The canopy expands, and increased light and moisture on the forest floor may stimulate the growth of understory shrubs and advanced regeneration of forest trees. Usually small gaps favor or stimulate the growth of tolerant species (Trimble et al., 1973), so that in a terminal community replacement species are likely to be those of the canopy.

If the gaps are large, such as would result from timber harvesting, insect damage, wind, or ice storms, the response may involve both vegetational reorganization and invasion by opportunistic species. In the Appalachian hardwood forests, for example, stump sprouts and suppressed seedlings may respond rapidly and fill in the gap in several years (Trimble, 1973). Invasion by opportunistic species is limited and the future composition of the forest will be determined in part by the competitive interactions of incoming growth. Intolerant species such as yellow poplar and black cherry may outcompete tolerant species such as sugar maple which will remain in the understory, capable of filling small gaps that might appear later in the established forest. Both yellow poplar and cherry would remain for some time as a component of the mature or equilibrium forest.

*Cyclic replacement.* Successional stages that appear to be directional are often in fact phases in a cycle of vegetational development. Such cyclic replacement results from destruction of vegetation by some characteristic of the dominant organisms or by periodic disturbances that start regeneration again at some particular stage. Such changes are a part of community dynamics, usually occur on a small scale within the community, and are repeated over the whole of the community. Each successive community or phase is related to the other by orderly changes in the upgrade and downgrade series. Such cyclic replacements contribute to community persistence.

These phasic cycles were recognized and described in the Scottish heaths (Watt, 1947). Scottish heather represents the peak of the upgrade series. After its death a lichen (*Cladonia silvatica*) becomes dominant and covers the dead heather stems. Eventually the lichen disintegrates to expose bare soil, the last of the downgrade series. The bare soil is colonized by bearberry to initiate the upgrade series. Heather then reclaims the area and dominates again. There are other shorter phasic cycles, one involving heather, lichen, bare soil and back to heather again.

A similar cycle, frequently initiated by ants or ground squirrels, occurs in old field communities in Michigan (Evans and Cain, 1952). Here bare areas at the bottom of the downgrade series are invaded by mosses to start the upgrade series (Figure 21-13). The mosses are invaded by Canada bluegrass and dock. The accumulation of dead culms of these plants is covered by lichens of several species. The organic debris and lichens crowd out the grass. Rain, frost, and wind destroy the lichens, and the bare soil is left open to invasion by mosses again.

Most grassland succession involves a form of cyclic replacement (Vogl, 1974). Dominance in grassland plants is less influenced by moisture and shade tolerances than by vegetative reproduction, prolific seed reproduction, and allelopathic effects. Grassland vegetation appears to be maintained as a vegetative cycle, usually controlled by fire or by some ecological equivalent, such as high rainfall and protracted wet periods, necessary to recycle nutrients, stimulate plant growth, and maintain vegetational composition.

Cyclic replacement is also important in the maintenance of marsh ecosystems. Although the classical concept of the marsh is a stage in the development of an open basin of water to a terrestrial community, vegetation cycles induced by changes in water availability prevent

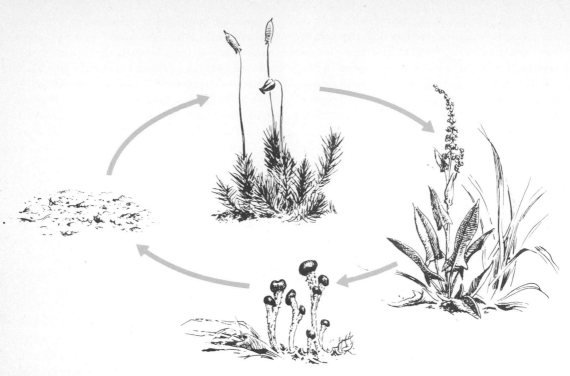

FIGURE 21-13.
*Cyclic replacement in an old field community. The cycle moves from dock to lichen to bare ground to moss and back to dock again.*

such a progression. During periods of drought (which in midwestern United States occur about every 5 to 20 years) shallow marshes dry up. Organic debris accumulated on the bottom decays rapidly releasing nutrients for recycling and the germination of seeds is stimulated to initiate the following cycle (Weller and Fredrickson, 1973): (1) germination phase, prevalent on mud flats or in water less than 2 inches deep; (2) newly flooded, with sparse, often well dispersed vegetation, annuals and immature perennials dominating; (3) flooded dense marsh dominated by perennials; (4) flooded hemi-marsh involving openings in dense emergents, usually created by muskrats; (5) deep open marsh rimmed with emergents. The cycle begins anew when the marsh dries (Figure 21-14). Although these short-term cycles give the shallow marsh the appearance of an unstable ecosystem, such cyclic replacements insure the long-term stability of the marsh ecosystem.

*Fluctuations.* Fluctuations are nonsuc-cessional or short-term reversible changes (Rabotnou, 1974). Fluctuations differ from succession in that floristic composition over time is stable, that is, no new species invade the site and changes in dominants may be reversible. These changes in floristic composition result from such environmental stresses as changes in soil moisture, wind, grazing, and the like.

An example is the fluctuation in populations of two species of grass, blue grama and buffalo grass, condominants in mixed prairie (Coupland, 1958). Blue grama has a greater physiological ability to resist dry conditions than buffalo grass. In western Kansas during the eight-year drought of the 1930s blue grama became two times as dense as buffalo grass. But when the rains came, buffalo grass quickly responded. In two years it quickly reversed its position and became five times as

FIGURE 21-14
*Cyclic replacement of vegetation and the input and output of seeds from the seed bank during the cycle of a prairie glacial marsh. The cycle is initiated by periods of drought followed by periods of normal rainfall. (After van der Valk and Davis, 1978.)*

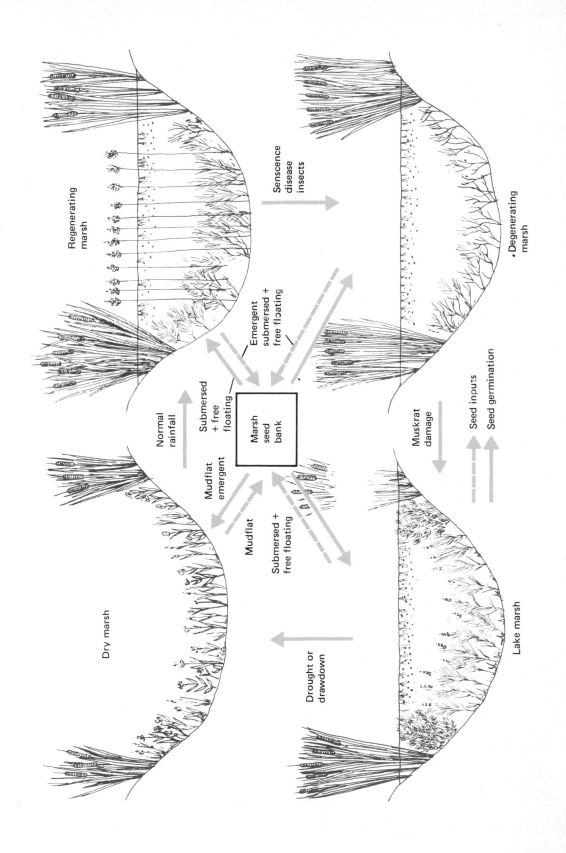

Regenerating marsh

Senscence
disease
insects

Degenerating marsh

Emergent
submersed +
free floating

Normal
rainfall

Submersed
+ free floating

Marsh
seed
bank

Muskrat
damage

Seed inputs

Seed germination

Mudflat
emergent

Mudflat

Submersed +
free floating

Dry marsh

Drought or
drawdown

Lake marsh

dense as blue grama. After 10 years of favorable conditions the two grasses regained their condominance.

Fluctuations in terminal forest communities may involve an alteration of species in small gap phase replacements. In such forests there appears to be a tendency for each species to be replaced by its competitor (Fox, 1977). If one species becomes moderately abundant in the canopy, saplings of alternate species may be abundant beneath it. Thus over time succession in the canopy may ultimately shift in favor of the temporarily disadvantaged species. For example, Fox (1977) found that in "virgin" northern hardwood stands sugar maple tends to replace beech in small openings and beech replaces sugar maple. In general the tendency is for a dominant tree to be replaced more than half the time by its competitor. Such alternation probably results for two reasons. (1) The dominant tree usurps the site concentrating the bulk of biomass at one particular place in a single tree. Its conspecifics are thinned out more severely than its competing or alternate species. (2) Because of its influence on nutrient regeneration and light and moisture regimes, the canopy tree creates a somewhat species-specific microhabitat for seeds and seedlings beneath it. Because an alternate species is favored in the understory, the terminal community is able to return to equilibrium after small-scale disturbances.

Fluctuations may also involve replacement of one age class by another within a species. Such fluctuations are important in maintaining certain forest ecosystems, particularly coniferous forest (Korchagin and Karpov, 1974). Sprugle (1976) describes a wave regeneration pattern in balsam fir *(Abies balsamae)* forests in northeastern United States. This fluctuation involves a type of disturbance in which

the trees continually die off at the edge of a "wave" and are replaced by vigorous stands of young balsam fir (Figure 21-15). The cycle is initiated when an opening occurs in the forest exposing trees on the leeward side of the opening to the wind. Dessication of the canopy foliage by winter winds, the loss of branches and needles in winter from rime ice forming on the needles, and decreased primary production due to the cooling of needles in summer cause the death of trees. Their death exposes the trees behind them to the same lethal conditions and they in turn die. This process continues so that a wave of dying trees through the forest is followed by a wave of vigorous reproduction.

These regeneration waves follow each other at intervals of about 60 years of age. The regularity of the process is such that all stages of degeneration and regeneration are found in the forest at all times, provided the stand is not cut. This phasic cycle results in a steady state because the degenerative changes in one part of the forest are balanced by regenerative stages in another. The wave regeneration process ensures the stability of the forest and prevents its advancement into a hardwoods stage.

Loucks (1970) proposed a similar pattern of response for the oak forests of Wisconsin. Oak

*FIGURE 21-15*
*Diagrammatic cross-section through a regeneration wave in a balsam fir forest. The wave is initiated at the location of standing dead trees, with mature trees beyond it and an area of vigorous regeneration below it. In the area where dead trees have fallen a crop of young fir seedlings is developing. Beyond these is a dense stand of fir saplings, followed by a mature fir forest and then by a second wave of dying trees. (After Sprugel, 1976.)*

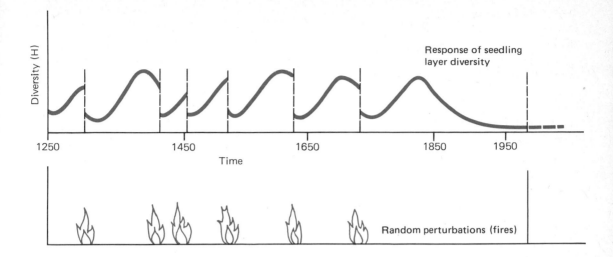

forests are maintained by a series of random perturbations in the form of fire at intervals of 30 to 200 years (Figure 21-16). When fire is excluded, forest succession proceeds to more tolerant sugar maple forests lower in species diversity and primary production.

## Influence of perturbations on succession

According to classical theory (exemplified by type I succession), succession is predictable and unidirectional and culminates in a terminal, stable, self-replicating community. But succession is not always directional nor does it necessarily end in a predictable terminal community. The course that succession takes is influenced by many biological and environmental factors. These include seed availability and dispersal, fire, soil drainage, and fertility, herbivory, disease and others.

### ROLE OF SITE AND EDAPHIC CONDITIONS

The outcome of succession is often determined by the nature of the site, past land use history, and soil conditions. Consider plant succession on two sites in the Canaan Valley of West Virginia: (1) a well-drained to moderately drained, medium acid to extremely acid abandoned pasture and (2) a wet, mineral, medium acid to extremely acid abandoned pasture. Canaan Valley is a high basin, 3200 feet above sea level. Hemmed in by higher mountains and boreal in climate, the valley once sup-

**FIGURE 21-16**
*Model of trends in diversity through succession as influenced by random natural disturbance across an extended period of time in southern Wisconsin oak forests. In this case the perturbation is fire. Note that the whole series of transient response phenomenona which makes up the stable system is restarted by each perturbation. With fire controlled by human interference with natural processes, the recycling process never occurs. Species diversity declines and the structure and composition of the forest changes. With fire intolerant species reach maximum production and development in an open environment and grow to a large size to dominate the site. They achieve this at the expense of survival in their own shade. These species can be maintained through time on a given site only by some strong perturbation such as fire that destroys the canopy, opens up the site, and allows seedlings of intolerant trees to gain possession of the site again. In the absence of perturbation shade-tolerant trees and associated species specialized for self-perpetuation in the shade possess the ability to saturate the environment with shade-tolerant progeny and claim the site. (After Loucks, 1970.)*

ported a dense forest of red spruce *(Picea rubens),* but logging, fires, and agricultural settlements resulted in the destruction of the spruce forest. The rugged slopes reverted to a northern hardwoods forest following a stage dominated by pin cherry, quaking aspen, and bigtooth aspen (Fortney, 1975). The valley,

**635**

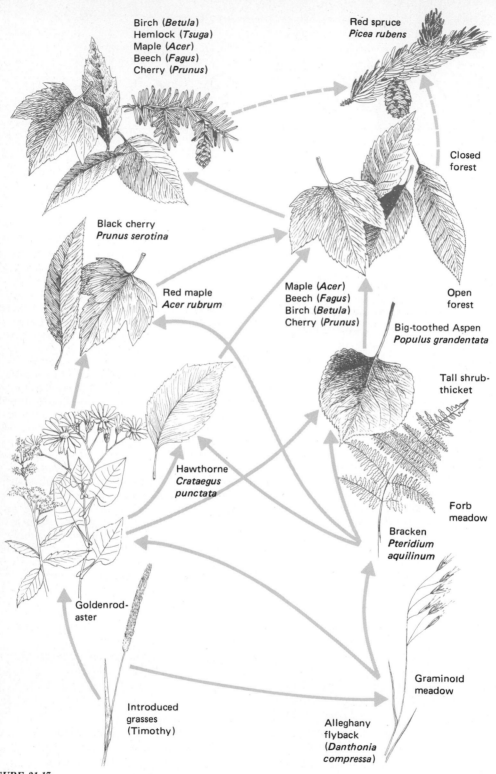

Birch (*Betula*)
Hemlock (*Tsuga*)
Maple (*Acer*)
Beech (*Fagus*)
Cherry (*Prunus*)

Red spruce
*Picea rubens*

Closed
forest

Black cherry
*Prunus serotina*

Red maple
*Acer rubrum*

Maple (*Acer*)
Beech (*Fagus*)
Birch (*Betula*)
Cherry (*Prunus*)

Open
forest

Big-toothed Aspen
*Populus grandentata*

Tall shrub-
thicket

Hawthorne
*Crataegus
punctata*

Forb
meadow

Bracken
*Pteridium
aquilinum*

Goldenrod-
aster

Graminoid
meadow

Introduced
grasses
(Timothy)

Alleghany
flyback
(*Danthonia
compressa*)

**FIGURE 21-17**
*Plant succession on well-drained to*
*moderately well-drained soils in Canaan*
*Valley, West Virginia. (After Fortney, 1975.)*

**636**

once converted to pastureland and now largely abandoned, is undergoing natural succession, the direction of which is affected by past grazing practices and current moisture conditions. In the reversion back to natural vegetation succession has been complex because of the wide soil variations that range from wet organic to moderately drained mineral soils and from medium acid to extremely acid soils. Although the theoretical climax should be red spruce, the successional directions on various sites are different, and in places the terminal vegetation is shrubs rather than trees.

Plant succession on the well-drained to moderately well-drained soils is illustrated in Figure 21-17. The starting point is introduced grasses or a native grass, Allegheny flyback *(Danthonia compressa)*. Succession advances to a forb meadow dominated by asters and goldenrod or by bracken fern. Succession may now take two directions: to thickets of dotted hawthorne or to open forest to bigtooth aspen or black cherry and red maple, and finally to a closed forest of sugar maple, beech, hemlock, black cherry and birch.

Plant succession on wet, mineral soil also starts with a graminoid meadow dominated either by sedges *(Carex)* or *Calamagrostris canadensis* (Figure 21-18). Succession may preceed to two terminal shrub communities, St. John's wort *(Hypericum densiflora)* or meadow sweet *(Spirea alba)*, which persist because their heavy dense growth effectively prevents invasion by trees. Or succession may proceed to alder, which also may persist for many years. On some sites the graminoid meadow may go directly to an open forest of quaking aspen *(Populus tremuloides)*. Eventually the aspen and alder will give way to a balsam fir, hemlock, red spruce, birch, and maple forest. In both situations there is only slight indication that succession might proceed to red spruce. A return to the climax forest is prevented by an unsuitable substrate for red spruce seedlings and by competition from hardwood species. The return to a climax red spruce forest will depend, at least in part, upon the regeneration of a moist soil with deep humus and litter to support spruce seedlings and a reduction in competition from hardwoods that produce heavy shade.

## FIRE

Fire is a major natural disturbance, influencing species composition and shaping the character of a community. It has long played an important role in vegetational development. Charcoal is universal in the organic layer of the prairie (Heinselman, 1971). Eighty to 90 percent of the virgin northern coniferous forests can be traced to a post-fire origin. More than 95 percent of the virgin forests of Wisconsin were burned during the five centuries before the land was settled by white Europeans (Curtis, 1959). These fires not only enabled such species as yellow birch, hemlock, pines, and oaks to persist, but also, from an ecological viewpoint, were normal and necessary to perpetuate those forests (Maisurow, 1941; Curtis, 1959). The open ponderosa pine forests of southwestern United States evolved under the influence of natural fires. Forty years of fire exclusion from these forests have resulted in a thick growth of young pines, stagnation of stands, elimination of grass, and detrimental changes in the forest community (C. F. Cooper, 1960). Even-aged stands of Douglas-fir, western white pine, and longleaf, red, and jack pine usually result from a fire-prepared seedbed. Chaparral communities are fire-controlled ecosystems. They are renewed by periodic fires that consume old plant cover and initiate new growth (see Biswell, 1974). Fire is a powerful selective and regulatory force on the evolution and maintenance of Mediterranean type ecosystems (Naneh, 1974). In Alaska, however, fires have converted white spruce stands into treeless herbaceous or shrub communities of fireweed and grass or dwarf birch and willow (Lutz, 1956), whose growth is so thick that forest trees cannot become established.

The effects of fire depend upon its intensity, the age of the plants affected, soil moisture at the time of burn, season of burn, health of herbaceous plants, and frequency of drought. Crown and severe surface fires can reestablish pioneer conditions and initiate secondary succession. Deep peat burns can convert forest to stagnant openings. Fire can inhibit the growth of shrubs in grasslands and stimulate the growth of grass. It can promote the growth of shrubs in forests and stimulate shrubby growth

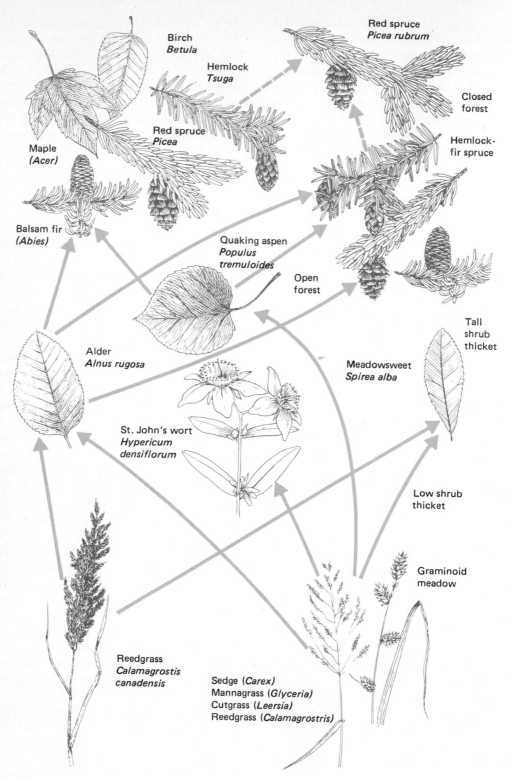

Red spruce
*Picea rubrum*

Birch
*Betula*

Hemlock
*Tsuga*

Closed
forest

Red spruce
*Picea*

Hemlock-
fir spruce

Maple
*(Acer)*

Balsam fir
*(Abies)*

Quaking aspen
*Populus
tremuloides*

Open
forest

Tall
shrub
thicket

Alder
*Alnus rugosa*

Meadowsweet
*Spirea alba*

St. John's wort
*Hypericum
densiflorum*

Low shrub
thicket

Graminoid
meadow

Reedgrass
*Calamagrostis
canadensis*

Sedge (*Carex*)
Mannagrass (*Glyceria*)
Cutgrass (*Leersia*)
Reedgrass (*Calamagrostris*)

**FIGURE 21-18**
*Plant succession wet mineral soil in*
*Canaan Valley, West Virginia. Note the*
*terminal shrub communities. (After*
*Fortney, 1975.)*

**638**

in chaparral. Fire can reduce the numbers of some plants and eliminate others altogether. It produces the vegetational pattern on which the animal component of ecosystems depends. Fire can produce successional stability by destroying the mature stages and initiating their redevelopment on the same site. In this manner fire recycles certain vegetational types such as chaparral and jack pine. Exclusion of fire from fire-dependent ecosystems will alter the type of vegetation that will ultimately claim the site. What type of vegetation might develop with fire exclusion is largely unknown to science and will be in a sense unnatural (Heinselman, 1971).

The plants that grow after a fire originate from several sources. Some are carried in by the wind or by animals. The former are well represented by aspens, paper birch, pine, and some herbs; the latter by fleshy-fruited plants. Others such as oaks, bracken fern, and some perennial grasses sprout from fire-resistant roots. Because the seeds of such trees as beech, birch, and hemlock are destroyed by heat, sprout trees, such as red oak, black oak, white oak, and scarlet oak, dominate the area and other species are reduced to a minimum. In this way fire can change the future composition of a forest (for a detailed discussion, see Ahlgren and Ahlgren, 1960; Kozlowski and Ahlgren, 1974).

In tropical regions cultivation associated with fire can permanently change vegetation. Kowal (1966) studied in detail the effects of fire and shifting cultivation in the Phillipines. Two types of cultivation are practiced: one is slash-and-burn, known as kaingin, and the other is the permanent garden. Practiced in the past on gentle slopes, kaingin caused few long-time environmental effects. Because of the topography, absence of hoeing, absorbent soil organic matter, and absence of fires, erosion was minimal. Once abandoned, the plots reverted to montane forest. But when kaingin was practiced on steep slopes soil erosion and slides so deteriorated the site that xerophytic pines and grasses seed in. Frequent fires, often only 1 to 5 years apart, encouraged continued erosion of the soil. Pines and grasses were never replaced by broad-leaved evergreen forests. Recurring fires in the pine forests slowly ate into the montane forest, which eventually became restricted largely to stream depressions and areas remote from cultivation. Today kaingin is replaced by permanent gardens that involve working the soil with hoe, intensive fertilization, and permanency, a result of increasing demands for truck crops by an expanding urban population. With permanent gardens the equilibrium between agriculture and the soil is broken. Severe erosion occurs even on gentle slopes. The destruction of the montane forest is even more extensive, and if and when abandoned, the land rarely returns to montane forest because of the deterioration of the soil and frequency of fires on the mountain slopes. Abandoned land is claimed by pines and grasses, which in turn may be cleared again for permanent gardens.

## TIMBER HARVESTING

Removal of a forest, especially by clear-cutting, turns the land back to an earlier stage of succession. Unless followed by fire or badly disturbed by erosion and logging activities, the cutover area fills in rapidly with herbs, shrubs, sprout growth, and seedlings of trees present as advanced regeneration in the understory. The area passes quickly through the shrub stage (tree and shrub species 2 to 10 ft. high) to an even-aged pole forest (trees 4 to 8 in. diameter breast height).

Humans often modify the forest in many ways to meet their requirements. Early in the life of a new forest, trees of species not desired for timber and of poor form can be removed. This improves, by economic but not necessarily ecologic standards, the composition of the stand and quality of the trees. The maximum growth of crop trees can be encouraged by thinning. Increased space between trees stimulates crown expansion and increases growth.

Many of the most valuable and desirable timber trees are pioneer species rather than species of late successional stages. To maintain and reproduce this seral stage is often a problem. Stands of pine, balsam fir, and some spruces, aspen, and yellow poplar are maintained by clear-cutting the mature trees to expose the ground to sunlight. Only under this condition will intolerant seedlings survive.

**639**

The other extreme in management is selection cutting. In this process single trees or groups of trees are removed based on their position in the stand and their possibilities of future growth. Because successional changes do not follow (see gap phase replacement and fluctuations), forest composition remains much the same. Intolerant trees are excluded, as well as wildlife species associated with them, and the proportion of tolerant trees is increased.

Mismanagement of the forest, such as clear-cutting excessively large areas, erosion initiated by poor layout of logging roads, and poor slash disposal, can limit the rate of succession and delay the return of the original vegetation. High-grading (taking the best and leaving the poorest) not only eliminates certain tree species from the regenerating forest, but also tends to leave genetically inferior trees to supply seed for the future. A more recent development, whole tree harvesting, in which the entire tree including the roots is removed from the site, can result in complete destruction of the natural forest including the associated understory plants and severely interfere with nutrient cycling.

## GRAZING AND BROWSING

Grazing and browsing by domestic and wild animals may arrest succession and even reverse it. Buffalo herds controlled the height of grass by grazing on the tall-grass species. This permitted short grasses to assume dominance. Plant clipping by prairie dogs tends to decrease the proportion of annual grasses and forbs (herbs other than grasses), to increase perennial forbs, and to influence the relative area of ground cover by each plant species (Koford, 1958). In prairies with both tall- and short-grass species, prairie dogs can develop and maintain a short-grass prairie. Overgrazing of grasslands by domestic stock results in denudation and erosion of the land. In the range lands of the southwestern United States overgrazing reduces the organic mat and thus the incidence of fire. Because of the reduced incidence of fire, reduced competition from grass, and dispersal of seeds through cattle droppings, mesquite and other unwanted shrubs rapidly invade the area (Phillips, 1963; Humphrey, 1958; Box et al., 1967).

The effects of overgrazing are well illustrated on the Wasatch Plateau in Utah. There the virgin subalpine meadows were so overgrazed by cattle and sheep in the 1880s and 1890s that the area was changed to a virtual desert. In spite of management efforts accelerated soil erosion still continues. The overgrazed meadows were taken over by annuals and early withering perennials. In moderately grazed meadows, however, vegetation reacted differently to sheep and cattle grazing. Grazing by sheep tended to favor development of grasses and low shrubs, whereas grazing by cattle resulted in dominance by forbs because grasses were suppressed (Ellison, 1958).

Wild grazing and browsing animals also influence community succession and development. In many parts of eastern North American white-tailed deer, overstocked because of the failure to harvest does as wells as bucks and because of the lack of natural predators, have destroyed natural reproduction and developed a browse line, the upper limits on the trees at which a deer can reach food. The effects of browsing can vary widely among forests depending upon the type and region. D. A. Marquis (1974) in his exclosure studies on the Allegheny Plateau in Pennsylvania found that white-tailed deer virtually eliminated pin cherry and selectively reduced sugar maple. Beech, birch, and striped maple, not preferred by deer, increased in the stands. Browsing by white-tailed deer resulted in regeneration failures in 25 to 40 percent of the areas studied. On some of the cutover areas forest regeneration was completely destroyed and the areas were dominated by grasses, ferns, goldenrods, and asters.

Mule deer and elk in the Rocky Mountain National Park prevent the invasion of forest openings by trees and shrubs. Extensive studies indicate that only in areas protected from deer by fencing did aspen, willow, and other woody invaders grow (Figure 21-19).

In Africa the elephant has a pronounced effect on the nature of the ecosystem. When elephant populations are in balance with forage supplies and their movements are not restricted, the elephant has an important role in creating and maintaining the forest. When elephants are overpopulated, their feeding habits combined with destructive fires can

(a)

(b)

**FIGURE 21-19**
*Influence of elk and deer on vegetation.
(a) When this exclosure was built, aspen
saplings were numerous on the hillside. Elk,
deer, and beaver have eliminated all those
that are not within the exclosure. (b) Old
scars on dying aspen resulted from elk
feeding on the bark.*

devastate flora, fauna, and soils. Elephant dep-
redation on trees acts as a catalyst to fires
which are the primary cause of converting
forest to grassland (Wing and Buss, 1970).

## Succession of human communities

No land-use change is more complete and final
than industrialization and urbanization, a
climax type in human succession. Natural
vegetation is destroyed by man and replaced
by ecologically permanent bare areas of concrete, asphalt, and steel. But even here a diversified group of animals and some plants are
able to exist. Norway rats, common rock
pigeons, starlings, English sparrows, cockroaches, and flies are common to this environment, as well as some grasses, algae, and
other plants able to gain a foothold in cracks in
concrete and vacant lots. Nighthawks have
substituted the artificial canyons created by
tall buildings for natural cliffs.

Fumes from factories, coke ovens, and smel-
ters destroy the vegetation of surrounding
areas. Even after the cause has been eliminated, many years are required before the
vegetation begins to return. Pollution of
streams by sewage, industrial wastes, and siltation eliminates oxygen-demanding fish like
trout; these are replaced by carp and bullheads, able to adapt to polluted conditions.
Dam construction for power drowns terrestrial
communities and converts part of the river
community to a deep lake. Migrant fish, particularly the salmon, may be blocked from
reaching their spawning grounds in the headwater streams.

Human settlement of an area from past to
present has undergone a sort of succession.
The first to live in or penetrate a region, the
pioneers, are hunters and trappers, who, aside
from harvesting animals, leave little mark on
the land. They are followed by a subsistence
farming or grazing culture, which can completely change a natural community. Some
plants and animals may be destroyed, succession set back to an earlier and more economically productive stage, and new animals introduced. Land too poor or too abused to support
human society economically may be abandoned and revert back to natural vegetation.
Traces of old settlements and abandoned land
can be found throughout the country. Ghost
towns, old stone and rail fences, house and
barn foundations, lilac bushes, wells and
springs hidden back in the woods, all attest to
former human occupancy of the land.

Industrial and urban settlements might be
considered climax stages of human succession.
The tremendous growth of suburban settlements onto fertile farmland marks well this

**641**

type of succession. Just as there are successional trends in the various later stages of forest development, so there are successional trends in the stages of urban development. The urban community begins as a small central core and grows outward into the surrounding countryside. Just as the forest invades an old field, so this outward expansion or invasion takes over the surrounding country.

Initially the center of the city is the most desirable place to locate. But because all resident and associated establishments cannot locate there, a sorting process takes place, resulting in the segregation of both functional units and social units based on socioeconomic status, culture, and race (Mayer, 1969). As the city grows, the pressure for other central locations forces outward expansion. This expansion or invasion does not take place at equal rates because of resistance (competition) from land-use functions, people, and transportation facilities. As one zone exerts pressure on the adjacent outer zone, it eventually replaces the outer zone. At the same time the outer zone tends to invade the next or colonize a new area. Thus one successional stage replaces another (Figure 21-20).

With the passage of time, the mature community deteriorates, and the zones about the central core become less desirable. Buildings deteriorate, industrial and commercial complexes become less efficient and far removed from the source of productivity—the young vigorous communities on the outer zones or the earlier stages of succession. There new central core areas develop, determined by transportation links, shopping centers, and industrial parks. As these new core areas become firmly established they in turn initiate their own successional forces and patterns. In time they will be absorbed by the expanding urban area.

Meanwhile, in the original central core, homes once occupied by higher socioeconomic classes have been abandoned to new immi-

FIGURE 21-20
*Expansion of urban areas is similar to succession in old fields. As the forest moves out into old fields, so do urban and suburban areas move into the countryside, their spread following avenues of transportation, as suggested by this model. (From Mayer, 1969.)*

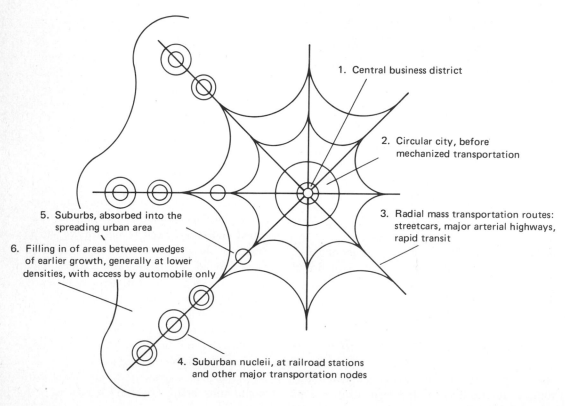

1. Central business district

2. Circular city, before mechanized transportation

3. Radial mass transportation routes: streetcars, major arterial highways, rapid transit

4. Suburban nucleii, at railroad stations and other major transportation nodes

5. Suburbs, absorbed into the spreading urban area

6. Filling in of areas between wedges of earlier growth, generally at lower densities, with access by automobile only

grants, the original residents having moved to outer zones. As the new inhabitants of the core better their socioeconomic status, they too move out to the adjacent zones, while their occupants move further out from the original city. Eventually the structure of the inner city become unattractive, the density of the population declines, and in time the core becomes a blighted area unable to pay any sort of economic rent. Subsidized public intervention becomes necessary. Just as fire or some other natural event tends to renew the aged, declining natural community, so urban renewal represents an attempt to put the core city back to an earlier, more productive stage of succession.

## Ecosystem development through Earth's history

The community changes we have been discussing take place over a relatively short period of time. Seen in terms of the history of the earth, they are changes that occur within one climatic or geological period. However, we can also examine community change from one glacial or geological period to another. The emergence of mountains, sinking and rising of seas, and glaciation, all influenced climate and other environmental conditions for a given area. Many plant and animal species became extinct and one type of climax was replaced by another.

Records of these past communities, their animals and plants, lie buried as fossils. These fossils enable us to determine plant and animal associations of the past and in a broad way the climatic changes that brought about the gradual destruction of one stable community and the emergence of another. Such interpretation is based on the assumption that organisms of the past possessed similar ecological requirements. For example, if modern palms and broadleaf evergreens are tropical plants, we assume that their ancient prototypes also lived in a tropical climate. The study of past relationships of ancient flora and fauna to their environment is *paleoecology*.

Earth, geologists estimate, is some 4600 million years old. Life formed in the late Precambrian era about 4000 million years ago and developed through geological ages with the evolution, emergence, and extinction of numerous groups of plants and animals (see Table 21-2).

### CONTINENTAL DRIFT

The distribution of life, as indicated by fossils, suggests that the distribution of land masses during early geological ages was much different from that today. The flora that covered the southern land masses during the age of plants, the Permian period of the Paleozoic era, differed from that of the northern land mass. This and the discovery that glaciers covered the southern continents could be explained only if one considered that the land masses of the earth were joined together in a manner different from what exists today.

In 1924 the German meteorologist and astronomer Alfred Wegener observed that if the continental shelves of South America and Africa on either side of the Atlantic Ocean were fitted together like pieces in a jigsaw puzzle, many geological features that ended abruptly on continental edges became continuous. Fossils of similar plants and animals once separated by oceans and realms of fossils of different ages were brought together again. Wegener proposed that in the Permian and earlier periods the land masses of the earth were compressed into one large continent, Pangaea, surrounded by one ocean, Panthalassia. Over eons of time this land mass broke apart and drifted into the positions the continents occupy today. Because Wegener was not a geologist and little was known of the earth's surface and ocean floors, Wegener's theory was rejected. But as additional evidence on the nature of the earth's interior and crust, fossils, and paleomagnetism (magnetic directions of ancient rocks) accumulated, the theory of continental drift was revived; it is widely accepted today.

Some evidence seems to indicate that in the lower Paleozoic three separate land masses existed: Asia, North America and Europe, and Gondwanaland, which included modern-day Africa, South America, Australia, New Zealand, and Antarctica. During the Permian these three blocks joined, raising the Caledonian, Ural, and Appalachian mountain ranges and forming the single land mass Pangaea. During the Paleozoic, 420 million years ago,

TABLE 21-2
*Geological time scale*

| Era | Period | Epoch | Age (millions of years) | Dominant life — Plants | Animals |
|---|---|---|---|---|---|
| CENOZOIC: The age of mammals | Quaternary | Recent | 0.01 | Agricultural plants | Domesticated animals |
| | | Pleistocene | 2 | | Ice Age—First true men; mixture and then thinning out of mammalian faunas |
| | Tertiary | Pliocene | 10 | Herbaceous plants rise; forests spread | Culmination of mammals; radiation of apes |
| | | Miocene | 25 | First extensive grass lands | |
| | | Oligocene | 35 | | Modernization of mammals; mammals become dominant |
| | | Eocene | 55 | | Mammals become conspicuous |
| | | Paleocene | 70 | | Expansion of mammals; extinction of dinosaurs |
| MESOZOIC: The age of reptiles | Cretaceous | | 135 | Angiosperms rise; gymnosperms decline | Dinosaurs reach peak; first snakes appear |
| | Jurassic | | 180 | Cycads prevalent | First birds and mammals appear |
| | Triassic | | 230 | Gymnosperms rise; seed ferns die out | First dinosaurs; reptiles prominent |
| PALEOZOIC | Permian | | 280 | Conifers become forest trees; cycads important | Great expansion of primitive reptiles |
| | Carboniferous Pennsylvanian | | 310 | Lepidodendron, sigillaria, and calamites dominant; the swamp forest | Age of cockroaches; first reptiles |
| | Mississippian | | 345 | Lycopods and seed ferns abundant | Peak of crinoids and bryozoans |
| | Devonian | | 405 | First spread of forests | First amphibians; insects and spiders |
| | Silurian | | 425 | First known land plants | First land animals (scorpions) |
| | Ordovician | | 500 | Algae, fungi, bacteria | Earliest known fishes; peak of trilobites |
| | Cambrian | | 600 | Algae, fungi, bacteria; lichens on land | Trilobites and brachipods; marine invertebrates |
| PRECAMBRIAN | Late | | | Algae, fungi, bacteria | First known fossils |
| | Early | | 4500 | Bacteria | No fossils found |

North America and Africa lay close together around the South Pole and the rest of Gondwanaland lay on the far side of the South Pole, pointing toward the equator (Figure 21-21). Slowly the land mass moved northward so that by the Carboniferous age, 340 million years ago, the whole of Africa had moved across the South Pole and Antarctica lay in the region of the South Pole. Glaciers covered southern South America, South Africa, India, and Australia; and Europe and North America lay along the equator.

As Pangaea moved northward it began to break apart slowly, 5 to 10 cm a year (Figure 21-22). The first break in the single land mass apparently took place in the mid-Mesozoic

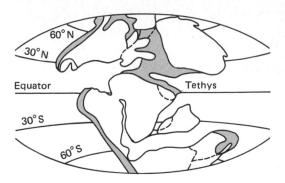

**FIGURE 21-22**
*Earth in the mid-Cretaceous, 105 million years ago. Pangaea not only moved northward, breaking way from Gondwanaland, but also began to break up into separate continents. Dotted lines indicate present-day continental coastlines; shaded areas indicate epicontinental seas. (From Cox et al, 1974.)*

**FIGURE 21-21**
*Map of Pangaea as it existed in the Permian and Triassic (without southeast Asia, because its position at this time is uncertain). Dashed lines indicate the outlines of present-day continental coast lines. Note that Laurasia, consisting of North America and Eurasia, is attached to Gondwanaland, the continents of Africa, South America, Anarctica, and Australia. The series of dark dots indicates the successive positions of the South Pole from the Cambrian to the Jurassic. (Redrawn from Cox et al., 1973.)*

Movement of South Pole

age, when North America and Africa separated to form the first narrow strip of the Atlantic. Africa and South America were still connected by the end of the Jurassic, but drift was taking place in the South Atlantic region. Africa was separated from the Antarctic between the middle Jurassic and middle Cretaceous. By the middle Cretaceous Africa and South America had split apart. By the late Cretaceous southern Greenland began to separate from the British Isles and to move northward. South America and Antarctica still clung together until sometime in the early Cenozoic, finally becoming separated in the Eocene.

### CLIMATIC CHANGES AND DISPERSAL OF SPECIES

The breakup and northward drift of continents resulted in broad climatic changes that affected the evolving plant and animal life. During the Carboniferous period Laurasia (North America and Europe), having moved north to the equator, had a warm, humid, and seasonless climate. A large part of the area was covered with swamps and tropical rainforests dominated by *Lepidodendron, Sigililaria,* and the forerunner of coniferous trees, *Cordaites.* Throughout the Mesozoic no distinct floral or faunal regions existed. For most of the Mesozoic the land, even though partially separated, was one. There were no effective barriers to the dispersal of plants and animals. Mountain ranges lay about the edges of Pangaea. Although shallow seas invaded the land, especially in the late Mesozoic, the invasions were short-lived, geologically speaking, and did not effectively disturb the distribution of animals. The climate, too, was warm and equitable, mostly tropical to subtropical, even to the coast of Alaska. These conditions allowed the great reptiles and early mammals to roam freely over the continental land masses.

But in the late Cretaceous conditions began to change as the continental land masses drifted apart. The warm, subtropical middle Cretaceous climate was replaced in the late Cretaceous by a warm, temperate climate. The cooling climate marked the end of the great reptiles. As the Cretaceous period moved into the Paleocene period of the Cenozoic, greater changes took place. Between the lower Cretaceous and the Eocene, the single, connected land mass inhabited by gymnosperms and reptiles was replaced by a divided land mass inhabited by flowering plants (angiosperms) and mammals. Continental drift effectively separated plant and animal populations and hastened the development of different floras and faunas.

The climate changed even more. Land areas along the seas experienced mild, less variable climates. New mountain ranges interrupted the pattern of rainfall, encouraging the development of deserts or grasslands on the lee side. The uplift of the Andes in South America resulted in a dry, cooler climate that favored the development of grassland on the eastern side of the continent, a situation that exists to the present day. As land masses moved northward the leading edges reached high latitudes and the land was covered by huge ice sheets. As Gondwanaland moved south, the southern limits of that land mass became buried under a layer of permanent ice. In the late Cretaceous Gondwanaland was broken up; the only major intact land mass was Laurasia.

Before Gondwanaland had broken up, the angiosperms experienced a sudden explosive evolutionary radiation. They supplanted the gymnosperms and, apparently before the breakup, spread throughout the super continent. As a result four families of angiosperms—Compositae, Graminae, Legumino-

sae, and Cyperaceae—are abundant throughout the world. When Gonwanaland broke apart, the resulting continental land masses of the southern hemisphere carried with them similar flora. Floral evolution in Laurasia was somewhat different. Laurasia retained the conifers that had arisen from the Cordaites. The conifers could not spread across the hotter equatorial regions, a barrier reinforced by the periodic invasion of shallow seas that covered southern Europe and northern Africa during the Jurassic and Cretaceous, separating Laurasia and Gondwanaland. Because of the colder climate in the northern part of Laurasia, a still different flora evolved in northern and central Eurasia. Later a shallow epicontinental sea, the Turgai Straits east of the Ural Mountains, separated eastern North America and central Europe from Asia and western North America, causing the development of two separate floral realms.

Although plants achieved worldwide distribution before the continents broke up, the mammals did not. They were not significant in the fauna until the dinosaurs disappeared, and by then the continents were well on their way to separation. One of the first groups of mammals, the marsupials, were confronted with competition from more advanced placental animals. Had the placental mammals achieved dominance before the breakup of the continents, the marsupials might never have survived. But marsupials apparently had spread to Antarctica and Australia before these land masses had separated. Antarctica moved south, where the cold eliminated terrestrial mammalian life, while Australia, separate from all other land masses, became a final refuge for marsupial life. Free from competing placental forms, marsupials were able to radiate into a variety of forms and to fill niches similar to those occupied by placental animals elsewhere. Marsupial mammals that existed in North America and South America were rapidly replaced by placental mammals, except for the opossum.

South America, isolated from North America during the lower Cenozoic (Tertiary period), supported diverse and unique forms of placental mammalian life, including the primitive ungulates *Thoatherium* and *Toxodon*. After a land bridge, the Isthmus of

Panama, became exposed between the two continents, these animals were unable to compete with the more advanced placentals that moved down from the north. While the competition resulted in the elimination of certain kinds of animals, the major influence on developing fauna was the rapidly changing climate of the northern land mass. The fauna of the Old World tropics remained free from the influences of great temperature changes, although the increasing aridity in Africa that began in the Oligocene and Miocene epochs of the Cenozoic era brought about the replacement of tropical forests by grasslands on which evolved the huge herds of diverse ungulate fauna and eventually man.

Europe, North America, and Asia were still connected in a manner that permitted animals to move from one to the other. Until the lower Eocene North America was still connected to Europe by Greenland and Scandinavia. At that time North America appeared to be the center of the early evolution of placental mammals. From there they spread to Europe. During that time Asia was separated from Europe by the Turgai Straits but was connected to North America by the Bering land bridge between Siberia and Alaska. However, the cold climate inhibited the movement of mammalian fauna from North America to Asia.

The mid-Eocene was a time of further changes. The North Atlantic joined the Arctic Ocean and separated Europe from North America (Figure 21-23). New mammalian groups that evolved in Europe could not cross to North America, but at this period in the earth's history the Turgai Straits dried up, allowing European mammals access to Asia. The same period saw the return of a warm, moist climate; a semitropical rain-forest extended to Alaska, and tropical conditions extended as far north as England. The climate of the Bering land bridge was benign enough to encourage the passage of mammals from Asia to North America. In the Oligocene the climate cooled again, restricting the migration of mammals to those tolerant of cold temperatures and eliminating tropical forms of plants from northern lands. From the Oligocene through the Pliocene the cooling trend continued. The movement of animals from Eurasia to North

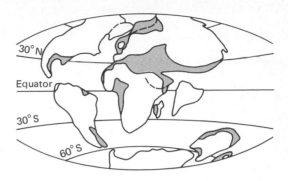

FIGURE 21-23
*Earth during lower Tertiary (upper Eocene), 50 million years ago. Dotted lines indicate present-day continental coastlines; shaded areas indicate shallow epicontinental seas. North America has separated from Eurasia, South America has moved well away from Africa, and Australia has separated from Antarctica. (From Cox et al., 1973.)*

America was restricted to such cold-tolerant species as mammoths and humans, and the flora of the continent was basically a modern one. The cooling of the Pliocene continued until the Pleistocene ushered in the Ice Age.

THE PLEISTOCENE EPOCH

Of particular interest to the paleoecologist are the climatic and vegetational changes that followed the advance and retreat of glaciers in the Pleistocene. As the glaciers moved south in several advances, vegetation was destroyed and the relief or physiognomy changed radically. The climate about the edges of the glacier supposedly was rather cold and optimal only for tundra and tiaga vegetation.

Changes in postglacial vegetation and climate are recorded in the bottoms of lakes and bogs. As the glaciers retreated, they left scooped out holes and dammed up rivers and streams, which filled with water to form lakes. Organic debris accumulated on the bottom to form peat, marl, and muds. Pollen, spores, and small invertebrates that blew in from adjacent vegetation settled on the water and sank. Microscopic examination of samples of organic bottom deposits obtained at regular intervals reveals the fossil remains of these organisms. Various genera of fossil pollen can be identified by comparisons with pollen growing to-

**647**

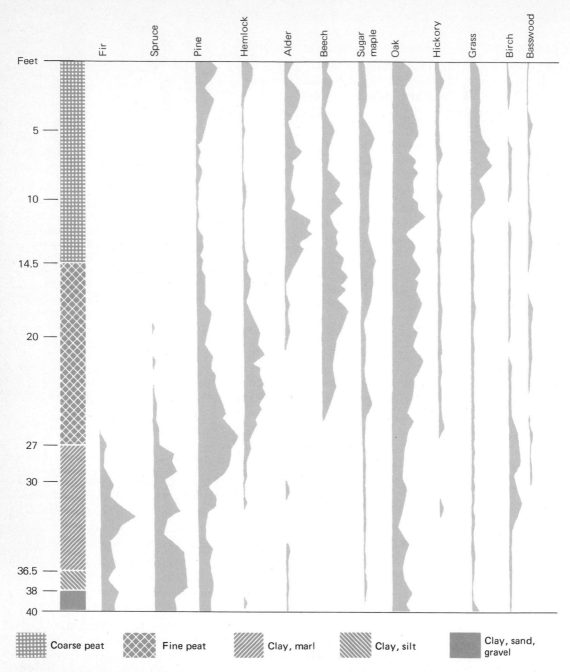

**FIGURE 21-24**
*Pollen diagram for Crystal Lake, Crawford County, Pennsylvania. The graphs indicate the percentage of various genera based on counts of 200 pollen grains for each spectrum level. Grass pollen counts are expressed as percentages of total free pollen. Note five major forest successions from bottom of diagram to top: (1) initial spruce-fir forest with oak and pine; (2) pine forest with oak and birch; (3) hemlock with codominant oak and accessory hickory and beech; (4) oak-hickory forest; (5) pine; hemlock increasing, oak remaining, beech and hickory decreasing. (Adapted from Walker and Hartman, 1960.)*

day. The relative abundance of pollen of several genera indicates the predominant vegetation at the specified depth of deposition (Figure 21-24).

Pollen investigation can indicate only trends in vegetation and climate through the past. At present it is impossible to determine the exact structure and composition of prevailing vegetation at any one time period. Tree species that produce more pollen than others may appear more abundant than they really were. Some pollen might have been carried some distance by the wind or perhaps buried deeper by soil invertebrates. Many pollen grains can be identified only to genera and not to species. For example, the pollen of different types of oak cannot be distinguished to give a clue to the particular type of oak forest existing at a particular time. However, modern paleoecological techniques that take these problems into account and improved identification procedures are enabling paleoecologists to provide a rather accurate picture of postglacial vegetation (see Dort and Jones, 1970).

The Pleistocene, which began some two million years ago, marked the end of the Tertiary period and the Pliocene epoch by ushering in the Ice Age. But recent studies of deep sea sediments, the geophysics of the ocean bottoms, and the Antarctic ice suggest that ice caps have been part of the earth's geological and thus ecological history at least since the Miocene and even earlier. Thus the Pleistocene is simply a stage on a continuum of ice and vegetation through Cenozoic. The present distribution of plants and animals can be appreciated only in the context of longer successional development in the past.

***Pre-Pleistocene Development.*** At the beginning of the Cenozoic era, some 65 million years ago, most of present-day continental North America and Europe was land. By the beginning of the Miocene epoch forests closely related to the present-day deciduous forest existed with little variation across the northern continents. Known as the Arcto-Tertiary forest, it was a mixture of broad-leafed and coniferous species roughly divided into boreal and temperate elements. The temperate or deciduous element was much like the mixed mesophytic forest of the central Appalachians.

The boreal element, consisting of pines, spruces, cypress, birches, and willows, bore little resemblance to today's uniform northern coniferous forests. Because tropical and subtropical climates existed far north of the present positions, neotropical and Paleotropical-tertiary forest, ancestors of today's tropical forests, covered most of central and southern Europe and central North America. Probably in the Eocene, a mixed woodland, the Madro-tertiary flora, developed on the Mexican plateau.

But from the Miocene on, the climate began to deteriorate. The western mountain system in North America rose; climatic zones and their biota were pushed southward and tropical forests were driven into central America. The American portion of the Arcto-tertiary forest was separated from that of Europe and Asia by continental drift, and certain species, such as *Metasequoia, Aliarthus,* and *Ginko* became extinct in North America. As the mountains rose the broad rain shadow on their lee side wiped out the Arcto-tertiary forest and caused the development of grasslands in the central part of North America. A relict Arcto-Tertiary forest, poor in species, but including the sequoias and redwoods, was left in the Pacific northwest. Elements of the Madro-tertiary forest moved northward to occupy dry lands vacated by the Arcto-tertiary forest and eastward to form a sclerophyllous-pine woodlands ancestral to the Southern oak-pine woodlands of today. In the late Pliocene a continuing climatic deterioration accompanied by mountain building resulted in the development of continental glaciers.

***Glacial periods and vegetation.*** The Pleistocene was an epoch of great climatic fluctuations throughout the world. Four times during the Pleistocene ice sheets advanced and retreated in North America (Figure 21-25) and at least three times they advanced and retreated in Europe (Figure 21-26). Four times in North America and three times in Eurasia the biota retreated and advanced, each advance having a somewhat different mix of species.

Each glacial period was followed by an interglacial period (Table 20-3 and Figure 21-27). The climate oscillations in each interglacial period had two major stages, cold and

*TABLE 21-3*
*Comparative glacial and interglacial stages*

| Britian | Northern Europe | North America | Climate |
|---|---|---|---|
| Flandrian (postglacial) | Weichselian | | temperate |
| Devensian (last glaciation) | Weichselian | Wisconsin | cold glacial |
| Ipswichian (last interglacial) | | Sangamon | temperate |
| Wolstonian | Saalian | Illinoian | cold glacial |
| Hoxnaian | Holstein | Yarmouth | temperate |
| Anglian | Elsterian | Kansas | cold glacial |
| Comerian | | Aftonian | temperate |
| Beestonian | | Nebraskan | cold |
| Pastonian | | | temperate |
| Baventian | | | cold |
| Older stages of cold and temperate climates | | | |

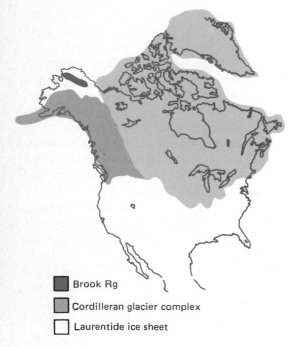

| | Brook Rg |
| | Cordilleran glacier complex |
| | Laurentide ice sheet |

*FIGURE 21-25*
*Glaciation in North America. In the Pleistocene northern North America was covered by four ice sheets, the limits of each usually marked by a terminal moraine. The last and perhaps most significant was the Wisconsin ice sheet.*

temperate. During the cold stage tundra-like vegetation dominated the landscape. As the glaciers retreated and the climate ameliorated, light-demanding forest trees such as birch and pine advanced. As the soil improved and the climate continued to warm, these trees were replaced by more shade-tolerant species as oak and ash. As the next glacial period began to develop, species such as firs and spruces dominated the forest, changing the soil from mull to acid mors. As both climate and soil began to deteriorate, heaths began to dominate the vegetation and forest species disappeared.

Major differences in vegetation bordering the glacier existed between Europe and North America. In Britain and Europe a wide belt of tundra edged the glacier. The Alps, with its large ice cap, created an air flow pattern in which the westerly flow of warm air was diverted southward. As a result the Arcto-tertiary flora was decimated by the early cold stages of the Pleistocene. Temperate genera were forced southward, but the southward retreat was blocked by mountains, deserts, and seas (Figure 21-28). Only the hardy boreal genera could survive. In North America no such barriers existed. The glacier extended further south into a warmer zone across which the flow of warm air was unimpeded (Figure 21-29). As a result spruce forests grew virtually to the edge of the ice and Tertiary precursors of the present-day deciduous forest survived to spread north.

The last great ice sheet, the Laurentian, reached its maximum advance about 12,000 to 14,000 years ago during what is known as the Wisconsin glaciation stage in North America. Spruce forest covered most of eastern and central United States as far west as northwestern Kansas. Canada was under ice, its forest cover eliminated. West of Kansas lay

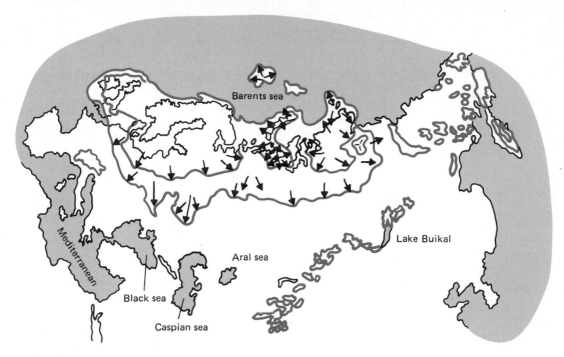

Barents sea

Lake Buikal

Mediterranean

Aral sea

Black sea

Caspian sea

**FIGURE 21-26**
*Glaciation in Europe. In the Pleistocene northern Eurasia was covered by ice sheets similar to those covering North America. The most important was the last, known as the Weischselian. Note the disjunct glacier in the region of the Alps. (After Flint, 1971.)*

uninterrupted sand dunes, a treeless landscape that resulted from intense winds generated by the nearby ice sheet. At the time of glacial maxima, tundra probably existed in the high Appalachians, a few relict examples of which still exist today, and in a narrow belt in New England and northern Minnesota (Wright, 1970, 1971). As the ice lobes retreated the winds decreased, the dunes stabilized, and spruce forest spread rapidly over them as far west as the Black Hills of South Dakota (Wright, 1970). South of the spruce, but still very much a part of the boreal forest, stands of spruce and pine stretched from Illinois eastward to the Virginian Appalachians. On the coastal plain extending from the southern Appalachians to Georgia were forests of jack pine and red pine (Whitehead, 1967).

The climatic changes that accelerated the retreat of the Laurentian ice sheet in the late Wisconsin also brought a sudden end to the spruce forest in unglaciated North America

over a period of perhaps 100 to 700 years depending upon the locality. In the western part of the glacial region spruce was replaced by prairie grasses. Further east, closer to the edge of the present prairie region, spruce gave way to pine and birch. In the southern Appalachians spruce was taken over by oak and pine, except for relict stands at high elevations. Pines moved northward rapidly from their Appalachian refugia in the Carolinas and dominated much of the region about the newly formed Great Lakes.

In the western Cordilla (parallel chains of mountains) during the Wisconsin glaciation the tree line and tundra vegetation moved down slope 800 to 1000 meters lower than today. Many of the modern desert basins were shrub steppe dominated in part by sagebrush.

**FIGURE 21-27 (page 652)**
*Schematic representation of the glacial-interglacial cycle in northwestern Europe. (After van der Hammen, Wijmstra, and Zagwijn, 1974.)*

**FIGURE 21-28 (page 653)**
*Generalized distribution of vegetation in Europe at the Weichsel/Wurm maximum. The Black Sea and Caspian Sea are interconnected lakes. (After Flint, 1971.)*

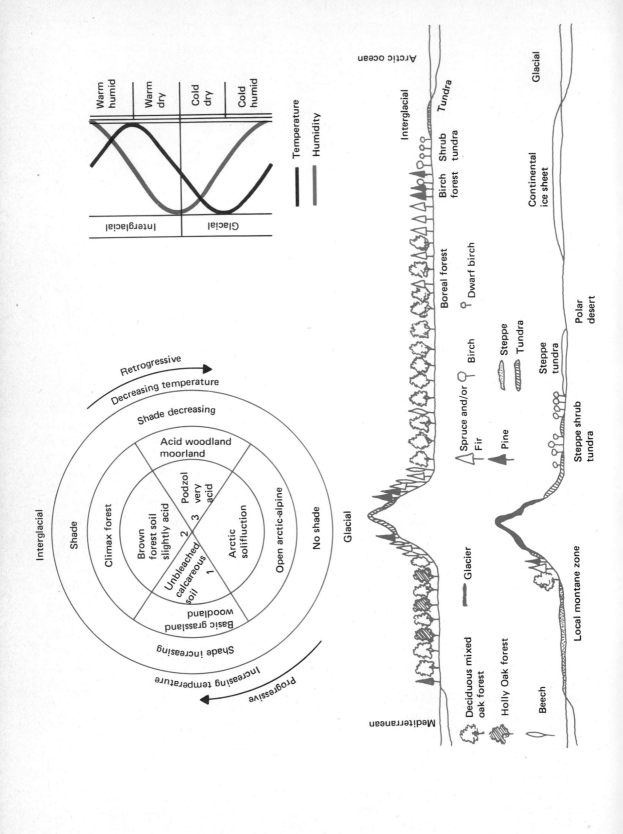

Tundra, mountain
vegetation

Prairie (long grass
with scattered
trees)

Boreal forest

Temperate mixed
forest chiefly
broadleaf

Mediterranean
vegetation, chiefly
coniferous

Mediterranean, chiefly
of dry steppe type

Water (lakes)

Glacier ice

Steepe

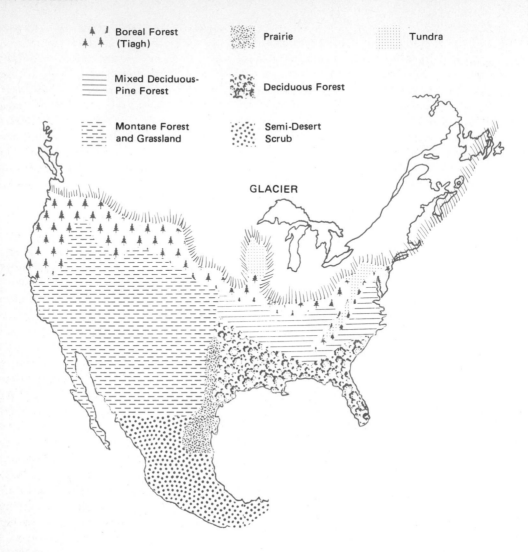

| ▲ ┛<br>▲ ▲ | Boreal Forest<br>(Tiagh) | | Prairie | | Tundra |
|---|---|---|---|---|---|
| ☰ | Mixed Deciduous-<br>Pine Forest | | Deciduous Forest | | |
| ┄ | Montane Forest<br>and Grassland | | Semi-Desert<br>Scrub | | |

GLACIER

**FIGURE 21-29**
*Probable distribution of vegetation in North
America during the glacial optimum of the
Wisconsian glacial period. (Adapted from
Flint, 1971, Ross, 1967.)*

At the end of the Wisconsin period the valley
glaciers melted, the climate warmed, and the
tundra and coniferous forest ascended the
mountains.

*Speciation.* The expansion of vegetation
during the interglacial periods and the destruction and isolation of large areas of vegetation during the full glacial periods undoubtedly set the stage for speciation of many forms
of life, such as insects (Ross, 1970), birds
(Mengel, 1964, 1970; Selander, 1965), mammals (Flerow, 1971), and amphibians (Blair,
1965). In North America much of this specia-

tion apparently took place in the forest regions, for they were subject to long periods of
geographical isolation during the glacial
periods. Little speciation took place on grasslands (Ross, 1971; Mengel, 1964) because the
grassland was not divided into isolated units
as were the forests. Rather the central grasslands acted as a persistent although changeable "sea" that separated the deciduous forest
"continent" from "islands" of western montane
forests and woodlands (Mengel, 1970). Acting
as a land bridge that periodically united the
two was the spruce forest, expansive during
interglacial stages and narrow and broken
during glacial periods (Figure 21-30). During
full glacial times species occupying the spruce
forest would be isolated in eastern and western
refugia and could begin to differentiate. Dur-

ing the interglacial periods the western forms would be further isolated as the montane forest moved upslope as the climate warmed.

R. Mengel (1964) developed a model to show how an ancestral species invaded the spruce after the first glaciation when the boreal Arcto-tertiary forest was forced south by glaciation. Upon retreat of the glacial ice and the northward expansion of the boreal forest, the species would spread throughout the transcontinental range. At the second glaciation the species would be disjoined and would subsequently differentiate not only into eastern and western species, but also into varying forms in the western montane islands. In the next interglacial stage the eastern species would expand again and the entire process of expansion and subsequent disjunction during the advance of the ice sheet would be repeated through the third and fourth glaciations.

A possible example that fits the model well is the black-throated green warbler complex (Figure 21-31). The parent species would be an ancestral black-throated green warbler *(Dendroica virens)*. Nearly continental in distribution the black-throated green warbler is found to a limited extent in the central Appalachian deciduous forest and more extensively in deciduous forest containing some hemlock *(Tsuga)*.

In the west are three and possibly four related species. Townsend's warbler *(D. Townsendi)*, an inhabitant of the coniferous forest of the northwest Pacific coast, is similar to the black-throated green in plumage pattern and in quality of song. It was probably derived from a westward invasion of the ancestral black-throated green warbler. Another species, the hermit warbler *(D. occidentalis)*, found in the tall conifers of the Cascades and Sierra Nevadas, also is only slightly different from the black-throated green, and may stem from the same invasion. Both Townsend's and the hermit warbler hybridize in a narrow zone of sympatry. Overlapping the ranges of both is the black-throated gray warbler *(D. nigrescens)* which occupies shrubby openings in the northwest coniferous forests, shrubby openings in western mixed woods, and dry slopes covered with chaparral. Small in size and completely lacking yellow pigments, the bird nevertheless is remarkably similar in color pattern to Townsend's warbler. It may have descended from a western colonization more remote in time than that postulated for the other two warblers. In the Edwards Plateau of Texas is the golden-cheeked warbler *(D. chrysoparia)*, which inhabits the relict deciduous forest. Appearing as a dark form of the black-throated green warbler and possessing a similar song, this species is regarded as an offshoot of the black-throated green warbler or its ancestor, which may have reached Texas during the Wisconsin glaciation.

*Extinction.* The Pleistocene was not only a time of widespread speciation; it was also a time of extinctions. At the end of the Pliocene a number of warm climate forms—southern species of deer, tapirs, elephants, and apes—disappeared from northern Asia and Europe. But as these forms disappeared, new groups evolved. By the middle of the Pleistocene a number of Arctic and Subartic genera of mammals evolved at the edges of the ice sheet—reindeer, musk-ox, wooly rhinocerous, long-horned bison, saber-toothed tiger, wolverine, and various lemmings. By the end of the Pleistocene most of the cold climate forms had disappeared and those that survived were limited to restricted ranges. The Pleistocene also was the expansion of *Homo sapiens* across the North American continent. Because of the lowered sea levels during glacial times humans were able to move across the land bridge from Asia to Alaska. Much of Alaska was unglaciated and during the glacial retreat 11,000 to 12,000 years ago, new immigrants were able to move into North America through an unglaciated corridor that led down through Canada. There human colonists found an abundance of large mammals. The spread of humans over the continent was spurred by a population increase and the need to move southward as resources were depleted (Martin, 1973).

Accompanying this explosive increase in humans was the extinction of many species, such as the wooly mammoth, the mastedon, wooly rhinocerous, camel, horse, and royal bison. Some students of the Quaternary Period believe that the reason for the extinction of certain large mammals was overkill by humans. P. S. Martin (1973) points out that overkill was most pronounced in North and

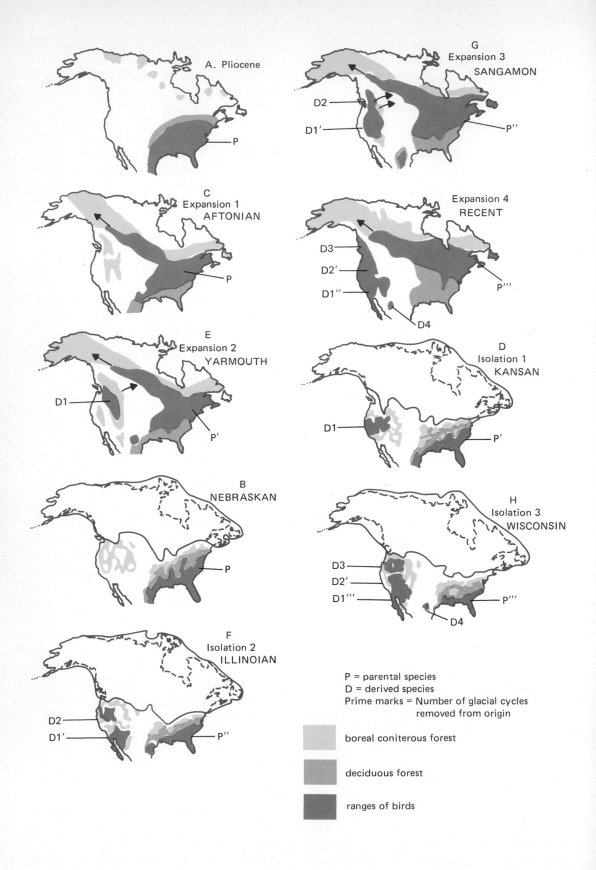

A. Pliocene

P

G
Expansion 3
SANGAMON

D2

D1'

P''

C
Expansion 1
AFTONIAN

P

Expansion 4
RECENT

D3

D2'

D1''

P'''

D4

E
Expansion 2
YARMOUTH

D1

P'

D
Isolation 1
KANSAN

D1

P'

B
NEBRASKAN

P

H
Isolation 3
WISCONSIN

D3

D2'

D1'''

P'''

D4

F
Isolation 2
ILLINOIAN

D2

D1'

P''

P = parental species
D = derived species
Prime marks = Number of glacial cycles
                        removed from origin

boreal coniterous forest

deciduous forest

ranges of birds

FIGURE 21-30

*FIGURE 21-30*

*Model of the influence of Pleistocene glaciation on speciation as exemplified by wood warblers in North America. With the first (Nebraskan) glaciation components of the boreal coniferous Arcto-Tertiary forest invaded deep into the Arcto-Tertiary deciduous forest of southeastern North America. This provided an opportunity for some warblers to adapt to coniferous forests or their seral stages. In the next interglacial period a broad transition zone of coniferous forest developed, permitting warblers to establish a continent-wide range and providing conditions for separation of stocks into eastern and western segments during subsequent glacial advances. In the west montane forest replaced glaciers during interglacial periods. This created islands of habitats in which differentiation could proceed. Repetition of the process during four glacial cycles resulted in the differentiation of the 12 endemic western species of warblers. (After Mengel, 1964.)*

South America. According to his hypothesis humans swept through North and South America in a series of advancing fronts of dense population. A front remained stationary for a decade or less until humans reduced or eliminated populations of big game. Then the front advanced to a new area and remained there until the animal resource was exhausted. In time as food resources were depleted, human populations, too, declined.

The hypothesis of mammalian extinctions by early human populations has its critics. Some argue that it is difficult to imagine that primitive humans could have destroyed a whole species by killing off individuals, even if the methods of taking animals were wasteful. (A common method of capturing or killing large game was to drive herds over a cliff or to surround them with fire.) Others believe that although hunting by humans was extensive, climatic changes and changes in food supply helped to cause extinctions (Bryson, 1970; Flerow, 1971). Climatic changes that replaced browse with short grass over large areas of the continent could have reduced the food supply for large browsers such as the wooly mammoth (note the location of the grasslands in the last glaciation in Figure 21-29). This argument is further supported by the fact that the mammoth became extinct in the Soviet Union about the same time it became extinct in North America (Flerow, 1971). If we were to assume that the highest density of population was associated with extinctions, we might have to conclude that Europe was less densely populated than Siberia because such Pleistocene species as big-horned deer and aurocks became extinct in Siberia, but the wisent or European bison still survives. This is highly improbable. Another descendent of the Pleistocene fauna, the American bison, did not face demise until white Europeans deliberately went about its extermination to subjugate the descendents of Pleistocene human settlers.

Extinction probably resulted from a number of causes, differing for each species. Extinction for one set of large grazing herbivores might have resulted from a combination of a declining food base and intense predation from both humans and large carnivores. There is considerable anthropological evidence that the Indians of central United States depended heavily on white-tailed deer for meat (Smith, 1975). So, too, did the wolf and the mountain lion. Wolves and to an extent mountain lions take mostly young of the year and old and sick individuals. Indians, as excavations of old settlements reveal, took mostly individuals 1 to 4 years old. Thus humans and wolves were noncompetitive complementary predators that together held deer herds to some level below the carrying capacity of the habitat. If humans, saber-toothed tigers, dire wolves, and others were complementary predators, all of them acting together could have exerted more predatory pressure than the large herbivores, faced with a changing vegetation because of a warming climate, could withstand. As a result the larger mammals disappeared, human populations sharply declined, and large predators became extinct.

## SUMMARY

With the passing of time natural communities change. Old fields of today return to forests tomorrow; weedy fields in prairie country revert to grassland.

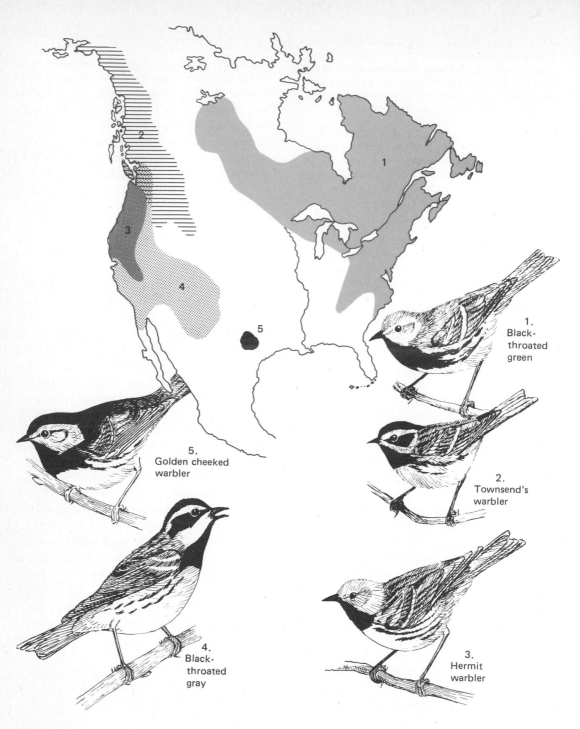

**FIGURE 21-31**
*Approximate breeding distribution of the black-throated green warbler group. (From Mengel, 1964.)*

1. Black-throated green

2. Townsend's warbler

3. Hermit warbler

4. Black-throated gray

5. Golden cheeked warbler

This gradual change in community composition is called succession. It is characterized by a replacement of opportunistic species by equilibrium species, a progressive change in species structure, an increase in biomass and organic matter accumulation, and a gradual balance between community production and community respiration.

Succession may come about because of changes caused by the organisms themselves. As they exploit the environment their life activities make the habitat unfavorable for their own continued survival and create an environment for a different group of organisms. Or succession may begin with any one or several groups of organisms capable of growing on the site successfully and arriving there early. They preempt the space and continue to exclude or inhibit the growth of others until the former colonists die or are damaged releasing resources and allowing new species to enter.

Eventually, however, these communities arrive at some form of steady state with the environment. This stage, usually called the climax, is more or less self-maintaining and usually long-lived (on a human time scale). If free from major disturbances, the climax experiences only small-scale changes as individuals die and are replaced. The degree of change in the climax depends on whether replacements are of the same or different species. If replacements are the same species, stability is assured. But because of conditions involving microorganisms that often develop in the soil around long-lived plants, offspring of the same species are usually less favored than those of other species. As a result the climax most often is a mosiac of regenerating patches in which the new growth may not be the same species as the individual being replaced. Thus stability of a system often requires an introduction of instability into the climax community. Although response to disturbance in the climax community may involve replacement of one species by another on a temporary or long-term basis, it does result in the maintenance of the community in general.

Small-scale disturbances may result in one of several kinds of responses. Gap phase replacement involves both vegetational reorganization and invasion by opportunistic or equilibrium species depending upon the nature of the environmental disturbance and biotic conditions. Cyclic replacement results from destruction of vegetation induced by some periodic biotic or environmental disturbance that starts regeneration again at some particular stage. Each successive phase or community is related to the other by orderly changes in the rise and decline of successive communities. Such cyclic successions aid in community persistence. Non-successional, reversible changes or fluctuations in community composition usually result from environmental stresses such as changes in soil moiture, wind, and grazing.

In the majority of natural communities succession is interrupted by major disturbances such as wind, fire, grazing, insect outbreaks, and the like. Disturbances such as grazing and fire can shape and modify the nature of the community and the direction of succession by favoring certain species and eliminating others. Overgrazing of forests and grasslands by both domestic and wild herbivores has resulted in denudation of grasslands and serious disturbance and even destruction of forests. In cutting forests for timber humans most often have tended either to remove trees completely or to take the best and leave the poorest to regenerate the future forest.

Perhaps the most outstanding characteristic of natural communities is their dynamic nature. They are constantly changing through time, rapidly in early stages of development, more slowing in later stages. Even the seemingly most stable natural communities slowly change through time.

**659**

That they have changed in the geological past is shown by fossil records. Some of the most pronounced changes, as evidenced by pollen profiles of bogs, occured during the Pleistocene when several advances of the ice sheet in the northern hemisphere eliminated vegetation and the several retreats allowed the northward invasion of vegetation. Isolation and environmental changes during the glacial and interglacial stages encouraged the speciation of some forms of plant and animal life and the extinction of others. The nature of vegetational and faunal communities today reflects the evolutionary impact of changing conditions during the Pleistocene.

# Appendixes

# An annotated bibliography of statistical methods
## APPENDIX A

Modern studies of plant and animal life must be quantified in contrast to the purely descriptive studies of an earlier day. Even the most elementary studies demand a quantified approach. Because much ecological work involves sampling of one sort or another, statistics are involved. But in the hands of those who know little about them statistics can be dangerous. They are often abused rather than used and misapplied to problems at hand. Too often data are collected and forced to fit some statistical procedure; instead research should be planned with some particular statistical approach in mind. Quantified approaches also may employ computers and computer programming. Because programming is not only specialized, but also constantly changing, I have not included any specific references on its use. Instead the investigator should consult local experts.

ALCHLEY, W. R., AND E. H. BRYANT (eds.). 1977. *Multivariate Statistical Methods: Among-Groups Covariance*. Benchmark® Papers. Dowden, Hutchinson & Ross, Stroudsburg, Pa.; distr. by Academic Press, New York. An important set of original papers on the use of multivariate analysis.

ANDREWARTHA, H. G. 1970. *Introduction to the Study of Animal Populations*. University of Chicago Press, Chicago. Contains helpful and easily understood discussions on the use and application of statistics to the study of animal populations: sampling, tests for nonrandomness, analysis of variance, etc.

BATSCHELET, E. 1965. *Statistical Methods for the Analysis of Problems in Animal Orientation and Certain Biological Problems*. American Institute of Biological Sciences, Washington, D.C. Prepared for the use of biologists working in the area of animal migration and homing, biological clocks, and periodic activity. Contains solved examples.

BROWER, J. E., AND J. H. ZAR. 1977. *Field and Laboratory Methods for General Ecology*. Brown, Dubuque, Iowa. An invaluable laboratory manual that provides an excellent introduction to elementary ecological statistics.

BRYANT, E. H., AND W. R. ALCHLEY (eds.). 1977. *Multivariate Statistical Methods: Within-Groups Convariance* (Benchmark Papers). Dowden, Hutchinson & Ross, Stroudsburg, Pa.; distr. by Academic Press, New York. Companion volume to Alchley and Bryant, 1977.

COCHRAN, W. C., AND G. M. COX. 1957 *Experimental Designs*. Wiley, New York. Application and interpretation of experimental design. Requires a background of statistics. Should be consulted if experiment requires experimental design.

COX, D. R. 1958. *Planning of Experiments*. Wiley, New York. Describes basic ideas underlying statistical aspects of experimental design with emphasis on planning. Probably the best general introduction to the design of experiments now available to nonstatisticians.

COX, G. W. 1976. *Laboratory Manual of General Ecology*. Brown, Dubuque, Iowa. A highly useful laboratory manual that provides an easy step-by-step introduction to ecological statistics.

DENENBERG, V. H. 1976. *Statistics and Experimental Design for Behavioral and Biological Researchers*. (Wiley,) Halsted Press, New York. Basic reference strong on randomization and experimental design.

DIXON, W. J., AND F. J. MASSEY, JR. 1969. *Introduction to Statistical Analysis*. McGraw-Hill, New York. Standard basic reference.

GREIG-SMITH, P. 1964. *Quantitative Plant Ecology*. Butterworth, Washington, D.C. Statistical approach to plant ecology. A must for all ecologists concerned with vegetation analysis.

HOGG, R. V., AND A. T. CRAIG. 1970. *An Introduction to Mathematical Statistics*. Macmillan, New York. Clear mathematical exposition of statistics.

KERSHAW, K. A. 1964. *Quantitative and Dynamic Ecology*. Elsevier, New York. Oriented toward plant ecology. Considers simpler statistical procedures and their application to vegetation studies. Strong on positive and negative associations between species, Poisson series, and the detection of nonrandomness and natural groupings.

LEHNER, P. 1978. *The Handbook of Ethological Measurements*. Garland STMP Press, New York. How to study animal behavior.

LEWIS, T., AND L. R. TAYLOR. 1967. *An Introduction to Experimental Ecology*. Academic Press, New York. Useful, but not as detailed as Southwood, 1966.

MUELLER-DOMBOIS, D., AND H. ELLENBERG. 1974. *Aims and Methods of Vegetation Ecology*. Wiley, New York. An important manual on statistical analysis of vegetation data. Provides detail not found in other texts and manuals.

ORLOCI, L. 1975. *Multivariate Analysis in Vegetation Research*. W. Junk, The Hague. Provides an introduction to concepts and procedures of multivariate analysis and a source of worked examples.

PATTEN, B. C. (ed.). 1977 on. *Systems Analysis and Simulation in Ecology*. Academic Press, New York. Multivolume work reviewing the application of systems science and computer technology to ecology. Invaluable reference on modeling.

PIELOU, E. C. 1974. *Population and Community Ecology: Principles and Methods*. Gorden and Breach, New York. Rather comprehensive examination of mathematical approach to population and community structure. Strong on mathematics of predation and competition.

PIELOU, E. C. 1975. *Ecological Diversity*. Wiley, New York. All you will want to know about mathematical analysis of diversity.

PIELOU, E. C. 1976. *Mathematical Ecology*. Wiley, New York. Excellent introduction to statistical and mathematical approach to population ecology. A graduate level book.

POOLE, R. W. 1974. *An Introduction to Mathematical Ecology*. McGraw-Hill, New York. Excellent introduction that should be in the reference library of every ecologist.

SCHULTZ, V. 1961. *An Annotated Bibliography on the Uses of Statistics in Ecology — A Search of 31 Periodicals*. Pub. Tid 3908. U.S. Atomic Energy Commission, Office of Technical Information, Environmental Science Branch, Division of Biology and Medicine, Washington, D.C. Useful compendium summarizing use of statistics in ecology to 1961.

SCHULTZ, V., et al. 1977. *A Bibliography of Quantitative Ecology*, Dowden, Hutchinson & Ross, Stroudsburg, Pa.; distr. by Academic Press, New York. Comprehensive bibliography arranged in subject matter categories. Supplements Schultz, 1961.

SEBER, G. A. F. 1973. *The Estimation of Animal Abundance and Related Parameters*. Griffin, London. Excellent manual on handling census and recapture data, etc.

SIEGEL, S. 1956. *Nonparametric Statistics for the Behavioral Sciences*. McGraw-Hill, New York. Collection of nonparametric tests in common use. Procedures explained and illustrated by mathematical examples.

SNEDECOR, G. W. 1957. *Statistical Methods*. Iowa State College Press, Ames. Excellent reference for those with some familiarity with statistics.

SOKAL, R. R., AND F. J. ROHLF. 1969. *Biometry, The Principles and Practices of Statistics in Biological Research*. Freeman, San Francisco. Perhaps the best introduction to biometrics and applied statistics.

SOKAL, R. R. AND F. J. ROHLF. 1973. *Introduction to Biostatistics*. Freeman, San Francisco. Excellent introduction. Requires minimal background in math and covers nonparametric tests.

SOUTHWOOD, T. R. 1966. *Ecological Methods: With Particular Reference to the Study of Insect Populations*. Barnes & Noble, New York. Excellent handbook on statistics in ecology. Especially good on population statistics.

STEEL, R. G. D., AND J. H. TORRIE. 1960. *Principle and Procedures of Statistics*. McGraw-Hill, New York. One of the standard texts on statistics. Good on procedures to problem solving.

TANNER, J. T. 1978. *Guide to the Study of Animal Populations*. University of Tennessee Press, Knoxville. Short but excellent guide to methods of population measurements with emphasis on statistical analysis.

ZAR, J. H. 1973. *Biostatistical Analysis*. Prentice-Hall, Englewood Cliffs, N.J. Best introductory text on biological statistics available. Requires no mathematics beyond elementary algebra and contains many worked out problems.

# Sampling plant and animal populations
## APPENDIX B

One of the major problems in ecology is the determination of population distribution, size, and changes in abundance. The problem involves sampling to estimate some characteristic of the population. By taking into account the variability within the population, one can make some general inferences about the population as a whole. To be valid the samples must be completely random, that is, all combinations of sampling units must have an equal probability of being selected. To characterize the population as a whole, certain *parameters* are used, for example, the proportion of males in a population or the mean value per plot.

The object of sampling is to estimate some parameter or a function of some parameter. The value of the parameter as estimated from a sample is the *estimate,* which is hoped to be *accurate,* that is, close to the true value. But often the estimate is *biased.* Bias is a systematic distortion due to some flaw in the measurement or to the method of collecting the sample. In some statistical procedures bias cannot be avoided, but it is always important to recognize the source of the bias and to take it into account. A biased estimate can never be accurate, although it may be *precise.* Precision refers to a clustering of sample values about their own mean (Figure B-1).

*FIGURE B-1*
*A graphic model of the various statistical concepts involved in estimating populations. The solid bull's-eye represents the parameter being estimated, in this instance population size. The dotted target represents the sampling estimates. The distance of the samples or shots from the true target or bull's-eye is the "deviation." Value of estimates gives one the mean value or expectation of the estimate. The distance of the shots from the mean point of impact are the deviations. The distance between the mean point of impact or expectation of the estimator and the true value, the bull's-eye (which one may never know) is the* bias. *A tight shot or a clumping of estimates around a mean value indicates a high precision but the estimates are not necessarily accurate. A shot group close to the parameter being estimated increases accuracy. The variance is the mean squared deviation about the mean point of impact. (From Giles, 1969,* **Wildlife Management Techniques.***)*

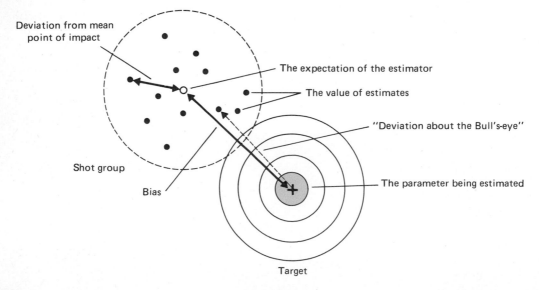

Deviation from mean point of impact

The expectation of the estimator

The value of estimates

"Deviation about the Bull's-eye"

Shot group

Bias

The parameter being estimated

Target

## SAMPLING PLANT POPULATIONS

## Methods

Methods of analyzing the vegetation occupying a given site are numerous and the literature discussing them, the underlying philosophies, and the statistical treatments is extensive. The basic references given at the end of this appendix should be consulted. Because a major decision in ecological studies involves the methods to be used, a number of ways of handling the vegetation are given here with some comments on their advantages and disadvantages. Again these are simply personal selections, based on experience, and they are not complete by any means.

### QUADRATS OR SAMPLE PLOTS

Strictly speaking, the quadrat applies to a square sample unit or plot. It may be a single sample unit or it may be divided into subplots. Quadrats may vary in size, shape, number, and arrangement, depending upon the nature of the vegetation and the objectives of the study.

The size of the quadrat must be adapted to the characteristics of the community. The richer the flora, the larger or more numerous the quadrats must be. To sample forest trees, the fifth-acre plot is a popular size, but it may be too large if trees are numerous or if many species are involved. Smaller plots can be used to study shrubs and understory trees. For grass and herbaceous plants, 1 m² is the usual size. The quadrat is usually square, but rectangular or circular ones may work better. Circular plots are the easiest to lay out because one needs only a center stake and string of desired length.

The number of sample units to be employed always presents some problems. The number will vary with the characteristics of the community, objectives of the investigation, degree of precision, and so on. The final number more often than not is arbitrary, but by using statistical methods the reliability of the sample and the number of samples needed for any desired degree of accuracy can be determined, once a normal distribution around a mean has been established.

*Species-area curves.* A second approach to this problem is the use of the species-area curve (Figure B-2), obtained by plotting the number of species found in plots of different sizes (vertical axis) against the sample size area (horizontal axis). The curve rises sharply at first because the number of new species found is large. As the sample plot size or number is increased, the quantity of new species added declines to a point of diminishing returns where there is little to be gained by continuing the sampling. This curve can be plotted on an arithmetic or a logarithmic base (see Vestal, 1949). The method can be employed to determine the largest size of a single plot (minimal area) needed to survey the community adequately. In this case the sampling should be done by using a geometric system of nested plots (Figure B-3). Or the curve can be used to determine the minimum number of small multiple plots needed for a satisfactory sample. In addition the species-area curve can be used to compare one community with another (Figure B-4).

*Kinds of quadrats.* Quadrats are often labeled according to the type of data recorded.

FIGURE B-2
*Species are curves: (a) for minimal area of quadrat; (b) for minimal number of quadrats. Arrows indicate minimal areas.*

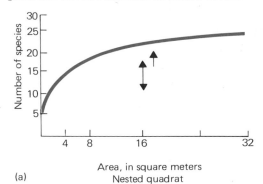

(a)

Area, in square meters
Nested quadrat

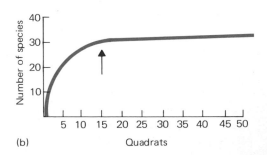

(b)

Quadrats

**665**

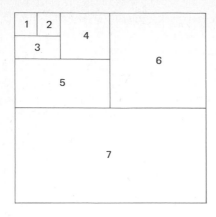

*FIGURE B-3*
*An example of nested quadrats.*

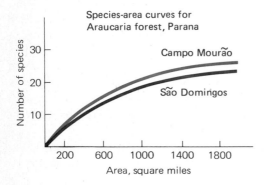

*FIGURE B-4*
*Species-area curve for a stand of* **Araucaria**
*forest on different sites near Campo Mourão,
Parana, showing the use of species-area
curves to compare one community with
another. (From Cain and Castro, 1959,*
**Manual of Vegetation Analysis,** *Harper &
Row, New York.)*

1. List quadrat. Organisms found are listed
   by name. A series of list quadrats gives a
   floristic analysis of the community and
   allows an assignment of a frequency index,
   but nothing else.
2. Count quadrat. Numbers as well as names
   of species encountered are recorded.
   Quadrats used in deer-browse studies fall
   into this category and it is widely used in
   forest survey work. In forest studies
   additional information, such as height,
   volume, basal area, etc., is also taken.
3. Cover quadrat. Actual or relative coverage
   is recorded, usually as a percentage of the
   area of the ground surface covered or
   shaded by vegetation.

4. Chart quadrat. A quadrat is mapped to
   scale to show the location of individual
   plants. This is a tedious job, but it
   provides an overall view especially useful
   in long-range studies of vegetation
   changes.
   ***Location of plots.*** For statistically reliable
estimates, the plots must be randomized. This
is rather easily done. Numbered grid lines are
drawn over an area photo or map of the study
area. The numbers of the vertical and horizontal grid lines are written on small squares of
paper. To draw random numbers for the two
lines, use a table of random numbers, available in statistical tables and statistics texts.

**THE BELT TRANSECT**

A variation of the quadrat method is the belt
transect. A *transect* is a cross section of an
area used as a sample for recording, mapping,
or studying vegetation. Because of its con-
tinuity through an area, the transect can be
used to relate changes in vegetation within it
to changes in the environment. As a sample
unit the measurements within a transect are
pooled and each transect is treated as a single
observation.

1. Determine the total area of the site to be
   sampled; then divide by 5 or 10 to obtain
   the total sample area.
2. Lay out a series of belt transects of a
   predetermined width and length, sufficient
   to embrace the area to be sampled. Then
   divide the belts into equal-sized segments.
   These are sometimes called quadrats or
   plots, but they differ from true quadrats in
   that each represents an observational unit
   rather than a sampling unit.
3. Measure the vegetation in each unit for
   some attribute, depending upon the
   problem at hand: abundance, sociability,
   frequency, stem counts, etc.
   A variation of the segmented-belt transect
consists of taking observations only on alternate segments. The precision seems to be affected very little (Oosting, 1956). For example,
10 quadrats alternately spaced on a 20-ft belt
are nearly twice as efficient statistically as 10
quadrats on a 10-ft belt.
   ***Advantages and Disadvantages.*** The quadrat method has its advantages and disadvantages. It is a popular method, easily employed.

If the individual organisms are randomly distributed, then the accuracy of the sample and the estimate of the density depend upon the size of the sample. But individuals seldom are randomly dispersed, so the accuracy of quadrat sampling may be low, unless a great number of plots are involved. The quadrat method is tedious and time-consuming.

The belt transect is well adapted to estimate abundance, frequency, and distribution. But for estimating the frequency index, it has the disadvantage that frequency by classes is related to the size of the plot. If one wishes to compare one area with another, the segment size used in sampling must be the same for both areas.

*Analysis.* There are several ways to record data from quadrats and belt transects. One is simply to record the presence of a species. This perhaps is the most objective method, but it limits analysis to frequency and relative frequency.

A second method is to record the number of individuals of the various species found in the quadrat. This gives a density figure. Because of the variations in growth forms among the various species, numbers mean little. Counts are most useful in certain situations such as counting the number of stems of shrubby plants available for deer browse or counting the amount of forest reproduction. The samples are broken down into classes, such as 1 ft high, 2 to 12 ft, or 1 to 3 in., 4 to 9 in., and so on.

A third method is that of Braun-Blanquet (1951). This involves a total estimate based on abundance and cover. If the number of individuals in a plant community is estimated but not counted, the data are referred to as *abundance*. Abundance implies a number of individuals, but number does not necessarily reflect dominance or *cover*. Cover is the result of both numbers and massiveness. Although abundance and coverage are separate and distinct, they can be combined in a community description as the total estimate. For many field studies this method works very well. But it is subjective and the data are difficult to handle statistically (see Mueller-Dumbois and Ellenberg, 1974). However, this method does provide a useful general picture of the plant community. The scales are given in Table B-1. Along with total estimates an estimate of

**TABLE B-1**
*Total estimate scale (abundance plus coverage)*

| | |
|---|---|
| + | Individuals of a species very sparsely present in the stand; coverage very small |
| 1 | Individuals plentiful, but coverage small |
| 2 | Individuals very numerous if small; if large, covering at least 5% of area |
| 3 | Individuals few or many, collectively covering 6–25% of the area |
| 4 | Individuals few or many, collectively covering 26–50% of the area |
| 5 | Plants cover 51–75% of the area |
| 6 | Plant species cover 76–100% of the area |

**TABLE B-2**
*Sociability classes of Braun-Blanquet*

| | |
|---|---|
| Class 1 | Shoots growing singly |
| Class 2 | Scattered groups or tufts of plants |
| Class 3 | Small, scattered patches or cushions |
| Class 4 | Large patches or broken mats |
| Class 5 | Very large mats of stands or nearly pure populations that almost blanket the area |

sociability of each species should be given (as in Table B-2)—whether the plant grows singly, in clumps, mats, and so on.

The total estimate and the sociability estimate can be expressed together to give a paired value for each species, for example, a plant species with the value 4.3, in which the first figure is the total estimate, the second the sociability. Once a number of stands have been surveyed, the community characteristics can be combined in a sort of an association table. The plant species usually are listed on the basis of fidelity or presence, the characteristic species of the community often heading the list. An example is given in Table B-3. It is far from being complete and is given here only to show how such tables are constructed.

Data so collected describe individual stands. By the use of another attribute, presence, one can compare stands of a community type or of related types. Presence refers to the degree of regularity with which a species recurs in different examples of a community type. It is commonly expressed as a percentage that can be assigned to one of a limited number of pre-

**667**

TABLE B-3
*Stand composition, Cumberland Plateau, West Virginia*

| Herbaceous species | 1 | 2 | 3 | 4 | 5 | 6 | 7 | 8 | 9 | 10 | Frequency (%) |
|---|---|---|---|---|---|---|---|---|---|---|---|
| *Polystichum aerostichoides* | 2·2 | +·1 | 1·2 | 1·2 | 2·2 | 2·2 | 2·2 | 2·2 | 2·1 | 1·1 | 100 |
| *Cimicifuga racemosa* | 3·2 | 2·2 | | | 2·2 | 3·2 | 2·2 | 1·2 | 2·2 | 2·2 | 80 |
| *Geranium maculatum* | | +·1 | +·1 | +·1 | 1·2 | 2·2 | 2·2 | +·2 | 1·1 | +·1 | 90 |
| *Disporum lanuginosum* | 3·2 | 3·3 | | 3·3 | 1·1 | 2·2 | 1·1 | +·1 | +·2 | 2·2 | 90 |
| *Galium circaezans* | +·2 | +·2 | | | 1·2 | +·1 | 2·2 | +·1 | | | 60 |
| *Thalictrum dioicum* | | +·2 | | | 1·2 | 1·1 | 2·1 | | | +·2 | 50 |
| *Sanicula canadensis* | +·1 | 2·2 | 1·1 | | +·1 | | | | | | 40 |

*Note:* Selected species only, to illustrate table construction.

TABLE B-4
*Presence classes*

| Presence class | Stands of one community type studied in which species occur (%) |
|---|---|
| 1 | 1–20 |
| 2 | 21–40 |
| 3 | 41–60 |
| 4 | 61–80 |
| 5 | 81–100 |

sence classes, as given in Table B-4. Presence is determined by dividing the total number of stands in which the species is found by the total number of stands investigated. Species that have a high percentage of presence or which fall within presence class 5 often are regarded as more or less characteristic of that community.

**LINE INTERCEPT**

The line intercept is one-dimensional. Most useful for sampling shrub stands and woody understory of the forest, the line intercept or line transect method consists of taking observations on a line or lines laid out randomly or systematically over the study area. The procedure is as follows.

1. Stretch a metric steel tape, steel chain, or a tape between two stakes 50 to 100 m or ft apart.
2. Subdivide the line into predetermined intervals such as 1 m, 5 m, etc.
3. Move along the line, and for each interval record the plant species found and the distance they cover along that portion of the line intercept. Consider only those plants touched by the line or lying under or over it. Treat each stratum of vegetation separately, if necessary,
   a. For grasses, rosettes, and dicot herbs, measure the distance along the line at ground level.
   b. For shrubs and tall dicot herbs, measure the shadow or distance covered by a downward projection of the foliage above.
4. Usually 20 to 30 such lines are sufficient. The data can be summarized as follows.
1. Number of intervals in which each species occurs along the line.
2. Frequency of occurrence for each species in relation to total intervals sampled.
3. Total linear distance covered by each species along the transect.
4. Total length of line covered by vegetation and total "open" length:
5. Total number of individuals, if they can be so recorded. Because of the nature of branching and size variations it is difficult to count individual plants on line transect.

*Advantages and disadvantages.* This method is rapid, objective, and relatively accurate. The area may be determined directly from recorded observations. The lines can be randomly placed and replicated to obtain the desired precision. The method is well adapted for measuring changes in vegetation if the ends of the lines are well marked. Generally it is more accurate in mixed plant communities than quadrat sampling and is especially well suited for measuring low vegetation.

On the debit side, the method is not well adapted for estimating frequency or abundance because the probability of an individual's being sampled is proportional to its size. Nor is it suited where vegetation types are intermingled and the boundaries indistinct.

*Analysis.* From the data the following measurements may be calculated. (Calculation of the relative density may not be possible if individual plants can not be identified.)

$$\frac{\text{relative}}{\text{density}} = \frac{\text{total individuals species A}}{\text{total individuals all species}} \times 100$$

$$\frac{\text{dominance}}{\text{(cover)}} = \frac{\text{total intercept length, species A}}{\text{total transect length}} \times 100$$

$$\frac{\text{relative}}{\text{dominance}} = \frac{\text{total intercept length, species A}}{\text{total intercept length, all species}} \times 100$$

$$\text{frequency} = \frac{\text{intervals in which species occurs}}{\text{total number of transect intervals}} \times 100$$

$$\frac{\text{relative}}{\text{frequency}} = \frac{\text{frequency value, species A}}{\text{total frequency value, all species}} \times 100$$

### POINT-QUARTER METHOD

Several variations of the variable plot or "plotless" method have been developed for ecological work. These methods arose from the variable radius method of forest sampling developed in Germany by Bitterlich. He used it to determine timber volume without establishing plot boundaries. The method was introduced into the United States by Grosenbaugh (1952, 1958).

One of the most useful of the plotless methods is the point-quarter method (see Cottam and Curtis, 1956; Greig-Smith, 1964). It is most useful in sampling communities in which individual plants are widely spaced or in which the dominant plants are large shrubs or trees.

The procedure is as follows.

1. Locate a series of random points within the stand to be sampled, or pick random points along a line transect passing through the stand.
2. At each station mark a point in the ground.
3. Divide the working area into four quarters or quadrants by visualizing a grid line, predetermined by compass bearing, and a line crossing it at right angles, both passing through the point (Figure B-5).
4. Select the tree (or plant) in each quarter which is closest to the point. Record its distance from the point, diameter breast height, and species. The tally sheet will thus contain data for four trees at each point, one from each quarter.
5. Tally at least 50 such points.

The computations are as follows.

1. Add all distances in the samples and divide the total distance by the number of distances to obtain a mean distance of point to plant.

$$\text{mean distance} = \frac{\text{total distance}}{\text{number of distances}}$$

**FIGURE B-5**
*The point-quarter method of sampling forest stands. See text for details.*

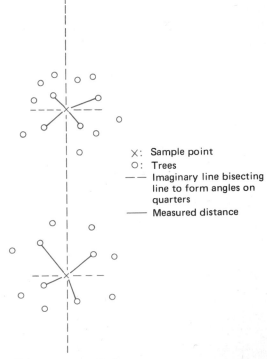

X: Sample point
O: Trees
– – Imaginary line bisecting line to form angles on quarters
—— Measured distance

Point quarter method

2. Square the mean distance to obtain the mean area covered on the ground per plant.

3. Divide the mean area per plant into the unit area on which density is expressed. If the area is in feet then divide the mean distance squared into 43,560 ft$^2$ to obtain the total density of trees per acre.

4. Determine basal area for each tree (see Table B-5 and B-6) from diameter measurements. (Basal area is the area of a plane passed through the stem of a tree at right angles to its longitudinal axis and at breast height. Because the cross section approximates a circle, its area can be computed from a standard formula.)

5. $$\text{relative density} = \frac{\text{individual of species A}}{\text{total individuals of all species}} \times 100$$

6. $$\text{density} = \frac{\text{relative density of species A}}{100} \times \text{total density of all species}$$

7. $$\text{relative dominance} = \frac{\text{total basal area of species A}}{\text{total basal area of all species}} \times 100$$

8. $$\text{frequency} = \frac{\text{number of points at which species A occurs}}{\text{total number of points sampled}} \times 100$$

9. $$\text{relative frequency} = \frac{\text{frequency value for species A}}{\text{total frequency value for all species}} \times 100$$

10. Absolute values for the number of trees per unit area of any species and the basal area per unit area of any species are determined by multiplying the relative figures for density by the total trees per acre to determine density and by the total basal area per acre to determine absolute dominance.

*Advantages and disadvantages.* The point-quarter method is simple and rapid and works quite well. The underlying assumption is that individuals of all species together are randomly dispersed. Although this assumption may not be true, it does not seem to produce significant error, except where a deviation from overall randomness is quite obvious. In spite of this, relative density and relative dominance are valid even if dispersion is not random. Error would appear in the calculation of absolute density and absolute dominance.

IMPORTANCE VALUES

In regions where the plant communities are highly heterogeneous, the classification of communities on the basis of dominants or codominants becomes impractical. Therefore, Curtis and McIntosh (1951) came up with the index "importance value" to develop a logical arrangement of the stands. This index is based on the fact that most species do not normally reach a high level of importance in the community, but those that do serve as an index, or guiding species.

Importance value (IV) is the sum of relative density, relative frequency, and relative dominance for each species involved. It may be expressed as a range from 0 to 3.00 or 300 percent or it may be divided by 3 to give importance percentage which will range from 0 to 1.0 or 100 percent.

Importance value provides an overall estimate of the influence or importance of a species in a community. Once importance values have been obtained for species within a stand, stands can be grouped by their leading dominants according to importance values, and the groups can be placed in a logical order based on the relationships of several predominant species. In Table B-7, for example, are four species that were the leading dominants in 80 of 95 forest stands in southern Wisconsin. Note that the dominants are arranged in order of decreasing importance value, from stands dominated by black oak to those dominated by sugar maple. Such an arrangement also shows increasing values for sugar maple. Trees intermediate in dominance can be handled in the same way.

CONTINUUM INDEX

The importance value can be expanded further into a continuum index, a composite figure that can be used to compare a large number of

## Areas of circles in square feet for diameters in inches

| Dia-meter | 0.0 | 0.1 | 0.2 | 0.3 | 0.4 | 0.5 | 0.6 | 0.7 | 0.8 | 0.9 |
|---|---|---|---|---|---|---|---|---|---|---|
| 0 | 0.000 | 0.000 | 0.000 | 0.000 | 0.001 | 0.001 | 0.002 | 0.003 | 0.003 | 0.004 |
| 1 | 0.005 | 0.007 | 0.008 | 0.009 | 0.011 | 0.012 | 0.014 | 0.016 | 0.018 | 0.020 |
| 2 | 0.022 | 0.024 | 0.026 | 0.029 | 0.031 | 0.034 | 0.037 | 0.040 | 0.043 | 0.046 |
| 3 | 0.049 | 0.052 | 0.056 | 0.059 | 0.063 | 0.067 | 0.071 | 0.075 | 0.079 | 0.083 |
| 4 | 0.087 | 0.092 | 0.096 | 0.101 | 0.106 | 0.110 | 0.115 | 0.120 | 0.126 | 0.131 |
| 5 | 0.136 | 0.142 | 0.147 | 0.153 | 0.159 | 0.165 | 0.171 | 0.177 | 0.183 | 0.190 |
| 6 | 0.196 | 0.203 | 0.210 | 0.216 | 0.223 | 0.230 | 0.238 | 0.245 | 0.252 | 0.260 |
| 7 | 0.267 | 0.275 | 0.283 | 0.291 | 0.299 | 0.307 | 0.315 | 0.328 | 0.332 | 0.340 |
| 8 | 0.349 | 0.358 | 0.367 | 0.376 | 0.385 | 0.394 | 0.403 | 0.413 | 0.422 | 0.432 |
| 9 | 0.442 | 0.452 | 0.462 | 0.472 | 0.482 | 0.492 | 0.503 | 0.513 | 0.524 | 0.535 |
| 10 | 0.545 | 0.556 | 0.567 | 0.579 | 0.590 | 0.601 | 0.613 | 0.624 | 0.636 | 0.648 |
| 11 | 0.660 | 0.672 | 0.684 | 0.696 | 0.709 | 0.721 | 0.734 | 0.747 | 0.759 | 0.772 |
| 12 | 0.785 | 0.799 | 0.812 | 0.825 | 0.839 | 0.852 | 0.866 | 0.880 | 0.894 | 0.908 |
| 13 | 0.922 | 0.936 | 0.950 | 0.965 | 0.979 | 0.994 | 1.009 | 1.024 | 1.039 | 1.054 |
| 14 | 1.069 | 1.084 | 1.100 | 1.115 | 1.131 | 1.147 | 1.163 | 1.179 | 1.195 | 1.211 |
| 15 | 1.227 | 1.244 | 1.260 | 1.277 | 1.294 | 1.310 | 1.327 | 1.344 | 1.362 | 1.379 |
| 16 | 1.396 | 1.414 | 1.431 | 1.449 | 1.467 | 1.485 | 1.503 | 1.521 | 1.539 | 1.558 |
| 17 | 1.576 | 1.595 | 1.614 | 1.632 | 1.651 | 1.670 | 1.689 | 1.709 | 1.728 | 1.748 |
| 18 | 1.767 | 1.787 | 1.807 | 1.827 | 1.847 | 1.867 | 1.887 | 1.907 | 1.928 | 1.948 |
| 19 | 1.969 | 1.990 | 2.011 | 2.032 | 2.054 | 2.074 | 2.095 | 2.117 | 2.138 | 2.160 |
| 20 | 2.182 | 2.204 | 2.226 | 2.248 | 2.270 | 2.292 | 2.315 | 2.337 | 2.360 | 2.382 |
| 21 | 2.405 | 2.428 | 2.451 | 2.474 | 2.498 | 2.521 | 2.545 | 2.568 | 2.592 | 2.616 |
| 22 | 2.640 | 2.664 | 2.688 | 2.712 | 2.737 | 2.761 | 2.786 | 2,810 | 2.835 | 2.860 |
| 23 | 2.885 | 2.910 | 2.936 | 2.961 | 2.986 | 3.012 | 3.038 | 3.064 | 3.089 | 3.115 |
| 24 | 3.142 | 3.168 | 3.194 | 3.221 | 3.247 | 3.274 | 3.301 | 3.328 | 3.355 | 3.382 |
| 25 | 3.409 | 3.436 | 3.464 | 3.491 | 3.519 | 3.547 | 3.574 | 3.602 | 3.631 | 3.659 |
| 26 | 3.687 | 3.715 | 3.744 | 3.773 | 3.801 | 3,830 | 3.859 | 3.888 | 3.917 | 3.947 |
| 27 | 3.976 | 4.006 | 4.035 | 4.065 | 4.095 | 4.125 | 4.155 | 4.185 | 4.215 | 4.246 |
| 28 | 4.276 | 4.307 | 4.337 | 4.368 | 4.399 | 4.430 | 4.461 | 4.493 | 4.524 | 4.555 |
| 29 | 4.587 | 4.619 | 4.650 | 4.682 | 4.714 | 4.746 | 4.779 | 4.811 | 4.844 | 4.876 |
| 30 | 4.909 | 4.942 | 4.974 | 5.007 | 4.041 | 5.074 | 5.107 | 5.140 | 5.174 | 5.208 |
| 31 | 5.241 | 4.275 | 5.309 | 5.343 | 5.378 | 5.412 | 5.446 | 5.481 | 5.515 | 5.550 |
| 32 | 5.585 | 5.620 | 5.655 | 5.690 | 5.726 | 5.761 | 5.796 | 5.832 | 5.868 | 5.904 |
| 33 | 5.940 | 5.976 | 6.012 | 6.048 | 6.084 | 6.121 | 6.158 | 6.194 | 6.231 | 6.268 |
| 34 | 6.305 | 6.342 | 6.379 | 6.417 | 6.454 | 6.492 | 6.529 | 6.567 | 6.605 | 6.643 |
| 35 | 6.681 | 6.720 | 6.758 | 6.796 | 6.835 | 6.874 | 6.912 | 6.951 | 6.990 | 7.029 |
| 36 | 7.069 | 7.108 | 7.147 | 7.187 | 7.227 | 7.266 | 7.306 | 7.346 | 7.286 | 7.426 |
| 37 | 7.467 | 7.507 | 7.548 | 7.588 | 7.629 | 7.670 | 7.711 | 7.752 | 7.793 | 7.834 |
| 38 | 7.876 | 7.917 | 7.959 | 8.001 | 8.042 | 8.084 | 8.126 | 8.169 | 8.211 | 8.253 |
| 39 | 8.296 | 8.338 | 8.381 | 8.424 | 8.467 | 8.510 | 8.553 | 8.596 | 8.640 | 8.683 |
| 40 | 8.727 | 8.770 | 8.814 | 8.858 | 8.902 | 8.946 | 8.990 | 9.035 | 9.079 | 9.124 |
| 41 | 9.168 | 9.213 | 9.258 | 9.303 | 9.348 | 9.393 | 9.439 | 9.484 | 9.530 | 9.575 |
| 42 | 9.621 | 9.667 | 9.713 | 9.759 | 9.805 | 9.852 | 9.898 | 9.945 | 9.991 | 10.038 |
| 43 | 10.085 | 10.132 | 10.179 | 10.226 | 10.273 | 10.321 | 10.368 | 10.416 | 10.463 | 10.511 |
| 44 | 10.559 | 10.607 | 10.655 | 10.704 | 10.752 | 10.801 | 10.849 | 10.898 | 10.947 | 10.996 |
| 45 | 11.045 | 11.094 | 11.143 | 11.192 | 11.242 | 11.291 | 11.341 | 11.391 | 11.441 | 11.491 |
| 46 | 11.541 | 11.591 | 11.642 | 11.692 | 11.743 | 11.793 | 11.844 | 11.895 | 11.946 | 11.997 |
| 47 | 12.048 | 12.100 | 12.151 | 12.203 | 12.254 | 12.306 | 12.358 | 12.410 | 12.462 | 12.514 |
| 48 | 12.566 | 12.619 | 12.671 | 12.724 | 12.777 | 12.830 | 12.882 | 12.936 | 12.989 | 13.042 |
| 49 | 13.095 | 13.149 | 13.203 | 13.256 | 13.310 | 13.364 | 13.418 | 13.472 | 13.527 | 13.581 |
| 50 | 13.635 | 13.690 | 13.745 | 13.800 | 13.854 | 13.909 | 13.965 | 14.020 | 14.075 | 14.131 |

*Area of circles in square feet for circumferences and diameters in inches and in square meters for circumferences and diameters in centimeters*

| Diameter (in./cm) | Circumference (in./cm) | Area (ft²) | (m²) | Diameter (in./cm) | Circumference (in./cm) | Area (ft²) | (m²) |
|---|---|---|---|---|---|---|---|
| 1 | 3.14 | 0.005 | — | 51 | 160.22 | 14.186 | 0.240 |
| 2 | 6.28 | 0.022 | — | 52 | 163.36 | 14.748 | 0.212 |
| 3 | 9.42 | 0.049 | 0.001 | 53 | 166.50 | 15.321 | 0.221 |
| 4 | 12.57 | 0.087 | 0.001 | 54 | 169.65 | 15.904 | 0.229 |
| 5 | 15.71 | 0.136 | 0.002 | 55 | 172.79 | 16.499 | 0.238 |
| 6 | 18.85 | 0.196 | 0.003 | 56 | 175.93 | 17.104 | 0.246 |
| 7 | 21.99 | 0.267 | 0.004 | 57 | 179.07 | 17.721 | 0.255 |
| 8 | 25.13 | 0.349 | 0.005 | 58 | 182.21 | 18.348 | 0.264 |
| 9 | 28.27 | 0.442 | 0.006 | 59 | 185.35 | 18.986 | 0.273 |
| 10 | 31.42 | 0.545 | 0.008 | 60 | 188.50 | 19.635 | 0.283 |
| 11 | 34.56 | 0.660 | 0.010 | 61 | 191.64 | 20.295 | 0.292 |
| 12 | 37.70 | 0.785 | 0.011 | 62 | 194.78 | 20.966 | 0.302 |
| 13 | 40.84 | 0.922 | 0.013 | 63 | 197.92 | 21.648 | 0.312 |
| 14 | 43.98 | 1.069 | 0.015 | 64 | 201.06 | 22.340 | 0.322 |
| 15 | 47.12 | 1.227 | 0.018 | 65 | 204.20 | 23.044 | 0.332 |
| 16 | 50.26 | 1.396 | 0.020 | 66 | 207.34 | 23.758 | 0.342 |
| 17 | 53.41 | 1.576 | 0.023 | 67 | 210.49 | 24.484 | 0.352 |
| 18 | 56.55 | 1.767 | 0.025 | 68 | 213.63 | 25.220 | 0.363 |
| 19 | 59.69 | 1.969 | 0.028 | 69 | 216.77 | 25.967 | 0.374 |
| 20 | 62.83 | 2.182 | 0.031 | 70 | 219.91 | 26.725 | 0.385 |
| 21 | 65.97 | 2.405 | 0.035 | 71 | 223.05 | 27.494 | 0.396 |
| 22 | 69.12 | 2.640 | 0.038 | 72 | 226.19 | 28.274 | 0.407 |
| 23 | 72.26 | 2.885 | 0.042 | 73 | 229.34 | 29.065 | 0.418 |
| 24 | 75.40 | 3.142 | 0.045 | 74 | 232.48 | 29.867 | 0.430 |
| 25 | 78.54 | 3.409 | 0.049 | 75 | 235.62 | 30.680 | 0.442 |
| 26 | 81.68 | 3.687 | 0.053 | 76 | 238.76 | 31.503 | 0.454 |
| 27 | 84.82 | 3.976 | 0.057 | 77 | 241.90 | 32.338 | 0.466 |
| 28 | 87.96 | 4.276 | 0.062 | 78 | 245.04 | 33.183 | 0.478 |
| 29 | 91.11 | 4.587 | 0.066 | 79 | 248.18 | 34.039 | 0.490 |
| 30 | 94.25 | 4.909 | 0.071 | 80 | 251.33 | 34.907 | 0.503 |
| 31 | 97.39 | 5.241 | 0.075 | 81 | 254.47 | 35.785 | 0.515 |
| 32 | 100.53 | 5.585 | 0.080 | 82 | 257.61 | 36.674 | 0.528 |
| 33 | 103.67 | 5.940 | 0.086 | 83 | 260.75 | 37.574 | 0.541 |
| 34 | 106.81 | 6.305 | 0.091 | 84 | 263.89 | 38.484 | 0.554 |
| 35 | 109.96 | 6.681 | 0.096 | 85 | 267.04 | 39.406 | 0.567 |
| 36 | 113.10 | 7.069 | 0.102 | 86 | 270.18 | 40.339 | 0.581 |
| 37 | 116.24 | 7.467 | 0.108 | 87 | 273.32 | 41.282 | 0.594 |
| 38 | 119.38 | 7.876 | 0.113 | 88 | 276.46 | 42.237 | 0.608 |
| 39 | 122.52 | 8.296 | 0.119 | 89 | 279.60 | 43.202 | 0.622 |
| 40 | 125.66 | 8.727 | 0.126 | 90 | 282.74 | 44.179 | 0.636 |
| 41 | 128.81 | 9.168 | 0.132 | 91 | 285.88 | 45.166 | 0.650 |
| 42 | 131.95 | 9.621 | 0.138 | 92 | 289.03 | 46.164 | 0.665 |
| 43 | 135.09 | 10.085 | 0.145 | 93 | 292.17 | 47.173 | 0.679 |
| 44 | 138.23 | 10.559 | 0.152 | 94 | 295.31 | 48.193 | 0.694 |
| 45 | 141.37 | 11.045 | 0.159 | 95 | 298.45 | 49.224 | 0.709 |
| 46 | 144.51 | 11.541 | 0.166 | 96 | 301.59 | 50.266 | 0.724 |
| 47 | 147.65 | 12.048 | 0.173 | 97 | 304.73 | 51.318 | 0.739 |
| 48 | 150.80 | 12.566 | 0.181 | 98 | 307.88 | 52.382 | 0.754 |
| 49 | 153.94 | 13.095 | 0.189 | 99 | 311.02 | 53.456 | 0.770 |
| 50 | 157.08 | 13.635 | 0.196 | 100 | 314.16 | 54.542 | 0.785 |

*The average importance value index of trees in stands with four species as the leading dominants*

| Species | Leading dominant in stand | | | | Ecological sequence number |
| | Quercus velutina | Quercus alba | Quercus rubra | Acer saccharum | |
| --- | --- | --- | --- | --- | --- |
| Black oak *(Quercus velutina)* | 165.1 | 39.6 | 13.6 | 0 | 2 |
| Shagbark hickory *(Carya ovata)* | 0.3 | 8.8 | 5.2 | 5.9 | 3.5 |
| White oak *(Quercus alba)* | 69.9 | 126.8 | 52.7 | 13.7 | 4 |
| Black walnut *(Juglans nigra)* | 1.5 | 1.2 | 2.2 | 1.9 | 5 |
| Red oak *(Quercus rubra)* | 3.6 | 39.2 | 152.3 | 37.2 | 6 |
| American basswood *(Tilia americana)* | 0.3 | 5.9 | 19.0 | 33.0 | 8 |
| Sugar maple *(Acer saccharum)* | 0 | 0.8 | 11.7 | 127.0 | 10 |

*Source:* Adapted from Curtis and McIntosh, 1951.

stands. Curtis and others in Wisconsin found in their study of importance values that each species reaches its best development in stands whose position bears a definite relationship to that of other species. In other words the stands varied continuously along a gradient, thus the word *continuum*. The continuum index is extremely useful because it can be employed to investigate environmental relationships of component communities, to designate the position of a stand on a gradient, and to provide a background for studies of other organisms.

The continuum index for a stand is obtained by first assigning a climax adaptation number to the various tree species involved. Curtis and McIntosh (1951) established a 10-part scale for Wisconsin, in which the highly tolerant sugar maple was given a value of 10, because it was best adapted to maintain itself in a stand. The intolerant bur oak and quaking aspen were given a value of 1 because they are early successional species. Cain and Castro (1959) suggest that the term *ecological sequence number* be used in place of the *climax adaptation number* to get rid of that questionable term *climax*. Next the importance value of each species is multiplied by its ecological sequence number, and the values for all the species in the stand are added. The sum is the stand continuum index. This is used to place that stand on a continuum scale that runs from 300 to 3000. After the stand indices have been calculated, the position of the individual species can be plotted in relation to the position of the stand on the index (Figure B-6).

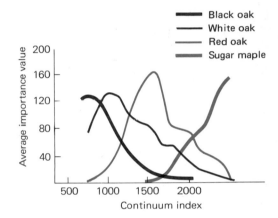

**FIGURE B-6**
*Distribution of four forest tree species on a continuum index. (From Curtis and McIntosh, 1951.)*

## Periphyton and phytoplankton

In aquatic communities algae are the dominant vegetation. Two kinds of growth are involved: the plankton suspended on the water and the periphyton growing attached to some substrate.

### PHYTOPLANKTON

The phytoplankton can be obtained by drawing water samples from several depths (see Appendix D). Cell counts of algae present in each sample, either normal or concentrated, can be made with a Sedgewich-Rafter counting chamber and a Whipple ocular. If necessary, the samples can be concentrated by centrifu-

**673**

gation in a Foerst plankton centrifuge. The centrifuged samples are then diluted to a suitable volume (100 to 200 ml) in a volumetric flask.

As cells are counted, a separate tally is kept for each species to permit an analysis of community structure at each station. The number of cells for single-celled forms and the number of colonies for colonial forms are recorded. The number of colonies is multiplied by an appropriate factor for each species to convert the colonies into cells. These factors are predetermined by averaging the cell counts from a large number of typical colonies from the area in question.

Another method of handling the phytoplankton is by filtration (see McNabb, 1960; Clark and Sigler, 1963). The organisms in the sample are first fixed by the addition of 4 parts 40 percent aqueous solution of formaldehyde to each 100 parts of the sample. The analysis is as follows.

1. Thoroughly agitate the sample. Withdraw a fraction with a pipette large enough to hold a sample that will provide an optimum quantity of suspended matter on the filter.
2. Place the sample in the tube of a filter apparatus designed to accommodate a 1-in.-diameter membrane filter. Draw the water through the filter with a vacuum pump.
3. Remove the filter and place it on a glass slide. Put two or three drops of immersion oil over the residue, and store the slide in the dark to dry (about 24 hours). The oil replaces the water in the pores of the filter and makes it transparent
4. Place a cover slip over the transparent filter.
5. Determine the most abundant species by scanning and then choose a quadrat size that will contain individuals of this species approximately 80 percent of the time.
6. Move the mechanical stage so that approximately 30 random quadrats are viewed. Note the presence or absence of individual species. There is no need to count.
7. When 30 quadrats have been surveyed, calculate the percentage frequency.

$$\text{frequency} \ (\%) = \frac{\text{total number of occurrences of a species}}{\text{total number of quadrats examined}} \times 100$$

## PERIPHYTON

The periphyton has not received quite the same attention from ecologists as the phytoplankton, particularly in a quantitative way. Methods for studying the periphyton are given in detail by Sladeckova, 1962.

Epiphyton, the periphyton growing on living plants and animals, can be observed in place on the organism if the substrate is thin or transparent enough to allow the transmission of light. If the leaves are thin and transparent, the task is relatively easy, but the growth on one side must be scraped away. If the leaf is opaque, the chlorophyll can be extracted by dipping the leaf in chloral hydrate. Small leaves can be examined over the whole area. Large leaves can be sampled in strips marked by grids on a slide or by an ocular micrometer. With large aquatic plants a square will have to be cut from the leaf or stem. If the leaf is too thick to handle under the microscope, scrape off the periphyton and mount in a counting cell for examination. The results can be related back to the total surface area.

Algae growing on such aquatic animals as turtles, mollusks, etc., and on stones must be removed for study. The quantitative scraping and transfer is difficult, but there are several techniques available.

One method employs a simple hollow, square instrument with a sharpened edge, which is pressed closely on or driven into the substrate. This separates out a small area of given size around which the periphyton is washed away. The instrument is then raised and the periphyton remaining in the sample square is scraped into a collecting bottle.

If the stones can be picked up from the bottom, then the periphyton can be removed with an apparatus consisting of a polyethylene bottle with the bottom cut out and a brush with nylon bristles. A section of the stone is delimited by the neck of the bottle held tightly on the surface. The periphyton is scraped loose by the brush and then washed into a collecting bottle with a fine-jet pipette.

The periphyton can be counted in a Sedgewich-Rafter cell recording a predetermined number, usually 100 to 1000, as they appear in the field of view. The results can be expressed as a percentage; or the algae can be checked for frequency in the field of view, using the Braun-Blanquet scale of total estimate.

Some of the difficulties can be avoided by growing the periphyton on an artificial substrate, usually glass or transparent plastic slides, attached in the water in a variety of ways. In lentic situations they can be placed on sand or stones in the water. In lotic situations they can be placed in saw-cuts on boards, in holes in bricks, clipped to a rope, attached to a wooden frame, or tucked into rubber corks.

In fresh water the glass slides are placed in a vertical or horizontal position. In shallow water the plates are usually laid horizontally, directly on the bottom, especially if the influence of light on the composition of periphyton is one of the objectives of a study. Or they can be hung vertically in the water; in this case both sides will be covered with periphyton. Plates placed horizontally collect true periphyton, detritus, etc., on the upper surface, while the bottom will be colonized by heterotrophic organisms. In running water the vertical position with plates parallel to the current is best since the surfaces will not become too badly filled with mud and debris.

For algae and protozoa the plates are exposed for one to two weeks; for hydras, sponges, and the like, about one month. At the end of this time, the slides are collected and placed in wide-mouthed jars filled with water from the locality. In the lab the periphyton can be observed directly on the slides under the microscope. For counting, the slides should be marked in a grid either by using a Whipple ocular micrometer or by cleaning the bottom of the slide and placing it on an auxiliary slide marked off in a grid. If the growth is very dense, it can be scraped off and examined in a Sedgewich-Rafter cell.

Biomass and production of periphyton can be determined by methods described in Appendix C: dry weight, loss through ignition to determine organic carbon, pigment extraction, light and dark bottles containing pieces of plant with and without periphyton attached.

## SAMPLING STREAM-BOTTOM ORGANISMS

Samples of stream-bottom organisms can be taken with a modified Surber bottom-fauna sampler (Figure B-7). The sampler consists of a box made of a brass frame and stainless steel side pieces and containing a current baffle. To this is attached on a removable brass frame two cone-shaped nets. A smaller cone, made of coarse net (19 meshes per linear inch), is fitted within a larger cone, made of fine net (74 meshes per linear inch). Flanges on the insert prevent the coarse net from being forced into the fine net.

This modified sampler picks up many small organisms that might otherwise be lost. In fact collection of virtually all macroorganisms is assured. In addition one obtains two subsamples with respect to size, and the small organisms are associated with fine detritus only.

The sampler encloses a specified area of stream (500 cm²), which is the sample unit. Organisms, detritus, and trash are scrubbed free from the substrate, and the current washes it into the net. The contents can be

*FIGURE B-7*
(a) *A modified Surber bottom sampler. (Redrawn from Withers and Benson, 1962.)*
(b) *Construction details for the Surber bottom sampler.*

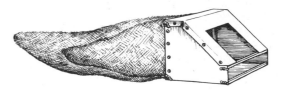

(a)

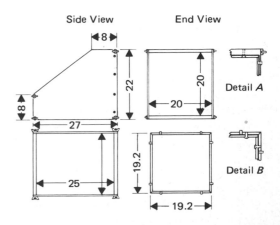

(b)

transferred to a container and taken back to the lab for examination and sorting.

*Artificial stream habitat.* As with pond and lake inhabitants, it is often desirable to observe stream organisms in an indoor aquarium. The problem of maintaining a stream population under artificial conditions is more difficult, but a setup that works fairly well can be constructed. It consists of a length of gutter, preferably wooden, set on an angle with water from an outlet running in at the top and draining into a sink at the bottom. Water depth can be controlled by inserting a series of wooden partitions, each notched in the top, through which water pours like little waterfalls from one compartment to the next. The bed of the "stream" should be lined with small stones and gravel.

Conditions in a slow stream or pool can be simulated by arranging a series of shallow pans, one above the other. Each is filled with an overflow pipe covered with a fine wire filter. If the overflow pipes are arranged about 9 in. above the next pan, the oxygen concentration will approach saturation (see Warren and Davis, 1971).

## Dendrochronology

Dendrochronology is the science of dating past events by the study of the aging of trees. As such it is a valuable tool for the ecologist. It has been used in a number of studies—to age trees for management information, to establish dates of past forest fires, insect outbreaks, glaze, periods of suppression and release in the life history of forest trees. It has been involved in hydrological and archaeological studies and even in legal cases involving boundary disputes in which specimens are taken from fence posts and witness trees. An outstanding example of growth-ring analysis in an ecological investigation is Spencer's study (1958) of porcupine fluctuations in the Mesa Verde National Park.

Dendrochronology is based on the variation of growth rings. Growth rings, despite popular belief, are not regular nor are they all necessarily laid down annually. Because of the failure of the cambium to form a sheath of xylem the entire length of the bole, rings may be omitted, especially near the base. This may be caused by the lack of food manufacture in the crown, by drought, fire, extreme cold, insects, and so on. At the other extreme are multiple rings produced by multiple waves of cambial activity during the growing season. These are caused by temporary interruptions in the normal growth, such as a late spring frost, or by regrowth taking place after normal seasonal growth had ended. Thus the growth rings reflect the interaction of woody plants and their environment as well as the passage of time.

The fundamental principle of dendrochronology is cross-dating, the correlation of distinctive patterns of growth between trees for a given sequence of years. Because no two plants have exactly the same growing conditions and life history (although the broad features are common to all trees involved), the similarities are relative rather than quantitative. The relative widths of corresponding rings are the same in relation to adjoining rings. By lining up these similarities, the investigator can establish the relative identity of any rings in sequence and aberrant rings in the individual specimen. A great number of specimens must be cross-dated before each ring with a sequence can be dated.

### COLLECTION OF MATERIAL

A recently logged-over area can provide an abundance of material, but new sections must be cut from the stump. Stump sections should be cut at a 30° angle. If the cut is clean enough, they may be ready for examination or they may require smoothing with a carpenter's plane or by machine sanding. If the study involves shrubs, cut sample stems close to the ground and use the entire cross section.

*Increment boring.* The usual method of obtaining samples from forest trees is increment boring. The increment borer, available from forestry supply houses, is an instrument designed to bore a core from a tree. It consists of a T handle, a hollow bit, and an extractor. Increment borers are fairly easy to use, but without care they can be damaged or broken. Here are a few hints.

1. For growth and age studies, remove the core as near to the base of the tree as the instrument handle will allow.

2. Coat the screw with heavy-grade oil.
3. To start the borer use a strong pushing and twisting motion until the borer is engaged in the wood.
4. Line the borer on the radius, keep the borer straight, and attempt to reach the center of the trunk.
5. When the core is drilled, insert the extractor and press firmly to cut the core from the trunk.
6. Remove the borer with reverse rotation.
7. Paint the wound with tree paint; a small cork can be inserted.
8. Store the cores in large diameter soda straws or polyethylene tubing. Be sure to label each sample fully, including the directional side of the tree from which it was removed.

To obtain a freshly cut edge for examination, the core is held firmly in a core holder. The groove in a plastic foot ruler is fine if the ruler is clamped to a table and the end is stopped. With a razor blade a transverse cut is made the length of the top of the core. It can then be brushed with water or kerosene to make the rings stand out better. When the core is ready, clip it to the stage of a microscope for examination. One-hundred-power magnification usually is sufficient.

The distance of each ring is measured with the use of a graduated mechanical stage, a stage micrometer, or a dial micrometer. The total distance included in the layers observed can be measured and then compared with the accumulated individual measurements. Any error should be distributed over the individual measurements. For serious research a dendrochronometer, a special device with a microscope and precise measuring devices, should be used.

CROSS-DATING

Although the more involved methods and problems of cross-dating cannot be described in detail here, the basic procedure is as follows.

1. On graph paper write down a series of numbers horizontally from left to right to represent growth layers. You can begin with one or with the years, the first number being the season preceding. This gives a series of numbers starting with the present and leading backward through the

tree's life. A number of such blanks should be made up.
2. Set up a scale on the graph in thousands-of-an-inch so that the largest bars represent the *narrowest* widths.
3. Make a small bar graph for each year of the tree's life.
4. Make such a coded summary of all wood samples available.
5. Compare these visually, two at a time, sliding them along each other. Keep looking for corresponding groups of years with the same pattern of ring sequences. By such a technique, multiple rings can be checked, or extremely narrow growth rings previously missed can be picked up.

A simpler, but less precise method is to draw a line under the year that has a ring slightly less than the rings adjacent to it; to draw two lines if the decrease is more pronounced; to draw three lines if very narrow; and to draw two lines above the year for very wide rings.

STATISTICAL ANALYSIS

The data can be reduced to average values, and then compared to weather data covering the principal growing season for the species involved. Comparisons can be made between rainfall and the current year's growth, rainfall and the previous year's growth, monthly evapotranspiration deficits and growth, with frost-free periods, and so on.

Data will have to be analyzed by simple or multiple regression, depending upon the variables, using partial correlations and standard errors for tests of significance. For sample analysis and interpretations, see Fretts, 1962.

## Palynology

Palynology is the study of past plants communities by the analysis of pollen profiles. These studies are especially enlightening if they are coupled with carbon-14 dating.

PEAT COLLECTION

Peat cores are bored at one to several stations in the bog. They are taken with a peat borer, available commercially. At each station two separate borings should be drilled, several feet apart. Then by taking successive samples from alternate borings (example, first foot sample in core number one; second foot sample from core number two) contamination of one sample with another can be prevented. Two 6-in. samples are collected at each boring, one from the lower part of the cylinder, the other from the upper, at each boring. The samples are placed in glass vials. If the vials are completely filled and tightly sealed, no preservative should be needed (Walker and Hartman, 1960).

### TREATMENT OF SAMPLES

Back in the laboratory, the samples can be treated as follows.

1. Thoroughly mix each 6-in. sample and remove a pea-size lump for deflocculation.
2. Boil the peat for a few minutes in a dilute solution of NaOH, gently breaking it apart with a wooden cocktail stirrer. Use only one stirrer for each sample to avoid contamination.
3. Add several drops of gentian-violet stain to the boiling mixture.
4. Stir vigorously and strain through fine wire mesh. Then stir again and draw up a 0.5-ml sample into a pipette.
5. Add a very small amount of warm glycerine jelly to the sample and mix.
6. Mount several drops on a slide and add cover slip.

### EXAMINATION

The samples, now transferred to slides, should be examined under a microscope equipped with a mechanical stage.

1. Tally 100 or 200 pollen grains as they are encountered by systematically moving the slide.
2. Identify each kind of pollen grain, if only by code, and tally the kinds separately. Identification should be made from a reference pollen collection made up beforehand.
3. Record the results from each slide directly as a percentage for each kind of pollen.

### PLOTTING THE POLLEN PROFILE

The pollen profile can be constructed by plotting a graph for each species or kind of pollen. The vertical scale is set up for depth in feet or meters; or the horizontal scale is percentage,

based on counts of 100 or 200 pollen grains for each spectrum level.

## SAMPLING ANIMAL POPULATIONS

The study of animals involves considerably more problems than the study of plants. Animals are harder to see and most are not stationary. When it comes to sampling, the animals have something to say about getting caught, and they are more liable to mortality than plants. The following methods of estimating animal numbers, determining age structure, mortality, home range, and so on enable the field biologist to make some measurements, however rough, of animal populations in the ecosystem.

## Trapping and collecting

The sampling of an animal population involves collecting animals, either alive, for marking and release, or dead. Because so much information on collecting and trapping is available in other publications (see references at the end of this appendix), I will deal with this topic very briefly.

### FLYING INSECTS

Diurnal insects are collected with aerial nets and heavy-duty sweep nets designed to withstand the hard wear encountered when put through grass and woody vegetation. Nocturnal insects may be collected by using traps containing ultraviolet light or a mercury-vapor light or an old sheet fitted on a slant against some support with a strong light above it. Insects can then be picked off the sheet. If the insects are to be killed they are placed in a killing jar containing a layer, either on the bottom or in a deep lid, of plaster of paris and potassium cyanide. Thin layers of tissue or light cloth in the jar prevent damage to moths and butterflies.

### AQUATIC ORGANISMS

Aquatic organisms may be collected with dip nets for organisms in the water, bottom nets for scraping along the bottom of ponds, wire-basket scraper nets, and plankton towing nets.

For collecting from the shore, aquatic throw nets are useful. A bottom dredge lowered from a boat can be used to collect bottom organisms in deep water. Fish, tadpoles, and large crustaceans can be collected with seines. A set of assorted widths will be necessary.

## SOIL ORGANISMS

The most difficult components of soil fauna to study are the soil arthropods. They are the most numerous, the most difficult to identify, and possibly the most difficult to sample accurately. Nematodes, white or pot worms, and protozoans require such specialized extraction techniques that they will not be discussed here. Those interested are referred to the book *Progress in Soil Zoology* (Murphy, 1962).

Soil arthropods can be extracted from the soil by means of a Tullgren funnel, an improved version of the Berlese funnel, the construction of which is simple (Figure B-8). Essentially it consists of a heat source, such as a light bulb, mounted above a funnel; a smooth funnel, preferably glass, fitted into a collecting vial; and a screen made of hardware cloth or a sieve meshing inside the funnel. The sample is placed on the screen; the heat and then dessication drive the arthropods downward, until they fall through the funnel into the collecting bottle.

The procedure is rather simple.

1. Place the sample of litter or soil on the hardware cloth so fitted in the funnel that air space is present between the wire and the wall of the funnel.
2. To begin extraction, open the lid of the funnel 90° and turn on the 100-watt bulb.
3. After about 16 hours, depending on sample size and moisture content, change to a 15-watt bulb and shut the lid. There will be two periods of arthropod exodus, the first wave due to heat, the second due to desiccation. The collecting bottle beneath the funnel may contain alcohol, formalin, or water. Water may be preferable, since it increases the humidity gradient toward which the animals move.
4. Sort and identify the animals under the microscope.

These funnels are adequate for introductory soil biology. For more efficient extraction, necessary for serious studies in soil zoology, a

(a)

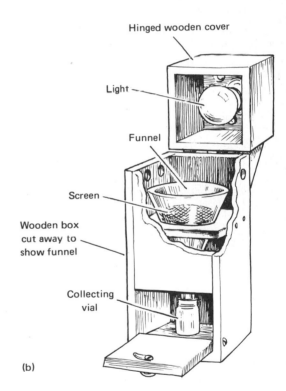

(b)

**FIGURE B-8**
*Although the Berlese and other types of funnels can be purchased, they can be constructed easily in the workshop: (a) relatively simple, sufficient for introductory work; (b) more elaborate and more efficient.*

better extractor is required. A new extractor for woodland litter has been described by Kempson, Lloyd, and Ghelardi (1963) in the journal *Pedobiologia*. The funnels are replaced by wide-mouthed bowls filled with an aqueous solution of picric acid. The acid not only preserves the specimens but also produces by evaporation a high humidity in the air just under the sample. The humid air is cooled by conduction from a cold-water bath in which the bowls are immersed.

Another method of extraction is flotation. Procedures, although relatively simple, are too lengthy to include here. Refer to Jackson and Raw (1966) and Andrews (1972).

Larger soil animals, such as spiders and beetles, can be taken in traps made from funnels and cans set in the soil to ground level. Boards placed on the ground may attract millipedes, centipedes, and slugs. Meat bait in small wire traps will attract scavenger insects.

Sampling earthworm populations presents some difficulties, for no really successful method has been devised to extract the animals from lower layers of the soil. One of the better methods is a combination of formalin and a shovel.

1. Apply a dilute solution of formalin (25 ml of 40 percent formalin to 1 gal of water) to a quadrat 2 ft². Within a few minutes worms will come to the surface.
2. After earthworm movement to the surface stops, pour on a second application.
3. When worms cease to come to the surface the second time, dig out the quadrat as deep as necessary.
4. Hand-sort the soil for maximum recovery. Earthworm cocoons can be extracted by the flotation method.

### SMALL ANIMALS IN VEGETATION

Sweep nets with stout frames to withstand sweeps close to the ground and in woody growth are useful for collecting many types of insects and even some arboreal amphibians and reptiles. Drag nets consisting of light tubular frames, to which are attached canvas bags, are useful on flat ground. Overhead vegetation can be sampled by beating the limbs with sticks to dislodge the animals, which should fall into canvas collecting trays beneath.

### BIRDS AND MAMMALS

Birds can be trapped for banding in specially constructed traps, cannon nets for larger game birds, and mist nets. Both federal and state permits are required for such work. Once a permit is granted, the operator of the banding station will be furnished with plans for suitable traps. For mammals, live traps of wood or wire and snap traps are used. Both are available commercially, but live traps are easily constructed (see references). Traps can be baited with natural foods, dripping water, etc. For small mammals, a mixture of peanut butter and oatmeal works well. Also useful are grain, apple, meat, and appropriate scents.

## Marking animals

Marking individuals in an animal population is necessary if you wish to distinguish certain members of a population at some future date, to recognize individuals from their neighbors, to study movements, or to estimate populations by the mark-recapture method.

Arthropods and snails are best marked with a quick-drying cellulose paint. It is easily applied with any pointed object. Marking butterflies is a two-man operation. One has to hold the wings together with a pair of forceps, while the other marks the side of the wing exposed at rest. For aquatic mollusks and insects better results are obtained through the use of ship-fouling paint because the acetate paints do not hold up well in water.

Fish are usually marked by tagging in several ways. Strap tags of monel metal may be attached to the jaw, the preopercle, or the operculum. Stream or pennant tags attached to various parts of the body, usually at the base of the dorsal fin, are used in some studies. Another method is to insert a plastic tag into the body cavity. The tag is inserted through a narrow incision made in the side of the abdominal wall. Once the incision heals, which it does quickly, the tag is carried by the fish for life and is recovered only when the fish is cleaned. Clipping the fins is still another way to mark fish, but it does not permit the individual recognition of a very large number of fish.

Frogs, toads, salamanders, and most lizards

can be marked by some system of toe clipping, which involves the removal of the distal part of one or more toes. One method worked out by Martof (1953) is as follows. The toes on the left hind foot are numbered 1 to 5, the toes on the right foot 10 to 50. The left forefoot toes are numbered 100 to 400, and the toes on the right forefoot 800, 1600, 2400, 3200. Thus one can mark up to 6399 individuals by clipping no more than 2 toes.

Snakes and lizards can be marked by removing scales or patches or scales in certain combinations.

Birds are usually marked by serially numbered aluminum bands and by cellulose and aluminum colored bands. The colored bands are necessary for individual recognition in the field. In some specialized studies, the plumage is dyed a conspicuous or contrasting color.

Small mammals may be marked by toe clipping in combinations similar to those given for amphibians or by notching the ear.

A number of other methods have been devised for marking mammals. Fur clipping and tattooing may be employed. Bear, deer, elk, moose, rabbits, and hares can be marked with strap tags or plastic discs attached to the ear. Aluminum bands similar to those used on birds can be attached to the forearm of bats. Dyes can be used to mark both large and small mammals. Small amphibians and reptiles can be marked by branding. For details, see Clark (1968).

### RADIOACTIVE TRACERS

The use of radioactive tracers is a particularly useful method for studying animals that are secretive in habits, live in dense cover, spend part or all of their lives underground, or have radically different phases in their life cycle, such as the moths and butterflies.

Animals are fed small traces of gamma-emitting radioactive material. The material is metabolically incorporated into the tissue and the tracer becomes a part of the animal. It is passed along to egg or offspring. Radioactive larvae remain so as they transform to adults; the material is passed on to the egg. The same is true for birds. This technique is useful for studying dispersal, for identification of specific broods or litters, for obtaining data on population dynamics and natural selection.

Another method involves the application of a radioactive tracer in or on an animal in such a way that the animal is not seriously injured and behaves in a normal way. This is usually done by fastening a radioactive wire to the animal or inserting it under the skin of the abdomen with a hypodermic needle. The movements of the tagged individual are then followed with a Geiger counter.

Although these techniques have their merits, they also have their disadvantages. The greatest is the potential radioactive hazard to the investigator himself, to other humans, and to the ecosystem. Another disadvantage is the impossibility of separating one animal from another. Because most work with radioactive tracers requires a federal license, details are not given here. Specific techniques can be reviewed in Tester, 1963; Godfrey, 1954; Pendleton, 1956; and Graham and Ambrose, 1967.

### Aging techniques

Information on the age structure of wild populations is not easily obtained. During the past several decades, a number of aging techniques have been developed, mostly for game and fish. Fish aging began first when Hoffbauer (1898) published his studies on the scale markings of known-age carp. Since then the technique has been further refined. It is based on the fact that a fish scale starts as a tiny plate and grows as the fish grows. A number of microscopic ridges, the circuli are laid down about the center of the scale each year (Figure B-9). When the fish is growing well in summer, the ridges are far apart. During winter, when growth slows down, the ridges are close together. This annual check on growth enables the biologist to determine the age of a fish by counting the number of areas of closed rings, the annuli. Salmon and some species of trout spend one or two years in streams before migrating out to sea or into lakes. Because stream growth is slower than lake or sea growth, the scales show when the fish migrate. When salmonid fish spawn, reabsorption of scales occurs, eroding the margins of the scales and interrupting the pattern of circular ridges. This erosion leaves a mark

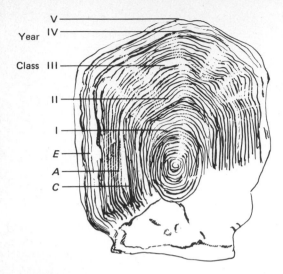

**FIGURE B-9**
*The age of fish can be determined by the growth rings on the scale. C is the circuli; A, the annuli; E, erosion of the scale from spawning.*

that can be detected in later years. Because the growth of a scale continues throughout the life of a fish, it also provides information on the growth rate. This is obtained by measuring the total radius of the scale, the radius to each year's growth ring, and the total body length of the fish. Then by simple proportion, the yearly growth rate can be determined.

Other techniques in aging fish include the length-frequency distribution, vertebral development, and rings or growth layer in the otolith or ear stone.

Because of the large number of year classes (animals born in a population during a particular year) that can be identified, one can determine dominant year classes, learn the age when fish reach sexual maturity, estimate production mortality, and estimate the effects of fish harvest.

Aging techniques for mammals and birds have developed more slowly and are not as refined as those for fish, but a number of methods are in common use. The age of the animal may be indicated by specific characteristics of body parts (Figure B-10). Among birds, plumage development is frequently used. Until molted, the tail feathers of juvenile waterfowl are notched at the tip, in contrast to the normally contoured feather of the winter plumage. The shape of the primary wing feather separates adults from young among many gallinaceous game birds. The presence or depth of the bursa of Fabricus, a blind pouch lying dorsal to the cecum and opening into the cloaca, indicates juvenile birds.

Among mammals the examination of reproductive organs is useful because the majority do not breed until the second year. This method can be used only during the breeding season. The presence of the epiphyseal cartilage in rabbits and squirrels identifies juveniles up to 6 or 7 months. Black bars on the pelage of the underside of the tail of juvenile gray squirrels separate the young from the adults. Primeness of pelt on the inside of skins is a good means of aging muskrat during the trapping season. Dark pigmentation on the flesh side of the pelt indicates areas of growing hair. This pigmentation in adults appears in irregular, scattered dark areas, whereas in immature animals it is more or less symmetrical and linear. Skull measurements are useful in beavers and muskrats. Annual growth rings on the roots of canine teeth indicate age for the first few years of life in the fur seal and other pinnipeds and in canids. Growth rings also show up in the horns of mountain sheep. The wear and replacement of teeth in deer and elk permit the determination of different age classes in these mammals (see Giles, 1969).

**LENS-WEIGHT TECHNIQUE**

Because the lens of the eye of most mammals (and possibly birds) grows continuously throughout life and because there is only slight variation among individuals in lens size and growth, the measurement of the lens is a feasible method for aging a number of mammals. It has been done successfully for the cottontail rabbit, racoon, black bear, and fur seal.

The technique involves comparing the weight of the dry lens with a chart of lens weights of known-age individuals (Figure B-11). The investigator may have to develop the chart by rearing young animals in captivity and sacrificing them week by week for their eyes. A table has been prepared for rabbits by Lord (1959, 1963).

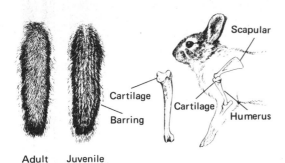

(a) Gray squirrel tail pelage    (b) Rabbit

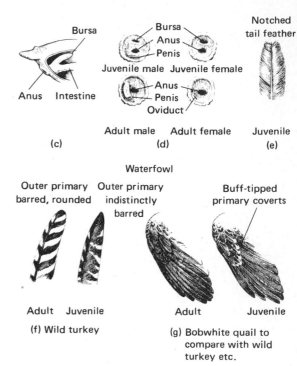

(c)    (d)    (e)

Waterfowl

(f) Wild turkey

(g) Bobwhite quail to compare with wild turkey etc.

The technique is as follows.

1. Remove eyes as soon as possible after the animal is killed and place in a solution of 10 percent formalin. The formalin will harden the lens so that it can be removed from the vitreous humor.

2. Fix for a minimum of 1 week, but the longer the better.

3. After fixing, remove the lens from the eye and roll it on a paper towel for a few minutes to remove excess moisture.

4. Place the lens in an oven to dry at 80° C.

5. Lenses are considered dry when repeated weighing after intervals of drying results in no additional loss of weight. This will usually require 24 to 36 hours.

6. Weigh immediately after removal from oven, since the dried lens are hygroscopic and take on water. Electric scales that read weights rapidly are preferred over other types of balances.

The lens-growth curve permits a rather close approximation of the age of the mammal. For cottontail rabbits, the method permits the determination of the month of birth of young rabbits and the year of birth of rabbits over one year of age.

## Sex determination

The sex of mammals in most instances can be determined by examining external genitalia, and the sex of birds by plumage differences (Figure B-12). For example, the male ruffed

**FIGURE B-10**

*Age determination in some game birds and mammals. (a) Regular barring on the underside of the tail distinguishes the juvenile gray squirrel from the adult. (b) Presence of the epiphyseal cartilage on the humerus of the juvenile cottontail rabbit separates that age class from the adult. This method is useful when the biologist wishes to collect data from a wide area by requesting that successful rabbit hunters turn in these bones. (c) The bursa of Fabricus (enlarged). Its presence or greater depth indicates a juvenile bird. The depths vary with the species. This method is useful in both waterfowl and some gallinaceous game birds. (d) A method of sexing and aging water fowl by examining the cloaca. Note the presence of the bursal opening on juvenile waterfowl, its absence on the adult. (e) The notched tail feather of juvenile waterfowl. (f) The number X (ten) primary in juvenile gallinaceous birds is sharply pointed; in adults it is rounded. The juvenile wild turkey in addition has its outer primary indistinctly barred. (g) The juvenile bobwhite quail, in addition to having a pointed number X primary, also possess buff-tipped primary coverts. (For a complete discussion on aging and sexing techniques, see R. Giles,* Wildlife Investigational Techniques.*)*

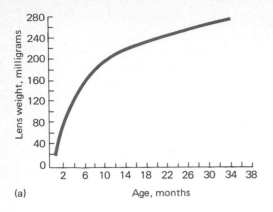

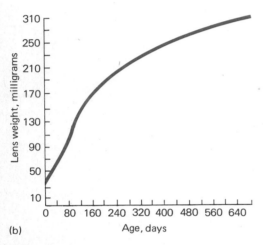

**FIGURE B-11**
*Growth rate curves for the lens of* (a) *cottontail rabbit (from Lord, 1961) and* (b) *black-tailed jack rabbits. (After Tiemeier and Plenert, 1964.)*

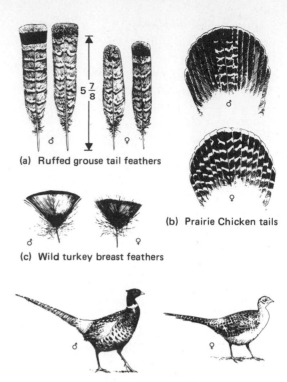

(a) Ruffed grouse tail feathers

(b) Prairie Chicken tails

(c) Wild turkey breast feathers

(d) Ring-necked pheasant

**FIGURE B-12**
*Sex determination in game birds.* (a) *Length of the central tail feather separates male and female ruffed grouse. The male tail feather is 5 7/8 in. long or more; the female tail is shorter.* (b) *The female prairie chicken has heavily barred tail feathers; barring is absent on the outer tail feather of the male.* (c) *The breast feathers of the male wild turkey are black tipped; those of the female are brown tipped.* (d) *Sexual dimorphism distinguishes males from females in many species of birds.*

grouse can be distinguished from the female by the length of the central tail feather; the female prairie chicken can be distinguished from the male by strong mottling or barring of all the tail feathers; the male wild turkey has black-tipped body feathers whereas the female has brown-tipped feathers. (For detailed information, see Giles, 1969.)

## Determination of home range and territory

There are a number of methods available for obtaining an approximation of the size of home range. Some methods are offered here.

### CALCULATING HOME RANGE

On a map, outline and measure the area that includes all the observations made on the movements of individuals. If the observations are obtained by trapping, then one must assume that the animal could have gone halfway toward an adjacent trap, especially if the traps were set in a regular grid.

### CENTER OF ACTIVITY

Arrange the recaptures on a grid and determine the values on an $XY$ axis. An average of these locations will give the center of activity (see Hayne, 1949*b*). This method has the advantages that the information is relatively easy to summarize and the calculations are not complicated. However, a map of the area must be made and many recaptures of the same individual are required before the extent of the home range can be obtained. As the number of recaptures increases, the area of the known range increases. At least 15 recaptures or more are necessary. This method is unsuited for mammals that follow paths or tunnel underground.

### FREQUENCY OF CAPTURE

Record the distances between captures in live traps set randomly or in grids. Record the number of captures as a frequency distribution according to the distance between them. Distances between captures are then tallied and proportions calculated for each distance by sex or age categories.

The distances can be measured in two ways. They can be taken from the place where the animal was first marked or observed or from each successive location.

This method has the advantage that the traps can be set out haphazardly, avoiding the labor of setting them out in grids. The recaptures of all individuals can be used. And information can be obtained during a short period of time because the data from animals captured only two or three times can be used. The disadvantages of this method are that short movements are favored and that no definite boundaries of home ranges can be given. Home range is described as a frequency of distances observed.

### DYED BAIT

The fact that small mammals come readily to bait can be exploited in a manner that does not involve trapping. Bait boxes made from quart-size paper milk containers are set out in a 50-ft grid over the study area. The boxes serve as containers for bait and as receptacles for feces. If food at certain stations is stained with a dye, then the visitation by animals that fed on the dyed bait to other bait boxes can be traced through colored droppings. The distance moved by a particular animal from a station with dyed bait can be determined when colored scats are recovered at various stations in the grid. The results with this method compare favorably with those obtained by other methods, and it does not involve the time and work of trapping. Complete details, as well as a list of suitable dyes, can be found in a paper by New (1958).

### TERRITORY

Because territorial boundaries are rather rigidly maintained by birds during the breeding season, territorial boundaries can be mapped by observing the movements of the birds during the day, by plotting singing perches, by observing locations of territorial disputes, and on occasion by chasing the bird. When the bird arrives at the boundary of its territory, it generally will double back.

### RADIOTRACKING

The development of transistors and other miniature electronic devices has made possible the construction of small transmitters that can be attached to animals usually by a specially designed collar or harness. Mercury cells are used as the source of power. The transmitter is a transistor-crystal-controlled oscillator with the tank coil for the oscillator acting as the magnetic-dipole transmitting antenna. The antenna is constructed of copper or aluminum and has a figure-8 directional pattern. The receiver is a portable battery-powered unit, whose basic components are the receiver, a radio range filter, and two transistorized radio-frequency converters. Positions of stationary animals can be obtained by a single portable direction-finding receiver. Running animals are best located by triangulation, using at least two direction-finding receivers. Radiotracking equipment is available commercially.

Although highly useful in obtaining data on animals that could not be acquired otherwise, radiotracking has its disadvantages for most investigators. The most serious disadvantage is the need for electronic expertise. Unless the investigator is an electronics whiz or has the assistance of an electronics technician, he should not attempt radiotracking. Other disadvantages are cost and the need for an FCC license. (See Giles, 1969.)

## Estimating the numbers of animals

Basic to the study of animal populations is the estimation of their numbers, no small task in wild populations. During the past several decades, much work has gone into the development of techniques and statistical methods to arrive at some estimates of animal populations. Basically the methods of estimating the numbers of animals can be put into three categories: *true census,* a count of all individuals on a given area; *sampling estimate,* derived from counts on sample plots; and *indices,* the use of different types of counts, such as roadside counts, animals signs, and call counts, to determine trends of populations from year to year or from area to area.

### TRUE CENSUS

A true census implies a direct count of all individuals in a given area. This is rather difficult to do for most wild populations, but there are situations where a total count can be made.

Many territorial species are easily seen and heard and can be located in their specific area. Such a census is regularly used for birds. The spot-map method is probably the best approach. A sample plot of at least 25 acres (10 hectares) is marked out in a grid with numbered stakes or tree tags placed at intervals of 50 m. Five or more daily counts are made throughout the breeding season. Each time a bird is observed, it is marked on a map of the plot. At the end of the census period all the spots at which a species is observed are placed on one map. The spots should fall into groups, with each group indicating the presence of a breeding pair. The groups for each species can then be counted in order to arrive at the total population for the given area. Results are usually expressed as animals per acre or per hectare.

Direct counts can be made in areas of concentration. Deer in open country, herds of elk and caribou, waterfowl on wintering grounds, rookeries, roosts, breeding colonies of birds and mammals permit direct counting usually either from the air from aerial photographs. Covies of bobwhite quail can be located and counted with the aid of a well-trained bird dog.

### SAMPLING ESTIMATE

The sample estimate of population size involves two basic assumptions: (1) the mortality and recruitment during the period the data are being taken are negligible or can be accounted for and (2) that all members have an equal probability of being counted—that the members of the population are not trap-shy or trap-addicted, that they are distributed randomly through the population if marked and released, and that they do not group by age, sex, or some other characteristic.

Sampling also involves one major general consideration. The method employed in taking the sample must be adapted to the particular species, time, place, and purpose.

*Sample plots.* Relatively immobile forms, such as barnacles, mollusks, and cicada emergence holes, can be estimated by the quadrat method, similar to that employed for plants. The data can be analyzed for presence, frequency, etc. or the results can be converted to a density per acre, etc. The size and shape of the quadrat will depend upon the density of the population, the diversity of the habitat, the nature of the organism involved. A few preliminary surveys are made before settling on a quadrat size.

Foliage arthropods may be sampled by a number of strokes with a standard sweep net over a 10-m² area. The number of strokes needed to secure the sample must be predetermined. It will vary with the type of vegetation.

Estimates of zooplankton, obtained by pulling a plankton net through a given distance of water at several depths, can be made by filtering a known volume of sample through a funnel using a filter pump. The filter paper should be marked off in equal squares. With the aid of a hand lens or a binocular microscope, the organisms in each square can be counted. The numbers then can be related back to the total volume of water sampled.

If the organisms are too small to be counted in this manner, a Rafter plankton-counting cell can be used. This consists of a microscope slide base plate ruled into ten 1-cm squares. The slides are made from strips of microscope glass slides cemented to the base with Canada balsam. This should hold 1 cc of liquid. After a small volume of water is introduced, the cell is covered with a long cover glass and placed under the microscope. The organisms are counted square by square and the number of each form recorded per square until at least 100 observations have been made. The occurrence of individual species can be recorded as percentage frequency. (Note: Plankton-counting cells and eyepiece micrometers can be purchased commercially, but they are expensive.)

*Mark-Recapture Method.* This method is based on trapping, banding, or marking, and then later recapturing sample individuals. A known number of marked animals are turned loose in the original area. After an appropriate interval of time (approximately 1 week for rabbits), a sample of the population is taken. An estimate of the total population is then computed from the ratio of marked to unmarked individuals.

$$N : T :: n : t$$

or

$$N = \frac{T}{t/n} \text{ or } \frac{nT}{t}$$

where

$T$ = number marked in the precensus period

$t$ = number of marked animals trapped in the census period

$n$ = total animals trapped in the census period

$N$ = the population estimate

Suppose that in a precensus period a biologist tags 39 rabbits. Then during the census period he traps 15 tagged rabbits and 19 unmarked ones, a total of 34. The following ratio is set up:

$$N : 39 :: 34 : 15$$

or

$$N = \frac{39}{15/34} = 88 \text{ rabbits}$$

The confidence limits at the 95 percent level may be calculated from

$$\text{S.E.} = N \sqrt{\frac{(N - T)(N - n)}{Tn(N - 1)}}$$

(S.E = Standard error.)

To determine the limits within which the population lies, add and subtract two standard errors from the estimate. A large standard error and rather wide confidence limits are the result of a small number of recaptures.

For this example

$$\text{S.E.} = 88 \sqrt{\frac{(88 - 39)(88 - 34)}{(34)(39)(87)}}$$
$$= (0.1513)(88) = 13.31$$

Upper limits: $88 + 26 = 114$

Lower limits: $88 - 26 = 62$

The chances are 95 out of 100 that the population of rabbits lies between 62 and 114. This wide spread is typical in wildlife studies.

The variation of this procedure is to accumulate the captures and recaptures. There are several ways in which this can be done, but only the Schnabel method will be illustrated here (for others, see Giles, 1969). All animals captured are tagged or marked and released daily. A record is kept of the total animals caught each day, the number of recaptures, and the number of animals newly tagged. An example is given in Table B-8. The method of calculating the population is the same as that already given. However, in the Schnabel method $T$ (the number marked in the precensus period) becomes progressively larger. Population estimates can be calculated daily later in the period or the season can be divided into periods of, say, a week, and the population computed for each period. True confidence limits cannot be determined, and the calculation of the standard error becomes rather involved. (For details, see Ricker, 1958.) An excellent approach using the regression method to estimate population size from mark-recapture data is described by Marten (1970).

Another variation of the mark-recapture method is the multiple recapture, in which all animals caught on any particular day are marked and released including the recaptures. Thus an animal caught on day 1 and again on day 2 will bear the marks of both days. By such a method one not only can keep account of total marks recaptured each day of trapping, but can also relate recaptures to the initial day of marking. Known as the Jolly method (Jolly, 1965), this technique is useful in following population trends. The basic equation is

TABLE B-8
*Schnabel method of estimating populations*

| Period (date) (P) | Number trapped (A) | Number marked | Marked animals in area (B) | (A) × (B) | (A) × (B) Sum | Recaptures | Sum of recaptures (C) | (A) × (B) / (C) Estimated population |
|---|---|---|---|---|---|---|---|---|
| 1 | 4 | 4 | — | 00 | 0 | — | — | — |
| 2 | 4 | 4 | 4 | 16 | 16 | 0 | 0 | — |
| 3 | 2 | 2 | 8 | 16 | 32 | 0 | — | — |
| 4 | 6 | 6 | 10 | 60 | 92 | 0 | — | — |
| 5 | 10 | 7 | 16 | 160 | 252 | 3 | 3 | — |
| 6 | 4 | 4 | 23 | 92 | 344 | 0 | 3 | — |
| 7 | 8 | 6 | 27 | 216 | 560 | 2 | 5 | — |
| 8 | 4 | 2 | 33 | 132 | 692 | 2 | 7 | — |
| 9 | 5 | 4 | 35 | 175 | 867 | 1 | 8 | — |
| 10 | 7 | 6 | 39 | 273 | 1140 | 1 | 9 | — |
| 11 | 7 | 6 | 45 | 315 | 1455 | 1 | 10 | 145 |
| 12 | 9 | 7 | 51 | 459 | 1914 | 2 | 12 | 159 |
| 13 | 6 | 3 | 58 | 348 | 2262 | 3 | 15 | 150 |
| 14 | 10 | 6 | 61 | 610 | 2872 | 4 | 19 | 151 |
| 15 | 8 | 5 | 67 | 536 | 3408 | 3 | 22 | 154 |
| 16 | 6 | 1 | 72 | 432 | 3840 | 5 | 27 | 142 |
| 17 | 4 | 2 | 73 | 292 | 4132 | 2 | 29 | 142 |
| 18 | 12 | 7 | 75 | 900 | 5032 | 5 | 34 | 148 |
| 19 | 8 | 4 | 82 | 656 | 5688 | 4 | 38 | 149 |

$$P_i = \frac{\hat{M}_i n_i}{r_i}$$

where

$P_i$ = estimate of population on day $i$

$\hat{M}_i$ = estimate of total number of marked animals in population on day $i$

$r_i$ = total number of marked animals on day $i$

$n_i$ = total number captured on day i

The formula is essentially that of the Lincoln index.

For bookkeeping, the trellis diagram is most convenient (Figure B-13). In the trellis diagram the marginal column on the left running downward from left to right contains the totals captured for each day. The marginal column on the right contains the totals released as marked animals each day. In the body of the table are the figures for the recaptures as they relate to the day they were originally marked. For example, on trapping day 5, 220 animals were caught, 214 released. Of the 220 captured, 30 had been marked on day 4, 13 on day 3, 8 on day 2, and 2 on day 1. The data can provide two different population estimates. A column starting at any date and running down-

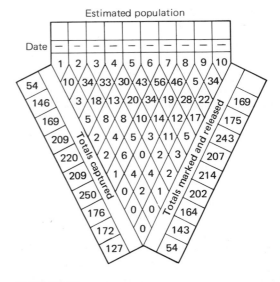

FIGURE B-13
*Trellis diagram.*

ward to the left gives the necessary information for determining the population on the day of recapture. Columns starting at any date and running downward to the right give the necessary information to determine population size on the day of release. To use this method, trapping does not have to be done on consecutive days, but it is necessary that it be done at equally spaced times.

Although new raw data can be used to estimate population (for details, see Southwood, 1966, pp. 83-87), it is more meaningful if the recapture values are corrected to the number of marked recaptures per 100 marked on day $i$ and per 100 in the recapture sample. This must be done for each recapture value in the body of the table. The corrected values are then substituted for the raw values. The formula is

$$y = \frac{X_i}{1} \times \frac{100}{M} \times \frac{100}{n_i}$$

where

$X_i$ = number recaptured on day $n_i$, sample $i$
$M$ = total number of marked animals initially released (prior to time of sample)
$n_i$ = total number of marked and unmarked animals in sample $i$
$y$ = corrected recaptures in sample $i$

For example, on day 5, there were four sets of recaptures: 30, 13, 18, and 2, out of 220 captures. To correct these values:

$$\text{recaptures from day } 4 : y = \frac{30}{1} \times \frac{100}{202} \times \frac{100}{220}$$
$$= 6.7$$
$$\text{recaptures day } 3 : y = \frac{13}{1} \times \frac{100}{164} \times \frac{100}{202}$$
$$= 3.6, \text{ and so on}$$

The next step is to calculate a weighted ratio, $r$, to show the rate of decrease of recapture values. Determine $r$ for each column running to the right, which gives you $r+$ values, size of the population on the day of release. Do the same for each column running to the left, which gives you $r-$ values for the size of the population on the day of recapture. For three or four values for $y$, the formula is

$$r = \frac{y_2 + y_3 + \cdots y_n}{y_1 + y_2 + \cdots y_{n-1}}$$

For more than four values for $y$, the formula is

$$r = \frac{y_3 + y_4 + \cdots y_n}{y_1 + y_2 + \cdots y_{n-2}}$$

With this weighted ratio for each method you can calculate the theoretical number of recap-

tures that would have been obtained on the day of release:

$$y_0 = \frac{y_1 + y_2 + \cdots y_{n-1}}{r} - (y_1 + y_2 + \cdots | y_{n-2})$$

With the theoretical values for recaptures at the time of release, you can estimate population size for each day by the following:

$$N \text{ (on day } i) = \frac{100}{1} \times \frac{100}{y_0}$$

For convenience and for comparison of population estimates by the two methods, positive and negative, you may tabulate your calculations in a table headed as follows.

| Day | $r+$ | $y_0$ | $N$ | $r-$ | $y_0$ | $N$ |
|-----|------|-------|-----|------|-------|-----|
|     |      |       |     |      |       |     |

This method works quite well where relatively large numbers of animals are involved, such as insect populations.

***Removal method.*** The removal method of estimating populations has been widely used in small-mammal studies, although the assumptions on which it is based are open to question. The method assumes (1) that the population is essentially stationary; (2) that the probability of capture during the trapping period is the same for each animal exposed to capture; (3) that the probability of capture remains constant from trapping to trapping (not trap-shy, etc.). The trapping program consists of setting the same number of traps for several nights. The number caught the first night is expected to exceed that caught the second, and the second night's catch should exceed the third. In other words, the population becomes depleted.

The field procedure is as follows.

1. Set the traps (snap) out either in a grid system, 3 traps to a station, or, preferably, in two parallel lines 50 ft apart, 20 stations to a line, 3 traps at each station. Space the stations 25 or 50 ft apart, depending upon the nature of the vegetation.
2. Trap for three nights. Prebait for best success.

There are several ways of handling the data to obtain the estimate. The most popular is to plot the size of the daily catch against the number of animals previously caught (Figure

**689**

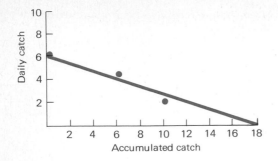

**FIGURE B-14**
*Daily catch on a 3-day removal trap line plotted against accumulated catch to estimate the exposed population.*

B-14). A straight line is drawn through the plotted points to cut the horizontal axis. The point at which the horizontal axis is cut represents the population estimate.

A simpler method of estimating the population based on two captures is suggested by Zippin (1956). The population $N$ equals the number caught the first day ($y_1$) squared, divided by the difference between the number caught the first day ($y_1$) and the number caught the second ($y_2$). Using the same data as on the graph,

$$N = \frac{(y_1)^2}{y_1 - y_2} = \frac{36}{2} = 18$$

The two methods give approximately the same answer.

A second variation of the removal method is the proposed IBP Standard Minimum method. This method involves a 16 by 16 grid of points or stations spaced 15 m apart, each station containing two snap traps. The stations are prebaited for 3 days, a procedure that attracts small mammals to the area and usually ensures the largest catch on the first day. The prebaiting period is followed by a 5-day trapping period. Again the daily catch is plotted against accumulated catch, and the population is estimated at the point where the axis crosses the horizontal line. But a better fit is obtained by calculating a simple regression line from the 5-day catch data after the method suggested by Hayne (1949b).

As with all trapping procedures, there is disagreement over the Standard Minimum method. Some argue for a smaller grid of, say 8 by 8. Traps on the outer stations usually capture more animals than those in the center of the grid because the animals in the border zone react to the sudden removal of the animals in the center. These immigrants are picked up by the border traps. Thus outside invaders can significantly contribute to the catch on the margin zones and influence population estimates. Some attempt should be made to assess the effect of movement into the grid by estimating the size of the border area. Although nearly all animals are removed in the 5-day period, some animals still avoid capture.

The removal method is useful where one desires a relative measure or index figure for small-mammal populations in order to compare one habitat with another. The data will be more valuable if details on vegetation and litter are recorded for each station. Often some association can be obtained between vegetation and trapping success at the various stations.

**INDICES**

Indices are estimates of animal populations derived from counts of animal signs, calls, roadside counts, and so on. In this type of estimating all data are relative and must be compared with data from other areas or other time periods. The results do not give estimates of absolute populations, but they do indicate trends of populations from year to year and from habitat to habitat. Often this type of information is all that is needed.

*Call counts.* Call counts are used chiefly to obtain population trends of certain game birds such as the mourning dove, bobwhite quail, woodcock, and pheasant. A predetermined route is established along country roads; it should be no longer than can be covered in 1 hour's time. Stations are located at quarter- or half-mile intervals, depending upon the terrain and the species involved. The route is run around sunrise for gallinaceous birds and doves and around sunset for woodcock. The exact time to start must be determined for each area by the investigator. The observer stops at each station, listens for a standardized period of time (a minute or two), records the calls heard, and goes on to the next station.

Routes should be run several times and an average taken. The number of calls divided by the number of stops gives a call-index figure.

*Roadside counts.* The roadside count is similar to a call count, with the exception that the number of animals observed along the route is recorded and the results divided by the number of miles. Other variations include counting of animal tracks, browse, signs, active dens and lodges, and so on.

*Pellet counts.* Counting pellet or fecal groups is widely used to estimate big-game populations. This method involves the counting of pellet groups in sample plots or transects located in the study area. It may be used for estimating the relative intensity of use of the range by one or more kinds of animals, to determine trends in animal populations, or in rarer cases to estimate the total population. The latter is possible only when an entire herd is known to occupy a given area for a definite period. Intensity of use is usually expressed as the number of pellets or pellet groups per unit area.

The accuracy of estimating populations by this method depends upon some knowledge of the rate of defecation by the animals involved. Herein lies the weakness of the technique, because pellet groups vary with the diet, season, age, sex, rate of decomposition, and the type of vegetation (that is, the plants can cover the pellet groups). Usually rates vary with the region, so some preliminary observations have to be made to arrive at some useful figure. For deer, a pellet-group figure of 15 (per deer) is satisfactory on good range, and 13 for poor range. Rabbits vary too widely in their pellet groups for the technique to have much value with that group.

The field procedure is as follows.

1. On randomly located transect lines, establish a number of rectangular plots of 1/50th acre, 12 × 72.6 ft divided in half longitudinally for ease in counting.
2. Count pellet groups at the most favorable time, when plant growth, leaf fall, and so on are least likely to interfere.
3. Mark the plots permanently and clear or paint the pellet groups at the beginning of the study where age determination of pellets is difficult.

4. Then let

$$t = \frac{1}{na'} y$$

where

  $y$ = sum of pellet groups counted over the plots
  $a'$ = area of one plot
  $n$ = number of plots
  $t$ = pellet groups per unit area

5. Determine the value for $t$. To translate $t$ to total deer days of use:
   a. Assume a defecation rate of 13 pellet groups/deer/day for poor range, 15/day for adequate range.
   b. Determine the period, the number of days, over which the pellet groups were deposited (for example, since the last count, time plots cleared, etc.)
   c. Divide $t$ by the defecation rate to obtain days of utilization by deer per acre.
   d. Divide the number of days of utilization by the number of days in the period to obtain the number of deer per acre (assuming a constant population).
6. Multiply result by 640 to obtain the number of deer per square mile.

*Dropping boards.* A modification of the pellet-group method for small mammals is the dropping-board method (see Emlen et al., 1957).

1. Set out 4-in. squares of weatherproof plywood (in natural color) in lines or grids. Use at least 100 boards. This will cover 1.2 acres if spaced 25 ft apart, 4.7 acres with a 50-ft spacing, and 18.8 acres with a 100-ft spacing. Be sure the squares are level and placed firmly on the ground.
2. Number each station.
3. The boards may or may not be baited depending upon local conditions.
4. Make a series of at least three visits. The time of day the visits are made and their frequency will depend upon local conditions, such as coprophagous insect activity. Daily visits may be necessary.
5. At each station record the presence of droppings by species, and brush the board

clean for the next visit. The droppings of small mammals are distinctive and with some experience can be identified (see Murie, 1954).

6. Results can be expressed as incidence of droppings for each species.

Figures obtained from the record of usage are indices of the population useful in comparative studies of interspecific, interseasonal, and interregional abundance. The dropping-board technique can be used in studies of population trends and fluctuations, local distribution, species association, activity, rhythms, effects of weather and environmental conditions on activity, and movements if the animal is tagged with a radioactive tracer or with dyed bait.

## Measuring mortality

The measurement of mortality in natural populations is imperative in the analysis of animal populations. Coupled with a knowledge of age structure, mortality data can be used to construct life tables, life equations, survivorship curves, growth curves, and the like (see Appendix C). Knowledge of the mortality rate is basic in the management of game and fish populations.

Several methods of determining mortality in wild populations will be discussed here. Further details on these methods and the description of others, including the construction of life tables and the conversion of monthly rates of probability of dying to annual rates and to death rates and estimations of life expectancy can be found in Giles, 1969; Davis, 1960; Davis and Golley, 1963; and Ricker, 1958.

**KNOWN DEATHS**

If a number of marked animals is found dead a known time after marking, the percentage dying can be plotted against time and a curve drawn through the points. The probability of dying can be read directly from the chart, or it can be calculated by dividing the number of individuals that die during a period of time by the initial population. At best, however, this is a very rough measurement.

If the animals can be readily aged, as is true with deer, rabbits, mountain sheep, and others, and if sufficient lower jaws, eye lenses, or horns can be obtained, then mortality and the probability of dying can be expressed on an age basis.

If the size of a population can be determined at two different times, if immigration and emigration are held to a minimum, and if there is no recruitment from births, then the difference in population size between time *A* and time *B* is the result of deaths. (Winter deer yards offer an excellent opportunity to determine mortality in this way.) The probability of dying during the period under observation can be calculated by dividing the number dying over the period (population time A − population time *B*) by the initial population. Davis and Golley (1963) give formulas for changing such data to an annual basis. However, because the probability of dying may vary with the season, as it does with deer, an annual probability of dying based on a seasonal figure may be erroneous.

## References

*Vegetation analysis*

BRAUN-BLANQUET, J. 1951. *Pflanzensoziologie: Grundzuge der Vegetationskunde,* 2d ed. Springer, Vienna.

BRAUN-BLANQUET, J. 1932. *Plant Sociology.* McGraw-Hill, New York. (An English translation of the 1st edition of above.)

CAIN, S. A., AND G. M. DE O. CASTRO. 1959. *Manual of Vegetation Analysis.* Harper & Row, New York.

CHAPMAN, S. B. (ED.). 1976. *Methods in Plant Ecology.* Halsted Press, Wiley, New York.

GOUNOT, M. 1969. *Methodes d'etude quantitative de la vegetation.* Masson, Paris.

GREIG-SMITH, P. 1964. *Quantitative Plant Ecology,* 2d ed. Academic Press, New York.

KERSHAW, K. A. 1973. *Quantitative and Dynamic Ecology,* 2nd ed. Edward Arnold, London; Elsevier, New York.

MUELLER-DOMBOIS, D., AND H. ELLENBERG. 1974. *Aims and Methods of Vegetation Ecology.* Wiley, New York.

OOSTING, H. J. 1956. *The Study of Plant Communities.* Freeman, San Francisco.

WHITTAKER, R. (ED.). 1973. *Ordination and Classification of Communities.* Vol. 5, *Handbook of Vegetation Science.* W. Junk, The Hague.

WOOD, R. D. 1975. *Hydrobotanical Methods.*
University Park Press, Baltimore.

## *Methodology*

ANDERSON, D. R., et al. 1976. *Guidelines for Line Transect Sampling of Biological Populations.* Utah Cooperative Wildlife Research Unit, Logan, Utah.

BECKER, D. A., AND J. J. CROCKETT. 1973. Evaluation of sampling techniques on tall-grass prairie. *J. Range Management,* 26:61–65.

BORMANN, F. H. 1953. The statistical efficiency of sample plot size shape in forest ecology. *Ecology* 34:474–487.

BRAY, R., AND J. T. CURTIS. 1957. An ordination of upland forest communities of southern Wisconsin. *Ecol. Monographs,* 27:325–349.

COTTAM, G., AND J. T. CURTIS. 1956. The use of distance measures in phytosociological sampling. *Ecology,* 37:451–460.

CURTIS., J. T., AND R. P. MCINTOSH. 1951. The upland forest continuum in the prairie-forest border region of Wisconsin. *Ecology,* 32:476–496.

GROSENBAUGH, L. R. 1952. Plotless timber estimates—new, fast, easy. *J. Forestry,* 50:32–37.

GROSENBAUGH, L. R. 1958. *Point-Sampling and Line-Sampling: Probability Theory, Geometric Implications, Synthesis.* Southern Forest Expt. Sta. Occ. Paper 160. U.S. Forest Service Experiment Station, New Orleans, LA.

HYDER, D. N., AND F. A. SNEVA. 1960. Bitterlich's plotless method for sampling basal ground cover of bunch grasses. *J. Range Management,* 13:6–9.

LONG, G. A., P. S. POISSONET, J. A. POISSONET, P. M. DAGET, AND M. P. GORDON. 1972. Improved needle point frame for exact line transects. *J. Range Management,* 25:228.

POISSONET, P. S., J. A. POISSONET, M. P. GORDON, AND G. A. LONG. 1973. A comparison of sampling methods in dense herbaceous pasture. *J. Range Management,* 26:65–67.

U.S. FOREST SERVICE. 1958. *Techniques and Methods of Measuring Understory Vegetation, a Symposium.* Southern Forest Experiment Station, Washington, D.C.

U.S. FOREST SERVICE EXPERIMENT STATION, New Orleans, LA. 1962. *Range Research Methods; A Symposium.* Misc. Publ. 940. U.S. Department of Agriculture, Washington, D.C.

VESTAL, A. G. 1949. Minimum areas for different vegetations. *Illinois Biol. Monographs,* 20:1–129.

## *Phytoplankton and Microorganisms*

CLARK, W. J., AND W. F. SIGLER. 1963. Method of concentrating phytoplankton samples using membrane filters. *Limnol. Oceanog.,* 8:127–129.

COWELL, R. R., AND R. Y. MORITA. 1975. *Marine and Estuarine Microbiology Laboratory Manual.* University Park Press, Baltimore.

HARTMAN, R. T. 1958. Studies of plankton centrifuge efficiency. *Ecology,* 39:374–376.

LIND, O. T. 1974. *Handbook of Common Methods in Limnology.* Mosby, St. Louis.

LUND, J. W. G., AND J. F. TALLING. 1957. Botanical limnological methods with special reference to algae. *Botan. Rev.,* 23:489–583.

MCNABB, C. D. 1960. Enumeration of fresh water phytoplankton concentrated on the membrane filter. *Limnol. Oceanog.,* 5:57–61.

PARKINSON, D., T. R. G. GRAY, AND S. T. WILLIAMS. 1971. *Methods for Studying the Ecology of Soil Microorganisms.* IBP Handbook no. 19. Blackwell, Oxford.

RODINA, A. G. 1971. *Methods in Aquatic Microbiology.* University Park Press, Baltimore.

SCHWOERBEL, J. 1970. *Methods of Hydrobiology (Freshwater Biology).* Pergamon Press, Elmsford, N.Y.

SLADECKOVA, ALENA. 1962. Limnological investigation methods for the periphyton (aufwuchs) community. *Botan. Rev.,* 28:286–350.

WELCH, P. S. 1948. *Limnological Methods.* McGraw-Hill—Blakiston, New York.

WOOD, E. J. F. 1962. A method for phytoplankton study. *Limnol. Oceanog.,* 7:32–35.

WOOD, R. D. 1975. *Hydrobotanical Methods.* University Park Press, Baltimore.

## *Dendrochronology*

FRETTS, H. C. 1960. Multiple regression analysis of radial growth in individual trees. *Forest Sci.,* 6:334–349.

FRETTS, H. C. 1962. An approach to dendrochronology—screening by means of multiple regression techniques. *J. Geophys. Res.,* 67:1413–1420.

ROUGHTON, R. D. 1962. A review of literature on dendrochronology and age determination of woody plants. Tech. Bull. 15. Colorado Department of Fish and Game, Denver.

SPENCER, D. A. 1958. Porcupine population fluctuations in past centuries revealed by dendochronology. PhD thesis, University of Colorado, Boulder.

TAYLOR, R. F. 1936. An inexpensive increment core holder. *J. Forestry,* 34:814–815.

## *Palynology*

BROWN, C. A. 1960. *Palynological Techniques.* Published by author, 1180 Standford Ave., Baton Rouge 8, La.

ERDTMANN, G. 1954. *An Introduction to Pollen Analysis.* Ronald Press, New York.

FAEGRI, K., AND J. IVERSON. 1964. *Textbook of Pollen Analysis,* 2d ed. Hafner, New York.

FELIX, C. F. 1961. An introduction to palynology. In H. N. Andrews (ed.), *Studies in Paleobotany.* Wiley, New York.

WALKER, P. C., AND R. T. HARTMAN. 1960. The forest sequence of the Hartstown bog area in western Pennsylvania. *Ecology,* 41:461–474.

WODEHOUSE, R. P. 1935. *Pollen Grains.* McGraw-Hill, New York.

### Animal populations, general

ANDREWARTHA, H. G. 1970. *An Introduction to the Study of Animal Populations,* 2d ed. University of Chicago Press, Chicago.

ANDREWS, W. A. 1972. *A Guide to the Study of Freshwater Ecology.* Prentice-Hall, Englewood Cliffs, N.J.

ANDREWS, W. A. 1972. *A Guide to the Study of Soil Ecology.* Prentice-Hall, Englewood Cliffs, N.J.

GILES, R. H., JR. 1969. *Wildlife Management Techniques.* The Wildlife Society, Washington, D.C.

HOLME, N. A., AND A. D. MCINTYRE. 1971. *Methods for the Study of Marine Benthos.* IPB Handbook no. 16. Blackwell, Oxford.

JACKSON, R. M., AND F. RAW. 1966. *Life in the Soil.* St. Martin's Press, New York.

MADSEN, R. M. 1967. *Age Determination of Wildlife, A Bibliography.* Biblio. no. 2. U.S. Department of the Interior, Department Library, Washington, D.C.

PARKINSON, D., T. R. G. GRAY, AND S. T. WILLIAMS. 1971. *Methods for Studying the Ecology of Soil Microorganisms.* IBP Handbook no. 19. Blackwell, Oxford.

PHILLIPSON, J. (ED.). 1970. *Methods of Study in Soil Ecology,* UNESCO, Paris.

PHILLIPSON, J. (ED.). 1971. *Methods of Study in Quantitative Soil Ecology.* IBP Handbook no. 18. Blackwell, Oxford.

RICKER, W. E. 1958. Handbook of computations for biological statistics of fish populations. *Fishery Res. Board Can. Bull.,* 119:1–300.

RICKER, W. E. 1971. *Methods for Assessment of Fish Production in Fresh Waters.* IBP Handbook no. 3. Blackwell, Oxford.

SCHWOERBEL, J. 1970. *Methods of Hydrobiology (Freshwater Biology).* Pergamon Press, Elmsford, N.Y.

SEBER, G. A. F. 1973. *The Estimation of Animal Abundance and Related Parameters.* Griffin, London.

SOUTHWOOD, T. R. E. 1966. *Ecological Methods.* Methuen, London.

TEPPER, E. E. 1967. *Statistical Methods in Using Mark-Recapture Data for Population Estimation (A Compilation).* Biblio. no. 4. U.S. Department of the Interior, Department Library, Washington, D.C.

WELCH, P. S. 1948. *Limnological Methods.* McGraw-Hill–Blakiston, New York.

### Animal populations articles

ADAMS, L. 1951. Confidence limits from the Petersen or Lincoln index in animal population studies. *J. Wildlife Management,* 15:13–19.

BALPH, M. H., ET AL. 1977. Simple techniques for analyzing bird transect counts. *Auk,* 94:606–607.

CLARK, D.R., JR. 1968. Branding as a marking technique for amphibians and reptiles. *Copia,* 1968:148–151.

COCHRAN, W. W., AND R. D. LORD, JR. 1963. A radio-tracking system for wild animals. *J. Wildlife Management,* 27:9–24.

DAVIS, D. E. 1960. A chart for estimation of life expectancy. *J. Wildlife Management,* 24:344–348.

DAVIS, D. E., AND F. B. GOLLEY. 1963. *Principles in Mammalogy.* Van Nostrand Reinhold, New York.

EBERHARDT, L. L. 1978. Transect methods for population studies. *J. Wildlife Management,* 42:1–31.

ELLIS, R. J. 1964. Tracking raccoon by radio. *J. Wildlife Management,* 28:363–368.

EMLEN, J. T. 1977. Estimating breeding season bird densities from transect counts. *Auk,* 94:455–468.

EMLEN, J. T., et al. 1957. Dropping boards for population studies of small mammals. *J. Wildlife Management,* 21:300–414.

FRANZREB, K. E. 1976. Comparison of various transect and spot map methods for censusing avian populations in a mixed-coniferous forest. *Condor,* 78:260–262.

GODFREY, G. K. 1954. Tracing field voles *(Microtus agrestis)* with a Geiger-Muller counter. *Ecology,* 35:5–10.

GRAHAM, W. J., AND H. W. AMBROSE III. 1967. A technique for continuously locating small mammals in field enclosures. *J Mammal.,* 48:639–642.

GRODZINSKI, W., Z. PUCEK, AND L. RYSZKOWSKI. 1966. Estimation of rodent numbers by means of prebaiting and intensive removal. *Acta Theriol.,* 11:297–314.

HAYNE, D. W. 1949*a.* An examination of the strip census method for estimating animal populations. *J. Wildlife Management,* 13:145–157.

HAYNE, D. W. 1949*b.* Two methods for estimating populations of mammals from trapping records. *J. Mammal.,* 30:399–411.

JARVINEN, O., AND R. A. VAISANEN. 1975. Estimating relative densities of breeding birds by the line transect method. *Oikos,* 26:316–322.

JOLLY, G. M. 1965. Explicit estimates from capture-recapture data with both death and immigration—stochastic model. *Biometrika,* 52:225–247.

KAYE, S. V. 1960. Gold-198 wires used to study movements of small mammals. *Science,* 13:824.

LORD, R. D., JR. 1959. The lens as an indicator of age in cottontail rabbits. *J. Wildlife Management,* 23:358–360.

LORD, R. D., JR. 1963. The cottontail rabbit in Illinois, *Illinois Dept. Conserv. Tech. Bull.,* no 3.

MACLULICH, D. A. 1951. New techniques of animal census, with examples. *J. Mammal.,* 32:318–328.

MARION, W. R., AND J. D. SHAMUS. 1977. An annotated bibliography of bird-marking techniques. *Bird-Banding,* 48:42–61.

MARTEN, G. C. 1970. A regression method for mark-recapture estimate of population size with unequal catchability. *Ecology,* 51:257–312.

MARTOF, B. S. 1953. Territoriality in the green frog *Rana clamitans. Ecology,* 34:165–174.

NEVILLE, A. C. 1963. Daily growth layers for determining the age of grasshopper populations. *Oikos,* 14:1–8.

NEW, J. G. 1958. Dyes for studying the movements of small mammals. *J. Mammal.,* 39:416–429.

NEW, J. G. 1959. Additional uses of dyes for studying the movements of small mammals. *J. Wildlife Management,* 23:348–351.

O'FARRELL, M. J., ET AL. 1977. Use of live-trapping with the assessment line method for density estimation. *J. Mammal.,* 55:575–582.

PENDLETON, R. C. 1956. Uses of marking animals in ecological studies: Labelling animals with radioisotopes. *Ecology,* 37:686–689.

ROBINETTE, W. L., R. B. FERGUSON, AND J. S. GASHWEILER. 1958. Problems involved in the use of deer pellet group counts. *Trans. North Am. Wildlife Conf.,* 23:411–425.

SCHULTZ, V. 1972. *Ecological Techniques Utilizing Radionuclides and Ionizing Radiation, A Selected Bibliography.* RLO-2213 (Suppl. 1). U.S. Atomic Energy Commission, Washington, D.C.

STICKEL, LUCILLE F. 1946. Experimental analysis of methods for measuring small mammal populations. *J. Wildlife Management,* 10:140–158.

SWIFT, D. M., ET AL. 1976. A technique for estimating small mammal population densities using a grid and assessment lines. *Acta Thierologica,* 21:471–480.

TABER, R. D. 1956. Marking of mammals: standard methods and new developments. *Ecology,* 37:681–685.

TESTER, J. R. 1963. Techniques for studying movements of vertebrates in the field. In *Radioecology.* Van Nostrand Reinhold, New York.

TIEMEIER, O. W., AND M. L. PLENERT. 1964. A comparison of three methods for determining the age of black-tailed jackrabbits. *J. Mammal.,* 45:409–416.

VAN ETTEN, R. C., AND C. L. BENNET, JR. 1965. Some sources of error in using pellet group counts for censusing deer. *J. Wildlife Management,* 29:723–729.

VERTS, B. J. 1963. Equipment and techniques for radio-tracking skunks. *J. Wildlife Management,* 27:325–339.

WILLIAMS, G. E., III. 1974. New technique to facilitate handpicking macrobenthos. *Trans. Amer. Microscop. Soc.,* 93:220–226.

ZIPPIN, C. 1958. The removal method of population estimation. *J. Wildlife Management,* 22:325–339.

### Collecting animals

ANDERSON, R. M. 1948. Methods of collecting and preserving vertebrate animals. *Nat. Museum Can. Bull.,* 69.

KNUDSEN, J. 1966. *Biological Techniques.* Harper & Row, New York.

NEEDHAM, J. G. (ED.). 1937. *Culture Methods for Invertebrate Animals.* Reprint, Dover, New York.

OMAN, P. W., AND A. D. CUSHMAN. 1948. *Collection and preservation of insects.* Misc. Publ. 60. U.S. Department of Agriculture, Washington, D.C.

PETERSEN, A. M. 1953. *A Manual of Entomological Techniques.* Published by the author, Ohio State University, Columbus.

WAGSTAFFE, R. J., AND J. H. FIDLER. 1955. *The Preservation of Natural History Specimens.* Vol. 1, *The Invertebrates.* Philosophical Library, New York.

# Measuring productivity and community structure
## APPENDIX C

Estimating the rates of energy fixation and storage in an ecosystem is difficult, and the task of measuring all aspects of energy flow is enormous. Several major techniques for determining primary production are in current use. All estimate energy fixation indirectly by relating the amounts of materials, oxygen, or carbon dioxide released or used.

## Methods of estimating production

### LIGHT AND DARK BOTTLES

The light-and-dark-bottle method, commonly used in aquatic environments, is based on the assumption that the amount of oxygen produced is proportional to gross production, since one molecule of oxygen is produced for each atom of carbon fixed. Two bottles containing a given concentration of phytoplankton are suspended at the level from which the samples were obtained. One bottle is black to exclude light; the other is clear. In the light bottle a quantity of oxygen proportional to the total organic matter fixed (gross production) is produced by photosynthesis. At the same time the phytoplankton is using some of the oxygen for respiration. Thus the amount of oxygen left is proportional to the amount of fixed organic matter remaining after respiration or net production. In the dark bottle oxygen is being utilized but is not being produced. Thus the quantity of oxygen utilized, obtained by subtracting the amount of oxygen left at the end of the run (usually 24 hours) from the quantity at the start, gives a measure of respiration. The amount of oxygen in the light bottle added to the amount used in the dark provides an estimate of total photosynthesis or gross production.

Pratt and Berkson (1959) point out two sources of error in this method. First, at a temperature range of 11° to 21° C, bacteria are responsible for 40 to 60 percent of the total respiration customarily attributed to the phytoplankton. Failure to adjust gives a low estimate of production by plants. Second, in a 2-day experiment large changes in the plankton population occur in the light bottle. Thus a difference exists in the concentration of phytoplankton inside and outside the bottle. The increase in the plankton inside the bottle is caused by an accelerated regeneration of nutrients by bacteria attached to the bottle walls.

There are other shortcomings. Estimations of production are confined to that portion of the plankton community contained within the sample bottles. The method fails to take into account the metabolism of the bottom community. And the procedure is based on the assumption that respiration in the dark is the same as that in the light, which is not necessarily true.

A modification of this method involves the whole aquatic ecosystem, which becomes the light and dark bottles, the daytime representing the light bottle, the nighttime the dark. The oxygen content of the water is taken every 2 to 3 hours during a 24-hour period. The rise and fall of the oxygen during the day and night can be plotted as a diurnal curve. To obtain a correct estimate for the oxygen production of plants, the oxygen exchanged between air and water and between the water and bottom must be estimated and deducted. Details for this method, adaptable to the study of flowing waters, can be found in Odum, 1956.

### CARBON DIOXIDE

The light-and-dark-bottle method is the most useful in aquatic ecosystems, while the measurement of the uptake of carbon dioxide and its release in respiration is better adapted to the study of terrestrial ecosystems. In this method a sample of the community, which may be a twig and its leaves, a segment of the tree stem, the ground cover and soil surface, or even a portion of the total community, is enclosed in a plastic tent. Air is drawn through the enclosure, and the carbon dioxide concentration of the incoming and outgoing air is measured with an infrared gas analyzer. The assumption is that any carbon dioxide removed from the incoming air has been incorporated into organic matter. A similar sample may be enclosed with a dark bag. The amount of carbon dioxide produced in the dark bag is a measure of respiration. In the light bag the quantity of carbon dioxide would be equivalent to photosynthesis minus respiration. The two results added together indicate gross production.

Because the enclosure of a segment of a community necessarily alters the environment of the sample, new methods that produce relatively little disturbance to the normal environment have been sought. One technique is the so-called aerodynamic method (Lemon, 1960; Monteith, 1960). It is based on the assumption that the reduction of carbon dioxide in a canopy of vegetation is equivalent to net photosynthesis. The method involves the periodic measurement of the vertical gradient of carbon dioxide concentration by means of sensors arranged in a series from the ground to a point above the canopy of the vegetation. The difference in the flux or flow between each layer represents the amount of carbon dioxide utilized by the foliage of each layer. The rate of diffusion or flux within the canopy is proportional to the gradient of carbon dioxide and to wind speed and turbulence within the canopy (see Inoue, 1968; Lemon, 1968). The technique in many ways is still in the developmental phase, particularly in the design of instruments sensitive to short-term responses. One shortcoming of this and other techniques is the inability to distinguish between soil and plant contributions to the carbon dioxide flux and to determine carbon dioxide uptake by the roots and its transport to the chloroplast site.

### CHLOROPHYLL

An estimate of the production of some ecosystems can be obtained from chlorophyll and light data. This technique evolved from the discoveries by plant physiologists that a close relationship exists between chlorophyll and photosynthesis at any given light intensity. This relationship remains constant for different species of plants and thus communities, even though the chlorophyll content of organisms varies widely, as a result of nutritional status and duration and intensity of the light to which the plant is exposed.

This method, best adapted to aquatic ecosystems, involves the determination of chlorophyll $a$ content of the plant per gram or per square meter, which under reasonably favorable conditions remains the same. Because the quantity of chlorophyll in aquatic (and terrestrial) communities tends to increase or decrease with the amount of photosynthesis (which varies at different light intensities), the chlorophyll per square meter indicates the food manufacturing potential at the time. This photosynthesis to chlorophyll ratio remains rather constant even in cells whose chlorophyll content varies widely because of nutrition or the duration of the intensity of light to which they were previously exposed.

Chlorophyll $a$ can be extracted by filtering natural water through a membrane filter and then extracting the pigments with acetone. The light absorption of the acetone extract is measured at selected wave lengths in the spectrophotometer; the chlorophyll $a$ content is computed from this information and expressed in grams per square meter. Nomographs for converting the plankton pigment into chlorophyll biomass can be found in a paper "Plankton pigment nomographs" by Duxbury and Yentsch (1956).

To estimate production one must know, in addition to the chlorophyll content of the water, the total daily solar radiation reaching the water's surface and the extinction coefficient of light. The following expression is used to determine daily photosynthesis.

$$P = \frac{R}{k} \times C \times p(\text{sat})$$

where

$P$ = photosynthesis of the phytoplankton population in grams of carbon per square meter per day

$R$ = relative photosynthesis determined from the curve in Figure C-1 for the appropriate value of surface radiation

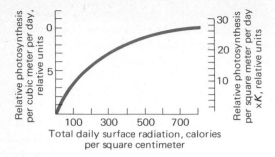

*FIGURE C-1*
*Radiation between total daily surface*
*radiation and daily relative photosynthesis*
(R) *beneath a unit of sea surface. (Adapted*
*from Ryther and Yentsch, 1957.)*

$k$ = extinction coefficient, per meter, as measured

$p$(sat) = photosynthesis of a sample of the population at 2000 foot-candles, as measured in grams of carbon per cubic meter per hour (a rough value for this is 3.7)

$C$ = grams of chlorophyll per cubic meter in a sample of a homogeneously distributed population

The chlorophyll method is well suited for survey work of aquatic ecosystems. It eliminates the need to enclose a sample of the community in artificial containers or to make time studies of photosynthesis. The method is less useful in terrestrial ecosystems. Productivity is limited more by the lack of nutrients than by the quantity of chlorophyll. The concentration of chlorophyll is greater in plants of shady habitats than in plants growing in sunlight. Yet the productivity of such plants is low, although the efficiency of the utilization of light is high. In terrestrial ecosystems the amount of chlorophyll per unit of leaf and stem is useful in determining the size of photosynthetic systems and in comparing one photosynthetic system with another (Newbould, 1967).

### HARVEST

Widely used in terrestrial ecosystems is the harvest method. It is most useful for estimating the production of cultivated land and range, where production starts from zero at seeding or planting time, becomes maximum at harvest, and is subject to minimal utiliza-

tion by consumers. A modification of this method is used to estimate forest productivity (see Newbould, 1967).

Briefly, the techniques involve the clipping or removal of vegetation at periodic intervals, drying to a constant weight, and expressing that weight as biomass in grams per square meter per year. Caloric value of the material can also be determined. The biomass is converted to calories and the harvested material is expressed as kilocalories per square meter per year.

To estimate community production from a sample taken only at the end of the growing season seriously underestimates production. In order to arrive at accurate estimates plant material must be sampled throughout the growing season and the contribution of each individual species determined. Different species of plants reach their peak production at different times during the growing season.

Some studies determine only the aboveground production and leave the root matter unsampled. Other studies attempt to estimate the root biomass. Sampling the root biomass is difficult at best. Although the roots of some annual and crop plants may be removed from the soil, the problem becomes more difficult with grass and herbaceous species and even more difficult with forest trees. In other than annual plants investigators face the almost impossible task of separating new roots from older ones. They have the added problem of estimating the turnover of short-lived small roots and root materials and the variability of the sample.

Because plants of different age, size, and species make up the forest and shrub ecosystems, a modified harvest technique known as dimension analysis is used. Dimension analysis involves the measurement of height, diameter at breast height (DBH), and diameter growth rate of trees in a sample plot. A set of sample trees is cut, weighed, and measured, usually at the end of the growing season. Height to the top of the tree, DBH, depth and diameter of the crown, and other parameters are taken. Total weight, both fresh and dry, of the leaves and branches is determined, as is the weight of the trunk and limbs. Roots are excavated and weighed. By various calculations, the net annual production of wood, bark, leaves, twigs, roots, flowers, and fruits is obtained. From this information the biomass and production of the trees in the sample unit are estimated and then summed for the whole forest. The biomass of the ground vegetation and litterfall are also estimated. By the use of such techniques, the biomass and nutrient content of various trees and forest stands can be obtained. Because energy utilized by plants and plant material consumed by animals and microorganisms is not accounted for, the harvest method estimates net community rather than net primary production.

### RADIOACTIVE TRACERS

The most recent method of determining production involves the measurement of the rate of uptake of radioactive carbon ($^{14}C$) by plants. This is the most sensitive technique now available to measure net photosynthesis under field conditions.

Basically the method involves the addition of a quantity of radioactive carbon as a carbonate ($^{14}CO_3$) to a sample of water containing its natural phytoplankton population. After a short period of time, to allow photosynthesis to take place, the plankton material is strained from the water, washed, and dried. Then radioactivity counts are taken, and from them calculations are made to estimate the amount of carbon dioxide fixed in photosynthesis. This estimate is based on the assumption that the ratio of activity of $^{14}CO_3$ added to the activity of phytoplankton is proportionate to the ratio of the total carbon available to that assimilated.

In common with other techniques the $^{14}C$ method has certain inherent weaknesses. The method does not adequately measure changes in the oxidative states of the carbon fixed. All of the carbon fixed is not retained by the producers. Some tends to seep out of algal cells as water-soluble organic compounds used by bacteria. The various primary producers have different abilities to utilize available light. The amount of carbon fixed is influenced by the species composition of the plant community.

Radioactive phosphorus, $^{32}P$, also has been used, but because phosphorus tends to be absorbed by sediments and organisms more rapidly than it is assimilated by organisms, estimates obtained from phosphorus are inaccurate. This tracer is much more useful for determining the direction of energy flow than for estimating the fixation of energy.

Odum and Kuenzler (1963) used $^{32}P$ rather successfully in studying energy flow in an old-field ecosystem. The investigators labeled three dominant species of plants, *Heterotheca*, *Rumex*, and *Sorgum*, by spraying a solution of $^{32}P$ on the crowns of the plants; the three were isolated in separate quadrats. The solution was soon absorbed and incorporated into the plant biomass. An animal eating the plant in turn became radioactive. The investigators found that the cricket *(Oecanthus)* and the ant *(Dorymyrmex)* were actively feeding on the plants. Radiation showed up in these two animals in 1 to 2 weeks, whereas it took from 2 to 5 weeks to appear in other grazing herbivores, including the grasshopper. Among the predators, such as the spider, the radioactive tracer did not reach its maximum level until 4 weeks after the initial labeling. Late, too, in showing any concentration of $^{32}P$ were the detritus- or litter-feeders, the snails. Thus radioactive tracers not only aid in the separation of animals of the community into their appropriate trophic levels but also make possible the determination of habitat niches. For example, in their experiment Odum and Kuenzler found that the most common grazing herbivores in the old-field community fed freely on several dominant species of plants without any marked preference, whereas some of the rarer species were quite selective.

## Estimating consumer production

Estimating consumer production also has its problems and difficulties. Methods involve the determination of food consumption, energy assimilation, heat production, maintenance requirements, and growth.

The first step involves some estimation of food consumption. This can be determined in the laboratory or estimated in the field. Laboratory determinations involve feeding the animal a known quantity of its natural foods, allowing it to eat over a period of time, usually 24 hours, then removing the food and weighing the remains. The amount of food consumed equals the amount fed minus the amount removed. The caloric value of the food consumed can be determined by burning a sample in the calorimeter or by obtaining the caloric value for the foods involved from a table (if one exists). If the activity periods of the animal and the weight of the food its stomach will hold are known, then consumption can be rather accurately determined by multiplying the activity periods by the mean weight of observed stomach contents from a sample of animals from the population. Activity periods are used because most animal activity usually is concerned with feeding.

Once consumed, the food must be assimilated. Assimilation can be determined by subtracting the energy voided in feces from energy consumed. The assimilation of natural foods by animals is still largely unknown.

The energy assimilated is used for maintenance and growth. Energy used for maintenance is lost. The cost of maintenance can be determined by confining the animal to a calorimeter and measuring the heat production directly, or the energy used in maintenance can be determined indirectly by placing the animal in a respirometer and measuring the oxygen consumed or the carbon dioxide produced. These results are then converted to calories of heat. But to do this one must know the respiratory quotient, the ratio of the volume of carbon dioxide produced to the oxygen consumed. The respiratory quotient varies with the type of food utilized in the body. To estimate accurately the heat production of a population from laboratory determinations, one must also know the daily activity periods, the weight distribution of the population, and the environmental temperature.

Production or storage of energy is estimated by weighing individuals fed on a natural diet in the laboratory or by weighing animals each successive time they are caught in the field. An indirect and usually more useful method is based on the age distribution of a population, the growth curve for the species, and the caloric value of the animal tissue. Growth curves must be obtained for each population under investigation and for each season under study. Once a growth curve is available and the age distribution of the population is known, the weight of the tissue produced in a given period can be estimated for each age category. The weight gain is then converted to caloric equivalents.

Radioisotopes are also used to determine secondary production, particularly of insects. The rate at which a radioisotope is ingested can be converted to food consumption as long as the concentration of the isotope in the food is known. The method is based on the fact that insects feeding on plant material tagged with a radioactive tracer $^{137}Cs$ accumulate the isotope in their tissues until there is a steady-state equilibrium. The intake through the consumption of plants is balanced by loss through biological elimination. The balance is given as

$$I = \frac{kQ}{a}$$

where

$I$ = feeding rate in units of radioactivity
$a$ = proportion of ingested isotope assimilated
$k$ = elimination coefficient (0.693/biological half-life)
$Q$ = equilibrium whole-body burden of the isotope

By using such a method, Reichle and Crossley (1969) determined that forest geometrid caterpillars in a tulip poplar *(Liriodendron)* forest consumed 3.39 mg of foliage/day or about 56 percent of their dry body weight; the tree cricket *(Oecanthus)* consumed 7.35 mg dry weight of foliage/individual/day or about 81 percent of its body weight. Maintenance energy flow amounted to 20.34 kcal/m²/year for the geometrids and 12.39 kcal/m²/year for the tree crickets.

## Community structure

### POPULATION DISPERSION

One of the problems associated with community structure is the spatial pattern of organisms in the community as it relates to the interaction of organisms with various aspects of the environment. Data collected from quadrats, point quadrats, etc., may be used to determine intrapopulation dispersion, as long as one remembers that the analysis can be influenced by the size of the sampling unit.

As pointed out in Chapter 13, population dispersion may be uniform, random, or clumped. In situations where the density of individuals is low relative to the available surface area or volume, the Poisson method is useful to determine types of dispersion. The Poisson distribution furnishes values expected on the basis of a random dispersion pattern and approximates an extremely asymmetrical distribution. Because the mean of the Poisson is equal to its variance, the Poisson distribution is completely specified by the mean. Thus the theoretical Poisson distribution corresponding to the observed distribution can be constructed from the sample mean alone.

To calculate the Poisson one must know the number of sample units, the number of organisms in each sample unit, and the probability or chance that the organism is located in the sample unit or area. The Poisson series is expressed as

$$P_x = e^{-x} \left( 1, x, \frac{x^2}{2!}, \cdots, \frac{x^i}{i!} \right)$$

The steps for setting up the Poisson distribution are as follows.

1. Determine the sample means obtained by the equation:

$$\bar{x} = \frac{fX}{N}$$

where

$f$ = observed frequencies
$X$ = frequency class
$N$ = total frequency

This is an estimate of $np$, which is then substituted into the general expression for Poisson probability, $e^{-\bar{x}}$, where $e$ is the base of natural logarithms.

2. Determine from a table of exponential functions (or by using a scientific calculator) the value of $e^{-\bar{x}}$.

3. Calculate the Poisson probabilities (see example below).

4. Multiply each probability distribution by the total frequency $N$ to convert it to absolute frequency, so that the probabilities are comparable with the observed distribution.

As an example we can use data from Gabutt (1961) on the distribution of fleas on mice. The null hypothesis is that fleas are randomly distributed through a population of mice. The information for *Microtis* is lumped into five classes as follows.

| Fleas per mouse | (X) | 0 | 1 | 2 | 3 | 4+ |
|---|---|---|---|---|---|---|
| Mice | (f) | 44 | 8 | 9 | 3 | 4 |

Calculating for the mean

$$x = \frac{\Sigma f(X)}{N}$$

$$= \frac{(0 \times 44) + (1 \times 8) + (2 \times 9) + (3 \times 3) + (4 + 4)}{68}$$

$$= 0.75$$

To determine the Poisson probability

$$X_0 = e^{-\bar{x}} = 0.472$$

where

$$e^{-\bar{x}} = e^{-.75} = 0.472$$
$$\bar{x} = 0.75$$

$$X_1 = \bar{x}e^{-\bar{x}} = 0.472 \times 0.75 = 0.354$$

$$X_2 = \frac{\bar{x}^2}{2!} e^{-\bar{x}} = 0.472 \times \frac{(0.75)^2}{2} = 0.133$$

$$X_3 = \frac{\bar{x}^3}{3!} e^{-\bar{x}} = 0.472 \times \frac{(0.75)^3}{6}$$
$$= 0.0345 \quad (0.04)$$

$$X_4 = \frac{\bar{x}^4}{4!} e^{-\bar{x}} = 0.472 \times \frac{(0.75)^4}{24}$$
$$= 0.00623 \quad (0.01)$$

To obtain the theoretical frequency or distribution multiply $N$, the total number of observations, by the Poisson probability for each class (Table C-1). For example, the theoretical distribution for class 0 is $68 \times 0.47 = 31.96$ or 32.

TABLE C-1
*Poisson probability and theoretical frequency*

| Number of fleas per mouse (X) | Observed frequency (f) | f(X) | Poisson probability | Theoretical frequency |
|---|---|---|---|---|
| 0 | 44 | 0 | .47 | 32.0 |
| 1 | 8 | 8 | .35 | 23.8 |
| 2 | 9 | 36 | .13 | 8.9 |
| 3 | 3 | 27 | .04 | 2.7 |
| 4+ | 4 | 64 | .01 | 0.6 |
| Total | 68 | 135 | | 68.0 |

Once the theoretical frequencies have been obtained, the next step is to determine how well the data fit the Poisson distribution. To do this we have to set up a chi-square table in which the observed distribution, $O$ (or $f$) can be compared with the expected value, $E$. This is done in Table C-2.

There are five classes after lumping the data and two constants, $\bar{x}$ and $N$. Thus the degrees of freedom are $5 - 2 = 3$. The high value of chi-square (see statistics books for tabled values of chi-square) with a probability lying well below 0.005 indicates that there is no agreement between the observed and expected distribution and the null hypothesis must be rejected. Thus the population of fleas is not randomly dispersed among the mice.

A further test for randomness is the ratio of the variance, $s^2$, to the mean, $\bar{x}$. This is based on the characteristic of the Poisson distribution in which the population mean, $\mu$, is equal to the population variance, $\sigma^2$. Thus a randomly distributed population would have a ratio of its variance to its mean equal to 1.0. A ratio much less than 1.0 would indicate a uniform distribution and a ratio much greater than 1.0 would indicate clumped distribution. In the example of the fleas on mice the ratio of 1.925 (see Table C-2) indicates contagious or clumped distribution.

Significance of departure from randomness may be assessed by the equation

$$t = \frac{s^2/\bar{x} - 1.0}{\sqrt{2/n - 1}}$$

Compare $t$ value to critical values for $t$ in $t$ table for $n - 1$ degrees of freedom.

Another method is to use the chi-square statistic:

$$\chi^2 = \frac{\text{sum of squares}}{\bar{x}}$$

The sum of squares may be computed by

$$ss = (n - 1)(s^2)$$

The statistical significance may be obtained by using chi-square tables. The degrees of freedom are $n - 1$.

## ASSOCIATION BETWEEN SPECIES

Some species in a community may occur together more frequently than by chance. This may result from symbiotic relationships, from food-chain coactions, or from similarities in adaptation and response to environmental conditions. Some measurement of this association provides an objective method for recognizing natural groupings of species. Negative associations may indicate interactions detrimental to one or both species, such as interspecific competition, or adaptations to different sets of environmental conditions.

The data are obtained by sampling quadrats or point-centered quadrats. Presence or ab-

TABLE C-2
*Comparison of observed distribution with expected distribution*

| Number of fleas per mouse (X) | Observed distribution (O or f) | Expected distribution (E) | O − E | (O − E)² | $\frac{(O - E)^2}{E}$ |
|---|---|---|---|---|---|
| 0 | 44 | 32.0 | 12.0 | 144 | 4.500 |
| 1 | 8 | 23.8 | −15.8 | 249.64 | 10.480 |
| 2 | 9 | 8.9 | 0.1 | .01 | .001 |
| 3 | 3 | 2.7 | 0.3 | .09 | .033 |
| 4 | 4 | 0.6 | 3.4 | 11.56 | 19.266 |
| Total | 68 | 68.0 | | | $\chi^2 = 34.280$ |

$\chi^2 = 34.28$

variance: $s^2 = \dfrac{f(X - \bar{x})^2}{N - 1} = \dfrac{\Sigma f(X^2) - N(\bar{x}^2)}{N - 1} = \dfrac{135 - 68(0.56225)}{67} = 1.444$

$\dfrac{s^2}{\bar{x}} = \dfrac{1.444}{0.75} = 1.925 \qquad P > .005$

sence data for pairs of species are arranged in a 2 × 2 contingency table:

**Species A**

|  |  | + | − |  |
|---|---|---|---|---|
| **Species B** | + | $a$ | $b$ | $a + b$ |
|  | − | $c$ | $d$ | $c + d$ |
|  |  | $a + c$ | $b + d$ | $a + b + c + d = n$ |

where

$a$ = samples containing both species A and B
$b$ = samples containing only species B
$c$ = samples containing only species A
$d$ = samples containing neither species

From these data a coefficient of association, $C$, can be calculated. It will vary from a +1.0 for maximum possible association to −1.0 for negative association. A value of 0 suggests that frequency of association is that expected by chance.

If $bc > ad$ and $d \geq a$, then

$$C = \frac{ad - bc}{(a + b)(a + c)}$$

If $bc > ad$ and $a > d$, then

$$C = \frac{ad - bc}{(b + d)(c + d)}$$

If $ad \geq bc$, then

$$C = \frac{ad - bc}{(a + b)(b + d)}$$

To determine whether the coefficient of association is significant, one can by a chi-square test determine the significant level of the deviations between the observed values of the contingency table and the expected based on chance association. This requires a slight modification of the contingency table above, as indicated in Table C-3. The expected number of samples containing both species can be found by

$$\frac{(a + b)(a + c)}{n}$$

The remaining expected values can be calculated in a similar manner or obtained by subtraction from the observed marginal totals.

The chi-square value, with one degree of freedom can be found from the formula

*TABLE C-3*
*Contingency table for determining the degree of association*

**Species A**

|  |  | + | | − | |
|---|---|---|---|---|---|
|  |  | Observed | Expected | Observed | Expected |
| **Species B** | + |  |  |  |  |
|  | − |  |  |  |  |
|  |  |  |  |  |  |

$$\text{chi square} = \sum \frac{(\text{observed} - \text{expected})^2}{\text{expected}}$$

**COMMUNITY SIMILARITY**

Community similarity can be measured from data as simple as presence and absence of species or as detailed as multivariate measures of species importance values. Probably the most widely used index of similarity between two stands or communities is the coefficient of community. This index ranges in value from 0 to indicate communities with no species in common to 100 to indicate two communities with identical species composition. (The proportion of each species is expressed as density, biomass, frequency, etc.). The index can be calculated using the equation

$$C = \frac{2W}{a + b} (100)$$

where

$a$ = sum of scores for one stand
$b$ = sum of scores for second stand
$W$ = sum of the lower scores for each species

An example of the calculation is given in Table C-4.

A second measure of community similarity is the index of percent similarity. The abundance of a species in a community is tabulated as a percentage of the total species present in that community (percent presence = number of individuals of a species/total number of individuals in a community). The lowest percentage for each species for the communities being compared is identified and used to calculate percent similarity using the equation

**703**

TABLE C-4
*Determination of coefficient of community*

| Species | Stand 1 | Stand 2 |
|---|---|---|
| Red oak *(Quercus rubra)* | 10.25 | 0[a] |
| Sugar maple *(Acer saccharum)* | 39.74 | 16.94[a] |
| Black maple *(Acer nigrum)* | 10.16[a] | 39.37 |
| Slippery elm *(Ulmus rubra)* | 20.17[a] | 78.87 |
| American elm *(Ulmus americana)* | 0[a] | 30.77 |
| Shagbark hickory *(Carya ovata)* | 26.14 | 0[a] |
| Ironwood *(Ostyra virginiata)* | 7.02 | 0[a] |
| Beech *(Fagus grandiflora)* | 29.39 | 0[a] |
| Black walnut *(Juglans nigra)* | 39.85 | 0[a] |
| Yellow poplar *(Liriodendron tulipifera)* | 10.44[a] | 21.49 |
| Redbud *(Cercis canadensis)* | 6.83[a] | 12.98 |
| Black locust *(Robinia pseudoacacia)* | 0[a] | 12.26 |
| Sum | 199.98 | 212.68 |

$$C = \frac{2W}{a + b}(100)$$

$a = 199.98$

$b = 212.68$

$W = 64.54$ (sum of lower scores)

$$C = \frac{2(64.54)}{199.98 + 212.68}(100) = 31.28$$

[a]Lower score for each species.

---

$PS = \Sigma$(lowest percentage for each species)

An example is considered in Table C-5.

## COMMUNITY ORDINATION

Ordination is the technique of arranging units (for example, forest stands) in a uni- or multidimensional order in such a manner that the position of each unit along the axis or axes conveys the maximum information about its composition or relationship with the other units involved.

Ordination is based on the assumption that community composition varies gradually over a continuum of environmental conditions. For this reason communities cannot be classified in discrete units; rather they form a continuum changing in composition and structure over environmental gradients (temperature, elevation, soil, etc.).

Community ordination may be accomplished by two different approaches in deriving the axes. The axes can be based on (1) change in environmental conditions or (2) change in community composition. When the axes represent change in environmental conditions, the position of communities along the axes reflects change in community composition influenced by environmental conditions (gradient analysis). When the axes are based on community composition, the configuration of communities in a geometric space reveals relationship between communities based on similarity in composition.

As an example we will use the Bray-Curtis method, based on the second approach to community ordination. The first step is to determine the degree of similarity among communities or stands using the coefficient of community. The stands involved are described in Table C-6 which gives stand composition and importance values for trees in the canopy layer.

When comparisons are being made among a number of communities, the results are usually presented as a matrix of values representing all pairwise comparisons between communities or stands (see Table C-7).

The next step in the ordination process is to convert the similarity coefficients (Table C-7) to values which express dissimilarity, because the distance between communities in ordination space represents the degree of difference rather than similarity. (Two stands with low values of dissimilarity will appear close together in the ordination arrangement. If the similarity coefficient were used, the greater the similarity the further apart the two communities would be positioned.) The coefficient of dissimilarity is obtained by substracting the coefficient of similarity from the highest value of similarity possible. In our example that value is 100, the highest value for the coefficient of community. The coefficient of dissimilarity for the stands described in Table C-6 is given in Table C-8.

The position of communities along the ordination axes is determined by calculating values for each stand along the $x$ and $y$ axes (and $z$ axis in the case of three-dimensional ordina-

TABLE C-5
**Determination of percent similarity**

| Species | STAND 1 | | STAND 2 | |
| --- | --- | --- | --- | --- |
| | *Number* | *% Presence* | *Number* | *% Presence* |
| Yellow poplar | | | | |
| (*Liriodendron tulipifera*) | 83 | 50.6 | 0 | 0 |
| Red oak | | | | |
| (*Quercus rubra*) | 25 | 15.3[a] | 55 | 34.4 |
| Red maple | | | | |
| (*Acer rubrum*) | 19 | 11.6 | 13 | 8.1 |
| Black cherry | | | | |
| (*Prunus serotina*) | 25 | 15.3[a] | 27 | 16.9 |
| Black birch | | | | |
| (*Betula lenta*) | 2 | 1.2[a] | 11 | 6.8 |
| Sugar maple | | | | |
| (*Acer saccharum*) | 1 | 0.6[a] | 2 | 1.3 |
| Sassafras | | | | |
| (*Sassafras albidum*) | 0 | 0[a] | 6 | 3.7 |
| Black locust | | | | |
| (*Robinia pseudoacacia*) | 1 | 0.6 | 0 | 0[a] |
| Black gum | | | | |
| (*Nyssa sylvatica*) | 3 | 1.9 | 0 | 0[a] |
| White ash | | | | |
| (*Fraxinus americana*) | 1 | 0.6 | 0 | 0[a] |
| Chestnut oak | | | | |
| (*Quercus prinus*) | 4 | 2.4[a] | 46 | 28.8 |
| Sum | 164 | | 160 | |

PS = Σ (lowest percentage for each species)
   = 0 + 15.3 + 8.1 + 15.3 + 1.2 + 0.6 + 0 + 0 + 0 + 0 + 2.4
   = 42.9

[a]Lowest percentage for each species.

tion). To position the stands along the $x$ axis, terminal points must first be chosen. The dissimilarity values between each stand and every other stand are summed. The stand with the highest total of dissimilarity values is placed at the 0 point along the $x$ axis and designated as stand A (Table C-9). In our example this happens to be stand 6. The stand with the greatest dissimilarity with A is chosen as the end point along the $x$ axis and designated as stand B. Note in Table C-6 that the greatest dissimilarity value with stand 6 is 100, for stands 2, 4, 5, and 7. Since the four values are equal, one is chosen arbitrarily as stand B. Thus stand 5 is made the end point. The remaining stands are placed along the $x$ axis a given distance, $D_x$, from stand A using the equation (based on the Pythagorean theorem)

$$D_x = \frac{L^2 + (DA^2) - (DB^2)}{2L}$$

where

$L$ = dissimilarity value between A and B
$DA$ = dissimilarity value between A and stand in question
$DB$ = dissimilarity value between B and stand in question

The calculation of the $y$ coordinate is designed to account for the greatest amount of remaining between-stand variation (Table C-10). First, the stand with the poorest fit along the $x$ axis is determined by calculating a poorness of fit value, $e$, for each stand using the equation

$$e = \sqrt{DA^2 - x^2}$$

The stand having the largest value of $e$ is designated as stand A' and given the value of 0 along the $y$ axis. The stand showing the greatest dissimilarity with A' and located within (.1) $L$ of A' along the $x$ axis is chosen as

*TABLE C-6*
**Stand composition and importance values for canopy layer**

| Species | Stand | | | | | | | |
|---|---|---|---|---|---|---|---|---|
| | *1* | *2* | *3* | *4* | *5* | *6* | *7* | *8* |
| Black oak (*Quercus velutina*) | | | | | | | | 42.90[a] |
| Red oak (*Quercus rubra*) | 10.25 | | | | 23.05 | | 29.98 | 7.81 |
| Chestnut oak (*Quercus prinus*) | | | | | 4.68 | | 114.46[a] | |
| Sugar maple (*Acer saccharum*) | 39.74[a] | 16.94 | 30.40 | 27.56 | 4.46 | | 25.56 | 21.02 |
| Black maple (*Acer nigrum*) | 10.16 | 39.37[a] | 28.95 | 28.95 | | | | |
| Slippery elm (*Ulmus americana*) | 20.17 | 78.87[a] | | | 13.60 | | | |
| American elm (*Ulmus americana*) | | 30.77 | | | | | | 11.59 |
| Shagbark hickory (*Carya ovata*) | 26.14[a] | | | | | | | 36.67[a] |
| Ironwood (*Ostrya virginiana*) | 7.02 | | | | 4.46 | | | 40.49[a] |
| American hornbeam (*Carpinus caroliniana*) | | | | | | 14.01 | | |
| Beech (*Fagus grandifolia*) | 29.39[a] | | 46.86[a] | | | [a]161.78 | | |
| Black walnut (*Juglans nigra*) | 39.85[a] | | 17.83 | | | | | |
| Yellow poplar (*Liriodendron tulipifera*) | 10.44 | 21.49 | 51.86[a] | 114.10[a] | | | | |
| White ash (*Fraxinus americana*) | | | 11.11 | | 106.44[a] | | | |
| Redbud (*Cercis canadensis*) | 6.83 | 12.98 | | | | | | 32.86[a] |
| Flowering dogwood (*Cornus florida*) | | | | | | | 24.21 | 6.66 |
| Black locust (*Robinia pseudoacacia*) | | 12.26 | | | | | | |
| Black cherry (*Prunus serotina*) | | | | 17.75 | 43.31 | | | |
| Sycamore (*Platanus occidentalis*) | | | | 4.64 | | | | |

[a]Dominant species.

*TABLE C-7*
**Similarity matrix for stands in Table C-6**

| Stand | *2* | *3* | *4* | *5* | *6* | *7* | *8* |
|---|---|---|---|---|---|---|---|
| 1 | 28.86 | 52.53 | 24.08 | 16.39 | 14.70 | 17.91 | 34.41 |
| 2 | | 33.84 | 33.84 | 9.03 | 0.00 | 8.47 | 14.27 |
| 3 | | | 54.19 | 7.79 | 23.43 | 12.78 | 17.00 |
| 4 | | | | 11.11 | 0.00 | 12.78 | 10.51 |
| 5 | | | | | 0.00 | 16.10 | 8.37 |
| 6 | | | | | | 0.00 | 3.33 |
| 7 | | | | | | | 14.42 |

TABLE C-8
**Dissimilarity matrix for stands in Table C-6**

| Stand | 2 | 3 | 4 | 5 | 6 | 7 | 8 |
|---|---|---|---|---|---|---|---|
| 1 | 71.14 | 47.47 | 75.92 | 83.61 | 85.30 | 82.09 | 65.59 |
| 2 | | 66.16 | 66.16 | 90.97 | 100.00 | 91.53 | 85.73 |
| 3 | | | 45.81 | 92.92 | 76.57 | 87.22 | 83.00 |
| 4 | | | | 88.89 | 100.00 | 87.22 | 89.49 |
| 5 | | | | | 100.00 | 83.90 | 91.63 |
| 6 | | | | | | 100.00 | 96.67 |
| 7 | | | | | | | 85.58 |

the end point along the $y$ axis and designated as stand B′. The value for stand B′ along the $y$ axis is its dissimilarity value with A′. The remaining stands are positioned at a given distance, $D_y$, from A′ (as with the x axis), using the equation

$$D_y = \frac{L^2 + (DA'^2) - (DB'^2)}{2L}$$

where

$L$ = dissimilarity value between A′ and B′
$DA'$ = dissimilarity value between A′ and stand in question
$DB'$ = dissimilarity value between B′ and stand in question

In our example stand A′ is stand 2 and B′ is stand 7.

With the points now determined, they can be plotted on an ordination graph, as illustrated in Figure C-2. (For an example of construction of a third axis for a three-dimensional ordination, see Mueller-Dombois and Ellenberg, 1976).

### SPECIES DIVERSITY

Species diversity implies both the number of species and the number of individuals in a community. But one also has to consider how the individuals are apportioned among the species. For example, a community consisting of 5 species and 100 individuals with the individuals equally divided among all 5 species would be more equitable than a community in which 80 individuals were of 1 species and the remaining 20 were allotted to the other 4 species.

Two approaches to species diversity are widely used today: Simpson's index (Simpson, 1949) and the Shannon-Wiener formula (Shannon and Wiener, 1963). Both are sensitive to changes in the number of species and to changes in the distribution of individuals among the species. However, the index value of both is influenced by sample size. Thus if the index is to be used to compare the diversity among communities, the sample sizes must be equal. Or if complete census data are used, the areas sampled must be of equal size.

Simpson's index of diversity is based on the number of samples of random pairs of individuals that must be drawn from a community to provide at least a 50 percent chance of obtaining a pair with both individuals of the same species. The index is calculated by the formula

$$D = \frac{N(N-1)}{\Sigma\, n(n-1)}$$

where

**FIGURE C-2**
*An ordination graph of communities used as an example in the text.*

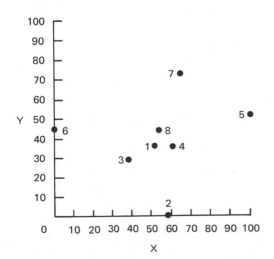

TABLE C-9
*Method of choosing terminal points and positions of stands along x axis*

1. Stand with highest total dissimilarity values is the 0 point on x axis, designated as stand A.

| Stand | Sum of dissimilarity values (from Table C-8) |
|---|---|
| 1 | 511.12 |
| 2 | 571.69 |
| 3 | 498.44 |
| 4 | 553.49 |
| 5 | 631.21 |
| 6 | 658.54 stand A |
| 7 | 617.54 |
| 8 | 579.69 |

2. Stand with greatest dissimilarity with stand A is end point on x axis, designated as stand B.

| Stand | Dissimilarity with A (stand 6) |
|---|---|
| 1 | 85.30 |
| 2 | 100.00 |
| 3 | 76.57 |
| 4 | 100.00 |
| 5 | 100.00 stand B (arbitrary |
| 7 | 100.00 choice since 2, 4, 5, |
| 8 | 96.97 and 7 are same) |

3. Distance *(Dx)* from stand A along x axis is determined by equation

$$D_x = \frac{L^2 + (DA^2) - (DB^2)}{2L}$$

For stand 1:

$$D_x = \frac{(100)^2 + (85.30)^2 - (83.61)^2}{2(100)}$$

$$D_x = 51.43$$

Values for each stand are determined in same fashion:

| Stand | Value along x |
|---|---|
| 1 | 51.43 |
| 2 | 58.62 |
| 3 | 36.80 |
| 4 | 60.49 |
| 5 | 100 |
| 6 | 0 |
| 7 | 64.80 |
| 8 | 54.75 |

TABLE C-10
*Method of choosing terminal points and positions of stands along y axis.*

1. Stand with the largest poorness of fit value, e, is the 0 point on the y axis, designated as A'. Values for each stand are calculated by the equation

$$e = \sqrt{DA^2 - X^2}$$

For stand 1:

$$e = \sqrt{(85.30)^2 - (51.43)^2}$$

Values for each stand are determined in the same fashion:

| Stand | e |
|---|---|
| 1 | 68.05 |
| 2 | 81.02 stand A' |
| 3 | 67.15 |
| 4 | 79.63 |
| 5 | (would be 0) |
| 6 | (not applicable) |
| 7 | 76.16 B' |
| 8 | 79.67 |

2. Stand with the greatest dissimilarity with A' (see Table C-8) and located within (.1)L of A' on the x axis is the end point of y axis, designated as B'. Stand 6 has the greatest dissimilarity value with stand 2 (100.00), but is not within (.1)L of stand 2 along the x axis. Stand 7 has the next greatest value (91.53) and is within (.1)L of stand 2, so it becomes the end point along the y axis, designated as B'. It now has a value along the y axis, equal to its dissimilarity value.

3. Distance *(Dy)* from stand A' along the y axis is determined by the equation

$$D_y = \frac{L^2 + (DA'^2) - (DB'^2)}{2L}$$

| Stand | Value along y axis |
|---|---|
| 1 | 36.60 |
| 2 | 0 |
| 3 | 28.12 |
| 4 | 28.12 |
| 5 | 52.52 |
| 6 | 45.77 |
| 7 | 91.53 |
| 8 | 45.91 |

$D$ = the diversity index
$N$ = total number of individuals of all species
$n$ = number of individuals of a species

A community containing only one species would have a value of 1.0. From this values would increase to an infinite one in which every individual belongs to a different species.

The Shannon-Wiener formula comes from information theory. In ecological use the function describes the degree of uncertainty of predicting the species of a given individual picked at random from the community. As the number of species increases and as the individuals are more equally distributed among the species present, the more the uncertainty increases. This function has been criticized as being biologically meaningless (Hurlbutt, 1971). The Shannon-Wiener formula in a general form is

$$H' = -\Sigma \, p_i \log p_i$$

where $p_i$ is the decimal fraction of total individuals belonging to the $i$th species.

In words the formula states that the probability that any one individual belongs to species $i$ is $p_i$. This in turn is equal to the ratio $n_i/n$, where $n_i$ is the number of individuals in the $i$th species and $n$ is the total number of individuals of all species. Diversity is greatest if each individual belongs to a different species; the least if all individuals belong to one species. A working formula is

$$H' = -\sum_{i=1}^{s} \left(\frac{n_i}{N}\right) \log_2 \left(\frac{n_i}{N}\right)$$

where

$s$  = the total number of species collected
$\log_2 = 3.322 \log_{10}$

For calculation of the index the equation used is

$$H = 3.322 \, [\log_{10}n - (1/n \, \Sigma n_i \, \log_{10}n_i)]$$

To compute:
1. Obtain appropriate $\log_{10}n$.
2. For each species calculate $\log_{10} n_i$.
3. Calculate $n_i\log_{10}n_i$, which means multiply the number of individuals in each species by $\log_{10}n_i$.

4. Sum the $n_i\log_{10}n_i$ and divide the value by $n$, the total number of individuals of all the species.
5. Subtract this value from $\log_{10}n$ and multiply by 3.322 to convert, index value to $\log_2$.

Any log base may be used to calculate diversity as long as it is used consistently. Most commonly used logarithmic bases are 10, e, and 2. Because the Shannon-Weiner formula comes from communication engineering, base 2 is commonly employed and is used here.

If species abundance data are sampled in a nonrandom fashion or if data have been collected for a whole community (such as a total census of organisms), a measure of diversity better suited than Shannon-Weiner is Brillouin's index:

$$H - \log_{10} \frac{N!}{n_i!}$$

where

$N$ = total number of individuals in all species
$n_i$ — number of individuals in the $i$th species

For calculation the equation used is

$$H = \frac{c}{N} \, (\log_{10}N! - \Sigma \log_{10}n_i!)$$

where $c$ is a constant for conversion of logarithms from the base 10 to the base chosen for measure. If the base is 2, $c$ is 3.3219; if the base is $e$, $c$ is 2.3026.

### NICHE BREADTH AND NICHE OVERLAP

Needed for the analysis of niche breadth and niche overlap are quantative measurements on resource utilization (food, habitat, etc.) for a group of species over a range of values for a particular resource. To eliminate the need for calculus, the resource spectrum can be divided into discrete categories $(n)$; species use of the resource in each category can then be recorded (Table C-11).

Numerous methods of quantifying niche breadth have been proposed. (Levins, 1968; Pielou, 1972; Cowell and Futuyma, 1971; Culver, 1972). The most widely adopted of these indices is based on the Shannon-Weiner index of diversity, described above. Niche width is calculated by the equation

TABLE C-11
*Prey utilization by size categories*

| Species | Prey size | | | |
|---------|-----------|---|---|---|
| | 0–10 | 11–20 | 21–30 | 31–40 |
| Least flycatcher *(Empidonax minimus)* | 11 | 2 | 0 | 0 |
| Wood pewee *(Contopis virens)* | 5 | 2 | 1 | 0 |
| Great crested flycatcher *(Myiarchus crinitus)* | 3 | 1 | 1 | 3 |

*Source:* Data from Johnston, 1971.

$$B_i = \frac{\log_{10} \Sigma N_{ij} - (1/\Sigma N_{ij}) \, \Sigma N_{ij} \log_{10} N_{ij}}{\log_{10} r}$$

where

$N_{ij}$ = value of species $i$ in niche category $j$
$r$ = number of niche categories

This value ranges from close to 0 for species in only one or very few categories to 1.0 for species represented equally in all resource categories.

As an example we can use the data from Table C-11.

$$B_1 =$$
$$\frac{\log_{10} 13 - (1/13)[(11 \log_{10} 11) + (2 \log_{10} 22)]}{\log_{10} r}$$
$$= .310$$

$B_2$ and $B_3$ calculated in a similar manner give the results

$B_2 = .694$
$B_3 = .604$

Another widely used index is that of Levins (1968):

$$B_i = \frac{(\underset{j}{\Sigma} N_{ij})^2}{\underset{j}{\Sigma} N_{ij}^2}$$

where $N_{ij}$ is value for species $i$ in niche category $j$

It is a much simpler index, but it has been criticized because it is difficult to standardize procedures so that measurements of resource utilization are comparable for different species

(for discussion, see Colwell and Futuyma, 1971). Using data from Table C-11, we have

$$B_1 = \frac{(11 + 2)^2}{(11^2 + 2^2)} = 1.352$$

$B_2 = 2.133$
$B_3 = 3.200$

As with niche breadth, a number of indices have been proposed to quantify niche overlap (Levins, 1968; Morisita, 1959; Pielow, 1972; Horn, 1966; Colwell and Futuyma, 1971). The data used to determine niche breadth can also be used to determine niche overlap between species. The simplest of these indices is a measure of proportional overlap (Cowell and Futuyma, 1971):

$$C_{ik} = 1 - \frac{1}{2} \Sigma \left[ \frac{N_{ij}}{N_i} - \frac{N_{hj}}{N_h} \right]$$

where

$N_{ij}$ = value for species $i$ in niche category $j$
$N_i$ = total of values for species $i$ in all niche categories
$N_{hj}$ = value for species $h$ in niche category $j$
$N_h$ = total of values for species $h$ in all niche categories

This index ranges from 0 for no overlap to 1.0 for complete resource overlap.

Using data from Table C-11,

$$C_{12} = 1 - \frac{1}{2} \left[ \left( \frac{11}{13} - \frac{5}{8} \right) + \left( \frac{2}{13} - \frac{2}{8} \right) + \left( \frac{0}{13} - \frac{1}{8} \right) \right]$$
$$= .779$$

$C_{13} = .500$
$C_{23} = .750$

A second method (Horn, 1966) is based on information theory:

$$C_{ih} = \frac{\Sigma\left[(N_{ij} + N_{hj})\log_{10}(N_{ij} + N_{hj})\right] - \Sigma N_{ij}\log_{10}N_{ij} - \Sigma N_{hj}\log_{10}N_{hj}}{(N_i + N_h)\log_{10}(N_i + N_h) - N_i\log_{10}N_i - N_h\log_{10}N_h}$$

where

$N_{ij}$ = value for species $i$ in niche category $j$
$N_{hj}$ = value for species $h$ in niche category $j$
$N_i$ = total of values for species $i$ in all niche categories
$N_h$ = total of values for species $h$ in all niche categories

This index also ranges from 0 to 1.0.

Using data from Table C-11,

$$C_{12} = \frac{\begin{array}{c}\left[(11 + 5)\log_{10}(11 + 5) + (2 + 2)\log_{10}(2 + 2) + (0 + 1)\log_{10}(0 + 1)\right] - \\ \left[(11\ \log_{10}11) + (2\ \log_{10}2)\right] - \left[(5\ \log_{10}5) + (2\ \log_{10}2) + (1\ \log_{10}1)\right]\end{array}}{\left[(13 + 8)\log_{10}(13 + 8)\right] - (13\ \log_{10}13) - (8\ \log_{10}8)} = .91078$$

$C_{13} = 0.65809$
$C_{23} = 0.77440$

## *Population structure*

### LIFE TABLES

The life table has been explained in some depth in Chapter 13. Procedures for constructing a life table are considered here.

To construct a life table it must be possible to age the organism in some manner to establish age classes or age intervals. For small rodents or rabbits the age intervals may be 1 month, for deer 1 year, for humans 5 years. Information is needed on survival, mortality, or rate of mortality by age classes for the given population. Data on survivorship in each age class provides the information needed for the survivorship column, $l_x$. Age specific mortality data provides information for the mortality column, $d_x$. The sum of the mortality for each age class over time can be used as the size of the initial population. Thus given any one column the rest can be calculated.

To demonstrate the construction of a life table, we will use the life table of an unexploited (unhunted) gray squirrel population, Table C-12. Data for the life table was obtained by noting the survivorship of gray squirrels marked as nestlings over an 8-year period (A). The nestlings for each period were pooled to comprise a single cohort (hense a dynamic-composite life table).

1. Construct a table with the following columns:

$x$ = age interval or age class
$n_x$ = number of survivors at start of age interval $x$ (raw field data)
$l_x$ = proportion of organisms surviving to start age interval $x$
$d_x$ = number or proportion dying during age interval $x$ to $x + 1$
$q_x$ = rate of mortality during the age interval $x$ to $x + 1$
$L_x$ = number of individuals alive on the average during the age interval $x$ to $x + 1$
$T_x$ = total years lived by individuals in the population
$e_x$ = mean expectation of life for individuals alive at start of age interval $x$

2. Tally raw data for survivorship in $n_x$ column; adjust this survivorship on basis of 1000 and record in $l_x$ column. In the case of the gray squirrels, the proportion of the number known to be alive was divided by the number marked and adjusted to 1000 marked (B).

3. Determine mortality by subtracting $l_{x+1}$ from $l_x$. This gives $d_x$. For example, in the squirrel life table initial population $l_{x0} - l_{x1} = 1000 - 247.3 = 752.7$ ($d_x$); for age class 1–2 247.3 – 112.4 = 134.9, etc.

*TABLE C-12*
**Dynamic-composite life table for squirrels marked as nestlings, 1956–1964 (sexes combined)**
A. *Returns of marked squirrels: basis for survivorship data*

| Year marked | Nestlings marked | Year of return | | | | | | | |
|---|---|---|---|---|---|---|---|---|---|
| | | *1957* | *1958* | *1959* | *1960* | *1961* | *1962* | *1963* | *1964* |
| 1956 | 40 | 8 | 4 | 3 | 2 | 0 | 0 | 0 | 0 |
| 1957 | 138 | | 60 | 30 | 28 | 13 | 9 | 4 | 3 |
| 1958 | 229 | | | 61 | 26 | 12 | 10 | 7 | 3 |
| 1959 | 193 | | | | 58 | 26 | 19 | 12 | 9 |
| 1960 | 162 | | | | | 19 | 13 | 8 | 6 |
| 1961 | 99 | | | | | | 4 | 1 | 1 |
| 1962 | 82 | | | | | | | 18 | 6 |
| 1963 | 80 | | | | | | | | 25 |

B. *Determination of $l_x$ from survivorship data*

| Age (x) | Total known alive ($n_x$) | Maximum available for recapture | Known live per 1,000 available ($l_x$) |
|---|---|---|---|
| 0–1 | 1,023 | 1,023 | 1,000.0 |
| 1–2 | 253 | 1,023 | 247.3 |
| 2–3 | 106 | 943 | 112.4 |
| 3–4 | 71 | 861 | 82.5 |
| 4–5 | 43 | 762 | 56.4 |
| 5–6 | 25 | 600 | 41.7 |
| 6–7 | 7 | 407 | 17.2 |
| 7–8 | 3 | 178 | 16.9 |

C. *Life table*

| Age (x) | Known survival ($l_x$) | Apparent mortality ($d_x$) | Proportional mortality ($q_x$) | Average years lived ($L_x$) | Total years lived ($T_x$) | Live expectancy ($e_x$) |
|---|---|---|---|---|---|---|
| 0–1 | 1,000.0 | 752.7 | .753 | 538.9[a] | 989.6 | .99 |
| 1–2 | 247.3 | 134.9 | .545 | 179.9[b] | 450.7 | 1.82 |
| 2–3 | 112.4 | 29.9 | .266 | 97.4 | 270.8 | 2.41 |
| 3–4 | 82.5 | 26.1 | .316 | 69.5 | 173.4 | 2.10 |
| 4–5 | 56.4 | 14.7 | .261 | 49.0 | 103.9 | 1.84 |
| 5–6 | 41.7 | 24.5 | .588 | 29.4 | 54.9 | 1.32 |
| 6–7 | 17.2 | .3 | .017 | 17.1 | 25.5 | 1.48 |
| 7–8 | 16.9 | 16.9 | 1.000 | 8.4 | 8.4 | .50 |

[a] $n = \dfrac{1,000\,(l - e^{l})}{i}$; $\overline{qx} = 0.635$.

[b] $n = \dfrac{l_x + (l_{x+1})}{2}$.

4. Calculate $q_x$ by dividing $d_x$ by $l_x$. For example, for age class 1–2.

$$q_{x2} = \frac{d_{x2}}{l_{x2}} = \frac{134.9}{247.3} = .545$$

5. Calculation of life expectancy in column $e_x$ requires additional information:

$$L_x = \frac{l_x + l_{x+1}}{2}$$

For example, for age class 2–3

$$L_{x3} = \frac{112.4 + 82.5}{2} = 97.45$$

Do this for each age interval.

TABLE C-13
*Survivorship and fecundity table, unexploited gray squirrel population*

| Age (x) | Age structure | Survival ($l_x$) | Productivity ($m_x$) | $l_x m_x$ | $x l_x m_x$ |
|---|---|---|---|---|---|
| 0–1 (0.5)[a] | 530 | 1.000 | 0.05 | .050 | .025 |
| 1–2 (1.5) | 134 | .253 | 1.28 | .324 | .486 |
| 2–3 (2.5) | 56 | .116 | 2.28 | .264 | .660 |
| 3–4 (3.5) | 39 | .089 | 2.28 | .203 | .710 |
| 4–5 (4.5) | 23 | .058 | 2.28 | .132 | .594 |
| 5–6 (5.5) | 12 | .039 | 2.28 | .089 | .489 |
| 6–7 (6.5) | 5 | .025 | 2.28 | .057 | .370 |
| 7–8 (7.5) | 2 | .022 | 2.28 | .050 | .375 |

$$\Sigma\, l_x m_x = R_o = 1.169^{b}$$
$$\Sigma x l_x m_x = 3.709$$

[a]Age categories have been converted to half years, 0.5, 1.5, etc., to permit correct calculation of $x l_x m_x$ for year class 0 and to take into consideration the productivity of year class 0–1. Some female gray squirrels may produce a litter late in the year of their birth. If age category 0 (the 1st year class) was listed as 0, the production of that age class would not be considered in the later calculations of mean generation time and the rate of increase $r$.

[b]The value of $R_o$ (1.169) is just above replacement rate which would be expected in an unhunted gray squirrel population. A hunted population should have a higher reproductive rate and few if any individuals in the 5 to 7 year age classes.

*Source:* Data from Barkalow, 1970.

6. Sum the $L_x$ column cumulatively from the bottom up to obtain $T_x$. Thus for age class 3–4,

$$T_{x4} = L_8 + L_7 + L_6 + L_5 + L_4$$
$$= 8.4 + 17.1 + 29.4 + 49.0 = 103.9$$

7. Obtain life expectancy for each age class by

$$e_x = \frac{T_x}{l_x}$$

For example, life expectancy for age class 4–5 is

$$e_x = \frac{T_{x5}}{l_{x5}} = \frac{54.9}{41.7} = 1.316 \ (1.32)$$

**RATE OF INCREASE**

Given a life table and a fecundity table, the rate of increase, $r_m$, for a population in a particular environment can be determined.

To the data in the life and fecundity tables is added another column, $x l_x m_x$ which records the sums obtained by multiplying the values in the $l_x m_x$ column by the appropriate age, $x$. For the gray squirrel population the rate of increase can be estimated by using survival and fecundity data for the female segment of the population. Necessary data are provided in Table C-13.

Once the values for the $x l_x m_x$ column have been determined, the mean generation time, $T_c$, can be computed by the equation

$$T_c = \frac{x l_x m_x}{R_0}$$

where $R_0$ is the net reproductive rate, the sum of the $l_x m_x$ column. For the unexploited gray squirrel population

$$T_c = \frac{3.708}{1.169} = 3.173$$

This mean generation time for the unexploited gray squirrel population is a rather long 3.173 years. A hunted population would have a shorter mean generation time because hunting pressure would decrease survivorship and life expectancy.

To find an approximate value for $r_m$, the following formula applies

$$r_m = \frac{\log_e R_0}{T_c} = \frac{\log_e 1.169}{3.173} = \frac{0.1561}{3.173} = 0.049$$

Thus an estimate of $r_m$ for the unexploited gray squirrel population is .049.

A more accurate assessment of the rate of increase, also based on life tables and fecundity tables at any given density with a stable age distribution, can be obtained by the formula derived by the eighteenth-century French mathematician, Leonard Euler:

$$\Sigma\, l_x m_x e^{-rx} = 1$$

Because there is no real way to solve for *r*, it can be obtained only by substituting values for *r* in the equation by trial and error until the right side balances with the left. The gray squirrel population again can serve as an example, although we may be erroneously assuming that it has a stable age distribution.

The first step is to obtain some approximate estimate of *r* to be used in the Euler equation. The estimate of $r = 0.049$, obtained by using net reproductive rate $R_0$ and mean generation time $T_c$, can be used as the first estimate substituted in the Euler equation (see Table C-14). The procedure is as follows.

1. Multiply the estimated *r* value by age as indicated in the formula to obtain *rx*. Thus for the first estimate value 0.049 is multiplied by 0.5, 1.5, 2.5, 3.5, 4.5, 5.5, 6.5, and 7.5 to give the following for values for *rx*: .025, .074, .110, .172, .221, .269, .319, and .368.

2. In a table of functions look up the tabled value for each of the above values of $e^{-rx}$ in the $e^{-x}$ column or, preferably, determine the value on a scientific electronic calculator. Record the values as done in Table C-14.

3. Multiply the values of $e^{-rx}$ by the

appropriate $l_x m_x$ and record. Sum the column. For the *r* value 0.049, the sum is 1.0068, close, but not quite the required value of 1.0.

4. Proceed by trial and error to come closer. In Table C-14 the next step is to try an *r* value of 0.051. The resulting sum of 0.9980 is still not quite close enough. A trial of $r = 0.050$ gives a sum of 1.0008, which is a close enough approximation.

### AGE DISTRIBUTION

A population growing geometrically with constant age-specific mortality and fecundity rates assumes and maintains a stable age distribution (see Chapter 14). From any set of life tables and fecundity tables for a population, the stable age distribution can be determined by the equation

$$C_x = \frac{\lambda^{-x} l_x}{\sum\limits_{i=0}^{\infty} \lambda^{-i} l_i}$$

where

$C_x$ = proportion of organisms in age category $x$ to $x + l$ in a population increasing geometrically

$\lambda$ = $e^r m$ = finite rate of increase

$l_x$ = survivorship function from life table

$x, i$ = subscripts indicating age

The gray squirrel population again may serve as the example as detailed in Table C-15. In this case $\lambda = e^r = e^{.05} = 1.05$.

Calculating the proportion of squirrels in each age category gives, for example,

TABLE C-14
**Determination of $r_m$**

| Age (x) | $l_x m_x$ | r = 0.049 | | | r = 0.051 | | | r = 0.050 | | |
| | | rx | $e^{-rx}$ | $e^{-rx} l_x m_x$ | rx | $e^{-r=0.051}$ | $e^{-rx} l_x m_x$ | rx | $e^{-rx}$ | $e^{-rx} l_x m_x$ |
|---|---|---|---|---|---|---|---|---|---|---|
| 0.5 | .050 | .025 | .9753 | .0487 | .025 | .9753 | .0478 | .025 | .9753 | .0478 |
| 1.5 | .324 | .074 | .9286 | .3008 | .0765 | .9263 | .3001 | .075 | .9277 | .3005 |
| 2.5 | .264 | .110 | .8958 | .2365 | .1275 | .8802 | .2323 | .125 | .8825 | .2329 |
| 3.5 | .203 | .172 | .8419 | .1709 | .1785 | .8365 | .1698 | .175 | .8394 | .1703 |
| 4.5 | .132 | .221 | .8017 | .1058 | .2295 | .7949 | .0149 | .225 | .7985 | .1054 |
| 5.5 | .089 | .269 | .7641 | .0677 | .2805 | .7554 | .0672 | .275 | .7595 | .0675 |
| 6.5 | .057 | .319 | .7334 | .0418 | .3315 | .7178 | .0409 | .325 | .7225 | .0412 |
| 7.5 | .050 | .368 | .6921 | .0346 | .3825 | .6822 | .0341 | .375 | .6873 | .0343 |
| | | | $\Sigma e^{-rx} l_x m_x =$ | 1.0068 | | | 0.9980 | | | 1.0008 |

TABLE C-15
*Determination of stable age distribution,*
*unexploited gray squirrel population*

| Age (x) | $l_x$ | $\lambda^{-x}$ | $\lambda^{-x}l_x$ | $C_x$ |
|---|---|---|---|---|
| 0 | 1.000 | 1.0000 | 1.000 | 0.652 |
| 1 | .253 | .9524 | .241 | .157 |
| 2 | .116 | .9070 | .105 | .068 |
| 3 | .089 | .8638 | .077 | .050 |
| 4 | .058 | .7921 | .046 | .030 |
| 5 | .039 | .7835 | .031 | .020 |
| 6 | .025 | .7462 | .018 | .012 |
| 7 | .022 | .7107 | .015 | .009 |

$$\Sigma\ \lambda^{-x}l_x = 1.533$$

$$\lambda = e^r = e^{.05} = 1.05$$

$$C_0 = \frac{1.000}{1.533} = .652$$

$$C_5 = \frac{.013}{1.533} = .020$$

Compare these calculated age distribution values with the actual age structure of the population given in Table C-13.

## References

### Productivity

*In the past few years a number of excellent books on methods of estimating productivity have appeared. Outstanding are the International Biological Program (IBP) handbooks, which belong in all ecological libraries.*

ECKARDT, F. E. (ED.). 1968. *Functioning of Terrestrial Ecosystems at the Primary Production Level.* UNESCO, Paris.

ELLENBERG, H. (ED.). 1971. *Integrated Experimental Ecology.* Springer-Verlag, New York.

INOUE, E. 1968. The $CO_2$ concentration profile within crop canopies and its significance for the productivity of plant communities. In F. E. Eckardt (ed.), *Functioning of Terrestrial Ecoysystems at the Primary Production Level.* UNESCO, Paris.

KALBER, F. A., AND F. G. WATSON. 1974. *Handbook of Marine Science.* Vol. 2, *Biological and Primary Productivity.* CRC Press, West Palm Beach, Fla.

LEITH, H., AND R. H. WHITTAKER (EDS.). 1975. *Primary Productivity in the Biosphere.* Springer-Verlag, New York.

LEMON, E. R. 1960. Photosynthesis under field conditions, II. An aerodynamic method for determining the turbulent carbon dioxide exchange between the atmosphere and a corn field. *Argon. J.,* 52:697–703.

LEMON, E. R. 1968. The measurement of height distribution of plant community activity using the energy and momentum balances approaches. In F. E. Eckardt (ed.), *Functioning of Terrestrial Ecosystems at the Primary Production Level.* UNESCO, Paris.

MILNER, C., AND R. E. HUGHES. 1968. *Methods for the Measurement of the Primary Production of Grassland.* IBP Handbook no. 6. Blackwell, Oxford.

MONTEITH, J. L., AND G. SZEICZ. 1960. The carbon-dioxide flux over a field of sugar beets. *Quart. J. R. Met. Soc.,* 86:206–214.

NEWBOULD, P. J. 1967. *Methods in Estimating the Primary Production of Forests.* IBP Handbook no. 2. Blackwell, Oxford.

ODUM, H. T. 1956. Primary production in flowing waters. *Limnol. Oceanog.,* 1:102–117.

PRATT, D. M., AND BERKSON. 1959. Two sources of error in the oxygen light and dark bottle method. *Limnol. Oceanog.,* 4:328–334.

RYTHER, J. H., AND C. S. YENTSCH. 1957. The estimation of phytoplankton production in the ocean from chlorophyll and light data. *Limnol. Oceanog.,* 3:381–386.

U.S. FOREST SERVICE. 1962. *Range research methods: a symposium.* Misc. Publ 940. U.S. Department of Agriculture, Washington, D.C.

VOLLENWEIDER, R. A. 1969. *A Manual on Methods for Measuring Primary Production in Aquatic Environments.* IBP Handbook no. 12. Blackwell, Oxford.

### Secondary production

EDMONDSON, W. T., AND G. G. WINBERG. 1971. *A Manual on Methods for the Assessment of Secondary Production in Fresh Waters.* IBP Handbook no. 17. Blackwell, Oxford.

ODUM, E. P., AND E. J. KUENZLER. 1963. Experimental isolation of food chains in old-field ecosystem with the use of phosphorus-32. In V. Schultz and A. W. Klements (eds.), *Radioecology.* Proceedings of the First National Symposium on Radioecology. Van Nostrand Reinhold, New York.

PETRUSEWICZ, K. (ED.). 1967. *Secondary Productivity of Terrestrial Ecosystems,* 2 vols. Pantsworve Wydawnictwo Naukowe, Warsaw.

PETRUSEWICZ, K., AND A. MACFADYEN. 1970. *Productivity of Terrestrial Animals — Principles and Methods.* IBP Handbook no. 13. Blackwell, Oxford.

PHILLIPSON, J. (ED.). 1970. *Methods of Study in Soil Ecology.* UNESCO, Paris.

PHILLIPSON, J. (ED.). 1971. *Methods of Study in Quantitative Soil Ecology.* IBP Handbook no. 18. Blackwell, Oxford.

REICHLE, D. E., AND D. A. CROSSLEY, JR. 1969. Trophic level concentrations of cesium-137, sodium, and potassium in forest arthropods. In D. J. Nelson and F. E. Evans (eds.), *Symposium on Radioecology.* Proceedings of the Second National

Symposium on Radioecology, CONF-670503. National Technical Information Service, Springfield, Va.

SCHWOERBEL, J. 1970. *Methods of Hydrobiology: Freshwater Biology.* Pergamon Press, New York.

SOROKIN, Y., AND H. KADOTA (EDS.) 1972. *Handbook on Microbial Production.* IBP Handbook no. 23. Blackwell, Oxford.

UNESCO. 1969. *Soil Biology.* UNESCO, Paris.

WINBERG, G. G. (ED.). 1971. *Methods for the Estimation of Production of Aquatic Animals.* Academic Press, New York.

*Community structure*

ANDREWARTHA, H. G. 1970. *Introduction to the Study of Animal Populations.* University of Chicago Press, Chicago.

BARKALOW, F. S., ET AL. 1970. The vital statistics of an unexploited gray squirrel population. *J. Wildlife Management,* 39:489–500.

BRILLOUIN, L. 1962. *Science and Information Theory,* 2nd ed. Academic Press, New York.

BROWER, J. E., AND J. H. ZAR. 1977. *Field and Laboratory Methods for General Ecology.* Brown, Dubuque, Iowa.

COLWELL, R. K., AND D. J. FUTUYMA. 1971. On the measurement of niche breadth and overlap. *Ecology,* 52:567–576.

COX, G. W. 1976. *Laboratory Manual of General Ecology,* 3rd ed. Brown, Dubuque, Iowa.

CULVER, D. C. 1972. A niche analysis of Colorado ants. *Ecology,* 53:126–131.

GARBUTT, P. D. 1961. The distribution of some small mammals and their associated fleas from central Labrador. *Ecology,* 42:518–525.

GREIG-SMITH, P. 1964. *Quantitative Plant Ecology.* Butterworth, London.

HURLBUTT, S. H. 1971. The nonconcept of species diversity: a critique and alternative parameters. *Ecology,* 52:577–586.

HORN, H. S. 1966. Measurement of "overlap" in comparative ecological studies. *Amer. Natur.,* 100:419–424.

JOHNSON, D. W. 1971. Niche relationships among some decidious forest flycatchers. *Auk,* 88:796–804.

KREBS, C. J. 1978. *Ecology: The Experimental Analysis of Distribution and Abundance,* 2nd ed. Harper & Row, New York.

LEVINS, R. 1968. *Evolution in Changing Environments.* Princeton University Press, Princeton, N.J.

LLOYD, M., AND R. J. GHELARDI. 1964. A table for calculating the "equitability" component of species diversity. *J. Animal Ecology,* 33:217–225.

MACARTHUR, R. H., AND J. W. MACARTHUR. 1961. On bird species diversity. *Ecology,* 42:594–598.

MORISITA. M. 1959. Measuring of interspecific association and similarity between communities. *Memoirs of the Faculty of Science, Kyushu University,* Series E (Biology), 3:66–80.

MUELLER-DOMBOIS, D., AND H. ELLENBERG. 1974. *Aims and Methods of Vegetation Ecology.* Wiley, New York.

PEET, R. K. 1974. The measurement of diversity. *Ann. Rev. Ecology and Systematics,* 5:285–307.

PIELOU, E. C. 1975. *Ecological Diversity.* Wiley, New York.

POOLE, R. W. 1974. *An Introduction to Quantative Ecology.* McGraw-Hill, New York.

SHANNON, C. E., AND W. WIENER. 1963. *The Mathematical Theory of Communication.* University of Illinois Press, Urbana.

SIMPSOM, E. H. 1949. Measurement of diversity. *Nature,* 163:688.

TRAMER, E. J. 1969. Bird species diversity: Components of Shannon's formula. *Ecology,* 50:927–929.

# *Environmental measurements*
## *APPENDIX D*

The methods suggested here are for the most part the simplest and least expensive for the field student to use. Some of them have a little of the homemade quality about them. They may not give the precise results of techniques using more refined instrumentation, but if performed carefully, these methods will give results accurate enough for most studies. Fortunately, there are now on the market some pre-packaged kits for analyzing water and air pollution. For air pollution testing equipment and material write:

> National Environmental Instruments
> P.O. Box 590 Pilgrim Station
> Warwick, R. I. 02888

For excellent water analysis kits write:

> Hach Chemical Co.
> Box 907
> Ames, Iowa 50010

> Lamotte Chemical Co.
> P.O. Box 329
> Chestertown, Md. 21620

Lamotte also supplies a good soil testing kit for field use.

For graphic analysis of the aquatic environment, see Kaill and Frey (1973) for plotting a unique environmental profile.

## *Collecting water samples*

Collecting samples of water from various depths of lakes, ponds, and estuaries can be a difficult task. Water samplers designed to collect waters from various depths can be purchased from scientific apparatus companies or they can be made by assembling a Meyer sampler. The Meyer sampler consists of a heavy bottle attached to a wire harness and fitted with a weight heavy enough to sink it when full of air. A stopper with an eyebolt attached is loosely inserted. A cord is attached to the eyebolt and to the wire harness. When the bottle is lowered to the desired depth (indicated by a marked line), the cord is jerked to remove the cork. Sufficient time for the bottle to fill is allowed and then it is pulled to the surface. Because the bottle cannot be recorked, there is some admixture with water from other layers, but it is minimal.

Water samples must be obtained in 200-ml or 250-cc glass bottles with tight-fitting glass stoppers and taken so that no bubbling or splashing occurs. The sample bottle should be completely filled, so that there is no air space between the neck and the stopper. Special water samplers are available that make it possible to secure water without introducing atmospheric oxygen. If a sampler is used, then when the water is transferred to the bottle, it should be allowed to overflow the container two or three times to eliminate atmospheric oxygen. The water is then ready for analysis in the field by the rapid Winkler method.

## *Oxygen determination*

### REAGENTS NEEDED

The following reagents are needed.
1. Alkaline sodium solution. Dissolve 500 g of NaOH and 135 g of NaI to 1 liter of water.
2. Manganous sulfate. Dissolve 480 g $MnSO_4 \cdot 4H_2O$ in distilled water; dilute to 1 liter. Where water is organically enriched, add to the solution 10 g of sodium azide, $NaN_3$, dissolved in 40 ml of water. This effectively removes interference from nitrites in the water sample.

3. Sodium thiosulfate. Prepare a solution (compute first) of 0.025 $N$ $Na_2S_2O_3 \cdot 5H_2O$. When dissolved, add a good drop of chloroform to act as a stabilizer or add 3 g borax/liter, since this solution does not store. Next standardize a 250-ml sample by adding 25 ml of 0.025 $N$ potassium dichromate, $K_2Cr_2O_7$. Then add 1 ml of the alkaline NaI solution and 5 ml of concentrated hydrochloric acid (HCl). Titrate this solution with the thiosulfate until the solution turns a faint yellow. Add 10 drops of starch solution (see next paragraph) and continue titration until the blue color disappears. The solution is now ready for use; it should be stored in a brown bottle.

4. Starch reagent. Into 350 ml of water stir 2 g of powdered potato starch. To this add 30 ml of 20 percent NaOH, mix, and let stand 1 hour. Neutralize with HCL, using litmus as an indicator. Then add 1 ml of glacial acetic acid.

### DETERMINATION

To determine oxygen follow this procedure.
1. Add 1 ml of the manganous sulfate solution so that it sinks to the bottom. (Slowly and carefully add all reagents deep into the sample with a narrow pipette. When all reagents have been added, the bottle will overflow.) Then with a clean pipette add 1 ml of alkaline sodium iodine solution. Close the bottle, quickly upend, and shake for 20 seconds. A glass bead added to the bottle speeds up mixing. A yellow precipitate will form and settle to the bottom.

2. Add 2 ml of concentrated sulfuric acid ($H_2SO_4$) and invert several times to mix the acid with the solution. This will dissolve the precipitate (manganese hydroxide, in which is fixed the oxygen in the solution) and leave a yellowish-brown color, the deepness of which is proportional to the amount of dissolved oxygen in the sample.

3. After inverting the bottle several times, place the sample into a 400-ml beaker and titrate with the 0.025 $N$ sodium thiosulfate solution until a pale straw yellow is reached. Before doing this, *be sure* to read

the level of the sodium thiosulfate in the burette and record.

4. Place a white background against the beaker. Then add 10 drops of the starch indicator. Continue titration until the blue color disappears. Read the new level of the sodium thiosulfate in the burette. Record and then calculate the amount used in titration. The number of milliliters used represents the quantity of dissolved oxygen in parts per million.

## Free carbon dioxide concentration

### REAGENTS NEEDED

1. Phenolphthalein indicator.
2. 0.02 N sodium hydroxide (NaOH)

### DETERMINATION

In general the amount of carbon dioxide varies inversely with the dissolved oxygen content. Free carbon dioxide occurs only in acid waters.
1. Obtain a water sample and from it fill a Nessler tube to the 100-cc mark. Do not agitate or splash, because $CO_2$ will easily diffuse into the air.

2. Add 10 drops of phenolphthalein solution. This can be made by dissolving 5 g of phenolphthalein in 1 liter of 50 percent alcohol, then neutralizing with 0.02 $N$ NaOH.

3. Titrate this with NaOH, but be sure to record the level of the solution in the burette before starting. When a pink color appears for a few seconds under agitation, stop. Read the new level and record the amount of NaOH used. Multiply amount used by 10. The answer equals the amount of $CO_2$ in parts per million in the sample.

## Temporary hardness

Temporary hardness of water is caused by the carbonates of calcium and magnesium. Because any carbon dioxide present in alkaline and neutral waters will be in the form of bicarbonate, the determination of temporary hardness and the determination of combined carbon dioxide are one and the same.

1. Hydrochloric acid, 0.01 $N$.
2. Indicator solution consisting of a mixture of 0.02 percent methyl red and 0.1 percent brome cresol green mixed in 95 percent alcohol.

### DETERMINATION

1. To the water sample (100 ml) add a couple of drops of indicator.
2. Titrate the HCL into solution until the first appearance of pink.
3. Record the amount in cubic centimeters of HCl needed per 100 ml of water.
4. Multiply this by 5 and the answer will be the amount of $CO_2$ expressed as parts per million of $CaCO_3$.

## Total hardness

Total or permanent hardness generally is caused by the chlorides and sulfates of calcium and magnesium.

There are several ways of determining total hardness, an old-fashioned way, with the soap method, and a colorimetric method. The latter is the most accurate and once the stock solution is mixed, the easiest. The colorimetric method is based on the ability of sodium versenate (sodium diethylenediamine tetracetate, $Na_2C_{10}H_{14}O_8N_2 \cdot 2H_2O$) to form un-ionized complexes with calcium and magnesium. If eriochrome black $T$, a dark blue dye, is added to this solution containing $Ca^{2+}$ and $Mg^{2+}$ ions, a complex, pink in color, is formed. By adding sodium versenate solution to this complex, the solution can be turned back to blue again by removing $Ca^{2+}$ $Mg^{2+}$ from the dye complex to the versenate complex again. The end point— when the pink changes back to blue—can be used as a measure of total hardness.

### REAGENTS NEEDED

1. Standard calcium chloride solution. Dissolve 55 g of $CaCl_2 \cdot 6H_2O$ in distilled water; then add enough to make 1 liter of solution.
2. Sodium versenate solution. This can be purchased as a standard solution or it can be mixed as follows. Dissolve 2.5 g of sodium versenate in 2 liters of distilled water. To this add 13.5 cc of $N$ NaOH solution, prepared by mixing 40 g in 1 liter of water, and make up to 2500 cc. This solution must be adjusted against the standard chloride solution so that 1 cc equals 0.1 mg of $Ca^{2+}$. Use eriochrome black $T$ as the indicator.
3. Eriochrome black $T$ *indicator* solution. To 30 cc of distilled water add 1 g of the eriochrome black $T$ and 1 cc of $N$ $Na_2CO_3$. Mix and make up to 100 cc with isopropyl alcohol.
4. Buffer solution. Dissolve 40 g of borax in 800 cc of water and 10 g of NaOH and 5 g of sodium sulfide ($Na_2S_2 \cdot 9H_2O$) in 100 cc of distilled water. Mix the two solutions together and dilute to 1 liter. This is used to control pH at the 8 to 10 level and to eliminate the effects of copper, iron, and manganese ions.

### DETERMINATION

1. Take a 100-cc sample of water and slightly acidify with 0.1 $N$ HCL; boil for a few minutes.
2. Add 0.5 cc of the buffer solution.
3. Add 5 drops of eriochrome black $T$ indicator solution.
4. Titrate with the standard sodium versenate, the end point being reached when the blue color appears. Each cubic centimeter of sodium versenate needed to titrate the 100-cc sample equals 1 part per million of $Ca^{2+}$ and $Mg^{2+}$.

## Chlorides: salinity

Salinity of coastal waters varies little, but in the estuary it varies considerably, both vertically and horizontally. Because of its biological importance, the determination of chlorides is necessary in any estuarine study. The most accurate method is titration.

### REAGENTS NEEDED

1. 0.01 $N$ sodium chloride (NaCl)
2. 0.01 $N$ silver nitrate ($AgNO_3$)
3. Potassium chromate indicator.

## DETERMINATION

1. Add a few drops of potassium chromate solution to a 100-cc sample of water.
2. Titrate with 0.01 $N$ silver nitrate solution, stirring constantly. When a faint red color of silver chromate appears, the end point has been reached. Determine the number of cubic centimeters of 0.01 $N$ silver nitrate used and multiply this by 0.000355. The answer equals chlorides in parts per million.

## Acidity

Colorimetric determination of pH is the quickest and simplest method in the field and fairly accurate. For aquatic work there are the narrow-range pH papers. For soil work there are a number of soil-acidity test kits on the market that use color indicators, but be sure to obtain a kit with a selection of narrow-range indicators.

For more sophisticated measurement and research, the electronic pH meter should be used. Most of these are expensive, although moderately priced pocket meters are available.

## Current

### STREAM

Estimates of stream flow are necessary in any study of the flowing-water community. Fortunately, a short, accurate method for estimating the volume of flow is available without resorting to current meters (Robins and Crawford, 1954). It works very well for most stream studies.

1. Choose a cross section of the stream where the current and depth are most uniform, and measure the width.
2. Divide this width into three equal segments, marked or separated by pushing sticks into the bottom, coloring the surface of stones, and so on.
3. Record the depth at the midpoint of each segment and determine the velocity of the surface current. The velocity can be found by dropping a fisherman's float (without projecting arms) attached to 5 ft of limp, monofilament nylon fishing line

(0.005-0.01-in. diameter). Record with a stopwatch the time required for the float to travel the 5 ft. This should be done several times and the average figure recorded and converted into feet per second.

4. Determine the volume of flow $R$ for each segment of the cross section by the formula

$$R = WDaV$$

where

$a$ = a bottom factor constant (0.8 for rocks and coarse gravel; 0.9 for mud, sand, hardpan, bedrock)
$W$ = width of the segment
$D$ = depth at the midpoint of the segment
$V$ = surface current velocity taken at the midpoint of the segment

5. Determine the total flow by adding the flows for the three segments.

### ESTUARY

The study of currents in the estuary is more complicated because of the outgoing flow of fresh water to the sea and the incoming flow of seawater at high tide. A knowledge of the circulation set up by the incoming and outgoing flow is of major importance in the study of both the physical and biological aspects of the estuary.

A rapid technique for obtaining current profiles in estuarine waters has been developed by Prichard and Burt (1951). It involves the use of a current indicator consisting of a submerged biplane-shaped drag and a device for reading the angle made by the suspending wire and the vertical. The device is made from two half-inch (five ply) fir plywood panels 4 ft long and 3 ft high, assembled so that each panel bisects the other. This gives four planes 2 by 3 ft. To this is added a 30-lb weight, and the entire biplane is attached to hydrographic wire and suspended into the water from a block extending as far out as possible from the anchored ship. The current will swing this drag out in the direction of the current, whose speed of flow is then computed by the angle of the wire from the vertical. The velocity of flow is proportional to the square root of the tangent of the angle made by the supporting line with the vertical, so the current velocity can be solved by the formula

$$v = k \sqrt{\tan \theta}$$

The value of $k$, a proportionality factor to take care of the opposing forces of frictional drag of water flowing past the drag and the restoring force of gravity, for this device is 1.04. For those interested, the formula for $k$ can be found in the original paper. This method works well for the determination of current velocity down to 50 ft, even where the velocity is as low as 1 cm/sec (less than 0.02 knot).

## Light

### AQUATIC

A commonly employed method of estimating the depth of light penetration into a body of water is the Secchi disk, which is easily constructed.

1. Take a firm metal disk about 8 in. in diameter and paint it with several coats of white enamel. Divide the disk into quadrants and paint the alternate sectors black or red.
2. Attach the disk by the center to a rod, cord, or chain calibrated for depth. Be sure the disk is fitted to ride level when it is lowered into the water.
3. Lower the disk over the side of the boat and note the depth at which the disk disappears. Sink the disk several more feet, then raise it, noting the depth at which the disk reappears.
4. Average the two observations to obtain a single light-penetration reading. This method gives an *estimate* of the compensation level.

For precision readings of light in aquatic environments photometers adapted for underwater work are necessary.

### TERRESTRIAL

Light-intensity data in terrestrial communities is most commonly obtained by the use of a photometer or exposure meter. Among the variety of types available today, photometers calibrated in foot-candles are preferable for this application.

Although the photometer works satisfactorily in open situations, its use in the forest community leaves a good deal to be desired. Because the photometer measures light intensity at a given spot at a given minute, it fails to yield information on the changing intensity of light on the forest floor. The only alternative with this instrument is to take a large number of readings over a period of time, which is time-consuming, inaccurate, and even impossible when more than a few locations are involved.

Superior to the exposure meter is the chemical light meter described briefly in a paper by W. G. Dore (1958). It offers several advantages: (1) it measures the cumulative amount of light reaching a particular location during a period of time; (2) it is comparatively inexpensive; (3) a great number of locations can be sampled at the same time; and (4) measurement can be made in almost any location where a small vial of the light-sensitive chemical can be placed.

The principle of the method is based on the fact that anthracene ($C_{14}H_{10}$) in benzene will polymerize into insoluble dianthracene ($C_{14}H_{10})_2$ upon exposure to light. This property can be used to measure the amount of light that enters an environment over a period of time. In use, vials of the anthracene-benzene solution are exposed to the light of a particular environment for a period of time, then are analyzed to determine the amount of unconverted anthracene remaining in the solution. The analysis is made with a spectrophotometer. A standardization curve that relates the percent transmittance from the spectrophotometer to concentration of anthracene and a calibration curve that relates the concentration of anthracene to light exposure are required to convert the chemical reaction into standard units of light.

Details of the procedure are too lengthy to describe here, but they are given in step-by-step pictorial detail by Marquis and Yelenosky, 1962.

## Temperature

The temperature of air and water at any one given time can be recorded with a standard mercury or dial thermometer. The latter is preferable for ecological work. The cost is moderate, they are not easily broken, and they can be used to take temperatures in a variety of locations, such as streams, crevices in bark, underneath litter, and in the soil.

A semirecording instrument is the maximum-minimum thermometer. It consists of two thermometers, each with a metallic float that in the one thermometer lodges at the highest temperature, in the other at the lowest. For new readings the floats are reset at the top of the mercury columns, preferably with a small magnet. Temperatures usually are recorded on a 24-hour basis. For ecological purposes any time for resetting can be chosen, preferably at some time away from the extremes. In terrestrial communities, maximum-minimum thermometers should be protected in slotted wooden shelters and should be placed at several levels, including the ground level. In animal studies the ground level is best.

The maximum-minimum thermometer can be used to record the water temperatures at different depths in lakes and ponds. Attach one end of the thermometer to a measured line and to the other end attach a weight. Take readings vertically from the surface to the bottom at 0.5- or 1.0-m intervals.

In the absence of a maximum-minimum thermometer, the temperatures of deep water can be taken as soon as the water samples are raised to the surface.

Electronic temperature recording devices with battery power and with a sensing probe or lead especially designed to read instantaneously the temperature of water at various depths are now available at reasonable cost. Even some mail-order houses have them available as fish-finding devices.

Highly accurate temperature measurements of microlocations—soil, leaf litter, tree wood, bark, leaf, etc.—can be taken with resistance thermometers or thermistors and microprobes that are buried in the material or object. The thermistors have leads projecting above the surface. To take a reading, the wires are attached to the terminals of a potentiometer, on which temperatures are read directly. Compact sensory units are available with which one can take simultaneous readings of both the soil-moisture content and the soil temperature.

Available for continuous recording of temperatures are recording thermographs. A number of types are available, but all work with a simple clock within a cylinder drum upon which temperatures are recorded in graph form by pen and ink.

## Moisture

*Atmospheric Moisture* The usual method of determining atmospheric moisture involves the use of the sling psychrometer. This instrument can be purchased rather reasonably from scientific supply houses. When readings are to be taken, the fabric over the wet bulb is soaked in distilled water. Then the instrument is rapidly swung so that the evaporation cools the wet bulb. When the readings of the two bulbs, wet and dry, are constant, the temperatures on both are recorded. Relative humidity is obtained from the dry-bulb reading and the difference between the dry-bulb reading and the wet-bulb reading. The readings are converted to relative humidity by consulting a conversion table. Vapor-pressure deficits can also be obtained from the original readings with the aid of suitable tables.

Humidity also can be recorded continuously with the use of a recording hygrograph, which plots the humidity on a graph attached to a revolving drum. In fact the thermograph and hygrograph are often combined into one (rather expensive) instrument, the hygrothermograph.

### RATE OF EVAPORATION

For certain studies the natural rates of evaporation may mean more than the relative humidity.

A useful instrument for measuring total evaporative power is the atmometer or, more specifically, the Livingston (or James) atmometer, a standardized and calibrated wet surface constant in size, shape, and texture. It consists of a hollow unglazed sphere connected to a reservoir or graduated cylinder.

For time measurements over a period of hours or days the atmometer is connected to a large reservoir of distilled water (about 500 ml in a flask or bottle) by ordinary glass tubing sealed at the end but with a small hole in the end about 1 in. from the top. This in turn is covered with tightly fitting rubber tubing, which acts as a valve to allow the water to pass from the reservoir into the porous chamber but not back again. The tube is inserted into the reservoir through one hole in a two-hole stopper. The bottom of the tube should reach nearly to the bottom of the reservoir. A hooked vent pipe is placed in the other hole, with the external opening pointing downward. The whole apparatus is weighed at the beginning and at the end of the procedure to determine the loss of weight over a period of time. The raw data is then multiplied by the conversion factor on the glazed neck of each atmometer to give a measure of evaporative power.

A simple short-term atmometer can be assembled by filling an inverted atmometer with distilled water and inserting a one-hole rubber stopper in the neck. A 1-ml graduated pipette is inserted through the stopper and into the atmometer far enough to cause the water to rise in the graduated pipette. The atmometer is exposed for a constant period of time in the habitat under study and the amount of water evaporated is noted.

### SOIL MOISTURE

Soil moisture is best determined through the use of moisture blocks, thermistors, and a potentiometer. This has been discussed briefly under temperature. The whole setup can be purchased from agricultural and forestry supply houses.

An alternate but less accurate method is to collect soil samples in the field, place them in plastic bags, and bring them back to the laboratory. If quite wet, the samples should be allowed to drain. Soil moisture content is determined by comparing the weight of the samples at the time of collection with their weight in oven-dry condition. A sample of more than 10g is weighed, placed on an evaporating dish, and put in an oven at 105° C until the weight is constant. It is cooled in the desiccator and then weighed again. The loss from fresh weight to ovendry weight (OD) represents the moisture content (MC) or actual amount of water lost. The loss in weight expressed as a percentage, the percent moisture content (PMC), is found as follows:

$$\frac{MC \times 100}{OD} = PMC$$

### Soil atmosphere

The soil atmosphere, its fluctuations in composition between oxygen and carbon dioxide, and its influence on animal distribution in the soil have received little attention. It is a fertile field for investigation, although its study requires some equipment and is rather time-consuming. The collection of the soil gases is relatively simple. An ingenious procedure for obtaining samples of soil atmosphere from the litter has been devised by Haley and Brierley (1953). This procedure might be modified to study atmosphere of deeper soil layers. The apparatus used (Figure D-1) consists of polyvinyl plastic tubing 30 cm long and 3 cm in diameter. Each end is fitted with a rubber stopper. One is fitted with a small piece of glass capillary tubing, which in turn is fitted with a stopper made by fitting a piece of glass rod inside a short rubber tube. The other stopper is fitted with wider glass tubing, to which is attached a short piece of rubber tubing with a screw clamp attached. A length of wire is soldered to the clamping screw.

These tubes, which will hold two 25-cc gas samples, are then buried in the surface soil or the litter with the location wire projecting upward. The nozzles of the tubes are left closed in location for several days until the disturbed soil conditions have returned to normal. Then the tubes are opened by turning the wire han-

**FIGURE D-1**
*Tube for sampling the soil atmosphere.*

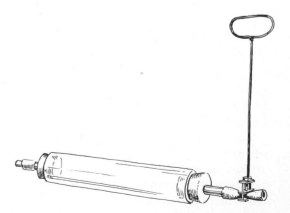

dles on the screw clamps. Gaseous diffusion between the tubes and the soil atmosphere will take place until equilibrium is established, usually in about 24 hours although best results are achieved by waiting a week. The nozzle is then closed and the tubes removed.

The gases within the tube are then tested in the laboratory for oxygen and carbon dioxide. Analysis procedure can be obtained in analytical chemistry books. A required piece of equipment is the Bancroft-Haldane or Haldane-Gutherie gas-analysis apparatus, obtainable from scientific supply houses. A more accurate field method involving more equipment is described by Wallis and Wilde, 1957.

## Soil organic matter

Accurate estimate of soil organic matter is difficult without special techniques, but a reasonable estimate can be obtained by loss of weight on ignition.

1. Take about 5 g of oven-dry soil, weigh carefully, and then heat red hot in a crucible for about a half hour. Cool in a desiccator, reweigh, and heat again.
2. Repeat this process until weight becomes constant. The loss in weight, expressed as a percentage, represents the amount of oxidizable organic matter present.
3. To correct for accompanying decomposition of carbonates, add some ammonia carbonate to the cooled sample and then heat in the oven to 105° C to drive off the excess solution. Cool and reweigh. The gain of weight indicates the amount of $CO_2$ lost by the carbonates during the previous heating. This gain in weight can then be subtracted from the initial weight loss before percentage of organic matter is determined.

## Soil texture and composition

1. Examine samples of sand, silt, clay, and gravel with a hand lens. Moisten each of these samples and feel them between your fingers. Note the differences as to their feel and the way they behave when moist.

2. Collect samples of soils from various areas—forest, fields, and gardens. Rub a sample of each between your fingers. If gritty, sand particles are present. If some or all of the soil feels like flour or talcum powder, then it is silty. Clay soils have enough clay particles to give them a harsh feeling when dry.
3. Dampen the samples by sprinkling them with water until they have the consistency of workable putty. From this sample make a ball about half an inch in diameter. Hold the ball between thumb and forefinger. Gradually press your thumb forward, forming the soil into a ribbon. If the ribbon forms easily and remains long and flexible, the soil is probably a clay or silty clay; if it breaks easily under its own weight, it probably is a clay loam or a silty clay loam. If a ribbon is not formed, the soil is probably a silt loam, sandy loam, or sand.

Soil textures are best determined in the laboratory by passing a known weight of oven-dry soil from several layers through a series of sieves of different size (U.S. Department of Agriculture standard sieves with decreasing aperture sizes), separating the various components. By determining the proportion of each, the soil texture can be classified.

A similar though less accurate method can be used to demonstrate soil texture.

1. Collect samples of various soils, enough to fill a pint jar about half full of each sample. Add water to fill the jar to the shoulder. Put the cap on and shake the jar. Then let the soil settle out. Allow plenty of time, for small particles will be slow in settling. The heavier particles will settle out first, the silt next, and the clay last. These components will build up several layers.
2. Hold up a card or heavy piece of paper against the side of the jar and draw a diagram showing the different layers. Label each layer. The relative amounts of each help determine the texture of the soil.

The ecologist is more concerned with combined effects than with detailed particle analysis, so this rough estimate may be sufficient.

## Sampling the soil

The most efficient device for sampling the soil is a soil augur, which enables the investigator to extract soil from known depths and horizons. After extraction the soil is laid out to dry before determining its composition.

## Soil profile

Because the distribution of soil inhabitants and vegetation is influenced by the nature of the soil, some attention must be given to the soil profile. To become familiar with a soil profile

1. Make a vertical cut through a well-drained soil. A good place to do this is along a roadside cut, but be careful to select an area where no debris or fill has been added to the surface soil. Observe the depth of the various horizons, their texture, color, and pH. Measure the pH of the three major horizons with a soil-pH test kit. Make a sketch of the soil profile, indicating the horizons, their depth, color, and pH.
2. Make a similar cut in a wooded section and note the difference between disturbed and undisturbed soil.
3. Sample the profile of the soil in an area where water stands most of the year. Note the difference between the vegetation growing on the well-drained soil and that growing on the wet areas.
4. Select an area where the slope of the land is variable and possesses a number of drainage conditions. Dig into the subsoil of each drainage class and compare the color of the subsoil on the top of the ridge with that lower on the slope and at the bottom of the slope. Notice the differences in the type or condition of the vegetation growing on these various drainage situations.
5. Dig a pit to expose $A$ and $B$ horizons under a deciduous and a coniferous woodland or under an old, established grassland and a woods. Compare the distribution of organic matter.
6. Learn from your local soil conservation service office the major soil groups found in your area and their location.

## Mapping the study area

In some studies an accurate map is desirable and often necessary. It may be needed to show the location of sample plots, animal concentrations and movements, vegetative cover, and the like. The details of the map and what it should show depend upon the nature of the study.

### AERIAL PHOTOGRAPHY

The simplest method of mapping is with the use of aerial photographs. If the area to be mapped is located within the center of the photograph and the terrain is not too rough, the major vegetational units and features of the landscape can be traced directly from a photograph enlarged to a scale of 660 ft to 1 in. (Photographs of most areas are available from the Agricultural Stabilization and Conservation Service, U.S. Department of Agriculture, Washington, D.C.) Scale can be corrected by pacing a given distance along a road, field boundary, or other straight-line feature, comparing it with the distance scale of the photograph, and making any necessary adjustments. For rough or hilly country or for areas requiring several photographs, the map should be prepared by radial triangulation. (For details, see American Society of Photogrammetry, 1966; Avery, 1957, 1962; Spurr, 1960.)

### TAPE AND COMPASS

The area can also be mapped by tape and compass. This will give accurate enough maps for most work. More precise surveying instruments require knowledge of land-surveying techniques.

The box-type compass is the most satisfactory for mapping work. It requires the use of a tripod, so that the compass can be leveled at each survey point.

The face of the compass is graduated into quadrants. The graduations at the North and South Poles are 0 and increase through 90° from each pole to east and west. The greater the number of degrees, the more the bearing approaches east or west. The bearing is always read from north or south to east or west. For example, S50E means a point 50° away from south toward the east. If the needle points directly east or west, it is read so. Or the compass may be graduated into 360°. North is 0 and the degrees increase clockwise. East is 90, south 180, west 270, and north 360. Both 0 and 360 are the same—north.

The compass needle on a central pivot always points to magnetic north. The angle this line makes with a line pointing to true north is known as the angle of declination. This declination varies in different parts of the country. The degrees of declination are usually indicated on geologic survey topographic maps. If the survey lines are to be changed to true north, they must be corrected by degrees of declination. This is best done by adjusting the dial of the compass to true north before mapping.

The east and west directions on the box-type compass are printed on opposite sides. In other words, east is where one normally would expect to find west and vice versa. There is a reason for this. The direction is indicated by means of a sight or line on the compass lid. The bearing is always read on the graduated dial at the point where the north end of the needle comes to rest. If the sighting is to the magnetic north, the needle will rest at north; if the compass is shifted to the west, the needle will still point north, but the dial is shifted so that west is at the tip of the needle.

Distance can be measured by using a standard measuring device, such as a tape or surveyor's chain, marked in meters, feet, or chains (66 ft), or by pacing.

Pacing is faster, but not as accurate, and it is more difficult to correct for slope. However, if the terrain is fairly level, it will suffice for most mapping. First, one must know the distance of one's average step or pace. Distance by pacing is measured by counting the paces and multiplying the number by the distance covered in each pace. Usually only one foot, the left, is counted. Thus the pace is the average distance taken in two steps. Pacing should be practiced by stepping off measured distances in different terrains and dividing the distance by the number of paces or steps taken.

### THE SURVEY TRAVERSE

A traverse is a survey line or a series of connected survey lines. A boundary traverse begins at some established point, follows the exterior limits of the area, and closes at the starting point. A meandering traverse follows the direction of some interior feature of the area, such as a road or stream.

The compass is set up at the starting point, A, and a sight is taken on the next station, B. The bearing is read and entered in the notebook. The distance between A and B is then measured or paced and the distance recorded. When the line AB is completed, the compass is set up at B, a bearing taken on C, and the distance to C measured. The work progresses in this manner until completed. Accurate and complete notes must be taken as the survey progresses. From these notes the map will be drawn. Notes must be entered so they can be interpreted accurately.

After the exterior lines have been surveyed, interior details can be determined by meandering traverses, as the course of a road or stream. The starting point is located at the exact point on the exterior boundary where the road or stream enters the area. The course is determined by plotting the bearings and distances. The point on the opposite boundary traverse where the meandering traverse leaves the area is also determined.

*Drawing the map.* A map is plotted from the traverse notes. A protractor, a ruler or engineer's scale, and a straight edge are needed. Before transferring the notes to the map paper, the scale must be determined. The scale will depend upon the extent to which the area can be reduced to fit the size of the map paper. For small areas, the scale of 1 in. to 660 ft is satisfactory. For larger areas, the scale can be reduced; for smaller areas it can be increased.

The protractor is used to transfer the bearings of the individual lines of the map. The protractor is graduated in degrees, used in plotting the angles.

A point is selected on the map to represent the starting point of the survey, A. A straight line is drawn through the point for the meridian, or north-south line. The protractor is centered over the initial point and carefully aligned to the meridian. The bearing of the initial line, AB, is marked by a pencil dot according to the number of degrees it bears east or west from north or south. A line is drawn between the starting station and the point located by the protractor. The appropriate distance from A to B is then scaled off on this line. The protractor is then centered at B and the same operation repeated to locate station *C*. The plotting is continued until all stations and lines are located on the map.

*Adjusting the errors.* More often than not, errors in plotting the bearings and distance will be great enough so that the map will not close. The error of closure is determined by measuring the distance connecting the last point plotted with the starting point. If the length of line or error of closure is 0.5 percent of the total length of the perimeter, the error is not excessive and will not seriously affect area determinations. If the error is larger than this, the field work should be checked.

Small errors must be corrected if the map is to close. This can be done as follows.

1. Draw and measure the line representing the closing error.
2. Through each angle, draw lines parallel to the closing error.
3. Divide the length of each line between stations by the total perimeter. Multiply the results by the length of the closing error.
4. Mark off on the parallel lines the respective amounts of closing error, increasing cumulatively at each successive corner.
5. Connect the newly located points and the adjusted boundary to the map is completed.

*Interior details.* Interior details are now added from the notes to the outline. Care must be taken to have the traverse of vegetation boundaries, roads, etc., end on or leave the area at a point on the map that corresponds to the measured distance in the field notes.

*Area determination.* If a planimeter is not available, the area can be determined as follows.

1. Divide the map into triangles.
2. Compute in square inches, square feet, or square meters the areas of each triangle by the formula: area = base × ½ altitude. If the calculations are made in feet, divide by 43,560 to obtain the area in acres; if in chains, divide by 10; if in meters, divide by 10,000 to obtain the area in hectares.

### MAPPING LAKES, PONDS, AND STREAMS

Mapping is essential to the study of any natural habitat, and lakes, ponds, and streams are no exception. One does not need to spend a long time on this, but something fairly accurate is necessary.

The map of a lake or pond needs to show its shape and the surrounding features, including the major zones of vegetation. In addition it should include the depth contours and the bottom vegetation (Figure D-2). Depth can be roughly determined by taking soundings with a weighted, marked line lowered from a boat. It can be determined more accurately with a hydrograph or sonar, if available. Mapping will require a number of transects across the pond. The contours can be plotted and drawn on the map, or depths can be indicated by inserting the figures at appropriate places on the map. The bottom vegetation (weed beds) should be mapped and emergents as well as submerged vegetation indicated. The result should be a base map on which can be placed transparent overlays of transects, temperature, and oxygen profiles, fish cover, spawning beds, waterfowl and other bird nests, etc.

**FIGURE D-2**
*A map of a lake showing depth profiles and a bed of aquatic vegetation.*

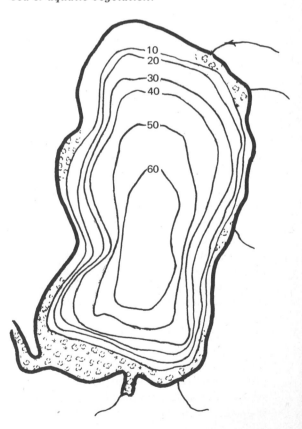

A section of stream, if not too wide, can be laid out in a grid and the surface plan of the bottom mapped on paper. It should show such major features as: current flow, large stones, current deflection around the stones, depth contours (far easier to determine than in lakes and ponds), distribution and nature of vegetation, and bottom substrate. By using transparent overlays, one can plot in the distribution of aquatic organisms in relation to current, depth, bottom substrate, etc. Line transects run across the stream will be helpful in determining the distribution of aquatic organisms.

## Remote sensing

Most techniques used in the study of environment involve the handling and measurement of some component of the ecosystem. Of growing importance is remote sensing, techniques of learning about ecosystems and the environment without direct contact. These methods involve the use of electromagnetic and sound waves, radiant energy, and ionizing radiation.

Remote sensing is not new. One method, the use of aerial photography, has been employed for years. Black and white stereoscopic photos have been used for some time to determine the volume of timber stands, to count wild animals, to record vegetational and man-made changes in the landscape, and to prepare topographical maps. More recently color photography has been widely employed, for its visual contrasts make much easier the identification of tree species, insect damage, plant disease, and details of aquatic environment.

One of the newer techniques is the use of infrared radiation. Infrared photography is very useful in the study of vegetation because the molecules of pigments in plants do not absorb infrared wavelengths. Instead the infrared is either transmitted through the leaf or is reflected by the cell walls. Cells of one species of plant have a different reflectivity than those of another, and normal cells have a different reflectivity than abnormal ones. Therefore, infrared photography can be used to distinguish species and to detect unhealthy plants. In fact diseases and environmental stresses in plants can be detected by infrared photography before they can be detected visually.

Infrared photography is also used to detect differences in environmental temperatures. Thermographs, pictures in which differences in temperature appear as contrasting bands of color, permit the detection of thermal pollution.

Because more information from aerial photography can be obtained if different types of photography are employed, a new technique called multispectral scanning has been developed. With this method several or many spectral bands from ultraviolet to infrared are recorded. The visual data are gathered by telemetry or are recorded on film or magnetic tape. The image data are fed into a multichanneled sensor, which sorts out the various displays or images. Because all the channels are recorded simultaneously, each image is in perfect register. This enables the interpreter to pick out and study an object of interest with different types of images taken at the same instant.

Because of its technical nature, remote sensing beyond the use of black and white and color aerial photographs is beyond the range of most field biologists and ecologists.

## References

*Books*

AMERICAN SOCIETY OF PHOTOGRAMMETRY. 1966. *Manual of Photogrammetry.* American Society of Photogrammetry, Falls Church, Va.

ANDREWS, W. A. (ED.). 1972. *A Guide to the Study of Environmental Pollution.* Prentice-Hall, Englewood Cliffs, N. J.

ANDREWS, W. A. (ED.). 1972. *A Guide to the Study of Fresh Water Ecology.* Prentice-Hall, Englewood Cliffs, N.J.

ANDREWS, W. A. (ED.). 1972. *A Guide to the Study of Soil Ecology.* Prentice-Hall, Englewood Cliffs, N.J.

ANDREWS, W. A. (ED.). 1972. *A Guide to the Study of Terrestrial Pollution.* Prentice-Hall, Englewood Cliffs, N.J.

AVERY, G. 1977. *Interpretation of Aerial Photographs,* 3rd ed. Burgess, Minneapolis.

BARNES, H. 1963. *Oceanography and Marine Biology: A Book of Techniques.* Macmillan, New York.

BLACK, C. A., ET AL. 1965. *Methods of Soil Analysis. Part 2, Chemical and Microbiological Properties.* American Society of Agronomy, Madison, Wis.

BROWER, J. E., AND J. H. ZAR. 1977. *Field and Laboratory Methods for General Ecology*. Brown, Dubuque, Iowa.

CAIRNS, J., JR., AND K. L. DICKSON (EDS.). 1973. *Biological Methods for the Assessment of Water Quality*. American Society for Testing and Materials, Philadelphia.

GOLTERMAN, H. L. (ED.). 1969. *Methods for Chemical Analysis of Fresh Water*. IBP Handbook no. 8. Blackwell, Oxford.

JACKSON, M. L. 1958. *Soil Chemical Analysis*. Prentice-Hall, Englewood Cliffs, N.J.

JOHNSON, P. L. 1969. *Remote Sensing in Ecology*. University of Georgia Press, Athens.

KAILL, W. M., AND J. K. FREY. 1973. *Environments in Profile*. Canfield, San Francisco.

KUCHLER, A. W. 1967. *Vegetation Mapping*. Ronald Press, New York.

LIND, O. T. 1974. *Handbook of Common Methods in Limnology*. Mosby, St. Louis.

PHILIPSON, J. 1971. *Methods of Study in Quantative Soil Ecology*. Blackwell, Oxford.

PLATT, R. B., AND J. F. GRIFFITHS. 1964. *Environmental Measurement and Interpretation*. Van Nostrand Reinhold, New York.

RAND, M. C., A. E. GREENBERG, AND M. J. TARAS (EDS.). 1975. *Standard Methods for the Examination of Water and Waste Water,* 14th ed. American Public Health Association, Washington, D.C.

SCHLIEFER, C. (ED.). 1968. *Research Methods in Marine Biology*. University of Washington Press, Seattle.

SCHWOERKEL, J. 1970. *Methods of Hydrobiology*. Pergamon Press, Elmsford, N.Y.

SOIL SURVEY STAFF. 1971. *Handbook of Soil Survey Investigations*. U.S. Government Printing Office, Washington, D.C.

STRICKLAND, J. D. H., AND T. R. PARSON. 1972. *A Practical Handbook of Seawater Analysis*. Bulletin 167. Fisheries Research Board of Canada, Toronto.

STROBBE, M. A. 1972. *Environmental Science Laboratory Manual*. Mosby, St. Louis.

WADSWORTH, R. M. (ED.). 1968. *The Measurement of Environmental Factors in Terrestrial Ecology*. Blackwell, Oxford.

WELCH, P. S. 1948. *Limnological Methods*. McGraw-Hill–Blakiston, New York.

WHITE, W., JR. (ED.). 1972. *North American Reference Encyclopedia of Ecology and Pollution*. North American, Philadelphia.

WOOD, R. D. 1975. *Hydrobotanical Methods*. University Park Press, Baltimore.

## Papers

AVERY, G. 1957. Foresters' guide to aerial photo interpretation. *Southern Forest Exp. Sta. Occ. Paper,* 156.

DORE, W. G. 1958. A simple chemical light meter. *Ecology,* 39:151–152.

GERKING, S. D. 1959. A method of sampling the littoral macrofauna and its application. *Ecology,* 38:219–226.

HALEY, J. L., AND J. K. BRIERLEY. 1953. A method of estimation of oxygen and carbon dioxide concentrations in the litter layer of beech woods. *J. Ecology,* 41:385–387.

JOHNSTON, R. 1964. Recent advances in the estimation of salinity. In H. Barnes (ed.), *Oceanography and Marine Biology,* ann. rev. no. 2. Allen and Unwin, London.

MARQUIS, D. A., AND G. YELENOSKY. 1962. A chemical light meter for forest research. *Northeast Forest Exp. Sta. Paper,* 165. U.S. Forest Service, Broomall, Pa. 19008.

PRITCHARD, D. W., AND W. V. BURT. 1951. An inexpensive and rapid technique for obtaining current profiles in estuarine waters. *J. Marine Res.,* 10:180–189.

ROBINS, C. R., AND R. W. CRAWFORD. 1954. A short, accurate method for estimating the volume of stream flow. *J. Wildlife Management,* 18:366–369.

WALLIS, G. W., AND S. A. WILDE. 1957. Rapid method for the determination of carbon dioxide evolved from forest soils. *Ecology,* 38:359–361.

# *Glossary*

**A horizon**   surface stratum of mineral soil characterized by maximum accumulation of organic matter, maximum biological activity, and loss of such materials as iron, aluminum oxides, and clays

**abyssal**   relating to bottom waters of oceans, usually below 1000 m

**acclimation**   alteration through long-term exposure to certain conditions of physiological rate or other capacity to perform a function

**acclimatization**   changes or differences in a physiological state that appear after exposure to different natural environments

**adaptation**   genetically determined characteristic (behavioral, morphological, physiological) that improves an organism's ability to survive and successfully reproduce under prevailing environmental conditions.

**adaptative radiation**   evolution from a common ancestor of divergent forms adapted to distinct ways of life

**aggressive mimicry**   resemblance of a predator or parasite to a harmless species to deceive potential prey

**agonistic behavior**   all types of hostile responses to other organisms ranging from overt attack to overt escape

**alfisol**   soil characterized by an accumulation of iron and aluminum in lower or B horizon

**allele**   one of a pair of characters that occupy the same relative position or locus on homologous chromosome and appear alternatively at that locus

**allelopathy**   effect of metabolic products of plants (excluding microorganisms) on the growth and development of other nearby plants

**allochthonous**   food material reaching an aquatic community from outside the community in the form of organic detritus

**allopatric**   having different areas of geographical distribution; possessing nonoverlapping ranges

**allopatric speciation**   separation of a population into two or more species by reproductive isolation brought about by geographical separation of the subpopulations

**alluvial soil**   soil developing from recent alluvium (material deposited by running water); exhibits no horizon development; typical of flood plains

**amensalism**   relationship between two species in which one is inhibited or harmed by the presence of another

**anadromous fish**   fish that typically inhabit seas or lakes but ascend streams to spawn

**anaerobic**   adapted to environmental conditions devoid of oxygen

**antibiotic**   substance produced by a living organism which is toxic to organisms of different species

**aposematism**   possession of warning coloration; conspicuous markings on animals that are poisonous, distasteful, or possess some unpleasant defensive mechanism

**assimilation**   transformation or incorporation of a substance by organisms; absorption and conversion of energy and nutrient intake into constitutents of an organism

**association**   a natural unit of vegetation characterized by a relatively uniform species composition and often dominated by particular species

**aufwuchs**   community of plants and animals attached to or moving about on submerged surfaces; also called *periphyton,* but this term more specifically applies to organisms attached to submerged plant stems and leaves

**autecology**   ecology of individuals species in response to environmental conditions

**autotrophy**   ability of an organism to produce organic material from inorganic chemicals and some source of energy

**B horizon**   soil stratum beneath the A horizon characterized by an accumulation of silica, clay, and iron and aluminum oxides and possessing blocky or prismatic structure

**balanced polymorphism**   maintenance of more than one allele in a population; results from the selective superiority of the heterozygote over the homozygote

**Batesian mimicry**   resemblance of a palatable species, the mimic, to an unpalatable species, the model

**benthos**   animals and plants living on the bottom of a lake or sea from high water mark to the deepest depths

**biennial**   plant that requires two years to complete a life cycle, with vegetative growth the first year and reproductive growth (flowers and seeds) the second

**biomass**   weight of living material, usually expressed as dry weight per unit area

**biome**   major regional ecological community of plants and animals; usually corresponds to plant ecologists' and European ecologists' classification of plant formation and classification of life zones

**biosphere**   thin layer surrounding the earth in which all living organisms exist

**bog**   wetland ecosystem characterized by accumulation of peat, acid conditions, and dominance of sphagnum moss

**boreal forest**   needle-leafed evergreen or coniferous forest bordering subpolar regions; also called tiaga.

**bryophyte**   member of the division in the plant kingdom of nonflowering plants comprising mosses (Musci), liverworts (Hepaticae), and hornworts (Anthocerotae)

**C horizon**   soil stratum beneath the solum (A and B horizons) relatively little affected by biological activity and soil forming process

**calcification**   process of soil formation characterized by accumulation of calcium in lower horizons

*calorie* amount of heat needed to raise 1 gram of water 1° C, usually from 15° C to 16° C

*carnivore* organism that feeds on animal tissue; taxonomically, a member of the order Carnivora

*carr* vegetation dominated by alder and willow and occupying eutrophic peat

*carrying capacity (K)* number of individual organisms the resources of a given area can support usually through the most unfavorable period of the year

*catadromous fish* fish that feed and grow in fresh water, but return to the sea to spawn

*chaparral* vegetation consisting of broad-leafed evergreen shrubs found in regions with a mediterranean climate of hot, dry summers and mild, wet winters

*character convergence* evolution of similar appearance or behavior in unrelated species

*character divergence* evolution of behavioral, physiological, or morphological differences among species occupying the same area brought about by selective pressures of competition

*circadian rhythm* endogenous rhythm of physiological or behavioral activity of approximately 24 hours duration

*climax* stable end community of succession that is capable of self-perpetuation under prevailing environmental conditions

*cline* gradual change in population characteristics over a geographical area, usually associated with changes in environmental conditions

*coevolution* joint evolution of two or more noninterbreeding species that have a close ecological relationship; through reciprocal selective pressures the evolution of one species in the relationship is partially dependent on the evolution of the other

*coexistence* two or more species living together in the same habitat, usually with some form of competitive interaction

*commensalism* relationship between species which is beneficial to one, but neutral or of no benefit to the other

*community* a group of interacting plants and animals inhabiting a given area

*compensation point* point at which photosynthesis and respiration balance each other so that net production is 0; in aquatic systems, usually the depth of light penetration at which oxygen utilized in respiration equals oxygen produced by photosynthesis

*competition* common use of a resource in short supply by a number of organisms of the same or different species; also, common use of a resource, regardless of supply, resulting in harm to one organism by another seeking the resource

*competitive exclusion* hypothesis that states that when two or more species coexist using the same resource, one must displace or exclude the other

*continuum index* measure of a position of a community on a gradient defined by species composition

*convergent evolution* development of similar characteristics in different species living in different areas but under similar environmental conditions

*critical thermal maximum* temperature at which an animal's capacity to move becomes so reduced it cannot escape from thermal conditions that will lead to death

*cryptic coloration* coloration of animals that makes them resemble or blend into their habitat or background

*cyclic replacement* type of succession in which the sequence of seral stages is repeated by imposition of some disturbance so that the sere never arrives at a climax or stable state

*death rate* percentage of individuals in a population dying in a specified time interval

*deciduous* (of leaves) shed during a certain season (winter in temperate regions, dry season in the tropics); (of trees) having deciduous parts

*decomposer* organism that obtains energy from breakdown of dead organic matter to more simple substances; most precisely refers to bacteria and fungi

*deme* local population, or interbreeding group within a larger population

*density dependent* varying in relation to population density

*density independent* unaffected by population density

*detritivore* organism that feeds on dead organic matter; usually applies to detritus-feeding organisms other than bacteria and fungi

*detritus* fresh to partly decomposed plant and animal matter

*diapause* period of arrested growth and activity in which metabolism is greatly decreased; usually seasonal in nature and common among insects

*dimorphism* existing in two forms, two color forms, two sexes, etc.

*diploid* having chromosomes in homologous pairs or twice the haploid number of chromosomes

*dispersal* leaving an area of birth of activity for another area

*dispersion* distribution of organism in given area

*disruptive selection* selection in which two extreme phenotypes in the population leave more offspring than the intermediate phenotype which has lower fitness

*diversity* abundance in number of species in a given location

*dominance* (ecological) control within a community over environmental conditions influencing associated species by condition one or several species, plant or animal, because of their number, density, or growth form; (social) behavioral, hierarchical order in a population that gives high-ranking individuals priority of access to essential requirements

**dominant** population possessing ecological dominance in a given community and thereby governing type and abundance of other species in the community

**dormant** state of cessation of growth and suspended biological activity during which life is maintained

**dystrophic** term applied to a body of water with a high content of humic organic matter, often with high littoral productivity and low planktonic productivity

**ecological efficiency** percentage of biomass produced by one trophic level that is incorporated into biomass of the next highest trophic level

**ecological release** expansion of habitat or increase in food availability resulting from release of a species from interspecific competition

**ecotone** transition zone between two structurally different communities; the edge

**ecotype** subspecies or race adapted to a particular set of environmental conditions

**ectothermy** determination of body temperature primarily by external thermal conditions

**edaphic** relating to soil

**egestion** elimination of undigested food material

**endemic** restricted to a given region

**endothermy** regulation of body temperature by internal heat production; allows maintenance of appreciable difference between body temperature and external temperature

**environment** total surroundings of an organism including other plants and animals including those of its own kind.

**ephemeral** organism with a short life cycle; lasting only a season or less

**epilimnion** upper, warm, oxygen-rich zone of a lake that lies above the thermocline

**epiphyte** organism that lives wholly on the surface of plants, deriving support but not nutrients from the plants

**equilibrium species** species whose population exists in equilibrium with resources and at a stable density

**equitability** evenness of distribution of species abundance patterns, the maximum equitability being the same number of individuals in all species

**ethology** study of animal behavior

**euphotic zone** surface layer of water to the depth of light penetration where photosynthetic production equals respiration

**eutrophic** term applied to body of water with high nutrient content and high productivity

**eutrophication** nutrient enrichment of a body of water; called cultural eutrophication when accelerated by introduction of massive amounts of nutrients by human activity

**evapotranspiration** loss of moisture by evaporation from land and water surfaces and by transpiration from plants

**evolution** change in gene frequency through time resulting from natural selection and producing cumulative changes in characteristics of a population

**fecundity** potential ability of an organism to produce eggs or young

**fen** wetlands dominated by sedges in which peat accumulates

**fitness** genetic contribution by an individual's descendents to future generations

**fixation** process in soil by which certain chemical elements essential for plant growth are converted from soluble or exchangeable form to a less soluble or nonexchangeable form

**flux** flow of energy from a source to a sink or receiver

**food chain** movement of energy and nutrients from one feeding group of organisms to another in a series that begins with plants and ends with carnivores

**food web** interlocking pattern formed by a series of interconnecting food chains

**forb** herbaceous plant other than grass, sedge, or rush

**formation** classification of vegetation based on dominant life forms

**free-running cycle** length of a circadian rhythm in absence of external time cues

**frugivore** organism that feeds on fruits

**fugitive species** species characteristic of temporary habitats

**functional response** change in rate of exploitation of a prey species by a predator in relation to changing prey density

**gap phase replacement** successional development in small disturbed areas within a stable plant community; filling in of a space left by a disturbance, not necessarily by the species eliminated by the disturbance

**gene** unit of material of inheritance.

**gene flow** exchange of genetic material between populations

**genetic feedback** evolutionary response of a population to adaptations of predators, parasites, or competitors

**genotype** genetic constitution of an organism

**geometric rate of increase** factor by which size of a population increases over a period of time

**gley soil** soil developed under conditions of poor drainage resulting in reduction of iron and other elements and in gray colors and mottles

**granivore** organism that feeds on seeds

**gross production** production, most precisely, photosynthetic production, before respiration losses are deducted

**growth form** morphological category of plants, such as tree, shrub, vine

**group selection** elimination of one group of individuals by another group of individuals possessing superior genetic traits (a not widely accepted hypothesis)

**guild**   group of species that share a resource in a community

**habitat**   place where a plant or animal lives

**haploid**   having a single set of unpaired chromosomes in each cell nucleus

**herbivore**   organism that feeds on plant tissue

**heterotrophic**   requiring a supply of organic matter or food from the environment

**heterozygous**   containing two different alleles of a gene, one from each parent, at the corresponding loci of a pair of chromosomes

**hibernation**   winter dormancy in animals characterized by a great decrease in metabolism

**histosol**   soil characterized by high organic matter content

**homeostasis**   maintenance of nearly constant conditions in function of an organism or in interaction among individuals in a population

**homeothermy**   regulation of body temperature by physiological means

**home range**   area over which an animal ranges throughout the year

**homozygous**   containing two identical alleles of a gene at the corresponding loci of a pair of chromosomes

**host**   organism that provides food or other benefit to another organism of a different species; usually refers to an organism exploited by a parasite

**humus**   organic material derived from partial decay of plant and animal material

**hybrid**   plant or animal resulting from a cross between genetically different parents

**hyperthermia**   rise in body temperature to reduce thermal differences between an animal and a hot environment, thus reducing rate of heat flow into body

**hypha**   filament of a fungus thallus or vegetative body

**hypolimnion**   cold, oxygen-poor zone of a lake that lies below the thermocline

**immobilization**   conversion of an element from inorganic to organic form in microbial or plant tissue rendering nutrient relatively unavailable to other organisms

**importance value**   sum of relative density, relative dominance, and relative frequency of a species in a community

**imprinting**   type of rapid learning at a particular and early stage of development in which the individual learns identifying characteristics of another individual or object

**incipient lethal temperature**   temperature at which a stated fraction of a population of poikilothermic animals (usually 50 percent) will die when brought rapidly to it from a different temperature

**infralittoral**   region below the littoral region of the sea

**innate capacity for increase** (r)   intrinsic growth rate of a population under ideal conditions

without inhibition from competition

**instar**   form of insect or arthropod between successive molts

**interspecific**   between individuals of different species

**intraspecific**   between individuals of the same species

**inversion**   (genetic) reversal of part of a chromosome so that genes within that part lie in reverse order; (meteorological) increase rather decrease in air temperature with height caused by radiational cooling of the earth (radiational inversion) or by compression and consequent heating of subsiding air masses from high pressure areas (subsidence inversion)

**krumholtz**   stunted form of trees characteristic of transition zone between alpine tundra and subalpine coniferous forest

**K-selection**   selection under carrying capacity conditions

**laterization**   soil-forming process in hot, humid climates characterized by intense oxidation resulting in loss of bases and in a deeply weathered soil composed of silica, sesquioxides of iron and aluminum, clays, and residual quartz

**lentic**   pertaining to standing water, as lakes and ponds

**life table**   tabulation of mortality and survivorship schedule of a population

**life zone**   major area of plant and animal life equivalent to a biome; transcontinental region or belt characterized by particular plants and animals and distinguished by temperature differences; applies best to mountainous regions where temperature changes accompany changes in altitude

**limnetic zone**   shallow water zone of lake or sea in which light penetrates to the bottom

**lithosol**   soil showing little or no evidence of soil development and consisting mainly of partly weathered mass of rock fragments or nearly barren rock

**littoral**   shallow water zone of lake in which light penetrates to the bottom permitting vegetative growth; also shore zone of tidal water between high water and low water marks

**logistic curve**   S-shaped curve of population growth which slows at first, steepens, and then flattens out at asymptote; determined by carrying capacity

**log-normal distribution**   frequency distribution of species abundance in which the $x$ axis is expressed in a logarithmic scale; with the $x$ axis representing the number of individuals in the sample and the $y$ axis the number of species

**lotic**   pertaining to flowing water

**marl**   earthy, unconsolidated deposit formed in fresh-water lakes that consists chiefly of calcium carbonate mixed with clay and other impurities

*meiosis*   two successive divisions by gametic cells, with only one duplication of chromosomes so that the number of chromosomes in daughter cells is one-half the diploid number

*mesic*   moderately moist habitat

*metalimnion*   transition zone in lake between hypolimnion and epilimnion; region of rapid temperature decline

*mimicry*   resemblance of one organism to another or to an object in the environment evolved to deceive predators

*mineralization*   microbial breakdown of humus and other organic matter in soil to inorganic substances

*mire*   wetland characterized by an accumulation of peat

*mitosis*   cell division involving chromosome duplication resulting in two daughter cells with full complement of chromosomes, genetically the same as parent cells

*mollisol*   soil formed by calcification characterized by accumulation of calcium carbonate in lower horizons and high organic content in upper horizons

*monogamy*   mating of an animal with only one member of the opposite sex at a time

*montane*   pertaining to mountains

*mor*   type of forest humus layer of unincorporated organic matter usually matted or compacted or both and distinct from mineral soil; poor in bases and acid in reaction

*morphology*   study of the form of organisms

*mull*   humus which contains appreciable amounts of mineral bases and forms a humus-rich layer of forested soil consisting of mixed organic and mineral matter; blends into upper mineral layer without abrupt changes in soil characteristics

*Mullerian mimicry*   resemblance of two or more conspicuously marked distasteful species which increases predator avoidance

*mutation*   transmissable change in structure of gene or chromosome

*mutualism*   relationship between two species in which both benefit

*mycelium*   mass of hyphae that make up the vegetative portion of fungus

*mycorrhiza*   association of fungus with roots of higher plants which improves the plants' uptake of nutrients from the soil

*natural selection*   differential reproduction and survival of individuals that results in elimination of maladaptive traits from a population

*net above-ground production*   accumulation of biomass in above-ground parts of plants over a given period of time

*net production*   accumulation of total biomass over a given period of time; energy left over after respiration deducted from gross production

*net reproductive rate*   number of young a female can be expected to produce during a lifetime

*niche*   functional role of a species in the community including activities and relationships

*numerical response*   change in size of a population of predators in response to change in density of its prey

*oligotrophic*   term applied to a body of water low in nutrients and in productivity

*opportunistic species*   organisms able to exploit temporary habitats or conditions

*optimum yield*   amount of material that can be removed from a population that will result in production of maximum amount of biomass on a sustained yield

*oscillation*   regular fluctuations with a fixed cycle in a population

*overturn*   vertical mixing of layers in a body of water brought about by seasonal changes in temperature

*oxisol*   soil formed by process of laterization

*paleoecology*   study of ecology of past communities by means of fossil record

*parapatric*   having ranges coming into contact but not overlapping by much more than the dispersal range of an individual within its lifetime

*parapatric speciation*   separation into different species of a parent population and a previously isolated subpopulation with a fixed gene mutation; if subpopulation expands range and comes in contact with parent population, hybrids are selected against, producing a tension zone that prevents each population from penetrating the other, ultimately resulting in formation of a new species

*parasitism*   relationship between two species in which one benefits while the other is harmed (although not usually killed directly)

*peat*   unconsolidated material consisting of undecomposed and only slightly decomposed organic matter under conditions of excessive moisture

*pedalfer*   podzolic soil possessing a layer of iron accumulation (hardpan) that impedes free circulation of air and water

*pelagic*   pertaining to the sea

*permafrost*   permanently frozen soil

*perturbation*   disturbance to a system

*phenology*   study of seasonal changes in plant and animal life and the relationship of these changes to weather and climate

*phenotype*   physical expression of a characteristic of an organism as determined by genetic constitution and environment

*pheromone*   chemical substance released by an animal that influences behavior of others of the same species

*photoperiodism*   response of plants and animals to changes in relative duration of light and dark

*photosynthesis*   synthesis of carbohydrates from carbon dioxide and water by chlorophyll using light as energy and releasing oxygen as a by-product

**physiological longevity** maximum lifespan of an individual in a population under given environmental conditions

**phytoplankton** small, floating plant life in aquatic ecosystems; planktonic plants

**plagioclimax** climax community maintained by a continuous human activity

**plankton** small, floating or weakly swimming plants and animals in fresh-water and marine ecosystems

**podzolization** soil-forming process resulting from acid leaching of the A horizon and accumulation of iron, aluminum, silica, and clays in lower horizon

**poikilothermy** variation of body temperature with external conditions

**polyandry** mating of one female with several males

**polygyny** mating of one male with several females

**polymorphism** occurrence of more than one distinct form of a species

**polyploid** having three or more times the haploid number of chromosomes

**primary production** production by green plants

**production** amount of energy formed by an individual, population, or community per unit time

**pycnocline** layer of water that exhibits a rapid change in density

**r-selection** selection under low population densities

**recombination** exchange of genetic material resulting from independent assortment of chromosomes and their genes during gamete production followed by a random mix of different sets of genes at fertilization

**recruitment** addition by reproduction of new individuals to a population

**respiration** metabolic assimilation of oxygen accompanied by production of carbon dioxide and water, release of energy, and breaking down of organic compounds

**rhizobia** bacteria capable of living mutualistically with higher plants

**rhizome** underground stem bearing buds in axils of reduced scale-like leaves

**riparian** along banks of rivers and streams; riverbank forests are often called gallery forests

**saprophyte** plant that obtains food from dead or decaying organic matter

**sclerophyll** woody plant with hard, leathery, evergreen leaves that prevent moisture loss

**secondary production** production by consumer organisms

**senescense** process of aging

**sere** series of stages that follow one another in succession

**serotinous cones** cones of pine that remain on the tree several years and require the heat of fire to open them and release the seeds

**sessile** not free to move about; permanently attached to a substrate

**sibling species** species with similar appearance but unable to interbreed

**sigmoid curve** S-shaped curve of logistic growth

**site** combination of biotic, climatic, and soil conditions that determine an area's capacity to produce vegetation

**soil horizon** developmental layer in the soil, with its own characteristics of thickness, color, texture, structure, acidity, nutrient concentration, and the like

**soil profile** distinctive layering of horizons in the soil

**soil series** basic unit of soil classification consisting of soils which are essentially alike in all major profile characteristics except texture of the A horizon

**soil type** lowest unit in the natural system of soil classification consisting of soils which are alike in all characteristics including texture of the A horizon

**speciation** separation of a population into two or more reproductively isolated populations

**species diversity** measurement which relates density of organisms of each type present in a habitat to the number of species in a habitat

**species packing** increase in species diversity within a comparatively narrow range of resource variation

**spodosol** soil characterized by the presence of a horizon in which organic matter and amorphous oxides of aluminum and iron have precipitated; includes podzols

**stability** ability of a system to resist change or to recover rapidly after a disturbance; absence of fluctuations in a population

**stable age distribution** constant proportion of individuals of various age classes in a population through population changes

**stand** unit of vegetation that is essentially homogeneous in all layers and differs from adjacent types qualitatively and quantitatively

**stratification** division of an aquatic or terrestrial community into distinguishable layers on the basis of temperature, moisture, light, vegetative structure, and other such factors creating zones for different plant and animal types

**sublittoral** lower division of sea from about 40 to 60 m to below 200 m

**subspecies** geographical unit of a species population distinguishable by certain morphological, behavioral, or physiological characteristics

**succession** replacement of one community by another; often progresses to a stable terminal community called the climax

**sustained yield** yield per unit time from an exploited population equal to production per unit time

**swamp** wooded wetlands in which water is near or above ground level

**symbiosis** living together of two or more species

**sympatric** living in the same area; usually refers to overlapping populations

**sympatric speciation** speciation without geographical isolation

**synecology** study of groups of organisms in relation to their environment; community ecology

**territory** area defended by an animal; varies among animals according to social behavior, social organization, and resource requirements of different species

**thermocline** layer in a thermally stratified body of water in which temperature changes rapidly relative to the remainder of the body

**thermogenesis** increase in production of metabolic body heat to counteract loss of heat to a colder environment

**tiaga** needle-leafed coniferous forest bordering the northern subpolar regions; boreal forest

**translocation** transport of materials within a plant; absorption of minerals from soil into roots and their movement throughout the plant

**transpiration** loss of water vapor by land plants

**trophic** related to feeding

**trophic level** functional classification of organisms in an ecosystem according to feeding relationships from first level autotrophs through succeeding levels of herbivores and carnivores

**tundra** area in arctic and alpine (high mountains) regions characterized by bare ground, absence of trees, and growth of mosses, lichens, sedges, forbs and low shrubs

**upwelling** upward or vertical movement of water currents in oceans that bring up nutrients from the deep

**vagile** free to move about

**wilting point** moisture content of soil on an oven-dry basis at which plants wilt and fail to recover their turgidity when placed in dark, humid atmosphere

**xerophyte** plants adapted to life in a dry or physiologically dry (saline) habitat

**zooplankton** floating or weakly swimming animals in fresh-water and marine ecosystems; planktonic animals

# General references

## AIDS TO IDENTIFICATION

*The following is a list of recommended guides to various groups of organisms. The list is not complete, but the volumes represent one man's opinion about what are the best and most acceptable guides to plants and animals. The regional works included are those applicable to a wider region than the state borne in the title. Except in rare instances, technical monographs are not included.*

## Lower plants

BODENBERG, E. T. 1954. *Mosses: A New Approach to the Identification of Common Species.* Burgess, Minneapolis.

COBB, B. 1956. *A Field Guide to the Ferns.* Houghton Mifflin, Boston.

CONRAD, H. J. 1956. *How to Know the Mosses and Liverworts.* Brown, Dubuque, Iowa.

DAWSON, E. Y. 1956. *How to Know the Seaweeds.* Brown, Dubuque, Iowa.

FINK, B. 1935. *The Lichen Flora of the United States.* University of Michigan Press, Ann Arbor.

FRYE, T., AND L. CLARK. 1937–1947. *Hepaticae of North America,* 2 vols. University of Washington Press, Seattle.

GRAHAM, V. O. 1970. *Mushrooms of the Great Lakes Region.* Reprint, Dover, New York.

GROOT, A. J. 1947. *Mosses with a Hand Lens,* 4th ed. Published by the author, 1 Vine Street, New Brighton, Staten Island, N.Y.

HALE, M. E. 1969. *How to Know the Lichens.* Brown, Dubuque, Iowa.

KAUFFMAN, C. H. 1971. *The Gilled Mushrooms (Agaricaceae) of Michigan and the Great Lakes Region,* 2 vols. Reprint of 1918 ed., Dover, New York.

KREIGER, L. C. C. 1967. *The Mushroom Handbook.* Reprint of 1947 ed., Dover, New York.

MILLER, O. K., JR. 1975. *Mushrooms of North America,* 2nd rev. ed. Dutton, New York.

PARSONS, F. T. 1961. *How to Know the Ferns: A Guide to Names, Haunts, and Habits.* Reprint, Dover, New York.

PRESCOTT, G. W. 1964. *How to Know the Fresh Water Algae.* Brown, Dubuque, Iowa.

SHAVER, J. M. 1970. *Ferns of the Eastern Central States, with Special Reference to Tennessee,* Reprint of 1954 ed., Dover, New York.

SMITH, A. H. 1951. *Puffballs and Their Allies in Michigan.* University of Michigan Press, Ann Arbor.

SMITH, A. H. 1963. *The Mushroom Hunter's Field Guide,* rev. ed. University of Michigan Press, Ann Arbor.

SMITH, G. 1950. *The Fresh Water Algae of the United States.* McGraw-Hill, New York.

SMITH, G. 1951. *Manual of Phycology: An Introduction to the Algae and Their Biology.* Chronica Botanica, Waltham, Mass.

SMITH, H. V., AND H. ALEXANDER. 1973. *The Non-gilled Fleshy Fungi.* Brown, Dubuque, Iowa.

TAYLOR, W. R. 1937. *Marine Algae of the Northeastern Coast of North America.* University of Michigan Science Series 13. University of Michigan Press, Ann Arbor.

TAYLOR, W. R. 1960. *Marine Algae of the Eastern Tropical and Subtropical Coasts of the Americas.* University of Michigan Press, Ann Arbor.

THOMAS, W. C. 1936. *Field Book of Common Mushrooms,* Putnam, New York.

WATSON, E. V. 1955. *British Mosses and Liverworts,* Cambridge University Press, London. (Most species treated also occur in northeastern North America.)

WHERRY. E. T. 1942. *Guide to Eastern Ferns,* 2d ed. Science Press, Lancaster, Pa.

## Grasses and wildflowers

BRITTON, N. L., AND J. N. ROSE. 1937. *The Cactaceae,* 2 vols. Reprint, Dover, New York.

CRAIGHEAD, J., F. C. CRAIGHEAD, AND R. J. DAVIS. 1963. *Field Guide to Rocky Mountain Wildflowers.* Houghton Mifflin, Boston.

CUTHBERT, M. 1943. *How to Know the Spring Wildflowers.* Brown, Dubuque, Iowa.

CUTHBERT, M. 1948. *How to Know the Fall Wildflowers.* Brown, Dubuque, Iowa.

DANA, WM. STARR. 1962. *How to Know the Wildflowers.* Rev. modernized ed. of 1900 volume. Dover, New York.

DAWSON, E. Y. 1963. *How to Know the Cacti.* Brown, Dubuque, Iowa.

DAYTON, W. A. 1960. *Notes on Western Range Forbs.* Handbook 161. U.S. Department of Agriculture, Washington, D.C.

DEGENER, O. 1946–1957. *New Illustrated Flora of the Hawaiian Islands.* Published by the author, Makuleia Beach, Waialua, Oahu, Hawaii.

FASSETT, N. C. 1940. *A Manual of Aquatic Plants.* McGraw-Hill, New York.

FERNALD, M. L. 1950. *Gray's Manual of Botany,* 8th ed. American Book, New York.

GLEASON, H. A. 1952. *The New Britton and Brown Illustrated Flora of the Northeastern United States and Adjacent Canada,* 3 vols. New York Botanical Garden, New York.

HITCHCOCK, A. S. 1950. *Manual of Grasses of the United States,* rev. Agnes Chase. Reprint, Dover, New York.

HITCHCOCK, C. L., A. CRONQUIST, M. OWNBEY, AND J. W. THOMPSON. 1964. *Vascular Plants of the Pacific Northwest,* 5 vols. University of Washington Press, Seattle.

HOTCHKISS, N. 1972. *Common Marsh, Underwater and Floating-leaved Plants.* Dover, New York.

HUTLEN, E. 1960. *Flora of the Aleutian Islands,* 2d ed. Hafner, New York.

JAQUES, H. E. 1949. *Plant Families, How to Know Them.* Brown, Dubuque, Iowa.

JAQUES, H. E. 1972. *How to Know the Weeds.* Brown, Dubuque, Iowa.

KUMMER, A. P. 1951. *Weed Seedlings.* University of Chicago Press, Chicago.

MARTIN, A. C., H. S. ZIN, AND A. L. NELSON. 1951. *American Wildlife and Food Plants.* McGraw-Hill, New York: Dover, New York, 1961.

MATTHEWS, F. S. 1955. *Field Book of American Wildflowers,* rev. ed. Norman Taylor. Putnam, New York.

MUENSCHER, W. C. 1944. *Aquatic Plants of the United States.* Comstock, Ithaca, N.Y.

MUENSCHER, W. C. 1955. *Weeds.* Macmillan, New York.

PETERSON, R. T., AND M. MCKENNY. 1968. *A Fieldguide to the Wildflowers of Northeastern and Northcentral North America.* Houghton Mifflin, Boston.

POHL, R. W. 1954. *How to Know the Grasses.* Brown, Dubuque, Iowa.

PORSILD, A. E. 1957. Illustrated flora of the Canadian Arctic Archipelago. *Nat. Museum Can. Bull.,* 146.

PRESCOTT, G. W. 1969. *How to Know the Aquatic Plants.* Brown, Dubuque, Iowa.

RICKETT, H. W. 1966-1971. *Wild Flowers of the United States.* Vol. 1, *The Northeastern States;* 2, *The Southeastern States;* 3, *Texas;* 4, *The Southwestern States;* 5, *The Northwestern States;* 6, *The Central Mountains and Plains.* McGraw Hill, New York.

RYDBERG, P. A. 1969. *Flora of the Rocky Mountains and Adjacent Plains.* Hafner, New York.

RYDBERG, P. A. 1971. *Flora of the Prairies and Plains of Central North America.* Reprint of 1932 ed., Dover, New York.

U. S. DEPARTMENT OF AGRICULTURE. 1971. *Common Weeds of the United States.* Reprint of *Selected Weeds of the U.S.,* 1970, Dover, New York.

WHERRY, E. T. 1948. *The Wildflower Guide.* Doubleday, Garden City, N. Y.

WIGGANS, I. L., AND J. H. THOMAS. 1962. *Flora of the Alaskan Arctic Slope.* University of Toronto Press, Toronto.

## Trees and shrubs

BAERS, H. 1955. *How to Know the Western Trees.* Brown, Dubuque, Iowa.

BENSON, L. D., AND R. A. DARROW. 1945. *A Manual of Southwestern Desert Trees and Shrubs.* University of Arizona Press, Tucson.

BROCKMAN, C. F. 1968. *Trees of North America.* Golden, New York.

GRAVES, A. H. 1956. *Illustrated Guide to Trees and Shrubs.* Harper & Row, New York.

GRIMM, W. C. 1960. *Recognizing Trees.* Stackpole Harrisburg, Pa.

GRIMM, W. C. 1966. *How to Recognize Shrubs (Recognizing Native Shrubs).* Stackpole, Harrisburg, Pa.; Castle, New York.

HARLOW, W. M. 1957. *Trees of Eastern and Central United States and Canada; Fruit Key and Twig Key to Trees and Shrubs.* Reprint, Dover, New York.

HARRAR, E. S., AND J. G. HARRAR. 1946. *Guide to Southern Trees.* Reprint, Dover, New York.

HAYES, DORIS W. 1960. *Key to Important Woody Plants of Eastern Oregon and Washington.* Handbook 148. U.S. Department of Agriculture, Washington, D.C.

HOSIE, R. C. 1973. *Native Trees of Canada,* 7th ed. Canadian Forestry Service, Department of Environment, Ottawa.

JAQUES, H. E. 1946. *How to Know the Trees,* rev. ed. Brown, Dubuque, Iowa.

KEELER, H. 1969. *Our Northern Shrubs and How To Identify Them.* Reprint, Dover, New York.

LITTLE, E. L., JR., AND F. H. WADSWORTH, 1964. *Common Trees of Puerto Rico and the Virgin Islands.* Forest Service Agricultural Handbook 249. U.S. Department of Agriculture, Washington D.C.

LITTLE, E. L., JR., R. O. WOODBURY, and F. H. WADSWORTH. 1974. *Trees of Puerto Rico and the Virgin Islands,* vol. 2. Forest Service Agricultural Handbook 449. U.S. Department of Agriculture, Washington, D.C.

MCMINN, H. E., AND E. MAINO. 1946. *An Illustrated Manual of Pacific Coast Trees.* University of California Press, Berkeley.

PEATTIE, D. C. 1953. *Natural History of Western Trees.* Houghton Mifflin, Boston.

PEATTIE, D. C. 1966. *Natural History of Trees of Eastern and Central North America.* Houghton Mifflin, Boston.

PETRIDIES, G. 1972. *A Field Guide to Trees and Shrubs.* Houghton Mifflin, Boston.

SARGENT, C. S. 1922. *Manual of Trees of North America,* 2 vols. Reprint, Dover, New York.

TRELEASE, W. 1967. *Winter Botany: An Identification Guide to Native Trees and Shrubs.* Reprint, Dover, New York.

VINES, R. A. 1963. *Trees, Shrubs and Woody Vines of the Southwest.* University of Texas Press, Austin.

## Invertebrates

### General

DAVIS, C. C. 1955. *The Marine and Fresh Water Plankton.* Michigan State University Press, East Lansing.

EDMONDSON, W. T. (ED.) 1959. *Fresh-water Biology,* 2d ed. Wiley, New York.

JOHNSON, M. E., AND H. J. SNOOK, 1967. *Seashore*

*Animals of the Pacific Coast.* Reprint, Dover, New York.

KLOTS, E. B. 1966. *New Field Book of Freshwater Life.* Putnam, New York.

LIGHT, S. F., R. I. SMITH, F. A. PITELKA, D. P. ABBOTT, AND F. M. NEESNER. 1957. *Intertidal Invertebrates of the Central California Coast.* University of California Press, Berkeley.

MINER, R. W. 1950. *Field Book of Seashore Life.* Putnam, New York.

NEEDHAM, J. G., AND P. R. NEEDHAM. 1962. *A Guide to the Study of Fresh Water Biology.* Holden-Day, San Francisco.

PENNAK, R. 1953. *Fresh Water Invertebrates of the United States.* Ronald Press, New York.

PRATT, H. S. 1953. *Manual of the Common Invertebrate Animals.* McGraw-Hill—Blakiston, New York.

### Protozoa

JAHN, T. L. 1949. *How to Know the Protozoa.* Brown, Dubuque, Iowa.

### Colenterata

SMITH, F. G. W. 1948. *Atlantic Reef Corals.* University of Miami Press, Coral Gables.

### Earthworms

EATON, T. H., JR. 1942. Earthworms of the northeastern United States. *J. Wash. Acad. Sci.,* 32:242-249

OLSON, H. W. 1928. The earthworms of Ohio. *Ohio Biol. Survey,* 4:45-90.

### Mollusca

ABBOTT, R. T. 1954. *American Seashells.* Van Nostrand Reinhold, New York.

BAKER, F. C. 1939. *Fieldbook of Illinois Land Snails, Manual 2.* Illinois Natural History Survey, Urbana.

BURCH, J. B. 1962. *How to Know the Eastern Land Snails.* Brown, Dubuque, Iowa.

KEEN, A. M. 1963. *Marine Molluscan Genera of Western North America.* Stanford University Press, Stanford, Calif.

KEEP, J. 1935. *West Coast Shells.* Rev. J. L. Bailey, Jr., Stanford University Press, Stanford, Calif.

MORRIS, P. A. 1966. *A Field Guide to the Shells of the Pacific Coast and Hawaii,* 2nd. ed. Houghton Mifflin, Boston.

MORRIS, P. A. 1973. *A Field Guide to the Shells of the Atlantic and Gulf Coasts and the West Indies,* 3rd ed. Ed. W. J. Clench. Houghton Mifflin, Boston.

OLIVER, A. P. H. 1975. *Guide to Shells.* Harper & Row/Quadrangle, New York.

PILSBRY, H. H. A. 1939-1948. *Land Mollusca of North America.* Monograph no. 3. Academy of Natural Sciences, Philadelphia, Pa.

### Insects

BORROR, D. J., AND R. E. WHITE. 1970. *A Field Guide to the Insects of North America.* Houghton Mifflin, Boston.

CHU, H. F. 1949. *How to Know the Immature Insects.* Brown, Dubuque, Iowa.

DILLON, E., AND L. S. DILLON. 1961. *A Manual of Common Beetles of Eastern North America.* Harper & Row, New York; reprint, Dover, New York.

EHRLICH, P. R. 1970. *How to Know the Butterflies.* Brown, Dubuque, Iowa.

ESSIG, E. O. 1958. *Insects and Mites of Western North America.* Macmillan, New York.

HELFER, J. R. 1963. *How to Know the Grasshoppers, Cockroaches, and Their Allies.* Brown, Dubuque, Iowa.

HOLLAND, W. J. 1949. *The Butterfly Book.* Doubleday, Garden City, N.Y.

HOLLAND, W. J. 1968. *The Moth Book.* Reprint, Dover, New York.

JAQUES, H. E. 1947. *How to Know the Insects.* Brown, Dubuque, Iowa.

JAQUES, H. E. 1951. *How to Know the Beetles.* Brown, Dubuque, Iowa.

KLOTS, A. K. 1951. *Field Guide to the Butterflies.* Houghton Mifflin, Boston.

LUTZ, F. E. 1935. *Field Book of Insects,* 3rd rev. ed. Putnam, New York.

NEEDHAM, J. G., AND M. J. WESTFALL, JR. 1955. *A Manual of the Dragonflies of North America.* University of California Press, Berkeley.

SWAIN, R. B. 1948. *The Insect Guide.* Doubleday, Garden City, N.Y.

USINGER, R. I. (ED.). 1956. *Aquatic Insects of California, with Keys to North American Genera and California Species.* University of California Press, Berkeley.

### Spiders

COMSTOCK, J. H. 1965. *The Spider Book,* 2nd rev. ed. Ed. W. J. Gertsch. Doubleday, Garden City, N.Y.

EMERTON, J. H. 1961. *The Common Spiders of the United States.* Reprint, Dover, New York.

KATSON, B. J., AND E. KATSON. 1972. *How to Know the Spiders.* Brown, Dubuque, Iowa.

## Vertebrates

### General

BLAIR, W. F., A. P. BLAIR, P. BRODKROB, F. R. CAGLE, AND G. A. MOORE. 1957. *Vertebrates of the United States.* McGraw-Hill, New York.

### Fishes

BIGELOW, H. B., AND W. C. SCHRODER. 1953. Fishes of the gulf coast of Maine. *U.S. Fish and Wildlife Service Fishery Bull,* No. 74.

BREDER, C. M. 1948. *Field Book of Marine Fishes.* Putnam, New York.

EDDY, S. 1957. *How to Know the Freshwater Fishes.* Brown, Dubuque, Iowa.

HARLAN. J. R., AND E. B. SPEAKER. 1956. *Iowa Fish and Fishing,* 3rd ed. Iowa Conservation Commission, Des Moines.

HUBBS, C., AND K. LAGLER. 1958. Fishes of the Great Lakes region. *Cranbook Institute of Science Bulletin,* no. 26 (Bloomfield, Ill.)

MCCLANE, A. J. 1978. *Field Guide to Freshwater Fishes of North America.* Holt, Rinehart and Winston, New York.

MCCLANE, A. J., 1978. *Field Guide to Saltwater Fishes of North America.* Holt, Rinehart and Winston, New York.

PERLMUTTER, A. 1961. *Guide to Marine Fishes.* New York University Press, New York.

SCOTT, W. B. 1967. *Freshwater Fishes of Eastern Canada.* University of Toronto Press, Toronto.

TRAUTMAN, M. B. 1957. *The Fishes of Ohio.* Ohio State University Press, Columbus.

### Amphibians and reptiles

BISHOP, S. 1943. *Handbook of Salamanders.* Comstock, Ithaca, N. Y.

CARR, A. 1952. *Handbook of Turtles.* Comstock, Ithaca, N.Y.

CONANT, R. 1958. *A Field Guide to the Reptiles and Amphibians of the United States and Canada East of the 100 Meridian.* Houghton Mifflin, Boston.

DITMARS, R. 1949. *Fieldbook of North American Snakes.* Doubleday, Garden City, N.Y.

ERNST, C. H., AND R. W. BARBOUR. 1972. *Turtles of the United States.* University of Kentucky Press, Lexington.

PICKWELL, G. 1972. *Amphibians and Reptiles of the Pacific States.* Reprint, Dover, New York.

POPE, C. 1939. *Turtles of the United States and Canada.* Knopf, New York.

SCHMIDT, K., AND D. D. DAVIS. 1941. *Field Book of Snakes of the United States and Canada,* Putnam, New York.

SMITH, H. M. 1946. *Handbook of Lizards of the United States and Canada.* Comstock, Ithaca, N.Y.

STEBBINS, R. C. 1954. *Amphibians and Reptiles of Western North America.* McGraw-Hill, New York.

STEBBINS, R. C. 1966. *A Field Guide to Western Reptiles and Amphibians.* Houghton Mifflin, Boston.

WRIGHT, A. H., AND A. WRIGHT. 1949. *Handbook of Frogs and Toads,* 3rd ed. Comstock, Ithaca, N.Y.

WRIGHT, A. H., AND A. A. WRIGHT. 1967. *Handbook of Snakes of the United States and Canada,* vols. 1, 2, Comstock, Ithaca, N.Y.

### Birds

BELLROSE, F. C. 1976. *Ducks, Geese, and Swans of North America.* Stackpole, Harrisburg, Pa.

BOND, J. 1971. *A Field Guide to the Birds of the West Indies.* Houghton Mifflin, Boston.

DE SCHAUENSEE, R. M. 1970. *A Guide to the Birds of South America.* Livingston, Wynnewood, Pa.

DE SCHAUENSEE, R. M., AND W. H. PHELPS, JR. 1978.

*A Guide to the Birds of Venezuela.* Princeton University Press, Princeton, N. J.

HARRISON, COLIN. 1978. *Field Guide to Nests, Eggs, and Nestlings of North American Birds.* Quadrangle, New York.

HARRISON, COLIN. 1976. *Field Guide to the Nests, Eggs, and Nestlings of British and European Birds.* Quadrangle, New York.

HARRISON, H. H. 1975. *A Field Guide to Bird Nests (Found East of the Mississippi River).* Houghton Mifflin, Boston.

HEADSTROM, R. 1979. *A Complete Field Guide to Nests in the United States.* Washburn, New York.

JAQUES, H. E. 1947. *How to Know the Land Birds.* Brown, Dubuque, Iowa.

PETERSON, R. T. 1947. *A Field Guide to the Birds.* Houghton Mifflin, Boston.

PETERSON, R. T. 1961. *A Field Guide to Western Birds.* Houghton Mifflin, Boston.

PETERSON, R. T. 1963. *A Field Guide to the Birds of Texas and Adjacent States.* Houghton Mifflin, Boston.

PETERSON, R. T., AND E. L. CHALIF. 1973. *A Field Guide to Mexican Birds.* Houghton Mifflin, Boston.

PETERSON, R. T., G. MOUNTFORT, AND P. A. D. HOLLOM, 1974. *A Field Guide to the Birds of Britain and Europe,* 3rd ed. Houghton Mifflin, Boston.

POUGH, R. H. 1946-1951. *Audubon Bird Guides.* Vol. 1., *Eastern Land Birds;* 2, *Water Birds,* Doubleday, Garden City, N.Y.

POUGH, R. H. 1957. *Audubon Western Bird Guide.* Doubleday, Garden City, N.Y.

RIDGELY, R. S. 1976. *A Guide to the Birds of Panama.* Princeton University Press, Princeton, N.J.

ROBBINS, C. S., B. BRUNN, AND H. S. ZIM. 1966. *Birds of North America.* Golden, New York.

SAUNDERS, A. A. 1959. *A Guide to Bird Song.* Doubleday, Garden City, N.Y.

SHARROCK, J. R. T. 1977. *The Atlas of Breeding Birds in Britain and Ireland.* Poyser, London.

*Note: For identification of bird song see the Peterson's Field Guide Records, Dover Records, and records produced by The Federation of Ontario Naturalists.*

### Mammals

BOOTH, E. S. 1961. *How to Know the Mammals.* Brown, Dubuque, Iowa.

BURT, W. H. 1957. *Mammals of the Great Lakes Region.* University of Michigan Press, Ann Arbor.

BURT, W. H., AND R. P. GROSSENHEIDER. 1963. *A Field Guide to the Mammals,* 2d ed. Houghton, Mifflin, Boston.

GODIN, A. J. 1977. *Wild Mammals of New England.* Johns Hopkins Press, Baltimore.

HALL, E. R., AND K. R. KELSON. 1959. *Mammals of North America.* Ronald Press, New York.

HAMILTON, W. J., JR. 1943. *The Mammals of*

*Eastern United States*. Comstock, Ithaca, N.Y.

INGLES, L. G. 1965. *Mammals of the Pacific States*. Stanford University Press, Stanford, Calif.

JACKSON, H. H. T. 1961. *The Mammals of Wisconsin*. University of Wisconsin Press, Madison.

LEOPOLD, A. S. 1959. *Wildlife in Mexico*. University of California Press, Berkeley.

MURIE, O. J. 1954. *A Field Guide to Animal Tracks*. Houghton Mifflin, Boston.

PALMER, R. S. 1954. *The Mammal Guide*. Doubleday, Garden City, N.Y.

SCHWARTZ, C. W., AND E. R. SCHWARTZ. 1960. *The Wild Mammals of Missouri*. University of Missouri Press, Columbia.

## *JOURNALS OF INTEREST TO ECOLOGISTS*

*The following list is by no means complete. The number of new journals has increased considerably during the past few years. The scope is international. A number of journals in foreign languages also publish many papers in English or have English abstracts. The date given is the first year of publication.*

*Acta Botanica Neerlandica,* 1952, Royal Botanical Society of the Netherlands. Wide diversity of papers, mostly in English. Important.

*American Birds,* 1946, formerly *Audubon Field Notes*. Devoted to the reporting of distribution and abundance of birds. Important information on continental trends in bird populations.

*American Journal of Botany,* 1914, Botanical Society of America. Technical, of interest chiefly to the professional.

*American Midland Naturalist,* 1909, University of Notre Dame, Notre Dame, Ind. Specializes in papers in field natural history.

*American Naturalist,* 1867 on, American Society of Naturalists. Concerned largely with morphology, evolution and physiology, but recent increase in papers on ecology.

*American Scientist,* 1913, Society of Sigma Xi. Contains in every volume several papers of importance to ecologists.

*American Zoologist,* 1916, Quarterly Publication of American Society of Zoologists. Most papers published originate in symposia of the society and its annual refresher course.

*Animal Behavior,* 1952, Baillière, London. Contains a wide range of papers on animal behavior.

*Annales Zoologici Fennici,* 1963, Societas Biologica Fennica Vanamo, Helsinki. An important journal; many papers on ecology.

*Annals of Botany,* 1887, Annals of Botany Co., Clarendon, Oxford. Heavy stress on physiological ecology.

*Annuals of the Entomological Society of America,* 1908, Entomological Society of America. The major entomological journal in the United States. Strong on taxonomy and morphology.

*Arctic,* 1948, Arctic Institute of North America, McGill-Queens University Press, Montreal. Papers heavily oriented toward natural and human ecology of the arctic region.

*The Auk,* 1884, American Ornithologists Union. Leading American ornithological journal.

*Australian Journal of Zoology,* 1958, Commonwealth, Scientific and Industrial Research Organization. General zoological papers, many of ecological interest.

*Behavior Genetics,* 1970, Plenum, New York. Papers on inheritance of behavior in animals and humans.

*Behavioral Ecology and Sociobiology,* 1976, Springer-Verlag, New York. New journal emphasizing ecological and field approach to behavior and sociobiology.

*Behavior,* 1947, Brill, Leiden, Netherlands. The journal of animal behavior; strongly ethological in viewpoint.

*Biological Bulletin,* 1927, Marine Biological Laboratory, Woods Hole, Mass. General papers with emphasis on marine biology.

*Biological Conservation,* 1970, Applied Science Publishers, Barking, Essex, England. Important new journal on scientific protection of plants and animals and management of natural resources.

*Bioscience,* 1951, American Institute of Biological Sciences. General biological journal with many papers devoted to ecology and environment.

*Bird-Banding,* 1930, Bird Banding Organizations. Small quarterly noted for its excellent coverage of current literature.

*Botanical Gazette,* 1862, University of Chicago Press, Chicago. Covers all departments of botanical science including plant ecology.

*Botanical Review,* 1935, New York Botanical Garden. Excellent review papers, often on plant ecology.

*British Birds,* 1897, Witherby, London. An important ornithological journal issued monthly. Contains some first-rate papers on the ecology of birds.

*Brittonia,* 1949, American Society of Plant Taxonomists. Papers deal largely with plant taxonomy.

*The Bryologist,* 1899, American Bryological and Lichenological Society. Only journal devoted to mosses and lichens. Many papers of ecological interest.

*Bulletin of Environmental Contamination and Toxicology,* 1966, Springer-Verlag, New York. Short, current papers dealing largely with pesticides and heavy metal pollution.

*California Fish and Game,* 1914, State of

California, Department of Fish and Game. Papers largely devoted to fish and game management. Contains many important papers on western wildlife.

*Cambridge Philosophical Society,* 1923, Biological Reviews, Cambridge University Press. General review papers, many of importance to ecology and behavior.

*Canadian Entomologist,* 1868, Entomological Society of Ontario. A major entomological journal devoted to insects of Canada.

*Canadian Field Naturalist,* 1887, Ottawa Field-Naturalist's Club. Papers on all phases of natural history. Of strong interest to the field biologist.

*Canadian Journal of Botany,* 1921, National Research Council of Canada, Ottawa. General botanical papers, a few of ecological interest.

*Canadian Journal of Zoology,* 1923, National Research Council of Canada, Ottawa. General zoological papers, but many of ecological interest.

*Chesapeake Science,* 1960, State of Maryland, Department of Research and Education, Chesapeake Biological Laboratory, Solomons. A regional publication, but papers of wide interest.

*Condor,* 1899, Cooper Ornithological Society. An ornithological journal with excellent scientific papers.

*Copeia,* 1913, American Society of Ichthyologists and Herpetologists. Papers on fishes, amphibians, and reptiles.

*East African Wildlife Journal,* 1963, East African Wildlife Society. Covers general ecology and wildlife and park management in East Africa.

*Ecological Entomology,* 1976, Royal Entomological Society of London, Blackwell, Oxford. Insect ecology.

*Ecological Modelling,* 1975, Elsevier, Amsterdam, The Netherlands. Deals with the use of mathematical models and systems analysis for the description of ecosystems and for the control of environmental pollution and resource development.

*Ecological Monographs,* 1930, Ecological Society of America. Contains papers too long for *Ecology.*

*Ecology,* 1920, Ecological Society of America. Indispensable to ecologists.

*Ecology Law Quarterly,* 1971, School of Law, University of California, Berkeley. Papers on law and the environment. Informative new journal.

*Economic Botany,* 1947, Society of Economic Botany. Many interesting papers on human and plant relationships. More ecology than economics.

*Environment,* 1958, formerly *Scientist and Citizen,* Scientists Institute for Public Information, St. Louis. Included among journals because it does contain significant, well-documented articles.

*Environment and Behavior,* 1969, Sage Publications, New York. Interesting papers on relationship between human environment and behavior.

*Environment and Planning,* 1969, Pion Ltd.,

London. Papers on environment and ecological approach to urban and suburban planning.

*Environment, Science and Technology,* 1967, American Chemical Society. Environmental articles written from the technological point of view.

*Environmental Affairs,* 1971, Boston College Environmental Law Center, Brighton, Mass. New journal devoted to environmental law. Similar to *Ecology Law Quarterly.*

*Environmental Biology of Fishes,* 1977, Dr. W. Junk. The Hague, Netherlands. Papers on environmental physiology of fish.

*Environmental Conservation,* 1974, Elsevier, Sequoia S.A., Lausanne, Switzerland. Articles range from case histories of resource use to environmental policy, education, and management.

*Environmental Law Reporter,* 1971, Environmental Law Institute, Washington, D.C. Analysis of laws concerning environmental protection and natural resource use, design of new institutional arrangements to carry out environmental policy.

*Environmental Letters,* 1970, Dekker, New York. Short papers quickly published on current environmental pollution problems.

*Environmental Management,* 1976, Springer-Verlag, New York. New journal focusing on real environmental problems and their solution. Articles deal with practical application of environmental management.

*Environmental Pollution,* 1970, Applied Science Publishers, Barking, Essex, England. Important journal with papers on pollution written from an ecological point of view.

*Evolution,* 1947, Society for the Study of Evolution. Papers on evolution and natural selection.

*Fishery Bulletin,* 1881, U.S. Department of Commerce. Wide range of well-edited papers on ecology and life history of commercial marine species, marine environment, and population dynamics.

*Forest Ecology and Management,* 1976, Elsevier, Amsterdam. Papers concerned with forest science and conservation and application of ecological knowledge to the management of man-made and natural forests. Covers all forest ecosystems.

*Forest Science,* 1955, Society of American Foresters. Technical, but occasionally contains papers of interest to ecologists.

*Forestry,* 1927, Society of Foresters of Great Britain, Oxford University Press, London. Frequently contains good papers on forest ecology.

*The Geographical Journal,* 1893, Royal Geographical Society. Many articles on human-relations. Geographical journals a good source of material on human ecology.

*Geographical Review,* 1916, American Geographical Society of New York. Outstanding journal with articles of strong ecological interest.

*Geography,* 1916, The Geographical Association, Sheffield, England. Another source of papers relating to ecology and man.

*Herpetologica,* 1945, Herpetologists League. Papers on all aspects of herpetology. Has improved with age.

*Holarctic Ecology,* 1978, Munksgaard, Copenhagen. New journal covering all aspects of the ecology of the Holarctic region and emphasizing descriptive and analytical studies. Complements *Oikos.*

*Human Ecology,* 1972, Plenum, New York. New journal devoted to the ecology of man.

*Hydrobiologica,* 1948, Dr. W. Junk, The Hague. Important journal on fresh-water biology.

*Hydrobiological Journal,* 1965, American Fisheries Society and Scripta Publishing Co., New York. Translation of Russian journal. Valuable source on fresh-water biology.

*Ibis,* 1859, British Ornithological Society. The leading British ornithological journal.

*Journal of Animal Ecology,* 1933, British Ecological Society, Blackwell, Oxford. Devoted exclusively to animal ecology. Contains many valuable papers.

*Journal of Applied Ecology,* 1966, British Ecological Society, Blackwell Oxford. Devoted to the applied aspects of ecology. Many useful papers on wildlife and human dominated ecosystems.

*Journal of Chemical Ecology,* 1975, Springer-Verlag, New York. Major reference for papers on chemical ecology of plants and animals: pheromones, chemical defense, attraction, orientation, etc.

*Journal of Ecology,* 1912, British Ecological Society, Blackwell Oxford. Papers on general ecology but with major emphasis on plants. An important reference journal.

*Journal of Environmental Quality,* 1972, American Society of Agronomy, Crop Science Society of America, and Soil Science Society of America. Madison, Wis. Papers on environmental quality in natural and agricultural ecosystems. A significant new journal.

*Journal of Environmental Systems,* 1971, Baywood. Papers on analysis, design, and management of environmental problems. Applied systems analysis.

*Journal of Experimental Marine Biology and Ecology,* 1963, North Holland. All aspects of marine ecology.

*Journal of Fish Biology,* 1969, Fisheries Society of the British Isles, Academic Press, New York. Papers devoted to fishery biology and fishery management.

*Journal of the Fisheries Research Board of Canada,* 1943, Fisheries Research Board of Canada. Very valuable source of papers on fish, fish management, and ecology.

*Journal of Forestry,* Society of American Foresters. Papers on general forestry. Some of interest to ecologists.

*Journal of Herpetology,* 1967, Society of the Study of Amphibians and Reptiles. General papers on herpetology.

*Journal of Mammalogy,* 1919, American Society of Mammalogists. Only journal in English devoted to mammals.

*Journal of the Marine Biological Association of the United Kingdom,* 1920, Marine Biological Association of the United Kingdom. Contains papers on marine biology.

*Journal of Marine Research,* 1937, Sears Foundation for Marine Research, Bingham Oceanographic Laboratory, Yale University Press, New Haven, Conn. A major source of papers on oceanography, heavy on the physical side.

*Journal of Parasitology,* 1914, American Society of Parasitology. All aspects of animal parasitism.

*Journal of Range Management,* 1947, American Society of Range Management. Advances in the science and art of grazing land management, understanding of practical and scientific range and pasture problems.

*Journal of Thermal Biology,* 1976, Papers on impact of thermal changes on organisms and thermal responses of organisms to temperature changes.

*Journal of Water Pollution Control Federation,* 1902, Water Pollution Control Federation. Important source of material on water pollution from causes and effects to solutions.

*Journal of Wildlife Management,* 1937, The Wildlife Society. Devoted to wildlife research and management. Contains excellent material for the field biologist.

*Limnology and Oceanography,* 1956, American Society of Limnology and Oceanography. The major journal in the field. All papers of strong interest to ecologists.

*Marine Biology,* 1967, Springer-Verlag, New York. International journal devoted to life in the oceans and coastal waters.

*Microbial Ecology,* 1975, Springer-Verlag, New York. Important new journal covering advances in microbiology of natural ecosystems. Contains papers in those branches of ecology in which microorganisms are involved.

*Natural Resources Journal,* 1961, The University of New Mexico School of Law. The original environmental law journal, in print long before a general interest in law and the environment developed.

*Nature,* 1869, Macmillan, London. Contains a number of important papers of interest to ecologists and field biologists.

*New York Fish and Game Journal,* 1953, State of New York, Conservation Department. Important regional publication with a number of major papers in fish and wildlife field.

*Oecologica,* 1963, Springer-Verlag, New York, in cooperation with International Association for

Ecology. International journal containing general ecological papers.

*Oikos, Acta Oecologica Scandinavica,* 1950, Munksgaard, Copenhagen. A major ecological journal emphasizing experimental and theoretical studies. Papers are in English.

*Pesticide Monitoring Journal,* 1967, Federal Workers Group on Pesticide Management, Environmental Protection Agency, Chamblee, Ga. Important current source on pesticide information.

*Physiological Zoology,* 1927, University of Chicago Press, Chicago. Contains many papers on physiological zoology.

*Proceedings of the Zoological Society of London,* 1822, Zoological Society of London. Contains a number of major papers on ecology and behavior of animals.

*Quarterly Review of Biology,* 1926, American Institute of Biological Sciences. A review occasionally publishing papers of interest to field biologists. Excellent book reviews.

*Radiation Botany,* 1961, Pergamon Press, Elmsford, N.Y. International journal with papers on effects of ionizing radiation on plants.

*Science,* 1833, American Association for the Advancement of Science. Covers the whole field of science, but some papers of interest to ecologists.

*Soviet Journal of Ecology,* 1970 (translated issues from 1972), Consultants Bureau, New York. The major ecological journal of the Soviet Union. Wide range of papers.

*Systematic Zoology,* 1952, Society of Systematic Zoologists. Papers mainly on taxonomy, but many of strong ecological interest.

*Taxon,* 1951, International Association for Plant Taxonomy, Utrecht, The Netherlands. The international journal on plant taxonomy. Papers mostly of a technical nature.

*Transactions of American Fisheries Society,* 1959. Contains a number of papers on fish and fresh-water ecology.

*Urban Ecology,* 1975, Elsevier, New York. International journal dealing with ecological processes and interactions within urban areas and between human settlements and surrounding natural systems that support them.

*Viltrevy, Swedish Wildlife,* 1964, Swedish Sportsman's Association, Stockholm. Long papers on wildlife covering both ecology and behavior.

*Wilson Bulletin,* 1888, Wilson Ornithological Society. Ornithological journal especially oriented to the field.

## GENERAL BIBLIOGRAPHIES

*In recent years a number of specialized bibliographies have appeared. However useful, they are too narrow to list here. The following list, also incomplete, is confined to current bibliographies. These are excellent gateways to the literature.*

*Applied Ecology Abstracts,* 1975 on, Information Retrieval Inc., 1911 Jefferson Davis Highway, Arlington, Va. 22202. Monthly abstracts of papers dealing with interactions between microbes, plants, and animals and their environments. Monitors a list of 4300 primary journals and other reference sources.

*Aquatic Sciences and Fishery Abstracts,* 1971 on, Information Retrieval Inc., Arlington, Va. Monthly abstracts of papers relating to studies on the aquatic environment, including physical and chemical aspects, pollution, utilization of resources, and biology of aquatic organisms.

*Bibliography of Agriculture,* 1942 on, U.S. Department of Agriculture. Monthly; covers wide variety of biological subjects. Especially useful for coverage of publications of state agricultural experiment stations.

*Biological Abstracts,* 1926 on, Philadelphia. Most comprehensive biological abstracting journal in North America. A primary reference, it includes brief abstracts available in groups on microfilm.

*Current Contents: Agriculture, Biology & Environmental Sciences; Life Sciences,* 1970 on. Weekly journal invaluable for scanning contents of current scientific journals.

*Entomology Abstracts,* 1970 on, Information Retrieval Inc., Arlington, Va. Covers systematics, physiology, biology and ecology, genetics and evolution, geography.

*Forestry Abstracts,* 1940 on, Commonwealth Agricultural Bureau, Oxford, England. Abstracts of forestry subjects compiled from world literature. Covers all phases of forestry including ecology, soils, general and systematic zoology, animal ecology, general botany, etc. A major bibliography for the ecologist.

KLEMENT, A. W., JR., AND V. SCHULTZ, *Terrestrial and Freshwater Radioecology. A Selected Bibliography.* (Report TID-3910 and 7 supplements), 1962–1971, U.S. Atomic Energy Commission, Division of Technical Information. Invaluable source of materials on radioecology.

SCHULTZ, V., *Ecological Techniques Utilizing Radionuclides and Ionizing Radiation. A Selected Bibliography* (RLO-2213, Suppl. 1), 1972, U.S. Atomic Energy Commission, Technical Information Center. Sourcebook for techniques.

SCHULTZ, V., AND F. W. WHICKER, *A Selected Bibliography of Terrestrial Freshwater and Marine Radiation Ecology* (TID-25650), 1971, U.S. Atomic Energy Commission. Available from National Technical Information Service, Springfield, Va. Highly useful and valuable bibliography for ecologists. Broken down into subject matter.

*Sport Fishery Abstracts,* 1956 on, U.S. Fish and Wildlife Service. Covers sport fishery research and management, limnology, ecology, and natural history of fishes. Indispensable to the fishery biologist, ecologist, and limnologist. Contains short abstracts of each paper.

*Water Resource Abstracts,* 1968 on, U.S. Department of Commerce. Comprehensive. Covers water-related aspects of life, physical and social sciences, as well as related engineering and legal aspects of water.

*Wildlife Abstracts,* 1935-1951; 1952-1955; 1956-1960. Annotated bibliography of the publications abstracted in *Wildlife Review.* A major reference source.

*Wildlife Review,* 1935 on, U.S. Fish and Wildlife Service. Coverage much wider than title indicates. Invaluable to field biologists; a necessity for those in wildlife management.

*Zoological Record,* 1864 on, Zoological Society of London. The major reference for zoologists. Worldwide coverage; essential.

# Selected references

Since 1970 there has been an avalanche of ecological literature: thousands of research papers, hundreds of books, and many new journals. To the student and professional alike, the literature seems overwhelming. While textbooks and research papers and the bibliographies they contain provide an introduction to the literature, reference books provide a more immediate introduction and access to specialized areas of ecology. Although some books become dated, they do provide a reference framework and foundation for further inquiry into a subject.

The reference books listed here are arranged by broad topics to fit the organization of the text. The selection is wholly personal and many books listed probably could be replaced by others. Many more titles could be added, but the list must necessarily be restricted. Recent titles are given preference over older ones which appeared in previous editions of this text. Older titles are retained when their merits and the concepts presented outweigh the dated material; classic works which ecologists should peruse are also retained. When titles are self-explanatory, notations have been omitted.

## General

### Reviews

In recent years a number of review volumes have come on the market. Among the most valuable are Annual Review of Entomology, Annual Review of Plant Physiology, and Annual Review of Ecology and Systematics, *published by Annual Reviews, Palo Alto, Calif., and* Advances in Ecological Research, *published by Academic Press, New York.*

### Readings

COBB, J. S., AND M. M. HARLIN (EDS.). 1976. *Marine Ecology, Selected Readings.* University Park Press, Baltimore. Wide ranging collection of recent studies in marine ecology.

CONNELL, J. H., D. B. MERTZ, AND W. W. MURDOCK (EDS.). 1970. *Readings in Ecology and Ecological Genetics.* Harper & Row, New York. Thirty selections in areas of community and ecosystems, population dynamics and genetics. Editors provide some illuminating commentary. Additional bibliographies.

DAWSON, P. S., AND C. E. KING (EDS.). 1971. *Readings in Population Biology.* Prentice-Hall, Englewood Cliffs, N.J. Covers a wide spectrum of populations from genetics to population dynamics, 41 papers.

FORD, R. F., AND W. E. HAZEN (EDS.). 1972, *Readings in Aquatic Ecology.* Saunders, Philadelphia. From physiology and behavior to aquatic pollution; 29 papers.

HAZEN, W. E. (ED.). 1970. *Readings in Population and Community Ecology.* Saunders, Philadelphia.

Contains a full set of papers of debate on balance of nature between Ehrlich and Birch, Slobodkin, etc.

NYBAKKEN, J. W. (ED.). *Readings in Marine Ecology.* Harper & Row, New York. Excellent collection of 32 papers on marine ecology; some overlap with Ford and Hazen.

SCHULTZ, V., AND F. W. WHICKER (EDS.). 1972. *Ecological Aspects of the Nuclear Age: Selected Readings in Radiation Ecology.* TID-25978. National Technical Information Service, Springfield, Va. The best general introduction to radiation ecology.

SMITH, R. L. (ED.): 1976. *Ecology of Man: An Ecosystem Approach,* 2nd ed. Harper & Row, New York. Papers and lengthy commentaries view man and the environment from an ecosystem point of view.

### Texts

ALEXANDER, M. 1971. *Microbial Ecology.* Wiley, New York. Excellent book on microbial ecology by a well-known soil microbiologist.

BROCK, T. 1966. *Principles of Microbial Ecology.* Prentice-Hall, Englewood Cliffs, N.J. The first book on the subject.

CAMPBELL, R. E. 1977. *Microbial Ecology.* Halsted Press, New York. Emphasizes role of microbes in biogeochemical cycling. Brief.

COLLIER, B. D., G. W. COX, A. W. JOHNSON, AND P. C. MILLER. 1973. *Dynamic Ecology,* Prentice-Hall, Englewood Cliffs, N.J. A graduate-level text in ecology, theoretically oriented.

EHRLICH, P. R., A. H. EHRLICH AND J. HOLDREN. 1978. *Population Resources, Environment,* 3rd ed. Freeman, San Francisco. An important sourcebook that considers all three items from an ecological viewpoint.

KREBS, C. J. 1977. *Ecology: Experimental Analysis of Distribution and Abundance,* 2nd ed. Harper &Row, New York. Experimental in approach, an intermediate text largely on population ecology.

MACARTHUR, R. H. 1972. *Geographical Ecology.* Harper & Row, New York. Largely theoretical, this is one of the most important ecology books of the 1970s.

MAY, R. B. (ed.). 1976. *Theoretical Ecology, Principles and Applications,* Saunders, Philadelphia, Pa. Selected topics in population biology. Advanced.

NAUMOV, N. P. 1972 (Russia, 1963). *The Ecology of Animals,* University of Illinois Press, Urbana Translation of the leading Russian text in ecology. Interesting contrast. Russian ecology is much more applied than American. Should be read by all ecologists for a different viewpoint.

ODUM, E. P. 1971. *Fundamentals of Ecology,* 3rd ed. Saunders, Philadelphia. The original ecology text updated. Chapters on remote sensing, modeling.

PIANKA, E. 1978. *Evolutionary Biology,* 2nd ed. Harper & Row, New York. Abstract, conceptual approach to ecology with heavy emphasis on population biology.

PRICE, P. 1975. *Insect Ecology*. Wiley, New York. More a general ecology than an insect ecology text. Good coverage of population biology.

RICKLEFS, R. 1979. *Ecology*. Chiron Press, Newton, MA. Heavily oriented to population biology and ecological genetics. Evolutionary approach.

STONEHOUSE, B. AND C. M. PERRINS. *Evolutionary Ecology*. University Park Press, Baltimore. Compendium of 21 chapters ranging from population regulation to breeding adaptation and behavior.

WHITTAKER, R. H. 1976. *Communities and Ecosystems,* 2nd ed. Macmillan, New York. Somewhat advanced, smaller text that is a rich storehouse of concepts in community ecology.

## Part I

### History and general

BATES, M. 1960. *The Forest and the Sea*. Random House, New York (several editions available). Probably the finest general approach written.

BATES, M. 1961. *The Nature of Natural History*. Scribner, New York. Well-written thought-provoking, humanistic approach to ecology.

BLAIR, W. F. 1977. *Big Biology: THE US/IBP,* Academic Press, New York. Historical account of the origin and evolution of U.S. participation in the International Biological program.

EGERTON, F. AND R. P. MCINTOSH (EDS.) 1977. *History of American Ecology*. Arno Press, New York.

FARB, P. 1964. *Ecology*. Life Nature Library, Time-Life Books, New York. Enjoyable and concise introduction to ecology, heavily illustrated with outstanding color photographs.

LEOPOLD, A. 1949. *A Sand County Almanac,* Oxford University Press, New York (several editions available). A classic in ecological writing, rediscovered in the 1970s.

MARGALEF, R. 1968. *Perspectives in Ecological Theory*. University of Chicago Press, Chicago. Thought-provoking ideas for advanced students.

SCIENTIFIC AMERICAN. 1970. *The Biosphere*. Freeman, San Francisco. Excellent series of articles on the ecosystem.

WATTS, M. T. 1976. *Reading the American Landscape*. Macmillan, New York. Delightful book that should be read by all ecologists. An updated version of the 1968 edition.

WORSTER, D. 1977. *Nature's Economy: The Roots of Ecology*. Sierra Club Books, San Francisco, Engaging history of ecology.

WORTHINGTON, E. B. 1975. *The Evolution of IBP*. Cambridge University Press, New York. IBP as it developed in Europe.

### Systems and processes

ANDERSON, M. M., AND A. MACFAYDEN (EDS.). 1977. *The Role of Terrestrial and Aquatic Organisms in Decomposition Processes*. Seventeenth Symposium, British Ecological Society. Blackwell, Oxford.

BARBERM, J. (ED.). 1977. *Primary Processes of Photosynthesis*. Vol. 2, *Topics in Photosynthesis*. Elsevier, New York. Primary reactions involved in capture of light energy.

BURRIS, R. H., AND C. C. BLACK (EDS.). 1976. $CO_2$ *Metabolism and Plant Productivity*. University Park Press, Baltimore. Advanced treatment of photosynthesis with special consideration of $C_3$ and $C_4$ pathways.

COOPER, J. P. 1975. *Photosynthesis and Productivity in Different Environments*. Cambridge University Press, New York. Comparative study of photosynthesis in different ecosystems.

DEVLIN, R. M., AND A. J. BARKER. 1971. *Photosynthesis*. Van Nostrand Reinhold, New York. Introductory text.

GOVINDJEE (ED.). 1975. *Bioenergetics of Photosynthesis*. Academic Press, New York.

GREGORY, R. P. F. 1977. *Biochemistry of Photosynthesis,* 2nd ed. Wiley, New York.

HALFON, E. (ED.). 1979. *Theoretical Systems Ecology*. Academic, New York.

HALL, C. A. S., AND J. W. DAY, JR. (EDS.). 1977. *Ecosystem Modeling in Theory and Practice*. Wiley, New York.

MATTSON, W. J. (ED.). 1977. *The Role of Arthropods in Forest Ecosystems*. Springer-Verlag, New York.

## Part II

### Environmental influences

CUSHING, C. E., JR. (ED.). 1976. *Radioecology and Energy Resources*. Dowden, Hutchinson & Ross, Stroudsburg, Pa. Proceedings of the Fourth National Symposium on Radioecology. Good source of most recent information.

GATES, D. M. 1962. *Energy Exchange in the Biosphere*. Harper & Row, New York. Read in particular Chap. 1.

GEIGER, R. 1957. *The Climate Near the Ground,* 2nd ed. Harvard University Press, Cambridge, Mass. Emphasizes the microclimate, which makes this the climatology book for ecologists.

GOLTERMAN, H. L., AND R. S. CLYMO. 1967. *Chemical Environment in Aquatic Habitats, An IBP Symposium*. Noord-Hollandsche Vitgevers, Maatsschappy, Amsterdam.

HENDERSON, L. J. 1913. *The Fitness of the Environment*. Reprint, Beacon, Boston. An old classic that ecologists still should read.

JOHNSON, C. G., AND L. P. SMITH (EDS.). 1965. *Biological Significance of Climatic Changes in*

*Britain.* Academic Press, New York. Some ecological information.

KOZLOWSKI, T. T., AND C. E. Ahlgren. 1974. *Fire and Ecosystems.* Academic Press, New York. Role of fire in all environments. Comprehensive.

LEE, R. 1978. *Forest Climatology.* Columbia University Press, New York. Concise introduction to the climate in the forest.

LOWERY, W. P. 1969. *Weather and Life,* Academic Press, New York. Introductory text on weather and its effects on plants and animals, including humans.

NELSON, D. J (ED.). 1973. *Radionuclides in Ecosystems.* Proceedings of the Third National Symposium on Radioecology, 2 vol., CONF-710501-Pl. National Technical Information Service, Springfield, Va. Wealth of information on radioecology up to 1971.

NELSON, D. J., AND F. C. EVANS (EDS.). 1969. *Symposium on Radioecology.* Proceedings of the Second National Symposium on Radioecology, CONF 670503. National Technical Information Service, Springfield, Va. Major reference source.

SCHROEDER, M. J., AND C. C. BUCK. 1970. *Fire Weather.* Agricultural Handbook 360. U.S. Department of Agriculture, Washington, D.C. Although written as a forest fire control handbook, this book is a superb reference on climatology.

SCHULTZ, V., AND A. W. KLEMENT (EDS.). 1963. *Radioecology.* Proceedings of the First National Symposium on Radioecology. Van Nostrand Reinhold, New York. Important sourcebook.

U.S. FOREST SERVICE. 1971. *Fire in the Northern Environment—A Symposium.* Northwest Forest and Range Experimental Station, Portland, Ore. The role of fire in the northern coniferous forest and tundra.

STUDY OF MAN'S IMPACT ON CLIMATE. 1971. *Inadvertent Climate Modification.* MIT Press, Cambridge, Mass. Excellent introduction to climate and man's impact on it.

### Physiological ecology

BANNISTER, P. 1977. *Introduction to Physiological Plant Ecology.* Halsted Press, New York. Plant ecology from the standpoint of plant response.

BENTLEY, P. 1971. *Endocrines and Osmoregulation.* Springer-Verlag, New York. Comparative account of regulation of water and salt in vertebrates.

BLIGHT, J. 1973. *Temperature Regulation in Mammals and Other Vertebrates.* Elsevier, New York. Comprehensive analyses of evolution and physiology of temperature regulation in vertebrates.

BROWN, F. A., ET AL. 1970. *Biological Clocks, Two Views.* Academic Press, New York. Discusses the controversy about exogenous and endogenous control over circadian rhythms.

BUNNING, E. 1976. *The Physiological Clock,* 3rd ed.

Academic Press, New York. Short book by an authority in the field of periodicity.

DECOURSEY, P. J. 1976. *Biological Rhythms in the Marine Environment.* University of South Carolina Press, Columbia.

ESCH, G. W., AND R. W. MCFARLAND (EDS.). 1976. *Thermal Ecology II.* National Technical Information Service, Springfield, Va. Reviews thermal ecology at individual, population, and ecosystem level.

ETHERINGTON, J. R. 1975. *Environment and Plant Ecology.* Wiley, New York. Examines interrelationships among plants, animals, and microbes and environmental interactions as they relate to plant physiological ecology.

FOLK, G. E. 1974. *Textbook of Environmental Physiology,* 2nd ed. Lea & Febiger, Philadelphia. Excellent introduction to environmental physiology of mammals.

GORDON, M. S. 1971. *Animal Functions, Principles and Adaptation.* Macmillan, New York. Valuable because of its ecological approach.

IRVING, L. 1972. *Arctic Life of Birds and Mammals, Including Man.* Springer-Verlag, New York. Review of adaptative physiology of arctic homeotherms.

LARCHER, W. 1975. *Physiological Plant Ecology.* Springer-Verlag, New York.

NEWELL, R. C. 1970. *Biology of Intertidal Animals.* Elsevier, New York. Thorough review of physiology of intertidal animals.

NICOL, J. A. C. 1967. *Biology of Marine Animals.* Wiley, New York. Somewhat similar to Newell, but considers animals of open water.

PALMER, J. D. 1976. *An Introduction to Biological Rhythms.* Academic Press, New York.

PENGELLEY, E. T. 1974. *Circannual Clocks: Annual Biological Rhythms.* Academic Press, New York.

PHILLIPS, J. G. (ED.). 1975. *Environmental Physiology.* Wiley, New York. Good elementary introduction.

SAUNDERS, D. S. 1977. *An Introduction to Biological Rhythm.* Halsted Press, New York.

SCHMIDT-NEILSEN, K. 1975. *Animal Physiology, Adaptation and Environment.* Cambridge University Press, New York.

STRAIN, B. R., AND W. B. BILLINGS (EDS.). 1975. *Vegetation and Environment.* Vol. 6, *Handbook of Vegetation Science.* W. Junk, The Hague.

TRESHOW, M. 1970. *Environment and Plant Response.* McGraw-Hill, New York. Discusses plants under environmental stress, especially pollution.

VERNBERG, F. J. (ED.). 1975. *Physiological Adaptation to the Environment.* Intext, New York.

VERNBERG, F. J., AND A. B. VERNBERG. 1970. *The Animal and the Environment.* Holt, Rinehart and Winston, New York. Animal ecology from a physiological point of view.

YOUSEF, M. K., R. W. BULLARD, AND S. M. HORVATH.

1972. *Physiological Adaptations in Desert and Mountains.* Academic Press, New York.

WITTOW, G. C. 1970–1971. *Comparative Physiology of Thermoregulation.* Vol. 1, *Invertebrates and Non-mammalian Vertebrates;* Vol. 2. *Mammals.* Thorough review of the literature.

### Energy and productivity

BROWN, L. R., AND G. W. FINSTERBUSCH. 1972. *Man and His Environment: Food.* Harper & Row, New York. Superb account of the environmental consequences of man's quest for food.

CRAWFORD, M. A. (ED.). 1968. Comparative Nutrition of Wild Animals. *Symposium Zoological Society, London, No. 21.* Academic Press, New York. Important reference on nutrition of wild animals; contains information often hard to find.

CRISP, D. J. (ED.). 1965. *Grazing in Terrestrial and Marine Environments.* Blackwell, Oxford. Energy flow at the herbivore level.

DUVIGNEAUD, P. (ED.). 1971. *Productivity of Forest Ecosystems.* UNESCO, Paris. Invaluable source of information on primary production in the forest. Worldwide in scope.

ECKARDT, F. E. (ED.). 1968. *Functioning of Terrestrial Ecosystems at the Primary Production Level.* UNESCO, Paris. Comprehensive look at primary production of terrestrial ecosystems. Includes research and methods.

GOLDMAN, C. R. (ED.). 1966. *Primary Productivity in Aquatic Environments.* University of California Press, Berkeley. Contains some technical, but basic papers.

GOLLEY, F. B., ET AL. (EDS.). 1975. *Small Mammals: Their Productivity and Population Dynamics.* Cambridge University Press, New York.

LEITH, H. (ED.). 1978. *Patterns of Primary Production in the Biosphere.* Benchmark® Papers. Dowden, Hutchinson & Ross, Stroudsburg, Pa. Collection of historical papers on primary production.

LEITH, H., AND R. H. WHITTAKER (EDS.). 1975. *Primary Productivity of the Biosphere.* Springer-Verlag, New York. Summary of current understanding and research methods.

MILLER, G. T., JR. 1971. *Energy, Kinetics, and Life.* Wadsworth, Belmont, Calif. Outstanding elementary approach to energy and ecology.

MOROWITZ, H. J. 1968. *Energy Flow in Biology.* Academic Press, New York. Physicist's view of energy in the biological world; advanced.

NATIONAL ACADEMY OF SCIENCE. 1975. *Productivity of World Ecosystems.* National Academy of Science, Washington, D.C. Symposium papers; good overview.

PETRUSEWICZ, K. (ED.). 1967. *Secondary Productivity of Terrestrial Ecosystems,* 2 vols. Institute of Ecology, Polish Academy of Science, International Biological Program, Warsaw. Wide range of papers on secondary production.

PHILLIPSON, J. 1966. *Ecological Energetics.* St.

Martin's Press, New York. Short, lucid introduction to the subject.

RAYMONT, J. E. C. 1963. *Plankton and Productivity in the Oceans.* Macmillan, New York. Primary production well covered in a readable fashion in chaps. 6–10.

REICHLE, D. E. (ED.). 1970. *Analysis of Temperate Forest Ecosystems.* Springer-Verlag, New York. Good synthesis of productivity in temperate forests.

RUSSELL-HUNTER, W. D. 1970. *Aquatic Productivity.* Macmillan, New York. Presents current knowledge of aquatic production from phytoplankton to human harvest of the sea.

STEELE, J. H. (ED.). 1970. *Marine Food Chains.* University of California Press, Berkeley. Secondary productivity in marine environment.

WIERGERT, R. G. 1976. *Ecological Energetics.* Benchmark® Papers. Dowden, Hutchinson & Ross, Stroudsburg, Pa; distr. by Academic Press, New York. Collection of papers surveying development of concept.

### Nutrient cycles

BARTHOLOMEW, W. V., AND F. F. CLARK (EDS.). 1963. *Soil Nitrogen.* American Society of Agronomists, Madison, Wis. Review of nitrogen in the soil and role in nitrogen cycle.

BROWN, A. W. A. 1978. *Ecology of Pesticides.* Wiley, New York. Up-to-date review of pesticide literature and impact of pesticides on the environment.

DREVER, J. E. (ED.). 1977. *Sea Water: Cycles of Major Elements.* Benchmark® Papers in Geology. Collection of papers on development of nutrients cycles in sea.

DUGAN, P. R. 1972. *Biochemical Ecology of Water Pollution.* Plenum, New York. Significant types and causes of water pollution.

EDWARDS, C. E. 1974. *Persistent Pesticides in the Environment.* CRC Press, West Palm Beach, Fla. State of knowledge of the problem.

FRIBERG, L. T. 1972. *Mercury in the Environment.* CRC Press, West Palm Beach, Fla.

FRIBERG, L. T., ET AL. 1974. *Cadium in the Environment,* 2nd ed. CRC Press, West Palm Beach, Fla.

HASLER, A D. (ED.). 1975. *Coupling of Land and Water Ecosystems.* Springer-Verlag, New York. Interrelationships of nutrient cycles between land and fresh-water ecosystems.

HOWELL, F. G., J. B. GENTRY, AND M. H. SMITH (EDS.). 1975. *Mineral Cycling in Southeastern Ecosystems.* National Technical Information Service, Springfield, Va. Compendium of information.

THE INSTITUTE OF ECOLOGY. 1971. *Man in the Living Environment.* University of Wisconsin Press, Madison. Excellent assessment of human impact on the environment. Considers effects on biogeochemical cycles, agriculture, and resource management.

LIKENS, G. E., ET AL. 1977. *Biogeochemistry of a Forested Ecosystem*. Springer-Verlag, New York. The quantified details of the Hubbard Brook study of forest functions.

NATIONAL ACADEMY OF SCIENCE. 1969. *Eutrophication: Causes, Consequences, and Corrections*. National Academy of Science, Washington, D.C. Excellent overview of eutrophication. Worldwide in scope.

PIMENTEL, D. D. 1971. *Ecological Effects of Pesticides on Non-Target Species*. Executive Office of the President, Office of Science and Technology, Washington, D.C. Excellent summary and bibliography of subject.

POMEROY, L. R. 1974. *Cycles of Essential Elements*. Benchmark® Papers in Ecology. Dowden, Hutchinson & Ross, Stroudsburg, Pa.; distr. by Academic Press, New York. Collection of important papers on mineral cycling.

SINGER, S. F. (ED.). 1970. *Global Effects of Environmental Pollution*. American Association for the Advancement of Science, Symposium. Springer-Verlag, New York. Contains a number of significant papers on pollution, especially atmospheric.

STERN, A. C. (ED.). 1968. *Air Pollution, A Comprehensive Treatise*, 2d ed., 3 vols. Academic Press New York. Covers all aspects of air pollution in detail.

STEWART, W. D. P. (ED.). 1975. *Nitrogen Fixation by Free-living Organisms*. Cambridge University Press, New York.

WOODWELL, G. M., AND E. PECAN (EDS.). 1973. *Carbon and the Biosphere*. Proceedings of Symposium, Brookhaven National Laboratory, May 1972. National Technical Information Service, Springfield, Va.

## Part III

*A popular, elementary, yet authoritative treatment of much of the material covered in this section is contained in the fourteen volumes of* Our Living World of Nature Series, *published by McGraw-Hill, New York. The volumes include forest, tropical forest, grasslands, ponds, streams, seashores, African plains, far north, and other ecosystems. All are clearly written and exceptionally well illustrated.*

### Biogeography

*An elementary treatment of biogeography for the beginning student and lay reader is contained in the six-volume series on land and wildlife which is part of the Life Nature Library, published by Time-Life Books, New York. The volumes cover South America, Australia, Eurasia, Africa, North America, and Tropical Asia.*

*The books below are more technical references.*

CAIN, S. A. 1944. *The Foundations of Plant Geography*. Harper & Row, New York.

COLLINSON, A. S. 1978. *Introduction to World Vegetation*. Allen and Unwin, Chicago.

DARLINGTON, P. J. JR. 1957. *Zoogeography: The Geographical Distribution of Animals*. Wiley, New York.

DAUBENMIRE, R. 1978. *Plant Geography with Special Reference to North America*. Academic Press, New York.

UDVARDY, M. D. F. 1969. *Dynamic Zoogeography, with Special Reference to Land Animals*. Van Nostrand Reinhold, New York.

WALTER, H. 1974. *Vegetation of the Earth*. Springer-Verlag, New York. Description of zonal vegetation from an ecophysiological viewpoint.

### Fresh-water ecosystems

AMOS, W. H. 1970. *The Infinite River*. Random House, New York. Story of the river from the headwaters to the sea.

BARDACH, J. E. 1964. *Downstream: A Natural History of the River*. Harper & Row, New York. Follows development of the river from headwaters, with practical considerations.

BENNETT, G. W. 1962. *Management of Artificial Lakes and Ponds*. Van Nostrand Reinhold, New York. Limnology and fishery biology from a practical approach. Deals only with artificial impoundments, especially small ponds, popular in ecology laboratory work.

BROWN, A. L. 1977. *Ecology of Fresh Water*. Harvard University Press, Cambridge, Mass. Introduction to the study of fresh-water biology.

COKER, R. E. 1954. *Streams, Lakes, and Ponds*. University of North Carolina Press, Chapel Hill. An account for the general reader.

DYKYJOVA, D., AND J. KVET (EDS.). 1978. *Pond Littoral Ecosystems, Structure and Function*. Springer-Verlag, New York. Emphasis on eastern European systems.

FREY, D. G. (ED.). 1963. *Limnology in North America*. University of Wisconsin Press, Madison. Thorough account.

HUTCHINSON, G. E. 1957–1967. *A Treatise on Limnology*. Vol. 1, *Geography, Physics and Chemistry;* Vol. 2, *Introduction to Lake Biology and Limnoplankton*. Wiley, New York. Invaluable reference on lake ecosystems.

HYNES, H. B. N. 1960. *The Biology of Polluted Waters*. Liverpool University Press, Liverpool. Effects of pollution on fresh-water ecosystems.

HYNES, H. B. N. 1970. *The Ecology of Running Water*. University of Toronto Press, Toronto. A major work on running-water ecosystems. Indispensable reference.

MACAN, T. T. 1963. *Fresh Water Ecology*. Longmans, Essex. Revised Edition of a major work on aquatic life.

MACAN, T. T. 1970. *Biological Studies of English Lakes*. Elsevier, New York. Good summary of lake biology.

MACAN, T. T. 1973. *Ponds and Lakes,* Crane, Russak, New York.

MACAN, T. T., AND E. B. WORTHINGTON. 1952. *Life in Lakes and Rivers.* Collins, London. Excellent limnology; the finest general book on the subject in spite of the date.

MOORE, P. D., AND D. J. BELLAMY. 1974. *Peatlands.* Springer-Verlag, New York. Major basic reference.

NEEDHAM, P. R. 1969. *Trout Streams.* Winchester, New York. Reprint of a classic published in 1938, with annotations to update the text.

OSVALT, H. 1970. *Vegetation and Stratigraphy of Peatlands of North America.* Nova Acta Regiase Societatis Scientarium Upsaliensis, ser. V:C, vol. 1. Royal Scientific Society, Upsala, Sweden.

POPHAM, E. J. 1961. *Life in Fresh Water.* Harvard University Press, Cambridge, Mass. Excellent introduction to fresh-water ecology. Contains account of a long-term survey of an actual pond that can serve as a model for practical field work.

REID, G. K. 1961. *Ecology of Inland Waters and Estuaries.* Van Nostrand Reinhold, New York. Good summary of fundamentals.

RUTTNER, F. 1963. *Fundamentals of Limnology,* rev. ed. University of Toronto Press, Toronto. Basic reference, but somewhat dated.

WETZEL, R. 1975. *Limnology.* Saunders, Philadelphia. Finest reference available. Strong on functioning of fresh-water ecosystems, especially nutrient cycles.

WITTON, B. A. (ED.). 1975. *River Ecology.* University of California Press, Berkeley. Valuable literature source.

## Estuaries

BARNES, R. 1976. *The Coastline.* Wiley, New York. Contribution to understanding of ecology and physiography in relation to land use and mánagement of coastal resources.

CHAPMAN, V. T. (ED.). 1977. *Wet Coastal Ecosystems.* Vol. 1, *Ecosystems of the World.* Elsevier, New York. Survey of world salt marshes and mangroves.

CLARK, J. 1974. *Coastal Ecosystems.* The Conservation Foundation, Washington, D.C. Emphasis on management.

COWELL, E. B. (ED.). 1971. *The Ecological Effects of Oil Pollution on Littoral Communities.* Institute of Petroleum, London. Papers dealing with all aspects of oil spills on life along the coast.

GREEN, J. 1968. *Biology of Estuarine Animals.* University of Washington Press, Seattle. From vegetation to vertebrates, an overview of the estuary.

LAUFF, G. (ED.). 1967. *Estuaries.* American Association for the Advancement of Science, Washington, D.C. Comprehensive look at estuaries. An important basic reference.

PERKINS, E. J. 1974. *The Biology of Estuaries and Coastal Waters.* Academic Press, New York. Describes interrelations between chemistry and biology of estuarine and coastal waters; applied approach.

RANWELL, D. S. 1972. *Ecology of Salt Marshes and Sand Dunes.* Chapman & Hall, London. Good introduction; more technical than Teal. Plant oriented, little on animal life.

REMANE, A., AND C. SCHLIEPER. 1971. *Biology of Brackish Water.* Wiley, New York. Important reference that covers not only the estuary, but all brackish water habitats.

STEVENSON, L. H., AND R. R. COLWELL (EDS.). 1976. *Estuarine Microbial Ecology.* University of South Carolina Press, Columbia. Major reference source.

TEAL, J., AND M. TEAL. 1969. *Life and Death of a Salt Marsh.* Little, Brown, Boston. Fine reference on salt marsh.

## Seashore

BERRILL, N. J. 1951. *The Living Tide.* Dodd, Mead, New York, Popular book on life in the tidal zone.

CAREFOOT, T. 1977. *Pacific Seashore: A Guide to Intertidal Ecology.* University of Washington Press, Seattle.

CARSON, R. 1955. *The Edge of the Sea.* Houghton Mifflin, Boston. Beautifully written book that deals with marine life along the eastern coast of North America.

ELTRINGHAM, S. K. 1971. *Life in Mud and Sand.* Crane, Russak, New York. Fresh, concise introduction to physical environment and life of muddy and sandy seashores.

MACGINITIE, G. E., AND N. MACGINITIE. 1968. *Natural History of Marine Animals,* 2d ed. McGraw-Hill, New York. Excellent introduction to animal life between the tide marks.

RICKETTS, E. F., J. CALVIN, AND J. HEDGPETH (EDS.). 1968. *Between Pacific Tides,* 4th ed. Stanford University Press, Stanford, Calif. Contains a wealth of information on Pacific coast intertidal life. A must.

RUDLOE, J. 1971. *The Erotic Ocean: Handbook for Beachcombers.* Harcourt Brace Jovanovich, New York. Excellent handbook for seaside ecologists also.

SILVERBERG, R. 1972. *The World Within the Tide-pool.* Weybright & Talley, New York. Well-illustrated popular account of the marine microcosm.

SOUTHWARD, A. J. 1976. *Life on the Seashore.* Harvard University Press, Cambridge, Mass. Basic descriptive and theoretical introduction to shore life.

STEPHENSON, T. A., AND A. STEPHENSON. 1972. *Life Between Tidemarks on Rocky Shores.* Freeman, San Francisco. Excellent descriptions of rocky shores over the world.

## Open sea

BARNES, H. (ED.). 1963. *Oceanography and Marine Biology.* Allen and Unwin, London. Annual review publication.

CARSON, R. 1961. *The Sea Around Us.* Oxford University Press, New York. A classic; best popular account of the sea.

COKER, R. E. 1947. *This Great and Wide Sea.* University of North Carolina Press, Chapel Hill. Brief, but comprehensive book on the ocean.

CUSHING, D. H., AND J. T. WALSH. 1977. *The Ecology of the Seas.* Saunders, Philadelphia.

DRAKE, C. L., ET AL. (EDS.). 1978. *Oceanography.* Holt, Rinehart and Winston, New York. Broad overview of oceanography: physical, chemical, biological, geological.

EKMAN, S. 1953. *Zoogeography of the Sea.* Sidgwick and Jackson, London. Indispensable reference on the "shelf fauna" and around the world.

ENGEL, L. 1961. *The Sea.* Life Nature Library. Time-Life, New York. Brief introduction to the sea with an excellent collection of illustrations.

FREDRICH, H. 1970. *Marine Biology.* University of Washington Press, Seattle. Technical introduction to marine world.

HARDY, A. 1971. *The Open Sea: Its Natural History.* Houghton Mifflin, Boston. The original two-volume work now in one volume. One of the great writings on the sea.

HEDGPETH, J. W. (ED.). 1957. *Treatise on Marine Ecology and Paleoecology.* Vol. 1, *Ecology.* Mem. 67. Geological Society of America, New York. Immense in scope—a bible for the marine ecologist. Covers the open sea, the sea shore, the estuary, etc., but unfortunately out-of-print.

HILL, M. N. (ED.). 1962–1964. *The Seas: Ideas and Observations on Progress in the Study of the Seas,* 3 vols. Wiley, New York. The three volumes cover physical oceanography, seawater, comparative and descriptive oceanography, etc. A major reference on the sea.

HOOD, D. W. (ED.). 1971. *Impingement of Man on the Oceans.* Wiley, New York. Ocean pollution and other impacts on the sea.

KINNE, O. (ED.). 1971–1976. Marine Ecology. Vol. 1, *Environmental Factors;* vol. 2, *Physiological Mechanisms.* Comprehensive, advanced treatment of life and environment in oceans and coastal waters. Part of a proposed five-volume treatise on marine ecology.

OLSEN, T. A., AND F. J. BURGESS (EDS.). 1967. *Pollution and Marine Ecology.* Wiley, New York. Studies on the effects of pollution on marine life and ecosystems.

RUSSEL, F. S. (ED.). 1963. *Advances in Marine Biology.* Academic Press, New York. Annual review volume.

SCIENTIFIC AMERICAN. 1971. *Oceanography.* Freeman, San Francisco. Collection of articles arranged to provide an excellent introduction to the ocean and its use by humans.

SEARS, M. (ED.). 1961. *Oceanography.* American Association for the Advancement of Science, Washington, D.C. A symposium volume.

STEELE, J. H. 1974. *The Structure of Marine Ecosystems.* Harvard University Press, Cambridge, Mass. Flow of energy through marine food webs.

TAIT, R. V., AND R. S. DESANTO. 1974. *Elements of Marine Ecology.* Plenum, New York. Elementary and readable introduction to marine ecology.

WEINS, H. J. 1962. *Atoll Environment and Ecology.* Yale University Press, New Haven, Conn. Excellent and complete book on ecology of coral reefs, but now somewhat dated.

## Soil

BRADY, N. C. 1974. *The Nature and Properties of Soils,* 8th ed. Macmillan, New York. Classic college textbook that covers most aspects of soil science.

BUOL, S. W., ET AL. 1973. *Soil Genesis and Classification.* Iowa State University Press, Ames. Next text on this subject.

EYRE, S. R. 1968. *Vegetation and Soils, A World Picture,* 2d ed. Aldine, Chicago. Broad picture of soils of the world and their development.

FARB, P. 1959. *The Living Soil.* Harper & Row, New York (in several editions). Popular account of life in the soil; scientifically sound and beautifully written.

KEVAN, D. K. MCE. 1962. *Soil Animals.* Philosophical Library, New York. Simplified introduction to life in the soil. Good on the ecology of soils.

KEVAN, D. K. MCE. (ED.). 1955. *Soil Zoology.* Butterworth, Washington, D.C. Symposium proceedings. Excellent; notable are the keys to soil invertebrates.

LUTZ, H. J., AND R. F. CHANDLER. 1946. *Forest Soils.* Wiley, New York. Out of print, but a basic reference on the subject.

RICHARDS, B. N. 1974. *Introduction to the Soil Ecosystem.* Longman, New York. Strong on the role of soil microbes in nutrient cycling and energy flow.

UNESCO. 1969. *Soil Biology.* UNESCO, Paris. Outstanding collection of papers on all aspects of soil biology.

U.S. DEPARTMENT OF AGRICULTURE. 1957. Soils. *Yearbook of Agriculture.* U.S. Department of Agriculture, Washington, D.C. Principles of soils, soil fertility, and soil management.

WALLWORK, J. A. 1970. *Ecology of Soil Animals.* McGraw-Hill, New York. Up-to-date reference with a good review of world literature.

## Grassland

DIX, R. L., AND R. G. BEIDLEMAN (EDS.). 1969. *The Grassland Ecosystem, A Preliminary Synthesis, Range Science Department, Science Series no. 2,* Colorado State University Fort Collins. The finest reference on grasslands. It is relatively unavailable.

CARPENTER, J. R. 1940. The grassland biome. *Ecol. Monographs,* 10:617–684. Included among the books because it is an excellent review of North American grasslands.

DUFFEY, E. 1974. *Grassland Ecology and Wildlife Management,* Chapman and Hall, London. An English approach to management of tame grasslands. Excellent reference.

HEAL, O. W., AND D. F. PERKINS (EDS.) 1978. *The Ecology of Some British Moore and Montane Grassland.* Springer-Verlag, New York.

HUMPHREY, R. R. 1962. *Range Ecology.* Ronald Press, New York. Ecology text with grassland orientation.

SPEDDING, C. R. W. 1971. *Grassland Ecology.* Clarendon, Oxford. Synthesis of ecology and agriculture. A fresh modern approach to grassland management.

SPRAGUE, H. E. (ED.). 1959. *Grasslands.* American Association for the Advancement of Science, Washington, D.C. Symposium volume with a number of papers on grassland ecology.

WEAVER, J. E. 1954. *North American Prairie.* Johnsen, Lincoln, Neb. Detailed description of the original prairie.

WEAVER, J. E. 1968. *Prairie Plants and Their Environment.* University of Nebraska Press, Lincoln. Summary of many years' work on prairie grasslands.

WEAVER, J. E., AND F. W. ALBERTSON. 1956. *Grasslands of the Great Plains.* Johnsen, Lincoln, Neb. Sound reference on the grasslands of the midcontinent.

### Shrubland

FRIEDLANDER, C. P. 1961. *Heathland Ecology.* Harvard University Press, Cambridge, Mass. Good introduction to the ecology of shrub communities, although oriented toward British heathland.

MCKELL, C. M., ET AL. (EDS.). 1972. *Wildland Shrubs, Their Biology and Utilization.* Forest Service General Tech. Rept. INT-1. Intermountain Forest and Range Experiment Station, U.S. Department of Agriculture, Washington, D.C. Definitive and much-needed reference on shrubs, a subject on which little literature exists. Good bibliography.

MOONEY, H. A. (ED.). 1977. *Convergent Evolution in Chile and California Mediterranean Climate Ecosystems.* US/IBP Synthesis Series. Academic Press, New York.

POLLARD, E., M. D. HOOPER, AND N. W. MOORE. 1969. *Hedges.* Collins, Glasgow, Scotland. Only major reference on a very important habitat.

### Desert

BROWN, G. W., JR. (ED.). 1976–1977. *Desert Biology,* 2 vols. Academic Press, New York. Encyclopedia; basic information on biological and physical features of world deserts.

BUXTON, P. A. 1923. *Animal Life in Deserts.* Edward Arnold, London. Old book on animals in the desert that still has its interest.

CLOUDSLEY-THOMPSON, J. L. 1954. *Biology of the Deserts.* Hafner, New York. General book on desert life.

GOODALL, D. W. (ED.). 1976. *Evolution of Desert Biota.* University of Texas Press, Austin. Survey of geological history of deserts, evolution of their plants and animals, and adaptations to environment.

HOWES, P. G. 1954. *The Giant Cactus Forest and Its World,* Duell, Sloan & Pearce, New York. Pleasant introduction to the southwestern desert.

JAEGER, E. C. 1957. *The North American Deserts.* Stanford University Press, Stanford, Calif. General survey; excellent reference.

KIRMIZ, J. P. 1962. *Adaptation to Desert Environment; A Study On the Jerboa, Rat and Man.* Butterworth, Washington, D.C. Detailed treatment of the adaptations of mammalian life in the desert.

KRUTCH, J. W. 1952. *The Desert Year.* Sloane, New York. A classic on the desert that should be read for both the description and the philosophy.

LEOPOLD, A. STARKER. 1961. *The Desert.* Life Nature Library. Time-Life, New York. Excellent, well-illustrated general account of the desert.

MABRY, T. J., ET AL. (EDS.). 1977. *Creosote Bush, Biology and Chemistry of Larrea in New World Deserts.* Dowden, Hutchinson & Ross, Stroudsburg, Pa.; distr. by Academic Press, New York.

ORIONS, G. H., AND O. T. SOLBRIG. 1977. *Convergent Evolution in Warm Deserts.* US/IBS Synthesis Series. Dowden, Hutchinson & Ross, Stroudsburg, Pa.; distr. by Academic Press, New York.

SHREVE, F., AND I. L. WIGGINS. 1963. *Vegetation and Flora of the Sonoran Desert,* 2 vols. Stanford University Press, Stanford, Calif. Monumental study of the desert vegetation of the Southwest.

WEST, N. E., AND J. SKUJINS (EDS.). 1978. *Nitrogen in Desert Ecosystems.* Dowden, Hutchinson & Ross, Stroudsburg, Pa.; distr. by Academic Press, New York.

### Forest

AUBERT, DE LA RUE, E. F. BOURLIERE, AND J. HARROY. 1957. *The Tropics.* Knopf, New York. Good introduction to the tropical world.

BRAUN, E. L. 1950. *Deciduous Forests of Eastern North America.* McGraw-Hill–Blakiston, New York. The ecology of the forests of eastern North America, their development, composition, and distribution.

COUSENS, J. 1974. *An Introduction to Woodland Ecology.* Oliver Boyd, Edinburgh, Scotland; distr. by Longman, New York. Excellent introduction to woodland ecology with emphasis on British woodlands.

CURTIS, J. T. 1959. *The Forest of Wisconsin.* University of Wisconsin Press, Madison. Sound regional ecology of forests that should be a model for future studies.

FARB, P. 1961. *The Forest.* Life Nature Library. Time-Life, New York. Excellent introduction to forest ecology; beautifully written and illustrated.

FARNSWORTH, E. G. AND F. B. GOLLEY: 1974. *Fragile Ecosystems: Evaluation of Research and Applications in the Neotropics.* Springer-Verlag, New York. Overview of tropical ecology and interactions of humans and tropical environments.

GOLLEY, F. B. 1975. *Mineral Cycling in Tropical Ecosystems.* University of Georgia Press, Athens.

GOLLEY, F. B., AND E. MEDINA (EDS.). 1975. *Tropical Ecological Systems.* Springer-Verlag, New York. Research trends, theoretical and applied, in the tropics.

HOLDRIDGE, L. R., ET AL. (EDS.). 1971. *Forest Environments in Tropical Life Zones: A Pilot Study.* Pergamon Press, Elmsford, N.Y.

MCCORMICK, J. 1959. *The Living Forest.* Harper & Row, New York. Short introduction to the forest—ecology, disease, insects, harvesting, types.

NEAL, E. 1958. *Woodland Ecology.* Harvard University Press, Cambridge, Mass. A gem; written with the British woodland in mind, but overall presentation applicable to all temperate forests.

PLATT, R. 1965. *The Great American Forest.* Prentice-Hall, Englewood Cliffs, N.J. Sweeping look at the forest in all its aspects. A popular introduction.

RICHARDS, P. W. 1952. *The Tropical Rain Forest.* Cambridge University Press, London. The book on the tropical rain forest.

U.S. DEPARTMENT OF AGRICULTURE. 1949. Trees. *Yearbook of Agriculture.* U.S. Department of Agriculture, Washington, D.C. Overall view of the forest—the trees, forest regions, forest management.

WHITMORE, T. C. 1975. *Tropical Rain Forests of the Far East.* Clarendon, Oxford. Excellent survey of Asian rain forests with emphasis on autecology, soils, diversity, and stratification.

### Tundra

BRITTON, M. E. 1957. Vegetation of the arctic tundra. *18th Biol. Colloq. Oregon State Chapter Phi Kappa Phi.* Oregon State College, Corvallis. Detailed summary.

BROOKS, M. B. 1967. *The Life of the Mountains.* Our Living World of Nature Series. McGraw-Hill, New York. Beautifully written survey of mountain life in North America.

BRUEMMER, F. 1974. *The Arctic.* Quadrangle, New York. Although for nonspecialists, a sophisticated introduction to the arctic ecosystem. Superbly illustrated.

FREUCHEN, P., AND F. SALMOSEN. 1958. *The Arctic Year.* Putnam, New York. Excellent reading, scientifically sound.

IVES, J. D., AND R. G. BARRY (EDS.). 1974. *Arctic and Alpine Environments.* Barnes & Noble Books, New York. A storehouse of information.

LEY, WILLY. 1962. *The Poles.* Life Nature Library. Time-Life, New York. Description of life and exploration in the polar regions. Well illustrated.

MILNE, L., AND M. MILNE. 1962. *The Mountains.* Life Nature Library. Time-Life, New York. Account of alpine life.

PEARSALL, W. H. 1960. *Mountains and Moorlands.* Collins, London. European mountains and moorlands well described; emphasis on Britain.

POLUNIN, N. 1948. *Botany of the Canadian Eastern Arctic; III, Vegetation and Ecology.* Bulletin 104. National Museum of Canada, Ottawa. Basic reference by an authority of the tundra.

P'YAVCHENKO, N. I. 1964. *Peat Bogs of the Russian Forest Steppe.* Israel Program for Scientific Translations, Jerusalem. Extensive survey of the vast peatlands of Eurasia, their development and ecology.

ROSSWALL, T., AND O. W. HEAL (EDS.). 1975. *Structure and Function of Tundra Ecosystems.* Ecological Bulletin no. 20. Swedish Natural Sciences Research Council, Stockholm.

STONEHOUSE, B. 1971. *Animals of the Arctic; the Ecology of the Far North.* Holt, Rinehart and Winston, New York. Colorful and informative book on arctic animal life.

WASHBURN, A. L. 1956. Classification of patterned ground and review of suggested origins. *Geol. Soc. Am. Bull.,* 67:823–865. Detailed discussion of patterned ground in the Arctic.

WIELGOLASKI, F. E. (ED.). 1975–1976. *Fennoscandian Tundra Ecosystems.* Part 1, *Plants and Microorganisms;* Part 2, *Animals and Systems Analysis.* Springer-Verlag, New York.

ZWINGER, A. H., AND B. E. WILLARD. 1972. *Land Above the Trees.* Harper & Row, New York. Excellent guide to American alpine tundra.

## Part IV

### Ecological genetics and natural selection

BEADLE, G., AND M. BEADLE. 1966. *The Language of Life.* Doubleday, Garden City, N.Y. Engaging and lucid account of genetics.

BLAIR, W. F. (ED.). 1961. *Vertebrate Speciation.* University of Texas Press, Austin. Symposium volume containing a tremendous amount of compressed information.

COLD SPRING HARBOR SYMPOSIA ON QUANTITATIVE BIOLOGY. 1955. *Population Genetics: The Nature and Causes of Genetic Variability in Populations.* No. 20. The Biological Laboratory, Cold Spring Harbor, N.Y. A volume that well summarizes the field up the the the mid-1950s.

CREED, R. (ED.). 1971. *Ecological Genetics and Evolution.* Blackwell, Oxford. Essays in honor of E. B. Ford that stress polymorphism and melanism.

DAWKINS, R. 1976. *The Selfish Gene.* Oxford University Press, New York. Interesting exposition of altruism, mortality, parental behavior, and parental care.

DOBZHANSKY, T. 1951. *Genetics and the Origin of Species,* 3rd ed. Columbia University Press, New York. Hard reading in places, but one of the most important books on evolution.

DOBZHANSKY, T., ET AL. 1968. *Evolutionary Biology.* Prentice-Hall, Englewood Cliffs, N.J. Review volumes on the subject.

DOBZHANSKY, T., F. J. AYALA, AND G. L. STEBBINS. 1977. *Evolution.* Freeman, San Francisco. Modern and substantial textbook.

ENDLER, J. A. 1977. *Geographic Variation, Speciation, and Clines.* Princeton University Press, Princeton, N.J.

FORD, E. B. 1970. *Ecological Genetics,* 3rd ed. Methuen, London. Ecological aspects well developed.

GRANT, V. 1971. *Plant Speciation.* Columbia University Press, New York. Speciation phenomena in higher plants.

JAMESON, D. (ED.). 1977. *Evolutionary Genetics.* Benchmark® Papers. Dowden, Hutchinson & Ross, Stroudsburg, Pa.; distr. by Academic Press, New York. Twenty-five papers covering significant contributions to early development of evolution from a theoretical viewpoint.

JAMESON, D. L. (ED.). 1977. *Genetics of Speciation.* Benchmark® Papers. Dowden, Hutchinson & Ross, Stroudsburg, Pa.; distr. by Academic Press, New York. Selected papers summarizing development of understanding of genetic basis of origin of species.

JOHNSON, C. 1976. *Introduction to Natural Selection.* University Park Press, Baltimore. Operation and Measurement of natural selection.

LACK, D. 1974. *Darwin's Finches: An Essay on the General Biological Theory of Evolution.* Cambridge University Press, London. A classic of biology. Probably the best introduction to the actual origin of new species.

LACK, D. 1971. *Ecological Isolation in Birds.* Blackwell, London. Valuable book on interspecific competition and isolating mechanisms.

LEVIN, R. 1968. *Evolution in a Changing Environment.* Princeton University Press, Princeton, N.J. Theoretical look at strategies of adaptation and evolution.

LEWONTIN, R. C. 1974. *The Genetic Basis of Evolutionary Change.* Columbia University Press, New York.

LOWE-MCCONNELL, R. H. (ED.). 1967. Speciation in tropical environments. *Biol. J. Linn. Soc. Lond.,* 1:1–2. Academic, New York. Excellent papers on speciation and species diversity in the tropics.

MAYR, E. 1970. *Population, Species, and Evolution.* Harvard University Press, Cambridge, Mass. An abridgement of *Animal Species and Evolution* and much more usable. Excellent reference.

SAVORY, T. 1963. *Naming the Living World.* Wiley, New York. Introduction to the principles of biological nomenclature.

SCIENTIFIC AMERICAN. 1978. *Evolution.* Special Issue, September 1978. Scientific American, New York.

SLOBODCHIKOFF, C. N. 1976. *Concepts of Species.* Benchmark® Papers. Dowden, Hutchinson & Ross, Stroudsburg, Pa.; distr. by Academic Press, New York. Historical papers on the subject.

STEBBINS, G. L. 1971. *Processes of Organic Evolution.* Prentice-Hall, Englewood Cliffs, N.J. Introductory text strong on speciation.

WHITE, M. J. D. 1978. *Modes of Speciation.* Freeman, San Francisco. Speciation from the viewpoint of animal cytology; excellent discussion of sympatric speciation.

WILSON. E. O. (ED.). 1974. *Ecology, Evolution, and Population Biology.* Freeman, San Francisco. Readings from *Scientific American.*

### *Behavior*

ALCOCK, J. 1978. *Animal Behavior: An Evolutionary Approach.* Sinauer, Sunderland, Mass. Popular, well-illustrated text.

ARMSTRONG, E. A. 1964. *Bird Display and Behavior,* rev. ed. Dover, New York. Overall survey of bird behavior. Dated, but still of interest.

BANKS, E. M. 1977. *Vertebrate Social Organization.* Benchmark® Papers. Dowden, Hutchinson & Ross, Stroudsburg, Pa., dist. by Academic, New York. Emphasis on ecological and evolutionary constraints on animal social behavior.

BURGHARDT, G. M., AND H. BEKOFF (EDS.). 1978. *The Development of Behavior: Comparative and Evolutionary Aspects.* Garland STMP Press, New York. Covers behavioral development throughtout animal kingdom.

COLLIAS, N. E., AND E. C. COLLIAS. 1976. *External Construction by Animals.* Benchmark® Papers. Dowden, Hutchinson & Ross, Stroudsburg, Pa., dist. by Academic, New York.

CROOK, J. H. (ED.). 1970. *Social Behavior in Birds and Mammals.* Academic Press, New York. Good on feeding dispersal of birds, breeding adaptations, and parental care.

DARLING, F. F. 1937. *A Herd of Red Deer.* Oxford University Press, New York. A classic in the study of animal behavior.

DAVIS, D. E. 1974. *Behavior as an Ecological Factor.* Benchmark® Papers. Dowden, Hutchinson & Ross, Stroudsburg, Pa., dist. by Academic, New York.

EIBL-EIBESFELDT, I. 1970. *Ethology, The Biology of Behavior.* Holt, Rinehart and Winston, New York. Excellent introductory text on behavior.

GEIST, V. 1971. *Mountain Sheep, A Study in Behavior and Evolution.* University of Chicago Press, Chicago. Good field study of behavior, well illustrated.

GRZIMEK, B.(ED.). 1977. *Grzimek's Encyclopedia of Ethology.* Van Nostrand Reinhold, New York. Reasonably thorough coverage of ethology.

HAFEZ, E. S. (ED.). 1962. *The Behavior of Domestic Animals.* Williams & Wilkins, Baltimore. A thorough reading of this book would be a valuable beginning for the study of big game animals.

HARBORNE, J. B. 1977. *Introduction to Ecological Chemistry.* Academic, New York. An excellent short introduction to basics of chemical defense and communication.

HESS, E. H., AND S. B. PETROVICH. 1977. *Imprinting.* Benchmark® Papers. Dowden, Hutchinson & Ross, Stroudsburg, Pa. Reference volume providing guide to original literature.

HINDE, R. A. 1970. *Animal Behavior,* 2nd ed. McGraw-Hill, New York. Synthesis of ethology and comparative psychology. Concentrates on the causes of behavior.

JEWELL, P. A., AND C. LORZOS (EDS.). 1966. *Play, Exploration, and Territory in Animals.* Symposium of the Zoological Society of London. Academic Press, New York. Contains a group of informative papers.

JOHNSGARD, P. A. 1965. *Handbook of Waterfowl Behavior.* Cornell University Press, Ithaca, N.Y. Excellent descriptions of waterfowl displays.

JOHNSGARD, P. A. 1971. *Animal Behavior,* 2nd ed. Brown, Dubuque, Iowa. Introduction to animal behavior.

KENDEIGH, S. C. 1952. Parental care and its evolution in birds. *Illinois University Biological Monographs 22.* University of Illinois Press, Urbana. Summarizes parental care by birds with special reference to the house wren.

KLOPFER, P. H. 1973. *Behavioral Aspects of Ecology,* 2nd. ed. Prentice-Hall, Englewood Cliffs, N.J. Behavioral Approach to ecology. Highly speculative in places, but a refreshing point of view.

KREBS, J. R., AND N. B. DAVIES (EDS.). 1978. *Behavioral Ecology: An Evolutionary Approach.* Blackwell, Oxford; distr. by Sinauer, Sunderland, Mass. Up-to-date reference covering optimal foraging, predator and prey, sex and mating signal, habitat selection, territory, optimal tactics of reproduction.

LORENZ, K. 1952. *King Solomon's Ring.* Crowell, New York. Popular book in which this great ethologist describes his experiences in studying animal behavior.

RHEINGOLD, H. 1963. *Maternal Behavior in Mammals.* Wiley, New York. The first general review of maternal behavior in mammals.

SCHEIN, M. W. (ED.). 1975. *Social Hierarchy and Dominance.* Benchmark® Papers. Dowden, Hutchinson & Ross, Stroudsburg, Pa., dist. by Academic, New York. Papers tracing the development of the concept of social hierarchy.

SEBEOK, T. (ED.). 1977. *How Animals Communicate.* Indiana University Press, Bloomington. Detailed spurcebook on phenomenon of animal communication.

SHOREY, H. H. 1976. *Animal Communication by Pheromones.* Academic Press, New York.

SILVER, R. (ED.). 1978. *Parental Behavior in Birds.*

Benchmark® Papers. Dowden, Hutchinson & Ross, Stroudsburg, Pa. Covers all aspects of parental behavior from breeding season to post hatching.

SMITH, W. J. 1977. *The Behavior of Communicating: An Ecological Approach.* Harvard University Press, Cambridge, Mass.

STOKES, A. W. (ED.). 1974. *Territory.* Benchmark® Papers. Dowden, Hutchinson & Ross, Stroudsburg, Pa. Survey of development of ideas on territory.

TAVOLGA, W. N. 1969. *Principles of Animal Behavior.* Harper & Row, New York.

TAVOLGA, W. N. 1976. *Sound Reception in Fishes.* Benchmark® Papers. Dowden, Hutchinson & Ross, Stroudsburg, Pa.

TAVOLGA, W. N. 1977. *Sound Production in Fishes.* Benchmark® Papers. Dowden, Hutchinson & Ross, Stroudsburg, Pa.

THIELCKE, G. A. 1976. *Bird Sounds.* University of Michigan Press, Ann Arbor. Short, informative work on all aspects of bird vocalizations.

THORPE, W. H. 1961. *Bird Song: The Biology of Vocal Communication and Expression in Birds.* Cambridge University Press, London. Important work on vocal communications in birds.

THORPE, W. H. 1963. *Learning and Instinct in Animals,* rev. ed. Methuen, London. Basic work in the field of ethology.

TINBERGEN, N. 1951. *The Study of Instinct.* Oxford University Press, New York. Major work in the comparative study of behavior.

TINBERGEN, N. 1960. *The Herring Gull's World.* Basic Books, New York. Tinbergen's finest book, an excellent example of how basic research can be presented as good literature.

TINBERGEN, N. 1965. *Animal Behavior.* Life Nature Library. Time-Life, New York. Highly readable and richly illustrated introduction to ethology.

WILSON, E. O. 1971. *Insect Societies.* Harvard University Press, Cambridge, Mass. Outstanding book on social insects and their behavior.

WILSON, E. O. 1975. *Sociobiology.* Harvard University Press, Cambridge, Mass. Important synthesis of biological principles that govern social behavior and social organization in all kinds of animals.

### Populations

ANDREWARTHA, H. G. 1970. *Introduction to the Study of Animal Populations.* University of Chicago Press, Chicago. Basically a summary of the following volume.

ANDREWARTHA, H. G., AND L. C. BIRCH. 1954. *The Distribution and Abundance of Animals.* University of Chicago Press, Chicago. Stimulating and controversial text.

CHENG, T. C. 1971. *Aspects of the Biology of Symbiosis.* University Park Press, Baltimore. Papers from AAAS Symposium. Wide range of topics.

CHRISTIANSEN, F. B., AND T. M. FENCHEL. 1977.

*Theories of Population and Biological Communities*. Springer-Verlag, New York. Advanced, highly theoretical text with emphasis on mathematical models.

CLARK, L. R., ET AL. (EDS.). 1967. *The Ecology of Insect Populations in Theory and Practice*. Methuen, London. Introduction to population ecology from entomological point of view. Pedantic, but important reference.

CURIO, E. 1976. *The Ethology of Predation*. Springer-Verlag, New York. Refreshingly different view of predation that emphasizes the individual rather than the population.

DORST, J. 1963. *The Migration of Birds*. Houghton Mifflin, Boston. Survey of bird migration the world over.

EASTWOOD, E. 1967. *Radar Ornithology*. Methuen, London. Account of bird movements as observed by radar.

ELTON, C. E. 1958. *Voles, Mice and Lemmings*. Methuen, London. A classic book on animal populations.

ERRINGTON, P. L. 1963. *Muskrat Populations*. Iowa State University Press, Ames. Results of a 20-year study of muskrats. A model of field investigations.

ERRINGTON, P. L. 1967. *Of Predation and Life*. Iowa State University Press, Ames. Philosophical view of predation.

FENNER, F., AND F. M. RATCHIFFE. 1965. *Myxomatosis*. Cambridge University Press, New York. The story of myxomatosis, rabbit populations, and the adaptations of hosts and parasites.

HARRISON, G. A., AND A. J. BOYCE (EDS.). 1971. *The Structure of Human Populations*. Clarendon, London. Demography, ecology, psychology, anthropology, and sociology brought together in an analysis of human populations.

HENRY, I. S. M. 1966. *Symbiosis*, vols. 1, 2. Academic Press, New York. Sweeping survey of symbiosis. Important reference.

HUTCHINSON, G. E. 1978. *An Introduction to Population Ecology*. Yale University Press, New Haven, Conn. Excellent introductory text by a world authority; ranges from historical past to current research.

JOHNSON, C. G. 1969. *Migration and Dispersal of Insects by Flight*. Methuen, London. Exhaustive review of subject. Basic reference.

JONES, F. R. H. 1968. *Fish Migration*. St. Martin's Press, New York. Exhaustive account of fish migration over the world.

KEYFITZ, N., AND W. FLIEGER. 1971. *Populations, Facts and Methods of Demography*. Freeman, San Francisco. Fundamentals of demography and a compendium of the world's human populations.

LACK, D. 1954. *The Natural Regulation of Animal Numbers*. Oxford University Press, New York. Emphasizes the role of food.

LACK, D. 1966. *Population Studies of Birds*. Oxford University Press, New York. Case history studies.

LE CREN. E. D., AND M. W. HOLDGATE (EDS.). 1962. *The Exploitation of Natural Animal Populations*. Wiley, New York. Symposium volume which explores the scientific basis of the response of natural animal populations to exploitation.

MACFAYDEN, A. 1963. *Animal Ecology*, 2nd ed. Pitman, London. Technical introduction with emphasis on soil organisms.

ORR, R. T. 1970. *Animals in Migration*. Macmillan, New York. Excellent popular account of migration among animals.

ROUGHGARDEN, J. 1969. *Theory of Population Genetics and Evolutionary Ecology: An Introduction*. Macmillan, New York.

SCHMIDT-KOENIG, K. 1975. *Migration and Homing in Animals*. Springer-Verlag, New York. Introduction to subject.

SCHMIDT-KOENING, K., AND W. T. KEETON (EDS.). 1978. *Animal Migration, Navigation, and Homing*. Springer-Verlag, New York. Papers cover hypotheses and experimental evidence of migratory behavior and mechanisms of nagivation.

SINCLAIR, A. R. E. 1977. *The African Buffalo: A Study of Resource Limitations of Populations*. University of Chicago Press, Chicago. Significant field study of population regulation through starvation and associated problems of malnutrition.

SLOBODKIN, L. B. 1962. *Growth and Regulation of Animal Populations*. Holt, Rinehart and Winston, New York. Population dynamics, laboratory rather than field oriented.

SMITH, J. M. 1974. *Models in Ecology*. Cambridge University Press, New York. Mathematical models aimed at understanding general properties of population ecology.

TAMARIN, R. H. (ED.). 1978. *Population Regulation*. Benchmark® Papers. Dowden, Hutchinson & Ross, Stroudsburg, Pa.; distr. by Academic Press, New York. Papers showing the historical development of this controversial subject and presenting areas of most active research.

VERNBERG, N. B. (ED.). 1977. *Symbiosis in the Sea*. University of South Carolina Press, Columbia. Summary of knowledge about certain aspects of marine symbiosis.

WATSON, A. 1970. *Animal Populations in Relation to Their Food Resources*. Blackwell, Oxford. Investigations into animal-food interactions. A number of interesting papers.

WICKLER, W. 1968. *Mimicry in Plants and Animals*. World University Library, McGraw-Hill, New York. Brief introduction to subject.

## Part V

### Community and succession

AGER, D. V. 1963. *Principles of Paleoecology*. McGraw-Hill, New York. Survey of the subject of communities in the past.

ASHBY, M. 1969. *Introduction to Plant Ecology*. St. Martin's Press, New York. Elementary text that is both interesting and stimulating.

BROOKHAVEN NATIONAL LABORATORY. 1969. *Diversity and Stability in Ecological Systems*. Brookhaven Symposia in Biology no. 22. National Technical Information Service, Springfield, Va. Summary of current status of theory.

CLAPHAM, A. R., AND B. N. NICHOLSON. 1975. *The Oxford Book of Trees*. Oxford University Press, London. Although basically aimed at identification, contains many beautiful illustrations of plant communities and vegetation development together with descriptions.

CODY, M., AND J. M. DIAMOND (EDS.). 1975. *Ecology and Evolution of Communities*. Harvard University Press, Cambridge, Mass. Stimulating theoretical approach to population ecology, but often highly speculative with minimal field data.

CUSHING, E. J., AND H. E. WRIGHT. 1967. *Quaternary Paleoecology*. Yale University Press, New Haven, Conn. Covers glacial sequences, methodology, regional studies, climatic history.

DAUBENMIRE, D. 1968. *Plant Communities: A Textbook of Synecology*. Harper & Row, New York. Strong on succession.

DICE, L. C. 1952. *Natural Communities*. University of Michigan Press, Ann Arbor. Good book that has never been fully appreciated.

DRURY, H. H., AND C. T. NISBET. 1975. Succession. *J. Arnold Arboretum*, 54:331–368. Paper included here to call attention to a significant contribution that for various reasons was eventually published in a relatively obscure journal, included in Golley, 1977.

ELTON, C. E. 1927. *Animal Ecology*. Sidgwick & Jackson, London.

ELTON, C. E. 1958. *The Ecology of Invasion by Plants and Animals*. Wiley, New York. This and the previous book, classics in the best sense of the word, are excellent introductions to community ecology. Short, clearly written.

GOLLEY, F. B. (ED.). *Ecological Succession*. Benchmark® Papers. Dowden, Hutchinson & Ross, Stroudsburg, Pa., distr. by Academic Press, New York. Classical papers in plant succession.

GRAHAM, E. H. 1944. *Natural Principles of Land Use*. Oxford University Press, New York (Greenwood). Excellent book stressing the biological concepts applied to land management. Should be required reading for all.

HAWKSWORTH, D. L. 1974. *The Changing Flora and Fauna of Britain*. Academic Press, New York. Important look at effects of environmental changes on flora and fauan.

KNAPP, R. (ED.). 1974. *Vegetation Dynamics*. Vol. 8. *Handbook of Vegetation Science*. W. Junk, The Hague. Classical view of plant succession; valuable for information on vegetation development in Europe.

KOMAREK, E. V. (ED.). 1962–1976. *Annual Tall Timbers Fire Ecology Conference*. Tallahassee, Fla. The finest source of information on forest fire ecology available.

MCINTOSH, R. H. (ED.). 1978. *Phytosociology*. Benchmark® Papers. Dowden, Hutchinson & Ross, Stroudsburg, Pa.; distr. by Academic Press, New York. Traces origin and development of plant sociology. Many hard-to-find papers reprinted.

MAY, R. M. 1973. *Stability and Complexity in Model Ecosystems*. Princeton University Press, Princeton, N.J. Theoretical examination of community structure, community interactions (feedbacks), environmental fluctuations, and stability in communities.

OOSTING, H. S. 1956. *The Study of Plant Communities*. Freeman, San Francisco. General introduction to plant ecology; something of a classic.

PENNINGTON, W. 1969. *History of British Vegetation*. The English University Presses, London.

PIELOW, E. C. 1975. *Population and Community Ecology*. Gordon and Breach, New York. Mathematical and theoretical approach.

RAUP, D. M., AND S. M. STANLEY. 1978. *Principles of Paleontology*. Freeman, San Francisco.

THOMAS, W. L., JR. (ED.). 1956. *Man's Role in Changing the Face of the Earth*. University of Chicago Press, Chicago. Symposium volume that presents a thorough discussion of human impact through the ages on the biosphere.

TUBBS, C. R. 1968. *The New Forest: An Ecological History*. David and Charles, Newton Abbot, England. Interesting discussion of vegetation disturbance and development on a tract of land through the centuries.

TUREKIAN, K. K. (ED.). 1971. *The Late Cenozoic Glacial Ages*. Yale University Press, New Haven, Conn. Description of worldwide tectonic and glacial events during the late Cenozoic.

WEST, R. G. 1968. *Pleistocene Geology and Biology*. Wiley, New York. General introduction of interest as background to paleoecology.

WHITTAKER, R. H., AND S. A. LEVIN (EDS.). 1975. *Niche: Theory and Application*. Benchmark® Papers. Halsted Press, New York. Survey of history of niche concept.

WINDLEY, B. F. 1977. *The Evolving Continents*. Wiley, New York. Continental drift.

WRIGHT, H. E., AND D. G. FREY (EDS.). 1965. *The Quaternary of the United States*. Princeton University Press, Princeton, N.J. A compendium on the Pleistocene in the United States covering all aspects. Dated, but still a major reference.

# Bibliography

ABEE, A., AND D. LAVENDER

1972 Nutrient cycling in throughfall and litterfall in a 450-year-old Douglas fir stand. In J. F. Franklin et al., 1972, pp. 133–143.

ABRAHAMSON, W. G., AND M. D. GADGIL

1973 Growth form and reproductive effort in goldenrod (Solidago, Compositae). Amer. Natur., 107:651–661.

ADKISSON, P. L.

1964 Action of the photoperiod in controlling insect diapause. Amer. Natur., 98:357–374.

1966 Internal clocks and insect diapause. Science, 154:234–241.

AHLGREN, C. E.

1974 Effects of fire on temperate forests: North Central United States. In T. Kozlowski and C. E. Ahlgren (eds.), 1974, pp. 195–223.

AHLGREN, I. F.

1974 The effects of fire on soil organisms. In T. Kozlowski and C. E. Ahlgren (eds.), 1974, pp. 47–72.

AHLGREN, I. F., AND C. E. AHLGREN

1960 Ecological effects of forest fires. Bot. Rev., 26:483–533.

AHMED, A. K.

1976 PCBs in the environment. Environment, 18(2):6–11.

ALCOCK, J.

1973 Cues used in searching for food by red-winged blackbirds (Agelaires phoeniceus). Behaviour, 46:174–188.

ALEXANDER, R. D.

1960 Sound communication in Orthoptera and Cicadidae. In W. E. Lanyon and W. E. Tavolga (eds.), 1960, pp. 38–96.

1961 Aggressiveness, territoriality and sexual behavior in field crickets (Orthoptera: Gryllidae). Behaviour, 17:130–223.

1962 The role of behavioral study in cricket classification. System. Zool., 11:53–72.

ALEXANDER, R. D., AND G. BORGIA

1978 Group selection, altruism, and levels of organization of life. Ann. Rev. Ecol. Syst., 9:449–474

ALLEE, W. C., B. GREENBERG, G. M. ROSENTHAL, AND P. FRANK

1948 Some effects of social organization on growth in the green sunfish Lepomis cyanellas. J. Exp. Zool., 108:1–19.

ALLEN L. N., AND E. R. LEMON

1976 Carbon dioxide exchange and turbulence in a Costa Rican tropical rainforest. In J. C. Monteith (ed.), 1976, pp. 265–308.

ALM, G.

1952 Year class fluctuations and span of life of perch. Rep. Inst. Freshwater Res. Drottningholm, 33:17–38.

ALTMANN, M.

1952 Social behavior of elk, Cervus canadensis nelsoni, in the Jackson Hole area of Wyoming. Behaviour, 4:116–143.

1960 The role of juvenile elk and moose in the social dynamics of their species. Zoologica, 45:35–40.

ALTSCHUL, A. M.

1967 Food proteins: new sources from seeds. Science, 158:221–226.

AMADON, D.

1947 Ecology and evolution of some Hawaiian birds. Evolution, 1:63–68.

1950 The Hawaiian honey creepers (Aves, Drepanididae). Bull. Amer. Mus. Natur. Hist., 100:397–451.

AMERICAN CHEMICAL SOCIETY

1969 Cleaning Our Environment: The Chemical Basis for Action, American Chemical Society, Washington.

AMMANN, G. A.

1957 The Prairie Grouse of Michigan, Michigan Dept. of Conservation, Lansing.

ANDERSON, D. W., J. J. HICKEY, R. W. RISEBROUGH, D. F. HUGHES, and R. E. CHRISTENSEN

1969 Significance of chlorinated hydrocarbon residues to breeding pelicans and cormorants. Can. Field Natur., 83:91–112.

ANDERSON, R. C.

1974 Seasonality in terrestrial primary production. In H. Leith, 1974, pp. 103–111.

ANDREWARTHA, H. G., AND L. C. BIRCH

1954 The Distribution and Abundance of Animals, University of Chicago Press.

ANDREWS, R., D. C. COLEMAN, J. E. ELLIS, AND J. S. SINGH

1974 Energy flow relationships in a short grass prairie ecosystem. Proc. 1st Inter. Cong. Ecol., 22–28.

ANDREWS, R., AND A. S. RAND

1974 Reproductive efforts in anoline lizards. Ecology, 55:1317–1327.

ANTOINE, L. H., JR.

1964 Drainage and best use of urban land. Public Works, 95:88–90.

ANTONOVICS, J.

1971 The effects of a heterogeneous environment on the genetics of natural populations. Amer. Sci., 59:593–599.

ASCHMANN, H.

1973 Distribution and peculiarity of mediterranean ecosystems. In F. di Castri and H. A. Mooney (eds.), 1973, pp. 11–19.

ASCHOFF, J.

1958 Tierische Periodik unter dem Einfluss von Zeitgebern. Z. F. Tierpsychol., 15:1–30.

1966 Circadian activity pattern with two peaks. Ecology, 47:657–662.

1969 Desynchronization and resynchronization of human rhythms. Aerospace Med., 40:844–849.

ASH, J. E., AND J. P. BARKHAM

1976 Changes and variability in the field layer of a coppiced woodland in Norfolk, England. J. Ecol., 64:697–712.

ATSATT, P. R., AND D. J. O'DOWD

1976 Plant defense guilds. Science, 193:24–29.

AUCLAIR, A. N., AND G. COTTAM

**771**

1971 Dynamics of black cherry (*Prunus serotina* Erbr.) in southern Wisconsin oak forests. *Ecol. Monogr.,* 41:153–177.

AUCLAIR, A. N., AND F. G. GOFF
1971 Diversity relations of upland forests in the western Great Lakes area. *Amer. Natur.,* 105:499–528.

AUERBACH, S. I., D. J. NELSON, AND E. G. STRUXNESS
1974 Environmental Sciences Division Annual Progress Report Period ending Sept. 30, 1973. *Environ. Sci. Div. Pub. No. 575,* Oak Ridge National Laboratory.

AUSMUS, B. S., N. T. EDWARDS, AND M. WITKAMP
1975 Microbial immobilization of carbon, nitrogen, phosphorus, and potassium: implications for forest ecosystem processes. *Proc. Brit. Ecol. Soc. Symp. on Decomposition,* Blackwell, Oxford.

AVERY, R. A.
1975 Clutch size and reproductive effort in the lizard *Lacerta vivipara* Jaquin. *Oecologia,* 19:165–170.

BAAKEN, A.
1959 Behavior of gray squirrels. *Symp. Gray Squirrel, Contr. 162,* Maryland Dept. Resource Education, Annapolis, pp. 393–407.

BACKIEL, T., AND E. D. LECREN
1967 Some density relationships for fish population parameters. In S. D. Gerking (ed.), *The Biological Basis of Freshwater Fish Production,* Wiley, New York, pp. 261–293.

BAER, J. G.
1951 *Ecology of Animal Parasites,* University Illinois Press, Urbana.

BAERENDS, G. P.
1959 The ethological analysis of incubation behavior. *Ibis,* 101:357–368.

BAERENDS, G. P., AND J. M. BAERENDS-VAN ROON
1950 An introduction to the study of the ethology of cichlid fishes. *Behaviour Suppl.,* 1:7–242.

BAES, C. F., JR., H. E. GOELLER, J. S. OLSON, AND R. M. ROTTY
1977 Carbon dioxide and the climate: the uncontrolled experiment. *Amer. Sci.,* 65:310–320.

BALL, R. C., AND F. F. HOOPER
1963 Translocation of phosphorus in a trout stream ecosystem. In V. Schultz and A. Klement (eds.), 1963, pp. 217–228.

BALSER, DONALD S., H. H. DILL, AND H. K. NELSON
1968 Effect of predator reduction on waterfowl nesting success. *J. Wild. Manage.,* 32:669–682.

BARBER, H. S.
1951 North American fireflies of the genus *Photuris. Smithsonian Inst. Misc. Coll.,* 117:1–58.

BARKALOW, F. S., JR., R. B. HAMILTON, AND R. F. SOOTS, JR.
1970 The vital statistics of an unexploited gray squirrel population, *J. Wild. Manage.,* 34:489–500.

BARLOCHER, F., AND B. KENDRICK
1974 Dynamics of the fungal population of leaves in a stream. *J. Ecology,* 62:761–791.

BARLOW, J. C., AND D. M. POWER
1970 An analysis of character variation in red-eyed and Philadelphia vireos (Aves: Vireonidae) in Canada. *Can. J. Zool.,* 48:673–678.

BARLOW, J. P.
1955 Physical and biological processes determining the distribution of zooplankton in a tidal estuary. *Biol. Bull.,* 109:211–225.
1956 Effect of wind on salinity distribution in an estuary. *J. Marine Res.,* 15:192–203.

BARSDALE, R. J., AND V. ALEXANDER
1979 Nitrogen balance of Arctic tundra pathways, rates, and environmental implications. *J. Env. Quality,* 4:111–117.

BARTHOLOMEW, B.
1970 Bare zone between California shrub and grassland communities: the role of animals. *Science,* 170:1210–1212.

BARTHOLOMEW, G. A.
1959 Mother-young relations and the maturation of pup behaviour in the Alaskan fur seal. *Anim. Behav.,* 7:163–171.

BARTHOLOMEW, G. A., AND T. J. CADE
1957 Temperature regulation, hibernation and aestivation in the little pocket mouse *Perognathus longimembris. J. Mammal.,* 38:60–72.

BARTHOLOMEW, G. A., T. R. HOWELL, AND T. J. CADE
1959 Torpidity in the white-throated swift, Anna hummingbird, and poor-will. *Condor,* 59:145–155.

BASSAM, J. A.
1965 Photosynthesis. In J. Bonner and J. E. Varnea (eds.), *Plant Biochemistry,* Academic, New York, pp. 875–902.
1977 Increasing crop production through more controlled photosynthesis. *Science,* 197:630–638

BASTOCK, M., D. MORRIS, AND M. MOYNIHAN
1953 Some comments on conflict and thwarting in animals. *Behaviour,* 6:66–84.

BATCHELDER, R. B.
1967 Spatial and temporal patterns of fire in the tropical world. *Proc. 6th Tall Timbers Fire Ecol. Conf.,* pp. 171–208.

BATESON, G.
1963 The role of somatic change in evolution. *Evolution,* 17:529–539.

BATZLI, G. O., AND F. A. PITELKA
1970 Influence of meadow mouse populations on California grassland. *Ecology,* 51:1027–1039.

BAUMGARTNER, A.
1968 Ecological significance of the vertical energy distribution in plant stands. In F. E. Eckardt (ed.), 1968, pp. 367–374.

BAZZAZ, F. A.
1975 Plant species diversity in old field successional ecosystems in southern Illinois. *Ecology,* 56:485–488.

BEALS, E. W.
1968 Spatial pattern of shrubs on a desert plain in Ethiopia. *Ecology,* 49:744–746.

BEAMISH, R. J.

1976 Acidification of lakes in Canada by acid precipitation and the resulting effects on fish. *Water Air Soil Pollut.*, 6(2, 3, 4):501–514.

BEATLEY, J. C.

1969 Dependence of desert rodents on winter annuals and precipitation. *Ecology,* 50:721–724.

1976 Rainfall and fluctuating plant populations in relation to distributions and numbers of desert rodents in southern Nevada. *Oecologia,* 24:21–42.

BECKWITH, S. L.

1954 Ecological succession on abandoned farm lands and its relationship to wildlife management. *Ecol. Monogr.,* 24:349–376.

BEDDINGTON, J. R., M. P. HASSELL, AND J. H. LAWTON

1976 The components of arthropod predation. II. The predator rate of increase. *J. Anim. Ecol.,* 45:165–185.

BEEBE, F. L.

1960 The marine peregrines of the northwest Pacific coast. *Condor,* 62:145–189.

BELL, R. H. V.

1974 A grazing ecosystem in the Serengeti. *Sci. Amer.,* 225(1): 86–93.

BELLINGER, P. F.

1954 Studies of soil fauna with special reference to the Collembola. *Conn. Agr. Exp. Sta. Bull. No. 583.*

BELLIS, U.

1974 Medium salinity plankton systems. In H. T. Odum, B. J. Copeland, and E. A. McMahan (eds.), 1974, Vol. 2, pp. 358–396.

BELLROSE, F. C., T. G. SCOTT, A. S. HAWKINS, AND J. B. LOW

1961 Sex ratios and age ratios in North American ducks. *Illinois Natur. Hist. Surv. Bull.,* 27:391–474.

BEMMISH, R. J.

1976 Acidification of lakes in Canada by acid precipitation and the resulting effects on fish. In L. Dochinger and T. Seliga (eds.), *Proc. 1st Int. Symp. on Acid Precipitation and the Forest Ecosystem, USDA Forest Serv. Gen. Tech. Rept. NE-23.*

BENDELL, J. F.

1974 Effects of fire on birds and mammals. In T. Kozlowski and C. E. Ahlgren (eds.), 1974, pp. 73–138.

BENNINGHOFF, W. S.

1952 Interaction of vegetation and soil frost phenomena. *Arctic,* 5:34–44.

BENSON, L.

1962 *Plant Taxonomy, Methods and Principles,* Ronald, New York.

BENTLEY, B. L.

1977 Extrafloral nectaries and protection by pugnacious bodyguards. *Ann. Rev. Ecol. Syst.,* 8:407–427.

BENTLEY, P. J.

1966 Adaptations of amphibians to arid environments. *Science,* 152:619–623.

BERG, A., S. KJELVICK, AND F. E. WIELGOLASKI

1975a Measurement of leaf areas and leaf angles

of plants at Hardangervidda, Norway. In F. E. Wielgolaski (ed.), 1975, pp. 103–110.

1975b Distribution of $^{14}C$ photosynthates in Norwegian alpine plants. In F. E. Wielgolaski (ed.), 1975, pp. 208–215.

BERG, W., A. JONNELS, B. SJOSTRAND, AND T. WESTERMARK

1966 Mercury content in feathers of Swedish birds from the past 100 years. *Oikos,* 17:71–83.

BERGERUD, A. T.

1971 The population dynamics of Newfoundland caribou. *Wildl. Monogr.,* No. 25.

BERGERUD, A. T., AND F. MANUEL

1969 Aerial census of moose in central Newfoundland. *J. Wildl. Manage.,* 33:910–916.

BERGH VAN DEN, J. P.

1969 Distribution of pasture plants in relation to chemical properties of the soil. In I. H. Rorison (ed.), 1969, pp. 11–23.

BERGH VAN DEN, J. P., AND C. T. DEWIT

1960 Concurrentie tussen Timothee en Reukgras. *Meded. Inst. biol. scherk. Onderz. Lanib. Gewass.,* 121:155–165.

BERGLUND, B. E.

1967 Vegetation and human influence on south Scandinavia during prehistoric times. *Oikos Suppl.,* 12:9–28.

BERNHARD REVERSAT, F.

1975 Nutrients in throughfall and their quantitative importance in rain forest mineral cycles. In F. Golley and E. Medina (eds.), 1975, pp. 153–159.

BIEL, E. R.

1961 Microclimate, bioclimatology, and notes on comparative dynamic climatology. *Amer. Sci.,* 49:326–357.

BILLINGS, W. D., AND H. A. MOONEY

1968 The ecology of arctic and alpine plants. *Biol. Rev.,* 43:481–529.

BIRCH, L. C., AND D. P. CLARK

1953 Forest soil as an ecological community with special reference to the fauna. *Quart. Rev. Biol.,* 28(1):13–36.

BISWELL, H. H.

1974 Effects of fire on chaparral. In T. Kozlowski and C. E. Ahlgren (eds.), 1974, pp. 321–364.

BITMAN, J.

1970 Hormonal and enzymatic activity of DDT. *Agr. Sci. Rev.,* 7(4):6–12.

BJORKMAN, O.

1973 Comparative studies on photosynthesis in higher plants. In A. C. Giese (ed.), *Photophysiology,* Vol. 8, Academic, New York, pp. 1–63.

BJORKMAN, O., AND J. BERRY

1973 High efficiency photosynthesis. *Sci. Amer.,* 229(4):80–93.

BLACK, C. C. JR.

1971 Ecological implications of dividing plants into groups with distinct photosynthetic production capacities. *Adv. Ecol. Res.,* 7:87–114.

1973 Photosynthetic carbon fixation in relation to

net $CO_2$ uptake. *Ann. Rev. Plant Physiol.,* 24:253–286.

BLAIR, A. P.

1942  Isolating mechanisms in a complex of four species of toad. *Biol. Symp.,* 6:235–249.

BLAIR, W. F.

1955  Mating call and stage of speciation in the *Microhyla olivacea–M. carolinensis* complex. *Evolution,* 9:469–480.

1965  Amphibian speciation. In H. E. Wright and D. G. Frey (eds.), 1965, pp. 543–556.

BLEST, A. D.

1957  The function of eyespot patterns in Lepidoptera. *Behaviour,* 11:209–256.

BLISS, L. C.

1956  A comparison of plant development in microenvironments of arctic and alpine tundras. *Ecol. Monogr.,* 26:303–337.

1963  Alpine plant communities of the Presidential Range, New Hampshire. *Ecology,* 44:678–697.

1975  Devon Island, Canada. In T. Rosswall and O. W. Heal (eds.), 1975, pp. 17–60.

BLUM, J. L.

1960  Algae populations in flowing waters. In *Ecology of Algae, Spec. Pub. No. 2,* Pymatuning Lab. of Field Biology, pp. 11–21.

BOGERT, C. M.

1960  The influence of sound on the behavior of amphibians and reptiles. In W. E. Lanyon and W. N. Tavolga (eds.), 1960, pp. 137–320.

BOND, R. R.

1957  Ecological distribution of breeding birds in the upland forests of southern Wisconsin. *Ecol. Monogr.,* 27:351–384.

BORG, K., H. WANNTROP, K. ERNE, AND E. HANKO

1969  Alkyl mercury poisoning in terrestrial Swedish wildlife. *Viltrevy,* 6(4):301–379.

BORMANN, F. H., G. E. LIKENS, AND J. H. MELILLO

1977  Nitrogen budget for an aggrading northern hardwood forest ecosystem. *Science,* 196:981–983.

BORMANN, F. H., G. E. LIKENS, T. H. SICCOMA, R. J. PIERCE, AND J. S. EATON

1974  The effect of deforestation on ecosystem export and steady state conditions at Hubbard Brook. *Ecol. Monogr.,* 44:255–277.

BORNEBUSCH, C. H.

1930  *The Fauna of Forest Soils,* Nielsen and Lydiche, Copenhagen.

BOTKIN, D. B.

1977  Forest, lakes, and the anthropogenic production of carbon dioxide. *Bioscience,* 27:325–331.

BOTKIN, D. B., AND C. R. MALONE

1968  Efficiency of net primary production based on light intercepted during the growing season. *Ecology,* 49:438–444.

BOTT, T. L., AND T. D. BROCK

1969  Bacterial growth rates above 90°C in Yellowstone hot springs. *Science,* 164:1411–1412.

BOX, T. W., J. POWELL, AND D. L. DRAWE

1967  Influence of fire on south Texas chaparral.

*Ecology,* 48:955–961.

BRADLEY, W. G., AND R. A. MAUER

1971  Reproduction and food habits of Merriam's kangaroo rat, *Dipodomys merriami. J. Mamm.,* 52:497–507.

BRAGG. A. N.

1950  Observations on *Microhyla* (Salientia: Microhylidae). *Wasmann J. Biol.,* 8:113–118.

BRANT, D. H.

1962  Measures of the movements and population densities of small rodents. *Univ. Calif. Pub. Zool.,* 62:105–184.

BRATTSTROM, B. H.

1963  A preliminary review of the thermal requirements of amphibians. *Ecology,* 44:238–255.

1965  Body temperature of reptiles. *Amer. Midl. Natur.,* 73:376–422.

BRAWN, V. M.

1961  Reproductive behaviour of the cod (*Gadus callarias* L.). *Behaviour,* 18:177–198.

BRAY, J. R., AND E. GORHAM

1964  Litter production in the forests of the world. *Adv. Ecol. Res.,* 2:101–157.

BROCK, T. R.

1966  *Principles of Microbial Ecology,* Prentice-Hall, Englewood Cliffs, N.J.

BROECKER, W. S.

1970  Man's oxygen reserves. *Science,* 168:1537–1538.

BRONSON, F. H.

1969  Pheromonal influences on mammalian reproduction. In M. Diamond (ed.), *Perspectives in Reproduction and Sexual Behaviour,* Indiana University Press, Bloomington.

BROOKS, J. L., AND S. I. DODSON

1965  Predation, body size, and composition of plankton. *Science,* 150:28–35.

BROOKS, M. G.

1940  The breeding warblers of the central Allegheny Mountain region. *Wilson Bull.,* 52:249–266.

1951  Effect of black walnut trees and their products on other vegetation. *W.Va. Univ. Agr. Exp. Stat. Bull.,* 347:1–31.

1955  An isolated population of the Virginia varying hare. *J. Wildl. Manage.,* 19:54–61.

BROWER, J. V. Z.

1958  Experimental studies of mimicry in some North American butterflies: 1. The monarch, *Danaus plexippus,* and Viceroy, *Limenitis archippus;* 2. *Battus philenor* and *Papilio troilus, P. polyxenes* and *P. glaucus;* 3. *Danaus gilippus berenice* and *Limenitis archippus floridensis. Evolution,* 12:32–47, 123–136, 273–285.

BROWER, J. V. Z., AND L. P. BROWER

1962  Experimental studies of mimicry: 6. The reaction of toads (*Bufo terrestris*) to honeybees (*Apis mellifera*) and their dronefly mimics (*Eristalis vinetorum*). *Amer. Natur.,* 97:297–307.

1969  Ecological chemistry. *Sci. Amer.,* 220(2):22–29.

BROWER, L. P., J. V. Z. BROWER, AND F. P. CRANSTON
1965   Courtship behaviour of the Queen butterfly, *Danaus gilippus berenice* Cramer. *Zoologica,* 50:1–39.

BROWER, L. P., J. V. Z. BROWER, AND P. W. WESTCOTT
1960   Experimental studies of mimicry: 5. The reaction of toads *(Bufo terrestris)* to bumblebees *(Bombus americanorum)* and their robberfly mimics *(Mallophora bomboides)* with a discussion of aggressive mimicry. *Amer. Natur.,* 94:343–355.

BROWN, E. R.
1961   The black-tailed deer of western Washington. *Washington State Game Dept. Biol. Bull. No. 13,* Olympia.

BROWN, F. A., JR.
1959   Living clocks. *Science,* 130:1535–1544.

BROWN, J. H.
1971   Mammals on mountaintops: nonequilibrium insular biogeography. *Amer. Natur.,* 105:467–478.

BROWN, J. H., AND D. W. DAVIDSON
1977   Competition between seed-eating rodents and ants in desert ecosystems. *Science,* 196:880–882.

BROWN, J. H., AND A. K. LEE
1969   Bergmann's rule and climate adaptation in woodrats *(Neotoma). Evolution,* 23:329–338.

BROWN, J. L.
1963   Aggressiveness, dominance, and social organization in the Stellar jay. *Condor,* 65:460–484.

1964   The evolution of diversity in avian territorial systems. *Wilson Bull.,* 76:160–169.

1969   Territorial behavior and population regulation in birds: a review and reevaluation. *Wilson Bull.,* 81:293–329.

1974   Alternate routes to sociality in jays—with a theory for altruism and communal breeding. *Amer. Zool.,* 14:63–80.

1975   *The Evolution of Behavior,* Norton, New York.

BROWN, L.
1976   *British Birds of Prey,* Collins, London.

BROWN, L. R., AND G. W. FINSTERBUSCH
1972   *Man and His Environment: Food,* Harper & Row, New York.

BROWN, R. J. E.
1970   *Permafrost in Canada,* University of Toronto Press.

BROWN, R. J. E., AND G. H. JOHNSON
1964   Permafrost and related engineering problems. *Endeavour,* 23:66–72

BROWN, W. L., AND E. O. WILSON
1956   Character displacement. *System. Zool.,* 5:49–64.

BRUCE, V. G.
1960   Environmental entrainment of circadian rhythms. *Cold Spring Harbor Symp. Quant. Biol.,* 25:29–47.

BRYSON, R. A., D. A. BAERRIS, AND W. M. WENDLAND
1970   The character of late glacial and post glacial climatic changes. In W. Dort and J. K. Jones (eds.), 1970, pp. 53–74.

BRYSON, R. A., AND R. A. RAGOTZKIE
1960   On internal waves in lakes. *Limnol. Oceanogr.,* 5:397–408.

BUCKNER, C. H., AND W. J. TURNOCK
1965   Avian predation on the larch sawfly, *Pristiphora erichsonii* (Hymenoptera: Tenthredinidae). *Ecology,* 46:223–236.

BUDYKO, M. I.
1963   *The Heat Budget of the Earth,* Hydrological Publishing House, Leningrad.

BUELL, M. F., AND R. E. WILBUR
1948   Life form spectra of the hardwood forests of the Itaska Park region, Minnesota. *Ecology,* 29:352–359.

BULLEN, F. T.
1966   Locusts and grasshoppers as pests of crops and pasture—a preliminary economic approach. *J. Appl. Ecol.,* 3:147–168.

BULMER, M. G.
1974   A statistical analysis of the 10 year cycle in Canada. *J. Anim. Ecol.,* 43:701–718.

1975   Phase relations in the 10 year cycle. *J. Anim. Ecol.,* 44:609–621.

BUNNELL F. L., S. F. MACLEAN, JR., AND J. BROWN
1975   Barrow, Alaska, U.S.A. In T. Rosswall and O. W. Heal (eds.), 1975, pp. 73–124.

BUNNING, E.
1964   *The Physiological Clock,* 2nd ed., Academic, New York.

BUNNING, E., AND G. JOERRENS
1960   Tagesperiodische antogonistische Schwankungen de Blauviolettund Gelbrot Empfindlichkeit als Grundlage der photoperiodischen Diapause-Induktion bei *Pieris brassicae. Z. Naturforsch.,* 15b:205–223.

BURBANCH, M. P., AND R. B. PLATT
1964   Granite outcrop communities of the Piedmont Plateau in Georgia. *Ecology,* 45:292–306.

BURDICK, G. E, E. J. HARRIS, H. J. DEAN, T. M. WALKER, J. SKEA, AND D. COLBY
1964   The accumulation of DDT in lake trout and the effect on reproduction. *Trans. Amer. Fish. Soc.,* 93:127–136.

BURGER, G. V.
1969   Response of gray squirrels to nest boxes at Remington Farms, Maryland. *J. Wild. Manage.,* 33:796–801.

BURGESS, R. L., AND R. V. O'NEILL
1976   Eastern Deciduous Forest Biome Progress Report September 1, 1974–August 31, 1975, *EDFB/IBP 76/5. Env. Sci. Div. Pub. No. 871,* Oak Ridge National Laboratory, Oak Ridge, Tenn.

BURT, W. V., AND J. QUEEN
1957   Tidal overmixing in estuaries. *Science,* 126:973–974.

BUSH, G. L.
1974   The mechanism of sympatric host race formation in the true fruit flies (Tephritidae). In M. J. D. White (ed.), 1974, pp. 3–23.

1975 Modes of animal speciation. *Ann. Rev. Ecol. Syst.* 6:339–364.

1975 Sympatric speciation in phytophagous insects. In P. W. Price (ed.), 1975, pp. 187–206.

BUSKIRK, R. E., AND W. H. BUSKIRK
1976 Changes in anthropod abundance in a highland Costa Rica forest. *Amer. Midl. Natur.,* 95:288–298.

BUTLER, P. A.
1964 Commercial fisheries investigations. In Pesticide-Wildlife Studies. *U.S. Fish and Wildl. Serv. Circ. 226,* pp. 11–25.

CAIN, A. J., AND P. M. SHEPPARD
1954 Natural selection in *Cepaea. Genetics,* 39:89–116.

CAIN, S. A., AND G. M. CASTRO
1959 *Manual of Vegetation Analysis,* Harper & Row, New York.

CAMERON, G. N.
1972 Analysis of insect trophic diversity in two salt marsh communities. *Ecology,* 53:58–73.

CAMPBELL, C. A., E. A. PAUL, D. A. RENNIE, AND K. J. MC CALLUM
1967 Applicability of the carbon-dating method of analysis to soil humus studies. *Soil Sci.,* 104:217–224.

CAMPBELL, R. W.
1969 Studies on gypsy moth population dynamics. In *Forest Insect Population Dynamics, USDA Res. Paper NE-125,* pp. 29–34.

CANTLON, J. E.
1953 Vegetation and microclimates on north and south slopes of Cushetunk Mt., New Jersey. *Ecol. Monogr.,* 23:241–270.

CARL, E.
1971 Population control in arctic ground squirrels. *Ecology,* 52:395–413.

CARLISLE, A., A. H. F. BROWN, AND E. J. WHITE
1966 The organic matter and nutrient elements in the precipitation beneath a sessile oak canopy. *J. Ecol.,* 54:87–98.

CARPENTER, C. R.
1934 A field study of the behavior and social relations of howling monkeys *(Aloatta palliata). Comp. Psychol. Monogr. No. 2.*

CARSON, H. L.
1968 The population flush and its genetic consequences. In R. C. Lewontin (ed.), *Population Biology and Evolution,* Syracuse University Press, New York, N.Y., pp. 123–137.

1973 Reorganization of the gene pool during speciation. *Population Gen. Monogr.,* 3:274–280.

1975 The genetics of speciation at the diploid level. *Amer. Natur.,* 109:83–92.

CASTRI, F. DI
1970 Les Grands problems qui se posent aux ecologistes pour l'etude des ecosystemes du sol. In J. Phillipson (ed.), *Methods of Study in Soil Biology,* UNESCO, Paris, pp. 15–31.

CASTRI, F. DI, AND H. A. MOONEY (EDS.)
1973 *Mediterranean Type Ecosystems, Origin and Structure,* Springer-Verlag, New York.

CASWELL, H., F. REED, S. N. STEPHENSON, AND P. A. WEAVER
1973 Photosynthetic pathways and selective herbivory: a hypothesis. *Amer. Natur.,* 107:465–480.

CAUGHLEY, G.
1966 Mortality patterns in mammals. *Ecology,* 47:906–918.

1970 Eruption of ungulate populations with emphasis on Himalayan thar in New Zealand. *Ecology,* 51:53–72.

1976a Wildlife management and the dynamics of ungulate populations. *Appl. Biol.,* 1:183–246.

1976b Plant and herbivore systems. In R. B. May (ed.), 1976, pp. 94–113.

CERNUSA, A.
1976 Energy exchange within individual layers of a meadow. *Oecologia* 23:141–149.

CHAFFEE, R. R. J., AND J. C. ROBERTS
1971 Temperature acclimation in birds and mammals. *Ann. Rev. of Physiol.,* 33:155–202.

CHANGNON, S. A.
1968 La Porte weather anomaly: Fact or fiction? *Bull. Amer. Meteorol. Soc.,* 49:4–11.

CHAPMAN, J. A. AND R. P. MORGAN III
1973 Systematic status of the cottontail complex in western Maryland nearby West Virginia. *Wildl. Monogr. 36,* Wildlife Society, Washington, D.C.

CHAPMAN, V. J.
1960 *Salt Marshes and Salt Deserts of the World,* Leonard Hill, London.

1976 *Coastal Vegetation,* 2nd ed., Pergamon Press, Oxford.

CHARLEY, J. C.
1972 The role of shrubs in nutrient cycling. In C. M. McKell et al. (eds.), 1972, pp. 182–203.

CHEATUM, E. L., AND C W. SEVERINGHAUS
1950 Variations in fertility of white-tailed deer related to range conditions. *Trans. North Amer. Wildl. Conf.,* 15:170–189.

CHERNOV, YU I. ET AL.
1975 Tareya, USSR. In T. Rosswall and O. W. Heal (eds.), 1975, pp. 159–181.

CHESTNUT, A. F.
1974 Oyster reefs. In H. T. Odum, B. J. Copeland, and E. A. McMahan (eds.), 1974, Vol. 2, pp. 171–203.

CHEVALIER, J. R.
1973 Cannibalism as a factor in the first year survival of walleye in Oneida Lake. *Trans Amer. Fish. Soc.,* 102:739–744.

CHEW, R. M., AND A. E. CHEW
1970 Energy relationships of the mammals of a desert shrub *(Larrae tridentatia)* community. *Ecol. Monogr.,* 40:1–21.

CHITTY, D.
1960 Population processes in the vole and their reference to general theory. *Can. J. Zool.,* 38:99–113.

CHITTY, D., AND E. PHIPPS
1966 Seasonal changes in survival in mixed

populations of two species of vole. *J. Anim. Ecol.,* 35:313–331.

CHOW, T. J.
1970 Lead accumulation in roadside soil and grass. *Nature,* 225:295–296.

CHOW, T. J., AND J. L. EARL
1970 Lead aerosols in the atmosphere: increasing concentrations. *Science,* 169:577–580.

CHRISTENSEN, A. M., AND J. J. MC DERMOTT
1958 Life history of the oyster crab *Pinnotheres ostreum. Biol. Bull.,* 114:146–179.

CHRISTIAN, J. J.
1963 Endocrine adaptive mechanisms and the physiologic regulation of population growth. In W. V. Mayer and R. G. Van Gelder (eds.), *Physiological Mammalogy,* Vol. 1, *Mammalian Populations,* Academic, New York, pp. 189–353.

CHRISTIAN, J. J., AND D. E. DAVIS
1964 Endocrines, behaviour and populations. *Science,* 146:1550–1560.

CHRISTIANSEN, F. B., AND T. M. FENCHEL
1977 *Theories of Population in Biological Communities,* Springer-Verlag, New York.

CLACK, L.
1975 Subspecific intergradation and zoogeography of the painted turtle *(Chrysemys picta)* in northern West Virginia. Unpublished MS Thesis, West Virginia University, Morgantown.

CLARK, C.
1967 *Population Growth and Land Use,* Macmillan, London, and St. Martins Press, New York.

CLARK, F. E.
1969a The microflora of grassland soils and some microbial influences on ecosystems functions. In R. L. Dix and R. G. Beidleman (eds.), 1969, pp. 361–376.
1969b Ecological associations among soil micro-organisms. In *Soil Biology,* UNESCO, Paris, pp. 125–161.

CLARK, F. W.
1972 Influence of jackrabbit density on coyote population change. *J. Wildl. Manage.,* 36:343–356.

CLARKE, B., AND J. MURRAY
1969 Ecological genetics and speciation in land snails in the genus *Paitula. Biol. J. Linn. Soc.,* 1:31–42.

CLARKE, J. F.
1969 Nocturnal urban boundary layer over Cincinnati, Ohio. *Monthly Weather Rev.,* 97:582–589.

CLAUSEN, J.
1965 Population studies of alpine and subalpine races of conifers and willows in the California High Sierra Nevada. *Evolution,* 19:56–68.

CLAUSEN, J., D. D. KECK, AND W. M. HIESEY
1948 Experimental studies on the nature of species: 3. Environmental responses of climatic races of *Achillea. Carnegie Inst. Wash. Pub. No. 581.*

CLAWSON, S. G.

1958–59 Fire ant eradication and quail. *Alabama Conser.,* 30(4)14–15, 25.

CLEMENTS, F. C.
1916 *Plant Succession, Carnegie Inst. Wash. Pub. No. 242.*

CLEMENTS, R. G., AND J. COLON
1975 The rainfall interception process and mineral cycling in a montane rain forest in eastern Puerto Rico. In F. G. Howell et al. (eds.), 1975, pp. 813–823.

CLIFFORD, H. F.
1966 The ecology of invertebrates in an intermittent stream. *Inv. Indiana Lakes and Streams,* 7(2):57–98.

CODY, M. L.
1966 A general theory of clutch size. *Evolution,* 20:174–184.
1969 Convergent characteristics in sympatric species: a possible relation to interspecific competition and aggression. *Condor,* 71:222–239.

CODY, M. L., AND J. M. DIAMOND (EDS.)
1975 *Ecology and Evolution of Communities,* Harvard University Press, Cambridge, Mass.

COKER, R. E.
1947 *This Great and Wide Sea,* University of North Carolina Press, Chapel Hill.

COLE, D. W., S. P. GESSEL, AND S. F. DICE
1967 Distribution and cycling of nitrogen, phosphorus, potassium and calcium in a second-growth Douglas-fir ecosystem. In *Symposium on Primary Productivity and Mineral Cycling in Natural Ecosystems,* University of Maine Press, Orono, pp. 197–232.

COLE, L. C.
1946 A study of the crypotozoa of an Illinois woodland. *Ecol. Monogr.,* 16:49–86.
1954a The population consequences of life history phenomena. *Quart. Rev. Biol.,* 29:103–137.
1954b Some features of random cycles. *J. Wildl. Manage.,* 18:107–109.
1960 Competitive exclusion. *Science,* 132:348–349.

COLLIAS, N. E., AND R. D. TABER
1951 A field study of some grouping and dominance relations in ring-necked pheasants. *Condor,* 53:265–275.

COMMONER, B.
1970 Threats to the integrity of the nitrogen cycle: nitrogen compounds in soil, water, atmosphere, and precipitation. In S. F. Singer (ed.), 1970, pp. 70–95.

CONLEY, W.
1976 Competition between microbes: a behavioral hypothesis. *Ecology,* 57:224–237.

CONNELL, J. H.
1961 The influence of interspecific competition and other factors on the distribution of the barnacle *Chthamalus stellatus. Ecology,* 42:710–723.
1975 Some mechanisms producing structure in natural communities: a model and evidence from field experiments. In M. Cody and J. Diamond

(eds.), 1975, pp. 460–480.

CONNELL, J. H., AND E. ORIAS
1964  The ecological regulation of species diversity. *Amer. Natur.*, 98:399–414.

CONNELL, J. H., AND R. O. SLATYER
1977  Mechanisms of succession in natural communities and their role in community stability and organization. *Amer. Natur.*, 111: 1119–1144.

COOKE, F., AND F. G. COOCH
1968  The genetics of polymorphism in the goose *Anser caerulescens*. *Evolution*, 22:289–300.

COOKE, F., AND J. R. RYDER
1971  The genetics of polymorphism in the Ross Goose *(Anser rossii)*. *Evolution*, 25:483–496.

COOKE, G. D.
1967  The pattern of autotrophic succession in laboratory microcosms. *Bioscience*, 17:717–721.

COOPER, A. W.
1974  Salt marshes. In H. T. Odum, B. J. Copeland, and E. A. McMahan (eds.), 1974, Vol. 2, pp. 55–98.

COOPER, C. F.
1960  Changes in vegetation, structure, and growth of southwestern pine forest since white settlement. *Ecol. Monogr.*, 30:129–164.

COPE, O. B.
1971  Interaction between pesticides and wildlife. *Ann. Rev. Entomol.*, 16:325–364.

COPELAND, B. J., K. R. TENORE, AND D. B. HORTON
1974  Oligohaline regime. In H. T. Odum, B. J. Copeland, and E. A. McMahan (eds.), 1974, Vol. 2, pp. 315–358.

CORBETT, E. S., AND R. P. CROUSE
1968  Rainfall interception by annual grass and chaparral. *USDA Forest Serv. Res. Paper PSW 48.*

CORNFORTH, I. S.
1970a  Leaf-fall in a tropical rain forest. *J. Appl. Ecol.*, 7:603–608.
1970b  Reafforestation and nutrient reserves in the humid tropics. *J. Appl. Ecol.*, 7:609–615.

CORRELL, D. L.
1978  Estuarine productivity. *Bioscience*, 28:646–650.

COULSON, J. C.
1968  Differences in the quality of birds nesting in the center and on the edges of a colony. *Nature*, 217:478–479.

COUPLAND, R. T.
1950  Ecology of mixed prairie in Canada. *Ecol. Monogr.*, 20:217–315.
1958  The effects of fluctuations in weather upon the grassland of the Great Plains. *Botan. Rev.*, 24:273–317.
1959  Effect of changes in weather conditions upon grassland in the northern Great Plains. In G. Sprague (ed.), *Grassland*, American Association Advancement Science, Washington, D.C., pp. 291–306.

COUTANT, C.
1970  Biological aspects of thermal pollution: 1. Entrainment and discharge canal effects. *CRC Critical Reviews in Environ. Control*, Nov. 1970, 341–381.

COWAN, I., AND V. GEIST
1961  Aggressive behavior in deer of the genus *Odocoileus*. *J. Mammal*, 42:522–526.

COWAN, R. L.
1962  Physiology of nutrition as related to deer. *Proc. 1st Natl. White-tailed Deer Disease Symp.*, pp. 1–8.

COWLES, H. C.
1899  The ecological relations of the vegetation on the sand dunes of Lake Michigan. *Botan. Gaz.*, 27:95–117, 167–202, 281–308, 361–391.

COWLES, R. B., AND C. M. BOGERT
1944  A preliminary study of the thermal requirements of desert reptiles. *Bull. Amer. Mus. Natur. Hist.*, 83:265–296.

COX, C. B., I. N. HEALEY, AND P. D. MOORE
1973  *Biogeography, An Ecological and Evolutionary Approach*, Blackwell, Oxford.

COX, G. W., AND R. E. RICKLEFS
1977  Species diversity and ecological release in Caribbean land bird faunas. *Oikos*, 28:113–122.

CRAGG, J. B. (ED.)
1962  *Advances in Ecological Research*, Vol. 1, Academic, New York.

CRITCHFIELD, W. B.
1971  Profiles of California vegetation. *USDA Forest Serv. Res. Paper PSW-76.*

CROCH, J. H., J. E. ELLIS, AND J. D. GOSS-CUSTARD
1976  Mammalian social systems: structure and function. *Anim. Behav.*, 24(2): 261–274.

CRONIN, L. E., AND A. J. MANSUETI
1971  The biology of the estuary. In P. A. Douglas and R. H. Stroud (eds.), *A Symposium on the Biological Significance of Estuaries*, Sport Fishing Institute, Washington, D.C., pp. 14–39.

CROSBY, G. T.
1972  Spread of the cattle egret in the Western hemisphere. *Bird-banding*, 43:205–212.

CROSSLEY, D. A., JR., AND K. K. BOHNSACK
1960  Long-term ecological study in the Oak Ridge area: 3. Oribatid mite fauna in pine litter. *Ecology*, 41:628–639.

CROSSMAN, E. J.
1959a  Distribution and movement of a predator, the rainbow trout, and its prey, the redside shiner, in Paul Lake, British Columbia. *J. Fish. Res. Board Can.*, 16:247–267.
1959b  A predator prey interaction in freshwater fish. *J. Fish. Res. Board Can.*, 16:269–281.

CROZE, H.
1970  *Searching Image in Carrion Crows*, Paul Parey, Berlin.

CUMMINGS, B. G., AND F. WAGNER
1968  Rhythmic processes in plants. *Ann. Rev. Plant Physiol.*, 19:381–416.

CUMMINS, K. W.
1974  Structure and function of stream ecosystems. *Bioscience*, 24:631–641.
1975  The importance of different energy sources in freshwater ecosystems. In National Academy

Sciences, 1975, pp. 50–59.

CUMMINS, K. W., W. P. COFFMAN, AND P. A. ROFF
1966  Trophic relationships in a small woodland stream. *Verheindlung der Internationalen Vereinigung fur Theoretische und Angewandte Limnologie,* 16:627–637.

CURIO, E.
1976  *The Ethology of Predation,* Springer-Verlag, New York.

CURTIS, J. T.
1959  *The Vegetation of Wisconsin,* University of Wisconsin Press, Madison.

CURTIS, J. T., AND R. P. MCINTOSH
1951  An upland forest continuum in the prairie-forest border region of Wisconsin. *Ecology,* 32:476–496.

CUSHING, E. J., AND H. E. WRIGHT, JR. (EDS.)
1967  *Quaternary Paleoecology,* Yale University Press, New Haven, Conn.

DAHL, E.
1953  Some aspects of the ecology and zonation of the fauna of sandy beaches. *Oikos,* 4:1–27.

DANSEREAU, P.
1945  Essae de correlation sociologique entre les plantes superieures et les poissons de la Beine du Lac Saint-Louis. *Rev. Can. Biol.,* 4:369–417.
1959  Vascular aquatic plant communities of southern Quebec: a preliminary analysis. *Trans. 10th Northeast Wild. Conf.,* pp. 27–54.

DANSEREAU, P., AND F. SEGADAS-VIANNA
1952  Ecological study of the peat bogs of eastern North America. *Can. J. Bot.,* 30:490–520.

DARLINGTON, P. J., JR.
1957  *Zoogeography: The Geographical Distribution of Animals,* Wiley, New York.

DARNELL, R. M.
1961  Trophic spectrum of an estuarine community based on studies of Lake Pontchartrain, Louisiana. *Ecology,* 42:553–568.

DARWIN, C.
1881  *The Formation of Vegetable Mould Through the Action of Worms, with Observations on Their Habits,* Murray, London.

DATON, J. S., G. E. LIKENS, AND F. H. BORMANN
1973  Throughfall and stemflow chemistry in a northern hardwood forest. *J. Ecol.,* 65:495–508.

DAUBENMIRE, R. F.
1959  *Plants and Environment: A Textbook of Plant Autecology,* Wiley, New York.
1966  Vegetation: identification of typal community. *Science,* 151:291–298.
1968a  Soil moisture in relation to vegetation distribution in the mountains of northern Idaho. *Ecology,* 49:431–438.
1968b  Ecology of the fire in grasslands. *Adv. Ecol. Res.,* 5:208–266.
1968c  *Plant Communities: A Textbook of Plant Synecology,* Harper & Row, New York.

DAVIES, N. B.
1976  Food flocking and territorial behavior of the pied wagtail (*Motacilla alba yarrellii* Gould) in winter. *J. Anim. Ecol.,* 45:235–252.

1977  Prey selection and social behaviour in wagtails (Aves: Moticillidae). *J. Anim. Ecol.,* 46:37–57.

DAWKINS, M.
1971  Perceptual change in chicks: another look at the "search image" concept. *Anim. Behav.,* 19:566–574.

DAWKINS, R.
1976  *The Selfish Gene,* Oxford, New York.

DAWSON, W. R., AND G. A. BARTHOLOMEW
1968  Temperature regulation and water economy of desert birds. In G. W. Brown (ed.), *Desert Biology,* Academic, New York.

DAY, F. P., JR., AND D. T. MCGINTY
1975  Mineral cycling strategies of two deciduous and two evergreen tree species on a southern Appalachian watershed. In F. C. Howell et al. (eds.), 1975, pp. 736–743.

DAY, P. R.
1974  *Genetics of Host-Parasite Interaction,* Freeman, San Francisco.

DEAN, R., J. E. ELLIS, R. W. RICE, AND R. E. BEMERET
1975  Nutrient removal by cattle from a short grass prairie. *J. App. Ecol.,* 12:25–29.

DECOURSEY, P. J.
1960a  Phase control of activity in a rodent. *Cold Spring Harbor Symp. Quant. Biol.,* 25:49–54.
1960b  Daily light sensitivity rhythm in a rodent. *Science,* 131:33–35.
1961  Effect of light on the circadian activity rhythm of the flying squirrel, *Glaucomys volans. Z. Vergleich. Physiol.,* 44:331–354.

DELACOUR, J., AND C. VAURIE
1950  Les mesanges charbonnieres (revision de l'espece *Parus major*). *L'Oiseau,* 20:91–121.

DELONG, K. T.
1966  Population ecology of feral house mice: interference by *Microtus. Ecology,* 47:481–484.

DEMPSTER, J.
1975  *Animal Population Ecology,* Academic, New York.

DETHIER, V. G.
1970  Chemical interactions between plants and insects. In E. Sondheimer and J. B. Simone (eds.), *Chemical Ecology,* Academic, New York, pp. 83–102.

DHYSTERHOUS, E. J., AND E. M. SCHMUTZ
1947  Natural mulches or "litter of grasslands"; with kinds and amounts on a southern prairie. *Ecology,* 28:163–179.

DIAMOND, J. M.
1975  The island dilemma: Lessons of modern biogeographic studies for the design of natural preserves. *Biol. Conserv.,* 7:129–146.

DICE, L. R.
1943  *The Biotic Provinces of North America,* University of Michigan Press, Ann Arbor.

DICKINSON, H., AND J. ANTONOVICS
1973  Theoretical considerations of sympatric divergence. *Amer. Natur.,* 107:256–274.

DIETZ, D. R.
1965  Deer nutrition research in range

management. *Trans. 30th N. Amer. Wildl. Conf.,*
pp. 274–285.

DIETZ, D. R., R. H. UDALL, AND L. E. YEAGER
1962   Chemical composition and digestibility by
mule deer of selected forage species, Cache le
Poudre Range, Colorado. *Colorado Dept. Game
and Fish Tech. Bull. 14.*

DILGER, W. C.
1956a   Adaptive modifications and ecological
isolating mechanisms in the thrush genera
*Catharus* and *Hylocichla. Wilson Bull.,*
68:171–199.
1956b   Hostile behavior and reproductive isolating
mechanisms in the avian genera *Catharus* and
*Hylocichla. Auk,* 73:313–353.
1960   Agonistic and social behavior of captive
redpolls. *Wilson Bull.,* 72:115–132.

D'ITRI, F.
1971   Mercury accumulation in the aquatic
environment. *Proc. 162nd Amer. Chem. Soc.
Meeting,* Washington, D.C.

DIX, R. L.
1960   The effects of burning on the mulch
structure and species composition of grassland in
western North Dakota. *Ecology,* 41:49–56.

DIX, R. L., AND R. G. BIEDLEMAN (EDS.)
1969   The grassland ecosystem: a preliminary
synthesis. *Range Sci. Dept. Sci. Ser. No. 2,*
Colorado State University, Fort Collins.

DIX, R. L. AND F. E. SMEINS
1967   The prairie, meadow, and marsh vegetation
of Nelson County, North Dakota. *Can. J. Bot.,*
45:21–58.

DIXON, A. F. G.
1970   Quality and availability of food for a
sycamore aphid population. In A. Watson (ed.),
*Animal Populations in Relation to Their Food
Resources,* Blackwell, Oxford.

DOBZHANSKY, T.
1947   Effectiveness of intraspecific and
interspecific matings in *Drosophila
pseudoobscura* and *D. persimilis. Amer. Natur.,*
81:66–73.
1950   Evolution in the tropics. *Amer. Sci.,*
38:209–221.
1951   *Genetics and the Origin of Species,* 3rd ed.,
Columbia University Press, New York.

DOLBEER, R. A.
1972   Population dynamics of the snowshoe hare
in Colorado. PhD thesis, Colorado State
University, Fort Collins.

DORNSTREICH, M. D., AND G. E. B. MORREN
1974   Does New Guinea cannibalism have
nutritional value? *Human Ecol.,* 2:1–12.

DORST, R.
1958   Uber die Ansiedlung vob jung ins Binnerien
verfrachteten Silbernowen *(Larus argentatus).
Vogelwarte,* 17:169–173.

DORT, W. JR., AND J. K. JONES, JR. (EDS.)
1970   *Pleistocene and Recent Environments of the
Central Great Plains. Spec. Publ. No. 3,* Dept.
Biology, University of Kansas, Lawrence.

DOTY, M. S.
1957   Rocky intertidal surfaces. In J. W. Hedgpeth
(ed.), 1957a, pp. 535–585.

DOWDY, W. W.
1944   The influence of temperature on vertical
migration of invertebrates inhabiting different
soil types. *Ecology,* 25:449–460.
1951   Further ecological studies on stratification of
the arthropods. *Ecology,* 32:37–52.

DRIFT, J. VAN DER
1951   Analysis of the animal community in a
beech forest floor. *Tijdschrift voor Entomologie,*
94:1–168.
1971   Production and decomposition of organic
matter in an oakwood in the Netherlands. In P.
Duvigneaud (ed.), 1971, pp. 631–634.

DRURY, W. H., JR.
1961   The breeding biology of shorebirds on Bylot
Island, Northwest Territories, Canada. *Auk,*
78:176–219.

DRURY, W. H., JR., AND I. C. T. NISBET
1973   Succession. *J. Arnold Arboretum,*
54:331–368

DRURY, W. H., JR., I. C. T. NISBET, AND R. E.
RICHARDSON
1961   The migration of "angels." *Natur. Hist.,*
70(8):11–16.

DUDDINGTON, C. L.
1955   Interrelations between soil microflora and
soil nematodes. In D. Kevan, 1955, pp. 284–301.

DUFFEY, E., ET AL.
1974   *Grassland Ecology and Wildlife
Management,* Chapman and Hall, London.

DUGDALE, R. C., AND J. J. GEORING
1967   Uptake of new and regenerated forms of
nitrogen in primary productivity. *Limnol. and
Oceanogr.,* 12:196–206.

DUVIGNEAUD, P. (ED.)
1971   *Productivity of Forest Ecosystems* (Proc.
Brussels Symposium 1969), UNESCO, Paris.

DUVIGNEAUD, P., AND S. DENAEYER-DESMET
1967   Biomass, productivity and mineral cycling
in deciduous forests in Belgium. In *Symposium
on Primary Productivity and Mineral Cycling in
Natural Ecosystems,* University of Maine Press,
Orono, pp. 167–186.
1970   Biological cycling of minerals in temperate
deciduous forests. In D. Reichle (ed.), 1970, pp.
199–225.
1975   Mineral cycling in terrestrial ecosystems. In
National Academy of Science, 1975, pp. 133–154.

DYBAS, H. S., AND M. LLOYD
1962   Isolation by habitat in two-synchronized
species of periodical cicadas (Homoptera,
Cicadidae, Magicada). *Ecology,* 43:444–459.

EARP, R.
1974   Tidepools. In H. T. Odum, B. J. Copeland,
and E. A. McMahan (eds.), 1974, Vol. 2, pp. 1–29.

EASTWOOD, E.
1971   *Radar Ornithology,* Methuen, London.

EATON, J. S., G. E. LIKENS, AND F. H. BORMANN
1973   Throughfall and stemflow chemistry in a

northern hardwood forest. *J. Ecol.,* 61:495–508.

ECKHARDT, F. E. (ED.)

1968  *Functioning of Terrestrial Ecosystems at the Primary Production Level,* Proc. Copenhagen Symposium, Natural Resources Research, UNESCO, Paris.

EDINGTON, J. M., AND M. A. EDINGTON

1972  Spatial patterns and habitat partition in the breeding birds of an upland wood. *J. Anim. Ecol.,* 41:331–357.

EDMONDSON, W. T.

1956  The relation of photosynthesis by phytoplankton to light in lakes. *Ecology,* 37:161–174.

1969  Eutrophication in North America. In *Eutrophication: Causes, Consequences, Correctives,* National Academy of Sciences, Washington, D.C., pp. 124–149.

1970  Phosphorus, nitrogen, and algae in Lake Washington after diversion of sewage. *Science,* 169:690–691.

EDWARDS, C. A., AND G. W. HEATH

1963  The role of soil animals in the breakdown of leaf material. In J. Doeksen and J. van der Drift (eds.), *Soil Organisms,* North Holland Publishing Co., Amsterdam, pp. 76–84.

EGLER, F. E.

1953  Vegetation management for rights-of-way and roadsides. *Smithsonian Inst. Ann. Rept. 1953,* pp. 299–322.

1954  Vegetation science concepts. 1. Initial floristic composition—a factor in old field vegetation development. *Vegetatio,* 4:412–417.

EHRLICH, P. R., AND P. H. RAVEN

1965  Butterflies and plants: A study in coevolution. *Evolution* 18:586–608.

EHRLICH, P. R., D. E. BREEDLOVE, P. F. BRUSSARD, AND M. A. SHARP

1972  Weather and the "regulation" of subalpine populations. *Ecology,* 53:243–247.

EISNER, E.

1960  The relationship of hormones to the reproductive behavior of birds, referring especially to parental behavior: a review. *Anim. Behav.,* 8:155–179.

EISNER, T.

1970  Chemical defense against predation in arthropods. In E. Sondheimer and J. B. Simeone (eds.), *Chemical Ecology,* Academic, New York, pp. 157–217.

EISNER, T., AND J. MEINWALD

1966  Defensive secretions of arthropods. *Science,* 153:1341–1350.

ELLIOTT, P. F.

1974  Evolutionary responses of plants to seed eaters: pine squirrels predation on lodgepole pine. *Evolution,* 28:221–231.

ELLISON, L.

1954  Subalpine vegetation of the Wasatch Plateau, Utah. *Ecol. Monogr.,* 24:89–184.

1960  Influences of grazing on plant succession on rangelands. *Bot. Rev.,* 26:1–78.

ELSTER, H. J.

1965  Absolute and relative assimilation rates in relation to phytoplankton populations. In C. R. Goldman (ed.), *Primary Productivity in Aquatic Environments,* Mem. 1st Ital. Idrobiol. 18 Suppl., University of California Press, Berkeley, pp. 79–103.

ELTON, C. S.

1927  *Animal Ecology,* Sidgwick & Jackson, London.

1958  *The Ecology of Invasions by Animals and Plants,* Methuen, London.

ELTRINGHAM, S. K.

1971  *Life in Mud and Sand,* Crane Russak, New York.

ELWOOD, J. W., AND G. S. HENDERSON

1975  Hydrologic and chemical budgets at Oak Ridge, TN. In A. D. Hasler (ed.), 1975, pp. 31–51.

EMANUELSSON, A., E. ERIKSSON, AND H. EGNER

1954  Composition of atmospheric precipitation in Sweden. *Tellus,* 3:261–267.

EMLEN, J. M.

1973  *Ecology: An Evolutionary Approach,* Addison-Wesley, Reading, Mass.

EMLEN, J. T., JR.

1940  Sex and age ratios in the survival of California quail. *J. Wildl. Manage.,* 4:2–99.

EMLEN, S. T., AND L. W. ORING

1977  Ecology, sexual selection, and the evolution of mating systems. *Science,* 197:215–223.

ENDERS, F.

1975  The influence of hunting manner on prey size, particularly on spiders with long attack distances (Araneidae, Linyphiidae, and Salticidae). *Amer. Natur.,* 109:737–763.

ENDLER, J. A.

1973  Gene flow and population differentiation. *Science,* 179:243–250.

1977  *Geographic Variation, Speciation, and Clines,* Princeton University Press, New Jersey.

ENGELMANN, M. D.

1961  The role of soil arthropods in the energetics of an old field community. *Ecol. Monogr.,* 31:221–238.

ENGLE, L. G.

1960  Yellow poplar seedfall pattern. *Central States Forest. Expt. Stat. Note 143.*

ENRIGHT, J. T.

1970  Ecological aspects of endogenous rhythmicity. *Ann. Rev. Ecol. Syst.,* 1:221–238.

1975  Orientation in time: endogenous clocks. In O. Kinne (ed.), *Marine Ecology: Vol. 2, Physiological Mechanisms, Part 2,* pp. 917–944.

EPLING, C.

1947  Natural hybridization of *Salvia apicna* and *S. mellifera. Evolution,* 1:69–78.

ERIKISSON, E.

1952  Composition of atmospheric precipitation: 1. Nitrogen compounds. *Tellus,* 4:214–232.

1963  The yearly circulation of sulfur in nature. *J. Geophys. Res.,* 68:4001–4008.

ERRINGTON, P. L.

1943 Analysis of mink predation upon muskrats in the north-central U.S. *Iowa Agr. Exp. Sta. Res. Bull.,* 320:797–924.

1946 Predation and vertebrate populations. *Quart. Rev. Biol.,* 21:144–177, 221–245.

1963 *Muskrat Populations,* Iowa State University Press, Ames.

ETHERINGTON, J. R.

1976 *Environment and Plant Ecology,* Wiley, New York.

EVANS, F. C., AND S. A. CAIN

1952 Preliminary studies on the vegetation of an old-field community in southeastern Michigan. *Contrib. Lab. Vert. Biol. University of Michigan* 51:1–17.

EWEL, J. T.

1976 Litterfall and leaf decomposition in a tropical forest succession in eastern Guatemala. *J. Ecol.,* 64:293–308.

EWER, R. F.

1968 *Ethology of Mammals,* Plenum, New York.

EYRE, S. R.

1963 *Vegetation and Soils: A World Picture,* Aldine, Chicago.

FARNER, D. S.

1955 The annual stimulus for migration: experimental and physiologic aspects. In A. Wolfson (ed.), *Recent Studies in Avian Biology,* pp. 198–237.

1964a The photoperiodic control of reproductive cycles in birds. *Amer. Sci.,* 52:137–156.

1964b Time measurement in vertebrate photoperiodism, *Amer. Natur.,* 98:375–386.

FARNSWORTH, E., AND F. GOLLEY (EDS.)

1973 *Fragile Ecosystems,* Springer-Verlag, New York.

FEENEY, P.

1969 Seasonal changes in oak leaf tannins and nutrients as a cause of spring feeding by winter moth caterpillars. *Ecology,* 51:565–581.

1975 Biochemical coevolution between plants and their insect herbivores. In L. E. Gilbert and P. H. Raven (eds.), 1975, pp. 3–19.

FELTON, P. M., AND H. W. LULL

1963 Suburban hydrology can improve watershed conditions. *Public Works,* 94:93–94.

FENCHEL, T.

1974 Character displacement and coexistence in mud snails (Hydrobridae). *Oecologia,* 20:19–32.

FENNER, F.

1953 Host-parasite relationships in myxomotosis of the Australian wild rabbit. *Cold Spring Harbor Symp. Quant. Biol.,* 18:291–294.

FICKEN, M. S.

1963 Courtship of the American redstart. *Auk,* 80:307–317.

FICKEN, R. W.

1963 Courtship and behavior of the common grackle *Quiscalus quiscula. Auk,* 80:52–72.

FISCHER, A. G.

1960 Latitudinal variation in organic diversity. *Evolution,* 14:64–81.

FISCHER, R. A.

1929 *The Genetical Theory of Natural Selection,* 2nd rev. ed., Dover, New York.

FISHER, S. G., AND G. E. LIKENS

1973 Energy flow in Bear Brook, New Hampshire: an integrative approach to stream ecosystem metabolism. *Ecol. Monogr.,* 43:421–439.

FITZPATRICK, L. C.

1973 Energy allocation in the Allegheny mountain salamander, *Desmognathus ochrophaeus. Ecol. Monogr.,* 43:43–58.

FLEROW, C. C.

1971 The evolution of certain mammals during the late Cenozoic. In K. K. Turekian (ed.), 1971, pp. 479–485.

FLINT, H. L.

1974 Phenology and genecology of woody plants. In H. Leith (ed.), 1974, pp. 83–97.

FLINT, R. F.

1970 *Glacial and Quaternary Geology,* Wiley, New York.

FLOOK, D. R.

1970 Causes and implications of an observed sex differential in the survival of wapiti. *Can. Wildl. Serv. Rept. Ser. No. 11.*

FONS, W. L.

1940 Influence of forest cover on wind velocity. *J. Forest.,* 38:481–486.

FORMAN, R. T. T., A. E. GALLI, AND C. F. LECK

1976 Forest size and avian diversity in New Jersey woodlots with some land use implications. *Oecologia,* 26:1–8.

FORTESQUE, J. A. C., AND G. C. MARTIN

1970 Micronutrients: forest ecology and systems analysis. In D. C. Reichle (ed.), 1970, pp. 173–198.

FORTNEY, J. L.

1974 Interactions between yellow perch abundance, walleye predation, and survival of alternate prey in Oneida Lake, New York. *Trans. Amer. Fish Soc.,* 103:15–24.

FORTNEY, R. H.

1975 The vegetation of Canaan Valley: a taxonomic and ecological study. PhD thesis, West Virginia University. Morgantown.

FOSTER, M. S.

1974 A model to explain molt, breeding overlap, and clutch size in some tropical birds. *Evolution,* 28:182–190.

FOWELLS, H. A.

1948 The temperature profile in a forest. *J. Forest.,* 46:897–899.

FOWELLS, H. A. (ED.)

1965 *Silvics of Forest Trees of the United States. USDA Agricultural Handbook 271.* Washington, D. C.

FOX, J. F.

1977 Alternation and coexistence of tree species. *Amer. Natur.,* 111:69–89.

FOX, L. R.

1975a Some demographic consequences of food shortage for the predator *Notonecta hoffmanni.*

*Ecology,* 56:868–880.

1975b   Factors influencing cannibalism, a mechanism of population limitation in the predator *Notonecta hoffmanni. Ecology,* 56:933–941.

1975c   Cannibalism in natural populations. *Ann. Rev. Ecol. Syst.,* 6:87–106.

FOX, M. W.

1969a   Ontogeny of prey-killing behavior in Canidae. *Behaviour,* 35:259–272.

1969b   The anatomy of aggression and its ritualization in Canidae: a developmental and comparative study. *Behaviour,* 35:242–258.

1970   A comparative study of the development of facial expressions in canids: wolf, coyote, and foxes. *Behaviour,* 36:4–73.

FRANCIS, W. J.

1970   The influence of weather on population fluctuations in California quail. *J. Wildl. Manage.,* 34:249–266.

FRANKIE, G. W., H. G. BAKER, AND P. A. OPLER

1974   Tropical plant phenology: applications for studies in community ecology. In H. Leith (ed.), 1974, pp. 287–296.

FRANKLIN, J. F., L. J. DEMPSER, AND R. H. WARING (EDS.)

1972   *Proceedings Research on Coniferous Forest Ecosystems, a Symposium.* Pacific Northwest Forest and Range Experiment Station, Portland, Oregon.

FRANKLIN, J. F., AND C. T. DRYNESS

1973   Natural Vegetation of Oregon and Washington. *USDA For. Serv. Gen. Tech. Rept. PNW-8.*

FREDRIKSEN, R. L.

1972   Nutrient budget of a Douglas-fir forest on an experimental watershed in western Oregon. In Franklin et al. (eds.), 1973, pp. 115–131.

FRENCH, C. E., L. C. MCEWEN, N. C. MAGRUDER, R. H. INGRAM, AND R. W. SWIFT

1955   Nutritional requirements of white-tailed deer for growth and antler development. *Penn. State Univ. Agr. Exp. Sta. Bull. No. 600.*

FRENCH, N. R. (ED.)

1971   *Preliminary analysis of structure and function in grasslands.* Range Sci. Dept., Sci. Ser. No. 10, Colorado State University, Fort Collins.

FRENCH, N. R., W. E. GRANT, W. E. GRODZINSKI, AND D. M. SWIFT

1976   Small mammal energetics in grassland ecosystems. *Ecol. Monogr.* 46:201–220.

FRENZEL, G.

1936   *Untersuchugen uber die Tierwe··· des Weisenbodens,* Gustav Fischer, Jena, E. Germany.

FRETWELL, S. D., AND H. L. LUCAS

1969   On territorial behavior and other factors influencing habitat distribution in birds. *Acta Biotheoretica,* 19:16–36.

FRINK, C. R.

1971   Plant nutrients and water quality. *Agr. Sci. Rev.,* 9(2):11–25.

FRY, F. E. J.

1947   Effects of the environment on animal activity. *Univ. Toronto Stud. Biol.,* 55:1–62.

FRY, F. E. J., J. S. HART, AND K. F. WALKER

1946   Lethal temperature relations for a sample of young speckled trout *Salvelinus fontinalis. Univ. Toronto Studies, Fish. Res. Lab. No. 66.*

GADGIL, M., AND W. H. BOSSERT

1970   The life historical consequences of natural selection. *Amer. Natur.,* 104:1–24.

GADGIL, M., AND O. T. SOLBRIG

1972   The concept of *r* and *K* selection: evidences from wild flowers and some theoretical considerations. *Amer. Natur,* 106:14–31.

GALLI, A. E., C. F. LECK, AND R. T. T. FORMAN

1976   Avian distribution patterns in forest islands of different sizes in central New Jersey. *Auk,* 93:356–364.

GASHWILER, J. S.

1970a   Plant and mammal changes on a clearcut in West Central Oregon. *Ecology,* 51:1018–1026.

1970b   Further study of conifer seed survival in a western Oregon clearcut. *Ecology,* 51:849–854

GATES, D.

1962   *Energy Exchange in the Biosphere,* Harper & Row, New York.

1965   Radiant energy: its receipt and disposal. *Meteorol. Monogr.,* 6:1–26.

1966   Spectral distribution of solar radiation at the earth's surface. *Science,* 151:523–528.

1968a   Toward understanding ecosystems. *Adv. Ecol. Res.,* 5:1–35.

1968b   Energy exchange between organisms and environment. In W. P. Lowry (ed.), *Biometeorology,* Proc. 28th Ann. Biol. Colloq., Oregon State University Press, Corvallis.

GAUSE, G. F.

1934   *The Struggle for Existence,* Williams & Wilkins, Baltimore.

GAY, L. W., AND K. R. KNOERR

1975   *The Forest Radiation Budget.* Bull. 19, Duke University School Forestry and Environmental Studies.

GEIS, A. D., R. I. SMITH, AND J. P. ROGERS

1971   Black duck distribution, harvest characteristics, and survival. *U. S. Fish and Wildl. Serv. Spec. Sci. Rept., Wildl. No. 139.*

GEIST, V.

1963   On the behaviour of the North American moose (*Alces alces andersoni* Peterson, 1950) in British Columbia. *Behaviour,* 20:377–416.

1971   *Mountain Sheep: A Study in Behaviour and Evolution,* University of Chicago Press.

1977   A comparison of social adaptations in relation to ecology in gallinaceous birds and ungulate societies. *Ann. Rev. Ecol. Syst.,* 8:193–207.

GERSPA, P. L., AND N. HOLOWAYCHUK

1971   Some effects of stemflow from forest canopy trees on chemical properties of soils. *Ecology,* 52:691–702.

GHILAROV, M. S.

1970 Soil biocoenosis. In J. Phillipson (ed.), *Methods of Study in Soil Ecology*, UNESCO, Paris, pp. 67–77.

GHISELIN, M. T.
1974 *The Economy of Nature and the Evolution of Sex*. University of California Press, Berkeley.

GIBSON, J. B., AND J. M. THODAY
1962 Effects of disruptive selection: 6. A second chromosome polymorphism. *Heredity*, 17:1–26.

GIESEL, J. T.
1976 Reproductive strategies as adaptations to life in temporarily heterogenous environments. *Ann. Rev. Ecol. Syst.*, 7:51–79.

GILBERT, L. E.
1975 Ecological consequences of a coevolved mutualism between butterflies and plants. In L. Gilbert and P. Raven (eds.), 1975, pp. 210–240.

GILBERT, L. E., AND P. H. RAVEN (EDS.)
1975 *Coevolution of Animals and Plants*. University of Texae Press, Austin.

GILBERTSON, C. B., ET AL.
1970 The effect of animal density and surface slope on the characteristics of runoff, solid waste, and nitrate movement on unpaved feedyards. *Nebraska Agr. Exp. Sta. Bull. 508.*

GILBERTSON, M., R. D. MORRIS, AND R. A. HUNTER
1976 Abnormal chicks and PCB residue levels in eggs of colonial birds on the lower Great Lakes (1971–1973). *Auk*, 93:434–442.

GILL, D. E.
1975 Spatial patterning of pines and oaks in the New Jersey pine barrens. *J. Ecol.*, 63:291–298.

GILL, F. B., AND L. L. WOLF
1975 Economics of feeding territoriality in the golden-winged sunbird. *Ecology*, 56:333–345.

GILL, L. S., AND F. G. HAWKSWORTH
1961 The mistletoes: a literature review. *U.S. Dept. Agr. Tech. Bull. No. 1242*, pp. 1–87.

GILPIN, M. E.
1975 *Group Selection in Predator Prey Communities*, Princeton University Press.

GILPIN, M. E., AND F. J. AYALA
1973 Global models of growth and competition. *Proc. Natur. Acad. Sci.*, 70:3590–3593.

GISBORNE, H. T.
1941 How the wind blows in the forest of northern Idaho. *Northern Rocky Mt. Forest Range Expt. Sta.*

GIST, C. S., AND D. A. CROSSLEY, JR.
1975 A model of mineral-element cycling for an invertebrate food web in a southeastern hardwood forest litter community. In F. Howell, et al. (eds.), 1975, pp. 84–106.

GLEASON, H. A.
1917 The structure and development of the plant association. *Bull. Torrey Bot. Club*, 44:463–481.
1926 The individualistic concept of the plant association. *Bull. Torrey Bot. Club*, 53:7–26.

GOKSØYR, J.
1975 Decomposition, microbiology, and ecosystem analysis. In F. E. Wielgolaski (ed.), 1975a, pp. 230–238.

GOLDSMITH, J. R.
1969 Epidemiological bases for possible air quality criteria for lead. *Air Poll. Contr. Assoc. J.*, 19:714–719.

GOLDSMITH, J. R., AND A. C. HEXTER
1967 Respiratory exposure to lead: epidemiological and experimental dose-response relationship. *Science*, 158:132–134.

GOLLEY, F. B.
1960 Energy dynamics of a food chain of an old field community. *Ecol. Monog.*, 30:187–206.
1972a Energy flux in ecosystems. In J. S. Wiens (ed.), 1972, pp. 69–90.
1972b Summary. In P. M. Golley and F. B. Golley (eds.), 1972, pp. 407–413.
1975 Productivity and mineral cycling in tropical forests. In National Academy Science, 1975, pp. 106–115.

GOLLEY, F. B., AND J. B. GENTRY
1966 A comparison of variety and standing crop of vegetation on a one year and a twelve year abandoned field. *Oikos*, 15:185–199.

GOLLEY, F. B., J. T. MCGINNIS, R. C. CLEMENTS, G. S. CHILD, AND M. J. DUEVER
1975 *Mineral Cycling in a Tropical Moist Forest Ecosystem*, University of Georgia Press, Athens.

GOLLEY, F. B., K. PETRUSEWICZ, AND L. RYSZKOWSKI
1975 *Small Mammals: Their Productivity and Population Dynamics*, Cambridge University Press, Cambridge, England.

GOLLEY, F. B., AND E. MEDINA (EDS.)
1975 *Tropical Ecological Systems Trends in Terrestrial and Aquatic Research*, Springer-Verlag, New York.

GOLLEY, P. M., AND F. B. GOLLEY (EDS.)
1972 *Tropical Ecology, with an Emphasis on Organic Production*, University of Georgia Press, Athens.

GOOD, R. E., D. F. WHIGHAM, AND R. L. SIMPSON (EDS.)
1978 *Freshwater Wetlands*, Academic, New York.

GORDON, M. S.
1972 *Animal Physiology*, 2nd ed., Macmillan, New York.

GOREAU, T. F.
1963 Calcium carbonate deposition by coralline algae and corals in relation to their role as reef builders. *Ann. N.Y. Acad. Sci.*, 109:127–167.

GOSS-CUSTARD, J. D.
1970 The responses of redshank *Tringa totanus* (L) to spatial variation in the density of their prey. *J. Anim. Ecol.*, 39:91–113.
1977a The energetics of prey selection by redshank *Tringa totanus* (L) and a preferred prey *Corophium volutator* (Pallas). *J. Anim. Ecol.*, 46:21–35.
1977b The energetics of prey selection by redshank *Tringa totanus* (L) in relation to prey density. *J. Anim. Ecol.*, 46:1–19.

GOSZ, J. R., E. LIKENS, AND F. H. BORMANN
1976 Organic matter and nutrient dynamics of the forest and forest floor in the Hubbard Brook Forest. *Oecologica*, 22:305–320.

GOSZ, J. R., G. E. LIKENS, J. S. EATON, AND F. H. BORMANN

1975 Leaching of nutrients from leaves of selected tree species in New Hampshire. In F. C. Howell et al. (eds.), 1975, pp. 630–641.

GOTTLIEB, G.

1963 "Imprinting" in nature. *Science,* 139:497–498.

GOULD, S. J., AND R. F. JOHNSTON

1972 Geographic variation. *Ann. Rev. Ecol. and Syst.,* 3:457–498.

GRANHALL, R., AND V. LID-TORSVIK

1975 Nitrogen fixation by bacteria and free-living blue-green algae in tundra ecosystem. In F. E. Wielgolaski (ed.), 1975a, pp. 305–315.

GRANT, P. R.

1972 Convergent and divergent character displacement. *Biol. J. Linn. Soc.,* 4:39–68.

1972 Interspecific competition among rodents. *Ann. Rev. Ecol. Syst.,* 3:79–106.

1975 The classical case of character displacement. *Evolutionary Biology,* 8:237–357.

GRANT, P. R., AND R. D. MORRIS

1971 The distribution of *Microtus pennsylvanicus* within grassland habitat. *Can. J. Zool.,* 49:1043–1052.

GRANT, V.

1963 *The Origins of Adaptations,* Columbia University Press, New York.

1971 *Plant Speciation,* Columbia University Press, New York.

GREENWOOD, P. H.

1974 The cichlid fishes of Lake Victoria, East Africa; the biology and evolution of a species flock. *Bull. Brit. Mus. (Natur. Hist.) Zool.,* 6:1–134.

GREIG-SMITH, P.

1964 *Quantitative Plant Ecology,* 2nd ed., Butterworth, Washington, D.C.

GRIER, C. C., AND D. W. COLE

1972 Elemental transport changes occurring during development of a second growth Douglas fir ecosystem. In J. L. Franklin, et al., 1972, pp. 103–113.

GRIGGS, R. F.

1946 The timberlines of northern America and their interpretation. *Ecology,* 27:275–289.

GRIME, J. P.

1966 Shade avoidance and shade tolerance in flowering plants. In F. Bainbridge et al. (eds.), *Light as an Ecological Factor,* Blackwell, Oxford, pp. 187–207.

1977 Evidence for the existence of three primary strategies in plants and its relevance to ecological and evolutionary theory. *Amer. Natur.,* 111:1169–1194.

GRIME, J. P., AND A. V. CURTIS

1976 The interaction of drought and mineral nutrient stress in calcareous grassland. *J. Ecol.,* 64:975–988.

GRINNELL, J.

1917 The niche relationships of the California thrasher. *Auk,* 34:427–433.

1924 Geography and evolution. *Ecology,* 5:225–229.

1928 Presence and absence of animals. *University of California Chronicle,* 30:429–450.

GROSS, M.

1972 *Oceanography,* Prentice-Hall, Englewood Cliffs, N.J.

GUHL, A. M., AND W. C. ALLEE

1944 Some measurable effects of social organization in flocks of hens. *Physiol. Zool.,* 17:320–347.

GULLION, G. W.

1970 Factors influencing ruffed grouse populations. *Trans. 35th N. Am. Wildl. and Natur. Resource Conf.,* pp. 93–105.

GUNTER, G.

1961 Some relations of estuarine organisms to salinity. *Limno. Oceanogr.,* 6:182–190.

GYSEL, L. W.

1951 Borders and openings of beech-maple woodlands in southern Michigan. *J. Forest.,* 49:13–19.

1960 An ecological study of the winter deer range of elk and mule deer in the Rocky Mountain National Park. *J. Forest.,* 58:696–703.

HAARLOV, N.

1960 Microarthropods from Danish soils. *Oikos, Suppl. No. 3,* pp. 1–176.

HAARTMAN, L. VON

1956 Territory in the pied flycatcher, *Muscicapa hypoleuca. Ibis,* 98:460–475.

HACSKAYLO, E.

1971 Metabolite exchanges in ectomycorrhizae. In E. Hacskaylo (ed.), *Mycorrhizae, USDA Forest Serv. Misc. Pub. No. 1189.*

HADLEY, E. B., AND B. J. KIECKHEFER

1963 Productivity of two prairie grasses in relation to fire frequency. *Ecology,* 44:389–395.

HAILMAN, J. P.

1977 Communication by reflected light. In T. Sebeok (ed.), 1978, pp. 184–210.

HAIRSTON, N. G.

1949 The local distribution and ecology of the plethodontid salamanders of the southern Appalachians. *Ecol. Monogr.,* 19:47–73.

1969 On the relative abundance of species. *Ecology,* 50:1091–1094.

HAIRSTON, N. G., AND C. H. POPE

1948 Geographic variation and speciation in Appalachian salamanders (*Plethodon jordani* group). *Evolution,* 2:266–278.

HAIRSTON, N. G., F. E. SMITH, AND L. B. SLOBODKIN

1960 Community structure, population control, and competition. *Amer. Natur.,* 94:421–425.

HALDANE, J. S. B.

1954 The measurement of natural selection. *Proc. 9th Int. Congr. Genetics.,* pp. 480–487.

HALE, N.

1971 *Biology of Lichens,* Edward Arnold, London.

HALL, D. J., W. E. COOPER, AND E. E. WERNER

1970 An experimental approach to the production

**785**

dynamics and structure of freshwater animal communities. *Limnol. Oceanogr.,* 15:839–928.

HAMILTON, W. D.

1972   Altruism and related phenomena, mainly in social insects. *Ann. Rev. Ecol. Syst.,* 3:193–232.

HAMILTON, W. J., III

1959   Aggressive behavior in migrant pectoral sandpipers. *Condor,* 61:161–179.

HAMMOND, A. L.

1972   Chemical pollution: polychlorinated biphenyls. *Science,* 175:155–156.

HAMMOND, J.

1953   Periodicity in animals: the role of darkness. *Science,* 177:389–390.

HAMNER, W. M.

1963   Diurnal rhythm and photoperiodism in testicular recrudescence of the house finch. *Science,* 142:1294–1295.

1968   The photorefractory period of the house finch. *Ecology,* 49:211–227.

HANSON, H. C.

1958   Principles concerned in the formation and classification of communities. *Bot. Rev.,* 24:65–125.

HANSON, W. C.

1971   Seasonal patterns in native residents of three contrasting Alaskan villages. *Health Physics,* 20:585–591.

HARDIN, G.

1960   The competitive exclusion principle. *Science,* 131:1292–1297.

HARLEY, J. L.

1959   *The Biology of Mycorrhiza,* Plant Science Monographs, Leonard Hill, London.

HARNER, E. J., AND R. C. WHITMORE

1977   Multivariate measures of niche overlap using discriminate analysis. *Theor. Pop. Biol.,* 12:21–36.

HARNER, M.

1977   The enigma of Aztec sacrifice. *Natur. Hist.,* 6(4):46–51.

HARPER, J. L.

1961   Approaches to the study of plant competition. In F. L. Milthorpe (ed.), *Mechanisms in Biological Competition, Symp. Soc. Exp. Biol.,* 15:1–39.

1967   A Darwinian approach to plant ecology. *J. Ecol.* 55:247–270.

1969   The role of predation in vegetational diversity. *Brookhaven Symposia in Biology No. 22,* pp. 48–62.

1977   *Population Biology of Plants,* Academic, New York.

HARPER, J. L., P. H. LOVELL, AND K. G. MOORE

1970   The shapes and sizes of seeds. *Ann. Rev. Ecol. Syst.,* 1:327–356.

HARPER, J. L., AND J. OGDEN

1970   The reproductive strategy of higher plants. 1. The conception of strategy with special reference to *Senecio vulgaris. J. Ecol.,* 58:681–698.

HARPER, J. L., AND J. WHITE

1974   The demography of plants. *Ann. Rev. Ecol. Syst.,* 5:419–463.

HARRINGTON, R. W., JR.

1959   Photoperiodism in fishes in relation to the annual sexual cycle. In R. Witherow (ed.), *Photoperiodism and Related Phenomena,* pp. 651–667.

HARRIS, V. T.

1952   An experimental study of habitat selection by prairie and forest races of the deermouse, *Peromyscus maniculatus. Contrib. Lab. Vert. Zool., University of Michigan,* 56:1–53.

HARRIS, W. F., P. SOLLINS, N. T. EDWARDS, B. E. DINGER, AND H. F. SHUGART

1975   Analysis of carbon flow and productivity in a temperate deciduous forest ecosystem. In National Academy Science, 1975, pp. 116–122.

HARRISON, J. L.

1962   Distribution of feeding habits among animals in a tropical rain forest, *J. Anim. Ecol.,* 31:53–63.

HART, J. S.

1951   Photoperiodicity in the female ferret. *J. Expt. Biol.,* 28:1–12.

HARTMAN, W. L.

1972   Lake Erie: effects of exploitation, introductions, and eutrophication on the salmonid community. *J. Fish. Res. Board Can.,* 29:899–912.

HARVEY, H. W.

1950   On the production of living matter in the sea off Plymouth. *J. Marine Biol. Assoc. U.K.,* n.s., 29:97–137.

HASLER, A. D.

1954   Odour perception and orientation in fishes. *J. Fish. Res. Board Can.,* 11:107–129.

1960   Guideposts of migrating fishes. *Science,* 132:785–792.

1969   Cultural eutrophication is reversible. *Bioscience,* 19:425–431.

HASLER, A. D. (ED.)

1975   *Coupling of Land and Water Systems,* Ecological Studies, Vol. 10, Springer-Verlag, New York.

HASLER, A.D., AND H. O. SCHWASSMAN

1960   Sun orientation of fish at different latitudes, *Cold Spring Harbor Symp. Quant. Biol.,* 25:429–441.

HASSELL, M. P.

1966   Evaluation of parasite or predator response. *J. Anim. Ecol.,* 35:65–75.

HASSELL, M. P., J. H. LAWTON, AND J. R. BEDDINGTON

1976   The components of arthropod predation. 1. The prey death rate. *J. Anim. Ecol.,* 45:135–164.

1977   Sigmoid functional responses by invertebrate predators and parasitoids. *J. Anim. Ecol.,* 46:249–262.

HASSELL, M. P., AND R. M. MAY

1973   Stability in insect host-parasite models. *J. Anim. Ecol.,* 42:693–726.

1974   Aggregation in predators and insect parasites and its effect on stability. *J. Anim.*

*Ecol.,* 43:567–597.

HASTINGS, J. W.

1959   Unicellular clocks. *Ann. Rev. Microbiol.,* 13:297–312.

1970   Cellular-biochemical clock hypothesis. In F. Brown et al., *The Biological Clock,* Academic, New York.

HAVEN, S. B.

1973   Competition for food between the intertidal gastropods, *Acmaea scabia* and *Acmaea digitalis. Ecology,* 54:143–151.

HAYNE, D. W., AND R. C. BALL

1956   Benthic productivity as influenced by fish predation. *Limnol. Oceanogr.,* 1:162–175.

HAYNES, C. U.

1970   Geochronology of Man-Mammoth sites and their bearing on the Llano Complex. In W. Cort and J. K. Jones, (eds.), 1970, pp. 77–92.

HAYS, H.

1972   Polyandry in the spotted sandpiper. *Living Bird,* 11:43–57.

HAYS, H., AND R. W. RISEBROUGH

1972   Pollutant concentrations in abnormal terns from Long Island Sound. *Auk,* 89.19–35.

HEAL, O. W., H. E. JONES, AND J. B. WHITTAKER

1975   Moore House, U.K. In T. Rosswall & O. W. Heal (eds.), 1975, 295–320.

HEATH, J. E.

1965   Temperature regulation and diurnal activity in horned lizards. *Univ. Cal. Pub. Zool.,* 64:97–136.

HEDGPETH, J. W. (ED.)

1957a   *Treatise in Marine Ecology and Paleoecology:* 1. *Ecology.* Memoir 67, Geological Soc. Amer.

1957b   Sandy beaches. In Hedgpeth (ed.), 1957a, pp. 587–608.

HEGGLESTAD, H. E.

1969   Consideration of air quality standards for vegetation with respect to ozone. *Air. Pollut. Assoc. Cont. J.,* 19:424–426.

HEINSELMAN, M. L.

1963   Forest sites, bog processes, and peatland types in the glacial Lake Agassiz region, Minnesota. *Ecol. Monogr.,* 33:327–374.

1970   Landscape evolution, peatland types, and the environment in Lake Agassiz peatlands natural area, Minnesota. *Ecol. Monogr.,* 33:327–374.

1971   The natural role of fire in northern coniferous forests. In *Fire in the Northern Environment: A Symposium,* Pacific Northwest Forest and Range Expt. Station, Portland, Oregon, pp. 61–72.

1975   Boreal peatlands in relation to environment. In A. D. Hasler (ed.), 1975, pp. 93–103.

HELLER, H. C.

1971   Altitudinal zonation of chipmunks *(Eutamias):* interspecific aggression. *Ecology,* 52:312–319.

HELLER, H. C., AND D. GATES

1971   Altitudinal zonation of chipmunks *(Eutamias):* energy budgets. *Ecology,* 52:424–433.

HENDERSON, N. E.

1963   Influence of light and temperature on the reproductive cycle of the eastern brook trout *Salvelinus fontinalis. J. Fish. Res. Board Can.,* 20:859–897.

HENDRIKSSON, E., AND B. SIMU

1971   Nitrogen fixation by lichens. *Oikos,* 22:119-121.

HENRY, S. M. (ED.)

1966   *Symbiosis,* Academic, New York.

HERREID, C. F., AND S. KINNEY

1967   Temperature and development of wood frog *Rana sylvatica* in Alaska. *Ecology,* 48:579–590.

HENSLEY, M. M., AND J. B. COPE

1951   Further data on removal and repopulation of the breeding birds in a spruce-fir forest community. *Auk,* 68:483–493.

HESPENHEIDE, H. A.

1973   Ecological inferences from morphological data. *Ann. Rev. Ecol. Syst.,* 4:213–229.

HESS, E. H.

1959   Imprinting. *Science,* 130:133–141.

1964   Imprinting in birds. *Science,* 146:1129–1139.

1971   The imprinting process. In *Topics in the Study of Life, BIOS: The Bio Source Book,* Harper & Row, New York, pp. 314–320.

HICKMAN, J. C.

1975   Environmental unpredictability and plastic energy allocation strategies in annual *Polygonum cascadense* (Polygonaceae). *J. Ecol.,* 63:689–701.

HICKS, K. P., AND J. O. TAHVANIANEN

1974   Niche differentiation by crucifer-feeding flea beetles Coleoptera: Chupomelida. *Amer. Midl. Natur.,* 91:406–423.

HILDEN, O.

1964   Ecology of duck populations in the island group of Valassarret, Gulf of Bothnia. *Ann. Zool. Fennica,* 1:153–279.

1965   Habitat selection in birds: A review. *Ann. Zool. Fennica,* 2:53–75.

HILLBRICHT-ILKOWSKA, A.

1974   Secondary productivity in freshwaters, its value and efficiencies in plankton food chain. In *Proc. 1st Inter. Congr. Ecol.:* 164–167.

HINDE, R. A.

1959   Behavior and speciation in birds and lower vertebrates. *Biol. Rev.,* 34:85–128.

1970   *Animal Behaviour: A Synthesis of Ethology and Comparative Psychology,* McGraw-Hill, New York.

HINNERI, S., M. SONESSON, AND A. K. VEUM

1975   Soils of Fennoscandian I.B.P. Tundra ecosystems. In F. E. Wielgolaski (ed.), 1975a, pp. 31–40.

HIRSHFIELD, M. F., AND D. W. TINKLE

1975   Natural selection and the evolution of reproductive effort. *Proc. Nat. Acad. Sci.,* 72(6):2227–2231.

HOBBIE, J. E. ET AL.

1972   Carbon flux through a tundra pond ecosystem at Barron, Alaska. *U.S. Tundra Biome*

*Rep.* 72-1.

HOCK, R. J.
1960  Seasonal variations in physiological functions of arctic ground squirrels and black bears. In C. P. Lyman and A. R. Dawe (eds.), 1960, pp. 155–169.

HOCKING, B.
1975  Ant-plant mutualism: evolution and energy. In L. E. Gilbert and P. H. Raven (eds.), 1975, pp. 78–90.

HOESE, H. D.
1960  Biotic changes in a bay associated with the end of a drought. *Limnol. Oceanogr.,* 5:326–336.

HOFFMAN, K.
1965  Clock-mechanisms in celestial orientation of animals. In J. Aschoff (ed.), *Circadian Clocks,* North Holland Publishing, Amsterdam, pp. 87–94.

HOLDEN, C.
1974  Fish flour: protein supplement has yet to fulfill expectations. *Science,* 173:410–412.

HOLDRIDGE, L. R.
1967  Determination of world plant formations from simple climatic data. *Science,* 130:572.

HOLDRIDGE, L. R., W. C. GRENKE, W. H. HATHEWAY, T. LIANG, AND J. A. TOSI, JR.
1971  *Forest Environments in Tropical Life Zones, A Pilot Study,* Pergamon, New York.

HOLLING, C. C.
1959  The components of predation as revealed by a study of small mammal predation of the European pine sawfly. *Can. Entomol.,* 91:293–320.
1961  Principles of insect predation. *Ann. Rev. Entomol.,* 6:163–182.
1965  The functional response of predators to prey density and its role in mimicry and population regulation. *Mem. Entomol. Soc. Can. No. 45.*
1966  The functional response of invertebrate predators to prey density. *Mem. Entomol. Soc. Can. No. 48.*
1973  Resilience and stability of ecological systems. *Ann. Rev. Ecol. Syst.,* 4:1–23.

HOLMGREN, R. C.
1956  Competition between annuals and young bitterbrush *(Purshia tridentata)* in Idaho. *Ecology,* 37:370–377.

HOPKINS, H. H.
1954  Effects of mulch upon certain factors of the grassland environment. *J. Range Manage.,* 7:255–258.

HOPKINS, S. H.
1958  The planktonic larvae of *Polydora websteri* Hartman (Annelida, Polychaeta) and their settling on oysters. *Bull. Marine Sci. of Gulf and Caribbean,* 8:268–277.

HORNOCKER, M. G.
1969  Winter territoriality in mountain lions. *J. Wildl. Manage.,* 33:457–464.
1970  An analysis of mountain lion predation upon mule deer and elk in the Idaho primitive area. *Wildl. Monogr. No. 21.*

HOWARD, W. E.
1960  Innate and environmental dispersal of individual vertebrates. *Amer. Mid. Natur.,* 63:152–161.

HOWE, H. F.
1976  Egg size, hatching asynchrony, sex, and brood reduction in the common grackle. *Ecology,* 57:1195–1207.

HOWELL, F. G., J. B. GENTRY, AND M. H. SMITH (EDS.)
1975  *Mineral Cycling in Southeastern Ecosystems* (ERDA Symposium Series), National Technical Information Science, U.S. Dept. Commerce.

HOWELL, R. K., AND D. F. KREMER
1970  Alfalfa yields as influence by air quality. *Phytopathology,* 60:1297.

HUBBELL, S., AND P. A. WERNER
1979  On measuring the intrinsic rate of increase of populations with heterogeneous life histories. *Amer. Natur.,* 113:277–293.

HUBER, B.
1952  Der Emfluss der Vegetation auf die Schwankungen des $CO_2$ Gehaltes der Atmosphare. *Arch. Met. Geophys. Bioklim.*

HUDSON, H. J., AND J. WEBSTER
1958  Succession of fungi on decaying stems of *Agropyron repens. Brit. Mycol. Soc. Trans.,* 41:165–177.

HUEY, R. B., E. R. PIANKA, M. E. EGAN, AND L. W. COONS
1974  Ecological shifts in sympatric Kalahari fossorial lizards (Typhlosaurus). *Ecology,* 58:304–16.

HUFFAKER, C. B.
1958  Experimental studies on predation: dispersion factors and predator-prey oscillations. *Hilgardia,* 27:343–383.

HUMPHREY, R. R.
1958  The desert grassland, a history of vegetational changes and an analysis of causes. *Bot. Rev.,* 24:193–252.

HUMPHRIES, D. A., AND P. M. DRIVER
1967  Erratic display as a device against predators. *Science,* 156:1767–1768.

HUNT, E. G., AND A. I. BISCHOFF
1960  Inimical effects on wildlife of periodic DDT applications to Clear Lake. *Calif. Fish Game,* 46:91–106.

HUNT, R. L.
1975  Use of terrestrial invertebrates as food by salmonids. In A. Hasler (ed.), 1975, pp. 137–151.

HURD, L. E., M. V. MELLINGER, L. L. WOLF, AND S. T. MCNAUGHTON
1971  Stability and diversity at three trophic levels in terrestrial successional ecosystems. *Science,* 173:134–136.

HURLBERT, S. H.
1971  The nonconcept of species diversity: a critique and alternative parameters. *Ecology,* 52:577–586.

HUTCHINSON, G. E.
1957  *A Treatise on Limnology, Vol. 1, Geography, Physics, Chemistry,* Wiley, New York.

1959   Homage to Santa Rosalia, or why are there so many kinds of animals? *Amer. Natur.,* 93:145–159.

1958   Concluding remarks. *Cold Spring Harbor Symp. Quant. Biol.,* 22:415–427.

1969   Eutrophication, past and present. In National Academy Science, 1969, *Eutrophication: Causes, Consequences, Correctives,* Washington, D.C., pp. 12–26.

1978   *An Introduction to Population Ecology,* Yale University Press, New Haven. Conn.

HYDER, D. N.
1969   The impact of domestic animals on the structure and function of grassland ecosystems. In R. L. Dix and R. G. Beidleman (eds.), 1969, pp. 243–260.

HUXLEY, J. S.
1934   A natural experiment on the territorial instinct. *Brit. Birds,* 27:270–277.

HYNES, H.
1970   *Biology of Running Water,* University of Toronto Press, Ontario.

IKUSIMA, I.
1965   Ecological studies on the productivity of aquatic plant communities: 1, Measurement of photosynthetic activity. *Bot. Mag. Tokyo,* 78:202–211.

ILLIES, J., AND L. BOTOSANEANU
1963   Problems et methodes de la classification et de la zonation ecologique des eaux courantes, considerees surtout du point de vue faunistique. *Mitt. int. Verein theor. Angew. Limnol.,* 12:1–57.

IMMELMANN, K.
1975   Ecological significance of imprinting and early learning. *Ann. Rev. Ecol. Syst.,* 6:15–37.

IRVING, I.
1960   *Birds of Anaktuviik Pass, Kovuk and Old Crow: a study in Arctic adaptation. U.S. Nat. Mus. Bull.,* 217.

IRVING, I., AND J. KROGH
1954   Body temperatures of arctic and subarctic birds and mammals, *J. Appl. Physiol.,* 6:667–680.

IVLEV, V. S.
1961   *Experimental Ecology of the Feeding of Fishes,* Yale University Press, New Haven, Conn.

IWAKI, H.
1974   Comparative productivity of terrestrial ecosystems in Japan, with emphasis on the comparison between natural and agricultural ecosystems. *Proc. 1st Int. Cong. of Ecol.,* pp. 40–45.

JACOBSON, J. S., AND A. C. HILL
1970   *Recognition of Air Pollution Injury to Vegetation: A Pictorial Atlas,* Air Pollution Control Association, Pittsburgh, Penn.

JACOT, A. P.
1930   Reduction of spruce and fir litter by minute animals, *J. Forestry,* 37:858–860.

1940   The fauna of the soil. *Quart. Rev. Biol.,* 15:38–58.

JAEGER, R. G.
1971   Competitive exclusion as a factor influencing the distribution of two species of terrestrial salamanders. *Ecology,* 52:632–637.

JAFFE, L. S.
1970   The global balance of carbon monoxide. In F. S. Singer (ed.), 1970, pp. 34–49.

JAMES, F. C., AND H. SHUGART
1974   The phenology of the nesting season of the American robin *(Turdus migratorius)* in the United States. *Condor,* 76:159–168.

JANSEN, D. H.
1966   Coevolution of mutualism between ants and acacias in Central America. *Evolution,* 20:249–275.

1967   Synchronization of sexual reproduction of trees within the dry season in Central America. *Evolution,* 21:620–637.

1969   Seed eaters versus seed size, number, toxicity and dispersal. *Evolution,* 23:1–27.

1970   Herbivores and the number of tree species in tropical forests. *Amer. Natur.,* 104:501–528.

1971   Seed predation by animals. *Ann. Rev. Ecol. Syst.,* 2:465–492.

JARVIS, P. G., G. B. JAMES, AND J. J. LANDSBERG
1976   Coniferous forest. In J. L. Monteith (ed.), 1976, pp. 171–240.

JAWORSKI, N. A., AND L. J. HELLING
1970   Relative contribution of nutrients to the Potomac River Basin from various sources. In *Relationship of Agriculture to Soil and Water Pollution,* Cornell University Press, Ithaca, N.Y.

JEFFERIES, R. L.
1972   Aspects of salt marsh ecology with particular reference to inorganic plant nutrients. In R. S. K. Barnes and J. Green (eds.), *The Estuarine Environment,* Applied Science Publishers, London, pp. 61–85.

JENKIN, J. F.
1975   Macquarie Island, Subantarctic. In T. Rosswall and O. W. Heal (eds.), 1975, pp. 375–397.

JENKINS, D., A. WATSON, AND G. R. MILLER
1963   Population studies on red grouse *Lagopus lagopus scoticus* (Lath) in northeast Scotland. *J. Anim. Ecol.,* 32:317–376.

JENKINS, D. W.
1944   Territory as a result of despotism and social organization in geese. *Auk,* 61:30–47.

JENNI, D. A.
1974   Evolution of polyandry in birds. *Amer. Zool.,* 14:129–144.

JENNI, D. A., AND G. COLLIER
1972   Polyandry in the American jacana *(Jacana spinosa). Auk,* 89:743–765.

JENNY, H.
1933   Soil fertility losses under Missouri conditions. *Missouri Agri. Exp. Sta. Bull.,* 324.

JENSEN, S.
1966   A new chemical hazard. *New Scientist,* 32:612.

JEPSEN, G. L., E. MAYR, AND G. G. SIMPSON (EDS.)
1949   *Genetics, Paleontology, and Evolution,*

Princeton University Press, New Jersey.

JOENSUU, O. I.

1971   Fossil fuels as a source of mercury pollution. *Science,* 172:1027–1028.

JOHANNES, R. E.

1964   Uptake and release of dissolved organic phosphorus by representatives of a coastal marine ecosystem. *Limnol. Oceanogr.,* 9:224–234.

1965   Influence of marine protozoa on nutrient regeneration. *Limnol. Oceanogr.,* 10:434–442.

1967   Ecology of organic aggregates in the vicinity of coral reef. *Limnol. Oceanogr.,* 12:189–195.

1968   Nutrient regeneration in lakes and oceans. In M. R. Droop and E. J. Ferguson (eds.), *Advances in the Microbiology of the Sea,* Vol. 1, Academic, New York, pp. 203–213.

JOHANNES, R. E., AND K. L. WEBB

1970   Release of dissolved organic compounds by marine and fresh water invertebrates. *Proc. Conf. on Organic Matter in Natural Waters, Occ. Pub. No. 1.,* Inst. Marine Sci., University of Alaska.

JOHANSSON, L. G., AND S. LINDER

1975   The seasonal pattern of photosynthesis of some vascular plants on a subarctic mire. In F. E. Wielgolaski (ed.), 1975, pp. 194–200.

JOHNSGARD, P. A.

1961   The sexual behavior and systematic position of the hooded merganser. *Wilson Bull.,* 73:227–236.

JOHNSON, C.

1976   *Introduction to Natural Selection,* University Park Press, Baltimore, Md.

JOHNSON, C. G.

1969   *Migration and Dispersal of Insects by Flight,* Methuen, London.

JOHNSON, D. W.

1968   Pesticides and fishes: a review of selected literature. *Trans. Amer. Fish. Soc.,* 97:398–424.

JOHNSON, F. S.

1970   The oxygen and carbon dioxide balance in the earth's atmosphere. In F. S. Singer (ed.), 1970, pp. 4–11.

JOHNSON, H. B.

1975   Plant pubescence: an ecological perspective. *Bot. Rev.,* 41:233–258.

JOHNSON, L.

1972   Keller Lake: characteristics of a culturally unstressed salmonid community. *J. Fish Res. Board Can.,* 29:731–740.

JOHNSON, M.

1970   Preliminary report on species composition, chemical composition, biomass, and production of marsh vegetation in the upper Patuxent estuary, Maryland. *Chesapeake Biol. Lab. Rept. Ref. No. 70-130,* pp. 164–178.

JOHNSON, N. K.

1975   Controls of number of bird species on montane islands in the Great Basin. *Evolution,* 29:545–567.

JOHNSON, N. M., R. C. REYNOLDS, AND G. E. LIKENS

1972   Atmospheric sulphur: its effect on the chemical weathering of New England. *Science,* 177:514–515.

JOHNSON, P. L., AND W. O. BILLINGS

1962   The alpine vegetation of the Beartooth Plateau in relation to cryopedogenic processes and patterns. *Ecol. Monogr.,* 32:105–135.

JOHNSON, P. L., AND W. T. SWANK

1973   Studies of cation budgets in the southern Appalachians on four experimental watersheds with contrasting vegetation. *Ecology,* 54:70–80.

JOHNSON, W. K., AND G. J. SCHROEPFER

1964   Nitrogen removal by nitrification and denitrification. *J. Water Pollut. Control Fed.,* 36:1011–1036.

JOHNSTON, D. W., AND E. P. ODUM

1956   Breeding bird populations in relation to plant succession on the Piedmont of Georgia. *Ecology,* 37:50–62.

JOHNSTON, J. W.

1936   The macrofauna of soils as affected by certain coniferous and hardwood types in the Harvard forest. PhD. Dissertation, Harvard University Library, Cambridge, Mass.

JOHNSTON, R. F.

1956a   Predation by short-eared owls in a *Salicornia* salt marsh. *Wilson Bull.,* 68:91–102.

1956b   Population structure in salt marsh song sparrows: Part 2, Density, age structure and maintenance. *Condor,* 58:254–272.

JOHNSTON, R. F., AND R. K. SELANDER

1971   Evolution in the house sparrow: 2. Adaptative differentiation in North American populations. *Evolution,* 25:1–28.

JONES, H. LEE, AND J. M. DIAMOND

1976   Short time base studies of turnover in breeding bird populations on the California channel islands. *Condor,* 78:526–549.

JONES, M. G.

1933   Grassland management and its influence on the sward. *J. Roy. Agr. Soc.,* 94:21–41.

JORDAN, C. F.

1971a   Productivity of a tropical forest and its relation to a world pattern of energy storage. *J. Ecol.,* 59:127–142.

1971b   A world pattern in plant energetics. *Amer. Sci.,* 59:425–433.

KABAT, C., N. E. COLLIAS, AND R. C. GUETTINGER

1953   *Some Winter Habits of White-tailed Deer and the Development of Census Methods in the Flag Yard of Northern Wisconsin. Tech. Wildl. Bull. No. 7,* Wisconsin Conserv. Dept., Madison.

KABAT, C., AND D. R. THOMPSON

1963   *Wisconsin Quail, 1834–1962:Population Dynamics and Habitat Management. Tech. Bull. No. 30,* Wisconsin Conserv. Dept., Madison.

KAHN, A. J., AND P. J. LEVITAN

1976   Effect of habitat complexity on population density and species richness in tropical intertidal predatory gastropod assemblages. *Oecologica,* 25:199–210.

KALININ, G. R., AND V. D. BYKOV

1969 The world's water resources, present and future. *Impact of Science on Society,* 19:135–150.

KALLE, K.

1971 Salinity: general introduction. In O. Kinne (ed.), *Marine Ecology: Vol. 1, Environmental Factors, Part 2,* Wiley, New York, pp. 683–688.

KALLIO, P.

1974 The ecology of the nitrogen fixation in subarctic lichens. *Oikos,* 25:194–198.

KALLIO, S., AND P. KALLIO

1975 Nitrogen fixation in lichens at Kevo, North Finland. In F. E. Wielgolaski (ed.), 1975, pp. 292–304.

KALLIO, P., AND N. VALANNE

1975 On the effect of continuous light on photosynthesis in mosses. In F. E. Wielgolaski (ed.), 1975, pp. 149–162.

KARPLUS, M.

1949 Bird activity in continuous day-light of arctic summer. *Bull. Ecol. Soc. Amer.,* 30:60.

KARR, J. R.

1971 Structure of avian communities in selected Panama and Illinois habitats. *Ecol. Monogr.,* 41:207–233.

1975 Production and energy pathways, and community diversity in forest birds. In F. Golley and E. Medina (eds.), 1975, pp. 161–176.

1976 Seasonality, resource availability and community diversity in tropical bird communities. *Amer. Natur.,* 110:973–994.

KARR, J. R., AND R. R. ROTH

1971 Vegetation structure and avian diversity in several new world areas. *Amer. Natur.,* 105:423–435.

KATZ, B.A., AND H. LEITH

1974 Seasonality of decomposers. In H. Leith (ed.), 1974, pp. 163–184.

KAYLL, A. J.

1974 Use of fire in land management. In T. F. Kozlowski and C. E. Ahlgren (eds.), 1974, pp. 483–511.

KAYS, S., AND J. L. HARPER

1974 The regulation of plant and tiller density in a grass sward. *J. Ecol.,* 62:97–105.

KEAST, A.

1959 Australian birds: their zoogeography and adaptation to an arid continent. *Biogeography and Ecol. in Australia,* 8:89–114.

KEEVER, C.

1950 Causes of succession in old fields of the Piedmont, North Carolina. *Ecol. Monogr.,* 20:229–250.

KEITH, L. B.

1963 *Wildlife's Ten-year Cycle,* University of Wisconsin Press, Madison.

1974 Some features of population dynamics in mammals. *Proc. Inter. Congr. Game Biol. Stockholm,* 11:17–58.

KEITH, L. B., AND L. A. WINDBERG

1978 A demographic analysis of the snowshoe hare cycle. *Wildl. Monogr. 58.*

KELLOGG, C. E.

1936 *Development and Significance of the Great Soil Groups of the United States. USDA Misc. Pub. 229.*

KELLOGG, W. W., ET AL.

1972 The sulfur cycle. *Science,* 175:587–599.

KEMP, G. A., AND L. B. KEITH

1970 Dynamics and regulation of red squirrels (*Tamiasciurus hudsonicus*) populations. *Ecology,* 51:765–779.

KERSTER, H. W.

1968 Population age structure in the prairie forb, *Liatris aspera. Bioscience,* 18:430–432.

KETCHUM, B. H.

1967 The phosphorus cycle and productivity of marine phytoplankton. In *Symposium on Primary Productivity and Mineral Cycling in Natural Ecosystems.* University of Maine Press, Orono, pp. 32–51.

KETCHUM, B. H., AND N. CORWIN

1965 The cycle of phosphorus in a plankton bloom in the Gulf of Maine. *Limnol. Oceanogr. Suppl. to Vol. 10.,* pp. R148–R161.

KETTLEWELL, H. B. D.

1961 The phenomenon of industrial melanism in Lepidoptera. *Ann. Rev. Entomol.,* 6:245–262.

1965 Insect survival and selection for pattern. *Science,* 148:1290–1296.

KEVAN, D. K. MCE.

1955 *Soil Zoology,* Butterworth, Washington, D.C.

1962 *Soil Animals,* Philosophical Library, New York.

KEY, K. H. L.

1974 Speciation in the Australian morabine grasshoppers—taxonomy and ecology. In M. J. D. White (ed.), 1974, pp. 43–56.

KIESTER, A. R.

1971 Species density of North American amphibians and reptiles. *Syst. Zool.,* 20:127–137.

KING, J. A.

1955 Social behavior, social organization and population dynamics in a black-tailed prairie dog town in the Black Hills of South Dakota. *Contrib. Lab. Vert. Biol., Univ. of Michigan, No. 67,* pp. 1–123.

1973 The ecology of aggressive behavior. *Ann. Rev. Ecol. and Syst.,* 4:117–139.

KINNE, O. (ED.)

1970 *Marine Ecology: A Comprehensive Integrated Treatise on Life in Oceans and Coastal Waters: Vol. 1, Environmental Factors,* Wiley, New York.

KIRA, T., AND T. SHIDEI

1967 Primary production and turnover of organic matter in different forest ecosystems of the western Pacific. *Japanese J. Ecol.,* 17:70–87.

KLEEREKOPER, H.

1953 The mineralization of plankton, *J. Fish. Res. Board Can.,* 10:283–291.

KLEIN, D. R.

1970 Food selection by North American deer and their response to overutilization of preferred plant species. In A. Watson (ed.), *Animal Populations in Relation to Their Food Resources,*

Blackwell, Oxford, pp. 25–46.

KLEIN, R. G.

1969    *Man and Climate in the Late Pleistocene: A Case Study,* Chandler, San Francisco.

KLOPFER, P. H.

1962    *Behavioral Aspects of Ecology,* Prentice-Hall, Englewood Cliffs, N.J.

1973    *An Introduction to Animal Behavior: Ethology's First Century,* 2nd ed., Prentice-Hall, Englewood Cliffs, N.J.

KNAPP, R. (ED.)

1974    *Vegetation Dynamics:* Part 8. *Handbook of Vegetation Science,* Junk, The Hague, Netherlands.

KNOX, E. G.

1952    *Jefferson County (N.Y.) Soils and Soil Map,* N.Y. State College Agr., Cornell University, Ithaca, N.Y.

KOBLENTZ-MISHKE, J. J., V. V. VOLKOVINSKY, AND J. G. KABANOVA

1970    Plankton primary production of the world ocean. In W. S. Wooster (ed.), *Scientific Exploration of the South Pacific,* National Academy of Sciences, Washington, D.C.

KOFORD, C. B.

1958    Prairie dogs, white faces, and blue grama. *Wildl. Monogr. No. 3.*

KOK, B.

1965    Photosynthesis: the pathway of energy. In J. Bonner and J. E. Varner (eds.), *Plant Biogeochemistry,* Academic, New York, pp. 904–960.

KOLENOSKY, G. B.

1972    Wolf predation on wintering deer in east central Ontario, *J. Wildl. Manage.,* 36:357–369.

KOPEC, R. J.

1970    Further observations of the urban heat island of a small city. *Bull. Amer. Meteorol. Soc.,* 51:602–606.

KORCHAGIN, A. A., AND V. G. KARPOV

1974    Fluctuations in coniferous tiaga communities. In R. Knapp (ed.), 1974, pp. 225–231.

KOWAL, N. E.

1966    Shifting agriculture, fire, and pine forest in the Cordillera Central, Luzon, Philippines. *Ecol. Monogr.,* 36:389–419.

KOZLOVSKY, D. C.

1968    A critical evaluation of the trophic level concept: 1. Ecological efficiencies. *Ecology* 49:48–60.

KOZLOWSKI, T. F., AND C. E. AHLGREN (EDS.)

1974    *Fire and Ecosystems,* Academic, New York.

KRAMM, K. R.

1975a   Entrainment of circadian activity rhythms in squirrels. *Amer. Natur.,* 109:379–389.

1975b   Circadian activity of the red squirrel *Tamiascuirius hudsonicus* in continuous darkness and continuous illumination. *Int. J. Biometeor.,* 19:232–245.

1976    Phase control of circadian activity in the antelope ground squirrel. *J. Interdicip. Cycle*

*Res.,* 7:127–138.

KRAUSE, R.

1963    Food habits of the yellow bass *Roccus mississippiensis,* Clear Lake, Iowa, summer 1962. *Proc. Iowa Acad. Natur. Sci.,* 70:209–215.

KREBS, C.

1963    Lemming cycle at Baker Lake, Canada, during 1959–62. *Science,* 146:1559–1560.

1964    The lemming cycle at Baker Lake, Northwest Territories, during 1959–1962. *Tech. Paper No. 15, Arctic Institute of North America.*

1978    *Ecology: The Experimental Analysis of Distribution and Abundance,* 2nd ed., Harper & Row, New York.

KREBS, C. J., M. S. GAINES, B. L. KELLER, J. H. MYERS, AND R. H. TAMARIN

1973    Population cycles in small rodents. *Science,* 179:35–41.

KREBS, C. J., B. L. KELLER, AND J. H. MYERS

1971    *Microtus* population densities and soil nutrients in southern Indiana grasslands. *Ecology,* 52:660–663.

KREBS, C. J., AND J. H. MYERS

1974    Population cycles in small mammals. *Adv. Ecol. Res.,* 8:267–399.

KREBS, J. R.

1971    Territory and breeding density in the great tit *Parus major. Ecology,* 52:2–22.

KRIEGER, R. I., P. P. FEENY, AND C. F. WILKINSON

1971    Detoxication enzymes in the guts of caterpillars: an evolutionary answer to plant defenses? *Science,* 172:579–581.

KROG, J.

1955    Notes on temperature measurements indicative of special organization in arctic and subarctic plants for utilization of radiated heat from the sun. *Physiol. Plantarum,* 8:836–839.

KROODSMA, R. L.

1974    Species recognition behavior of territorial male rose-breasted and black-headed grosbeaks *(Pheucticus). Auk,* 91:54–64.

KUCERA, C. L., R. C. DAHLMAN, AND M. R. KOELLING

1967    Total net productivity and turnover on an energy basis for tallgrass prairie. *Ecology,* 48:536–541.

KUENZLER, E. J.

1958    Niche relations of three species of Lycosid spiders. *Ecology,* 39:494–500.

1961    Phosphorus budget of a mussel population. *Limnol. Oceanogr.,* 6:400–415.

KUHNELT, W.

1950    *Bodenbiologie mit besonderer Berucksichtigung der Tierwelt,* Herold, Vienna.

1970    Structural aspects of soil-surface-dwelling biocoenoses. In J. Phillipson (ed.), *Methods of Study in Soil Biology,* UNESCO, Paris, pp. 45–56.

KURCHEVA, G. F.

1964    Wirbellose Tiere abd Faktor der Zersetzung von waldstreu. *Pedobiologia,* 4:8–30.

LACK, D. L.

1947    The significance of clutch size. *Ibis,*

89:30–52; 90:25–45.

1953  *The Life of the Robin,* Penguin Books, London.

1954  *The Natural Regulation of Animal Numbers,* Clarendon Press, Oxford.

1964  A long term study of the great tit *(Parus major). J. Anim. Ecol. Suppl.,* 33:159–173.

1966  *Population Studies of Birds,* Clarendon Press, Oxford.

1971  *Ecological Isolation in Birds,* Harvard University Press, Cambridge, Mass.

LACK, D. L., AND L. S. V. VENERABLES

1939  The habitat distribution of British woodland birds. *J. Anim. Ecol.,* 8:39–71.

LA MOTTE, M.

1975  The structure and function of a tropical savannah ecosystem. In F. B. Golley and E. Medina (eds.), 1975, pp. 179–222.

LANCE, A. N.

1978  Territories and the food plant of individual red grouse: 2. Territory size compared with an index of nutrient supply in heather. *J. Anim. Ecol.,* 47:307–313.

LANDAHL, J., AND R. B. ROOT

1969  Differences in the life tables of tropical and temperate milkweed bugs Genus *Oncopeltus* (Hemoptera Lygaerdae). *Ecology,* 50:734–737.

LANDSBERG, H. E.

1970  Man-made climatic changes. *Science,* 170:1265–1274.

LANG, G. E., W. A. REINERS, AND R. R. HEIER

1976  Potential alterations of precipitation chemistry by epiphytic lichens. *Oecologica,* 25:229–241.

LANGFORD, A. N., AND M. F. BUELL

1969  Integration, identity, and stability in the plant association. *Adv. Ecol. Res.,* 6:84–135.

LANYON, W. E., AND W. N. TAVOLGA (EDS.)

1960  *Animal Sounds and Communication. AIBS Symposium Series No. 7.*

LARCHER, W., ET AL.

1975  Mt. Patscherfofel, Austria. In T. Rosswall and O. W. Heal (eds.), 1975, pp. 125–139.

LASIEWSKI, R. C., AND G. K. SNYDER

1969  Responses to high temperatures in nestling double-crested and pelagic cormorants. *Auk,* 86:529–540.

LAUFF, G. (ED.)

1967  *Estuaries,* American Association Advancement Science, Washington, D.C.

LAVE, L. B., AND E. P. SESKIN

1970  Air pollution and human health. *Science,* 169:723–733.

LAWLOR, L. R.

1976  Molting, growth and reproductive strategies in the terrestrial isopod *Armadillidum vulgare. Ecology,* 57:1179–1194.

LAWRENCE, D. B.

1958  Glaciers and vegetation in southeastern Alaska. *Amer. Sci.,* 46:89–122.

LAWTON, J. H., M. P. HASSELL, AND J. R. BEDDINGTON

1975  Prey death rates and rate of increase of arthropod predator populations. *Nature,* 255:60–62.

LEAN, D. R. S.

1973a  Movements of phosphorus between its biologically important forms in lake water. *J. Fish. Res. Board Can.,* 30:1525–1536.

1973b  Phosphorus dynamics in lake water. *Science,* 179:678–680.

LEARY, M. G., JR., AND G. B. CRAIG, JR.

1967  Barriers to hybridization between *Aedes aegypti* and *Aedes albopictus* (Diptera Culicidae). *Evolution,* 21:41–58.

LEES, D. R., AND E. R. CREED

1975  Industrial melanism in *Biston betularia:* the role of selective predation. *J. Anim. Ecol.,* 44:67–83.

LEIGH, E. G., JR.

1975  Structure and climate in tropical rain forest. *Ann. Rev. Ecol. Syst.,* 6:67–86.

LEITH, H.

1960  Patterns of change within grassland communities. In J. L. Harper (ed.), *The Biology of Weeds,* Oxford Univ. Press, Oxford, pp. 27–39.

1963  The role of vegetation in the carbon dioxide content of the atmosphere. *Geophys. Res.,* 68:3887–3898.

1973  Primary production: terrestrial ecosystems. *Human Ecol.,* 1:303–332.

1975  Primary production of the major vegetation units of the world. In R. H. Whittaker and H. Leith, *Primary Productivity of the Biosphere,* Springer-Verlag, New York, pp. 203–205.

LEITH, H. (ED.)

1974  *Phenology and Seasonality Modeling,* Springer-Verlag, New York.

LEOPOLD, A.

1933  *Game Management,* Scribner, New York.

LEVIN, D. A.

1971  Plant phenolics, an ecological perspective. *Amer. Natur.,* 105:151–181.

1976  The chemical defenses of plants to pathogens and herbivores. *Ann. Rev. Ecol. Syst.,* 7:121–159.

LEVIN, D. A., AND H. W. KERSTER

1974  Gene flow in seed plants. *Evolutionary Biology,* 7:139–220.

LEVINS, R.

1968  *Evolution in Changing Environments,* Princeton University Press, New Jersey.

LEWIS, J. K.

1971  The grassland biome; a synthesis of structure and function, 1970. In N. R. French (ed.), (1971), pp. 317–387.

LIDICKER, W. Z., JR.

1966  Ecological observations on a feral house mouse population declining to extinction. *Ecol. Monogr.,* 36:27–50.

1973  Regulation of numbers in an island population of the California voles, a problem in community dynamics. *Ecol. Monogr.,* 43:271–302.

1975  The role of dispersal in the demography of small mammals. In F. B. Golley et al. (eds.), 1975, pp. 103–128.

LIGHT, L. E.
1967  Growth inhibition in crowded tadpoles: intraspecific and interspecific effects. *Ecology,* 48:736–745.

LIGNON, J. D.
1968  Sexual differences in foraging behavior in two species of *Dendrocopos* woodpeckers. *Auk,* 85:203–215.

LIKENS, G. E.
1975  Nutrient flux and cycling in freshwater ecosystems. In F. G. Howell et al. (eds.), 1975, pp. 314–348.

LIKENS, G. E., ET AL.
1977  *Biogeochemistry of a Forested Ecosystem,* Springer-Verlag, New York.

LIKENS, G. E., AND F. H. BORMANN
1974  Acid rain: a serious regional environmental problem. *Science,* 184:1176–1179.
1974  Effects of forest clearing on the northern hardwood forest ecosystem and its biochemistry. *Proc. 1st Int. Cong. Eco.,* pp. 330–335.
1974  Linkages between terrestrial and aquatic ecosystems. *Bioscience,* 24(8):447–456.
1975  Nutrient-hydrologic interactions (eastern United States). In A. D. Hasler (ed.), 1975, pp. 1–5.
1975  An experimental approach in New England landscapes. In A. D. Hasler (ed.), 1975, pp. 7–29.
1976  Effects of forest clearing on the northern hardwood forest ecosystem and its biogeochemistry. In *Proc. 1st Int. Cong. Ecol.: Structure, Functioning, and Management of Ecosystems,* Center for Agricultural Publ. & Documentation, Wageninger, Netherlands.

LIKENS, G. E., F. H. BORMANN, AND N. M. JOHNSON
1969  Nitrification: importance to nutrient losses from a cutover forest ecosystem. *Science,* 163:1205–1206.

LIKENS, G. E., F. H. BORMANN, N. M. JOHNSON, AND R. S. PIERCE
1967  The calcium, magnesium, potassium, and sodium budgets for a small forested ecosystem. *Ecology,* 38: 46–49.
1970  Effects of forest cutting and herbicide treatment on nutrient budgets in the Hubbard Brook watershed ecosystem. *Ecol. Monogr.,* 40:23–47.

LIKENS, G. E., F. H. BORMANN, R. S. PIERCE, AND D. W. FISHER
1971  Nutrient hydrologic cycle interaction in small forested watershed ecosystems. In P. Duvigneaud (ed.), 1971, pp. 553–563.

LILLYWHITE, H. B.
1970  Behavorial temperature regulation in the bullfrog, *Rana catesbeiana. Copeia,* 1970:158–168.

LINDQUIST, B.
1942  Experimentelle Untersuchingen uber die Bedeutung einiger Landmollusken fur die zersetgung der Waldstreu. *Kgl Fysiograf Sallskap Lund Forh.,* 11:144–156.

LINHART, Y. B.
1974  Intra-population differentiation in annual plants. 1. *Veronica peregrina* L. raised under non-competitive conditions. *Evolution,* 28:232–243.

LITTLE, E. L., JR.
1971  *Atlas of United States Trees: Vol. 1, Conifers and Important Hardwoods. USDA Misc. Pub. No. 1146.*

LLOYD, M., AND H. S. DYBAS
1966  The periodical cicada problem. 1. Population ecology. *Evolution,* 20:133–149.

LLOYD, M., AND R. J. GHERARDI
1964  A table for calculating the "equitability" component of species diversity. *J. Anim. Ecol.,* 33:217–225.

LOOMIS, R. S., ET AL.
1967  Community architecture and the productivity of terrestrial plant communities. In A. San Pietro et al. (eds.), *Harvesting the Sun,* Academic, New York, pp. 291–308.

LORD, R. D.
1960  Litter size and latitude in North American mammals. *Amer. Midl. Natur.,* 64:488–499.
1961a  A population study of the gray fox. *Amer. Midl. Natur.,* 66:87–109.
1961b  Magnitudes of reproduction in cottontail rabbits. *J. Wild. Manage.,* 25:28–33.
1961c  Mortality rates of cottontail rabbits. *J. Wildl. Manage.,* 25:33–40.

LORENZ, K.
1931  Beitrage zur Ethologie der sozialer corviden. *J. fur Ornith.,* 46:67–127.
1935  Der Kumpan in der Umwelt des Vogels. *J. fur Ornith.,* 83:137–213, 289–413.
1937  The companion in the bird's world. *Auk,* 54:245–273.

LOSSAINT, P.
1973  Soil-vegetation relationships in Mediterranean ecosystems of southern France. In F. di Castri and H. A. Mooney (eds.), 1973, pp. 199–210.

LOTKA, A. J.
1925  *Elements of Physical Biology,* Williams & Wilkins, Baltimore.

LOUCKS, O. L.
1970  Evolution of diversity, efficiency, and community stability. *Amer. Zool.,* 10:17–25.

LOWE, V. P. Q.
1969  Population dynamics of red deer *(Cervus elaphus L.)* on Rhum. *J. Anim. Ecol.,* 38:425–457.

LOWE-MCCONNELL, R. H.
1969  Speciation in tropical freshwater fishes. *Biol. J. Linn. Soc.,* 1:51–75.

LULL, H. W.
1967  Factors influencing water production from forested watersheds. *Municipal Watershed Mgmt. Symp. Proc. 1965, Univ. Mass Coop. Ext. Serv. Pub.,* 446:2–7.

LULL, H. W., AND W. E. SOPPER
1969  Hydrologic effects from urbanization of forested watersheds in the northeast. *USDA*

**794**

Forest Serv. Res. Paper, NE-146.

LUTZ, H. J.
1956 Ecological effects of forest fires in the interior of Alaska. *USDA Tech. Bull. No. 1133.*

LYMAN, C. P., AND A. R. DAWE (EDS.)
1960 *Mammalian hibernation,* Museum Comparative Zoology, Harvard University, Cambridge, Mass.

LYNCH, J. F., AND N. K. JOHNSON
1974 Turnover and equilibria in insular avifaunas, with special reference to the California Channel Islands. *Condor,* 76:370–384.

MACARTHUR, R. H.
1958 Population ecology of some warblers of northeastern coniferous forests. *Ecology,* 39:599–619.
1960 On the relative abundance of species. *Amer. Natur.,* 94:25–36.
1961 Population effects of natural selection. *Amer. Natur.,* 95:195–199.
1972 *Geographical Ecology,* Harper & Row, New York.

MACARTHUR, R. H., AND R. LEVINS
1967 The limiting similarity convergence and divergence of coexisting species. *Amer. Natur.,* 101:377–385.

MACARTHUR, R. H., AND J. W. MACARTHUR
1961 On bird species diversity. *Ecology,* 42:594–598.

MACARTHUR, R. H., and E. O. WILSON
1963 An equilibrium theory of insular zoogeography. *Evolution,* 17:373–387.
1967 *The Theory of Island Biogeography,* Princeton University Press, New Jersey.

MACCLINTOCH, L., R. F. WHITCOMB, AND B. L. WHITCOMB
1977 Evidence for the value of corridors and minimization of isolation in preservation of biotic diversity. *Amer. Birds,* 31(1):6–12.

MACFAYDEN, A.
1963 *Animal Ecology: Aims and Methods,* 2nd ed., Pitman, London.

MACLULICH, D. A.
1937 Fluctuations in the numbers of varying hare *(Lepus americanus). Univ. Toronto Biol. Serv. No. 43.*

MADGWICK, H. A. I., AND J. D. OVINGTON
1959 The chemical composition of precipitation in adjacent forest and open plots. *Forestry,* 32:14–22.

MAGUIRE, B., JR.
1971 Phytotelmata: biota and community structures determination in plant held waters. *Ann. Rev. Ecol. and Syst.,* 2:439–464.
1973 Niche response structure and the analytical potentials of its relationship to the habitat. *Amer. Natur.,* 107:213–246.

MAHALL, B. E., AND R. B. PORK
1976 The ecotone between *Spartina foliosa* Trin. and *Salicornia virginica* L. in salt marshes of northern San Francisco Bay. 1. Biomass and production. *J. Ecol.,* 64:421–433.

MAIO, J. J.
1958 Predatory fungi. *Sci. Amer.,* 199:67–72.

MAISUROW, D. K.
1941 The role of fire in the perpetuation of virgin forests of northern Wisconsin. *J. Fores.,* 39:201–207.

MALMER, N.
1975 Development of bog mires. In A. D. Hasler (ed.), 1975, pp. 85–92.

MALTHUS, T. R.
1798 *An Essay on Principles of Population,* Johnson, London (numerous reprints).

MALVIN, R. L., AND M. RAYNER
1968 Renal function and blood chemistry in Cetacea. *Amer. J. Physiol.,* 214:187–191.

MANNING, A.
1972 *An Introduction to Animal Behaviour,* 2nd ed., Edward Arnold, London.

MARCHAND, D. E.
1973 Edaphic control of plant distribution in the White Mountains, eastern California. *Ecology,* 54:233–250.

MARGALEF, R.
1963 On certain unifying principles in ecology. *Amer. Natur.,* 47:357–374.
1968 *Perspectives in Ecological Theory,* University of Chicago Press.

MARKS, P. L.
1974 The role of pin cherry (*Prunus pensylvanica* L.) in the maintenance of stability in northern hardwood ecosystems. *Ecol. Monogr.,* 44:73–88.

MARKS, P. L., AND F. H. BORMANN
1972 Revegetation following forest cutting: mechanisms for return to steady-state nutrient cycling. *Science,* 176:914–915.

MARQUIS, D. A.
1974 The impact of deer browsing on Allegheny hardwood regeneration. *USDA Forest. Serv. Res. Paper NE-308.*

MARSDEN, H. M., AND N. R. HOLLER
1964 Social behavior in confined populations of the cottontail and swamp rabbit. *Wild. Monogr. No. 13.*

MATHER, K.
1955 Polymorphism as an outcome of disruptive selection. *Evolution* 9:52–61.

MARTIN, M. M.
1970 The biochemical basis of the fungus-attine ant symbiosis. *Science,* 169:16–20.

MARTIN, P. S.
1973 The discovery of America. *Science,* 179:969–974.

MARX, D. H.
1971 Ectomycorrhizae as biological deterrents to pathogenic root infections. In E. Hacskaylo (ed.), *Mycorrhizae, USDA Misc. Pub. 1189,* pp. 81–96.

MATVEYEVA, N. U., O. M. PARINKINA, AND Y. I. CHERNOV
1975 Maria Pronchitsheva Bay, U.S.S.R. In T. Rosswall and O. W. Heal (eds.), 1975, pp. 61–72.

MAY, R.
1973 *Stability and Complexity in Model*

*Ecosystems.* Princeton University Press, New Jersey.

1976  Models for two interacting populations. In R. May (ed.), 1976, pp. 49–70.

MAY, R. (ED.)

1976  *Theoretical Ecology,* Saunders, Philadelphia.

MAYER, H. M.

1969  *The Spatial Expression of Urban Growth, Commission on College Geography Resource Paper No. 7,* Association of American Geographers, Washington, D.C.

MAYER, W. V.

1960  Histological changes during the hibernation cycle in the arctic ground squirrel, in C. P. Lyman and A. R. Dawe (eds.), 1960, pp. 131–148.

MAYFIELD, H. R.

1960  The Kirtland's warbler. *Cranbrook Inst. Sci. Bull. No. 40.*

MAYNARD SMITH, J.

1964  Group selection and kin selection. *Nature,* 201:1145–1147.

MAYR, E.

1942  *Systematics and the Origin of Species,* Columbia University Press, New York.

1963  *Animal Species and Evolution,* Harvard University Press, Cambridge, Mass.

1974  The definition of the term disruptive selection. *Heredity,* 32:404–406.

MCARDLE, R. E., W. H. MEYER, AND D. BRUCE

1949  The yield of Douglas-fir in the Pacific Northwest. *USDA Tech. Bull. No. 201* (rev.).

MCBEE, R. H.

1971  Significance of intestinal microflora in herbivory. *Ann. Rev. Ecol. Syst.,* 2:165–176.

MCBRIDE, G. I., P. PARE, AND F. FOENANDER

1969  The social organization and behavior of the feral domestic fowl. *Anim. Behav. Monogr.,* 2:127–181.

MCCALLA, T. M.

1943  Microbiological studies of the effects of straw used as a mulch. *Trans. Kansas Acad. Sci.,* 43:52–56.

MCGINNES, W. G.

1972  North America. In C. M. McKell et al. (eds.), 1972, pp. 55–66.

MCILROY, R. J.

1972  *An Introduction to Tropical Grassland Husbandry,* 2nd ed., Oxford University Press, London.

MCINTOSH, R. P.

1958  Plant communities. *Science,* 128:115–120.

1967  An index of diversity and the relation of concepts to diversity. *Ecology,* 48:392–403.

1975  H. A. Gleason—"Individualistic Ecologist" 1882–1975: His contributions to ecological theory. *Bull. Torrey Bot. Club,* 102(5):253–273.

1976  Ecology since 1900. In B. J. Taylor and T. J. White (eds.), 1976, *Issues and Ideas in America.* University Oklahoma Press, Norman.

MCKELL, C. M., ET AL. (EDS.)

1972  *Wildland Shrubs: Their Biology and Utilization. USDA Forest Serv. Gen. Tech. Rept.*

*INT-1.*

MCKEY, D.

1975  The ecology of coevolved seed dispersal systems. In L. E. Gilbert and P. H. Raven (eds.), 1975, pp. 159–191.

MCMILLAN, C.

1959  The role of ecotypic variation in the distribution of the Central Grassland of North America. *Ecol. Monogr.,* 29:285–308.

MCNAB, B. K.

1963  Bioenergetics and the determination of home range size. *Amer. Natur.,* 97:133–140.

1971  On the ecological significance of Bergmann's rule. *Ecology,* 52:845–854.

1978  The evolution of endothermy in the phylogeny of mammals. *Amer. Natur.,* 112:1–21.

MCNAUGHTON, S. J.

1968  Structure and function in California grassland. *Ecology,* 49:962–972.

1975  $r$ and $K$ selection in *Typha. Amer. Natur.,* 109:251–261.

MCNAUGHTON, S. J., AND L. L. WOLF

1970  Dominance and the niche in ecological systems. *Science,* 167:131–139.

MCPHERSON, J. K., AND C. H. MULLER

1969  Allelopathic effects of *Adenostoma fasciculatum* "chamise" in the California chaparral. *Ecol. Monogr.,* 39:177–179.

MEANLEY, B.

1957  Notes on the courtship behavior of the king rail. *Auk,* 74:433–440.

MELLINGER, M. V., AND S. J. MCNAUGHTON

1975  Structure and function of successional vascular plant communities in central New York. *Ecol. Monogr.,* 45:161–182.

MENAKER, M. (ED.)

1971  *Biochionometry, Proceedings Friday Harbor Symposium, 1969,* National Academy Science, Washington, D.C.

MENGE, B. A., AND J. P. SUNDERLAND

1976  Species diversity gradients: synthesis of the role of predation competition and temporal heterogeneity. *Amer. Natur.,* 110:351–369.

MENGEL, R. M.

1964  The probable history of species formation in some northern wood warblers (Parulidae). *Living Bird,* 3:9–43.

1970  The North American Central Plains as an isolating agent in bird speciation. In W. Dort and J. K. Jones (eds.), 1970, pp. 279–345.

MENZIE, C. M.

1969  Metabolism of pesticides. *USDI Fish and Wildl. Serv. Sp. Sci. Rept. Wildl. No. 127.*

MERRIAM, C. H.

1898  Life zones and crop zones of the United States. *Bull. U.S. Bureau Biol. Survey,* 10:1–79.

MESLOW, E. C., AND L. B. KEITH

1968  Demographic parameters of a snowshoe hare population. *J. Wildl. Manage.,* 32:812–835.

METTLER, L. E., AND T. G. GREGG

1969  *Population Genetics and Evolution,* Prentice-Hall, Englewood Cliffs, N.J.

MILLER, A. H.
1942   Habitat selection among higher vertebrates and its relation to intraspecific variation. *Amer. Natur.*, 76:25–35.

MILLER, G. R.
1968   Evidence for selective feeding on fertilized plots by red grouse, hares and rabbits. *J. Wildl. Manage.*, 32:849–853.

MILLER, G. R., AND A. WATSON
1978a   Territories and the food plants of individual red grouse: 1. Territory size, number of mates and brood size compared with the abundance, production, and diversity of heather. *J. Anim. Ecol.*, 47:293–305.

1978b   Heather productivity and its relevance to the regulation of red grouse populations. In O. W. Heal and D. F. Perkins (eds.), *Production Ecology of British Moors and Grasslands,* Springer-Verlag, New York, pp. 277–285.

MILLER, R. S.
1967   Pattern and process in competition. *Adv. Ecol. Res.*, 4:1–74.

MILLER, W. R., AND F. E. EGLER
1950   Vegetation of the Wequetequock-Pawcatuck tidal marshes, Connecticut. *Ecol. Monogr.* 20:141–172.

MILLICENT, E., AND J. M. THODAY
1960   Gene flow and divergence under disruptive selection. *Science,* 131:1311–1312.

1961   Effects of disruptive selection: 4. Gene flow and divergence. *Heredity,* 16:199–217.

MILNE, A.
1957   Theories of natural control of insect populations. *Cold Spring Harbor Symp. Quant. Biol.,* 22:253–271.

MILTON, W. E. J.
1940   The effect of manuring, grazing and cutting on the yield; botanical and chemical composition of natural hill pastures: 1. Yield and botanical composition. *J. Ecol.,* 28:326–356.

MINSHALL, G. W.
1967   Role of allochthonous detritus in the trophic structure of a woodland spring brook community. *Ecology,* 48:139–149.

1978   Autotrophy in stream ecosystems. *Bioscience,* 28:767–771.

MISHUSTIN, E. N., AND V. K. SHILNIKOVA
1969   The biological fixation of atmospheric nitrogen by free-living bacteria. In *Soil Biology,* UNESCO, Paris, pp. 65–124.

MOEN, A. M.
1968   The critical thermal environment. *Bioscience,* 18:1041–1043.

MOIR, W. H.
1969a   Energy fixation and the role of primary producers in the energy flux of grassland ecosystems. In R. L. Dix and R. G. Beidleman (eds.), 1969, pp. 125–147.

1969b   Steppe communities in the foothills of the Colorado Front Range and their relative productivities. *Amer. Midl. Natur.,* 81:331–340.

MOLCHANOV, A. A.

1960   *The Hydrological Cycle of the Forest,* Israel Program for Scientific Publication, Jerusalem.

MONK, C. A.
1967   Tree species diversity in the eastern deciduous forest with particular reference to northcentral Florida. *Amer. Natur.,* 101:173–187.

1970   An ecological significance of energetics. *Ecology,* 47:504–505.

MONRO, J.
1967   The exploitation and conservation of resources by populations of insects. *J. Anim. Ecol.,* 36:531–547.

MONSI, M.
1968   Mathematical models of plant communities. In F. E. Eckardt (ed.), 1968, pp. 131–149.

MONTEITH, J. L.
1962   Measurement and interpretation of carbon dioxide flues in the field. *Netherlands J. Agr. Sci.,* 10(sp. issue):334–346.

MONTEITH, J. L. (ED.)
1976   *Vegetation and the Atmosphere, Vol. 2, Case Studies,* Academic, London.

MOOK, L. J.
1963   Birds and spruce budworm. In R. Morris (ed.), 1963, pp. 244–248.

MOONEY, H. A. (ED.)
1977   *Convergent Evolution in Chile and California Mediterranean Climate Ecosystems,* Academic, New York.

MOONEY, H. A., AND E. L. DUNN
1970a   Convergent evolution of Mediterranean climate evergreen sclerophyll shrubs. *Evolution,* 24:292–303.

1970b   Photosynthetic systems of Mediterranean climate shrubs and trees of California and Chile. *Amer. Midl. Natur.,* 104:447–453.

MOONEY, H. A., AND D. J. PARSONS
1973   Structure and function of the California chaparral—an example from San Dimas. In F. di Castri and H. A. Mooney (eds.), 1973, pp. 83–112.

MOONEY, H. A., D. J. PARSONS, AND J. KUMMEROW
1974   Plant development in Mediterranean climates. In H. Leith (ed.), 1974, pp. 225–267.

MOORE, C. W. E.
1959   The competitive effect of *Danthonia* spp. on the establishment of *Bothriochloa ambigua. Ecology,* 40:141–143.

MOORE, H. B.
1958   *Marine Ecology,* Wiley, New York.

MOORE, J. A.
1949a   Geographic variation of adaptive characters in *Rana pipiens* Schreber. *Evolution,* 3:1–24.

1949b   Patterns of evolution in the genus *Rana.* In G. L. Jepsen, E. Mayr, and G. G. Simpson (eds.), 1949, pp. 315–338.

MOORE, J. J., P. DOUDING, AND B. HEALY
1975   Glenamoy, Ireland. In T. Rosswall and O. W. Heal (eds.), 1975, pp. 321–343.

MOORE, N. W., M. D. HOOPER, AND B. N. K. DAVIS
1967   Hedges: 1. Introduction and reconnaissance studies. *J. Appl. Ecol.,* 4:201–220.

MOORE, P. D., AND D. J. BELLAMY
1973   *Peatlands,* Springer-Verlag, New York,

MORRIS, R. F.
1963   Predictive population equations based on key factors. *Entomol. Soc. Can. Mem. 32,* pp. 16–21.

MORRIS, R. F. (ED.)
1963   The dynamics of epidemic spruce budworm populations. *Entomol. Soc. Can. Mem. 31.*

MORRIS, R. F., W. F. CHESHIRE, C. A. MILLER, AND D. G. MOTT
1958   The numerical response of avian and mammalian predators during a gradation of the spruce budworm. *Ecology, 39:*487–494.

MORRISON, R. C., AND G. A. YARRANTON
1974   Vegetational heterogeneity during a primary sand dune succession. *Can J. Bot.,* 52:397–410.

MORTIMER, G. H.
1952   Water movements in lakes during summer stratification: evidence from the distribution of temperature in Windermer. *Phil. Trans. Roy. Soc., Ser. B.,* London, pp. 236–355.

MOSS, R.
1972   Food selection by red grouse *(Lagopus lagopus scoticus* Lath.) in relation to chemical composition. *J. Anim. Ecol.,* 41:411–428.

MOSS, R., G. R. MILLER, AND S. F. ALLEN
1972   Selection of heather by captive red grouse in relation to the age of the plant. *J. Appl. Ecol.,* 9:771–781.

MOSSER, J. L., N. S. FISHER, T. TENG, AND C. F. WURSTER
1972   Polychlorinated biphenyls toxicity to certain phytoplankton. *Science,* 175:191–192.

MOULDER, B. C., AND D. E. REICHLE
1974   Significance of spider predation in the energy dynamics of forest floor arthropod communities. *Ecol. Monogr.,* 42:473–498.

MOYNIHAN, M.
1955a   Some aspects of reproductive behaviour in the blackheaded gull and related species. *Behaviour Suppl.,* 4:1–201.
1955b   Types of hostile display. *Auk,* 72:247–259.
1955c   Remarks on the origianl source of displays. *Auk.* 72:240–246.
1968   Social mimicry: character convergence versus character displacement. *Evolution,* 22:315–331.

MUIR, R. C.
1965   The effect of sprays on fauna of apple trees. *J. Appl. Ecol.,* 2:31–41, 43–57.

MULLER, C. H., R. B. HANAWALT, AND J. K. MCPHERSON
1968   Allelopathic control of herb growth in the fire cycle of California chaparral. *Bull. Torrey Bot. Club,* 95:225–231.

MULLER-SCHWARZE, D.
1971   Pheromones in black-tailed deer. *Anim. Behav.,* 19:141–152.

MURDOCH, W. W.
1969   Switching in general predators: experiments on predator specificity and stability of prey populations. *Ecol. Monogr.* 39:335–354.
1973   The functional response of predators. *J. Appl. Ecol.,* 10:335–342.

MURDOCH, W. W., AND A. OATEN
1975   Predation and population stability. *Adv. Ecol. Res.,* 9:1–131.

MURPHY, G. I.
1966   Population biology on the Pacific sardine. *Proc. Calif. Acad. Sci. 4th Series,* 34:1–84.
1967   Vital statistics of the Pacific sardine and the population consequences. *Ecology,* 48:731–736.

MURPHY, P. W.
1952   Soil faunal investigations. In *Report on Forest Research for the Year Ending March, 1951,* Forest. Comm., London, pp. 130–134.
1953   The biology of the forest soils with special reference to the mesofauna or meiofauna. *J. Soil Sci.,* 4:155–193.

MUUL, I.
1965   Daylength and food caches. *Natur. Hist.,* 74(3):22–27.
1969   Photoperiod and reproduction flying squirrels, *Glaucomys volans. J. Mammal.,* 50:542–549.

MYERS, J. H., AND C. J. KREBS
1971   Genetic, behavioral, and reproductive attributes of dispersing field voles *Microtus pennsylvanicus* and *Microtus ochrogaster. Ecol. Monogr.,* 41:53–78.

MYERS, K., AND W. E. POOLE
1967   A study of the biology of the wild rabbit *Oryctolagus cuniculus* L. in confined populations: 4. The effects of rabbit grazing on sown pastures. *J. Ecol.,* 55:435–451.

MYERS, K., C. S. HALE, R. MYKYTOWYCZ, AND R. L. HUGHS
1971   The effects of varying density and space on sociality and health in animals. In A. H. Esser, *Behavior and Environment: The Use of Space by Animals and Men,* Plenum, New York, pp. 148–187.

NACE, R. L.
1969   Human uses of ground water. In R. J. Chorley (ed.), *Water, Earth, and Man,* Methuen, London, pp. 285–294.

NAPIER, J. R.
1966   Stratification and primate ecology. *J. Anim. Ecol.,* 35:411–412.

NATIONAL ACADEMY OF SCIENCES
1970   *Land Use and Wildlife Resources,* National Academy of Sciences, Washington, D.C.
1974   *U.S. Participation in the International Biological Program, Rept. No. 6,* U.S. Committee for the International Biological Program, National Academy of Science, Washington, D.C.
1975   *Productivity of World Ecosystems,* National Academy Science, Washington, D.C.

NAVTH, Z.
1974   Effects of fire in the Mediterranean region. In T. F. Kozlowski and C. E. Ahlgren (eds.),

1974, pp. 401–434.

NEAVE, F.
1944 Racial characteristics and migratory habits in *Salmo gairdneri*. *J. Fish. Res. Board Can.,* 6:245–251.

NEEL, J. K.
1951 Interrelations of certain physical and chemical features in headwater limestone streams. *Ecology,* 32:368–391.

NEILL, W. E.
1975 Experimental studies of microcrustacean competition, community composition, and efficiency of resource utilization. *Ecology,* 56:809–826.

NELLIS, C. H., S. P. WETMORE, AND L. B. KIETH
1972 Lynx-prey interaction in central Alberta. *J. Wildl. Manage.,* 36:320–328.

NELSON, D. J.
1962 Clams as indicators of strontium-90. *Science,* 138:38–39.

NELSON, D. J., ET AL.
1971 Hazards of mercury. *Envir. Res.* 4:3–69.

NELSON, D. J., AND F. E. EVANS (EDS.)
1967 *Symposium on Radioecology, Conf. 670503,* National Technical Information Services, Springfield, Va.

NELSON, D. J., AND C. C. SCOTT
1962 Role of detritus in the productivity of a rock-outcrop community in a Piedmont stream. *Limnol. Oceanogr.,* 3:396 413.

NEUWIRTH, R.
1957 Some recent investigations into the chemistry of air and of precipitation and their significance for forestry. *Allg. Fost-v. Jagdztg.,* 128:147–150.

NEVO, E.
1969 Mole rat *Spalax ehrenbergi* mating behavior and its evolutionary significance. *Science,* 163:484–486.

NEVO, E., AND S. A. BLONDHEIM
1972 Acoustic isolation in the speciation of mole crickets. *Ann. Entomol. Soc. Amer.,* 65:980–981.

NEVO, E., Y. J. KIM, C. R. SHAW, AND C. S. THAELER, JR.
1974 Genetic variation, selection, and speciation in *Thomomys talpoides* pocket gophers. *Evolution,* 28:1–23.

NEVO, E., AND C. R. SHAW
1972 Genetic variation in a sub-terranian mammal *Spalax ehrenbergi. Biochem. Genet.,* 7:235–241.

NEWELL, R. C.
1965 The role of detritus on the nutrition of two marine deposit feeders, the prosobranch *Hydrobia ulvae* and the bivalva *Macoma balthica. Proc. Zool. Soc. London,* 144:25–45.

NICE, M. M.
1941 The role of territory in bird life. *Amer. Midl. Natur.,* 26:441–487.
1943 Studies in the life history of the song sparrow: 2. *Trans. Linn. Soc. New York,* 6:1–329.

1962 Development of behavior in precocial birds. *Trans. Linn. Soc. New York, Vol. 8.*

NICE, M. M., AND J. J. TER FELWYK
1941 Enemy recognition by the song sparrow. *Auk,* 58:195–214.

NICHOLSON, A. J.
1954 An outline of the dynamics of animal populations. *Aust. J. Zool.,* 2:9–65.
1957 The self-adjustment of populations to change. *Cold Spring Harbor Symp. Quant. Biol.,* 22:153–173.

NICHOLSON, A. J., AND V. A. BAILEY
1935 The balance of animal populations: Part 1. *Proc. Zool. Soc. London,* pp. 551–598.

NIELSEN, A.
1950 The torrential invertebrate fauna. *Oikos,* 2:176–196.

NIELSEN, C. O.
1961 Respiratory metabolism in some populations of enchytraeid worms and free-living nematodes. *Oikos,* 12:17–35.

NIERING, W. A., AND F. E. EGLER
1955 A shrub community of *Viburnum lentago* stable for twenty-five years. *Ecology,* 36:356–360.

NIERING, W., AND R. GOODWIN
1974 Creation of relatively stable shrublands with herbicides: arresting "succession" on rights-of-way and pasture land. *Ecology,* 55:784–795.

NIERING, W. A., R. H. WHITTAKER, AND C. H. LOWE
1963 The saguaro: a population in relation to environment. *Science,* 142:15–23.

NIXON, S. W., AND C. A. OVIATT
1973 Ecology of a New England salt marsh. *Ecol. Monogr.,* 43:463–498.

NOBLE, G. K.
1936 Courtship and sexual selection of the flicker. *Auk,* 53:269–383.

NOLAN, V., JR.
1958 Anticipatory food-bringing in the prairie warbler. *Auk,* 75:263–278.

NOMMIK, H.
1965 Ammonium fixation and other reactions involving nonenzymatic immobilization. In W. Bartholomew and F. Clark (eds.), *Soil Nitrogen,* American Society Agronomy, Madison, Wis., pp. 198–258.

NOY-MIER, I.
1973 Desert ecosystems: environment and producers. *Ann. Rev. Ecol. Syst.,* 4:25–51.
1974 Desert ecosystems: higher trophic levels. *Ann. Rev. Ecol. Syst.,* 5:195–214.
1975 Stability of grazing systems: an application of predator-prey graphs. *J. Ecol.,* 63:459–481.

ODEN, S., AND T. AHL
1970 Forsurningen av skandinaviska vatten (The acidification of Scandinavian lakes and rivers). *Ymer Arsbok,* pp. 103–122.

ODUM, E. P.
1969 The strategy of ecosystem development. *Science,* 164:262–270.
1971 *Fundamentals of Ecology,* 3rd ed., Saunders, Philadelphia.

ODUM, E. P., C. E. CONNELL, AND L. B. DAVENPORT
1962   Population energy flow of three primary
consumer components of old field ecosystems.
*Ecology,* 43:88–96.

ODUM, E. P., AND E. J. KUENZLER
1963   Experimental isolation of food chains in an
old-field ecosystem with the use of phosphorus-
32. In V. Schultz and A. Klement (eds.), 1963, pp.
113–120.

ODUM, H. T.
1956   Primary production in flowing water.
*Limnol. Oceanogr.,* 1(2):102–117.

1957a   Trophic structure and productivity of Silver
Springs, Florida. *Ecol. Monogr.,* 27:55–112.

1957b   Primary production measurements in
eleven Florida springs and a marine turtlegrass
community. *Limnol. Oceanogr.,* 2:85–97.

1970 Summary: An emerging view of the ecological
system at El Verde. In H. T. Odum and R. F.
Pigeon (eds.), *A Tropical Rain Forest,* pp. I/191-
I/218.

ODUM, H. T., B. J. COPELAND, AND E. A. MCMAHAN
1974   *Coastal Ecological Systems of the U.S.,* Vols.
1–4, Conservation Foundation, Washington, D.C.

OGDEN, J.
1967   Radiocarbon and pollen evidence for a
sudden change in climate in the Great Lakes
region approximately 10,000 years ago. In E. J.
Cushing and H. E. Wright (eds.), 1967, pp.
117–127.

1974   The reproductive strategy of higher plants:
2. The reproductive strategy of *Tussilago
furfara* L. *J. Ecol.,* 62:291–324.

OLSON, J. S.
1958   Rates of succession and soil changes on
southern Lake Michigan sand dunes. *Bot. Gaz.,*
119:125–170.

1970   Carbon cycles and temperate woodlands. In
D. E. Reichle (ed.), 1970, pp. 226–241.

O'NEILL, R. V.
1976   Ecosystem persistence and heterotrophic
regulations. *Ecology,* 57:1244–1253.

O'NEILL, R. V., ET AL.
1975   Theoretical basis for ecosystem analysis
with particular reference to element cycling. In
F. G. Howell et al. (eds.), 1975, pp. 38–40.

OPLER, P. A.
1974 . Oaks as evolutionary islands for leaf
mining insects. *Amer. Sci.,* 62:67–73.

ORIANS, G.
1961   The ecology of blackbird *(Agelaius)* social
systems. *Ecol. Monogr.,* 31:285–312.

ORIANS, G. H., AND O. T. SOLBRIG
1977   *Convergent Evolution in Warm Deserts,*
Academic, New York.

ORING, L. W., AND M. L. KNUDSON
1972   Monogamy and polygamy in the spotted
sandpiper. *Living Bird,* 11:59–73.

OSTBYE, E., ET AL.
1975   Hardangervidda, Norway. In T. Rosswall
and O. W. Heal (eds.), 1975, pp. 225–264.

OTTE, D.
1978   Communication in Orthoptera. In T. A.
Sebeok, (ed.), 1978, pp. 334–361.

OTTE, D., AND A. JOERN
1975   Insect territoriality and its evolution:
population studies of desert grasshoppers on
creosote bushes. *J. Anim. Ecol.,* 44:29–54.

OVINGTON, J. D.
1961   Some aspects of energy flow in plantation of
*Pinus sylvestris* L. *Ann. Bot., London, n.s.,*
25:12–20.

OVINGTON, J. D., D. HEITKAMP, AND D. B. LAWRENCE
1963   Plant biomass and productivity of prairie
savanna, oakwood and maize field ecosystems in
Central Minnesota. *Ecology,* 44:52–63.

OVINGTON, J. D., AND D. B. LAWRENCE
1967   Comparative chlorophyll and energy studies
of prairie, savanna, oakwood, and maize field
ecosystems. *Ecology,* 48:515–524.

OVINGTON, J. D., AND H. A. I. MADGWICK
1959   The growth and composition of natural
stands of birch: 1. Dry matter production. *Plant
Soil,* 10:271–283.

PAINE, R. T.
1966   Food web complexity and species diversity.
*Amer. Natur.,* 100:65–75.

1969   The *Pisaster-Tegula* interaction: Prey
patches, predator food preference and intertidal
community structure. *Ecology,* 50:950–961.

1974   Intertidal community structure:
experimental studies on the relationship between
a dominant competitor and its principal predator.
*Oecologica,* 15:93–120.

PAJUNEN, V. I.
1966   The influence of population density in the
territorial behavior of *Leucorrhinia rubicunda* L.
(Odon. Libellulidae). *Ann. Zool. Fenn,.* 3:40–52.

PALMER, H. E., W. C. HANSON, B. I. GRIFFIN, AND W. C.
ROESCH
1963   Cesium-137 in Alaskan Eskimos. *Science,*
142(3588):64–65.

PALMER, J. D.
1973   Tidal rhythms: the clock control of the
rhythmic physiology of marine organisms. *Biol.
Rev.,* 48:377–418.

1974   *Biological Clocks in Marine Organisms,*
Wiley, New York.

1976   *An Introduction to Biological Rhythms,*
Academic, New York.

PALMGREN, P.
1949   Some remarks on the short term
fluctuations in the numbers of northern birds
and mammals. *Oikos,* 1:114–121.

PALOHEIMO, J. E., AND L. M. DICKIE
1970   Production and food supply. In J. H. Steele
(ed.), *Marine Food Chains,* University of
California Press, Berkeley.

PARENTI, R. L., AND E. L. RICE
1969   Inhibitional affects of *Digitaria sanguinalis*
and possible role in old field succession. *Bull.
Torrey Bot. Club,* 96:70–78.

PARIS, O. H.
1969   The function of soil fauna in grassland

ecosystems. In R. L. Dix and R. G. Beidleman (eds.), 1969, pp. 331–360.

PARK, T.

1948   Experimental studies of interspecies competition: 1. Competition between populations of the flour beetles, *Trilobium confusum* Duval and *Trilobium castaneum* Herbst. *Ecol. Monogr.,* 18:265–308.

1954   Experimental studies of interspecies competition: 2. Temperature, humidity and competition in two species of *Trilobium. Physiol. Zool.,* 27:177–238.

1955   Experimental competition in beetles with some general implications. In J. B. Cragg and N. W. Pirie (eds.), *The Numbers of Man and Animals,* Oliver & Boyd, London.

PATE, V. S. L.

1933   Studies on fish food in selected areas: a biological survey of Raquette Watershed. *N.Y. State Conserv. Dept. Biol. Survey No. 8,* pp. 136–157.

PATTERSON, D. T.

1975   Nutrient return in stemflow and throughfall of individual trees in the Piedmont deciduous forest. In F. G. Howell et al. (eds.), 1975, pp. 800–812.

PATTON, J. L.

1972   Patterns of geographic variation in karotype in the pocket gopher *Thomomys bottae* (Eydoux and Gervais). *Evolution,* 26:574–586.

PAYNE, R.

1968   Among wild whales. *N.Y. Zool. Soc. Newsletter,* Nov. 1968.

PEAKALL, D. B.

1970   Pesticides and the reproduction of birds. *Sci. Amer.,* 222:72–78.

PEAKALL, D. B., AND J. L. LINGER

1970   Polychlorinated biphenyls: another long-life widespread chemical in the environment. *Bioscience,* 20:958–964.

PEARCE, R. B.

1967   Photosynthesis in plant communities as influenced by leaf angle. *Crop Sci.,* 7:321–326.

PEARL, R., AND L. J. REED

1920   On the rate of growth of the population of the U.S. since 1790 and its mathematical representation. *Proc. Natur. Acad. Sci.,* 6:275–288.

PEARSON, D. L.

1971   Vertical stratification of birds in a tropical dry forest. *Condor,* 73:46–55.

PEARSON, O. P.

1966   The prey of carnivores during ᴏ ᴏ cycle of mouse abundance. *J. Anim. Ecol.,* 35:217–233.

1971   Additional measurement of the impact of carnivores on California voles. *J. Mammal.,* 52:41–49.

PEET, R. K.

1974   The measurement of species diversity. *Ann. Rev. Ecol. Syst.,* 5:285–307.

PEMADASA, M. A., AND P. H. LOVELL

1974   Interference in populations of some dune animals. *J. Ecol.,* 62:855–868.

PENNAK, R. W.

1942   Ecology of some copepods inhabiting intertidal beaches near Woods Hole, Massachusetts. *Ecology,* 23:446–456.

1951   Comparative ecology of the interstitial fauna of fresh-water and marine beaches. *Ann. Biol.,* 27:217–248.

PENNAK, R. W., AND E. D. VAN GERPEN

1947   Bottom fauna production and physical nature of a substrate in a northern Colorado trout stream. *Ecology,* 28:42–48.

PEQUEGNAT, W. E.

1961   Life in the scuba zone: 2. *Nat. Hist.,* 70(5):46–54.

PERRINS, C. M.

1965   Population fluctuations and clutch size in the great tit *Parus major* L. *J. Anim. Ecol.,* 34:601–647.

PETERLE, T. J.

1969   DDT in Antarctic snow. *Nature,* 224:620.

PETRIDES, G. A., AND W. G. SWANK

1966   Estimating the productivity and energy relations of the African elephant. *Proc. 9th Inter. Grassland Conv.,* pp. 831–842.

PETROV, V. S.

1946   Aktevnaia reaktsiia pochvy pH kah faktor rasprpstaneniia dozhdevykh chorvei (Lumbricidae, Oligochaetae). *Zool. Zh.,* 25:107–110.

PETRUSEWICZ, K. (ED.)

1967   *Secondary Productivity of Terrestrial Ecosystems.* Polish Academy Sciences, Warsaw.

PETRUSEWICZ, K., AND H. L. GRODZINSKI

1975   The role of herbivore consumers in various ecosystems. In National Academy Science, 1975, pp. 64–70.

PFEIFFER, W.

1962   The fright reaction of fish. *Biol. Rev.,* 37:495–511.

PHILLIPS, J.

1965   Fire—as master and servant: its influence in the bioclimatic regions of trans-Saharan Africa. *Proc. 4th Tall Timbers Fire Ecol. Conf.,* pp. 7–109.

PHILLIPS, P., AND M. M. BARNES

1975   Host race formation among sympatric apple, walnut, and plum populations of the codling moth. *Ann. Entomol. Soc. Amer.,* 68:1053–1060.

PHILLIPS, R. C.

1974   Temperate grass flats. In H. T. Odum, B. J. Copeland and E. A. McMahan (eds.), 1974, Vol. 2, pp. 244–299.

PHILLIPS, W. S.

1963   Depth of roots in soil. *Ecology,* 44:242.

PIANKA, E. R.

1966   Latitudinal gradients in species diversity: a review of concepts. *Amer. Natur.,* 100:33–46.

1967   On lizard species diversity, North American flatlands desert. *Ecology,* 48:333–351.

1972   *r* and *K* selections or *b* and *d* selection? *Amer. Natur.,* 100:65–75.

1974   Niche overlap and diffuse competition.
    *Proc. Nat. Acad. Sci.,* 71:2141–2145.
1976   Competition and niche theory. In R. May
    (ed.), 1976, pp. 114–141.
1978   *Evolutionary Ecology,* 2nd ed., Harper &
    Row, New York.

PIANKA, E., AND W. S. PARKER
1975   Age specific reproductive tactics. *Amer.
    Natur.,* 109:453–464.

PICKETT, S. T. A.
1976   Succession: an evolutionary interpretation.
    *Amer. Natur.,* 110:107–119.

PIELOU, E. C.
1974   *Population and Community Ecology,* Gordon
    and Breach, New York.
1975   *Ecological Diversity,* Wiley, New York
1977   *Mathematical Ecology,* Wiley, New York.

PIERCE, R. S., C. W. MARTIN, C. C. REEVES, G. E.
    LIKENS, AND F. H. BORMANN
1972   Nutrient loss from clearcuttings in New
    Hampshire. In *Symposium on Watersheds in
    Transition,* American Water Resources
    Association and Colorado State University, Fort
    Collins, Colorado, pp. 285–296.

PIJL, L. VAN DER
1969   *Principles of Dispersal in Higher Plants,*
    Springer-Verlag, New York.

PIMENTEL, D.
1961   Animal population regulation by the genetic
    feedback mechanism. *Amer. Natur.,* 95:65–79.
1968   Population regulation and genetic feedback.
    *Science,* 159:1432–1437.
1971a  *Ecological Effects of Pesticides on Non-
    target Species,* Executive Office of the President,
    Office of Science and Technology, Washington,
    D.C.
1971b  Evolutionary and environmental impact of
    pesticides. *Bioscience,* 21:109.

PIMENTEL, D., J. E. DEWEY, AND H. H. SCHWARDT
1951   An increase in the duration of the life cycle
    of DDT-resistant strains of the house fly. *J. Econ.
    Entomol.,* 44:477–481.

PIMENTEL, D., E. H. FEINBERG, P. W. WOOD, AND
J. T. HAYES
1965   Selection, spacial distribution, and the
    coexistence of competing fly species. *Amer.
    Natur.,* 99:97–109.

PIMENTEL, D., W. P. NAGEL, AND J. L. MADDEN
1963   Space-time structure of the environment
    and the survival of the parasite-host system.
    *Amer. Natur.,* 97:141–167.

PIMLOTT, D. H.
1967   Wolf predation and ungulate populations.
    *Amer. Zool.,* 7:267–278.

PITELKA, F. A.
1957a  Some aspects of population structure in the
    short term cycle of the brown lemming in
    northern Alaska. *Cold Spring Harbor Symp.
    Quant. Biol.,* 22:237–251.
1957b  Some characteristics of microtine cycles in
    the Arctic. *Proc. 18th Biol. Coll.,* Oregon State
    College, Corvallis, pp. 73–88.

1964a  The nutrient recovery hypothesis for Arctic
    microtine cycles: 1. Introduction. In D. J. Crisp
    (ed.), *Grazing in Terrestrial and Marine
    Environments,* Blackwell, Oxford, pp. 55–56.
1964b  Predation in the lemming cycle at Barrow,
    Alaska. *Proc. 16th Inter. Cong. Zool.,* 1:265.

PLATT, W. J., AND I. M. WEIS
1977   Resource partitioning and competition
    within a guild of fugitive prairie plants. *Amer.
    Natur.,* 479–513.

PODOLER, H., AND D. ROGERS
1975   A new method for the identification of key
    factors from life table data. *J. Anim. Ecol.,*
    44:85–114.

POLT, J. M., AND E. H. HESS
1964   Following the imprinting effects of light and
    social experience. *Science,* 143:1185–1187.

POMEROY, L. R.
1959   Algae productivity in salt marshes of
    Georgia. *Limnol. Oceanogr.,* 4:386–397.

POMEROY, L. R., ET AL.
1969   The phosphorus and zinc cycles and
    productivity of a salt marsh. In D. J. Nelson and
    F. E. Evans (eds.), 1969, pp. 412–419.

POMEROY, L. R., AND E. J. KUENZLER
1969   Phosphorus turnover by coral reef animals.
    In D. J. Nelson and F. E. Evans (eds.), 1969, pp.
    478–483.

POMEROY, L. R., H. M. MATHEWS, AND H. SHIKMIN
1963   Excretion of phosphate and soluble organic
    phosphorus compounds by zooplankton. *Limnol.
    Oceanogr.,* 4:50–55.

POORE, M. E. D.
1962   The method of successive approximation in
    descriptive ecology. *Adv. Ecol. Res.,* 2:35–68.

PORTER, K. G.
1973   Selective grazing and differential digestion
    of algae by zooplankton. *Nature,* 244:179–180.

PORTER, W. P., AND D. M. GATES
1969   Thermodynamic equilibria of animals with
    environment. *Ecol. Monogr.,* 39:245–270.

POWER, J. F.
1970   Leaching of nitrate-nitrogen under dryland
    agriculture in the northern Great Plains. In
    *Relationship of Agriculture to Soil and Water
    Pollution,* Cornell University Press, Ithaca,
    N.Y., pp. 111–122.

PREISTER, L. E.
1965   The accumulation in metabolism of DDT,
    parathion, and endrin by aquatic food chain
    organisms. Ph.D. thesis, Clemson University.

PRESTON, F. W.
1948   The commonness and rarity of species.
    *Ecology,* 29:254–283.
1962   The canonical distribution of commonness
    and rarity: Parts 1 and 2. *Ecology,* 43:185–215,
    410–432.

PRICE, P. W.
1974   Strategies for egg production. *Evolution,*
    28:76–84.

PRICE, P. W. (ED.)
1975   *Evolutionary Strategies of Parasitic Insects,*

Plenum, New York.

PRITCHARD, D. W.
1952   Salinity distribution and circulation in the Chesapeake Bay estuarine system. *J. Marine Res.,* 11:106–123.

PROCTOR, M. C., AND P. F. YEO
1973   *The Pollination of Flowers,* Collins, London.

PROSSER, C. L. (ED.)
1958   *Physiological Adaptations,* American Physiological Society, Washington, D.C.

PRUITT, W. O., JR.
1960   Behavior of the barren-ground caribou. *Biol. Paper No. 3,* University of Alaska.
1970   Some aspects of interrelationships of permafrost and tundra biotic communities. In *Productivity and Conservation in Northern Circumpolar Lands,* IUCN Publ. n.s. 10:33-41.

PUTWAIN, P. D., AND J. L. HARPER
1970   Studies of dynamics of plant populations: 3. The influence of associated species on populations of *Rumex acetosa* L. and *R. acetosella* L. in grassland. *J. Ecol.,* 58:251–264.

PUTWAIN, P. D., D. MACHIN, AND J. L. HARPER
1968   Studies in the dynamics of plant population: 2. Components and regulation of a natural population of *Rumex acetosella* L. *J. Ecol.,* 56:421–431.

RABATNOV, T. A.
1974   Differences between fluctuations and successions. In R. Knapp (ed.), 1974, pp. 21–24.

RANDALL, J. E.
1965   Grazing effect on sea grasses by herbivorous reef fishes in the West Indies. *Ecology,* 46:255–260.

RANNEY, J. W.
1977   Forest island edges—their structure, development, and importance to regional forest ecosystem dynamics. *EDFB/IBP Cont. No. 77/1,* Oak Ridge National Laboratory, Oak Ridge, Tenn.

RANWELL, D. S.
1961   *Spartina* salt marshes in southern England: 1. The effects of sheep grazing at the upper limits or *Spartina* marsh in Bridgwater Bay. *J. Ecol.,* 49:325–340.

RAPOPORT, E. H.
1969   Gloger's rule and pigmentation of Collembola. *Evolution,* 23:622–626.

RAUNER, Y. L.
1972   *Heat Balance of the Vegetation Cover,* Gidrometeoizdat, Leningrad.

RAUNKIAER, C.
1934   *The Life Form of Plants and Statistical Plant Geography,* Clarendon Press, Oxford.

RAUSCH, R. A.
1967   Some aspects of the population ecology of wolves. *Amer. Zool.,* 7:253–256.

RAVEN, P. H.
1973   The evolution of Mediterranean flora. In F. di Castri and H. A. Mooney (eds.), 1973, pp. 213–224.

RAYMONT, J. E. G.
1963   *Plankton and Productivity in Ocean,* Pergamon, Elmsford, N.Y.

REDD, B. L., AND A. BENSON
1962   Utilization of bottom fauna by brook trout in a northern West Virginia stream. *Proc. West Va. Acad. Sci.,* 34:21–26.

REEMOLD, R. J.
1972   The movement of phosphorus through the salt marsh cord grass *Spartina alterniflora* Loisel. *Limnol. Oceanogr.,* 17:606–611.

REGIER, F. E., AND E. O. WILSON
1968   The alarm defense system of the ant, *Acanthomyops caviger. J. Insect Physiol.,* 14:955–970.

REGIER, H. A., AND K. H. LOFTUS
1972   Effects of fisheries exploitation on salmonid communities in oligotrophic lakes. *J. Fish. Res. Board Can.,* 29:959–968.

REICHLE, D. E.
1975   Advances in ecosystem analysis. *Bioscience,* 25:257–264.

REICHLE, D. E. (ED.)
1970   *Analysis of Temperate Forest Ecosystems, Ecological Studies 1,* Springer-Verlag, New York.

REICHLE, D. E., ET AL.
1973   Carbon flow and storage in a forest ecosystem. In G. M. Woodwell and E. V. Pecan (eds.), 1973, pp. 345–365.

REICHLE, D. E., AND D. A. CROSSLEY, JR.
1967   Investigation of heterotrophic productivity in forest insect communities. In K. Petrusewicz (ed.), *Secondary Productivity of Terrestrial Ecosystems,* Polish Academy of Sciences, Warsaw, pp. 563–587.

REICHLE, D. E., P. B. DUNAWAY, AND D. J. NELSON
1970   Turnover and concentration of radionuclides in food chains. *Nuclear Safety,* 11:43–56.

REICHLE, D. E., R. A. GOLDSTEIN, R. I. VAN HOOK, AND G. J. DODSON
1973   Analysis of insect consumption in a forest canopy. *Ecology,* 54:1076–1084.

REIFSNYDER, W. E., AND H. W. LULL
1965   Radiant energy in relation to forests. *USDA Tech. Bull. No. 1344.*

REIMOLD, R. J.
1972   Salt marsh ecology: the effects on marine food webs of direct harvest of marsh grass by man and the contribution of marsh grass to the food available to marine organisms. *Sea Grant Rept.,* University of Georgia, Athens.

REINERS, W. A.
1973   A summary of the world carbon cycle and recommendations for critical research. In G. M. Woodwell and E. Pecan (eds.), 1973, pp. 368–382.

REITEMEIER, R. F.
1957   Soil potassium and fertility. *Yearbook of Agriculture,* USDA, Washington, D.C., pp. 101–106.

RICE, E. L.
1964   Inhibition of nitrogen-fixing and nitrifying bacteria by seed plants. *Ecology,* 45:824–837.
1965   Inhibition of nitrogen-fixing and nitrifying

bacteria by seed plants: 2. Characterization and identification of inhibitors. *Physiol. Plant,* 18:255–268.

1972  Allelopathic effects of *Andropogon virginicus* and its persistence in old fields. *Amer. J. Bot.,* 59:752–755.

RICE, E. L., AND R. L. PARENTI

1967  Inhibition of nitrogen-fixing and nitrifying bacteria by seed plants. V. Inhibitors produced by *Bromus japonicus* Thunb. *Southwest Natur.,* 12:97–103

RICH, P. H., AND R. G. WETZEL

1978  Detritus in the lake ecosystem. *Amer. Natur.,* 112:57–71.

RICHARDS, C. M.

1958  The inhibition of growth in crowded *Rana pipiens* tadpoles. *Physiol. Zool.,* 31:138–151.

1962  The control of tadpole growth by algal-like cells. *Physiol. Zool.,* 35:285–296.

RICHARDS, P. W.

1952  *The Tropical Rain Forest,* Cambridge University Press, London.

RICHARDSON, C. J., AND J. A. LUND

1975  Effects of clearcutting on nutrient losses in aspen forests on three soil types in Michigan. In F. G. Howell et al. (ed.), 1975, pp. 673–680.

RICKER, W. E.

1940  On the origin of Kokanee, a freshwater type of sockeye salmon. *Trans. Roy. Soc. Can. Sect. V,* 34:121–135.

1954  Stock and recruitment. *J. Fish. Res. Board Can.,* 11:559–623.

1958a  Maximum sustained yields from fluctuating environments and mixed stocks. *J. Fish. Res. Board Can.,* 15:991–1006.

1958b  Handbook of computations for biological statistics of fish populations. *Bull. 119, J. Fish. Res. Board Can.,* pp. 1–300.

RICKLEFS, R.

1978  *Ecology,* 2nd ed., Chiron Press, New York.

RIGLER, F. H.

1956  A tracer study of the phosphorus cycle in lake water. *Ecology,* 37:550–562.

1964  The phosphorus fractions and turnover time of inorganic phosphorus in different types of lakes. *Limno. Oceanogr.,* 9:511–518.

1973.  A dynamic view of the phosphorus cycle in lakes. In E. J. Griffith et al. (eds.), *Environmental Phosphorus Handbook,* Wiley, New York, pp. 539–572.

RILEY, G. A.

1970  Particulate organic matter in sea water. *Adv. Marine Biol.,* 8:1–118.

RIPLEY, E. A., AND R. E. REDMANN

1976  Grassland. In J. L. Monteith (ed.), 1976, pp. 349–398.

ROBINSON, M. H., AND B. ROBINSON

1970  Prey caught by a sample population of the spider in Panama: a year's census data. *Zool. J. Linn. Soc.,* 49:345–358.

RODIN, L. E., AND N. I. BAZILEVIC

1964  The biological productivity of the main vegetation types in the Northern Hemisphere of the Old World. *Forest. Abst.,* 27:369–372.

1967  *Production and Mineral Cycling in Terrestrial Vegetation* (transl. from Russian by Scripta Technica), G. E. Fogg (ed.), Oliver & Boyd, London.

ROOT, R. B.

1967  The niche exploitation pattern of the blue gray gnatcatcher. *Ecol. Monogr.,* 37:317–350.

RORISON, I. H. (ED.)

1969  *Ecological Aspects of Mineral Nutrition of Plants,* Blackwell Scientific Publ., Oxford.

ROSENZWEIG, M. L., AND R. H. MACARTHUR

1963  Graphical representation and stability conditions of predator-prey interactions. *Amer. Natur.,* 97:209–223.

ROSS, B. A., J. R. BRAY, AND W. H. MARSHALL

1970  Effects of a long-term deer exclusion on a *Pinus resinosa* forest in north-central Minnesota. *Ecology,* 51:1088–1093.

ROSS, H. H.

1970  The ecological history of the Great Plains— evidence from grassland insects. In W. Dort and J. K. Jones (eds.), 1970, pp. 225–240.

ROSSWALL, T., ET AL.

1975  Stordalen (Abisko) Sweden. In T. Rosswall and O. W. Heal (eds.), 1975, pp. 265–294.

ROSSWALL, T., AND O. W. HEAL (EDS.)

1975  *Structure and Function of Tundra Ecosystems,* Swedish Natural Science Research Council, Stockholm.

ROSSWALL, T. AND U. GRANHALL

1980  Nitrogen cycling in a subarctic ombrotrophic mire. In M. Sonesson (ed.), *Ecology of A Subarctic Mire Ecol. Bull. 30.* Swedish Natural Science Research Council, Stockholm.

ROTHSCHILD, H.

1975  Remarks on carotenoids on the evolution of signals. In L. E. Gilbert and P. H. Raven (eds.), 1975, pp. 20–50.

ROTHSCHILD, M.

1972  Secondary plant substances and warning coloration in insects. In H. F. Van Emden (ed.), *Insect-Plant Relationships,* Blackwell Scientific Publ., Oxford, pp. 59–83.

ROUGHGARDEN, J.

1974  Species packing and the competition function with illustrations from coral reef fish. *Theor. Pop. Biol.,* 5:163–186.

ROUGHGARDEN, J., AND M. FELDMAN

1975  Species packing and predation pressure. *Ecology,* 56:489–492.

ROWAN, W. R.

1925  Relation of flight to bird migration and developmental changes. *Nature,* 115:494–495.

1929  Experiments in bird migration: 1. Manipulation of the reproductive cycle, seasonal histological changes in the gonads. *Proc. Boston Soc. Natur. Hist.,* 39:151–208.

ROWELL, C. H. F.

1961  Displacement grooming in the chaffinch. *Anim. Behav.,* 9:38–63.

ROWLAND, F. S.
1973 Mercury levels in swordfish and tuna. *Biol. Conser.,* 5:52–53.

ROYAMA, T.
1970 Factors governing the hunting behavior and selection of food by the great tit. *J. Anim. Ecol.,* 39:619–668.
1971 A comparative study of models for predation and parasitism. *Res. Pop. Ecol.,* 13 (Suppl. 1):1–91.

RUDD, R. L.
1964 *Pesticides and the Living Landscape,* University of Wisconsin Press, Madison.

RUDD, R. L., AND R. E. GENELLY
1956 Pesticides: their use and toxicity in relation to wildlife. *Calif. Fish and Game Bull. No. 7.*

RUSCH, D. H., E. C. MESLOW, P. D. DOERR, AND L. B. KEITH
1972 Response of great horned owl populations to changed prey densities. *J. Wildl. Manage.,* 36:282–296.

RYTHER, J. H.
1969 Photosynthesis and fish production in the sea. *Science,* 166:72–75.

SABINE, W. S.
1959 The winter society of the Oregon junco: intolerance, dominance, and the pecking order. *Condor,* 61:110–135.

SAGAR, G. R., AND A. L. MORTIMER
1976 An approach to the study of the population dynamics of plants with special reference to weeds. *Appl. Biol.,* 1:1–47.

SALT, G. W.
1957 An analysis of avifaunas in the Teton Mountains and Jackson Hole, Wyoming. *Condor,* 59:373–393.

SANDERS, H. L.
1968 Marine benthic diversity: a comparative study. *Amer. Natur.,* 102:243–283.

SARUKHAN, J.
1974 Studies on plant demography: *Ranunculus repens* L., *R. bulbosus* L., and R. *acris* L.: 2. Reproductive strategies and seed population dynamics. *J. Ecol.,* 62:151–177.

SARUKHAN, J., AND J. HARPER
1974 Studies on plant demography: 1. Population flux and survivorship. *J. Ecol.,* 61:676–716.

SATHER, J. H. (ED.)
1976 *Proceedings of National Wetland Classification and Inventory Workshop,* U.S. Fish and Wildlife Service, Washington, D.C.

SCHAEFER, R.
1973 Microbial activity under seasonal conditions of drought in Mediterranean climates. In F. di Castri and H. A. Mooney (eds.), 1973, pp. 191–198.

SCHAFFER, W. G.
1974 Selection for optimal life histories: The effects of age structure. *Ecology,* 55:291–303.

SCHALLER, G. B.
1972 *Serengeti: A Kingdom of Predators,* Knopf, New York.

SCHARITZ, R. R., AND J. F. MCCORMICK
1973 Population dynamics of two competing annual plant species. *Ecology,* 54:723–740.

SCHEFFER, V. C.
1951 The rise and fall of a reindeer herd. *Sci. Month.,* 73:356–362.

SCHELDERUP-EBBE, T.
1922 Beitrage zur Socialpsychologie des Haushuhns. *Zeitschr. Psychol.,* 88:225–252.

SCHENKEL, R.
1948 Ausdruckstudien an Wolfen. *Behaviour,* 1:81–130.

SCHINDLER, D. W., R. W. NEWBURY, K. G. BEATY, AND P. CAMPBELL
1976 Natural water and chemical budgets for a small Precambrian lake basin in central Canada. *J. Fish. Res. Board Can.,* 33:2526–2543.

SCHLESINGER, W. H.
1977 Carbon balance in terrestrial detritus. *Ann. Rev. Ecol. Syst.,* 8:51–81.

SCHMIDT-NIELSEN, K.
1960 The salt secreting gland of marine birds. *Circulation,* 21:955–967.
1964 *Desert Animals: Physiological Problems of Heat and Water,* Oxford University Press, London.

SCHOENER, T. W.
1968 Sizes of feeding territories among birds. *Ecology,* 49:123–141.

SCHOLANDER, P. F., R. HOCK, V. WALTERS, F. JOHNSON, AND L. IRVING
1950 Heat regulation in some arctic and tropical birds and mammals. *Biol. Bull.,* 99:237–258.

SCHOLANDER, P. F., V. WALTERS, R. HOCK, L. IRVING, AND F. JOHNSON
1950 Body insulation of some arctic and tropical mammals and birds. *Biol. Bull.,* 99:225–236.

SCHULTZ, V., AND A. W. KLEMENT (EDS.)
1963 *Radioecology,* Van Nostrand Rheinhold, New York.

SCIDENSLICKER, J. C., IV, M. G. HORNOCKER, W. V. WILES, AND J. P. MESSICK
1973 Mountain lion social organization in the Idaho primitive area. *Wildl. Monogr.* 35:1–60.

SCLATER, P. L.
1858 On the general geographical distribution of the members of the class Aves. *J. Proc. Limnol. Soc. (Zool.),* 2:130–145.

SCOTT, D., AND W. O. BILLINGS
1964 Effects of environmental factors on standing crop and productivity of an alpine tundra. *Ecol. Monogr.,* 34:243–370.

SCOTT, J. P.
1958 *Animal Behavior,* University Chicago Press.
1962 Critical periods in behavioral development. *Science,* 138:949–958.

SCOTT, T. C.
1943 Some food coactions of the northern plains red fox. *Ecol. Monogr.,* 13:427–479.
1955 An evaluation of the red fox. *Ill. Natur. Hist. Surv., Biol. Notes No. 35,* pp. 1–16.

SEBEOK, T. A. (ED.)

1978  *How Animals Communicate,* Indiana University Press, Bloomington.

SEGERSTRALE, S. G.

1947  New observations on the distribution and morphology of the amphipod *Gammarus zaddachi* Sexon, with notes on related species. *J. Marine Biol. Assoc. U.K.,* 27:219–244.

SELANDER, R. K., AND R. F. JOHNSTON

1967  Evolution in the house sparrow: 1. Intrapopulation variation in North America. *Condor,* 69:217–258.

SELLECK, G. W.

1960  The climax concept. *Bot. Rev.,* 26:534–545.

SERVENTY, D. L.

1971  Biology of desert birds. In D. S. Farner and J. R. King (eds.), *Avian Biology,* Academic, New York, pp. 287–339.

SEVENSTER, P.

1961  *A Causal Analysis of Displacement Activity in Fannings in* Gasterosteus aculeatus *L.,* E. Brill, Leiden.

SEVERINGHAUS, C. W.

1972  Weather and the deer population. *The Conservationist,* 27(2):28–31.

SEVERINGHAUS, C. W., AND R. GOTTLEIB

1959  Big deer vs. little deer. *N.Y. State Conservationist,* 14(2):30–31.

SHAPIRO, A. M.

1970  The role of sexual behavior in density related dispersal of pierid butterflies. *Amer. Natur.,* 104:367–372.

SHAW, S. P., AND C. G. FREDINE

1956  Wetlands of the United States. *U.S. Fish and Wildl. Circ. 39.*

SHELDON, W. G.

1967  *The Book of the American Woodcock,* University of Massachusetts Press, Amherst.

SHELFORD, V. E., AND A. C. TWOMEY

1941  Tundra animals in the vicinity of Churchill, Manitoba. *Ecology,* 22:47-69.

SHEPPARD, P. M.

1959  *Natural Selection and Heredity,* Hutchinson, London.

SHERVIS, L. H., G. M. BUSH, AND C. F. KOVAL

1970  Infestation of sour cherries by apple maggot: confirmation of a previously uncertain host status. *J. Econ. Ent.,* 63:294–295.

SHOREY, H. H.

1978  Pheromones. In T. A. Sebeok (ed.), 1978, pp. 137–163.

SHURE, D. J., AND H. S. RAGSDALE

1977  Patterns of primary succession on granite outcrop surfaces. *Ecology,* 58:993–1006.

SHUSTER, C. N., JR.

1966  The nature of a tidal marsh. *The Conservationist,* 21(1):22–29, 36.

SIBLEY, C. G.

1957  The evolutionary and taxonomic significance of sexual dimorphism and hybridization in birds. *Condor,* 59:166–191.

1960  The electrophoretic patterns of avian egg-white proteins as taxonomic characters. *Ibis,* 102:215–284.

SIEBURTH, J. MCN., AND A. JENSEN

1970  Production and transformation of extracellular organic matter from marine littoral algae: a resume. In D. E. Hood (ed.), *Organic Matter in Natural Waters,* University of Alaska, pp. 203–223.

SIEGENTHALER, U., AND H. OESCHGER

1978  Predicting future atmospheric carbon dioxide levels. *Science,* 199:388–395.

SIMBERLOFF, D. S.

1974  Equilibrium theory of island biogeography and ecology. *Ann. Rev. Ecol. Sys.,* 5:161–182.

1976  Species turnover and equilibrium island biogeography. *Science,* 194:572–578.

SIMBERLOFF, D. S., AND L. A. ABELE

1976  Island biogeographic theory and conservation practice. *Science,* 191:285–286.

SIMBERLOFF, D. S., AND E. O. WILSON

1969  Experimental zoogeography of islands. The colonization of empty islands. *Ecology,* 50:278–296.

1970  Experimental zoogeography of islands. A two-year record of colonization. *Ecology,* 50:278–296.

SIMMS, E.

1965  The effects of cold weather of 1962/63 on the blackbird population of Dallis Hill. *Brit. Birds.* 58:33–43.

SIMONS, S., AND J. ALCOCK

1971  Learning and foraging persistence of white-crowned sparrows *Zonotricha leucophrys. Ibis,* 111:477–482.

SIMPSON, G. G.

1964  Species density of North American recent mammals. *Syst. Zool.,* 13:57–73.

SIMS, P. L., AND J. S. SINGH

1971  Herbage dynamics and net primary production in certain grazed and ungrazed grasslands in North America. In N. R. French (ed.), 1971, pp. 59–124.

SINGER, F. S. (ED.)

1970  *Global Effects of Environmental Pollution,* Springer-Verlag, New York.

SINGH, J. S., AND R. MISRA

1968  Diversity, dominance, stability, and net production in the grasslands at Varanasi, India. *Can. J. Bot.,* 47:425–427.

SINGH, R. N.

1961  *Role of Blue-Green Algae in the Nitrogen Economy of Indian Agriculture,* Indian Council of Agricultural Research, New Delhi.

SLUSS, R. R.

1967  Population dynamics of the walnut aphid *Chronophes juglandicola* (Kalt.) in northern California. *Ecology,* 48:41–58.

SMALLEY, A. E.

1960  Energy flow of a salt marsh grasshopper population. *Ecology,* 41:672–677.

SMITH, B. D.

1974 Predator and prey relationships in southeastern Ozarks, A.D. 1300. *Human Ecol.,* 2:31–43.

SMITH, C. C.

1968 The adaptative nature of social organization in the genus of tree squirrels *Tamiasciurus. Ecol. Monogr.,* 38:31–63.

1970 The coevolution of pine squirrels and conifers. *Ecol. Monogr.,* 40:349–371.

1975 The coevolution of plants and seed predators. In L. E. Gilbert and P. H. Raven (eds.), 1975, pp. 53–77.

SMITH, C. C., AND S. D. FRETWELL

1974 The optimal balance between size and number of offspring. *Amer. Natur.,* 108:499–506.

SMITH, F. E.

1972 Spatial heterogeneity, stability and diversity in ecosystems. *Trans. Conn. Acad. Arts and Science,* 44:309–355.

SMITH, J. N. M.

1974a The food searching behaviour of two European thrushes: 1. Description and analysis of search paths. *Behaviour,* 48:276–302.

1974b The food searching behaviour of two European thrushes: 2. The adaptiveness of the search patterns. *Behaviour,* 49:1–61.

SMITH, J. N. M., AND R. DAWKINS

1971 The hunting behaviour of individual great tits in relation to spatial variations in their food density. *Anim. Behav.,* 19:695–706.

SMITH, J. N. M., AND H. P. A. SWEATMAN

1974 Food searching behavior of titmice in patchy environments. *Ecology,* 55:1216–1232.

SMITH, M.

1974 Seasonality in mammals. In H. Leith (ed.), 1974, pp. 149–162.

SMITH, N.

1968 The advantage of being parasitized. *Nature,* 219:690–694.

SMITH, R. E., AND B. A. HORWITZ

1969 Brown fat and thermogenesis. *Physiol. Rev.,* 49:330–425.

SMITH, R. H., AND S. GALLIZIOLI

1965 Predicting hunter success by means of a spring call count of Gambel quail. *J. Wildl. Manage.,* 29:806–813.

SMITH, R. L.

1959a Conifer plantations as wildlife habitat. *N.Y. Fish Game J.,* 5:101–132.

1959b The songs of the grasshopper sparrow. *Wilson Bull.,* 71:141–152.

1962 Acorn consumption by white-footed mice *(Peromyscus leucopus). Bull. 482T, W. Va. Univ. Agr. Expt. Sta.*

1963 Some ecological notes on the grasshopper sparrow. *Wilson Bull.,* 75:159–165.

1966 Animals and the vegetation of West Virginia. In E. L. Core, *Vegetation of West Virginia,* McClain, Parsons, W.Va., pp. 17–24.

1972 *The Ecology of Man: An Ecosystem Approach,* Harper & Row, New York.

1976 Socio-ecological evolution in the hill country of southwestern West Virginia. In J. Luchok et al. (eds.), 1976, *Hill Lands. Proc. International Symposium,* pp. 198–202.

1977 Ecological genesis of endangered species: the philosophy of preservation. *Ann. Rev. Ecol. Syst.,* 7:33–55.

SMITH, S. M.

1978 The "underworld" in a territorial adaptive strategy for floaters. *Amer. Natur.,* 112:570–582.

SMYTH, M., AND G. A. BARTHOLOMEW

1966 The water economy of the black-throated sparrow and the rock wren. *Condor,* 68:447–458.

SMYTHE, N.

1970 Relationships between fruiting seasons and seed dispersal methods in a neotropical forest. *Amer. Natur.,* 104:25–35.

SNOW, D. W.

1958 *A Study of Blackbirds,* Allen-Unwin, London.

1966 A possible selective factor in the evaluation of fruiting seasons in tropical forest. *Oikos,* 15:274–381.

1971 Evolutionary aspects of fruit-eating by birds. *Ibis,* 113:194–202.

1976 *The Web of Adaptation: Bird Studies in the American Tropics,* Quadrangle, New York.

SOLBRIG, O.

1970 *Principles and Methods of Plant Biosystematics,* Macmillan, New York.

SOLBRIG, O., AND B. B. SIMPSON

1974 Components of regulation of a population of dandelion in Michigan. *J. Ecol.,* 62:473–486.

SOLOMON, M. E.

1949 The natural control of animal populations. *J. Anim. Ecol.,* 18:1–32.

1957 Dynamics of insect populations. *Ann. Rev. Entomol.,* 2:121–142.

1964 Analysis of processes involved in the natural control of insects. *Adv. Ecol. Res.,* 2:1–58.

SORENSEN, T.

1954 Adaptation of small plants to deficient nutrition and short growing season. *Botan. Tidsskr.,* 51:339–361.

SOWLS, L. K.

1960 Results of a banding study of Gambel's quail in southern Arizona. *J. Wildl. Manage.,* 24:185–190.

SPECHT, R. L.

1973 Structure and functional response of ecosystems in the Mediterranean climate of Australia. In F. di Castri and H. A. Mooney (eds.), 1973, pp. 113–120.

SPIGHT, T. M., AND J. EMLEN

1976 Clutch sizes of two marine snails with a changing food supply. *Ecology,* 57:1162–1178.

SPOONER, G. M.

1947 The distribution of *Gammarus* species in estuaries: Part 1. *J. Marine Biol. Assoc. U. K.,* 27:1–52.

SPRUGEL, D. G.

1976 Dynamic structure of wave generated *Abies balsamea* forests in north-eastern United States. *J. Ecol.*, 64:889–911.

SPURR, S. H.
1952 Origin of the concept of forest succession. *Ecology*, 33:426–427.
1957 Local climate in the Harvard Forest. *Ecology*, 38:37–56.

STARK. N.
1972 Nutrient cycling pathways and litter fungi. *Bioscience*, 22:355–360.
1973 *Nutrient Cycling in a Jeffey Pine Forest Ecosystem*, Montana Forest and Conservation Experiment Station, Missoula.

STEARNS, S. C.
1976 Life history tactics: a review of ideas. *Quart. Rev. Biol.*, 51:3–47.
1977 The evolution of life history traits. *Ann. Rev. Ecol. Syst.*, 8:145–171.

STEBBINS, G. L., JR.
1950 *Variation and Evolution in Plants*, Columbia University Press, New York.
1972 Evolution and diversity of arid-land shrubs. In C. McKell et al. (eds.), 1972, pp. 111–116.

STEELE, J. H.
1974 *The Structure of Marine Ecosystems*, Harvard University Press, Cambridge, Mass.

STEGEMAN, L. C.
1960 A preliminary survey of earthworms of the Tully Forest in central New York. *Ecology*, 41:779–782.

STEHLI, F. R., R. G. DOUGLAS, AND N. D. NEWELL
1969 Generation and maintenance of gradients in taxonomic diversity. *Science*, 164:947–949.

STENGER, J., AND J. B. FALLS
1959 The utilized territory of the ovenbird. *Wilson Bull.*, 71:125–140.

STEPHENSON, T. A., AND A. STEPHENSON
1952 Life between tide-marks in North America: 2. North Florida and the Carolinas, *J. Ecol.*, 40:1–49
1954 Life between the tide-marks in North America: 3A. Nova Scotia and Prince Edward Island: The geographical features of the region. *J. Ecol.*, 42:14–45, 46–70.
1971 *Life Between the Tide-marks on Rocky Shores*, Freeman, San Francisco.

STERN, W. L., AND M. F. BUELL
1951 Life-form spectra of New Jersey pine barren forest and Minnesota jack pine forest. *Bull. Torrey Bot. Club*, 78:61–65.

STEWARD, G. A.
1970 High potential productivity of the tropics for cereal crops, grass forage crops, and beef. *J. Aust. Inst. Agr. Sci.*, 36:85.

STEWART, W. D. P.
1967 Nitrogen-fixing plants. *Science*, 158:1426–1432.

STILES, F. G.
1975 Ecology, flowering phenology, and hummingbird pollination of some Costa Rican *Heliconia* species. *Ecology*, 56:285–301.

STILES, F. G., AND L. L. WOLF
1970 Hummingbird territoriality at a tropical flowering tree. *Auk*, 87:469–491.

STOECKLER, J. H.
1962 Shelterbelt influence on Great Plains field environment and crops. *USDA Prod. Res. Rept. No. 62.*

STRAHLER, A.
1971 *The Earth Sciences*, Harper & Row, New York.

STRICKLAND, J. D. H..
1965a Production of organic matter in the primary stages of the marine food chain. In J. P. Riley and G. Skirrow (eds.), *Chemical Oceanography, Vol. 1*, Academic, New York, pp. 477–610.
1965b Phytoplankton and marine primary production. *Ann. Rev. Microbiol.*, 19:127–162.

STROUD, L. M.
1969 Color-infrared aerial photographic interpretation and net primary productivity of a regularly flooded North Carolina salt marsh. M.S. thesis, North Carolina State University, Raleigh.

STRUMWASSER, F.
1960 Some physiological principles governing hibernation in *Citellus beecheyi*. In C. P. Lyman and A. R. Dawe (eds.), 1960, pp. 285–318.

STUDY OF CRITICAL ENVIRONMENTAL PROBLEMS
1970 *Man's Impact on the Global Environment*, MIT Press, Cambridge, Mass.

STUDY OF MAN'S IMPACT ON CLIMATE
1971 *Inadvertent Climate Modification*, MIT Press, Cambridge, Mass.

STUIVER, M.
1978 Atmospheric carbon dioxide and carbon reservoir changes. *Science*, 199:253–258.

STURKIE, P. D. (ED.)
1976 *Avian Physiology*, Springer-Verlag, New York.

SUTTON, G. M.
1951 Dispersal of mistletoe by birds. *Wilson Bull.*, 63:235–237.

TABER, R. D., AND R. F. DASMANN
1957 The dynamics of three natural populations of the deer *Odocoileus hemionus columbianus*. *Ecology*, 38:233–246.
1958 The black-tailed deer of the chaparral. *Calif. Dept. Fish Game, Game Bull. No. 8.*

TAIT, R. V.
1968 *Elements of Marine Ecology*, Plenum, New York.

TAMARIN, R. H.
1978 Dispersal, population regulation, and K-selection in field mice. *Amer. Natur.*, 112:545–555.

TAMARIN, R. H., AND C. J. KREBS
1969 Microtus population biology: 2. Genetic changes at the transferrin locus in fluctuating populations of two vole species. *Evolution*, 23:183–211.

TAMM, C. O.

1951 Removal of plant nutrients from tree crowns by rain. *Physiol. Plant.*, 4:184–188.

TANNER, J. T.

1966 Effects of population density on growth rates of animal populations. *Ecology*, 47:733–745.

1975 The stability and intrinsic growth rates of prey and predator populations. *Ecology*, 56:855–867.

TANSLEY, A. G.

1935 The use and abuse of vegetational concepts and terms. *Ecology*, 16:284–307.

TARRANT, R. F.

1971 Persistence of some chemicals in Pacific Northwest forest. In *Pesticides, Pest Control, and Safety on Forest Lands*, Continuing Education Books, Corvallis, Oregon, pp. 133–141.

TATUM, L. A.

1971 The southern corn leaf blight epidemic. *Science*, 171:1113–1116.

TAYLOR, C. R.

1969 The eland and the oryx, *Sci. Amer.*, 220(1):88–95.

1970a Strategies of temperature regulation: effect of evaporation on East African ungulates. *Amer. J. Physiol.*, 219:1131–1135.

1970b Dehydration and heat: effects on temperature regulation of East African ungulates. *Amer. J. Physiol.*, 219:1136–1139.

TAYLOR, F. G., JR.

1974 Phenodynamics of production in a mesic deciduous forest. In H. Leith (ed.), 1974, pp. 237–254.

TAYLOR, K.

1971 Biological flora of the British Isles. *Rubus chamaemorus* L., *J. Ecol.*, 59:293.

TEAL, J. M.

1957 Community metabolism in a temperate cold spring. *Ecol. Monogr.*, 27:283–302.

1962 Energy flow in the salt marsh ecosystem of Georgia. *Ecology*, 43:614–624.

TEAL, J. M., AND J. KANWISHER

1961 Gas exchange in a Georgia salt marsh. *Limnol. Oceanogr.*, 6:388–399.

TENAZA, R.

1971 Behavior and nesting success relative to nest location in Adelie penguins *(Pygoscelis adeliae)*. *Condor*, 73:81–92.

TERBORGH, J. W.

1974 Preservation of natural diversity: the problem of extinction prone species. *Bioscience*, 24:715–722.

1975 Faunal equilibria and the design of wildlife preserves. In F. B. Golley and E. Medina (eds.), 1975, pp. 269–380.

TERRI, J., AND L. STOWE

1976 Climate patterns and distribution of $C_4$ grasses in North America. *Oecologia*, 23:1–12.

TESTER, J. R., AND W. H. MARSHALL

1961 A study of certain plant and animal interrelations on a native prairie in Northwestern Minnesota. *Minn. Mus. Natur. Hist. Occasional Paper No. 8.*

THAILER, C. S., JR.

1974 Four contacts between ranges of different chromosome forms of the *Thomomys talpoides* complex. (Rodentia, Geomyidae). *Syst. Zool.*, 23:343–354.

THIELCKE, G. A.

1966 Ritualized distinctiveness of song in closely related sympatric species. *Phil. Trans. Roy. Soc. London, B*, 251:493–497.

1976 *Bird Sounds*, University of Michigan Press, Ann Arbor.

THODAY, J. M.

1959 Effects of disruptive selection: 1. Genetic flexibility. *Heredity*, 13:187–203.

1960 Effects of disruptive selection: 3. Coupling and repulsion. *Heredity*, 14:35–49.

THODAY, J. M., AND T. G. BOAM

1959 Effects of disruptive selection: 2. Polymorphism and divergence without isolation. *Heredity*, 13:205–218.

THODAY, J. M., AND J. B. GIBSON

1962 Isolation by disruptive selection. *Nature*, 193:1164–1166.

1970 The probability of isolation by disruptive selection. *Amer. Natur.*, 104:219–230.

THOMAS, J. W., C. MASER, AND J. E. RODICK

1978 Edges—their interspersion, resulting diversity, and its measurement. In *Proceedings Workshop on Nongame Bird Habitat in Coniferous Forests of Western United States*, USDA Forest Ser, Gen. Tech. Rept. PNW-64, pp. 91–100.

THOMPSON, D. Q., AND R. H. SMITH

1970 The forest primeval in the northeast: a great myth? *Proc. 10th Tall Timbers Fire Ecol. Conf.*, pp. 255–265.

THOMPSON, H. V.

1953 The grazing behavior of the wild rabbit. *Brit. J. Anim. Behav.*, 1:16–20.

1954 The rabbit disease, myxomatosis. *Ann. Appl. Biol.*, 41:358–366.

THOMPSON, W. L.

1960 Agonistic behavior in the house finch: Part 1. Annual cycle and display patterns. *Condor*, 62:245–271.

THOREAU, H. D.

1860 Succession of forest trees: Address read to the Middlesex Agricultural Society, Sept. 1860. In *Excursion*, 1891, Houghton Mifflin, Boston, Mass.

THORSON, G.

1950 Reproductive and larval ecology of marine bottom invertebrates. *Biol. Rev.*, 25:1–45.

1957 Bottom communities in J. Hedgpeth (ed.), 1957a, pp. 461–534.

THURSTON, J. M.

1969 The effect of liming and fertilizers on the botanical composition of permanent grassland and on the yield of hay. In I. H. Rorison (ed.), 1969, pp. 3–10.

TILLY, L. J.

1968 The structure and dynamics of Cone Spring.

*Ecol. Monogr.,* 28:169–197.

TINBERGEN, L.

1960  The natural control of insects in pinewoods: 1. Factors influencing the intensity of predation by songbirds. *Arch. Neerl. Zool.,* 13:265–343.

TINBERGEN, N.

1951  *The Study of Instinct,* Oxford University Press, New York.

1952  Derived activities, their causation, biological significance, origin and emancipation during evolution. *Quart. Rev. Biol.,* 27:1–32.

1953  *The Herring Gull's World,* Collins, London.

1960  Comparative studies of the behaviour of gulls (Laridae): a progress report. *Behaviour,* 15:1–70.

TINBERGEN, N., AND M. MOYNIHAN

1952  Head flagging in the black-headed gull: its function and origin. *Brit. Birds,* 45:19–22.

TINKLE, D. W.

1969  The concept of reproductive effort and its relation to the evolution of life histories of lizards. *Amer. Natur.,* 103:501–516.

TINKLE, D. W., AND R. E. BALLINGER

1972  *Sceloporus undulatus,* a study of the intraspecific comparative demography of a lizard. *Ecology,* 53:570–585.

TODD, R. L., J. B. WARDE, AND B. W. CORNABY

1975  Significance of biological nitrogen fixation and dentifrication in a deciduous forest ecosystem. In F. G. Howell et al. (eds.), 1975, pp. 729–735.

TOMANEK, G. W.

1969  Dynamics of mulch layer in a grassland ecosystem. In R. L. Dix and R. G. Beidleman (eds.), 1969, pp. 225–240.

TORDOFF, H. B.

1954  Social organization and behaviour in a flock of captive, non-breeding red crossbills. *Condor,* 36:346–358.

TORNABENE, T. G., AND H. W. EDWARDS

1972  Microbial uptake of lead. *Science,* 176:1334–1335.

TOTH, R. S., AND R. M. CHEN

1972  Development and energetics of *Notonecta undulata* during predation in *Culex tarsalis.* *Ann. Ent. Soc. Amer.,* 65:1270–1279.

TRACY, C. R.

1976  A model of the dynamic exchanges of water and energy between a terrestrial amphibian and its environment. *Ecol. Monogr.,* 46:293–326.

TRESHOW, M.

1970  *Environment and Plant Response,* McGraw-Hill, New York.

TRIMBLE, G. R., JR.

1973  The regeneration of central Appalachian hardwoods with emphasis on the effects of site quality and harvesting practice. *USDA Forest Serv. Res. Paper NE-282.*

TRIMBLE, G. R., JR., ET AL.

1974  Some options for managing forest land in the central Appalachians. *USDA Forest Serv. Gen. Tech. Rept. NE-12.*

TRIMBLE, C. R., JR., AND S. WEITZMAN

1954  Effect of a hardwood forest canopy on rainfall intensities. *Trans. Amer. Geophys. Union,* 35:226–234.

TRIVERS, R. L.

1971  The evolution of reciprocal altruism. *Quart. Rev. Biol.,* 46:34–57.

TRUE, R. P., ET AL.

1960  Oak wilt in West Virginia. *West Va. Univ. Agr. Expt. Sta. Bull. 448T.*

TURCEK, F. J.

1951  On the stratification of the avian populations of the Querceto-Carpinetum forest communities in southern Slovakia (English summary). *Sylvia,* 13:71–86.

TUREKIAN, K. K. (ED.)

1971  *The Late Cenozoic Glacial Ages.* Yale University Press, New Haven, Conn.

TYRTIKOV, A. P.

1959  Perennially frozen ground. In *Principles of Geocryology,* Part I, *General Geocryology* (trans. from Russian by R. E. Brown). Nat. Res. Coun. Canada. Tech Trans., 1163(1964):399–421.

UDVARDY, M. D. F.

1958  Ecological and distributional analysis of North American birds. *Condor,* 60:50–66.

URBAN, D.

1970  Raccoon populations, movement patterns, and predation on a managed waterfowl marsh. *J. Wildl. Manage.,* 34:372–382.

VALLENTYNE, J. R.

1957  The principles of modern limnology. *Amer. Sci.,* 45:218–244.

1974  *The Algal Bowl—Lakes and Man.* Misc. Spec. Publ. 22, Department of Environment, Ottawa, Canada.

VANCE, B. D., AND W. DRUMMOND

1969  Biological concentration of pesticides by algae. *J. Amer. Water Works Assoc.,* 61:360–362.

VANDERMEER, J. H.

1972  Niche theory. *Amer. Rev. Ecol. Syst.,* 3:107–132.

VAN DER HAMMER, T., T. A. WIJMSTRA, AND W. H. ZAGWIGN

1971  The floral record of late Cenozoic of Europe. In K. K. Turekian (ed.), 1971, pp. 391–424.

VAN DER VALK, A. G., AND C. B. DAVIS

1978  The role of seed banks in the vegetation dynamics of prairie glacial marshes. *Ecology,* 59:322–335.

VAN HOOKE, R. I.

1971  Energy and nutrient dynamics of spider and orthropteran populations in a grassland ecosystem. *Ecol. Monogr.,* 41:1–26.

VAN IERSEL, J. J. A.

1953  An analysis of the parental behaviour of the three-spined stickleback. *Behaviour, Suppl. No. 3.*

VAN IERSEL, J. J. A., AND A. C. A. BOL

1958  Preening of two tern species: a study on displacement activities. *Behaviour,* 13:1–88.

VAN VALEN, L.

1965 Morphological variation and the width of the ecological niche. *Amer. Natur.*, 99:377–389.

VAURIE, C. S. O.

1950 Rock nuthatches, adaptive differences between two sympatric species of nuthatches (*Sitta*). In *Proc. 10th Inter. Ornithol. Congr.*, pp. 163–166.

VEALE, P. T., AND H. L. WASCHER

1956 Henderson County soils. *Illinois Univ. Agr. Expt. Stat. Soil Rept. No. 77.*

VEITH, G. D., D. W. KUEHL, F. A. PUGLISI, G. F. GLASS, AND J. G. EATON

1977 Residues of PCB's and DDT in the western Lake Superior Ecosystem. *Arch. Environ. Contam. Toxicol.*, 5(4):487–499.

VIESSMAN, W., JR.

1966 The hydrology of small impervious areas. *Water Resources Res.*, 2:405–412.

VIRO, P. J.

1953 Loss of nutrients and the natural nutrient balance of the soil in Finland. *Comm. Inst. Forest. Fenn.*, 42:1–50.

VITOUSEK, P. M., AND W. A. REINERS

1975 Ecosystem succession and nutrient retention: a hypothesis. *Bioscience*, 25:376–381.

VOGL, R. J.

1969 One hundred and thirty years of plant succession in a southeastern Wisconsin lowland. *Ecology*, 50:248–255.

1974 Effect of fire on grasslands. In T. F. Kozwolski and C. E. Ahlgren (eds.), 1974, pp. 139–194.

VOIGHT, J. W., AND J. E. WEAVER

1951 Range condition classes of mature midwestern pasture, an ecological analysis. *Ecol. Monogr.*, 21:39–60.

VOIGT, G. K.

1960 Distribution of rain under forest stands. *For. Sci.*, 6:2–10.

1971 Mycorrhizae and nutrient mobilization. In E. Hacskaylo (ed.), *Mycorrhizae, USDA Forest Serv. Misc. Pub. No. 1189,* pp. 122–131.

VOLTERRA, V.

1926 Variazione e fluttazioni de numero d'individiu in specie animali conviventi. Translated in R. N. Chapman, 1931, *Animal Ecology,* McGraw-Hill, New York.

WADDINGTON, C. H.

1957 *The Strategy of the Genes,* Allen, London.

WAGNER, F. H., AND L. C. STODDART

1972 Influence of coyote predation on black-tailed jack rabbit populations in Utah. *J. Wildl. Manage.*, 36:329–342.

WAHRMAN, J. R. GOITEIN, AND E. NEVO

1969 Mole rat *Spalax*. Evolutionary significance of chromosone variation. *Science,* 164:82–84.

WALES, B. A.

1972 Vegetation analysis of north and south edges in a mature oak-hickory forest. *Ecol. Monogr.*, 42:451–471.

WALKER, P. C., AND R. T. HARTMAN

1960 Forest sequence of the Hartstown bog area

in western Pennsylvania. *Ecology,* 41:461–474.

WALKER, T. J., AND A. D. HASLER

1949 Detection and discrimination of odors of aquatic plants by the bluntnosed minnow (*Hyborhynchus notatus*). *Physiol. Zool.*, 22:45–63.

WALLACE, A. R.

1876 *The Geographical Distribution of Animals,* 2 vols., Macmillan, London.

WALLACE, B.

1968 *Topics in Population Genetics,* Norton, New York.

WALTER, E.

1934 Grundlagen der allegmeinen fisherielichen Produktionslehre. *Handb. Binnenfisch. Mitteleur.*, 14:480–662.

WALTER, H.

1971 *Ecology of Tropical and Subtropical Vegetation,* Oliver and Boyd, Edinburgh.

WALTHER, F. R.

1969 Flight behaviour and avoidance of predators in Thomson's gazelle (*Gazella thomsoni* Guenther, 1884). *Behaviour,* 34:184–221.

WARNER, R. E.

1968 The role of introduced diseases in the extinction of the endemic Hawaiian avifauna. *Condor,* 70:101–120.

WARREN, C. E., AND G. E. DAVIS

1971 Laboratory stream research: objectives, possibilities, constraints. *Ann. Rev. Ecol. Syst.*, 2:111–144.

WASSINK, E. C.

1959 Efficiency of light energy conversion in plant growth. *Plant Physiol.*, 34:356–361.

1968 Light energy conversion in photosynthesis and growth of plants. In F. E. Eckhardt (ed.), 1968, pp. 53–66.

WATERHOUSE, F. L.

1955 Microclimatological profiles in grass cover in relation to biological problems. *Quart. J. Roy. Meteorol. Soc.*, 81:63–71.

WATERS, T. F.

1961 Standing crop and drift of stream bottom organisms. *Ecology,* 42:532–537.

1964 Recolonization of denuded stream bottom areas by drift. *Trans. Amer. Fish. Soc.*, 91:243–250.

1965 Interpretation of invertebrate drift of day-active stream invertebrates. *Ecology,* 46:327–334.

1968 Diurnal periodicity in the drift of day-active stream invertebrates. *Ecology,* 49:152–153.

1972 The drift of stream insects. *Ann. Rev. Entomol.*, 17:253–272.

WATSON, A., AND D. JENKINS

1968 Experiments on population control by territorial behaviour in red grouse. *J. Anim. Ecol.*, 37:595–614.

WATSON, A., AND R. MOSS

1970 Dominance, spacing behaviour and aggression in relation to population limitation in vertebrates. In A. Watson (ed.), *Animal*

*Populations in Relation to Their Food Resources,* Blackwell, Oxford, pp. 167–218.

1971 Spacing as affected by territorial behavior, habitat, and nutrition in red grouse (*Lagopus l. scoticus*). In A. E. Esser (ed.), 1971, *Behavior and Environment,* Plenum, New York, pp. 92–111.

1972 A current model of population dynamics in red grouse. In *Proc. 15th Inter. Ornithol. Congr.,* pp. 134–149.

WATSON, G. E.

1962a Three sibling species of *Alectoris* partridge, *Ibis,* 104:353–367.

1962b Sympatry in Palearctic *Alectoris* partridges. *Evolution,* 16:11–19.

WATT, A. S.

1947 Pattern and process in the plant community. *J. Ecol.,* 35:1–22.

WATT, K. E. F.

1968 *Ecology and Resource Management: A Quantitative Approach,* McGraw-Hill, New York.

WEAVER, G. T.

1975 The quantity and distribution of four nutrient elements in high elevation forest ecosystems, Balsam Mountains, North Carolina. In F. G. Howell et al. (eds.), 1975, pp. 715–728.

WEAVER, J. A., AND F. C. CLEMENTS

1938 *Plant Ecology,* McGraw-Hill, New York.

WEAVER, J. E.

1954 *North American Prairie,* Johnson, Lincoln, Nebr.

WEAVER, J. E., AND F. W. ALBERTSON

1956 *Grasslands of the Great Plains: Their Nature and Use,* Johnson, Lincoln, Nebr.

WEAVER, J. E., AND N. W. ROWLAND

1952 Effects of excessive natural mulch on development, yield and structure of native grassland. *Botan. Gaz.,* 114:1–19.

WEBSTER, J.

1956–1957 Succession of fungi on decaying cocksfoot culms: Parts 1 and 2. *J. Ecol.,* 44:517–544; 45:1–30.

WEBSTER, J. R., J. B. WAIDE, AND B. C. PATTEN

1975 Nutrient recycling and the stability of ecosystems. In F. G. Howell et al. (eds.), 1975, pp. 1–27.

WECKER, S. C.

1963 The role of early experience in habitat selection by the prairie deer mouse, *Peromyscus maniculatus bairdi. Ecol. Monogr.,* 33:307–325.

WEIR, B. J., AND I. W. ROWLANDS

1973 Reproductive strategies in mammals. *Ann. Rev. Ecol. Syst.,* 4:139–163.

WEISE, C. M.

1974 Seasonality in birds. In H. Leith (ed.), 1974, pp. 139–147.

WEIS-FOGH, T.

1948 Ecological investigations of mites and collembola in the soil. *Nat. Jutland,* 1:135–270.

WEISS, H. V., M. KORDE, AND E. D. GOLDBERG

1971 Mercury in a Greenland ice sheet: evidence of recent input by man. *Science,* 174:692–694.

WELCH, P. S.

1952 *Limnology,* McGraw-Hill, New York.

WELLER, M. W.

1959 Parasitic egg laying the redhead (*Aythya americana*) and other North American Anatidae. *Ecol. Monogr.,* 29:333–365.

WELLER, M. W., AND L. H. FREDRICKSON

1973 Avian ecology of a managed glacial marsh. *Living Bird,* 12:269–291.

WELLINGTON, W. G.

1960 Qualitative changes in natural populations during changes in abundance. *Can. J. Zool.,* 38:238–314.

WELLS, H. W.

1961 The fauna of oyster beds, with special reference to the salinity factor, *Ecol. Monogr.,* 31:239–266.

WELLS, L.

1960 Seasonal abundance and vertical movements of planktonic crustaceans in Lake Michigan. *USDI Fish. Bull.,* 60(172):343–369.

WENT, F. W.

1955 The ecology of desert plants. *Sci. Amer.,* 192:68–75.

WENT, F. W., AND L. O. SHEPS

1969 Environmental factors in regulation of growth and development: ecological factors. In F. Steward (ed.), *Plant Physiology: A Treatise,* Academic, New York, pp. 299–406.

WENT, F. W., AND N. STARK

1968 Mycorrhiza. *Bioscience,* 18(11):1035–1039.

WERNER, E. E., AND D. J. HALL

1976 Niche shift in sunfishes: experimental evidence and significance. *Science,* 191:404–406.

WESSEL, M. A., AND A. DOMINSKI

1977 The children's daily lead. *Amer. Sci.,* 65:294–298.

WEST, D. A.

1962 Hybridization in grosbeaks (*Pheucticus*) of the Great Plains. *Auk,* 79:339–424.

WEST, G. C.

1965 Shivering and heat production in wild birds. *Physiol. Zool.,* 38:111–120.

WEST, N. E., AND P. T. TUELLER

1972 Special approaches to studies of competition and succession in shrub communities. In C. M. McKell et al. (eds.), 1972, pp. 172–181.

WESTLAKE, D. F.

1959 The effects of organisms on pollution. *Proc. Linn. Soc., London,* 170:171–172.

WETZEL, R.

1975 *Limnology,* Saunders, Philadelphia.

WETZEL, R. G., AND H. L. ALLEN

1970 Function and interactions of dissolved organic matter and the littoral zone in lake metabolism and eutrophication. In Z. Kabap and A. Hillbricht-Ilkowska (eds.), *Productivity Problems of Freshwater,* PWN Polish Sci. Publ., Warsaw, pp. 333–347.

WETZEL, R. G., P. H. RICH, M. C. MILLER, AND H. L. ALLEN

1972 Metabolism of dissolved and particulate

detrital carbon in a temperate hardwater lake. *Mem. 1st Ital. Idrobiol. 29 (Supp.),* pp. 185– 243.

WHITCOMB, R. F., J. F. LYNCH, P. A. OPLER, AND C. S. ROBBINS

1976 Island biogeography and conservation: strategy and limitations. *Science,* 193:1030– 1032.

1977 Long term turnover and effects of selective logging on the avifauna of forest fragments. *Amer. Birds,* 31(1):17– 23.

WHITE, E. J., AND F. TURNER

1970 A method of estimating income of nutrients in a catch of airborne particles by a woodland canopy. *J. Appl. Ecol.,* 7:441– 461.

WHITE, M. J. D.

1968 Models of speciation. *Science,* 159:1065– 1070.

1973 *Animal Cytology and Evolution,* 3rd ed., Cambridge University Press.

1974 Speciation in Australian morabine grasshoppers. The cytogenic evidence. In M. J. D. White (ed.), 1974, pp. 1– 42.

WHITE, M. J. D. (ED.)

1974 *Genetic Mechanisms of Speciation in Insects,* Australian and New Zealand Book Company, Sydney.

WHITEHEAD, D. R.

1967 Studies of full-glacial vegetation and climate in southeastern United States. In E. J. Cushing and H. E. Wright (eds.), 1967, pp. 237– 248.

WHITMORE, T. C.

1975 *Tropical Rain Forest of the Far East,* Oxford University Press, London.

WHITTAKER, R. H.

1951 A criticism of the plant association and climatic climax concept. *Northwest Sci.,* 25:17– 31.

1952 A study of summer foliage insect communities in the Great Smoky Mountains. *Ecol. Monogr.,* 22:1– 44.

1953 A consideration of the climax theory: the climax as a population and pattern. *Ecol. Monogr.,* 23:41– 78.

1956 Vegetation of the Great Smoky Mountains. *Ecol. Monogr.,* 26:1– 80.

1960 Vegetation of the Siskiyou Mountains, Oregon and California. *Ecol. Monogr.,* 30:279– 338.

1961 Estimation of net primary production of forest and shrub communities. *Ecology,* 42:177– 183.

1962 Classification of natural communities. *Bot. Rev.,* 28:1– 239.

1963 Net production of heath balds and forest heaths in the Great Smoky Mountains. *Ecology,* 44:176– 182.

1965 Dominance and diversity in land plant communities. *Science,* 147:250– 260.

1966 Forest dimensions and production in the Great Smoky Mountains. *Ecology,* 47:103– 121.

1967 Gradient analysis of vegetation. *Biol. Rev.,* 42:207– 264.

1970a *Communities and Ecosystems,* Macmillan, New York.

1970b The biochemical ecology of higher plants. In E. Sondheimer and J. B. Simeone (eds.), *Chemical Ecology,* Academic, New York, pp. 43– 70.

WHITTAKER, R. H., F. H. BORMANN, G. E. LIKENS, AND T. G. SICCAMA

1974 The Hubbard Brook ecosystem study: forest biomass and production. *Ecol. Monogr.,* 44:233– 252.

1975 The biosphere and man. In R. H. Whittaker and H. Leith (eds.), *Primary Productivity of the Biosphere,* Springer-Verlag, New York, pp. 305– 328.

WHITTAKER, R. H., AND P. R. FEENEY

1971 Allelochemics: chemical interactions between species. *Science,* 171:757– 770.

WHITTAKER, R. H., S. A. LEVIN, AND R. B. ROOT

1973 Niche, habitat, and ecotope. *Amer. Natur.,* 107:321– 338.

WHITTAKER, R. H., AND G. E. LIKENS

1973 Carbon in the biota. In G. M. Woodwell and E. V. Pecan (eds.), 1973, pp. 281– 300.

WHITTAKER, R. H., AND W. NIERING

1975 Vegetation of the Santa Catalina Mountains, Arizona. V. Biomass production and diversity along the elevation gradient. *Ecology,* 56:771– 790.

WHITTAKER, R. H., R. B. WALKER, AND A. R. KRUCKEBERG

1954 The ecology of serpentine soils. *Ecology,* 35:258– 288.

WHITTAKER, R. H., AND G. M. WOODWELL

1969 Structure, production, and diversity of the oak-pine forest at Brookhaven, New York. *J. Ecol.,* 57:155– 174.

WHYTE, R. O.

1968 *Grasslands of the Monsoon,* Praeger, New York.

WIEGERT, R. G.

1964 Population energetics of meadow spittlebugs as affected by migration and habitat. *Ecol. Monogr.,* 34:217– 241.

1965 Energy dynamics of the grasshopper populations in old field and alfalfa field ecosystems. *Oikos,* 16:161– 176.

WIEGERT, R. G., AND F. C. EVANS

1967 Investigations of secondary productivity in grasslands. In K. Petrusewicz, pp. 499– 518.

WIEGERT, R. G., AND D. F. OWEN

1971 Trophic structure, available resources, and population density in terrestrial vs. aquatic ecosystems. *J. Theoret. Biol.,* 30:69– 81.

WIELAND, N. K., AND F. A. BAZZAZ

1975 Physiological ecology of three codominant successional annuals. *Ecology,* 56:681– 688.

WIELGOLASKI, F. E.

1975a Productivity of tundra ecosystems. In National Academy of Science, 1975, pp. 1– 12.

1975b Primary productivity of alpine meadow communities. In F. E. Wielgolaski (ed.), 1975, pp. 121– 128.

WIELGOLASKI, F. E. (ED.)

1975 *Fennoscandian Tundra Ecosystems: Part 1, Plants and Microorganisms,* Springer-Verlag, New York.

WIELGOLASSI, F. E., AND S. KJELVIK
1975 Energy content and use of solar radiation of Fennoscandian tundra plants. In F. E. Wielgolaski (ed.), 1975, pp. 201–207.

WIELGOLASKI, F. E., S. KJELVIK, AND P. KALHIO
1975 Mineral content of tundra and forest tundra plants in Fennoscandia. In F. E. Wielgolaski (ed.), 1975, pp. 316–332.

WIENS, J. A.
1972 *Ecosystem Structure and Function, Proc. 31st Ann. Biol. Colloq.,* Oregon University Press, Salem.
1973 Pattern and process in grassland bird communities. *Ecol. Monogr.,* 43:237–270.
1974 Climatic instability and the "ecological saturation" of bird communities in North American grasslands. *Condor,* 76:385–400.
1975 Avian communities, energetics, and functions in coniferous forest habitats. In *Proc. Symp. Manage. Forest Range Habitats for Nongame Birds. USDA Forest Serv. Gen. Tech. Rept. WO-1.*
1976 Populations responses to patchy environments. *Ann. Rev. Ecol. Syst.,* 7:81–120.
1977 On competition and variable environments. *Amer. Sci.,* 65:590–597.

WIGGINS, I. L., AND J. H. THOMAS
1962 A flora of the Alaskan arctic slope. *Publ. Arctic Inst. N. Amer. No. 4,* University of Toronto Press, Toronto.

WILEY, R. H.
1974 Evolution of social organization and life history patterns among grouse (Aves: Tetraonidae). *Quart. Rev. Biol.,* 49:207–227.

WILLIAMS, C. B.
1964 *Patterns in the Balance of Nature,* Academic, New York.
1966 *Adaptation and Natural Selection,* Princeton University Press, New Jersey.

WILLIAMSON, P.
1971 Feeding ecology of the red-eyed vireo and associated foliage-gleaning birds. *Ecol. Monogr.,* 41:129–152.

WILLIS, A. J.
1963 Braunton burrows: the effects on vegetation of the addition of mineral nutrients to the dune soils. *J. Ecol.,* 51:353–374.

WILSON, D.
1975 A theory of group selection. *Proc. Nat. Acad. Sci.,* 72:143–146.

WILSON, D. S.
1976 Deducing the energy available in the environment: an application of optimal foraging theory. *Biotropica,* 8(2):96–103.

WILSON, E. O.
1971 Competitive and aggressive behavior. In J. Eisenberg and W. Dillon (eds.), *Man and Beast: Comparative Social Behavior,* Smithsonian Institute Press, Washington, D.C., pp. 183–217.

1975 *Sociobiology,* Harvard University Press, Cambridge, Mass.

WILSON, E. O., AND W. BOSSERT
1977 *A Primer of Population Ecology,* Sinauer Associates, Sunderland, Mass.

WILSON, E. O., AND E. O. WILLIS
1975 Applied biogeography. In M. L. Cody and J. M. Diamond (eds.), 1975, pp. 522–533.

WILSON, R. E., AND E. H. RICE
1968 Allelopathy as expressed by *Helianthus annus* and its role in old field succession. *Bull. Torrey Bot. Club,* 95:432–448.

WING, L. D., AND I. D. BUSS
1970 Elephants and forests. *Wildl. Monogr. No. 19.*

WINSTON, F. W.
1956 The acorn microsere with special reference to arthropods. *Ecology,* 37:120–132.

WIT, C. T. DE
1968 Photosynthesis: its relationship to overpopulation. In A. San Pietro et al. (eds.), *Harvesting the Sun,* Academic, New York, pp. 315–320.

WITHERSPOON, J. P., S. I. AVERBACH, AND J. S. OLSON
1962 Cycling of Cesium-134 in white oak trees on sites of contrasting soil type and moisture. *Oak Ridge Natur. Lab.,* 3328:1–143.

WITHROW, R. B. (ED.)
1959 *Photoperiodism and Related Phenomena in Plants and Animals,* Pub. No. 55, American Association for the Advancement of Science, Washington, D.C.

WITKAMP, M.
1963 Microbial populations of leaf litter in relation to environmental conditions and decomposition. *Ecology,* 44:370–377.

WITKAMP, M., AND D. A. CROSSLEY
1966 The role of arthropods and microflora on the breakdown of white oak litter. *Pedabiologia,* 6:293–303.

WITKAMP, M., AND J. S. OLSON
1963 Breakdown of confined and nonconfined oak litter. *Oikos,* 14:138–147.

WOLF, L. L., F. G. STILES, AND F. R. HAINSWORTH
1976 Ecological organization of a highland tropical hummingbird community. *J. Anim. Ecol.,* 45:349–379.

WOLFE, J. N., R. T. WAREHAM, AND H. T. SCOFIELD
1949 Microclimates and macroclimates of Neotoma, a small valley in central Ohio. *Ohio Biol. Survey Bull. No. 41.*

WOLFSON, A.
1959 The role of light and darkness in the regulation of spring migration and reproductive cycles in birds. In R. B. Withrow (ed.), 1959, pp. 679–716.
1960 Regulation of annual periodicity in the migration and reproduction of birds. *Cold Spring Harbor Symp. Quant. Biol.,* 25:507–514.

WOOD-GUSH, D. G. M.
1955 The behaviour of the domestic chicken: a review. *Brit. J. Anim. Behav.,* 3:81–110.

WOODWELL, G. M., AND D. B. BOTKIN
1970   Metabolism of terrestrial ecosystems by gas exchange techniques: The Brookhaven approach. In D. E. Reichle (ed.), 1970, pp. 73–85.

WOODWELL, G. M., P. P. CRAIG, AND H. A. JOHNSON
1971   DDT in the biosphere: where does it go? *Science,* 174:1101–1107.

WOODWELL, G. M., AND W. R. DYKEMAN
1966   Respiration of a forest measured by carbon dioxide accumulations during temperature inversions. *Science,* 154:1031–1034.

WOODWELL, G. M., AND T. G. MARPLES
1968   The influence of chronic gamma radiation on the production and decay of litter and humus in an oak-pine forest. *Ecology,* 49:456–465.

WOODWELL, G. M., AND E. V. PECAN (EDS.)
1973   *Carbon and the Biosphere, Conf. 72501,* National Technical Information Service, Springfield, Va.

WOODWELL, G. M., AND R. H. WHITTAKER
1968   Primary productivity in terrestrial ecosystems. *Amer. Zool.,* 8:19–30.

WOODWELL, G. M., R. H. WHITTAKER, W. A. REINERS, G. E. LIKENS, C. C. DELWICHE, AND D. B. BOTKIN
1978   The biota and the world carbon budget. *Science,* 199:141–146.

WOODWELL, G. M., C. F. WORSTER, JR., AND P. A. ISAACSON
1967   DDT residues in an east coast estuary: a case of biological concentration of a persistent pesticide. *Science,* 156:821–823.

WOOLFENDEN, G.
1973   Nesting and survival in a population of Florida scrub jays. *Living Bird,* 12:25–49.
1975   Florida scrub jay helpers at the nest. *Auk,* 92:1–15.

WOOLHOUSE, W. W. (ED.)
1969   *Dormancy and Survival, Symp. Soc. Expt. Biol. No. 23,* Academic, New York.

WOOLPY, J. H.
1968   The social organization of wolves, *Natur. Hist.,* 77(5):46–55.

WRIGHT, H. E., JR.
1970   Vegetational history of the Great Plains. In W. Dort and J. K. Jones (eds.), 1970, pp. 157–172.

WRIGHT, H. E., JR., AND D. G. FREY (EDS.)
1965   *The Quaternary of the U.S.,* Princeton University Press, New Jersey.

WRIGHT, R. F., ET AL.
1976   Impact of acid precipitation on freshwater systems in Norway. *Water, Air, Soil Pollution* 6(2,3,4,):483–499.

WRIGHT, R. T.
1970   Glycollic acid uptake by plankton bacteria. In D. Wood (ed.), *Organic Matter in Natural Waters, Occ. Pub. No. 1,* Institute Marine Science, University of Alaska.

WRIGHT, S.
1931a   Evolution in Mendelian populations. *Genetics,* 16:97–159.
1931b   Statistical theory of evolution. *Amer. Statist. J.,* March Suppl., pp. 201–208.
1935   Evolution in population in approximate equilibrium. *J. Genet.,* 30:243–256.

WURSTER, C. F., JR.
1968   DDT reduces photosynthesis by marine plankton. *Science,* 159:1477–1475.
1969   Chlorinated hydrocarbon insecticides and the world ecosystem. *Biol. Conser.,* 1:123–129.

WYNNE-EDWARDS, V. C.
1962   *Animal Dispersion in Relation to Social Behavior,* Hafner, New York.
1963   Intergroup selection in the evolution of social systems. *Nature,* 200:623–628.
1965   Self-regulating system in populations of animals. *Science,* 147:1543–1548.

YARRANTON, G. A., AND R. G. MORRISON
1974   Spatial dynamics of a primary succession: nucleation. *J. Ecol.,* 62:417–428.

YOM-TOV, Y.
1974   The effect of food and predation on breeding density and success, clutch size and laying date of the crow. *J. Anim. Ecol.,* 43:479–498.

YONGE, C. M.
1949   *The Sea Shore,* Collins, London.

ZACH, R., AND J. B. FALLS
1976a   Ovenbird hunting behavior in a patchy environment: an experimental study. *Can. J. Zool.,* 54:1863–1879.
1976b   Foraging behavior, learning, and exploration by captive ovenbirds. *Can. J. Zool.,* 54:1880–1893.
1976c   Do ovenbirds hunt by expectation? *Can. J. Zool.,* 54:1894–1903.

ZAK, B.
1964   Role of mycorrhizae in root disease. *Ann. Rev. Phytopathol.,* 2:377–392.

ZARET, T. M., AND R. T. PAINE
1973   Species introduction in a tropical lake. *Science,* 182:449–455.

ZELLER, D.
1961   Certain mulch and soil characteristics of major range sites in western North Dakota as related to range conditions. M.A. thesis, North Dakota State University, Fargo.

ZIMMERMAN, J. L.
1971   The territory and its density dependent effect in *Spiza americana. Auk,* 88:591–612.

# Index

Abundance, 667
Acclimation, 54–61
Acclimatization, 55
Acid rainfall, 178, 179
Acidity, determination of, 720
Accumulator plant, 327–328, 331
Active transport, 65
Adaptation, 6, 7, 19, 36, 358. *See also* Evolution,
    Natural Selection, Succession
  to aridity, 63–67
  coevolution, 558–575
  light, 95–97
  moisture, 64–67
  photoperiod, 97, 104–106
  predation, 562–567
  to salinity, 64, 67, 238, 250, 260, 262
  to temperature, 52–61
Adaptive radiation, 393–395
Adiabatic process, 82
ADP, 21, 38
Aerial photography, in mapping, 725
Age
  determination of, 681–683
  pyramids, 453–457
  stable age distribution, 462
    determination of, 714
Aggregations, 435
Aggregative response to prey density, 527
Agonistic display, 407–408
Agriculture
  food production, 147–149
  as pollution source, 173, 182
Algae, 217, 281, 284
  blue-green, 37, 38
    nitrogen fixation by, 232
Allele, 361
Allelopathy, 515–516, 583
Allen's rule, 375–376, 377
Allomones, 415
Allopolyploid, 388
Alternate prey, 528–529, 531
Altricial young, 425
Altruism, 370
Aluminum, 40, 41, 42, 43, 45, 46
Amensalism, 504
Ammonification, 169
Amphibians
  chemical defense in, 562
  cline in, 374
  diversity gradient in, 591–592
  effects of density on, 486
  marking of, 681
  parental behavior in, 423–424
  reproductive strategies in, 468
  sound production in, 416–417
  temperature regulation in, 56
  tundra, 342
Animals. *See also* Amphibians, Birds, Insects,
    Mammals, Reptiles
  aging of, 681
  distribution and continental drift, 646–647
  and fire, 94

  marking of, 680–681
  response to temperature, 54–60
  trapping and collecting, 678–680
  water balance, 64–66
Ants, 284
  competition with rodents, 515
  grassland, 295
  social parasitism, 573
Appeasement display, 408–409
Apple maggot, sympatric speciation in, 385–387
Aquatic organisms. *See also* Lentic Ecosystem,
    Lotic Ecosystem, Wetlands
  collecting, 678–679
Aquatic succession, 632–633, 614–615
Atmosphere
  DDT in, 185, 186
  lead pollution in, 183
  and sulfur cycle, 175, 176
  water cycle in, 155, 159
ATP, 21, 38, 161
Aufwuchs, 209
Auklet, Cassin's *(Ptychoramphus aleuticus)*,
    territoriality in, 488–489
Autecology, 8
Autopolyploid, 387
Autotrophs, 16–17, 578

Bacteria, 26–28, 30, 31, 37, 41, 42, 133, 169, 247,
    248, 250, 267, 281, 284, 335, 349. *See also*
    Decomposers
  in aquatic ecosystems, 30–31
  decomposition, 27–28
  and disease, 217–218, 483
  nitrogen-fixing, 38
  sulfur, 175–176
Bajadas, 304
Barnacles, 554
  interspecific competition, 512
Bathypelagic zone, 239, 266
Behavior, 6, 36
  aggressive, 491
    and competition, 513–514
  appeasement display, 408–410
  appetitive, 400
  attack, 407–408
  care of young, 425–426
  conflict, 402
  courtship, 401, 410–413
  distraction display, 414–415
  food storage, 103–104
  habitat selection, 405–406
  hierarchial, 417–419
  imprinting, 404–405
  incubation, 424–425
  innate, 399–403
  as isolating mechanism, 389–390
  parental, 423–427
  and population regulation, 487–491
  redirection activity, 402
  social, 399
  social dominance, 417–419
  stereotyped, 399–403

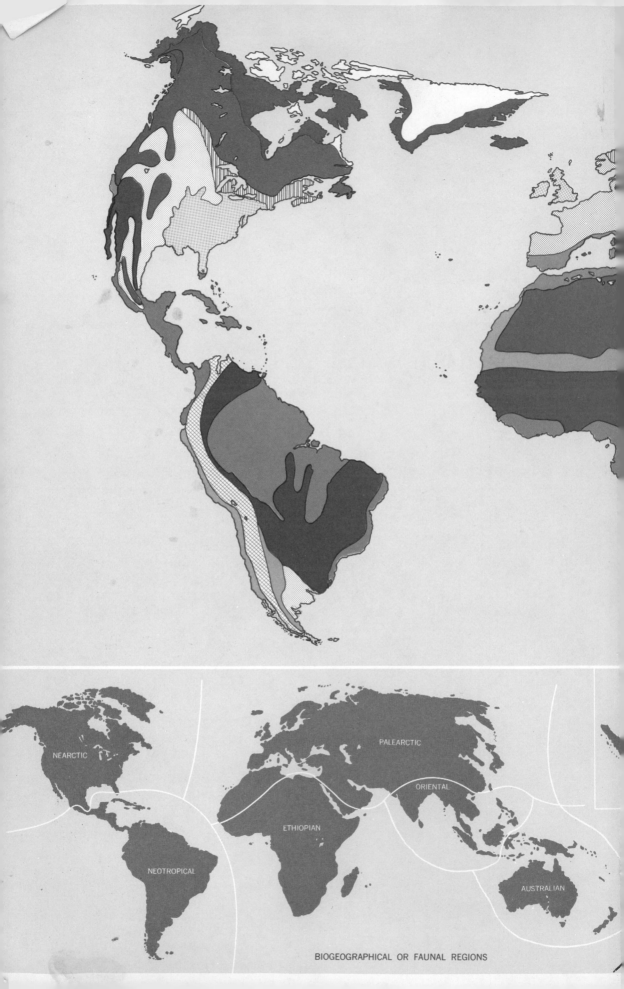

BIOGEOGRAPHICAL OR FAUNAL REGIONS

NEARCTIC

PALEARCTIC

ORIENTAL

ETHIOPIAN

NEOTROPICAL

AUSTRALIAN